Regulation of Photosynthesis

Advances in Photosynthesis and Respiration

VOLUME 11

The scope of our series, beginning with volume 11, reflects the concept that photosynthesis and respiration are intertwined with respect to both the protein complexes involved and to the entire bioenergetic machinery of all life. *Advances in Photosynthesis and Respiration* is a book series that provides a comprehensive and state-of-the-art account of research in photosynthesis and respiration. Photosynthesis is the process by which higher plants, algae, and certain species of bacteria transform and store solar energy in the form of energy-rich organic molecules. These compounds are in turn used as the energy source for all growth and reproduction in these and almost all other organisms. As such, virtually all life on the planet ultimately depends on photosynthetic energy conversion. Respiration, which occurs in mitochondrial and bacterial membranes, utilizes energy present in organic molecules to fuel a wide range of metabolic reactions critical for cell growth and development. In addition, many photosynthetic organisms engage in energetically wasteful photorespiration that begins in the chloroplast with an oxygenation reaction catalyzed by the same enzyme responsible for capturing carbon dioxide in photosynthesis. This series of books spans topics from physics to agronomy and medicine, from femtosecond processes to season long production, from the photophysics of reaction centers, through the electrochemistry of intermediate electron transfer, to the physiology of whole orgamisms, and from X-ray christallography of proteins to the morphology or organelles and intact organisms. The goal of the series is to offer beginning researchers, advanced undergraduate students, graduate students, and even research specialists, a comprehensive, up-to-date picture of the remarkable advances across the full scope of research on photosynthesis, respiration and related processes.

The titles published in this series are listed at the end of this volume and those of forthcoming volumes on the back cover.

Regulation of Photosynthesis

Edited by

Eva-Mari Aro
University of Turku, Finland

and

Bertil Andersson
University of Stockholm/Linköping, Sweden

KLUWER ACADEMIC PUBLISHERS
DORDRECHT / BOSTON / LONDON

A C.I.P. Catalogue record for this book is available from the Library of Congress.

ISBN 0-7923-6332-9

Published by Kluwer Academic Publishers,
P.O. Box 17, 3300 AA Dordrecht, The Netherlands.

Sold and distributed in North, Central and South America
by Kluwer Academic Publishers,
101 Philip Drive, Norwell, MA 02061, U.S.A.

In all other countries, sold and distributed
by Kluwer Academic Publishers,
P.O. Box 322, 3300 AH Dordrecht, The Netherlands.

The camera ready text was prepared
by Lawrence A. Orr, Center for the Study of Early Events in Photosynthesis
Arizona State University, Tempe, Arizona 85287-1604, U.S.A.

The photograph for the cover
was kindly provided by Dr. Alexander V. Vener

Printed on acid-free paper

Printed in the Netherlands.

Editorial

Advances in Photosynthesis and Respiration

I am proud to announce the publication of Volume 11, *Regulation of Photosynthesis*, edited by Eva-Mari Aro and Bertil Andersson, in our Series. This volume is the first one to appear under the new title of *Advances in Photosynthesis and Respiration*. Further, a new beginning is made here with the appointment of new members of the Board of Consulting Editors. They are: Christine Foyer, UK; Elisabeth Gantt, USA; John H. Golbeck, USA; Susan Golden, USA; Wolfgang Junge, Germany; Hartmut Michel, Germany; and Kimiyuki Satoh, Japan. I take this opportunity to welcome them and to thank the members of the previous Board of Consulting Editors: Jan Amesz*, the Netherlands, Eva-Mari Aro, Finland; James Barber, UK; Robert E. Blankenship, USA; Norio Murata, Japan; and Donald R. Ort, USA. Eva-Mari has kindly agreed to stay with the Board until Volume 12, after which we will appoint someone devoted solely to the area of "Respiration."

Published Volumes

The present volume is a sequel to the following ten volumes in the *Advances in Photosynthesis* (AIPH) series.

(1) *Molecular Biology of Cyanobacteria* (D.A. Bryant, editor, 1994);
(2) *Anoxygenic Photosynthetic Bacteria* (R.E. Blankenship, M.T. Madigan and C.E. Bauer, editors, 1995);
(3) *Biophysical Techniques in Photosynthesis* (J. Amesz and A.J. Hoff, editors, 1996);
(4) *Oxygenic Photosynthesis: The Light Reactions* (D.R. Ort and C.F. Yocum, editors, 1996);
(5) *Photosynthesis and the Environment* (N.R. Baker, editor, 1996);
(6) *Lipids in Photosynthesis: Structure, Function and Genetics* (P.-A. Siegenthaler and N. Murata, editors, 1998);
(7) *The Molecular Biology of Chloroplasts and Mitochondria in Chlamydomonas* (J.-D. Rochaix, M. Goldschmidt-Clermont and Sabeeha Merchant, editors, 1998);
(8) *The Photochemistry of Carotenoids* (H.A. Frank, A.J. Young, G. Britton and R.J. Cogdell, editors, 1999);
(9) *Photosynthesis: Physiology and Metabolism* (R.C. Leegood, T.D. Sharkey and Susanne von Caemmerer, editors, 2000); and
(10) *Photosynthesis: Photobiochemistry and Photobiophysics* (Bacon Ke, author, 2001)

See <http://www.wkap.nl/series.htm/AIPH> for further information and to order these books. Please note that the members of the International Society of Photosynthesis Research, ISPR (<http://www.photosynthesisresearch.org>) receive special discounts.

Regulatory Aspects of Photosynthesis

This new volume, Volume 11, is devoted to a critical area of photosynthesis, regulation. Regulation is the key for the optimum functioning of photosynthesis, and for dealing with both the abiotic (temperature, CO_2-deficiency, drought, and even light) and biotic stresses that are imposed on the plants. Eva-Mari Aro and Bertil Andersson have provided us with an authoritative book that has a broad perspective of the regulatory processes of photosynthesis. I have read the entire book and am impressed with the thoroughness of each chapter; it is a unique book in a rapidly growing field of plant biology. I predict that this book will remain a major resource in the area of 'regulatory processes' for decades to come. It provides the genetic and biochemical basis of: regulation of chloroplast gene expression and signal transduction; how nuclear and chloroplast genomes cooperate; how photosynthesis controls its own functions through redox regulation and through metabolism of the end-products; how proteins are correctly placed in the membranes; biogenesis and assembly of photosynthetic apparatus; and, how a myriad of enzymes, present at low levels, control and regulate many processes in the photosynthetic cells, among other topics. This book is highly suitable for advanced undergraduate and graduate students, beginning researchers, and even experts in the areas of plant biochemistry; plant physiology; plant molecular

* Deceased January 29, 2001

biology; molecular and cellular biology; integrative biology; microbiology; and plant biology.

The Scope of the Series

We expect that the scope of our series, beginning with this volume, will reflect the concept that photosynthesis and respiration are intertwined with respect to both the protein complexes involved and to the entire bioenergetic machinery of all life. *Advances in Photosynthesis and Respiration* is a book series providing a comprehensive and state-of-the-art account of research in photosynthesis and respiration. Photosynthesis is the process by which higher plants, algae, and certain species of bacteria transform and store solar energy in the form of energy-rich organic molecules. These compounds are in turn used as the energy source for all growth and reproduction in these and almost all other organisms. As such, virtually all life on the planet ultimately depends on photosynthetic energy conversion. Respiration, which occurs in mitochondrial and bacterial membranes, utilizes energy present in organic molecules to fuel a wide range of metabolic reactions critical for cell growth and development. In addition, many photosynthetic organisms engage in energetically wasteful *photorespiration* that begins in the chloroplast with an oxygenation reaction catalyzed by the same enzyme responsible for capturing carbon dioxide in photosynthesis. This series of books spans topics from physics to agronomy and medicine, from femtosecond processes to season long production, from the photophysics of reaction centers, through the electrochemistry of intermediate electron transfer, to the physiology of whole organisms, and from X-ray crystallography of proteins to the morphology of organelles and intact organisms. The intent of the series is to offer beginning researchers, advanced undergraduate students, graduate students, and even research specialists, a comprehensive, up-to-date picture of the remarkable advances across the full scope of research on bioenergetics and carbon metabolism.

Future Books

The readers of the current series are encouraged to watch for the publication of the forthcoming books:

(1) *Light-harvesting Antennas in Photosynthesis* (Editors: B.R. Green and W.W. Parson);
(2) *Photosynthesis in Algae* (Editors: A.W.D. Larkum, S. Douglas, and J.A. Raven);
(3) *Photosynthetic Nitrogen Assimilation and Associated Organic Acid Metabolism* (Editors: C.H. Foyer and G. Noctor);
(4) *Chlorophyll a Fluorescence: A Signature of Photosynthesis* (Editors: G. Papageorgiou and Govindjee); and,
(5) *Photosystem II: The Water/Plastoquinone Oxido-reductase in Photosynthesis* (Editors: T. Wydrzynski and K. Satoh).

In addition to these contracted books, invitations are out for several books. Topics planned are: Archael, Bacterial and Plant Respiration; Protein Complexes of Photosynthesis and Respiration; Photoinhibition and Photoprotection; Photosystem I; Protonation and ATP Synthesis; Global Aspects of Photosynthesis; Functional Genomics; History of Photosynthesis; The Chlorophylls; The Cytochromes; The Chloroplast; Laboratory Methods for Studying Leaves and Whole Plants; and C-3 and C-4 Plants. In view of the interdisciplinary character of research in photosynthesis and respiration, it is my earnest hope that this series of books will be used in educating students and researchers not only in Plant Sciences, Molecular and Cell Biology, Integrative Biology, Biotechnology, Agricultural Sciences, Microbiology, Biochemistry, and Biophysics, but also in Bioengineering, Chemistry, and Physics.

I take this opportunity to thank Eva-Mari Aro; Bertil Andersson; all the authors of volume 11; Larry Orr; Jacco Flipsen, Gloria Verhey; and my wife Rajni Govindjee for their valuable help and support that made the publication of *Regulatory Aspects of Photosynthesis* possible.

Readers are requested to send their suggestions for future volumes, authors or editors to me by E-mail (gov@uiuc.edu) or fax (1-217-244-7246).

May 14, 2001

Govindjee
Series Editor,
Advances in Photosynthesis and Respiration
University of Illinois at Urbana-Champaign
Department of Plant Biology
265 Morrill Hall
505 South Goodwin Avenue
Urbana, IL 61801-3707, U.S.A.

Email: gov@uiuc.edu
URL: http://www.life.uiuc.edu/govindjee

Contents

Part III: Biogenesis, Turnover and Senescence

Part IV: Regulation of Carbon Metabolism

Part V: Acclimation and Stress Responses

Preface

The progress in photosynthesis research has been quite dramatic during the last two decades. The Nobel prizes awarded to Peter Mitchel (1978), to Johannes Deisenhofer, Hartmut Michel and Robert Huber (1988), to Rudolf Marcus (1992) and to Paul Boyer and John Walker (1997) have recognized—directly or indirectly—the structural or mechanistic discoveries related to the photosynthetic energy conversion. Actually, photosynthesis may be the first biological process described, not only in molecular terms, but even in atomic terms.

Much of the excitement around photosynthesis is based upon the connection between light and life. Light is an elusive 'substrate' that cannot be handled in the same way as conventional chemical substrates in biological metabolic reactions. Thus, during the last several years a new branch of photosynthesis research has evolved that not only deals with the actual process of energy conversion but also how the photosynthetic process is regulated. This new volume (Volume 11) in the series *Advances in Photosynthesis and Respiration* is devoted to this topic—the regulation of photosynthesis. Many chapters in the book describe this regulation from the view of gene expression and signal transduction. These are essential and topical issues of life sciences in general but the function of two cooperating genomes, the nuclear and plastidic ones, as well as light as a factor in signaling, make these central issues even more complicated when it comes to regulation of photosynthesis. The book covers a wide range of topics about regulation of photosynthesis, occurring under natural daily conditions or when plants are exposed to stress conditions. The 32 chapters, contributed by authors who are leading experts in each particular field, have been divided into six distinct parts in this book.

Part I—*Evolution, complexity and regulation of photosynthetic structures*—gives a comprehensive overview to the different topics in the book and emphasizes the complex evolution and genetics the of photosynthetic machinery.

Part II—*Gene expression and signal transduction*—is devoted to specific light receptors, signal transduction and expression of photosynthetic genes which occur both in the nuclear/cytosol compartment and in the chloroplast. Several chapters are devoted to the transcription machinery and the two plastid RNA-polymerase complexes, to the regulation of photosynthesis genes by redox signaling both in chloroplasts and in the prokaryotic systems, as well as to the sugar sensing mechanisms. Chapters also cover important regulatory aspects imposed by post-transcriptional modifications and degradation of mRNA molecules, and the translational regulation mechanisms operating in chloroplasts.

Part III—*Biogenesis, turnover and senescence*—is closely connected to the question of regulation. The chapters included emphasize how the complicated membrane structures, composed of both nuclear and chloroplast encoded proteins, are synthesized and put together with a correct orientation in a membrane and how they bind an array of different redox ligands and pigments. Several of the chapters are devoted to genetic and biochemical aspects on lipid and pigment biosynthesis, on membrane transporters and on a group of relatively 'new' photosynthetic enzymes that assist in the assembly and turnover of membrane protein complexes. The latter group is represented by an extremely complicated set up of auxiliary enzymes present in very low abundance in connection with the photosynthetic machinery such as desaturases, kinases, phosphatases, isomerases, chaperones and proteases. This section also covers regulation of the latest phases of development in chloroplasts, also called senescence.

Part IV—*Regulation of Carbon metabolism*—Although the previous Volume 9 (*Photosynthesis: Physiology and Metabolism*, eds. RC Leegood, TD Sharkey and S von Caemmerer) in Advances in Photosynthesis already gave a comprehensive view of carbon metabolism and its regulation in plants, some strategically important aspects are also included in this volume. These include chapters that particularly emphasize the use of mutants in studies of photosynthetic regulation and the role of carbonic anhydrases in CO_2 fixation. Two chapters are devoted to the structure and regulatory mechanisms mediated by thioredoxins, universal regulators of photosynthetic enzymes in carbon fixation and which more recently have been discovered to be involved in regulation of a vast number of other photosynthetic processes as well. Posttranslational modification of phosphoenolpyruvate carboxylase by reversible phosphorylation provides an example of a stringent regulatory mechanism of C4 metabolism, a mechanism highly used also in the regulation of other photosynthetic processes.

Part V—*Acclimation and stress responses*—Photosynthetic organisms experience a number of abiotic stress factors related to temperature, CO_2-deficiency and drought. Even light—an absolute necessity for photosynthesis—can be harmful to oxygenic photosynthetic organisms leading to photoinhibitory damages to the photosystems. Several chapters address questions of how stress induced damages can be minimized or repaired. These include the enzymatic mechanisms of reversible thylakoid protein phosphorylation, zeaxanthin-violaxanthin interconversion, redox sensing of reactive oxygen species and consequent acclimation processes, upregulation of stress-related proteins and regulatory modifications in the electron transport pathways of the thylakoid membrane.

Part VI—*Photosynthetic regulation and genomics—methodological implications for the future*—takes into consideration the conceptual and methodological development that is influencing photosynthesis research by the dramatic revolution driven by genomics. Two chapters are devoted to the functional genomics of *Arabidopsis* and *Synechocystis* 6803, the model organisms with a complete genome sequence available and also frequently used in photosynthesis research. Certainly we are only in the beginning of an era where we will address questions in photosynthesis and its regulation by new and extremely efficient approaches. The book sets the scene for this new era of photosynthesis research.

This new volume in *Advances in Photosynthesis and Respiration* is the first that takes a broad perspective on regulatory aspects of photosynthesis and will provide a unique and updated frame of information in a rapidly growing area of photosynthesis research.

As editors of this volume 11 of *Advances in Photosynthesis and Respiration*, we want to express our sincere thanks to all the authors for their time and expertise in contributing their chapters on various regulatory aspects of photosynthesis. Special thanks are also due to Drs. Torill Hundal and Eira Kanervo for their excellent assistance, support and sense of humor in various phases and moments involved in the preparation of this volume. Govindjee is particularly acknowledged for his enthusiasm and encouragement, and Larry Orr for all possible technical help and patience with electronic communication which did not always work that perfectly.

Eva-Mari Aro
University of Turku, Finland
Email: evaaro@utu.fi

Bertil Andersson
University of Linköping, Sweden
Email: bertil.andersson@rek.liu.se

Color Plates

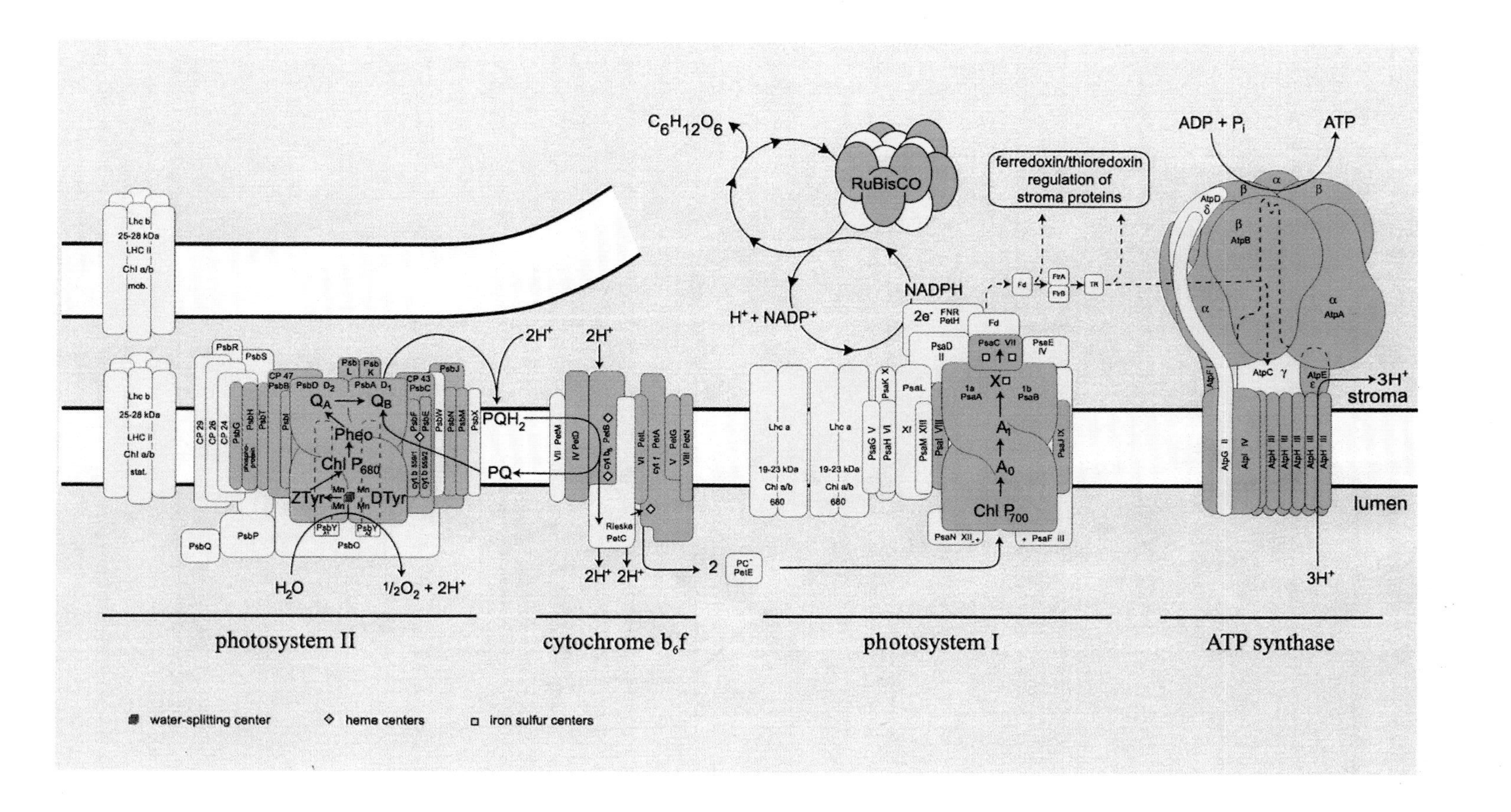

Color Plate 1. Model of the thylakoid membrane and of ribulose bisphosphate carboxylase (RuBisCo) from organisms of the chlorophyll *a/b* lineage of plants, illustrating the composition and the intracellular genetic origin of intrinsic and peripheral polypeptides of the membrane. Plastid-encoded components are presented in green, those of nuclear origin in yellow. PC = plastocyanin, Fd = ferredoxin, PQ/PQH$_2$ = plastoquinon/plastoquinol, FNR = ferredoxin oxidoreductase, Ftr = ferredoxin-thioredoxin oxidoreductase, TR = thioredoxines. Modified from Herrmann (1996). (See Chapter 1, p. 3.)

Eva-Mari Aro and Bertil Andersson (eds): Regulation of Photosynthesis, pp. CP-1–CP-7.

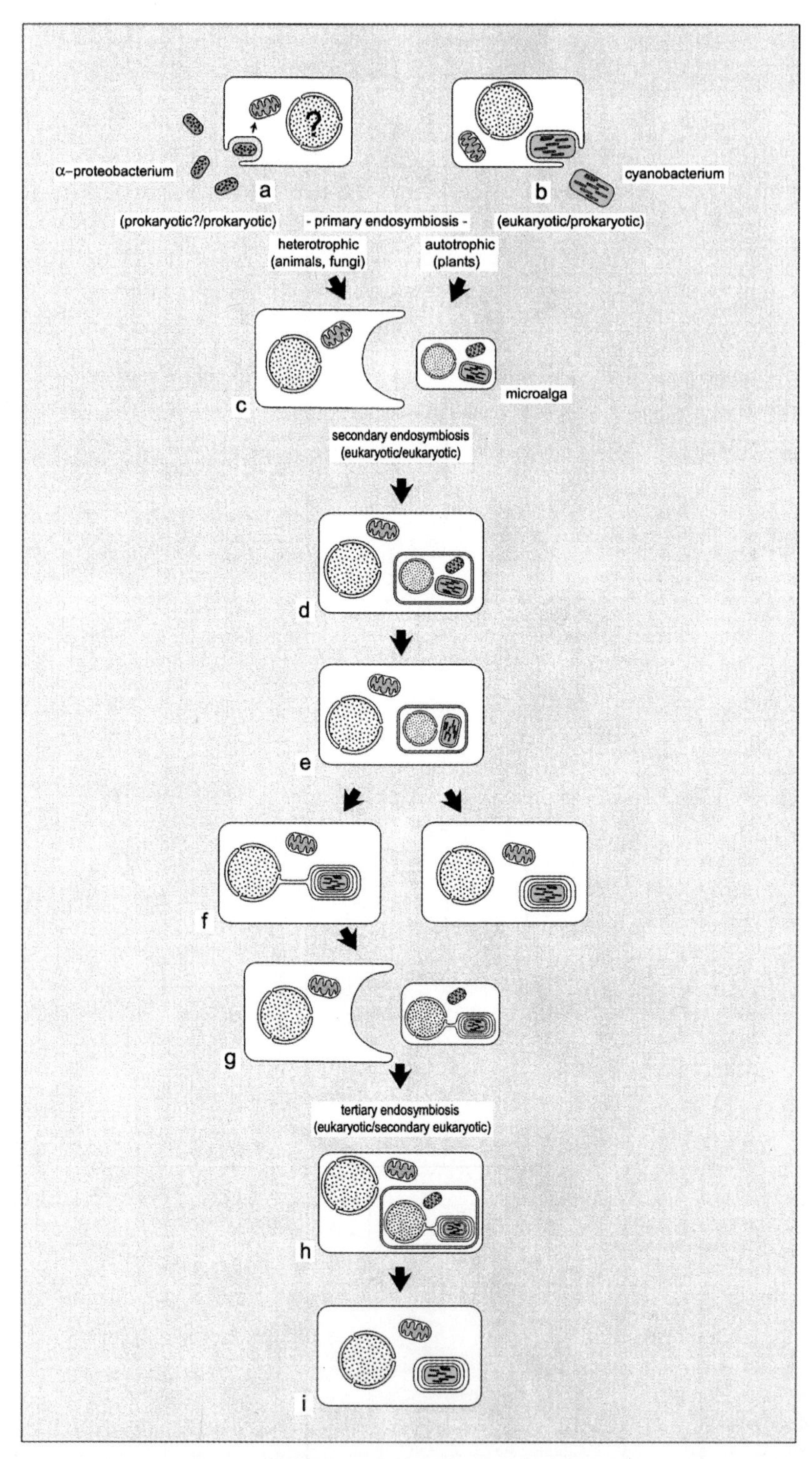

Color Plate 2. Simplified scheme illustrating the principal steps in the origin and evolution of photosynthetic (and respiratory) organelles and of plant cells. Primary mitochondria and plastids resulting from phagocytosis of (a) a heterotrophic prokaryote (α-proteobacterium) into a prokaryotic or protoeukaryotic (see text) host cell, or (b) of a photoautotrophic prokaryote (cyanobacterium) into a eukaryotic cell. (c-f) Complex plastids generated by secondary endosymbiosis, i.e. by engulfment of photoautotrophic microbes (microalgae) into heterotrophic (or autotrophic) cells. Stage (d) is hypothetical. (g-h) Plastids generated by tertiary endosymbiosis. Modified from Herrmann (1997). (See Chapter 1, p. 6.)

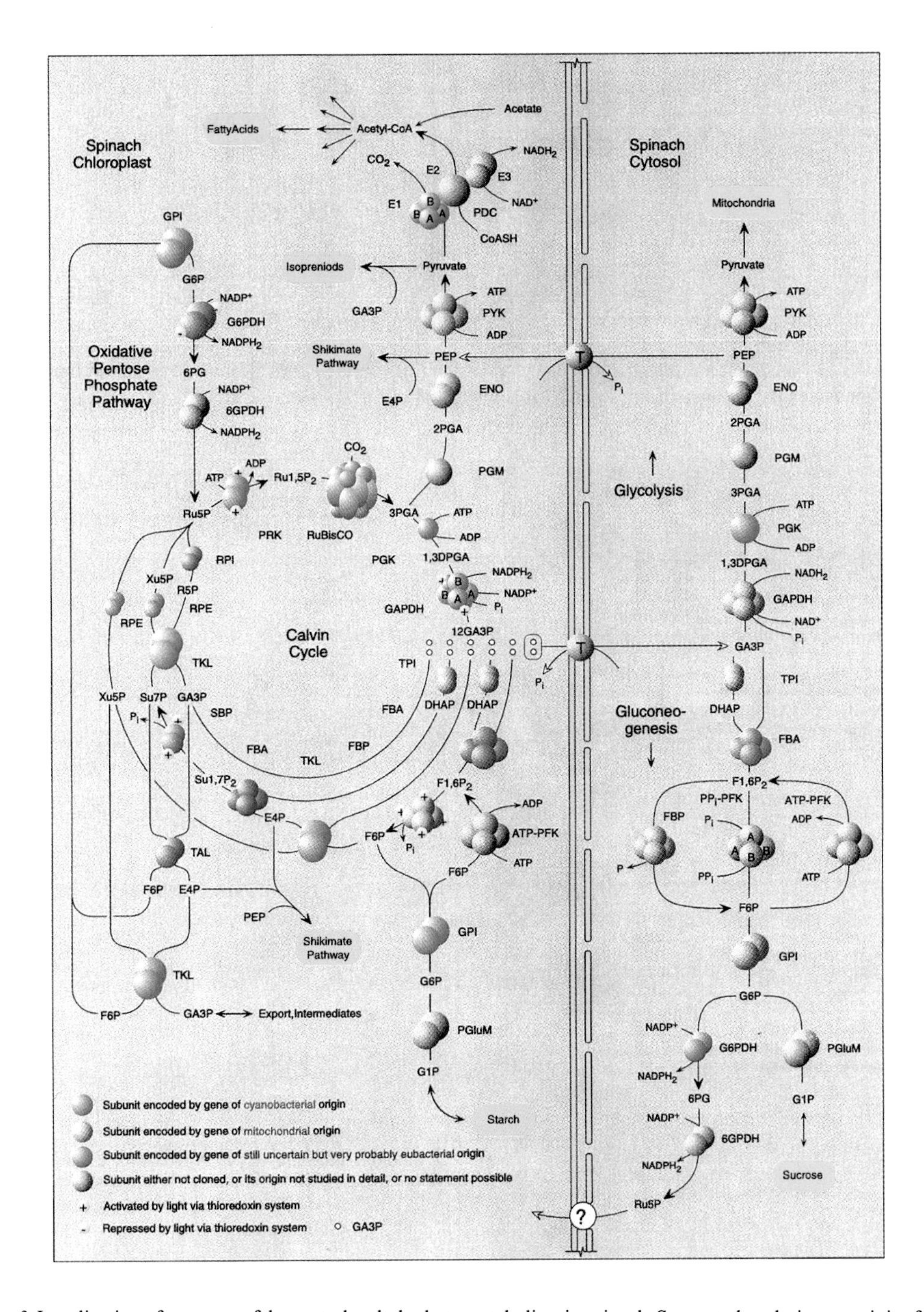

Color Plate 3. Localization of enzymes of the central carbohydrate metabolism in spinach. Suggested evolutionary origins for the nuclear genes are color coded. Enzymes regulated through the thioredoxin system (Fig. 1) are indicated. Enzyme abbreviations are: FBA, fructose-1,6-biophosphate aldolase; FBP, fructose-1,6-bisphosphatase; GAPDH, glyceraldehyde-3-P dehydrogenase; PGK, 3-phosphoglycerate kinase; PRI, ribulose-5-P isomerase; PRK, phosphoribulokinase; RPE, ribulose-5-P 3-epimerase; SBP, sedoheptulose-1,7-bisphosphatase; TLK, transketolase; TPI, triosephosphate isomerase; TAL, transaldolase; GPI, glucose-6-P isomerase; G6PDH, glucose-6-P dehydrogenase; 6GPDH, 6-phosphogluconate dehydrogenase; pGluM, phosphoglucomutase; PGM, phosphoglyceromutase; PFK, phosphofructokinase (pyrophosphate and ATP-dependent); ENO, enolase; PYK, pyruvate kinase; PDC, pyruvate dehydrogenase complex (E1, E2, E3 components); and T, translocator. PDC is a multienzyme complex, but only one set of components is drawn here. Open arrowheads indicate transport rather than conversion. Solid arrowheads indicate physiologically irreversible reactions. For details, Martin and Herrmann (1998). (See Chapter 1, p. 9.)

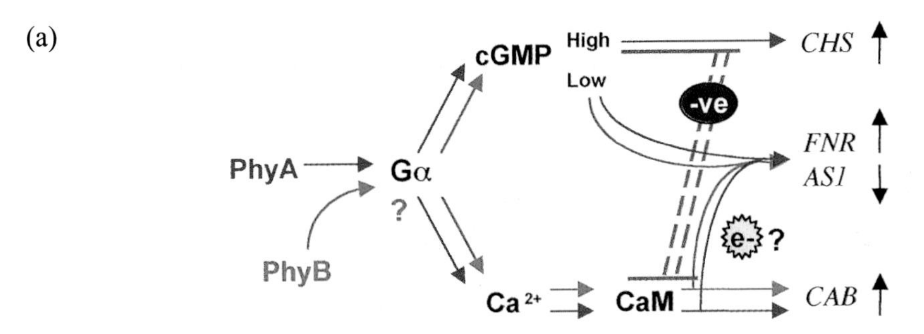

Color Plate 4a. Phytochrome signaling pathways identified by microinjection approaches. PhyA pathways are indicated in red, PhyB in green. Reciprocal negative regulation (denoted -ve) between the cGMP- and calcium-dependent pathways is indicated by dashed lines. Redox signals (denoted e^-) arising from plastoquinone are proposed to regulate the relative activities of the two calcium-dependent pathways. For further details see Sections III and IV. Abbreviations: *CHS*, chalcone synthase; *FNR*, ferredoxin oxidoreductase; *CAB*, chlorophyll *a,b*-binding proteins; *AS1*, asparagine synthetase. (See Chapter 3, p. 54.)

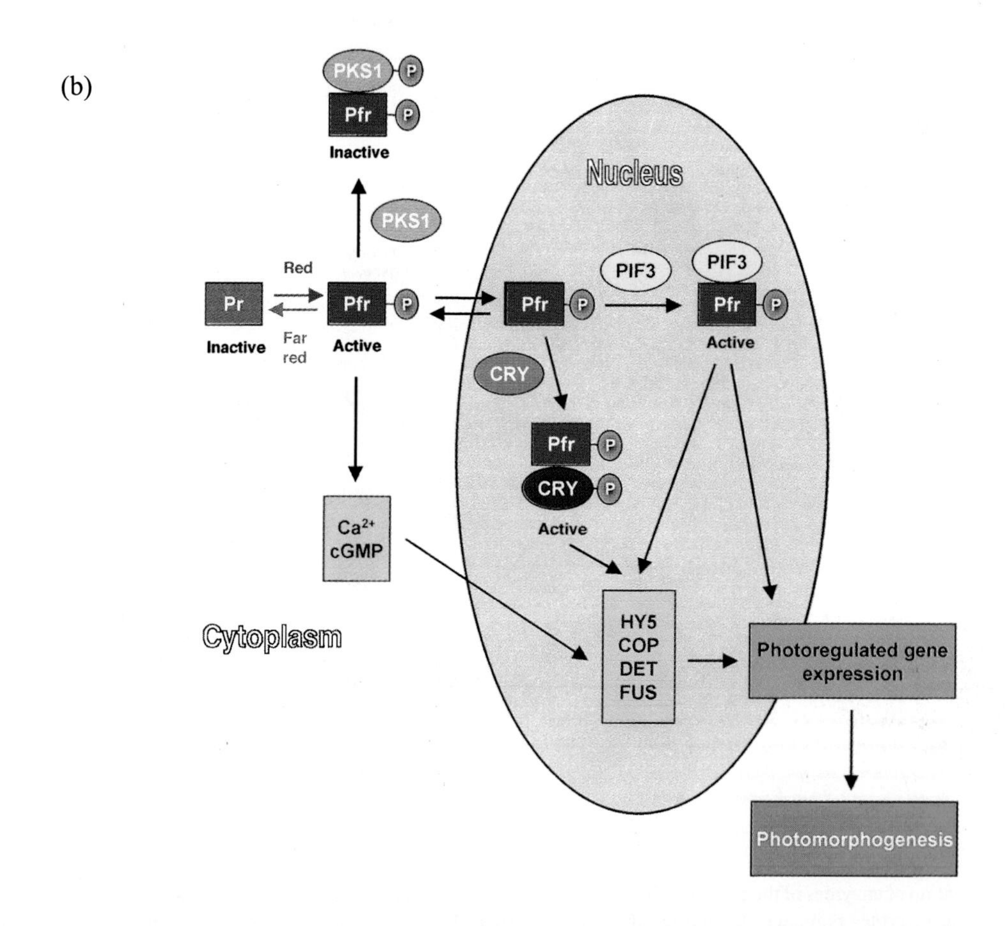

Color Plate 4b. A Simplified Model of Phytochrome Signal Transduction. Activated phytochrome (Pfr) is proposed to be phosphorylated and subsequently (a) to be translocated to the nucleus, (b) to activate calcium- and cGMP-dependent signaling pathways, or (c) to be sequestered away from the active pool by phytochrome kinase substrate 1 (PKS1), which in turn becomes phosphorylated by Pfr. Once inside the nucleus, Pfr can interact with PIF3 and cryptochomes (CRY) to control photoregulated gene expression. Additional nuclear factors such as HY5 and the COP/DET/FUS proteins may act as intermediates for photomorphogenesis either downstream of the cytoplasmic calcium and cGMP signals or downstream of the nuclear-localized phytochromes. The model is simplistic in that it does not account for differences between PhyA and PhyB signaling, e.g. the nuclear SPA1 and FAR1 proteins, that are specific for PhyA signaling (Hoecker et al., 1999; Hudson et al., 1999), are not shown. (See Chapter 3, p. 60.)

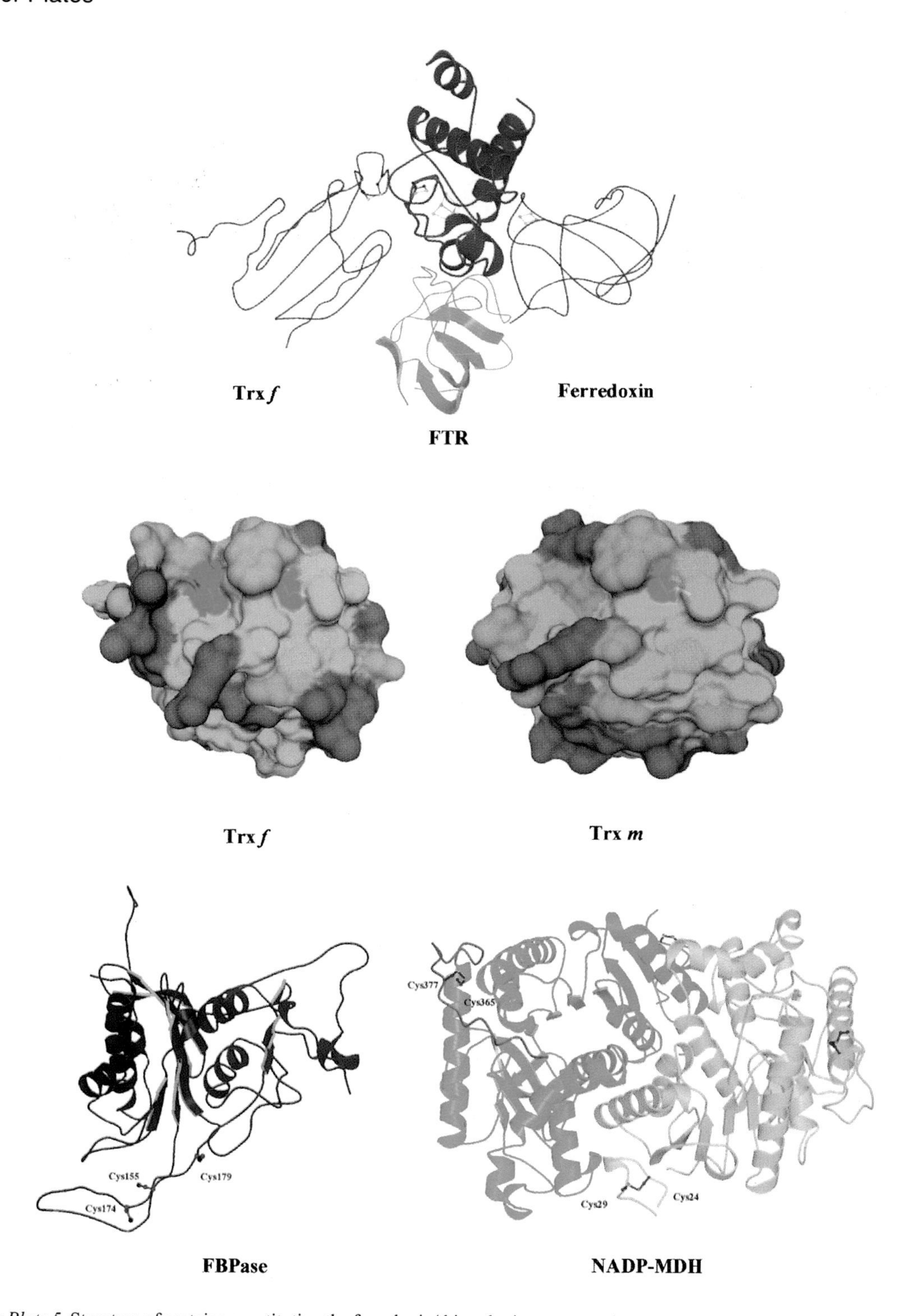

Color Plate 5. Structure of proteins constituting the ferredoxin/thioredoxin system and two target enzymes. The top of the figure shows the FTR (*Synechocystis)* from the side with the variable subunit in green and the catalytic subunit in red. The concave disk shape of the heterodimer allows simultaneous docking of a thioredoxin (on the left) and of a ferredoxin molecule (on the right). Two spinach chloroplast thioredoxins are given in surface view in the middle . The colors (as shown in the color-plate section of this book) represent the following residue type: green—Cys; red—charged (+ or -); blue—polar; yellow—polar; gray—backbone. The bottom structures represent spinach FBPase (only one subunit), with the regulatory loop carrying the Cys, extending out of the core structure, and oxidized sorghum NADP-MDH dimer with the disulfide bonds implicated in regulation. (See Chapter 20, p. 335.)

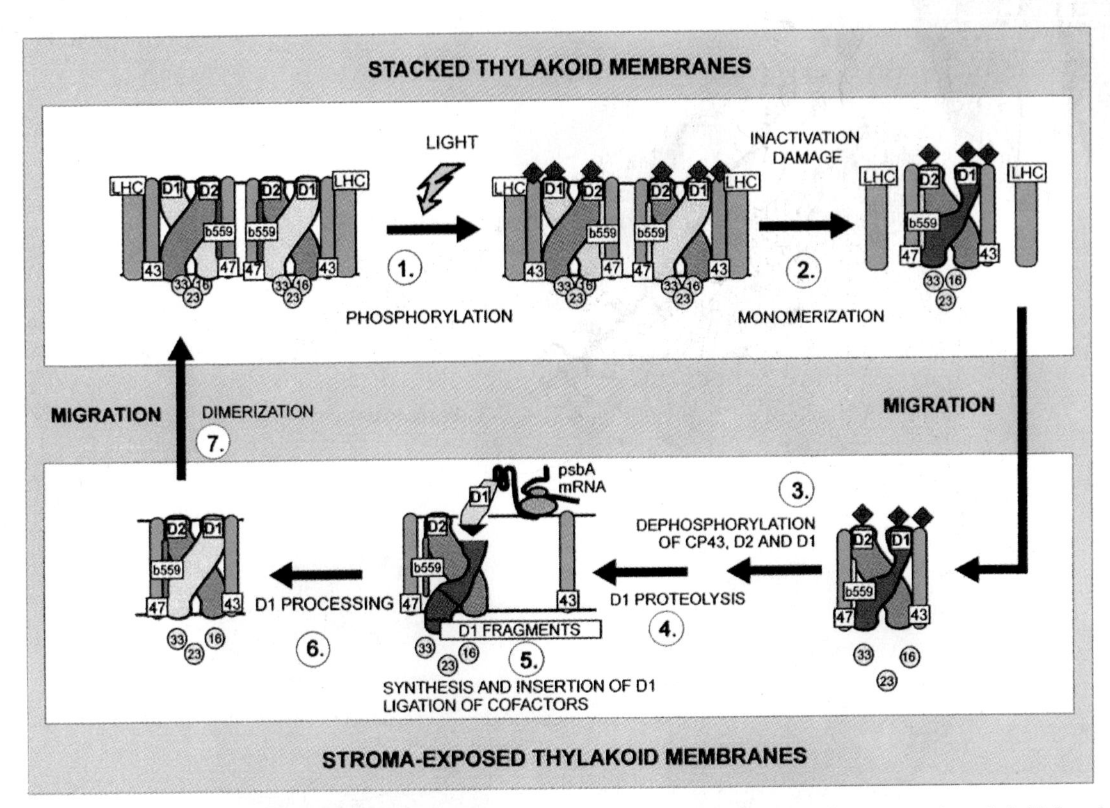

Color Plate 6. Phosphorylation of thylakoid proteins during various stages of Photosystem II photodamage-repair cycle in higher plants. Phosphorylation and migration of the photodamaged PS II from grana region of thylakoids to stroma-exposed membranes for repair are presented. LHCII phosphorylation occurs mainly at low light and is not shown in the figure. The various steps of the cycle, marked from 1 to 7, are described in the text. (See Chapter 23, p. 409.)

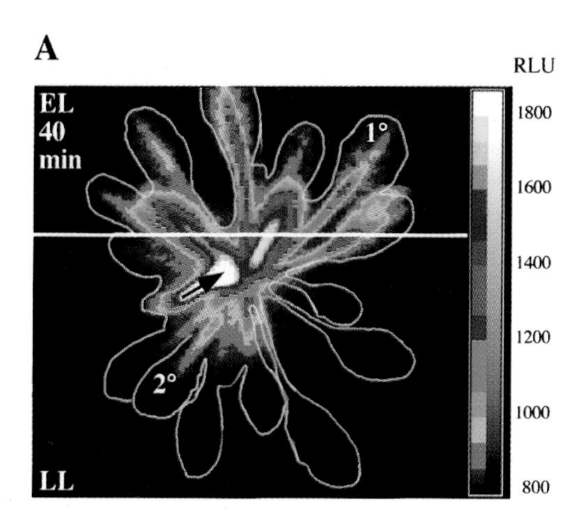

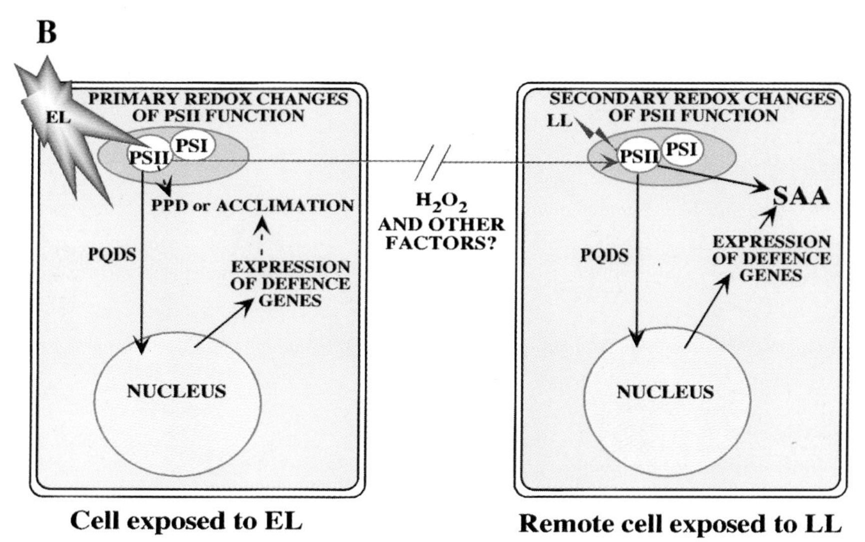

Color Plate 7. Systemic induction of *APX2-LUC* expression in transgenic *Arabidopsis* leaf tissue and the scheme illustrating the systemic acquired acclimation (SAA) mechanism. (A) Image of luciferase activity in relative light units (RLU). A part of the whole rosette (as shown) exposed to EL for 40 min, the arrow indicates the apical region of the rosette. A typical primary (1°) EL-exposed leaf and secondary (2°) LL-exposed leaf are shown (after Karpinski et al., 1999). (B) EL-induced permanent photodamage (PPD) and the SAA mechanism in plants. Induction of expression of the defense genes in the cell is controlled by plastoquinone dependent signaling (PQDS). SAA = systemic acquired acclimation at 200 μmol of photons $m^{-2} s^{-1}$, EL = 2500 μmol of photons $m^{-2} s^{-1}$, LL = 200 μmol of photons $m^{-2} s^{-1}$. (See Chapter 27, p. 481.)

Chapter 1

Thylakoid Biogenesis and Dynamics: The Result of a Complex Phylogenetic Puzzle

Reinhold. G. Herrmann*
Botanisches Institut der Ludwig-Maximilians-Universität, Menzinger Str. 67, D-80638 München, Germany

Peter Westhoff
Institut für Entwicklungs- und Molekularbiologie der Pflanzen der Heinrich-Heine-Universität, Universitätsstr. 1, D-40226 Düsseldorf, Germany

Summary

Biogenesis, maintenance, and adaptation of the thylakoid system in photosynthetic organelles are embodied in the genetic machinery of the plant cell. Plant cells are descendants of endosymbioses. Their genomes are compartmentalized and the result of a complex restructurating of the genetic potentials of (depending on the organism) three to five symbiotic partner cells during evolution that led to a common metabolism and a common inheritance in the entity. It appears that much of the thylakoid biogenesis and dynamics can be understood from the history of the plant cell. Inter-endosymbiotic genome restructuration included loss, gain, and intracellular transfer of genetic material. It was accompanied (i) by the generation of an integrated genetic system (rather

*Email: herrmann@biologie.botanik.uni-muenchen .de

Eva-Mari Aro and Bertil Andersson (eds): Regulation of Photosynthesis, pp. 1–28.

than a nucleus and semiautonomous organelles) that in its entirety is spatiotemporally regulated, (ii) by a massive intermixing of structural genes, (iii) a fundamental change in expression signals in the entire system which included the evolution of an exquisite set of regulatory mechanisms that operate in concert with basically ancient regulatory circuits originating in the organelle ancestors, and (iv) by the establishment of nuclear regulatory dominance that is found at all levels of regulation. Most conspicuous are fundamental changes of the transcription machinery in all three subgenomes, in the establishment of an elaborate posttranscriptional RNA modification system in chloroplasts that prokaryotes are largely lacking, and in sophisticated protein trafficking and assembly devices. The genetics of the photosynthetic machinery was fitted into the respective genetic programs for multicellular and terrestrial plants that developed new biochemistries including those for a spatiotemporal morphogenetic potential.

I. Introduction

Life on earth ultimately depends on energy derived from the sun. Photosynthesis is globally the only process of biological importance that can harvest this energy. It provides energy, organic matter, in its oxygenic version also oxygen, for nearly all biotic processes, and represents the only renewable energy source of our planet. The photosynthetic process is remarkably effective since it captures almost all the energy of the light it absorbs. Life would have been different, or probably even extinct, without the utilization of the external energy source through biological photosynthesis. If we can trust fossil records, this process was invented as early as ≥3.85 billion years ago, no more than 600–700 million years after the creation of the planet Earth. Photosynthesis has undergone crucial changes with far-reaching global consequences, notably the generation of the Photosystem II assembly which is capable to use solar energy to catalyze the breakdown of water to reducing equivalents and molecular oxygen. This caused the change from a reducing to an oxidizing environment. To date, the amount of energy stored by photosynthesis is enormous. Estimates suggest that more than 10.000 million tons of carbon are transformed annually into carbohydrate and other forms of organic matter.

In plants, photosynthesis takes place in a distinct cellular organelle, the chloroplast (rhodoplast, cyanoplast, phaeoplast etc.). Typical photoautotrophic organelles are double membrane-bound, those of secondary or tertiary endosymbiotic origin (Section II) triple- or quadruple membrane-bound entities. Their thylakoid system and stroma house the functional components required both for the light-driven and enzymatic reactions of the process. Thylakoid membranes are unique biomembranes. In oxygenic photosynthesis, they are specialized of converting inorganic substrates (CO_2 and H_2O) with solar energy to utilizable high-energy chemical metabolites, initially in the forms of ATP and NADPH. These compounds in turn provide the necessary energy and reduction equivalents for the 'dark' reactions, i.e. the fixation of atmospheric CO_2 in carbohydrates. Several different multisubunit protein complexes are inserted in the lipid bilayers of thylakoid membranes in a vectorially oriented way (Fig. 1; reviewed in Herrmann et al., 1991; Andersson and Barber, 1994). These supramolecular protein assemblies, the Photosystems I and II, each with an ensemble of light-collecting antenna, the cytochrome b_6f complex, the ATP synthase, and the NAD(P)H dehydrogenase, bind and organize pigments and other non-proteinaceous cofactors. Together with mobile electron carriers in the stroma (ferredoxin, FNR), the lumen (plastocyanin), and the lipid bilayer (plastoquinone) they cooperate in the conversion of radiant energy, and provide the link to the stromal metabolism (Fig. 1).

The photosynthetic process is anchored in the genetic system of the plant cell. Plant cells and their genomes are beyond doubt the result of endocytobioses (Fig. 2; Woese, 1987; Woese et al., 1990). Chloroplasts, as mitochondria, are descendants of endosymbiotic cells and have preserved remnants of the ancestral genomes which is a second fundamental feature of photoautotrophic organelles. They contain not only DNA, but also the mechanisms to maintain this information and to convert it into function.

Abbreviations: CFo – coupling factor membrane moiety of ATP synthase; DPE – downstream promoter element; ER – endoplasmatic reticulum; EST – expressed sequence tag; FNR – ferredoxin NAD(P)H oxidoreductase; GAPDH – glyceraldehyde-3 phosphate dehydrogenase; Inr – initiator; NEP – nuclear-encoded DNA-dependent RNA polymerase in chloroplasts; PCR – polymerase chain reaction; PEP – plastid-encoded DNA-dependent RNA polymerase; RT-PCR – reverse transcriptase PCR; SRP – signal receptor particle

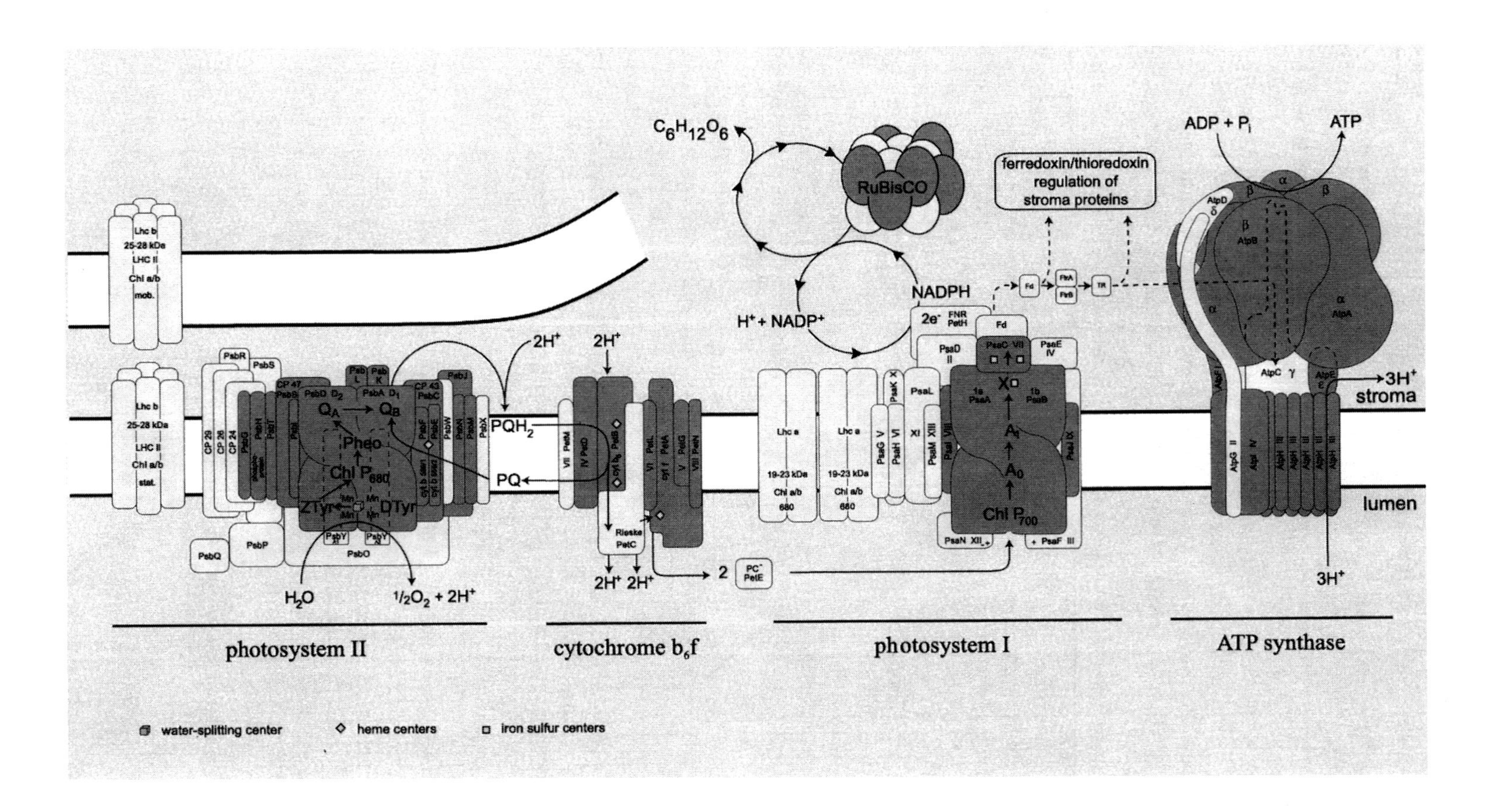

Fig. 1. Model of the thylakoid membrane and of ribulose bisphosphate carboxylase (RuBisCo) from organisms of the chlorophyll *a/b* lineage of plants, illustrating the composition and the intracellular genetic origin of intrinsic and peripheral polypeptides of the membrane. Plastid-encoded components are presented in green, those of nuclear origin in yellow. PC = plastocyanin, Fd = ferredoxin, PQ/PQH_2 = plastoquinone/plastoquinol, FNR = ferredoxin oxidoreductase, Ftr = ferredoxin-thioredoxin oxidoreductase, TR = thioredoxines. For gene nomenclature see Legen et al. (2001). Modified from Herrmann (1996). See also Color Plate 1.

Molecular data have verified that oxygenic photosynthesis originated in the eubacterial domain (Woese et al., 1990), initially probably in an anoxygenic form (Blankenship, 1992). Different from photoautotrophic prokaryotic cells, therefore, plant cells contain a *compartmentalized genetic system*. This set-up developed through a massive intracellular restructuring of the ancient endosymbiotic genomes (summarized in Herrmann, 1997, 2000). The rearrangements led to an integrated genome and to a dispersal of genes for complex chloroplast structures, such as the thylakoid membrane (Fig. 1), among two cellular subgenomes in the organelle and the nucleus. The operation of the partite genetic machinery in the plant cell and the dual genetic origin of organelle structures add substantially to the regulatory complexity of the system. Different from corresponding (photoautotrophic) prokaryotes, it required the establishment of intercompartmental regulatory circuitries, since the delivery of the components from two subgenomes and sources of protein synthesis in the cytosol and plastid respectively, has to be coordinated during biogenesis (Sections II and III).

Symbiosis is usually not considered to be a major driving force in evolution. However, molecular approaches have indisputably established that it was one of *the* driving forces that shaped the evolution of photosynthetic eukaryotes. Symbiotic events not only generated one of the three domains of Life (Woese, 1987; Woese et al., 1990). They permitted also the development of true multicellular life, in combination with oxygenic photosynthesis also of terrestrial life (Herrmann, 1997). Multicellular organisms display a division of labor, in that they are organized as complex societies of cells and tissues, each with distinct functions. Metabolism and ontogeny of the different parts of such organisms are tightly interconnected. For instance, in plants, the photosynthetic products that are generated in the green ('source') tissue have to be efficiently translocated to other parts of the individuum, the non-green ('sink') organs, where they are used for growth or stored as e.g. sucrose, fats, starch or proteins. Effective transport of the photosynthate to sink organs maintains high photosynthesis rates. Multicellularity arose approximately 1.3 billion years ago, first probably with thallophytic plants (red algae), and developed gradually. Again, this step and later the transition from the aqueous to a gaseous environment, approximately 600 million years ago, were accompanied by the generation of new biochemistries and regulation schemes (Section IV). Genetic programs for the synthesis, maintenance and adaptation of the photosynthetic apparatus in multicellular organisms again require an additional quality of regulatory information since the integrated genetic system has not only to be expressed in time and quantity, but also spatially to account for organ and tissue specificity.

Historically our understanding of the photosynthetic process has progressed continuously, often connected with technical progress. During the past two decades, photosynthetic research entered a new area with the intensified application of the technologies of molecular biology as well as X-ray and two-D ultrastructural analysis of bacterial photosynthetic reaction centers and antenna systems. These approaches have extended its basis from the level of biochemical, biophysical and physiological aspects to that of molecular characteristics. Their principal focus has been on functional, structural and biogenetic aspects onto which the energy-transducing process is based, notably on the analysis of (1) structures, (2) structure/function relationships, and (3) on the biogenesis of the photosynthetic machinery. The progress has been significant. The photosynthetic protein complexes can now be described at a highly refined structural and functional level, and basic outlines for the controlling levels of their biogenesis emerged. It turned out that their biogenesis is one to two orders of magnitude more complex than their mere structures in terms of requirement of genetic information, since regulation occurs at all levels studied, and these are intermingled. The functional consequences of genome compartmentation and of multicellularity in the biogenesis of the photosynthetic membrane are obvious at all levels of regulation. It is no overstatement to say that the thylakoid system belongs to the best characterized complex biological structures.

During the past decade molecular approaches have again generated two further areas of photosynthesis research that (4) on membrane dynamics and (5) on the evolution of the photosynthetic process. These fields have not yet received the scientific attention they deserve, reminiscent to the hesitant application of molecular biology in the field in the late seventieth, although both are of comparable importance. Thylakoid membranes are highly dynamic and vulnerable structures. To cope with a changing environment and to avoid serious consequences for their life due to impairment of photosynthesis, photoautotrophic organisms have evolved sophisticated protective systems to ensure high photosynthetic efficiency, despite highly fluctuating and even stressful conditions. These are often accompanied by marked changes in gene expression, membrane structure, and basic metabolism. The ways how this is achieved differ from those of animals last not least because plants cannot escape unfavorable milieus (Weis and Berry, 1988; Schöffl et al., 1998). The molecular details of plant signal transduction pathways, and which elements involved resemble or differ from those in animals and prokaryotic cells, are not yet clearly understood. However, one of the principal features of maintenance, acclimation, and protection of the thylakoid system that appeared is the complexity and sophistication of its regulation (Section V). The identification of the individual processes, the underlying mechanisms, the components involved, and ultimately their interaction and integration into the photosynthetic and cellular metabolom belong therefore to the central challenges of current photosynthetic research and of organelle biology in general.

Comparably, the static, and more recently the functional molecular phylogenetic-comparative approach have caused a paradigm change in the way we view living matter, its genomes and the evolution of biogenetic potentials, such as that for photoautotrophy. The approach is based on the assumption that each change in an information-storing molecule and its perpetuation or relationship to other changes is a document for evolution (Zuckerkandl and Pauling, 1965; Woese et al., 1990). Unlike phenotypic approaches, 'molecular phylogeny' provides therefore a direct means to understand how genomes, biological structure, and pathways evolved, and will permit to deduce a natural and global genealogical system. The massive changes that occurred during the evolution of plant genomes had profound consequences for the photosynthetic machinery. Intriguing phylogenetic aspects of photosynthesis that lead to a 'functional molecular phylogeny' have emerged, and it becomes more and more evident that the history of the plant genome explains much of the design, biogenesis and regulation of photosynthetic structure. The insight into the phylogeny of photosynthesis, especially with regard to biogenetic and functional aspects, represents one of the most fascinating chapters in the field presently.

In discussing the outlined aspects, we will focus on general principles rather than on details and integrate some dispersed material from our previous reviews in order to present a unified view of the biogenetic complexity of the photosynthetic machinery and its evolution.

II. Aspects of Chloroplast and Plant Genome Evolution—Plant Genome Structure

Biogenesis, function and modification of the thylakoid system, as of the entire chloroplast, are embedded into the genetic context of the respective plant or plant cell. Although the basic oxygenic photosynthetic machinery is phylogenetically conserved and structural and functional comparisons suggest that both photosystems share a common ancestry (Nitschke and Rutherford, 1991) as well as ancestry with the purple bacterial, green sulfur bacterial and heliobacterial photosynthetic reaction centers (Blankenship, 1992), the biogeneses of the photosynthetic machineries in different organisms differ markedly. Before considering the generation and dynamics of that device, it appears therefore appropriate to outline relevant basic elements in the structure of the plant genome and its evolution. The birth of the plant cell as a new cellular entity appeared with the symbiotic incorporation of a (photosynthetic, oxygen-producing) cyanobacterial cell into a mitochondria-containing, heterotrophic cell (eukaryotic/prokaryotic endosymbiosis; Fig. 2b). This event followed the α-proteobacterial endocytobiosis probably some 400 or 500 million years later that was ancestral to mitochondria and hydrogenosomes, their DNA-less derivatives in anaerobic eukaryotic microbes (Fig. 2a).

Available evidence suggests that the evolution of photosynthetic organelles, plant genomes and plant cells was by far more complex than generally assumed, although all plastids, as mitochondria, appear to have originated in a single, primary symbiosis. A monophyletic ancestry of plastids has been inferred primarily from the remarkable similarity between corresponding cyanobacterial and plastid gene and operon structures, nuclear genes encoding chloroplast proteins that plastid operons are missing, a high degree of sequence conservation, and the almost identical design of the basic photosynthetic machinery (but not of antenna systems) in the various plant lineages (Kowallik, 1997; Martin et al., 1998). The resulting autotrophic eukaryote diverged into Glaucocystophyta, red algae, and 'green algae' and their derivatives (Martin et al., 1998). Different from

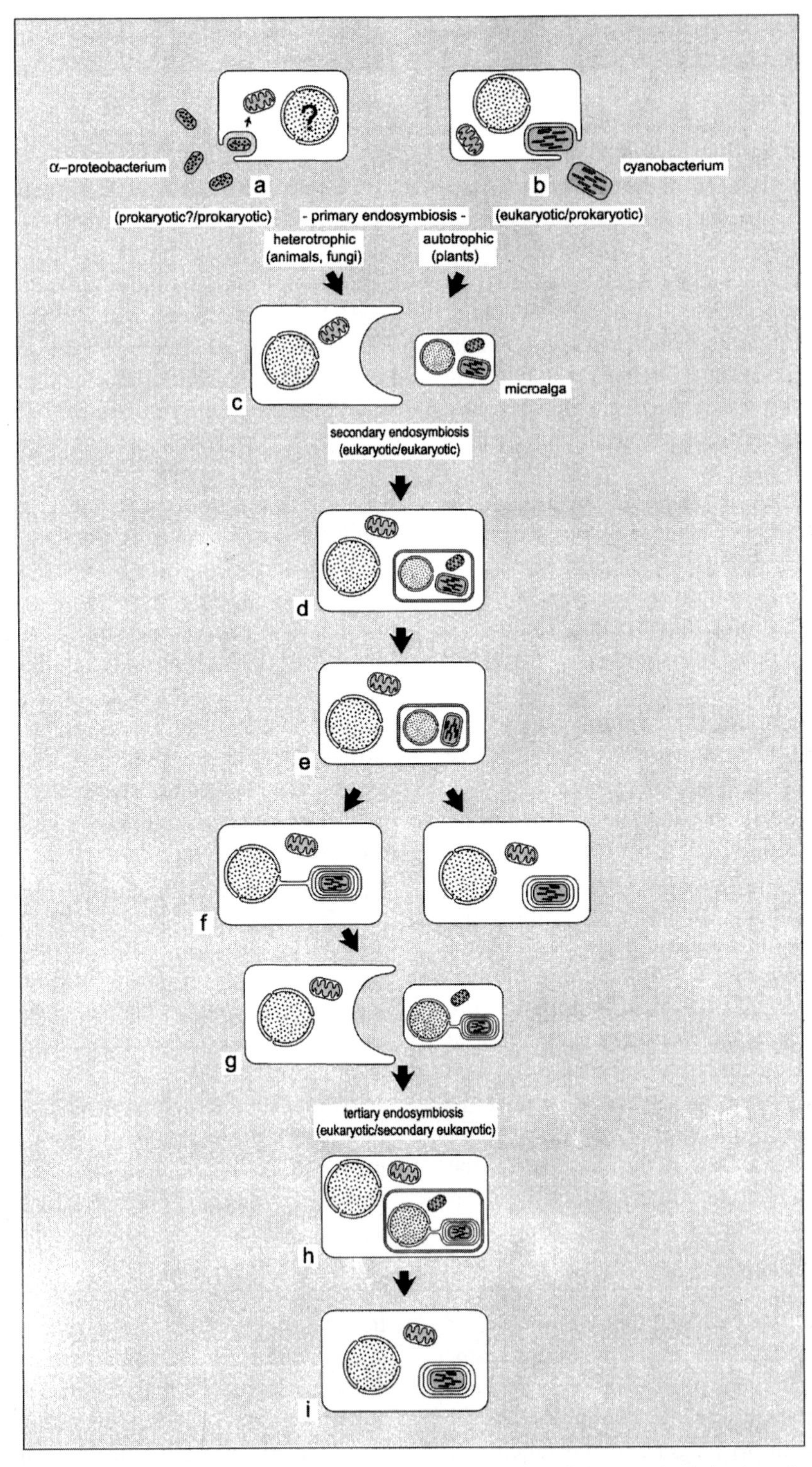

Fig. 2. Simplified scheme illustrating the principal steps in the origin and evolution of photosynthetic (and respiratory) organelles and of plant cells. Primary mitochondria and plastids resulting from phagocytosis of (a) a heterotrophic prokaryote (α-proteobacterium) into a prokaryotic or protoeukaryotic (see text) host cell, or (b) of a photoautotrophic prokaryote (cyanobacterium) into a eukaryotic cell. (c-f) Complex plastids generated by secondary endosymbiosis, i.e. by engulfment of photoautotrophic microbes (microalgae) into heterotrophic (or autotrophic) cells. Stage (d) is hypothetical. (g-h) Plastids generated by tertiary endosymbiosis. Modified from Herrmann (1997). See also Color Plate 2.

animal and fungal genomes, plant genomes may however trace back to associations of four or even five cells, because photosynthetic organelles, different from mitochondria, have also arisen through the incorporation of unicellular eukaryotic cells into heterotrophic (Fig. 2c) or perhaps even autotrophic host cells. Unlike primary eukaryotes, organisms with such complex photosynthetic organelles, which are surrounded by three or four envelope membranes (Fig. 2e and f), each probably of different origin and history, arose *polyphyletically*, at least seven times with both different, non-photosynthetic hosts and different engulfed microalgae. The majority of the photoautotrophic protophytic and thallophytic photoautotrophs, such as Cryptophyta, Chlorarachniophyta, Euglenophyta, brown algae, diatoms, Dinophyta, Haptophyta, Xanthophyta and Chrysophyta are the result from such serial, secondary (Fig. 2c and d, eukaryotic/eukaryotic), in the case of some fucoxanthin-containing Dinophyta even from tertiary (Fig. 2g–i, eukaryotic/secondary eukaryotic) endosymbioses. It is commensurate with this deduction that three remotely related lineages, Cryptophyta, Chloroarachniophyta, and some Dinophyta, (still) contain a vestigial nucleus of the endocytobiotic microalga, the nucleomorph, with three residual chromosomes in the known cases, and a residual cytosol with an additional category of functional 80S ribosomes, as expected in the space between the second and the third envelope layer (Sitte, 1987; Maier et al., 1991; McFadden and Gilson, 1995; Maier et al., 2000). Evidently, the integration of the eukaryotic endosymbionts and their conversion to photosynthetic organelles caused an additional biogenetic complexity.

The initially free-living three to five cells, each possessing an own metabolism and an own inheritance, developed into an entity with a tri- or quadripartite genetic system, represented by nucleus/cytosol, the two energy-transducing organelles and, where existing, the nucleomorph. This partite genome encodes a *common* metabolism and a *common* inheritance. *Consequently, a principal aspect in the evolution of plant genomes resides in the streamlining of the symbiotic genetic potentials, that is, in an intriguing capacity of restructuring genetic matter.* Besides genetic compartmentation, this is a second fundamental feature in the evolution of the plant kingdom with relevance to photosynthesis. As will be outlined below, work on the photosynthetic membrane has shown that the functional consequences of genome rearrangement are massive and obvious at all levels of regulation, with different mechanisms and over a wide range of time scales.

DNA rearrangements in the symbiotic cellular entity must have been frequent, with diverse kinetics and time scales, and unexpectedly complex (Herrmann, 1996, 1997; Martin and Schnarrenberger, 1997; Martin and Herrmann, 1998; Douglas et al., 2001). Remarkably, intracellular DNA translocations occurred between all genetic compartments, although the DNA net flow was preferentially from the organelles to the nucleus. A possible exception is the direct DNA translocation from mitochondria to plastids, which has not yet been demonstrated, perhaps because the changes in mitochondrial genomes had already too much progressed when the symbiotic cyanobacterium entered the cell. Indicative are repeated convergent losses and functional transfers of genes in individual plant lineages (Martin et al., 1998), or a substantial fraction of organelle-derived promiscuous DNA (Ellis, 1982), generally non-functional DNA that has been detected to a considerable extent in the nucleus, or of plastid and nuclear DNA sequences in mitochondria (Stern and Lonsdale, 1982; Brennicke et al., 1993; summarized in Herrmann, 1997). For example, plastid *ndh* genes, encoding proteins with homology to subunits of complex I of the mitochondrial respiratory chain, have been repeatedly lost, e.g. in the red alga *Prophyra* with its relatively large plastid chromosome and in the unrelated gymnosperms. The latter differ in this respect from other plants of the chlorophyll *a/b* lineage. That genomes can naturally incorporate foreign DNA has been amply demonstrated for cyanobacteria (Shestakov and Khyen, 1970).

Although the causes for genome restructuring and the underlying mechanisms are not yet well understood, the outlines what happened emerged during the past few years (Herrmann, 1997; Martin and Herrmann, 1998). The genome changes were substantial and rest on *four* principal processes that occur simultaneously: loss of function, intercompartmental DNA transfer, and gain of function, either by restructuring or functional re-dedication of existing genetic information, or by lateral transfer of genes (Herrmann, 1997). All these processes are well established. They caused not only the complete restructuring and intermixing of the genetic potential of the partner cells but also a fundamental change in expression signals in all three cellular subgenomes. This was accompanied by the establishment of nuclear regulatory dominance that operates at almost all levels of regulation, transcriptionally, posttranscriptionally, translationally, and posttranslationally (Sections III and V). It is evident already in the mere fact of gene translocations. Today, plastid chromosomes encode only a fraction, generally in the order of 5% or less, of their ancestral genome that may have been in the order of 2,000–16,000 genes based on genome sizes, e.g. of the cyanobacterium *Synechocystis* (Kaneko et al., 1996). They specify relatively few proteins, between approximately 50 (in *Euglena*) and 200 (in red algae; Martin et al., 1998). On the other hand, chloroplasts house probably in the order of 2000–3500 proteins (Martin and Herrmann, 1998; The Arabidopsis Genome Initiative, 2000). The overwhelming majority of the polypeptides that reside in the organelle are therefore nuclear gene products.

A particularly interesting aspect of nucleomorph-containing plant cells is the fate of the genes that were translocated from the primary endosymbiont to the nucleomorph ancestor in the once free-living microalgae (Fig. 2c and e; Herrmann, 1997; Deane et al., 2000). The now available sequence of the entire nucleomorph genome of the cryptomonad *Guillardia theta* establishes that this genome contains only few, in the order of 30, genes for structural components of the primary organelle (Douglas et al., 2001) and suggests that gene translocations could have been sequential (Deane et al., 2000), i.e. from the primary endosymbiont to the (microalgal) nucleus ancestral to the nucleomorph, and subsequently from the nucleomorph (initially from the nucleus of the secondary endosymbiont) to the nucleus of the secondary host cell. This would preclude the alternative that the secondary endosymbiont replaced a preceding plastid and adapted the genetic information translocated to the final host nucleus from that hypothetical primary organelle.

The approach of molecular phylogeny has shown that important genome changes occurred predominantly at the unicellular level, often repeatedly and gradually (Martin et al., 1998). This allows several fundamental inferences. For instance, much information for spatial regulation in multicellular plants had to be generated *after* gene translocation (Sections III and IV). In general, genes for single- or simple-chain enzymes of the organelle should not provide special problems for translocation. Consequently, they were probably transferred early from the organelles and are all of nuclear origin today. The massive protein import from the cytosol into the organelle is a consequence of gene translocation, since the functionality of an organelle depended upon the return of the product of the translocated gene (Weeden, 1981) or of an equivalent gene. This includes the generation of import machineries. For an organelle location this required also the acquisition of appropriate sorting signals, generally N-terminal extensions (Section III).

A substantial body of data establishes that the product-return-hypothesis is in fact only part of the story and hence provides an only incomplete picture of what happened. In principle, a gene, when functionally translocated from an organelle, can have three fates. It may gain a transit peptide or not, and in the former case it may acquire a transit peptide with correct, but also with a different sorting specificity. Without a sorting sequence, a polypeptide would end up in the cytosol, provided its gene is properly expressed. Three relevant observations can be mentioned in this context. Several nuclear-coded enzymes of the chloroplast are of α-proteobacterial (mitochondrial) origin (the reverse has not been demonstrated yet), and functionally related isozymes that reside in different cellular spaces, e.g. of the Calvin cycle or glycolysis, such as GAPDH, may be of different phylogenetic origin (Fig. 3; Martin and Schnarrenberger, 1997; Martin and Herrmann, 1998). Furthermore, cladistic analysis established that the majority of the enzymes in the cytosol and some of the chloroplast stroma that are nuclear-coded are of α-proteobacterial or cyanobacterial origin (Fig. 3). If the endosymbiotic host cell was an archaebacterium, as the nuclear system and more recent data suggest, it did not contribute equivalently to the cell and organelle metabolism. The eubacterial basic metabolism of the eukaryotic cell indicates that principal gene losses occurred predominantly from the host cell (Martin and Schnarrenberger, 1997). This would mean that gene translocations have replaced most of the host metabolic set-up, and that the plant cell contains more of the endosymbiont (plastid and mitochondrial) genomes than is reflected in the size of the residual organelle genomes.

Due to biogenetic complexity and different from stromal and cytosolic enzymes, the transfer of genes for complex organelle structures had to occur gradually. This is the most likely explanation that such structures are of dual genetic origin (Herrmann, 1997). In this instance, the gain of transit peptides had to be compartment-specific, different from those

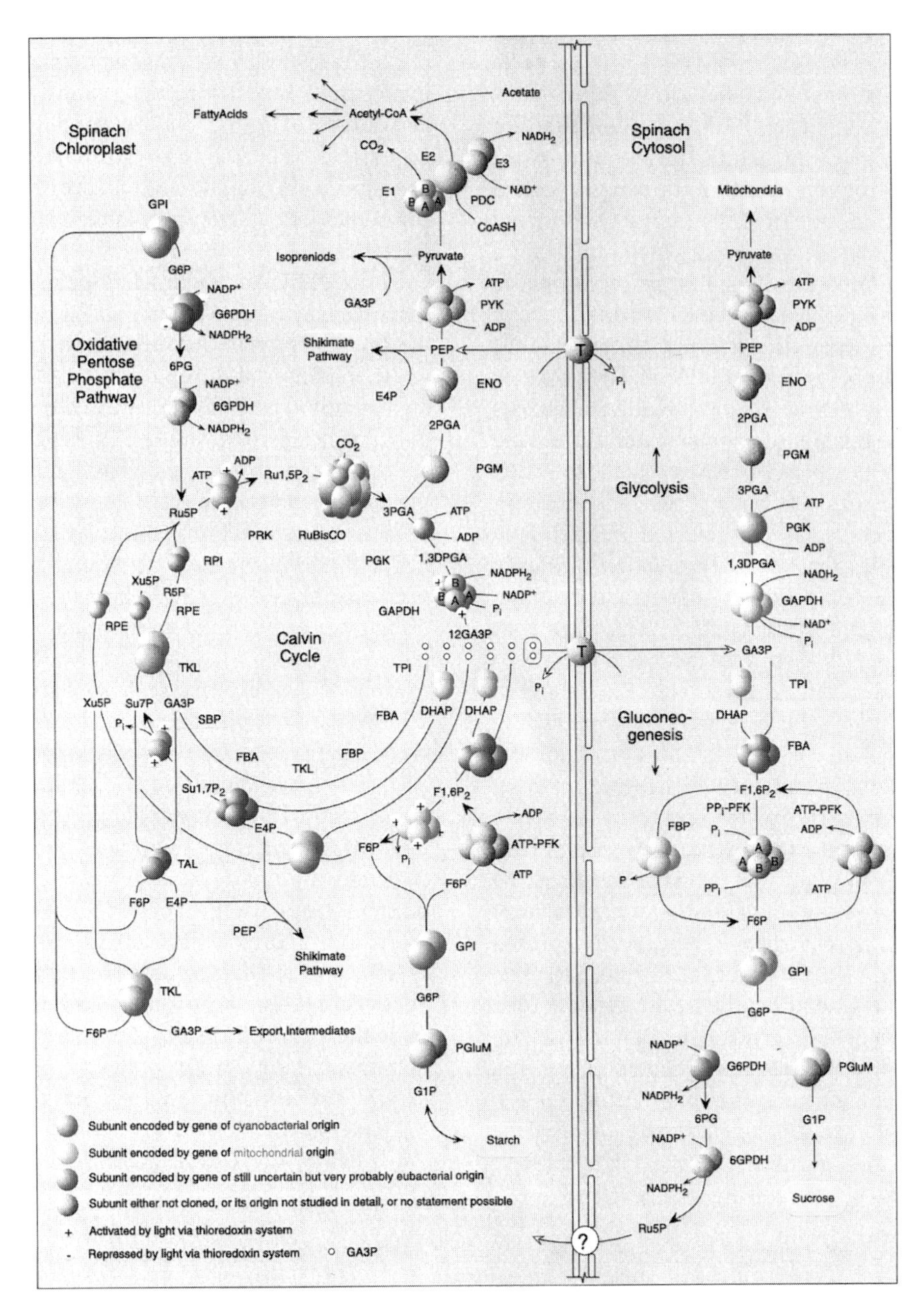

Fig. 3. Localization of enzymes of the central carbohydrate metabolism in spinach. Suggested evolutionary origins for the nuclear genes are color coded. Enzymes regulated through the thioredoxin system (Fig. 1) are indicated. Enzyme abbreviations are: FBA, fructose-1,6-biophosphate aldolase; FBP, fructose-1,6-bisphosphatase; GAPDH, glyceraldehyde-3-P dehydrogenase; PGK, 3-phosphoglycerate kinase; PRI, ribulose-5-P isomerase; PRK, phosphoribulokinase; RPE, ribulose-5-P 3-epimerase; SBP, sedoheptulose-1,7-bisphosphatase; TLK, transketolase; TPI, triosephosphate isomerase; TAL, transaldolase; GPI, glucose-6-P isomerase; G6PDH, glucose-6-P dehydrogenase; 6GPDH, 6-phosphogluconate dehydrogenase; pGluM, phosphoglucomutase; PGM, phosphoglyceromutase; PFK, phosphofructokinase (pyrophosphate and ATP-dependent); ENO, enolase; PYK, pyruvate kinase; PDC, pyruvate dehydrogenase complex (E1, E2, E3 components); and T, translocator. PDC is a multienzyme complex, but only one set of components is drawn here. Open arrowheads indicate transport rather than conversion. Solid arrowheads indicate physiologically irreversible reactions. For details see Martin and Herrmann (1998). See also Color Plate 3.

of soluble cytosolic or stromal enzymes. The comparison of genomes and structures of chloroplasts and mitochondria illustrates that the genetic integration of the photosynthetic organelle is less advanced, consistent with their endosymbiotic succession. Different from 'phylogenetically advanced' mitochondria that house only two genetically hybrid structures, the energy-transducing respiratory membrane and the organelle translation machinery required for its synthesis, chloroplasts contain at least eight compounds of dual genetic origin in the organelle stroma, the thylakoid membrane and inner envelope (Table 1). The existence of large amounts of promiscuous DNA in the nucleus reveals that the transfer of genetic material from the organelles may have been stochastic, but the selection of functional genes was not. Genes for regulatory components were apparently selected preferentially (Herrmann, 1997), for thylakoid membranes probably also peripheral compounds (Kapoor and Sugiura, 1998; Martin and Herrmann, 1998). The core components of the supramolecular thylakoid complexes that organize all cofactors involved in redox and catalytic processes are usually organelle-coded, much more so than in the corresponding complexes of the respiratory membrane in mitochondria. Comparably, the patterns of gene translocations from such structures may differ between different plant lineages. For instance, the transfer of genes for thylakoid proteins is most progressed in the chlorophyll *a/b* lineage (Martin and Herrmann, 1998; Martin et al., 1998). Together, this reflects the gradual nature of the process (that can be used to deduce genealogical relationships), and tells that the partite plant genome is not in a phylogenetic equilibrium. *All available data suggest that the ultimate aim of genome restructuring in the plant cell, as in the eukaryotic cell in general, is the elimination of genome compartmentation while retaining physiological compartmentation.* It is interesting to note in this context that hydrogenosmes, derivatives of mitochondria in obligate anaerobic protists, may lack DNA simply because both the respiratory membrane and consequently organelle ribosomes are dispensable under such conditions (Herrmann, 1997).

Collectively, the outlined aspects reinforce the massive intercompartmental DNA restructuring in the post-endosymbiotic cell, most impressively in nucleomorph-containing plastids of secondary endosymbiotic origin. They illustrate the complexity of changes, the intermixing of genetic material from the initial endosymbiotic cell association, as well as the present status of the cell. The selective advantage of the eubacterial metabolic machinery is not understood. *In any case, the result of restructuring the symbiotic genomes is an integrated genetic system, rather than semiautonomous organelles and nucleus* (Herrmann, 1996, 1997). Evidently, the elimination of redundant or dispensable genetic information from the cellular conglomerate, for instance to avoid competition between the homologous basic metabolisms of the symbiotic partners, as well as the intermixing of the endosymbiotic genomes, abolished the independence of the initially individual partner cells and of *all* of the cellular genetic compartments derived from them. This includes the nucleus. Although nuclear regulatory dominance was established, to date the compartments are mutually dependent upon each other, and it is the *entire* integrated genetic system that is regulated in time, quantity, in multicellular organisms also spatially to ensure proper synthesis of the photosynthetic machinery.

Table 1: Structures of dual genetic origin in chloroplasts and plant *mitochondria*

thylakoid membrane–*respiratory membrane*	(Herrmann, 1996; 1997)
organelle stroma	
ribulose bisphosphate carboxylase/oxygenase	
70S ribosome–*70S/60S ribosome*	
eubacterial-type RNA polymerase	(Link, 1996)
acteyl-CoA carboxylase	(Sasaki et al., 1993)
Clp protease	(Sokolenko et al., 1998)
transcript processing machinery	(Rochaix, 1996)
inner envelope membrane	(Rolland et al., 1997)

III. Functional Consequences of Genome Rearrangement—Regulatory Levels

The functional impact of the outlined genome restructuring upon the photosynthetic machinery and plant cell is profound and apparent at all levels of regulation with a multitude of mechanisms, over a wide range of time scales, and with nuclear regulatory dominance. Most obvious are a fundamental change of the transcription machinery of the entire cell, that is, in all three subgenomes, the establishment of an elaborate posttranscriptional RNA modification system in chloroplasts, and sophisticated protein trafficking and assembly devices. If the host genome was a sink for eubacterial, i.e. endosymbiotic, genetic information, genome expansion may not only have generated the chimeric archaebacterial/eukaryotic biochemistry. The partite nuclear genome and the lack of operon-like transcription units in eukaryotic genomes may as well be a consequence of genome restructuring, and hence a derived feature (Herrmann, 2000). Then, the host cell may well have been archaebacterial, as various data suggest now, and the nuclear genome would not trace back to a genome derived from the progenote, the hypothetical, last common ancestor for the three domains of life (Woese; 1998), from a comparable cell type, or from a chimerical archaebacterial/eubacterial hybrid host. Such a change would necessarily have to be accompanied with an alteration of the expression machinery.

Supramolecular thylakoid complexes represent appealing models for studying the nucleo-chloroplast interactions and their evolution in photosynthesis due to their hybrid genetic nature and the fact that operon design in plastid chromosomes often resembles that of the prokaryotic ancestors allowing comparisons between the organelle and its ancestral cell type. Their assembly into photosynthetically functional structure in chloroplasts results from a cooperation between genes and their products derived from both plastid and nuclear genomes. The chosen examples illustrate the complexity and subtlety of the changes at major levels of regulation that have to serve to coordinate the integrated genetic system spatiotemporally, while allowing cellular input into the functional and developmental decision. Little is known about intercompartmental regulatory circuitries or the evolution of spatial or ecophysiological adaptation of expression programs in multicellular plants.

A. Nuclear Gene Expression

1. Promoter Structure and Evolution

The symbiont's photosynthesis genes that were transferred to the host nucleus during chloroplast evolution required the acquisition of eukaryotic promoters, transit peptides either during the translocation process or by rearrangement after translocation (both cases exist), and integration into the signal transduction pathways of the host cells (Herrmann, 1997; Martin and Herrmann, 1998; Gray, 1999). The dispersal of genes for thylakoid proteins in the nuclear genome (Legen et al., 2001), the dissimilarity of their promoters and transit peptide sequences, even of genes that encode different subunits of the same membrane assembly, and the similarity of promoters of homologous genes in different higher plants, are all consistent with phylogenetically individual transfer events of originally organelle-encoded genes.

The application of transgenic approaches with modified promoter sequences, especially from single copy genes, and appropriate reporter genes, of gel retardation assays, the use of gene machines, and sequence comparisons have been the primary sources of information on the expression of nuclear genes for thylakoid proteins. Although genes encoding structural components of the photosynthetic machinery are regulated in a similar way, comparative promoter studies have shown an unexpected complexity of transcriptional control. Promoters of genes for thylakoid proteins generally lack common *cis*-elements, and probably common *trans*-acting factors, that are known from other collectively expressed groups of genes (e.g. heat shock genes; Schöffl et al., 1998). Each of the promoters studied possesses a distinctively different design (Lübberstedt et al., 1994). Furthermore, the initially acquired DNA segments of functionally translocated genes have been substantially changed at least in three ways. They had (i) to be integrated into signal transduction chains in order to ensure a coordinated, often induced (light, hormones) polypeptide synthesis; (ii) their expression profiles had to be adapted for plants that grow in different environments, and (iii) they had to be altered for tissue- and organ-specific expression in multicellular plants (Section IV). Most of these adjustments occurred after gene translocation, since transfers took place predominantly at the unicellular level (Martin et al., 1998). A parade example is found in the promoters of genes of close relatives involved in C3 or C4 photosynthesis (Stockhaus et al., 1997) which indicates that at least some information to ensure spatial expression in C4 background has been acquired with the separation of the two lineages.

In eukaryotic protein-coding genes, promoter elements are generally classified into two categories: common core promoter elements for basal transcription near the transcription initiation site and gene-specific elements located usually, but not always, upstream (Roeder, 1998). Gene-specific elements influence expression quantitatively (enhancer or silencer sequences) or qualitatively. The latter, although crucial, can display only residual activity to the overall expression pattern which may render their detection difficult. Gene-specific elements are arranged and operate in a highly complex way. They may be scrambled into super-elements or more or less spread individually. For instance, a 40 nucleotide pair segment in the promoter of the *psa*F gene, which encodes the plastocyanin-docking subunit of Photosystem I, directs quantitative, light-responsive and developmental expression (Flieger et al., 1993). Gene-specific *cis* elements are often located in the first 200–300 upstream nucleotides, but may be scattered within a sequence interval of more than 1000 nucleotides (Lübberstedt et al., 1994). Each gene appears to operate with a unique ensemble of regulatory elements which generally interact with nuclear proteins, often yielding complex patterns. The diversity of promoters probably finds an equivalent in *trans*-acting factors.

Core promoters can complicate the expression of nuclear genes. As upstream regulatory elements, they can influence gene expression qualitatively and differentially. In plants, for instance, they can contribute to light-responsive transcription (Yamamoto et al., 1997). Three basic elements have been defined in eukaryotic protein-coding loci that serve as basal core promoter modules and direct transcription initiation, notably TATA box, Inr, and DPE (Burke et al., 1998; Smale and Baltimore, 1989; Smale et al., 1998). They reside at equivalent positions and operate alternatively per se or in various combinations (Azizkhan et al., 1993; Roeder, 1998), for instance with TATA$^-$Inr$^+$ (Smale and Baltimore, 1989), TATA$^+$Inr$^+$ (Roy *et al.*, 1993), or Inr$^+$DPE$^+$ (Burke and Kadonaga, 1996). Degenerate consensus motifs have been deduced for vertebrates and invertebrates, PyPyAN(T/A)PyPy for Inr and (A/G)G(A/T)CGTG for DPE (Burke et al., 1998; Smale et al., 1998). TATA box-less promoters are not frequent, but have also been noted for plant genes, first in one of the light-regulated *psa*D genes of *Nicotiana sylvestris* which encodes the ferredoxin-binding subunit of Photosystem I (Yamamoto et al., 1997). Its promoter lacks a TATA-like motif near –30 but houses a functional Inr segment with a typical pyrimidine-rich motif at an appropriate position. *In silico* searches show that this promoter type appears to be remarkably common for nuclear genes encoding components of the photosynthetic machinery (J. Obokata, personal communication). Although this finding suggests that core promoter heterogeneity reflects group-specific gene expression, work predominantly in the animal field did not substantiate such ideas, but whether such core promoter elements are interchangeable or fulfil differential functions (Ohtsuki *et al.*, 1998; Smale *et al.*, 1998), and whether they are related to phylogenetic events, are intriguing questions. In spite of some knowledge in detail, the phylogenetic integration of the individually acquired promoter sequences into the signal transduction chains, the principles of transcriptional changes as a means of physiological adaptation of thylakoids remain to be explored, as the link to the complex network of interacting signaling components and cascades. A coherent picture of the spatiotemporal operation of an individual promoter or of transcriptional coordination of gene sets has not yet emerged.

2. Protein Import into Chloroplasts—Assembly Processes

An intracellular gene transfer requires the return of the respective or of a related protein into the chloroplast. Nuclear-coded plastid proteins therefore operate with import sequences that direct translocation into and across the envelope membranes. In some cases, they even target proteins subsequently to their suborganelle destination. Import and stepwise transport to and assembly at various locations within the chloroplast depend on the operation of specific targeting and sorting signals, of protein folding/unfolding catalysts (e.g. chaperones) as well as on highly specific, soluble and membrane-bound receptor systems that decode targeting signals and initiate integration or translocation through protein conducting channels (Chapters 10 and 11). Some of these proteins act globally, whereas others exert more specific effects. Those of the former category include members of chaperones, proteases, in plastids also protein kinases and phosphatases (Waegemann and Soll, 1996). These transport processes are especially complex with multilayered plastids of secondary endosymbiotic origin, with their envelopes of different origin (Fig. 2e and f; Section II). For complex plastids, even a vesicle transport step had been postulated, first from ultrastructural work (Gibbs, 1979; Hofmann et al., 1994). Using protein fusion with the green fluorescent protein (GFP; Haseloff, 1999) as a reporter, protein trafficking into the plastids of *Toxoplasma gondii* and *Plasmodium falicparum* which are surrounded by four envelope membranes was shown to include steps of the secretory pathway (Waller et al., 1998; 2000).

The thylakoid polypeptides encoded by nuclear and plastid genes follow two fundamentally different pathways to their sites of residence within the organelle. The former are translated on cytosolic ribosomes and are imported by chloroplasts posttranslationally, whereas proteins produced in the organelle are translated on endogenous ribosomes that may be bound to thylakoids. They may be inserted into it cotranslationally. These two pathways must converge at some point into an assembly pathway for multisubunit structures that are genetic hybrids. How proteins that originate in two compartments assemble into multimeric complexes is not known, although available evidence suggests that subunits assemble sequentially and may interact with several chaperones successively. Both genetic and biochemical strategies are available to address this problem (Section V). Major aspects in the field have been reviewed recently and will not be covered here (Robinson and Klösgen, 1994; Heins et al., 1998; Robinson et al., 1998; Schnell, 1998; Dalbey and Robinson, 1999; Keegstra and Cline, 1999; Cline, 2000; Schleiff and Soll, 2000). However, two points to note in the context are that the machineries for these pathways are composed entirely of nuclear-coded proteins and that their phylogeny is intriguing.

In the plant cell, the most important primary protein sorting events lead to the secretory pathway, to chloroplasts and mitochondria, to the nucleus and to other membrane embodied cellular differentiations. While strong homologies exist between components of the protein translocation systems found in the inner membrane of bacteria, the ER membrane, and the thylakoid membrane (Schatz and Dobberstein, 1996), suggesting a common origin, so far only few of the components identified in the chloroplast outer and inner membrane translocation machineries have known homologues in other organelles or cells (Heins et al., 1998; Robinson et al., 1998; Schnell, 1998; Keegstra and Cline, 1999; Schleiff and Soll, 2000). The chloroplast protein import machinery thus seems to be unique, and appears to be largely a phylogenetic gain. This applies to the equivalent mitochondrial machinery as well. One link between protein import machineries of plastid and mitochondria (and ER) are the proteases that cleave sorting sites (Nunnari et al., 1993; VanderVere et al., 1995). In contrast, an unexpected variety of pathways for targeting of proteins into and across thylakoid membranes has been deciphered that regulates the generation and modification of the photosynthetic machinery at the thylakoid membrane. These multiple integration and translocation machineries function non-competitively with specific subclasses of membrane proteins (Robinson and Klösgen, 1994; Schnell, 1998; Robinson et al., 1998, Keegstra and Cline, 1999; Cline, 2000), and include the Sec-, SRP-like, ΔpH- (or Tat-) and the so-called spontaneous routes. All these pathways possess ancestry in translocation systems that operate in bacteria. Initially, the ΔpH route was thought to be a phylogenetic gain, since it was found only with nuclear-coded, lumenal thylakoid components, such as the 23 and 16 kDa polypeptides of the oxygen-evolving system, that cyanobacteria are lacking. This turned out not to be correct (Berks, 1996). The route specialized for the translocation of co-factor containing, tightly folded proteins exists in a spectrum of biological membranes (Dalbey and Robinson, 1999; Summer et al., 2000). A further example is the recent discovery of a new translocase (or part of it) in *E. coli* for the so-called spontaneous route (Samuelson et al., 2000) that appears to be used in mitochondria and probably in thylakoids as well (see below). It is thus clear that the bacterial inner membrane translocation system can be viewed as an evolutionary precursor to the thylakoid system as well as to the ER (the Sec61 complex of the ER is derived from the Sec machinery). However, studies of the energetic requirements suggest that additions or variations on the pathway not found in bacteria exist in thylakoids (Robinson and Klösgen, 1994).

Plastid targeting pre-sequences are distinct from secretory signal peptides and mitochondrial targeting sequences, both in terms of overall design and in terms of cleavage sites (von Heijne et al., 1989). Expectedly, as promoters, they show no obvious homology at the level of primary sequence. They fall into two principal classes, stroma- and stroma-/thylakoid-targeting presequences. In the latter, two modules arranged in tandem operate successively with two distinct translocation and processing systems residing in the envelope/stroma and thylakoids, respectively. Plastid proteins of nucleo/cytosolic origin usually operate with a stroma targeting transit peptide. This module has been acquired during evolution, in all known instances. The probability of picking up appropriate DNA segments for polypeptide transport and/or gene expression is not low (Baker and Schatz, 1987), in the order of 1–3%. In contrast, thylakoid-targeting signals of composite transit peptides that follow the stroma-targeting moiety resemble prokaryotic signal peptides (von Heijne et al., 1989), and possess a prokaryotic ancestry in most, but not all, instances. Well known examples are plastocyanin or the 33 kDa polypeptide of the oxygen-evolving complex. The terminal processing of bipartite transit peptides occurs in the thylakoid lumen with an endopeptidase that again resembles bacterial leader peptidases. Plastid-encoded thylakoid proteins [as nuclear-encoded proteins of the outer envelope membrane; Steinkamp et al. (2000); Soll et al. (2000)] possess generally intrinsic thylakoid-targeting and assembly signals, but in rare cases even those can possess a classical signal peptide that targets to the thylakoid. For instance, cytochrome *f* is made as an apoprotein precursor molecule with a transitory, N-terminal signal peptide-like extension. Most of these epitopes and the corresponding machinery appear to be inherited from the prokaryotic progenitor of chloroplasts.

There are several exceptions to this scheme, found, for instance, in some of the second domains of bipartite transit peptides. These reinforce the diversity and subtlety of phylogenetic changes. An instructive example that bears on several aspects has been found in the ATP synthase subunits CFo-I (b) and -II (b′) (Michl et al., 1994). These two thylakoid proteins originated by an ancient gene duplication event at the prokaryotic stage which gave rise to the paralogous *atp*F and *atp*G genes. This event is not found in the respiratory enzymes of bacteria and mitochondria, but appears to predate the divergence of photosynthetic F-type ATP synthases. In the chlorophyll *a/b* lineage of plants, one of the genes (*atp*G) has been transferred to the nucleus, unlike in plants with plastids of secondary symbiotic origin. Remarkably, both the plastid and nuclear coded proteins integrate with different epitopes and mechanisms into the thylakoid membrane. CFo-I operates with a single hydrophobic domain for both targeting to and anchoring within the membrane (probably via the Sec route), while CFo-II requires an additional hydrophobic domain in its transit peptide for integration *via* the so-called spontaneous route (Michl et al., 1994, see also above). The CFo-II transit peptide displays all characteristics of bipartite presequences and it is cleaved off by the lumenal endopeptidase, but differs from those, since this second domain is not found in the corresponding plastome-encoded and prokaryotic subunits. This verifies that this domain is of a different phylogenetic origin, that is, it does not have a phylogenetic precedent and thus was not inherited from the prokaryotic progenitor of chloroplasts. *The membrane location, the bitopic nature of the two related proteins and the different intracellular location of their genes was favorable since they served as an example that genome rearrangements had substantial consequences for later biogenetic processes as well.* The change in mechanism is not caused by the necessity of a posttranslational *vs.* cotranslational integration due to gene translocation (Michl et al., 1999). It is worth noting in this context that all bitopic polypeptides of nuclear (but not plastid) origin known use this integration mode as a general route. Again, phylogeny, complexity and subtlety of regulation is evident at all stages of the system.

An often noted feature of transit peptides is their low degree of sequence conservation. For instance, the plastocyanin polypeptide is conserved in the order of 90% among vascular plants, unlike the corresponding bipartite transit peptides (approximately 60%). Although stroma-targeting presequences are distinct from other presequences, enriched in hydroxylated and basic amino acids, and generally lack acidic residues (von Heijne et al., 1989), no obvious sequence nor structural elements could be deduced, and whether a particular structure is induced upon contact with the chloroplast surface is unclear. On the other hand, defined functional epitopes, typical of prokaryotic signal peptides, have been noted for the second domain of bipartite transit peptides, notably a charged N-terminus, a hydrophobic domain and a cleavage site motif with small side chain residues at positions –1 and –3, generally A-X-A, that is preceded by helix-breaking and turn-inducing residues. No evidence exists that the presence or absence of intermediate cleavage sites in composite transit peptides is of functional or evolutionary relevance. It is conceivable that the dissimilarity of transit peptides between different polypeptides and the divergence of a given transit peptide reflect a selective advantage, because some flexibility may be initially advantageous to allow for import with the more or less randomly acquired sorting signals.

The discovery of diverse protein transport routes and the finding that import sequences, at least some of the chloroplast import machinery and sorting epitopes may be gains have contributed significantly to our understanding of assembly processes of the photosynthetic machinery. It has provided new insights into evolutionary relationships, differences and novelties between corresponding pro- and eukaryotic structures and between membrane-targeting systems of different organelles.

B. Gene Expression in Plastids

1. Transcriptional Control

It was only in recent years that transcription turned out to be a prime level of regulation in plastids, as well as to be highly complex, especially when compared with that in photoautotrophic prokaryotes. It is now firmly established that chloroplast genomes operate with more than one, possibly with as many as three RNA polymerases, comparably to nuclear genomes (Chapters 2 and 5): The eubacterial multisubunit $2\alpha\beta\beta'\sigma$-type of enzyme (PEP) tracing back to the endosymbiont, a phage-type RNA polymerase (NEP) that was probably acquired by lateral gene transfer (Hess and Börner, 1999) and a recently discovered third enzyme of unknown ancestry (Bligny et al., 2000). Prokaryotes usually house only one polymerase. As in prokaryotes, the coding potential of plastid chromosomes is generally organized in operons, in an economically not yet intelligible way. Genes for thylakoid membranes are dispersed throughout the chromosome and are often cotranscribed with genes for other multisubunit assemblies. This has important consequences for gene expression, especially with regard to posttranscriptional processes (see below). The existence of three polymerases and of an elaborate post-transcriptional RNA modification machinery are distinguishing features to prokaryotes and indicative of the changes towards the eukaryotic level. These findings again reflect one of the most intriguing recent aspects in organelle biology, as in the biology of eukaryotes in general, with a fundamental functional, phylogenetically caused impact to the photosynthetic machinery.

Several lines of molecular genetic and biochemical evidence suggested the existence of more than one enzyme to operate in the organelle, (i) the persisting transcription of a rudimentary plastid chromosome in the non-photosynthetic higher plant parasite *Epifagus virginiana* that has lost the entire *rpo*B/C1/C2 operon, encoding subunits of the PEP polymerase (Morden et al., 1991), although translocation of *rpo* genes to the nucleus is a conceivable possibility in this case; (ii) evidence for a phage-type RNA polymerase of low abundance in chloroplast fractions (Lerbs-Mache 1993), although contamination was not rigorously excluded, since the material used was not axenically grown; (iii) work on ribosome-deficient plastids (Falk et al., 1993; Hess et al., 1993); and (iv) the maintenance of plastid transcription in Δ*rpo*A, B, C1 and C2 mutant background (Allison et al., 1996; Serino and Maliga, 1998; De Santis-Maciossek et al., 1999). All this and other biochemical evidence was more or less indirect, and hence discussions on this point were controversial. The isolation of a nuclear-coded, second plastid-located RNA polymerase resembling phage-type and mitochondrial enzymes finally settled that point (Hedtke et al., 1997).

Sequence comparison has shown that the NEP gene and enzyme are homologous to the mitochondrial enzyme(s). Gene anatomy and sequences, even of the various introns of the genes, are highly conserved, except those for transit peptides and promoters (Hess and Börner, 1999). This has suggested that the chloroplast enzyme is the result of a cell-internal gene duplication that was accompanied by the acquisition of appropriate promoter and transit peptide sequences. The eubacterial and phage-type enzymes do not appear to share essential common subunits (Serino and Maliga, 1998). They require different promoter types (Kapoor et al., 1997; Liere and Maliga, 1999), presumably also different terminators. However, core promoters of the phage-type polymerase in plastids and mitochondria expectedly resemble each other (Kapoor et al., 1997; Liere and Maliga, 1999). The establishment of the novel, additional RNA polymerase in both organelles had therefore profound consequences for gene expression. Expression signals in the organelle (sub)genomes had to be and were replaced, completely in mitochondria where the phage-type enzyme is the sole transcriptase, and partially in the case of plastids. That the mitochondrial genome of *Reclinomonas americana*, a protist discovered only ten years ago, harbors an ancient, eubacterial-type RNA polymerase, different from all other mitochondrial genomes studied (Lang et al., 1997), is consistent with the replacement hypothesis of the ancient RNA polymerase in plants, animals and fungi.

In higher plant plastid chromosomes promoter changes and promoters are complex due to multiple transcriptional initiation sites. A few operons seem to be solely transcribed by either the plastid-coded or the nuclear-coded RNA-polymerase, the majority of them is transcribed by both (Hajdukiewicz et al., 1997, Kapoor et al., 1997). This is indicative of gradual changes. It is tempting to assume therefore that the complexity of the transcriptional system in plastids rests again in an intermediate stage, as the degree of gene translocations deduced above. Thus, during chloroplast evolution, not only the many of the symbiont's photosynthesis genes transferred to the host nucleus, required fundamental changes of expression signals due to the acquisition of eukaryotic promoters, transit peptides, and integration into the signal transduction pathways of the host cells (Herrmann, 1997; Martin and Herrmann, 1998; Gray, 1999).

The question of whether PEP and NEP are both involved in the expression of only a distinct set of genes, the latter in that of house-keeping genes contributing to establish the organelle's genetic system, the former predominantly in those of the photosynthetic machinery (Hajdukiewicz et al., 1997), is a controversial issue, as is the idea that the nuclear-coded enzyme precedes and thus controls the expression of the plastid-coded PEP genes in ontogeny. Proof of this regulatory cascade model is lacking. That promoters of distinct operons are transcribed by only one of the RNA polymerases, either NEP or PEP, does not necessarily mean that these operons are not read by the other enzyme. Comparison of transcript patterns between wild-type and Δ*rpo* material demonstrates that not only promoters but also terminator signals (and implicitly operon structures) for the two enzymes are different, and that an operon not read by one enzyme, may well be transcribed by the other via read-through starting from an upstream gene cluster. In fact, each of the two polymerases appears to transcribe the entire chromosome (Krause et al., 2000).

The analysis of the overall biogenetic potential of the residual organelle after plastid *rpo* gene disruption shows that the functional consequences of NEP integration into the genome and cellular context are highly complex (De Santis-Maciossek et al., 1999; Krause et al., 2000). Material lacking the eubacterial-type RNA polymerase (Δ*rpo*A, B, C1 or C2 plastome mutants) is capable of chlorophyll and lipid biosynthesis similar in composition to the wildtype. Also, fully processed nuclear-coded stroma and thylakoid polypeptides are present, in amounts similar to those found in the wild type. These findings and the formation of starch grains in the mutant plastids establish that the nuclear-coded components and the energy supply most likely from a respiratory source are basically provided by the cell. In Δ*rpo* material, different from etioplasts (Herrmann et al., 1992), thylakoid proteins encoded by plastid genes could not be detected, although the corresponding operons are transcribed, functional organelle ribosomes are present and at least some of the thylakoid proteins appear to be made in the organelle, but these are not stable. These observations demonstrate that the Δ*rpo* phenotype is not merely based on transcriptional regulatory control. They are indicative of genetic integration and instructive with regard to the cell status and the complexity of changes. We now need to know how the two plastid RNA polymerases interact in the tissue- and development-specific expression of plastid genes, how they communicate with the nucleo/cytosolic system, what is the phylogenetic origin of the nuclear-coded enzyme and the moment of the duplication event of its gene that generated the isozymes operating in different organelles, and what is the reason why two apparently opposite processes appear to be superimposed upon each other: Is it only the generation of nuclear regulatory dominance why a second enzyme has been established lately in spite of the reductive evolution of organelle genomes (Section II)?

2. Posttranscriptional Regulation

Chloroplast transcription units are usually multigenic, and primary transcripts derived from them are extensively processed in one or more ways (Barkan, 1988; Westhoff and Herrmann, 1988). Posttranscriptional RNA changes are complex (Chapter 7). Nearly a dozen enzymatic machineries have been deduced including tRNA excision, 5′ and 3′ terminal shortage by endo- and exonucleolytic activities, group I and group II intron excision and splicing, endonucleolytic cleavage, 3′ polyadenylation of mRNA fragments, 3′ CCA addition to tRNAs or nucleotide substitutions and nucleotide modifications. In addition, specific mechanisms for stabilization of certain transcripts have been documented. Collectively, these give rise to complex sets of overlapping RNA species.

As transcriptional and translational control, modification of RNA molecules represents a key level of regulation in the plastid and a prime example for complexity and network regulation. Some recent reviews discuss distinct aspects of RNA modification in plastids (Brennicke et al., 1999; Hayes et al., 1999; Barkan and Goldschmidt-Clermont, 2000; Bock, 2000; Monde et al., 2000). Biochemical work has shown that polycistronic mRNAs, e.g. of the *psb*B operon, can be decoded (Barkan, 1988), but that monocistronic *pet*D mRNA that is excised from that polycistronic RNA may be more efficiently translated than its precursors (Barkan et al., 1994). Similarly, antisense RNAs in instances where the complementary strand encodes genes, e.g. for *psb*N in the *psb*B operon or for *trn*S in the *psb*C/D operon in plastid chromosomes of higher plants, are selectively removed from the overlapping transcripts, but only after processing (Herrmann et al., 1992). Also, quite serious phenotypes appear with impaired processing patterns (summarized in Barkan and Goldschmidt-Clermont, 2000). With the exception of some components involved in the 3′ processing of plastid encoded mRNAs and RNA stability (Hayes et al., 1996), the enzymes and factors involved in posttranscriptional processing and transcript stability are not known. Nevertheless, it is known that the RNA modification machinery appears to be of dual genetic origin (Table 1, Rochaix, 1996), although the overwhelming majority of regulatory components required for processing plastid RNA are likely to be coded in nuclear DNA (Maier et al., 1996, Barkan and Stern, 1998, Brennicke et al., 1999; Barkan and Goldschmidt-Clermont, 2000; Monde et al., 2000). Relevant as well in the context is that the extensive transcript processing in chloroplasts and some of its mechanisms are phylogenetically derived traits (Malek et al. 1996; Brennicke et al., 1999).

RNA editing is a particularly instructive process in this context. The enzymatic activity changes the nucleotide sequence of a transcript so that the RNA sequence differs from the genomic template from which it is transcribed. In plastids and mitochondria, RNA editing generally changes genomically encoded C residues to nucleotides which behave like uridines in RT-PCR analyses (Maier et al., 1996). The nature of the underlying enzymatic activity is not known but circumstantial evidence suggests that de- or transamination of cytidines is the most likely mechanism (Blanc et al., 1995; Yu and Schuster, 1995). C-to-U conversions and the reverse process are often essential for the decoding and/or for the structural and functional fidelity of a protein, or change the expression patterns and/or RNA stability. Editing in plastid transcripts is found in reading frames as well as in nontranslated RNA segments. It can restore initiation codons, termination codons, or codons for amino acid residues that are conformationally crucial, frequently residues conserved in other organisms, and hence represents an important means for nuclear regulatory control, as the other posttranslational modifications. To give an example: The editing of a distinct site in cytochrome b_6 mRNA into a proline codon is imperative for heme ligation, correct assembly, and hence photosynthetic electron transport (Freyer et al., 1993; Zito et al., 1997). Editing is not known from prokaryotes, protophytes and thallophytes. It is found to some extent in bryophyta. It thus appears to be a phylogenetic innovation that spread out (Malek et al., 1996, Brennicke et al., 1999) both within the organelle and within the cell. In plastids, it may have been acquired from the mitochondrial compartment as judged from the parallel occurrence of the process in both compartments and its higher frequency in that organelle. If correct, this would be an interesting analogy to the RNA polymerase NEP, and imply that the nucleus, which itself operates with RNA processing in gene expression, converts analogous complexity to the regulation of gene expression to the organelle.

3. Translation

Although relatively little is known of how mechanistically plastid mRNAs are recognized by ribosomes and their translation is regulated, chloroplast and eubacterial translation shows many similarities, it also exhibits notable differences (Chapter 8; Hirose and Sugiura, 1996; Fargo et al., 1998; summarized in Sugiura et al. 1998; Zerges, 2000). One of the differences lies in the mechanism of translational initiation. In eubacteria, the initiation codon is usually selected as a consequence of interactions between mRNA and 16S rRNA. The best understood of these interactions occurs at the Shine-Dalgarno sequence. In plastid translation, mRNAs basically fall into at least two classes, those that follow that eubacterial mode, and others that lack a Shine-Dalgarno-like sequence at the expected position. In the plastid chromosome of tobacco, approximately one-third do not contain a typical Shine-Dalgarno-like sequence (Hirose and Sugiura, 1996). Their mechanism of initiation codon selection is quite different from that in eubacteria and integrates also elements known from cytosolic (eukaryotic) mRNA translation (Preiss and Hentze, 1999). It may require RNA processing and editing as well as a cooperative interaction of several, 5′ but also 3′ UTR-located *cis*-elements and *trans*-factor(s) for efficient translation. In *Chlamydomonas* chloroplasts interaction of mRNA *cis*-elements with *trans*-factors appears to be controlled through the redox potential and/or ATP/ADP ratio (Danon and Mayfield, 1994 a,b). Also differences have been noted between different organisms, even for the same gene or operon. The possibility that nuclear-coded factors affect chloroplast transcriptional elongation has not yet been seriously addressed. Advancement in these areas may be expected by chloroplast transformation and the recent development of an in vitro translation system from chloroplasts (Hirose and Sugiura, 1996).

4. Nuclear Regulatory Control

A striking feature of chloroplast gene expression is the fact that the expression of each gene for a photosynthetic compound is controlled by multiple gene- or RNA-specific *trans*-factors. This is obvious from work at all levels of regulation (Herrmann 1996, 1997; Rochaix, 1996; Allison, 2000; Barkan and Goldschmidt-Clermont, 2000). The majority of these factors appears to be gene- and operon-specific (a consequence of individual gene translocations and product integration?), rather than operate in a general way. The relatively large number of the gene-specific factors identified so far that are required for expression at posttranscriptional steps, such as processing of organelle pre-mRNAs, stabilization of transcripts, translation of RNAs, or assembly of polypeptides into functional structures suggest that probably more than one order of magnitude more loci are required to manage the expression of a single thylakoid gene and its product during biogenesis. Assuming that approximately 1.5–2% of the cellular coding potential is required for the photosynthetic structures (The Arabidopsis Genome Initiative, 2000), this would imply that approximately 25% or more of the nuclear coding potential is involved in the management of the energy-transducing organelles (Herrmann, 1997; Section V). How the different regulatory components are co-ordinated per se and integrated to ensure correct thylakoid biogenesis is one of the future prime aspects to solve in photosynthetic research.

IV. The Impact of Multicellularity and Terrestrial Life upon Thylakoid Biogenesis

To establish a functional photosynthetic apparatus in complex forms of life required new qualities of genetic information, notably information for spatial and for an extended temporal aspect of regulation. Multicellularity and the step to land, more than 600 million years ago, exposed plants to new physiological conditions. The latter was a comparably critical period for evolution as the radical step from an anaerobic to aerobic environment due to oxygenic photosynthesis, in the period from three to two billion years ago. Both, multicellularity and the step to land mark intervals of unparalleled innovation for genomes and phenotypes. Both included new biochemistries, novel biochemical pathways, such as for lignin, flavonoid, wax and hormone synthesis as well as unique genetic programs and biochemistries for form. The enormous spatiotemporal morphogenetic potential distinguishes plants from prokaryotic autotrophs. The novel structural potential includes the differentiation of an extraordinary array of approximately 70 cell types, tissues and organs, an internalization of vital functions, such as fluid transportation systems (vascular tissue), gas exchange and protecting surfaces with stomata and lenticells, cell and tissue systems for stabilization, storage facilities etc. Creation of form can also be seen with subcellular structures. Chloroplasts diversified into various photosynthetic and non-photosynthetic modifications with different functions, such as bundle sheath and mesophyll chloroplasts with structurally and functionally different thylakoid membranes, leucoplasts or chromoplasts, each one in distinct tissues. Division of labor in plants included the generation of programs for the differentiation of photosynthetically competent and non-competent cells and tissues. Those for the latter are probably different from the well known de-etiolation processes, when chloroplasts of higher plants develop from etioplasts and a new gene progamme is induced. All these developments resulted in significantly more highly differentiated plants. In the plant kingdom, the structural potential including that for complex life cycles is evident in the three major morphological categories, from protophytes (protists), through thallophytes (algae) to kormophytes (vascular plants). Some relevant aspects emerge in this context.

Development of form, plant diversification and adaptation to life on land, were gradual processes. This may mean that principal steps of the phylogenetic development and the gene sets encoding them would still be deposited in the genomes. Ontogenies should therefore be under a similar kind of developmental control in all principal plant groups. Much morphological diversity should be reducible to the gradual acquisition of novel genes or gene combinations, and their integration into existing regulatory networks. The basic steps may involve no more than two to three dozen key genes that include regulation of (i) cell communication, (ii) unequal cell division, (iii) information for spindle arrangement in mitosis in one plane which resulted in a thread-like, trichal (filamentous) body (this stage is reached in all thallophytic groups, probably convergently), (iv) the change from intercalate growth to tip cell growth, (v) the periodic 45° shift of the cell division plane which resulted in dichotomic or subapical (monopodial) branching, (vi) changes in the sectoring of tip cells, (vii) the integration of sectoring changes into spatial and developmental processes, obvious for instance in mosses, and (viii) the generation of meristems including (ix) the establishment of lateral meristems (cambia) combined with the potential to built decay-resistant lignin which was critical to the evolution of large terrestrial plants.

This enumeration, of course, is not complete and does not explain all topographic diversity. 'Fundamental' innovations are complemented by modulating genes or gene combinations, or by re-shuffling genetic information. Much 'ornamental' difference can be interpreted in terms of changes of basic schemes or by dormancy/activation and abortion of meristematic, branching or cell expansion systems with specifically regulated genetic programs. An example is found in the organ 'leaf', with its functionally quite different photosynthetic-, floral- and storage-type structures.

The outlined features can appear successively or in parallel, with varying combinations. In each instance, however, the biogenesis of the photosynthetic machinery has to be embedded into the respective developmental programs that, again, are predominantly nuclear-guided. Most of the two principal gene sets mentioned were probably generated at the eukaryotic level, and most of them perhaps *after* the principal genome rearrangements. The relatively few gene translocations from organelles to the nucleus in multicellular plants (Martin et al., 1998) suggest that multicellularity poses 'spatial' constraints to genome restructuring. Furthermore, present knowledge indicates that plants and animals use the same, largely transcription factor-based strategy for their body plans, but both evolved independently, consistent with the different gene families that were established in both groups, *mads* box and *hox* genes, respectively. It will be highly interesting to explore and compare the gene spectrum for this convergent evolution. No systematic research exists to investigate the gradual evolution of biochemical potential at the eukaryotic level and the genetic integration of the energy-providing system into the respective body plans.

V. Maintenance and Acclimation of Thylakoids

The list of auxiliary compounds that are essential for the biogenesis, maintenance and acclimation of the thylakoid system is wide (Chapters 9–16). In addition to the approximately 80 major polypeptide species that constitute the major supramolecular membrane assemblies (Fig. 1), at least twice as many minor polypeptide species can be estimated to be associated with the photosynthetic membrane, and a relatively large number is found in the soluble phase surrounding it. Disregarding compounds that are involved in processes such as plastid transcription, transcript processing and modification, translation, or in insertion and translocation of polypeptides into or across membranes (Section III), these operate in the assembly, modification and degradation of thylakoid proteins or the production, transport and ligation of cofactors and pigments. Various polypeptides of this sort have been identified in the thylakoid membrane. They include chaperon-like proteins (Meurer et al., 1998), ion transporters (Bartsevich et al., 1995), thylakoid-associated protein kinases and phosphatases (Snyders and Kohorn, 1999; Vener et al., 1998; 1999; Weber et al., 2001), immunophilins (Fulgosi et al. 1998), or proteases

(Chapter 15; Sokolenko et al., 1997; Adam, 2000). Other proteins are involved in plastid lipid synthesis or transport (Márechal et al., 1997; Kroll et al., 2001) and cofactor synthesis (Chapter 14; Nakamoto et al., 2000).

To grow and develop optimally, plants need to perceive and process information from their surroundings. A particularly important factor is light which is used as a source for energy, orientation and morphogenetically. The diverse responses of the photosynthetic machinery to light require sophisticated sensing of its intensity, spectral distribution, direction and duration. At least three photoreceptor systems, the phytochromes absorbing in the red/far red (Chapter 3; reviewed in Smith, 2000), the cryptochromes in the blue/near-UV (reviewed in Briggs and Huala, 1999), and a putative receptor system operating in the UV spectral range, have been uncovered by action spectra of light responses, and in the first two instances verified by molecular identification of the receptors. These photoreceptors ultimately feed into terminal response pathways which may also affect the expression of genes encoding thylakoid proteins. However, a complete transduction cascade, linking early and late processes or biochemical and genetic data has not been achieved.

Photosynthesis is not only fueled by light, in excess, light can also cause problems, and even harm a plant (Long et al., 1994; Niyogi, 1999). In their natural habitat, plants are exposed to light intensity and quality changes that can lead to an imbalance between the excitation rates of the two photosystems, and thus diminish photosynthetic efficiency. Light can also induce photoinactivation of electron transport and damage both photosystems (Prasil et al., 1992; Aro et al., 1993; Sonoike 1995). Comparably, high temperatures dramatically inhibit carbon dioxide fixation (Berry and Björkman, 1980; Feller et al., 1998), and the heat tolerance limits of leaves, which possess surprisingly little heat capacity, can depend on the thermal sensitivity of photochemical reactions in thylakoid membranes (Berry and Björkman, 1980; Weis and Berry, 1988; Havaux et al., 1996). Even short exposure to increasing temperatures in the range of 40 °C can cause a progressive destacking of stacked thylakoids (Gounaris et al., 1984), a lateral redistribution of the membrane protein complexes that constitute this membrane (Sundby et al., 1986), an increase in the ionic permeability and fluidity of the membrane (Havaux et al., 1996; Tardy and Havaux, 1997), rapid dephosphorylation and degradation of the Photosystem II core protein D1 (Rokka et al., 2000), and a release of extrinsic proteins of the Photosystem II oxygen-evolving system (Enami et al., 1994).

To avoid such serious consequences, plants have evolved mechanisms to adapt and protect their photosynthetic system (Chapters 22–29). In fact, thylakoid membranes are highly dynamic structures that can record and transduce signals themselves. Although the molecular details of the underlying signal transduction pathways are not yet clearly understood, reversible phosphorylation of thylakoid proteins and proteases are now recognized to be key players (Allen, 1992; Rintamäki et al., 1997; 2000; Vener et al., 1998; 1999). Protein phosphorylation/dephosphorylation appears to be involved in at least three major processes: (i) in adjusting the balance of the light-driven electron transfer between the two photosystems through reversible association of the light-harvesting chlorophyll *a/b* antenna (LHCII), thus modulating the relative light absorption cross section of the two photosystems via an intriguing redox-controlled membrane protein kinase (State I/State II transitions; Chapter 23); (ii) in the repair and photodamage of Photosystem II which appears to be a particularly vulnerable structure (Chapter 22), and (iii) presumably in the control of chloroplast gene expression. Proteolytic activities, in turn, can be involved in short-term processes such as in D1 degradation, but are frequently operating in long-term acclimation (Adam, 2000; Andersson and Aro, 1997).

The inventory of nuclear and plastid coded proteins required to assemble and maintain a functional thylakoid system demonstrates clearly that plastid and nuclear genomes interact in at least two ways. First, genes of both compartments contribute to chloroplast protein function. This level is obvious from the mere fact that plastid gene products do not act alone. They are components of multimeric protein complexes that include nuclear-encoded components as well. At least eight of these genetic chimeric complexes are common to most photosynthetic plastid types (Table 1). Second, both cellular subgenomes interact to affect the synthesis, assembly, repair, acclimation, and controlled senescence during cellular apoptosis (Chapter 16) of thylakoid structure. To an overwhelming degree the principal genes for these processes reside in the nucleus.

VI. Outlook—New Approaches

In this article we have depicted thylakoid biogenesis in a phylogenetic context with a broad brush, de-emphasizing details and minor exceptions in order to highlight common principles and their evolution. The uniformity of the structural design of thylakoids, but not of the basic mechanisms underlying their biogenesis, maintenance and adaptation, that has emerged from recent work places the photosynthetic process firmly within the framework of other complex biological structures of fundamental importance. Of course, some of the principles described here may not hold the test of time. It is still prudent to seek simplicity, and then to distrust it.

It is obvious that plant genomes and the principal energy providing system of the plant cell evolved in a highly complex manner. The plant cell may be considered as a 'genomic laboratory' with an enormous potential of restructuring genetic material. The integrated genetic system of the eukaryotic cell must possess an enormous genetic and functional plasticity, since the entire system remains functional in spite of often substantial individual changes. It is also obvious that features central for regulation have been changed in all compartments probably to account for the 'eukaryotic' situation. The regulation in and between cellular genetic compartments does not merely follow the expression characters of the initial endosymbiotic cells, and this holds for all compartments. The generation of complex regulation schemes for the photosynthetic machinery may lie in the history and design of the eukaryotic cell with its increased demand for regulation. This rests in nuclear genome expansion, genome and operon fragmentation, but also in the establishment of nuclear regulatory dominance in the partite genome (Herrmann, 1997). The molecular phylogenetic analysis to determine how complex plants have developed from cell associations, maintained genetic integrity over time, while keeping the energy-transducing photosynthetic and repiratory machineries fully functional is one of the most amazing and fascinating chapters in modern biology.

The diversity of cases and our still fragmentary knowledge render it difficult to formulate a comprehensive concept that describes events, genetic make-up and evolution of the photosynthetic process. Even the origin of the water oxidation system is enigmatic. It does not seem to reflect a structural 'learning process' by replacing H_2S by H_2O (Blankenship, 1992). The oxygen-evolving system was established at the prokaryotic level and transferred to all photosynthetic organelles. It is presumably older than 3 billion years and required that two, possibly related (Nitschke and Rutherford, 1991) photosystems had to be combined. Evidently, as in plant cells, genome rearrangements have occurred in phototrophic prokaryotes, as may be inferred from the evolution of *atp* operon(s) for instance, encoding subunits of photosynthetic or respiratory ATP synthases. Advances in the techniques of molecular and cell biology during the past two decades have generated unprecedented opportunities for exploring such new horizons. Their potential has not yet been fully applied in the field, last not least because relevant information is scattered among various disciplines, often with little relationship to each other. Because of their impact expected for future work in photosynthesis, a few of these techniques will be mentioned.

A. Genetic Approaches

Genetic approaches have proven to be extremely powerful to dissect, for the first time, complex regulatory, developmental or metabolic processes into individual or groups of loci and to identify the genes involved. This requires genetically or environmentally appropriately defined materials and tools such as molecular marker, EST, PCR, megabase and array technologies or genome saturating random mutagenesis by chemical and/or biological mutagens, to isolate the affected gene(s) and study their impact. With respect to thylakoid membrane biogenesis the cyanobacterium *Synechocystis* PCC6803, the unicellular green alga *Chlamydomonas reinhardtii* and the two angiosperms *Arabidopsis thaliana* and *Zea mays* may be considered the most advanced models. Complete genome sequences, e.g. of *Synechocystis* (Kaneko et al., 1996) or *Arabidopsis thaliana* (The Arabidopsis Genome Initiative, 2000), are available and genome and EST sequencing efforts are in progress for both *Chlamydomonas* (http://www.biology.duke.edu/chlamy_genome/) and *Zea mays* (http://www.zmdb.iastate.edu/).

In *Arabidopsis*, extensive mutagenesis programs are being carried out both in the academia and the private sector that use the T-DNA of *Agrobacterium tumefaciens* (Azpiroz-Leehan and Feldmann, 1997) or heterologous transposable elements (Wisman et al., 1998) as gene inactivation tools. In addition, a multitude of molecular markers and the availability of genome-covering BAC libraries allow positional cloning even of chemically mutagenized genes (Lukowitz et al., 2000). Selection for genes associated with thylakoid membrane biogenesis may rely on the high chlorophyll fluorescence phenotype as an easily scorable marker (Meurer et al., 1996). More sophisticated screening methods use time-resolved fluorescence imaging methods (Niyogi et al., 1998).

In addition to forward genetics, reversed genetics approaches are almost routine for *Arabidopsis* as well. They may be based on PCR (Baumann et al., 1998) or *in silico* screening (Parinov et al., 1999) of T-DNA- or transposon-mutagenized populations of that plant. Alternatively, gene inactivation by RNA interference is possible (Chuang and Meyerowitz, 2000). Reversed genetic approaches are particularly rewarding when considering the extensive duplications found in the *Arabidopsis* genome. Mutations of redundant genes may only lead to a mutant phenotype if all redundant genes have been mutated and the resulting alleles been combined in a single plant (cf. Pelaz et al., 2000).

Two further strategies to probe into the genetic anchoring of the photosynthetic process are chloroplast and compartment-alien (allotopic) transformation, and cybrid technology, respectively. In the tripartite plant genome, in principle the target of transformation can be any of the three organelles, with compartment-specific but also with compartment-alien transformation, that is, placing a gene equipped with appropriate regulatory and targeting sequences in a 'wrong' compartment. The latter approach would allow a search for constraints in (artificial) gene translocations, for instance. To date, the application of chloroplast transformation technology is restricted to two materials: the green alga *Chlamydomonas* and some Solanacean species, notably tobacco. Transformation of higher plant mitochondria has not been technically solved. Both a biolistic approach, originally developed for the unicellular alga *Chlamydomonas reinhardtii,* and a PEG-based strategy are available to transform chloroplasts of higher plants (Svab et al., 1990; Koop and Kofer, 1995). Despite of improvements for high transformation efficiency, rapid regeneration (Dovzhenko et al., 1998) and plastome segregation within 3–4 months, a more widespread application has been hampered from the lack of appropriate tissue or protoplast culture protocols and, in particular, by the lack of appropriate selection makers. The *aad*A cassette, conferring spectinomycin and streptomycin resistance and the only marker gene available todate, cannot be applied to various dicotyledonous species, including *Arabidopsis*, because both drugs display pseudo-auxin effects so that transformed material is being overgrown from non-transformed. The development of additional selective markers for plastid transformation is one of the more pressing questions in chloroplast biology.

Cybrid and hybrid technologies, in turn, are instrumental in investigating novel aspects in the compartmental interaction in photosynthesis and its impact for evolution. The partite plant genome reveals its integrated character in combinations of non-naturally occurring cellular subgenomes. Already early formal genetics has convincingly demonstrated that plastids and nuclei cannot be arbitrarily exchanged between even closely related species without the risk of impairing harmoniously balanced growth. Such an exchange either by interspecific crosses (Renner, 1934; Stubbe, 1959) or by somatic fusion of protoplasts with (nuclear-free) micro- or cytoplasts (summarized in Medgyesy, 1990) can greatly impair the photosynthetic machinery; the resulting hybrids are frequently bleached. 'Hybrid bleaching' reflects genetic incompatibility between alien genomes and plastomes combined in one cell and illustrates the specificity of the interaction between genetic compartments and, implicitly, their co-evolution. This represents a general principle and an integral factor in the evolution of plants (Kirk and Tilney-Bassett, 1978). The underlying mechanisms are not known. 'Hybrid bleaching' is a reversible and temporary feature, since impaired plastids from disharmonic combinations restore their original morphology and functionality when recombined with their natural (or another compatible) genome even after many generations under the influence of an incompatible nucleus. The fundamental role of both organelles in energy metabolism, genetic data and equivalent recent findings of degenerative human diseases caused by mitochondrial genomes (Wallace 1999) reinforce that organelle genomes are not only adjuncts in the genetic design of the cell.

B. Transcriptomics and Proteomics

New developments in techniques and instrumentation allow as well the simultaneous and high throughput determination of transcript (Chapters 31 and 32; Duggan et al., 1999) and protein (Pandey and Mann,

2000) patterns. Knowing the changes in transcript and protein amounts in time and space allows to establish a comprehensive expression catalogue of all the genes involved in a particular biological process (Chu et al., 1998). With this information available regulatory and functional networks can be characterized (Thieffry, 1999; Aharoni et al., 2000) and the underlying *cis-* and *trans*-regulatory signaling elements (Roth et al., 1998) may be identified. Genome-wide expression profiling may be particularly rewarding for the discovery of regulatory and functional networks, provided genetically well characterized material is available (Hughes et al., 2000). Unfortunately, these techniques have not been applied yet in a systematic way to study thylakoid membrane biogenesis in both pro- and eukaryotic photosynthetic organisms.

Acknowledgments

We would like to thank Dr. Rainer Maier and Holger Hupfer for assistance in the preparation of this manuscript, and Prof. Dr. W. Martin for agreement to include Fig. 3. The help of Ms. Renate Reichinger with the preparation of the figures is greatly appreciated. This work was supported by the Deutsche Forschungsgemeinschaft (SFB TR1) and the Fonds der Chemischen Industrie.

References

Adam Z (2000) Chloroplast proteases: Possible regulators of gene expression? Biochimie 82: 647–654

Aharoni A, Keizer LCP, Bouwmeester HJ, Sun ZK, Alvarez-Huerta M, Verhoeven HA, Blaas J, Van Houwelingen AMML, De Vos RCH, Van der Voet H, Jansen RC, Guis M, Mol J, Davis RW, Schena M, Van Tunen AJ and O'Connell AP (2000) Identification of the *SAAT* gene involved in strawberry flavor biogenesis by use of DNA microarrays. Plant Cell 12: 647–661

Allen JF (1992) Protein phosphorylation in regulation of photosynthesis. Biochim Biophys Acta 1098: 275–335

Allison LA (2000) The role of sigma factors in plastid transcription. Biochimie 82: 537–548

Allison LA, Simon LD and Maliga P (1996) Deletion of *rpoB* reveals a second distinct transcription system in plastids of higher plants. EMBO J 15: 2802–2809

Andersson B and Aro E-M (1997) Proteolytic activities and proteases of plant chloroplasts. Physiol Plant 100: 780–793

Andersson B and Barber J (1994) Composition, organization and dynamics of thylakoid membranes. In: E.E. Bittar (ed) Advances in molecular and cell biology (eds), pp 1–53. JAI press

Aro E-M, Virgin I and Andersson B (1993) Photoinhibition of Photosystem II. Inactivation, protein damage and turnover. Biochim Biophys Acta 1143: 113–134

Azizkhan JC, Jensen DE, Pierce AJ and Wade M (1993) Transcription from TATA-less promoters: Dihydrofolate reductase as a model. Crit Rev Eukaryot Gene Express 3: 229–254

Azpiroz-Leehan R and Feldmann KA (1997) T-DNA insertion mutagenesis in *Arabidopsis*: Going back and forth. Trends Genet 13: 152–156

Baker A, Schatz G (1987) Sequences from a prokaryotic genome or the mouse dihydrofolate reductase gene can restore the import of a truncated precursor protein into yeast mitochondria. Proc Natl Acad Sci USA 84: 3117–3121

Barkan A (1988) Proteins encoded by a complex chloroplast transcription unit are each translated from both monocistronic and polycistronic mRNAs. EMBO J 7: 2637–2644

Barkan A and Goldschmidt-Clermont M (2000) Participation of nuclear genes in chloroplast gene expression. Biochimie 82: 559–572

Barkan A and Stern D (1998) Chloroplast mRNA processing: intron splicing and 3′-end metabolism. In: Bailey-Serres J and Gallie DR (eds) A Look beyond transcription: Mechanisms determining mRNA stability and translation in plants, pp 162–173. Amer Soc Plant Physiol, Rockeville

Barkan A, Walker M, Nolasco M and Johnson D (1994) A nuclear mutation in maize blocks the processing and translation of several chloroplast mRNAs and provides evidence for the differential translation of alternative mRNA forms. EMBO J 13: 3179–3181

Bartsevich VV and Pakrasi HB (1995) Molecular identification of an ABC transporter complex for manganese: Analysis of a cyanobacterial mutant strain impaired in the photosynthetic oxygen evolution process. EMBO J 14: 1845–1853

Baumann E, Lewald J, Saedler H, Schulz B and Wisman E (1998) Successful PCR-based reverse genetic screens using an En-1-mutagenised *Arabidopsis thaliana* population generated via single-seed descent. Theor Appl Genet 97: 729–734

Berks BC (1996) A common export pathway for proteins binding complex redox cofactors? Mol Microbiol 22: 393–404

Berry J and Björkman O (1980) Photosynthetic response and adaptation to temperature in higher plants. Annu Rev Plant Physiol 31: 491–543

Blanc V, Litvak S and Araya A (1995) RNA editing in wheat mitochondria proceeds by a deamination reaction. FEBS Lett 373: 56–60

Blankenship RE (1992) Origin and early evolution of photosynthesis. Photosynth Res 33: 91–111

Bligny M, Courtois F, Thaminy S, Chang C-C, Lagrange T, Baruah-Wolff J, Stern D and Lerbs-Mache (2000) Regulation of plastid rDNA transcription by interaction of CDF2 with two different RNA polymerases. EMBO J 19: 1851–1860

Bock R (2000) Sense from nonsense: How the genetic information of chloroplasts is altered by RNA editing. Biochimie 82: 549–557

Brennicke A, Grohmann L, Hiesel R, Knoop V and Schuster W (1993) The mitochondrial genome on its way to the nucleus: Different stages of gene transfer in higher plants. FEBS Lett 325: 140–145

Brennicke A, Marchfelder A and Binder S (1999) RNA editing. FEMS Microbiol Rev 23: 297–316

Briggs WR and Huala E (1999) Blue-light photoreceptors in higher plants. Annu Rev Cell Dev Biol 15: 33–62

Burke TW and Kadonaga JT (1996) *Drosophila* TFIID binds to a conserved downstream basal promoter element that is present in many TATA-box-deficient promoters. Genes Dev 10: 711–724

Burke TW, Willy PJ, Kutach AK, Butler JEF and Kadonaga JT (1998) The DPE, a conserved downstream core promoter element that is functionally analogous to the TATA-box. Cold Spring Harb Symp Quant Biol 63: 75–82

Chu S, DeRisi J, Eisen M, Mulholland J, Botstein D, Brown PO and Herskowitz I (1998) The transcriptional program of sporulation in budding yeast. Science 282: 699–705

Chuang CF and Meyerowitz EM (2000) Specific and heritable genetic interference by double-stranded RNA in *Arabidopsis thaliana*. Proc Natl Acad Sci USA 97: 4985–4990

Cline K (2000) Gateway to the chloroplast. Nature 403: 148–149

Dalbey RE and Robinson C (1999) Protein translocation into and across the bacterial plasma membrane and the plant zhylakoid membrane. Trends Biochem Sci 24: 17–21

Danon A and Mayfield SPY (1994a) ADP dependent phosphorylation regulates RNA-binding in vitro: Implications in light-modulated translation. EMBO J 13: 2227–2235

Danon A and Mayfield SPY (1994b) Light regulated translation of chloroplasts messenger RNAs through redox potential. Science 266: 1717–1719

Deane JA, Fraunholz M, Su V, Maier U-G, Martin W, Durnford DG and McFadden GI (2000) Evidence for nucleomorph to host nucleus gene transfer: light-harvesting complex proteins from cryptomonads and chlorarachniophytes. Protist 151: 239–252

DeSantis-Maciossek G, Kofer W, Bock A, Schoch S, Maier RM, Wanner G, Rüdiger W, Koop H-U and Herrmann RG (1999) Targeted disruption of the plastid RNA polymerase genes *rpo*A, B and C1: Molecular biology, biochemistry and ultrastructure. Plant J 18: 477–490

Douglas S, Zauner S, Fraunholz M, Beaton M, Penny S, Deng L-T, Reith M, Cavalier-Smith T and Maier U-G (2001) The highly reduced genome of an enslaved algal nucleus. Nature 410: 1091–1096

Dovzhenko A, Bergen U and Koop H-U (1998) Thin-alginate-layer technique for protoplast culture of tobacco leaf protoplasts: shoot formation in less than two weeks. Protoplasma 204: 114–118

Duggan DJ, Bittner M, Chen YD, Meltzer P and Trent JM (1999) Expression profiling using cDNA microarrays. Nature Genet 21 Suppl: 10–14

Ellis J (1982) Promiscuous DNA—chloroplast genes inside plant mitochondria. Nature 299: 678–679

Enami I, Kitamura M, Tomo T, Isokawa Y, Ohta H and Katoh S (1994) Is the primary cause of thermal inactivation of oxygen evolution in spinach PS II membranes release of the extrinsic 33 kDa protein or of Mn? Biochim Biophys Acta 1186: 52–58

Falk J, Schmidt A and Krupinska K (1993) Characterization of plastid DNA transcription in ribosome-deficient plastids of heat-bleached barley leaves. J Plant Physiol 141: 176–180

Fargo DC, Zhang M, Gillham NW and Boynton JE (1998) Shine-Dalgarno-like sequences are not required for translation of chloroplast mRNAs in *Chlamydomonas reinhardtii* chloroplasts or in *Escherichia coli*. Mol Gen Genet 257: 271–282

Feller U, Crafts-Brandner SJ and Salvucci ME (1998) Moderately high temperatures inhibit ribulose-1,5-bisphosphate carboxylase/oxygenase (Rubisco) activase-mediated activation of Rubisco. Plant Physiol 116: 539–546

Flieger K, Tyagi A, Sopory S, Cséplö A, Herrmann RG, Oelmüller R (1993) A 42 bp promotor fragment of the gene for subunit III of Photosystem I (*psaF*) is crucial for its activity. Plant J 4: 9–17

Freyer R, Hoch B, Neckermann K, Maier RM and Kössel H (1993) RNA editing in maize chloroplasts is a processing step independent of splicing and cleavage to monocistronic mRNAs. Plant J 4: 621–629

Fulgosi H, Vener AV, Altschmied L, Herrmann RG and Andersson B (1998) A novel multi-functional chloroplast protein: identification of a 40 kDa immunophilin-like protein located in the thylakoid lumen. EMBO J 17: 1577–1587

Gibbs SP (1979) The route of entry of cytoplasmatically synthesized proteins into chloroplasts of algae possessing chloroplast ER. J Cell Sci 35: 253–266

Gounaris K, Brain ARR, Quinn PJ and Williams WP (1984) Structural reorganisation of chloroplast thylakoid membranes in response to heat-stress. Biochim Biophys Acta 766: 198–208

Gray MW (1999) Evolution of organellar genomes. Curr Opinion Genet Devel 9: 678–687

Hajdukiewicz P T J, Allison L A and Maliga P (1997) The two RNA polymerases encoded by the nuclear and plastid compartments transcribe distinct groups of genes in tobacco plastids. EMBO J 16: 4041–4048

Haseloff J (1999) GFP variants for multispectral imaging of living cells. Methods Cell Biol 58: 139–151

Havaux M, Tardy F, Ravenel J, Chanu D and Parot P (1996) Thylakoid membrane stability to heat stress studied by flash spectroscopic measurements of the electrochimic shift in intact potato leaves: Influence of xanthophyll content. Plant Cell Environ 19: 1359–1368

Hayes R, Kudla J, Schuster G, Gabay L, Maliga P and Gruissem W (1996) Chloroplast mRNA 3′-end processing by a high molecular weight protein complex is regulated by nuclear encoded RNA binding proteins. EMBO J 15:1132–1141

Hayes R, Kudla J and Gruissem W (1999) Degrading chloroplast mRNA: The role of polyadenylation. Trends Biochem Sci 24: 199–202

Hedtke B, Börner T and Weihe A (1997) Mitochondrial and chloroplast phage-like RNA polymerases in *Arabidopsis*. Science 277: 809–811

Heins L, Collison I and Soll J (1998) The protein translocation apparatus of chloroplast envelopes. Trends Plant Sci 3: 56–61

Herrmann RG (1996) Photosynthesis research: Aspects and perspectives. In: Andersson B, Salter HA and Barber J (eds) Frontiers of Molecular Biology. Molecular Genetics in Photosynthesis, pp 1–44. IRL Press, Oxford

Herrmann RG (1997) Eukaryotism, towards a new interpretation. In: Schenk HEA, Herrmann RG, Jeon KW, Müller NE and Schwemmler W (eds) Eukaryotism and Symbiosis, pp 73–118. Springer, Heidelberg, New York

Herrmann RG (2000) Organelle genetics—Part of the integrated plant genome. Vortrg Pflanzenzücht 48: 279–296

Herrmann RG, Oelmüller R, Bichler J, Schneiderbauer A, Steppuhn J, Wedel N, Tyagi AK and Westhoff P (1991) The thylakoid membrane of higher plants: Genes, their expression and interaction. In: RG Herrmann and BA Larkins (eds) Plant

Molecular Biology 2, NATO ASI Series A: Life Sciences, Vol 212, pp 411–427. Plenum Publ Corp, New York

Herrmann RG, Westhoff P and Link G (1992) Biogenesis of plastids in higher plants. In: Herrmann RG (ed) Plant Gene Research, Vol 6, pp 275–349. Springer, Wien, New York

Hess WR and Börner T (1999) Organellar RNA polymerases of higher plants. Int Rev Cytol 190: 1–59

Hess WR, Prombona A, Fieder B, Subramanian AR and Börner T (1993) Chloroplasts *rps15* and the *rpoB/C1/C2* gene cluster are strongly transcribed in ribosome-deficient plastids: Evidence for a functioning non-chloroplast-encoded RNA polymerase. EMBO J 12: 563–571

Hirose T and Sugiura M (1996) *Cis*-acting elements and *trans*-acting factors for accurate translation of chloroplast *psbA* mRNAs: Development of an in vitro translation system from tobacco chloroplast. EMBO J 15: 1687–1695

Hofmann CJB, Rensing SA, Häuber MM, Martin WF, Müller SB, Couch J, McFadden GI, Igloi GL and Maier U-G (1994) The smallest known eukaryotic genomes encode a protein gene: Towards an understanding of nucleomorph functions. Mol Gen Genet 243: 600–604

Hughes TR, Marton MJ, Jones AR, Roberts CJ, Stoughton R, Armour CD, Bennett HA, Coffey E, Dai HY, He YDD, Kidd MJ, King AM, Meyer MR, Slade D, Lum PY, Stepaniants SB, Shoemaker DD, Gachotte D, Chakraburtty K, Simon J, Bard M and Friend SH (2000) Functional discovery via a compendium of expression profiles. Cell 102: 109–126

Kaneko T, Sato S, Kotani H, Tanaka A, Asamizu E, Nakamura Y, Miyajima N, Hirosawa M, Sugiura M, Sasamoto S, Kimura T, Hosouchi T, Matsuno A, Muraki A, Nakazaki N, Naruo K, Okumura S, Shimpo S, Takeuchi C, Wada T, Watanabe A, Yamada M, Yasuda M and Tabata S (1996) Sequence analysis of the genome of the unicellular cyanobacterium *Synechocystis* sp. strain PCC6803. II. Sequence determination of the entire genome and assignment of potential protein-coding regions. DNA Res 3: 109–136

Kapoor S and Sugiura M (1998) Expression and regulation of plastid genes. In: Raghavendra AS (ed) Photosynthesis, pp 72–86. Cambridge University Press, Cambridge

Kapoor S, Suzuki Y-Y and Sugiura M (1997) Identification and functional significance of a new class of non-consensus-type plastid promoters. Plant J 11: 327–337

Keegstra K and Cline K (1999) Protein import and routing systems of chloroplasts. Plant Cell 11: 557–570

Kirk JTO and Tilney-Bassett RAE (1978) The Plastids. Their chemistry, structure, growth and inheritance. Elsevier North Holland, Amsterdam

Koop H-U and Kofer W (1995) Plastid transformation by polyethylene glycol treatment of protoplasts and regeneration of transplastomic tobacco plants. In: Potrykus I and Spangenberg G (eds) Gene transfer to plants, pp. 75–82. Springer Verlag, Berlin

Kowallik KV (1997) Origin and evolution of chloroplasts: Current status and future perspectives. In: Schenk HEA, Herrmann, RG, Jeon KE and Schwemmler W (eds) Eukaryotism and Symbiosis, pp 3–23. Springer, Heidelberg

Krause K, Maier RM, Kofer W, Krupinska K and Herrmann RG (2000) Disruption of plastid-encoded RNA polymerase genes in tobacco: Expression of only a distinct set of genes is not based on selective transcription of the plastid chromosome. Mol Gen Genet 263: 1022–1030

Kroll D, Meierhoff K, Bechtold N, Kinoshita M, Westphal S, Vothknecht U, Soll J and Westhoff P (2001) VIPP1, a nuclear gene of *Arabidopsis thaliana* essential for thylakoid membrane formation. Proc Natl Acad Sci USA 98: 4238–4242

Lang BF, Burger G, O'Kelly CJ, Cedergren R, Golding GB, Lemieux C, Sankoff D, Turmel M and Gray MW (1997) An ancestral mitochondrial DNA resembling a eubacterial genome in miniature. Nature 387: 493–497

Legen J, Meurer J, Herrmann RG and Miséra S (2001) Map positions of 70 *Arabidopsis thaliana* genes that encode all known constituent polypeptides and various regulatory factors of the photosynthetic machinery: A case study. DNA Res 8: 53–60

Lerbs-Mache S (1993) The 110-kDa polypeptide of spinach plastid DNA-dependent RNA polymerase: Single-subunit enzyme or catalytic core of multimeric enzyme complexes? Proc Natl Acad Sci USA 90: 5509–5513

Liere K and Maliga P (1999) In vitro characterization of the tobacco *rpo*B promoter reveals a core sequence motif conserved between phage-type plastid and plant motochondrial promoters. EMBO J 18: 249–257

Link G (1996) Green life: Control of chloroplast gene transcription. BioEssays 18: 465–471

Long SP, Humphries S and Falkowski PG (1994) Photoinhibition of photosynthesis in nature. Annu Rev Plant Physiol Plant Mol Biol 45: 633–662

Lübberstedt T, Bolle C E H, Sopory S, Flieger K, Herrmann R G and Oelmüller R (1994) Promoters from genes for plastid proteins possess regions with different sensitivities toward red and blue light. Plant Physiol 104: 997–1006

Lukowitz W, Gillmor CS and Scheible WR (2000) Positional cloning in *Arabidopsis*. Why it feels good to have a genome initiative working for you. Plant Physiol 123: 795–805

Maier RM, Zeltz P, Kössel H, Bonnared G, Gualberto JM and Grienenberger JM (1996) RNA editing in plant mitochondria and chloroplasts. Plant Mol Biol 32: 343–365

Maier UG, Hofmann CJB, Eschbach S, Wolters J and Igloi GL (1991) Demonstration of nucleomorph-encoded eukaryotic ribosomal ssu RNA in cryptomonads. Mol Gen Genet 230: 155—160

Maier U-G, Douglas SE and Cavalier-Smith T (2000) The nucleomorph genomes of cryptopyhtes and chlorarachniophytes. Protist 151: 103-109

Malek O, Lättig K, Hiesel R, Brennicke A and Knoop V (1996) RNA editing in bryophytes and a molecular phylogeny of land plants. EMBO J 15: 1403–1411

Maréchal E, Block MA, Dorne AJ, Douce R and Joyard J (1997) Lipid synthesis and metabolism in the plastid envelope. Physiol Plant 100: 65–77

Martin W and Herrmann RG (1998) Gene transfer from organelles to the nucleus: how much, what happens, and why? Plant Physiol 118: 9–17

Martin W and Schnarrenberger C (1997) The evolution of the Calvin cycle from prokaryotic to eukaryotic chromosomes: A case study of functional redundancy in ancient pathways through endosymbiosis. Curr Genet 32: 1–8

Martin W, Stoebe B, Goremykin V, Hansmann S, Hasegawa M and Kowallik KV (1998) Gene transfer to the nucleus and the evolution of chloroplasts. Nature 393: 162–165

McFadden GI and Gilson P (1995) Something borrowed, something green: Lateral transfer of chloroplasts by secondary endosymbiosis. Trends Ecol Evol 10: 12–17

Medgyesy P (1990) Selection and Analysis of Cytoplasmatic Hybrids. In: Dix PJ (ed) Plant Cell Line Selection, pp 287–316. VCH Verlagsgesellschaft mbH, Weinheim, New York, Basel Cambridge

Meurer J, Meierhoff K and Westhoff P (1996) Isolation of high-chlorophyll-fluorescence mutants of *Arabidopsis thaliana* and their characterisation by spectroscopy, immunoblotting and Northern hybridisation. Planta 198: 385–396

Meurer J, Plücken H, Kowallik KV and Westhoff P (1998) A nuclear-encoded protein of prokaryotic origin is essential for the stability of Photosystem II in *Arabidopsis thaliana*. EMBO J 17: 5286–5297

Michl D, Robinson C, Shackleton J B, Herrmann R G and Klösgen R B (1994) Targeting of proteins to the thylakoids by bipartite presequences: CFoII is imported by a novel, third pathway. EMBO J 13: 1310—1317

Michl D, Karnauchov I, Berghöfer J, Herrmann RG and Klösgen RB (1999) Phylogenetic transfer of organelle genes to the nucleus can lead to new mechanisms of protein integration into membranes. Plant J 17: 31–40

Monde RA, Schuster G and Stern DB (2000) Processing and degradation of chloroplast mRNA. Biochimie 82: 573–582

Morden CW, Wolfe KH, dePamphilis CW and Palmer JD (1991) Plastid translation and transcription genes in a non-photosynthetic plant: Intact, missing and pseudo genes. EMBO J 10: 3281–3288

Nakamoto SS, Hamel P and Merchant S (2000) Assembly of chloroplast cytochromes *b* and *c*. Biochimie 82: 603–614

Nitschke W and Rutherford AW (1991) Photosynthetic reaction centres: Variations on a common structural theme? TIBS 16: 241–245

Niyogi KK (1999) Photoprotection revisited: Genetic and molecular approaches. Annu Rev Plant Physiol Plant Mol Biol 50: 333–359

Niyogi KK, Grossman AR and Björkman O (1998) *Arabidopsis* mutants define a central role for the xanthophyll cycle in the regulation of photosynthetic energy conversion. Plant Cell 10: 1121–1134

Nunnari J, Fox TD, Walter P (1993) A mitochondrial protease with two catalytic subunits of nonoverlapping specifities. Science 262: 1997–2004

Ohtsuki S, Levine M and Cai HN (1998) Different core promoters possess distinct regulatory activities in the *Drosophila* embryo. Genes Dev 12: 547–556

Pandey A and Mann M (2000) Proteomics to study genes and genomes. Nature 405: 837–846.

Parinov S, Sevugan M, Ye D, Yang WC, Kumaran M and Sundaresan V (1999) Analysis of flanking sequences from *Dissociation* insertion lines: A database for reverse genetics in Arabidopsis. Plant Cell 11: 2263–2270

Pelaz S, Ditta GS, Baumann E, Wisman E and Yanofsky MF (2000) B and C floral organ identity functions require *SEPALLATA* MADS-box genes. Nature *405*: 200–203

Prasil O, Adir N and Ohad I (1992) Dynamics of Photosystem II: Mechanism of photoinhibition and recovery processes. In: Barber J (ed) The Photosystems: Structure, Function and Molecular Biology, pp 295–348. Elsevier, Amsterdam

Preiss T and Hentze MW (1999) Form factors to mechanisms: translation and translational control in eukaryotes. Curr Opin Genet Dev 9: 515–521

Renner O (1934) Die pflanzlichen Plastiden als selbständige Elemente der genetischen Konstitution. Ber Sächs Akad Wiss, Math-Phys Kl 86: 214–266

Rintamäki E, Salonen M, Suoranta UM, Carlberg I, Andersson B and Aro E-M (1997) Phosphorylation of light-harvesting complex II and Photosystem II core proteins shows different irradiance-dependent regulation in vivo. Application of phosphothreonine antibodies to analysis of thylakoid phosphoproteins. J Biol Chem 272: 30476–30482

Rintamäki E, Martinsuo P, Pursiheimo S and Aro EM (2000) Cooperative regulation of light-harvesting complex II phosphorylation via the plastoquinol and ferredoxin-thioredoxin system in chloroplasts. Proc Natl Acad Sci USA 97: 11644–11649.

Robinson C and Klösgen RB (1994) Targeting of proteins into and across the thylakoid membrane—a multitude of mechanisms. Plant Mol Biol 26: 15–24

Robinson C, Hynds P J, Robinson D and Mant A (1998) Multiple pathways for the targeting of thylakoid proteins in chloroplasts. Plant Mol Biol 38: 209–221

Rochaix J D (1996) Post-transcriptional regulation of chloroplast gene expression in *Chlamydomonas reinhardtii*. Plant Mol Biol 32: 327–341

Roeder RG (1998) Role of general and gene-specific cofactors in the regulation of eukaryotic transcription. Cold Spring Harb Symp Quant Biol 63: 201–218

Rokka A, Aro E-M, Herrmann RG, Andersson B and Vener AV (2000) Dephosphorylation of Photosystem II reaction center proteins in plant photosynthetic membranes as an immediate response to abrupt elevation of temperature. Plant Physiol 123: 1525–1535

Rolland N, Dorne AJ, Amoroso G, Sultemeyer DF, Joyard J and Rochaix JD (1997) Disruption of the plastid *ycf*10 open reading frame affects uptake of inorganic carbon in the chloroplast of *Chlamydomonas*. EMBO J 16: 6713–6726

Roth FP, Hughes JD, Estep PW and Church GM (1998) Finding DNA regulatory motifs within unaligned noncoding sequences clustered by whole-genome mRNA quantitation. Nature Biotechnol 16: 939–945

Roy AL, Malik S, Meisterernst M and Roeder RG (1993) An alternative pathway for transcription initiation involving TFII-I. Nature 365, 355–359

Samuelson JC, Chen M, Jiang F, Möller I, Wiedmann M, Kuhn A, Philips GJ and Dalbey RE (2000) YidC mediates membrane protein insertion in bacteria. Nature 406: 637–641

Sasaki Y, Sekiguchi K, Nagano Y and Matsuno R (1993) Chloroplast envelope protein encoded by chloroplast genome. *FEBS Lett* 316: 93–98

Schatz G and Dobberstein B (1996) Common principles of protein translocation across membranes. Science 271: 1519–1526

Schleiff E and Soll J (2000) Travelling of proteins through membranes: translocation into chloroplasts. Planta 211: 449–456

Schnell D J (1998) Protein targeting to the thylakoid membrane. Annu Rev Plant Physiol Plant Mol Biol 49: 97–126

Schöffl F, Prandl R and Reindl A (1998) Regulation of the heat-shock response. Plant Physiol 117: 1135–1141

Serino G and Maliga P (1998) RNA polymerase subunits encoded by the plastid rpo genes are not shared with the nucleus-encoded plastid enzyme. Plant Physiol 117: 1165–1170

Shestakov S and Khyen N (1970) Evidence for genetic transformation in blue-green algae. Mol Gen Genet 107: 372–375

Sitte P (1987) Zellen in Zellen: Endocytobiose und die Folgen. In: Lüst R (ed) Experiment und Theorie in Naturwissenschaft und Medizin, pp 431–446. Wissenschaftliche Verlagsgesellschaft, Stuttgart

Smale ST and Baltimore D (1989) The 'Initiator' as a transcription control element. Cell 57: 103–113

Smale ST, Jain A, Kaufmann J, Emami KH, Lo K and Garraway IP (1998) The initiator element: A paradigm for core promoter heterogeneity within metazoan protein-coding genes. Cold Spring Harb Symp Quant Biol 63: 21–31

Smith H (2000) Phytochromes and light signal perception by plants—an emerging synthesis. Nature 407: 585–591

Snyders S and Kohorn BD (1999) TAKs, thylakoid membrane protein kinases associated with energy transduction. J Biol Chem 274: 9137—9140

Sokolenko A, Altschmied L and Herrmann RG (1997) Sodium dodecylsulfate-stable proteases in chloroplasts. Plant Physiol 115: 827–832

Sokolenko A, Lerbs-Mache S, Altschmied L and Herrmann RG (1998) Clp protease complexes and their diversity in chloroplasts. Planta 207: 286–295

Soll J, Bolter B, Wagner R and Hinnah SC (2000) ...Response: The chloroplast outer envelope: A molecular sieve? Trends Plant Sci 5: 137–138

Sonoike K, Terashima I, Iwaki M and Itoh S (1995) Destruction of Photosystem I iron-sulfur centres in leaves of *Cucumis sativus* L. by weak illumination at chilling temperatures. FEBS Lett 362: 235–238

Steinkamp T, Hill K, Hinnah SC, Wagner R, Rohl T, Pohlmeyer K and Soll J (2000) Identification of the pore-forming region of the outer chloroplast envelope protein OEP16. J Biol Chem 275: 11758–11764

Stern DB and Lonsdale DM (1982) Mitochondrial and chloroplast genomes of maize have a 12-kilobase DNA sequence in common. Nature 299: 698–702

Stockhaus J, Schlue U, Koczor M, Chitty JA, Taylor WC and Westhoff P (1997) The promoter of the gene encoding the C_4 form of phosphoenolpyruvate carboxylase directs mesophyll specific expression in transgenic C_4 *Flaveria* spp. Plant Cell 9: 479–489

Stubbe W (1959) Genetische Analyse des Zusammenwirkens von Genom und Plastom bei Oenothera. Z Vererbungslehre 90: 288–298

Sugiura M, Hirose T and Sugita M (1998) Evolution and mechanisms of translation in chloroplasts. Annu Rev Genet 32: 437–459

Summer EJ, Mori H, Settles AM and Cline K (2000) The thylakoid delta pH-dependent pathway machinery facilitates RR-independent N-tail protein integration. J Biol Chem 275: 23483–23490

Sundby C, Melis A, Mäenpää P and Andersson B (1986) Temperature-dependent changes in antenna size of Photosystem II. Reversible conversion of Photosystem IIα to Photosystem IIβ. Biochim Biophys Acta 851: 475–483

Svab Z, Hajdukiewicz P and Maliga P (1990) Stable transformation of plastids in higher plants. Proc Natl Acad Sci USA 87: 8526–8530

Tardy F and Havaux M (1997) Thylakoid membrane fluidity and thermostability during the operation of the xanthophyll cycle in higher-plant chloroplasts. Biochim Biophys Acta 1330: 179–193

The *Arabidopsis* Genome Initiative (2000) Analysis of the genome sequence of the flowering plant *Arabidopsis thaliana*. Nature 408: 796–815

Thieffry D (1999) From global expression data to gene networks. BioEssays 21: 895–899

VanderVere PS, Bennet TM, Oblong JE and Lampp GK (1995) A chloroplast processing enzyme involved in precursor maturation shares a zinc-motif with a recently recognized family of metallopeptidases. Proc. Natl. Acad. Sdci. USA 92: 7177–7181

Vener A, Ohad I and Andersson B (1998) Protein phosphorylation and redox sensing in chloroplast thylakoids. Curr Opin Plant Biol 1: 217–223

Vener AV, Rokka A, Fulgosi H, Andersson B and Herrmann RG (1999) A cyclophilin-regulated PP2A-like protein phosphatase in thylakoid membranes of plant chloroplasts. Biochemistry 38: 14955–14965

von Heijne G, Steppuhn J and Herrmann RG (1989) Domain structure of mitochondrial and chloroplast targeting peptides. Eur J Biochem 180: 535–545

Waegemann K and Soll J (1996) Phosphorylation of the transit sequence of chloroplast precursor proteins. J Biol Chem 271: 6545–6554

Wallace DC (1999) Mitochondrial disease in man and mouse. Science 283: 1482–1488

Waller RF, Keeling PJ, Donald RG, Striepen B, Handman E, Lang-Unnasch N, Cowman AF, Besra GS, Roos DS and McFadden GI (1998) Nuclear-encoded proteins target to the plastid in *Toxoplasma gondii* and *Plasmodium falciparum*. Proc Natl Acad Sci USA 95: 12352–12357

Waller RF, Reed MB, Cowman AF and McFadden GI (2000) Protein trafficking to the plastid of *Plasmodium falciparum* is via the secretory pathway. EMBO J 19: 1794–1802

Weber P, Sokolenko A, Eshaghi S, Fulgosi H, Vener AV, Andersson B, Ohad I and Herrmann RG (2001) Elements of signal transduction involved in thylakoid membrane dynamics. In: Sopory SK, Oelmüller R and Maheshwari SC (eds) Signal Transduction in Plants, Current Advances, pp 222–227. Kluwer Academic/Plenum Publishers, Dordrecht

Weeden N W (1981) Genetic and biochemical implications of the endosymbiotic origin of the chloroplast. J Mol Evol 17: 133–139

Weis E and Berry JA (1988) Plants and high temperature stress. Symp Soc Exp Biol 42: 329–346

Westhoff P and Herrmann RG (1988) Complex RNA maturation in chloroplasts: The *psb*B operon from spinach. Eur J Biochem 171: 551–564

Wisman E, Cardon GH, Fransz P and Saedler H (1998) The behaviour of the autonomous maize transposable element *En/Spm* in *Arabidopsis thaliana* allows efficient mutagenesis. Plant Mol Biol 37: 989–999

Woese CR (1987) Bacterial evolution. Microbiol Rev 51: 221–271

Woese C (1998) The universal ancestor. Proc Natl Acad Sci USA 95: 6854–6859

Woese CR, Kandler O and Wheelis ML (1990) Towards a natural system of organisms: Proposal for the domains Archaea, Bacteria, and Eucarya. Proc Natl Acad Sci USA 87: 4576–4579

Yamamoto YY, Kondo Y, Kato A, Tsuji H and Obokata J (1997) Light-responsive elements of the tobacco PSI-D gene are located both upstream and within the transcribed region. Plant J 12: 255–265

Yu W and Schuster W (1995) Evidence for a site-specific cytidine deamination reaction involved in C-to-U RNA editing of plant mitochondria. J Biol Chem 270: 18227–18233

Zerges W (2000) Translation in chloroplasts. Biochimie 82: 583–601

Zito F, Kuras R, Choquet Y, Kossel H and Wollman FA (1997) Mutations of cytochrome b6 in *Chlamydomonas reinhardtii* disclose the functional significance for a proline to leucine conversion by petB editing in maize and tobacco. Plant Mol Biol 33: 79–86

Zuckerkandl E and Pauling L (1965) Molecules as documents of evolutionary history. J Theor Biol 8: 357–366

Chapter 2

Plastid RNA Polymerases in Higher Plants

Karsten Liere* and Pal Maliga†
Waksman Institute, Rutgers, the State University of New Jersey,
190 Frelinghuysen Road, Piscataway, NJ, 08854-8020, U.S.A.

Summary

Plastids evolved from ancestral cyanobacteria through gradual conversion of an endosymbiont to a plant organelle. Plastids maintained a cyanobacterium-like (eubacterial) transcription machinery. The eubacterial core-enzyme consists of four plastid-encoded subunits (α_2, β, β' and β''), and may associate with multiple, nuclear-encoded σ^{70}-type specificity factors. This holo-enzyme is the plastid-encoded plastid RNA polymerase (PEP). The promoters recognized by the PEP are of σ^{70}-type with conserved –10 (TATAAT) and –35 (TTGACA) elements. In addition, species-specific cis-elements and trans-factors regulate *psbA*, *psbD* and *rrn16* promoter activity. The PEP in chloroplasts associates with up to eight auxiliary proteins. One of them is the plastid transcription kinase (PTK), an enzyme which regulates PEP transcription by σ factor phosphorylation. PTK activity itself is regulated by phosphorylation and the redox state of plastids.

In addition to the eubacterial enzyme, plastids have acquired a second, phage-type RNA polymerase (NEP, nuclear-encoded plastid RNA polymerase). NEP probably evolved by duplication of the mitochondrial transcription machinery. A nuclear gene encodes the NEP catalytic core with a plastid targeting N-terminal sequence. The NEP subunit composition is likely to be similar to the mitochondrial enzyme, which associates

*Current address: Humboldt-University Berlin, Institute of Biology. †Author for correspondence, email: maliga@wakeman.rutgers.edu

Eva-Mari Aro and Bertil Andersson (eds): Regulation of Photosynthesis, pp. 29–49.

with at least two specificity factors. NEP recognizes two distinct promoters. Type-I NEP promoters are ~15 nt AT-rich region upstream (−14 to +1) of the transcription initiation site (+1) with a conserved YRTA core, a feature shared with plant mitochondrial promoters. Type-II NEP promoters are mainly downstream (−5 to +25) of the transcription initiation site.

There is a division of labor between the two plastid RNA polymerases. Photosynthetic genes and operons have PEP promoters, whereas most non-photosynthetic genes involved in housekeeping functions, such as transcription and translation, have promoters for both RNA polymerases. The NEP promoter(s) of these genes are, with a few exceptions, silent in chloroplasts. Only a few genes are transcribed exclusively from a NEP promoter. One of these is the *rpoB* operon encoding three of the four PEP core subunits. Through transcription of the PEP genes by the NEP the nucleus indirectly controls transcription of plastid genes, thereby integrating the endosymbiont-turned-organelle into the developmental network of multicellular plants.

I. Introduction

The plant cell has three DNA-containing compartments: the nucleus, the mitochondria and the plastids. Plastids evolved from ancestral cyanobacteria through gradual conversion of the endosymbiont to a plant organelle. Although plastids have their own genome and prokaryotic-type gene expression system, they are genetically semi-autonomous and require nuclear gene products for biogenesis and function (Sugiura, 1992; Gray, 1993).

Since the early 1970s, plastid transcription has been extensively studied. Sequence analysis of plastid genomes revealed *rpo* genes encoding homologues of the eubacterial RNA polymerase (RNAP) α, β and β' subunits in the unicellular alga *Euglena gracilis* (Little and Hallick, 1988), and in maize (Hu and Bogorad, 1990, 1991). Biochemical characterization of sigma factor homologues was first accomplished in *Chlamydomonas reinhardtii* (Surzycki and Shellenbarger, 1976), followed by similar work in spinach (Lerbs et al., 1983) and mustard (Bülow and Link, 1988). Cloning of the genes for the sigma factors awaited the discovery of PCR (polymerase chain reaction). Homologues of the *E. coli* σ^{70} factor were reported only a few years ago in the red alga *Cyanidium caldarium* (Liu and Troxler, 1996; Tanaka et al., 1996), then in *Arabidopsis thaliana* (Isono et al., 1997b; Tanaka et al., 1997). Promoters for the *E. coli*-like enzyme were first characterized in chloroplast extracts in vitro (Jolly and Bogorad, 1980; Orozco et al., 1985) then in vivo using a transgenic approach (Allison and Maliga, 1995).

The role of a phage-type plastid RNA polymerase in plastid gene transcription has been recognized only recently. The catalytic subunit of the plastid enzyme was fortuitously cloned while searching for the catalytic subunit of the mitochondrial RNA polymerase (Hedtke et al., 1997). Characterization of promoters for the phage-type plastid enzyme was made feasible by plastid transformation, the method used to obtain tobacco plants lacking the *E. coli*-like enzyme by targeted deletion of the plastid *rpo* genes (Allison et al., 1996; Serino and Maliga, 1998). The two plastid RNA polymerases are distinguished by the localization of the genes for their catalytic cores: the *E. coli*-like enzyme is referred to as plastid-encoded plastid RNA polymerase or PEP, whereas the phage-type enzyme is termed NEP for nucleus-encoded plastid RNA polymerase (Hajdukiewicz et al., 1997).

Data on plastid gene transcription were previously viewed as universally applicable from algae to higher plants. As more information accumulates on the plastid transcription machinery, it is becoming increasingly clear that there are significant differences between the plastids of unicellular algae and higher plants. For example *Chlamydomonas reinhardtii*, the model unicellular alga, apparently lacks the phage-type plastid transcription machinery (Rochaix, 1995). Furthermore, there are significant differences between monocots and dicots, or even within dicots, with respect to specific details of transcription. Earlier work on the plastid transcription machinery has been summarized by Bogorad (1991), Igloi and Kössel (1992), Sugiura (1992), Mullet (1993), and Link (1996). For more recent developments see reviews

Abbreviations: AGF – AAG-box binding factor; BLRP – blue-light responsive promoter; CDF – chloroplast DNA-binding factor; NEP – nuclear-encoded plastid RNA polymerase; PCR – polymerase chain reaction; PEP – plastid-encoded plastid RNA polymerase; PGTF – PGT-box binding factor; PTK – plastid transcription kinase; RNAP – RNA polymerase; SLF – sigma-like factor

by Stern et al. (1997), Maliga (1998), and Hess and Börner (1999). This chapter will limit itself to an assessment of data published on the plastid transcription machinery of higher plants.

II. The Plastid-Encoded Plastid RNA Polymerase (PEP)

A. The PEP Enzyme

PEP has been purified as a soluble or DNA-bound polymerase activity. The latter was also termed transcriptionally active chromosome (TAC). There has been much discussion in the literature concerning the composition and role of the TAC in plastid gene expression (Hallick et al., 1976; Briat et al., 1979; Gruissem et al., 1983a,b; 1986b; Greenberg et al., 1984; Narita et al., 1985; Reiss and Link, 1985; for reviews see Igloi and Kössel, 1992; Lakhani et al., 1992; Link, 1994). It appears that both soluble and TAC preparations contain PEP, at least they share the core α and β PEP subunits (Suck et al., 1996; K. Krause and K. Krupinska, unpublished). Therefore, it is most likely that the complete PEP is integrated into the TAC complex, which might be a functional super-complex combining transcriptional, post-transcriptional and even translational activity. Whether NEP is also part of the TAC complex remains to be determined.

Soluble plastid transcription extracts prepared from maize contain polypeptides encoded by the plastid *rpo* genes, which correspond to the core subunits of the *E. coli* RNA polymerase (Hu and Bogorad, 1990; Hu et al., 1991). However, the peptide pattern of plastid RNA polymerase preparations is more complex than that of the bacterial enzyme. Most illuminating was the study about the peptide composition of mustard plastid RNA polymerases at different developmental stages (Pfannschmidt and Link, 1994). Etioplasts, the plastid type in dark-grown leaves, contain a plastid RNAP similar in composition to the *E. coli* enzyme. This simple enzyme (peak B) consists of the four *E. coli*-like core subunits. When dark-grown plants are transferred to light, etioplasts develop into photosynthetically active chloroplasts. In mustard chloroplasts a more complex form (peak A) is present consisting of at least 13 major polypeptides. Nevertheless, both the *E. coli*-like and the more complex form of the enzyme recognize the same set of plastid promoters, suggesting conversion of the simple form (in etioplasts) to the complex form (in chloroplasts) by recruiting additional components (Fig. 1; Pfannschmidt and Link, 1997). One of the additional components is a heterotrimeric kinase (Baginsky et al., 1997; 1999). The identity of most of the other polypeptides in the complex form remains to be determined.

There are conflicting reports on the composition of plastid RNA polymerases (Table 1). These may be attributable to the characterization of the simple or complex forms from different species. Alternatively, the differences may reflect genuine species-specific

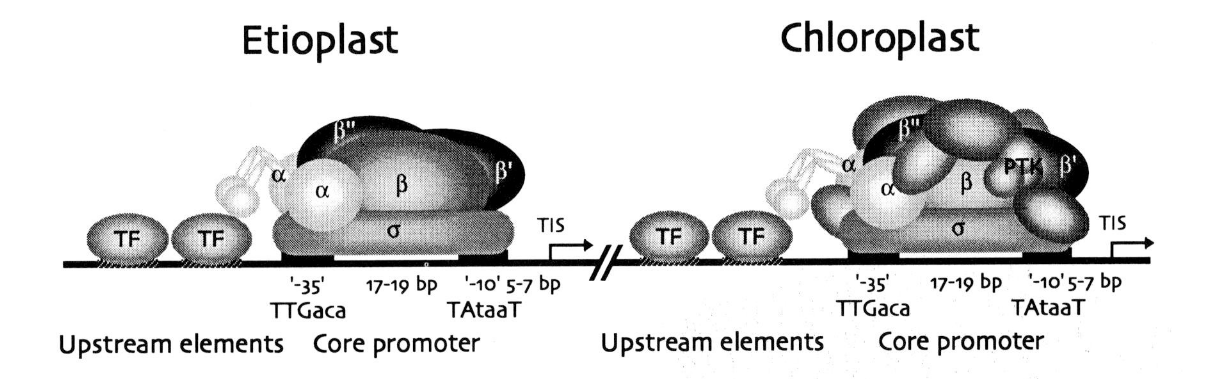

Fig. 1. The plastid-encoded plastid RNA polymerase (PEP): the simple form of PEP exists in etioplasts and the complex form in chloroplasts. In etioplasts, PEP consists of the core α_2, β, β', and β'' subunits and the nuclear-encoded σ-like factor required for promoter recognition. In chloroplasts, PEP associates with additional nuclear-encoded auxiliary factors (unmarked proteins), including the plastid transcription kinase (PTK) (Table 1). Upstream cis-elements recognized by specific trans-factors (TF) may enhance PEP transcription. Conserved hexamers of the plastid σ^{70}-type core promoter are also shown. Highly conserved nucleotides of the –10/–35 promoter elements are in capital letters. Horizontal arrows mark transcription initiation sites (TIS).

Table 1. Subunit composition of chloroplast RNA polymerases from maize, pea, mustard and spinach

Maize[a]	Maize[a*]	Pea[b]	Pea[b*]	Mustard 'B'[c]	Mustard 'A'[c]	Spinach[d]
180 = β''	180	180	150	154 β''[5]	141	150 {β,β' -like}[3]
120 = β	140	140	130	120 β[5]	110	145 {β,β' -like}[3]
85	120	110	115	93	107	110
78 = β' [a**]	110	95	110	85	78	102
64[a**]	100	65	95	72 β'[5]	72 [K][c*]	90 {σ}[4, d*]
61[a**] = BP[1]	95	47	(90)	61	(70)	80 {β,β' -like}[3] [DB][d**]
55[a**] {RbcL}[2]	85	27	85	42	48 54 [K][c*]	75 [DB][d**]
38 = α	75		75	38 α[5]	(35/6) 41 [K][c*]	64
	70		68[b**]	33	29	58
	55		60[b**]		(26)	53
	70		48		(16)	48
	42		44		(13)	38
	40		39			34
	38		(34)			33 {σ}[4, d*]
	27		(32)			27
			(27)			17
			15[b**]			

= identity to chloroplast proteins by peptide sequence
[1] BP, Rubisco binding protein
{} identified by antibody cross-reactivity to: [2] RubisCo large subunit
[3] *E. coli* RNA polymerase β, β' subunits
[4] *E. coli* σ^{70}
[5] predicted identity by authors
[ ] function determined experimentally: [DB] DNA-binding protein
[K] Plastid Transcription Kinase (PTK)
( ) minor proteins
[a] (Hu and Bogorad, 1990); [a*](Kidd and Bogorad, 1980); [a**](Hu et al., 1991)
[b] (Tewari and Goel, 1983); [b*](Rajasekhar et al., 1991); [b**](Rajasekhar and Tewari, 1995)
[c] (Pfannschmidt and Link, 1994); [c*](Baginsky et al., 1997); 'A' 'B' refer to the complex and the simple (*E. coli*-like) RNAP forms, respectively
[d] (Lerbs et al., 1985); [d*](Lerbs et al., 1988); [d**](Lerbs et al., 1983)

differences in size and subunit composition, differential protein modification, or differences in the purification protocols.

The *E. coli* RNA polymerase is sensitive to rifampicin. Interestingly, the simple plastid PEP is rifampicin sensitive while the complex form is rifampicin resistant in vitro, although the two enzymes share the same catalytic core. Lack of rifampicin sensitivity of the complex form may be explained by inaccessibility of the rifampicin target site in the β subunit (Pfannschmidt and Link, 1994).

B. The PEP Catalytic Core

Plastids, in agreement with their endosymbiotic origin, have maintained a cyanobacterium-like (eubacterial) transcription machinery. Eubacterial RNA polymerases, for example the *E. coli* enzyme, consist of four subunits (α_2, β, β'), which form the core-enzyme. An additional protein factor (σ) is needed to obtain the holo-enzyme, capable of specific promoter recognition and transcription (Busby and Ebright, 1994). Genes encoding homologues of the *E. coli* core subunits were found on plastid genomes of all photosynthetic plants and algae investigated to date. The α subunit homologue is encoded by *rpoA*, the β subunit homologue by *rpoB*. The eubacterial β' subunit is split in plastids: the amino-terminal domain (β') is encoded by *rpoC1* and the carboxy-terminal domain (β'') is encoded by *rpoC2*. The same subunit structure is found in cyanobacteria (Kaneko et al., 1996). The *rpo* gene organization is conserved between eubacteria and plastids: *rpoB, rpoC1* and *rpoC2* form one operon, whereas *rpoA* is co-transcribed with genes for ribosomal proteins. An additional feature is the occurrence of an intron in the plastid *rpo* genes: in *rpoC1* in dicots, and in *rpoC2* in monocots (Sugiura, 1992).

C. The PEP Specificity Factors

In eubacteria, accurate transcription initiation requires a transcription factor (σ), which is responsible for promoter recognition and contributes to DNA melting around the initiation site (Wösten, 1998). Sigma-like activities in plant plastids were reported early on in *Chlamydomonas* (Surzycki and Shellenbarger, 1976), spinach (Lerbs et al., 1983) and mustard (Bülow and Link, 1988). From mustard, three different sigma-like factors (SLFs) were purified (SLF67, SLF52, SLF29; Tiller and Link, 1993b). SLFs were detected immunologically in chloroplast RNA polymerase preparations of maize, rice, *Chlamydomonas* and *Cyanidium* (Troxler et al., 1994). The *Cyanidium* plastid SLF is similar to eubacterial σ^{70}-factors (Liu and Troxler, 1996; Tanaka et al., 1996). Genes encoding eubacterial σ^{70}-type factors were also cloned from *Arabidopsis*, mustard, rice and wheat (Isono et al., 1997b; Tanaka et al., 1997; Kestermann et al., 1998; Tozawa et al., 1998; Morikawa et al., 1999).

Bacterial σ^{70}-factors have conserved functional regions and are commonly divided into three different groups (Lonetto et al., 1992; Gruber and Bryant, 1997): primary factors, responsible for essential RNA synthesis (group 1) and alternative or nonessential factors (group 2 and 3). Group 2 factors are closely related to group 1 polypeptides, except that they lack subregion 1.1, whereas group 3 factors are less conserved. An additional σ^{70}-factor subfamily regulates genes for proteins with extracytoplasmic function (Lonetto et al., 1994). The putative cloned SLFs of higher plants are similar in size (53–65 kDa) and lack subregion 1.1. Therefore, in higher plants, presence or absence of subregion 1.1 does not distinguish essential and non-essential sigma factors. In this regard, higher plants are similar to *Anabaena* (Brahamsha and Haselkorn, 1991) in which both essential (SigA) and nonessential sigma factors (SigB and SigC) lack subregion 1.1. The only sigma factor in *Cyanidium* also lacks subregion 1.1. (Troxler et al., 1994; Liu and Troxler, 1996),

In *Arabidopsis*, the three SLFs were named SigA (56 kDa), SigB (64 kDa) and SigC (65 kDa; Tanaka et al., 1997) or SIG2, SIG1, and SIG3 (Isono et al., 1997b), respectively. The *Arabidopsis* SigA (SIG2) shows the highest conservation with SIG1 from mustard (Kestermann et al., 1998), with SigA from rice (Tozawa et al., 1998) as well as with SigA from wheat (Morikawa et al., 1999). Since the *Arabidopsis* SigA homologues are highly expressed during chloroplast biogenesis and this coincides with the highest levels of PEP activity it is assumed, that the SigA genes encode the principal chloroplast SLF (Tanaka et al., 1997; Kestermann et al., 1998; Tozawa et al., 1998; Morikawa et al., 1999). It appears that the role of different SLFs is to selectively modify promoter activity in response to developmental and environmental cues.

The relationship between the cloned mustard SaSIG1 gene and the biochemically identified SLFs from mustard remains to be defined. SLF52 is the most abundant of the three biochemically-identified factors. It is within the size range of SIG1, hence a candidate for the native SIG1 gene product (Kestermann et al., 1998). Relatedness of the mustard SIG1 polypeptide and eubacterial σ-factors was shown by reconstitution of a functional heterologous enzyme with the *E. coli* core RNA polymerase (Kestermann et al., 1998).

D. Cis-*Elements and* Trans-*Factors Regulating Individual Promoter Activity*

The plastid RNA polymerase has evolved from an eubacterial-type RNA polymerase. In agreement with this, plastid promoters contain a variant of the –10 (TAtaaT) and –35 (TTGaca) consensus sequences of typical σ^{70}-type *E. coli* promoters (Reznikov et al., 1985; for reviews see Gruissem et al., 1988; Sugiura, 1992; Gruissem and Tonkyn, 1993; Link, 1994). Nucleotides in capital letters are highly conserved in plastid promoters. Indeed, the *E. coli* RNA polymerase faithfully recognizes plastid σ^{70}-type promoters (Gatenby et al., 1981; Bradley and Gatenby, 1985; Boyer and Mullet, 1988; Eisermann et al., 1990).

Some plastid σ^{70}-type promoters contain additional *cis*-elements, which have a regulatory significance: a TATA-box localized between the conserved –10/–35 promoter sequences, a TGn motif located one base pair upstream of the –10 region (extended –10 promoter) and an AAG-box upstream of the conserved

–35 hexamer. The role of these promoter motifs and of cognate *trans*-factors in regulating individual promoter activity is discussed in this section.

1. *The* psbA *Promoter*

During proplastid to chloroplast development, transcription of genes required for plastid maintenance is activated first, followed by transcription of photosynthetic genes (Mullet, 1993). The *psbA* gene encodes the D1 polypeptide of the photosystem II reaction center. *psbA* transcription is developmentally timed and activated by light (Klein and Mullet, 1990; Schrubar et al., 1990; Baumgartner et al., 1993).

Early data derived from in vitro transcription experiments established that the *psbA* gene is transcribed from a σ^{70}-type promoter with the typical conserved –10 and –35 elements. Data were obtained in mustard (Link, 1984), spinach (Gruissem and Zurawski, 1985), pea (Boyer and Mullet, 1986) and barley (Boyer and Mullet, 1988). Detailed characterization of the *psbA* promoter was carried out in vitro in mustard, a dicotyledonous plant (Eisermann et al., 1990). This study identified a TATATA promoter element between the –10 and –35 hexamers resembling the TATA-box of nuclear genes transcribed by RNA polymerase II (Link, 1994). In vitro transcription extracts were prepared from dark-grown and light-grown plants containing etioplasts and chloroplasts, respectively. The TATATA element and –10 region were sufficient to obtain basic transcription levels in plastid extracts prepared from both dark- and light-grown plants. However, enhanced rates of *psbA* transcription, characteristic of chloroplasts of light-grown plants, were obtained only if the –35-element was present in the tested promoter (Link, 1984; Eisermann et al., 1990). The role of –10/–35 sequences in mustard *psbA* promoter recognition is shown in Fig. 2a.

The *psbA* promoter was also characterized in two monocotyledonous plants, wheat and barley. The barley *psbA* promoter also contains the TATATA motif between the –10/–35-elements. In barley, as in mustard, the TATA-box is important for full promoter activity. However, unlike in mustard, the –35 sequence is absolutely required for transcription in vitro (Fig. 2a; Kim et al., 1999b).

The wheat *psbA* promoter also has the TATA-box and, in addition, the TGn motif upstream of the –10 element which is referred to as extended –10 sequence (Bown et al., 1997; Satoh et al., 1999). Interestingly, the TATA-box in the wheat *psbA* promoter does not

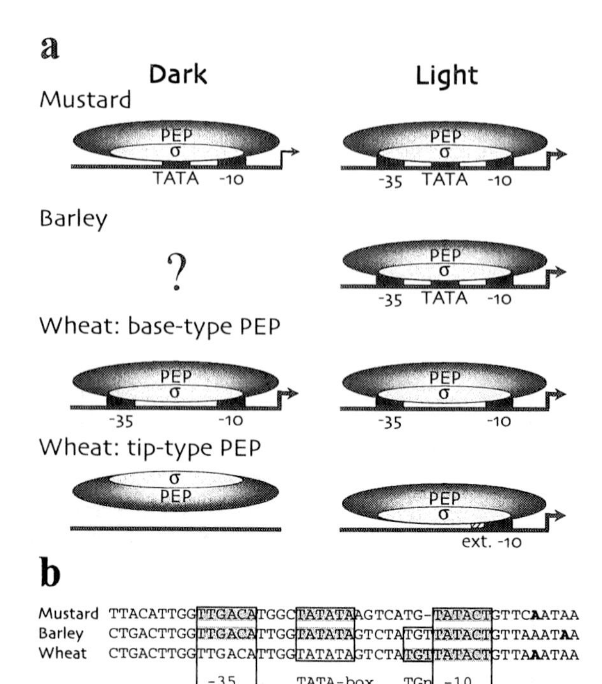

Fig. 2. Species-specific utilization of *psbA* promoter elements. (a) Elements important for promoter activity. TATA-box, –10/–35 consensus sequences and extended –10 sequence (TGn) are shown only when important for promoter function. (b) Alignment of *psbA* promoter sequences of mustard (Link and Langridge, 1984), barley (Boyer and Mullet, 1988) and wheat (Hanley-Bowdoin and Chua, 1988). Conserved promoter elements are boxed. Sequences required for full *psbA* promoter function in extracts of light-grown plants are shaded. Nucleotides at which transcription initiates are in bold.

seem to be important. The wheat *psbA* study was carried out using extracts from two developmentally distinct leaf parts. The basal leaf segments contain young chloroplasts whereas the leaf tip contains mature chloroplasts. The study identified two PEP types. Constitutive (light-independent) transcription by the base-type PEP (Fig. 2a) required both the –10 and –35 elements for promoter activity. However, transcription by the tip-type PEP (Fig. 2a) was dependent only on the –10 region. It was proposed, that the base-type and tip-type PEPs differ by the associated sigma factor, and that the TGn motif (in this case TGT) of the extended –10 region may be involved in recognition by the tip-type PEP (Fig. 2a; Satoh et al., 1999).

The mustard, barley and wheat *psbA* promoter sequences are highly conserved (Fig. 2b). Differences in the utilization of cis-elements probably are the result of divergent evolution of trans-factors in these species.

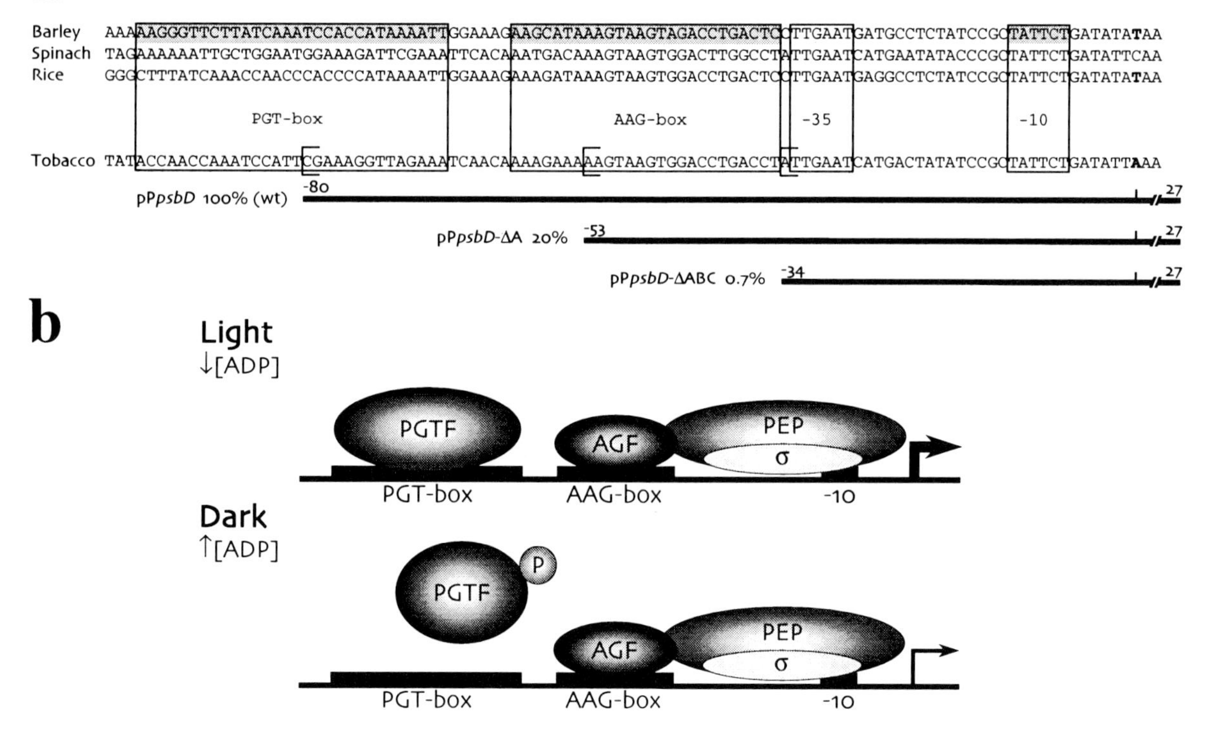

Fig. 3. Upstream activators regulate *psbD* promoter activity. (a) Alignment of *psbD* promoter regions from tobacco (Allison and Maliga, 1995), spinach (Holschuh et al., 1984), rice (To et al., 1996) and barley (Kim et al., 1999b). Promoter elements are boxed and sequences required for full *psbD* promoter activity are shaded. Promoter fragments tested in vivo are shown below. Transcription initiation sites are marked in bold. (b) Role of AGF and PGTF factors in light-induced psbD transcription. AGF constitutively binds to the AAG-box; AGF-PEP interaction substitutes for PEP and –35-element interaction. Transcription is enhanced by PGTF binding in the light. ADP-dependent PGTF phosphorylation results in loss of affinity for the PGT element, one mechanism proposed to explain reduced transcription in the dark.

2. *The* psbD *Promoter*

The rate of transcription of photosynthetic genes declines during chloroplast development in mature chloroplasts, with the exception of the *psbD-psbC* genes (Klein and Mullet, 1990; Baumgartner et al., 1993; DuBell and Mullet, 1995). High-levels of *psbD* transcripts in mature chloroplasts are due to activation of the blue light-responsive promoter (BLRP; Sexton et al., 1990). The *psbD* BLRP architecture was studied in a number of species (Christopher et al., 1992), including barley (Kim and Mullet, 1995; Kim et al., 1999b), tobacco (Allison and Maliga, 1995), wheat (Wada et al., 1994), rice (To et al., 1996) and *Arabidopsis* (Hoffer and Christopher, 1997). The *psbD* promoter in each of these species has a similar architecture and is thus likely to be regulated similarly.

The *psbD* promoter is σ^{70}-type, with poorly conserved –10/–35-elements which are spaced relatively close (15 nucleotides instead of the usual 17 to 19; Fig. 3a). Architecture of the *psbD* BLRP was studied in vivo, using a transgenic approach (Allison and Maliga, 1995). The 107-bp region studied as the promoter contained the –10/–35-elements, the AAG-box and part of the PGT-box (Fig. 3). Activity of this truncated promoter was taken as 100%. Deletion of part of the PGT-box reduced mRNA levels 5-fold. Deletion of a DNA segment containing the AAG-box sequences further reduced transcript levels 30-fold. Transcription initiation from the remaining promoter core was inefficient and could be detected only in vivo. Thus, light-induced (150-fold) transcript accumulation in the in vivo study was mediated by sequences directly upstream of the promoter core.

Transcription from the *psbD* promoter depends on the –10, but not on the –35 promoter element in rice, wheat and barley (To et al., 1996; Satoh et al., 1997; Kim et al., 1999b). Upstream sequences involved in transcriptional activation have been characterized in vitro. The AAG-box located between nucleotides –36 to –64 of the barley promoter is the binding site for the nuclear-encoded AAG-binding factor or AGF (Kim and Mullet, 1995). The PGT-box located between nucleotides –71 to –100 is the binding site for the PGT-binding factor (PGTF), the activity of which is regulated by an ADP-dependent kinase (Kim et al., 1999a). The model, based on experiments in barley is shown in Fig. 3b. Entry of the PEP is facilitated by AGF constitutively binding to the upstream AAG-element. Transcriptional activation in light is mediated by PGTF bound to the PGT site. Phosphorylation of PGTF results in loss of its affinity for the PGT element, a mechanism proposed to explain lack of transcription from this promoter in the dark. It is assumed, that the *Arabidopsis* DET1 gene product regulates the activity of these DNA-binding complexes via a photosensoric pathway (Christopher and Hoffer, 1998).

3. The rbcL Promoter

The plastid *rbcL* gene encodes the large subunit of ribulose-1,5-bisphosphate carboxylase (EC 4.1.1.39). The *rbcL* gene is transcribed by the PEP from a single promoter which has been mapped in tobacco (Shinozaki and Sugiura, 1982), maize, spinach, pea (Mullet et al., 1985), barley (Reinbothe et al., 1993), and *Arabidopsis* (Isono et al., 1997a). The tobacco *rbcL* promoter has well conserved –10 and –35 elements (TACAAT, TTGCGC) and canonical spacing (18 nucleotides). In vitro studies confirmed the importance of the –10 and –35 box spacing and sequence for *rbcL* promoter strength (Gruissem and Zurawski, 1985; Hanley-Bowdoin et al., 1985).

Sequences positioned between nucleotides –16 and –102 relative to the *rbcL* transcription initiation site were proposed to function as a binding site for the chloroplast DNA-binding factor 1 (CDF1) in maize. Segments of the CDF1 binding site are conserved between maize, pea, spinach, and tobacco (Lam et al., 1988). Conserved regions in the tobacco promoter are shown in Fig. 4. One of these, CDF1 region II, is reminiscent of the AT-rich UP element, which stimulates transcription by a factor of 30 in *E. coli* (Ross et al., 1993).

A study of chimeric *uidA* genes confirmed that the *rbcL* core promoter is sufficient to obtain wild-type rates of transcription (Shiina et al., 1998). Therefore, no sequences outside the promoter core significantly contribute to *rbcL* promoter function. Furthermore, the rates of *rbcL* transcription were significantly reduced in dark-adapted plants. Reduced rates of transcription resulted in lower steady-state *uidA* mRNA levels in the dark, unless the 5′ segment of the *rbcL* 5′ untranslated region (5′ UTR) was included in the chimeric construct. This finding indicates that accumulation of *rbcL* mRNA in a light-independent manner is due to stabilization of the mRNA via its 5′ UTR, compensating for reduced rates of transcription in the dark.

4. The rrn Operon Promoters

The plastid *rrn* operon in maize, pea and tobacco is transcribed by PEP from σ^{70}-type promoters (P1 or Nt-Prrn-114; Strittmatter et al., 1985; Sun et al.,

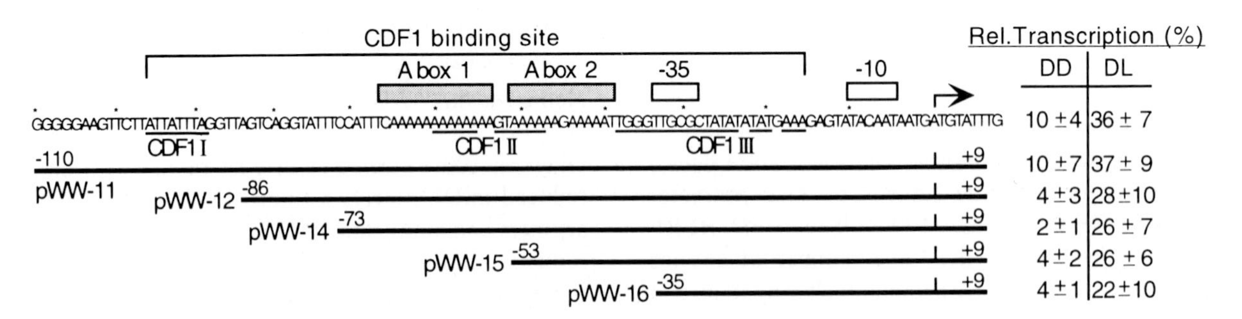

Fig. 4. The *rbcL* promoter core is sufficient for light-induced transcription. The tobacco *rbcL* promoter region is shown with the transcription initiation site (horizontal arrow), the –10/–35 promoter elements, the two A-rich boxes (A box 1, A box 2) and the conserved CDF1 regions (I, II, III; underlined). Note that these elements were found to have no effect in vivo by transgenic testing in tobacco. Deletion derivatives studied in vivo are identified by their pWW plasmid number. Numbers indicate endpoints in nucleotides relative to the transcription initiation site. The Table on the right lists the relative transcription rate (in percentage of *rrn16* transcription) of *rbcL* deletion derivatives from dark-adapted (96h) plants (DD) and dark-adapted (72h), illuminated (24h) plants (DL). Adopted from Shiina et al. (1998).

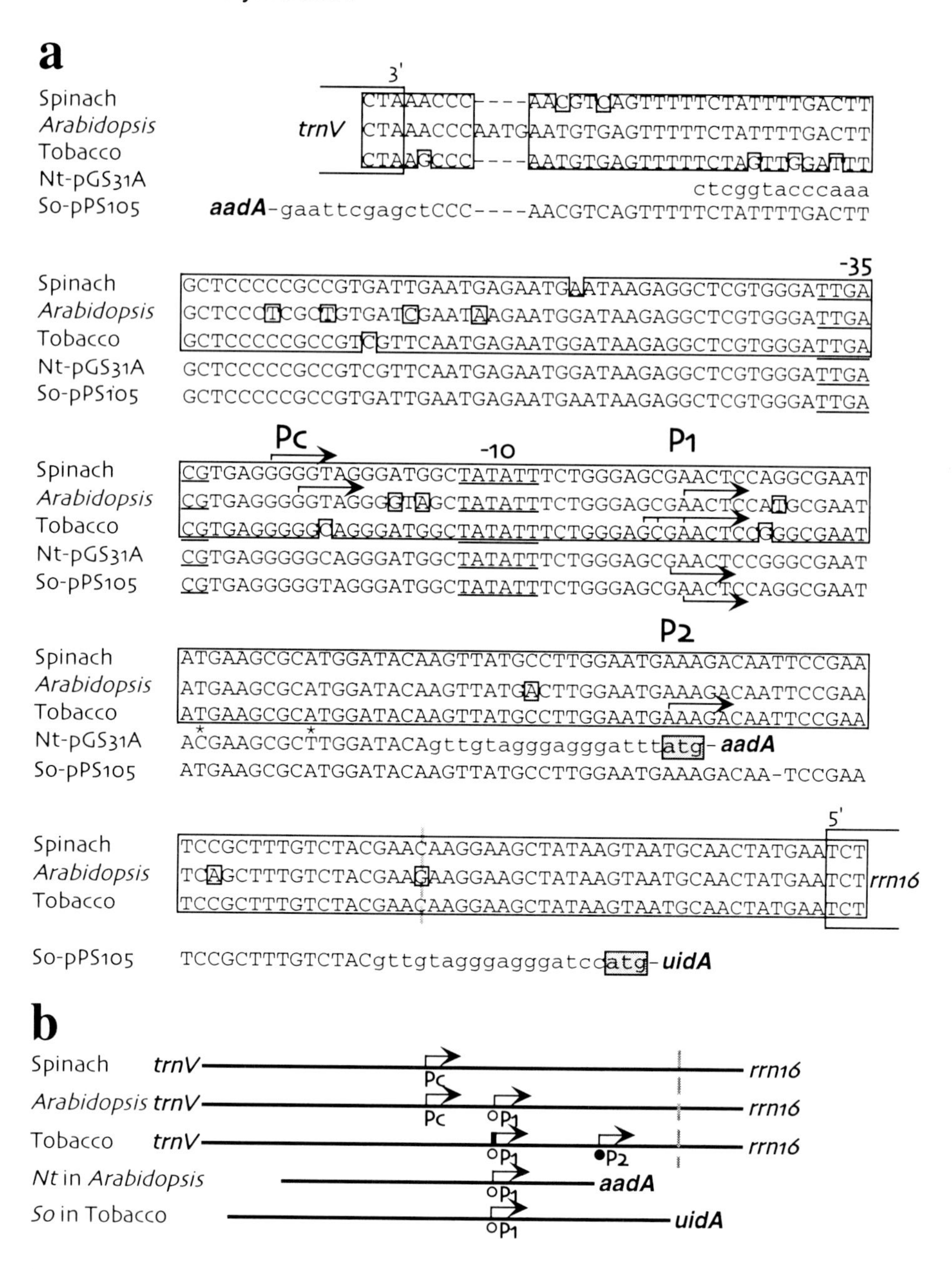

Fig. 5. The rRNA operon promoters of spinach, *Arabidopsis*, and tobacco. (a) Alignment of the DNA sequences between *trnV* and *rrn16*. The sequences of the tobacco *rrn16* promoter region in transplastomic *Arabidopsis* (Nt-pGS31A) and the spinach *rrn16* promoter sequences in transplastomic tobacco (So-pPS105) are also shown. P1 and P2 mark transcription initiation by PEP and NEP respectively. Pc marks transcription initiation in spinach from a yet uncharacterized promoter. Conserved sequences are boxed and transcription initiation sites marked with horizontal arrows. A dashed vertical line indicates the RNA processing site. (b) Schematic representation of transcription initiation sites shown in Fig. 5a. PEP and NEP transcription initiation sites are represented by open and filled circles, respectively. Based on Sriraman et al. (1998b).

1989; Vera and Sugiura, 1995; Allison et al., 1996). In addition to P1, the *rrn* operon in tobacco has a second promoter (P2 or Nt-Prrn-62; Fig. 5) recognized by NEP which is inactive in chloroplasts, and functions only in BY2 tissue culture cells (Vera and Sugiura, 1995) and in plants lacking PEP (Allison et al., 1996). In maize plastids there is no active NEP promoter directly upstream of the *rrn* operon (Silhavy and Maliga, 1998a).

Transcription of the *rrn* operon in spinach plastids initiates in a region, which is highly conserved in tobacco (Fig. 5a). This region contains typical –10/–35-elements active as *rrn* operon promoters in other species. However, in spinach the σ^{70}-type promoter sequences are not utilized by plastid RNA polymerases in vivo. Instead, transcripts initiate from a site between the conserved –10/–35 hexamers (Pc promoter; Baeza et al., 1991; Iratni et al., 1997). Sequences relevant for transcription initiation from Pc have yet to be identified. The Pc promoter is active in chloroplasts but not in roots. The Pc site appears to be faithfully recognized by the mustard PEP in vitro (Pfannschmidt and Link, 1997). A good candidate for the Pc activating factor is CDF2, a protein, the binding of which is correlated with transcription from Pc in spinach (Iratni et al., 1997).

In *Arabidopsis*, *rrn* operon transcripts were mapped to both the major tobacco PEP P1 and the spinach Pc initiation sites. A study of promoters in heterologous plastids indicates that tobacco plastids lack the factor required for transcription from Pc, while spinach has an intact P1 promoter but lacks the cognate P1 activator (Sriraman et al., 1998b).

E. Regulation of PEP Activity by Sigma Factor Modification

PEP activity depends on the developmental stage of the plastids: it is down regulated in etioplasts and is active in chloroplasts (Rapp et al., 1992; DuBell and Mullet, 1995). Furthermore, rates of PEP transcription are higher in the light than in the dark (Shiina et al., 1998). Changes in PEP transcription activity were explained by changes in the phosphorylation state of SLFs. In vitro studies in mustard have shown that SLFs in their active (non-phosphorylated) state, bind relatively loosely to the PEP core enzyme, and are efficiently released subsequent to transcription initiation. In contrast, the phosphorylated SLF binds tightly to the PEP core. As a consequence, subsequent release of the SLF is hindered, which overall leads to inefficient recycling of SLF and hence reduced transcription efficiency (Fig. 6a; Tiller et al., 1991; Tiller and Link, 1993a; 1993b; reviewed in Link, 1996).

An in vivo role for the regulation of PEP activity by SLF phosphorylation was supported by the characterization of a heterotrimeric complex with serine-type kinase activity from mustard chloroplasts. This kinase, designated plastid transcription kinase (PTK), is either associated with the PEP or is present in a free form (Baginsky et al., 1997). Decreased in vitro transcription by the phosphorylated PEP is consistent with recent in vivo findings in barley, where the presence of kinase inhibitors enhanced and phosphatase inhibitors decreased transcription activity from the light-responsive *psbD-C* operon promoter (Christopher et al., 1997).

PTK was proposed to be part of the signaling pathway controlling PEP activity based on the observation that PTK itself is differentially regulated by phosphorylation (Fig. 6a) and redox state (Fig 6b; Baginsky et al., 1999). The unphosphorylated PTK was shown to be more active than the phosphorylated form. Interestingly, reduced glutathione (GSH) in vitro negatively affected unphosphorylated PTK, but enhanced the kinase activity of the phosphorylated form. Activation and inactivation of PTK affects

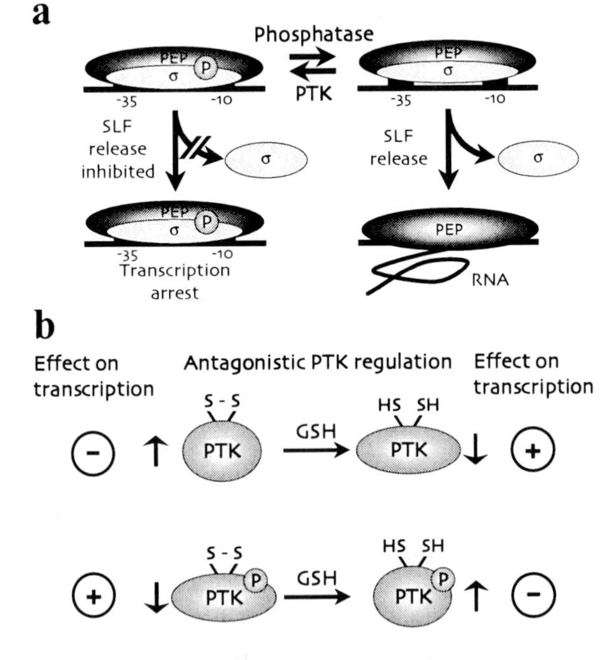

Fig. 6. Regulation of PEP activity by phosphorylation and the redox state. (a) Reversible SLF phosphorylation regulates PEP activity. PTK is the plastid transcription kinase. SLF phosphorylation arrests its subsequent release after initiation of transcription, which results in lower transcription efficiency. Based on Tiller and Link (1993a). (b) Antagonistic regulation of PTK activity by glutathione (GSH). Changes in PTK activity are indicated by arrows, increase (↑) and decrease (↓). Active (unphosphorylated) PTK (↑) is negatively affected by GSH (↓), therefore transcription remains high (+). Inactive (phosphorylated) PTK (↓) is positively affected by GSH (↑) which leads to more efficient SLF phosphorylation and thereby reduced transcription (-). Based on Baginsky et al. (1999).

PEP activity via phosphorylation of SLFs, as summarized in Fig. 6a. Direct redox control of plastid gene transcription was also shown via an as yet unknown mechanism by the redox state of the plastoquinone pool (Pfannschmidt et al., 1999). This topic is the subject of Chapter 5 (Link).

III. The Nuclear-Encoded Plastid RNA Polymerase (NEP)

Multiple clues have pointed to the existence of a nuclear-encoded plastid RNA polymerase (NEP). The oldest reports came from studies on transcription in ribosome-deficient plastids of heat-bleached rye leaves and the barley mutant *albostrians* (Bünger and Feierabend, 1980; Siemenroth et al., 1981). Since these plants have no translation of plastid proteins on plastid ribosomes, the synthesis of RNA was taken as evidence for the nuclear localization of genes for a plastid RNA polymerase. Further analysis of these (Falk et al., 1993; Hess et al., 1993) and other similar mutants, i.e. the maize *iojap* mutant (Han et al., 1993), and heat-bleached barley (Hess et al., 1993) revealed transcription of distinct sets of plastid genes.

An antibody-linked polymerase assay of polypeptides translated from spinach polyA$^+$-RNA suggested a nuclear origin of the plastid RNA polymerase (Lerbs et al., 1985). Further purification of plastid RNA polymerase activities identified a single 110-kDa protein with non-specific transcription from plastid DNA templates. The 110-kDa spinach protein recognized the T7-phage promoter. Based on these findings a nuclear-encoded phage-type RNA polymerase was proposed to exist in plastids (Lerbs-Mache, 1993).

Further support for the existence of NEP in plastids was obtained by analysis of the plastid genome of a non-photosynthetic parasitic plant, *Epifagus virginiana*. The reduced plastid genome of this plant lacks the *rpoB*, *rpoC1* and *rpoC2* genes, while *rpoA* is present as a pseudogene (Morden et al., 1991; Wolfe et al., 1992a,b,c). Still, in the non-green *Epifagus* plastids transcription of a subset of plastid genes is maintained (DePamphilis and Palmer, 1990; Ems et al., 1995). The promoters for these transcripts have not been identified. Therefore, the *Epifagus* plastid genes may be transcribed by the NEP, or by an eubacterial-type RNA polymerase, for which the subunit genes have been transferred to the nucleus as part of adaptation to the parasitic lifestyle.

Currently, the existence of NEP has been established by targeted deletion of *rpo* genes from the tobacco plastid genome (Allison et al., 1996; Serino and Maliga, 1998). Although the knockout plastids lack PEP, a subset of plastid genes is still transcribed (Allison et al., 1996; Hajdukiewicz et al., 1997).

A. The NEP gene(s)

Screening of an *Arabidopsis* genomic library with a *Chenopodium album* mitochondrial RNA polymerase probe identified three genes. Two of these are targeted to mitochondria and one is targeted to plastids (Hedtke et al., 1997; 1999). The latter is a candidate for the putative NEP catalytic subunit. This *Arabidopsis* gene was initially called *RpoZ* (Hedtke et al., 1997), then later re-named *RpoT;3* and its product RPOT;3 (Hess and Börner, 1999). The maize and wheat NEP genes are designated *RpoTp* and their product RpoTp (Chang et al., 1999; Ikeda and Gray, 1999a). The genes targeted to mitochondria in *Arabidopsis* are *RpoT;1* (formerly *RpoY*; Hedtke et al., 1997; 1999) and *RpoT;2* (Hedtke et al., 1999). The maize and wheat mitochondrial homologues have also been cloned (*RpoTm*; Chang et al., 1999; Ikeda and Gray, 1999a).

The plastid-targeted *Arabidopsis* RPOT;3 and maize RPOTP have a molecular mass of ~113 kDa, which is in good agreement with the size of phage RNA polymerases (~95 kDa) and with biochemical data obtained for the putative NEP catalytic subunit from spinach chloroplasts (110 kDa; Lerbs-Mache, 1993). This size estimate includes the plastid targeting transit peptide. The cleavage site for the transit peptides has not been determined.

The catalytic subunit of the plastid phage-type RNA polymerase probably evolved by duplication of the mitochondrial gene. Conservation of the genomic structure of the plastid and mitochondrial genes supports this hypothesis. The *Arabidopsis* and maize *RpoT* genes share the same organization of 19 exons and 18 introns, which are inserted at identical amino acid positions (Gray and Lang, 1998; Hess and Börner, 1999). The evolutionary time of gene duplication is uncertain. DNA sequences related to phage-type RNA polymerases are widespread throughout organisms with and without plastids in the eukaryotic lineage (Cermakian et al., 1996). The evolutionary study did not distinguish between potential mitochondrial and plastid targeted genes.

The earliest known example for a functional plastid NEP promoter was found in *Marchantia polymorpha*, a liverwort (Sriraman et al., 1998a). In lower eukaryotes, the only methodical study so far was carried out in the unicellular alga *Chlamydomonas*. Experiments to knock out *rpo* genes from the *Chlamydomonas* plastid genome failed (Rochaix, 1995; Fischer et al., 1996). One explanation is that *Chlamydomonas* altogether lacks the phage-type plastid RNA polymerase, and the lack of translation due to the lack of ribosomes and mRNAs is lethal. Alternatively, unlike in tobacco (Allison et al., 1996), the *Chlamydomonas* PEP may solely transcribe genes required to maintain essential functions, and thus it is not dispensable.

B. The NEP Promoters

Transcription initiation sites for NEP were unambiguously identified in plants lacking PEP due to nuclear mutations leading to the loss of plastid ribosomes, and hence translation, in the monocots barley and maize (Hübschmann and Börner, 1998; Silhavy and Maliga, 1998a). In dicots, unambiguous identification of NEP promoters was carried out in tobacco plants, which lack PEP due to targeted deletion of the plastid *rpo* genes. In these plants, however, translation of plastid mRNAs is likely to be maintained (Allison et al., 1996; Hajdukiewicz et al., 1997; Serino and Maliga, 1998). NEP promoters were also mapped in tobacco BY2 tissue culture cells, with elevated levels of NEP activity. The BY2 cells, unlike the tobacco *rpo*-deletion mutants, have both NEP and PEP enzymes. Classification of NEP promoters in this system was based on the lack of eubacterial σ^{70}-type promoter elements, and enhanced activity relative to photosynthetically active leaf cells (Kapoor et al., 1997; Miyagi et al., 1998).

Identification of NEP promoters revealed that photosynthetic genes and operons have PEP promoters, whereas most non-photosynthetic genes involved in housekeeping functions such as transcription and translation have promoters for both RNA polymerases. The NEP promoter(s) of these genes are, with a few exceptions, silent in chloroplasts. Only a few genes are known to be transcribed exclusively from a NEP promoter. These are *accD*, encoding a subunit of the acetyl-CoA carboxylase in dicots (Hajdukiewicz et al., 1997); *clpP* encoding the proteolytic subunit of the Clp ATP-dependent protease in monocots; and the *rpoB* operon encoding three of the four PEP core subunits in all higher plants (Silhavy and Maliga, 1998a).

The first clues for NEP promoter architecture came from aligning 11 tobacco NEP transcription initiation sites. The alignment identified a tentative, loose ~10-nt consensus directly upstream of the transcription initiation site (Hajdukiewicz et al., 1997). Additional NEP promoters were mapped in monocots (Hübschmann and Börner, 1998; Silhavy and Maliga, 1998a). Clearly, none of the NEP promoters have the conserved σ^{70}-type promoter elements. Instead, they resemble mitochondrial and phage promoters, characterized by conservation of sequences around the transcription initiation site and an essential core motif which may be at a variable distance from the transcription initiation site (Maliga et al., 1999; Weihe and Börner, 1999).

There are two distinct promoter types active in plants lacking PEP (Fig. 7). The Type-I NEP promoters are characterized by a conserved YRTA-motif embedded in a ~15 nt AT-rich region upstream of the transcription initiation site. Systematic dissection and point mutational analysis of NEP promoters in vitro verified the importance of promoter elements identified by sequence alignment. The Type-I promoters analyzed thus far are included in a small DNA fragment (–15 to +5) upstream of the transcription initiation site (+1) (PatpB-289; Kapoor and Sugiura, 1999; S. Popescu, K. Liere and P. Maliga, unpublished; PrpoB-345; Liere and Maliga, 1999a; PaccD-129; Liere and Maliga, 1999b). The CRT-motif is critical for *rpoB* promoter recognition defining it as the NEP promoter core. The maize *rpoB* promoter, which also contains a core CAT-motif (Silhavy and Maliga, 1998a), is faithfully recognized by the tobacco extract, further substantiating the importance of this element (Liere and Maliga, 1999a). The YRTA sequence motif is also present in the conserved plant mitochondrial YRTAT promoter motif (Brown et al., 1991; Covello and Gray, 1991; Mulligan et al., 1991; Rapp and Stern, 1992; Binder and Brennicke, 1993; Rapp et al., 1993; Caoile and Stern, 1997; Yan and Pring, 1997). Conservation of the core element between the plastid Type-I NEP and mitochondrial promoters supports the common origin of phage-type organellar RNA polymerases.

A subset of Type-I NEP promoters possesses a second, conserved sequence motif (ATAN$_{0-1}$GAA) ~18 to 20 bp upstream of the YRTA-motif, designated box II or GAA-box (Kapoor et al., 1997; Silhavy and

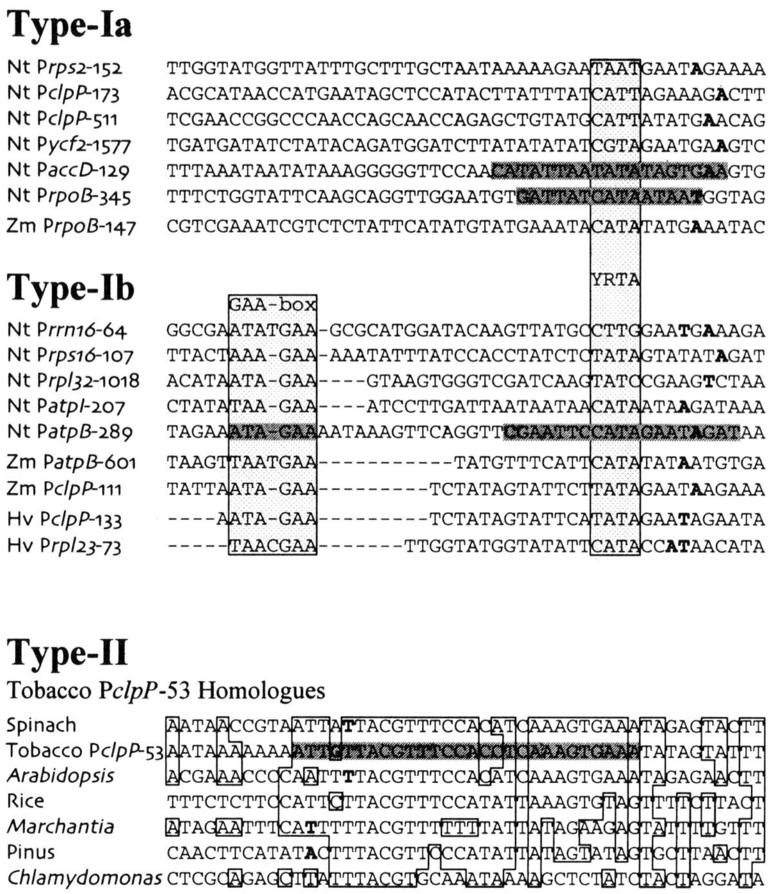

Type-Ia

Nt Prps2-152 TTGGTATGGTTATTTGCTTTGCTAATAAAAAGAATAATGAATAGAAAA
Nt PclpP-173 ACGCATAACCATGAATAGCTCCATACTTATTTATCATTAGAAAGACTT
Nt PclpP-511 TCGAACCGGCCCAACCAGCAACCAGAGCTGTATGCATTATATGAACAG
Nt Pycf2-1577 TGATGATATCTATACAGATGGATCTTATATATATCGTAGAATGAAGTC
Nt PaccD-129 TTTAAATAATATAAAGGGGTTCCAACATATTAATATATAGTGAAGTG
Nt PrpoB-345 TTTCTGGTATTCAAGCAGGTTGGAATGTGATTATCATAATAATGGTAG
Zm PrpoB-147 CGTCGAAATCGTCTCTATTCATATGTATGAAATACATATATGAAATAC

YRTA

Type-Ib

GAA-box

Nt Prrn16-64 GGCGAATATGAA-GCGCATGGATACAAGTTATGCCTTGGAATGAAAGA
Nt Prps16-107 TTACTAAA-GAA-AAATATTTATCCACCTATCTCTATAGTATATAGAT
Nt Prpl32-1018 ACATAATA-GAA----GTAAGTGGGTCGATCAAGTATCCGAAGTCTAA
Nt PatpI-207 CTATATAA-GAA----ATCCTTGATTAATAATAACATAATAAGATAAA
Nt PatpB-289 TAGAAATA-GAAAATAAAGTTCAGGTTCGAATTCCATAGAATAGATAA
Zm PatpB-601 TAAGTTAATGAA-----------TATGTTTCATTCATATATAATGTGA
Zm PclpP-111 TATTAATA-GAA---------TCTATAGTATTCTTATAGAATAAGAAA
Hv PclpP-133 ----AATA-GAA---------TCTATAGTATTCATATAGAATAGAATA
Hv Prpl23-73 -----TAACGAA-------TTGGTATGGTATATTCATACCATAACATA

Type-II

Tobacco P*clpP*-53 Homologues

Spinach AATAACCGTAATTATTACGTTTCCACATCAAAGTGAAATAGAGTACTT
Tobacco PclpP-53 AATAAAAAAAATTGTTACGTTTCCACCTCAAAGTGAAATATAGTATTT
Arabidopsis ACGAAACCCCAATTTTTACGTTTCCACATCAAAGTGAAATAGAGAACTT
Rice TTTCTCTTCCATTCTTACGTTTCCATATTAAAGTGTAGTTTTCTTACT
Marchantia ATAGAATTTCATTTTTACGTTTTTTTATTATAGAAGAGTATTTTGTTT
Pinus CAACTTCATATACTTTACGTTCCCATATTATAGTATAGTGCTTAACTT
Chlamydomonas CTCGCAGAGCTTATTTACGTGCAAATAAAAGCTCTATCTACTAGGATA

Fig. 7. Type-I and Type-II plastid NEP promoters. The sequence of experimentally defined promoters is highlighted, including PaccD-129, PrpoB-345, PatpB-289 (Type-I) and PclpP-53 (Type-II) from tobacco. Nucleotides at the 5'-end of the primary transcripts are in bold. The YRTA promoter core and GAA-box is marked in Type-I promoters. Conserved regions of Type-II promoters are boxed.

Maliga, 1998a). Preliminary data obtained by mutational analyses of the GAA-box in in vitro transcription experiments suggest, at least for the tobacco P*atpB*-289 promoter, a functional role of this element in promoter recognition (Kapoor and Sugiura, 1999; S. Popescu, K. Liere and P. Maliga, unpublished). Therefore, Type-I promoters could be divided into two subclasses: Ia, with only the YRTA-motif, and Ib, carrying both YRTA- and GAA-box (Fig. 7; Weihe and Börner, 1999).

Sequence conservation in the Type-II NEP promoters is downstream of the transcription initiation site (Fig. 7). So far this class is represented by a single example, P*clpP*-53, a promoter of the ClpP protease subunit gene. P*clpP*-53 was characterized using a transplastomic in vivo approach, since this promoter is well expressed in leaves. Primer extension analysis of *uidA* transcripts established that P*clpP*-53 is contained in sequences mainly downstream of the transcription initiation site (–5 to +25) (Sriraman et al., 1998a). The *clpP*-53 promoter sequences and the transcription initiation site are conserved among monocots, dicots, conifers and liverworts, indicating the appearance of the NEP transcription machinery early in evolution. Although the tobacco PclpP-53 sequence motif is present in rice and *Chlamydomonas*, it is not used as a promoter. Interestingly, if the rice sequence is introduced into tobacco plastids, it is recognized by the tobacco NEP suggesting that the lack of transcription in rice from the PclpP-53 homologue is due to the lack of the Type-II specificity factor (Sriraman et al., 1998b).

C. The NEP Enzyme and its Regulation

Genes encoding the tentative NEP catalytic subunit

were identified by their similarity to the phage-type mitochondrial RNA polymerases. Evidence that the genes may play a role in plastid gene transcription was obtained by identification of N-terminal extensions targeting the protein to plastids. NEP transcription activity has been characterized in vitro in plastid extracts prepared from tobacco plants lacking PEP activity due to targeted deletion of the PEP α subunit gene (Liere and Maliga, 1999a). The pattern of plastid NEP drug resistance was similar to that of the phage T7 RNA polymerase, hence biochemically corroborating the assignment of NEP as a phage-type RNA polymerase. NEP transcription is sensitive to actinomycin, a non-specific RNA polymerase inhibitor, and is partially resistant to rifampicin and rifamycin SV, inhibitors of eubacterial RNA polymerases (Hartmann et al., 1967). Furthermore, NEP is resistant to α-amanitin, an inhibitor of nuclear Type-II RNA polymerases and to tagetin, a selective inhibitor of nuclear RNA polymerase III, the *E. coli* RNA polymerase and the related PEP (Lukens et al., 1987; Mathews and Durbin, 1990; Steinberg et al., 1990). Thus far, there is no direct evidence that the *RpoT;3* or *RpoTp* gene products are the catalytic subunit of the polymerase characterized in plastid extracts. It is encouraging, however, that an antibody prepared against the RpoTp C-terminus recognized a spinach RNA polymerase activity in an antibody-linked transcription assay (Chang et al., 1999).

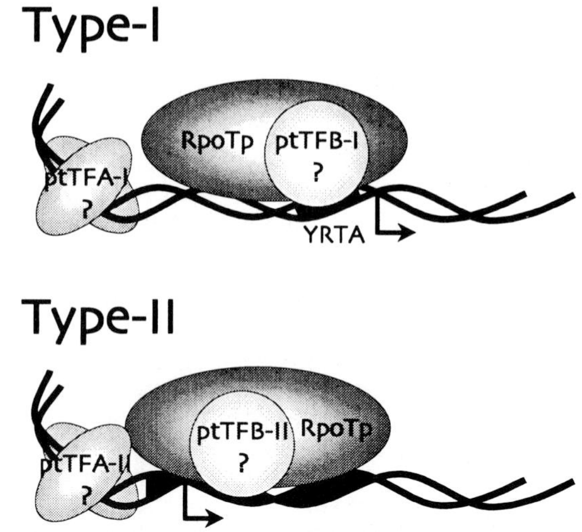

As to the NEP subunit composition, we can only speculate based on information on the related mitochondrial RNA polymerases (Fig. 8). Mitochondrial transcription complexes from human, mouse, *Xenopus laevis* and yeast consist of a minimum of two components: the catalytic core enzyme (mtRPO, ~ 120–150 kDa), and a specificity factor (mtTFB, ~ 40–45 kDa), which confers promoter recognition. An additional component (mtTFA, 20–25 kDa) which binds the DNA further upstream of the transcription initiation site, enhances mitochondrial transcription in vitro. This DNA-binding protein belongs to the HMG (high mobility group) family and may also facilitate the interaction with other *trans*-acting factors (reviewed in: Jaehning, 1993; Shadel and Clayton, 1993; Tracy and Stern, 1995; Hess and Börner, 1999). We expect that NEP promoter recognition is dependent on the plastid homologues of yeast and mammalian mitochondrial mtTFA and mtTFB factors.

In plants, candidates for mitochondrial transcription factors have been identified. One of them is MCT, a maize nuclear gene. Its target is the mitochondrial *cox2* promoter, which is active only when the dominant MCT allele is present (Newton et al., 1995). Additional candidates are two pea polypeptides (32 and 44 kDa) with affinity to the mitochondrial *atp9* promoter region (Hatzack et al., 1998). In wheat, a cDNA for a putative mitochondrial specificity factor has been cloned (Ikeda and Gray, 1999b). The encoded 69-kDa protein shows similarities to regions 2 and 3 of bacterial σ factors and the yeast mtTFB (MTF1), displays promoter selective binding and enhances transcription in vitro.

Since the genes for the catalytic subunits of the plastid and mitochondrial phage-type RNA polymerases are related, we expect that the organellar specificity factors are also encoded by related genes. Interestingly, accurate in vitro transcription in mitochondrial extracts requires purification (Rapp and Stern, 1992; Rapp et al., 1993; Caoile and Stern, 1997; Chang et al., 1999; Ikeda and Gray, 1999a), whereas purification of plastid extracts leads to reduction or loss of transcription activity (Liere and

Fig. 8. Speculative model for the nuclear-encoded plastid RNA polymerase (NEP). RpoTp is the NEP catalytic subunit; ptTFA and ptTFB are homologues of the mitochondrial mtTFA and mtTFB factors, respectively. YRTA is the conserved Type-I NEP promoter core. Model is based on Shadel and Clayton (1993). We assume that the Type-I and Type-II promoters are recognized by distinct ptTFA and ptTFB factors. Note that some Type-I NEP promoters are active and others inactive in the same plastid type. This could be explained by assuming a family of ptTFA-I and/or ptTFB-I factors with different sequence recognition specificity.

Maliga, 1999a). Therefore, the tobacco NEP transcription factors may have been lost during purification, indicating differences in the biochemical properties of the mitochondrial versus the plastid specificity factors.

A factor that may be involved in NEP transcription is the plastid ribosomal protein L4 (RPL4; encoded by the nuclear *Rpl4* gene). A role for RPL4 in transcription was proposed, as it co-purifies with the T7-like transcription complex in spinach (Trifa et al., 1998). The ribosomal protein L4 was shown to have extraribosomal functions in transcriptional regulation in prokaryotes (Zengel et al., 1980). The spinach and *Arabidopsis RPL4* genes have acquired a remarkable 3′ extension during their evolutionary transfer to the nuclear genome, which resembles highly acidic C-terminal ends of some transcription factors. A function for this protein in NEP or PEP transcription, however, has yet to be demonstrated.

IV. The Role of NEP and PEP in Plastid Gene Expression

The fact that plastid genes are transcribed by at least two different RNA polymerases, the phage-type NEP and the *E. coli*-type PEP, raises questions as to why plastids need these different RNA polymerases and what is their role and function in plastid development and gene expression. It is assumed that the phage-type plastid RNA polymerase evolved by duplication of the nuclear gene encoding the mitochondrial enzyme, and re-targeting of the gene product to plastids (Hedtke et al., 1997). After accidental targeting of the nuclear gene product into chloroplasts, transcription of plastid genes by NEP probably occurred from spurious promoters, generating additional sets of RNAs for plastid genes. NEP probably became indispensable when it took over transcription of essential genes, such as the *rpoB* operon encoding the subunits for PEP, the RNA polymerase derived from the ancestral cyanobacterium. Transcription of PEP genes by NEP was probably a critical step of the nucleus indirectly taking control of the transcription of plastid genes, thereby fully integrating the photosynthetic endosymbiont-turned organelle into the developmental network of multicellular plants.

Transcription of plastid genes during chloroplast development is timed: housekeeping genes are transcribed first, followed by photosynthetic genes. In mature chloroplasts transcription of photosynthetic genes declines, with the exception of *psbA* and *psbD*. Developmentally timed expression of housekeeping and photosynthetic genes was proposed to be based on sequential transcription of the relevant genes by NEP and PEP, respectively (Mullet, 1993; DuBell and Mullet, 1995). Developmental timing was assumed to be due to the availability of the RNA polymerases: NEP in proplastids and PEP in chloroplasts. Although the model was attractive, reality turned out to be more complex. It appears that NEP is present in all plastid types, as there are essential genes expressed from a single NEP promoter. An example is the Type-I NEP promoter of *clpP* in monocots, which is expressed in all plastid types, including mature chloroplasts (Silhavy and Maliga, 1998a; 1998b). In mature chloroplasts the Type-I *rpoB* NEP promoter is poorly expressed, indicating regulation of transcription from Type-I NEP promoters via promoter-specific factors, and not through the availability of NEP. The ribosomal RNA (*rrn*) operon in monocots is expressed only from a PEP promoter (Silhavy and Maliga, 1998a; 1998b). This again suggests that PEP should be available in all plastid types for the expression of the essential *rrn* operon. These findings argue against a cascade-type promoter switch in the expression of genes with promoters for both RNA polymerases and suggest a parallel rather than a hierarchical role of the two transcription systems in plastid development.

V. Unsolved Mystery: tRNA Transcription

Most plastid tRNAs are transcribed by PEP from upstream σ^{70}-type promoters. However, a few tRNA genes such as the spinach *trnS*, *trnR* and *trnT* (Gruissem et al., 1986a; Cheng et al., 1997) as well as *trnS*, *trnH* and *trnR* from mustard (Neuhaus and Link, 1990; Nickelsen and Link, 1990; Liere and Link, 1994), and the *Chlamydomonas trnE* gene (Jahn, 1992) seem to be transcribed from internal promoters. Best characterized is transcription and processing of the spinach *trnS* gene (Wu et al., 1997). The in vitro transcription start site was mapped 12 nucleotides upstream of the mature tRNA coding region. The *trnS* coding region contains sequences resembling the A and B blocks of nuclear tRNA promoters transcribed by the eukaryotic RNA polymerase III (Galli et al., 1981). The coding region

(+1/+93) containing the A and B blocks promoted basal levels (8%) of transcription. Inclusion of the AT-rich region between –31 and –11 immediately upstream of the transcription start site was necessary to restore wild-type promoter strength. This promoter structure is reminiscent of the promoter of yeast tRNA genes (Geiduschek et al., 1995), suggesting that the plastid tRNAs may be transcribed by an RNA polymerase III-type enzyme in plastids. However, the biochemical properties and enzyme composition of this transcription activity remains to be determined. Alternatively, tRNAs may be transcribed from internal promoters by a specialized NEP or PEP associated with distinct transcription factors.

Acknowledgments

We thank T.M. Ikeda and M.W. Gray, W.R. Hess and T. Börner, S. Kapoor and M. Sugiura, K. Krause and K. Krupinska for information on their unpublished work, and J. Suzuki for providing Table 1. KL was a recipient of a postdoctoral fellowship from the Deutsche Forschungsgemeinschaft. Research in this laboratory was supported by grants from the National Science Foundation MCB 96-30763, the Rockefeller Foundation and Monsanto Co. to PM.

References

Allison LA and Maliga P (1995) Light-responsive and transcription-enhancing elements regulate the plastid *psbD* core promoter. EMBO J 14: 3721–3730

Allison LA, Simon LD and Maliga P (1996) Deletion of *rpoB* reveals a second distinct transcription system in plastids of higher plants. EMBO J 15: 2802–2809

Baeza L, Bertrand A, Mache R and Lerbs-Mache S (1991) Characterization of a protein binding sequence in the promoter region of the 16S rRNA gene of the spinach chloroplast genome. Nucleic Acids Res 19: 3577–3581

Baginsky S, Tiller K and Link G (1997) Transcription factor phosphorylation by a protein kinase associated with chloroplast RNA polymerase from mustard (*Sinapis alba*). Plant Mol Biol 34: 181–189

Baginsky S, Tiller K, Pfannschmidt T and Link G (1999) PTK, the chloroplast RNA polymerase-associated protein kinase from mustard (*Sinapis alba*), mediates redox control of plastid *in vitro* transcription. Plant Mol Biol 39: 1013–1023

Baumgartner BJ, Rapp JC and Mullet JE (1993) Plastid genes encoding the transcription/translation apparatus are differentially transcribed early in barley (*Hordeum vulgare*) chloroplast development: Evidence for selective stabilization of *psbA* mRNA. Plant Physiol 101: 781–791

Binder S and Brennicke A (1993) Transcription initiation sites in mitochondria of *Oenothera berteriana*. J Biol Chem 268: 7849–7855

Bogorad L (1991) Replication and transcription of plastid DNA. In: Bogorad L and Vasil IK (eds) The Molecular Biology of Plastids, pp 93–124. Academic Press, San Diego

Bown J, Barne K, Minchin S and Busby S (1997) Extended –10 promoters. Nucleic Acids Mol. Biol. 11: 41–52

Boyer SK and Mullet JE (1986) Characterization of *P. sativum* chloroplast *psbA* transcripts produced in vivo and in vitro and in *E. coli*. Plant Mol Biol 6: 229–243

Boyer SK and Mullet JE (1988) Sequence and transcript map of barley chloroplast *psbA* gene. Nucleic Acids Res 16: 8184

Bradley D and Gatenby AA (1985) Mutational analysis of the maize chloroplast ATPase-beta subunit gene promoter: The isolation of promoter mutants in *E. coli* and their characterization in a chloroplast in vitro transcription system. EMBO J 4: 3641–3648

Brahamsha B and Haselkorn R (1991) Isolation and characterization of the gene encoding the principal sigma factor of the vegetative cell RNA polymerase from the cyanobacterium *Anabaena* sp. strain PCC 7120. J Bacteriol 173: 2442–2450

Briat JF, Laulhere JP and Mache R (1979) Transcription activity of a DNA-protein complex isolated from spinach plastids. Eur J Biochem 98: 285–292

Brown GG, Auchincloss AH, Covello PS, Gray MW, Menassa R and Singh M (1991) Characterization of transcription initiation sites in the soybean mitochondrial genome allows identification of a transcription-associated sequence motif. Mol Gen Genet 228: 345–355

Bülow S and Link G (1988) Sigma-like activity from mustard (*Sinapis alba* L.) chloroplasts confering DNA-binding and transcription specificity to *E. coli* core RNA polymerase. Plant Mol Biol 10: 349–357

Bünger W and Feierabend J (1980) Capacity for RNA synthesis in 70S ribosome-deficient plastids of heat-bleached rye leaves. Planta 149: 163–169

Busby S and Ebright RH (1994) Promoter structure, promoter recognition, and transcription activation in prokaryotes. Cell 79: 743–746

Caoile AGFS and Stern DB (1997) A conserved core element is functionally important for maize mitochondrial promoter activity in vitro. Nucleic Acids Res 25: 4055–4060

Cermakian N, Ikeda TM, Cedergren R and Gray MW (1996) Sequences homologous to yeast mitochondrial and bacteriophage T3 and T7 RNA polymerases are widespread throughout the eukaryotic lineage. Nucleic Acids Res 24: 648–654

Chang C-C, Sheen J, Bligny M, Niwa Y, Lerbs-Mache S and Stern DB (1999) Functional analysis of two maize cDNAs encoding T7-like RNA polymerases. Plant Cell 11: 911–926

Cheng YS, Lin CH and Chen LJ (1997) Transcription and processing of the gene for spinach chloroplast threonine tRNA in a homologous in vitro system. Biochem Biophys Res Commun 233: 380–385

Christopher DA and Hoffer PH (1998) DET1 represses a chloroplast blue light-responsive promoter in a developmental and tissue-specific manner in *Arabidopsis thaliana*. Plant J 14: 1–11

Christopher DA, Kim M and Mullet JE (1992) A novel light-regulated promoter is conserved in cereal and dicot chloroplasts. Plant Cell 4: 785–798

Christopher DA, Li XL, Kim M and Mullet JE (1997) Involvement of protein kinase and extraplastidic Serine/Threonine protein phosphatases in signaling pathways regulating plastid transcription and the *psbD* blue light-responsive promoter in barley. Plant Physiol 113: 1273–1282

Covello PS and Gray MW (1991) Sequence analysis of wheat mitochondrial transcripts capped in vitro: Definitive identification of transcription initiation sites. Curr Genet 20: 245–252

DePamphilis CW and Palmer JD (1990) Loss of photosynthetic and chlororespiratory genes from the plastid genome of a parasitic flowering plant. Nature 348: 337–339

DuBell AN and Mullet JE (1995) Differential transcription of pea chloroplast genes during light-induced leaf development. Plant Physiol 109: 105–112

Eisermann A, Tiller K and Link G (1990) In vitro transcription and DNA binding characteristics of chloroplast and etioplast extracts from mustard (*Sinapis alba*) indicate differential usage of the *psbA* promoter. EMBO J 9: 3981–3987

Ems SC, Morden CW, Dixon CK, Wolfe KH, DePamphilis CW and Palmer JD (1995) Transcription, splicing and editing of plastid RNAs in the nonphotosynthetic plant *Epifagus virginiana*. Plant Mol Biol 29: 721–733

Falk J, Schmidt A and Krupinska K (1993) Characterization of plastid DNA transcription in ribosome deficient plastids of heat-bleached barley leaves. J. Plant Physiol. 141: 176–181

Fischer N, Stampacchia O, Redding K and Rochaix JD (1996) Selectable Marker Recycling In the Chloroplast. Mol Gen Genet 251: 373–380

Galli G, Hofstetter H and Birnstil ML (1981) Two conserved blocks within eukaryotic tRNA genes are major promoter elements. Nature 294: 626–631

Gatenby AA, Castleton JA and Saul MW (1981) Expression in *E. coli* of maize and wheat chloroplast genes for large subunit of ribulose bisphosphate carboxylase. Nature 291: 117–121

Geiduschek EP, Bardeleben C, Joazeiro CA, Kassavetis GA and Whitehall S (1995) Yeast RNA polymerase III: Transcription complexes and RNA synthesis. Braz J Med Biol Res 28: 147–159

Gray MW (1993) Origin and evolution of organelle genomes. Curr Opin Genet Dev 3: 884–890

Gray MW and Lang BF (1998) Transcription in chloroplasts and mitochondria: A tale of two polymerases. Trends Microbiol 6: 1–3

Greenberg BM, Narita JO, DeLuca-Flaherty C, Gruissem W, Rushlow KA and Hallick RB (1984) Evidence for two RNA polymerase activities in *Euglena gracilis* chloroplasts. J Biol Chem 259: 14880–14887

Gruber TM and Bryant DA (1997) Molecular systematic studies of eubacteria, using sigma70-type sigma factors of group 1 and group 2. J Bacteriol 179: 1734–747

Gruissem W and Tonkyn JC (1993) Control mechanisms of plastid gene expression. Crit Rev Plant Sci 12: 19–55

Gruissem W and Zurawski G (1985) Analysis of promoter regions for the spinach chloroplast *rbcL*, *atpB* and *psbA* genes. EMBO J 4: 3375–3383

Gruissem W, Greenberg BM, Zurawski G, Prescott DM and Hallick RB (1983a) Biosynthesis of chloroplast transfer RNA in a spinach chloroplast transcription system. Cell 35: 815–828

Gruissem W, Narita JO, Greenberg BM, Prescott DM and Hallick RB (1983b) Selective in vitro transcription of chloroplast genes. J Cell Biochem 22: 31–46

Gruissem W, Elsner-Menzel C, Latshaw S, Narita JO, Schaffer MA and Zurawski G (1986a) A subpopulation of spinach chloroplast tRNA genes does not require upstream promoter elements for transcription. Nucleic Acids Res 14: 7541–7556

Gruissem W, Greenberg BM, Zurawski G and Hallick RB (1986b) Chloroplast gene expression and promoter identification in chloroplast extracts. Meth Enzymol 118: 253–270

Gruissem W, Barkan A, Deng XW and Stern D (1988) Transcriptional and post-transcriptional control of plastid mRNA levels in higher plants. Trends Genet 4: 258–263

Hajdukiewicz PTJ, Allison LA and Maliga P (1997) The two RNA polymerases encoded by the nuclear and the plastid compartments transcribe distinct groups of genes in tobacco plastids. EMBO J 16: 4041–4048

Hallick RB, Lipper C, Richards OC and Rutter WJ (1976) Isolation of a transcriptionally active chromosome from chloroplasts of *Euglena gracilis*. Biochemistry 15: 3039–3045

Han CD, Patrie W, Polacco M and Coe EH (1993) Aberrations in plastid transcripts and deficiency of plastid DNA in striped and albino mutants in maize. Planta 191: 552–563

Hanley-Bowdoin L and Chua N-H (1988) Transcription of the wheat chloroplast gene that encodes the 32 kd polypeptide. Plant Mol Biol 10: 303–310

Hanley-Bowdoin L, Orozco EMJ and Chua NH (1985) In vitro synthesis and processing of a maize chloroplast transcript encoded by the ribulose 1,5-bisphosphate carboxylase large subunit gene. Mol Cell Biol 5: 2733–2745

Hartmann G, Honikel KO, Knusel F and Nuesch J (1967) The specific inhibition of the DNA-directed RNA synthesis by rifamycin. Biochim Biophys Acta 145: 843–844

Hatzack F, Dombrowski S, Brennicke A and Binder S (1998) Characterization of DNA-binding proteins from pea mitochondria. Plant Physiol 116: 519–527

Hedtke B, Börner T and Weihe A (1997) Mitochondrial and chloroplast phage-type RNA polymerases in *Arabidopsis*. Science 277: 809–811

Hedtke B, Meixner M, Gillandt S, Richter E, Börner T and Weihe A (1999) Green fluorescent protein as a marker to investigate targeting of organellar RNA polymerases of higher plants in vivo. Plant J 17: 557–561

Hess WR and Börner T (1999) Organellar RNA polymerases of higher plants. Int Rev Cytol 190: 1–59

Hess WR, Prombona A, Fieder B, Subramanian AR and Börner T (1993) Chloroplast *rps15* and the *rpoB/C1/C2* gene cluster are strongly transcribed in ribosome-deficient plastids: Evidence for a functioning non-chloroplast-encoded RNA polymerase. EMBO J 12: 563–571

Hoffer PH and Christopher DA (1997) Structure and blue-light-responsive transcription of a chloroplast *psbD* promoter from Arabidopsis thaliana. Plant Physiol 115: 213–222

Holschuh K, Bottomley W and Whitfeld PR (1984) Structure of the spinach chloroplast genes for the D2 and 44 kd reaction-centre proteins of photosystem II and for tRNASer (UGA). Nucleic Acids Res 12: 8819–8834

Hu J and Bogorad L (1990) Maize chloroplast RNA polymerase: The 180-, 120-, and 38-kilodalton polypeptides are encoded in chloroplast genes. Proc Natl Acad Sci USA 87: 1531–1535

Hu J, Troxler RF and Bogorad L (1991) Maize chloroplast RNA polymerase: The 78-kilodalton polypeptide is encoded by the

plastid *rpoC*1 gene. Nucleic Acids Res 19: 3431–3434
Hübschmann T and Börner T (1998) Characterisation of transcript initiation sites in ribosome-deficient barley plastids. Plant Mol Biol 36: 493–496
Igloi GL and Kössel H (1992) The transcriptional apparatus of chloroplasts. Crit Rev Plant Sci 10: 525–558
Ikeda TM and Gray MW (1999a) Identification and characterization of T7/T3 bacteriophage-like RNA polymerase sequences in wheat. Plant Mol Biol 40: 567–578
Ikeda TM and Gray MW (1999b) Characterization of a DNA-binding protein implicated in transcription in wheat mitochondria. Mol Cell Biol 19: 8113–8122
Iratni R, Diederich L, Harrak H, Bligny M and Lerbs-Mache S (1997) Organ-specific transcription of the *rrn* operon in spinach plastids. J Biol Chem 272: 13676–13682
Isono K, Niwa Y, Satoh K and Kobayashi H (1997a) Evidence for transcriptional regulation of plastid photosynthesis genes in *Arabidopsis thaliana* roots. Plant Physiol 114: 623–630
Isono K, Shimizu M, Yoshimoto K, Niwa Y, Satoh K, Yokota A and Kobayashi H (1997b) Leaf-specifically expressed genes for polypeptides destined for chloroplasts with domains of sigma70 factors of bacterial RNA polymerases in *Arabidopsis thaliana*. Proc Natl Acad Sci USA 94: 14948–14953
Jaehning JA (1993) Mitochondrial transcription: Is a pattern emerging? Mol Microbiol 8: 1–4
Jahn D (1992) Expression of the *Chlamydomonas reinhardtii* chloroplast tRNA(Glu) gene in a homologous in vitro transcription system is independent of upstream promoter elements. Arch Biochem Biophys 298: 505–513
Jolly SO and Bogorad L (1980) Preferential transcription of cloned maize chloroplast DNA sequences by maize chloroplast RNA polymerase. Proc Natl Acad Sci USA 77: 822–826
Kaneko T, Sato S, Kotani H, Tanaka A, Asamizu E, Nakamura Y, Miyajima N, Hirosawa M, Sugiura M, Sasamoto S, Kimura T, Hosouchi T, Matsuno A, Muraki A, Nakazaki N, Naruo K, Okumura S, Shimpo S, Takeuchi C, Wada T, Watanabe A, Yamada M, Yasuda M and Tabata S (1996) Sequence analysis of the genome of the unicellular cyanobacterium *Synechocystis* sp. strain PCC6803. II. Sequence determination of the entire genome and assignment of potential protein-coding regions. DNA Res 3: 109–136
Kapoor S and Sugiura M (1999) Identification of two essential sequence elements in the nonconsensus Type II PatpB-290 plastid promoter by using plastid transcription extracts from cultured tobacco BY-2 cells. Plant Cell 11: 1799–1810
Kapoor S, Suzuki JY and Sugiura M (1997) Identification and functional significance of a new class of non-consensus-type plastid promoters. Plant J 11: 327–337
Kestermann M, Neukirchen S, Kloppstech K and Link G (1998) Sequence and expression characteristics of a nuclear-encoded chloroplast sigma factor from mustard (*Sinapis alba*). Nucleic Acids Res 26: 2747–2753
Kidd GH and Bogorad L (1980) A facile procedure for purifying maize chloroplast RNA polymerase from whole cell homogenates. Biochim Biophys Acta 609: 14–30
Kim M and Mullet JE (1995) Identification of a sequence-specific DNA binding factor required for transcription of the barley chloroplast blue light-responsive *psbD-psbC* promoter. Plant Cell 7: 1445–1457
Kim M, Christopher DA and Mullet JE (1999a) ADP-Dependent phosphorylation regulates association of a DNA-binding complex with the barley chloroplast *psbD* blue-light-responsive promoter. Plant Physiol 119: 663–670
Kim M, Thum KE, Morishige DT and Mullet JE (1999b) Detailed architecture of the barley chloroplast *psbD-psbC* blue light-responsive promoter. J Biol Chem 274: 4684–4692
Klein RR and Mullet JE (1990) Light-induced transcription of chloroplast genes. *psbA* transcription is differentially enhanced in illuminated barley. J Biol Chem 265: 1895–1902
Lakhani S, Khanna NC and Tewari KK (1992) Two distinct transcriptional activities of pea (*Pisum sativum*) chloroplasts share immunochemically related functional polypeptides. Biochem J 286: 833–841
Lam E, Hanley-Bowdoin L and Chua NH (1988) Characterization of a chloroplast sequence-specific DNA binding factor. J Biol Chem 263: 8288–8293
Lerbs S, Briat JF and Mache R (1983) Chloroplast RNA polymerase from spinach: Purification and DNA-binding proteins. Plant Mol Biol 2: 67–74
Lerbs S, Bräutigam E and Parthier B (1985) Polypeptides in DNA-dependent RNA polymerase of spinach chloroplasts: characterization by antibody-linked polymerase assay and determination of sites of synthesis. EMBO J 4: 1661–1666
Lerbs S, Bräutigam E and Mache R (1988) DNA-dependent RNA polymerase of spinach chloroplasts: characterization of α-like and σ-like polypeptides. Mol Gen Genet 211: 459–464
Lerbs-Mache S (1993) The 110-kDa polypeptide of spinach plastid DNA-dependent RNA polymerase: Single-subunit enzyme or catalytic core of multimeric enzyme complexes? Proc Natl Acad Sci USA 90: 5509–5513
Liere K and Link G (1994) Structure and expression characteristics of the chloroplast DNA region containing the split gene for tRNA(Gly) (UCC) from mustard (*Sinapis alba* L.). Curr Genet 26: 557–563
Liere K and Maliga P (1999a) In vitro characterization of the tobacco *rpoB* promoter reveals a core sequence motif conserved between phage-type plastid and plant mitochondrial promoters. EMBO J 18: 249–257
Liere K and Maliga P (1999b) Novel in vitro transcription assay indicates that the *accD* NEP promoter is contained in a 19 bp fragment. In: Argyroudi-Akoyunoglou JH and Senger H (eds) The Chloroplast: From Molecular Biology to Biotechnology, pp 79–84. Kluwer Academic Publishers, Dordrecht
Link G (1984) DNA sequence requirements for the accurate transcription of a protein-coding plastid gene in a plastid in vitro transcription system from mustard (*Sinapis alba* L.). EMBO J 3: 1697–1704
Link G (1994) Plastid differentiation: Organelle promoters and transcription factors. In: Nover L (ed) Plant Promoters and Transcription Factors—Results and Problems in Cell Differentiation, Vol 20, pp 65–85. Springer Verlag, Berlin
Link G (1996) Green life: Control of chloroplast gene transcription. Bioessays 18: 465–471
Link G and Langridge U (1984) Structure of the chloroplast gene for the precursor of the Mr 32,000 Photosystem II protein from mustard (*Sinapis alba* L.). Nucleic Acids Res 12: 945–958
Little MC and Hallick RB (1988) Chloroplast *rpoA*, *rpoB*, and *rpoC* genes specify at least three components of a chloroplast DNA-dependent RNA polymerase active in tRNA and mRNA transcription. J Biol Chem 263: 14302–14307
Liu B and Troxler RF (1996) Molecular characterization of a positively photoregulated nuclear gene for a chloroplast RNA

polymerase sigma factor in *Cyanidium caldarium*. Proc Natl Acad Sci USA 93: 3313–3318

Lonetto M, Gribskov M and Gross CA (1992) The sigma 70 family: Sequence conservation and evolutionary relationships. J Bacteriol 174: 3843–3849

Lonetto MA, Brown KL, Rudd KE and Buttner MJ (1994) Analysis of the Streptomyces coelicolor *sigE* gene reveals the existence of a subfamily of eubacterial RNA polymerase sigma factors involved in the regulation of extracytoplasmic functions. Proc Natl Acad Sci USA 91: 7573–7577

Lukens JH, Mathews DE and Durbin RD (1987) Effect of tagetitoxin on the levels of ribulose 1,5-bisphosphate carboxylase, ribosomes, and RNA in plastids of wheat leaves. Plant Physiol 84: 808–813

Maliga P (1998) Two plastid polymerases of higher plants: An evolving story. Trends Plant Sci 3: 4–6

Maliga P, Liere K, Sriraman P and Svab Z (1999) A transgenic approach to characterize the plastid transcription machinery in higher plants. In: Argyroudi-Akoyunoglou JH and Senger H (eds) The Chloroplast: From Molecular Biology to Biotechnology, pp 317–323. Kluwer Academic Publishers, Dordrecht

Mathews DE and Durbin RD (1990) Tagetitoxin inhibits RNA synthesis directed by RNA polymerases from chloroplasts and *Escherichia coli*. J Biol Chem 265: 493–498

Miyagi T, Kapoor S, Sugita M and Sugiura M (1998) Transcript analysis of the tobacco plastid operon *rps2/atpI/H/F/A* reveals the existence of a non-consensus type II (NCII) promoter upstream of the *atpI* coding sequence. Mol Gen Genet 257: 299–307

Morden CW, Wolfe KH, dePamphilis CW and Palmer JD (1991) Plastid translation and transcription genes in a non-photosynthetic plant: Intact, missing and pseudo genes. EMBO J 10: 3281–3288

Morikawa K, Ito S, Tsunoyama Y, Nakahira Y, Shiina T and Toyoshima Y (1999) Circadian-regulated expression of a nuclear encoded plastid sigma factor gene (*sigA*) in wheat seedlings. FEBS Lett 451: 275–278

Mullet JE (1993) Dynamic regulation of chloroplast transcription. Plant Physiol 103: 309–313

Mullet JE, Orozco EM and Chua N-H (1985) Multiple transcripts for higher plant *rbcL* and *atpB* genes and localization of the transcription initiation sites of the *rbcL* gene. Plant Mol Biol 4: 39–54

Mulligan RM, Leon P and Walbot V (1991) Transcription and posttranscriptional regulation of maize mitochondrial gene expression. Mol Cell Biol 11: 533–543

Narita JO, Rushlow KE and Hallick RB (1985) Characterization of a *Euglena gracilis* chloroplast RNA polymerase specific for ribosomal RNA genes. J Biol Chem 260: 11194–11199

Neuhaus H and Link G (1990) The chloroplast *psbK* operon from mustard (*Sinapis alba* L.): Multiple transcripts during seedling development and evidence for divergent overlapping transcription. Curr Genet 18: 377–383

Newton KJ, Winberg B, Yamato K, Lupold S and Stern DB (1995) Evidence for a novel mitochondrial promoter preceding the *cox2* gene of perennial teosintes. EMBO J 14: 585–593

Nickelsen J and Link G (1990) Nucleotide sequence of the mustard chloroplast genes *trnH* and *rps19′*. Nucleic Acids Res 18: 1051

Orozco EM, Jr., Mullet JE and Chua NH (1985) An in vitro system for accurate transcription initiation of chloroplast protein genes. Nucleic Acids Res 13: 1283–1302

Pfannschmidt T and Link G (1994) Separation of two classes of plastid DNA-dependent RNA polymerases that are differentially expressed in mustard (*Sinapis alba* L.) seedlings. Plant Mol Biol 25: 69–81

Pfannschmidt T and Link G (1997) The A and B forms of plastid DNA-dependent RNA polymerase from mustard (*Sinapis alba* L.) transcribe the same genes in a different developmental context. Mol Gen Genet 257: 35–44

Pfannschmidt T, Nilsson A and Allen JF (1999) Photosynthetic control of chloroplast gene expression. Nature 397: 625–628

Rajasekhar VK and Tewari KK (1995) Analyses of the extent of immunological relatedness between a highly purified pea chloroplast functional RNA polymerase and *Escherichia coli* RNA polymerase. J. Plant Physiol. 145: 427–463

Rajasekhar VK, Sun E, Meeker R, Wu BW and Tewari KK (1991) Highly purified pea chloroplast RNA polymerase transcribes both rRNA and mRNA genes. Eur J Biochem 195: 215–228

Rapp WD and Stern DB (1992) A conserved 11 nucleotide sequence contains an essential promoter element of the maize mitochondrial *atp1* gene. EMBO J 11: 1065–1073

Rapp JC, Baumgartner BJ and Mullet J (1992) Quantitative analysis of transcription and RNA levels of 15 barley chloroplast genes. Transcription rates and mRNA levels vary over 300-fold; predicted mRNA stabilities vary 30-fold. J Biol Chem 267: 21404–21411

Rapp WD, Lupold DS, Mack S and Stern DB (1993) Architecture of the maize mitochondrial *atp1* promoter as determined by linker-scanning and point mutagenesis. Mol Cell Biol 13: 7232–7238

Reinbothe S, Reinbothe C, Heintzen C, Seidenbecher C and Parthier B (1993) A methyl jasmonate-induced shift in the length of the 5′ untranslated region impairs translation of the plastid *rbcL* transcript in barley. EMBO J 12: 1505–1512

Reiss T and Link G (1985) Characterization of transcriptionally active DNA-protein complexes from chloroplasts and etioplasts of mustard (*Sinapis alba* L.). Eur J Biochem 148: 207–212

Reznikov W, Siegle DA, Cowing DW and Gross CA (1985) The regulation of transcription initiation in bacteria. Annu Rev Genet 19: 355–387

Rochaix JD (1995) *Chlamydomonas reinhardtii* as the photosynthetic yeast. Annu Rev Genet 29: 209–230

Ross W, Gosink KK, Salomon J, Igarashi K, Zou C, Ishihama A, Severinov K and Gourse RL (1993) A third recognition element in bacterial promoters: DNA binding by the alpha subunit of RNA polymerase. Science 262: 1407–1413

Satoh J, Baba K, Nakahira Y, Shiina T and Toyoshima Y (1997) Characterization of dynamics of the *psbD* light-induced transcription in mature wheat chloroplasts. Plant Mol Biol 33: 267–278

Satoh J, Baba K, Nakahira Y, Tsunoyama Y, Shiina T and Toyoshima Y (1999) Developmental stage-specific multi-subunit plastid RNA polymerases (PEP) in wheat. Plant J 18: 407–416

Schrubar H, Wanner G and Westhoff P (1990) Transcriptional control of plastid gene expression in greening *sorghum* seedlings. Planta 183: 101–111

Serino G and Maliga P (1998) RNA polymerase subunits encoded by the plastid *rpo* genes are not shared with the nucleus-encoded plastid enzyme. Plant Physiol 117: 1165–1170

Sexton TB, Christopher DA and Mullet JE (1990) Light-induced switch in barley *psbD-psbC* promoter utilization: A novel mechanism regulating chloroplast gene expression. EMBO J 9: 4485–4494

Shadel GS and Clayton DA (1993) Mitochondrial Transcription. J Biol Chem 268: 16083–16086

Shiina T, Allison L and Maliga P (1998) *rbcL* transcript levels in tobacco plastids are independent of light: reduced dark transcription rate is compensated by increased mRNA stability. Plant Cell 10: 1713–1722

Shinozaki K and Sugiura M (1982) The nucleotide sequence of the tobacco chloroplast gene for the large subunit of ribulose-1,5-bisphosphate carboxylase/oxygenase. Gene 20: 91–102

Siemenroth A, Wollgiehn R, Neumann D and Börner T (1981) Synthesis of ribosomal RNA in ribosome-deficient plastids of the mutant 'albostrians' of *Hordeum vulgare* L. Planta 153: 547–555

Silhavy D and Maliga P (1998a) Mapping of the promoters for the nucleus-encoded plastid RNA polymerase (NEP) in the *iojap* maize mutant. Curr Genet 33: 340–344

Silhavy D and Maliga P (1998b) Plastid promoter utilization in a rice embryonic cell culture. Curr Genet 34: 67–70

Sriraman P, Silhavy D and Maliga P (1998a) The phage-type PclpP-53 plastid promoter comprises sequences downstream of the transcription initiation site. Nucleic Acids Res 26: 4874–4879

Sriraman P, Silhavy D and Maliga P (1998b) Transcription from heterologous rRNA operon promoters in chloroplasts reveals requirement for specific activating factors. Plant Physiol 117: 1495–1499

Steinberg TH, Mathews DE, Durbin RD and Burgess RR (1990) Tagetitoxin: A new inhibitor of transcription by RNA polymerase III. J Biol Chem 265: 499–505

Stern DB, Higgs DC and Yang JJ (1997) Transcription and translation in chloroplasts. Trends Plant Sci 2: 308–315

Strittmatter G, Godzicka-Josefiak A and Kössel H (1985) Identification of an rRNA operon promoter from *Zea mays* chloroplast which excludes the proximal tRNAVal from the primary transcript. EMBO J 4: 599–604

Suck R, Zeltz P, Falk J, Acker A, Kössel H and Krupinska K (1996) Transcriptionally active chromosomes (TACs) of barley chloroplasts contain the α-subunit of plastome encoded RNA polymerase. Curr Genet 30: 515–521

Sugiura M (1992) The Chloroplast Genome. Plant Mol Biol 19: 149–168

Sun E, Wu BW and Tewari KK (1989) In vitro analysis of the pea chloroplast 16S rRNA gene promoter. Mol Cell Biol 9: 5650–5659

Surzycki SJ and Shellenbarger DL (1976) Purification and characterization of a putative sigma factor from *Chlamydomonas reinhardtii*. Proc Natl Acad Sci USA 73: 3961–3965

Tanaka K, Oikawa K, Ohta N, Kuroiwa H, Kuroiwa T and Takahashi H (1996) Nuclear encoding of a chloroplast RNA polymerase sigma subunit in a red alga. Science 272: 1932–1935

Tanaka K, Tozawa Y, Mochizuki N, Shinozaki K, Nagatani A, Wakasa K and Takahashi H (1997) Characterization of three cDNA species encoding plastid RNA polymerase sigma factors in *Arabidopsis thaliana*: evidence for the sigma factor heterogeneity in higher plant plastids. FEBS Lett 413: 309–313

Tewari KK and Goel A (1983) Solubilization and partial purification of RNA polymerase from pea chloroplasts. Biochemistry 22: 2142–2148

Tiller K and Link G (1993a) Phosphorylation and dephosphorylation affect functional characteristics of chloroplast and etioplast transcription systems from mustard (*Sinapis alba* L.). EMBO J 12: 1745–1753

Tiller K and Link G (1993b) Sigma-like transcription factors from mustard (*Sinapis alba* L.) etioplast are similar in size to, but functionally distinct from, their chloroplast counterparts. Plant Mol Biol 21: 503–513

Tiller K, Eisermann A and Link G (1991) The chloroplast transcription apparatus from mustard (*Sinapis alba* L.). Evidence for three different transcription factors which resemble bacterial sigma factors. Eur J Biochem 198: 93–99

To KY, Cheng MC, Suen DF, Mon DP, Chen LFO and Chen SCG (1996) Characterization of the light-responsive promoter of rice chloroplast *psbD-C* operon and the sequence-specific DNA binding factor. Plant & Cell Physiology 37: 660–666

Tozawa Y, Tanaka K, Takahashi H and Wakasa K (1998) Nuclear encoding of a plastid sigma factor in rice and its tissue- and light-dependent expression. Nucleic Acids Res 26: 415–419

Tracy RL and Stern DB (1995) Mitochondrial transcription initiation: promoter structures and RNA polymerases. Curr Genet 28: 205–216

Trifa Y, Privat I, Gagnon J, Baeza L and Lerbs-Mache S (1998) The nuclear *RPL4* gene encodes a chloroplast protein that co-purifies with the T7-like transcription complex as well as plastid ribosomes. J Biol Chem 273: 3980–3985

Troxler RF, Zhang F, Hu J and Bogorad L (1994) Evidence that sigma factors are components of chloroplast RNA polymerase. Plant Physiol 104: 753–759

Vera A and Sugiura M (1995) Chloroplast rRNA transcription from structurally different tandem promoters: an additional novel-type promoter. Curr Genet 27: 280–284

Wada T, Tunoyama Y, Shiina T and Toyoshima Y (1994) In vitro analysis of light-induced transcription in the wheat *psbD/C* gene cluster using plastid extracts from dark-frown and short-term-illuminated seedlings. Plant Physiol 104: 1259–1267

Weihe A and Börner T (1999) Transcription and the architecture of promoters in chloroplasts. Trends Plant Sci 4: 169–170

Wolfe KH, Katz-Downie DS, Morden CW and Palmer JD (1992a) Evolution of the plastid ribosomal RNA operon in a nongreen parasitic plant: Accelerated sequence evolution, altered promoter structure, and tRNA pseudogenes. Plant Mol Biol 18: 1037–1048

Wolfe KH, Morden CW, Ems SC and Palmer JD (1992b) Rapid evolution of the plastid translational apparatus in a nonphotosynthetic plant: Loss or accelerated sequence evolution of tRNA and ribosomal protein genes. J Mol Evol 35: 304–317

Wolfe KH, Morden CW and Palmer JD (1992c) Function and evolution of a minimal plastid genome from a nonphotosynthetic parasitic plant. Proc Natl Acad Sci USA 89: 10648–10652

Wösten MM (1998) Eubacterial sigma-factors. FEMS Microbiol Rev 22: 127–150

Wu CY, Lin CH and Chen LJ (1997) Identification of the transcription start site for the spinach chloroplast serine tRNA gene. FEBS Lett 418: 157–161

Yan B and Pring DR (1997) Transcriptional initiation sites in *sorghum* mitochondrial DNA indicate conserved and variable features. Curr Genet 32: 287–295

Zengel JM, Mueckl D and Lindahl L (1980) Protein L4 of the *E. coli* ribosome regulates an eleven gene r protein operon. Cell 21: 523–535

Chapter 3

Phytochrome and Regulation Of Photosynthetic Gene Expression

Michael Malakhov* and Chris Bowler†
Molecular Plant Biology Laboratory, Stazione Zoologica 'Anton Dohrn,' Villa Comunale, I-80121 Naples, Italy

Summary

Phytochromes are the best characterized plant photoreceptors. They are responsible for a wide range of photomorphogenic events ranging from seed germination, de-etiolation, and shade-avoidance responses to flowering. Many of these responses include the induction and subsequent regulation of genes encoding the photosynthetic components. The availability of mutants defective in individual phytochromes has now revealed which phytochromes are responsible for specific physiological responses. New molecular methods have revolutionized our understanding of the precise mode of action of the phytochromes, which have shattered many of the traditional dogmas. For example, phytochrome has now been demonstrated to have serine/threonine (Ser/Thr) protein kinase activity and to be translocated from the cytoplasm to the nucleus in a light-regulated manner. Several phytochrome-interacting proteins have now been identified, most of which are nuclear localized. Integration of these new downstream-interacting proteins into the signaling pathways, previously proposed by biochemical and genetic studies, reveals a highly sophisticated signal transduction circuitry that can allow the parallel function of both cytoplasmic- and nuclear-localized components.

*Present address: Scripps Research Institute, 10550 N. Torrey Pines Road, La Jolla, CA 92037, U.S.A.
†Author for correspondence, email: chris@alpha.szn.it

Eva-Mari Aro and Bertil Andersson (eds): Regulation of Photosynthesis, pp. 51–66.

I. Introduction

Light is the source of energy for plant cells, but also an important source of information about the environment. Dark-grown (etiolated) seedlings display an apical hook, closed and unexpanded cotyledons and elongated hypocotyl cells (skotomorphogenesis). This is necessary for the seedlings to grow through soil or fallen leaves and reach the light. Upon light exposure seedlings undergo de-etiolation: cotyledons open and expand, hypocotyl elongation is inhibited, and cell differentiation is initiated (photomorphogenesis). Another characteristic feature of de-etiolation is the pigmentation of seedlings due to the assembly of the photosynthetic complexes in the chloroplast. Photosynthetic pigments such as chlorophyll are synthesized, as well as general photoprotectants such as carotenoids and flavonoids (e.g. anthocyanins). These changes are mediated by the induction of specific sets of genes. The best studied of these are the *CAB* (chlorophyll *a/b* binding protein), also known as *LHCB* (light-harvesting chlorophyll binding protein) family of nuclear genes encoding the chlorophyll *a/b*-binding proteins, the *CHS* (chalcone synthase) family of genes encoding an enzyme catalyzing the first committed step of anthocyanin biosynthesis, and *RBCS* encoding the small subunit of ribulose bisphosphate carboxylase.

At least three groups of photoreceptors control photomorphogenesis: the ultraviolet (UV)-B light photoreceptors, UV-A and blue light photoreceptors (cryptochrome and phototropin), and red/far red light photoreceptors (phytochrome). Phytochrome is the best characterized and is responsible for initiating a wide range of photomorphogenic events including the expression of nuclear and chloroplastic genes encoding photosynthetic components (Kendrick and Kronenberg, 1994).

II. Activation of Phytochrome and Other Photoreceptors

Plant phytochrome was discovered 40 years ago and its molecular structure has been extensively studied (Quail et al., 1995; Chory, 1997). Phytochromes are encoded by small gene families. In *Arabidopsis*, for example, five types of phytochromes (PhyA-PhyE) have been identified. They form homodimers with a covalently attached photosensitive linear tetrapyrrole chromophore that, most classically, mediates responses to red and far red light. These responses occur through the photointerconversion of the chromophore between two stable isomers, the red light-absorbing form, termed Pr, and the far red light-absorbing form termed Pfr (Kendrick and Kronenberg, 1994). On the basis of physiological, genetic, and biochemical studies, Pfr is thought to be the active form, although Pr may play a role in seed germination in certain light conditions (Reed et al., 1994; Shinomura et al., 1994).

Two types of phytochromes have been defined on the basis of their lability in light (Table 1). Type I phytochrome (PhyA) is abundant in seed and etiolated seedlings, but its level drops significantly in green plants. As early as 1975 it was demonstrated (by use of immunolabeling) that upon exposure to light Type I phytochrome becomes associated with discrete regions of the cell. This event was given the term 'sequestering.' The regions in question did not appear to be nuclei, plastids, or mitochondria and did not contain any membranes (Mackenzie et al., 1975; McCurdy and Pratt, 1986). These phytochrome-containing structures have been proposed to represent 26S proteasomes aggregated with ubiquitinated PhyA destined for degradation (Clough and Vierstra, 1997). Type II phytochromes (e.g., PhyB and PhyC) are not subject to sequestering and are consequently the most abundant phytochromes in light-grown plants. Further classification of phytochromes can be made on the basis of the fluence required for the response; individual phytochrome responses can be classified as either Very Low Fluence Responses (VLFRs), Low Fluence Responses (LFRs), and High Irradiance Responses (HIRs), according to their fluence or fluence rate requirements (Furuya and Schäfer, 1996; Mustilli and Bowler, 1997). In some cases the

---

Abbreviations: CAB – chlorophyll *a,b*-binding protein; CCA – circadian clock associated; CHS – chalcone synthase; COP – constitutive photomorphogenic; CRY – cryptochrome; DET – de-etiolated; EST– expressed sequence tag; FRc – constant far red light irradiation; HIR – high irradiance response; LFR – low fluence response; LHY – late elongated hypocotyl; NPH – non-phototropic hypocotyl; NDPK – nucleotide diphosphate kinase; Pfr – far red light-absorbing form of phytochrome; PhyA – phytochrome A; PhyB – phytochrome B; PIF – phytochrome interacting factor; PKS – phytochrome kinase substrate; POR – protochlorophyllide oxidoreductase; Pr – red light-absorbing form of phytochrome; PS I – Photosystem I; PS II – Photosystem II; Rc – constant red light irradiation; SPA – suppressor of PhyA; VLFR – very low fluence response

Table 1. Functions of the different phytochromes

Name	Designation	Light lability	Function	Primary site of function
PhyA	Type I	Yes	Far red HIR, VLFR	Seeds, Etiolated seedlings
PhyB-E	Type II	No	Red/far red-reversible LFR, Red HIR, Shade-avoidance response, Flowering time	Seeds, Etiolated seedlings, Mature green plants

Abbreviations: HIR, high irradiance response; VLFR, very low fluence response

classification can be made according to a particular response mediated by the phytochrome (Table 1).

It should be emphasized that most of the work on individual photoreceptors and their roles in the regulation of gene expression has been done with seeds or in etiolated seedlings but not in photosynthetically active mature plants. As a consequence, only limited information is available about the contribution of photoreceptors in the regulation of structure and performance of the photosynthetic apparatus. It is likely that in mature plants phytochromes and other photoreceptors affect photosynthesis primarily through long term changes such as the regulation of PS I/PS II ratio and via the circadian clock (see Section VI).

Type II phytochrome is known to be responsible for the 'shade-avoidance syndrome' (Table 1). Because chlorophyll absorbs preferentially red light, tall plants create far-red-light- enriched environments in the understory (Kendrick and Kronenberg, 1994). This serves as information for seedlings and shade-avoiding plants to grow taller.

The exact mechanism of signal transfer from phytochrome to the downstream signal transduction components is not yet known. However, phytochrome has now been demonstrated to have Ser/Thr protein kinase activity. The first strong evidence for this came from the discovery of cyanobacterial phytochrome and from the demonstration of its histidine kinase activity (Kehoe and Grossman, 1996; Yeh et al., 1997). Plant PhyA has subsequently been shown to be a phosphoprotein in vivo. It does not, however, display the prokaryotic type of histidine phosphorylation but eukaryotic-type phosphorylation of serine/threonine residues (Lapko et al., 1997, 1999; Yeh and Lagarias, 1998). Phosphorylation is a light- and chromophore-regulated process, and it is believed that after photoconversion of Pr into Pfr, the phytochrome kinase activity initiates light signaling either by phosphorylating downstream elements of the signaling cascade and/or by phosphospecific interactions with them (Section V)(Yeh and Lagarias, 1998; Fankhauser and Chory, 1999).

Phytochrome is not the only photoreceptor with kinase activity. The flavoprotein phototropin (initially designated NPH1 (non-phototropic hypocotyl)), which senses UV-A and blue light and mediates phototropism (Liscum and Briggs, 1996), also exhibits a serine/threonine protein kinase activity (Christie et al., 1998).

Similarly to NPH1, cryptochromes (encoded by *CRY1* and *CRY2* in *Arabidopsis*) are flavoprotein photoreceptors that detect UV-A and blue-light. Cryptochromes are involved in the regulation of circadian rhythms, inhibition of hypocotyl elongation, expansion of cotyledons, and transition to flowering. These processes clearly overlap with phytochrome-mediated responses (Ahmad and Cashmore, 1996). It has been shown recently using the yeast two-hybrid system that phytochromes and cryptochromes can interact physically (Section V)(Ahmad et al., 1998).

III. Second Messengers in Phytochrome Signal Transduction

Pharmacological approaches, coupled with microinjection techniques, have indicated that membrane-bound heterotrimeric G-proteins are involved in amplification of PhyA signal transduction. Three distinct signaling cascades were found to control PhyA responses downstream of the G-protein (Fig. 1)(Neuhaus et al., 1993, 1997; Bowler et al., 1994a,b). First, cGMP can activate *CHS* genes and stimulate anthocyanin biosynthesis. Second, calcium (and calmodulin) can activate synthesis of light harvesting complexes, Photosystem II (PS II), ribulose bisphosphate carboxylase, and ATPase and, thus, partial chloroplast development. Third, a

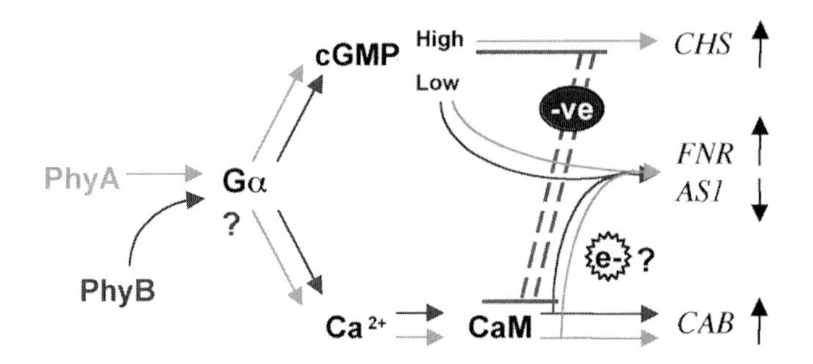

Fig. 1. Phytochrome signaling pathways identified by microinjection approaches. PhyA pathways are indicated in red, PhyB in green. Reciprocal negative regulation (denoted -ve) between the cGMP- and calcium-dependent pathways is indicated by dashed lines. Redox signals (denoted e^-) arising from plastoquinone are proposed to regulate the relative activities of the two calcium-dependent pathways. For further details see Sections III and IV. Abbreviations: *CHS*, chalcone synthase; *FNR*, ferredoxin oxidoreductase; *CAB*, chlorophyll *a,b*-binding proteins; *AS1*, asparagine synthetase. See also Color Plate 4a.

pathway requiring both calcium and cGMP can activate synthesis of Photosystem I (PS I) and cytochrome b_6f complexes to produce fully mature chloroplasts. Furthermore, it is likely that the latter pathway can mediate light-induced repression of the *AS1* gene (encoding asparagine synthetase).

It is known that to retain a high quantum efficiency of photosynthesis when plants are grown under different light conditions, their PS II:PS I ratios are modified (Anderson, 1986; Melis, 1991). Hence, a plant must be able to alter the stoichiometry between structurally and functionally distinct thylakoid complexes, while maintaining the polypeptide composition within these complexes absolutely constant. With this in mind, it is not surprising that cGMP-, Ca^{2+}- and cGMP/Ca^{2+}- requiring signal transduction pathways are divided along functional lines, and that synthesis of PS II and PS I is separately regulated. Nonetheless, an absolute requirement for the maintenance of photosynthetic efficiency must be the ability to make long-term adjustments in photosystem stoichiometry in response to varying light conditions during growth. These changes could be coordinated e.g. via reciprocal control between the calcium- and the cGMP-stimulated signal transduction pathways. Indeed, high cGMP concentrations cause negative regulation of both the Ca^{2+}- and the Ca^{2+}/cGMP-dependent pathways. Conversely, high activity of the Ca^{2+}–dependent pathway can negatively regulate the cGMP-dependent pathway (Fig. 1)(Bowler et al., 1994b).

The molecular details of calcium/calmodulin function in light signaling have not been studied. Other responses mediated by Ca^{2+}, however, have been well documented in plants, and several plant genes coding for calmodulins and calmodulin-dependent enzymes have been characterized (Trewavas and Malho, 1998). cGMP-mediated signaling is extremely poorly explored. Although the subject is now drawing increasing attention with the discovery of the involvement of cGMP in pathogen defense responses (Delledonne et al., 1998; Durner et al., 1998), so far no convincing data is available on plant genes/enzymes that are involved in the metabolism of cGMP. Recently a family of six genes encoding putative cyclic nucleotide-gated channels have been identified in *Arabidopsis* (Köhler et al., 1999). These genes, similarly to their animal counterparts, share a high degree of homology and contain potential cyclic nucleotide binding domains as well as calmodulin binding sites. It is possible that (some of) these channels mediate phytochrome signaling.

IV. Genetic Approaches to Dissect Phytochrome Signaling

Important knowledge about phytochrome signaling has been derived from the isolation and analysis of photomorphogenic mutants in *Arabidopsis*. These mutants can be classified either as insensitive or constitutive regarding their light responses (Table 2).

A. Insensitive Mutants

Insensitive mutants display a light-blind phenotype, i.e. even after exposure to light the seedlings show elongated hypocotyls, and only partially expanded, unfolded cotyledons. In addition, light-grown mutant

Table 2. Summary of light signaling mutants

Name	Gene	Phenotype	Function	Cellular localization
hy1	*HY1*	Insens. in WL	Chromophore biosynthesis (heme oxygenase)	P
hy2	?	Insens. in WL	Chromophore biosynthesis	?
hy3	*PHYB*	Insens. in Rc	Photoreceptor	N/C
hy4	*CRY1*	Insens. in B	Photoreceptor	N
hy5	*HY5*	Insens. in WL	Transcription factor	N
hy8	*PHYA*	Insens. in FRc	Photoreceptor	N/C
fhy1	?	Insens. in FRc	PhyA signaling?	?
fhy3	?	Insens. in FRc	PhyA signaling?	?
far1	*FAR1*	Insens. in FRc	PhyA signaling	N
red1	?	Insens. in Rc	PhyB signaling?	?
cue1	*PPT*	CAB underexpressed	Photosynthesis	P
127-4	*Toc33*	Pale green	Plastid import machinery	P
det1	*DET1*	Constitutive de-etiolation	?	N
cop1	*COP1*	Constitutive de-etiolation	?	C/N
cop9	*COP9*	Constitutive de-etiolation	COP9 complex	N
fus4	*FUS4*	Constitutive de-etiolation	COP9 complex	N
fus5	*FUS5*	Constitutive de-etiolation	COP9 complex	N
fus6	*GPS1*	Constitutive de-etiolation	COP9 complex	N
fus8	?	Constitutive de-etiolation	?	?
fus9	?	Constitutive de-etiolation	?	?
fus11	*FUS11*	Constitutive de-etiolation	COP9 complex	N
fus12	?	Constitutive de-etiolation	?	?
gun1, 2, 3	?	Constitutive *CAB*	?	?
doc1,2,3		Dark-expressed CAB	?	?
psi2	?	Up-regulated *CAB*	?	?
spa1	*SPA1*	Suppressor of *hy8*	PhyA signaling?	N
shy1	?	Suppressor of *hy2*	?	?
shy2	*IAA3*	Suppressor of *hy2*	Auxin metabolism	N
ted1,2,3	?	Suppressor of *det1*	?	?
ted4	*HY1*	Suppressor of *det1*	Allelic to *hy1*	P
ted5	*HY5*	Suppressor of *det1*	Transcription factor	N
det2	*DET2*	Partial de-etiolation	Brassinosteroid biosynthesis	C
cpd	*CPD*	Partial de-etiolation	Brassinosteroid biosynthesis	C
dim	*DIM*	Partial de-etiolation	Brassinosteroid biosynthesis	C

Abbreviations: PPT, phosphoenolpyruvat/phosphate translocator; N, nucleus; C, cytoplasm; P, plastid; B, blue light; WL, white light

plants are often characterized by early flowering. Some of these mutants, denoted *hy* for long hypocotyl, have defects in photoreceptor function [e.g. *hy3* (defective in *phyB*), *hy8* (defective in *phyA*), *hy4* (defective in *cry1*), *hy1* and *hy2* (defective in phytochrome chromophore biosynthesis)](Table 2) (Whitelam and Harberd, 1994; Chamovitz and Deng, 1996; Chory et al., 1996), while others encode positive regulators of light signal transduction pathways e.g. *hy5* (a bZIP transcription factor)(Oyama et al., 1997). Some insensitive mutants appear to have defects specific to PhyA or PhyB signaling, e.g. *fhy1*, *fhy3*, and *far1* for PhyA (Barnes et al., 1996a; Hudson et al., 1999) and *red1* for PhyB (Wagner et al., 1997) suggesting that different phytochromes may use independent signal transduction pathways.

Another group of insensitive mutants, designated *cue*, was isolated in *Arabidopsis* with reduced levels of *CAB3* gene expression as well as defects in plastid biogenesis. *cue* mutants display pale-green mesophyll cells and dark-green bundle sheath cells aligning the veins (Li et al., 1995; Lopez-Juez et al., 1998). Analysis of another mutant with a similar phenotype, denoted *127-4*, revealed that the mutated gene had strong homology with a component of the general protein import apparatus (Toc33) through which the majority of plastid-destined proteins are believed to pass (Jarvis et al., 1998). This indicates that at least some of the *cue* mutants cannot be classified as signaling mutants, even though expression patterns of light-responsive genes are similar to those of true light-insensitive regulatory mutants.

B. Constitutive Mutants

Constitutive mutants, e.g. *det/cop/fus* (de-etiolated/ constitutive photomorphogenic/ fusca) splay a light-grown phenotype when grown in the dark: short hypocotyl, cotyledon opening and expansion, epidermal cell differentiation, synthesis of anthocyanins and partial plastid development. Constitutive mutants grown in light are usually shorter than the dark-grown ones and strong alleles cause seedling lethality (Chory et al., 1989; 1996; Deng et al., 1991; Chamovitz and Deng, 1996). The mutants have been identified in different screens in several laboratories. After thorough genetic work and unifying the name for each locus (e.g. *COP11* = *FUS6*), 10 loci have been assigned to the *DET/COP/FUS* group (*DET1, COP1, COP9, FUS4, FUS5, FUS6, FUS8, FUS9, FUS11,* and *FUS12*)(Kwok et al., 1996)(Table 2). The recessive nature of these mutations suggests that they are loss-of-function mutations and that the wild-type genes are repressors of photomorphogenesis in darkness. However, many of these mutations are not specific for phytochrome signal transduction as they display many pleiotropic effects (e.g. Chory and Peto, 1990; Millar et al., 1994; Mayer et al., 1996; Szekeres et al., 1996). For example, light grown *cop1*, *det1* and *cop9* mutants display elevated expression of genes involved in stress responses (Mayer et al., 1996).

Another three groups of mutants, *gun* (including *gun1*, *gun2*, and *gun3;* Susek et al., 1993), *doc* (comprising *doc1*, *doc2*, and *doc3*; Li et al., 1994), and *psi2* (Genoud et al., 1998) have been isolated by screening for elevated or constitutive *CAB* gene expression. Whether these mutants are mutated in loci directly related to light signal transduction has not yet been reported.

C. Suppressor Mutants

To identify mutants potentially defective in signaling intermediates specific to PhyA, a screen was performed for extragenic mutations (i.e. lying in other genes) that suppress the morphological phenotype of the *phyA* mutant (Hoecker et al., 1998). Analysis of a recessive mutant, designated *spa1* (for suppressor of phyA) identified a component that acts in the phyA-specific signaling pathway. Sequence analysis demonstrated that SPA1 shows some homology to COP1 (Section V)(Hoecker et al., 1999); however, *spa1* mutations only affect seedlings grown in light whereas the *cop1* mutant phenotype is most apparent in the dark. A similar screen for *hy2* suppressors resulted in the isolation of two dominant mutants *shy1* and *shy2* (for suppressor of hy; Kim et al., 1996). *shy2*, which has been characterized in most detail, has pleiotropic effects and interacts not only with phytochrome but also with auxin-controlled developmental pathways (Kim et al, 1998; Tian and Reed, 1999).

A '*ted*' group of mutants has been isolated in the screens for reversal of *det1* phenotype (Pepper and Chory, 1997). Mutations in *ted4* and *ted5* loci identified new alleles of the previously described *hy1* and *hy5*, respectively, whereas three other loci (*ted1*, *ted2*, and *ted3*) may define new genes regulating photomorphogenesis.

D. Species-Specific Differences

There is an indication that phytochrome signaling pathways are not identical in different plants. It was recently discovered that mutation of the *DET1* homolog of tomato causes substantially different phenotypes as compared to *Arabidopsis*. This finding came from the analysis of one of the high pigment mutants (*hp-2*) of tomato which display exaggerated light responses. The mutation was localized in a gene highly homologous to *Arabidopsis DET1* and the gene was denoted *TDET1* (Mustilli et al., 1999).

Comparison of *Arabidopsis det1* and tomato *hp-2*

mutant phenotypes has revealed some interesting differences. In the dark, *hp-2* mutant seedlings are almost identical to wild-type plants except that the mutants show partial plastid development. After transfer of the plants to light, an exaggerated inhibition of hypocotyl elongation occurs in the *hp-2* mutant, together with an enhanced anthocyanin production. This is different from the *Arabidopsis det1* mutant, which shows a light-grown phenotype in darkness. The fact that *TDET1* mutation in tomato is not strongly manifested in the dark or in the absence of functional photoreceptors may be related to the observation that in tomato anthocyanin biosynthesis is strictly light-dependent, whereas in *Arabidopsis* it can be induced in the absence of light (Chory et al, 1989; Deng et al., 1991; Peters et al., 1992; Misera et al., 1994; Pepper et al., 1994; Kwok et al., 1996, Kerckhofs et al., 1997). This may also explain why no constitutive de-etiolated mutants, such as *cop* and *det*, have ever been reported in tomato.

Light grown tomato *hp-2* mutant plants appear shorter and more deeply pigmented than the wild-type, whereas *Arabidopsis det1* mutant plants, although short, appear pale-green. Moreover, *hp-2* mutants of tomato are able to enhance light responses, whereas *det1* mutants of *Arabidopsis* cannot. For example, light inducibility of *CHS* gene expression in *hp-2* mutants is more than two times higher as compared to wild-type seedlings. In *det1* mutants, on the contrary, this photoregulatory step is essentially lost, which results from both higher expression levels and ectopic expression of the *CHS* gene in all cell types in the dark (Chory and Peto, 1990; Misera et al., 1994; Mustilli et al., 1999).

These results therefore imply that DET1 plays an important role in regulating light responses, and that its activity may be regulated differently in tomato and *Arabidopsis*. Plant species existing in different environments and ecological niches should have evolved different mechanisms to cope with varying intensities and spectral qualities of light. It thus seems reasonable that regulatory circuits will not be identical in different species.

V. Nuclear-Localized Components of the Light Signaling Machinery

Studies of the molecular mechanisms of light signal transduction have recently made substantial progress and the most exciting results have revealed the importance of events within the nucleus. These studies have not yet corroborated the microinjection-based studies that have implicated G-proteins, calcium, and cGMP in phytochrome signaling but rather have identified a range of novel nuclear-localized proteins, all of which are involved in the regulation of expression of photosynthetic genes.

A. Phytochrome in the Nucleus and Nuclear-Localized Interacting Partners

It has now been shown that both PhyA and PhyB are translocated to the nucleus from the cytoplasm in a light-dependent fashion (Sakamoto and Nagatani, 1996; Kircher et al., 1999; Yamaguchi et al., 1999). While translocation is dependent upon the presence of the chromophore, the influence of phytochrome phosphorylation on nuclear import and subsequent export is not yet known.

Using a yeast two-hybrid screen, phytochrome has been demonstrated to interact directly with a constitutively nuclear-localized protein denoted PIF3 (Phytochrome Interacting Factor 3) (Ni et al., 1998). Physiological responses to light and expression of light-responsive genes are altered in transgenic lines that overexpress sense or antisense *PIF3* RNA, indicating that the gene product is important for appropriate responsiveness to red and far red light (Ni et al., 1998; Halliday et al., 1999). PIF3 is a basic helix-loop-helix transcription factor that contains sequence features that would allow it to interact directly with DNA (possibly with G-box sequences). Whether binding of PIF3 to DNA actually takes place, or not, has to be investigated but such a direct pathway from light receptor to DNA may provide a very short and rapid regulatory pathway.

Furthermore, binding of phytochrome to PIF3 is red/far red-light reversible, at least for PhyB (Ni et al., 1999). Hence, the classical signature of phytochrome action is manifested at the level of the Phy:PIF3 binding, implicating the significance of this interaction.

In addition to PIF3, phytochrome can also bind and phosphorylate CRY1, the blue light photoreceptor that is also localized in the nucleus (Ahmad et al., 1998). Although this interaction is not influenced by the phosphorylation status of phytochrome, such an interaction between two different classes of photoreceptors is likely to be of extreme physiological significance.

B. Components Identified by Mutant Approaches

The *COP9* gene encodes a protein present exclusively in a nuclear-localized high molecular weight protein complex designated the COP9 complex or signalosome (Wei et al., 1994). Purification of COP9 in *Arabidopsis* revealed a multisubunit complex with a total molecular mass of 550 kDa, also including the FUS6/COP11 protein (Castle and Meinke, 1994; Chamovitz et al., 1996; Staub et al., 1996), FUS5 (Karniol et al., 1999) and FUS4/COP8 (Serino et al., 1999). A mammalian counterpart of the plant COP9 complex has been purified and characterized (Seeger et al., 1998; Wei and Deng, 1998). It has been shown that both mammalian and plant COP9 complexes consist of nine highly homologous subunits (Wei et al., 1998). AJH1 and AJH2 in *Arabidopsis* are homologues of the mammalian transcriptional activator JAB1 (Kwok et al., 1998) and FUS6/COP11 is a homolog of the mammalian negative regulator GPS1 (Wei et al., 1998). Interestingly, the yeast genome does not contain any homologues of the COP9 complex subunits and it has been suggested that the action of the complex is specific to the functions of multicellular eukaryotes (Wei et al., 1998).

The *Arabidopsis* AJH proteins are present in two forms in wild-type cells. The monomeric form is preferentially localized in the cytoplasm while inside the nucleus it is associated with the COP9 complex. In some of the *cop/det/fus* mutants (*cop1* and *det1*) the AJH protein is present in the complexed nuclear form, whereas in other mutants (*fus4* (*cop8*), *cop9* and various other *fus* mutants) AJH1 is only found in the cytoplasm (Kwok et al., 1998). This suggests that FUS4 and most of the other FUS proteins are either integral parts of the COP9 complex or affect its assembly and/or stability.

Furthermore, three proteins have been identified that are not integral components of the COP9 complex but may associate with it and interact with FUS6/COP11 in two-hybrid screens (Karniol et al., 1998; Kwok et al., 1999). This result therefore indicates another level of functioning of the COP9 complex. Sequence analysis of the core components and associated proteins of the COP9 complex suggests evolutionary relationship of the complex with the 26S proteasome. Therefore, it is possible that the COP9 complex plays some role in degradation of nuclear proteins. A mechanism proposed by Deng and coworkers (Wei et al., 1998) involves selective phosphorylation/dephosphorylation of specific regulatory proteins by the COP9 complex followed by proteasome-mediated proteolysis. However, it is likely that the processes regulated by the COP9 complex are not specific to the photomorphogenic response because of the pleiotropic phenotypes of the *cop*/*fus* mutants affected in their function.

The DET1 and COP1 proteins, contrary to the COP9 complex, seem to have more specific roles in regulating photomorphogenesis. The *DET1* gene has been cloned but, to date, no data is available on the mechanism of action of this protein (Pepper et al., 1994; Mustilli et al., 1999). The translation product of the *DET1* gene has certain homology to mammalian expressed sequence tags (ESTs) of yet unknown function (Mustilli et al., 1999). The DET1 protein is, like the COP9 complex, nuclear-localized, but no DNA-binding activity has been detected (Chory et al., 1996). Importance of the DET1 protein in light signaling is inferred from the light hypersensitive phenotypes of tomato *hp-2* mutants, which are mutated in the tomato *DET1* gene (Mustilli et al., 1999; Section IV). DET1 does not appear to be associated with the COP9 complex.

The COP1 protein is not intrinsically associated with the COP9 complex either. The protein is a homodimer and contains a RING finger zinc-binding domain, a coiled-coil domain (dimerization module), a WD-40 repeat motif (repressor module), a bipartite nuclear localization signal, and a novel nuclear exclusion signal (Deng et al., 1992; Torii et al., 1998; Stacey et al., 1999). The total amount of cellular COP1 protein is not affected by light. However, in darkness the protein is predominantly localized in the nucleus, whereas light causes a specific redistribution and exclusion of this protein from the nucleus (von Arnim and Deng, 1994). Overexpression of the *COP1* gene results in partial inhibition of photomorphogenesis, thus providing evidence that COP1 acts as a molecular repressor of this process (McNellis et al., 1994). When the RING-finger and coiled-coil-containing domains are overexpressed in *Arabidopsis,* a dominant-negative phenotype similar to that of the loss-of-function *cop1* mutants, can be observed (McNellis et al., 1996). Further studies have demonstrated that the COP1 protein interacts directly with the transcriptional activator HY5 (Ang et al., 1998) and CIP7 (Yamamoto et al., 1998).

Two possible mechanisms of COP1 action have been proposed (Jarillo and Cashmore, 1998). In etiolated cells COP1 (localizing in the nucleus) may bind transcriptional activators thereby preventing them from binding to the promoters of light-regulated genes. Light may then cause a displacement of COP1 from the nucleus and liberation of transcriptional activators. Alternatively, a hypothetical protein activated by light may break the interaction between COP1 and an activator of transcription. COP1 was found to have (in its WD-repeat region) strong similarity to the recently cloned *SPA1* gene. Moreover, the SPA1 gene product was found to be localized in the nucleus. The authors therefore suggested that SPA1 and COP1 may act through potentially related mechanisms but at different points of the light signaling pathway (Hoecker et al., 1999). A distinctive feature of SPA1 is the presence of an N-terminal sequence with similarity to protein kinases; however data on its activity and specificity are not yet available.

Nuclear-cytoplasmic partitioning of COP1 is defective in all of the mutants that define the pleiotropic *cop/det/fus* loci (Chamovitz et al., 1996; von Arnim et al., 1997). It is thus possible that interaction between COP1 and the COP9 complex may be essential for COP1's stability or retention in the nucleus, although there is no direct evidence for this. On the other hand, in *cop1* and *det1* mutants, where integrity of the COP9 complex is not affected, the monomeric form of the AJH protein could not be detected (Kwok et al., 1998). This may indicate that there is an interdependence between DET1, COP1 and the COP9 complex, although identification and function of the links are not yet clear.

C. Cis-*Elements and Transcription Factors*

Considerable effort has been made to identify *cis*-elements and transcription factors involved in the regulation of light-inducible genes (Terzaghi and Cashmore, 1995). Besides PIF3, several other transcription factors have been proposed to be important. HY5 was initially identified from the light insensitive mutant *hy5* (Oyama et al., 1997). HY5 is a bZIP transcription factor which acts as a positive regulator of light-inducible gene expression and photomorphogenesis. HY5 interacts with the G-box motif present in many light-responsive promoters (Ang and Deng 1994; Chattopadhyay et al., 1998).

CIP7 (COP1 interaction protein 7) is a novel protein isolated by screening of an expression library with labeled COP1. CIP7 does not contain any recognizable DNA-binding motif but is nuclear-localized and can activate transcription (Yamamoto et al., 1998). Transgenic plants expressing antisense CIP7 RNA display defects in light-dependent accumulation of anthocyanin and chlorophyll as well as reduced expression of light-inducible genes. In wild-type plants expression of CIP7 was observed only in the light, whereas in *cop1* mutants the gene was partially derepressed in the dark.

A homeobox gene *ATH1* is not expressed in the dark but is induced by light. In the *det1* and *cop1* mutants the *ATH1* gene was also expressed in the dark (Quaedvlieg et al., 1995). It is thus possible that *ATH1* is another transcription factor interacting with the light regulated transcriptional machinery.

Different regulation levels are involved in the nuclear control of the light signaling machinery. Because, in most cases, light does not modulate the quantity of transcription factors, post translational control mechanisms must be important (Terzaghi and Cashmore, 1995). With the discovery of light-induced changes in the distribution pattern of COP1, one can hypothesize that regulation of transcription may be mediated via the light-dependent release of transcriptional activators from COP1 followed by their recruitment to the promoters (Jarillo and Cashmore, 1998). The COP9 complex may represent another level of regulation which involves the controlled proteolysis of transcriptional regulators (Kwok et al., 1999). It is tempting to speculate that DET1 represents yet a different level of regulation which involves chromatin remodeling or modification of the general transcriptional machinery (Chory et al., 1996). The latter two systems are well described and work hand in hand in *Drosophila* and yeast to control many developmental processes and environmental responses. The *Arabidopsis* genome sequencing project has revealed that homologues to some of these *Drosophila* and yeast proteins are present in plants; however, whether such systems are involved in photomorphogenic responses in plants remains to be determined.

D. Other Phytochrome-Interacting Proteins

In addition to the nuclear-localized PIF3, several cytoplasmically-localized kinases have been shown in yeast two-hybrid assays to interact directly with phytochrome. One of these, denoted Phytochrome Kinase Substrate 1 (PKS1) (Fankhauser et al., 1999) is, like phytochrome, a Ser/Thr protein kinase. Although PKS1 binds to both inactive (Pr) and active (Pfr) forms of phytochrome it only phosphorylates

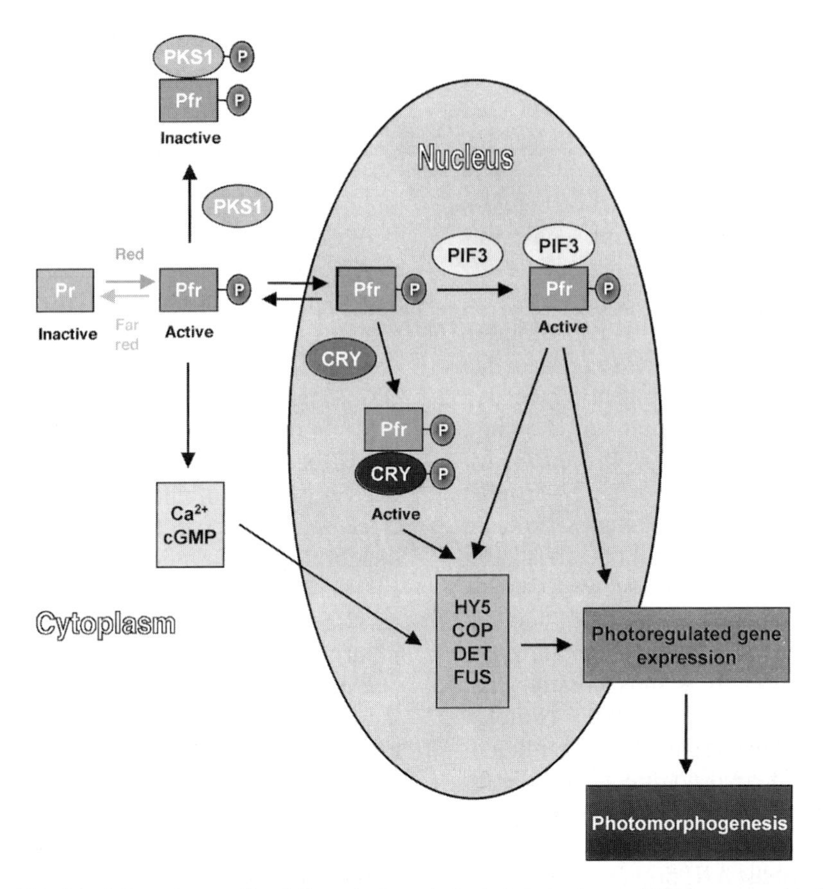

Fig. 2. A Simplified Model of Phytochrome Signal Transduction. Activated phytochrome (Pfr) is proposed to be phosphorylated and subsequently (a) to be translocated to the nucleus, (b) to activate calcium- and cGMP-dependent signaling pathways, or (c) to be sequestered away from the active pool by phytochrome kinase substrate 1 (PKS1), which in turn becomes phosphorylated by Pfr. Once inside the nucleus, Pfr can interact with PIF3 and cryptochomes (CRY) to control photoregulated gene expression. Additional nuclear factors such as HY5 and the COP/DET/FUS proteins may act as intermediates for photomorphogenesis either downstream of the cytoplasmic calcium and cGMP signals or downstream of the nuclear-localized phytochromes. The model is simplistic in that it does not account for differences between PhyA and PhyB signaling, e.g. the nuclear SPA1 and FAR1 proteins, that are specific for PhyA signaling (Hoecker et al., 1999; Hudson et al., 1999), are not shown. See also Color Plate 4b.

the Pfr form. Overexpression of PKS1 in transgenic plants leads to a suppression of light signaling, suggesting that it is a negative regulator. It has been proposed that one function of PKS1 may be to retain phytochrome in the cytoplasm, thereby preventing nuclear translocation and subsequent activation of some signaling pathways.

Nucleoside diphosphate kinases 1 and 2 (NDPK1 and NDPK2) have been identified as other phytochrome-interacting proteins (Choi et al., 1999). They have been localized in both nuclear and cytoplasmic compartments. Like PIF3, binding of phytochrome to NDPK1 and NDPK2 is red/far-red photoreversible, and furthermore, Pfr binding stimulates NDPK activity. As for PKS1 and PIF3, modulation of NDPK activities in transgenic plants results in altered light responses, providing direct evidence for the importance of this interaction.

E. A Model for Phytochrome Signaling

At first sight, our knowledge of phytochrome signaling summarized on the preceding pages comprises a series of conflicting data; genetic analyses have not corroborated the microinjection data, and results from yeast two-hybrid experiments have revealed an apparently unrelated series of phytochrome-interacting proteins. How can this data be summarized in a unifying model? An attempt to do so is presented in Fig. 2. Most probably, phytochrome activates several signaling pathways which lead to different or overlapping responses. For example, while phytochrome translocation to the nucleus is likely to be a key event, the slow translocation kinetics suggests that phytochrome is also likely to initiate signal transduction from the cytoplasm. Calcium and cGMP may be the key second messengers involved in activating these responses. PKS1 may sequester Pfr away from these pathways, as well as prevent nuclear import. Once inside the nucleus it is possible that phytochrome interacts with PIF3 to control directly a subset of photoregulated genes, as well as with the cryptochromes, to fine tune the overall output in different light conditions.

VI. Interactions Between Phytochrome and Other Signaling Pathways

As discussed above, the majority of *cop/det/fus* mutations do not specifically affect the regulation of photosynthetic genes but modulate various developmental processes in which light and phytochrome play crucial roles. Furthermore, besides phytochrome, many other factors contribute to the regulation of light-responsive genes. These include the circadian clock (Millar and Kay, 1996), phytohormones (Flores and Tobin, 1986; Bartholomew et al., 1991), metabolites such as sugars (Dijkwel et al., 1997), temperature, and light intensity. Interactions between these factors are complex. However, it is becoming clear that some of these pathways (e.g. circadian clock, hormones, and sugars) are directly coupled to the phytochrome signaling pathways (Mustilli and Bowler, 1997).

A. Phytochrome and the Circadian Clock

Millar and Kay (1996) have demonstrated that transcription of the *CAB* genes is upregulated at dawn and downregulated before sunset. This is an elegant example of energy saving at the transcriptional level. Processes affected by circadian rhythms are widespread in living organisms and in a range of prokaryotes and eukaryotes the clock can be reset by light. In plants phytochromes and cryptochrome are the photoreceptors known to synchronize the circadian clock (Somers et al., 1998) and, importantly, cryptochromes have also been found to reset clocks in insects and mammals (Thresher et al., 1998).

The Circadian Clock Associated (*CCA1*) gene encodes a MYB-related transcription factor involved in the phytochrome-mediated induction of the *CAB* gene. Expression of *CCA1* is transiently induced by phytochrome and oscillates with a circadian rhythm (Wang and Tobin, 1998). The lack of this transcription factor also affects the phytochrome regulation of gene expression, suggesting that CCA1 has additional roles in light signal transduction, possibly acting at the point of integration between phytochrome and the circadian clock (Green and Tobin, 1999). Another mutation of *Arabidopsis*, designated late elongated hypocotyl (*lhy*), disrupts circadian clock regulation of gene expression and leaf movements and causes flowering to occur independently of photoperiod. *LHY* encodes a protein highly homologous to CCA1 (Schaffer et al., 1998). It has been suggested that CCA1 and LHY perform similar roles and represent central oscillator controls for the plant cell (McClung, 1998).

B. Phytochromes and Phytohormones

The photomorphogenesis-related *SHY2* gene (Section IV) was identified as the *IAA3* gene, encoding a nuclear localized protein involved in auxin regulation (Tian and Reed, 1999). This result demonstrates that auxin and phytochrome pathways share common components. Addition of cytokinin to dark-grown wild-type seedlings results in a de-etiolated morphology, partial plastid development, and expression of some light-regulated genes, thus mimicking the phenotype of *det* mutants (Chory et al., 1996; Mustilli et al., 1999). Furthermore, a group of constitutive photomorphogenic mutants (*det2*, *cpd*, and *dim*) (Table 2) was found to be mutated in the brassinosteroid biosynthesis pathway. The effects of the mutations could be alleviated by exogeneous application of brassinolide (Chory et al., 1996; Szekeres et al., 1996). This indicates that brassinolide is involved in repressing light- and stress-regulated gene expression and in promoting cell expansion, leaf senescence, and flowering.

C. Phytochrome and Sugar Sensing

One can expect that sugars, the metabolic products of photosynthesis, may also play feedback regulatory roles in the phytochrome control of photosynthetic gene expression. Sucrose and glucose can repress the cotyledon opening and inhibition of hypocotyl elongation that occur in continuous far red light (a PhyA-mediated response). Barnes et al. (1996b) demonstrated that far red light-grown seedlings fail to green upon subsequent white light illumination. Microscopic analysis showed further that this far red block of greening correlates with both the failure of plastids to accumulate prolamellar bodies and the formation of vesicles in the stroma. Continuous far red irradiation causes severe repression of protochlorophyllide oxidoreductase (POR) genes coupled with irreversible plastid damage. This resulted from the temporal separation of PhyA-mediated POR repression from light-dependent protochlorophyllide reduction, two processes that are normally coordinated under white light. The far red response which results in block of greening could be suppressed by exogeneous sucrose indicating interaction of the two pathways.

Some sugar-sensing signaling systems have been proposed, most notably some studies have implicated hexokinase as a sensor and signal transmitter of sugar repression in higher plants (Jang et al., 1997). Another sensing mechanism, possibly employing sugar transporters, may interact with light signaling (Rook et al., 1998a; Lalonde et al., 1999). Conversely, expression of the *ATB2* gene (a sucrose-repressed bZIP transcription factor) is derepressed in *cop1* and *det1* mutants (Rook et al., 1998b).

Some mutants with sucrose-uncoupled (*sun*) far red light responses have been isolated indicating that the pathways of light signaling and sugar repression closely interact (Dijkwel et al., 1997). Similarly to *sun* mutants, the glucose-insensitive (*gin*) mutant de-etiolates regardless of the presence or absence of glucose. Exogeneous application of ethylene precursors can phenocopy the *gin* mutation suggesting a further interaction of pathways between phytochrome, ethylene, and glucose (Zhou et al., 1998).

D. Phytochrome and Environmental Cues

The contribution of light intensity and temperature to the expression of phytochrome-regulated genes has been well documented. For example, the synthesis of anthocyanin is activated by low temperature (Mol et al., 1996), and expression of the *CAB* genes is affected by temperature and light intensity (Escoubas et al., 1995). It has been suggested that temperature and light intensity operate through the same sensing mechanism, which involves the plastoquinone pool (Huner et al., 1998). Increase in light intensity or downshift of temperature induce an increase in the redox state of the plastoquinone pool which consequently leads to downregulation of *CAB* genes. It thus appears that chloroplast-derived signals efficiently regulate expression of nuclear genes. Nevertheless, it is not known whether redox and phytochrome signaling pathways cross-regulate each other. If they do, one might expect an interaction of redox and Type II phytochrome signaling pathways, since these are the dominant phytochromes in photosynthetically active mature plants (Fig. 1).

VII. Concluding Remarks

Significant progress has been made in the identification of factors affecting expression of the nuclear-encoded photosynthetic genes. The most recent results draw attention to the important role of nuclear factors such as COP1, DET1, PIF3, SPA1, the COP9 complex, and nuclear-localized Pfr. However, much work remains to be done to understand the precise functions of these different components and to define their interactions.

Additional questions that must be addressed include the cross-talk between phytochrome, circadian clock, and phytohormone signaling pathways. It seems probable that different options will be available to plant cells to allow integration between different pathways. For example PIF3 may provide a direct link between phytochrome and gene expression that bypasses other signals. Other pathways, for example those involving calcium and cGMP, may provide more options for the integration of different stimuli. Future research will undoubtedly continue to reveal the exquisite control of phytochrome signaling and will provide fundamental information for understanding how photosynthetic gene expression is regulated within the developmental circuitry that controls plant growth and development.

Acknowledgments

We thank Anna-Chiara Mustilli for critical reading of the manuscript. Signal transduction research in our laboratory has been financed by the Human Frontier Science Program (RG362/95), the Italian Ministry of Agriculture (MIRAAF PM356), the European Commission (BIO4-CT96-0101), and the Consiglio Nazionale delle Ricerche Target Project on Biotechnology (No. 97.01259.PF49) to C.B.

References

Ahmad M and Cashmore AR (1996) Seeing blue: The discovery of cryptochrome. Plant Mol Biol 30: 851–861

Ahmad M, Jarillo JA, Smirnova O and Cashmore AR (1998) The CRY1 blue light photoreceptor of *Arabidopsis* interacts with phytochrome A in vitro. Mol Cell 1: 939–948

Anderson, JM (1986) Photoregulation of the composition, function, and structure of thylakoid membranes. Annu Rev Plant Physiol 37: 93–136

Ang LH and Deng XW (1994) Regulatory hierarchy of photomorphogenic loci: Allele specific and light dependent interaction between *HY5* and *COP1* loci. Plant Cell 6: 613–628

Ang LH, Chattopadhyay S, Wei N, Oyama T, Okada K, Batschauer A and Deng XW (1998) Molecular interaction between COP1 and HY5 defines a regulatory switch for light control of *Arabidopsis* development. Mol Cell 1: 213–222

Barnes SA, Quaggio RB, Whitelam GC and Chua NH (1996a) *fhy1* defines a branch point in phytochrome A signal transduction pathways for gene expression. Plant J 10: 1155–1161

Barnes SA, Nishizawa NK, Quaggio RB, Whitelam GC and Chua NH (1996b) Far red light blocks greening of *Arabidopsis* seedlings via a phytochrome A-mediated change in plastid development. Plant Cell 8: 601–615

Bartholomew DM, Bartley GE and Scolnik PA (1991) Abscisic acid control of *RBCS* and *Cab* transcription in tomato leaves. Plant Physiol 96: 291–296

Bowler C, Neuhaus G, Yamagata H and Chua NH (1994a) Cyclic GMP and calcium mediate phytochrome phototransduction. Cell 77: 73–81.

Bowler C, Yamagata H, Neuhaus G and Chua NH (1994b) Phytochrome signal transduction pathways are regulated by reciprocal control mechanisms. Genes Dev 8: 2188–2202

Castle LA and Meinke DW (1994) A *FUSCA* gene of *Arabidopsis* encodes a novel protein essential for plant development. Plant Cell 6: 25–41

Chamovitz DA and Deng XW (1996) Light signaling in plants. Crit Rev Plant Sci 15: 455–478

Chamovitz DA, Wei N, Osterlund MT, von Arnim AG, Staub JM, Matsui M and Deng XW (1996) The COP9 complex, a novel multisubunit nuclear regulator involved in light control of a plant developmental switch. Cell 86: 115–121

Chattopadhyay S, Ang LH, Puente P, Deng XW and Wei N (1998) *Arabidopsis* bZIP protein HY5 directly interacts with light-responsive promoters in mediating light control of gene expression. Plant Cell 10: 673–683

Choi G, Yi H, Lee J, Kwon Y-K, Soh MS, Shin B, Luka Z, Hahn T-R, Song P-S (1999) Phytochrome signalling is mediated through nucleoside diphosphate kinase 2. Nature 401: 610–613

Chory J (1997) Light modulation of vegetative development. Plant Cell 9: 1225–1234

Chory J and Peto CA (1990) Mutations in the *DET1* gene affect cell-type-specific expression of light-regulated genes and chloroplast development in *Arabidopsis*. Proc Natl Acad Sci USA 87: 8776–8780

Chory J, Peto C, Feinbaum R, Pratt L and Ausubel F (1989) *Arabidopsis thaliana* mutant that develops as a light-grown plant in the absence of light. Cell 58: 991–999

Chory J, Chattergee M, Cook RK, Elich T, Fankhauser C, Li J, Nagpal P, Neff M, Pepper A, Poole D, Reed J and Vitart V (1996) From seed germination to flowering, light controls plant development via the pigment phytochrome. Proc Natl Acad Sci USA 93: 12066–12071

Christie JM, Reymond P, Powell GK, Bernasconi P, Raibekas AA, Liscum E and Briggs WR (1998) *Arabidopsis* NPH1: A flavoprotein with the properties of a photoreceptor for phototropism. Science 282: 1698–1701

Clough RC and Vierstra RD (1997) Phytochrome degradation. Plant Cell Environ 20: 713–721

Delledonne M, Xia Y, Dickson RA and Lamb C (1998) Nitric oxide functions as a signal in plant disease resistance. Nature 394: 585–588

Deng XW, Caspar T and Quail PH (1991) *cop1*: A regulatory locus involved in light-controlled development and gene expression in *Arabidopsis*. Genes Dev 5: 1172–1182

Deng XW, Matsui M, Wei N, Wagner D, Chu AM, Feldman KA and Quail PH (1992) COP1, an *Arabidopsis* regulatory gene, encodes a protein with both a zinc-binding motif and a G_β homologous domain. Cell 71: 791–801

Dijkwel PP, Huijser C, Weisbeek PJ, Chua NH and Smeekens SC (1997) Sucrose control of phytochrome A signaling in *Arabidopsis*. Plant Cell 9: 583–595

Durner J, Wendehenne D and Klessig DF (1998) Defence gene induction in tobacco by nitric oxide, cyclic-GMP, and cyclic ADP-ribose. Proc Natl Acad Sci USA 95: 10328–10333

Escoubas JM, Lomas M, LaRoche J and Falkowski PG (1995) Light intensity regulation of *cab* gene transcription is signaled by the redox state of the plastoquinone pool. Proc Natl Acad Sci USA 92: 10237–10241

Fankhauser C and Chory J (1999) Photomorphogenesis: Light receptor kinases in plants! Curr Biol 9: R123–R126

Fankhauser C, Yeh K-C, Lagarias JC, Zhang H, Elich TD and Chory J (1999) PKS1, a substrate phosphorylated by phytochrome that modulates light signaling in *Arabidopsis*. Science 284: 1539–1541

Flores S and Tobin EM (1986) Cytokinin modulation of *LHCP* messenger-RNA levels—the involvement of post-transcriptional regulation. Plant Mol Biol 11: 409–415

Furuya M and Schäfer E (1996) Photoperception and signalling of induction reactions by different phytochromes. Trends Plant Sci 9: 301–307

Genoud T, Millar AJ, Nishizawa N, Kay SA, Schäfer E, Nagatani A and Chua NH (1998) An *Arabidopsis* mutant hypersensitive to red and far red light signals. Plant Cell 10: 889–904

Green RM and Tobin EM (1999) Loss of the Circadian Clock-Associated protein 1 in *Arabidopsis* results in altered clock-regulated gene expression. Proc Natl Acad Sci USA 96: 4176–4179

Halliday KJ, Hudson M, Ni M, Qin M and Quail PH (1999) *poc1*: An *Arabidopsis* mutant perturbed in phytochrome signaling because of a T DNA insertion in the promoter of *PIF3*, a gene encoding a phytochrome-interacting bHLH protein. Proc Natl Acad Sci USA 96: 5832–5837

Hoecker U, Xu Y and Quail PH (1998) SPA1: A new genetic locus involved in phytochrome A-specific signal transduction. Plant Cell 10: 19–33

Hoecker U, Tepperman JM and Quail PH (1999) SPA1 a WD-repeat protein specific to phytochrome signal transduction. Science 284: 496–499

Hudson M, Ringli C, Boylan MT and Quail PH (1999) The *FAR1* locus encodes a novel nuclear protein specific to phytochrome A signaling. Genes Dev 13: 2017–2027

Huner NPA, Öquist G and Sarhan F (1998) Energy balance and acclimation to light and cold. Trends Plant Sci 3: 224–230

Jang JC, Leon P, Zhou L and Sheen J (1997) Hexokinase as a sugar sensor in higher plants. Plant Cell 9: 5–19

Jarillo JA and Cashmore AR (1998) Enlightenment of the COP1-HY5 complex in photomorphogenesis. Trends in Plant Sci 3: 161–163

Jarvis P, Chen L-J, Li H-M, Peto CA, Fankhauser C and Chory J (1998) An *Arabidopsis* mutant defective in the plastid general protein import apparatus. Science 282: 100–103

Karniol B, Yahalom A, Kwok S, Tsuge T, Matsui M, Deng XW and Chamovitz DA (1998) The *Arabidopsis* homologue of an eIF3 complex subunit associates with the COP9 complex. FEBS Lett 439: 173–179

Karniol B, Malec P and Chamovitz DA (1999) *Arabidopsis FUSCA5* encodes a novel phosphoprotein that is a component of the COP9 complex. Plant Cell 11: 839–848

Kendrick RE and Kronenberg GHM (1994) Photomorphogenesis in Plants. 2nd ed. Kluwer Academic Publishers, Dordrecht

Kehoe DM and Grossman AR (1996) Similarity of a chromatic adaptation sensor to phytochrome and ethylene receptors. Science 273: 1409–1412

Kerckhoffs LHJ, Schreuder MEL, Van Tuinen A, Koornneef M and Kendrick RE (1997) Phytochrome control of anthocyanin synthesis in tomato seedlings: Analysis using photomorphogenic mutants. Photochem Photobiol 65: 374–381

Kim BC, Soh MC, Kang BJ, Furuya M and Nam HG (1996) Two dominant photomorphogenic mutations of *Arabidopsis thaliana* identified as suppressor mutations of *hy2*. Plant J 9: 441–456

Kim BC, Soh MS, Hong SH, Furuya M and Nam HG (1998) Photomorphogenic development of the *Arabidopsis* shy2-1D mutation and its interaction with phytochromes in darkness. Plant J 15: 61–68

Kircher S, Kozma-Berger L, Kim L, Adam E, Harter K, Schafer E and Nagy F (1999) Light quality-dependent nuclear import of the plant photoreceptors phytochrome A and B. Plant Cell 11: 1445–1456

Köhler C, Merkle T and Neuhaus G (1999) Characterisation of a novel gene family of putative cyclic nucleotide and calmodulin-regulated ion channels in *Arabidopsis thaliana*. Plant J 18: 97–104

Kwok SF, Peikos B, Misera S and Deng X-W (1996) A complement of ten essential and pleiotropic *Arabidopsis COP/DET/FUS* genes is necessary for repression of photomorphogenesis in darkness. Plant Physiol 110, 731–741

Kwok SF, Solano R, Tsuge T, Chamovitz DA, Ecker JR, Matsui M and Deng XW (1998) *Arabidopsis* homologs of a c-Jun coactivator are present both in monomeric form and in the COP9 complex, and their abundance is differentially affected by the pleiotropic *cop/det/fus* mutations. Plant Cell 10: 1779–1790

Kwok SF, Staub JM and Deng XW (1999) Characterization of two subunits of *Arabidopsis* 19S proteasome regulatory complex and its possible interaction with the COP9 complex. J Mol Biol 285: 85–95

Lalonde S, Boles E, Hellman H, Barker L, Patrick JW, Frommer WB and Ward JM (1999) The dual function of sugar carriers. Transport and sugar sensing. Plant Cell 11: 707–726

Lapko VN, Jiang XY, Smith DL and Song PS (1997) Posttranslational modification of oat phytochrome A: Phosphorylation of a specific serine in a multiple serine cluster. Biochemistry 36: 10595–10599

Lapko VN, Jiang XY, Smith DL and Song PS (1999) Mass spectrometric characterization of oat phytochrome A: Isoforms and posttranslational modifications. Prot Sci in press

Li H-M, Altschmied L and Chory J (1994) *Arabidopsis* mutants define downstream branches in the phototransduction pathway. Genes Dev 8: 339–349

Li H-M, Culligan K, Dixon RA and Chory J (1995) *CUE1*: A mesophyll cell-specific positive regulator of light-controlled gene expression in *Arabidopsis*. Plant Cell 7: 1599–1610

Liscum E and Briggs WR (1996) Mutations of *Arabidopsis* in potential transduction and response components of the phototropic signaling pathway. Plant Physiol 112: 291–296

Lopez-Juez E, Jarvis RP, Takeuchi A, Page AM and Chory J (1998) New *Arabidopsis cue* mutants suggest a close connection between plastid- and phytochrome regulation of nuclear gene expression. Plant Physiol 118: 803–815

Mackenzie JM Jr, Coleman RA, Briggs WR and Pratt LH (1975) Reversible redistribution of phytochrome within the cell upon conversion to its physiologically active form. Proc Natl Acad Sci USA 72: 799–803

Mayer R, Raventos D and Chua N-H (1996) *det1*, *cop1*, and *cop9* mutations cause inappropriate expression of several gene sets. Plant Cell 8: 1951–1959

McClung CR (1998) It's about time: Putative components of an *Arabidopsis* circadian clock. Trends Plant Sci 3: 454–456

McCurdy DW and Pratt LH (1986) Immunogold electron microscopy of phytochrome in *Avena*: Identification of intracellular sites responsible for phytochrome sequestering and enhanced pelletability. J Cell Biol 103: 2541–2550

McNellis TW, von Arnim AG and Deng XW (1994) Overexpression of *Arabidopsis* COP1 results in partial suppression of light-mediated development: Evidence for a light-inactivable repressor of photomorphogenesis. Plant Cell 6: 1391–1400

McNellis TW, Torii KU and Deng XW (1996) Expression of an N-terminal fragment of COP1 confers a dominant-negative effect on light-regulated seedling development in *Arabidopsis*. Plant Cell 8: 1491–1503

Melis A (1991) Dynamics of photosynthetic membrane composition and function. Biochim Biophys Acta 1058: 87–106

Millar AJ and Kay S (1996) Integration of circadian and phototransduction pathways in the network controlling *CAB* gene transcription in *Arabidopsis*. Proc Natl Acad Sci USA 93: 15491–15496

Millar AJ, McGrath RB and Chua NH (1994) Phytochrome phototransduction pathways. Annu Rev Genet 28: 325–349

Misera S, Muller AJ, Weiland-Heidecker U and Jurgens G (1994) The *FUSCA* genes of *Arabidopsis*: Negative regulators of light responses. Mol Gen Genet 244: 242–252

Mol J, Jenkins G, Schaefer E and Weiss D (1996) Signal perception, transduction and gene expression involved in anthocyanin biosynthesis. Crit Rev Plant Sci 15: 525–557

Mustilli AC and Bowler C (1997) Tuning in to the signals controlling photoregulated gene expression in plants. EMBO J 16: 5801–5806

Mustilli AC, Fenzi F, Ciliento R, Alfano F and Bowler C (1999) Phenotype of the tomato high pigment-2 mutant is caused by a mutation in the tomato homolog of DEETIOLATED1. Plant Cell 11: 145–157

Neuhaus G, Bowler C, Kern R and Chua NH (1993) Calcium/

calmodulin-dependent and -independent phytochrome signal transduction pathways. Cell 73: 937–952

Neuhaus G, Bowler C, Hiratsuka K, Yamagata H and Chua NH (1997) Phytochrome-regulated repression of gene expression requires calcium and cGMP. EMBO J 16:2554–2564

Ni M, Tepperman JP and Quail PH (1998) PIF3, a phytochrome-interacting factor necessary for normal photoinduced signal transduction, is a novel basic helix-loop-helix protein. Cell 95: 657–667

Ni M, Teppermann JM and Quail PH (1999) Binding of phytochrome B to its nuclear signalling partner PIF3 is reversibly induced by light. Nature 400: 781–784

Oyama T, Shimura Y and Okada K (1997) The *Arabidopsis HY5* gene encodes a bZIP protein that regulates stimulus-induced development of root and hypocotyl. Genes Dev 11: 2983–2995

Pepper AE and Chory J (1997) Extragenic suppressors of the *Arabidopsis det1* mutant identify elements of flowering-time and light-response regulatory pathways. Genetics 145: 1125–1137

Pepper A, Delaney T, Washburn T, Poole D and Chory J (1994) *DET1*, a negative regulator of light-mediated development and gene expression in *Arabidopsis*, encodes a novel nuclear-localized protein. Cell 78: 109–116

Peters JL, Schreuder MEL, Verduin SJW and Kendrick RE (1992) Physiological characterization of a high pigment mutant of tomato. Photochem Photobiol 56: 75–82

Quaedvlieg N, Dockx J, Rook F, Weisbeek P and Smeekens S (1995) The homeobox gene *ATH1* of *Arabidopsis* is derepressed in the photomorphogenic mutants *cop1* and *det1*. Plant Cell 7: 117–129

Quail PH, Boylan MT, Parks BM, Short TW, Xu Y and Wagner D (1995) Phytochromes: Photosensory perception and signal transduction. Science 268: 675–680

Reed JW, Nagatani A, Elich TD, Fagan M and Chory J (1994) Phytochrome A and phytochrome B have overlapping but distinct functions in *Arabidopsis* development. Plant Physiol 104, 1139–1149

Rook F, Gerrits N, Kortstee A, van Kampen M, Borrias M, Weisbeek P and Smeekens S (1998a) Sucrose-specific signalling represses translation of the *Arabidopsis ATB2* bZIP transcription factor gene. Plant J 15: 253–263

Rook F, Weisbeek P and Smeekens S (1998b) The light regulated *Arabidopsis* bZIP transcription factor gene *ATB2* encodes a protein with an unusually long leucine zipper domain. Plant Mol Biol 37: 171–178

Sakamoto K and Nagatani A (1996) Nuclear localization activity of phytochrome B. Plant J 10: 859–68

Schaffer R, Ramsay N, Samach A, Corden S, Putterill J, Carre JA and Coupland G (1998) *LATE ELONGATED HYPOCOTYL*, an *Arabidopsis* gene encoding a MYB transcription factor, regulates circadian rhythmicity and photoperiodic responses. Cell 93: 1219–1229

Seeger M, Kraft R, Ferrell K, Bech-Otschir D, Dumdey R, Schade R, Gordon C, Naumann M and Dubiel W (1998) A novel protein complex involved in signal transduction possessing similarities to 26S proteasome subunits. FASEB J 12: 469–478

Serino G, Tsuge T, Kwok SF, Matsui M, Wei W and Deng X-W (1999) *Arabidopsis cop8* and *fus4* mutations define the same locus that encodes subunit 4 of the COP9 signalosome. Plant Cell 11: 1967–1979

Shinomura T, Nagatani A, Chory J and Furuya M (1994) The induction of seed germination in *Arabidopsis thaliana* is regulated principally by phytochrome B and secondarily by phytochrome A. Plant Physiol 104: 363–371

Somers DE, Devlin PF and Kay SA (1998) Phytochromes and cryptochromes in the entrainment of the *Arabidopsis* circadian clock. Science 282: 1488–1450

Stacey MG, Hicks SN and von Arnim AG (1999) Discrete domains mediate the light-responsive nuclear and cytoplasmic localization of *Arabidopsis* COP1. Plant Cell 11: 349–364

Staub JM, Wei N and Deng XW (1996) Evidence for FUS6 as a component of the nuclear-localized COP9 complex in *Arabidopsis*. Plant Cell 8: 2047–2056

Susek RE, Ausubel FM and Chory J (1993) Signal transduction mutants of *Arabidopsis* uncouple nuclear *CAB* and *RBCS* gene expression from chloroplast development. Cell 74: 787–799

Szekeres M, Nemeth K, Koncz-Kalman Z, Mathur J, Kauschmann A, Altmann T, Redei GP, Nagy F, Schell J and Koncz C (1996) Brassinosteroids rescue the deficiency of CYP90, a cytochrome P450, controlling cell elongation and de-etiolation in *Arabidopsis*. Cell 85: 171–182

Terzaghi WB and Cashmore AR (1995) Light-regulated transcription. Annu RevPlant Physiol Plant Mol Biol 46: 445–474

Thresher RJ, Vitaterna MH, Miyamoto Y, Kazantsev A, Hsu DS, Petit C, Selby CP, Dawut L, Smithies O, Takahashi JS and Sancar A (1998) Role of mouse cryptochrome blue-light photoreceptor in circadian photoresponses. Science 282: 1490–1494

Tian Q and Reed JW (1999) Control of auxin-regulated root development by the *Arabidopsis thaliana SHY2/IAA3* gene. Development 126: 711–721

Torii KU, McNellis TW and Deng XW (1998) Functional dissection of *Arabidopsis* COP1 reveals specific roles of its three structural modules in light control of seedling development. EMBO J 17: 5577–5587

Trewavas AJ and Malho R (1998) Ca^{2+} signaling in plant cells: the big network! Curr Opin Plant Biol 1: 428–433

von Arnim AG and Deng XW (1994) Light inactivation of *Arabidopsis* photomorphogenic repressor COP1 involves a cell-specific regulation of its nucleocytoplasmic partitioning. Cell 79: 1035–1045

von Arnim AG, Osterlund MT, Kwok SF and Deng XW (1997) Genetic and developmental control of nuclear accumulation of COP1, a repressor of photomorphogenesis in *Arabidopsis*. Plant Physiol 114: 779–788

Wagner D, Hoecker U and Quail PH (1997) *RED1* is necessary for phytochrome B-mediated red light-specific signal transduction in *Arabidopsis*. Plant Cell 9: 731–743

Wang ZY and Tobin EM (1998) Constitutive expression of the CIRCADIAN CLOCK ASSOCIATED 1 (CCA1) gene disrupts circadian rhythms and suppresses its own expression. Cell 93: 1207–1217

Wei N and Deng XW (1998) Characterization and purification of the mammalian COP9 complex, a conserved nuclear regulator initially identified as a repressor of photomorphogenesis in higher plants. Photochem Photobiol 68: 237–241

Wei N, Chamovitz DA and Deng XW (1994) *Arabidopsis* COP9 is a component of a novel signaling complex mediating light control of development. Cell 78: 117–124

Wei N, Tsuge T, Serino G, Dohmae N, Takio K, Matsui M and Deng X-W (1998) The COP9 complex is conserved between plants and mammals and is related to the 26S proteasome regulatory complex. Curr Biol 8: 919–922

Whitelam GC and Harberd NP (1994) Action and function of phytochrome family members revealed through the study of mutant and transgenic plants. Plant Cell Environ 17: 615–625

Yamaguchi R, Nakamura M, Mochizuki N, Kay, SA and Nagatani A (1999) Light-dependent translocation of a phytochrome B-GFP fusion protein to the nucleus in transgenic *Arabidopsis*. J Cell Biol 145: 437–445

Yamamoto YY, Matsui M, Ang LH and Deng XW (1998) Role of a COP1 interactive protein in mediating light-regulated gene expression in *Arabidopsis*. Plant Cell 10: 1083–1094

Yeh KC and Lagarias JC (1998) Eukaryotic phytochromes: Light-regulated serine/threonine protein kinases with histidine kinase ancestry. Proc Natl Acad Sci USA 95: 13976–13981

Yeh KC, Wu SH, Murphy JT and Lagarias JC (1997) A cyanobacterial phytochrome two-component light sensory system. Science 277: 1505–1508

Zhou L, Jang JC, Jones TL and Sheen J (1998) Glucose and ethylene signal transduction crosstalk revealed by an *Arabidopsis* glucose-insensitive mutant. Proc Natl Acad Sci USA 95: 10294–10299

Chapter 4

Regulating Synthesis of the Purple Bacterial Photosystem

Carl E. Bauer*
Department of Biology, Indiana University, Bloomington, IN 47405 U.S.A.

Summary

The photosystem synthesized by purple bacteria is composed of several membrane spanning polypeptides that act as scaffolding to bind cofactors bacteriochlorophyll, carotenoids, and quinones. Production of the photosystem is suppressed in aerobically growing cells which utilize oxidative phosphorylation (respiration) as their main energy generating system. Under anaerobic conditions, the cells synthesize copious amounts of a photosystem that is housed in intracellular membrane invaginations. Anaerobic synthesis of the pigment and polypeptide components of the photosystem are tightly coordinated with the level of synthesis modulated according to light intensity. To maintain stoichiometric amounts of the pigment and polypeptide components of the photosystem, the cells must coordinately regulate the pigment biosynthetic pathways with synthesis of the structural polypeptides that ultimately bind the pigments. Studies have demonstrated that regulating synthesis of components of the purple bacterial photosystem involves a complex set of transcriptional and post-transcriptional processes. This chapter summarizes our current understanding of these regulatory events.

*Email: cbauer@bio.indiana.edu

Eva-Mari Aro and Bertil Andersson (eds): Regulation of Photosynthesis, pp. 67–83.

I. Introduction

No organisms exhibit more versatility in energy metabolism than the purple non-sulfur photosynthetic bacteria. Living predominantly in soil or aquatic environments, most species can grow in the dark using either inorganic or organic compounds as an energy source (chemolithotrophic or chemoheterotrophic growth, respectively) or photosynthetically in the light in the presence or absence of organic compounds (photoautotrophic or photoheterotrophic growth). In most species where it has been examined, the synthesis of a photosystem that is responsible for absorbing and utilizing light as an energy source, occurs only during anaerobic growth. Aerobically grown cells are essentially colorless and exhibit a typical gram negative membrane architecture. In contrast, when growth is shifted to an anaerobic environment, the cells undergo rapid differentiation involving inward folding of the cytoplasmic membrane to form numerous intracellular invaginations known as intracytoplasmic membranes that house the photosystem (Cohen-Bazier et al., 1957). This chapter briefly describes components of the purple bacterial photosystem and summarizes what is currently known about the control of its synthesis. Studies of the purple non-sulfur bacterium *Rhodobacter capsulatus* are the main focus of this chapter since there is a deeper understanding of how this bacterium controls photosynthesis gene expression than in any other bacterial species. Our current understanding is that there exists a complex set of regulatory events, including transcriptional and post-transcriptional mechanisms that control synthesis of the photosystem. The described regulatory circuits ensure that stoichiometric amounts of photopigment and pigment-binding polypeptides are maintained even during transitions between aerobic and photosynthetic modes of growth. A minireview that covers aspects of this topic has also recently been published by Pemberton et al., (1998).

II. The Purple Bacterial Photosystem

The purple bacterial photosystem is composed of three distinct pigment-protein complexes designated light harvesting-I (LH-I), light harvesting-II (LH-II) and the reaction center (RC) (Fig. 1). The LH-I and LH-II complexes use bacteriochlorophyll and carotenoids to absorb light energy. Energy captured by these antenna complexes is transmitted to the RC which initiates electron transfer, essentially converting light energy into a useful source of cellular energy.

Structural analyses have provided new insights into the organization of bacterial photosystems. It has been established that LH-I and LH-II complexes are each composed of two small membrane spanning polypeptides designated α and β (reviewed by Zuber and Cogdell, 1995). These heterodimers create a polypeptide framework that non-covalently binds bacteriochlorophylls and carotenoids. Individual protein/pigment heterodimers then aggregate into larger supercomplexes comprised of 16 and 9 subunits for LH-I and LH-II, respectively. The LH-I and LH-II supercomplexes form ring structures that are embedded within the intracytoplasmic membrane (McDermott et al., 1995). The LH-I ring is thought to enclose a single RC comprising a 'core' unit that is fully capable of supporting photosynthetic growth (Karrasch et al., 1995). The photosystem core unit, in turn, interacts with arrays of LH-II rings that operate as dim light auxiliary antennae. X-ray diffraction study of crystallized LH-II isolated from *Rhodopseudomonas acidophila* has detailed the configuration of subunits within the complex (McDermott et al., 1995). Each LH-II ring contains 27 bacteriochlorophyll molecules that are precisely oriented to absorb photons of specific wavelengths (primarily infrared). The overlapping arrangement of bacteriochlorophyll molecules also facilitates efficient and extremely rapid transfer of LH-II absorbed energy to LH-I and ultimately to the RC complex.

The RC of purple bacteria, which is structurally and functionally related to the type II photosystem synthesized by cyanobacteria and plant chloroplasts, is composed of three small integral membrane proteins designated as H, M and L (reviewed by Lancaster et al., 1995). The M and L subunits form

Abbreviations: *bch* – bacteriochlorophyll biosynthesis genes; *crt* – carotenoid biosynthesis genes; *crtJ* – gene coding for the CrtJ aerobic repressor of *bch*, *crt* and *puc* genes; H-NS – bacterial nucleoid protein; HvrA – light regulated transcription factor; *lacZ* – β-galactosidase gene; LH-I – light harvesting-I; LH-II – light harvesting-II; *puc* – light harvesting-II structural genes; *puf* – light harvesting-I and reaction center L and M structural genes; *puh* – reaction center H structural gene; PYP – photoactive yellow protein; *R. capsulatus* – *Rhodobacter capsulatus*; *R. sphaeroides* – *Rhodobacter sphaeroides*; RC – reaction center; *regA* – gene coding for the transcription activator RegA that is phosphorylated by RegB; *regB* – gene coding for the histidine sensor kinase RegB that phosphorylates RegA; RegB″ – deletion derivative of RegB

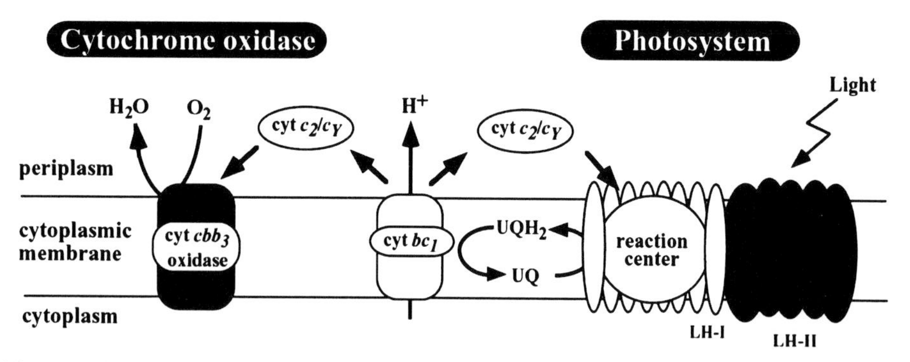

Fig 1. Structural features of the photosystem and the respiratory cytochrome oxidase of purple photosynthetic bacteria. Light energy absorbed by the reaction center and the light harvesting-I (LH-I) and light harvesting–II (LH-II) complexes is used to form reduced ubiquinone (UQH_2). Electrons are passed through the cytochrome bc_1 complex back to the reaction center, or to cytochrome oxidase, by cytochromes c_2 and c_y. All of these components are controlled by the RegB-RegA signal transduction cascade.

many of the noncovalent interactions with photopigments (one carotenoid, two bacteriochlorophylls, and two bacteriopheophytins) as well as with two quinones that act as primary electron acceptors. The H subunit has a role in providing interactions with soluble and membrane bound electron carriers such as cytochrome c_2 and c_y as well as a role in donating protons to the reduced quinone. Excitation of the RC by light energy converts a 'special pair' of bacteriochlorophyll molecules into strong reductants that supply high energy electrons ultimately to the quinone pool. Electrons subsequently flow from quinone back to the RC via cytochrome bc_1 and cytochrome c_y (or cytochrome c_2). The shuttling of electrons through cytochrome bc_1 is coupled to a translocation of protons across the membrane which generates an electrochemical gradient that is ultimately used to drive ATP synthesis. Since electron flow in purple bacteria is cyclic, there is no requirement for exogenous electron donors to the process. As a consequence of cyclic electron flow, purple and green bacterial photosynthesis does not evolve oxygen and are thus commonly referred to as 'anoxygenic photosynthesis.' Anoxygenic photosynthesis is contrasted by the photosystem of cyanobacteria and chloroplasts which evolve oxygen as a byproduct of oxidizing water.

III. The Photosynthesis Gene Cluster

Early generalized transduction experiments established that genes for carotenoid (*crt*) and bacteriochlorophyll (*bchl*) biosynthesis are clustered in the *R. capsulatus* chromosome (Yen et al., 1976) (Fig. 2). Demonstration of tight linkage of pigment biosynthesis genes was followed by the isolation of an R′ plasmid by Marrs (1981) that complemented an extensive collection of point mutations that were defective in photosynthesis. Physical mapping of photosynthesis genes carried by the R′, through complementation analysis of subcloned fragments (Taylor et al., 1983) and transposition mapping techniques (Biel and Marrs, 1983; Zsebo and Hearst, 1984), demonstrated that a 'photosynthesis gene cluster' contained an assembly of *bch* and *crt* genes (hatched and shaded boxes in Fig. 2, respectively) flanked by the LH-I and RC structural genes encoded by the *puf* and *puh* operons (solid boxes) (Taylor et al., 1983; Youvan et al., 1984; Zsebo and Hearst, 1984). A contiguous 45kb region of the *R. capsulatus* genome thus contains all of the essential genetic information needed to synthesize a core photosynthetic unit comprised of the pigment and polypeptide components of LH-I and the RC. Only the *puc* operon, containing the LH-II structural loci, is unlinked to the photosynthesis gene cluster (Youvan and Ismail, 1985).

A detailed understanding of the photosynthesis gene cluster was obtained by Hearst and coworkers who completed sequence analysis of the entire 45 kb region (Youvan et al., 1984; Armstrong et al., 1989; Alberti et al., 1995). Analysis of the function of individual loci was accomplished by the construction of defined interposon mutations within each of the identified open reading frames (Giuliano et al., 1988; Young et al., 1989, 1992; Yang and Bauer 1990; Bollivar et al., 1994ab). This provided the first comprehensive understanding of genes involved in carotenoid and Mg-tetrapyrrole (chlorophyll and bacteriochlorophyll) biosynthesis in photosynthetic organisms. Indeed, detailed information of pigment

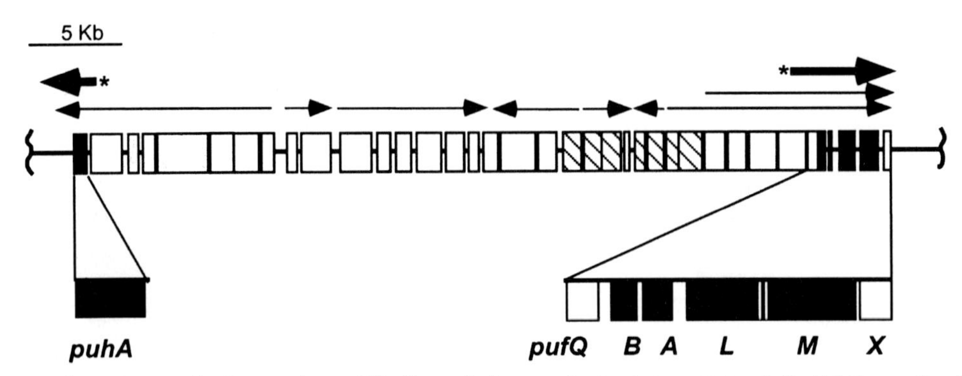

Fig 2. The *R. capsulatus* photosynthesis gene cluster. Filled boxes indicate *puf* and *puh* genes that code for light harvesting-I and reaction center apoproteins. Shaded boxes code for enzymes involved in syntheses of bacteriochlorophyll. Hatched boxes code for enzymes involved in synthesis of carotenoid. Open boxes are those involved in regulatory functions as well as those with unknown functions. The arrows denote unprocessed transcripts with the major RegB-RegA regulated *puf* and *puh* transcripts highlighted with a star. Thickness of the arrows indicates relative transcript abundance.

biosynthesis genes in *R. capsulatus* was a 'guiding light' for the characterization of chlorophyll and carotenoid biosynthesis genes in a variety of diverse experimental systems ranging from photosynthetic bacteria to plants (Bauer et al., 1993; Armstrong 1994; Kaneko et al., 1996; Suzuki et al., 1997; Xiong et al., 1998, 2000).

A similar photosynthesis gene cluster has also been found in other purple bacterial species such as *Rhodobacter sphaeroides* (Coomber et al., 1990; Naylor et al., 1999), *Rhodospirillum centenum* (Yildiz et al., 1992), *Rhodospirillum rubrum* (Sagesser, 1992), *Rhodopseudomonas viridis* (Wiessner et al., 1990) and *Rhodoferax gelatinosus* (Nagashima et al., 1993). There is also a 'photosynthesis gene cluster' present in the gram positive bacterium *Heliobacillus mobilus* which is closely related to cyanobacteria (Xiong et al., 1998). Curiously, no photosynthesis gene cluster exists in sequenced genomes of the cyanobacterium *Synechocystis* sp. PCC 6803 (Kaneko et al., 1996), the green sulfur bacterium *Chlorobium tepidum* (Xiong et al., 2000) or in the green non-sulfur bacterium *Chloroflexus aurantiacus* (Xiong et al., 2000). Thus, only three of the five photosynthetic bacterial lineages contain a tight clustering of photosynthesis genes.

IV. Regulating Photosystem Synthesis

Restricting synthesis of the photosystem to anaerobic conditions requires that the cell have sensory mechanisms to monitor the environment as well as the means to adjust photosynthesis gene expression according to changes in the environment. The combination of both fixed and varied structural components of the bacterial photosystem also imposes additional constraints on the regulation of photosynthesis gene expression. The incorporation of a RC complex into a LH-I ring requires that the expression of RC and LH-I structural genes (encoded by the *puh* and *puf* operons) be coordinated such that an appropriate stoichiometry of each component is maintained under all photosynthetic growth conditions. This coordinated expression is not extended to the *puc* operon that encodes LH-II structural genes (*puc*) since the ratio of LH-II to core photosynthetic units is known to be an inverse function of light intensity (Drews, 1995; Schumacher and Drews, 1979). This allows for optimization of energy absorption by the photosystem under varying light intensities. Superimposed over the regulation of LH and RC apoproteins is the synthesis of photopigments that must be adjusted to a level stoichiometric with LH and RC apoproteins. Consequently, the cell must either regulate the synthesis of biosynthetic enzymes and/or regulate enzymatic activity to adjust the pool of free photopigments in response to changes in the levels of structural proteins that bind them.

The remainder of this chapter will focus on the regulatory mechanisms that have been integrated into an elaborate network that controls the synthesis and activity of the photosystem. These systems include a strategic linkage of photosynthesis genes, transcriptional control, and differential mRNA decay as mechanisms that act in concert to influence production of the photosystem.

A. Photosynthesis Gene Cluster Superoperons

Promoter mapping studies initially revealed that the *R. capsulatus* photosynthesis gene cluster contains several large 'superoperons' such that transcripts originating from upstream pigment biosynthesis operons often include the downstream *puf* and *puh* coding regions (Beatty, 1995; Wellington et al., 1992) (Fig. 2). Specifically, RNA end-mapping experiments (Wellington and Beatty, 1989, 1991), and polar disruption studies (Young et al., 1989), demonstrated that the carotenoid *crtEF* and bacteriochlorophyll *bchCXYZ* (formerly called the *bchCA*) transcripts overlapped and that these pigment biosynthesis transcripts extended into the *pufQBALMX* operon that encodes the LH-I and reaction center structural polypeptides. The *pufQBALMX* operon also has its own endogenous promoter (Bauer et al., 1988; Adams et al., 1989) so it appears that transcription of the *puf* operon is initiated from three different sites. A similar arrangement was found to exist at the other end of the photosynthesis gene cluster where the *puhA* (reaction center H) gene was shown to be expressed as a shorter *puhA* specific transcript and as part of all kb *bchFBNHLM* transcript that coded for numerous upstream bacteriochlorophyll biosynthesis genes (Bauer et al., 1991).

One consequence of organizing the photosynthesis gene cluster into superoperons is that it couples the synthesis of the LH/RC structural polypeptides with that of the pigment biosynthetic enzymes (Bauer et al., 1991; Wellington and Beatty, 1991). Furthermore, because transcription of upstream pigment biosynthesis genes is only weakly repressed by oxygen, a low but significant level of LH and RC (*puh* and *puf*) gene expression is maintained even during aerobic growth. This basal level of expression is eclipsed by vigorous transcription from the *puh* and *puf* promoters, which are induced by anaerobiosis. Mutations that uncouple the overlapping transcriptional units caused substantial growth lags in cultures that are shifted from aerobic to anaerobic conditions (Bauer et al., 1991; Wellington and Beatty, 1991). This has been interpreted as evidence that coupled low level aerobic synthesis of photopigments and pigment-binding polypeptides may facilitate an efficient transition from respiration to photosynthetic growth. Furthermore, similarities in the photosynthesis gene clusters from such diverse species as *Rhodobacter sphaeroides* (Coomber et al., 1990; Naylor et al., 1999), *Rhodospirillum rubrum* (Sagesser, 1992), *Rhodoferax gelatinosus* (Nagashima et al., 1993), *Rhodospirillum centenum* (Yildiz et al., 1992) and *Rhodopseudomonas viridis* (Wiessner et al., 1990) imply that an important selective advantage is conferred by this arrangement.

B. Control of Transcription Initiation

1. Regulation in Response to Oxygen

The first comprehensive investigation into the effect of oxygen on purple non-sulfur bacteria was reported by Cohen-Bazire et al., (1957) who noted that the amounts of photopigments in cultures grown under 'really rigorous aerobiosis' were barely detectable. They concluded that when photosynthetic cells were exposed to oxygen, carotenoid and bacteriochlorophyll biosynthesis was abruptly halted. Years later, Biel and Marrs (1983) used *lacZ* transcriptional fusions to various carotenoid (*crt*) and bacteriochlorophyll (*bch*) genes to show that expression of these genes was two-fold higher in cells grown in low (3%) oxygen tension when compared to cells grown in high (23%) oxygen tension. In contrast to the two-fold regulation of pigment biosynthesis genes, *lacZ* translational fusions to the *puf* and *puh* operons, that code for LH and RC structural genes, exhibited much more dramatic 15- to 30-fold induction upon a reduction in oxygen tension (Bauer et al., 1988, 1991). RNA hybridization experiments also revealed that LH and RC operons were transcribed at much higher levels than were the bacteriochlorophyll or carotenoid genes (Zhu and Kaplan, 1985; Zhu and Hearst, 1986; Zhu et al., 1986). These studies also concluded that LH and RC genes were induced 30- to 100-fold in response to oxygen deprivation which is significantly higher then the two-fold induction seen with most bacteriochlorophyll and carotenoid genes. Moreover, it became obvious that separate regulatory systems controlled pigment biosynthetic and structural polypeptide gene expression. For example, pigment biosynthesis gene expression was found to be subject to aerobic repression, while synthesis of the LH and RC structural polypeptides are shown to rely on anaerobic activation of expression. Below are details of these two different types of regulation.

a. Anaerobic Activation

The first description of a *trans*-acting regulator that

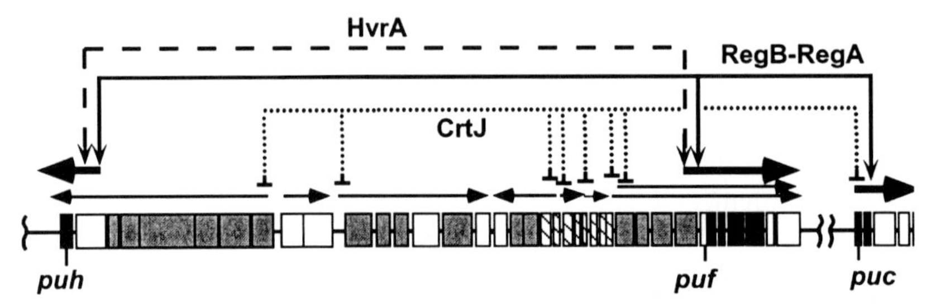

Fig 3. Regulatory circuits as discussed in the text that are responsible for activating or repressing photosynthesis gene expression. RegA and RegB (solid line) function as anaerobic transcriptional activators of the highly expressed *puf, puc* and *puh* operon promoters. HvrA (dashed line) functions as a dim light activator of the *puf* and *puh* promoters. CrtJ (dotted line) functions as an aerobic repressor of bacteriochlorophyll, carotenoid and *puc* (light harvesting II) genes. Boxes are as in Fig. 2 with thickness of the arrows indicating relative transcript abundance.

controlled expression of bacterial photosynthesis genes was a report by Sganga and Bauer (1992) describing the isolation of mutants of *R. capsulatus* that failed to induce synthesis of the photosystem when growth was shifted from aerobic to anaerobic environments. Mutants obtained through this procedure were screened on the basis of poor colony pigmentation as well as a reduction in β-galactosidase activity derived from a *puf::lacZ* translational fusion. Several mutants were mapped to a single locus, designated *regA* that was found to encode a response regulator class of DNA-binding proteins (Sganga and Bauer, 1992). Response regulators are the DNA-binding components of a two-component signal transduction system, the other member being a histidine sensor kinase that activates the response regulator by phosphorylation of a conserved aspartate.

Disruption of the *regA* gene severely inhibits photosynthetic growth, particularly under low light intensities, but has no observable effect on the growth of aerobic cultures (Sganga and Bauer, 1992). Electron microscopy of *regA* cells from anaerobic cultures contain poorly developed intracytoplasmic membranes while spectral analyses indicate that the amount of photosystem synthesized is greatly reduced relative to wild-type cells (Sganga and Bauer, 1992). *lacZ* fusions to various photosynthesis genes demonstrated that *regA* strains fail to anaerobically induce light harvesting and reaction center gene expression (*puf, puc* and *puh*) above the basal levels observed under aerobic conditions (Sganga and Bauer, 1992) (Fig. 3). In contrast, expression of bacteriochlorophyll and carotenoid (*bch* and *crt*) genes are essentially identical in *regA* and wild-type backgrounds, indicating that these loci are excluded from the RegA regulon (Sganga and Bauer, 1992).

A second oxygen regulator of photosynthesis gene expression, designated *regB*, was discovered using a modified version of the procedure used to isolate *regA* mutants (Mosley et al., 1994). *regB* mapped to a region upstream and in opposite orientation of the operon containing *regA*. The deduced amino acid sequence established RegB as a histidine sensor/kinase with six hydrophobic regions near the amino-terminus that anchor the protein to the cytoplasmic membrane (Chen et al., 2000). As for *regA*, disruptions in *regB* reduce the level of *puf, puc* and *puh* expression but do not affect expression of pigment biosynthesis genes (Mosley et al., 1994). Genetic analysis of the phenotypes associated with *regA* and *regB* mutants have been interpreted as evidence that RegA and RegB constitute a signal transduction system that regulates photosynthesis gene expression in response to alterations in oxygen tension.

Biochemical analysis of RegB has been undertaken using a soluble form of RegB that lacks most of its transmembrane domains (RegB″) (Inoue et al., 1995; Bird et al., 1999). In vitro experiments have demonstrated that the RegB″ autophosphorylates when mixed with [γ-P^{32}]ATP and that the phosphate moiety can be transferred from RegB″~P to RegA. The results of chemical treatments of RegB″~P and RegA~P are consistent with the predictions that these proteins are phosphorylated on histidine and aspartate residues, respectively. Phosphorylation of RegA has been shown to increase its binding affinity to target promoters from 10- to 20-fold (Du et al., 1998; Bird et al., 1999). Phosphorylated RegA has also been shown to stimulate in vitro transcription of *puf* and *puc* using purified *R. capsulatus* RNA polymerase (Bowman et al., 1999).

Based on genetic and molecular analysis, it is believed that RegB senses changes in oxygen tension either directly or indirectly by monitoring the redox state of the cell. Although the exact nature of the sensory mechanism is still unknown, it has been demonstrated that mutations in the respiratory terminal electron acceptor cytochrome cbb_3 oxidase significantly increase expression of genes in the RegB-RegA regulon (Buggy and Bauer, 1995; O'Gara, et al., 1998; Oh et al., 2000). Thus, it is possible that RegB is capable of monitoring the electron flow through the oxidase. Presumably an absence of a substrate for the oxidase (oxygen) triggers autophosphorylation of RegB and subsequent phosphorylation of RegA (Fig. 4a).

Recently, it has been established that RegA and RegB control expression of many additional genes beyond just the light harvesting and reaction center apoproteins. We now know that the RegB and RegA regulon includes genes involved in carbon fixation (Johsi and Tabita, 1996; Qian and Tabita 1996; Dubbs et al., 2000; Vichivanives et al., 2000), nitrogen fixation (Johsi and Tabita, 1996; Elsen et al., 2000), hydrogen utilization (Elsen et al., 2000), several terminal oxidases, as well as expression of cytochromes c_2, c_y and the cytochrome bc_1 complex (Swem and Bauer, 2001) (Fig. 4a). Thus, RegB and RegA constitute a signal transduction system that is not limited to photosynthesis, but rather functions as a global regulator of many metabolic pathways. One common theme of genes that are controlled by RegB and RegA is that they all affect the redox state of the cell. Specifically, photosynthesis and hydrogenase both generate reducing power in the form of reduced ubiquinone. In contrast, carbon fixation, nitrogen fixation and respiration, all utilize reducing power. So one can envision that the main role of RegB and RegA is to mediate the balance of the generation of reducing power with the utilization of excess reducing equivalents. In many respects, RegB and RegA appear to be functionally similar to the ArcB-ArcA system in *Escherichia coli* which are also global regulators of numerous anaerobic processes (Bauer et al., 1999).

Another interesting aspect of RegB and RegA is the fact that homologs are present in a number of different photosynthetic bacteria such as *R. capsulatus* (Sganga and Bauer, 1992; Mosley et al., 1994), *R. sphaeroides* (Eraso and Kaplan 1994, 1995; Phillips-Jones and Hunter, 1994), *Rhodovulvum sulfidophulum* (Masuda et al., 1999), *Roseobacter denitrificans* (Masuda et al., 1999) as well as in non-photosynthetic bacteria such as *Rhizobium meliloti* (Tiwari et al., 1996) and *Bradyrhizobium japonicum* (Bauer et al., 1998). Furthermore, the RegB and RegA sequences from these different genera are the most highly conserved members of the large sensor kinase and response regulator family (Masuda et al., 1999). Curiously, the putative helix-turn-helix motif of RegA is conserved in all RegA homologs that have been sequenced. This high level of sequence conservation suggests that there are significant constraints in the ability of RegA to genetically drift among these different species. It also suggests that RegB and RegA may play important roles in regulating anaerobic metabolisms in these species.

b. Aerobic Repression

Mutational analyses of the *R. capsulatus* and *R. sphaeroides* photosynthesis cluster suggested that an open reading frame termed *crtJ* codes for a *trans-acting* repressor of photopigment biosynthesis genes (Penfold and Pemberton, 1991, 1994; Bollivar et al., 1994a). Spectral analyses of aerobically grown cultures demonstrated that the steady-state level of bacteriochlorophyll is elevated 2.6 fold in a *crtJ* mutant as compared to wild-type cells (Bollivar et al., 1994a). Aerobic expression of various photopigment biosynthesis genes is also elevated two-fold higher in a *crtJ* mutant strain suggesting that CrtJ functions as an aerobic repressor of *bch* and *crt* genes (Ponnampalam et al., 1995; Gomelsky and Kaplan, 1995) (Fig. 3). Interestingly, the effect of CrtJ is not limited to the pigment biosynthesis genes. Aerobic expression of the *puc* operon, which codes for the light harvesting II structural apoproteins, is also two-fold higher in a *crtJ* mutant despite the fact that *puf* and *puh* expression levels are essentially the same as in wild-type (Ponnampalam et al., 1995). Thus, *puc* operon expression is subject to activation by RegB-RegA as well as repression by CrtJ.

CrtJ is a soluble protein (52 kDa) with a deduced amino acid sequence that exhibits a putative helix-turn-helix DNA binding domain (Penfold and Pemberton, 1991; 1994; Alberti et al., 1995). An examination of the DNA sequences of promoters that are regulated by CrtJ reveals a consensus sequence of TGT-N_{12}-ACA (Alberti et al., 1995; Elsen et al., 1998). Gel mobility shift and DNA footprint experiments have demonstrated that CrtJ binds to this palindrome's sequence (Ponnampalam and Bauer, 1997; Elsen et al., 1998; Ponnampalam et al., 1998). In the *bchC* promoter region there are two CrtJ binding palindromes separated 8 bp apart, one that straddles the –35 region and the other the –10

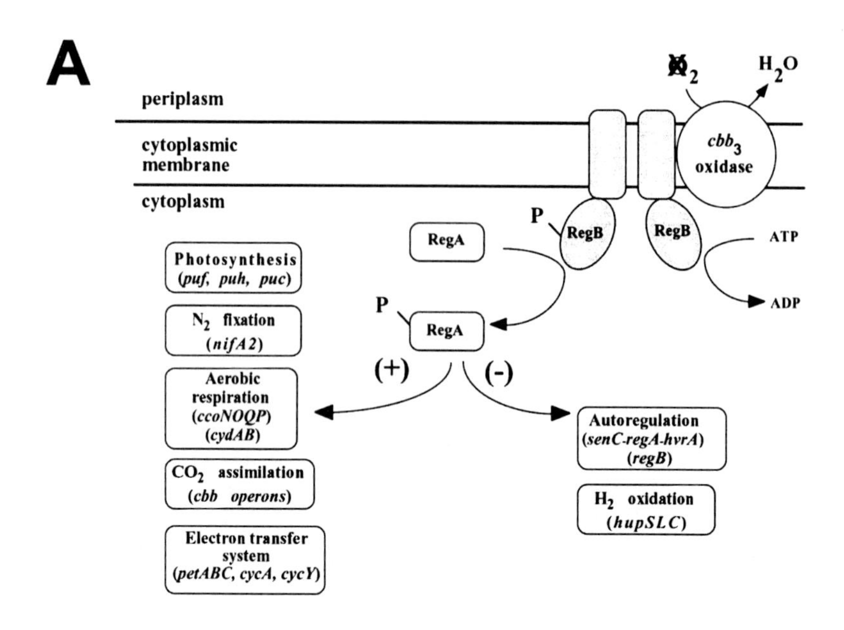

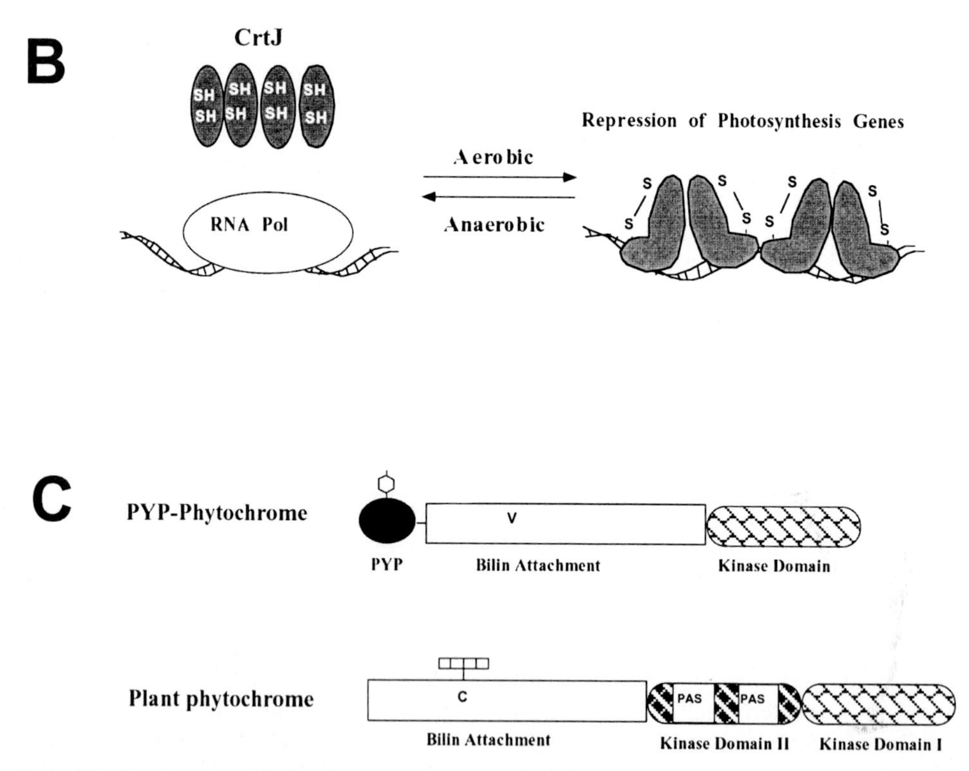

Fig. 4. Molecular details of various transcription factors. (A) RegB is a membrane bound histidine sensor kinase that is thought to monitor respiratory activity of cbb_3 oxidase. Upon anaerobiosis, RegB autophosphorylates and subsequently transfers a phosphate to RegA. RegA~P then activates or represses expression of the various indicated target genes. (B) Under aerobic conditions, CrtJ forms an intramolecular disulfide bond that stimulates DNA-binding and subsequent repression of gene expression by CrtJ. Under anaerobic conditions, the disulfide bond is reduced which inhibits DNA binding. (C) PYP-phytochrome has structural features very similar to plant phytochromes. Both contain sensor kinase domains as well as bilin attachment domains, the difference being that plant phytochrome contains a conserved Cys used for attachment of a liner tetrapyrrole as a red light absorbing chromophore while PYP-phytochrome lacks this Cys. PYP-phytochrome has an additional PYP domain that attaches a blue light absorbing chromophore, *p*-hydroxycinnamic acid.

region (Ponnampalam and Bauer, 1997). Binding of CrtJ to the *bchC* promoter is cooperative, with protein bound to the –35 palindrome interacting with protein bound to the –10 palindrome (Ponnampalam et al., 1998). In vitro and in vivo experiments have demonstrated that CrtJ is unable to effectively bind to the *bchC* promoter if the spacing between the two *bchC* palindromes is altered by adding or removing just a few bps (Ponnampalam et al., 1998).

Unlike the *bchC* operon, other CrtJ repressed genes in *R. capsulatus* contain only a single palindrome within the –10 to –35 promoter region (Alberti et al., 1995; Elsen et al., 1998). However, in every case that has been examined, an additional palindrome has been found within close proximity, usually 100 to 150 bps upstream of the transcription start-site. In vivo and in vitro experiments have demonstrated that CrtJ bound to the 'distant palindrome' cooperatively interacts with CrtJ bound to the –10 or –35 promoter region (Elsen et al., 1998). This indicates that DNA looping must be occurring that permits CrtJ interactions between palindromes that are separated by 100 or more bp apart. For the *puc* operon, which is activated by RegA and repressed by CrtJ, it has been demonstrated that the CrtJ binding site overlaps that of the RegA binding site (Du et al., 1998; Elsen et al., 1998; Bowman et al., 1999). It has also been demonstrated that these two sites cannot be co-occupied by RegA and CrtJ invoking competition for binding as the mechanism of *puc* repression by CrtJ (Bowman et al., 1999).

In vitro gel mobility shift experiments have demonstrated the CrtJ effectively binds to the target palindrome under oxidizing but not under reducing conditions (Ponnampalam and Bauer, 1997; Dong and Bauer, 2000). Studies have failed to detect the presence of metal as a redox responding cofactor which has been found in other redox sensing proteins (Ponnampalam and Bauer, 1997). Recently, it has been demonstrated that oxidizing conditions promote the formation of an intramolecular disulfide bond in CrtJ (Dong and Bauer, unpublished) (Fig. 4b). Disulfide bond formation in CrtJ was also shown to occur in vivo when *R. capsulatus* cells are grown under oxidizing conditions but not when grown under reducing conditions. Presumably, the formation of the disulfide bond causes CrtJ to adopt a conformational form that permits it to effectively bind DNA. Interestingly, one of the Cys that is involved in disulfide bond formation is located very near the CrtJ helix -turn-helix motif (Dong and Bauer, unpublished).

2. Light Regulation

Early physiological studies of purple bacteria demonstrated that when photosynthetic cultures are shifted from high to low light conditions they exhibit a distinct growth lag that correlated with increased rates of photopigment production (Cohen-Bazire et al., 1957). This indicates that cells compensate for a decline in light intensity by increasing synthesis of the photosystem to maintain an adequate supply of energy. Careful investigations into this phenomenon revealed that cells adapt to a drop in light intensity by increasing the number of core photosynthetic units per cell (up to five-fold) while simultaneously increasing the ratio of LH-II antenna complexes to reaction centers (up to two-fold) (Shumacher et al. 1979; Golecki et al., 1980; Drews, 1985). Molecular and genetic investigations later confirmed that high light intensities caused a reduction in photopigment production and led to diminished levels of *puf, puh* and *puc* transcripts (Zhu and Kaplan, 1985; Zhu and Hearst, 1986; Zhu et al., 1986; Buggy et al., 1994).

Shimada et al., (1992) used Northern hybridization experiments to examine the dependence of *puf* and *puc* gene expression on specific light wavelengths and found that the largest inhibition occurred under blue light (approximately 450 nm). This suggested that a transcription factor, possibly in combination with a chromophore, might sense light intensity and modulate photosynthesis gene expression accordingly. The authors later reported that a factor present in cell extracts of cultures grown photosynthetically under various light conditions bound to the *puf* promoter at a site previously identified as a *cis*-acting oxygen control element (Shimada et al. 1993; Shimada et al., 1996). Gel mobility shift and footprint experiments indicated that binding appeared to be enhanced when cell extracts were made from cultures grown semi-aerobically under blue light indicating that the factor might be a repressor of *puf* transcription.

Difficulties in identifying transcriptional regulators of light responsive genes through classical genetics have been compounded by the fact that the effect is small (two-fold) compared to activation by anaerobiosis (>30-fold). In fact, the identification of the only gene known to influence expression of the photosystem in response to changes in light intensity occurred because it is transcriptionally linked to the oxygen regulator, *regA*. Disruption of the *hvrA* locus, located immediately downstream of *regA*, produced a strain with an impaired capacity for photosynthetic growth under low light (Buggy et al., 1994). Although growth under saturating light intensity (100 μmol photons m^{-2} s^{-1}) is virtually unaffected, the doubling times of *hvrA* mutants are substantially longer than for wild-type when grown under low intensities (2 μmol photons m^{-2} s^{-1}). Furthermore, the absorption spectra of membrane fractions made from the mutant strain indicate that the ability to increase the amount of photosystem in cells grown under low light intensity was significantly impaired. The role of *hvrA* in photosynthesis gene expression was examined using translational *lacZ* fusions to promoters from various photosynthesis genes. These experiments showed that expression of the *crt, bch* and *puc* operons were essentially the same in wild-type and *hvrA* backgrounds. However, the two-fold elevation in *puf* and *puh* expression in response to low light intensity was greatly reduced in the *hvrA* strain relative to the wild-type indicating that HvrA may be a light responsive *trans*-activator of *puf* and *puh* expression (Buggy et al., 1994).

Inspection of the HvrA deduced amino acid sequence revealed that it is a small polypeptide (11.5 kDa) with a positive net charge (isoelectric point of 10.2) and a putative helix-turn-helix DNA binding domain. The protein has been purified and used in DNA footprinting experiments where it was observed to bind to the promoter regions (approximately 90 to 120 bps upstream of the transcription start-sites) of both the *puf* and *puh* operons (Kouadio and Bauer, unpublished). The sites protected by HvrA do not exhibit any obvious similarities in DNA sequence but are relatively A/T rich suggesting that the protein may bind to a common structure such as an intrinsic curve in the DNA. Similar DNA-binding properties have been proposed for H-NS, a bacterial nucleoid protein that has been shown to affect transcription from several disparate genes in response to various environmental stimuli. H-NS is known to constrain negative supercoils in the *E. coli* chromosome and is believed to work in conjunction with additional transcription factors to influence gene expression. HvrA shares 35% sequence identity to H-NS fueling speculation that it too may act in concert with other regulatory factors to stimulate *puf* and *puh* expression (Kouadio and Bauer, unpublished).

The mechanism that controls HvrA activity is not known although several hypotheses have been considered. The possibility that HvrA senses light intensity through binding of a chromophore has been explored by examining the absorption properties of the native protein isolated from *R. capsulatus*. However, these experiments have provided no evidence to suggest that HvrA contains a light sensitive cofactor (Kouadio and Bauer, unpublished). A second prospect is that HvrA does not sense light intensity directly, but rather monitors ATP levels or 'energy charge' within the cell. It has been postulated that energy charge is an important factor in determining the activity levels of bacteriochlorophyll biosynthetic enzymes and it has been shown that reductions in light intensity coincide with a drop in the rate of photophosphorylation. However, beyond the observation that the protein contains a short amino acid sequence with similarity to an ATP binding motif, no evidence has been obtained thus far to suggest that HvrA monitors cellular ATP concentrations. Lastly, HvrA may have no intrinsic sensory capability but rather, function as an accessory factor to other *trans*-activators of photosynthesis genes. If so, this activity could be controlled if the concentration of HvrA varied.

It has also been reported that the accumulation of LH-II polypeptides in cells that have experienced a shift to low light intensity lags behind increased expression of *puf* and *puh* encoded proteins (Klug et al., 1985). This would indicate the presence of a specific light-regulated transcription factor for the *puc* operon. So far, no light sensitive transcriptional regulators of the *puc* operon have yet been discovered.

An interesting light regulator of gene expression has recently been described for the bacterium *Rhodospirillum centenum.* This species has been shown to contain a hybrid polypeptide comprised of photoactive yellow protein (PYP) at the amino terminus and a plant phytochrome like carboxyl terminus (Jiang et al., 1999) (Fig. 4c). PYP's have been described in several species of purple photosynthetic bacteria (Meyer 1985; Kort et al., 1996). PYP is a small 120 amino acid protein containing a blue light absorbing chromophore, *p*-hydroxycinnamic acid attached by a covalent thioester linkage to a conserved cysteine. The chromophore undergoes a photocycle with the ground state having a maximum absorbance at 446 nm and intermediates at 465 nm and 355 nm. The role of PYP is unclear since mutations in the *Rhodobacter capsulatus* PYP have no noticeable phenotype (Jiang and Bauer, unpublished data). However, the *R. centenum* PYP-phytochrome hybrid was shown to be involved in blue light regulated expression of chalcoln synthase (Jiang et al., 1999). Interestingly, chalcoln synthase expression in higher plants is also under red-light control that is meditated by phytochrome (Martin, 1993). Plant phytochrome has a liner tetrapyrrole (bilin) attached to a conserved Cys that absorbs red and far red light (Quail, 1991). In the *R. centenum* PYP-phytochrome hybrid, the conserved bilin attachment Cys is replaced with a Val (Jiang et al., 1999). This suggests that PYP-phytochrome has only the blue light chromophore *p*-hydroxycinnamic acid associated with this photoreceptor. However, isolation of full length PYP-phytochrome from *R. centenum* has not been reported so it remains to be determined if a second red light absorbing chromophore is also present in this photoreceptor.

C. mRNA Processing Control

In addition to the control of transcription initiation, mRNA processing has been shown to be an important factor in regulating light harvesting and reaction center gene expression. A series of mRNA processing events occurs that give rise to mRNA segments with differential mRNA decay rates. These mRNA processing events are important for maintaining the stoichiometry of the photosystem structural polypeptides. The sections below discuss mRNA processing events that are known to occur in *R. capsulatus*. This topic has also been reviewed elsewhere (Klug, 1993, 1995)

1. The puf mRNA Processing Paradigm

Northern blot analysis of the *R. capsulatus puf* operon initially revealed two mRNAs 2.7 kb and 0.5 kb in length (Belasco et al., 1985) (Fig. 5). Hybridization studies demonstrated that the larger message contained the *pufBALMX* coding region while the smaller contained only the *pufBA* region (Belasco, 1985). Analysis of mRNA decay rates indicated that the smaller message was generated by a $3' \rightarrow 5'$ degradation of the *pufBALMX* transcript which terminated at a stable stem loop structure located between the *pufA* and *pufL* genes. The half-lives of the smaller and larger transcripts were calculated to be 33 and 8 min, respectively, that resulted in a steady-state accumulation of 10-fold more *pufBA* message than *pufBALMX* message. This difference correlates well with the observed 15-fold excess of LH-I (*pufBA* gene products) to RC polypeptides (*pufLM* gene products).

Several experiments have been carried out to

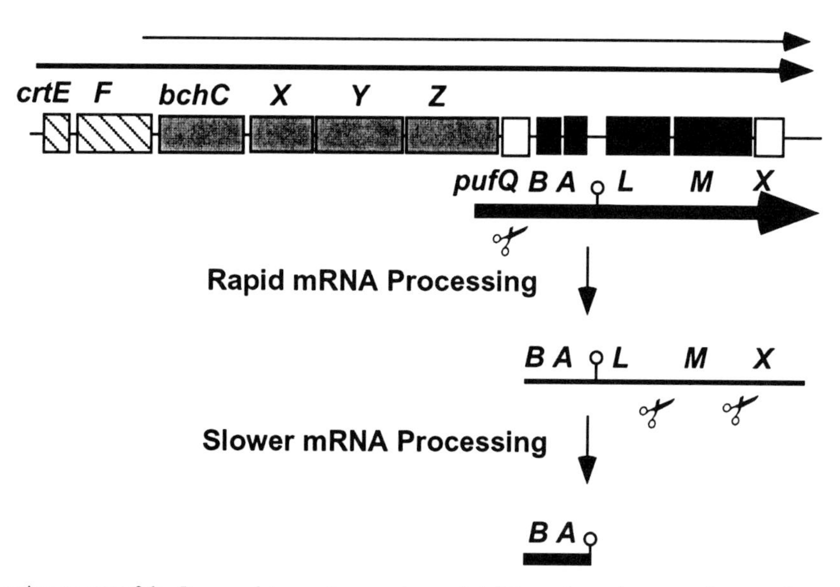

Fig. 5. mRNA processing events of the *R. capsulatus puf* operon transcript. It is unclear what type of processing events occur with the large upstream *crt* and *bch* transcripts that extend into the *puf* operon (thin arrows). However, the highly expressed RegA~P regulated transcript (thick arrow) undergoes rapid (<0.5 min half life) processing from a 3.4 kb *pufQBALMX* transcript to a 2.7 kb *pufBALMX* transcript. The *pufBALMX* transcript, which has a 8 min half life, is further processed to a *pufBA* transcript that has a 33 min half life.

investigate the role of secondary structure in attenuating mRNA decay of the *pufBALMX* transcript. Deletion of the hairpin between the *pufA* and *pufL* coding regions resulted in loss of the otherwise stable 0.5 kb *pufBA* transcript and had little effect on decay of the larger *pufBALMX* transcript (Klug et al., 1987). Similarly, a mutation that eliminated the stop codon of the *pufA* gene, which allowed translocating ribosomes to proceed through the 3′ hairpin structure, significantly destabilized the 0.5 kb *pufBA* mRNA (Klug and Cohen, 1991). These data support the view that the stem loop protects the *pufBA* message from 3′ to 5′ exonuclease degradation.

It became evident that the *puf* mRNA decay pathway was even more complex when the *puf* promoter was located several hundred base pairs upstream of the stable 5′ end of the *pufBALMX* and *pufBA* mRNAs (Bauer et al., 1988; Adams et al., 1989). In fact, an additional gene, *pufQ*, was found to reside within the operon upstream of *pufB* which is not present in the processed message (Bauer and Marrs, 1988). A 3.4 kb *pufQBALMX* transcript was subsequently shown to undergo rapid processing (half life of <0.5 min.) creating the smaller 2.7 kb *pufBALMX* transcript. Thus, the mRNA processing model, supported by the data, involves rapid processing of the ~3.4 kb *pufQBALMX* message to a 2.7 kb *pufBALMX* message followed by a slower decay of *pufBALMX* by 3′ → 5′ degradation to the stable *pufBA* transcript (Fig. 5). Superimposed over the this model is the observation that the half-life of the *pufBALMX* transcript was reduced from 8 to 3 minutes in cells undergoing aerobic growth while stability of the *pufBA* transcript was unaffected (Klug, 1991).

Insights into the mechanisms that control *puf* mRNA decay suggest that processing of the 3.4 kb *pufQBALMX* may be similar to that of the *lacZ* mRNA in *E. coli*. *lacZ* mRNA processing is known to involve an initial cleavage at an internal RNaseE or RNaseK type processing site followed by rapid exoribonuclease degradation (Lundberg et al., 1990; Ehretsmann et al., 1992). Internal cleavage sites within the *pufBALMX* transcript have been identified which apparently contribute to mRNA instability (Klug and Cohn, 1990). Furthermore, studies have shown that while the kinetics of 2.7 kb *puf* mRNA decay in *R. capsulatus* is essentially the same as when it is expressed in *E. coli*, the message is stabilized significantly in an RNaseE mutant strain (Klug et al., 1992). These data suggest two possible mechanisms for oxygen control of *puf* mRNA stability; (i) directly through oxygen regulated RNase activity (Klug, 1991, 1993), or (ii) indirectly by growth condition dependent alterations in transcription or translation initiation/ elongation rates which expose RNase sites on the mRNA.

While there is no precedent for oxygen control of RNase activity, studies of *lacZ* mRNA stability in *E. coli* indicate that coordination of transcription and translation are critical to control turnover rates (Lost and Dreyfus, 1995). Presumably, uncoupling of transcription and translation processes alters mRNA decay by exposing internal RNase cleavage sites that are otherwise masked (Petersen, 1992). Klug and Cohen (1991) have reported that the introduction of stop codons near the start of *pufL* in *R. capsulatus* decreased stability of the *pufBALMX* mRNA. If translation was allowed to proceed at least 107 nucleotides into the *pufL* gene, normal stability of the message was restored even though the ribosomes failed to reach the crucial endonucleolytic cleavage sites. The authors concluded that functional interaction of the translational apparatus with internal processing sites must be affecting *puf* mRNA stability through a mechanism that does not require physical protection mediated by ribosomes.

2. puh *and* puc *mRNA Processing Events*

While not studied as extensively as *puf* mRNA processing, there is evidence that transcript processing also plays an important role in regulating expression of the RC-H and LH-II structural proteins encoded by the *puh* and *puc* operons, respectively. Northern blot analysis has revealed that three mRNA species of 0.9, 1.1 and 11 kb in size are derived from the *puh* operon in *R. capsulatus* (Bauer et al., 1991). Introduction of a transcription terminator 4 kb upstream of *puhA* resulted in the loss of the 11 and 1.1 kb mRNAs but did not affect the 0.9 kb transcript (Bauer et al., 1991). These results were interpreted as evidence that the 11 kb transcript originated from a superoperon that included the upstream *bchFBNHLM*-orf444 region and that the 1.1 kb mRNA was a product of the larger transcript. This hypothesis is supported by the observation that the stable 5′ end of the 1.1 kb segment contains a hairpin structure that resembles an RNaseIII cleavage site. Mutational analysis indicated that the 0.9 kb fragment was generated from a RegA/RegB dependent promoter element designated as the *puhA* promoter.

Experiments to determine the relative stabilities of the *puhA* mRNAs have not yet been undertaken. Therefore, it is uncertain how each of the *puh* transcripts contribute to the synthesis of RC complexes. It has been demonstrated that the presence of the 0.9 kb transcript alone is sufficient to maintain a wild-type growth rate under steady-state photosynthetic conditions but cannot provide for a rapid transition from aerobic to photosynthetic conditions without a significant growth lag (Bauer et al., 1991). This would suggest that the 11 kb and 1.1 kb transcripts are responsible for promoting a basal level of *puhA* expression during aerobic growth that is required for a rapid transition to photosynthetic growth.

Northern blot experiments have also indicated that the *pucBACDE* operon produces multiple mRNA species of 2.4, 0.5, and 1.0 kb in length. The 2.4 kb mRNA appears to encode the entire *puc* operon whereas the 0.5 kb and 1.0 kb transcripts have been shown to contain the *pucBA* and *pucDE* genes, respectively (Tichy et al., 1991; Leblanc and Beatty, 1993). A putative hairpin structure located within the *pucA-pucC* intercistronic region is thought to stabilize the 0.5 kb *pucBA* segment as described for the *pufBA* mRNA (Zucconi and Beatty, 1988, Tichy et al., 1991). *R. sphaeroides* has similar *puc* mRNA species to *R. capsulatus* and also appears to contain a stem loop secondary structure between *pucA* and *pucC* (Lee et al., 1989). Moreover, it has been demonstrated that the 2.3 kb mRNA from this species has a half-life of less than five minutes whereas the 0.5 kb mRNA has a half-life of greater than 20 minutes. As for the *puf* mRNA in *R. capsulatus*, these turnover rates have been found to be dependent upon oxygen tension (Lee et al., 1989).

V. Concluding Statements

During the past decade, there has been much progress in the identification of numerous transcription factors that control synthesis of the purple bacterial photosystem. At this stage, we have a rough sketch of the various players that are involved. What is needed is a better understanding of how these regulatory proteins function at a molecular level. For example, we need a better understanding of how RegB senses alterations in redox and the mechanism of transcription activation that phosphorylated RegA provides. We also need to understand how light affects HvrA and PYP-phytochrome's activity. Other aspects, that are just now emerging, are how the cell uses various transcription factors to control a variety of different cellular processes. For example, we have recently determined that RegA is used in conjunction with other transcription factors to control photosynthesis, nitrogen fixation, carbon fixation, hydrogen utilization and respiration. Each of these processes have their own dedicated transcription factors in addition to RegB-RegA, so it will be interesting to see if any interactions occur between RegA~P and other dedicated transcription factors. It appears that the cell carefully balances physiological processes that generate reducing power with those that utilize reducing power. A major challenge in the future will be to determine the molecular mechanisms behind this control.

Acknowledgment

Research for the C. Bauer laboratory has been supported by the National Institutes of Health.

References

Adams CW, Forrest ME, Cohen SN and Beatty JT (1989) Structural and functional analysis of transcriptional control of the *Rhodobacter capsulatus puf* operon. J Bacteriol 171: 473–482

Alberti M, Burke DH and Hearst JE (1995) Structure and sequence of the photosynthesis gene cluster. In: Blankenship RE, Madigan MT and Bauer CE (eds)Anoxygenic Photosynthetic Bacteria, pp 1083–1106. Kluwer Academic Publishers, Dordrecht

Armstrong GA (1994) Eubacteria show their true colors: Genetics of carotenoid pigment biosynthesis from microbes to plants. J Bacteriol 176: 4795–802

Armstrong GA, Alberti M, Leach F and Hearst JE (1989) Nucleotide sequence, organization and nature of protein products of the carotenoid biosynthesis gene cluster of *Rhodobacter capsulatus*. Mol Gen Genet 216: 254–268

Bauer CE and Marrs BL (1988) *Rhodobacter capsulatus puf* operon encodes a regulatory protein (PufQ) for bacteriochlorophyll biosynthesis. Proc Natl Acad Sci USA 85: 7074–7078

Bauer CE, Young DY and Marrs BL (1988) Analysis of the *Rhodobacter capsulatus puf* operon: Location of the oxygen-regulated promoter region and the identification of an additional *puf*-encoded gene. J Biol Chem 263: 4820–4827

Bauer CE, Buggy JJ, Yang Z and Marrs BL (1991) The superoperonal organization of genes for pigment biosynthesis and reaction center proteins is a conserved feature in *Rhodobacter capsulatus*: analysis of overlapping *bchB* and *puhA* transcripts. Mol Gen Genet 228: 438–444

Bauer CE, Bollivar DW and Suzuki JY (1993) Genetic analysis of photopigment biosynthesis in eubacteria: A guiding light for algae and plants. J Bacteriol 175: 39193925

Bauer E, Kaspar T, Fischer HM and Hennecke H (1998) Expression of the *fixR-nifA* operon in *Bradyrhizobium japonicum* depends on a new response regulator, RegR. J Bacteriol 180: 3853–3863

Bauer CE, Elsen S and Bird T (1999) Mechanisms that control gene expression by redox. Ann Rev Microbiol 53: 495–523

Beatty JT (1995) Organization of photosynthesis gene transcripts. In: Blankenship RE, Madigan MT and Bauer CE (eds) Anoxygenic Photosynthetic Bacteria, pp 1209–1219. Kluwer Academic Publishers, Dordrecht

Belasco JG, Beatty TJ, Adams CW, Von Gabain A and Cohen SN (1985) Differential expression of photosynthesis genes in *R. capsulata* results from segmental differences in stability within the polycistronic *rxc* transcript. Cell 40: 171–181

Biel AJ and Marrs BL (1983) Transcriptional regulation of several genes for bacteriochlorophyll synthesis in *Rhodopseudomonas capsulata* in response to oxygen. J Bacteriol 156: 686–694

Bird TH, Du S and Bauer CE (1999) Autophosphorylation, phosphotransfer and DNA binding characteristics of the RegB-RegA two component regulon involved in controlling anaerobic gene expression in *Rhodobacter capsulatus*. J Biol Chem 274: 16343–16348

Bollivar DW, Suzuki JY, Beatty JT, Dobrowlski J and Bauer CE (1994a) Directed mutational analysis of bacteriochlorophyll *a* biosynthesis in *Rhodobacter capsulatus*. J Mol Biol 237: 622–640

Bollivar DW, Wang S, Allen JP and Bauer CE (1994b) Molecular genetic analysis of terminal steps in bacteriochlorophyll *a* biosynthesis: Characterization of a *Rhodobacter capsulatus* strain that synthesizes geranylgeraniol esterified bacteriochlorophyll *a*. Biochemistry 33: 12,763–12,768

Bowman WC, Du S, Bauer CE and Kranz RG (1999) In vitro activation and repression of photosynthesis gene transcription in *Rhodobacter capsulatus*. Mol Microbiol 33: 429–437

Buggy JJ and Bauer CE (1995) Cloning and characterization of *senC*, a gene involved in both aerobic respiration and photosynthesis gene expression in *Rhodobacter capsulatus*. J Bacteriol 177: 6958–6965

Buggy JJ, Sganga MW and Bauer CE (1994) Characterization of a light-responding trans-activator responsible for differentially controlling reaction center and light-harvesting-I gene expression in *Rhodobacter capsulatus*. J Bacteriol 176: 6936–6943

Chen W, Jager A and Klug G (2000) Correction of the DNA sequence of the *regB* gene of *Rhodobacter capsulatus* with implications for the membrane topology of the sensor kinase RegB. J Bacteriol 182: 818–820

Cohen-Bazire GW, Sistrom WR and Stanier RY (1957) Kinetic studies of pigment synthesis by non-sulfur purple bacteria. J Cellular Comp Physiol 49: 25–68

Coomber SA, Chaudhri M, Conner M, Britton G and Hunter CN (1990) Localized transposon Tn5 mutagenesis of the photosynthetic gene cluster of *Rhodobacter sphaeroides*. Mol Microbiol 4: 977–989

Drews G (1985) Structure and functional organsization of light-harvesting complexes and photochemical reaction centers in membranes of phototrophic bacteria. Microbial Rev 49: 59–70

Du S, Bird TH and Bauer CE (1998) DNA binding characteristics of RegA. A constitutively active anaerobic activator of photosynthesis gene expression in *Rhodobacter capsulatus*. J Biol Chem 273: 18509–18513

Dubbs JM, Bird TH, Bauer CE and Tabita FR (2000) Binding of CbbR and a constitutively active mutant form of the activator RegA with the *Rhodobacter sphaeroides cbbI* promoter-operator region. J Biol Chem 275: 19224–19230

Ehretsmann CP, Agamemnon JC, and Krisch HM (1992). Specificity of *Escherichia coli* endoribonuclease E: In vivo and in vitro analysis of mutants in a bacteriophage T4 mRNA processing site. Genes Develop 6:149–159

Elsen S, Ponnampalam SN and Bauer CE (1998) CrtJ bound to distant binding sites interacts cooperatively to aerobically repress photopigment biosynthesis and light harvesting II gene expression in *Rhodobacter capsulatus*. J Biol Chem 273: 30762–30769

Elsen S, Dischert W, Colbeau A and Bauer CE (2000) Expression of uptake hydrogenase and molybdenum nitrogenase in *Rhodobacter capsulatus* is coregulated by the RegB-RegA two-component regulatory system. J Bacteriol 182: 2831–2837

Eraso JM and Kaplan S (1994) *prrA*, a putative response regulator involved in oxygen regulation of photosynthesis gene expression in *Rhodobacter sphaeroides*. J Bacteriol 176: 32–43

Eraso JM and Kaplan S (1995) Oxygen-insensitive synthesis of the photosynthetic membranes of *Rhodobacter sphaeroides:* A mutant histidine kinase. J Bacteriol 177: 2695–2706

Giuliano G, Pollock D, Stapp H and Scolnik PA (1988) A genetic-physical map of the *Rhodobacter capsulatus* carotenoid biosynthesis cluster. Mol Gen Genet 213: 78–83

Golecki JR, Schumacher A and Drews G (1980) The differentiation of the photosynthetic apparatus and the intracytoplasmic membrane in cells of *Rhodopseudomonas capsulata* upon variation of light intensity. Eur J Cell Biol 23:1–5

Gomelsky M and Kaplan S (1995) Genetic evidence that PpsR from *Rhodobacter sphaeroides* 2.4.1 functions as a repressor of *puc* and *bch*F expression. J Bacteriol 177: 1634–1637

Inoue K, Kouadio JL, Mosley CS and Bauer CE (1995) Isolation in vitro phosphorylation of sensory transduction components controlling anaerobic induction of light harvesting and reaction center gene expression in *Rhodobacter capsulatus*. Biochemistry 34: 391–396

Jiang Z-Y, Swem L, Rushing B, Devanathan S, Tollin G and Bauer CE (1999) Bacterial photoreceptor with similarity to both photoactive yellow protein and plant phytochromes. Science 285: 406–409

Joshi HM, and Tabita FR (1996) A global two component signal transduction system that integrates the control of photosynthesis, carbon dioxide assimilation, and nitrogen fixation. Proc Natl Acad Sci USA 93:14515–14520

Kaneko T, Sato S, Kotani H, Tanaka A, Asamizu E, Nakamura Y, Miyajima N, Hirosawa M, Sugiura M, Sasamoto S, Kimura T, Hosouchi T, Matsuno A, Muraki A, Nakazaki N, Naruo K, Okumura S, Shimpo S, Takeuchi C, Wada T, Watanabe A, Yamada M, Yasuda M, and Tabata S (1996) Sequence analysis of the genome of the unicellular cyanobacterium *Synechocystis* sp. strain PCC6803. II. Sequence determination of the entire genome and assignment of potential protein-coding regions. DNA Res 3: 109–136

Karrasch S, Bullough PA and Ghosh R (1995) The 8.5 A projection map of the light-harvesting complex I from *Rhodospirillum rubrum* reveals a ring composed of 16 subunits. EMBO J 14: 631–638

Klug G (1991) Endonucleolytic degradation of *puf* mRNA in *Rhodobacter capsulatus* is influenced by oxygen. Proc Natl Acad Sci USA 88: 1765–1769

Klug G (1993) The role of mRNA degradation in the regulated expression of bacterial photosynthesis genes. Mol Microbiol 9: 1–7

Klug G (1995) Post-transcriptional control of photosynthesis gene expression. In: Blankenship RE, Madigan MT and Bauer CE (eds)Anoxygenic Photosynthetic Bacteria, pp 1235–1244. Kluwer Academic Publishers, Dordrecht

Klug G and Cohen SN (1990) Combined action of multiple hairpin loop structures and sites of rate-limiting endonucleolytic cleavage determine differential degradation rates of individual segments within polycistronic *puf* operon mRNA. J Bacteriol 172: 5140–5146

Klug G and Cohen SN (1991) Effects of translation on degradation of mRNA segments transcribed from the polycistronic *puf* operon of *Rhodobacter capsulatus*. J Bacteriol 173: 1478–1484

Klug G, Kaufmann N and Drews G (1985) Gene expression of pigment-binding proteins of the bacterial photosynthetic apparatus: transcription and assembly in the membrane of *Rhodopseudomonas capsulata*. Proc Natl Acad Sci USA 82: 6485–6489

Klug G, Adams CW, Belasco JG, Dörge B and Cohen SN (1987) Biological consequences of segmental alterations in mRNA stability: Effects of the intercistronic hairpin loop region of the *Rhodobacter capsulatus puf* operon. EMBO J 6: 3515–3520

Klug G, Jock S and Rothfuchs R (1992) The rate of decay of *Rhodobacter capsulatus*specific *puf* mRNA segments is differentially affected by RNase E activity in *Escherichia coli*. Gene 121: 95–102

Kort R, Hoff WD, West MV, Kroon AR, Hoffer SM, Vlieg KH, Crielaard W, Van Beeumen JJ and Hellingwerf KJ (1996) The xanthopsins: A new family of eubacterial blue-light photoreceptors EMBO J 15: 3209–3218

Lancaster CRD, Ermler U and Michel H (1995) The Structure of photosynthetic reaction centers from purple bacteria reveled by x-ray crystallography. In: Blankenship RE, Madigan MT and Bauer CE (eds)Anoxygenic Photosynthetic Bacteria, pp 503–526. Kluwer Academic Publishers, Dordrecht

LeBlanc NH and Beatty JT (1993) *Rhodobacter capsulatus puc* operon: Promoter location, transcript sizes and effects of deletions on photosynthetic growth. J Gen Microbiol 139: 101–109

Lee JK, Kiley PJ and Kaplan S (1989) Post-transcriptional control of *puc* operon expression of B800-850 light harvesting complex formation in *Rhodobacter sphaeroides*. J Bacteriol 171: 3391–3405

Lost I and Dreyfus M (1995) The stability of *Escherichia coli lacZ* mRNA depends upon the simultaneity of its synthesis and translation. EMBO J 14: 3252–3261

Lundberg U, von Gabin A and Meleförs O (1990) Cleavages in the 5′ region of the *omp*A and *bla* mRNA control stability: Studies with an *E. coli* mutant altering mRNA stability and a novel ribonuclease. EMBO J 9: 2731–2741

Marrs BL (1981) Mobilization of the genes for photosynthesis from *Rhodopseudomonas capsulata* by a promiscuous plasmid. J Bacteriol 146: 1003–1012

Martin CR (1993) Structure, function, and regulation of the chalcone synthase. Int Rev of Cytol 147: 233–284

Masuda S, Matsumoto Y, Nagashima KV, Shimada K, Inoue K, Bauer CE and Matsuura K (1999) Structural and functional analyses of photosynthetic regulatory genes *regA* and *regB* from *Rhodovulum sulfidophilum, Roseobacter denitrificans*, and *Rhodobacter capsulatus*. J Bacteriol 181: 4205–4215

McDermott G, Prince SM, Freer AA, Hawthornthwaite-Lawless AM, Papiz MZ, Cogdell RJ and Isaacs NW (1995) Crystal structure of an integral membrane light-harvesting complex from photosynthetic bacteria. Nature 374: 517–521

Meyer TE (1985) Isolation and characterization of soluble cytochromes, ferredoxins and other chromophore proteins from the halphilic phototrophic bacterium *Ectothiorhodospira halophila*. Biochim Biophys Acta 806: 175–183

Mosley CS, Suzuki JY and Bauer CE (1994) Identification and molecular genetic characterization of a sensor kinase responsible for coordinately regulating light harvesting and reaction center gene expression in response to anaerobiosis. J Bacteriol 176: 7566–7573

Nagashima KVP, Shimada K and Matsuura K (1993) Phylogenetic analysis of photosynthetic genes of *Rhodocyclus gelatinosus*: Possibility of horizontal gene transfer in purple bacteria. Photosynth Res 36: 185–191

Naylor GW, Addlesee HA, Gibson LCD and Hunter CN (1999) The photosynthesis gene cluster of *Rhodobacter sphaeroides*. Photosynth Res 62: 124–139

O'Gara JP, Eraso JM and Kaplan S (1998) A redox-responsive pathway for aerobic regulation of photosynthesis gene expression in *Rhodobacter sphaeroides* 2.4.1. J Bacteriol 175: 2292–2302

Oh J-H, Eraso JM and Kaplan S (2000) Interacting regulator circuits involved in orderly control of photosynthesis gene expression in *Rhodobacter sphaeroides* 2.4.1. J Bacteriol 182: 3081–3087

Pemberton JM, Horne IM and McEwin AG (1998) Regulation of photosynthetic gene expressions in purple bacteria. Microbiology 144: 267–278

Penfold RJ and Pemberton JM (1991) A gene from the photosynthetic gene cluster of *Rhodobacter sphaeroides* induces *trans*-suppression of bacteriochlorophyll and carotenoid levels in *R. sphaeroides* and *R. capsulatus*. Curr Microbiol 23: 259–263

Penfold RJ and Pemberton JM (1994) Sequencing, chromosomal inactivation and functional expression in *E. coli* of *pps*, a gene which represses carotenoid and bacteriochlorophyll synthesis in *Rhodobacter sphaeroides*. J Bacteriol 176: 2869–2876

Petersen C (1992) Control of functional mRNA stability in bacteria: Multiple mechanisms of nucleolytic and non-nucleolytic inactivation. Mol Microbiol 6: 277–282

Phillips-Jones MK and Hunter CN (1994) Cloning and nucleotide sequence of *regA*, a putative response regulator gene of *Rhodobacter sphaeroides*. FEMS Lett 116: 269–275

Ponnampalam SN and Bauer CE (1997) DNA binding

characteristics of CrtJ. A redox-responding repressor of bacteriochlorophyll, carotenoid, and light harvesting-II gene expression in *Rhodobacter capsulatus*. J Biol Chem 272: 18391–18396

Ponnampalam SN, Buggy JJ and Bauer CE (1995) Characterization of an aerobic repressor that coordinately regulates bacteriochlorophyll, carotenoid, and light harvesting-II expression in *Rhodobacter capsulatus*. J Bacteriol 177: 2990–2997

Ponnampalam SN, Elsen S and Bauer CE (1998) Aerobic repression of the *Rhodobacter capsulatus bchC* promoter involves cooperative interactions between CrtJ bound to neighboring palindromes. J Biol Chem 273: 30757–30761

Qian Y and Tabita FR (1996) A global signal transduction system regulates aerobic and anaerobic CO_2 fixation in *Rhodobacter sphaeroides*. J Bacteriol 178: 12–18

Quail PH (1991) Phytochrome: a light-activated molecular switch that regulates plant gene expression. Annu Rev Genet 25: 389–409

Sagesser R (1992) Identifikation und charakterisierung des photosynthese-genclusters von *Rhodospirillum rubrum*. Ph.D. Thesis, Universität Zürich

Schumacher A and Drews G (1979) Effects of light intensity on membrane differentiation in *Rhodopseudomonas capsulata*. Biochim Biophys Acta 501: 417–428

Sganga MW and Bauer CE (1992) Regulatory factors controlling photosynthetic reaction center and light harvesting gene expression in *Rhodobacter capsulatus*. Cell 68: 945–954

Shimada H, Iba K and Takamiya K-I (1992) Blue-light irradiation reduces the expression of *puf* and *puc* operons of *Rhodobacter sphaeroides* under semi-aerobic conditions. Plant Cell Physiol 33: 471–475

Shimada H, Ohta H, Masuda T, Shioi Y and Takamiya K-I (1993) A putative transcription factor binding to the upstream region of the *puf* operon in *Rhodobacter sphaeroides*. FEBS Letters 328: 41–44

Shimada H, Wada T, Handa H, Ohta H, Mizoguchi H, Nishimura K, Masuda T, Shioi Y and Takamiya K-I (1996) A transcription factor with a leucine-zipper motif involved in light- dependent inhibition of expression of the *puf* operon in the photosynthetic bacterium *Rhodobacter sphaeroides*. Plant Cell Physiol 37: 515–522

Suzuki JY, Bollivar DW and Bauer CE (1997) Genetic analysis of the chlorophyll biosynthesis. Ann Review of Genetics 31: 61–89

Taylor DP, Cohen SN, Clark WG, and Marrs BL (1983) Alignment of genetic and restriction maps of the photosynthesis region of the *Rhodopseudomonas capsulata* chromosome by a conjugation-mediated marker rescue technique. J Bacteriol 154: 580–590

Tichy H-V, Oberle' B, Stiehl H, Schlitz E and Drews G (1991) Analysis of the *Rhodobacter capsulatus puc* operon: The *pucC* gene plays a central role in the regulation of LHII (B800-850 complex) expression. EMBO J 10: 2949–2956

Tiwari RP, Reeve WG, Dilworth MJ and Glenn AR (1996) Acid tolerance in *Rhizobium meliloti* strain WSM419 involves a two-component sensor-regulator system. Microbiology 142: 1693–1704

Vichivanives P, Bird TH, Bauer CE and Tabita FR (2000) Multiple regulators and their interactions in vivo and in vitro with the *cbb* regulons of *Rhodobacter capsulatus*. J Mol Biol 300: 1079–1099

Wellington CL and Beatty JT (1989) Promoter mapping and nucleotide sequence of the *bchC* bacteriochlorophyll biosynthesis gene from *Rhodobacter capsulatus*. Gene 83: 251–261

Wellington CL and Beatty JT (1991) Overlapping mRNA transcripts of photosynthesis gene operons in *Rhodobacter capsulatus*. J Bacteriol 173: 1432–1443

Wellington CL, Bauer CE and Beatty JT (1992) Photosynthesis gene superoperons in purple nonsulfur bacteria: The tip of the iceberg? Can J Micro 38: 20–27

Wiessner C, Dunger I and Michel H (1990) Structure and transcription of the genes encoding the B1015 light harvesting complex β and α subunits and the photosynthetic reaction center L, M, and cytochrome *c* subunits from *Rhodopseudomonas viridis*. J Bacteriol 172: 2877–2887

Xiong J, Inoue K and Bauer CE (1998) Tracking molecular evolution of photosynthesis by characterization of a major photosynthesis gene cluster from *Heliobacillus mobilis*. Proc Natl Acad Sci USA 95: 14851–14856

Xiong J, Fisher WM, Inoue K, Nakahara M and Bauer CE (2000) Molecular evidence for the early evolution of photosynthesis. Science 289: 1724–1730

Yang ZY and Bauer CE (1990) *Rhodobacter capsulatus* genes involved in early steps of the bacteriochlorophyll biosynthetic pathway. J Bacteriol 126: 619–629

Yen H-C, Hu NT and Marrs BL (1976) Map of genes for carotenoid and bacteriochlorophyll biosynthesis in *Rhodopseudomonas capsulata*. J Bacteriol 126: 619–629

Yildiz FH, Gest H and Bauer CE (1992) Conservation of the photosynthesis gene cluster in *Rhodospirillum centenum* Mol Microbiol 6: 2683–2691

Young DA, Bauer CE, Williams JC, and Marrs BL (1989) Genetic evidence for superoperonal organization of genes for photosynthetic pigments and pigment-binding proteins in *Rhodobacter capsulatus*. Mol Gen Genet 218: 1–12

Young DA, Rudzik MB and Marrs BL (1992) An overlap between operons involved in carotenoid and bacteriochlorophyll biosynthesis in *Rhodobacter capsulatus*. FEMS Microbiol Lett 95: 213–218

Youvan DC and Ismail S (1985) Light-harvesting II (B800-B850 complex) structural genes from *Rhodopseudomonas capsulata*. Proc Natl Acad Sci USA 82: 58–62

Youvan DC, Bylina EJ, Alberti M, Begusch H and Hearst JE (1984) Nucleotide and deduced polypeptide sequence of the photosynthetic reaction center, B870 antenna, and flanking polypeptides from *Rhodopseudomonas capsulata*. Cell 37: 949–957

Zhu YS and Kaplan S (1985) Effects of light, oxygen and substrates on steady-state levels of mRNA coding for ribulose-1, 5-bisphosphate carboxylase and light harvesting and reaction center polypeptides in *Rhodopseudomonas sphaeroides*. J Bacteriol 162: 925–932

Zhu YS and Hearst JE (1986) Regulation of expression of genes for light-harvesting antenna proteins LH-I and LH-II; reaction center polypeptides RC-L, RC-M, and RC-H; and enzymes of bacteriochlorophyll and carotenoid biosynthesis in *Rhodobacter capsulatus* by light and oxygen. Proc Natl Acad Sci U S A 83: 7613–7617

Zhu YS, Cook DN, Leach F, Armstrong GA, Alberti M and Hearst JE (1986) Oxygen-regulated mRNAs for light-harvesting and reaction center complexes and for bacteriochlorophyll and carotenoid biosynthesis in *Rhodobacter capsulatus* during the shift from anaerobic to aerobic growth. J Bacteriol 168: 1180–1188

Zsebo KM and Hearst JE (1984) Genetic physical mapping of a photosynthetic gene cluster from *R. capsulata*. Cell 37: 937–947

Zuber H, and Cogdell RJ (1995) Structure and organization of purple bacterial antenna complexes. In: Blankenship RE, Madigan MT and Bauer CE (eds) Anoxygenic Photosynthetic Bacteria, pp 315–348. Kluwer Academic Publishers, Dordrecht

Zucconi AP and Beatty JT (1988) Post-transcriptional regulation by light of the steady-state levels of mature B880-850 light harvesting complexes in *Rhodobacter capsulatus*. J Bacteriol 170: 877–882

Chapter 5

Redox Regulation of Photosynthetic Genes

Gerhard Link*
Ruhr-Universität Bochum, Pflanzliche Zellphysiologie und Molekularbiologie, Universitätsstr. 150, D-44780 Bochum, Germany

Summary

Chloroplasts are the sites of photosynthesis and several other major biosynthetic pathways. A large number of proteins, comprising of products from both chloroplast and nuclear genes, participates in the biogenesis and function of the mature organelle. The dual genetic origin of the photosynthetic apparatus and other chloroplastidic supramolecular complexes requires mechanisms of intracellular integration, involving mutual exchange of information between the chloroplast and nucleo-cytoplasmic compartments. An important part of the signaling network that integrates cellular gene expression is the functional two-way connection between the photosynthetic apparatus and the gene expression machineries, both inside and outside the chloroplast. Apart from its primary role in light energy capture, photosynthetic electron transport acts as a source of redox signal(s) that initiates signal transduction events and gene expression responses at various levels and intracellular locations. The photosynthetic apparatus can adapt to environmental changes, including those in light quality and intensity,

*Email: Gerhard.Link@ruhr-uni-bochum.de

Eva-Mari Aro and Bertil Andersson (eds): Regulation of Photosynthesis, pp. 85–107.

with associated changes in redox state of electron carriers. Photostress is known to accelerate deleterious effects that are part of normal oxygenic photosynthesis, i.e. degradation and replenishment of (reaction center) proteins. This inherent feature of photosynthesis requires a balanced supply of gene products in response to redox state. Accumulating evidence suggests that both transcriptional and posttranscriptional steps in gene expression are subject to control by photosynthetic electron flow, and gene expression both inside and outside the chloroplast is affected. Identified key mechanisms that provide regulatory feed-back connections between photosynthetic electron transport and gene expression are the phosphorylation and SH-group redox control of proteins. Despite this unifying picture, it has become increasingly clear, however, that the details can be quite variable, suggesting a possible existence of multiple and/or split signal transduction pathways.

I. Introduction

A. The Dual Role of Photosynthetic Electron Transport

Redox (reduction-oxidation) chemistry plays a fundamental role in living organisms. According to the Nernst equation (Fig. 1), the redox potential of a cellular electron carrier is given by the relative concentrations of its reduced versus oxidized forms. Metabolic changes in their concentrations will shift the redox potential, which can sequentially affect the redox state of a second, third, fourth etc. electron carrier connected in series. The power of biological electron transfer chains is highlighted by the photosynthetic apparatus, which consists of four multisubunit pigment-protein complexes that act together in the photoproduction of oxygen, ATP, and NADPH—the basis for autotrophic plant growth and, in general, most of life on earth (for review, see Simpson and von Wettstein, 1989; Bogorad and Vasil, 1991; Pakrasi, 1995; Allen and Williams 1998).

In addition to capture of light energy, a second function can be assigned to photosynthetic electrochemistry. The efficiency of photosynthesis is dependent on the environmental conditions (most notably light intensity, quality and temperature), leading to both short and long-term responses. Resulting changes in the steady-state levels of reducing equivalents can affect the properties of many chloroplast proteins, as has been well-established e.g. for the Calvin cycle enzymes with known redox-regulated activity (Buchanan, 1991; Scheibe, 1991; Schürmann, 1995). Likewise, evidence has become available that plant gene expression—both inside and outside the organelle—can be modulated according to the energy state of the photosynthetic electron transport chain (Allen, 1993; Vener et al., 1998; Foyer and Noctor, 1999). Hence, in this respect the photosynthetic apparatus acts like a (photo)sensor, which by way of intermediate electron carriers initiates signal transduction pathways for downstream metabolic and/or gene regulatory responses.

B. Photosynthetic Redox Regulation of Gene Expression: How, When and Where?

Coupling of photosynthetic electron transport with mechanisms involved in both cellular maintenance and differentiation appears useful and necessary for several reasons: (i) The photosynthetic apparatus is

$$E = E_m + \frac{RT}{nF} \cdot \ln \frac{[ox]}{[red]}$$

E	redox potential
E_m	midpoint (50%) potential
R	gas constant
T	abs. temperature
n	number of electrons
F	Faraday constant
[ox]	conc. of oxidized form
[red]	conc. of reduced form

Fig. 1. Concentration dependence of the redox potential.

Abbreviations: APX – ascorbate peroxidase; bromanil – tetrabromo-1,4-benzoquinone; CCCP – carbonal cyanide-m-chlorophenylhydrazone; DBMIB – 2,5-dibromo-3-methyl-6-isopropyl-1,4-benzoquinone; DCMU – 3-(3,4-dichlorophenyl)-1,1-dimethyl urea; DCPIP – 2,6-dichlorophenolindophenol; DTT – dithiothreitol; FNR – ferredoxin-$NADP^+$ reductase; GBF – G-box binding factor; GR – glutathione reductase; GSH – reduced dithiol form of glutathione; GSSG – oxidized disulfide form of glutathione; LHC – light-harvesting complex; NEP – nuclear-encoded plastid RNA polymerase; PEP – plastid-encoded RNA polymerase; PS I – Photosystem I; PS II – Photosystem II; PTK – plastid transcription kinase; ROI – reactive oxygen intermediate; SLF – sigma-like factor; SOD – superoxide dismutase

formed de-novo during the plant development and is restricted to only a fraction of the cells that make up the plant body (Boyer et al., 1989). (ii) A close coordination both inside and outside the organelle is required during the chloroplast biogenesis. Both structural and regulatory genes are involved in the assembly of the photosynthetic machinery, many of which are nuclear genes for proteins that are imported into the chloroplast. (iii) Even in fully functional chloroplasts the efficiency of photosynthetic electron transport can vary considerably, depending on the species, age, cell position, and environmental conditions. (iv) Moreover, there is a constant requirement for a balanced supply of newly-synthesized photosynthetic proteins due to the inherent instability of reaction center proteins during active photosynthesis and even more so, under photostress conditions (photoinhibition). All these complex tasks could be best achieved if both the chloroplast and nucleo-cytoplasmic genetic systems would be 'informed' of the state of photosynthesis in an integrated manner. Redox poise would seem to be a suitable intracellular signal to play this role.

Various aspects of photosynthetic redox chemistry have been reviewed in depth previously (see Hipkins and Baker, 1986) and also in other chapters of this book. The same holds true for the molecular biology of 'photosynthetic' genes, i.e. both the chloroplast-localized and nuclear sets of genes that encode components of the photosynthetic apparatus (Bogorad and Vasil, 1991; Herrmann et al., 1992; Rochaix, 1995). In this chapter, attempts will be made to bring these two areas together by focusing on the question of what might be the mechanism(s) responsible for redox regulation of photosynthetic gene expression. As both chloroplast and nuclear genes participate in the biogenesis and maintenance of the photosynthetic machinery (Goldschmidt-Clermont, 1998), we will look at both the plastid and nucleo-cytoplasmic compartments. Pertinent questions to be discussed will include: Which steps during the flow of genetic information from the DNA to the mature functional protein are controlled by photosynthetic redox poise? Is there a single one-step control mechanism, or are multiple stages of gene expression affected in the same or different ways? Where are the 'outlets' of the photosynthetic apparatus that act as redox-sensors and signal transmitters to the gene expression systems? What are the components and mechanisms that receive the signal and initiate gene expression responses? How closely is photosynthetic electron transport coupled to gene expression, and what are the connecting signal transduction mechanisms? In going through the various aspects of the topic of this chapter, we will probably come to the conclusion that many of these questions have just begun to be addressed experimentally. Nevertheless, several general lines have already become visible and can be expected to have ongoing impact on both photosynthesis research and chloroplast molecular biology.

C. Transcriptional versus Posttranscriptional (Redox) Regulation

There has long been a debate on whether chloroplast gene expression in general is regulated 'transcriptionally' or 'posttranscriptionally'. The former of these two terms relates to the process of transcription, i.e. de novo synthesis (not: accumulation) of RNA; the latter refers to everything beyond the transcription stage. At the RNA level, posttranscriptional regulation includes processing, modification and degradation of the newly-synthesized transcription products (transcripts). At the protein level, at least as many steps are involved, ranging from (initiation of) translation, i.e. the mRNA-directed de novo synthesis of (precursor) polypeptides, to protein processing and posttranslational modification(s) into mature functional gene products.

Considering the complex gene expression machinery of the chloroplast (Table 1), it has become increasingly clear that often it is neither possible nor useful to assign a single regulatory event. Instead, there seems to be a close physical and functional interaction among the various submechanisms along the gene expression pathway. For instance, an RNA molecule that is not properly processed or edited (Sugiura et al., 1998) may no longer be an efficient template for protein synthesis. Such 'non-functional' transcripts may be present in the same quantities, and even may have the same size, as their functional counterparts that are generated in different developmental or environmental contexts (Hirose et al., 1996). As a result, using the common blot transfer techniques (northern, dot blot, western), one would detect different amounts of protein, but not RNA, in these two hypothetical situations; yet the conclusion that this is 'translational' control would obviously be incomplete, if not misleading.

Even under circumstances where the steady-state concentrations of a specific transcript remains virtually constant, the expression of that particular

Table 1. Regulatory mechanisms in chloroplast gene expression

Stage	Mechanism	Variables
(A) DNA		
	Gene dosis (copy number)	Plastids per cell, DNA molecules per plastid, genes per DNA molecule
	DNA superstructure, topology, accessibility	Degree of condensation, packing conformation, local superstructure, membrane association, DNA modification (methylation)
(B) RNA synthesis (transcription)		
	Promoter architecture	Single promoter for single gene, multiple (alternative) promoters for single gene, single promoter for several genes (operon), closely spaced promoters for overlapping or divergent genes (promoter occlusion) multiple promoter elements
	Promotor recognition, binding, initiation	Promoter strength, differential usage of promoter elements, multiple RNA polymerases, multiple initiation factors
	Elongation, termination	Elongation factors, termination factors, abortive transcription, antitermination
(C) RNA maturation		
	Polycistronic primary transcripts	Intergenic cutting (trimming)
	Internal RNA processing	Intron class I - III splicing (cis-splicing)
		Transon splicing (trans-splicing)
		Twintron (intron within intron) splicing
		Alternative splicing
	Terminal (5′/3′) processing	Exo- and endonucleases
	Terminal modifications	3′ polyadenylation
	Internal modifications	RNA editing (coding region, start/stop codons)
	RNA stability	Sequence elements and factors affecting RNA turnover
	Antisense RNA	Interference with RNA function
(D) Protein synthesis (translation) / posttranslational regulation		
	Mono-, poly(ribo)somes	
	Free/membrane-bound polysomes	
	Translation efficiency	Precurser transcripts, mature mRNAs, translation arrest
	Translation factors	Initiation, elongation, termination factors
	Protein modifications	Acetylation, methylation, phosphorylation, glycosylation, palmitoylation
	Processing	Proteolysis, precursor removal, protein splicing (?)
	Assembly	Proteins, pigments

gene may be highly regulated, e.g. if changes in transcription rates are compensated by changes in RNA stability (Shiina et al. 1998). Despite these difficulties, it seems appropriate to operationally assign observed changes in RNA or protein concentration to '(post-)transcriptional' or 'translational' control, yet without excluding mechanisms at other control levels that were not investigated rigorously.

As more and more comparative data has become available, another relevant point has emerged: one and the same gene seems to be regulated quite differently in different organisms, developmental stages, and environmental conditions. Considering this, many of the conflicting conclusions reached by different groups regarding the control level(s) of light-regulated plastid gene expression can be resolved. For instance, it is not unexpected to find regulatory details in studies on a single-celled green alga like *Chlamydomonas reinhardtii* to be different from those found in tissues of multi-cellular higher plants. The same holds true, if one compares light/

dark regulation in young seedlings to fully mature, dark-adapted and re-illuminated plants. Furthermore, the gene expression program in a dicotyledonous seedling will not necessarily be the same as that along the axis of a monocotyledonous leaf with its basal meristem. As molecular biologists have become increasingly aware of these differences in basic botany, it is widely accepted today that gene regulation is quite flexible in a specific cellular and environmental context. One of the best examples is perhaps the differential gene expression program in the mesophyll and bundle sheath cells of the C4-plants such as maize (Hatch, 1992). There is not only an opposite cell-specific expression of the genes for the principal 'C4-type' carbon dioxide fixing enzymes PEPCase and Rubisco (for reviews, see Langdale and Nelson, 1991), but also photosynthetic genes for the PS II proteins are differentially expressed within the two photosynthetic cell types (Meierhoff and Westhoff, 1993).

In concluding this part of the review, it is important to be aware of the extreme complexity and interdependent regulation of the gene expression system(s) for photosynthetic proteins, and to take into consideration the cellular and environmental context. In the next section we will discuss recent work showing that (redox) regulation of the photosynthetic genes is evident both inside and outside the chloroplast, and it involves both transcriptional and posttranscriptional regulatory mechanisms.

II. Redox Regulation of Nuclear Genes for Photosynthetic Proteins

A. The Chloroplast Redox Signal—in Brief

A number of groups have investigated nuclear gene expression under conditions where the photosynthetic apparatus is severely affected by physical (heat, high irradiation) or chemical (70S ribosome-specific antibiotics, norflurazon) treatments (for review see Taylor, 1989), or by genetic defects in chloroplast traits (Susek et al., 1993; Hess et al., 1994; Lopez-Juez et al., 1998). Typically, it was found for multi-component systems such as Rubisco and the thylakoid protein complexes that the absence of chloroplast-encoded constituents correlated with a negative effect on the expression of their nuclear-encoded counterparts. This has led to the concept of a chloroplast 'signal' that somehow informs the nuclear system and thus brings about the coordination of gene expression within the two compartments (Taylor, 1989; Goldschmidt-Clermont, 1998). Elucidation of both the identity of this signal and the exact transduction mechanisms across the compartment borders are, however, just started (Kropat et al., 1997). Clearly, there could be various participating metabolites and macromolecules, including redox carrier systems.

One limitation of using antibiotics (or other 'general' inhibitors) to eliminate plastid functions is the lack of a specific causal relationship between the photosynthetic electron transport and nuclear gene expression. If plastid protein synthesis, in general, is inhibited by a 70S ribosome-specific inhibitor such as chloramphenicol or lincomycin, this has dramatic effects on the organelle and it would thus seem likely that metabolic feedback inhibition takes place if proper transport across the envelope of a damaged plastid is no longer possible. What would be needed, therefore, is a controlled up- or down-regulation of the photosynthetic electron transport without general destruction of plastid ultrastructure, metabolism and gene expression. Several approaches have been taken to test the dependence of nuclear gene expression mechanisms on photosynthetic electron transport activity. These experiments typically involve use of photosynthetically active radiation of variable intensity (or spectral quality) in combination with specific inhibitors of photosynthetic electron transport. The most commonly used inhibitors are DCMU and DBMIB, i.e. chemicals that interfere with the flow of electrons through the plastoquinone pool (Trebst, 1980). Effects of these compounds on gene expression can thus be taken as an operational criteria for photosynthetic electron transport being involved.

B. Light Intensity Regulation of Nuclear Gene Expression

1. Transcriptional Control during Photoacclimation

Many plant and algal cells are known to compensate for changes in photon flux density by altering the abundance of chlorophyll and photosynthetic proteins in a physiological process known as acclimation. Among the most dramatic changes are perhaps those in the LHCII proteins, i.e. the major chlorophyll *a/b* binding light harvesting apoproteins associated with PS II, which are more abundant under conditions of low growth irradiance than at higher light intensity (Green et al., 1991). The LHCII apoproteins are encoded by the nuclear *cab* gene family, consisting of members that are expressed in a light-responsive manner in different plant species and at various developmental stages (Cashmore, 1984). Photoacclimation can be distinguished, however, from light-induced developmental processes (such as greening during seedling development) both by the criteria of reversibility and by its presence in fully differentiated cells.

The unicellular green algae *Dunaliella tertiolecta* (Escoubas et al., 1995) and *D. salina* (Maxwell et al., 1995) were chosen for studies of *cab* gene expression during photoacclimation because of their high degree of physiological plasticity and several technically useful features for in vivo labeling experiments. It was found that the light intensity-related changes in LHCII apoprotein concentration were reflected by altered *cab* transcript abundance in this system. Based on nuclear run-on transcription assays and measurements of *cab* RNA stability, the authors concluded that the *cab* gene expression in *Dunaliella* seems to be mainly controlled at the level of transcription under the experimental conditions used. Interestingly, the effects of varying light intensity could partially be mimicked by the presence of inhibitors that affect the redox state of the photosynthetic plastoquinone pool. Under high-light conditions, appropriate sublethal concentrations of DCMU in the culture medium resulted in an increase in the total amount of *cab* transcripts, thus resembling the situation where a high-light culture (in the absence of DCMU) was transferred to low-light conditions. Conversely, in the presence of DBMIB in a low-light culture, a decreased steady-state concentration of *cab* transcripts was observed. In contrast to the effects of DCMU and DBMIB, neither an inhibitor of water splitting (hydroxylamine) nor uncouplers of photophosphorylation (CCCP and methylamine) were found to have any effect on the cellular chlorophyll-*a* content. As DCMU inhibits plastoquinone reduction, and DBMIB prevents oxidation of plastoquinol, these data are consistent with the notion that this redox carrier system is part of a signal transduction mechanism that accounts for the *cab* transcript fluctuations in response to altered photosynthetic electron flow during photoacclimation (Escoubas et al., 1995; Durnford and Falkowski, 1997). Although DBMIB can, in principle, also inhibit the mitochondrial electron transport, Escoubas et al. (1995) concluded from the control experiments with dark-grown cells that the mitochondrial inhibition was not involved in the signal transduction in their studies.

In addition to suggesting a role for the plastoquinone pool in redox signaling, these studies also addressed the question of possible downstream signal transduction mechanisms. Earlier work had established the phosphorylation of photosynthetic proteins, including LHCII, D1 and other PS II proteins, as well as the existence of redox-responsive photosynthesis-related protein kinase(s) that could play a role in signal transduction (for recent reviews see Gal et al., 1997; Vener et al., 1998). Using inhibitors of cytosolic phosphatase (okadaic acid, microcystin-LR, tautomycin), Escoubas et al. (1995) noted a partial inhibition of the normal photoacclimatory response in *Dunaliella*, which could mean the existence of a phosphorylation cascade involving component(s) within the nucleo-cytoplasmic compartment. Consistent with this notion, the authors could show a high light intensity-dependent formation of DNA-protein complexes within the 5′ upstream region of a cloned *cab* (*LhcII*) gene from *Dunaliella*. This DNA region contains a sequence element that resembles the G-box motif which is present in many light-regulated plant promoters (Menkens et al., 1995). The cognate G-box binding factors (GBFs) are well-characterized nuclear transcription factors of the bZIP (basic domain/leucine zipper) type (Meshi and Iwabushi, 1995). Phosphorylation has been shown to affect both the binding activity and intracellular (cytoplasmic versus nuclear) location of factors in this group (Harter et al., 1994). It is thus possible that a phosphorylated GBF-like factor acts as a negative transcription regulator under high-light conditions, and the unphosphorylated factor as a positive regulator under low-light conditions.

2. A Working Model and Open Questions

Molecular studies on photosynthetic redox effects on nuclear gene transcription in *Dunaliella* (Escoubas et al., 1995; Maxwell et al., 1995) has provided important initial observations and a working model on how the complex intracellular signal transduction might operate (Durnford and Falkowski, 1997) (Fig. 2). It should be emphasized, however, that the details of this model await to be further addressed

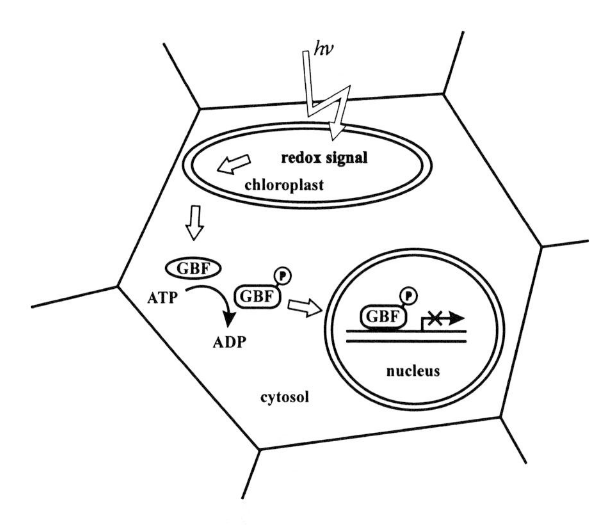

Fig. 2. Regulation of nuclear gene transcription by photosynthetic redox state during photoacclimation in the green alga *Dunaliella tertiolecta* (Escoubas et al., 1995; Durnford and Falkowski, 1997). Only the situation following transfer of the cells to high light is shown. Under these conditions (reduced plastoquinone pool), a redox signal is generated, which may involve the LHCII kinase or another redox-regulated protein kinase and a downstream phosphorylation cascade. The signal is then transmitted by unknown mechanisms across the envelope. One subsequent element in the signal transduction chain is the phosphorylation of a cytoplasmically localized form of a nuclear transcription factor. This factor, which seems to be a member of the GBF class (Meshi and Iwabushi, 1995), then migrates to the nucleus, where it binds tightly to G-box promoter elements (Menkens et al., 1995), acting as a negative transcription regulator of *cab* (*LhcII*) genes.

experimentally: (i) The proposed mechanism connecting the plastoquinone pool directly to the LHCII kinase (Escoubas et al., 1995) is conceivable, but there could be other explanations. A number of distinct kinase activities are known to be associated with thylakoid preparations (Gal et al., 1997; Vener et al., 1998), any or all of which could be involved in light intensity modulation of signal transduction (Rintamäki et al., 1997). (ii) Modulation of gene expression by DCMU and DBMIB points to a role for electron flow through the plastoquinone pool. This does not necessarily mean, however, that this part of the photosynthetic electron transport chain is the direct 'outlet' for a redox signal. Recent evidence supports the role of reducing equivalents generated by PS I that could directly link the photosynthetic apparatus to signal transduction (Carlberg et al., 1999). (iii) How the signal transduction across the chloroplast envelope might operate, remains to be established. (iv) Phosphorylation/dephosphorylation is known to play a role in numerous intracellular signaling pathways (Hunter, 1995). In experiments using protein kinase and/or phosphatase inhibitors in vivo one has to take into consideration that these chemicals might inhibit general metabolic pathways and could have pleiotropic effects. It thus seems important to show that the expression of one particular group of genes (e.g. nuclear-encoded photosynthetic genes) responds to the externally applied drugs, whereas other genes remain unaffected.

3. Posttranscriptional Control of Nuclear Gene Expression

One question that seems pertinent to ask is, whether or not the photosynthetic redox regulation of nuclear genes is entirely transcriptional, or whether posttranscriptional mechanisms are involved. This was addressed in work using transgenic tobacco plants that contain the Ferredoxin-1 (*Fed-1*) gene from pea (Petracek et al., 1997; 1998). It was shown that *Fed-1* transcripts accumulated in a light-dependent manner, i.e. following re-illumination of dark-adapted plants using (non-stress) growth-light conditions, and this accumulation was inhibited by DCMU. *Fed-1* transcripts were found to be polysome-associated in the re-illuminated control plants, but they were mostly found within the non-polysomal fraction in dark-adapted plants and in re-illuminated plants exposed to DCMU (Petracek et al., 1997). In an elegant approach using a repressible promoter system for controlled down-regulation of *Fed-1* transcription, evidence was provided that *Fed-1* mRNA has a reduced half-life both in the dark and in the presence of DCMU in the light (Petracek et al., 1998). Together, this work demonstrates that changes in photosynthetic electron transport can have dramatic effects on posttranscriptional steps in the expression of nuclear genes for photosynthetic proteins. Although the details of the signal transduction pathway(s) involved remain to be established, it is already clear that a short sequence element within the 5′ untranslated region of *Fed-1* mRNA is involved (Dickey et al., 1998).

4. Nuclear Gene Regulation under Photostress Conditions

Photoacclimation, i.e. the dynamic regulation of the antenna size, is an important mechanism to adjust the photosynthetic electron transport according to the environmental conditions at low and 'normal' growth light intensities. In addition, photoacclimation can be viewed as part of a protective system, including also other mechanisms such as thermal dissipation, which together prevent that too much absorbed light energy is transferred to the reaction centers. At excessive light intensities, particularly if accompanied by other adverse stress factors such as high or low temperature, drought, or low carbon dioxide supply, the capacity of these protective mechanisms is exceeded, causing symptoms of light stress (Barber and Andersson, 1992; Aro et al., 1993). The main target of light stress is PS II with its light-driven breakdown-repair cycle that involves degradation and subsequent replacement of the reaction center protein D1, occurring even under normal growth-light conditions (Mattoo et al., 1984). If the rate of photodamage to PS II exceeds the rate of resynthesis and repair, as e.g. under high-light conditions, a decrease in photochemical activity of PS II can be observed, which is designated photoinhibition.

Although PS II has been identified as an important target for light stress, it has become increasingly clear that photodamage can occur in other parts of the photosynthetic apparatus as well, most notably at PS I (Sonoike et al., 1997; Tjus et al., 1998). The reducing side of PS I is a major source of photodamage, since molecular oxygen can serve as an electron acceptor in a reaction mainly mediated by ferredoxin (Mehler reaction) and this photoreduction of O_2 yields reactive oxygen radical anions and H_2O_2. These and other reactive oxygen intermediates (ROIs) can cause considerable damage to membrane lipids, induce base mutations and DNA strand breakage, as well as protein degradation. ROIs are normally scavenged by effective detoxification systems of the chloroplast, including superoxide dismutases (SOD), ascorbate peroxidase (APX) and other enzymes of the ascorbate-glutathione cycle (Bowler et al., 1992; Noctor and Foyer, 1998). However, under conditions of excessive electron transport in combination with other stress factors (high or low temperature, drought or salinity), the capacity of the antioxidative defense mechanisms is exceeded, resulting in oxidative stress symptoms.

Photooxidative stress is not restricted to the chloroplast but it also affects the cytosolic compartment which contains its own defense system against ROIs. Again, a number of enzymes are involved, including cytosolic forms of ascorbate peroxidase, catalase, dehydroascorbate reductases, glutathione reductase and superoxide dismutases (Bowler et al., 1992; Noctor and Foyer, 1998). Karpinski et al. (1997) addressed the question of whether the photosynthetic electron transport has an effect on the expression of the nuclear genes for this cytosolic defense system. The approach taken was to expose *Arabidopsis* plants to either normal growth light or to excessive light intensity causing photoinhibition, both in the absence or presence of DCMU or DBMIB. Using the cloned cDNAs for the cytosolic defense proteins as probes, the steady-state concentrations of their transcripts were assessed using total RNA from leaves or leaf discs that had been subjected to various light and inhibitor treatments. The authors could provide conclusive evidence that the levels of transcripts for cytosolic ascorbate peroxidases (APX1 and APX2) increase within 15 min after exposure of leaves to high light. Using leaf discs incubated with the photosynthetic electron transport inhibitors, it was found that DBMIB had a stimulatory effect on both APX1 and APX2 transcripts, either under low-light or high-light conditions, whereas DCMU treatments resulted in constant or even reduced transcript concentrations. Transcripts of a third ascorbate peroxidase gene *APX3* remained constant under all light conditions and inhibitor treatments that were tested (Karpinski et al., 1997). It was concluded that the redox state of the plastoquinone pool initiates signaling mechanisms that result in the induction of the cytosolic defense system. In a recent extension of this work, these findings were further substantiated (Karpinski et al., 1999). Using partial high-light exposure of *Arabidopsis* plants, it was shown that the cytoplasmic defense mechanisms to excessive excitation energy are systemic responses, i.e. they take place even in leaves that were not exposed to light stress. Furthermore, the systemic response was found to reflect the redox status of the plastoquinone pool, and H_2O_2 was implicated as a systemic signal (Karpinski et al., 1999; Foyer and Noctor, 1999).

5. Glutathione as a Redox Regulator

One key result in the work of Karpinski et al. (1997) is that the induction of cytosolic defense gene expression was abolished upon treatment of leaf disks with reduced glutathione (GSH). The intracellular ratio of GSH versus GSSG (the oxidized disulfide form of glutathione) decreased significantly during the photostress period, and it increased during the post-stress recovery phase when plants were transferred back to growth light conditions. Furthermore, transcript levels for cytosolic glutathione reductase (GOR2) increased significantly during the latter phase. These data support the idea that the glutathione redox state possibly acts as an intracellular signal in this stress response. This is consistent with the previous results obtained with transgenic tobacco (Broadbent et al., 1995) and poplar explants (Foyer et al., 1995), in which a foreign (*gor*) gene for glutathione reductase (GR) was overexpressed. Although the antioxidant capacity of the transgenic lines differed considerable from each other, a number of these lines showed both elevated glutathione pool sizes and enhanced GR activity as well as revealed greater tolerance to oxidative stress.

As in other organisms, glutathione is the most abundant non-protein thiol compound in plant cells (Bergmann and Rennenberg, 1993), and it can be considered to have a universal role as a main antioxidant molecule and redox buffer system (Meister, 1995). Moreover, in concert with other biothiols it is involved in the reversible reduction and oxidation of proteins through their disulfide (S-S) and sulfhydryl (SH) groups, thus influencing their activity via changes in redox state. Thiol redox reactions can hence be considered as a key mechanism for signaling both within single cellular compartments and across compartmental boundaries (Garcia-Olmedo et al., 1994). Although S-S/SH exchange reactions differ substantially from phosphorylation/dephosphorylation and other reversible protein modifications, both sets of reactions act as switches that can positively or negatively modulate enzymatic activity and can participate in interconnected cascade-type signaling chains. Indeed, both SH-group reactions and (de)phosphorylation can be part of one and the same regulatory step in a signal transduction pathway, operating either synergistically or antagonistically. We will see examples of this when discussing the redox regulation of photosynthetic genes inside the chloroplast.

III. Redox Regulation of Gene Expression Inside the Chloroplast

A. Posttranscriptional Regulation of Chloroplast Gene Expression

Photoregulation is a key issue in chloroplast molecular biology, and there have been numerous reports on light effects at various stages in plastid gene expression (for review see Gruissem, 1989; Link, 1991; Mullet, 1993). By using appropriate (red/far-red or blue light) irradiation programs, a number of these effects have been traced to the phytochrome and/or cryptochrome photosensory systems residing outside the chloroplast (Batschauer, 1998), rather than to changes in photosynthetic electron flow. In vivo studies with (photosynthetically active) white light usually require additional criteria in order to be decided whether the observed effects at the RNA or protein level are specific responses to photosynthetic redox poise, or whether they merely reflect a general metabolic or developmental state initiated by light. As in the case of nuclear gene regulation (Escoubas et al., 1995; Karpinski et al., 1997; Petracek et al., 1997), the use of specific inhibitors of photosynthetic electron transport can provide clues as to the existence of redox-regulated gene expression inside the chloroplast. In addition, as the photosynthetic apparatus and the organellar gene expression system both operate in close physical proximity, with many of their constituents being attached to the thylakoid membranes (Bogorad, 1991), complementary information can be obtained by using isolated chloroplasts, thylakoid preparations, and solubilized in vitro systems.

Plastid gene expression is thought to be primarily regulated posttranscriptionally via RNA stability and translation, although this view is based on results obtained by techniques that do not necessarily detect functionally important details from other stages of gene expression. Despite these notes of caution, however, clear-cut evidence is now available for translational control of plastid gene expression both in algae and higher plants. For instance, accumulation of specific proteins can increase dramatically after illumination, and this is often not reflected by a similar magnitude of changes in the steady-state concentrations of the corresponding transcripts (for review see Gillham et al., 1994; Mayfield et al., 1995; Rochaix, 1995; Danon, 1997; Stern et al., 1997).

1. Translational Control by a Redox-responsive RNA Binding Complex

The role of photoregulation of chloroplast translation was addressed by Danon and Mayfield (1991; 1994a,b) in the studies done with *Chlamydomonas*. They purified a multi-component protein complex that specifically binds to the 5′-noncoding region of *psbA* RNA, i.e. the messenger RNA for the D1 reaction center protein of PS II. A 47-kDa polypeptide (RB47) was assigned as the major binding protein, although other constituents were implicated to be functionally involved in determining the specificity and intensity of RNA binding. This complex was equally abundant in dark-grown and in light-grown cells, but its RNA binding activity was higher in preparations obtained under light-grown conditions, suggesting that protein activation/inactivation rather than differential gene expression might be involved (Danon and Mayfield, 1991).

Important clues as to what mechanism(s) might be involved in the modulation of the *psbA* 5′-mRNA binding complex came from the observation that the binding activity depends on the phosphorylation state of a component in the protein complex. It was shown that a serine/threonine-type protein kinase is responsible for phosphorylation of a 60 kDa polypeptide (RB60) in the binding complex. Interestingly, unlike most of the other known serine/threonine kinases, this activity was found to utilize ADP (rather than ATP) as the phosphate donor. Based on the observation that phosphorylation of the purified complex inhibited its RNA binding activity, it was proposed that this might reflect in vivo regulation by changes in ADP/ATP ratios in dark-grown versus light-grown cells. A relatively high ADP level in the dark could inhibit RNA binding and hence *psbA* mRNA translation. Under conditions of photosynthetic ATP production in the light, inhibition of the RNA binding activity would seem to be released allowing translation (Danon and Mayfield, 1994a).

Although this work provided strong arguments for the phosphorylation state being an important determinant of *psbA* 5′ mRNA binding in *Chlamydomonas*, it did not exclude the possibility that other mechanisms may also be involved in RNA binding and chloroplast translation. This was indeed shown to be the case by Danon and Mayfield (1994b) in their studies on redox regulation of the purified RNA binding complex. The oxidant dithionitrobenzoic acid (DTNB) completely abolished the RNA binding activity that could be restored by the dithiol reductant DTT, but not by the monothiol β-mercaptoethanol. Further biochemical experiments indicated that reactivation of the RNA binding activity resulted from the breakage of a disulfide bond and formation of vicinal sulfhydryl groups. Reduced thioredoxin from *E. coli* was found to enhance the reactivation of *psbA* mRNA binding beyond the level obtained with DTT alone, suggesting that chloroplast thioredoxin might be a factor involved in redox regulation of the binding complex in vivo.

To test the possible role of photosynthetic redox poise in vivo, Danon and Mayfield (1994b) took advantage of a *Chlamydomonas* mutant strain (cc703) that is deficient in the PS I reaction center. As PS I is the primary reducer of ferredoxin, and hence of thioredoxin, it was reasoned that the *psbA* 5′ binding complex isolated from the mutant should have reduced RNA binding activity. Indeed, gel mobility shift assays showed substantially decreased RNA-protein complex formation in preparations from the light-grown mutant cells compared to those from the wild-type. Furthermore, experiments using in vivo pulse labeling of proteins, followed by isolation of thylakoids and SDS-PAGE, indicated that the synthesis of D1 and other proteins (D2, CP47 and CP43) known to be light-regulated in the wild-type was affected in the cc703 mutant. In contrast, synthesis of the (non-photoregulated) alpha and beta subunits of the ATPase remained unaffected. Transcript levels for all proteins that were investigated remained at the wild-type level, suggesting that chloroplast translation is a key step affected in the mutant (Danon and Mayfield, 1994b).

Together, these data support the role of redox poise in the posttranscriptional regulation of chloroplast genes for photosynthetic proteins and provides evidence of the reducing power from PS I being transmitted via ferredoxin, ferredoxin-thioredoxin reductase, and thioredoxin to the organellar gene expression system. Clearly, further clarifying the details involved would require both genetic information from mutants defective in the ferredoxin-thioredoxin system and techniques to demonstrate physical and functional interactions of redox signaling proteins with the RNA binding complex. Other interesting points include the questions of whether the SH-group redox regulation described for the *psbA* 5′ mRNA binding complex is unique to this gene (as compared to other photosynthetic and non-photosynthetic chloroplast genes), and whether or not the same mechanism as described for *Chlamydomonas* operates in higher plant systems as well. Despite these ongoing questions, the work by Danon and Mayfield provides an attractive model of how the synthesis of D1 (and perhaps other photosynthetic proteins) might be regulated in response to changing light conditions (Fig. 3).

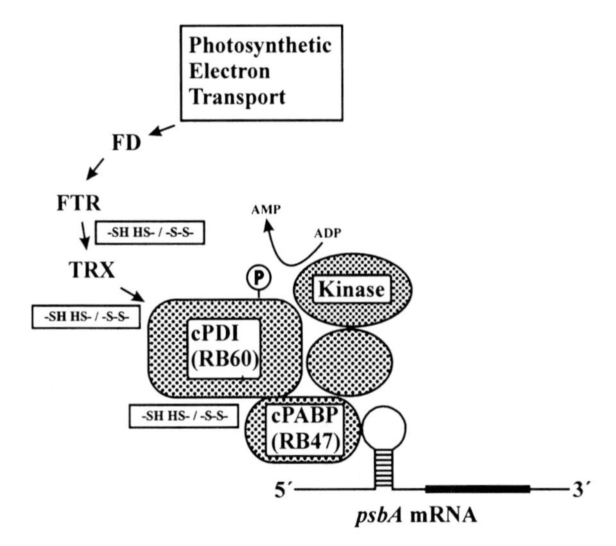

Fig. 3. Scheme summarizing the translational activation model of the light-controlled *psbA* gene expression in *Chlamydomonas* (Mayfield et al., 1995; Danon, 1997). As a result of photosynthetic electron flow through PS II and PS I, reducing equivalents are generated, which can be transferred from ferredoxin (FD) via ferredoxin-thioredoxin reductase (FTR) and thioredoxin (TRX) to redox-responsive target proteins in a sequence of thiol/disulfide exchange reactions (indicated by boxed symbols). The redox-responsive protein disulfide isomerase (cPDI, RB60) is part of a multimeric complex that binds to the 5′ non-coding region of the mature *psbA* transcript, i.e. the messenger RNA for the D1 reaction center protein of PS II. Within this complex, cPDI transmits the SH-group redox signal to an RNA-binding protein that resembles poly(A)-binding proteins (RB47, cPABP). RB47 confers strong RNA binding on the complex when it is in the reduced (thiol) form in the light, but not when it is in the oxidized (disulfide) form thought to prevail in the dark. Binding of the protein complex to sequence and structural elements within the *psbA* 5′ region is required for efficient translation. As indicated, the binding complex contains additional proteins, including an ADP-dependent protein kinase (Kinase). The latter can phosphorylate cPDI on a threonine residue, which results in a decrease in RNA binding activity. As chloroplast ADP levels are lower in the light than in the dark, it seems that phosphorylation and SH-group redox control act synergistically in this system, i.e. they both favor RNA binding and translation in the light, and they inhibit these processes in the dark.

In subsequent work of Mayfield's group, cDNAs representing the nuclear genes for two components of the 5′-RNA binding complex, RB60 and RB47, were cloned, sequenced and used for overexpression of the recombinant proteins. Based on conserved sequence characteristics, RB60 was assigned as a protein disulfide isomerase (PDI) and the over-expressed protein had properties very similar to the authentic chloroplast RB60 protein, i.e. it affected the RNA binding activity of RB47 in a redox- and phosphorylation-dependent manner (Kim and Mayfield, 1997). RB47 was shown to reveal sequence similarity to poly(A)-binding proteins from various organisms (Yohn et al., 1998), despite the facts that it binds to 5′ RNA sequences and that the *Chlamydomonas psbA* message is not known to contain a poly(A) tail at its 3′ end. However, using tobacco *psbA* mRNA in a chloroplast-derived in vitro translation system, Hirose and Sugiura (1996) identified a short stretch of adenosine residues within the 5′ region to be critical for translation. The 5′-untranslated region of the *psbA* gene was shown to direct the light-dependent expression of a fused reporter gene in transgenic tobacco (Staub and Maliga, 1993). It thus appears likely that short internal sequence element(s) may represent the recognition site for RB47. Interestingly, at least two nuclear mutants of *Chlamydomonas* that are defective in *psbA* translation were recently identified to be deficient in RB47 (Yohn et al., 1998). This work brings together genetic and biochemical data required for a more detailed understanding of light-activated translational regulation in chloroplasts and should open up new avenues for testing of RB47 functions by 'knock-in' strategies.

2. Initiation versus Elongation Control of Chloroplast Translation

The activation/inactivation of the 5′ RNA binding complex (Danon and Mayfield, 1994a,b) is consistent with a translational control mechanism that involves initiation of newly-synthesized polypeptides (Danon, 1997). On the other hand, evidence has become available for redox control affecting the elongation of nascent translation products. As was shown by Kim et al. (1991) in barley, ribosome pausing at specific sites during D1 synthesis results in the accumulation of intermediates, possibly as a consequence of delayed binding of essential cofactor(s), and this delay might facilitate the integration of the nascent polypeptide into the thylakoid membrane. Although initial evidence pointed at the ATP/ADP ratio to be critical (Michaels and Herrin, 1990), work by Kuroda et al. (1996), using protein synthesis in isolated pea chloroplasts in the presence of photosynthesis inhibitors, suggested that the crucial factor is a redox-reactive component generated by the operation of photosynthetic electron transport.

Similar conclusions were also reached by

Mühlbauer and Eichacker (1998) in their studies on barley chloroplasts. They also noted that inhibition of translation elongation in the light by DCMU was overcome by artificial electron donors that allow formation of a proton gradient, either by proton release into the lumen (PMS, DCPIP) or by allowing proton transport via the cytochrome b_6f complex (reduced plastoquinone and duroquinone). A working model was presented, in which the photosynthetic proton gradient serves as a sensitive monitor, thus coupling the rate of translational elongation to the pH gradient by so far unknown component(s). This mechanism seems to be superimposed on the SH-group redox mechanism thought to control translation initiation (Danon and Mayfield, 1994a,b; Danon, 1997), as it neither seems to involve phosphorylation, nor is restricted to D1 synthesis (Mühlbauer and Eichacker, 1998). The latter observation is supported by a study on protein synthesis in pea chloroplasts in the presence of redox-reactive reagents, which shows positive or negative effects on a number of synthesis products in SDS-PAGE gels (Allen et al., 1995).

Independent evidence for control of chloroplast protein synthesis by photosynthetic electron flow was obtained by Kettunen et al. (1997). The D1 protein synthesis was investigated under growth light, high light and in darkness, using in vivo pulse-labeling, in organello protein synthesis, and run-off translation on isolated thylakoid membrane preparations in the presence of lincomycin (an inhibitor of translation initiation). Although the rates of synthesis were accelerated during photoinhibition, the increased in vivo accumulation and in organello synthesis of D1 were not reflected by similar changes in the thylakoid run-off assays. This could mean that only a fraction of the thylakoid-associated *psbA* mRNA is committed to translation at high irradiance. It was proposed that there is an extra pool of mRNA that might function as a reserve under conditions of recovery from photoinhibition (Kettunen et al., 1997).

3. Chloroplast RNA Splicing, a Redox-regulated Process?

The 5′ RNA binding complex for translational control (Danon and Mayfield, 1994a,b; Levings and Siedow, 1995) represents only a fraction of the proteins that interact with chloroplast RNAs and have various roles in the synthesis, processing, and function of these key molecules during plastid gene expression (for review see Rochaix, 1996; Sugita and Sugiura, 1996; Stern et al., 1997; Hagemann et al., 1998). It was reported that the internal processing (splicing) of the intron-containing pre-mRNA of the *psbA* gene from *Chlamydomonas* is accelerated in light as compared to dark. This effect was abolished by inhibitors of photosynthetic electron transport (DCMU, DBMIB), but not by an inhibitor of ATP synthesis (CCCP). Furthermore, photosynthesis deficient mutants did not reveal the light/dark difference in the pre-*psbA* mRNA splicing. One mutant known to be defective in the *tscA* locus, i.e. the chloroplast gene for a small RNA that acts as an essential splicing factor (Rochaix, 1997), was transformed with the wild-type gene. The complementation restored photoautotrophic growth and resulted in light-accelerated splicing of the *psbA* pre-mRNA. Together, these results provide evidence that chloroplast RNA splicing can be under redox control operationally defined by the dependence on photosynthetic electron transport (Deshpande et al., 1997). It will be interesting to learn exactly which splicing factor(s) are the mediators and/or targets of this control.

4. The End of the Message—Redox Regulation of a 3′ RNA Binding Protein

At least some of the organellar proteins that assemble at the RNA 3′ end are known to be subject to phosphorylation control (Kanekatsu et al., 1993; Lisitsky and Schuster, 1995; Hayes et al., 1996). A 3′ RNA binding protein, which was identified as a sequence-specific endoribonuclease from mustard (*Sinapis alba*) (Nickelsen and Link, 1993), responds to both phosphorylation and redox reagents in vitro. The purified endonuclease, termed p54, was found to be activated by phosphorylation or oxidation by glutathione disulfide (GSSG), and was inhibited by either dephosphorylation or reduction by GSH. Kinase pretreatment of p54 prior to GSSG resulted in the highest levels of activation, suggesting that phosphorylation and redox state act together to control p54 activity in vitro and possibly also in vivo (Liere and Link, 1997). The antagonistic effects of phosphorylation and redox state are reminiscent of the situation for the 5′ mRNA binding complex from *Chlamydomonas* (Danon and Mayfield,1994). Binding activity of the latter, however, was found to be affected in an opposite direction, with phosphorylation resulting in inhibition and reducing reagents resulting in activation of the protein. In addition, p54 activity was shown to be modulated by GSH/GSSG but not by DTT (Liere and Link, 1997), whereas the 5′ binding complex responds to DTT and thioredoxin (Danon and Mayfield, 1994a,b). Hence, the mechanistic details of the two redox-regulated enzymatic complexes at the 5′ and 3′ ends of plastid RNAs seem to be quite different.

It is notable that p54 is capable of cleaving a RNA precursor that contain sequences of the two neighboring genes *trnK* and *psbA* (Nickelsen and Link, 1991). As a result, two classes of *psbA* RNAs exist, which differ in the length of their 5′ untranslated region. The longer one is generated by this cleavage, the shorter one (much more abundant) by transcription initiation at the *psbA* promoter. RNA 5′ regions of variable lengths have a potential for alternative folding, which can possibly result in a differential assembly of binding proteins, thus forming starting points for translation control. Furthermore, it was shown in animal systems that a functional interaction between 5′ and 3′ mRNA sequences can be a prerequisite for regulation of translation (Gunkel et al., 1998). The details of such regulatory interactions involved in (post)transcriptional chloroplast gene expression have only begun to be investigated, leaving room for multiple and perhaps unexpected controls, including those by redox state.

B. Redox Regulation of Chloroplast Transcription

1. Complexity and Photoregulation of the Plastid Transcription Apparatus

As in bacteria and eukaryotic nuclear systems, RNA synthesis in chloroplasts is a complex enzymatic process that is controlled by a variety of mechanisms, including multiple RNA polymerases, transcription factors and protein modifications (Gruissem, 1989; Igloi and Kössel, 1992; Gruissem and Tonkyn, 1993; Mullet, 1993; Link, 1994, 1996; Stern et al., 1997; Maliga, 1998). Environmental cues such as light can affect the plastid transcription apparatus, as was initially reported by Apel and Bogorad (1976), who could show that the activity of RNA polymerase was higher in preparations from maize chloroplasts than in those from etioplasts. Although organellar run-on transcription experiments (Deng and Gruissem, 1987; Mullet and Klein, 1987) pointed at posttranscriptional mechanisms as principal control levels, evidence was also obtained for transcriptional regulation in light-grown versus dark-adapted leaf tissue (Klein and Mullet, 1990; Krupinska, 1992; Rapp et al., 1992; Baumgartner et al., 1993), developing cotyledons (Schrubar et al., 1991) as well as in plastids from green versus non-green tissues (Isono et al., 1997; Sakai et al., 1998).

The notion that plastid transcription can indeed be subject to regulation was further supported by the findings that multiple RNA polymerases exist in the organelle, each having a specific role in a distinct developmental context (Mullet, 1993; Maliga, 1998). According to current nomenclature (Maliga, 1998), two principal types of transcription enzymes can be distinguished. One of them, termed Plastid-Encoded Polymerase (PEP) because of the organellar coding site of its multiple core subunits, resembles the bacterial RNA polymerase (for a review see Igloi and Kössel, 1992). The other enzyme, termed Nuclear-Encoded Polymerase (NEP), because its single catalytic subunit (Lerbs-Mache, 1993) is encoded by a nuclear gene (Hedtke et al., 1997), resembles those of T3/T7 bacteriophages and mitochondria. Furthermore, it was found that the PEP enzyme can exist in two different forms, 'A' and 'B', each of which have distinct biochemical characteristics and are differentially expressed in plastids (Pfannschmidt and Link, 1994; 1997). The 'B' enzyme most closely resembles bacterial RNA polymerase by several criteria, including its subunit structure and sensitivity against the transcription inhibitor rifampicin. This enzyme form predominates in etioplasts and perhaps in other non-photosynthetic plastid types. The 'A' enzyme is larger than B, containing a number of accessory polypeptides and is not affected by rifampicin. It is the major RNA polymerase in chloroplasts, where it is responsible for the regulated transcription of photosynthetic genes (Link, 1996) (Fig. 4).

In many experiments on plastid gene expression it has been difficult to decide whether the observed differences are direct consequences of light or dark conditions, or whether they reflect the general metabolic and developmental state of the plastid type. In at least one case, however, evidence was presented suggesting that blue/UV-light (acting through cryptochrome) (Batschauer, 1998) is capable of activating synthesis of a nuclear-encoded transcription factor, AGF (Kim and Mullet, 1995), which binds to a light-responsive element in front of the *psbD-psbC* operon on barley chloroplast DNA (Christopher et al., 1992; Wada et al., 1994; Allison et Maliga, 1995; To et al., 1996; Hoffer and

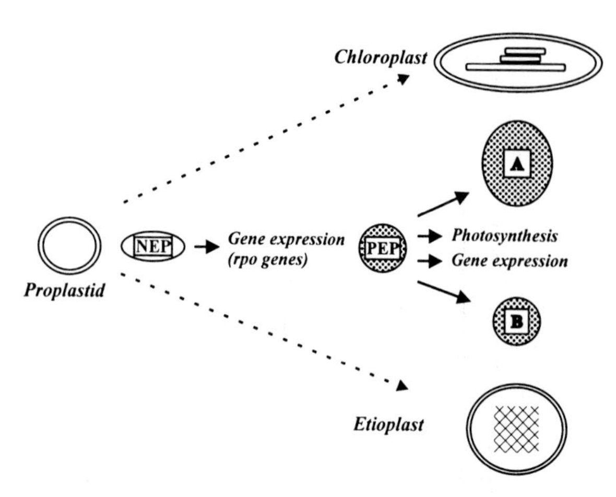

Fig. 4. Current view of the role of multiple DNA-dependent RNA polymerases during plastid development and function. Two principally different types of polymerases are involved, i.e. the nuclear-encoded single-subunit enzyme (NEP) and the plastid-encoded multi-subunit enzyme (PEP). The latter occurs in two different forms, A and B, that can be distinguished by several biochemical criteria (see text). NEP is considered as an 'early' protein that may be exclusively responsible for transcription in proplastids and at early stages during their differentiation into mature plastid forms. Once PEP has been synthesized and assembled, it successively takes over the functions of NEP. In mature plastids, PEP is the major polymerase and seems almost exclusively responsible for the transcription of photosynthetic genes. The A and B forms of PEP seem to coexist, but their relative abundance varies in a plastid type-specific manner. Etioplasts contain mostly the B form, whereas the A form is the dominating enzyme in chloroplast. The A form contains a number of accessory proteins not associated with the B enzyme, including a redox-responsive protein kinase that confers redox control on chloroplast transcription.

Christopher, 1997; Satoh et al., 1997; Kim et al., 1999). Likewise, evidence for red/far-red (phytochrome-mediated) control of plastid gene expression was obtained, although the transcriptional mechanism(s) involved remain to be established (DuBell and Mullet, 1995; for review, see Link, 1991). In view of only a limited number of data pointing directly at the photoreceptor-mediated plastid gene transcription, it is reasonable to assume that effects observed under (white) light versus dark conditions are either indirect consequences of processes occurring outside the organelle or, alternatively, they are related to changes in photosynthetic electron transport, thus representing redox-regulated responses.

2. Redox-Regulated Transcription in Non-Chloroplast Systems—a Paradigm?

One argument about the existence of redox-regulated transcription in chloroplasts is based on the increasing number of reports on redox-responsive transcriptional regulators in bacteria (Campbell et al., 1993; Bauer and Bird, 1996; Demple, 1998; Zeilstra-Ryalls et al., 1998; Zheng et al., 1998) and eukaryotic nuclear systems (Abate et al., 1990; Toledano and Leonard, 1991; Pahl and Baeuerle, 1994; Sen and Packer, 1996; Tell et al., 1998;). For instance, it was suggested (Allen, 1993) that chloroplast transcription may be controlled by a mechanism involving a redox sensor and response regulator(s) analogous to the prototype two-component systems that are widespread in prokaryotes (Stock et al., 1990; Parkinson and Kofoid, 1992). In its classical form, the (membrane-associated) sensor component contains a kinase activity that uses ATP for autophosphorylation at a histidine residue on its so-called transmitter domain. The phosphoryl group is then transferred to an aspartic acid residue on the receiver domain of the response regulator, which often seems to be a sequence-specific transcriptional regulator. With regard to the possibility of two-component regulatory system(s) in chloroplasts, it may be worth considering arguments discussed in the following.

Although several protein kinases are known to be associated with thylakoid preparations (Gal et al., 1997; Vener et al., 1998), none of them has been identified as a histidine kinase. On the other hand, serine/threonine kinases are thought to act in concert with, and even to substitute for, the histidine kinase moiety of 'classical' two-component systems (Zhang, 1996). Well-known examples related to this issue are the phytochromes, i.e. the chromoproteins acting as light sensors both in plants and cyanobacteria. Prokaryotic phytochrome was shown to reveal histidine kinase activity as well as other features that identify it as the sensor moiety of a two-component system (Kehoe and Grossman, 1996; Hughes et al., 1997; Yeh et al., 1997). Higher plant phytochrome has significant sequence similarity to the bacterial histidine kinase region but reveals serine/threonine kinase activity. This led Elich and Chory (1997) to suggest that plant phytochrome may represent a 'nonorthodox' member of the sensor kinase family, which was recently confirmed to be the case (Cashmore, 1998; Yeh and Lagarias, 1998). It thus seems possible that one of the known thylakoid-associated kinases might have a role as a chloroplast sensor kinase. Unfortunately, none of the chloroplast DNA binding proteins has yet been characterized as a (redox) response regulator, and hence the suggestion that the 'classical' two-component regulation may play a role in chloroplasts still awaits experimental support.

3. Effects of Electron Transport Inhibitors on Chloroplast RNA Synthesis

The possibility of redox regulation of chloroplast transcription was experimentally addressed by Pearson et al. (1993) in studies on RNA synthesis in isolated lettuce chloroplasts. They could show that dark-incubated chloroplasts were more active in RNA synthesis than light-incubated ones. They also found that inhibitors of electron flow through PS II (DCMU, DBMIB) or the Rieske iron-sulfur protein of the cytochrome b_6f complex (bromanil) resulted in enhanced RNA-synthesis in the light but not in the dark. It was concluded that the incorporation of radiolabel under the experimental conditions was favored if the cytochrome b_6f complex was in the oxidized state. In these experiments, ^{3}H-NAD served as the primary source of radioactive label, which was found to be rapidly converted to 5′-AMP and adenosine and then incorporated into RNA. The authors noted, however, that RNA synthesis was only partially inhibited by actinomycin D, leaving room for the interpretation that only a fraction of the incorporation of label represented template-dependent RNA synthesis (transcription). Considering that chloroplasts contain poly(A) RNA, it is conceivable that the synthesis observed by Pearson et al. (1993) mainly represents terminal addition of NAD-derived adenine residues.

Using a different approach, Pfannschmidt et al. (1999) recently provided evidence for control of specific plastid gene transcription by photosynthetic electron flow during light quality adaptation. They preincubated isolated chloroplasts in the absence or presence of DCMU or DBMIB under conditions known to favor the intactness and general activity of the organelles. This preincubation was then followed by in organello run-on transcription reactions and the formation of newly-synthesized transcripts from single chloroplast genes was assessed using gene-specific hybridization probes (Mullet and Klein, 1987). The mustard seedlings that served as a source for chloroplast isolation were grown under two spectrally different light regimes, which are known to favor either PS I or PS II function, thus resulting in an imbalance of electron flow between the two photosystems (Glick et al., 1986). Using a *psaAB* probe for the genes encoding the two major reaction center proteins of PSI, lower transcriptional activity was detected in plastids from seedlings grown under 'red' light, primarily absorbed by PS I, than in plastids from plants grown under 'yellow' PS II-sensitizing light. In plastids from the latter seedlings, the negative effect of PSI-sensitizing light could be mimicked by DCMU. Conversely, the positive effect of the PS II-light was partially mimicked by DBMIB in plastids from PSI-light seedlings. Considering the sites of action of the two photosynthesis inhibitors, it was concluded that *psaAB* transcription is downregulated when the plastoquinone pool is oxidized (PSI-light or DCMU), and it is enhanced when the pool is reduced (PS II-light or DBMIB) (Pfannschmidt et al., 1999).

These results confirm and extend those obtained in earlier work on light quality adaptation (Glick et al., 1986; Deng et al., 1989). It was noted by Deng et al. (1989) in work on spinach that among 10 different plastid genes tested, only *psaA* revealed light quality-dependent changes in transcription rate, and this could not fully account for the in vivo effects at the RNA level. This again emphasizes the importance of considering species-specific and context-specific differences and the integrated regulation of plastid gene expression at various levels. Furthermore, it is interesting to note that certain cyanobacterial strains (*Calothrix*) undergo cellular differentiation in response to changes in light spectral quality in a manner reminiscent of light quality adaptation in higher plants (Campbell et al., 1993). Using different light conditions in the presence of DCMU and DBMIB, results were obtained suggesting that red light excitation of PS I and the oxidized plastoquinone pool favor hormogonia differentiation, whereas green light excitation of PS II and net reduction of plastoquinone stimulate heterocyst formation.

4. Mechanisms and Factors in Redox Regulation of Chloroplast Transcription

What might be the mechanisms that transduce the redox signal to the chloroplast transcription apparatus? One possibility, for which experimental evidence has become available, is the phosphorylation/dephosphorylation of regulatory proteins that resemble bacterial sigma factors, i.e. proteins that confer promoter selection and transcription initiation specificity on the core RNA polymerase (Helmann and Chamberlin, 1988; Lonetto et al., 1992). Three different 'Sigma Like Factors' (SLFs) were purified from mustard chloroplasts and, based on their sizes, they were named SLF67, SLF52 and SLF29 (Tiller et al., 1991). Factors of the same sizes were also present in etioplasts, but they revealed enzymatic characteristics that differed from those of their chloroplast counterparts (Tiller and Link, 1993a). In vitro phosphorylation of chloroplast SLFs by using an animal protein kinase (catalytic subunit of bovine heart PKA) resulted in conversion of chloroplast SLF52 and SLF29 into 'etioplast-type' factors and, vice versa, treatment of the etioplast factors with calf intestinal alkaline phosphatase (CIAP) led to a recovery of chloroplast-type properties (Tiller and Link, 1993b). Based on the results of these vitro experiments, a model was presented, suggesting that plastid transcription in vivo might indeed be regulated by the phosphorylation state of sigma factors and perhaps other regulatory proteins.

Support for this model came from the purification of the chloroplast protein kinase that was shown to be responsible for sigma factor phosphorylation. This serine-specific kinase was associated with chloroplast RNA polymerase and hence was termed Plastid Transcription Kinase (PTK) (Baginsky et al., 1997). It could be purified as a 'free' heterotrimeric factor and one (54 kDa) of its polypeptides was identified as the catalytic subunit. Furthermore, the activity of PTK itself was found to be controlled by phosphorylation, suggesting that this kinase might be part of a signaling cascade that controls chloroplast transcription in vivo.

Subsequent work showed that PTK not only phosphorylates chloroplast sigma factors, but also some of the core polypeptides of the major chloroplast RNA polymerase, and PTK activity was found to determine the extent of faithful transcription from the *psbA* promoter in a homologous plastid in vitro system. Furthermore, the kinase activity in vitro responded to both phosphorylation and the SH-group redox reagent glutathione (GSH), yet in an antagonistic manner. The unphoshorylated (active) enzyme was inhibited by GSH, whereas the phosphorylated (inactive) PTK form was activated by the redox reagent (Baginsky et al., 1999). Together, this could mean that PTK is a target for both phosphorylation and redox control in vivo, thus representing a key element in a transduction pathway that might connect a sensor of photosynthetic electron flow (and ATP/ADP ratio) to chloroplast gene expression at the transcriptional level (Link et al., 1997) (Fig. 5).

PTK is only one of several accessory factors that can be purified as part of the major chloroplast RNA polymerase. Biochemical characterization of the large form 'A' complex of the PEP enzyme has provided evidence e.g. for a helicase-type activity (K. Ogrzewalla and G. Link, unpublished). Likewise, a proteomics approach involving 2D-separation and subsequent sequencing of the polymerase-associated polypeptides has revealed the existence of a component with sequence similarity to known RNA-binding proteins. Another polypeptide was tentatively identified as a chloroplast SOD (T. Pfannschmidt and G. Link, unpublished). These components of highly purified chloroplast RNA polymerase A could be involved in connecting transcription to posttranscriptional processes (RNA-binding proteins, helicase) as well as to signal transduction (SOD). It is interesting to note that the flavoprotein ferredoxin-$NADP^+$ reductase (FNR) is released from the thylakoids into the stroma under oxidative stress conditions, and is thereby converted from an NADPH-producing to an NADPH-consuming form (Palatnik et al., 1997). PTK exists both in a 'free' and a polymerase-associated form (Baginsky et al., 1997). It is conceivable that reversible association and dissociation of this and other factors with chloroplast RNA polymerase is part of an immediate response mechanism to changing environmental conditions.

IV. Outlook

For reasons of space, this review can only be focused on a limited number of reports, rather than providing an overview of all aspects of the entire field. Clearly, many other important approaches and findings should be discussed in detail. With regard to the photobiology side of the topic, we have mentioned photo-acclimation, light stress, and light quality adaptation as tools to manipulate the redox state and investigate subsequent changes in the expression of photo-synthetic genes. Other approaches that have been successfully applied include e.g. single turnover light flashes (Tyystjärvi et al., 1998) as well as short transient acidification of thylakoid membranes in

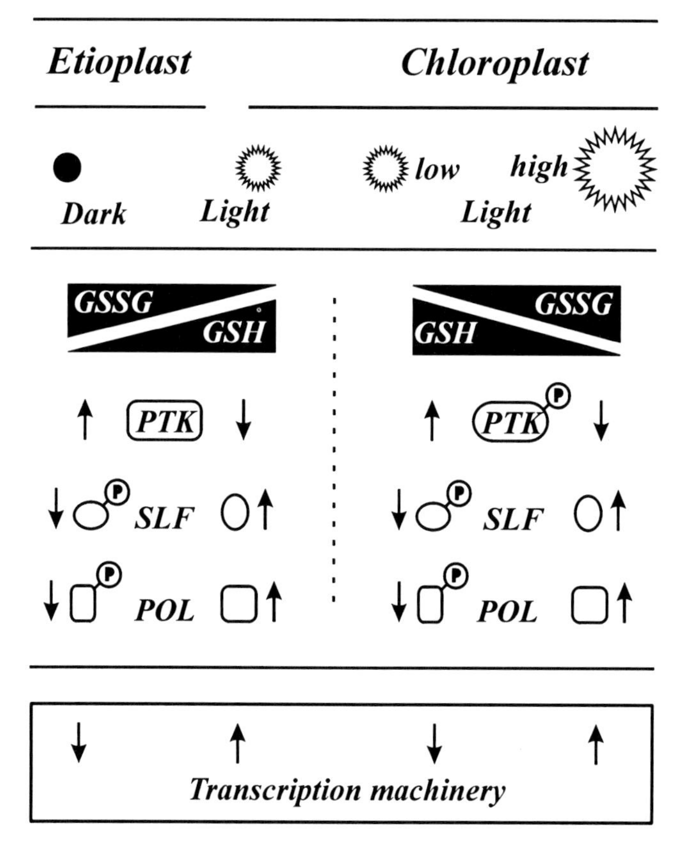

Fig. 5. Working model of regulation of plastid transcription in vivo by the redox and phosphorylation state. The hypothetical scheme is mostly based on data obtained by in vitro studies of the A and B forms of the PEP RNA polymerase and accessory factors, including the Sigma-Like Factors (SLFs) and the Plastid Transcription Kinase (PTK). Some of the predictions regarding transcription activity under various physiological conditions are consistent with the available in vivo data, whereas others remain to be tested. The basic in vitro findings are that PTK is inhibited either following SH-group reduction by glutathione or by phosphorylation. The kinase in its phosphorylated form, however, is activated by GSH. PTK is capable of phosphorylating both the core polymerase (POL) and SLFs, resulting in downregulation of transcription (Tiller and Link, 1993b; Baginsky et al., 1997; 1999). Considering changing ratios of the oxidized (GSSG) versus reduced (GSH) forms of glutathione in vivo, the following predictions can be made. In etioplasts (or other non-photosynthetic plastid forms), the GSSG/GSH ratio is thought to be higher than under reducing conditions in chloroplasts. Accordingly, PTK would be active and both SLF and POL would be phosphorylated, resulting in low transcription activity (left half of the figure). During chloroplast formation under (low) growth-light conditions, when GSH levels increase, PTK would be inactivated, thus releasing the transcriptional inhibition. On the other hand, PTK itself is a substrate for phosphorylation in vitro (Baginsky et al., 1997), and this is likely to occur during active photosynthesis (right half of the figure). As a result, there would be a gradual increase in the concentration of the kinase in its phosphorylated form, which is then activated (rather than inhibited) by GSH and results in downregulation of transcription. During 'normal' photosynthesis (low light), there would thus be a balanced equilibrium of active versus inactive PTK forms, which provides a constant supply of photosynthetic gene products needed under these conditions. Upon transfer to photostress (high light) conditions, however, GSH levels will transiently decrease relative to GSSG (Karpinski et al., 1997). Under these conditions, one might expect a high proportion of phosphorylated and inactivated PTK molecules to accumulate, which would result in upregulation of transcription and further, replenishment of the destroyed photosynthetic proteins. Arrows indicate activity (↑) and inactivity (↓) of the components.

darkness to shift the redox equilibrium of the plastoquinone pool (Vener et al., 1995, 1997).

What can be expected to help clarify the redox signaling mechanisms that connect photosynthetic electron transport with gene expression at various levels (Fig. 6)? The answers will likely be obtained from the combined efforts in molecular genetics and biochemistry. Both naturally occurring mutants (Barkan et al., 1994; Lopez-Juez et al., 1998; Meurer et al., 1998) and transgenic systems (Allison and Maliga, 1995; Rochaix, 1997; Shiina et al., 1998) will be important approaches to carry out functional

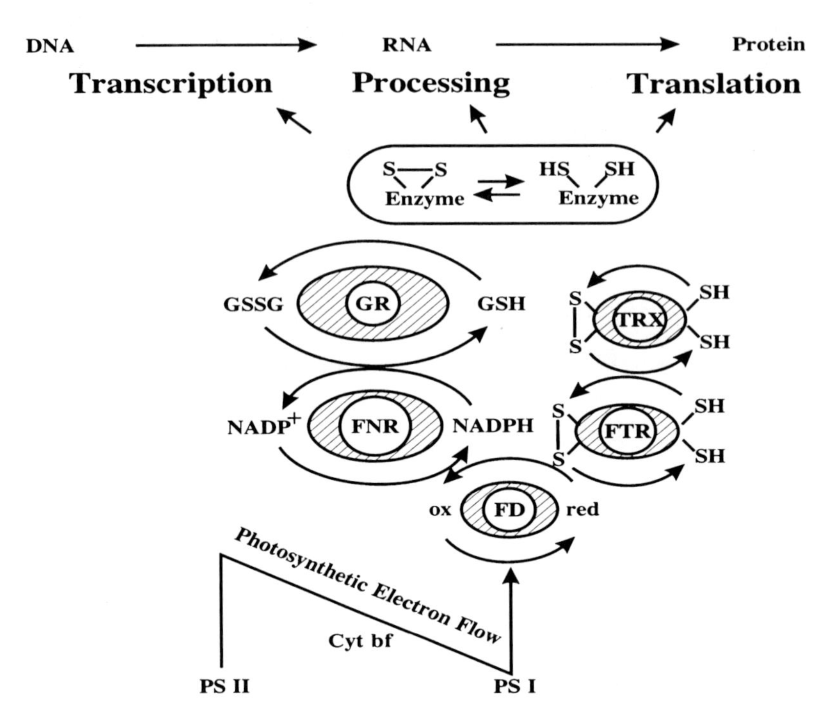

Fig. 6. Scheme presenting a unifying view of photosynthetic redox regulation of chloroplast gene expression. Ferredoxin (FD) is depicted as an outlet for the photosynthetic redox signal. Reducing equivalents include NADPH, GSH, ascorbate (not shown) as well as the SH-groups of ferredoxin-thioredoxin reductase (FTR) and thioredoxin (TRX). Also ferredoxin-NADP$^+$ reductase (FNR) and glutathione reductase (GR) have a central role in the transmission of the redox signal. The connection between these enzymes and the various steps in gene expression could be direct, or could involve intermediate cascade mechanisms. In any case, they seem to operate via SH-group redox (and phosphorylation) control. As the details, reported for the in vitro redox regulation of translation versus transcription and RNA maturation, are different (see text), is seems likely that portions of the signal transduction pathway(s) are branched.

analyses. The cloning of most, if not all, genes for photosynthesis proteins has been achieved (for review see Pakrasi, 1995), and the same holds true for the genes encoding the core components of the organellar gene expression machinery (reviewed by Sugita and Sugiura, 1996; Maliga, 1998). Cloning of the genes for sigma factors is currently in progress (Liu and Troxler, 1996; Tanaka et al., 1996; 1997; Allison, 1997; Isono et al., 1997; Kestermann et al., 1998), as also cloning of the genes for other accessory factors involved in chloroplast transcription and posttranscriptional processes. Together, this will open up functional reconstitution using both wild-type and mutant proteins. Their interaction will be shown by chemical crosslinking techniques, and the search for interacting components will certainly be complemented by 'n-hybrid' screening techniques in vivo. This way, we will probably see that photosynthesis and chloroplast gene expression are two sides of the same coin.

Acknowledgments

I apologize that, because of space constrains, only a fraction of all relevant work could be cited. I am indebted to members of my group for comments and discussion, in particular to Sacha Baginsky, Karsten Ogrzewalla and Thomas Pfannschmidt, who contributed unpublished results. Work of the group at Bochum was funded by the Deutsche Forschungsgemeinschaft and the Fonds der Chemischen Industrie, Germany.

References

Abate C, Patel L, Rauscher FJ,III and Curran T (1990) Redox regulation of Fos and Jun DNA-binding activity in vitro. Science 249: 1157–1161

Allen CA, Hakansson G and Allen JF (1995) Redox conditions specify the proteins synthesized by isolated chloroplasts and mitochondria. Redox Rep 1: 119–123

Allen JF (1993) Redox control of transcription: Sensors, response

regulators, activators and repressors. FEBS Lett 332: 203–207
Allen JP and Williams JC (1998) Photosynthetic reaction centers. FEBS Lett 438: 5–9
Allison LA (1997) Multiple sigma factors provide a mechanism for nuclear control over chloroplast gene expression. Plant Physiol 114, Supplement: 65
Allison LA and Maliga P (1995) Light-responsive and transcription-enhancing elements regulate the plastid *psbD* core promoter. EMBO J 14: 3721–3730
Apel K and Bogorad L (1976) Light-induced increase in the activity of maize plastid DNA-dependent RNA polymerase. Eur J Biochem 67: 615–620
Aro E-M, Virgin I and Andersson B (1993) Photoinhibition of Photosystem II. Inactivation, protein damage and turnover. Biochim Biophys Acta 1143: 113–134
Baginsky S, Tiller K and Link G (1997) Transcription factor phosphorylation by a protein kinase associated with chloroplast RNA polymerase from mustard (*Sinapis alba*). Plant Mol Biol 34: 181–189
Baginsky S, Tiller K, Pfannschmidt T and Link G (1999) PTK, the chloroplast RNA polymerase-associated protein kinase from mustard (*Sinapis alba*), mediates redox control of plastid in vitro transcription. Plant Mol Biol 39: 1013–1023
Barber J and Andersson B (1992) Too much of a good thing: Light can be bad for photosynthesis. Trends Biochem Sci 17: 61–66
Barkan A, Walker M, Nolasco M and Johnson D (1994) A nuclear mutation in maize blocks the processing and translation of several chloroplast mRNAs and provides evidence for the differential translation of alternative mRNA forms. EMBO J 13: 3170–3181
Batschauer A (1998) Photoreceptors of higher plants. Planta 206: 479–492
Bauer CE and Bird TH (1996) Regulatory circuits controlling photosynthesis gene expression. Cell 85: 5–8
Baumgartner BJ, Rapp JC and Mullet JE (1993) Plastid genes encoding the transcription/translation apparatus are differentially transcribed early in barley (*Hordeum vulgare*) chloroplast development. Evidence for selective stabilization of *psbA* mRNA. Plant Physiol 101: 781–791
Bergmann L and Rennenberg H (1993) Glutathione metabolism in plants. In: De Kok LJ, Stulen I, Rennenberg H, Brunold C and Rauser WE (eds) Sulfur nutrition and assimilation in higher plants, pp 109–123. SPB Academic Publishers, The Hague
Bogorad L (1991) Replication and transcription of plastid DNA. In: Bogorad L and Vasil IK (eds) The molecular biology of plastids, pp 93–124. Academic Press, San Diego
Bogorad L and Vasil IK (1991) The photosynthetic apparatus: Molecular biology and operation. Academic Press, San Diego
Bowler C, Van Montagu M and Inzé D (1992) Superoxide dismutase and stress tolerance. Annu Rev Plant Physiol Plant Mol Biol 43: 83–116
Boyer CT, Shannon JC and Hardison RC (1989) Physiology, biochemistry, and genetics of nongreen plastids. American Society of Plant Physiologists, Rockville
Broadbent P, Creissen GP, Kular B, Wellburn AR and Mullineaux PM (1995) Oxidative stress responses in transgenic tobacco containing altered levels of glutathione reductase activity. Plant J 8: 247–256
Buchanan BB (1991) Regulation of CO_2 assimilation in oxygenic photosynthesis: The ferredoxin/thioredoxin system: Perspective on its discovery, present status, and future development. Arch Biochem Biophys 288: 1–9
Campbell D, Houmard J and Tandeau de Marsac N (1993) Electron transport regulates cellular differentiation in the filamentous cyanobacterium *Calothrix*. Plant Cell 5: 451–463
Carlberg I, Rintamäki E, Aro E-M and Andersson B (1999) Thylakoid protein phosphorylation and the thiol redox state. Biochemistry 9: 3197–3204
Cashmore AR (1984) Structure and expression of a pea nuclear gene encoding a chlorophyll *a/b*-binding polypeptide. Proc Natl Acad Sci USA 81: 2960–2964
Cashmore AR (1998) Higher-plant phytochrome: 'I used to date histidine, but now I prefer serine'. Proc Natl Acad Sci USA 95: 13358–13360
Christopher DA, Kim M and Mullet JE (1992) A novel light-regulated promoter is conserved in cereal and dicot chloroplasts. Plant Cell 4: 785–798
Danon A (1997) Translational regulation in the chloroplast. Plant Physiol 115: 1293–1298
Danon A and Mayfield SP (1991) Light regulated translational activators: Identification of chloroplast gene specific mRNA binding proteins. EMBO J 10: 3993–4001
Danon A and Mayfield SP (1994a) ADP-dependent phosphorylation regulates RNA-binding in vitro: Implications in light-modulated translation. EMBO J 13: 2227–2235
Danon A and Mayfield SP (1994b) Light-regulated translation of chloroplast messenger RNAs through redox potential. Science 266: 1717–1719
Demple B (1998) Signal transduction—a bridge to control. Science 279: 1655–1656
Deng X-W and Gruissem W (1987) Control of plastid gene expression during development: the limited role of transcriptional regulation. Cell 49: 379–387
Deng X-W, Tonkyn JC, Peter GF, Thornber JP and Gruissem W (1989) Posttranscriptional control of plastid mRNA accumulation during adaptation of chloroplasts to different light quality environments. Plant Cell 1: 645–654
Deshpande NN, Bao Y and Herrin DL (1997) Evidence for light/redox-regulated splicing of *psbA* pre-RNAs in *Chlamydomonas* chloroplasts. RNA 3: 37–48
Dickey LF, Petracek ME, Nguyen TT, Hansen ER and Thompson WF (1998) Light regulation of *Fed-1* mRNA requires an element in the 5′ untranslated region and correlates with differential polyribosome association. Plant Cell 10: 475–484
DuBell AN and Mullet JE (1995) Differential transcription of pea chloroplast genes during light-induced leaf development transcription—Continuous far-red light activates chloroplast transcription. Plant Physiol 109: 105–112
Durnford DG and Falkowski PG (1997) Chloroplast redox regulation of nuclear gene transcription during photoacclimation. Photosynth Res 53: 229–241
Elich TD and Chory J (1997) Phytochrome: If it looks and smells like a histidine kinase, is it a histidine kinase? Cell 91: 713–716
Escoubas JM, Lomas M, LaRoche J and Falkowski PG (1995) Light intensity regulation of *cab* gene transcription is signaled by the redox state of the plastoquinone pool. Proc Natl Acad Sci USA 92: 10237–10241
Foyer CH and Noctor G (1999) Leaves in the dark see the light. Science 284: 599–601
Foyer CH, Souriau N, Perret S, Lelandais M, Kunert K-J, Pruvost

C and Jouanin L (1995) Overexpression of glutathione reductase but not glutathione synthetase leads to increases in antioxidant capacity and resistance to photoinhibition in poplar trees. Plant Physiol 109: 1047–1057

Gal A, Zer H and Ohad I (1997) Redox-controlled thylakoid protein phosphorylation. News and views. Physiol Plant 265: 869–885

Garcia-Olmedo F, Pineiro M and Diaz I (1994) Dances to a redox tune. Plant Mol Biol 26: 11–13

Gillham NW, Boynton JE and Hauser CR (1994) Translational regulation of gene expression in chloroplasts and mitochondria. Annu Rev Genet 28: 71–93

Glick RE, McCauly SW, Gruissem W and Melis A (1986) Light quality regulates expression of chloroplast genes and assembly of photosynthetic membrane complexes. Proc Natl Acad Sci USA 83: 4287–4291

Goldschmidt-Clermont M (1998) Coordination of nuclear and chloroplast gene expression in plant cells. Int Rev Cytol 177: 115–180

Green BR, Pichersky E and Kloppstech K (1991) Chlorophyll *a*/*b*-binding proteins: An extended family. Trends Biochem Sci 16: 181–186

Gruissem W (1989) Chloroplast gene expression: how plants turn their plastids on. Cell 56: 161–170

Gruissem W and Tonkyn JC (1993) Control mechanisms of plastid gene expression. Crit Rev Plant Sci 12: 19–55

Gunkel N, Yano T, Markussen FH, Olsen LC and Ephrussi A (1998) Localization-dependent translation requires a functional interaction between the 5′ and 3′ ends of *oskar* mRNA. Genes Dev 12: 1652–1664

Hagemann R, Hagemann MM and Bock R (1998) Genetic extranuclear inheritance: Plastid genetics. Progr Bot 59: 108–130

Harter K, Kircher S, Frohnmeyer H, Krenz M, Nagy F and Schäfer E (1994) Light-regulated modification and nuclear translocation of cytosolic G-box binding factors in parsley. Plant Cell 6: 545–559

Hatch MD (1992) C_4 photosynthesis: An unlikely process full of surprises. Plant Cell Physiol 33: 333–342

Hayes R, Kudla J, Schuster G, Gabay L, Maliga P and Gruissem W (1996) Chloroplast mRNA 3′-end processing by a high molecular weight protein complex is regulated by nuclear encoded RNA binding proteins. EMBO J 15: 1132–1141

Hedtke B, Börner T and Weihe A (1997) Mitochondrial and chloroplast phage-type RNA polymerases in *Arabidopsis*. Science 277: 809–811

Helmann JD and Chamberlin MJ (1988) Structure and function of bacterial sigma factors. Annu Rev Biochem 57: 839–872

Herrmann RG, Westhoff P and Link G (1992) Chloroplast biogenesis in higher plants. In: Herrmann RG (ed) Cell organelles, pp 275–349. Springer-Verlag, New York

Hess WR, Müller A, Nagy F and Börner T (1994) Ribosome-deficient plastids affect transcription of light-induced nuclear genes: Genetic evidence for a plastid-derived signal. Mol Gen Genet 242: 305–312

Hipkins MF and Baker NR (1986) Photosynthesis energy transduction—a practical approach. IRL Press, Oxford

Hirose T and Sugiura M (1996) Cis-acting elements and trans-acting factors for accurate translation of chloroplast *psbA* mRNAs: Development of an in vitro translation system from tobacco chloroplasts. EMBO J 15: 1687–1695

Hirose T, Fan H, Suzuki JY, Wakasugi T, Tsudzuki T, Kössel H and Sugiura M (1996) Occurrence of silent RNA editing in chloroplasts: Its species specificity and the influence of environmental and developmental conditions. Plant Mol Biol 30: 667–672

Hoffer PH and Christopher DA (1997) Structure and blue-light-responsive transcription of a chloroplast *psbD* promoter from *Arabidopsis thaliana*. Plant Physiol 115: 213–222

Hughes J, Lamparter T, Mittmann F, Hartmann E, Gärtner W, Wilde A and Börner T (1997) A prokaryotic phytochrome. Nature 386: 663

Hunter T (1995) Protein kinases and phosphatases: The yin and yang of protein phosphorylation and signaling. Cell 80: 225–236

Igloi GL and Kössel H (1992) The transcriptional apparatus of chloroplasts. Crit Rev Plant Sci 10: 525–558

Isono K, Niwa Y, Satoh K and Kobayashi H (1997) Evidence for transcriptional regulation of plastid photosynthesis genes in *Arabidopsis thaliana* roots. Plant Physiol 114: 623–630

Isono K, Shimizu M, Yoshimoto K, Niwa Y, Satoh K, Yokota A and Kobayashi H (1997) Leaf-specifically expressed genes for polypeptides destined for chloroplasts with domains of sigma-70 factors of bacterial RNA polymerases in *Arabidopsis thaliana*. Proc Natl Acad Sci USA 94: 14948–14953

Kanekatsu M, Munakata H, Furuzono K and Ohtsuki K (1993) Biochemical characterization of a 34 kDa ribonucleoprotein (p34) purified from the spinach chloroplast fraction as an effective phosphate acceptor for casein kinase II. FEBS Lett 335: 176–180

Karpinski S, Escobar C, Karpinska B, Creissen G and Mullineaux PM (1997) Photosynthetic electron transport regulates the expression of cytosolic ascorbate peroxidase genes in *Arabidopsis* during excess light stress. Plant Cell 9: 627–640

Karpinski S, Reynolds H, Karpinska B, Wingsle G, Creissen G and Mullineaux P (1999) Systemic signaling and acclimation in response to excess excitation energy in *Arabidopsis*. Science 284: 654–757

Kehoe DM and Grossman AR (1996) Similarity of a chromatic adaptation sensor to phytochrome and ethylene receptors. Science 273: 1409–1412

Kestermann M, Neukirchen S, Kloppstech K and Link G (1998) Sequence and expression characteristics of a nuclear-encoded chloroplast sigma factor from mustard (*Sinapis alba*). Nucleic Acids Res 26: 2747–2753

Kettunen R, Pursiheimo S, Rintamäki E, Van Wijk KJ and Aro EM (1997) Transcriptional and translational adjustments of *psbA* gene expression in mature chloroplasts during photoinhibition and subsequent repair of Photosystem II. Eur J Biochem 247: 441–448

Kim J, Klein PG and Mullet JE (1991) Ribosomes pause at specific sites during synthesis of membrane-bound chloroplast reaction center protein D1. J Biol Chem 266: 14931–14938

Kim JM and Mayfield SP (1997) Protein disulfide isomerase as a regulator of chloroplast translational activation. Science 278: 1954–1957

Kim M and Mullet JE (1995) Identification of a sequence-specific DNA binding factor required for transcription of the barley chloroplast blue light-responsive *psbD-psbC* promoter. Plant Cell 7: 1445–1457

Kim M, Christopher DA and Mullet JE (1999) ADP-dependent phosphorylation regulates association of a DNA-binding complex with the barley chloroplast *psbD* blue-light-responsive promoter. Plant Physiol 119: 663–670

Klein RR and Mullet JE (1990) Light-induced transcription of chloroplast genes. *psbA* transcription is differentially enhanced in illuminated barley. J Biol Chem 265: 1895–1902

Kropat J, Oster U, Rüdiger W and Beck CF (1997) Chlorophyll precursors are signals of chloroplast origin involved in light induction of nuclear heat-shock genes. Proc Natl Acad Sci USA 94: 14168–14172

Krupinska K (1992) Transcriptional control of plastid gene expression during development of primary foliage leaves of barley grown under a daily light-dark regime. Planta 186: 294–303

Kuroda H, Kobashi K, Kaseyama H and Satoh K (1996) Possible involvement of a low redox potential component(s) downstream of Photosystem I in the translational regulation of the D1 subunit of the Photosystem II reaction center in isolated pea chloroplasts. Plant Cell Physiol 37: 754–761

Langdale JA and Nelson T (1991) Spatial regulation of photosynthetic development in C4 plants. Trends Genet 7: 191–196

Lerbs-Mache S (1993) The 110-kDa polypeptide of spinach plastid DNA-dependent RNA polymerase: Single-subunit enzyme or catalytic core of multimeric enzyme complexes. Proc Natl Acad Sci USA 90: 5509–5513

Levings CS III and Siedow JN (1995) Regulation by redox poise in chloroplasts. Science 268: 695–696

Liere K and Link G (1997) Chloroplast endoribonuclease p54 involved in RNA 3′-end processing is regulated by phosphorylation and redox state. Nucleic Acids Res 25: 2403–2408

Link G (1991) Photoregulated development of chloroplasts. In: Bogorad L and Vasil IK (eds) The photosynthetic apparatus: Molecular biology and operation, pp 365–394. Academic Press, San Diego

Link G (1994) Plastid differentiation: organelle promoters and transcription factors. In: Nover L (ed) Plant promoters and transcription factors—Results and problems in cell differentiation, Vol. 20, pp 65–85. Springer-Verlag, Berlin

Link G (1996) Green life: control of chloroplast gene transcription. BioEssays 18: 465–471

Link G, Tiller K and Baginsky S (1997) Glutathione, a regulator of chloroplast transcription. In: Hatzios KK (ed) Regulation of Enzymatic Systems Detoxifying Xenobiotics in Plants, pp 125–137. Kluwer Academic Publishers, Dordrecht,

Lisitsky I and Schuster G (1995) Phosphorylation of a chloroplast RNA-binding protein changes its affinity to RNA. Nucleic Acids Res 23: 2506–2511

Liu B and Troxler RF (1996) Molecular characterization of a positively photoregulated nuclear gene for a chloroplast RNA polymerase sigma factor in *Cyanidium caldarium*. Proc Natl Acad Sci USA 93: 3313–3318

Lonetto M, Gribskov M and Gross CA (1992) The sigma-70 family: Sequence conservation and evolutionary relationships. J Bacteriol 174: 3843–3849

Lopez-Juez E, Jarvis RP, Takeuchi A, Page AM and Chory J (1998) New *Arabidopsis cue* mutants suggest a close connection between plastid- and phytochrome regulation of nuclear gene expression. Plant Physiol 118: 803–815

Maliga P (1998) Two plastid RNA polymerases of higher plants: An evolving story. Trends Plant Sci 3: 4–6

Mattoo AK, Hoffman-Falk H, Marder JB and Edelman M (1984) Regulation and protein metbolism: coupling of photosynthetic electron transport to in vivo degradation of the rapidly metabolized 32-kilodalton protein of the chloroplast membranes. Proc Natl Acad Sci USA 81: 1380–1384

Maxwell DP, Laudenbach DE and Huner NPA (1995) Redox regulation of light-harvesting complex II and *cab* mRNA abundance in *Dunaliella salina*. Plant Physiol 109: 787–795

Mayfield SP, Yohn CB, Cohen A and Danon A (1995) Regulation of chloroplast gene expression. Annu Rev Plant Physiol Plant Mol Biol 46: 147–166

Meierhoff K and Westhoff P (1993) Differential biogenesis of Photosystem II in mesophyll and bundle-sheath cells of monocotyledonous NADP-malic enzyme- type C_4 plants: The non-stoichiometric abundance of the subunits of Photosystem II in the bundle-sheath chloroplasts and the translational activity of the plastome-encoded genes. Planta 191: 23–33

Meister A (1995) Glutathione metabolism. Methods Enzymol 251: 3–7

Menkens AE, Schindler U and Cashmore AR (1995) The G-box: A ubiquitous regulatory DNA element in plants bound by the GBF family of bZIP proteins. Trends Biochem Sci 20: 506–510

Meshi T and Iwabuchi M (1995) Plant transcription factors. Plant Cell Physiol 36: 1405–1420

Meurer J, Grevelding C, Westhoff P and Reiss B (1998) The PAC protein affects the maturation of specific chloroplast mRNAs in *Arabidopsis thaliana*. Mol Gen Genet 258: 342–351

Michaels A and Herrin DL (1990) Translational regulation of chloroplast gene expression during the light-dark cell cycle of *Chlamydomonas*: Evidence for control by ATP/energy supply. Biochem Biophys Res Commun 170: 1082–1088

Mullet JE (1993) Dynamic regulation of chloroplast transcription. Plant Physiol 103: 309–313

Mullet JE and Klein RR (1987) Transcription and RNA stability are important determinants of higher plant chloroplast RNA levels. EMBO J 6: 1571–1579

Mühlbauer SK and Eichacker LA (1998) Light-dependent formation of the photosynthetic proton gradient regulates translation elongation in chloroplasts. J Biol Chem 273: 20935–20940

Nickelsen J and Link G (1991) RNA-protein interactions at transcript 3′ ends and evidence for *trnK-psbA* cotranscription in mustard chloroplasts. Mol Gen Genet 228: 89–96

Nickelsen J and Link G (1993) The 54 kDa RNA-binding protein from mustard chloroplasts mediates endonucleolytic transcript 3′ end formation in vitro. Plant J 3: 537–544

Noctor G and Foyer CH (1998) Ascorbate and glutathione: Keeping active oxygen under control. Annu Rev Plant Physiol Plant Mol Biol 49: 249–279

Pahl HL and Baeuerle PA (1994) Oxygen and the control of gene expression. BioEssays 16: 497–502

Pakrasi HB (1995) Genetic analysis of the form and function of Photosystem I and Photosystem II. Annu Rev Genet 29: 755–776

Palatnik JF, Valle EM and Carrillo N (1997) Oxidative stress causes ferredoxin $NADP^+$ reductase solubilization from the thylakoid membranes in methyl viologen treated plants. Plant Physiol 115: 1721–1727

Parkinson JS and Kofoid EC (1992) Communication modules in

bacterial signaling proteins. Annu Rev Genet 26: 71–112
Pearson CK, Wilson SB, Schaffer R and Ross AW (1993) NAD turnover and utilisation of metabolites for RNA synthesis in a reaction sensing the redox state of the cytochrome b_6f complex in isolated chloroplasts. Eur J Biochem 218: 397–404
Petracek ME, Dickey LF, Huber SC and Thompson WF (1997) Light-regulated changes in abundance and polyribosome association of ferredoxin mRNA are dependent on photosynthesis. Plant Cell 9: 2291–2300
Petracek ME, Dickey LF, Nguyen TT, Gatz C, Sowinski DA, Allen GC and Thompson WF (1998) Ferredoxin-1 mRNA is destabilized by changes in photosynthetic electron transport. Proc Natl Acad Sci USA 95: 9009–9013
Pfannschmidt T and Link G (1994) Separation of two classes of plastid DNA-dependent RNA polymerases that are differentially expressed in mustard (*Sinapis alba* L.) seedlings. Plant Mol Biol 25: 69–81
Pfannschmidt T and Link G (1997) The A and B forms of plastid DNA-dependent RNA polymerase from mustard (*Sinapis alba* L.) transcribe the same genes in a different developmental context. Mol Gen Genet 257: 35–44
Pfannschmidt T, Nilsson A and Allen JF (1999) Photosynthetic control of chloroplast gene expression. Nature 397: 625–628
Rapp JC, Baumgartner BJ and Mullet J (1992) Quantitative analysis of transcription and RNA levels of 15 barley chloroplast genes. Transcription rates and mRNA levels vary over 300-fold; Predicted mRNA stabilities vary 30-fold. J Biol Chem 267: 21404–21411
Rintamäki E, Salonen M, Suoranta UM, Carlberg I, Andersson B and Aro EM (1997) Phosphorylation of light-harvesting complex II and Photosystem II core proteins shows different irradiance-dependent regulation in vivo—Application of phosphothreonine antibodies to analysis of thylakoid phosphoproteins. J Biol Chem 272: 30476–30482
Rochaix J-D (1995) *Chlamydomonas reinhardtii* as the photosynthetic yeast. Annu Rev Genet 29: 209–230
Rochaix J-D (1996) Post-transcriptional regulation of chloroplast gene expression in *Chlamydomonas reinhardtii.* Plant Mol Biol 32: 327–341
Rochaix J-D (1997) Chloroplast reverse genetics: New insights into the function of plastid genes. Trends Plant Sci 2: 419–425
Sakai A, Suzuki T, Miyazawa Y, Kawano S, Nagata T and Kuroiwa T (1998) Comparative analysis of plastid gene expression in tobacco chloroplasts and proplastids: Relationship between transcription and transcript accumulation. Plant Cell Physiol 39: 581–589
Satoh J, Baba K, Nakahira Y, Shiina T and Toyoshima Y (1997) Characterization of dynamics of the *psb*D light-induced transcription in mature wheat chloroplasts. Plant Mol Biol 33: 267–278
Scheibe R (1991) Redox-modulation of chloroplast enzymes. A common principle for individual control. Plant Physiol 96: 1–3
Schrubar H, Wanner G and Westhoff P (1991) Transcriptional control of plastid gene expression in greening *Sorghum* seedlings. Planta 183: 101–111
Schürmann P (1995) Ferredoxin: Thioredoxin system. Methods Enzymol 252: 274–283
Sen CK and Packer L (1996) Antioxidant and redox regulation of gene transcription. FASEB J 10: 709–720
Shiina T, Allison L and Maliga P (1998) rbcL transcript levels in tobacco plastids are independent of light: Reduced dark transcription rate is compensated by increased mRNA stability. Plant Cell 10: 1713–1722
Simpson DJ and Von Wettstein D (1989) The structure and function of the thylakoid membrane. Carlsberg Res Commun 54: 55–65
Sonoike K, Kamo M, Hihara Y, Hiyama T and Enami I (1997) The mechanism of the degradation of *psaB* gene product, one of the photosynthetic reaction center subunits of Photosystem I, upon photoinhibition. Photosynth Res 53: 55–63
Staub JM and Maliga P (1993) Accumulation of D1 polypeptide in tobacco plastids is regulated via the untranslated region of the *psbA* mRNA. EMBO J 12: 601–606
Stern DB, Higgs DC and Yang J (1997) Transcription and translation in chloroplasts. Trends Plant Sci 2: 308–315
Stock JB, Stock AM and Mottonen JM (1990) Signal transduction in bacteria. Nature 344: 395–400
Sugita M and Sugiura M (1996) Regulation of gene expression in chloroplasts of higher plants. Plant Mol Biol 32: 315–326
Sugiura M, Hirose T and Sugita M (1998) Evolution and mechanism of translation in chloroplasts. Annu Rev Genet 32: 437–459
Susek RE, Ausubel FM and Chory J (1993) Signal transduction mutants of *Arabidopsis* uncouple nuclear *CAB* and *RBCS* gene expression from chloroplast development. Cell 74: 787–799
Tanaka K, Oikawa K, Ohta N, Kuroiwa H, Kuroiwa T and Takahashi H (1996) Nuclear encoding of a chloroplast RNA polymerase sigma subunit in a red alga. Science 272: 1932–1935
Tanaka K, Tozawa Y, Mochizuki N, Shinozaki K, Nagatani A, Wakasa K and Takahashi H (1997) Characterization of three cDNA species encoding plastid RNA polymerase sigma factors in *Arabidopsis thaliana*: Evidence for the sigma factor heterogeneity in higher plant plastids. FEBS Lett 413: 309–313
Taylor WC (1989) Regulatory interactions between nuclear and plastid genomes. Annu Rev Plant Physiol Plant Mol Biol 40: 211–233
Tell G, Scaloni A, Pellizzari L, Formisano S, Pucillo C and Damante G (1998) Redox potential controls the structure and DNA binding activity of the paired domain. J Biol Chem 273: 25062–25072
Tiller K and Link G (1993a) Sigma-like transcription factors from mustard (*Sinapis alba* L.) etioplast are similar in size to, but functionally distinct from, their chloroplast counterparts. Plant Mol Biol 21: 503–513
Tiller K and Link G (1993b) Phosphorylation and dephosphorylation affect functional characteristics of chloroplast and etioplast transcription systems from mustard (*Sinapis alba* L.). EMBO J 12: 1745–1753
Tiller K, Eisermann A and Link G (1991b) The chloroplast transcription apparatus from mustard (*Sinapis alba* L.)—Evidence for three different transcription factors which resemble bacterial sigma factors. Eur J Biochem 198: 93–99
Tjus SE, Moller BL and Scheller HV (1998) Photosystem I is an early target of photoinhibition in barley illuminated at chilling temperatures. Plant Physiol 116: 755–764
To K-Y, Cheng M-C, Suen D-F, Mon D-P, Chen L-FO and Chen S-CG (1996) Characterization of the light-responsive promoter of rice chloroplast *psbD-C* operon and the sequence-specific DNA binding factor. Plant Cell Physiol 37: 660–666
Toledano MB and Leonard WJ (1991) Modulation of transcription

factor NF-kappaB binding activity by oxidation-reduction in vitro. Proc Natl Acad Sci USA 88: 4328–4332
Trebst A (1980) Inhibitors in electron flow. Methods Enzymol 69: 675–715
Tyystjärvi T, Tyystjärvi E, Ohad I and Aro EM (1998) Exposure of *Synechocystis* 6803 cells to series of single turnover flashes increases the *psbA* transcript level by activating transcription and down-regulating *psbA* mRNA degradation. FEBS Lett 436: 483–487
Vener AV, Van Kan PJM, Gal A, Andersson B and Ohad I (1995) Activation/deactivation cycle of redox-controlled thylakoid protein phosphorylation—Role of plastoquinol bound to the reduced cytochrome bf complex. J Biol Chem 270: 25225–25232
Vener AV, Van Kan PJM, Rich PR, Ohad I and Andersson B (1997) Plastoquinol at the quinol oxidation site of reduced cytochrome bf mediates signal transduction between light and protein phosphorylation: Thylakoid protein kinase deactivation by a single-turnover flash. Proc Natl Acad Sci USA 94: 1585–1590
Vener AV, Ohad I and Andersson B (1998) Protein phosphorylation and redox sensing in chloroplast thylakoids. Curr Opinion Plant Biol 1: 217–223
Wada T, Tunoyama Y, Shiina T and Toyoshima Y (1994) In vitro analysis of light-induced transcription in the wheat *psbD/C* gene cluster using plastid extracts from dark-grown and short-term-illuminated seedlings. Plant Physiol 104: 1259–1267
Yeh KC and Lagarias JC (1998) Eukaryotic phytochromes: Light-regulated serine/threonine protein kinases with histidine kinase ancestry. Proc Natl Acad Sci USA 95: 13976–13981
Yeh KC, Wu SH, Murphy JT and Lagarias JC (1997) A cyanobacterial phytochrome two-component light sensory system. Science 277: 1505–1508
Yohn CB, Cohen A, Danon A and Mayfield SP (1998) A poly(A) binding protein functions in the chloroplast as a message-specific translation factor. Proc Natl Acad Sci USA 95: 2238–2243
Yohn CB, Cohen A, Rosch C, Kuchka MR and Mayfield SP (1998) Translation of the chloroplast *psbA* mRNA requires the nuclear-encoded poly(A)-binding protein, RB47. J Cell Biol 142: 435–442
Zeilstra-Ryalls J, Gomelsky M, Eraso JM, Yeliseev A, O'Gara J and Kaplan S (1998) Control of photosystem formation in *Rhodobacter sphaeroides*. J Bacteriol 180: 2801–2809
Zhang C-C (1996) Bacterial signaling involving eukaryotic-type protein kinases. Mol Microbiol 20: 9–15
Zheng M, Aslund F and Storz G (1998) Activation of the OxyR transcription factor by reversible disulfide bond formation. Science 279: 1718–1721

Chapter 6

Sugar Sensing and Regulation of Photosynthetic Carbon Metabolism

Uwe Sonnewald*
Institut für Pflanzengenetik und Kulturpflanzenforschung, Corrensstrasse 3, D-06466, Gatersleben, Germany

Summary

Regulation of photosynthetic carbon metabolism is central for plant growth and development. Photosynthesis occurs mainly in source leaves, which produce carbohydrates in excess to support growth of sink tissues, unable to produce assimilates. During plant development, sink to source ratios change, which implies that assimilate production must be adjusted to changing needs of distant tissues. In the past, single-site manipulations of metabolic pathways have been conducted to adopt plant metabolism to human needs. Although interesting results have been obtained, most of these changes have not lead to the expected changes in crop productivity. This is due to the enormous flexibility of plant metabolism allowing the induction of biosynthetic by-passes in many cases. One prominent example is the inhibition of the triose-phosphate translocator in the chloroplast envelope, which did not cause any growth retardation of the transgenic plants. This was due to a switch from triose-phosphate export during the light period to hexose and/or hexose-phosphate export at night, circumventing the triose-phosphate translocator. This flexibility is based on complex regulatory networks, not well understood until now. Among others, sugars play a pivotal role in this network complicating attempts to manipulate carbon metabolism and allocation. This chapter summarizes the biosynthetic pathway of photosynthetic carbon metabolism in C3 plants, focusing mainly on sucrose. Thereafter, the chapter reviews current hypotheses on possible mechanisms underlying source to sink interactions with a possible role of sugars as signal molecules.

*Email: uwe@ipk-gatersleben.de

Eva-Mari Aro and Bertil Andersson (eds): Regulation of Photosynthesis, pp. 109–120.

I. Photosynthetic Carbon Metabolism in C3 Plants

A. General Aspects

During photosynthesis light energy is converted into chemical energy, which is used to reduce atmospheric CO_2 to carbohydrates via the reductive pentose phosphate cycle (RPP or Calvin Cycle). Following carboxylation of ribulose-1,5-bisphosphate (RuBP) by the enzyme ribulose-1,5-bisphosphate carboxylase /oxygenase (Rubisco), 3-phosphoglycerate is formed which is subsequently reduced to triose-phosphates (glyceraldehydes-3-phosphate and dihydroxyacetone-3-phosphate). This reduction is catalysed by the successive action of two enzymes, 3-phosphoglycerate kinase (PGK) and NADP:glyceraldehyde-3-phosphate dehydrogenase (NADP:GAPDH). Continued CO_2 assimilation requires that the CO_2 acceptor, RuBP, is constantly regenerated. Therefore, part of the newly formed triose-phosphates is used to regenerate RuBP. Excessive triose-phosphates can be channeled into starch biosynthesis, which takes place in the stroma of chloroplasts, or they can be exported into the cytosol as precursors for sucrose synthesis.

Besides triose-phosphates, also erythrose-4-phosphate (E4P) and ribose-5-phosphate (R5P) are allocated to other metabolic pathways (Fig. 1). Condensation of E4P and phosphoenolpyruvate (PEP) represents the starting point of the shikimate pathway providing intermediates for aromatic compounds. Since chloroplasts do not contain a complete glycolytic pathway, PEP has to be imported into plastids via the phosphoenolpyruvate/phosphate translocator, which recently has been shown to be essential for the synthesis of aromatic amino acids in *Arabidopsis* leaves (Streatfield et al., 1999). R5P can be converted to 5-phosphoribosyl 1-pyrophosphate (PRPP), required for a number of metabolic pathways including nucleotide and histidine biosynthesis.

Flux through the different pathways is not static but responds to environmental and developmental changes. To guarantee efficient regeneration of RuBP, distribution of RPP cycle intermediates has to be under rigorous control, which requires extensive cross talk between the individual pathways. Especially communication between chloroplasts and cytosol is essential for adjusting the rate of photosynthesis to the demands of various parts of the plant for photoassimilates. This communication is supposed to occur at the membrane system of the chloroplasts. Two membranes enclose the plastidic stroma, an outer membrane that is permeable with an exclusion limit of 10 kDa, and an inner membrane that controls the substances entering and leaving the plastid. A specific transport system, the triose-phosphate translocator (TPT), which is located in the inner membrane of the chloroplast envelope catalyses the export of triose-phosphates out of the chloroplasts. The TPT mediates the transport of triose-phosphate, 3-phosphoglycerate (3PGA), and inorganic phosphate (P_i) in a strict counter-exchange mode. In the cytosol, inorganic phosphate is released and is made available as the counter-ion for triose-phosphate export during the biosynthetic processes (Flügge and Heldt, 1991).

Starting from triose-phosphate, the sucrose biosynthetic pathway consists of seven enzymatic steps. cDNA or genomic clones of all but one enzyme have been isolated. The remaining enzyme to be cloned is the last enzyme of the pathway, sucrose-6-phosphate phosphatase (SPP). The key regulatory steps of sucrose biosynthesis are supposed to be the interconversion of fructose-1,6-bisphosphate (FBP) and fructose-6-phosphate (F6P), and the formation of sucrose-6-phosphate (S6P) from UDP-glucose (UDPG) and F6P (Huber et al., 1985; Stitt and Quick 1989; Daie, 1993). The interconversion of FBP and F6P involves three enzymatic activities: the cytosolic fructose-1,6-bisphosphatase (cyFBPase) catalyses the forward reaction; the backward reaction is catalyzed by phosphofructokinase (PFK); the third enzyme activity, pyrophosphate:fructose-6-phosphate-1-phosphotransferase (PFP) is able to catalyze the reaction in both directions. In contrast only one

Abbreviations: 3PGA – 3-phosphoglycerate; AGPase – adenine-5′-diphosphate glucose pyrophosphorylase; CyFBPase – cytosolic fructose-1,6-bisphosphatase; E4P – erythrose-4-phosphate; F2,6BP – fructose-2,6-bisphosphate; F2,6Pase – fructose-2,6-bisphosphatase; F6P – fructose-6-phosphate; FBP – fructose-1,6-bisphosphate; GA – gibberellic acid; GAPDH – glyceraldehyde 3-phosphate dehydrogenase; HK – hexokinase; HT – hexose transporter; NADP – nicotinamide adenine dinukleotide phosphate; PEP – phosphoenolpyruvate; PF2K – 6-phosphofructo-2-Kinase; PFK – ATP-dependent phosphofructokinase; PFP – pyrophosphate:fructose-6-phosphate 1-phosphotransferase; PGK – 3-phosphoglycerate Kinase; Pi – inorganic phosphate; PRPP – 5-phosphoribosyl 1-pyrophosphate; R5P – ribose-5-phosphate; RPP – reductive pentose phosphate; RuBP – ribulose-1,5-bisphosphate; S6P – sucrose-6-phosphate; SNF1 kinase – sucrose nonfermenting 1-kinase; SPP – sucrose-6-phosphate phosphatase; SPS – sucrose-6-phosphate synthase; SUT – sucrose transporter; TPT – triose-phosphate translocator; UDPG – Uridine-5′-diphosphate glucose

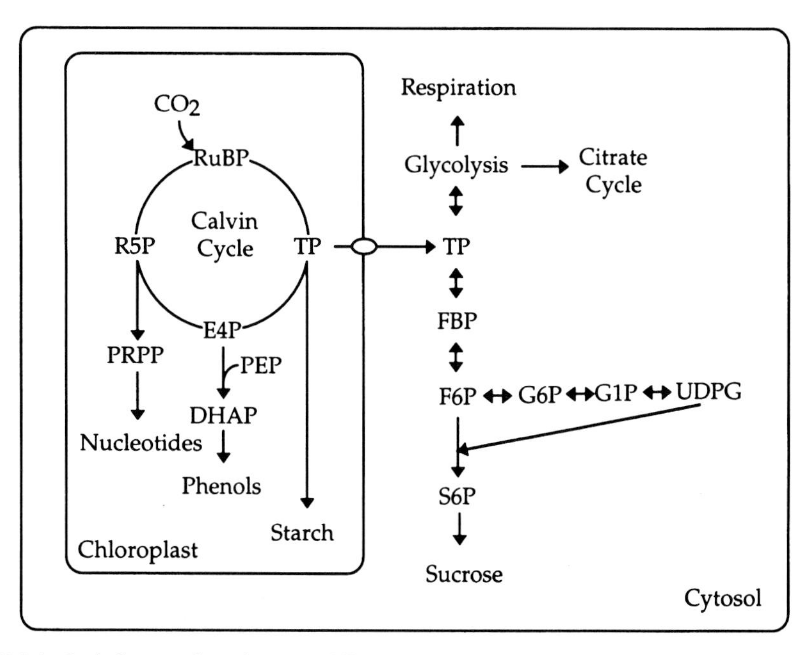

Fig. 1. Allocation of Calvin Cycle intermediates between different metabolic pathways. RuBP, Ribulose-1,5-bisphosphate; TP, triose-phosphate; E4P, erythrose-4-phosphate; PEP, phosphoenolpyruvate; DHAP, 3-deoxy-D-arabinoheptulosonic acid-7-phosphate; PRPP, 5-phosphoribosyl 1-pyrophoshate; R5P, ribose-5-phosphate; FBP, fructose-1,6-bisphosphate; F6P, fructose-6-phosphate; G6P, glucose-6-phosphate; G1P, glucose-1-phosphate; UDPG, UDP-glucose; S6P, sucrose-6-phosphate.

enzyme sucrose-phosphate synthase (SPS) is responsible for the formation of S6P, which is converted to the final product sucrose via SPP.

The activities of the enzymes cyFBPase and PFP are subject to allosteric control by fructose-2,6-bisphosphate (F2,6BP), which stimulates PFP and inhibits cyFBPase (Stitt, 1990). The level of F2,6BP in vivo is determined by the relative activities of 6-Phosphofructo-2-kinase (PF2K) and fructose-2,6-bisposphatase (F2,6Pase). Bifunctional PF2K/F2,6Pase enzymes exist in plants, like in animals. A corresponding cDNA clone has recently been isolated from potato (Draborg et al., 1999). To assess the extent to which F2,6BP regulates carbon metabolism a modified rat liver enzyme possessing only PF2K activity was introduced into tobacco plants (Scott and Kruger, 1995; Scott et al., 1995). Analysis of the transgenic plants revealed that F2,6BP inhibits flux of assimilates towards sucrose and stimulates flux to starch during photosynthesis which is consistent with the proposed model of F2,6BP action.

B. Regulation of Sucrose-6-Phosphate Synthase

The activity of SPS is modulated by several mechanisms, which act at different levels. In spinach SPS has been shown to be regulated by metabolites (Doehlert and Huber, 1983), by posttranslational protein modification (Huber et al., 1989), and changes in the amount of protein (Walker and Huber, 1989). The enzyme is activated by glucose-6-phosphate and inhibited by inorganic phosphate. Application of gibberellic acid (GA) to spinach and soybean leaves resulted in increased levels of SPS protein, indicating that GA is one of the endogenous hormonal factors regulating the steady-state level of SPS protein (Cheikh and Brenner, 1992; Cheikh et al., 1992). In addition, GA application has been shown to increase the photosynthetic rate of maize (Wareing et al., 1968). Thus, it is tempting to speculate that GA is part of the signaling pathway regulating the rate of photosynthetic carbon assimilation. SPS is subject to regulation by multi-site-seryl-phosphorylation (McMichael et al., 1993; Toroser and Huber, 1997; Toroser et al., 1998). Phosphorylation of serine residue 158 (Ser158) of the spinach enzyme lead to a reduced substrate (fructose-6-phosphate) and effector (glucose-6-phosphate) affinity and most likely is responsible for the light/dark regulation of the enzyme (McMichael et al., 1993). In vitro it could be shown that two SNF1-related protein kinases from spinach are able to phosphorylate Ser158, suggesting a possible role of these protein kinases in light/dark regulation of SPS (Sugden et al., 1999). The role of Ser158 has recently been analyzed by site-directed mutagenesis (Toroser et al., 1999). Changing Ser158 to alanine abolished phosphorylation and thereby dark inactivation of spinach SPS in transgenic tobacco plants. Interestingly, exchanging Ser158 to glutamate lead to a constitutive de-activated form, suggesting that the negative charge is responsible for the regulation of the enzyme activity. Phosphorylation of Ser424 activates the enzyme and might be involved in the activation of SPS during osmotic stress (Toroser and Huber, 1997).

Besides the two phosphorylation sites discussed above, additional phosphorylation site(s) are proposed to be involved in the binding of 14-3-3 proteins to SPS. 14-3-3 proteins were first identified as abundant brain proteins and named according to their mobility in protein gels. They bind target proteins in a sequence-specific and phosphorylation-dependent manner. In plants, the 14-3-3 proteins are encoded by gene families but the function of isoforms is still unknown (for recent review see, MacKintosh, 1998; Finnie et al., 1999). Moorhead and co-workers (1999) found a number of plant proteins interacting with digoxygenin labeled yeast 14-3-3 proteins. Among those, SPS could be identified. This observation is in agreement with the analysis of Toroser et al. (1998) who co-purified 14-3-3 proteins with spinach SPS. Although binding of the 14-3-3 proteins to SPS was found in both studies, the consequence of binding was different. Moorhead et al. (1999) reported that 14-3-3 binding activates SPS, whereas Toroser et al. (1998) suggest inactivation of SPS upon 14-3-3 binding. Multi-site phosphorylation and / or isoform-specific interactions may be responsible for the differences observed. Future experiments will have to clarify the function of 14-3-3 binding to SPS, and whether binding is involved in signal transduction processes regulating sucrose synthesis in source leaves according to sink demand.

Despite the numerous activities associated with 14-3-3 proteins, a major role as 'molecular clamp' facilitating protein:protein interactions is becoming apparent (Fig. 2). Assuming that SPS activity is regulated by the interaction with different polypeptides the contradictory results obtained may be explained by the differences in the physiological state of the tissues analyzed. One possible binding partner for SPS has recently been speculated to be UDP-glucose pyrophosphorylase (Toroser et al., 1998), other partners may be protein kinases regulating SPS activity.

II. Sink Regulation of Photosynthesis

A. The Role of Leaf Starch

Under normal growth conditions photosynthesis will

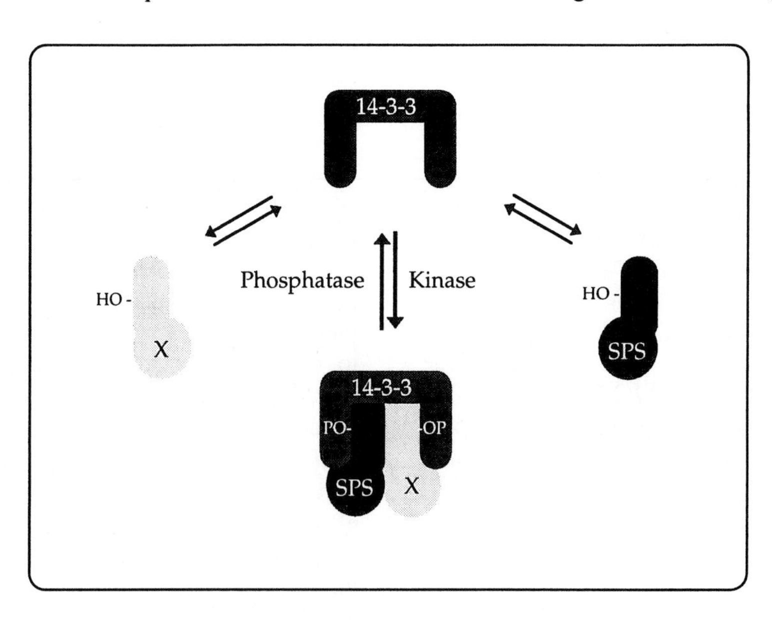

Fig. 2. Interaction of phospho-SPS with 14-3-3 proteins. Following seryl- phosphorylation 14-3-3 proteins bind to SPS and facilitate binding of interacting proteins. X, unknown binding partner; SPS, sucrose-6-phosphate synthase.

mainly be limited by light and CO_2 availability. Under saturating light and CO_2 concentration, however, photosynthesis may be controlled by synthesis and utilization of endproducts. The major endproducts of C3 photosynthesis are starch and sucrose. Sucrose is synthesized in the cytosol and serves as the major export form of carbohydrates from source leaves. Among others, sucrose biosynthesis will be controlled by the efficiency of assimilate export. In case of limited sink demand sucrose export and thereby biosynthesis will cease which consequently results in a reduced triose-phosphate export from the chloroplasts. Thus, plastidic 3PGA levels will raise and starch biosynthesis will be favored due to activation of ADP-glucose pyrophosphorylase (AGPase). In the case of limited carbon export, leaf starch can be considered as the first immediate sink. Thus, the rate of starch biosynthesis could be an important determinant to maintain a high photosynthetic capacity under conditions of limited sink demand.

To investigate the role of leaf starch in the regulation of photosynthesis, two starchless *Arabidopsis thaliana* mutants (TL25 and TL46) have been characterized for their photosynthetic capacity under varying light and CO_2 conditions (Sun et al., 1999). Compared to the wild type, starchless *Arabidopsis* mutants showed a reduced photosynthetic capacity. The rate of photosynthesis correlated with the rate of starch synthesis. Since the inhibition of starch accumulation in TL25 and TL46 was not restricted to leaves, a feedback inhibition of photosynthesis due to reduced sink strength cannot be excluded. To distinguish between immediate and distant sinks Ludewig et al. (1998) analyzed transgenic potato plants with a leaf-specific reduction in AGPase activity. Leaf-specific inhibition of AGPase activity was achieved by expressing an AGP-B antisense RNA under control of the leaf/stem-specific ST-LS1 promoter (Stockhaus et al., 1989). Biochemical analysis revealed that the transgenic potato plants had a strong reduction in transitory leaf starch but starch accumulation in tubers was indistinguishable from the wild type. Consistent with the data obtained with starchless *Arabidopsis* mutants, a positive correlation between starch accumulation and photosynthetic capacity was observed. Usually a negative correlation between the amount of soluble sugars in leaves and the rate of photosynthesis is expected. Interestingly, expression of photosynthetic genes did not decline under elevated atmospheric CO_2 concentration although the level of soluble sugars increased in transgenic plants with reduced leaf starch content. Similarly, long term exposure of *Arabidopsis* plants to low temperatures lead to an increase of soluble sugars in leaves without a concomitant decrease in the rate of photosynthesis and expression of the photosynthetic genes (Strand et al., 1997).

B. Strategies to Alter Source to Sink Relations

Changes in source to sink relations have been shown to alter the rate of photosynthesis in many experimental systems. Partial defoliation of bean plants, for instance, lead to an increased photosynthetic rate of the remaining leaves (Wareing et al., 1968). The increase in photosynthetic rate could be prevented by simultaneous removal of half of the root system. De-graining of wheat is another example in which decreased sink size lead to decreased photosynthetic rates (Blum et al., 1988). Besides the removal of organs hot and cold girdles have been applied to disconnect sink and source tissues. A significant disadvantage of the techniques discussed is the simultaneous impairment of all transport processes including sucrose, amino acid, mineral, and phytohormon transport. To circumvent this problem transgenic plants with either decreased levels of the sucrose transporter (Riesmeier et al., 1994) or elevated levels of a yeast-derived cell wall invertase (von Schwaewen et al., 1990; Stitt et al., 1990) have been constructed. Although, these plants clearly demonstrated the importance of sucrose export form source leaves, strong pleiotropic effects, caused by the constitutive expression of the export block, rendered these plants unsuitable for a deeper analysis of source to sink interactions. Problems associated with the constitutive expression of the transgene have recently been overcome by using a chimeric ethanol inducible promoter system to drive expression of yeast invertase in transgenic plants (Caddick et al., 1998).

C. Models of 'Sink Signals' in the Regulation of Photosynthesis

Based on the data accumulated it has been hypothesized that sink signals, adjusting source metabolism according to sink demand, must exist in plants (Herold, 1980). Mainly two concepts have been put forward to explain the so-called 'sink signal' (Fig. 3): hormonal regulation of photosynthesis and endproduct inhibition of photosynthesis. According

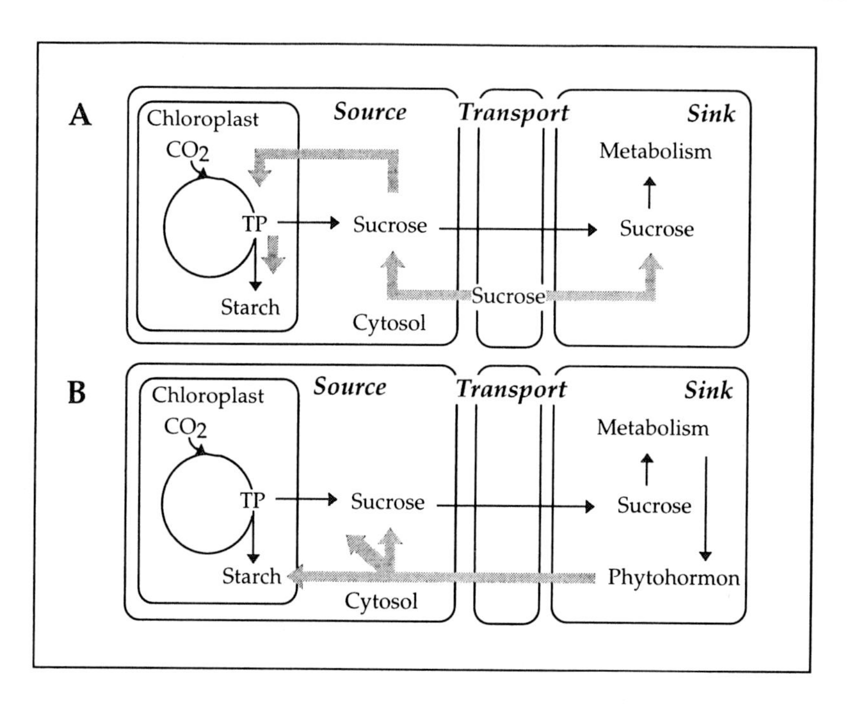

Fig. 3. Hypothetical sink signals (bold, gray arrows) regulating photosynthetic carbon metabolism in plant tissues. A. Surcose or derived sugars not utilized in sink tissues lead to down-regulation of photosynthesis. B. Hormonal signals are produced in sink tissues which regulate assimilate production in source leaves. TP, triose phosphate.

to the model of hormonal regulation, variation in the activity of sink tissues will be associated with changes in the content of phytohormons. These altered phytohormon levels will transmit appropriate messages to photosynthetic cells adjusting the rate of photosynthesis to sink demand. This concept is supported by the observation that application of cytokinin can partially overcome down-regulation of photosynthesis caused by the removal of sinks (Wareing et al., 1968). In addition to cytokinins, abscisic acid and gibberellins have been implicated to be involved in linking sink and source tissues.

According to the second model assimilate accumulation in source tissues is supposed to be responsible for inhibition of photosynthesis. This hypothesis is supported by a negative correlation between assimilate content and leaf photosynthetic rates found in most plant species investigated (Goldschmidt and Huber, 1992; Martinez-Carrasco et al., 1993). Down-regulation of photosynthesis may be achieved by fine regulation of biosynthetic enzymes or the adjustment of gene expression. Fine regulation of photosynthetic sucrose biosynthesis may be initiated by the sucrose-mediated inhibition of the activity of SPS. As a consequence of reduced SPS activity, cytosolic phosphate levels will decline and hexose-phosphates and triose-phosphates levels will increase. Accumulation of triose-phosphates in turn will activate F2,6BP kinase which leads to the accumulation of the signal metabolite F2,6BP. F2,6BP is a potent inhibitor of cyFBPase slowing down the channeling of assimilates towards the sucrose biosynthetic pathway. As a consequence of reduced sucrose biosynthesis and decreased cytosolic phosphate content, triose-phosphate export from chloroplasts will cease. Consequently, plastidic triose-phosphate and 3PGA levels will increase which activates AGPase, the key regulatory enzyme of starch biosynthesis. Thus, more photosynthates are stored as starch in chloroplasts finally leading to the down-regulation of photosynthesis (Fig. 4).

III. Sugar Regulation of Gene Expression

A. General Considerations

Fine regulation of carbon metabolism by metabolites can be considered as short term response to changing demands for assimilates. In case of continuous changes of assimilate demand altered gene expression will accompany metabolic adjustments. The

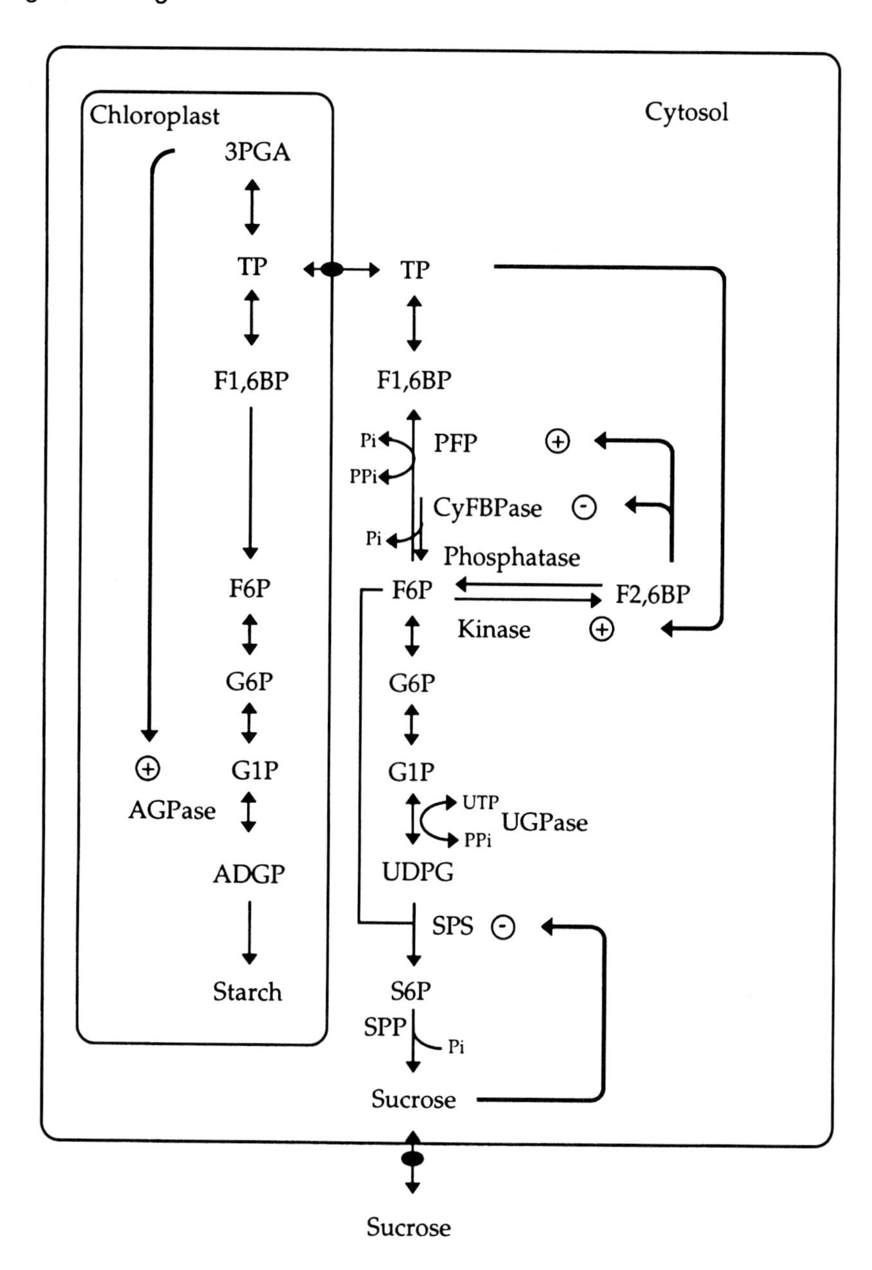

Fig. 4. Fine regulation of sucrose and starch synthesis in mesophyll cells. 3PGA, 3-phopshoglycerate; F1,6BP, fructose-1,6-bisphosphate; Pi, inorganic phosphate; Ppi, inorganic pyrophosphate; AGPase, ADP-glucose pyrophosphorylase; ADPG, ADP-glucose, cyFBP, cytosolic fructose-1,6-bisphosphatase; PFP, pyrophosphate:fructose-6-phopshate 1-phosphotransferase; UGPase, UDP-glucose pyrophosphorylase; SPS, sucrose-6-phosphate synthase; SPP, sucrose-6-phopshate phosphatase. +, up-regulation by indicated metabolite; –, down-regulation by indicated metabolite.

underlying molecular mechanism is largely unknown, but several lines of evidence suggest a prominent role of sugars in the down-regulation of photosynthetic genes. In agreement with this hypothesis several genes have been shown to be regulated by external sugars (for recent review see Koch, 1996; Smeekens, 1998; Roitsch, 1999; Sheen et al., 1999; Yu, 1999; Graham and Martin, 2000). Depending on their response to sugars plant genes have been grouped into three classes: down-regulated, up-regulated, and non-responding genes. Photosynthetic genes belong to the class of down-regulated genes, which is in agreement with the hypothesis that sugars are involved in sink regulation of photosynthesis. Theoretically,

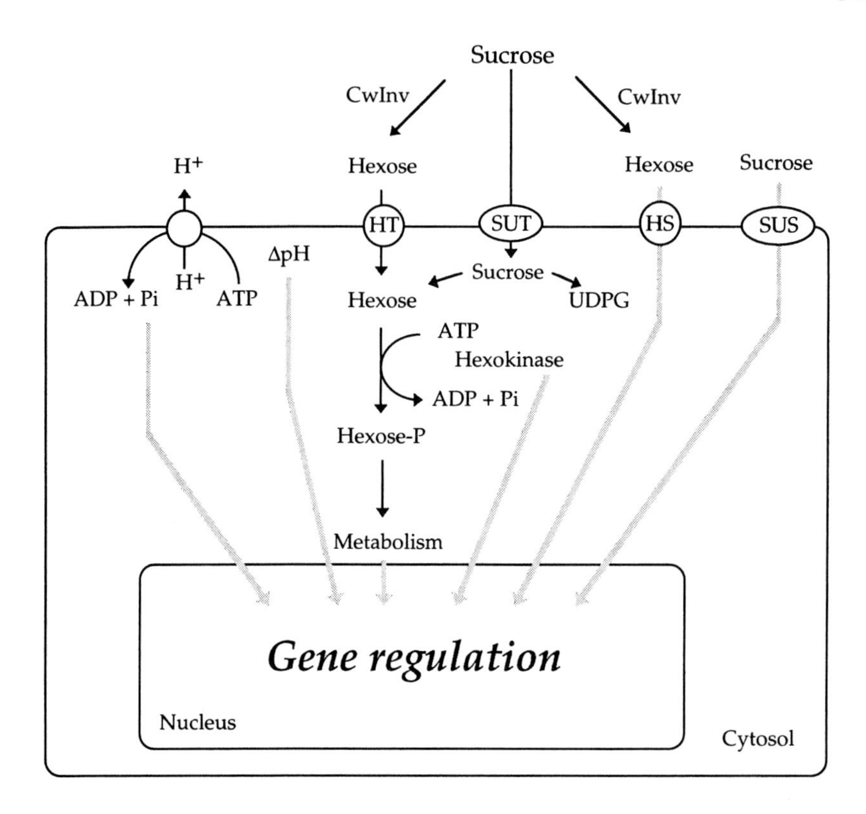

Fig.5. Possible signal transduction pathways (bold, gray arrows) of sugar mediated gene regulation in plants. CwInv, cell wall-bound invertase; HT, Hexose transporter; SUT, sucrose transporter; HS, trans-membrane hexose sensor; SUS, transmembrane sucrose sensor; ATPase, plasma-membrane bound ATPase; UDPG, UDG-glucose.

several concepts involved in sugar sensing can be conceived (Fig. 5).

Membrane-located receptors could monitor external sugar concentrations. If the external sugar concentration exceeds a given threshold a signal is generated in the absence of further transport and/or metabolism. Or, sensing might occur by means of (limited) metabolism of the respective sugar, for instance mediated by hexokinases. Signaling in this case could be an integrational event brought about by a sugar-sensing complex between hexokinases and transporter molecules as has been suggested by Jang and Sheen (1997). Alternatively, flux through respective metabolic pathways might be required. In addition, changes in the carbohydrate status might give rise to metabolic signals such as altered cellular redox states, cellular phosphate content, or changed intracellular pH-values, which could then modulate sugar-responsive gene expression. In soybean, sugar-inducible gene expression could be modulated by phosphate, suggesting that due to the accumulation of hexose-phosphates a concomitant reduction of cellular phosphate levels occurs which leads to altered gene expression (Sadka et al., 1994). In general, mechanisms underlying sugar sensing and signal transduction in plants are largely unknown. As a model system, mainly yeast has been employed to unravel possible mechanisms in plants. In yeast hexokinase, the enzyme that catalyses the ATP-dependent phosphorylation of hexoses, has been implicated as a hexose sensor (Rose et al., 1991). In addition sugar sensing occurs via a hexokinase-independent pathway mediated by two glucose receptors. (Özcan et al., 1996, 1998).

B. Role of Hexokinase in Sugar Sensing

Several lines of evidences suggest a prominent role of hexokinases in sugar sensing. Despite the results obtained from the application of different sugar analogs, which have been questioned in a recent publication (Klein and Stitt, 1998), analysis of transgenic *Arabidopsis* plants with altered hexokinase activities strongly suggest hexokinase being involved in sugar-sensing. In *Arabidopsis* two hexokinase isoforms (AtHK1 and AtHK2) have been isolated

(Jang et al., 1997). Transgenic *Arabidopsis* plants with reduced and elevated levels of AtHK1 or AtHK2 mRNAs have been constructed and the sugar response of seedlings investigated. Overexpression of hexokinase resulted in enhanced sensitivity to glucose whereas reduced levels of hexokinase led to a reduced glucose sensitivity of seedlings (Jang et al., 1997). This is in agreement with the signaling function (Jang and Sheen, 1997) of hexokinases. The experiments were carried out with seedlings grown on medium supplemented with external sugar sources. Therefore it remained unclear whether hexokinase(s) would also be involved in the sink regulation of photosynthesis.

To investigate whether hexokinases are involved in sensing endogenous sugar levels in leaves Dai et al. (1999) overexpressed AtHK1 in transgenic tomato plants. Increasing the activity of the hexokinase isoform 1 (HK1) lead to a reduced chlorophyll content, a reduced photosynthetic rate and accelerated senescence of the transgenic tomato plants. This lead the authors to conclude that HK1 is a regulatory enzyme involved in sugar sensing. In a similar study, Veramendi et al. (1999) transformed potato plants with *Solanum tuberosum* hexokinase-1 (StHK1) in sense and antisense orientation. Despite a wide range of hexokinase activities, no evidence could be found that this StHK1 is involved in sugar sensing in potato. Interestingly, reduced StHK1 activity in leaves of potato plants lead to a starch-excess phenotype, indicating that StHK1 plays a role in transitory starch degradation. A possible explanation for the observed phenotype comes from the biochemical and molecular characterization of spinach HK1 (Wiese et al., 1999). Subcellular localization of HK1 revealed its association with the outer envelope membrane of chloroplasts. Therefore the authors suggested that HK1 is involved in the energization of glucose export from plastids rather than in hexose sensing. The contradictory results may be explained by the different plant species investigated or by different subcellular localizations of the individual hexokinase isoforms. Biochemical data suggest plastidic, mitochondrial and cytosolic hexokinase isoforms, but only in the case of spinach the localization (outer envelope) has been shown.

C. Hexokinase-Independent Sugar Sensing

In addition to a hexokinase-mediated sensing mechanism there is good evidence for hexokinase-independent sensing involving monosaccharide transporter(s) or receptor proteins (Fig. 5). Godt et al. (1995) and Roitsch et al. (1995) presented data showing that sugar induction of cell wall invertase and sucrose synthase could be mimicked by the addition of 3-O-methyl-glucose and 6-deoxy-glucose. Similarly, the sugar-responsive patatin class I promoter was shown to be equally inducible by these analogs in *Arabidopsis* plants (Martin et al., 1997). Induction of sucrose synthase expression might also be caused by changes in the osmotic potential of the cells. This has been shown for one sucrose synthase gene (*Sus1*) in detached *Arabidopsis* leaves fed with either sucrose or non-metabolizable osmotica such as sorbitol and mannitol. Sucrose as well as non-metabolizable sugars led to the induction of *Sus1* mRNA accumulation in an ABA-independent way, as demonstrated by the use of ABA-deficient mutants (Dejardin et al., 1999). More recently, sucrose-specific regulation of a sucrose transport protein has been shown (Chiou and Bush, 1998). Feeding sucrose via the xylem transpiration stream revealed a decline in the amount of sucrose transport protein and its mRNA. Hexoses did not elicit the response, and manno-heptulose, a potent hexokinase inhibitor, did not block the sucrose effect. Furthermore translation of the ATB2 bZIP transcription factor from *Arabidopsis* has been shown to be specifically repressed by sucrose (Rook et al., 1998).

In maize regulation of cell wall invertase has been shown to occur at the transcriptional and post-transcriptional level. The *Incw*1 gene encodes two transcripts with differences in the length of the 3′-untranslated region. The shorter transcript is inducible by metabolizable sugars which results in the accumulation of active protein. The larger transcript is inducible by nonmetabolizable sugars including 2-deoxy-glucose, excluding a role of hexokinase in signal transduction (Cheng et al., 1999). Analyzing transgenic plants expressing invertase either in the cell wall, the vacuolar compartment, or the cytosol, Herbers et al. (1996) could demonstrate that down-regulation of photosynthetic genes requires sugar accumulation either in the vacuole or the cell wall. Based on this result it was speculated that the putative sugar sensor is membrane-bound.

D. Sugar Sensing Mutants

As evident multiple pathways will be responsible for sugar sensing and signaling in plants. To elucidate the underlying regulatory pathways *Arabidopsis* has been used as model system to isolate sugar sensing mutants. Basically two strategies to obtain sugar sensing mutants have been followed. In the first approach sugar-modulated gene expression has been taken as screening system. To this end promoter sequences of sugar-responsive genes have been fused to marker genes allowing easy detection of gene expression (Dijkwel et al., 1997; Martin et al., 1997; Mita et al., 1997). Constructs of either sugar induced (beta-amylase, patatin) or sugar repressed genes (plastocyanin) have been transformed into *Arabidopsis*. Homozygous plants expressing the respective chimeric gene constructs in a sugar regulated manner have been mutagenized and altered transgene expression of seedlings grown on sugar containing medium has been analyzed. The second approach is based on the observation that germination of *Arabidopsis* seeds is delayed on sugar containing medium. Therefore, mutants could be selected for sugar insensitive germination. Analysis of the first mutants have indicated possible cross-talk between sugar, light, and hormonal signals, which needs further clarification.

E. Summary of Possible Sugar Sensing Pathways

A summary of the current models of sugar sensing pathways is given in Fig. 5. Sucrose, synthesized in photosynthetically active tissues is translocated to sinks. In sink tissues sucrose can either be hydrolyzed by cell wall-bound invertases (cwInv) to glucose and fructose or it can be transported into sink cells by sucrose transport proteins (SUT) or through plasmodesmata. By analogy to yeast, members of the sucrose transporter family (SUS) have been proposed to be involved in sugar sensing (Lalonde et al., 1999). Direct effects of sucrose have been demonstrated for the expression of the sucrose transporter *BvSUT1* from sugar beet (Chiou and Bush, 1998). Alternatively, hexoses derived from sucrose hydrolysis can be translocated into sink cells via hexose transporters (HT). Subsequently, cytoplasmic hexoses can be phosphorylated by hexokinases which are hypothesized to transmit sugar signals to the nucleus (Jang and Sheen, 1997) leading to altered gene regulation. In addition to the hexokinase sensing pathway alternative pathways exist. Examples include sensing of extracellular sugars by putative transmembrane receptor proteins, the requirement for further metabolism of sugars, changes in cytoplasmic pH, and altered cytoplasmic phosphate content (Fig. 5).

F. Concluding Remarks

Regulation of photosynthesis is a complex phenomenon, which involves many interacting networks coordinating the biosynthetic capacity of leaves at the whole plant level. Regulation involves biochemical adjustments and changes in gene expression. Studies to unravel the underlying mechanisms usually involve artificial manipulation of sink or source tissues. Based on feeding experiments, there is no doubt that sugars are important regulators of gene expression. It is clear that multiple sugar-sensing pathways operate in plants and that sucrose as well as hexoses can initiate independent responses. Whether they are involved in the regulation of photosynthesis according to sink demand needs to be demonstrated. Analysis of the sugar-sensing mutants isolated should provide evidences for the role of sugars in the regulation of source to sink interactions.

Acknowledgment

The work has been supported by the Deutsche Forschungsgemeinschaft SFB 363.

References

Blum A, Mayer J, and Golan G (1988) The effect of grain number per ear (sink size) on source activity and its water-relations in wheat. J Exp Bot 39: 106–114

Caddick MX, Greenland AJ, Jepson I, Krause K-P, Qu N, Riddell M, Salter MG, Schuch W, Sonnewald U and Tomsett AB (1998) An ethanol inducible gene switch for plants used to manipulate carbon metabolism. Nat Biotech 16: 177–180

Cheikh N and Brenner ML (1992) Regulation of key enzymes of sucrose biosynthesis in soybean leaves. Plant Physiol 100: 1230–1237

Cheikh N, Brenner ML, Huber JL and Huber SC (1992) Regulation of sucrose phosphate synthase by gibberellins in soybean and spinach plants. Plant Physiol 100: 1238–1242

Cheng W-H, Taliecio EW and Chourey PS (1999) Sugars modulate an unusual mode of control of the cell-wall invertase gene (Incw1) through its 3′untranslated region in cell suspension culture of maize. Proc Natl Acad Sci USA 96: 10512–10517

Chiou T-J and Bush DR (1998) Sucrose is a signal molecule in assimilate partitioning. Proc Natl Acad Sci USA 95: 4784–4788

Dai N, Schaffer A, Petreikov M, Shahak Y, Giller Y, Ratner K,

Levine A and Granot D (1999) Overexpression of *Arabidopsis* hexokinase in tomato plants inhibits growth, reduces photosynthesis, and induces rapid senescence. Plant Cell 11: 1253–1266

Daie J (1993) Cytosolic fructose-1,6-bisphosphatase: A key enzyme in the sucrose biosynthetic pathway. Photosynth Res 38: 5–14

Dejardin A, Sokolov LN and Kleczkowski A (1999) Sugar/osmoticum levels modulate differential abscisic acid-independent expression of two stress-responsive sucrose synthase genes in *Arabidopsis*. Biochem J 344: 503–509

Dijkwel PP, Kock PAM, Bezemer R, Weisbeek PJ and Smeekens SCM (1996) Sucrose represses the developmentally controlled transient activation of the plastocyanin gene in *Arabidopsis thaliana* seedlings. Plant Physiol 110: 455–463

Doehlert DC and Huber SC (1983) Regulation of spinach leaf sucrose-phosphate synthase by Glc6P, inorganic phosphate and pH. Plant Physiol 73: 989–994

Draborg H, Villadsen D and Nielsen TH (1999) Cloning and expression of a bifunctional fructose-6-phosphate, 2-kinase/fructose-2,6-bisphosphatase from potato. Plant Mol Biol 39: 709–720

Finnie C, Borch J and Collinge DB (1999) 14-3-3 proteins: Eukaryotic regulatory proteins with many functions. Plant Mol Biol 40: 545–554

Flügge UI and Heldt HW (1991) Metabolite translocators of the chloroplast envelope. Annu Rev Plant Physiol Plant Mol Biol 42: 129–144

Godt DE, Riegel A and Roitsch T (1995) Regulation of sucrose synthase expression in *Chenopodium rubrum*: Characterisation of sugar-induced expression in photoautotrophic suspension cultures and sink tissue specific expression in plants. J Plant Physiol 146: 231–238

Goldschmidt EE and Huber SC (1992) Regulation of photosynthesis by endproduct accumulation in leaves of plants storing starch, sucrose, and hexose sugars. Plant Physiol 99: 1443–1448

Graham IA and Martin T (2000) Control of photosynthesis, allocation and partitioning by sugar regulated gene expression. In: Leegood RC, Sharkey TD and von Caemmerer S (eds) Photosynthesis: Physiology and Metabolism, pp 233–248. Kluwer Academic Publishers, Dordrecht

Herbers K, Meuwly P, Frommer WB, Métraux J-P and Sonnewald U (1996) Systemic acquired resistance mediated by the ectopic expression of invertase: Possible hexose sensing in the secretory pathway. Plant Cell 8: 793–803

Herold A (1980) Regulation of photosynthesis by sink activity—the missing link. New Phytol 86: 131–144

Huber SC, Kerr PS and Torres WK (1985) Regulation of sucrose synthesis and movement. In: Heath RL and Preiss J (eds) Regulation of Carbon Partitioning in Photosynthetic Tissue, pp 199–214. Williams and Wilkins, Baltimore

Huber SC, Nielsen TH, Huber JLA and Pharr DM (1989) Variation among species in light activation of sucrose-phosphate synthase. Plant Cell Physiol 30: 277–285

Jang JC and Sheen J (1997) Sugar sensing in higher plants. Trends Plant Sci 2: 208–213

Jang JC, Leon P, Zhou L and Sheen J (1997) Hexokinase as a sugar sensor in higher plants. Plant Cell 9: 5–19

Klein D and Stitt M (1998) Effects of 2-deoxyglucose on the expression of rbcS and the metabolism of *Chenopodium rubrum* cell-suspension cultures. Planta 205: 223–234

Koch KE (1996) Carbohydrate-modulated gene expression in plants. Annu Rev Plant Physiol Plant Mol Biol 47: 509–540

Lalonde S, Boles E, Hellmann H, Barker L, Patrick JW and Frommer WB (1999) The dual function of sugar carriers: Transport and sugar sensing. Plant Cell 11: 707–726

Ludewig F, Sonnewald U, Kauder F, Heineke D, Geiger M, Stitt M, Müller-Röber BT, Gillissen B, Kühn C and Frommer WB (1998) The role of starch in acclimation to elevated atmospheric CO_2. FEBS Lett 429: 147–151

MacKintosh C (1998) Regulation of cytosolic enzymes in primary metabolism by reversible protein phosphorylation. Curr Opin Plant Biol 1: 224–229

Martin T, Hellmann H, Schmidt R, Willmitzer L and Frommer WB (1997) Identification of mutants in metabolically regulated gene expression. Plant J 11: 53–62

Martinez-Carrasco R, Cervantes E, Perez P, Morcuende R and Martin del Molino IM (1993) Effect of sink size on photosynthesis and carbohydrate content of leaves of three spring wheat varieties. Physiol Plant 89: 453–459

McMichael RW Jr, Klein RR, Salvucci ME and Huber SC (1993) Identification of the major regulatory phosphorylation site in sucrose-phosphate synthase. Arch Biochem Biophys 307: 248–252

Mita S, Hirano H and Nakamura K (1997) Negative regulation in the expression of a sugar-inducible gene in *Arabidopsis thaliana*. A recessive mutation causing enhanced expression of a gene for beta-amylase. Plant Physiol 114: 575–582

Moorhead G, Douglas P, Cotelle V, Harthill J, Morrice N, Meek S, Deiting U, Stitt M, Scarabel M, Aitken A and MacKintosh C (1999) Phosphorylation-dependent interaction between enzymes of plant metabolism and 14-3-3 proteins. Plant J 18: 1–12

Özcan S, Dover J, Rosenwald AG, Wölf S and Johnston M (1996) Two glucose transporters in *Saccharomyces cerevisiae* are glucose sensors that generate a signal for induction of gene expression. Proc Natl Acad Sci USA 93: 12428–12432

Özcan S, Dover J and Johnston M (1998) Glucose sensing and signalling by two glucose receptors in the yeast *Saccharomyces cerevisiae*. EMBO J 17: 2566–2573

Riesmeier JW, Frommer WB and Willmitzer L (1994) Evidence for an essential role of the sucrose transporter in phloem loading and assimilate partitioning. EMBO J 13: 1–7

Roitsch T (1999) Source-sink regulation by sugars and stress. Curr Opin Plant Biol 2: 198–206

Roitsch T, Bittner M and Godt DE (1995) Induction of apoplastic invertase by D-glucose and a glucose analog and tissue specific expression suggest a role in sink-source regulation. Plant Physiol 108: 285–294

Rook F, Gerrits N, Kortstee A, van Kampen M, Borrias M, Weisbeek P and Smeekens S (1998) Sucrose-specific signalling represses translation of *Arabidopsis* ATB2 bZIP transcription factor gene. Plant J 15: 253–263

Rose M, Albig W and Entian KD (1991) Glucose repression in *Saccharomyces cerevisiae* is directly associated with hexose phosphorylation by hexokinase PI and PII. Eur J Biochem 199: 511–518

Sadka A, DeWald DB, May GD, Park WD and Mullet J.E. (1994) Phosphate modulates transcription of soybean VspB and other sugar-inducible genes. Plant Cell 6: 737–749

von Schaewen A, Stitt M, Schmidt R, Sonnewald U and Willmitzer

L (1990) Expression of a yeast-derived invertase in the cell wall of tobacco and *Arabidopsis* plants leads to accumulation of carbohydrate and inhibition of photosynthesis and strongly influences growth and phenotype of transgenic tobacco plants. EMBO J 9: 3033–3044

Scott P and Kruger NJ (1995) Influence of elevated fructose-2,6-bisphosphate levels on starch mobilization in transgenic tobacco leaves in the dark. Plant Physiol 108: 1569–1577

Scott P, Lange AJ, Pilkis SJ and Kruger NJ (1995) Carbon metabolism in leaves of transgenic tobacco (*Nicotiana tabacum* L.) containing elevated fructose 2,6-bisphosphate levels. Plant J 7: 461–469

Sheen J, Zhou L and Jang J-C (1999) Sugars as signalling molecules. Curr Opin Plant Biol 2: 410–418

Smeekens S (1998) Sugar regulation of gene expression in plants. Curr Opin Plant Biol 3: 230–234

Stitt M (1990) Fructose-2,6-bisphosphate as a regulatory molecule in plants. Annu Rev Plant Physiol Plant Mol Biol 41: 153–185

Stitt M and Quick WP (1989) Photosynthetic carbon partitioning: Its regulation and possibilities for manipulation. Physiol Plant 77: 633–641

Stitt M, von Schaewen A and Willmitzer L (1990) 'Sink' regulation of photosynthetic metabolism in transgenic tobacco plants expressing yeast invertase in their cell wall involves a decrease of Calvin-cycle enzymes and an increase of glycolytic enzymes. Planta 183: 40–50

Stockhaus J, Schell J and Willmitzer L (1989) Correlation of the expression of the nuclear photosynthetic gene ST-LS1 with the presence of chloroplasts. EMBO J 8: 2445–2451

Strand A, Hurry V, Gustafsson P and Gardström P (1997) Development of *Arabidopsis thaliana* leaves at low temperatures releases the suppression of photosynthesis and photosynthetic gene expression despite the accumulation of soluble carbohydrates. Plant J 12: 605–614

Streatfield SJ, Weber A, Kinsman EA, Hausler RE, Li J, Post-Beittenmiller D, Kaiser WM, Pyke KA, Flügge UI and Chory L (1999) The phosphoenolpyruvate/phosphate translocator is required for phenolic metabolism, palisade cell development, and plastid-dependent nuclear gene expression. Plant Cell 11: 1609–1622

Sugden C, Donaghy PG, Halford NG and Hardie DG (1999) Two SNF1-related protein kinases from spinach leaf phosphorylate and inactivate 3-hydroxy-3-methylglutaryl-coenzyme A reductase, nitrate reductase, and sucrose phosphate synthase in vitro. Plant Physiol 120: 257–274

Sun J, Okita TW and Edwards GE (1999) Modification of carbon partitioning, photosynthetic capacity, and O_2 sensitivity in *Arabidopsis* plants with low ADP-glucose pyrophosphorylase activity. Plant Physiol 119: 267–276

Toroser D and Huber SC (1997) Protein phosphorylation as a mechanism for regulation of sucrose-phosphate synthase under osmotic stress conditions. Plant Physiol 114: 947–955

Toroser D, Athwal GS and Huber SC (1998) Site-specific regulatory interaction between spinach leaf sucrose-phosphate synthase and 14-3-3 proteins. FEBS Lett 435: 110–114

Toroser D, McMichael R Jr, Krause K-P, Kurreck J, Sonnewald U, Stitt M and Huber SC (1999) Site-directed mutagenesis of serine 158 demonstrates its role in spinach leaf sucrose-phosphate synthase modulation. Plant J 17: 407–413

Veramendi J, Roessner U, Renz A, Willmitzer L and Trethewey RN (1999) Antisense repression of hexokinase 1 leads to an over-accumulation of starch in leaves of transgenic potato plants but not to significant changes in tuber carbohydrate metabolism. Plant Physiol 121: 123–133

Walker JLA and Huber SC (1989) Regulation of sucrose-phosphate synthase activity in spinach leaves by protein level and covalent modification. Planta 117: 116–120

Wareing PF, Khalifa MM and Treharne KJ (1968) Rate-limiting processes in photosynthesis at saturating light intensities. Nature 220: 453–457

Wiese A, Gröner F, Sonnewald U, Deppner H, Lerchl J, Hebbeker U, Flügge U-I and Weber A (1999) Spinach hexokinase 1 is located in the outer envelope membrane of plastids. FEBS Lett 461: 13–18

Yu S-M (1999) Cellular and genetic responses of plants to sugar starvation. Plant Physiol 121: 687–693

Chapter 7

Editing, Polyadenylation and Degradation of mRNA in the Chloroplast

Gadi Schuster*
Department of Biology, Technion - Israel Institute of Technology, Haifa 32000, Israel

Ralph Bock
Institut für Biologie III, Universität Freiburg, Schänzlestrasse 1, D-79104 Freiburg, Germany

Summary

In the chloroplasts of higher plants and green algae, mRNAs are transcribed as precursor RNAs which undergo a variety of maturation events, including *cis*- and *trans*-splicing, cleavage of polycistronic messages, processing of 5′ and 3′ ends, and in higher plants, also RNA editing. These post-transcriptional processes play an important role in regulating gene expression and they may be controlled mainly by nuclear-encoded genes. In the first part of this chapter we describe our present knowledge of the process of RNA editing in the chloroplast. Characterization of the editing sites identified in different plants with respect to their possible function, involvement of editing in the control of gene expression and some evolutionary aspects, are discussed. In the second part, we describe recent discoveries on RNA degradation processes in the chloroplast. Several of the proteins involved have been purified and identified. It was found that RNA polyadenylation is required for the mRNA degradation process. This enables us to suggest a model describing the molecular mechanism of mRNA degradation in the chloroplast.

*Author for correspondence, email: gadis@ts.techion.ac.il

Eva-Mari Aro and Bertil Andersson (eds): Regulation of Photosynthesis, pp. 121–136.

I. Introduction

In chloroplasts of higher plants and green algae, mRNAs are transcribed as precursor RNAs which undergo a variety of maturation events, including *cis*- and *trans*-splicing, cleavage of polycistronic messages, processing of 5′ and 3′ ends, and in higher plants, also RNA editing. These post-transcriptional processes play an important role in regulating gene expression and they may be controlled mainly by nuclear-encoded genes. For example, mutants of nuclear genes which affect the accumulation and translation of specific chloroplast transcripts have been characterized in maize (Barkan et al., 1995), *Arabidopsis* (Meurer et al., 1988, 1996) and the green alga *Chlamydomonas reinhardtii* (reviewed in Rochaix, 1996; Goldschmidt-Clermont, 1998). Different aspects of chloroplast mRNA processing and stability have been the subject of recently published reviews (Gruissem and Schuster, 1993; Gruissem and Tonkyn, 1993; Mayfield et al., 1995; Rochaix, 1996; Barkan and Stern, 1998; Drager and Stern, 1998; Goldschmidt-Clermont, 1998; Schuster et al., 1999) and will not be discussed here. In this chapter, we focus on recent developments in two of the post-transcriptional processes in plastid gene expression: RNA editing and the molecular mechanism of mRNA degradation.

II. RNA Editing in the Chloroplast

A. Revision of Genetic Information by RNA Editing

For several decades, the colinear and unidirectional flow of genetic information from DNA via RNA to protein has been viewed as the 'central dogma' of molecular biology, until several unexpected discoveries demanded extensive revisions of this chapter in the textbooks. The isolation of reverse transcriptases capable of converting RNA templates into DNA firmly planted the seed of uncertainty about the unidirectionality of the information flow as a universal principle. The second part of the dogma, the colinearity of DNA, RNA and protein, was first questioned when intervening sequences (introns) were found to interrupt coding regions. More recently, the discovery of protein splicing and RNA editing uncovered additional mechanisms violating the colinearity in the flow of genetic information.

RNA editing was first detected in mitochondrial transcripts of trypanosomes (Benne et al., 1986). These protozoans carry a single large mitochondrion, termed kinetoplast. In spite of good homology with mitochondrial genes from other organisms, most of the genes encoded by the kinetoplast DNA were found not to contain continuous reading frames to specify functional polypeptides. Surprisingly, it turned out that the genuine reading frames are restored at the mRNA level by post-transcriptional insertion or deletion of uridine residues at highly specific sites. The term 'RNA editing' was coined for such post-transcriptional alterations at the single nucleotide level. In the kinetoplast editing system, the site specificity of the nucleotide insertions or deletions is conferred by small *trans*-acting RNA molecules (so-called guide RNAs) which hybridize to the unedited precursor RNA. With the aid of several enzymatic activities organized as a complex called editosome, mismatches between the two RNA molecules are subsequently removed by U insertions or deletions in the mRNA strand of the duplex (for review see Stuart et al., 1997; Alfonzo et al., 1997). Soon after the initial discovery of editing in trypanosomatid mitochondria, editing processes were found to be present also in many other genetic systems (reviewed in Benne, 1996): in the nucleo-cytoplasmic compartment of animal cells (Powell et al., 1987; Sommer et al., 1991), in mitochondria of slime molds (Mahendran et al., 1991), in viral RNAs (Cattaneo et al., 1989; Polson et al., 1996), as well as in plant mitochondria (Covello and Gray, 1989; Gualberto et al., 1989; Hiesel et al., 1989) and chloroplasts (Hoch et al., 1991; Kudla et al., 1992). In the past few years, it has become quite clear that the editing processes in different genetic systems employ widely different mechanisms implying that editing activities may have evolved several times independently. In spite of this diversity, it appears useful to formally distinguish between the two major types of RNA editing: insertional/deletional editing and convertional editing (Benne, 1996; Smith et al., 1997). Whereas the aforementioned kinetoplastid editing is of the insertional/deletional type, the editing processes in plant mitochondria and chloroplasts fall into the convertional type of editing systems.

Abbreviations: PNPase – polynucleotidephosphorylase; PS II – Photosystem II; RNP – RNA binding protein; RT-PCR – reverse transcriptase-polymerase chain reaction; UTR – untranslated region

B. Editing of Chloroplast Messenger RNAs

Chloroplast RNA editing proceeds by pyrimidine-to-pyrimidine conversions at highly specific sites. In plastids of higher plants, the changes introduced into mRNAs by editing are restricted to conversions from C to U. Recently, 'reverse' editing by U-to-C alterations was discovered in chloroplasts of the hornwort *Anthoceros formosae* (Yoshinaga et al., 1996; Yoshinaga et al., 1997). Interestingly, predominance of C-to-U changes with the exception of rare reverse editing events is also characteristic of the editing system in plant mitochondria (Gualberto et al., 1990; Schuster et al., 1990). This is just one of the numerous parallels between the editing systems of the two plant cell organelles (see also below) suggesting that the two systems are possibly related and may share common evolutionary roots. The only significant difference between the plant mitochondrial and chloroplast RNA editing known to date lies in the editing frequencies: more than a thousand editing sites in the mitochondrial transcripts contrast altogether only about 25 sites encoded in the entire maize plastid genome (Maier et al., 1995; reviewed in Maier et al., 1996; Bock et al., 1997b; Mulligan and Maliga, 1997).

The vast majority of editing sites is located within the coding regions and is functionally significant in that the editing results in changes of the coding properties of the affected triplets. In most instances, these alterations lead to restoration of conserved amino acid residues (Fig. 1). Editing can also be involved in the creation of chloroplast reading frames by introducing initiation codons (by editing of an ACG into an AUG start codon; Hoch et al., 1991; Kudla et al., 1992; Neckermann et al., 1994; Fig. 1A) or termination codons for translation (e. g. by converting a CAA into a UAA stop codon; Wakasugi et al., 1996). All these editing-mediated changes in the coding properties of plastid RNAs strongly point to a functional relevance of RNA editing and suggest that editing is an essential processing step in the generation of mature mRNAs which then serve as faithful templates for the organellar protein biosynthesis machinery. The biological relevance of plastid RNA editing was directly demonstrated by the generation of tobacco plants carrying the spinach *psbF* editing site (Fig. 1A, B; Bock et al., 1993, 1994). In tobacco plastids, the *psbF* mRNA does not undergo RNA editing and the correct amino acid sequence is already specified by the genomic sequence (Fig. 1B). When the spinach *psbF* editing site was created within the tobacco *psbF* gene by chloroplast transformation, the tobacco plastids turned out to be incapable of editing this site resulting in the incorporation of a serine instead of a phenylalanine residue in the *psbF* gene product, the β subunit of cytochrome b_{559}. The mutant tobacco plants displayed a photosynthetically deficient phenotype confirming that the change of a serine codon into phenylalanine by RNA editing is a prerequisite for proper function of the cytochrome b_{559} (Bock et al., 1994).

Chloroplast RNA editing exhibits a strong preference for certain codon transitions as well as a strong bias towards transitions affecting the second codon position, again paralleled by a very similar codon transition preference and codon positions bias in plant mitochondria. 87% of the editing sites encoded by the chloroplast genomes of maize, tobacco and black pine are located in the second codon position, 12% affects the first and only 1% affects the third codon position. Some codon transitions are found particularly frequently, such as UCA (Ser) → UUA (Leu) or CCA (Pro) → CUA (Leu), whereas others occur rather rarely, or are not found at all. In general, codon transitions with a pyrimidine nucleotide in the first position predominate greatly, whereas those with a purine in the first codon position are rather rare. These preferences in the relationship of RNA editing with the genetic code are statistically highly significant and, therefore, most likely reflect functional or evolutionary constraints. The nature of these constraints, however, remains enigmatic.

RNA editing in plastid transcripts is usually very efficient: at most sites, the cDNA population shows a virtually complete transition from C to U. Editing occurs in transcripts of both of the two major classes of plastid-encoded genes: the genetic-system genes and the photosynthesis-related genes (Maier et al., 1995). Solely structural RNAs of the plastid genetic apparatus, such as tRNAs and rRNAs, seem to be exempt from processing by RNA editing. Transcripts for the subunits of all major protein complexes of the thylakoid membrane can be affected by RNA editing. Table 1 provides an overview of the editing-site-containing photosynthesis genes encoded by the maize chloroplast genome. There is no edited photosystem II (PSII) gene transcript in maize, but in other species, such as tobacco and spinach, some PSII mRNAs are subject to editing (Kudla et al., 1992; Bock et al., 1993; Fig. 1). As seen in Table 1, it

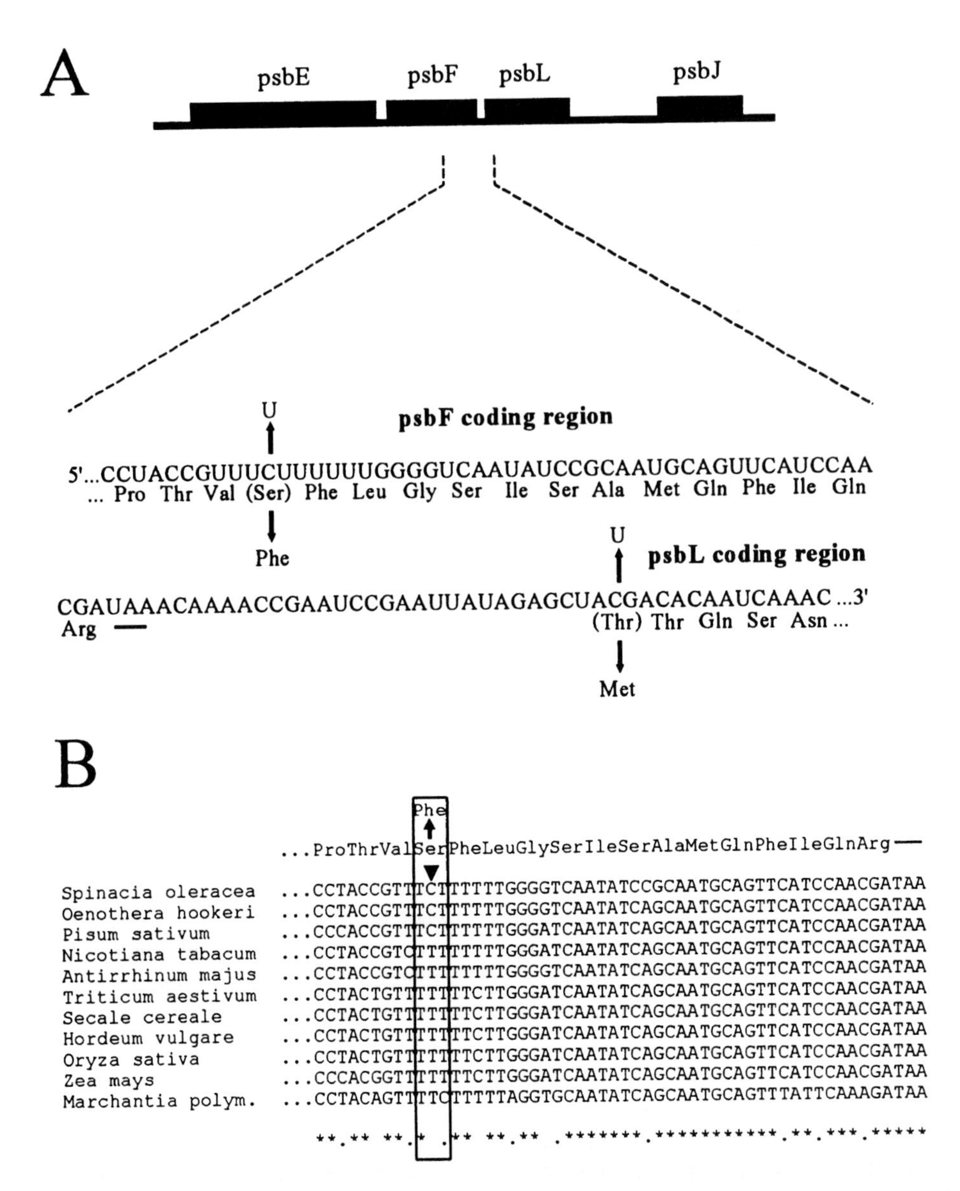

Fig. 1. RNA editing in transcripts of a chloroplast operon containing four genes encoding photosystem II subunits in spinach. (A) Structure of the spinach *psbE* operon and editing sites in the *psbF* and *psbL* coding regions. The *psbE* operon forms a tetracistronic transcription unit encompassing the *psbE*, *psbF*, *psbL* and *psbJ* genes. The *psbE* and *psbF* genes encode the α and β subunits of the cytochrome b_{559}, *psbL* and *psbJ* specify small polypeptides of the photosystem II complex with unknown function. A first RNA editing event changes a genomically encoded UCU serine triplet within the *psbF* coding region into a UUU phenylalanine codon by C-to-U editing in the second codon position. The second editing event creates the initiation codon for *psbL* translation by converting an ACG triplet into a standard AUG start codon (Bock et al., 1993). (B) Partial nucleotide sequence alignment of the 3′ portion of *psbF* genes from various plant species. The boxed amino acid position denotes a phenylalanine residue which is conserved in the *psbF* gene product (β subunit of cytochrome b_{559}) of all photosynthetically active organisms. A TTT or TTC phenylalanine triplet is encoded in the chloroplast DNAs from tobacco, snapdragon, wheat, rye, barley, rice, maize and the liverwort *Marchantia polymorpha*. The chloroplast DNAs from spinach, evening primrose and pea, however, carry a TCT serine triplet in this position which is post-transcriptionally converted into a phenylalanine codon by C-to-U editing in the second codon position (▼). Nucleotide positions identical in all species listed here are marked by asterisks, positions identical in at least nine out of the eleven species are indicated by dots (Bock et al., 1993).

Table 1. RNA editing sites in transcripts of photosynthesis genes encoded in the maize plastid genome (Maier et al., 1995). Asterisks denote intron-containing genes. Note that the *psbF* and *psbL* transcripts edited in spinach (Fig. 1) are not subject to editing in maize.

Protein complex	Plastid-encoded genes	Genes with editing sites	Number of editing sites
Photosystem II	*psbA, psbB, psbC, psbD, psbE, psbF, psbH, psbI, psbJ, psbK, psbL, psbM, psbN, psbT*	–	–
Cytochrome b_6f complex	*petA, petB**, *petD**, *petG, petL, petN*	*petB*	1
Photosystem I	*psaA, psaB, psaC, psaI, psaJ, ycf3**, *ycf4*	*ycf3*	2
NADH dehydrogenase	*ndhA**, *ndhB**, *ndhC, ndhD, ndhE, ndhF, ndhG, ndhH, ndhI, ndhJ, ndhK*	*ndhA*	4
		ndhB	6
		ndhD	1
		ndhF	1
ATPase	*atpA, atpB, atpE, atpF**, *atpH, atpI*	*atpA*	1

is also evident that RNA editing sites are not distributed evenly across the chloroplast genome. Whereas the maize *ndhB* gene encodes as many as six editing sites, many other plastid transcripts do not undergo editing at all (Table 1). It is currently unclear whether or not this uneven distribution of editing sites reflects some functional constraints.

C. Evolutionary Considerations

The discovery of RNA editing has also provided a wealth of interesting evolutionary puzzles: Is editing an evolutionary relic or a recent acquisition? Why is editing maintained and why are editing sites not simply lost by genomic C-to-T mutations? Does editing confer any selective advantage? There is presently no definitive answer to most of these questions. However, phylogenetic studies suggest that chloroplast RNA editing is absent from certain primitive moss lineages as well as from all algae and cyanobacteria (Freyer et al., 1997), the presumptive ancestors of present-day plastids (Gray, 1993). This could suggest that plastid editing is not an ancient trait but rather evolved with or shortly after the plants' conquest of land habitats. In addition, the absence of chloroplast RNA editing from certain lower plant lineages as well as the lack of evolutionary conservation of most (if not all) plastid editing sites (Freyer et al., 1997; Fig. 1B) may indicate that RNA processing by editing does not confer a strong selective advantage over encoding the 'correct' sequence already at the DNA level.

Remarkably, the evolution of plastid RNA editing seems to be paralleled by the evolution of editing in plant mitochondria and those taxa that exhibit plastid RNA editing usually also show editing in their mitochondrial compartment (Hiesel et al., 1994; Malek et al., 1996; Freyer et al., 1997). In turn, species like the liverwort *Marchantia polymorpha* show neither plastid nor mitochondrial editing activities. Interestingly, the absence of editing in the plastids of *Marchantia polymorpha* has been very instrumental in the identification of editing sites in higher plant plastids. Since the plastid genome from this bryophyte has been completely sequenced (Ohyama et al., 1986) and the plastid editing events generally result in the restoration of conserved amino acid residues, putative editing sites can be identified by simply aligning higher plant plastid DNA and derived protein sequences with the corresponding sequences from *Marchantia polymorpha*. One then searches for positions where conservation at the protein level can be improved by C-to-U changes in the higher plant mRNA sequence (Fig. 1B). In this way, candidate editing sites are identified which subsequently can be subjected to experimental testing by cDNA sequence analyses.

D. RNA Editing and Plastid Gene Expression

A number of studies have been performed on the relationship between RNA editing and other steps in plastid gene expression as well as other forms of RNA processing. Polycistronic primary transcripts (those not yet processed into monocistronic units) and unspliced precursor mRNAs were investigated with respect to their editing state. In most cases, these pre-mRNAs were found to be already completely, or at least partially edited indicating that RNA editing is a post-transcriptional event occurring rather early in plastid mRNA maturation (Freyer et al., 1993; Ruf et al., 1994). Surprisingly, the relationship between editing and translation appears to be somewhat more complex: whereas the editing of some sites turned out to be entirely independent of plastid translation (Zeltz et al., 1993), the editing of other sites was found to depend on plastid translation in an unknown (direct or indirect) manner (Karcher and Bock, 1998).

Practically every single step in the expression of plastid-encoded photosynthesis genes is influenced by environmental cues, the most important of which is light (reviewed in Gruissem and Tonkyn, 1993; Mayfield et al., 1995). In addition, plastid gene expression undergoes dramatic changes in response to developmental programs resulting in differentiation of plastids into specialized organelles like chloroplasts, chromoplasts or amyloplasts. Although there is most probably no qualitative influence of these developmental or environmental factors on the editing patterns of plastid transcripts, the efficiency of editing at individual sites can be subject to tissue-specific and/or developmental stage-specific changes (Bock et al., 1993; Ruf and Kössel, 1997; Karcher and Bock, 1998). However, the functional significance of these quantitative differences remains to be established and it is currently unclear whether they reflect regulation *by* editing or solely regulation *of* editing.

E. Mechanistic Aspects of Plastid RNA Editing

The biochemistry of plastid RNA editing is not yet resolved. Theoretically, a cytidine residue can be most simply converted into a uridine by oxidative deamination at the C4 atom of the pyrimidine ring. In fact, an editing system is already known to exist which makes use of a site-specific cytidine deaminase to convert a genomically encoded cytidine into a uridine: the editing of a single C in the mammalian apolipoprotein B mRNA is carried out by a multiprotein complex ('editosome') containing such a deamination activity. Recent results obtained for plant mitochondria suggest that plant mitochondrial editing may also proceed by cytidine deamination (Rajasekhar and Mulligan, 1993; Blanc et al., 1995; Yu and Schuster, 1995).

An intriguing question concerning plant organellar RNA editing is, how to achieve the extraordinarily high specificity with which the editing apparatus selects individual cytosine residues for modification. The sequences surrounding plastid editing sites lack any apparent conserved, consensus sequence-like elements at the primary or at the secondary structural levels. A number of in vivo studies employing chloroplast transformation technologies have demonstrated that mRNA sequences flanking the editing site are directly involved in plastid RNA editing site recognition (Chaudhuri et al., 1995; Bock et al., 1996, 1997a; Chaudhuri and Maliga, 1996). These studies have also indicated that the major *cis*-acting recognition elements at the mRNA level reside in the 5′ upstream region whereas only little contribution comes from the 3′ flanking region (Bock et al., 1996; Chaudhuri and Maliga, 1996). Therefore, it seems reasonable to assume that the essential 5′ upstream sequence element serves as the binding site for a *trans*-acting factor which then mediates the downstream C-to-U change. Evidence for the participation of *trans*-factors in plastid RNA editing has come from two different sets of in vivo experiments (Chaudhuri et al., 1995; Bock and Koop, 1997). Interestingly, at least the primary *trans*-acting recognition factor appears to be strictly site-specific and responsible only for a single editing site (Chaudhuri et al., 1995; Bock and Koop, 1997). Whether or not additional general *trans*-factors participate in the editing reaction is still unknown.

To date, no in vitro system for chloroplast RNA editing is available. In vivo experiments involving the generation of plants with transgenic chloroplasts have been the only successful approach to study functional and mechanistic aspects of editing. Unfortunately, these transgenic technologies are extremely time-consuming and laborious. To develop faithful in vitro systems for chloroplast RNA editing or to devise suitable genetic screens for editing factors, therefore, represent a major challenge for the future of this field.

III. Polyadenylation and Degradation of mRNA in the Chloroplast

A. Changes in mRNA Stability during Chloroplast Development

During leaf development and plastid differentiation, the levels of many plastid mRNAs vary dramatically. RNA processing and differential RNA stability are important factors contributing to the accumulation of specific mRNAs during chloroplast maturation. In order to study the half-life of mRNAs in the chloroplast of higher plants in vivo, spinach and barley plants were treated with transcriptional inhibitors. The rate of decay of chloroplast-encoded mRNAs was determined by quantitative northern analyses using gene-specific probes. The half-life of several chloroplast mRNAs, such as *psbA,* changed during development from young to mature leaves, whereas no changes occurred in the half-life of others, such as *rbcL,* (Klaff and Gruissem, 1991). Analysis of the transcription, translation and mRNA levels of 15 plastid genes during barley chloroplast development revealed a highly dynamic modulation of gene expression and mRNA stability (Rapp et al., 1992; Baumgartner et al., 1993). Enhanced levels of *psbA* mRNA in mature chloroplasts were shown to be due primarily to its selective stabilization. Nevertheless, the precise mechanism regulating the stability of a specific chloroplast mRNA during plant development and in response to physiological changes (such as light intensity or quality) is not yet understood. Moreover, the prokaryotic nature of the chloroplast gene expression system, in which transcription and translation can theoretically be coupled, requires a mechanism that rapidly degrades immature RNA in order to prevent the synthesis of aberrant proteins.

B. In vitro Chloroplast mRNA Degradation Systems

To study the pathways of mRNA decay, in vitro degradation systems from several organisms have been established. A system based on polysomal fractions has been used for mammalian cells, yeast and bacteria (Ross and Kobs, 1986; Vreken et al., 1992; Ingle and Kushner, 1996). A polysome-based mRNA degradation system also has been established for higher plants. Thus, degradation of the soybean SRS4 mRNA, which encodes the small subunit of RuBP-carboxylase, was analyzed in vitro and a set of proximal and distal endonucleolytic degradation products could be identified (Tanzer and Meagher, 1995). Soluble protein extracts from chloroplasts of higher plants and *Chlamydomonas reinhardtii* have been used to study the 3′ end processing, stability and degradation of in vitro-transcribed RNAs (Stern et al., 1989; Stern and Gruissem, 1989; Nickelsen and Link, 1993; Stern and Kindle, 1993; Nickelsen et al., 1994; Lisitsky et al., 1995; Hayes et al., 1996; Lisitsky et al., 1996, 1997b). In addition, an in vitro degradation system for the analysis of internal mRNA decay in the chloroplast has been described (Klaff, 1995). When lysed chloroplasts were incubated in the presence of yeast tRNA, which inhibits exonuclease activity, the accumulation of distinct endonucleolytic cleavage products was observed (Klaff, 1995). The endonucleolytic cleavage sites were mapped by primer extension and found to be scattered over the mRNA molecule. Some of these sites perfectly matched polyadenylation sites obtained by RT-PCR (reverse transcriptase-polymerase chain reaction) of RNA isolated from intact chloroplasts, indicating that most, if not all, of these endonucleolytic cleavage sites are used in vivo (Lisitsky et al., 1996; see below). By sequence comparison, no obvious motif could be identified which may determine the recognition of an endonucleolytic cleavage site by the processing enzymes.

C. Chloroplast RNA-binding Proteins

Several RNA-binding proteins (RNP) which may be involved in the processing, maturation and degradation of chloroplast RNAs have been characterized over the last few years. A family of nuclear-encoded chloroplast proteins having an RNA-binding recognition sequence motif (Burd and Dreyfuss, 1994) was described from several plants. Among these, the spinach 28RNP was characterized in detail: immunodepletion of this protein from soluble chloroplast extract, as well as addition of the recombinant protein made in bacteria, interfered with the in vitro 3′ end processing of chloroplast RNAs (Schuster and Gruissem, 1991; Lisitsky et al., 1995). Analysis of the sequence-specific RNA-binding properties of 28RNP revealed enhanced affinities to poly(U) or poly(G) sequences. Furthermore, the affinity of the protein to RNA was changed by phosphorylation of a serine residue (Lisitsky et al., 1994, 1995; Lisitsky and Schuster, 1995). Two additional proteins, 33 and 24 kDa, were also characterized from spinach. They are composed of four domains: a transit peptide, an amino-terminal acidic domain and two 80 amino acid RNA-binding domains (Gruissem and Schuster, 1993). In tobacco, five similar proteins were identified and characterized (Li and Sugiura, 1990; Li et al., 1991; Ye et al., 1991; Ye and Sugiura, 1992). Homologous proteins have been identified also in other plants and in cyanobacteria (reviewed in Sugita and Sugiura, 1996). Interestingly, the characterization of a similar protein from *Arabidopsis* (Atrbp33) disclosed the presence of multiple mRNA species that potentially encode different polypeptides, one of which lacks a chloroplast transit peptide and therefore may function outside the chloroplast (Cheng et al., 1994). RNPs in the range of 24 to 33 kDa are relatively abundant in the chloroplast. However, neither the 28RNP nor any of the tobacco RNPs were found to be associated with thylakoid-bound or soluble polysomes (Lisitsky et al., 1995; Sugita and Sugiura, 1996). It was suggested that the chloroplast RNPs, by analogy to hnRNP particles, are associated with non-polysomal pre-mRNAs or with mRNAs that are excluded from the ribosomes (Sugita and Sugiura, 1996).

Recently, two proteins, 37 and 38 kDa, were described which bind specifically to the barley *psbA* 3′ UTR (untranslated region)(Memon et al., 1996). Their binding site was mapped in the 3′ UTR between the translational stop codon and the stem-loop structure and was flanked by U-rich sequences. AU-rich sequences were also identified as the targets of the 55 to 57 kDa RNA-binding protein which possibly possesses an endoribonuclease activity (Chen and Stern, 1991b; Nickelsen and Link, 1993). Further characterization of the 37/38 kDa RNA-binding proteins should reveal their precise biological function.

1. The 100RNP/PNPase and the Chloroplast Degradosome

The search for a ribonuclease that is involved in the 3′ end processing of chloroplast mRNAs yielded a 100 kDa RNA-binding protein (Hayes et al., 1996). The purified 100 kDa protein had biochemical properties similar to those of one of the two exonucleases discovered to date in bacterial cells, the polynucleotide phosphorylase (PNPase) (Soreq and Littauer, 1982; Table 2). Furthermore, the deduced amino-acid sequence of the chloroplast 100RNP cDNA was highly homologous to the bacterial PNPase (Hayes et al., 1996). The chloroplast RNA processing and degradation system, therefore, appears to be similar to the recently discovered mechanisms in *E. coli* (Nierlich and Murakawa, 1996; Carpousis et al., 1999). However, in contrast to bacteria, the mRNA metabolism and associated enzymes in plastids are also controlled by the nucleus and may be regulated by light (Hayes et al., 1996).

The *E. coli* RNA degradosome is a multi-enzyme complex consisting of the exoribonuclease PNPase, the endonuclease RNase E, a DEAD-box ATP-dependent RNA helicase and the enzyme enolase (Carpousis et al., 1999). This high molecular weight protein complex is important in RNA processing and mRNA degradation in the bacterial cell and two of its components, PNPase and RNase E, have been shown to act as key elements in RNA metabolism. The chloroplast 100RNP/PNPase was isolated as part of a high molecular weight complex of about 600 kDa. A 67 kDa protein that crossreacts with antibodies prepared against RNase E of *E. coli* and displays endoribonuclease activity was copurified with this complex (Hayes et al., 1996). It is therefore tempting to suggest that a complex, similar to the bacterial degradosome exists in the chloroplast, preserving its

Table 2. Comparison between *a bacterial* PNPase and a chloroplastic 100RNP/PNPase

	Bacteria (*E. coli*)	Chloroplast (spinach)
MW	86 kDa	100 kDa
Structure	trimer	trimer/tetramer
RNA degradation	3′ to 5′ exonuclease working processively	3′ to 5′ exonuclease working processively
Degradation products	diphosphonucleosides	diphosphonucleosides
Phosphate stimulation	yes	yes
Polymerase activity	yes	yes
High binding affinity	poly(A) and poly(U)	poly(A) and poly(U)
Amino acid sequence homology	43% identity, 63% similarity	
Association with other proteins	endonuclease RNase E, DEAD-box RNA-helicase, enolase	an endonuclease probably related to RNase E plus possibly other proteins

ancestral prokaryotic origin.

In both bacteria and chloroplasts, it seems that not all the PNPase population is associated with the degradosome (Carpousis et al., 1994; Lisitsky et al., 1997b). Therefore, the question of whether different forms of the 100RNP/PNPase are involved in distinct RNA-metabolizing activities, such as 3′ end processing or degradation, remains open. In vitro experiments using synthetic RNAs and purified 100RNP/PNPase have shown a much higher activity of the enzyme when supplied with polyadenylated RNA substrates (Lisitsky et al., 1997b). This preference for polyadenylated RNA resulted from the high binding affinity of the 100RNP/PNPase to poly(A) sequences. Interestingly, a similar function and mode of action has been suggested for the *E. coli* PNPase (Lisitsky and Schuster, 1999), and also for another bacterial exoribonuclease, RNase II (Coburn and Mackie, 1996). Therefore, identification, isolation and characterization of the other chloroplast exoribonucleases will reveal whether the preference for poly(A)-rich RNAs is only intrinsic to the 100RNP/PNPase, or is shared by several or all of the chloroplast exonucleases.

2. Other Chloroplast Ribonucleases

A mustard 54 kDa protein was isolated as an endonuclease that may be involved in the 3′ end processing of mRNAs (Nickelsen and Link, 1993). This protein may be similar to the spinach 57 kDa RNA-binding protein which has been postulated to be responsible for endonucleolytic activity and was termed Endo C1 (Chen and Stern, 1991a; 1991b). A stable RNA-protein complex was shown to be formed in vitro between the spinach *petD* pre-mRNA and three proteins of 29, 41 and 55 kDa (Chen et al., 1995). These proteins bind specifically to the stem-loop structure and to a downstream AU-rich sequence element termed box II (Chen et al., 1995). Isolation and cDNA cloning of the 41 kDa protein (CSP41) revealed that it is encoded by a single nuclear gene exhibiting no homology to other RNA-binding proteins or ribonucleases (Yang et al., 1996). Interestingly, computer searches disclosed that CSP41 is homologous to nucleotide-sugar epimerases and hydroxysteroid dehydrogenases as well as to the *Synechocystis* open reading frame accession no. 1652543 (Baker et al., 1998). The biological significance of these homologies remains to be elucidated. Nevertheless, this protein, either purified from chloroplasts or expressed in *E. coli*, exhibited high endoribonuclease activity that cleaved RNA molecules non-specifically into small oligoribonucleotides (Yang et al., 1996; Yang and Stern, 1997). Further analysis of the 29-, 41-, 55 kDa protein complex and the in vivo activity of CSP41 are required to uncover their mode of operation in vivo.

Evidence for the activity of a 5′ to 3′ exoribonuclease in the chloroplast of *Chlamydomonas reinhardtii* has been obtained recently (Drager et al., 1998). Since no such activity was identified in bacteria, it will be interesting to characterize this enzyme and reveal its function and the mode of action in the chloroplast.

D. Polyadenylation of mRNA in the Chloroplast

1. Detection of Polyadenylated mRNA in the Chloroplast

Post-transcriptional addition of poly(A) tails to the 3′ end of mRNAs was first identified and characterized in eukaryotic cells for viral and nuclear-encoded mRNAs (Jackson and Standart, 1990; Sachs and Wahle, 1993; Wahle and Keller, 1996). These poly(A) tails are formed by the addition of about 250 adenylate residues to a 3′ end generated by endonucleolytic cleavage of the precursor RNA (Wahle and Keller, 1996). A complex assembly of proteins is required for 3′ end maturation, including the activity of a poly(A) polymerase. The long poly(A) tails of eukaryotic nuclear-encoded mRNAs are important determinants of mRNA stability and, moreover, have been implicated in the initiation of translation. Deadenylation, in addition to decapping, may be a major step in the degradation pathway of nuclear-encoded mRNAs.

Poly(A) tails have also been described in bacteria for several mRNA 3′ ends (Cohen, 1995; Sarkar, 1997). In the *rpsO* mRNA of *E. coli*, they are found at endonucleolytic and exonucleolytic degradation sites (Haugel-Nielsen et al., 1996). Contrary to the nucleus and cytoplasm of eukaryotic cells, where deadenylation of the long poly(A) tail is part of the mRNA degradation pathway of nuclear-encoded mRNAs, the addition of poly(A) tails to bacterial mRNAs promotes their degradation.

It is interesting to note that poly(A) RNA was detected in the chloroplast already more than 20 years ago (Burkard and Keller, 1974; Haff and

Bogorad, 1976). By hybridization experiments with chloroplast DNA and ^{125}I-labeled RNA from maize seedlings it was determined that about 6% of the poly(A)-containing RNA hybridized to chloroplast DNA and that the chloroplast poly(A) tracts averaged about 45 nucleotides in length (Haff and Bogorad, 1976). However, these findings have been largely ignored for more than two decades and polyadenylation was commonly regarded as a characteristic feature of eukaryotic nuclear and viral mRNAs. Nevertheless, polyadenylation of prokaryotic and organellar mRNAs has recently returned to the focus of research when its role in the mRNA degradation pathway became obvious.

The post-transcriptional addition of poly(A)-rich sequences to the plastid-encoded *psbA* mRNA, as well as to several other transcripts, has recently been described using RT-PCR of oligo-dT-primed cDNA from spinach chloroplasts (Kudla et al., 1996; Lisitsky et al., 1996). Unlike in eukaryotic nuclear-encoded and bacterial RNAs, the poly(A) moiety in the chloroplast was not found to be a ribohomopolymer of adenosine residues but rather consists of clusters of adenosines interrupted mostly by guanosines and, in rare cases, by cytidines or uridines. The tails can be several hundreds of nucleotides long (Lisitsky et al., 1996). Most of the polyadenylation sites found by RT-PCR of oligo-dT-primed chloroplast cDNA were localized within the amino acid-coding region or the 3′ UTR of the *psbA* mRNA. RT-PCR clones of *psbA* mRNA that were polyadenylated at the 3′ end were also obtained, but with at least 50 times lower frequency (Lisitsky et al., 1996). This result indicated that mostly truncated mRNAs are polyadenylated. The truncation may originate from either early transcription termination or cleavage of full-length transcripts. Two observations suggest that in vivo polyadenylation occurs subsequent to the cleavage of an mRNA as part of a specific degradation pathway. First, three of the polyadenylation sites mapped by RT-PCR in spinach *psbA* mRNA perfectly matched endonucleolytic cleavage sites that were mapped earlier by primer extension in the mRNA-degradation system from lysed chloroplasts described above (Lisitsky et al., 1996). Mapping the endonucleolytic cleavage site by primer extension identifies the first nucleotide at the 5′ end of the distal cleavage product, whereas the poly(A) tail is added to the 3′ end of the proximal cleavage product. Consequently, the nucleotide mapped by primer extension was the 3′ neighboring nucleotide to the one that was found to precede the poly(A) tail (Lisitsky et al., 1996). Second, polyadenylation sites located in the 3′ UTR of the *petD* mRNA determined by RT-PCR were mapped to the positions observed earlier as cleavage sites of the purified endoribonuclease p67 on synthetic, in vitro-transcribed RNA corresponding to the *petD* 3′ UTR (Kudla et al., 1996). These results strongly suggest that most of the polyadenylated mRNAs result from endonucleolytic cleavage of full-length transcripts. Nevertheless, the possibility that some of the polyadenylated RNA molecules in the chloroplast reflect the degradation of molecules resulting from premature transcription termination cannot be ruled out. Such molecules should enter the degradation pathway immediately, because otherwise they might serve as templates for translation of truncated, defective proteins.

2. Polyadenylation is Required for Degradation of Chloroplast mRNA

The question of whether polyadenylation is required for mRNA degradation or whether other decay pathways also exist for chloroplast mRNAs, was approached in studies using the lysed-chloroplast system and the polyadenylation inhibitor 3′-dATP (cordycepin triphosphate). Blocking the polyadenylation of RNA inhibits RNA degradation and has similar, if not identical, effects as direct blocking of the exoribonucleases by addition of an excess of yeast tRNA (Lisitsky et al., 1997a). In both treatments, the full-length mRNA was endonucleolytically cleaved yielding distinct degradation products that accumulated instead of being exonucleolytically degraded. Hence, the addition of poly(A)-rich sequences to endonucleolytic cleavage products of mRNAs is required to target these molecules for rapid exonucleolytic degradation in the chloroplast.

3. In vitro Polyadenylation of Chloroplast mRNAs

An in vitro polyadenylation system was used to elucidate the biochemistry of mRNA polyadenylation and degradation activities, to isolate the proteins involved, and to reconstitute their activities. In vitro transcribed RNAs corresponding to chloroplast mRNAs could be polyadenylated at their 3′ ends using a soluble chloroplast protein extract complemented by the addition of ATP (Kudla et al., 1996; Lisitsky et al., 1996). In vitro analysis of chloroplast polyadenylation activity revealed specific incorporation of ATP and GTP, reflecting in vivo composition of the poly(A)-rich tails. In this respect, it is interesting to note that poly(A)- and poly(G)-polymerase activities were purified from wheat chloroplasts more than 20 years ago (Burkard and Keller, 1974). Furthermore, the in vitro polyadenylation activity is dependent on the substrate structure. Unstructured RNAs were polyadenylated with high efficiency as compared to molecules forming the stable stem-loop structure characteristic of mature plastid mRNA 3′ ends.(Lisitsky et al., 1996).

When incubated in a soluble chloroplast protein extract, it was found that in vitro transcribed synthetic polyadenylated RNA was degraded more rapidly than the same but non-polyadenylated RNA. Competition experiments revealed that polyadenylated RNA molecules are more efficient competitors for the degradation machinery than their non-polyadenylated counterparts (Lisitsky et al., 1997b). These results suggest that poly(A)-rich tails play a major role in the rapid degradation of intermediate products of mRNA decay in the chloroplast, most probably by targeting the cleavage products for rapid degradation due to their high affinity to chloroplast exoribonuclease(s) (Kudla et al., 1996; Lisitsky et al., 1996).

The above discussed 100RNP/PNPase, like the bacterial PNPase, is a processive exoribonuclease that binds to the 3′ end and digests the substrate RNA, nucleotide by nucleotide, without dissociating from the molecule. In competition experiments using isolated, purified 100RNP/PNPase, the polyadenylated RNA competed with the non-polyadenylated RNA for the exonuclease from the soluble protein extract (Lisitsky et al., 1997b). Affinity-binding assays of the 100RNP/PNPase to poly(A) RNA as well as to other RNA sequences revealed higher binding affinity of this protein to poly(A) than to other RNA molecules (Lisitsky et al., 1997b). These results indicate that the preferential degradation of polyadenylated RNA in the chloroplast is based on the high binding affinity of the exoribonuclease 100RNP/PNPase to poly(A) sequences. However, the possibility that another, as yet unidentified chloroplast exoribonuclease(s) also binds polyadenylated RNA with higher affinity than non-polyadenylated RNA currently cannot be ruled out. In fact, higher in vitro degradation activity of bacterial RNase II with polyadenylated RNA as a substrate

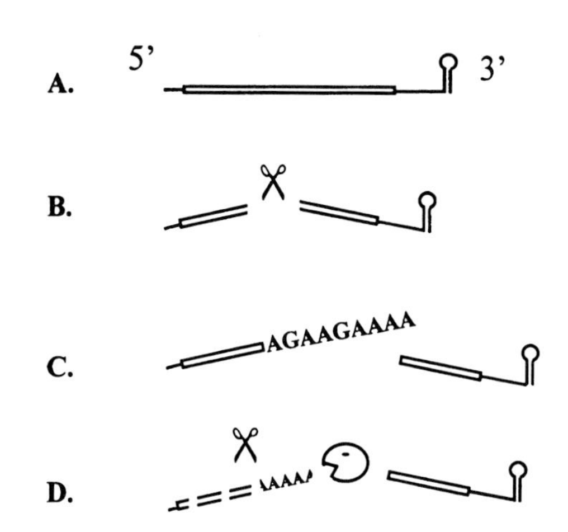

Fig. 2. A model for the degradation pathway of mRNA in the chloroplast. (A) Schematic representation of a *psbA* mRNA molecule. The open box represents the protein-coding region and the stem-loop structure represents inverted repeats in the 3′ untranslated region which potentially form an RNA secondary structure. (B) The initial step in the mRNA degradation process is suggested to be an endonucleolytic cleavage by an as yet unidentified endonuclease(s). The endonuclease is symbolized by scissors. (C) A poly(A)-rich tail, which can be up to several hundred nucleotides long, is then added to the 3′ end of the 5′ endonucleolytic cleavage product. (D) The polyadenylated RNA molecule is rapidly degraded by an exonuclease(s), possibly the 100RNP/PNPase.

has been detected recently (Coburn and Mackie, 1996). Also the bacterial PNPase was found to display high binding affinity to poly(A) and poly(U) ribohomopolymers (Lisitsky and Schuster, 1999).

E. The Molecular Mechanism of mRNA Degradation in the Chloroplast

Our current model of the mRNA degradation pathway in the chloroplast is presented in Fig. 2. The initial event is an endonucleolytic cleavage(s) that produces RNA molecules lacking the stem-loop at their 3′ end (B in Fig. 2). RNAs terminating with a stem-loop structure are poorly polyadenylated in vitro, and RT-PCR clones with poly(A)-rich sequences at the end of stem-loop-containing mRNA molecules are detected with much lower frequency than those having the polyadenylation site inside the coding region (Lisitsky et al., 1996). Therefore, we suggest that the stem-loop structures being characteristic of most chloroplast mRNAs and shown to be effective 3′ end processing signals (Stern and Gruissem, 1987; Stern et al., 1989, 1991; Stern and Kindle, 1993; Rott et al., 1996, 1998) serve only as poor polyadenylation sites thereby preventing exonucleolytic degradation of functional RNA molecules. Following the endonucleolytic cleavage(s), the proximal fragments are polyadenylated by the addition of poly(A)-rich sequences (C in Fig. 2). This step can be inhibited experimentally by 3′-dATP (cordycepin). Polyadenylation is then required for the continuation of mRNA degradation in the chloroplast, because only polyadenylated RNAs are rapidly digested by exonuclease(s) due to the higher affinity of this enzyme(s) to poly(A)-rich sequences (D in Fig. 2). This last stage can be slowed down by addition of yeast tRNA (Klaff and Gruissem, 1991; Lisitsky et al., 1997a).

The model described above leaves two possibilities for how the half-life time of a particular RNA molecule is determined. First, once the mRNA molecule has been endonucleolytically cleaved, it is subject to rapid degradation. In this case, the rate-limiting step is the initial endonucleolytic cleavage; once it has occurred, cleaved mRNA molecules will be rapidly polyadenylated and exonucleolytically degraded. Consequently, the nature, specificity and modulation of activity and/or expression of the endoribonuclease(s) would determine the half-life of a particular mRNA molecule, whereas the activities of the poly(A) polymerase and the exoribonuclease(s) are not rate limiting. Second, it is possible that other steps in the degradation pathway, not the endonucleolytic cleavage, limit the degradation rate. For example, the length of the poly(A) tail, which can amount to several hundreds of nucleotides (Lisitsky et al., 1996), could theoretically influence the rate of degradation. Furthermore, the guanosine residues that are characteristic of the poly(A) tails of chloroplast *psbA* mRNA may be involved in modulating activities of the degradation enzyme(s). Alternatively, these residues may simply reflect the specificity properties of the poly(A) polymerase(s) (Burkard and Keller, 1974; Lisitsky et al., 1996). In vitro experiments, in which synthetic RNAs carrying poly(A) tails of different lengths and different proportions of adenosine vs. guanosine residues were incubated with chloroplast protein extract, disclosed remarkable differences in degradation rate (Lisitsky and Schuster, unpublished). The significance of these results to the in vivo situation is still unclear and awaits further investigation.

So far, our model does not explain how the distal endonucleolytic cleavage product is degraded. One possibility is that several endonucleolytic cleavages occur along the mRNA until the small RNA fragment representing the 3′ UTR is degraded by the polyadenylation-dependent pathway. Endonucleolytic damage to the 3′ stem-loop structure potentially could then enable polyadenylation. A low degree of polyadenylation of the mature 3′ end has already been detected both in vivo and in vitro (Lisitsky et al., 1996). Alternatively, additional exonucleolytic degradation pathways which are not dependent on polyadenylation, may exist. However, the results of the experiments with the polyadenylation inhibitor 3′-dATP (cordycepin) indicated that at least the *rbcL* and *psbA* mRNAs are exonucleolytically degraded only in the polyadenylation-dependent degradation pathway. To understand the role of this degradation pathway in the developmental regulation of plastid mRNA stability, the requirement for and the extent of polyadenylation in etioplasts and root amyloplasts should be analyzed. In these tissues and developmental stages, the *psbA* and *rbcL* RNAs have short half-lives and are degraded rapidly after their synthesis (Deng and Gruissem, 1987).

During the past few years, our understanding of the chloroplast mRNA degradation pathway has progressed significantly. For certain mRNAs, such as *psbA,* the different steps of their specific decay have been disclosed. The succession of endonucleolytic-degradation events is clearly reminiscent of the RNA decay mechanisms operating in the prokaryotic (eubacterial) ancestors of present-day chloroplasts. This evolutionary relationship is also supported by the observation that one of the chloroplast RNA-degrading enzymes shares structural and functional homology with its prokaryotic counterpart. Chloroplast gene expression, however, is predominantly regulated by the nucleus and in response to external stimuli, such as light. This as well as the relative longevity of chloroplast mRNAs seem to reflect eukaryotic properties rather than bacterial ones. Therefore, the chloroplast may have adopted an intermediate position by combining both prokaryotic and eukaryotic features. Following the biochemical pathway of plastid mRNA degradation, future research will focus on the regulation of mRNA stability as part of the regulatory network that determines plant development and adaptation to different environmental conditions.

Acknowledgments

Work in G. S.'s laboratory was supported by the

Israel Science Foundation, administered by the Israel Academy of Science and Humanities, and by a United States-Israel Binational Agricultural Research and Development Fund grant. Work in R. B.'s laboratory is supported by grants from the Deutsche Forschungsgemeinschaft (DFG).

References

Alfonzo JD, Thiemann O and Simpson L (1997) The mechanism of U insertion/deletion RNA editing in kinetoplast mitochondria. Nucl Acids Res 25: 3751–3759

Baker ME, Grundy WN, and Elkan CP (1998) Spinach CSP41, an mRNA-binding protein and ribonuclease, is homologous to nucleotide-sugar epimerases and hydroxysteroid dehydrogenases. Biochem Biophys Res Commun 248: 250–254

Barkan A and Stern D (1998). Chloroplast mRNA processing: intron splicing and 3′-end metabolism. In: Bailey-Serres J and Gallie DR (eds) A Look Beyond Transcription: Mechanisms Determining mRNA Stability and Translation in Plants, pp 162–173. American Society of Plant Physiologists, Rockville

Barkan A, Voelker R, Mendel-Hartvig J, Johnson D, and Walker M (1995) Genetic analysis of chloroplast biogenesis in higher plants. Physiol Plant 93: 163–170

Baumgartner BJ, Rapp JC, and Mullet JE (1993) Plastid genes encoding the transcription-translation apparatus are differentially transcribed early in barley (*Hordeum-vulgare*) chloroplast development: Evidence for selective stabilization of *psbA* mRNA. Plant Physiol 101: 781–791

Benne R (1996) RNA editing: How a message is changed. Curr Opinion Genet Dev 6: 221–231

Benne R, van den Burg J, Brakenhoff JPJ, Sloof P, van Boom JH and Tromp MC (1986) Major transcript from the frameshifted coxII gene from trypanosome mitochondria contains four nucleotides that are not encoded in the DNA. Cell 46: 819–826

Blanc V, Litvak S and Araya A (1995) RNA editing in wheat mitochondria proceeds by a deamination mechanism. FEBS Lett 373: 56–60

Bock R and Koop H-U (1997) Extraplastidic site-specific factors mediate RNA editing in chloroplasts. EMBO J 16: 3282–3288

Bock R, Hagemann R, Kössel H and Kudla J (1993) Tissue- and stage-specific modulation of RNA editing of the *psbF* and *psbL* transcript from spinach plastids—a new regulatory mechanism? Mol Gen Genet 240: 238–244

Bock R, Kössel H and Maliga P (1994) Introduction of a heterologous editing site into the tobacco plastid genome: The lack of RNA editing leads to a mutant phenotype. EMBO J 13: 4623–4628

Bock R, Hermann M and Kössel H (1996) In vivo dissection of *cis*-acting determinants for plastid RNA editing. EMBO J 15: 5052–5059

Bock R, Hermann M and Fuchs M (1997a) Identification of critical nucleotide positions for plastid RNA editing site recognition. RNA 3: 1194–1200

Bock R, Albertazzi F, Freyer R, Fuchs M, Ruf S, Zeltz P and Maier RM (1997b) Transcript editing in chloroplasts of higher plants. In: Schenk HEA, Herrmann R, Jeon KW, Müller NE, Schwemmler W (eds) Eukaryotism and Symbiosis, pp 123–137. Springer-Verlag, Berlin

Burd CG, and Dreyfuss G (1994) Conserved structures and diversity of functions of RNA-binding proteins. Science 265: 615–621

Burkard G, and Keller EB (1974) Poly(A) polymerase and poly(G) polymerase in wheat chloroplasts. Proc Natl Acad Sci USA 71: 389–393

Carpousis AJ, Van Houwe G, Ehretsmann C, and Krisch HM (1994) Copurification of *E. coli* RNAase E and PNPase: Evidence for a specific association between two enzymes important in RNA processing and degradation. Cell 76: 889–900

Carpousis AJ, Vanzo NF and Raynal LC (1999) mRNA degradation, a tale of poly(A) and multiprotein machines. Trends Genet 15: 24–28

Cattaneo R, Kaelin K, Baczko K and Billeter MA (1989) Measles virus editing provides an additional cysteine-rich protein. Cell 56: 759–764

Chaudhuri S and Maliga P (1996) Sequences directing C to U editing of the plastid psbL mRNA are located within a 22 nucleotide segment spanning the editing site. EMBO J 15: 5958–5964

Chaudhuri S, Carrer H and Maliga P (1995) Site-specific factor involved in the editing of the *psbL* mRNA in tobacco plastids. EMBO J 14: 2951–2957

Chen H and Stern DB (1991a) Specific ribonuclease activities in spinach chloroplasts promote mRNA maturation and degradation. J Biol Chem 266: 24205–24211

Chen H and Stern DB (1991b) Specific binding of chloroplast proteins in vitro to the 3′ untranslated region of spinach chloroplast *petD* messenger RNA. Mol Cell Biol 11: 4380–4388

Chen Q, Adams CC, Usack L, Yang J, Monde R and Stern DB (1995) An AU-rich element in the 3′ untranslated region of the spinach chloroplast *petD* gene participates in sequence-specific RNA-protein complex formation. Mol Cell Biol 15: 2010–2018

Cheng SH, Cline K and Delisle AJ (1994) An *Arabidopsis* chloroplast RNA-binding protein gene encodes multiple mRNAs with different 5′ ends. Plant Physiol 106: 303–311

Coburn GA and Mackie GA (1996) Differential sensitivities of portions of the mRNA for ribosomal protein S20 to 3′-exonucleases dependent on oligoadenylation and RNA secondary structure. J Biol Chem 271: 15776–15781

Cohen SN (1995) Surprises at the 3′ end of prokaryotic RNA. Cell 80: 829–832

Covello PS and Gray MW (1989) RNA editing in plant mitochondria. Nature 341: 662–666

Deng XW and Gruissem W. (1987) Control of plastid gene expression during development: The limited role of transcriptional regulation. Cell 49: 379–387

Drager RG and Stern DB (1998) Chloroplast RNA Synthesis and Processing. In: Rochaix J-D, Goldschmidt-Clermont M and Merchant S (eds) Molecular Biology of *Chlamydomonas*: Chloroplasts and Mitochondria, Vol 7, pp 165–140. Kluwer Academic Publishers, Dordrecht

Drager RG, Girard-Bascou J, Choquet Y, Kindle KL and Stern DB (1998) In vivo evidence for 5′→ 3′ exoribonuclease degradation of an unstable chloroplast mRNA. Plant J 13: 85–96

Freyer R, Hoch B, Neckermann K, Maier RM and Kössel H

(1993) RNA editing in maize chloroplasts is a processing step independent of splicing and cleavage to monocistronic mRNAs. Plant J 4: 621–629

Freyer R, Kiefer-Meyer M-C and Kössel H (1997) Occurrence of plastid RNA editing in all major lineages of land plants. Proc Natl Acad Sci USA 94: 6285–6290

Goldschmidt-Clermont M (1998) Coordination of nuclear and chloroplast gene expression in plant cells. Int Rev Cyt 177: 115–180

Gray MW (1993) Origin and evolution of organelle genomes. Curr Opinion Genet Dev 3: 884–890

Gruissem W and Schuster G (1993) Control of mRNA degradation in organelles. In: Brawerman G and Belasco J (eds) Control of Messenger RNA Stability, pp 329–365. Academic Press, Orlando

Gruissem W and Tonkyn JC (1993) Control mechanisms of plastid gene expression. Crit Rev Plant Sci 12: 19–55

Gualberto JM, Lamattina L, Bonnard G, Weil J-H and Grienenberger J-M (1989) RNA editing in wheat mitochondria results in the conservation of protein sequences. Nature 341: 660–662

Gualberto JM, Weil J-H and Grienenberger JM (1990) Editing of the wheat coxIII transcript: evidence for twelve C to U and one U to C conversions and for sequence similarities around editing sites. Nucl Acids Res 18: 3771–3776

Haff LA and Bogorad L (1976) Poly(adenylic acid)-containing RNA from plastids of maize. Biochemistry 15: 4110–4115

Haugel-Nielsen J, Hajnsdorf E and Regnier P (1996) The *rpsO* mRNA of *Escherichia coli* is polyadenylated at multiple sites resulting from endonucleolytic processing and exonucleolytic degradation. EMBO J 15: 3144–3152

Hayes R, Kudla J, Schuster G, Gabay L, Maliga P and Gruissem W (1996) Chloroplast mRNA 3′-end processing by a high molecular weight protein complex is regulated by nuclear encoded RNA binding proteins. EMBO J 15: 1132–1141

Hiesel R, Wissinger B, Schuster W and Brennicke A (1989) RNA editing in plant mitochondria. Science 246: 1632–1634

Hiesel R, Combettes B and Brennicke A (1994) Evidence for RNA editing in mitochondria of all major groups of land plants except the Bryophyta. Proc Natl Acad Sci USA 91: 629–633

Hoch B, Maier RM, Appel K, Igloi GL and Kössel H (1991) Editing of a chloroplast mRNA by creation of an initiation codon. Nature 353: 178–180

Ingle CA and Kushner SR (1996) Development of an in vitro mRNA decay system for *Eschericia coli*: Poly(A) polymerase I is necessary to trigger degradation. Proc Natl Acad Sci USA 93: 12926–12931

Jackson RJ and Standart N (1990) Do the poly(A) tail and 3′ untranslated region control mRNA translation? Cell 62: 15–24

Karcher D and Bock R (1998) Site-selective inhibition of plastid RNA editing by heat shock and antibiotics: a role for plastid translation in RNA editing. Nucl Acids Res 26: 1185–1190

Klaff P (1995) mRNA decay in spinach chloroplasts: *psbA* mRNA degradation is initiated by endonucleolytic cleavages within the coding region. Nucl Acids Res 23: 4885–4892

Klaff P and Gruissem W (1991) Changes in chloroplast mRNA stability during leaf development. Plant Cell 3: 517–530

Kudla J, Igloi GL, Metzlaff M, Hagemann R and Kössel H (1992) RNA editing in tobacco chloroplasts leads to the formation of a translatable *psbL* mRNA by a C to U substitution within the initiation codon. EMBO J 11: 1099–1103

Kudla J, Hayes R and Gruissem W (1996) Polyadenylation accelerates degradation of chloroplast mRNA. EMBO J 15: 7137–7146

Li Y and Sugiura M (1990) Three distinct ribonucleoproteins from tobacco chloroplasts: Each contains a unique amino terminal acidic domain and two ribonucleoprotein consensus motifs. EMBO J 9: 3059–3066

Li Y, Ye L, Sugita M and Sugiura M (1991) Tobacco nuclear gene for the 31 kd chloroplast ribonucleoprotein: Genomic organization, sequence analysis and expression. Nucl Acids Res 19: 2987–2992

Lisitsky I and Schuster G (1995) Phosphorylation of a chloroplast RNA-binding protein changes the affinity to RNA. Nucl Acids Res 23: 2506–2511

Lisitsky I and Schuster G (1999) Preferential degradation of polyadenylated and polyuridinylated RNAs by bacterial exoribonuclease polynucleotide phosphorylase (PNPase). Eur J Biochem 261: 468–474

Lisitsky I, Liveanu V and Schuster G (1994) RNA-binding activities of the different domains of a spinach chloroplast ribonucleoprotein. Nucl Acids Res 22: 4719–4724

Lisitsky I, Liveanu V and Schuster G (1995) RNA-binding characteristics of a ribonucleoprotein from spinach chloroplast. Plant Physiol 107: 933–941

Lisitsky I, Klaff P and Schuster G (1996) Addition of poly(A)-rich sequences to endonucleolytic cleavage sites in the degradation of spinach chloroplast mRNA. Proc Natl Acad Sci USA 93: 13398–13403

Lisitsky I, Klaff P and Schuster G (1997a) Blocking polyadenylation of mRNA in the chloroplast inhibits its degradation. Plant J 12: 1173–1178

Lisitsky I, Kotler A and Schuster G (1997b) The mechanism of preferential degradation of polyadenylated RNA in the chloroplast: the exoribonuclease 100RNP/PNPase displays high binding affinity for poly(A) sequence. J Biol Chem 272: 17648–17653

Mahendran R, Spottswood MR and Miller DL (1991) RNA editing by cytidine insertion in mitochondria of Physarum polycephalum. Nature 349: 434–438

Maier RM, Neckermann K, Igloi GL and Kössel H (1995) Complete sequence of the maize chloroplast genome: Gene content, hotspots of divergence and fine tuning of genetic information by transcript editing. J Mol Biol 251: 614–628

Maier RM, Zeltz P, Kössel H, Bonnard G, Gualberto JM and Grienenberger JM (1996) RNA editing in plant mitochondria and chloroplasts. Plant Mol Biol 32: 343–365

Malek O, Lättig K, Hiesel R, Brennicke A and Knoop V (1996) RNA editing in bryophytes and a molecular phylogeny of land plants. EMBO J 15: 1403–1411

Mayfield SP, Yohn CB, Cohen A and Danon A (1995) Regulation of chloroplast gene expression. Annu Rev Plant Physiol Plant Mol Biol 46: 147–166

Memon AR, Meng B and Mullet JE (1996) RNA-binding proteins of 37/38 kDa bind specifically to the barley chloroplast *psbA* 3′ end untranslated RNA. Plant Mol Biol 30: 1195–1205

Meurer J, Berger A and Westhoff P (1996) A nuclear mutant of Arabidopsis with impaired stability on distinct transcripts of the plastid *psbB*, *psbD/C*, *ndhH*, and *ndhC* operons. Plant Cell 8: 1193–1207

Meurer J, Grevelding C, Westhoff P and Reiss B (1988) The PAC protein affects the maturation of specific chloroplast mRNAs in *Arabidopsis thaliana*. Mol Gen Genet 258: 342–351

Mulligan RM and Maliga P (1997) RNA editing in mitochondria and plastids. In: Bailey-Serres J and Gallie D (eds) Look Beyond Transcription: Mechanisms Determining mRNA Stability and Translation in Plants, pp 153–161. American Society of Plant Physiologists, Rockville

Neckermann K, Zeltz P, Igloi GL, Kössel H and Maier RM (1994) The role of RNA editing in conservation of start codons in chloroplast genomes. Gene 146: 177–182

Nickelsen J and Link G (1993) The 54 kDa RNA-binding protein from mustard chloroplasts mediates endonucleolytic transcript 3′ end formation in vitro. Plant J 3: 537–544

Nickelsen J, Van-Dillewijn J, Rahire M and Rochaix JD (1994) Determinants for stability of the chloroplast *psbD* RNA are located with in its short leader region in *Chlamydomonas reinhardtii*. EMBO J 13: 3182–3191

Nierlich DP and Murakawa GJ (1996) The decay of bacterial messenger RNA. Prog Nucl Acids Res Mol Biol 52: 153–216

Ohyama K, Fukuzawa H, Kohchi T, Shirai H, Sano T, Sano S, Umesono K, Shiki Y, Takeuchi M, Chang Z, Aota S-i, Inokuchi H and Ozeki H (1986) Chloroplast gene organization deduced from complete sequence of liverwort Marchantia polymorpha chloroplast DNA. Nature 322: 572–574

Polson AG, Bass BL and Casey JL (1996) RNA editing of hepatitis delta virus antigenome by dsRNA-adenosine deaminase. Nature 380: 454–456

Powell LM, Wallis SC, Pease RJ, Edwards YH, Knott TJ and Scott J (1987) A novel form of tissue-specific RNA processing produces apolipoprotein-B48 in intestine. Cell 50: 831–840

Rajasekhar VK and Mulligan RM (1993) RNA editing in plant mitochondria: α-phosphate is retained during C-to-U conversion in mRNAs. Plant Cell 5: 1843–1852

Rapp JC, Baumgartner BJ and Mullet J (1992) Quantitative analysis of transcription and RNA levels of 15 barley chloroplast genes. Transcription rates and mRNA levels vary over 300-fold; predicted mRNA stabilities vary 30-fold. J Biol Chem 267: 21404–21411

Rochaix J-D (1996) Post-transcriptional regulation of chloroplast gene expression in *Chlamydomonas reinhardtii*. Plant Mol Biol 32: 327–341

Ross J and Kobs G (1986) H4 histone messenger RNA decay in cell-free extracts initiates at or near the 3′ terminus and proceeds 3′ to 5′. J Mol Biol 267: 579–573

Rott R, Drager RG, Stern DB and Schuster G (1996) The 3′ untranslated regions of chloroplast genes in *Chlamydomonas reinhardtii* do not serve as efficient transcriptional terminators. Mol Gen Genet 252: 676–683

Rott R, Liveanu V, Drager RG, Stern DB and Schuster G (1998) The sequence and structure of the 3′-untranslated regions of chloroplast transcripts are important determinants of mRNA accumulation and stability. Plant Mol Biol 36: 307–314

Ruf S and Kössel H (1997) Tissue-specific and differential editing of the two ycf3 editing sites in maize plastids. Curr Genet 32: 19–23

Ruf S, Zeltz P and Kössel H (1994) Complete RNA editing of unspliced and dicistronic transcripts of the intron-containing reading frame IRF170 from maize chloroplasts. Proc Natl Acad Sci USA 91: 2295–2299

Sachs AB and Wahle E (1993) Poly(A) tail metabolism and function in eukaryotes. J Biol Chem 268: 22955–22958

Sarkar N (1997) Polyadenylation of mRNA in prokaryotes. Annu Rev Biochem 66: 173–197

Schuster G and Gruissem W (1991) Chloroplast mRNA 3′ end processing requires a nuclear-encoded RNA-binding protein. EMBO J 10: 1493–1502

Schuster G, Lisitsky I and Klaff P (1999) Update on chloroplast molecular biology: Polyadenylation and degradation of mRNA in the chloroplast. Plant Physiol 120: 937–944

Schuster W, Hiesel R, Wissinger B and Brennicke A (1990) RNA editing in the cytochrome *b* locus of the higher plant Oenothera berteriana includes a U-to-C transition. Mol Cell Biol 10: 2428–2431

Smith HC, Gott JM and Hanson MR (1997) A guide to RNA editing. RNA 3: 1105–1123

Sommer B, Köhler M, Sprengel R and Seeburg PH (1991) RNA editing in brain controls a determinant of ion flow in glutamate-gated channels. Cell 67: 11–19

Soreq H and Littauer UZ (1982) Polynucleotide phosphorylase. In: Boyer PD (ed) The Enzymes, Vol 15, pp 517–553. Academic Press, New York

Stern DB and Gruissem W (1987) Control of plastid gene expression: 3′ inverted repeats act as mRNA processing and stabilizing elements, but do not terminate transcription. Cell 51: 1145–1157

Stern DB and Gruissem W (1989) Chloroplast mRNA 3′ end maturation is biochemically distinct from prokaryotic mRNA processing. Plant Mol Biol 13: 615–625

Stern DB and Kindle KL (1993) 3′ end maturation of the *Chlamydomonas reinhardtii* chloroplast *atpB* mRNA is a two-step process. Mol Cell Biol 13: 2277–2285

Stern DB, Jones H and Gruissem W (1989) Function of plastid mRNA 3′ inverted repeats: RNA stabilization and gene-specific protein binding. J Biol Chem 264: 18742–18750

Stern D B, Radwanski ER and Kindle KL (1991) A 3′ stem/loop structure of the *Chlamydomonas* chloroplast *atpB* gene regulates mRNA accumulation in vivo. Plant Cell 3: 285–297

Stuart K, Allen TE, Heidmann S and Seiwert SD (1997) RNA editing in kinetoplast protozoa. Microbiol Mol Biol Rev 61: 105–120

Sugita M and Sugiura M (1996) Regulation of gene expression in chloroplasts of higher plants. Plant Mol Biol 32: 315–326

Tanzer MM and Meagher RB (1995) Degradation of the soybean ribulose-1,5-bisphosphate carboxylase small- subunit mRNA, SRS4, initiates with endonucleolytic cleavage. Mol Cell Biol 15: 6641–6652

Vreken P, Buddelmejer N and Raue HA (1992) A cell free extract from yeast cells for studying mRNA turnover. Nucl Acids Res 20: 2503–2510

Wahle E and Keller W (1996) The biochemistry of poly-adenylation. Trends Biochem Sci 21: 247–250

Wakasugi T, Hirose T, Horihata M, Tsudzuki T, Kössel H and Sugiura M (1996) Creation of a novel protein-coding region at the RNA level in black pine chloroplasts: The pattern of RNA editing in the gymnosperm chloroplast is different from that in angiosperms. Proc Natl Acad Sci USA 93: 8766–8770

Yang J, Schuster G and Stern DB (1996) CSP41, a sequence-specific chloroplast mRNA binding protein, is an endo-ribonuclease. Plant Cell 8: 1409–1420

Yang J and Stern DB (1997) The spinach chloroplast endoribonuclease CSP41 cleaves the 3′-untranslated region of *petD* mRNA primarily within its terminal stem-loop structure.

J Biol Chem 272: 12874–12880

Ye L and Sugiura M (1992) Domains required for nucleic acid binding activities in chloroplast ribonucleoproteins. Nucl Acids Res 20: 6275–6279

Ye L, Li Y, Fukami-Kobayashi K, Go M, Konishi T, Watanabe A and Sugiura M (1991) Diversity of a ribonucleoprotein family in tobacco chloroplasts: Two new chloroplast ribonucleoproteins and a phylogenetic tree of ten chloroplast RNA-binding domains. Nucl Acids Res 19: 6485–6490

Yoshinaga K, Iinuma H, Masuzawa T and Ueda K (1996) Extensive RNA editing of U to C in addition to C to U substitution in the rbcL transcripts of hornwort chloroplasts and the origin of RNA editing in green plants. Nucl Acids Res 24: 1008–1014

Yoshinaga K, Kakehi T, Shima Y, Iinuma H, Masuzawa T and Ueno M (1997) Extensive RNA editing and possible double-stranded structures determining editing sites in the atpB transcripts of hornwort chloroplasts. Nucl Acids Res 25: 4830–4834

Yu W and Schuster W (1995) Evidence for a site-specific cytidine deamination reaction involved in C to U RNA editing of plant mitochondria. J Biol Chem 270: 18227–18233

Zeltz P, Hess WR, Neckermann K, Burner T and Kössel H (1993) Editing of the chloroplast *rpoB* transcript is independent of chloroplast translation and shows different patterns in barley and maize. EMBO J 12: 4291–4296

Chapter 8

Regulation of Chloroplast Translation

Aravind Somanchi* and Stephen P. Mayfield

Department of Cell Biology and The Skaggs Institute for Chemical Biology, The Scripps Research Institute, 10550 North Torrey Pines Road, La Jolla, CA 92037 U.S.A.

Summary

Plastid gene expression is highly regulated in response to environmental parameters, such as light, during plant growth and development. The integrated regulation of post-transcriptional events, such as mRNA stability, mRNA processing, and translation, has been shown to play a major role in chloroplast gene expression. Biosynthesis of many chloroplast proteins shows a requirement for nuclear-encoded proteins and hence a coordination of the two genomes. Genetic analysis of mutants and biochemical analysis of proteins involved in gene expression have begun to reveal mechanisms of plastid gene expression. Analyses of mutants in green algae and plants have identified *trans*-acting nuclear encoded proteins, and *cis*-elements in the messenger RNAs that are involved in the expression of specific chloroplast genes. Biochemical studies have identified interactions of these *trans*-acting proteins with *cis*-elements found in both the 5′ and 3′ untranslated regions of plastid mRNAs. Translation initiation in chloroplast mRNAs has both prokaryotic and eukaryotic features indicating that chloroplast translational regulation is a hybrid between the two systems. An emerging theme suggests that translational regulation relies on specific RNA-protein and protein-protein interactions that influence the ability of the ribosome to correctly initiate translation at the start codon. Understanding the involvement of nuclear gene products in the regulation of chloroplast translation should allow for the identification of the mechanisms of chloroplast gene expression that facilitate the coordination of a prokaryotic-like organelle with the eukaryotic nuclear host.

*Author for correspondence, email: aravind@scripps.edu

Eva-Mari Aro and Bertil Andersson (eds): Regulation of Photosynthesis, pp. 137–151.

I. Introduction

Chloroplasts are thought to have arisen by endosymbiosis of a photosynthetic unicellular prokaryote into an eukaryotic host. This process of integration of the endosymbiont genome with that of the host involved gene translocation from the plastid to the host nucleus. This gene transfer necessitates a coordination of the two genomes for the regulation of photosynthetic protein expression.

Photosynthesis, the primary function of the chloroplast, is driven by light energy captured by protein complexes localized in the thylakoid membranes. The products of the light reactions, ATP and NADPH, are used for a number of important metabolic functions such as carbon and nitrogen fixation, amino acid biosynthesis, and lipid biosynthesis. Therefore, any change in the availability and intensity of light has pronounced effects on chloroplast metabolism. Fluctuations in light intensity and availability also have a direct influence on gene expression. An outstanding example of such an effect on chloroplast protein accumulation is observed during light induced chloroplast biogenesis, as some chloroplast proteins have been shown to increase more than 1000-fold (Gillham et al., 1994; Rochaix, 1996). Exposure of dark-adapted plants to light results in a rapid increase in the synthesis of chloroplast proteins (Malnoe et al., 1988). Although some modulation of transcription in response to environmental cues has been identified (Christopher et al., 1992; Chen et al., 1994; Wada et al., 1994; Kim et al., 1999), chloroplast gene expression is predominantly regulated post transcriptionally (Berry et al., 1986; Klein and Mullet, 1986; Malnoe et al., 1988; Kim and Mullet, 1994; Rochaix, 1996; Sugita and Sugiura, 1996). While many post transcriptional processes such as RNA splicing, editing, processing, and stability, are capable of modulating chloroplast protein synthesis, translation appears to provide the most efficient mechanism for regulating protein levels within the chloroplast. Translational modulation is thus a key component in the regulation of protein synthesis in the plastids.

The endosymbiotic nature of the prokaryotic chloroplast within the eukaryotic host, while drawing upon prokaryotic and eukaryotic features of the two genomes, demands the emergence of novel set of regulatory mechanisms. The elucidation of the mechanism of regulation of chloroplast translation is thus crucial for an understanding of gene expression in the chloroplast, and ultimately in understanding plant development.

II. Translation in the Chloroplast—An Overview

The basic tenets of translation, engaging the ribosomes to recognize and bind at the correct start codon, elongation of the nascent polypeptide chain, and termination of the event by disassembly of the ribosomal complex, have been recognized in the chloroplast. The components involved in translation of chloroplast messages show unique characteristics along with features that are analogous to both prokaryotes and eukaryotes (Table 1). In eukaryotes, a protein complex (eIF-4F, Mader and Sonenberg, 1995) facilitates binding of the translation initiation complex to the m^7G capped 5′ terminus of the mRNA. After binding to the 5′ terminus of the mRNA, the initiation complex scans the 5′ untranslated region (UTR) to identify the correct initiator codon (Jackson, 1996). The scanning process is dependent on helicase activity intrinsic to the initiation complex and requires energy in the form of ATP. Secondary structures in the 5′ UTR, and binding of specific proteins to the 5′ UTR and the polyadenylated tail, affect the rate of initiation of translation (Kozak, 1992). In contrast, the initiation complex in prokaryotes directly binds to a consensus ribosome binding site (RBS) known as a Shine-Dalgarno (SD) sequence (Gold, 1988). Binding of the initiation complex is facilitated by complementary base pairing of the SD and sequences in the 3′ terminus of the 16S rRNA of the 30 S small ribosomal subunit. The SD is typically located about five to nine nucleotides upstream of the initiator codon. Because of the close proximity between the SD and the initiator codon, binding to the SD localizes the initiation complex to the correct initiator codon. Prokaryotic translation initiation does not require ATP hydrolysis nor is it dependent on helicase activity. The direct binding of the ribosomes to SD allows internal initiation of translation in prokaryotes

Abbreviations: bp – base pairs; cPABP – chloroplast-localized PABP; eIF-4F – eukaryotic initiation factor 4F complex ; GUS – glucuronidase synthase; IR – inverted repeat; kDa – kilodaltons; ORF – open reading frame; PABP – poly(A) binding protein; PDI – protein disulfide isomerase; PS – photosystem; RB – RNA binding protein; RBS – ribosome binding site; RNP – ribonucleic protein; SD – Shine-Dalgarno sequence ; UTR – untranslated region

Table 1. Comparison of translation in prokaryotes, eukaryotes and chloroplast

Property	Prokaryotes	Eukaryotes	Chloroplast
Shine-Dalgarno sequence	–3 to –9 nt upstream of initiation codon	Not present of initiation codon	–27 to –30 nt upstream
Ribosome binding	SD sequence	5′-Cap	SD-like sequence and *cis*-elements
Initiation	Complimentary base pairing of SD and 16S rRNA	Initiation factor eIF-4F	Interaction of SD/*cis*-element with initiation complex
Ribosomal scanning	Not seen	Required	Not identified thus far
Helicase activity	Not required	Important	Not identified thus far
Energy requirement	None	ATP hydrolysis	Not known

independent of upstream sequence, thus allowing for the simultaneous translation of several open reading frames (ORFs) from a single polycistronic transcript (Gold, 1988).

Chloroplasts contain 70 S ribosomes and transcripts devoid of 5′ caps and polyadenylated tails. Expression of polycistronic transcripts in the chloroplast suggests that the chloroplast translation initiation complex binds in a prokaryotic fashion internal to the chloroplast transcript and not to its 5′ terminus (Barkan, 1988). Translation in the chloroplast is not coupled to transcription, and chloroplast mRNAs have been shown to exist as stable ribonucleic protein (RNP) complexes with half-lives up to 40 hours (Klaff and Gruissem, 1991; Kim et al., 1993). Neither scanning nor helicase activity have been identified in the chloroplasts. The 5′ and 3′ UTRs of several chloroplast sequences have been shown to associate with nuclear-encoded chloroplast-localized protein factors. This interaction appears to influence the stability of the message and also facilitate the recruitment of the ribosomes to the correct initiation codon. Chloroplast transcripts generally lack consensus prokaryotic SD sequences, but biochemical and genetic evidence supports the presence of SD like elements in many mRNAs, although these elements are much farther upstream of the initiation codon as compared to the prokaryotic consensus. These differences indicate that translation in the chloroplast diverged from that of prokaryotes and has incorporated components, and perhaps regulatory pathways, of eukaryotic translation.

III. Translational Regulation in the Chloroplast

The general components of prokaryotic and eukaryotic translation have been relatively well characterized, while the components of chloroplast translation are just being identified. As such, the understanding of the mechanism of translation of chloroplast mRNAs, and the involvement of nuclear encoded factors in this process is still in its initial stages. Translation initiation, the most complex step of the translation process has been shown to be an important site of regulation of gene expression in many organisms (Kozak, 1999). Studies done thus far suggest that the interaction of the nuclear encoded protein factors with sequence elements in specific mRNAs regulate translation initiation, and thereby protein synthesis in the chloroplast. The identification of distinct RNA elements in individual mRNAs, the characterization of specific proteins that interact with these mRNA sequences, and the characterization of nuclear gene products that impact this process, have begun to define these key regulators of chloroplast translation.

A. RNA Elements

The binding of the ribosome to a specific initiation site in the 5′ UTR of mRNAs (RBS), the first step in translation initiation, is often the rate limiting step of translation. In prokaryotes, this process is controlled by a SD sequence, secondary structural elements affecting accessibility of the start codon, and other

RNA enhancer elements (Gold, 1988). Translation initiation in the chloroplasts also appears to be impacted by such translation enhancers and SD-like sequences.

1. Cis *elements Enhancing Translation*

Studies on 5′UTRs of chloroplast mRNAs show that they are AU-rich and often contain potential stem-loop structures. In vitro and in vivo assays have shown that several regions of chloroplast mRNAs upstream of the translational initiation codon affect translation in tobacco chloroplasts (Hirose and Sugiura, 1996). Analysis of GUS reporter gene expression fused downstream of 5′ UTRs of various *Chlamydomonas reinhardtii* genes, shows a large variation in expression, suggesting that modulation of RNA translation efficiency by each of these 5′UTRs is quite unique (Bolle et al., 1996; Eibl et al., 1999). A stretch of 97 nucleotides located 236 nucleotides from the initiation codon in the *C. reinhardtii psbC* 5′ UTR is required for activation of translation initiation in this message (Zerges et al., 1997). Mutations leading to the loss of a potential secondary structure element in this 97-nucleotide region resulted in the loss of reporter gene expression. Regions approximately 150 nucleotides upstream of the initiation codon in the 5′ UTR of *C. reinhardtii petD* have been shown to enhance translation as well as message stability (Sakamoto et al., 1994a). Analysis of these long chloroplast 5′ UTRs suggests that these RNAs have extensive secondary structures that are predicted to form stem-loop structures directly upstream of the initiation codon (Hauser et al., 1998). Such secondary structures have also been detected in relatively short 5′ UTRs, as is seen in the case of the chloroplast *atpH* mRNA of *Euglena gracilis* (Betts and Spremulli, 1994).

Site-directed mutations affecting stem-loop elements directly upstream of potential SD sequences greatly reduce both D1 and D2 protein synthesis from the *psbA* and *psbD* mRNAs (Bruick and Mayfield, 1998; Nickelsen et al., 1999). Mutations in the 5′ UTR of *C. reinhardtii rps7* mRNA, that eliminate a predicted stem-loop structure, reduced the expression of a downstream aminoglycoside adenyltransferase (AAD) reporter gene, while complementary nucleotide changes that reconstitute the stem-loop structure restored the expression to wild type levels (Fargo et al., 1999). The 5′ UTR of *psbA* mRNA, has been shown to be processed by the removal of a stem-loop structure (Bruick and Mayfield, 1998). Loss of the ribosome binding site by mutation in the 5′ UTR of the message, or nuclear mutations blocking ribosome association, both result in the absence of mRNA processing and translation of the *psbA* mRNA. In vivo studies show that the stem-loop elements are removed by processing in the majority of mRNAs, making these mutant mRNAs indistinguishable from the wild type mRNAs in vivo. These data suggest a role for the stem-loop element in translation that must occur prior to the processing event.

Many chloroplast mRNAs have been shown to contain inverted repeats (IR) in their 3′ UTRs that fold into stem-loop structures and function in mRNA processing and stability (Gruissem and Schuster, 1993; Drager et al., 1996; Yang and Stern, 1997). Deletion of the IR in the 3′ UTR of the *C. reinhardtii atpB* gene reduced accumulation of this transcript, suggesting that this IR plays some role in stabilizing the transcripts (Rott et al., 1998). Deletion of 3′ IRs in several *C. reinhardtii* mRNAs resulted in altered transcript size, but not in a dramatic change in mRNA accumulation, suggesting that these 3′ IR are more important for 3′ end formation than for mRNA stability (Blowers et al., 1993). Deletions in the 3′ IR of the *C. reinhardtii atpB* mRNA showed a strong correlation between translation and mRNA processing at the 3′ UTR, suggesting that translation and 3′ end formation are coupled (Rott et al., 1998). Each of these data suggest that 3′ UTRs are functional components of translation that affect mRNA stability, perhaps through their role in translation indirectly.

Although the available data does not provide a clear role for these 5′ and 3′ RNA elements, it is clear they are required for normal translation. It is possible that the 5′ upstream sequence elements might affect events early in translation initiation, perhaps providing sites for interaction with *trans*-acting factors that direct the mRNA to a pre-initiation complex, after which the RNA element is removed by processing. The IRs in the 3′ UTR are required for correct mRNA processing, and may influence mRNA stability and translation directly or through interactions with other factors. These data also suggest that mRNA processing may be a consequence of translation but not necessarily a prerequisite for it, demonstrating that mRNA processing, stability and translation are intimately interconnected.

2. Shine-Dalgarno Sequences

Shine-Dalgarno sequences in prokaryotes consist of a stretch of 3-9 nucleotides, located 5-15 nucleotides upstream of the initiation codon. The SD sequence base pairs with a complimentary sequence on the 3′ terminus of the 16S rRNA (Kozak, 1999). Analysis of the 5′ UTRs of chloroplast mRNAs confirmed the presence of potential SD sequences in close proximity upstream of the initiation codons of many chloroplast mRNAs. Nonetheless, few of these potential SD elements are located at the prokaryotic consensus in relation to the initiation codon (Ruf and Kossel, 1988; Bonham-Smith and Bourque, 1989). Mutations in potential SD sequences significantly affected translation of the *psbA* and *rps14* mRNAs of tobacco and *C. reinhardtii* as well as the *atpH* mRNA of *E. gracilis* (Betts and Spremulli, 1994; Mayfield et al., 1994; Hirose and Sugiura, 1996; Hirose et al., 1998). However, similar mutations did not show any effect on the translation of *C. reinhardtii petD, atpB, atpE, rps4, rps7* and *E. gracilis rbcL* mRNAs (Koo and Spremulli, 1994; Sakamoto et al., 1994b; Fargo et al.,1998). In vitro analysis of barley chloroplast mRNAs showed that binding of the initiation complex is mediated by the SD-like sequence in messages having an upstream SD-like element, and that additional factors, unique to the chloroplast, are required in the remainder of chloroplast messages (Kim and Mullet, 1994).

Mutation analysis of a putative SD sequence located 27 nucleotides upstream of the initiation codon of the *C. reinhardtii psbA* and *psbD* mRNAs showed that an effective and accessible SD sequence is required for translation (Bruick and Mayfield, 1998; Nickelsen et al., 1999). Deletions placing the SD sequence at a prokaryotic consensus position result in loss of *psbA* translation in *C. reinhardtii,* but showed high translation efficiency in *Escherichia coli* (Bruick, unpublished). Mutations in a 7 nucleotide SD-like element present 30 nucleotides upstream of the start codon of the *C. reinhardtii psbD* mRNA, resulted in loss of translation of the D2 polypeptide (Nickelsen et al., 1999). These results suggest that some chloroplast messages use a SD-independent mechanism for translation, while other chloroplast mRNAs use a SD-dependent initiation mechanism. Both mechanisms of chloroplast translation have adapted to facilitate binding of *trans*-acting protein factors, in a message-specific manner, as a requirement to recognize the correct initiation codon. The determination of translation initiation by position rather than initiation codon context (Chen et al., 1995) also strengthens the above argument. Finding additional *cis* elements in the 5′ UTR may help in identifying the mechanism of chloroplast translation initiation, especially where messages lack an obvious SD element.

B. RNA Binding Proteins

Gene expression studies in *E coli* and bacteriophages provided the first evidence for translational regulation mediated by RNA binding proteins targeted to specific motifs in the mRNA (McCarthy and Gualerzi, 1990). Proteins bound to the mRNA have been shown to introduce conformational changes that alter the accessibility of the initiation codon to the ribosomes. Positive and negative regulatory effects of such RNA bound proteins have been shown in both prokaryotic and eukaryotic translation mechanisms (Kenan et al., 1991; Wulczyn and Kahmann, 1991; Goossen and Hentze, 1992; Draper, 1995).

Affinity chromatography studies have assisted in the identification of several proteins associated with chloroplast mRNAs (Danon, 1997; Bruick and Mayfield, 1998). The RNA binding property of many of these proteins does not necessarily suggest they recognize individual mRNAs or specific sequences within messages. These proteins are likely to play a variety of roles during translation, such as maintaining ribosomal accessibility by preventing formation of secondary structure, assist in unwinding the mRNA, increasing the stability of the initiation complex, or perhaps promote the movement of the complex along the RNA. Characterization of a number of such protein factors from *C. reinhardtii* showed that many of them are nuclear encoded and that a majority of them associate with the untranslated regions (UTRs) present on the 5′ and the 3′ end of chloroplast messages.

1. 3′ Untranslated Region Associating Proteins

The inverted repeat sequences in the 3′ UTRs of *C. reinhardtii* have been shown to associate with protein factors that allow processing of the mRNA (Blowers et al., 1993). It has been proposed that these RNA binding proteins influence the efficiency of endonucleolytic cleavage or the exonucleolytic trimming in the formation of a proper 3′ end (Stern and Kindle, 1993). Mutations in the 3′ UTR of *C. reinhardtii atpB* gene show that this 3′ end processing is not sequence specific (Rott et al., 1999). For example, some 3′ UTR binding proteins have been found to interact with the 5′ UTR or 5′ UTR binding proteins to control RNA degradation (Lisitsky and Schuster, 1995; Memon et al., 1996; Goldschmidt-Clermont, 1998). A 28 kDa protein isolated from spinach chloroplast has been shown to interact with an IR in the 3′ UTR of the mRNA as well as with sequence elements in the 5′ UTR (Lisitsky and Schuster, 1995). A 67 kDa protein has been isolated that associates with a high molecular weight complex, and is thought to be involved in 3′ end processing of spinach mRNAs (Hayes et al., 1996). This interaction is modulated by phosphorylation and has been shown to be required for 3′ end processing and message stability in vitro (Lisitsky et al., 1996; Goldschmidt-Clermont, 1998). Monitoring the expression of a GUS reporter gene fused with the UTRs of the tobacco *psbA, rbcL* and *rpl32* mRNAs, showed that 3′ UTRs influence mRNA processing and accumulation, but do not significantly impact translation (Eibl et al., 1999). Reporter gene expression in constructs with 5′ and 3′ UTRs of the same gene and constructs with UTRs from different genes shows distinct differences, suggesting that interactions between the two UTRs play a role in translation efficiency. The identification of proteins that bind to both the UTRs of *psbA* and *rbcL* (Stern et al., 1989) shows that interaction between 5′ and 3′ UTR often involves RNA binding proteins that recognize both ends. While the proteins binding 3′ UTR have not been shown to influence translation directly, they appear to play a significant role in mRNA processing and stability, which have been shown to be important for translation initiation and efficiency

2. 5′ Untranslated Region Associating Proteins

A biochemical approach used to identify factors regulating translation of chloroplast messages has been to isolate specific proteins that bind the 5′ UTR of chloroplast messages. While a number of proteins have been identified by RNA affinity chromatography, little is known regarding their specificity and function (Danon and Mayfield, 1991; Hauser et al., 1993; Zerges and Rochaix, 1994). A comparison of the 5′ UTRs, of *psbA, rps12, rbcL,* and *atpB,* led to the identification of at least seven proteins, of 15, 36, 38, 47, 56, 62, and 81 kDa, associated with the RNA (Hotchkiss and Hollingsworth, 1995, 1999; Danon, 1997). Multiple forms of the 47 kDa and 81 kDa proteins have been identified (Hauser et al., 1996). Three of these, the 38 kDa, 47 kDa, and 81 kDa proteins, have been shown to associate with the 5′ UTRs of all the checked RNAs. Cells, where mRNAs for chloroplast encoded ribosomal proteins are preferentially translated, show reduced chloroplast protein synthesis when a 36 kDa protein present in wild type cells is missing or greatly reduced. This suggests that this 36 kDa protein is required for translation of mRNAs encoding photosynthetic proteins (Hauser et al., 1996).

The 5′ UTR of the *C. reinhardtii psbD* mRNA has been shown to associate with a number of proteins. A nuclear mutation with reduced abundance of a 47 kDa protein showed decreased accumulation of the *psbD* mRNA, suggesting that this protein might influence message stability (Nickelsen et al., 1994). Studies on the proteins associating with the 5′ UTR of *psbC* resolved three proteins of 95, 65 and 40 kDa. A 46 kDa protein was detected binding to the 5′ UTR of *psbC* mRNA only in the nuclear mutant F64, which lacks translation of that mRNA leading the authors to hypothesize that the binding of the 46 kDa protein may inhibit translation of *psbC* (Zerges and Rochaix, 1994, 1998).

A protein complex consisting of four polypeptides of 38, 47, 55 and 60 kDa that binds to the 5′ UTR of *psbA* mRNA has been isolated (Danon and Mayfield, 1991). There is a strong correlation between the RNA binding activity of the complex and *psbA* mRNA translation under light and dark growth conditions (Danon and Mayfield, 1991). The binding activity of this complex is lower in the dark though there is little variation in accumulation of these proteins between the dark and light grown cells. This suggests that modulation of specific RNA binding activity is used to regulate mRNA translation (Danon and Mayfield, 1991). Two of these RNA binding (RB) proteins, RB47 and RB60, have been cloned and sequenced. The RB47 protein shows homology to poly(A) binding proteins (PABP; Yohn et al., 1998a) and RB60 shows homology to protein disulfide isomerases (PDI; Kim and Mayfield, 1997).

Cross-linking studies of RB47 indicate that this protein is directly in contact with the *psbA* 5′ UTR (Danon and Mayfield, 1991). In vitro studies showed that this protein is imported into the chloroplasts (Yohn et al., 1998b), and associates with a low-density membrane fraction that may be part of the envelope membrane (Zerges and Rochaix, 1998). Nuclear mutants that affect the levels and activity of this chloroplast localized PABP (cPABP) affect light activated translation initiation of the *psbA* mRNA (Yohn et al., 1996).

C. Nuclear Mutations Affecting Translation

Many nuclear mutations with defects in chloroplast translation have been identified. Some of these mutations affect components of the general translation machinery such as ribosomal RNA or ribosomal proteins. Others are more specific and affect a single or a small subset of chloroplast proteins. Nuclear mutants, *iojap* in maize and *albostrians* in barley, have been characterized to be ribosome deficient (Taylor, 1989). Other nuclear mutants of maize with deficiencies in photosynthesis, *cps1, cps2,* and *hcf7*, have general defects in chloroplast translation with reduction in polysome associated mRNAs and defective 16S rRNA processing (Barkan, 1993). A number of *C. reinhardtii* mutants with translational deficiencies, such as reduced numbers of ribosomes (Harris et al., 1994), impaired intron splicing in 23S rRNA or in *psbA* (Herrin et al., 1990, 1991) and unstable 50S subunits (Myers et al., 1984), have also been characterized.

Analysis of *C. reinhardtii*, maize, barley, spinach, tobacco and *Arabidopsis* mutants demonstrates the universal requirement of nuclear factors for translation of specific chloroplast mRNAs (Table 2). Genetic and molecular analyses have identified mutations of nuclear gene products, each of which causes translational deficiency of a specific chloroplast mRNA. While it is not obvious that each of these gene products is a component of the chloroplast translation machinery, more than one nuclear gene product has been shown to be necessary for the translation of single chloroplast mRNAs (Gillham et al., 1994; Rochaix 1996; Zerges et al., 1997).

Nuclear mutations F34 and F64 have been shown to affect the translation of the *psbC* gene encoding the 43 kDa subunit of Photosystem II (PS II). Monitoring the expression of the reporter gene *aadA* fused to the 5′ UTR of *psbC* has shown that these gene products interact with the 5′ UTR of the *psbC* mRNA (Rochaix et al., 1989; Zerges and Rochaix, 1994). Deletions in a region of the 5′ UTR relieve the

Table 2. Nuclear mutants affecting translation of chloroplast genes

Chloroplast gene	Nuclear mutant	Reference
psbA	F35	Yohn et al. (1996)
	hf149, hf233, hf261	Yohn et al. (1998)
psbB	*222E; GE2-10*	Monod et al. (1992)
	hcf109	Meurer et al. (1996)
psbC	F34 *(TBC1)*	Rochaix et al. (1989)
	F64 *(TBC2)*	Zerges and Rochaix (1994)
	TBC3	Zerges et al. (1997)
psbD	*nac 2*	Kuchka et al. (1989)
	ac-115	Kuchka et al. (1988)
	sup4b	Wu and Kuchka (1995)
atpA	F54	Drapier et al. (1992)
atpB	*thm24*	Drapier et al. (1992)
	atp 1-1	McCormac and Barkan (1999)
psaB	F15 *(TAB1)*	Stampacchia et al. (1997)
psaC	*hcf109*	Meurer et al. (1996)
petA	MΦ11	Gumpel et al. (1995)
petB	*crp1*	Barkan et al. (1994)
	MΦ37	Gumpel et al. (1995)
petD	F16	Drager et al. (1998)

need for the nuclear gene product F34 suggesting an interaction of this protein with this potential stem-loop region. Another nuclear mutant of *C. reinhardtii* (F15) in which the mutation has been mapped to the nuclear TAB1 locus, fails to translate the *psaB* mRNA, in spite of accumulating this mRNA to wild type levels. A chloroplast suppressor of this mutation was mapped to a region of the mRNA adjacent to a potential SD sequence within the 5′ UTR (Stampacchia et al., 1997). The suppressor mutation was hypothesized to have a destabilizing effect on the stem-loop structure in the 5′ UTR of the *psaB* mRNA. This suggests that the TAB1 factor might be required for alleviating structural constraints around the SD sequence of *psaB*, enhancing accessibility to a ribosome-binding site.

C. reinhardtii nuclear mutants, *nac1-18* and *ac-115* have defects in the synthesis of the D2 protein from the *psbD* mRNA (Kuchka et al., 1988). The *nac1* mutation has been shown to block translational elongation of the *psbD* mRNA and has been predicted to have a secondary effect on translation initiation of *psbA* (Rochaix, 1996). A mutation in a third nuclear locus, *sup4b*, has been shown to suppress both mutations (Wu and Kuchka, 1995). While this suppressor shows an ability to overcome three different mutations in these two loci, it does not recover nuclear mutations affecting translation of other chloroplast messages. This suggests that this suppressor encodes a factor specific to the *psbD* mRNA. The *ac115* gene containing a nonsense mutation that causes the mutant phenotype has been cloned (Rattanachaikunsopon et al., 1999). The novel gene product, predicted to be a 100 amino acid protein with a thylakoid membrane spanning sequence, has been presumed to function after translation initiation of D2 to assist in co-translational folding of D2 into the thylakoid membrane or co-translational binding of a D2 specific ligand. The *psbA* mRNA of the *vir-115* nuclear mutant of barley is quite stable despite the loss of D1 synthesis (Kim et al., 1994) suggesting a post initiation effect of the mutation.

Nuclear mutations spanning at least five nuclear loci have been shown to specifically effect *C. reinhardtii psbA* translation (Yohn et al., 1996; Yohn et al., 1998a,b). Several of these mutations have also been shown to affect the accumulation of *psbA* mRNA binding proteins (Yohn et al., 1998b). Only 4% of the *psbA* mRNA is associated with polysomes in the nuclear mutant F35, compared to nearly 35% in the wild type. This block in translation initiation and the subsequent absence of D1 synthesis coincides with a 90% reduction in cPABP accumulation and a corresponding loss of *psbA*-specific binding activity (Yohn et al., 1996). The nuclear mutant *hf149* also fails to synthesize the D1 protein, in conjunction with a 95% reduction in polysome association of the *psbA* mRNA, and this mutant completely lacks accumulation of the cPABP (Yohn et al., 1998a). Other nuclear mutants, lacking D1 synthesis (*hf261*, *hf859* and *hf1085*), have a reduced *psbA* RNA binding activity and show reduced cPABP accumulation (Yohn et al., 1998b). These data demonstrate that cPABP accumulation and activity is required for normal *psbA*-mRNA ribosome association. PABPs of eukaryotes have been shown to bind cytoplasmic mRNAs at the 3′ poly (A) tail and play a role in translational activation through interactions with the initiation complex (Tarun and Sachs, 1996). Though chloroplast messages lack poly (A) tails, RB47 has been shown to bind to the 5′ UTR of *psbA* mRNA and may interact with chloroplast initiation factors in a manner similar to eukaryotic PABPs.

A nuclear mutation, *hcf136* in *Arabidopsis thaliana* results in the complete loss of all the core subunits of PS II and the oxygen evolving complex, suggesting that this mutation affects a very central step in the biogenesis of the PS II complex (Meurer et al., 1998). The HCF136 protein has been speculated to be necessary for the translation of one of the proteins involved in the stability of the core complex. Mutations affecting nuclear genes required for translation of *atpB/E, petA* and *petD* in maize have been identified (Fisk et al., 1999; McCormac and Barkan, 1999). Mutations in a maize nuclear gene *crp1* lead to the absence of cytochrome b_6f and a decrease in PS I complex (Barkan et al., 1994). These mutations have been shown to affect the translation of *petA* mRNA and the synthesis of *petD* mRNA. Cloning and analysis of this gene suggests CRP1 to be a dual-function protein that influences the translation of *petA* and *petD* mRNAs, and *petD* mRNA processing independently. Nuclear mutants affecting the translation of *C. reinhardtii* proteins involved in the formation of the cytochrome b_6f complex have also been identified (Gumpel et al., 1995). These mutants have been proposed to affect mRNA stability and hence translation. A *C. reinhardtii* nuclear mutant (F16) harboring a mutation in *mcd1* gene, shows instability of *petD* mRNA, and consequently does not accumulate the cytochrome

b_6f complex. The degradation of *petD* mRNA by a 5′ to 3′ exoribonuclease activity has been attributed to the loss of a nuclear encoded protein, MCD1, providing evidence that this protein protects RNA from degradation by interacting with the 5′ UTR of the *petD* mRNA (Drager et al., 1998; 1999).

IV. Mechanism of Translational Activation

Protein synthesis in the chloroplast is regulated with high specificity. Light induces a 50- to 100-fold enhancement of synthesis of specific chloroplast proteins (Inamine et al., 1985; Laing et al., 1988; Malnoe et al., 1988; Berry et al., 1990; Keller et al., 1991) with no equivalent increase of their mRNAs. Recruitment of mRNAs into polysomes following illumination is observed for both *rbcL* (Berry et al., 1990) and *psbA*, suggesting that translation initiation is a key component of light activated translation. While translation initiation appears to play a role in light activated translation, light dependent formation of a proton gradient across the thylakoid membranes has also been shown to activate translational elongation of D1 (Edhofer et al., 1998; Muhlbauer and Eichacker, 1998), suggesting that both initiation and elongation are light activated processes.

In vitro, binding of proteins to the *psbA* 5′ UTR can be modulated by the addition of oxidizing and reducing agents to *C. reinhardtii* cell extracts (Danon and Mayfield, 1991, 1994a). The RNA binding activity abolished by incubation with the oxidizing agent dithionitrobenzoic acid (DTNB), can be restored by treatment with dithiothreitol (DTT), but not with β mercaptoethanol, indicating that the inhibition may be due to the formation of a disulfide bond between vicinal thiols (Danon and Mayfield, 1994a). In vitro, the binding activity of oxidized cPABP can be restored by a protein disulfide isomerase (PDI) and DTT (Kim and Mayfield, 1997). The regulatory switch of the binding of this complex in vivo, appears to operate through RB60, a chloroplast localized homolog of PDI (cPDI, Kim and Mayfield, 1997). The oxidation potential of cPDI could allow for the oxidation of cPABP when photosynthetic activity and consequently reducing potential of the chloroplast is low thus allowing for a reversible switch for cPABP RNA binding activity.

A threshold level of ADP (higher than 0.3 mM) was also found to inhibit the binding of *psbA* translational activators to the RNA by phosphorylation of RB60 (Danon and Mayfield, 1994b). Levels of ADP high enough to activate this phosphorylation are found in chloroplasts following transfer into the dark (Stitt et al., 1980; Hampp et al., 1982). This suggests that translation of *psbA* mRNA may be inhibited in the dark by the inactivation of the *psbA* RNA binding activators in response to a rise in stromal ADP concentrations (Danon and Mayfield, 1994b). Upon illumination, translation may be reactivated by reduction of the RNA binding proteins resulting from the production of reduced thioredoxin. The in vivo effect of chloroplast redox potential on D1 expression examined in a PS I deficient strain in which thioredoxin is poorly reduced (Buchanan, 1991) shows reduced binding of the complex to *psbA* mRNA and reduced D1 synthesis. This data suggests that in vivo the chloroplast redox potential generated by photosynthetic activity can be used to modulate the binding affinity of the protein complex to the *psbA* mRNA (Fig. 1) and hence translation of the D1 protein.

A protein complex that recognizes a specific AU rich region in the *psbA* 5′UTR has been identified in tobacco (Hirose and Sugiura, 1996). This element has been shown to confer light dependent translational regulation of a reporter gene (Staub and Maliga, 1994). A 43 kDa chloroplast homolog of the spinach ribosomal S1 protein has also been shown to associate with the *psbA* 5′ UTR of seedlings grown in the light (Klaff and Gruissem, 1995; Alexander et al., 1998). The ribosomal S1 protein of prokaryotes is thought to assist complex formation between the 30S ribosomal subunit and mRNAs, particularly in the absence of a SD sequence (Tzareva et al., 1994). Hence this chloroplast homolog may be involved in mediating light affected translation. Another example of a chloroplast counterpart of a prokaryotic regulatory mechanisms is seen in the feedback pathway for cytochrome *f* synthesis, where the cytochrome *f* protein interacts with its own 5′ UTR to inhibit translation (Choquet et al., 1998). Similarly to *E. coli* systems, excess cytochrome *f* of *C. reinhardtii* that fails to incorporate into cytochrome b_6f complex attenuates translation of its own *petA* mRNA, directly or indirectly through interactions with the *petA* 5′ UTR.

From the above examples a model can be proposed where the major step required to facilitate translation initiation involves the association of an initiation complex with the 5′ UTR of chloroplast mRNAs followed by the recognition of the correct initiator

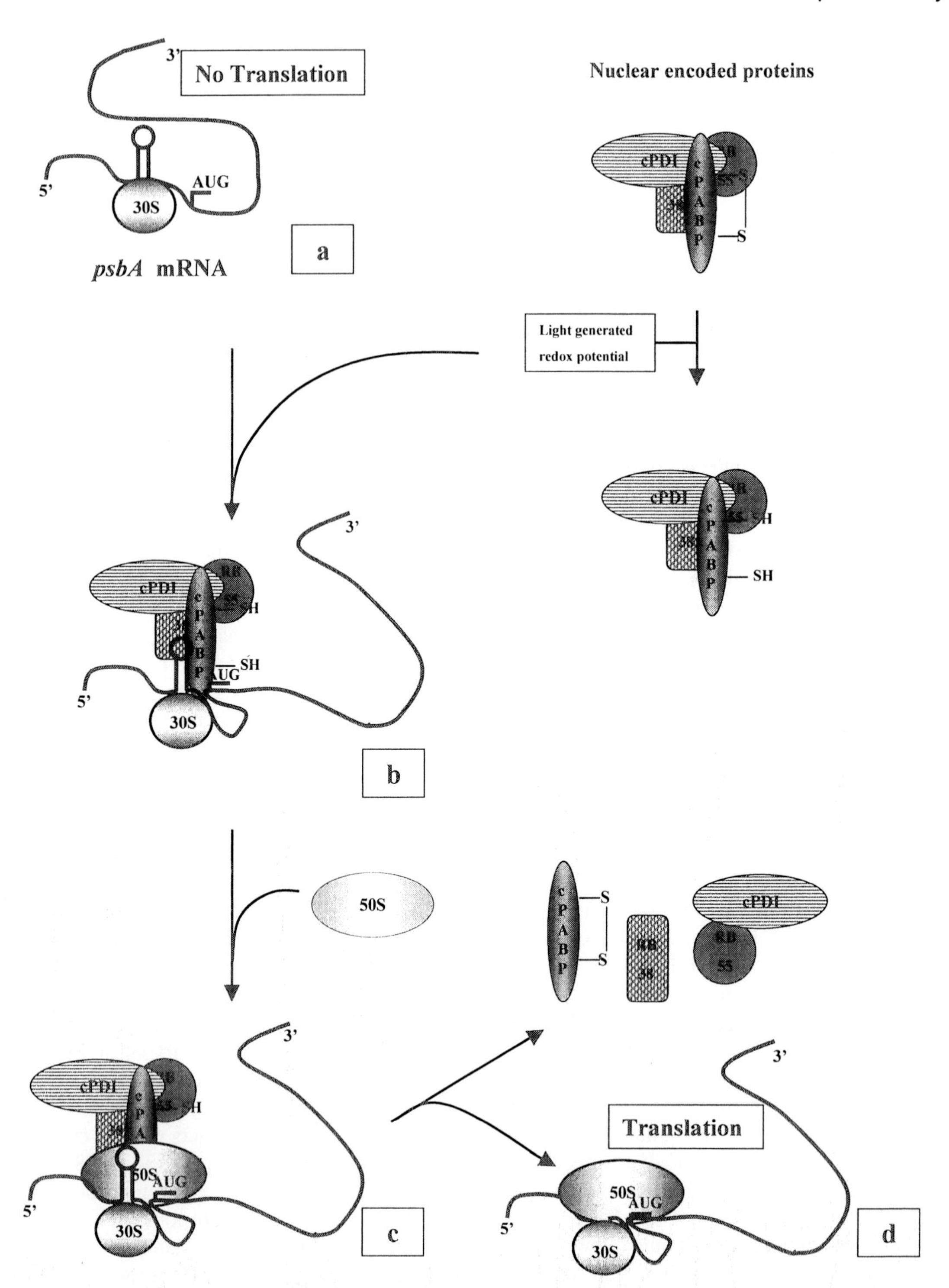

Fig. 1. Model for translation of the *psbA* mRNA. Though the 30S subunit of ribosomes associates with the *psbA* mRNA, translation is not initiated as the ribosome bound to the SD sequence cannot access the initiation codon (a). Four proteins, cPABP, cPDI, RB38 and RB55, encoded by the nucleus, are imported into the chloroplast. Reducing equivalents generated by the light reactions of PSI reduce cPDI, which reduces cPABP. This protein complex associates with the *psbA* mRNA and 30S subunit by binding between the stem-loop structure and the AUG initiation codon (b). Binding of the protein complex makes the AUG codon accessible to the 30S subunit following which the 50 S ribosomal subunit is recruited (c). The protein complex dissociates from the *psbA* mRNA as translation is initiated (d).

codon. The physical and functional interactions suggested between the UTRs and the associating protein factors allow the proposal of a 'closed-loop' mode of regulation analogous to the one proposed for eukaryotes (Preiss and Hentze, 1999). The association of the regulatory protein complexes to the 5′ and 3′ UTRs and the structural changes imparted by the ensuing RNA-protein and protein-protein interactions imply an intricate mechanism for translational regulation. The interaction of nuclear encoded proteins with specific elements in the 5′ and 3′ UTRs of chloroplast mRNAs may control the accessibility of the initiation site to the ribosomes, and modulate release of the nascent polypeptide thereby enhancing or repressing translation (Fig. 2). The observation that light affects both initiation and elongation of chloroplast translation provides support for the proposed mechanism of light generated changes in RNA-protein interactions regulating translational activation.

V. Conclusions and Perspectives

Analysis of nuclear mutants with deficiencies in photosynthesis suggests an intricate interaction between the nuclear and chloroplast genomes in photosynthetic protein synthesis. The large number of nuclear mutations that affect chloroplast translation suggests that many of these nuclear genes encode proteins with potentially novel mechanisms of regulating chloroplast translation. The interaction of nuclear encoded factors with *cis* elements in the 5′ and 3′ UTR emerges as key to the regulation of chloroplast translation. Some chloroplast messages appear to use uncharacteristic elements upstream of

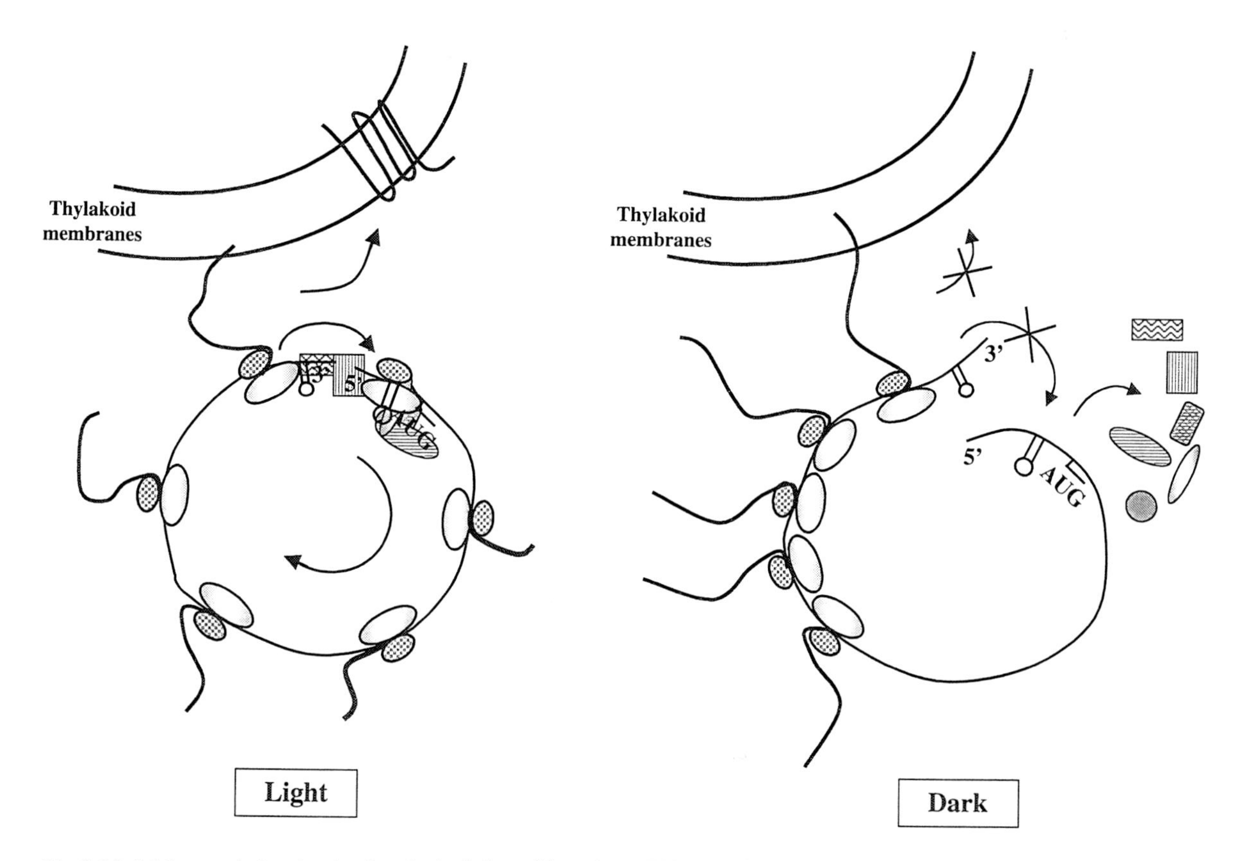

Fig. 2. Model for translational activation. In the light, a chloroplast mRNA may form a closed loop as protein factors bind to the 5' and 3' UTRs activating translation initiation and elongation. These protein-protein and RNA-protein interactions make the initiation codon accessible to ribosomes, and promote the mobility of ribosomes from the 3' end to the 5' end to reinitiate translation. The interaction of protein factors with the 3' UTR may be required to facilitate the release of the translated polypeptide from ribosomes. In the dark, these protein factors dissociate from the RNA making the initiation codon inaccessible to the ribosomes. This dissociation also inhibits the ability of the ribosomes to reinitiate translation and may block the release of the translated polypeptides from the ribosomes, which in turn inhibits elongation as the movement of ribosomes along the RNA is blocked

the initiation codon to assist in ribosome binding, while others use prokaryotic-like SD elements. Chloroplast redox potential and phosphorylation have been shown to be able to fine-tune the regulation of translation for some chloroplast mRNAs. Further characterization is required to assign specific roles in translation for individual RNA elements and *trans*-acting proteins. While available data allows the predicted working model, identification of homologs is required before this paradigm can be extrapolated to other chloroplast messages or from *C. reinhardtii* to higher plants.

While many *trans* acting factors have been proposed based on mutant analysis, few of them have been biochemically characterized. Identification of factors affecting translation of many chloroplast messages will be required to identify the general translational factors that must be present in the chloroplast. It remains unclear whether the numerous nuclear factors constitute specific groups of proteins for individual messages or general factors with specificity imparted by some specific modification or by interaction with other proteins. The identification of conserved features in higher plants is necessary before general rules for chloroplast translation can be instituted. Development of reliable in vitro translation systems, and technologies that facilitate study of RNA-protein, and protein-protein interactions should go a long way in the understanding of the mechanism of chloroplast translation.

References

Alexander C, Faber N and Klaff P (1998) Characterization of protein-binding to the spinach chloroplast *psbA* mRNA 5′ untranslated region. Nucleic Acids Res 26: 2265–72

Barkan A (1988) Proteins encoded by a complex chloroplast transcription unit are each translated from both monocistronic and polycistronic mRNAs. EMBO J 9: 2637–44

Barkan A (1993) Nuclear mutants of maize with defects in chloroplast polysome assembly have altered chloroplast RNA metabolism. Plant Cell 5: 389–402

Barkan A, Walker M, Nolasco M and Johnson D (1994) A nuclear mutation in maize blocks the processing and translation of several chloroplast mRNAs and provides evidence for the differential translation of alternative mRNA forms. EMBO J 13: 3170–3181

Berry JO, Nikolau BJ, Carr JP and Klessig DF (1986) Translational regulation of light-induced ribulose 1, 5-bisphosphate carboxylase gene expression in amaranth. Mol Cell Biol 6: 2347–2353

Berry JO, Breiding DE and Klessig DF (1990) Light-mediated control of translational initiation of ribulose-1, 5-bisphosphate carboxylase in amaranth cotyledons. Plant Cell 2: 795–803

Betts L and Spremulli LL (1994) Analysis of the role of the Shine-Dalgarno sequence and mRNA secondary structure on the efficiency of translational initiation in the *Euglena gracilis* chloroplast *atpH* mRNA. J Biol Chem 269: 26456–26463

Blowers AD, Klein U, Ellmore GS and Bogorad (1993) Functional in vivo analyses of the 3′ flanking sequences of the *Chlamydomonas* chloroplast *rbcL* and *psaB* genes. Mol Gen Genet 238: 339–349

Bolle C, Herrmann RG and Oelmuller R (1996) Different sequences for 5′-untranslated leaders of nuclear genes for plastid proteins affect the expression of the beta-glucuronidase gene. Plant Mol Biol 32: 861–868

Bonham-Smith PC and Bourque DP (1989) Translation of chloroplast-encoded mRNA: Potential initiation and termination signals. Nucleic Acids Res 17: 2057–2080

Bruick RK and Mayfield SP (1998) Processing of the *psbA* 5′ untranslated region in *Chlamydomonas reinhardtii* depends upon factors mediating ribosome association. J Cell Biol 143: 1145–1153

Buchanan BB (1991) Regulation of CO_2 assimilation in oxygenic photosynthesis—the ferredoxin thioredoxin system—perspective on its discovery, present status, and future development. Arch Biochem Biophys 288: 1–9

Chen Q, Osteryoung K and Vierling E (1994) A 21-kDa chloroplast heat shock protein assembles into high molecular weight complexes in vivo and in organelle. J Biol Chem 269: 13216–13223

Chen X, Kindle KL and Stern DB (1995) The initiation codon determines the efficiency but not the site of translation initiation in *Chlamydomonas* chloroplasts. Plant Cell 8: 1295–1305

Choquet Y, Stern DB, Wostrikoff K, Kuras R, Girard-Bascou J and Wollman FA (1998) Translation of cytochrome *f* is autoregulated through the 5′ untranslated region of *petA* mRNA in *Chlamydomonas* chloroplasts. Proc Natl Acad Sci USA 95: 4380–4385

Christopher DA, Kim M and Mullet JE (1992) A novel light-regulated promoter is conserved in cereal and dicot chloroplasts. Plant Cell 4: 785–798

Danon A (1997) Translational regulation in the chloroplast. Plant Physiol 115: 1293–8

Danon A and Mayfield SP (1991) Light regulated translational activators: identification of chloroplast gene specific mRNA binding proteins. EMBO J 10: 3993–4001

Danon A and Mayfield SP (1994a) Light-regulated translation of chloroplast messenger RNAs through redox potential. Science 266: 1717–1719

Danon A and Mayfield SP (1994b) ADP-dependent phosphorylation regulates RNA-binding in vitro: Implications in light-modulated translation. EMBO J 13: 2227-2235

Drager RG, Zeidler M, Simpson CL and Stern DB (1996) A chloroplast transcript lacking the 3′ inverted repeat is degraded by 3′→5′ exoribonuclease activity. RNA 2: 652–63

Drager RG, Girard-Basou J, Choquet Y, Kindle KL and Stern DB (1998) In vivo evidence for 5′-3′ exoribonuclease degradation of an unstable chloroplast mRNA. Plant J 13: 85–96

Drager RG, Higgs DC, Kindle KL and Stern DB (1999) 5′ to 3′ exoribonucleolytic activity is a normal component of chloroplast mRNA decay pathways. Plant J 19: 521–531

Draper DE (1995) Protein-RNA recognition. Annu Rev Biochem 64: 593–620

Drapier D, Girard-Bascou J and Wollman FA (1992) Evidence for nuclear control of the gene expression of the *atpA* and *atpB* chloroplast genes in *Chlamydomonas reinhardtii*. Curr Genet 22: 47–52

Edhofer I, Muhlbauer SK and Eichacker LA (1998) Light regulates the rate of translation elongation of chloroplast reaction center protein D1. Eur J Biochem 257: 78–84

Eibl C, Zou Z, Beck A, Kim M, Mullet J and Koop HU (1999) In vivo analysis of plastid *psbA*, *rbcL* and *rpl32* UTR elements by chloroplast transformation: Tobacco plastid gene expression is controlled by modulation of transcript levels and translation efficiency. Plant J 19: 333–345

Fargo DC, Zhang M, Gillham NW and Boynton JE (1998) Shine-Dalgarno-like sequences are not required for translation of chloroplast mRNAs in *Chlamydomonas reinhardtii* chloroplasts or in *Escherichia coli*. Mol Gen Genet 257: 271–82

Fargo DC, Boynton JE and Gillham NW (1999) Mutations altering the predicted secondary structure of a chloroplast 5′ untranslated region affect its physical and biochemical properties as well as its ability to promote translation of reporter mRNAs both in the *Chlamydomonas reinhardtii* chloroplast and in *Escherichia coli*. Mol Cell Biol 19: 6980–6990

Fisk DG, Walker MB and Barkan A (1999) Molecular cloning of the maize gene *crp1* reveals similarity between regulators of mitochondrial and chloroplast gene expression. EMBO J 18: 2621–2630

Gillham NW, Boynton JE and Hauser CR (1994) Translational regulation of gene expression in chloroplasts and mitochondria. Annu Rev Genet 28: 71–93

Gold L (1988) Post transcriptional regulatory mechanisms in *Escherichia coli*. Annu Rev Biochem 57: 199–233

Goldschmidt-Clermont M (1998) Coordination of nuclear and chloroplast gene expression in plant cells. Int Rev Cyt 177: 115–180

Goossen B and Hentze MW (1992) Position is the critical determinant for function of iron-responsive elements as translational regulators. Mol Cell Biol 12:1959–1966

Gruissem W and Schuster G (1993) Control of mRNA degradation in organelles. In: Belasco JG and Brawermann G (eds) Control of Messenger RNA Stability, pp 329–365. Academic Press, San Diego

Gumpel NJ, Ralley L, Girard-Bascou J, Wollman FA, Nugent JH and Purton S (1995) Nuclear mutants of *Chlamydomonas reinhardtii* defective in the biogenesis of the cytochrome b_6f complex. Plant Mol Biol 29: 921–32

Hampp R, Goller M and Ziegler H (1982) Adenylate levels, energy charge, and phosphorylation potential during dark-light and light-dark transition in chloroplasts, mitochondria and cytosol of mesophyll protoplasts from *Avena sativa* L. Plant Physiol 69: 448–455

Harris EH, Boynton JE and Gillham NW (1994) Chloroplast ribosomes and protein synthesis. Microbiol Rev 58: 700–754

Hauser CR, Randolph-Anderson BL, Hohl TM, Harris EH, Boynton JE and Gillham NW (1993) Molecular genetics of chloroplast ribosomes in *Chlamydomonas reinhardtii*. In: Nierhaus KH, Fraceschi F, Subramanian AR, Erdmann VA and Wittmann-Liebold (eds) The Translational Apparatus, pp 545–554. Plenum, New York

Hauser CR, Gillham NW and Boynton JE (1996) Translational regulation of chloroplast genes. Proteins binding to the 5′-untranslated regions of chloroplast mRNAs in *Chlamydomonas reinhardtii*. J Biol Chem 271: 1486–1497

Hauser CR, Gillham NW and Boynton JE (1998) Regulation of chloroplast translation. In: Rochaix J-D, Goldschmidt-Clermont M and Merchant S (eds) Molecular Biology of *Chlamydomonas*: Chloroplasts and Mitochondria. Kluwer Academic Publishers, Dordrecht

Hayes R, Kudla J, Schuster G, Gabay L, Maliga P and Gruissem W (1996) Chloroplast mRNA 3′-end processing by a high molecular weight protein complex is regulated by nuclear encoded RNA binding proteins. EMBO J 15: 1132–1141

Herrin DL, Chen YF and Schmidt GW (1990) RNA splicing in *Chlamydomonas* chloroplasts. Self-splicing of 23 S preRNA. J Biol Chem 265: 21134–21140

Herrin DL and Bao Y, Thompson AJ and Chen YF (1991) Self-splicing of the *Chlamydomonas* chloroplast *psbA* introns. Plant Cell 3: 1095–1107

Hirose T and Sugiura M (1996) *Cis*-acting elements and *trans*-acting factors for accurate translation of chloroplast *psbA* mRNAs: Development of an in vitro translation system from tobacco chloroplasts. EMBO J 15: 1687–1695

Hirose T, Kusumegi T and Sugiura M (1998) Translation of tobacco chloroplast *rps14* mRNA depends on a Shine-Dalgarno-like sequence in the 5′-untranslated region but not on internal RNA editing in the coding region. FEBS Lett 430: 257–260

Hotchkiss TL and Hollingsworth MJ (1995) Factors in a chloroplast extract specifically bind to the 5′ untranslated regions of chloroplast mRNAs. Nucleic Acids Symp Ser 33: 207—208

Hotchkiss TL and Hollingsworth MJ (1999) ATP synthase 5′ untranslated regions are specifically bound by chloroplast polypeptides. Curr Genet 35: 512–520

Inamine G, Nash B, Weissbach H and Brot N (1985) Light regulation of the synthesis of the large subunit of ribulose-1, 5-bisphosphate carboxylase in peas: Evidence for translational control. Proc Natl Acad Sci USA 82: 5690–5694

Jackson RJ (1996) A comparative view of initiation site selection mechanisms. In: Hershey JWB, Matthews MB and Sonenberg N (eds) Translational Control, pp 71–112. Cold Spring Harbor Laboratory Press, Cold Spring Harbor

Keller M, Chan RL, Tessier LH, Weil JH and Imbault P (1991) Post-transcriptional regulation by light of the biosynthesis of *Euglena* ribulose-1, 5-bisphosphate carboxylase/oxygenase small subunit. Plant Mol Biol 17: 73–82

Kenan DJ, Query CC and Keene JD (1991) RNA recognition:Towards identifying determinants of specificity. Trends Biochem Sci 16: 214–220

Kim J and Mullet JE (1994) Ribosome-binding sites on chloroplast *rbcL* and *psbA* mRNAs and light-induced initiation of D1 translation. Plant Mol Biol 25: 437–448

Kim J and Mayfield SP (1997) Protein disulfide isomerase as a regulator of chloroplast translational activation. Science 278: 1954–1957

Kim J, Klein PG and Mullet JE (1994) *Vir-115* gene product is required to stabilize D1 translation intermediates in chloroplasts. Plant Mol Biol 25: 459–467

Kim M, Christopher DA and Mullet JE (1993) Direct evidence for selective modulation of *psbA*, *rpoA*, *rbcL* and 16S RNA stability during barley chloroplast development. Plant Mol Biol 22: 447–463

Kim M, Thum KE, Morishige DT and Mullet JE (1999) Detailed architecture of the barley chloroplast *psbD-psbC* blue light-responsive promoter. J Biol Chem 274: 4684–4692

Klaff P and Gruissem W (1991) Changes in chloroplast mRNA stability during leaf development. Plant Cell 3: 517–529

Klaff P and Gruissem W (1995) A 43 kD light-regulated chloroplast RNA-binding protein interacts with the *psbA* non-translated leader RNA. Photosynth Res 46: 235–248

Klein RR and Mullet JE (1986) Regulation of chloroplast-encoded chlorophyll-binding protein translation during higher plant chloroplast biogenesis. J Biol Chem 261: 11138–11145

Koo JS and Spremulli LL (1994) Analysis of the translational initiation region on the *Euglena gracilis* chloroplast ribulose-bisphosphate carboxylase/oxygenase (*rbcL*) messenger RNA. J Biol Chem 269: 7494–7500

Kozak M (1992) Regulation of translation in eukaryotic systems. Annu Rev Cell Biol 8: 197–225

Kozak M (1999) Initiation of translation in prokaryotes and eukaryotes. Gene 234: 187–208

Kuchka MR, Mayfield SP and Rochaix JD (1988) Nuclear mutations specifically affect the synthesis and/or degradation of the chloroplast-encoded D2 polypeptide of Photosystem II in *Chlamydomonas reinhardtii.* EMBO J 7: 319–324

Kuchka MR, Goldschmidt-Clermont M, van Dillewijn J and Rochaix JD (1989) Mutations at the *Chlamydomonas* nuclear *Nac2* locus specifically affects stability of the chloroplast *psbD* transcript encoding polypeptide D2 of PS II. Cell 58: 869–876

Laing W, Kruenz K and Apel K (1988) Light dependent, but phytochrome-independent, translational control of the accumulation of the P700 chlorophyll–*a* protein of Photosystem I in barley (*Hordeum vulgare* L.). Planta 176: 269–276

Lisitsky I and Schuster G (1995) Phosphorylation of a chloroplast RNA-binding protein changes its affinity to RNA. Nucleic Acids Res 23: 2506–2511

Lisitsky I, Klaff P and Schuster G (1996) Addition of destabilizing poly (A)-rich sequences to endonuclease cleavage sites during the degradation of chloroplast mRNA. Proc Natl Acad Sci USA 93: 13398–13403

Mader S and Sonenberg N (1995) Cap binding complexes and cellular growth control. Biochimie 77: 40–44

Malnoe P, Mayfield SP and Rochaix JD (1988) Comparative analysis of the biogenesis of Photosystem II in the wild-type and Y-1 mutant of *Chlamydomonas reinhardtii.* J Cell Biol 106: 609–616

Mayfield SP, Cohen A, Danon A and Yohn CB (1994) Translation of the *psbA* mRNA of *Chlamydomonas reinhardtii* requires a structured RNA element contained within the 5′ untranslated region. J Cell Biol 127:1537–1545

McCarthy JEG and Gualerzi C (1990) Translational control of prokaryotic gene expression. Trends Genet 6: 78–85

McCormac DJ and Barkan A (1999) A Nuclear gene in maize required for the translation of the Chloroplast *atpB/E* mRNA. Plant Cell 11: 1709–1716

Memon AR, Meng B and Mullet JE (1996) RNA-binding proteins of 37/38 kDa bind specifically to the barley *psbA* 3′-end untranslated RNA. Plant Mol Biol 30: 1195–1205

Meurer J, Berger A and Westhoff P (1996) A nuclear mutant of Arabidopsis with impaired stability on distinct transcripts of the plastid *psbB*, *psbD/C*, *ndhH* and *ndhC* operons. Plant Cell 7: 1749–1761

Meurer J, Plucken H, Kowallik KV and Westhoff P (1998) A nuclear-encoded protein of prokaryotic origin is essential for the stability of Photosystem II in *Arabidopsis thaliana.* EMBO J 17: 5286–5297

Monod C, Goldschmidt-Clermont M and Rochaix, JD (1992) Accumulation of psbB RNA requires a nuclear factor in *Chlamydomonas reinhardtii.* Mol Gen Genet 231: 449–459

Muhlbauer SK and Eichacker LA (1998) Light-dependent formation of the photosynthetic proton gradient regulates translation elongation in chloroplasts. J Biol Chem 273: 20935–20940

Myers AM, Harris EH, Gillham NW and Boynton JE (1984) Mutations in a nuclear gene of *Chlamydomonas* cause the loss of two chloroplast ribosomal proteins, one synthesized in the chloroplast and the other in the cytoplasm. Curr Genet 8: 369–378

Nickelsen J, van Dillewijn J, Rahire M and Rochaix JD (1994) Determinants for stability of the chloroplast *psbD* RNA are located within its short leader region in *Chlamydomonas reinhardtii.* EMBO J 13: 3182–3191

Nickelsen J, Fleischmann M, Boudreau E, Rahire M and Rochaix JD (1999) Identification of *cis*-acting RNA leader elements required for chloroplast *psbD* gene expression in *Chlamydomonas.* Plant Cell 11: 957–970

Preiss T and Hentze MW (1999) From factors to mechanisms: Translation and translational control in eukaryotes. Curr Opin Genet Dev 9: 515–521

Rattanachaikunsopon P, Rosch C and Kuchka MR (1999) Cloning and characterization of the nuclear *AC115* gene of *Chlamydomonas reinhardtii.* Plant Mol Biol 39: 1–10

Rochaix JD (1996) Post-transcriptional regulation of chloroplast gene expression in *Chlamydomonas reinhardtii.* Plant Mol Biol 32: 327–341

Rochaix JD, Kuchka M, Mayfield S, Schirmer-Rahire M, Girard-Bascou J and Bennoun P (1989) Nuclear and chloroplast mutations affect the synthesis or stability of the chloroplast *psbC* gene product in *Chlamydomonas reinhardtii.* EMBO J 8: 1013–1021

Rott R, Levy H, Drager RG, Stern DB and Schuster G (1998) 3′-Processed mRNA is preferentially translated in *Chlamydomonas reinhardtii* chloroplasts. Mol Cell Biol 18: 4605–4611

Rott R, Liveanu V, Drager RG, Higgs D, Stern DB and Schuster G (1999) Altering the 3 UTR endonucleolytic cleavage site of a *Chlamydomonas* chloroplast mRNA affects 3-end maturation in vitro but not in vivo. Plant Mol Biol 40: 679–686

Ruf M and Kossel H (1988) Structure and expression of the gene coding for the alpha-subunit of DNA-dependent RNA polymerase from the chloroplast genome of *Zea mays.* Nucleic Acids Res 16: 5741–5754

Sakamoto W, Chen X, Kindle KL and Stern DB (1994a) Function of the *Chlamydomonas reinhardtii petD* 5′ untranslated region in regulating the accumulation of subunit IV of the cytochrome *b6/f* complex. Plant J 6: 503–512

Sakamoto W, Chen X, Kindle KL and Stern DB(1994b) *petD* mRNA maturation in *Chlamydomonas reinhardtii* chloroplasts: Role of 5′ endonucleolytic processing. Mol Cell Biol. 14: 6180–6186

Stampacchia O, Girard-Bascou J, Zanasco JL, Zerges W, Bennoun P and Rochaix JD (1997) A nuclear-encoded function essential for translation of the chloroplast *psaB* mRNA in *Chlamy-*

domonas. Plant Cell 9: 773–782
Staub JM and Maliga P (1994) Translation of *psbA* mRNA is regulated by light via the 5′-untranslated region in tobacco plastids. Plant J 6: 547–553
Stern DB, Jones H and Gruissem W (1989) Function of plastid mRNA 3′ inverted repeats. RNA stabilization and gene-specific protein binding. J Biol Chem 264: 18742–18750
Stern DB and Kindle KL (1993) 3′ end maturation of the *Chlamydomonas reinhardtii* chloroplast *atpB* mRNA is a two-step process. Mol Cell Biol 13: 2277–2285
Stitt M, Wirtz W and Heldt HW (1980) Metabolite levels during induction in the chloroplast and extrachloroplast compartments of spinach protoplasts. Biochim Biophys Acta 593: 85–102
Sugita M and Sugiura M (1996) Regulation of gene expression in chloroplasts of higher plants. Plant Mol Biol 32: 315–326
Sugiura M, Hirose T and Sugita M (1998) Evolution and mechanism of translation in chloroplasts. Annu Rev Genet 32: 437–459
Tarun SZ and Sachs AB (1996) Association of the yeast poly(A) tail binding protein with translation initiation factor eIF-4G. EMBO J 15: 7168–7177
Taylor WC (1989) Regulatory interactions between nuclear and plastid genomes. Annu Rev Plant Physiol Plant Mol Biol 406: 211–233
Tzareva NV, Makhno VI and Boni IV (1994) Ribosome-messenger recognition in the absence of the Shine-Dalgarno interactions. FEBS Lett 337: 189–194
Yang J and Stern DB (1997) The spinach chloroplast endoribonuclease CSP41 cleaves the 3′-untranslated region of *petD* mRNA primarily within its terminal stem-loop structure. J Biol Chem 272: 12874–12880
Yohn CB, Cohen A, Danon A and Mayfield SP (1996) Altered mRNA binding activity and decreased translational initiation in a nuclear mutant lacking translation of the chloroplast *psbA* mRNA. Mol Cell Biol 16: 3560–3566
Yohn CB, Cohen A, Danon A and Mayfield SP (1998a) A poly(A) binding protein functions in the chloroplast as a message-specific translation factor. Proc Natl Acad Sci USA 95: 2238–2243
Yohn CB, Cohen A, Rosch C, Kuchka MR and Mayfield SP (1998b) Translation of the chloroplast *psbA* mRNA requires the nuclear-encoded poly(A)-binding protein, RB47. J Cell Biol 142: 435–442
Wada T, Tunoyama Y, Shiina T and Toyoshima Y (1994) In vitro analysis of light-induced transcription in the wheat *psbD/C* gene cluster using plastid extracts from dark-grown and short-term-illuminated seedlings. Plant Physiol 104: 1259–1267
Wu HY and Kuchka MR (1995) A nuclear suppressor overcomes defects in the synthesis of the chloroplast *psbD* gene product caused by mutations in two distinct nuclear genes of *Chlamydomonas*. Curr Genet 27: 263–269
Wulczyn FG and Kahmann R (1991) Translational stimulation: RNA sequence and structural requirements for binding of Com protein. Cell 65: 259–269
Zerges W and Rochaix JD (1994) The 5′ leader of a chloroplast mRNA mediates the translational requirements for two nucleus-encoded functions in *Chlamydomonas reinhardtii*. Mol Cell Biol 14: 5268–5277
Zerges W and Rochaix JD (1998) Low density membranes are associated with RNA-binding proteins and thylakoids in the chloroplast of *Chlamydomonas reinhardtii*. J Cell Biol 140: 101–110
Zerges W, Girard-Bascou J and Rochaix JD (1997) Translation of the chloroplast psbC mRNA is controlled by interactions between its 5′ leader and the nuclear loci *TBC1* and *TBC3* in *Chlamydomonas reinhardtii*. Mol Cell Biol 17: 3440–3448

Chapter 9

Proteins Involved in Biogenesis of the Thylakoid Membrane

Klaas Jan van Wijk*

Department of Biochemistry, Arrhenius Laboratories for Natural Sciences, Stockholm University, S-10691 Stockholm, Sweden

Summary

Assembly of the four major complexes in the thylakoid membranes of green algae and higher plants, involves many processes and requires regulation at different levels. In this chapter, proteins directly involved in processing, targeting, insertion and assembly of components of the four photosynthetic complexes will be discussed. Since targeting and insertion of the polytopic chloroplast-encoded thylakoid proteins is tightly intertwined with translation, some aspects of translational regulation are discussed but only where it is directly relevant for targeting, insertion and assembly. The four photosynthetic complexes have many different cofactors. Thus to assemble and maintain functional photosynthetic complexes, these cofactors need to be transported, ligated and possibly transiently stored. Information on a number of (putative) proteins involved in cofactor storage/assembly is summarized. Observations of vesicles flow from the inner chloroplast envelope

* Present address: Department of Plant Biology, Cornell University, Ithaca, NY 14853, U.S.A., email: kv35@cornell.edu

Eva-Mari Aro and Bertil Andersson (eds): Regulation of Photosynthesis, pp. 153–175.

membrane, presence of tubules emerging from the chloroplast envelope, and low density membranes in the chloroplast and their possible involvement in biogenesis of the thylakoid membrane are summarized. It is likely that many more proteins involved in thylakoid biogenesis remain to be discovered. Proteomics will be discussed as a way to systematically identify these proteins.

I. Introduction

Chloroplasts in green algae and higher plants contain photosynthetic thylakoid membranes with four multi-subunit protein complexes (PS I, PS II, the ATP-synthase and the cytochrome b_6f complex). These four complexes comprise together at least 66 different proteins and carry out the photosynthetic reactions (Ort and Yocum, 1996; Rochaix et al., 1998; Wollman et al., 1999). Assembly of these complexes involves many processes and requires regulation at several levels, varying from communication between nucleus and chloroplast, transcriptional, translational and post-translational control mechanisms, regulation of biosynthesis and transport of cofactors, etc. (Mullet, 1988; Rochaix, 1996; Sugita and Sugiura, 1996; Kranz et al., 1998; Wollman et al., 1999).

In this chapter, we will focus on proteins directly involved in processing, targeting, insertion and assembly of proteins in the four photosynthetic complexes in the thylakoid membrane (Sections II–V). Since targeting and insertion of the polytopic chloroplast-encoded thylakoid proteins is tightly intertwined with translation, we will briefly address some aspects of translational regulation for these proteins, but only where it is directly relevant for targeting, insertion and assembly (Section II.C). The four photosynthetic complexes contain many different cofactors (manganese, calcium, chloride, iron, copper, chlorophylls, carotenoids, quinones). To assemble and maintain functional photosynthetic complexes, these cofactors need to be transported, ligated and possibly (transiently) stored. In many cases it is not clear if cofactors are ligated prior, during or after assembly of the complexes, but it is quite likely that soluble and membrane-bound chaperones are required. A number of proteins with (putative) functions in cofactor storage/assembly are discussed in Section V.E.

After biogenesis of the thylakoid membrane system, the protein complexes and the lipid bilayer have to be maintained in an active state and damaged components have to be replaced or repaired (Aro et al., 1993). Oxidative damage to cofactors, proteins and lipids does occur, despite different anti-oxidative protective systems and acclimation strategies, as described in several of the chapters in this book (Chapters 22–30, Acclimation and Stress Responses). It is quite likely that specific proteins, such as proteases, are required in these repair processes, as will be discussed in Section III. Enzymes involved in biosynthesis and conversion of pigments (e.g. the xanthophyll cycle) (Chapter 25, Eskling et al.) and anti-oxidative defense (Chapter 27, Karpinski et al.) will not be addressed in this chapter.

Finally, the lipids of the thylakoid membrane need to be synthesized and accumulated to form the lipid bilayer. Several lines of evidence have been presented suggesting active ATP-dependent vesicle flow from the inner chloroplast envelope to the thylakoid membrane. These and related studies, as well as their possible functional significance in term of thylakoid biogenesis, will be briefly summarized in Section VI.

Although it can be postulated that at least 50–75 proteins involved in biogenesis and maintenance should be present in the thylakoid membrane, only relatively few have been identified so far (see Tables 1,2,3). The biochemical identification process has been slow, most likely due to their low abundance. Assignment of a function in thylakoid biogenesis based on primary sequence information is often difficult. A number of biogenesis mutants have been collected by different laboratories, but for many of these mutants the identification of the disrupted genes and their precise functional role remains to be established. At the end of this chapter, proteomics is reviewed as a way to accelerate the identification of these (low abundance) proteins and to establish their functional role in biogenesis (Section VII). Proteins without assigned function, that have either been predicted or biochemically determined to be present in the thylakoid, will not be mentioned in this review.

Abbreviations: CF – coupling factor; cp – chloroplast; HCF – high chlorophyll fluorescence; LHC – light harvesting complex; OEC – oxygen evolving complex I; PS I, PS IIPhotosystem I and II, respectively; RNC – ribosome nascent chain; SPP – stromal processing peptidase; SRP – signal recognition particle; TAT – twin arginine translocation; TPP – thylakoid processing peptidase; UTL – untranslated leader

II. Chloroplast Proteins Involved in Targeting and Insertion into the Thylakoid Membrane

The four photosynthetic complexes in thylakoids from higher plants contain in total at least 66 different proteins, with 14 subunits in PS I (five chloroplast-encoded), 29 subunits in PS II (14 chloroplast-encoded), nine subunits in the cytochrome b_6f complex (six chloroplast-encoded) and nine subunits in the ATP-synthase (six chloroplast-encoded). Thus thirty-one of those proteins are encoded by the chloroplast genome, whereas the other thirty-five proteins are encoded by the nucleus and are synthesized in the cytosol.

The nuclear-encoded proteins are synthesized as precursors on cytosolic ribosomes and subsequently targeted to the chloroplast via a N-terminal transit peptide, which is proteolytically removed after import into the chloroplast (Section III) (Richter and Lamppa, 1998; Chen and Schnell, 1999; Dalbey and Robinson, 1999; Keegstra and Cline, 1999). Once inside the chloroplast, (at least) four pathways operate to target the proteins to the thylakoid membrane and possibly into the lumenal space of the thylakoid membrane (Dalbey and Robinson, 1999; Keegstra and Cline, 1999). The presequences of these nuclear-encoded chloroplasts share common features, which can be used to predict localization with moderate confidence (Emanuelsson et al., 1999; Nakai and Horton, 1999) and for discussion see (Peltier et al., 2000). However, their lack of conserved domains prevents a systematic PCR based screening of all chloroplast localized proteins and can therefore not be used to rapidly identify all nuclear genes encoding for chloroplast proteins.

Targeting of the 31 chloroplast-encoded thylakoid proteins is poorly understood and occurs both co- and post-translationally (Section II.C). The information to direct these proteins to the thylakoid membrane and subsequently into the lumen or into the thylakoid lipid bilayer, must be present within these (precursor) proteins. In addition, several soluble and membrane proteins are involved in these processes.

Below, I will discuss the identified soluble stromal proteins (Section II.A) and membrane proteins (Section II.B) and their functional significance in targeting and insertion of thylakoid proteins. The discussion is kept brief and focused on the most recent findings since this field has been thoroughly reviewed over the last years (Robinson et al., 1998; Settles and Martienssen, 1998; Dalbey and Robinson, 1999; Keegstra and Cline, 1999). mRNA binding proteins involved in 'directed synthesis' of the chloroplast-encoded proteins are discussed in Section II.C. Passage of nuclear-encoded precursor proteins through the chloroplast envelope is beyond the scope of this chapter and is discussed in recent reviews (Chen and Schnell, 1999; Dalbey and Robinson, 1999; Keegstra and Cline, 1999).

A. Soluble Stromal Proteins

Six different soluble stromal chaperones have been identified to date and a few of those play (possibly) a direct or indirect role in protein targeting to the thylakoid membrane (Table 1). The chaperones HSP70β (a homologue to bacterial DnaK) (Wang et al., 1993) and Cpn60α or Cpn60β (homologues to bacterial GroEL) (Hemmingsen et al., 1988; Nishio et al., 1999) have been shown to interact transiently with the nuclear-encoded Ferredoxin-NADP$^+$ reductase upon import into the chloroplast in an ATP-dependent manner (Tsugeki and Nishimura, 1993). The association of Ferredoxin-NADP$^+$ reductase with HSP70 preceded that of Cpn60, suggesting that these two chaperones sequentially assist in the maturation of newly imported thylakoid protein. Also during import of the Rieske Fe-S protein, Cpn60 and HSP70 interacted sequentially, but in this case interaction with Cpn60 preceded interaction with HSP70 (Madueno et al., 1993). When the hydrophobic D1 reaction center protein, fused to a chloroplast targeting transit peptide (Wu and Watanabe, 1997), was imported into intact chloroplasts in vitro, the fusion protein interacted tightly with Cpn60 (J. Larsson and K.J. van Wijk, unpublished). The chaperone Cpn21, a homologue to GroES (Bertsch et al., 1992) has also been identified in the chloroplast, but its role might be confined to assembly of Rubisco and other soluble complexes. Relatively large amounts of Cpn60, HSP70 and Cpn21 were found in thylakoid preparations either associated with membranes at the stromal surface or located in the thylakoid lumen (Schlicher and Soll, 1996; Peltier et al., 2000). At this point it is interesting to mention that HSP70β has also been reported to play a, so far undefined, role in PS II repair after light stress (Schroda et al., 1999). Two other chaperone-like proteins have been reported and they are homologous to bacterial GrpE and DnaJ (Schlicher and Soll, 1997) (Table 1), but a functional role for

Table 1: Proteins involved in transport, targeting, insertion, processing, folding and proteolysis

Functional class	Protein	Molecular mass (kDa) or No. of amino acid residues	Del. mutant
SPPs	CPE	1265/1210[1]	–
Chaperones	HSP70β	706/623[1]	–
	Cpn60α	587/540[2]	–
	Cpn60*a*	595/545[2]	–
	Cpn21	253/203[1]	–
	SRP43	376/317[1]	*A. thaliana*
	SRP54	564/490[1]	*A. thaliana*
	DnaJ	499/93[1]	–
	GrpE	311/279[1]	–
Thylakoid receptors	FtsY	366/326[1]	–
Protein translocators	SecA	1021/595[1]	Maize
	SecY	556/509[1]	Maize
	SecE	177/138[1]	–
	Tha4	170/109[3]	Maize
	Tha9	169/117[3]	–
	HCF106	243/176[3]	Maize
	TatC homologue	340/310[1]	–
	Albino3	462/407[1]	*A. thaliana*
Lumenal TPPs	CtpA	539/494[4]	*Spirodela*
	TPP	340/288[1]	–
General proteases	FtsH	709/661[1]	–
	DegP	437/395[1]	–
Isomerases	TLP–40	449/387/345[4]	–
	FKBP isomerase	ca 21 kDa[1]	–
	Rotamase (Roc4)	260/183[1]	–
Kinase	TAK1,2,3	ca 490/–[1]	–

The number of amino acid residues of the predicted precursor and mature protein or the apparent molecular mass in kDa are indicated between brackets, if available. Sequences are from [1] *Arabidopsis thaliana*, [2] pea or [3] maize, [4] spinach, # *Chlamydomonas reinhardtii or * Synechocystis PCC6803*. Cleavage site prediction was carried out using ChloroP (chloroplast) and SignalP (lumen). In several cases more than one sequence could be found for each protein but only one is listed. Availability of disruption or deletion mutants are indicated.

these chloroplast proteins has not been established.

Two subunits of the chloroplast signal recognition particle, the GTP binding protein cpSRP54 (homologous to Ffh in *Escherichia coli* and SRP54 in the endoplasmatic reticulum) and cpSRP43, have been identified (Franklin and Hoffman, 1993; Klimyuk et al., 1999). cpSRP54 was initially identified as a chloroplast envelope protein, but is predominantly localized as a soluble protein in the stroma. In *E. coli*, the signal recognition particle is a ribonucleoprotein complex consisting of one protein (referred to as P48 or fifty-four-homologue, Ffh) and a 4.5S RNA. In the endoplasmatic reticulum, SRP consists of 6 proteins and a 7S RNA. Eukaryotic SRP54 and prokaryotic P48 have shown to bind tightly to N-terminal signal sequences or signal anchor sequences of nascent proteins when they are emerging from the ribosome and SRP was shown to assist in targeting of the ribosome to the membrane (reviewed in Rapoport et al., 1996; Fekkes and Driessen, 1999). In chloroplasts, about 50% of the cpSRP54 population can be found associated with chloroplast 70S ribosomes and is likely to be involved in targeting of chloroplast-encoded thylakoid proteins (Franklin and

Hoffman, 1993; High et al., 1997; Nilsson et al., 1999). Recent data show that the nascent chains from the D1 protein could be photocrosslinked to cpSRP54 but only when the D1 nascent chain was still attached to the ribosome. The interaction was dependent on the length of the nascent chain that emerged from the ribosome (Nilsson et al., 1999).

Unexpectedly, cpSRP54 was also found to be involved in post-translational targeting of nuclear encoded LHCII (Li et al., 1995), when assembled into a 170 kDa complex together with the cpSRP43 (Schuenemann et al., 1998). cpSRP43 has four ankyrin repeats and two chromatin binding domain (chromodomains), which have both been implicated in mediating protein-protein interactions. It was suggested that the chromatin domains of cpSRP43 facilitates dimerization of two cpSRP43 subunits whereas the ankyrin domain interacts with cpSRP54 (Schuenemann et al., 1998; Klimyuk et al., 1999). An 18-aa sequence motif was determined in LHC (L18) that, along with a hydrophobic domain, is required for cpSRP interaction (DeLille et al., 2000). Additional data indicated that cpSRP43 binds to the L18 domain, that cpSRP54 binds to the hydrophobic domain, and that LHC and cpSRP54 independently bind to cpSRP43 (Tu et al., 2000). *Arabidopsis thaliana* mutants lacking (functional) cpSRP54 have a pleiotropic disturbance in chloroplast biogenesis (Pilgrim et al., 1998; Amin et al., 1999) suggesting a broad role of cpSRP54 in protein targeting. In contrast, *Arabidopsis* mutants lacking cpSRP43 have low levels of a number of chlorophyll *a/b* binding proteins, whereas other thylakoid proteins accumulate at wild-type levels (Amin et al., 1999; Klimyuk et al., 1999). Thus the cpSRP43 mutant phenotype points to a specific role of cpSRP43 for targeting of nuclear-encoded chlorophyll a/b binding proteins. So far, no homologues of cpSRP43 have been found in non-chloroplast SRPs.

B. Thylakoid-Bound Proteins

Currently four, fairly independent, translocation and insertion pathways into the thylakoid (SecA/Y, TAT/ΔpH, SRP/FTSY/Albino3 and 'spontaneous insertion') for nuclear-encoded proteins have been established through contributions from a number of laboratories. It was only in 1994 that the first protein involved in thylakoid membrane insertion, SecA, was identified through a biochemical approach by the laboratory of K. Cline (Yuan et al., 1994). From then on, progress has been very rapid and more proteins were found through screening of chloroplast biogenesis mutants in *Arabidopsis* and maize. Sequencing of the *Arabidopsis* genome has further accelerated the identification of a number of other components based on homology or similarity to proteins in prokaryotes , such as *E. coli*.

Currently, nine thylakoid-bound proteins involved in insertion and translocation of thylakoid proteins are described in the literature (Table 1). Those proteins can be classified in three groups corresponding to the Sec and ΔpH/TAT and SRP/FTSY/Albino3 translocation pathways.

1. The Sec Machinery

In the last five years homologues of different bacterial secretory (Sec) proteins have been identified in chloroplasts. In *E. coli*, the Sec translocation pathway is comprised of two trimeric membrane-embedded complexes consisting of SecY, SecE, SecG (the SecYEG complex), SecD, SecF and YajC (the SecDFYajC) and the peripherally bound ATPase, SecA. Together, they function in translocation and insertion of a large number of cytoplasmic, periplasmic and outer membrane proteins. SecYE is the conserved functional 'core' and channel, while SecDFYajC and SecG serve to promote SecYE function and stabilize SecA in its active conformation. The energy of ATP binding and hydrolysis promotes cycles of membrane insertion and deinsertion of SecA and catalyzes the movement of the preproteins across the membrane (Duong and Wickner, 1997; Fekkes and Driessen, 1999).

Chloroplast SecA, assigned cpSecA, was identified in 1994 through biochemical purification by using a peptide antibody against a conserved SecA domain in bacterial and algal sequences (Yuan et al., 1994). cpSecA was shown to be an azide sensitive component involved in insertion and translocation of a number of nuclear-encoded thylakoid proteins, such OEC33. Purified cpSecA supported post-translational translocation of a number of nuclear-encoded lumenal proteins (Nakai et al., 1994; Yuan et al., 1994). Import experiments with OEC33 using isolated chloroplasts showed that cpSecA can be crosslinked at the membrane surface to the translocation intermediate of OEC33 when translocation is blocked by azide (Haward et al., 1997).

A transposon-tagged deletion mutant of cpSecA in maize (*tha1*) further confirmed the role of SecA in post-translational targeting of a specific group of nuclear-encoded proteins (e.g. OEC33, plastocyanin). Interestingly, the mutant also showed an effect on at least one chloroplast-encoded protein, cytochrome *f* (Voelker and Barkan, 1995b; Voelker et al., 1997) (Section II.C). This is in agreement with the observation that the N-terminal signal peptide of cytochrome *f* can target β-galactosidase to the *E. coli* inner membrane in a SecA dependent manner (Rothstein et al., 1985). Cytochrome *f* is possibly co-translationally inserted into the thylakoid membrane, since *petA* mRNA is located on thylakoid-bound ribosomes (Friemann and Hachtel, 1988). In vitro studies using intact chloroplasts and import of chimeric constructs of cytochrome *f* showed that this fusion protein can also be post-translationally inserted into the thylakoid membrane in dependence of cpSecA and the N-terminal signal peptide (Nohara et al., 1996; Mould et al., 1997; Zak et al., 1997). cpSRP54 (Section II.A) could be crosslinked to cytochrome *f* nascent chains attached to 80S ribosomes from wheat germ extracts (High et al., 1997). However, in a homologous chloroplast translation system with 70S ribosomes, cytochrome *f* ribosome nascent chains could only be crosslinked to cpSecA and not to cpSRP54 (T. Röhl and K.J. van Wijk, unpublished).

SecY and SecE have also been identified in chloroplasts after their genomic sequence appeared in the *Arabidopsis* genome database (Laidler et al., 1995; Schuenemann et al., 1999). A transposon-tagged maize mutant showed the central importance of cpSecY in thylakoid biogenesis. Disruption of the cpSecY gene resulted in a non-photosynthetic mutant with an extremely poorly developed chloroplast (Roy and Barkan, 1998). In vitro experiments with isolated thylakoids showed that cpSecY antibodies inhibited translocation of the Sec pathway substrate OEC33 but not of substrates of the ΔpH (OEC17, OEC23) or the FTSY/Albino3 pathway (LHC) (Mori et al., 1999; Schuenemann et al., 1999). cpSecE could be co-purified with cpSecY as a complex of at least 180 kDa (Schuenemann et al., 1999), indicating that cpSecE is part of the Sec translocon in the thylakoid. It is quite possible that homologues of the other bacterial Sec components, SecG, D and F, and possibly YajC, are also present in thylakoids. It is likely that the SecY/E/A translocon is also involved in the co-translational insertion of a number of integral thylakoid proteins, such as D1, PsaA (Section II.C).

2. The TAT or ΔpH Pathway

In the last few years protein components of the so-called ΔpH or TAT targeting and translocation pathway in the thylakoid membrane have been identified. This pathway requires the trans-thylakoid proton gradient (the ΔpH) but neither soluble stromal components nor ATP or GTP, and was originally thought to be unique to the chloroplast.

Two transposon tagged mutants which disrupted insertion of several thylakoid membrane proteins, Tha4 and HCF106, have been described (Das and Martienssen, 1995; Voelker and Barkan, 1995a; Settles et al., 1997; Walker et al., 1999). The *Hcf106* mutation interfered with targeting of the nuclear-encoded proteins OEC23, which is known to be targeted via the ΔpH pathway. After cloning of the *Hcf106* gene, bacterial homologues were identified and were named TAT proteins, after twin arginine translocation, because the substrates always contain two arginines upstream of the hydrophobic region in the signal peptide (Sargent et al., 1998). Currently three proteins in the thylakoid TAT pathway have been experimentally identified: the *Hcf106* gene product (a TatB homologue) (Settles et al., 1997), Tha4 and Tha9 (TatA homologues) (Mori et al., 1999). A fourth TAT protein homologous to the *E. coli* protein TatC is also present in the *Arabidopsis* database but no experimental data about this protein have been presented. The TAT pathway is important for thylakoids, since 50% of the 26 non-redundant higher plant thylakoid proteins with experimentally determined lumenal transit peptides, contain a typical Twin Arginine motif (Peltier et al., 2000).

3. Albino3 and the cpSRP receptor

In 1997, an integral membrane protein, named Albino3, was identified in the thylakoid membrane by analysis of a transposon tagged *Arabidopsis* mutant (Sundberg et al., 1997). The mutant has an albino phenotype, is non-photosynthetic and sterile. The Alb3 protein is encoded by the nuclear genome and was shown to be targeted and inserted into the thylakoid membrane (Sundberg et al., 1997). Although it is clear that the protein is essential for chloroplast biogenesis, its precise functional role is unclear. However, antisera raised against the N-terminus of Albino3 (a predicted stromal domain) could prevent insertion of LHC into the thylakoid membrane, whereas antibodies against SecY or HCF106 did not effect insertion (Moore et al., 2000).

The SRP receptor, FtsY, is a 32 kDa polypeptide bound peripherally on the outer surface of thylakoids and has been identified in *Arabidopsis* (Kogata et al., 1999; Tu et al., 1999). The functional role of FtsY is most likely to interact with cpSRP54 or the cpSRP54/cpSRP43 complex (Section II.A). When chloroplast FtsY was combined with cpSRP and GTP, the three factors promoted efficient LHC integration into thylakoid membranes in the absence of stroma, demonstrating that they are all required for reconstituting the soluble phase of LHC transport (Tu et al., 1999). In an independent study (Kogata et al., 1999), antibodies raised against the FtsY chloroplast homologue inhibited the cpSRP-dependent insertion of the light-harvesting chlorophyll *a/b*-binding protein into thylakoid membranes, again suggesting that the chloroplast FtsY homologue is involved in the cpSRP-dependent protein targeting to the thylakoid membranes.

Homologues of Albino3 are present in the inner membrane of yeast mitochondria (Altamura et al., 1996; He and Fox, 1997; Hell et al., 1997; Kermorgant et al., 1997) and *E. coli* (Sääf et al., 1998) and are named Oxa1p and YidC, respectively. Oxa1p was initially detected for its requirement for correct assembly of the cytochrome *c* oxidase and the ATP synthase complex (Altamura et al., 1996). Oxa1p was shown to physically interact with imported nuclear encoded N-tail proteins. Furthermore, Oxa1p interacted with nascent polypeptide chains synthesized in mitochondria, including the fully synthesized pCoxII and CoxIII species and was required to export pCoxII to the intermembrane space (He and Fox, 1997; Hell et al., 1997, 1998; Kermorgant et al., 1997). Thus, Oxa1p represents a component of a general export machinery of the mitochondrial inner membrane. Interestingly, the bacterial homologue, YidC, has been reported to co-purify with the Sec translocon and crosslinks between YidC and nascent chains have been observed in an *E. coli* translation system (Scotti et al., 2000). Taking the strong phenotype of the *Arabidopsis* Albino3 mutant and the results from *E. coli* and yeast mitochondria together, it is quite likely that Albino3 is not only involved in translocation of nuclear-encoded LHC but that it has a broader function in protein targeting and translocation. It remains to be determined if Albino3 is associated with the Sec translocon.

C. Proteins involved in 'Directed Synthesis' of Chloroplast-Encoded Proteins

Chloroplasts in higher plants possess circular DNA, containing approximately 120 genes. About 36 genes encode for proteins located in the thylakoid membranes and the majority is essential for photosynthesis. Based on a number of observations in chloroplasts, such as run-off translations of thylakoids with bound ribosomes ('rough thylakoids') and detection of translation intermediates in the membrane, it can be postulated that insertion of at least the polytopic chloroplast-encoded thylakoid proteins occurs co-translationally (Margulies and Michaels, 1975; Margulies et al., 1975; Michaels and Margulies, 1975; Herrin and Michaels, 1985; Klein et al., 1988; van Wijk et al., 1996; Zhang et al., 1999).

Further evidence for synthesis of chloroplast proteins at the thylakoid surface (but not necessarily co-translational membrane insertion) came from observations that translation initiation inhibitors decreased the amount of bound mRNA, while translation elongation chain inhibitors (such as lincomycin and chloramphenicol) prevented most of the loss of bound mRNA (Boschetti et al., 1990; Jagendorf and Michaels, 1990). It should be noted that also the peripheral thylakoid membrane proteins CF1α,β (Bhaya and Jagendorf, 1985; Herrin and Michaels, 1985), the large subunit of Rubisco (Breidenbach et al., 1988; Mühlbauer and Eichacker, 1999) and possibly other soluble proteins are translated on ribosomes bound to the thylakoid surface (reviewed in Boschetti et al., 1990; Jagendorf and Michaels, 1990). Insertion of many smaller proteins with a molecular mass less than 4 kDa is necessarily post-translational, since approximately 35 amino acids are buried in the ribosome during translation.

In yeast mitochondria it has been shown that synthesis and assembly of the seven mitochondria-encoded hydrophobic membrane proteins is strongly dependent on nuclear-encoded, mRNA-specific translational activators that recognize the 5′-untranslated leaders (UTLs) of their target mRNAs (Costanzo and Fox, 1988; Sanchirico et al., 1998). These translational activators are themselves membrane-associated and could therefore tether translation to the inner membrane (Michaelis et al., 1991). Chimeric mRNAs of Cox2p and Cox3p with the UTL of the mRNA encoding the soluble mitochondrial protein Var1p, can not lead to functional expression of the integral membrane proteins Cox2p and Cox3p. Although cells expressing these chimeric mRNAs actively synthesized both membrane proteins, they were severely deficient in cytochrome *c* oxidase activity and in the accumulation of Cox2p and Cox3p, respectively (Sanchirico et al., 1998). It was thus suggested that the UTLs are important to obtain synthesis at the correct locations and this was therefore defined as 'directed synthesis.'

In chloroplasts, the 5′UTL has also been shown to be important in coupling synthesis and assembly of at least one of the chloroplast-encoded thylakoid membrane proteins. In an elegant series of papers, Wollman and collegues showed that the 5′UTL of *pet A*, encoding cytochrome *f*, played an important role in regulation of translation and assembly (Kuras and Wollman, 1994; Choquet et al., 1998). Using in vivo studies with *Chlamydomonas* mutants, they showed that negative autoregulation of cytochrome *f* translation occurs if the subunits of the cytochrome b_6f complex are absent (Kuras and Wollman, 1994). When expressed from a chimeric mRNA containing the *atpA* 5′ UTL, cytochrome *f* no longer showed an assembly-dependent regulation of translation. The precise molecular mechanism for this regulation is not understood but could possibly involve mRNA binding proteins (see below) and direct interaction between the C-terminus with the translating ribosome (Choquet et al., 1998).

Another example of coupling between synthesis and assembly, is the polytopic D1 reaction center protein of PS II, although neither regulatory proteins components nor a role for the 5′UTL of *psbA* mRNA have been demonstrated. Pulse-chase experiments with intact chloroplasts, followed by subfractionation, suggested that the D1 protein can assemble co-translationally (van Wijk et al., 1996; Zhang et al., 1999). Immunoprecipitation showed that assembly of the newly synthesized D1 protein into PS II involves direct interaction of the nascent D1 chains with the D2 protein (Zhang et al., 1999).

Initiation and translation of many (if not all) chloroplast-encoded proteins is regulated by nuclear-encoded mRNA binding proteins (Rochaix, 1996; Sugita and Sugiura, 1996; Cohen and Mayfield, 1997; Hippler et al., 1998) and several chapters in Rochaix et al. (1998). Several of such proteins regulating translation of *psbA*, (Kim and Mayfield, 1997; Yohn et al., 1998) or *psbC* (Zerges and Rochaix, 1994; Zerges et al., 1997) have been identified in *Chlamydomonas reinhardtii* through genetic and biochemical approaches (Rochaix, 1996; Cohen and Mayfield, 1997) and are listed in Table 1. Many of these mRNA binding proteins seem to be transcript specific, whereas a few others are involved in translational regulation of several transcripts.

So far only two mRNA binding proteins, involved either in regulation of translation of *petA/petD* and CF1β, have been identified in higher plants (Fisk et al., 1999; McCormac and Barkan, 1999) (Table 1). However, it is unclear to what extent (if any) these proteins are involved in directing synthesis to the thylakoid membrane. Interestingly, (Zerges and Rochaix, 1998) isolated a low density membrane fraction from the chloroplast (Section VI.B) which were enriched in the chloroplast localized poly(A) binding protein RB47 and other mRNA binding proteins. RB47 has been reported to be a specific activator of *psbA* mRNA translation (Yohn et al., 1998a, 1998b). These results indicate that initial steps in translation of chloroplast mRNAs encoding thylakoid proteins could possibly occur at either a subfraction of the chloroplast inner envelope membrane or an uncharacterized intra-chloroplast compartment, which is physically associated with thylakoids (Zerges and Rochaix, 1998) (Section VI.B). It is well known that several of the chloroplast mRNAs are processed prior to translation and ribosome nascent chain complexes usually contain processed transcripts, suggesting that processing is a prerequisite for translation. Since translation takes place mostly at the thylakoid membrane surface, it has been postulated that in chloroplasts the unprocessed transcript contains information needed for membrane association (Rochaix, 1996).

To reconstitute the targeting and insertion process, a homologous chloroplast in vitro initiation and translation system is required in which plasmid derived transcripts can be faithfully translated. The development of a translation system isolated from tobacco chloroplasts has opened up novel possibilities to address these important processes at a molecular level (Hirose and Sugiura, 1996). Recently, co-translational insertion into the thylakoid has been achieved with Leader peptidase I as a model protein, using a chloroplast translation extract, exogenous mRNA and nuclease treated thylakoids (Houben et al., 1999). The encoding region for Leader peptidase I was fused to the 5′UTL *psbA* and in vitro synthesized mRNA of this construct gave good translation rates of both truncated and full length Leader peptidase.

Co-translational insertion into the thylakoid membrane was strongly inhibited by azide, suggesting a requirement for SecA activity (Houben et al., 1999), similarly as in *E. coli*.

In conclusion, in vitro essays and mutant analysis of *Chlamydomonas*, tobacco and maize have shown that mRNA processing, translation initiation and elongation are tightly connected and possibly further linked to thylakoid membrane insertion and assembly. Synthesis of the chloroplast-encoded proteins might lead to functional products only if it takes place at the correct location; i.e. at the thylakoid membrane surface, possibly in the vicinity of the final complex. All these processes could be coordinated through mRNA binding proteins and other interactions involving 5′UTLs, and interaction of the nascent chains with other membrane components. The mRNA binding proteins or genetic loci, obtained by mutant screening, required for translation are listed in Table 1. mRNA binding proteins providing RNAse protection at the 3′UTL or 5′UTL of the mRNAs are not included in the table.

III. Peptidases and Proteases Responsible for Processing and Turnover

A. Stromal Processing Peptidases

The majority of the chloroplast proteins are encoded by the nuclear genome and are synthesized in the cytosol as precursor proteins with an N-terminal transit peptide that directs the import of the polypeptide into chloroplasts (Chen and Schnell, 1999; Dalbey and Robinson, 1999; Keegstra and Cline, 1999). The transit peptide of most precursors, is removed by a stromal processing activity that is sensitive to metal chelators. So far one stromal processing peptidase (SPP), a metallo-endopeptidase assigned CPE, has been identified (Oblong and Lamppa, 1992; VanderVere et al., 1995; Richter and Lamppa, 1999). CPE contains a zinc-binding motif (His-X-X-Glu-His), characteristic of the pitrilysin family. CPE overexpressed in *E. coli* was able to correctly cleave 10 different nuclear-encoded stromal and thylakoid precursor proteins. The cleaved signal peptide was subsequently degraded in an ATP dependent step, independent of SPP (Richter and Lamppa, 1999). The identity of the secondary protease has not been determined.

Based on different sensitivities and partial purifications it proposed that several processing peptidases are located in the chloroplast stroma of *Chlamydomonas reinhardtii* (Su and Boschetti, 1993, 1994) However, sequence information from those putative SPPs has not been obtained. Also in other studies several stromal proteins with precursor processing activities have been partially purified without successful identification (Creighton et al., 1993; Koussevitzky et al., 1998).

B. Thylakoid Processing Peptidases

A number of thylakoid proteins have cleavable lumenal transit peptides and processing of these transit peptides takes place at the lumenal side of the thylakoid membrane, immediately upon translocation of the N-terminus. A semi-conserved consensus for TPP was determined to be AXA↓X with a high frequency of alanines as well as leucines upstream of the cleavage side (Dalbey and Robinson, 1999). After aligning 26 non-redundant sequences of proteins with lumenal transit peptides according to their cleavage site in a logoplot, a nearly complete conservation (25 out of 26) for the –1 position (alanine) was observed (Peltier et al., 2000), in agreement with site-directed mutagenesis studies (Shackleton and Robinson, 1991). At the –3 position a preference for alanine as well as valine, serine (small and neutral residues), and unexpectedly aspartic acid, was noticeable. After the alanine/leucine-rich hydrophobic region, a fairly high frequency of prolines at the –6, –5 and –4 position could be seen; This is likely to stimulate helix-breaking of the hydrophobic region and possibly to ensure interaction with the thylakoid processing peptidase. Further downstream (+2, +4), a preference for the negatively charged glutamic acid was observed (Peltier et al., 2000) which is not found in Gram-positive and Gram-negative bacteria (Nielsen et al., 1997; Cristobal et al., 1999).

In several studies it was shown that the processing specificity closely resembled bacterial Leader peptidase I from *E. coli* (Halpin et al., 1989; Anderson and Gray, 1991; Shackleton and Robinson, 1991). A 36 kDa TPP has been identified in *Arabidopsis* both from cDNA and genomic sequences (Chaal et al., 1998). The predicted amino acid sequence of the protein includes regions highly conserved among Type I Leader peptidases and indicates that the enzyme uses a serine-lysine catalytic dyad mechanism. When the catalytic domain was overexpressed in *E. coli*, it generated a product capable of cleaving the thylakoid-transfer domain from a chloroplast protein (Chaal et al., 1998). So far, there is no reason to believe that there is more than one general TPP.

One other processing peptidase has been identified which is exclusively functioning to process the C-terminus of the D1 protein of PS II (Oelmuller et al., 1996). Several in vitro studies were carried out to determine the cleavage site in the D1 protein (Bowyer et al., 1992; Yamamoto and Satoh, 1998). In addition, a homologue with a similar function, assigned ctpA, was identified in *Synechocystis sp.* PCC6803 (Anbudurai et al., 1994; Mitchell and Minnick, 1997). The predicted sequence of the encoded protein showed significant similarity to that of the Prc protein, a carboxyl-terminal processing protease in *E. coli*. Interestingly, the localization of ctpA was suggested not to be the thylakoid membrane but the cytoplasmic membrane (B. Norling, B. Andersson, E. Zak, R. Maitra, H.P. Pakrasi, unpublished), raising questions about the site of biogenesis of thylakoid complexes in cyanobacteria.

C. General Thylakoid Proteases

Two more general proteases, FtsH and DegP, have been identified in the thylakoid (Adam, 1996). Membrane bound FtsH, is an ATP and Zn^{2+} dependent protease which has the catalytic domain on the stromal side (Lindahl et al., 1996). FtsH has been implied in degradation of unassembled Rieske Fe-S protein and photodamaged D1 protein (Ostersetzer and Adam, 1997; Spetea et al., 1999; Lindahl et al., 2000). DegP is a lumenal serine type protease homologous to the DegP localized in the periplasm of *E. coli* (Itzhaki et al., 1998). In *E. coli*, DegP has been shown to be involved in the degradation of several proteins during oxidative stress and other environmental stresses (Spiess et al., 1999). To my knowledge, no lumenal substrate for DegP has yet been identified and the cleavage site for bacterial or plant DegP is unknown (Chapter 15, Adam).

IV. Proteins involved in Folding and Post-translational Modifications

Several thylakoid proteins, including the major LHCs and the D1 and D2 reaction center proteins of PS II, are phosphorylated and dephosphorylated, in dependence of the redox state and of the trans-thylakoid proton gradient of the chloroplast. These reversible post-translational modifications help to regulate electron transport and also have a role in biogenesis; as this topic is extensively covered in Chapters 23 (Rintamäki and Aro) and 27 (Karpinski et al.), we will only briefly review the modifying enzymes that have been identified.

So far, one thylakoid bound kinase of 55 kDa has been purified and its gene as well as two other closely related kinases were identified (Table 1) (Snyders and Kohorn, 1999). No genes for thylakoid-bound phosphatases have been identified, although protein fractions with strongly enriched phosphatase activity have been obtained (Vener et al., 1999). A stromal phosphatase of 29 kDa has been purified and this protein was able to dephosphorylate synthetic phosphopeptides (both phosphothreionine and phosphoserine), mimicking the N-terminus of LHC-II, as well as LHC-II in situ (Hammer et al., 1997). So far, the corresponding gene has not been identified.

Several thylakoid proteins with isomerase function have been identified (Table 1). A lumenal 40 kDa protein, assigned TLP40, is characterized by a cyclophilin-like C-terminal segment of 20 kDa, a predicted N-terminal leucine zipper and a potential phosphatase-binding domain (Fulgosi et al., 1998). The isolated protein possesses peptidyl-prolyl cis-trans isomerase protein folding activity characteristic of immunophilins and TLP40 also exerts an effect on dephosphorylation of several key proteins of Photosystem II, probably as a constituent of a transmembrane signal transduction chain (Chapter 10, Vener). Systematic analysis of lumenal and peripheral thylakoid proteins by two-dimensional electrophoresis and mass spectrometry (Peltier et al., 2000), indicated the presence of a lumenal isomerase with a twin arginine motif in the lumenal transit sequence (Table 1). At present, only a partial N-terminal cDNA clone can be found in the database. A third isomerase was found in these two-dimensional maps of the peripheral thylakoid protein fractions and was earlier reported to be a stromal protein (Lippuner et al., 1994). This protein, a cyclophilin (assigned ROC4) is a protein with a peptidyl-prolyl *cis-trans* isomerase activity, and was found to be expressed only in photosynthetic organs. Import experiments with radiolabeled precursor proteins Indicated a location in chloroplast stroma (Lippuner et al., 1994), but the protein was fairly abundant on the two-dimensional maps suggesting that it must have a high affinity for the thylakoid membrane, and it could therefore be involved in folding of thylakoid proteins.

V. Proteins Assisting in Protein and Cofactor Transport, Storage and Ligation

Relatively few thylakoid proteins with a role in protein assembly or cofactor storage, transport and ligation have been identified (Table 1), as it is difficult to infer such functions from protein sequences present in the database. An additional reason is that disruption of the genes for such assembly factors will often lead to pleiotropic phenotypes, prohibiting direct conclusions about molecular mechanisms. Despite these potential difficulties, several laboratories have initiated a systematic search for proteins involved in thylakoid biogenesis, using different functional genomics tools.

So far, studies with the green algea *Chlamydomonas reinhardtii* have been most successful for several experimental reasons (Hippler et al., 1998; Rochaix et al., 1998). Studies with the photosynthetic cyanobacterium *Synechocystis* PCC6803 are also beginning to reveal a number of components, and even though this review deals mostly with chloroplasts from green algae and higher plants, these bacterial proteins are discussed and summarized in Table 1.

Progress in the identification of genes involved in thylakoid assembly in higher plants such as maize and *Arabidopsis* has accelerated through the systematic search for chloroplast biogenesis mutants by screening large numbers of tagged mutants (transposon or t-DNA) (Meurer et al., 1996; Roy and Barkan, 1998). The availability of full genome sequences or high coverage with EST sequences of *Arabidopsis* and other plant species in the near future, combined with different functional genomic strategies [such as proteomics (Section VII) and transcriptomics (Chapter 32, Scheible et al.)], will most likely lead to rapid progress in the identification of these assembly components. The experimental determination of the precise molecular function or biochemical mechanisms will then become the bottleneck.

In this section, protein components with (putative) roles in protein assembly, cofactor transport, storage or ligation are discussed. A number of assembly mutants for which the disrupted genes have not been identified are also highlighted.

A. Photosystem II

So far only one protein has been determined with a specific role in assembly of PS II (Table 2). The gene, *hcf136*, was identified by screening of a T-DNA tagged collection of *Arabidopsis* mutants (Meurer et al., 1998) and homologues are found in *Synechocystis sp.* PCC6803 (slr2034) and in the cyanelle genome of Cyanophora paradoxa (ORF333). The mutant is devoid of any Photosystem II activity, and none of the nuclear- and plastome-encoded subunits of PS II accumulate to significant levels. Protein labeling studies in the presence of cycloheximide showed that the plastid-encoded PS II subunits are synthesized but do not accumulate. The HCF136 protein is produced already in dark-grown seedlings and its levels do not increase dramatically during light-induced greening. This accumulation profile suggests that the HCF136 protein must be present when PS II complexes are made and the authors concluded that HCF136 is a stability and/or assembly factor of PS II.

B. Photosystem I

Three specific PS I assembly mutants have been isolated from *Chlamydomonas reinhardtii* and *Synechocystis sp. PPC6803* and the responsible disrupted genes have been identified (Table 2).

Transformants of *Chlamydomonas* lacking the chloroplast genes Ycf3 or Ycf4 did not stably accumulate Photosystem I (Boudreau et al., 1997). Ycf3 and Ycf4 are localized on thylakoid membranes but did not stable associate with PS I and both proteins accumulated to wild-type levels in mutants lacking PS I. The deduced amino acid sequences of Ycf4 and Ycf3 display high sequence identity with their homologues from plants and cyanobacteria. Transcripts of the chloroplast-encoded PS I proteins, PsaA, PsaB and PsaC, accumulated normally in Ycf3 or Ycf4 mutants and use of chimeric reporter genes revealed that Ycf3 is not required for initiation of translation of *psaA* and *psaB* mRNA. The authors concluded that Ycf3 and Ycf4 are required for stable accumulation of the PS I complex (Boudreau et al., 1997).

Disruption of the gene *btpA* in *Synechocystis sp.* PCC6803, leads to a photosynthesis-deficient mutant strain with a severely reduced PS I content, whereas PS II was present in normal amounts. Northern blot analysis revealed that the steady-state levels of the

Table 2. Proteins involved in assembly of thylakoid membrane complexes or in translational regulation

Functional class	Protein or nuclear locus No. of amino acid residues	Molecular mass (kDa) or	del. mutant
mRNA binding proteins in translational control of			
cyt*f*	CRP1	668/603[3]	maize
CF1β/CF1ε	ATP1	–	maize
D1	RB38, 47, 55, 60#	38–60 kDa	*C. reinhardtii*
CP43	TBC1, TBC2#	–	*C. reinhardtii*
D2	NAC2#	—	*C. reinhardtii*
PSI–B	F15 #		*C. reinhardtii*
specific assembly factors of			
PSII	HCF136	403/433[1]	*A. thaliana*
PSII	BtpA	287*	*Synechocystis*
PSI	Ycf3	–/126[1]	*C. reinhardtii*
PSI	Ycf4	–/184[1]	*C. reinhardtii*
Cyt*b6*	CCB1,2,3,4[#]	-	*C. reinhardtii*
Cyt*f*/cyt*c6*	CCSA	–/328[1]	*C. reinhardtii*
Cyt*f*/cyt*c6*	CCS1,2,3,4,[#]	-	*C. reinhardtii*

The number of amino acid residues of the predicted precursor and mature protein or the apparent molecular mass in kDa are indicated between brackets, if available. Sequences are from [1]*Arabidopsis thaliana*, [3]maize, [#] *Chlamydomonas reinhardtii or * Synechocystis* spp. PCC6803. Cleavage site prediction was carried out using ChloroP. In several cases more than one sequence could be found for each protein but only one is listed. Availability of disruption or deletion mutants are indicated.

transcripts from the *psaAB* operon remained unaltered in the mutant strain (Bartsevich and Pakrasi, 1997). BtpA is a 30 kDa peripheral thylakoid membrane protein, exposed to the cytoplasmic face (Zak et al., 1999). It was thus concluded that the BtpA protein regulates a post-transcriptional process during the life cycle of the PS I protein complex such as 1) translation of the psaAB mRNA, 2) assembly of the PsaA/PsaB polypeptides and their associated cofactors into a functional complex, or 3) degradation of the protein complex. Close relatives of the BtpA protein have been found in non-photosynthetic organisms, such as the archaebacterium *Methanococcus jannaschii*, the eubacterium *E. coli*, and the nematode *Caenorhabditis elegans*, suggesting that these proteins may regulate biogenesis of other protein complexes in these evolutionarily distant organisms (Bartsevich and Pakrasi, 1997).

C. Cytochrome b_6f Complex and Cytochrome c_6

Mutations in five genetic loci in *Chlamydomas reinhardtii,* leading to defects in the accumulation of the two chloroplast localized *c*-type cytochrome containing complexes, cytb_6f and cytc_6, have been identified (Xie and Merchant, 1996; Inoue et al., 1997; Kuras et al., 1997) and reviewed in (Xie and Merchant, 1998; Xie et al., 1998; Wollman et al., 1999). These genes were named *Ccs* genes, for *c*-type **c**ytochrome **s**ynthesis. One of the genes, Ccs*A*, is chloroplast open reading frame *ycf5*, whereas the others are nuclear genes (*Ccs1-Ccs4*). Homologues for several of these genes in bacteria as well as higher plants can be found. The gene products of the *CcsA* and *Ccs1-Ccs4* loci are not required for translocation or processing of the preproteins but they are required for *c*-heme attachment during assembly of both holocytochrome *f* and holocytochrome c_6, respectively. The *Ccs* genes show similarity to *Ccm* genes in mitochondria and also to the *Dsb* system in *E. coli* (Xie et al., 1998). Some of the experimental approaches that lead to identification of these mutants and genes are discussed below.

Chloroplast ORF *ycf5* displays limited sequence identity to bacterial genes (*ccl1/cycK*) required for the biogenesis of c-type cytochromes (Ritz et al., 1995), and was therefore tested for its function in chloroplast cytochrome biogenesis in *Chlamydomonas reinhardtii* (Xie and Merchant, 1996). Homologues can also be found in higher plants such as *Pinus*, maize, tobacco and *Arabidopsis*. Ycf5 is predicted to encode for a 353 amino acid protein and a specific antiserum detected a 29 kDa protein in the chloroplast. Targeted inactivation of the *ycf5* gene resulted in a non-photosynthetic phenotype attributable to the absence of c-type cytochromes. The cloned *ycf5* gene also complemented the phototrophic growth deficiency in strain B6 which is unable to synthesize functional forms of cytochromes *f* and c_6. The complementing gene, then renamed *Ccsa*, is expressed in wild-type and B6 cells but is non-functional in B6 due to a frameshift mutation (Xie and Merchant, 1996). Additional mutants in *Ccsa* (strains ct34 and ct59) were later identified by screening for acetate-requiring mutants of *Chlamydomonas reinhardtii* (Xie et al., 1998).

Mutations leading to similar deficiencies in an additional set of strains were localized to the nuclear genome (Xie et al., 1998). Complementation tests of these strains indicated that they defined four nuclear loci, *Ccs1-Ccs4*. One of the other strains, *abf3*, also fails to accumulate holocytochrome c_6, and was disrupted in a gene, *Ccs1*, predicted to encode for a 65 kDa protein. Homologues can be found in *Arabidopsis* as well as several algae and *Synechocystis*. On the basis of the pleiotropic *c*-type cytochrome deficiency in the *Ccs1* mutant, the predicted plastid localization of the protein, and its relationship to proteins in Gram-positive bacteria, it was concluded that *Ccs1* encodes a protein which is required for chloroplast c-type holocytochrome formation (Xie et al., 1998). Interestingly, the *ccsA* gene is transcribed in each of the *Ccs* nuclear mutants, but its protein product is absent in *Ccs1* mutants, and it appears to be susceptible to degradation in the *ccs3* and *ccs4* strains. It was therefore suggested that CCSl may be associated with CCSA in a multisubunit 'holocytochrome *c* assembly complex,' and the products of the other *Ccs* loci may correspond to other subunits (Xie et al., 1998).

Five nuclear mutants (*ccb* strains) were identified that are defective in holocytochrome b_6 formation (Kuras et al., 1997). The defect is specific for cytochrome b_6 assembly, because the *ccb* strains can synthesize other b cytochromes and all *c*-type cytochromes. The *ccb* strains, which define four nuclear loci (*Ccb1, Ccb2, Ccb3, and Ccb4*), provide the first evidence that a *b*-type cytochrome requires trans-acting factors for its heme association (Kuras et al., 1997).

Four transposon-induced maize mutants, were identified that lack cytochrome b_6f proteins but contain normal levels of other photosynthetic complexes (Voelker and Barkan, 1995a). The four mutations defined two nuclear genes. In each mutant the mRNAs encoding the known subunits of the complex were normal in size and abundance and the major subunits were synthesized at normal rates. Thus, it was concluded that these mutations block the biogenesis of the cytochrome b_6f complex at a post-translational step. The two nuclear genes identified by these mutations may encode previously unknown subunits, involved in prosthetic group synthesis or ligation, or facilitate assembly of the complex (Voelker and Barkan, 1995a).

D. ATP synthase

No mutants have been identified with disruptions in genes involved in the assembly of the ATP synthase. However, in vitro reconstitution experiments of a catalytically active CF1 core revealed a dependence on a crude mixture of chloroplast molecular chaperones (Chen and Jagendorf, 1994). The reconstitution was achieved by using purified subunits overexpressed in *E. coli*, combined in the presence of MgATP, K^+, and a crude mixture of several chloroplast molecular chaperones (including HSP70, Cpn60 and Cpn21). The combination of Cpn60 and Cpn21 alone failed to reconstitute the active CF1 core, as did the GroEL/GroES pair (*E. coli* chaperonin 60/10 homologues) (Chen and Jagendorf, 1994).

E. Cofactor Storage, Transport and Ligation

Together, the four photosynthetic complexes contain many cofactors, including different pigments such as chlorophylls and carotenoids (Chapters 12, Paulsen and 13, Apel). Biosynthesis of the pigments takes place in the chloroplast at the inner envelope membrane with the last step(s) at the thylakoid surface. Since the pigments are molecules with large hydrophobic tails, it is likely that they are assisted to the thylakoid membrane; this could be achieved through vesicle flow or by chaperone-like molecules.

A water soluble chlorophyll binding protein, named WSCP, has recently been identified in cauliflower (*Brassica oleracea*) and appeared to be identical to an earlier reported drought stress-induced protein, named BnD22 (Nishio and Satoh, 1997; Satoh et al., 1998) (Table 3). Curiously, it was reported that the BnD22 protein is related to the Kunitz family of protease inhibitors (Downing et al., 1992). When

Table 3. Proteins involved in post–translational modifications, vesicle flow or ion transport

Functional class	Protein	No. of amino acid residues	Del. mutant
Cofactor binding/transport	Pilin IV–like	168*	*Synechocystis*[8]
	ScpA	387*	*Synechocystis*[8]
	ScpB,C,D,E	58–70*	–
	Elip1	195/151[1]	–
	Elip2	191/150[1]	–
	Sep1	103/60[1]	–
	Sep2	181/160[1]	–
	Sep3	223/184[1]	–
	WSCP/bnd22	218/–[5]	–
	Ferritin	225/178[1]	*A. thaliana*[9]
	ORFslr0399	326*	*Synechocystis*[8]
Vesicle formation	Pftf	710/695[6]	–
	ALD1	610/–[1]	*A. thaliana*[7]
	ALD2	808/743[1]	–
	ALD3	836/–[1]	–
ABC transporter (localization unclear)	MNTA	260*	*Synechocystis*[8]
	MNTB	306*	*Synechocystis*[8]
	MNTC	330*	*Synechocystis*[8]

The number of amino acid residues of the predicted precursor and mature protein are indicated, if available. Sequences are from [1] *Arabidopsis thaliana*, [5] *Brassica oleraceae*, [6] *Capsicum annuum*, [#] *Chlamydomonas reinhardtii or* * *Synechocystis* spp. PCC6803. Cleavage site prediction was carried out using ChloroP. In several cases more than one sequence could be found for each protein but only one is listed. Availability of dominant negative mutant[7], deletion mutant[8] or antisense mutant[9] is indicated.

recombinant WSCP fused to maltose binding protein was incubated with thylakoid membranes, the MBP-WSCP removed chlorophylls from these membranes. During this process, the monomer of the apo-MBP-WSCP successfully bound chlorophylls and was converted into tetrameric holo-MBP-WSCP. Reconstituted MBP-WSCP exhibited absorption and fluorescent spectra identical to those of native WSCP purified from cauliflower leaves having a high chlorophyll *a/b* ratio. It was concluded that WSCP is a hydrophilic protein that can transfer chlorophylls from hydrophobic thylakoid proteins (Nishio and Satoh, 1997; Satoh et al., 1998).

In higher plants a group of light stress induced proteins, named Elips (early light induced proteins), were postulated to be involved in transient storage of pigments during light stress (Table 3). Additionally, another group of genes with a conserved chlorophyll binding domain has been identified, and were named *Seps* (stress enhanced proteins), since their expression is enhanced under different stress conditions (Table 3). Sep and Elip proteins have two or three transmembrane domains, respectively. They will not be further discussed in this chapter, as details are given in Chapter 28 (Adamska).

Recently several putative transient chlorophyll binders have also been reported in *Synechocystis* sp PCC6803. Upon non-denaturing gel electrophoresis of thylakoid extracts from *Synechocystis* a Type IV pilin-like protein (ORFsll1694), was found in chlorophyll-containing bands (Table 3). The authors proposed that this pilin is involved in (without being essential) delivering chlorophyll to nascent photosystems and antennae proteins (He and Vermaas, 1999). Five other genes were identified with significant sequence similarity to members of the Cab gene family (Table 3). Four of those were named *ScpA,B,C,D* (for small Cab-like proteins) and were predicted to have a single transmembrane helix (Funk and Vermaas, 1999). The fifth Cab-like gene is much longer and codes for a protein of which the N-terminus resembles ferrochelatase but the C-terminal domain has similarity to Cab regions. The SCPs were suggested to represent a new group of cyanobacterial proteins that, in view of their primary structure, are likely to be involved in transient pigment binding (Funk and Vermaas, 1999).

A putative chaperone-like protein (ORF Slr0399) for delivering quinones to the QA binding pocket of PS II has been identified in *Synechocystis* (Ermakova-

Gerdes and Vermaas, 1999) (Table 3). The protein is similar to hypothetical proteins in the cyanelle genome of *Cyanophora paradoxa*, in the chloroplast genomes of diatoms, dinoflagellates, red algae and in the nuclear genome of *Arabidopsis*. Slr0399 has a NAD(P)H binding motif near the N-terminus and has some similarity to isoflavone reductase-like proteins and to a subunit of the eukaryotic NADH dehydrogenase complex I. The authors suggested that ORF Slr0399 is a chaperone-like protein that aids in, but is not essential for, quinone insertion and protein folding around QA in PS II and possibly other photosynthetic and respiratory complexes (Ermakova-Gerdes and Vermaas, 1999).

An iron storage protein, ferritin, has been identified in plastids of higher plants, as well as in other tissues. Ferritin concentrates and stores cellular iron, to approximately 10^{11} times the solubility of the free ion and in the plastid it has shown to be also located at the stromal side of thylakoid membranes (Waldo et al., 1995). Ferritin was also found as a fairly abundant protein on two-dimensional gels of peripheral thylakoid proteins (Peltier et al., 2000). Ferritin protein accumulation is positively correlated to iron loading of the plant, and its expression is under developmental control (Lobreaux and Briat, 1991).

Manganese is a cofactor in the oxygen-evolving complex on the lumenal side of PS II and is absolutely required for water splitting activity. It was therefore interesting that high and low affinity Mn uptake systems were discovered in *Synechocystis* sp. PCC6803. One of the high affinity import system is an ABC transporter encoded by three genes, *mntA*, *mntB* and *mntC* (Table 3) which were induced under manganese starvation conditions (Bartsevich and Pakrasi, 1995, 1996). Unfortunately, no data are available on the location of the ABC transporter and to my knowledge, no studies on an ABC transporter, or any other Mn uptake systems, in the thylakoid membrane have been presented.

Finally, biophysical evidence for a cation selective ion channel, a calcium activated channel and a calcium proton antiporter in the thylakoid membrane have been reported but no genes have been identified (Vambutas et al., 1994; Pottosin and Schönknecht, 1996; Ettinger et al., 1999).

VI. Vesicles Formation, Low Density Membranes and Tubules

Biogenesis of the thylakoid membrane requires not only targeting, insertion and assembly of proteins and cofactors but also involves formation of the lipid bilayer. Over the last decades, thylakoid membrane formation has received a fluctuating level of attention, and in recent years several interesting leads have opened up. In this section three phenomena involving membranes and membrane vesicles are discussed.

A. Vesicle Flow from Inner Envelope to Thylakoid

The most abundant thylakoid lipids, monogalactosyldialcylglycerol (MGDG) and digalactosyldiacylglycerol (DGDG) are synthesized in the chloroplast envelope and transferred rapidly to the thylakoid membrane, as was demonstrated by radiolabeling studies (Marechal et al., 1997; Joyard et al., 1998). When leaf discs of different dicotelydonous species where incubated at suboptimal, non-freezing temperature, membrane vesicles accumulated in the chloroplast stroma within 30–60 min and these membranes were frequently continuous with the chloroplast envelope (Morré et al., 1991b). Reconstitution in a cell-free system prepared from isolated pea or spinach chloroplasts, indicated an ATP and temperature dependent transfer of monogalactosylglycerides from the chloroplast envelope to the chloroplast thylakoid (Morré et al., 1991a). Ultrastructural studies in the 1970's and 1980's revealed that vesicles can form at the inner chloroplast envelope and it was suggested that vesicle flow could transfer lipids from the inner envelope to the thylakoids.

Only recently, two proteins were identified, that support active vesicle formation within the chloroplast (Hugueney et al., 1995; Park et al., 1997, 1998) (Table 3). In 1995, a 72 kDa protein, designated plastid fusion and/or translocation factor (Pftf), was purified and identified. cDNA cloning revealed that mature Ptft has significant homology to yeast and animal proteins, as well as bacterial FtsH proteins, involved in vesicle fusion or membrane protein translocation. Using in vitro assays, the authors showed fusion of chromoplast membrane vesicles with dependence upon an ATP-requiring protein that is sensitive to N-ethylmaleimide.

A dynamin-like protein, named ALD1, was identified in the chloroplast of *Arabidopsis* and transgenic plants harboring various deletion mutant genes of ALD1 had a yellow leaf phenotype and the cells had very few chloroplasts (Park et al., 1997,1998). In addition, the remaining chloroplasts appeared morphologically not fully developed, whereas the accumulation of several thylakoid membrane proteins, such as LHC and CP29, was greatly reduced. Currently at least three more dynamin-like proteins can be found in the *Arabidopsis* database, and are named ALD2, ALD3 and phragmoplastin. Analysis of the transit peptides by Psort and CloroP, indicated chloroplast localization for ALD2. Also ALD3 was predicted to be chloroplast localized by the programs Psort (http://www.cbs.dtu.dk/services/ChloroP/) but not by ChloroP (http://www.cbs.dtu.dk/services/ChloroP/). Phragmoplastin was shown to be localized in the root tips in dividing soybean and data suggested that this protein may be associated with exocytic vesicles that are depositing cell plate material during cytokinesis in the plant cell (Gu and Verma, 1996).

B. Low Density Membranes—An Alternative Biogenesis Site?

In a recent report, a low density chloroplast membrane fraction was isolated from *Chlamydomonas* cells by differential centrifugation after mechanical disruption of crude thylakoid membranes in the absence of Mg^{2+} ions (Zerges and Rochaix, 1998). Under those specific conditions small amounts of membranes with a density lower than thylakoid membranes could be purified. The acyl lipid composition was similar to inner envelopes and thylakoid membranes. Whereas the possible presence of chloroplast inner envelope proteins was not tested, these membranes did not contain PS I or PS II core subunits. However, the poly(A) binding protein RB47 and other chloroplast localized mRNA binding proteins were found to be strongly enriched. RB47 has been reported to be a specific activator of *psbA* mRNA translation (Yohn et al., 1998 a,b), whereas some of the other mRNA binding proteins are interacting with the *psbC* 5′UTL (Zerges and Rochaix, 1998). These results suggest that translation of chloroplast mRNAs encoding thylakoid proteins could possibly occur at either a subfraction of the chloroplast inner envelope membrane or a previously uncharacterized intra-chloroplast compartment, which is physically associated with the thylakoids (Zerges and Rochaix, 1998). No nascent chains of chloroplast-encoded thylakoid proteins could be detected in radiolabeling experiments. To elucidate their role in biogenesis, these intriguing membranes need to be further characterized, including an analysis for marker proteins of the envelope. The possible presence of such low density membranes in higher plant chloroplasts should also be investigated.

C. Stromules

Two years ago, a study was published showing with high resolution that tube-like structures are protruding out of the chloroplast envelope (Köhler et al., 1997), and strongly confirming several reports from the 1960–1980s (Wildman et al., 1962). These tube-like structures were named stromules, for stroma filled tubules, and were detected by using transgenic plants in which green fluorescent protein (GFP) was targeted to the chloroplast stroma. The stromules were found to be highly dynamic, variable in length and shape and sometimes interconnecting different plastids. Subsequent studies also showed that they were much more abundant in chlorophyll-free proplastids, etioplasts, chromoplasts, and amyloplasts (Tirlapur et al., 1999; Köhler and Hanson, 2000). Surprisingly, these plastids surrounded nuclei and mitochondria, suggesting a direct contact to these organelles (Köhler and Hanson, 2000).

The role of stromules is so far unknown but they could allow plastid derived molecules to be delivered throughout the cell or they could assist uptake of cytoplasmic molecules. The observed close association of stromules with other organelles, including mitochondria, nuclei and other plastids, could help to enhance the transfer from one organelle to the other and possibly facilitate the coordination of plastid activities. Although possibly not directly relevant for thylakoid biogenesis, these structures are highly intriguing and modify the classical picture of the plastid as an oval-shaped organelle drifting freely in the cytoplasm.

VII. Proteomics as a Tool for Identification of Proteins involved in Thylakoid Biogenesis

As was mentioned in the introduction, it can be postulated that the chloroplast contains a large number of unidentified proteins which are involved in the biogenesis and maintenance of the chloroplast (auxiliary enzymes). These proteins should carry out processes such as biosynthesis and ligation of cofactors, insertion, folding and degradation of proteins, and signal transduction. Based on preliminary information and expected functions, it can be expected that at least 100 proteins involved in such processes should exist and that most of these proteins have a very low abundance.

The improvements of two-dimensional electrophoresis through the development of immobiline drystrip gels (IPGs) (Görg et al., 1988) and optimization of solubilization techniques (Rabilloud et al., 1997; Molloy et al., 1998), now allows the reproducible separation of up to 5000 proteins on a single two-dimensional gel. Such gel-separated proteins can be rapidly identified by mass spectrometry if genomic information is available, permitting the systematic identification of the protein complement of genomes ('the proteome') (Shevchenko et al., 1996; Dainese et al., 1997; Roepstorff, 1997; Yates, 1998). In addition, mass spectrometry is a powerful tool to analyze isoforms, secondary modifications of proteins (such as glycosylation, phosphorylation, isoprenylation, etc.) and proteolysis using low amounts (picomoles–attomoles) of proteins (Burlingame et al., 1998; Kuster and Mann, 1998; Wilkins et al., 1999). A proteomics approach allows therefore to bridge the gap between the genomic sequence information and the actual protein population in a cell, specific tissue or cellular compartment. Two-dimensional protein reference maps of subfractions of various organisms are expected to become a central tool for organizing and understanding proteome data.

Proteomics has already become an important tool for drug discovery and the analysis of yeast and *E. coli*. However, at the time of writing of this chapter, proteomics is not much in use yet in the field of plant biology possibly because no complete plant genome is available yet. It is however expected that with the completion of the *Arabidopsis* genome and EST sequencing of several plant species, proteomics will become an important tool for the plant research community. A number of plant proteomics studies have been published in recent years and were reviewed in (Thiellement et al., 1999). Mass spectrometry has been used to analyze purified PS II complexes (Zheleva et al., 1998; Whitelegge et al., 1999) but no systematic analysis of chloroplast proteins has been carried out to date. However, recently we have initiated a systematic analysis of thylakoid proteins (Peltier et al., 2000). We constructed high resolution two-dimensional maps of lumenal and peripheral proteins of the thylakoid membrane and showed that, after correction for possible isoforms and post-translational modifications, at least 200–230 different lumenal and peripheral proteins are present in the thylakoid. More than 60 proteins were identified by mass spectrometry and Edman sequencing (Peltier et al., 2000). Construction of high resolution maps of *Arabidopsis* chloroplast fractions are now also well underway. These maps will provide a database which will allow thorough and systematic studies of chloroplast biogenesis, using a reverse genetics approach or using other functional genomics tools Chapter 32, Scheible et al.).

Clearly, proteomics will prove valuable to identify and localize many known and unknown, low abundance proteins. Reference two-dimensional maps will also be used to follow up- and down-regulation of proteins under different environmental conditions, during different developmental stages or in mutants affected in chloroplast function.

VIII. Conclusions and Perspectives

Which proteins remain to be discovered in the years to come? It can be expected that translocators for cofactors are present in the thylakoid membrane. More isomerases, proteases, and possibly chaperones in the thylakoid, including the lumen, and possibly, membrane bound chaperones involved in ligation, transport or storage of cofactors could be present. Importantly, very few proteins, such as kinases and phosphatases, involved in signaling or regulatory processes have been identified so far; one could expect more of such enzymes, albeit at low abundance. Finally, post-translational modification of thylakoid proteins has been shown to occur and protein modifying enzymes might therefore be located at the thylakoid surface or within the membrane.

Many molecular mechanisms of thylakoid biogenesis are still largely unknown. Clearly, the understanding of thylakoid biogenesis and the identification of the gene products involved are still in its infancy. Most likely, disruption mutants are required to identify many of the proteins that are involved in biogenesis of the thylakoid. However, in many cases, the phenotypes of the mutants will be pleiotropic and detailed biochemical studies will be further needed to determine the precise molecular function of the different genes. Several of the components involved in biogenesis have been suggested to function in a larger complex; identification of such complexes will be a biochemical challenge for the near future. Once a sufficient number of gene products with a function in biogenesis of different membrane complexes in plants and other organisms have been identified and the molecular mechanism understood, it might ultimately become possible to predict function from primary, secondary or tertiary structure.

Acknowledgments

I gratefully acknowledge Jean-Benoit Peltier, Thomas Röhl and other members of my laboratory for numerous stimulating discussions and critically reading the manuscript. Iwona Adamska, Eva-Marie Aro, Neil Hoffman, Bertil Andersson and many other colleagues are greatly acknowledged for discussions on different aspects of chloroplast biogenesis over the last years. Funding was provided by the Swedish National Research Council (NFR), the Swedish Strategic Funds (SSF) and the Carl Trygger Foundation.

References

Adam Z (1996) Protein stability and degradation in chloroplasts. Plant Mol Biol 32: 773–783

Altamura N, Capitanio N, Bonnefoy N, Papa S and Dujardin G (1996) The *Saccharomyces cerevisiae* OXA1 gene is required for the correct assembly of cytochrome *c* oxidase and oligomycin-sensitive ATP synthase. FEBS Lett 382: 111–115

Amin P, Sy DA, Pilgrim ML, Parry DH, Nussaume L and Hoffman NE (1999) *Arabidopsis* mutants lacking the 43- and 54-kilodalton subunits of the chloroplast signal recognition particle have distinct phenotypes. Plant Physiol 121: 61–70

Anbudurai PR, Mor TS, Ohad I, Shestakov SV and Pakrasi HB (1994) The ctpA gene encodes the C-terminal processing protease for the D1 protein of the Photosystem II reaction center complex. Proc Natl Acad Sci USA 91: 8082–8086

Anderson CM and Gray J (1991) Cleavage of the precursor of pea chloroplast cytochrome *f* by leader peptidase from *Escherichia coli*. FEBS Lett 280: 383–386

Aro EM, Virgin I and Andersson B (1993) Photoinhibition of Photosystem II. Inactivation, protein damage and turnover. Biochim Biophys Acta 1143: 113–134

Bartsevich VV and Pakrasi HB (1995) Molecular identification of an ABC transporter complex for manganese: Analysis of a cyanobacterial mutant strain impaired in the photosynthetic oxygen evolution process. EMBO J 14: 1845–1853

Bartsevich VV and Pakrasi HB (1996) Manganese transport in the cyanobacterium *Synechocystis* sp. PCC 6803. J Biol Chem 271: 26057–26061

Bartsevich VV and Pakrasi HB (1997) Molecular identification of a novel protein that regulates biogenesis of Photosystem I, a membrane protein complex. J Biol Chem 272: 6382–6387

Bertsch U, Soll J, Seetharam R and Viitanen PV (1992) Identification, characterization, and DNA sequence of a functional 'double' groES-like chaperonin from chloroplasts of higher plants. Proc Natl Acad Sci USA 89: 8696–8700

Bhaya D and Jagendorf AT (1985) Synthesis of the alpha and beta subunits of coupling factor 1 by polysomes from pea chloroplasts. Arch Biochem Biophys 237: 217–223

Boschetti A, Breidenbach E and Blätter R (1990) Control of protein formation in chloroplasts. Plant Science 68: 131–149

Boudreau E, Takahashi Y, Lemieux C, Turmel M and Rochaix JD (1997) The chloroplast ycf3 and ycf4 open reading frames of *Chlamydomonas reinhardtii* are required for the accumulation of the Photosystem I complex. EMBO J 16: 6095–6104

Bowyer JR, Packer JC, McCormack BA, Whitelegge JP, Robinson C and Taylor MA (1992) Carboxyl-terminal processing of the D1 protein and photoactivation of water-splitting in Photosystem II. Partial purification and characterization of the processing enzyme from *Scenedesmus obliquus* and *Pisum sativum*. J Biol Chem 267: 5424–5433

Breidenbach E, Jenni E and Boschetti A (1988) Synthesis of two proteins in chloroplasts and mRNA distribution between thylakoids and stroma during the cell cycle of *Chlamydomonas reinhardii*. Eur J Biochem 177: 225–232

Burlingame AL, Boyd RK and Gaskell S (1998) Mass spectrometry. Anal Chem 70: 647R–716R.

Chaal BK, Mould RM, Barbrook AC, Gray JC and Howe CJ (1998) Characterization of a cDNA encoding the thylakoidal processing peptidase from *Arabidopsis thaliana*. Implications for the origin and catalytic mechanism of the enzyme. J Biol Chem 273: 689–692

Chen GG and Jagendorf AT (1994) Chloroplast molecular chaperone-assisted refolding and reconstitution of an active multisubunit coupling factor CF1 core. Proc Natl Acad Sci USA 91: 11497–11501

Chen X and Schnell DJ (1999) Protein import into chloroplasts. Trends Cell Biol 9: 222–227

Choquet Y, Stern DB, Wostrikoff K, Kuras R, Girard-Bascou J and Wollman FA (1998) Translation of cytochrome *f* is autoregulated through the 5′ untranslated region of petA mRNA in *Chlamydomonas* chloroplasts. Proc Natl Acad Sci USA 95: 4380–4385

Cohen A and Mayfield SP (1997) Translational regulation of gene expression in plants. Curr Opin Biotechnol 8: 189–194

Costanzo MC and Fox TD (1988) Specific translational activation by nuclear gene products occurs in the 5′ untranslated leader of a yeast mitochondrial mRNA. Proc Natl Acad Sci USA 85: 2677–2681

Creighton AM, Bassham DC and Robinson C (1993) The stromal processing peptidase activities from *Chlamydomonas reinhardtii* and *Pisum sativum:* Unexpected similarities in reaction specificity. Plant Mol Biol 23: 1291–1296

Cristobal S, de Gier JW, Nielsen H and von Heijne G (1999) Competition between Sec- and TAT-dependent protein translocation in *Escherichia coli*. EMBO J 18: 2982–2990

Dainese P, Staudenmann W, Quadroni M, Korostensky C, Gonnet

G, Kertesz M and James P (1997) Probing protein function using a combination of gene knockout and proteome analysis by mass spectrometry. Electrophoresis 18: 432–442
Dalbey RE and Robinson C (1999) Protein translocation into and across the bacterial plasma membrane and the plant thylakoid membrane. Trends Biochem Sci 24: 17–22
Das L and Martienssen R (1995) Site-selected transposon mutagenesis at the hcf106 locus in maize. Plant Cell 7: 287–294
DeLille J, Peterson EC, Johnson T, Moore M, Kight A and Henry R (2000) A novel precursor recognition element facilitates posttranslational binding to the signal recognition particle in chloroplasts. Proc Natl Acad Sci USA 97: 1926–1931
Downing WL, Mauxion F, Fauvarque MO, Reviron MP, de Vienne D, Vartanian N and Giraudat J (1992) A *Brassica napus* transcript encoding a protein related to the Kunitz protease inhibitor family accumulates upon water stress in leaves, not in seeds. Plant J 2: 685–693
Duong F and Wickner W (1997) Distinct catalytic roles of the SecYE, SecG and SecDFyajC subunits of preprotein translocase holoenzyme. EMBO J 16: 2756–2768
Emanuelsson O, Nielsen H and von Heijne G (1999) ChloroP, a neural network-based method for predicting chloroplast transit peptides and their cleavage sites. Protein Sci 8: 978–984
Ermakova-Gerdes S and Vermaas W (1999) Inactivation of the open reading frame slr0399 in *Synechocystis* sp. PCC 6803 functionally complements mutations near the Q(A) niche of Photosystem II. A possible role of Slr0399 as a chaperone for quinone binding. J Biol Chem 274: 30540–30549
Ettinger WF, Clear AM, Fanning KJ and Peck ML (1999) Identification of a Ca^{2+}/H^+ antiport in the plant chloroplast thylakoid membrane. Plant Physiol 119: 1379–1386
Fekkes P and Driessen AJ (1999) Protein targeting to the bacterial cytoplasmic membrane. Microbiol Mol Biol Rev 63: 161–173
Fisk DG, Walker MB and Barkan A (1999) Molecular cloning of the maize gene crp1 reveals similarity between regulators of mitochondrial and chloroplast gene expression. EMBO J 18: 2621–2630
Franklin AE and Hoffman NE (1993) Characterization of a chloroplast homologue of the 54-kDa subunit of the signal recognition particle. J Biol Chem 268: 22175–22180
Friemann A and Hachtel W (1988) Chloroplast messenger RNAs of free and thylakoid-bound polysomes from *Vicia faba* L. Planta 175: 50–59
Fulgosi H, Vener AV, Altschmied L, Herrmann RG and Andersson B (1998) A novel multi-functional chloroplast protein: Identification of a 40 kDa immunophilin-like protein located in the thylakoid lumen. EMBO J 17: 1577–1587
Funk C and Vermaas W (1999) A cyanobacterial gene family coding for single-helix proteins resembling part of the light-harvesting proteins from higher plants. Biochemistry 38: 9397–9404
Görg A, Postel W and Gunther S (1988) The current state of two-dimensional electrophoresis with immobilized pH gradients. Electrophoresis 9: 531–546
Gu X and Verma DP (1996) Phragmoplastin, a dynamin-like protein associated with cell plate formation in plants. EMBO J 15: 695–704
Halpin C, Elderfield PD, James HE, Zimmermann R, Dunbar B and Robinson C (1989) The reaction specificities of the thylakoidal processing peptidase and *Escherichia coli* leader peptidase are identical. EMBO J 8: 3917–3921
Hammer MF, Markwell J and Sarath G (1997) Purification of a protein phosphatase from chloroplast stroma capable of dephosphorylating the light-harvesting complex-II. Plant Physiol 113: 227–233
Haward SR, Napier JA and Gray JC (1997) Chloroplast SecA functions as a membrane-associated component of the Sec-like protein translocase of pea chloroplasts. Eur J Biochem 248: 724–730
He Q and Vermaas W (1999) Genetic deletion of proteins resembling Type IV pilins in *Synechocystis* sp. PCC 6803: Their role in binding or transfer of newly synthesized chlorophyll. Plant Mol Biol 39: 1175–1188
He S and Fox TD (1997) Membrane translocation of mitochondrially coded Cox2p: Distinct requirements for export of N and C termini and dependence on the conserved protein Oxa1p. Mol Biol Cell 8: 1449–1460
Hell K, Herrmann JM, Pratje E, Neupert W and Stuart RA (1997) Oxa1p mediates the export of the N- and C-termini of pCoxII from the mitochondrial matrix to the intermembrane space. FEBS Lett 418: 367–370
Hell K, Herrmann JM, Pratje E, Neupert W and Stuart RA (1998) Oxa1p, an essential component of the N-tail protein export machinery in mitochondria. Proc Natl Acad Sci USA 95: 2250–2255
Hemmingsen SM, Woolford C, van der Vies SM, Tilly K, Dennis DT, Georgopoulos CP, Hendrix RW and Ellis RJ (1988) Homologous plant and bacterial proteins chaperone oligomeric protein assembly. Nature 333: 330–334
Herrin D and Michaels A (1985) In vitro synthesis and assembly of the peripheral subunits of coupling factor CF1 (alpha and beta) by thylakoid-bound ribosomes. Arch Biochem Biophys 237: 224–236
High S, Henry R, Mould RM, Valent Q, Meacock S, Cline K, Gray JC and Luirink J (1997) Chloroplast SRP54 interacts with a specific subset of thylakoid precursor proteins. J Biol Chem 272: 11622–11628
Hippler M, Redding K and Rochaix JD (1998) *Chlamydomonas* genetics, a tool for the study of bioenergetic pathways. Biochim Biophys Acta 1367: 1–62
Hirose T and Sugiura M (1996) *Cis*-acting elements and *trans*-acting factors for accurate translation of chloroplast psbA mRNAs: Development of an in vitro translation system from tobacco chloroplasts. EMBO J 15: 1687–1695
Houben E, de Gier JW and van Wijk KJ (1999) Insertion of leader peptidase into the thylakoid membrane during synthesis in a chloroplast translation system. Plant Cell 11: 1553–1564
Hugueney P, Bouvier F, Badillo A, d'Harlingue A, Kuntz M and Camara B (1995) Identification of a plastid protein involved in vesicle fusion and/or membrane protein translocation. Proc Natl Acad Sci USA 92: 5630–5634
Inoue K, Dreyfuss BW, Kindle KL, Stern DB, Merchant S and Sodeinde OA (1997) Ccs1, a nuclear gene required for the post-translational assembly of chloroplast *c*-type cytochromes. J Biol Chem 272: 31747–31754
Itzhaki H, Naveh L, Lindahl M, Cook M and Adam Z (1998) Identification and characterization of DegP, a serine protease associated with the luminal side of the thylakoid membrane. J Biol Chem 273: 7094–7098
Jagendorf A and Michaels A (1990) Rough thylakoids: Translation on photosynthetic membranes. Plant Science: 137–145

Joyard J, Teyssier E, Miege C, Berny-Seigneurin D, Marechal E, Block MA, Dorne AJ, Rolland N, Ajlani G and Douce R (1998) The biochemical machinery of plastid envelope membranes. Plant Physiol 118: 715–723

Keegstra K and Cline K (1999) Protein import and routing systems of chloroplasts. Plant Cell 11: 557–570

Kermorgant M, Bonnefoy N and Dujardin G (1997) Oxa1p, which is required for cytochrome *c* oxidase and ATP synthase complex formation, is embedded in the mitochondrial inner membrane. Curr Genet 31: 302–307

Kim J and Mayfield SP (1997) Protein disulfide isomerase as a regulator of chloroplast translational activation. Science 278: 1954–1957

Klein RR, Mason HS and Mullet JE (1988) Light-regulated translation of chloroplast proteins. I. Transcripts of psaA-psaB, psbA, and rbcL are associated with polysomes in dark-grown and illuminated barley seedlings. J Cell Biol 106: 289–301

Klimyuk VI, Persello-Cartieaux F, Havaux M, Contard-David P, Schuenemann D, Meiherhoff K, Gouet P, Jones JD, Hoffman NE and Nussaume L (1999) A chromodomain protein encoded by the *Arabidopsis* CAO gene is a plant- specific component of the chloroplast signal recognition particle pathway that is involved in LHCP targeting. Plant Cell 11: 87–100

Kogata N, Nishio K, Hirohashi T, Kikuchi S and Nakai M (1999) Involvement of a chloroplast homologue of the signal recognition particle receptor protein, FtsY, in protein targeting to thylakoids. FEBS Lett 447: 329–333

Köhler RH and Hanson MR (2000) Plastid tubules of higher plants are tissue-specific and developmentally regulated. J Cell Science 113: 81–89

Köhler RH, Cao J, Zipfel WR, Webb WW and Hanson MR (1997) Exchange of protein molecules through connections between higher plant plastids. Science 276: 2039–2042

Koussevitzky S, Ne'eman E, Sommer A, Steffens JC and Harel E (1998) Purification and properties of a novel chloroplast stromal peptidase. Processing of polyphenol oxidase and other imported precursors. J Biol Chem 273: 27064–27069

Kranz R, Lill R, Goldman B, Bonnard G and Merchant S (1998) Molecular mechanisms of cytochrome *c* biogenesis: Three distinct systems. Mol Microbiol 29: 383–396

Kuras R and Wollman FA (1994) The assembly of cytochrome b_6/f complexes: An approach using genetic transformation of the green alga *Chlamydomonas reinhardtii*. EMBO J 13: 1019–1027

Kuras R, de Vitry C, Choquet Y, Girard-Bascou J, Culler D, Buschlen S, Merchant S and Wollman FA (1997) Molecular genetic identification of a pathway for heme binding to cytochrome b_6. J Biol Chem 272: 32427–32435

Kuster B and Mann M (1998) Identifying proteins and post-translational modifications by mass spectrometry. Curr Opin Struct Biol 8: 393–400

Laidler V, Chaddock AM, Knott TG, Walker D and Robinson C (1995) A SecY homolog in Arabidopsis thaliana. Sequence of a full-length cDNA clone and import of the precursor protein into chloroplasts. J Biol Chem 270: 17664–17667

Li X, Henry R, Yuan J, Cline K and Hoffman NE (1995) A chloroplast homologue of the signal recognition particle subunit SRP54 is involved in the posttranslational integration of a protein into thylakoid membranes. Proc Natl Acad Sci USA 92: 3789–3793

Lindahl M, Tabak S, Cseke L, Pichersky E, Andersson B and Adam Z (1996) Identification, characterization, and molecular cloning of a homologue of the bacterial FtsH protease in chloroplasts of higher plants. J Biol Chem 271: 29329–29334

Lindahl M, Spetea C, Hundal T, Oppenheim AB, Adam Z and Andersson B (2000) The thylakoid FtsH protease plays a role in the light-induced turnover of the Photosystem II D1 protein. Plant Cell 12: 419–432

Lippuner V, Chou IT, Scott SV, Ettinger WF, Theg SM and Gasser CS (1994) Cloning and characterization of chloroplast and cytosolic forms of cyclophilin from *Arabidopsis thaliana*. J Biol Chem 269: 7863–7868

Lobreaux S and Briat JF (1991) Ferritin accumulation and degradation in different organs of pea (*Pisum sativum*) during development. Biochem J 274: 601–606

Madueno F, Napier JA and Gray JC (1993) Newly imported Rieske iron-sulfur protein associates with both Cpn60 and Hsp70 in the chloroplast stroma. Plant Cell 5: 1865–1876

Maréchal E, Block MA, Dorne AJ, Douce R and Joyard J (1997) Lipid synthesis and metabolism in the plastid envelope. Physiol Plant 100: 65–77

Margulies MM and Michaels A (1975) Free and membrane-bound chloroplast polyribosomes in *Chlamydomonas reinhardtii*. Biochim Biophys Acta 402: 297–308

Margulies MM, Tiffany HL and Michaels A (1975) Vectorial discharge of nascent polypeptides attached to chloroplast thylakoid membranes. Biochem Biophys Res Commun 64: 735–739

McCormac DJ and Barkan A (1999) A nuclear gene in maize required for the translation of the chloroplast atpB/E mRNA. Plant Cell 11: 1709–1716

Meurer J, Meierhoff K and Westhoff P (1996) Isolation of high-chlorophyll-fluorescence mutants of *Arabidopsis thaliana* and their characterisation by spectroscopy, immunoblotting and Northern hybridisation. Planta 198: 385–396

Meurer J, Plucken H, Kowallik KV and Westhoff P (1998) A nuclear-encoded protein of prokaryotic origin is essential for the stability of Photosystem II in *Arabidopsis thaliana*. EMBO J 17: 5286–5297

Michaelis U, Korte A and Rodel G (1991) Association of cytochrome *b* translational activator proteins with the mitochondrial membrane: Implications for cytochrome *b* expression in yeast. Mol Gen Genet 230: 177–185

Michaels A and Margulies MM (1975) Amino acid incorporation into protein by ribosomes bound to chloroplast thylakoid membranes: formation of discrete products. Biochim Biophys Acta 390: 352–362

Mitchell SJ and Minnick MF (1997) A carboxy-terminal processing protease gene is located immediately upstream of the invasion-associated locus from *Bartonella bacilliformis*. Microbiology Uk 143: 1221–1233

Molloy MP, Herbert BR, Walsh BJ, Tyler MI, Traini M, Sanchez JC, Hochstrasser DF, Williams KL and Gooley AA (1998) Extraction of membrane proteins by differential solubilization for separation using two-dimensional gel electrophoresis. Electrophoresis 19: 837–844

Moore M, Harrison MS, Peterson EC and Henry R (2000) Chloroplast Oxa1p homolog Albino3 is required for post-translational integration of the light harvesting chlorophyll-binding protein into thylakoid membranes. J Biol Chem 275: 1529–1532

Mori H, Summer EJ, Ma X and Cline K (1999) Component specificity for the thylakoidal Sec and Delta pH-dependent protein transport pathways. J Cell Biol 146: 45–56

Morré DJ, Morre JT, Morre SR, Sundqvist C and Sandelius AS (1991a) Chloroplast biogenesis. Cell-free transfer of envelope monogalactosylglycerides to thylakoids. Biochim Biophys Acta 1070: 437–445

Morré DJ, Sellden G, Sundqvist C and Sandelius AS (1991b) Stromal Low temperature compartment derived from inner membrane of the chloroplast envelope. Plant Physiol 97: 1558–1564

Mould RM, Knight JS, Bogsch E and Gray JC (1997) Azide-sensitive thylakoid membrane insertion of chimeric cytochrome *f* polypeptides imported by isolated pea chloroplasts. Plant J 11: 1051–1058

Mühlbauer SK and Eichacker LA (1999) The stromal protein large subunit of ribulose–1,5-bisphosphate carboxylase is translated by membrane-bound ribosomes. Eur J Biochem 261: 784–788

Mullet JE (1988) Chloroplast development and gene expression. Ann Rev Plant Physiol Plant Mol Biol 39: 475–502

Nakai K and Horton P (1999) PSORT: A program for detecting sorting signals in proteins and predicting their subcellular localization. Trends Biochem Sci 24: 34–36

Nakai M, Goto A, Nohara T, Sugita D and Endo T (1994) Identification of the SecA protein homolog in pea chloroplasts and its possible involvement in thylakoidal protein transport. J Biol Chem 269: 31338–31341

Nielsen H, Engelbrecht J, Brunak S and vonHeijne G (1997) Identification of prokaryotic and eukaryotic signal peptides and prediction of their cleavage sites. Protein Eng 10: 1–6

Nilsson R, Brunner J, Hoffman NE and van Wijk KJ (1999) Interactions of ribosome nascent chain complexes of the chloroplast-encoded D1 thylakoid membrane protein with cpSRP54. EMBO J 18: 733–742

Nishio K, Hirohashi T and Nakai M (1999) Chloroplast chaperonins: Evidence for heterogeneous assembly of alpha and beta cpn60 polypeptides into a chaperonin oligomer. Biochem Biophys Res Commun 266: 584–587

Nishio N and Satoh H (1997) A water-soluble chlorophyll protein in cauliflower may be identical to BnD22, a drought-induced, 22-kilodalton protein in rapeseed. Plant Physiol 115: 841–846

Nohara T, Asai T, Nakai M, Sugiura M and Endo T (1996) Cytochrome *f* encoded by the chloroplast genome is imported into thylakoids via the SecA-dependent pathway. Biochem Biophys Res Commun 224: 474–478

Oblong JE and Lamppa GK (1992) Identification of two structurally related proteins involved in proteolytic processing of precursors targeted to the chloroplast. EMBO J 11: 4401–4409

Oelmuller R, Herrmann RG and Pakrasi HB (1996) Molecular studies of CtpA, the carboxyl-terminal processing protease for the D1 protein of the Photosystem II reaction center in higher plants. J Biol Chem 271: 21848–21852

Ort DR and Yocum CF (eds) (1996) Oxygenic Photosynthesis: The Light Reactions. Kluwer Academic Publishers, Dordrecht

Ostersetzer O and Adam Z (1997) Light-stimulated degradation of an unassembled Rieske FeS protein by a thylakoid-bound protease: The possible role of the FtsH protease. Plant Cell 9: 957–965

Park JM, Cho JH, Kang SG, Jang HJ, Pih KT, Piao HL, Cho MJ and Hwang I (1998) A dynamin-like protein in *Arabidopsis thaliana* is involved in biogenesis of thylakoid membranes. EMBO J 17: 859–867

Park JM, Kang SG, Pih KT, Jang HJ, Piao HL, Yoon HW, Cho MJ and Hwang I (1997) A dynamin-like protein, ADL1, is present in membranes as a high- molecular-mass complex in *Arabidopsis thaliana*. Plant Physiol 115: 763–771

Peltier JB, Friso G, Kalume DE, Roepstorff P, Nilsson F, Adamska I and van Wijk KJ (2000) Proteomics of the Chloroplast. Systematic identification and targeting analysis of lumenal and peripheral thylakoid proteins. Plant Cell 12: 319–342

Pilgrim ML, van Wijk KJ, Parry DH, Sy DA and Hoffman NE (1998) Expression of a dominant negative form of cpSRP54 inhibits chloroplast biogenesis in *Arabidopsis*. Plant J 13: 177–186

Pottosin, II and Schönknecht G (1996) Ion channel permeable for divalent and monovalent cations in native spinach thylakoid membranes. J Membr Biol 152: 223–233

Rabilloud T, Adessi C, Giraudel A and Lunardi J (1997) Improvement of the solubilization of proteins in two-dimensional electrophoresis with immobilized pH gradients. Electrophoresis 18: 307–316

Rapoport TA, Jungnickel B and Kutay U (1996) Protein transport across the eukaryotic endoplasmic reticulum and bacterial inner membranes. Annu Rev Biochem 65: 271–303

Richter S and Lamppa GK (1998) A chloroplast processing enzyme functions as the general stromal processing peptidase. Proc Natl Acad Sci USA 95: 7463–7468

Richter S and Lamppa GK (1999) Stromal processing peptidase binds transit peptides and initiates their ATP-dependent turnover in chloroplasts. J Cell Biol 147: 33–44

Ritz D, Thony-Meyer L and Hennecke H (1995) The cycHJKL gene cluster plays an essential role in the biogenesis of *c*- type cytochromes in *Bradyrhizobium japonicum*. Mol Gen Genet 247: 27–38

Robinson C, Hynds PJ, Robinson D and Mant A (1998) Multiple pathways for the targeting of thylakoid proteins in chloroplasts. Plant Mol Biol 38: 209–221

Rochaix J-D (1996) Post-transcriptional regulation of chloroplast gene expression in *Chlamydomonas reinhardtii*. Plant Mol Biol 32: 327–341

Rochaix J-D, Goldschmidt-Clermont M and Merchant S (eds) (1998) The Molecular Biology of Chloroplasts and Mitochondria in Chlamydomonas. Kluwer Academic Publishers, Dordrecht

Roepstorff P (1997) Mass spectrometry in protein studies from genome to function. Curr Opin Biotechnol 8: 6–13

Rothstein SJ, Gatenby AA, Willey DL and Gray JC (1985) Binding of pea cytochrome *f* to the inner membrane of *Escherichia coli* requires the bacterial secA gene product. Proc Natl Acad Sci USA 82: 7955–7959

Roy LM and Barkan A (1998) A SecY homologue is required for the elaboration of the chloroplast thylakoid membrane and for normal chloroplast gene expression. J Cell Biol 141: 385–395

Sääf A, Monne M, de Gier JW and von Heijne G (1998) Membrane topology of the 60-kDa Oxa1p homologue from *Escherichia coli*. J Biol Chem 273: 30415–30418

Sanchirico ME, Fox TD and Mason TL (1998) Accumulation of mitochondrially synthesized Saccharomyces cerevisiae Cox2p and Cox3p depends on targeting information in untranslated portions of their mRNAs. EMBO J 17: 5796–5804

Sargent F, Bogsch EG, Stanley NR, Wexler M, Robinson C, Berks BC and Palmer T (1998) Overlapping functions of components of a bacterial Sec-independent protein export pathway. EMBO J 17: 3640–3650

Satoh H, Nakayama K and Okada M (1998) Molecular cloning and functional expression of a water-soluble chlorophyll protein, a putative carrier of chlorophyll molecules in cauliflower. J Biol Chem 273: 30568–30575

Schlicher T and Soll J (1996) Molecular chaperones are present in the thylakoid lumen of pea chloroplasts. FEBS Lett 379: 302–304

Schlicher T and Soll J (1997) Chloroplastic isoforms of DnaJ and GrpE in pea. Plant Mol Biol 33: 181–185

Schroda M, Vallon O, Wollman FA and Beck CF (1999) A chloroplast-targeted heat shock protein 70 (HSP70) contributes to the photoprotection and repair of Photosystem II during and after photoinhibition. Plant Cell 11: 1165–1178

Schuenemann D, Gupta S, Persello-Cartieaux F, Klimyuk VI, Jones JDG, Nussaume L and Hoffman NE (1998) A novel signal recognition particle targets light-harvesting proteins to the thylakoid membranes. Proc Natl Acad Sci USA 95: 10312–10316

Schuenemann D, Amin P, Hartmann E and Hoffman NE (1999) Chloroplast SecY is complexed to SecE and involved in the translocation of the 33-kDa but not the 23-kDa subunit of the oxygen-evolving complex. J Biol Chem 274: 12177–12182

Scotti PA, Urbanus ML, Brunner J, de Gier JWL, von Heijne G, van der Does C, Driessen AJM, Oudega B and Luirink J (2000) YidC, the *E. coli* homologue of mitochondrial Oxa1p, is a component of the Sec-translocase. EMBO J 19, 542–549

Settles AM and Martienssen R (1998) Old and new pathways of protein export in chloroplasts and bacteria. Trends Cell Biol 8: 494–501

Settles AM, Yonetani A, Baron A, Bush DR, Cline K and Martienssen R (1997) Sec-independent protein translocation by the maize Hcf106 protein. Science 278: 1467–1470

Shackleton JB and Robinson C (1991) Transport of proteins into chloroplasts. The thylakoidal processing peptidase is a signal-type peptidase with stringent substrate requirements at the –3 and –1 positions. J Biol Chem 266: 12152–12156

Shevchenko A, Wilm M, Vorm O, Jensen ON, Podtelejnikov AV, Neubauer G, Mortensen P and Mann M (1996) A strategy for identifying gel-separated proteins in sequence databases by MS alone. Biochem Soc Trans 24: 893–896

Snyders S and Kohorn BD (1999) TAKs, thylakoid membrane protein kinases associated with energy transduction. J Biol Chem 274: 9137–9140

Spetea C, Hundal T, Lohman F and Andersson B (1999) GTP bound to chloroplast thylakoid membranes is required for light-induced, multienzyme degradation of the Photosystem II D1 protein. Proc Natl Acad Sci USA 96: 6547–6552

Spiess C, Beil A and Ehrmann M (1999) A temperature-dependent switch from chaperone to protease in a widely conserved heat shock protein. Cell 97: 339–347

Su Q and Boschetti A (1993) Partial purification and properties of enzymes involved in the processing of a chloroplast import protein from *Chlamydomonas reinhardii*. Eur J Biochem 217: 1039–1047

Su Q and Boschetti A (1994) Substrate- and species-specific processing enzymes for chloroplast precursor proteins. Biochem J 300: 787–792

Sugita M and Sugiura M (1996) Regulation of gene expression in chloroplasts of higher plants. Plant Mol Biol 32: 315–326

Sundberg E, Slagter JG, Fridborg I, Cleary SP, Robinson C and Coupland G (1997) ALBINO3, an *Arabidopsis* nuclear gene essential for chloroplast differentiation, encodes a chloroplast protein that shows homology to proteins present in bacterial membranes and yeast mitochondria. Plant Cell 9: 717–730

Thiellement H, Bahrman N, Damerval C, Plomion C, Rossignol M, Santoni V, de Vienne D and Zivy M (1999) Proteomics for genetic and physiological studies in plants. Electrophoresis 20: 2013–2026

Tirlapur UK, Dahse I, Reiss B, Meurer J and Oelmuller R (1999) Characterization of the activity of a plastid-targeted green fluorescent protein in *Arabidopsis*. Eur J Cell Biol 78: 233–240

Tsugeki R and Nishimura M (1993) Interaction of homologues of Hsp70 and Cpn60 with ferredoxin-NADP$^+$ reductase upon its import into chloroplasts. FEBS Lett 320: 198–202

Tu CJ, Schuenemann D and Hoffman NE (1999) Chloroplast FtsY, chloroplast signal recognition particle, and GTP are required to reconstitute the soluble phase of light-harvesting chlorophyll protein transport into thylakoid membranes. J Biol Chem 274: 27219–27224

Tu CJ, Peterson EC, Henry R and Hoffman NE (2000) The L18 domain of light-harvesting chlorophyll proteins binds to cpSRP43. J Biol Chem 275, 13187–13190

Vambutas V, Tamir H and Beattie DS (1994) Isolation and partial characterization of calcium-activated chloride ion channels from thylakoids. Arch Biochem Biophys 312: 401–406

van Wijk KJ, Andersson B and Aro EM (1996) Kinetic resolution of the incorporation of the D1 protein into Photosystem II and localization of assembly intermediates in thylakoid membranes of spinach chloroplasts. J Biol Chem 271: 9627–9636

VanderVere PS, Bennett TM, Oblong JE and Lamppa GK (1995) A chloroplast processing enzyme involved in precursor maturation shares a zinc-binding motif with a recently recognized family of metalloendopeptidases. Proc Natl Acad Sci USA 92: 7177–7181

Vener AV, Rokka A, Fulgosi H, Andersson B and Herrmann RG (1999) A cyclophilin-regulated PP2A-like protein phosphatase in thylakoid membranes of plant chloroplasts. Biochemistry 38: 14955–14965

Voelker R and Barkan A (1995a) Nuclear genes required for post-translational steps in the biogenesis of the chloroplast cytochrome b_6f complex in maize. Mol Gen Genet 249: 507–514

Voelker R and Barkan A (1995b) Two nuclear mutations disrupt distinct pathways for targeting proteins to the chloroplast thylakoid. EMBO J 14: 3905–3914

Voelker R, MendelHartvig J and Barkan A (1997) Transposon-disruption of a maize nuclear gene, tha1, encoding a chloroplast SecA homologue: In vivo role of cp-SecA in thylakoid protein targeting. Genetics 145: 467–478

Waldo GS, Wright E, Whang ZH, Briat JF, Theil EC and Sayers DE (1995) Formation of the ferritin iron mineral occurs in plastids. Plant Physiol 109: 797–802

Walker MB, Roy LM, Coleman E, Voelker R and Barkan A

(1999) The maize tha4 gene functions in sec-independent protein transport in chloroplasts and is related to hcf106, tatA, and tatB. J Cell Biol 147: 267–276

Wang H, Goffreda M and Leustek T (1993) Characteristics of an Hsp70 homolog localized in higher plant chloroplasts that is similar to DnaK, the Hsp70 of prokaryotes. Plant Physiol 102: 843–850

Whitelegge JP, le Coutre J, Lee JC, Engel CK, Prive GG, Faull KF and Kaback HR (1999) Toward the bilayer proteome, electrospray ionization-mass spectrometry of large, intact transmembrane proteins. Proc Natl Acad Sci USA 96: 10695–10698

Wildman SG, Hongladarom T and Honda SI (1962) Chloroplast and mitochondria in living plant cells: Cinematographic studies. Science 138: 434–436

Wilkins MR, Gasteiger E, Gooley AA, Herbert BR, Molloy MP, Binz PA, Ou K, Sanchez JC, Bairoch A, Williams KL and Hochstrasser DF (1999) High-throughput mass spectrometric discovery of protein post-translational modifications. J Mol Biol 289: 645–657

Wollman FA, Minai L and Nechushtai R (1999) The biogenesis and assembly of photosynthetic proteins in thylakoid membranes. Biochim Biophys Acta 1411: 21–85

Wu GJ and Watanabe A (1997) Import of modified D1 protein and its assembly into Photosystem II by isolated chloroplasts. Plant Cell Physiol 38: 243–247

Xie Z, Culler D, Dreyfuss BW, Kuras R, Wollman FA, Girard-Bascou J and Merchant S (1998) Genetic analysis of chloroplast *c*-type cytochrome assembly in *Chlamydomonas reinhardtii*: One chloroplast locus and at least four nuclear loci are required for heme attachment. Genetics 148: 681–692

Xie Z and Merchant S (1996) The plastid-encoded ccsA gene is required for heme attachment to chloroplast *c*-type cytochromes. J Biol Chem 271: 4632–4639

Xie Z and Merchant S (1998) A novel pathway for cytochromes *c* biogenesis in chloroplasts. Biochim Biophys Acta 1365: 309–318

Yamamoto Y and Satoh K (1998) Competitive inhibition analysis of the enzyme-substrate interaction in the carboxy-terminal processing of the precursor D1 protein of Photosystem II reaction center using substituted oligopeptides. FEBS Lett 430: 261–265

Yates JR, 3rd (1998) Mass spectrometry and the age of the proteome. J Mass Spectrom 33: 1–19

Yohn CB, Cohen A, Danon A and Mayfield SP (1998a) A poly(A) binding protein functions in the chloroplast as a message- specific translation factor. Proc Natl Acad Sci USA 95: 2238–2243

Yohn CB, Cohen A, Rosch C, Kuchka MR and Mayfield SP (1998b) Translation of the chloroplast psbA mRNA requires the nuclear-encoded poly(A)-binding protein, RB47. J Cell Biol 142: 435–442

Yuan J, Henry R, McCaffery M and Cline K (1994) SecA homolog in protein transport within chloroplasts: Evidence for endosymbiont-derived sorting. Science 266: 796–798

Zak E, Norling B, Andersson B and Pakrasi HB (1999) Subcellular localization of the BtpA protein in the cyanobacterium *Synechocystis* sp. PCC 6803. Eur J Biochem 261: 311–316

Zak E, Sokolenko A, Unterholzner G, Altschmied L and Herrmann RG (1997) On the mode of integration of plastid-encoded components of the cytochrome *bf* complex into thylakoid membranes. Planta 201: 334–341

Zerges W and Rochaix JD (1994) The 5′ leader of a chloroplast mRNA mediates the translational requirements for two nucleus-encoded functions in *Chlamydomonas reinhardtii*. Mol Cell Biol 14: 5268–5277

Zerges W and Rochaix JD (1998) Low density membranes are associated with RNA-binding proteins and thylakoids in the chloroplast of *Chlamydomonas reinhardtii*. J Cell Biol 140: 101–110

Zerges W, Girard-Bascou J and Rochaix JD (1997) Translation of the chloroplast psbC mRNA is controlled by interactions between its 5′ leader and the nuclear loci TBC1 and TBC3 in *Chlamydomonas reinhardtii*. Mol Cell Biol 17: 3440–3448

Zhang L, Paakkarinen V, van Wijk KJ and Aro EM (1999) Co-translational assembly of the D1 protein into Photosystem II. J Biol Chem 274: 16062–16067

Zheleva D, Sharma J, Panico M, Morris HR and Barber J (1998) Isolation and characterization of monomeric and dimeric CP47-reaction center Photosystem II complexes. J Biol Chem 273: 16122–16127

Chapter 10

Peptidyl-Prolyl Isomerases and Regulation of Photosynthetic Functions

Alexander V. Vener*
Department of Horticulture, University of Wisconsin, Madison, WI 53706, USA

Summary

Peptidyl-prolyl *cis-trans* isomerase (PPIase) are catalysts involved in protein folding functioning by accelerating the *cis-trans* isomerization of proline peptide bonds. The first PPIase was discovered over 15 years ago. Since then, almost two hundred enzymes of this type have been identified. PPlases are ubiquitous in all living organisms and belong to three distinct families: cyclophilins, FK506 binding proteins (FKBPs) and parvulins. PPIases are found as soluble, membrane-bound or associated with chaperone complexes, receptors, ribosomes, membrane channels or pores in cells. They are involved in a variety of cellular processes apart from protein folding, for example: protein biogenesis, intracellular signaling, protein trafficking, cell cycle control, heat shock responses, regulation of protein phosphorylation, pathogen-host interactions, RNA maturation and DNA degradation. The functional diversity of these proteins is based on the multiplicity of binding modules linked to the catalytic (isomerase) domains of various PPIases. Numerous PPIase genes have been identified in the plant genome and several complex PPIases with multi-domain structure have been characterized in plants. According to genomic analyses, about ten putative PPIases are localized in plant chloroplasts; however, only three chloroplast PPIases have been characterized to date. One of them, TLP40 (thylakoid lumen PPIase of 40 kDa), has a multi-modular structure; a cyclophilin-like C-terminal PPIase domain, an acidic linker region and an N-terminal leucine zipper flanked by phosphatase binding modules. TLP40 is localized in the lumen of

*Present Address: Division of Cell Biology, Linköping University, SE-581 85, Linköping, Sweden. Email: aleve@ibk.liu.se; avvener@facstaff.wisc.edu

Eva-Mari Aro and Bertil Andersson (eds): Regulation of Photosynthesis, pp. 177–193.

chloroplast thylakoids and binds reversibly to the inner membrane surface, regulating a membrane-bound protein phosphatase. This phosphatase functions to specifically dephosphorylate Photosystem II core phosphoproteins. This regulation is responsible for the fast dephosphorylation of Photosystem II during heat shock directly controlling the turnover rate of photosynthetic proteins. The role of additional putative chloroplast PPIases in regulation of photosynthesis has yet to be revealed.

I. Introduction

Among the canonical amino acids, proline is unique in that its side chain forms a ring structure with its own α-amino group. This leads to an exclusive structure of peptide bonds formed by proline residues in proteins. As illustrated in Fig. 1, this peptide bond can acquire two conformations; *cis* and *trans*. Conversions between these conformations lead to significant changes in the protein's tertiary structure. While both *cis*- and *trans*- Xaa-Pro conformers are equally represented in proteins with known tertiary structures, usually only one conformer is correct ensuring the structural integrity of a given protein. The transitions between conformations of proline residues are slow reactions with half times in the range of seconds to several minutes (Fischer et al., 1998). For a long time prolyl isomerization has been recognized as the rate limiting step in both folding and unfolding of proteins (Brandts et al., 1975; Schmid and Baldwin, 1978; Jackson and Fersht, 1991; Koide et al., 1993). Proline *cis-trans* isomerization in protein folding ensures the requirement for enzyme-assisted catalysis of this reaction in vivo. Until recently, it was not anticipated that enzymes catalyzing prolyl isomerization could be involved in the large number of cellular processes as it appears today. The multifunctional character of isomerases has only recently been demonstrated in the field of photosynthesis.

The first enzyme accelerating prolyl isomerization: peptidyl-prolyl *cis-trans* isomerase (PPIase, EC 5.1.2.8) was isolated in the mid-eighties by Fischer et al. (1984). Following this original work almost two hundred PPIases belonging to three different families have been identified. As expected many of these proteins have been found to be important catalysts involved in protein folding (Matouschek et al., 1995; Rassow et al., 1995; Hesterkamp et al., 1996; Lazar and Kolter, 1996; Dartigalongue and Raina, 1998). A large number of PPIases have also been identified as multifunctional proteins and functionally active components involved in many different cellular processes. PPIases appear to be constituents of: receptors (Owens-Grillo et al., 1996; Pratt et al., 1996); chaperone complexes (Freeman et al., 1996; Pratt, 1998); calcium channels (Marks, 1996; Marks, 1997); and mitochondrial membrane pores (Nicolli et al., 1996). Furthermore, PPIases have been found to participate in: intracellular signaling and protein trafficking (Schreiber, 1991; Walsh et al., 1992; Maleszka et al., 1997); RNA maturation (Mi et al., 1996; Hani et al., 1999); DNA degradation during apoptosis (Montague et al., 1997); pathogen-host interactions (Braaten et al., 1997; Hoerauf et al., 1997; Pissavin and Hugouvieux-Cotte-Pattat, 1997; Deng et al., 1998; Yin et al., 1998); cell

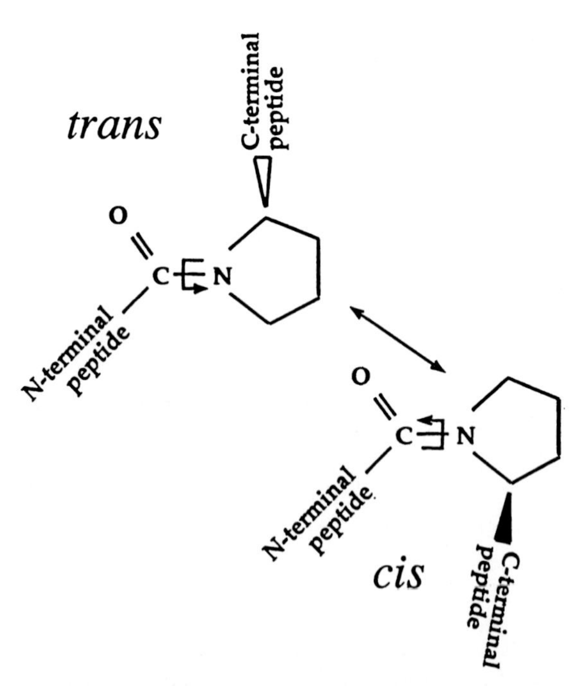

Fig. 1. Cis-*trans* isomerization of a peptidyl-prolyl bond. The peptide bond formed by proline residue may be present in two conformations: *cis* and *trans*. Conversions between these conformations of proline residues in proteins lead to significant changes in protein structure.

Abbreviations: CsA – cyclosporin A; CypA – cyclophilin A; FKBP – FK506 binding protein; PP2A – protein phosphatase 2A; PP2B – protein phosphatase 2B; PPIase – peptidyl-prolyl *cis-trans* isomerase; TLP40 – thylakoid lumen PPIase of 40 kDa; TPR – tetratricopeptide repeat

cycle control (Lu et al., 1996; Crenshaw et al., 1998; Shen et al., 1998); heat-shock-protein-complex signaling pathways under both stress and non-stress conditions (Duina et al., 1996, 1998a); as well as in protein biogenesis involving the proteasome pathway (Ferreira et al., 1998). Several of these newly discovered functions of PPIases are described in recent reviews (Maleszka et al., 1997; Pahl et al., 1997; Hunter, 1998; Gothel and Marahiel, 1999).

Analysis of PPIases in this review provides a clue towards understanding their functional diversity arising from the modular structure. The catalytic domains of many PPIases are linked to distinct binding modules and protein-protein interaction domains, which ensure specific targeting of the PPIase catalytic function. The first part of this review describes the three families of PPIases giving special attention to their structural organization. This provides a connection between the modular structure of PPIases and the diversity of their function in various organisms. The second part of this review addresses the structure and function of PPIases from plants and specifically, chloroplasts. Finally, TLP40, the first characterized PPIase associated with plant photosynthetic membranes, will be described comprehensively followed by a discussion on the possible roles PPIases play in the regulation of photosynthesis.

II. Structure and Function of Peptidyl-Prolyl Isomerases (PPlases)

A. Three Families of PPlases

The enzymes responsible for the catalysis of prolyl *cis-trans* isomerization reactions of proteins have been identified in all prokaryotic, eukaryotic and archea organisms. These ubiquitous enzymes belong to three distinct families: cyclophilins; FKBPs; and parvulins (Fig. 2). Their catalytic domains are relatively compact with molecular masses of the prototypic members ranging between 10 and 18 kDa. Despite a common catalytic activity, the members of the three families of PPIases have no significant protein sequence homology. Nevertheless, PPIases of all the three families are similar in their prolyl-specific binding affinity and in the hydrophobic nature of the prolyl-binding pockets of their active sites (Fischer et al., 1998). Notably, cyclophilins, FKBPs and parvulins do not catalyze isomerization of non-prolyl peptide bonds (Scholz et al., 1998). In addition, PPIases of different families have been found to have similar stereospecific recognition with respect to the prolyl-neighboring amino acid residues of the substrate peptides (Schiene et al., 1998). Surprisingly, deletion of cyclophilins and FKBPs in a number of organisms revealed that these knockout mutants are not lethal, while a group of parvulins, which function as mitotic regulators, is crucial for the survival of eukaryotic cells (Lu et al., 1996; Hunter, 1998; Gothel and Marahiel, 1999).

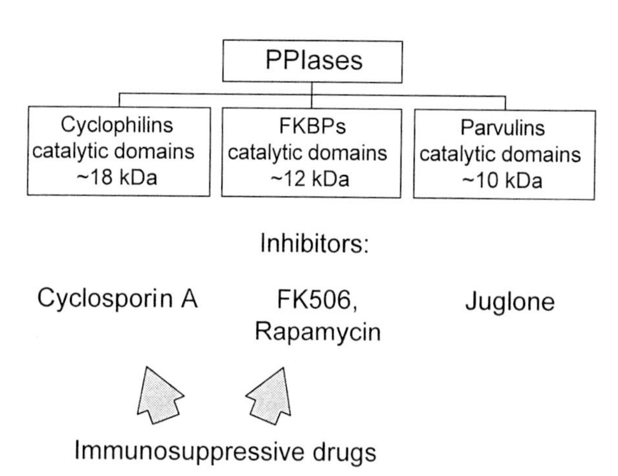

Fig. 2. Three families of PPIases. Cyclophilin and FKBP families are also collectively named immunophilins because they bind immunosuppressive drugs. Parvulins do not belong to immunophilins since they do not bind immunosuppressive drugs. The PPIases belonging to different families do not have significant sequence similarity.

A number of PPIases were found to maintain specific substrate-dependence activity (Kofron et al., 1991; Janowski et al., 1997; Scholz et al., 1997a). This can be explained by the presence of an additional substrate-binding domain in these enzymes (Fischer et al., 1998). The most recently discovered family of parvulins (Rahfeld et al., 1994) illustrates the importance of the specific substrate affinity in the prolyl-isomerization reaction. Some parvulins possess PPIase activity that specifically recognizes phosphoserine-proline or phosphothreonine-proline bonds in mitotic phosphoproteins (Ranganathan et al., 1997; Yaffe et al., 1997). This phosphorylation-dependent specificity is mediated by a small phosphate-specific binding domain present in parvulins, not through selectivity of their PPIase catalytic domains (Yaffe et al., 1997).

The elucidation of additional biological activities of PPIases, in addition to their enzymatic activity, followed soon after the identification of the first PPIase in a publication entitled 'Cyclophilin and peptidyl-prolyl *cis-trans* isomerase are probably identical proteins' (Fischer et al., 1989). Prior to that report, cyclophilins were known as proteins binding the immunosuppressive drug cyclosporin A (CsA), which is a cyclic undecapeptide produced in a variety of fungi. In the early 1980s CsA revolutionized the science of transplantation of human organs and tissues due to its powerful ability to suppress immunoresponses (Schreiber, 1991). When it was revealed that intracellular receptors for CsA were in fact PPIases, the family of CsA-binding PPIases was subsequently defined as cyclophilins. Two other immunosuppressive drugs, FK506 and rapamycin, unrelated to cyclophilins, were also found to bind with high affinity to PPIases. The latter family of PPIases has been designated FKBPs (FK506 binding proteins). Collectively, cyclophilins and FKBPs were designated immunophilins, which stands for immunosuppressant-binding proteins (Schreiber, 1991). Immunosuppressive drugs inhibit the specific enzymatic activity of the corresponding PPIases (Fig. 2). Immunosuppression in mammals occurs via protein-protein interactions involving immunophilins in mammalian T cells, not through the inhibition of PPIase activity. (Schreiber, 1992; Walsh et al., 1992). The complexes cyclophilin-CsA and FKBP-FK506 (but not cyclophilin, FKBP, FKBP-rapamycin, or FKBP-506BD) competitively bind to and inhibit Ca^{2+}- and calmodulin-dependent protein phosphatase calcineurin (Liu et al., 1991). As a result, the phosphatase is deactivated, unable to dephosphorylate substrate proteins, including a transcription factor responsible for expression of proteins required for T cell activation (Fig. 3). The interference of these immunophilin-drug complexes disrupting the signaling pathway in the immune system ultimately leads to the blockade of immunoresponse. The characterization of these PPIases interactions with the protein phosphatase calcineurin was the first demonstrated examples of the multiple cellular functions of these proteins.

Parvulins, the third PPIases family, are not immunophilins since they do not bind the immunosuppressive drugs CsA, FK506 or rapamycin. The recently discovered inhibitor of parvulins, juglone (Fig. 2), has no connection to immunosuppression (Hennig et al., 1998). Parvulins are also multifunctional proteins. The diverse cellular functions of all PPIases are reviewed in the following sections of this chapter.

B. Cyclophilins

Nature has selected a 'mix and match' strategy when creating the cyclophilins. While prototypic cytoplasmic CypA consists of only a PPIase catalytic core of 18 kDa, other members of the family have complex structural organization with a variety of other functional modules linked to PPIase domains. Emphasizing this complex organization, different cyclophilins have molecular masses ranging from 20 kDa to 360 kDa (Fig. 4). Representatives of modular cyclophilins have N-terminal pre-sequences targeting their specific cellular localization to: the periplasm of bacterium (Liu and Walsh, 1990); to mitochondrion (Connern and Halestrap, 1992); endoplasmic

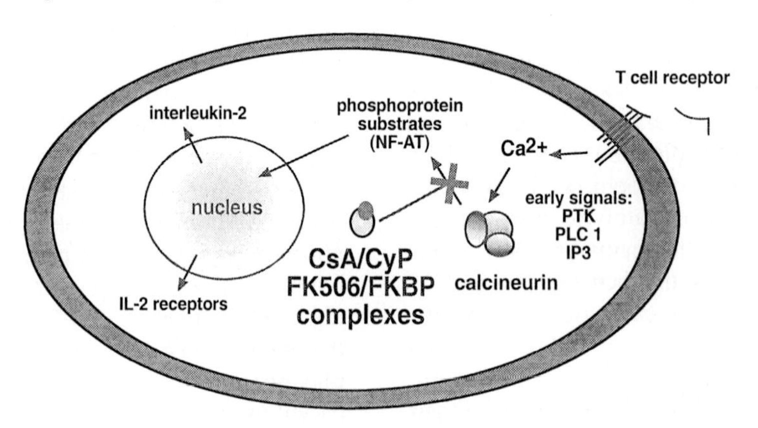

Fig. 3. Cartoon illustrating the mechanism of immunosuppressive drug action in mammalian T cell. CsA forms complex with cyclophilin and FK506 form complexes with FKBP. The complexes cyclophilin-CsA or FKBP-FK506 bind to and inhibit the Ca^{2+}- and calmodulin-dependent protein phosphatase calcineurin. The inhibited phosphatase does not dephosphorylate substrate proteins, including a transcription factor (NF-AT) responsible for expression of proteins required for T cell activation, which consequently leads to immunosuppression.

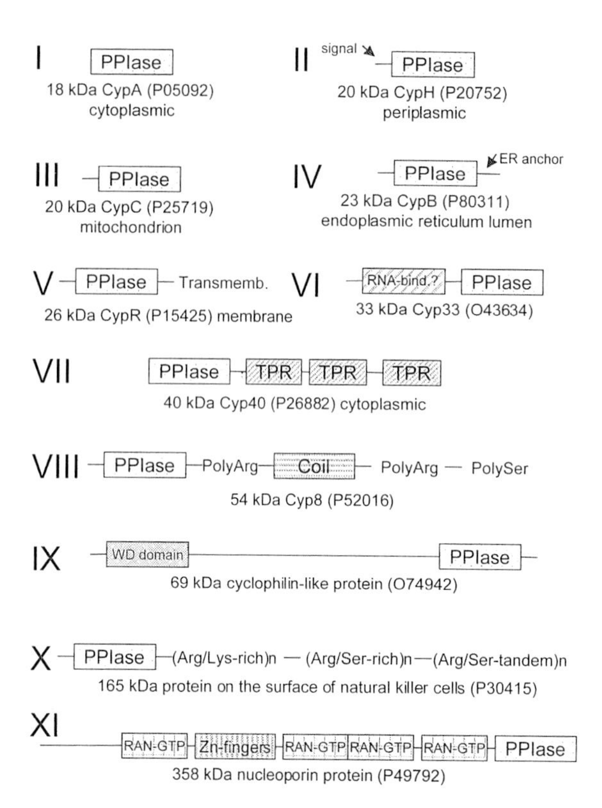

Fig. 4. Schematic representation of modular cyclophilins, showing fusion of the PPIase domain to different binding modules. The accession numbers of individual representatives of the protein family are shown in the brackets.

reticulum (Chen et al., 1995) or chloroplasts (Breiman et al., 1992; Lippuner et al., 1994) in eukaryotic cells. A cyclophilin identified in the matrix of yeast mitochondria, has been observed to accelerate the refolding of a fusion protein that was synthesized in reticulocyte lysate and imported into isolated mitochondria (Matouschek et al., 1995). Similarly, *Neurospora crassa* mitochondria lacking Cyp20, efficiently imported pre-proteins into the matrix, but correct folding of the imported proteins was significantly delayed. Folding intermediates of the accumulated imported proteins were found associated to the molecular chaperones Hsp70 and Hsp60 in the matrix, indicating that the cyclophilin is a component of the mitochondrial protein folding machinery (Rassow et al., 1995). Mammalian mitochondrial matrix cyclophilin PPIase has a short unique N-terminal sequence and reversibly binds to the membrane permeability transition pore, which in turn regulates pore closure and opening (Nicolli et al., 1996; Scorrano et al., 1997).

A number of cyclophilins function specifically to accelerate the folding of defined proteins, for example, photoreceptors. In *Drosophila*, a 26-kDa cyclophilin has a putative signal sequence and a predicted transmembrane domain, unlike most cyclophilins, which are soluble proteins (Fig. 4). This protein is expressed only in the fly eye and mutations in the cyclophilin gene severely reduce the amount of rhodopsin bound in the photoreceptors, suggesting that it is an opsin-specific PPIase required for correct folding and stability of rhodopsin (Schneuwly et al., 1989). Similarly, in vertebrate retina there are two cyclophilin isoforms, which are expressed at high levels in cone photoreceptors. The larger of the isoforms binds to Ran GTPase and thus is called Ran-binding protein-2 (Ferreira et al., 1996). The Ran-binding domain and the cyclophilin PPIase domain act in concert as chaperones for the opsin molecule of the red/green-sensitive visual pigment. The cyclophilin domain does not bind opsin directly but rather augments and stabilizes the interaction between the opsin and the Ran-binding domain. This involves a cyclophilin-mediated modification of the red/green opsin, possibly a proline isomerization reaction (Ferreira et al., 1996). In addition, the cyclophilin domain of Ran-binding protein-2 could specifically associate with the 19 S regulatory complex of the 26 S proteasome in the neuroretina. This complex association is believed to mediate the limited proteolysis of Ran-binding protein-2 into the smaller second isoform of retinal cyclophilin (Ferreira et al., 1997, 1998).

Specific domains in complex cyclophilins direct their binding to distinct protein complexes. The cytoplasmic Cyp40 has three TPR protein-protein binding modules downstream the PPIase catalytic domain (Fig. 4). Along with the molecular chaperones Hsp90 and Hsp70, Cyp40 has been observed to be a component of a steroid apo-receptor complex thought to maintain substrates in an intermediate folded state (Freeman et al., 1996). In *Saccharomyces cerevisiae,* mutations that knockout Cpr7 function, a Cyp40-type cyclophilin required for full Hsp90 activity, result in increased gene expression. This increased gene expression is dependent on heat shock transcription factors and the acquisition of a thermo-tolerant phenotype. Genetic assays also reveal that Hsp90 and Cpr7 function synergistically and are required for negative regulation of heat shock response under both stress, and non-stress conditions (Duina et al., 1998a). The TPR domains of Cpr7 mediate the binding to Hsp90, but no requirement for the PPIase domain has been defined. A mutation in the catalytic domain, altering a conserved site predicted to be critical for isomerase activity, does not compromise Cpr7 function. Furthermore, deletion of the entire PPIase domain does not significantly affect growth or Hsp90-mediated steroid receptor activity. These results indicate that the TPR-containing carboxyl terminus of Cpr7 is sufficient for fundamental Cpr7-dependent activity (Duina et al., 1998b).

The TPR domains are common modules mediating protein-protein interactions. In addition to the Hsp90-binding cyclophilins and FKBPs (see below), TPR-containing proteins include the anaphase promoting complex subunits cdc16, cdc23 and cdc27; the NADPH oxidase subunit p67 phox; transcription factors; the PKR protein kinase inhibitor; peroxisomal and mitochondrial import proteins; and protein phosphatase PP5 (Das et al., 1998). Besides binding to Hsp90, the TPR domains of Cyp40 have been shown to bind to the c-Myb transcription factor, which negatively regulates its DNA-binding activity (Hunter, 1998).

A nuclear cyclophilin Cyp33 (Fig. 4) from human T-cells has an RNA-binding domain. This protein proved to be the first example of a combined RNA-binding and protein folding function in a single protein (Mi et al., 1996). In support of this finding, recently, a PPIase of the parvulin family was implicated in 3'-end formation of a pre-mRNA (see below in II D). These findings expand the interactions of multi-domain PPIases beyond only protein-protein interplay, but implicate PPIase-RNA interactions.

While the binding domains of modular cyclophilins are important for interactions with numerous proteins, the catalytic PPIase domains can complex proteins as well. The binding involves prolyl-affinity in the hydrophobic pocket of the cyclophilin active site. Cytoplasmic cyclophilins have been observed in host-parasite interactions via direct binding to the pathogen protein, as have been shown in *Leishmania major* infected macrophages (Hoerauf et al., 1997) and human immunodeficiency virus type 1 (HIV-1) (Braaten et al., 1997). Completion of an early stage in HIV-1 life cycle requires incorporation of the cellular cyclophilin A (CypA) into virons. The hydrophobic pocket in the CypA active site was shown to interact directly with HIV-1 polyprotein. Notably, a critical proline residue in HIV-1 polyprotein was observed to mediate this binding (Braaten et al., 1997; Yin et al., 1998).

Besides PPIase activity, human recombinant cyclophilins A, B, and C have been shown to be multifunctional, having nucleolytic and DNA degradation activity. Nuclease activity inherent to cyclophilins is distinct from PPIase activity and is similar to that described for apoptotic nucleases. This suggests that cyclophilins are involved in the degradation of the genome during apoptosis (Montague et al., 1997).

C. FKBPs

Like cyclophilins, the FKBP PPIases comprise a family of proteins assembled with diverse functional domains (Fig. 5). In a number of cases, the additional modules bound to PPIase catalytic domains perform functions similar to analogous cyclophilins. For example, the TPR-containing FKBPs bind to Hsp90 like Cyp40 (Pratt, 1998), whereas other FKBPs participate in cellular processes such as co-translational protein folding, regulation of calcium channels and protein tyrosine phosphorylation. Cyclophilins and parvulins have not to date been observed in the latter processes.

The *Escherichia coli* trigger factor (Fig. 5) is a PPIase, which is specifically associated with the 50S ribosome sub-unit and catalyzes efficient proline-limited protein folding (Stoller et al., 1995). The catalytic domain of this PPIase is homologous to FKBP; however, its activity is not inhibited by FK506. Extended C- and N-terminal domains of the trigger factor mediate association of both the translating ribosome and the protein nascent chain during translation (Hesterkamp et al., 1996; Zarnt et al., 1997). These binding modules ensure two orders of magnitude higher protein folding activity of the trigger factor compared to other FKBPs, cyclophilins and parvulins (Scholz et al., 1997b; Zarnt et al., 1997). The isolated catalytic domain of the trigger factor retains full catalytic activity directed at short peptides, but in protein folding assays, its activity is reduced 800-fold (Scholz et al., 1997b).

The FKBP12.6 (Fig. 5) has been observed to associate with intracellular Ca^{2+}-release channels on the muscular sarcoplasmic reticulum (ryanodine receptors (RyRs)) (reviewed in Marks, 1996, 1997). RyR channels are tetramers composed of four subunits each with a molecular mass of approximately 560 kDa. The tetrameric structural organization of the channels is stabilized by one molecule of FKBP12.6 that associates with each subunit of the

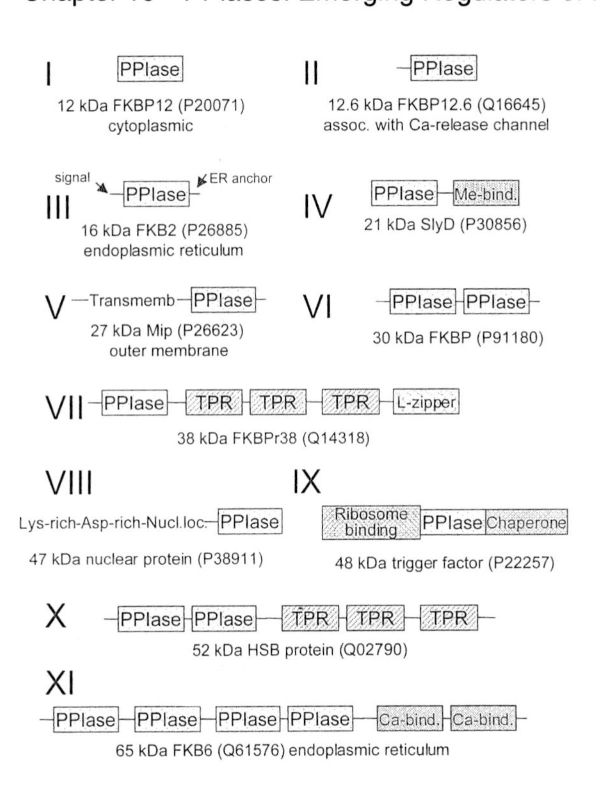

Fig. 5. Schematic representation of modular FKBPs. The accession numbers of individual representatives of the protein family are shown in the brackets.

channel complex. FK506 and rapamycin inhibit PPIase activity of FKBP12.6 and can dissociate FKBP12.6 from the channels. The role of prolyl isomerase activity for Ca^{2+}-release channels remains undefined. A mutant mouse deficient in FKBP12.6 has been observed to have severe heart disorder and altered RyRs single-channel properties (Shou et al., 1998).

FKBP12 was found to modulate tyrosine kinase activity of the epidermal growth factor (EGF) receptor from plasma membranes of a human epidermal cell line (Lopez-Ilasaca et al., 1998). Whereas FKBP12 inhibited autophosphorylation of EGF, the cyclophilins screened did not affect this activity. In contrast to the wild-type enzyme, several variants of FKBP12 with greatly reduced PPIase activity were unable to suppress the EGF receptor tyrosine kinase activity. Pervanadate, an inhibitor of protein tyrosine phosphatases, abolishes the effect of FKBP12 on EGF receptor autophosphorylation. Finally, FK506 and rapamycin induced significant stimulation of EGF receptor autophosphorylation in intact cells, suggesting suppression of EGF receptor autophosphorylation by intracellular FKBP12 in vivo. These data point to an inhibitory function of FKBP12 in EGF receptor signaling, possibly induced by the stimulation of a protein tyrosine phosphatase coupled to the EGF receptor. Both PPIase activity and substrate specificity of FKBP12 seem to be indispensable for this effect (Lopez-Ilasaca et al., 1998). Thus, in addition to inhibition of the Ser/Thr-protein phosphatase calcineurin by the FKBP-FK506 complex, FKBP12 is also capable of regulating the activity of a protein tyrosine phosphatase.

D. Parvulins

Parvulins are the most recently discovered family of PPIases. The name parvulins is based on the small size of the catalytic domain (Latin: *parvulus*, very small) (Rahfeld et al., 1994). While knockout mutants of cyclophilins and FKBPs in many organisms are not lethal, a number of parvulins were found to be crucial for the survival of both eukaryotic and prokaryotic cells (Lu et al., 1996; Dartigalongue and Raina, 1998; Hunter, 1998; Gothel and Marahiel, 1999). Specific binding modules fused to the PPIase catalytic domain were found to define many of the critical functions of these essential parvulins.

The PPIase Pin1 (Fig. 6) is a regulator of mitosis that is highly conserved from yeast to humans (Crenshaw et al., 1998; Shen et al., 1998). Pin1 is required for the cell's progression through mitosis via binding and regulation of a series of mitosis-specific proteins in response to their phosphorylation state (Yaffe et al., 1997; Shen et al., 1998). Pin1 is a phosphorylation-dependent PPIase that isomerizes proline residues preceded by a phosphorylated serine or threonine in substrates with up to 1300-fold selectivity compared with non-phosphorylated peptides (Yaffe et al., 1997). The effect of phosphorylated serine or threonine residues is specific, since neither phosphorylated tyrosine nor glutamic acid is able to accelerate the prolyl isomerization to a similar extent (Schutkowski et al., 1998). Pin1 interacts with specific mitotic proteins, including the phosphatase Cdc25 and its regulator protein (Crenshaw et al., 1998). This parvulin binds only to the phosphorylated, mitotically active form of Cdc25, which leads to inhibition of phosphatase activity (Shen et al., 1998). Pin1 acts as a general regulator of mitotic proteins phosphorylated by mitotic kinases and regulates mitotic progression by catalyzing sequence-specific and phosphorylation- dependent prolyl isomerization.

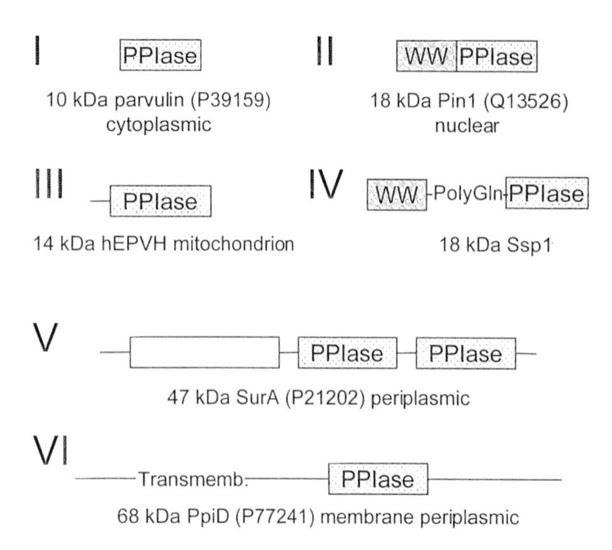

Fig. 6. Schematic representation of modular parvulins. The accession numbers of individual representatives of the protein family are shown in the brackets.

The phosphoserine- or phosphothreonine-binding specificity of Pin1 is determined by its N-terminal WW protein-interacting module (Fig. 6) (Lu et al., 1999b). WW modules are small domains observed in proteins that participate in cell signaling or regulation (Lu et al., 1999a,b; Wang et al., 1999). The WW domain functions as a phosphoserine- or phosphothreonine-binding module, allowing Pin1 to interact with its substrate in vitro and to perform phosphorylation-dependent functions in vivo (Lu et al., 1999b). Parvulins devoid of the WW domain do not have phosphorylation-dependent PPIase specificity.

The yeast homologue of Pin1 was found in genetic screens aimed at identifying factors involved in mRNA 3′-end processing in budding yeast (Hani et al., 1999). Mutations leading to defects in mRNA 3′-end formation caused by amino acid substitutions of highly conserved amino acid residues within the PPIase domain resulted in a marked decrease in activity of the mutant enzyme (Hani et al., 1999). Furthermore, the parvulin interacted specifically with the phosphorylated but not unphosphorylated carboxyl-terminal domain of the yeast kinase I and RNA polymerase II (Morris et al., 1999). The WW domain of the parvulin mediates this interaction. It is suggested that the WW domain interacts with the phosphorylated form of RNA polymerase II during elongation, while the PPIase domain of the parvulin modifies the substrate conformation by isomerization of proline residues (Morris et al., 1999).

A parvulin homologue from *Neurospora crassa*, Ssp1, contains a polyglutamine stretch between the N-terminal WW domain and the C- terminal PPIase domain (Fig. 6). Ssp1 is a site-specific PPIase: peptides with glutamate, phosphoserine, or phosphothreonine in the -1-position proved to be the best substrates. Ssp1 isomerizes small peptides and functions in protein folding, as shown with mouse dihydrofolate reductase (Kops et al., 1998). A complex parvulin protein found in yeast, *Drosophila* and humans is comprised of four modules: a WW domain, a parvulin PPIase domain, a nuclear localization motif and a long, surface-exposed α-helix likely to be involved in binding to cell cycle serine/threonine kinases (Maleszka et al., 1997).

The two largest PPIases of the parvulin family, SurA and PpiD (Fig. 6), are folding catalysts of outer membrane proteins in the periplasm of *Escherichia coli*. SurA contains two parvulin-like domains and assists in the folding of specific secreted proteins (Lazar and Kolter, 1996). The PPIase PpiD is anchored to the inner membrane via a single transmembrane spanning helix, and its catalytic domain is directed towards the periplasm (Dartigalongue and Raina, 1998). A null mutation in PpiD leads to an overall reduction in the level and folding of outer membrane proteins and to the induction of periplasmic stress response. The combination of PpiD and SurA null mutations is lethal (Dartigalongue and Raina, 1998).

III. Plant PPIases

A. Structure of Plant PPIases

The above general description of PPIases provides a framework for the classification of plant PPIases, as well as anticipating their putative regulatory functions in plant cells. Both prolyl isomerase activity and PPIase enzymes have been identified in all plant tissues to date (Gasser et al., 1990; Breiman et al., 1992; Luan et al., 1996; Chou and Gasser, 1997; Saito et al., 1999b); however, the understanding of their cellular functions and roles are incomplete. For example, there are representatives of all three PPIase families in plants (Fig. 7), but not a single parvulin has yet been characterized.

Plant cyclophilins are encoded by a number of different genes (Buchholz et al., 1994; Chou and Gasser, 1997). The majority of identified and

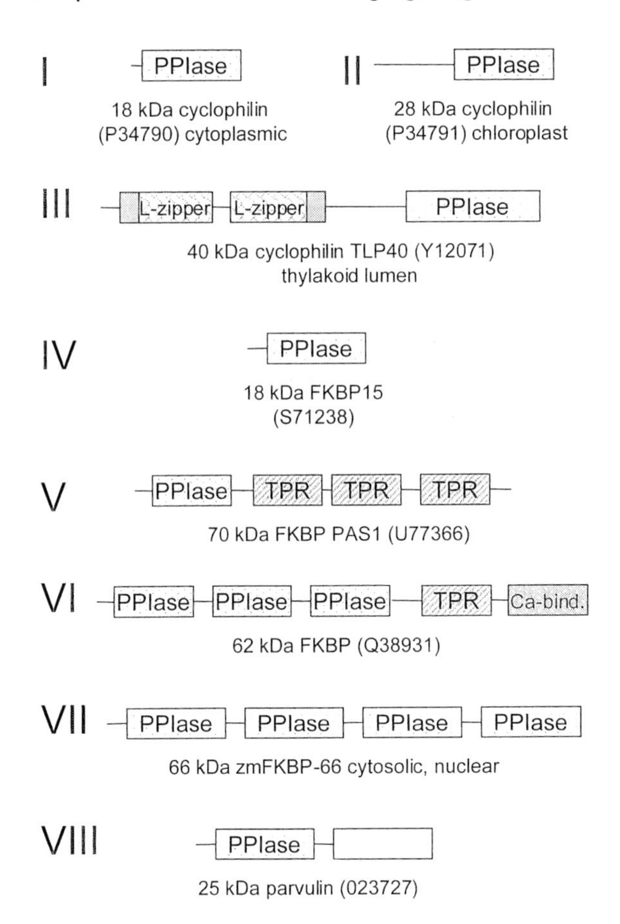

Fig. 7. Schematic representation of known modular PPlases from plants. The accession numbers of individual PPlases are shown in the brackets.

characterized cyclophilins from plants belong to simple, low-molecular weight PPlases (Fig. 7). Cyclophilins are localized to several compartments within the plant cell such as the cytosol (Breiman et al., 1992; Lippuner et al., 1994), mitochondria, chloroplasts (Breiman et al., 1992; Fulgosi et al., 1998; Mattoo, 1998), and endoplasmic reticulum (Sheldon and Venis, 1996; Saito et al., 1999a). Plant cytosolic cyclophilins contain an insertion of additional seven amino acids at the N-termini, as compared to mammalian enzymes (Buchholz et al., 1994; Chou and Gasser, 1997). Mitochondrial and chloroplast cyclophilins are synthesized as precursor proteins in the cytoplasm with corresponding transit peptides and imported (Chou and Gasser, 1997; Fulgosi et al., 1998; Saito et al., 1999a). Up to now, the only complex cyclophilin characterized from plants is TLP40, located in chloroplast thylakoid lumen (see below).

Like cyclophilins, plants FKBPs are present in all cellular compartments (Breiman et al., 1992; Luan et al., 1994a; Luan et al., 1996; Hueros et al., 1998; Mattoo, 1998). In addition to single domain PPlases several complex FKBPs have been characterized from different plant species. FKBP52, encoded by the PAS1 gene in *Arabidopsis thaliana* is characterized by an FKBP-like domain and three TPR units (Vittorioso et al., 1998). The majority of high molecular weight FKBP proteins present in plants have multiple FKBP-like domains in a single polypeptide chain (Fig. 7). *Arabidopsis thaliana* and wheat FKBPs of 62, 73 and 77 kDa each have an N-terminal FKBP domain followed by a TPR motif and a calmodulin binding site (Blecher et al., 1996; Vucich and Gasser, 1996; Reddy et al., 1998; Kurek et al., 1999). In maize, FKBP-66 carries four tandem copies of an FKBP-like binding domain (Hueros et al., 1998).

A comprehensive search in the *Arabidopsis thaliana* database for parvulin-like sequences has revealed parvulin-related protein, both of simple and complex organization. Due to the lack of experimental data on these proteins, a discussion of plant parvulins is beyond the scope of this review.

B. Functions of Plant PPlases

The ubiquitous distribution of PPlases in all plant tissues and the presence of PPlase activities in the majority of plant cell organelles is consistent with their essential role in protein folding and other cellular regulatory functions including plant specific roles. In the latter case, data on the expression and regulation of specific PPlases at different developmental stages in plants, and in response to hormones and various stress conditions, are of particular interest.

In developing bean plants, cyclophilin transcripts were first identified in all tissues three days following onset of germination (Marivet et al., 1994). High levels of cyclophilin mRNA transcripts were detected in developing plant tissues compared to mature tissues. Cyclophilin mRNAs were also observed in mosaic virus-infected bean leaves following ethephon and salicylic acid treatments (Marivet et al., 1994). A comparative study of bean and maize cyclophilin mRNA levels revealed striking differences between the two species in response to various external stimuli. In maize plants, cyclophilin mRNA was shown to accumulate in response to heat shock, whereas no cyclophilin mRNA was observed in bean plants hours following heat stress. Differences in mRNA accumulation were also observed upon salt stress, inducing an earlier response in maize than bean, whereas the opposite effect was observed when these two plants were subjected to cold-stress (Marivet et al., 1994). While pointing towards the physiological importance of cyclophilins, these findings also demonstrate the need for studies on particular PPIase isoforms, taking into account their diversity among different species.

The first described PPIase knockout mutant in plants involved the inactivation of a complex FKBP-like gene, PAS1 in *Arabidopsis thaliana* (Fig. 7). This mutant was defective in development and showed ectopic cell proliferation in cotyledons, extra layers of cells in the hypocotyl, and an abnormal apical meristem (Vittorioso et al., 1998). This phenotype was correlated to both cell division and elongation defects. PAS1 gene expression is increased in the presence of cytokinins, a class of phytohormones known to stimulate cell division. In addition, expression of another complex FKBP of 73 kDa, has been shown to be restricted mainly to young wheat tissues (Blecher et al., 1996).

Enhanced levels of several FKBPs have been observed in plants during a series of stress conditions. The expression of a 62 kDa FKBP, ROF1 (Fig. 7), increased several fold upon wounding of *Arabidopsis thaliana* or exposure to elevated NaCl levels (Vucich and Gasser, 1996). The wheat FKBP77, possessing three FKBP-like domains (Fig. 7), was found to be a heat-induced PPIase (Kurek et al., 1999). The mRNA steady-state level for this protein increased significantly during heat shock and returned to normal levels following recovery at the standard growth temperature. mRNA levels of FKBP15, a PPIase in plant endoplasmic reticulum, have also been shown to increase during heat shock (Luan et al., 1996). Analysis on a genomic clone of a maize cyclophilin revealed the presence of two putative heat shock elements in the promoter region, a metal regulatory element and a third heat shock element located in the 5' untranslated leader sequence. Increasing levels of two cytosolic cyclophilins were induced in proembryonic tissues of *Digitalis lanata* by cold or osmotic stress, which increased the freezing tolerance during cryopreservation (Kullertz et al., 1999). Transcription of cytosolic cyclophilins is also enhanced by hormone (salicylic acid) treatment (Marivet et al., 1995; Kullertz et al., 1999).

Elucidation of protein-protein interactions can provide valuable insights towards understanding the function of specific proteins. Using patch-clamp techniques, it was demonstrated that CsA and FK506 blocked Ca^{2+}-induced inactivation of K^+ channels in *Vicia faba* guard cells (Luan et al., 1993). The cyclophilin-CsA and FKBP-FK506 complexes inhibit the activity of endogenous Ca2+-dependent protein phosphatase. By analogy with mammalian PP2B (calcineurin) signaling system (see above in IIA), the authors proposed a hypothesis involving protein phosphatase in the Ca^{2+} signal transduction pathway of higher plants (Luan et al., 1993). In a recent study, a full-length cDNA sequence encoding a cyclophilin of guard cell protoplasts from *Vicia faba* was shown to have strong sequence homology to a previously described cytosolic plant PPIases (Kinoshita and Shimazaki, 1999). A complex of recombinant vfCyp and CsA inhibited phosphatase activity of bovine calcineurin. The endogenous protein phosphatase activity in the cytosolic fraction of guard cell protoplasts was increased by physiological concentrations of Ca^{2+}. Moreover, Ca^{2+}-stimulated activity was inhibited by CsA, suggesting that both cytosolic cyclophilin and calcineurin-like protein phosphatase are present in guard cells (Kinoshita and Shimazaki, 1999). The cloning of the FKBP12 gene and characterization of its gene product from *Vicia faba* provided opposing data. The VfFKBP12 protein failed to mediate the typical interaction with calcineurin in the presence of FK506 and rapamycin (Xu et al., 1998).

Immuno-affinity chromatography with immobilized maize FKBP-66, containing four FKBP domains was used to search for plant proteins interacting with FKBP (Fig. 7). This approach permitted the isolation and identification of an unknown 36 kDa polypeptide associated with FKBP-66 and elucidating a calmodulin-binding site in FKBP-66 (Hueros et al., 1998). In a yeast two-hybrid screen, a 37 kDa protein from *Arabidopsis thaliana* was identified to interact with FKBP12 (Faure et al., 1998). This 37 kDa protein was homologous to a mammalian protein, FAP48 that also binds to FKBP12 and abolishes binding in the presence of FK506, but not CsA.

The knowledge of animal multi-protein complexes involving Hsp90, Hsp70, p60, p23 and TPR-bearing PPIases stimulated studies on the involvement of TPR-containing FKBPs in corresponding plant complexes. The plant chaperone system can interact with animal chaperones: wheat Hsp70 protein functions in the rabbit reticulocyte lysate hetero-complex assembly system, and human p23 protein functions in the wheat germ lysate assembly system. The ATP-dependent formation of an animal p23/plant hsp90 complex has been identified (Owens-Grillo et al., 1996). Furthermore, the FKBP73 and FKBP77 PPIases (Fig. 7) have been detected in wheat chaperone hetero-complexes and observed to bind hsp90 via their TPR domains (Reddy et al., 1998). Binding of the plant FKBPs to plant Hsp90 protein was disrupted by adding purified TPR domains of human cyclophilin-40 to wheat germ lysate assay system (Owens-Grillo et al., 1996). These data demonstrate that the binding of PPIases to Hsp90 via TPR domains is conserved in both the plant and animal kingdoms.

Similar to mammalian hosts (see above), PPIases are involved in plant-pathogen interactions. For example, *Agrobacterium tumefaciens* induces crown gall tumors on plants by transferring a nucleoprotein complex, the T-complex, to the plant cell. The T-complex consists of a T-DNA and an endonuclease, VirD2, covalently bound to the 5' end of the T-DNA. The yeast two-hybrid system was used to screen for proteins in *Arabidopsis thaliana* that interact with the T-complex (Deng et al., 1998). The interaction of VirD2 with two isoforms of *Arabidopsis thaliana* cyclophilins was identified using this analysis. The VirD2 domain responsible for the interaction with the cyclophilin is a distinct sub-domain within the endonuclease. The VirD2-cyclophilin interactions were disrupted in vitro by CsA, which inhibited Agrobacterium-mediated transformation of *Arabidopsis thaliana* and tobacco. These data strongly suggest that host cyclophilins play a role in T-DNA transfer (Deng et al., 1998).

IV. PPIases and Photosynthetic Function

A. Chloroplast PPIases

In the first dedicated study on PPIase activities in plant organelles (performed by the group of A. Mattoo) (Breiman et al., 1992), prolyl isomerase activity was identified in pea chloroplasts. The major PPIase activity was present in the soluble (stroma) fraction of the purified chloroplasts. This activity was inhibited by CsA, indicating the presence of a cyclophilin-like PPIase (Breiman et al., 1992). Thylakoid-associated PPIase activity of the cyclophilin type was also detected (Breiman et al., 1992; Mattoo, 1998). Furthermore, both stroma and thylakoid fractions of pea chloroplasts bound radioactively labeled CsA (Breiman et al., 1992; Mattoo, 1998). However, the hydrophobic nature of this cyclic undecapeptide may have led to unspecific labeling effects. Neither cross-reactivity with anti-rapamycin-binding protein from yeast, nor binding of radioactive rapamycin to the chloroplast proteins was observed in these early studies (Breiman et al., 1992; Mattoo, 1998). Following this, Breiman et al. (1992) concluded that PPIases are involved in folding of newly synthesized prolyl-containing soluble and membrane proteins in plant plastids.

The gene of a chloroplast stroma cyclophilin from *Arabidopsis thaliana*, ROC4, has been cloned (Lippuner et al., 1994). This gene is expressed only in photosynthetic organs and encodes a precursor protein including an N-terminal extension typical for chloroplast transit peptides. PPIase activity measurements and immunoblot analyses of sub-cellular fractions performed by Lippuner et al. (1994), observed the presence of cyclophilin only in the chloroplast stroma, neither in the thylakoid membrane nor lumen. Expression of chloroplastic cyclophilin ROC4 is strongly induced with light (Chou and Gasser, 1997). A similar gene was also cloned from *Fava bean* (Luan et al., 1994b). The mRNA transcript levels of this cyclophilin are regulated by light and are induced by heat shock (Luan et al., 1994b).

In addition to the stromal cyclophilin discussed above, an FKBP-like PPIase (FKBP13) has been identified in chloroplasts (Luan et al., 1994a). Etiolated leaves produce detectable levels of cyclophilin but do not express FKBP13. Illumination of etiolated plants dramatically increases the expression of both FKBP13 and cyclophilin. Interestingly, light-induced expression of FKBP13 is directly correlated to the accumulation of chlorophyll (Luan et al., 1994a).

B. Structure and Function of TLP40

The first complex plant cyclophilin was isolated accidentally in an attempt to purify a protein phosphatase from spinach thylakoid membranes. A 40 kDa protein that was persistently co-purified with the membrane phosphatase was subjected to micro-sequencing revealing a previously unidentified gene product. This result was confirmed by cloning and sequencing of the corresponding gene (Fulgosi et al.,

1998). The protein was shown to be encoded by a nuclear gene and synthesized as a 49 kDa precursor with a 104 amino acid, N-terminal pre-sequence. This pre-sequence displayed all the attributes of a bipartite transit peptide typical of lumenal and bitopic thylakoid proteins. The pre-protein was shown to be imported into the chloroplast and subsequently into the thylakoid lumen, where its was processed as a mature protein of 40 kDa (Fulgosi et al., 1998). This protein was also identified as a lumenal component in a biochemical screening study of thylakoid lumen proteins (Kieselbach et al., 1998). Thus, this protein has been designated TLP40 (thylakoid lumen PPIase of 40 kDa) (Fulgosi et al., 1998).

Sequence analysis of the mature full length form of the TLP40 protein unexpectedly revealed that it belongs to a cyclophilin family of complex PPIases with a catalytic domain at the C-terminus (Fig. 8). The PPIase domain of TLP40 has approximately 25% amino acid sequence identity to human cyclophilin and possess PPIase activity comparable to other cyclophilins. While all five typical cyclophilin fingerprint motifs are present in this domain, the fourth motif, crucial for high affinity binding of CsA, is the least conserved and does not contain the signature tryptophan (Fulgosi et al., 1998). This explains the low affinity of TLP40 to CsA (IC_{50} is equal to 0.18 mM) (Vener et al., 1999). The N-terminal half of TLP40, upstream the PPIase domain, contains several predicted protein-protein interaction modules (Fig. 8). These modules include two leucine zipper elements, two phosphatase-binding modules, flanking the leucine zipper and an acidic stretch of 90 amino acid residues, which separates the leucine zippers, and the phosphatase binding regions from the PPIase domain. The N-terminal binding modules of TLP40 are likely to be responsible for the reversible binding of the protein to the inner surface of thylakoid membranes and regulation of a membrane protein phosphatase (Vener et al., 1999).

The existence of phosphatase binding domains in TLP40 explained co-purification results with the thylakoid membrane protein phosphatase. Furthermore, the protein phosphatase activity could be co-immunoprecipitated with antibodies raised against the TLP40 (Fulgosi et al., 1998).

TLP40 present in highly purified phosphatase preparations was removed with a high salt wash from a PP2A-like catalytic subunit of the membrane phosphatase fixed to microcystin-agarose (Vener et al., 1999). Most importantly, the activity of the membrane phosphatase was found to be regulated by TLP40: it was activated upon binding of CsA to TLP40 and suppressed upon interaction of TLP40 with peptide substrates containing proline residues. The phosphatase activation occurred only when CsA had access to TLP40, and at high CsA concentrations inhibiting the purified TLP40. Inhibition of the protein phosphatase activity by PPIase substrates, prolyl-containing peptides, could be observed only in ruptured thylakoids where the peptides had access to the active site of TLP40 (Vener et al., 1999). A dramatic acceleration of Photosystem II protein dephosphorylation under heat shock conditions coincided with a release of TLP40 from the inner thylakoid surface (Rokka et al., 1998, 2000; Vener at al., 1998).

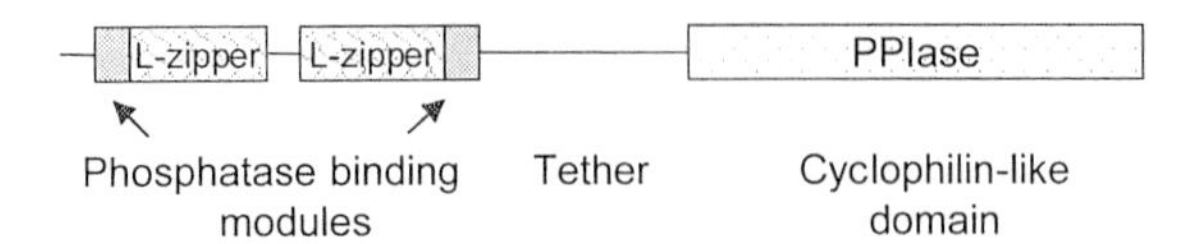

Fig. 8. Modular structure of TLP40. The N-terminal part of the protein contains a leucine zipper that is flanked by the phosphatase-binding modules. The central part contains a rather acidic stretch of about 90 amino acid residues (tether), which connects the leucine zipper to the PPIase domain. The catalytic cyclophilin-like domain is situated in the C-terminal part of TLP40.

Before the realization that TLP40 regulates thylakoid protein dephosphorylation, only calcineurin, a phosphatase of the PP2B type, has been known to bind cyclophilins (Liu et al., 1991; Walsh et al., 1992). PP2B enzymes interact with cyclophilins only in the presence of the exogenous drug CsA, while the interaction between the thylakoid phosphatase and TLP40 occurs in the absence of such a drug. Besides this, the protein phosphatase in thylakoid membranes is a PP2A-like enzyme (Vener et al., 1999). Typical cyclophilin-CsA complexes bind directly to the calcineurin active site inhibiting its dephosphorylating activity, while TLP40 appears to regulate the phosphatase activity across the thylakoid membrane (Vener et al., 1999). Such a form of regulation presents a novel-signaling pathway across the thylakoid membrane, directed from the inside to the outside of the lipid bilayer (Fig. 9). TLP40 was shown to be in dynamic association with the inner surface of thylakoid membrane (Vener et al., 1999; Rokka et al., 2000). Experimental data support a mechanism for which the activity of the thylakoid phosphatase at the outer face of the membrane is suppressed when TLP40 binds to a site

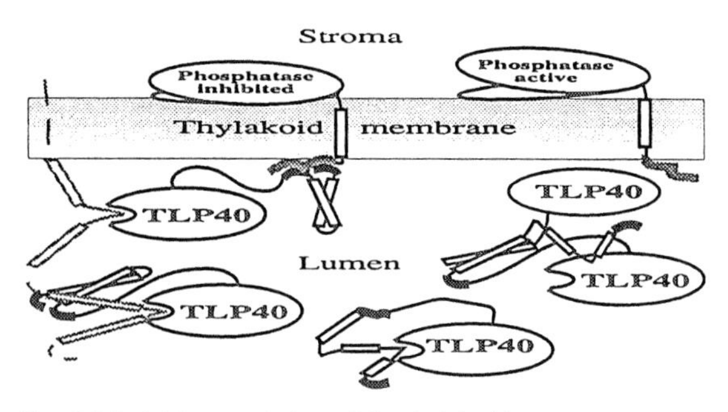

Fig. 9. Model for regulation of the thylakoid membrane protein phosphatase by reversible interaction with the complex cyclophilin TLP40. TLP40 binds to and inhibits the activity of the phosphatase when the leucine zipper of TLP40 is in a closed/assemble conformation and the phosphatase binding modules are brought together. Autocatalytic isomerization of proline residues in the first element of the leucine zipper by PPIase activity disrupts assembled conformation of the leucine zipper. This leads to spatial separation of the phosphatase binding modules and release of TLP40 from the membrane with concomitant activation of the phosphatase.

at the inner thylakoid surface and stimulated when TLP40 is released into the thylakoid lumen. An analogous system is observed in the regulation of the transmembrane mitochondrial permeability transition pore by reversible binding of a matrix cyclophilin (Cyp-M) (Nicolli et al., 1996). Binding of the Cyp-M to the membrane is involved in pore opening while dissociation of the Cyp-M to the matrix of mitochondrion inhibits this process. The Cyp-M release from the mitochondrial pore is induced by CsA (Nicolli et al., 1996; Halestrap et al., 1997).

The phosphatase binding sites, flanking the leucine zipper of TLP40 (Fig. 8), posses a high homology to well characterized epitopes of proteins known to bind the catalytic subunits of PP2A and PP2B (Vener et al., 1999). It is hypothesized that in an α-helical conformation the two leucine zipper elements of TLP40 make intermolecular anti-parallel 'zippering' bringing together the phosphatase anchoring epitopes. This causes binding of TLP40 to the membrane protein phosphatase (Fig. 9). The homologous epitopes of the regulatory subunit A of PP2A have been shown to form loops which are brought together by tightly packed α-helices in the core of the subunit (Ruediger et al., 1994). Leucine zippers are generally considered as highly ordered structures, however, only a few of them are entirely without discontinuities (reviewed in Lupas, 1996). Consistently, the first leucine zipper element of TLP40 contains two proline residues, which in their *cis* conformation could readily facilitate the distortion of a α-helix and ultimately lead to spatial separation of the phosphatase binding sites and subsequent release of TLP40 from the membrane (Fig. 9). These proline residues may be isomerized by the catalytic domain of TLP40 itself resulting in an autocatalytic conformation change in the structure of TLP40 leucine zipper. Such an auto-isomerization of the N-terminal proline residues could be influenced by competition with other prolyl-containing proteins in the lumen requiring folding assistance from TLP40. The proposed model for TLP40-phosphatase interaction (Fig. 9) implies coordination of prolyl isomerization and dephosphorylation of proteins on opposing sides of the photosynthetic membrane.

In addition to its proposed role in the regulation of protein phosphorylation, TLP40 is expected to catalyze the folding of proteins newly inserted in the thylakoid membrane or translocated into the thylakoid lumen (Fulgosi et al., 1998). The restriction of TLP40 to the inner membrane surface of the stroma-exposed thylakoids (Vener et al., 1999), the principal entry site of newly synthesized proteins (Cohen et al., 1995; van Wijk et al., 1996), is consistent with the role during biogenesis and protein turnover. The connection between the presence of prolyl-containing peptides in the lumen and protein dephosphorylation on the stroma side of thylakoid membranes suggests a dual physiological role of TLP40. During photoinhibitory stress conditions, as a part of the repair cycle, there is a rapid turnover of Photosystem II reaction center subunits, specifically the D1 protein (Aro et al., 1993; Andersson and Aro, 1997). It has been observed that this repair process is controlled by protein dephosphorylation (Andersson and Aro, 1997). In order to be replaced, the photodamaged D1 protein has to be dephosphorylated. Following dephosphorylation it can be recognized by the specific protease and degraded to leave space for a new protein copy to be inserted into the Photosystem II complex thereby restoring photosynthetic activity. It is believed that D1 protein turnover should require coordination between dephosphorylation and protein folding. These processes may be modulated by TLP40 - phosphatase transmembrane signaling based upon the hypothesis that dissociation of TLP40 from the inner thylakoid membrane activates the protein phosphatase. For example, release of TLP40 to the thylakoid lumen at elevated temperatures causes rapid dephosphorylation of the D1 protein (Fig. 10).

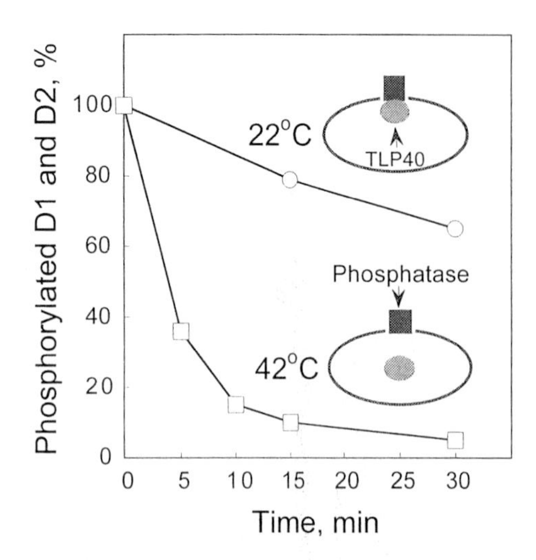

Fig. 10. Model for heat shock induced dephosphorylation of the D1 and D2 reaction center proteins of Photosystem II. The curves show dephosphorylation of these proteins at two different temperatures (22 °C and 42 °C). The cartoons illustrate regulation of the protein phosphatase by reversible binding of TLP40 to the inner membrane surface. At 22 °C the major part of TLP40 is bound to the membrane and the activity of the phosphatase is suppressed. At 42 °C TLP40 is released from the membrane and the phosphatase is activated.

TLP40 is the first identified PPIase that regulates photosynthetic functions. Predicted proteome analysis of *Arabidopsis thaliana* chloroplasts (Abdallah et al., 2000) reveals that more than ten different PPIases may be localized to chloroplasts. Recent proteomic approaches have identified additional PPIases in plant thylakoids (Kieselbach et al., 1998; Peltier et al., 2000; A. Andersson and A. Vener, unpublished). These findings permit speculation concerning the diverse involvement of different PPIases in regulation of plant photosynthesis. This review of the known PPIases and their structures and functions should facilitate characterization of novel PPIases in chloroplasts and understanding of the complex protein machinery dedicated for maintenance and control of photosynthesis.

References

Abdallah F, Salamini F and Leister D (2000) A prediction of the size and evolutionary origin of the proteome of chloroplasts of *Arabidopsis*. Trends Plant Sci 5:141–142

Andersson B and Aro E-M (1997) Proteolytic activities and proteases of plant chloroplasts. Physiol Plant 100: 780–793

Aro E-M, Virgin I and Andersson B (1993) Photoinhibition of Photosystem II. Inactivation, protein damage and turnover. Biochim Biophys Acta 1143: 113–134

Berry J and Björkman O (1980) Photosynthetic response and adaptation to temperature in higher plants. Annu Rev Plant Physiol 31: 491–543

Blecher O, Erel N, Callebaut I, Aviezer K and Breiman A (1996) A novel plant peptidyl-prolyl-*cis-trans*-isomerase (PPIase): cDNA cloning, structural analysis, enzymatic activity and expression. Plant Mol Biol 32: 493–504

Braaten D, Ansari H and Luban J (1997) The hydrophobic pocket of cyclophilin is the binding site for the human immunodeficiency virus type 1 Gag polyprotein. J Virol 71: 2107–2113

Brandts JF, Halvorson HR and Brennan M (1975) Consideration of the Possibility that the slow step in protein denaturation reactions is due to *cis-trans* isomerism of proline residues. Biochemistry 14: 4953–4963

Breiman A, Fawcett TW, Ghirardi ML and Mattoo AK (1992) Plant organelles contain distinct peptidylprolyl *cis,trans*-isomerases. J Biol Chem 267: 21293–21296

Buchholz WG, Harris-Haller L, DeRose RT and Hall TC (1994) Cyclophilins are encoded by a small gene family in rice. Plant Mol Biol 25: 837–843

Chen H, Li XL and Ljungdahl LG (1995) A cyclophilin from the polycentric anaerobic rumen fungus *Orpinomyces* sp. strain PC-2 is highly homologous to vertebrate cyclophilin B. Proc Natl Acad Sci USA 92: 2587–2591

Chou IT and Gasser CS (1997) Characterization of the cyclophilin gene family of *Arabidopsis thaliana* and phylogenetic analysis of known cyclophilin proteins. Plant Mol Biol 35: 873–892

Cohen Y, Yalovsky S and Nechushtai R (1995) Integration and assembly of photosynthetic protein complexes in chloroplast thylakoid membranes. Biochim Biophys Acta 1241: 1–30

Connern CP and Halestrap AP (1992) Purification and N-terminal sequencing of peptidyl-prolyl *cis-trans*- isomerase from rat liver mitochondrial matrix reveals the existence of a distinct mitochondrial cyclophilin. Biochem J 284: 381–385

Crenshaw DG, Yang J, Means AR and Kornbluth S (1998) The mitotic peptidyl-prolyl isomerase, Pin1, interacts with Cdc25 and Plx1. EMBO J 17: 1315–1327

Dartigalongue C and Raina S (1998) A new heat-shock gene, ppiD, encodes a peptidyl-prolyl isomerase required for folding of outer membrane proteins in *Escherichia coli*. EMBO J 17: 3968–3980

Das AK, Cohen PW and Barford D (1998) The structure of the tetratricopeptide repeats of protein phosphatase 5: Implications for TPR-mediated protein-protein interactions. EMBO J 17: 1192–1199

Deng W, Chen L, Wood DW, Metcalfe T, Liang X, Gordon MP, Comai L and Nester EW (1998) Agrobacterium VirD2 protein interacts with plant host cyclophilins. Proc Natl Acad Sci USA 95: 7040–7045

Duina AA, Chang HC, Marsh JA, Lindquist S and Gaber RF (1996) A cyclophilin function in Hsp90-dependent signal transduction. Science 274: 1713–1715

Duina AA, Kalton HM and Gaber RF (1998a) Requirement for Hsp90 and a CyP-40-type cyclophilin in negative regulation of the heat shock response. J Biol Chem 273: 18974–18978

Duina AA, Marsh JA, Kurtz RB, Chang HC, Lindquist S and Gaber RF (1998b) The peptidyl-prolyl isomerase domain of the CyP-40 cyclophilin homolog Cpr7 is not required to support growth or glucocorticoid receptor activity in *Saccharomyces cerevisiae*. J Biol Chem 273: 10819–10822

Faure JD, Gingerich D and Howell SH (1998) An *Arabidopsis* immunophilin, AtFKBP12, binds to AtFIP37 (FKBP interacting protein) in an interaction that is disrupted by FK506. Plant J 15: 783–789

Ferreira PA, Nakayama TA, Pak WL and Travis GH (1996) Cyclophilin-related protein RanBP2 acts as chaperone for red/green opsin. Nature 383: 637–640

Ferreira PA, Nakayama TA and Travis GH (1997) Interconversion of red opsin isoforms by the cyclophilin-related chaperone protein Ran-binding protein 2. Proc Natl Acad Sci USA 94: 1556–1561

Ferreira PA, Yunfei C, Schick D and Roepman R (1998) The cyclophilin-like domain mediates the association of Ran-binding protein 2 with subunits of the 19 S regulatory complex of the proteasome. J Biol Chem 273: 24676–24682

Fischer G, Bang H and Mech H (1984) Detarmination of enzymatic catalysis of the *cis-trans* isomerisation of peptide bound in proline containing peptides. Biomed Biochem Acta 43: 1101–1111

Fischer G, Wittmann-Liebold B, Lang K, Kiefhaber T and Schmid FX (1989) Cyclophilin and peptidyl-prolyl *cis-trans* isomerase are probably identical proteins. Nature 337: 476–478

Fischer G, Tradler T and Zarnt T (1998) The mode of action of peptidyl prolyl *cis/trans* isomerases in vivo: Binding vs. catalysis. FEBS Lett 426: 17–20

Freeman BC, Toft DO and Morimoto RI (1996) Molecular chaperone machines: Chaperone activities of the cyclophilin Cyp-40 and the steroid aporeceptor-associated protein p23. Science 274: 1718–1720

Fulgosi H, Vener AV, Altschmied L, Herrmann RG and Andersson B (1998) A novel multi-functional chloroplast protein: Identification of a 40 kDa immunophilin-like protein located in the thylakoid lumen. EMBO J 17: 1577–1587

Gasser CS, Gunning DA, Budelier KA and Brown SM (1990) Structure and expression of cytosolic cyclophilin/peptidyl-prolyl *cis-trans* isomerase of higher plants and production of active tomato cyclophilin in Escherichia coli. Proc Natl Acad Sci USA 87: 9519–9523

Gothel SF and Marahiel MA (1999) Peptidyl-prolyl *cis-trans* isomerases, a superfamily of ubiquitous folding catalysts. Cell Mol Life Sci 55: 423–436

Halestrap AP, Connern CP, Griffiths EJ and Kerr PM (1997) Cyclosporin A binding to mitochondrial cyclophilin inhibits the permeability transition pore and protects hearts from ischaemia/reperfusion injury. Mol Cell Biochem 174: 167–172

Hani J, Schelbert B, Bernhardt A, Domdey H, Fischer G, Wiebauer K and Rahfeld JU (1999) Mutations in a peptidylprolyl-*cis/trans*-isomerase gene lead to a defect in 3′-end formation of a pre-mRNA in *Saccharomyces cerevisiae*. J Biol Chem 274: 108–116

Havaux M and Tardy F (1996) Temperature-dependent adjustment of the thermal stability of Photosystem II in vivo: Possible involvement of xanthophyll-cycle pigments. Planta 198: 324–333

Hennig L, Christner C, Kipping M, Schelbert B, Rucknagel KP, Grabley S, Kullertz G and Fischer G (1998) Selective inactivation of parvulin-like peptidyl-prolyl *cis/trans* isomerases by juglone. Biochemistry 37: 5953–5960

Hesterkamp T, Hauser S, Lutcke H and Bukau B (1996) *Escherichia coli* trigger factor is a prolyl isomerase that associates with nascent polypeptide chains. Proc Natl Acad Sci USA 93: 4437–4441

Hoerauf A, Rascher C, Bang R, Pahl A, Solbach W, Brune K, Rollinghoff M and Bang H (1997) Host-cell cyclophilin is important for the intracellular replication of Leishmania major. Mol Microbiol 24: 421–429

Hueros G, Rahfeld J, Salamini F and Thompson R (1998) A maize FK506-sensitive immunophilin, mzFKBP-66, is a peptidylproline *cis-trans*-isomerase that interacts with calmodulin and a 36-kDa cytoplasmic protein. Planta 205: 121–131

Hunter T (1998) Prolyl isomerases and nuclear function. Cell 92: 141–143

Jackson SE and Fersht AR (1991) Folding of chymotrypsin inhibitor 2. 2. Influence of proline isomerization on the folding kinetics and thermodynamic characterization of the transition state of folding. Biochemistry 30: 10436–10443

Janowski B, Wollner S, Schutkowski M and Fischer G (1997) A protease-free assay for peptidyl prolyl *cis/trans* isomerases using standard peptide substrates. Anal Biochem 252: 299–307

Kieselbach T, Hagman, Andersson B and Schroder WP (1998) The thylakoid lumen of chloroplasts. Isolation and characterization. J Biol Chem 273: 6710–6716

Kinoshita T and Shimazaki K (1999) Characterization of cytosolic cyclophilin from guard cells of Vicia faba L. Plant Cell Physiol 40: 53–59

Kofron JL, Kuzmic P, Kishore V, Colon-Bonilla E and Rich DH (1991) Determination of kinetic constants for peptidyl prolyl *cis-trans* isomerases by an improved spectrophotometric assay. Biochemistry 30: 6127–6134

Koide S, Dyson HJ and Wright PE (1993) Characterization of a folding intermediate of apoplastocyanin trapped by proline isomerization. Biochemistry 32: 12299–12310

Kops O, Eckerskorn C, Hottenrott S, Fischer G, Mi H and Tropschug M (1998) Ssp1, a site-specific parvulin homolog from *Neurospora crassa* active in protein folding. J Biol Chem 273: 31971–31976

Kullertz G, Liebau A, Rucknagel P, Schierhorn A, Diettrich B, Fischer G and Luckner M (1999) Stress-induced expression of cyclophilins in proembryonic masses of *Digitalis lanata* does not protect against freezing/thawing stress. Planta 208: 599–605

Kurek I, Aviezer K, Erel N, Herman E and Breiman A (1999) The wheat peptidyl prolyl *cis-trans*-isomerase FKBP77 is heat induced and developmentally regulated. Plant Physiol 119: 693–704

Lazar SW and Kolter R (1996) SurA assists the folding of *Escherichia coli* outer membrane proteins. J Bacteriol 178: 1770–1773

Lippuner V, Chou IT, Scott SV, Ettinger WF, Theg SM and Gasser CS (1994) Cloning and characterization of chloroplast and cytosolic forms of cyclophilin from *Arabidopsis thaliana*. J Biol Chem 269: 7863–7868

Liu J and Walsh CT (1990) Peptidyl-prolyl *cis-trans*-isomerase from *Escherichia coli*: A periplasmic homolog of cyclophilin that is not inhibited by cyclosporin A. Proc Natl Acad Sci USA 87: 4028–4032

Liu J, Farmer JD, Jr., Lane WS, Friedman J, Weissman I and Schreiber SL (1991) Calcineurin is a common target of cyclophilin-cyclosporin A and FKBP- FK506 complexes. Cell 66: 807–815

Lopez-Ilasaca M, Schiene C, Kullertz G, Tradler T, Fischer G and Wetzker R (1998) Effects of FK506-binding protein 12 and FK506 on autophosphorylation of epidermal growth factor receptor. J Biol Chem 273: 9430–9434

Lu KP, Hanes SD and Hunter T (1996) A human peptidyl-prolyl isomerase essential for regulation of mitosis. Nature 380: 544–547

Lu PJ, Wulf G, Zhou XZ, Davies P and Lu KP (1999a) The prolyl isomerase Pin1 restores the function of Alzheimer-associated phosphorylated tau protein. Nature 399: 784–788

Lu PJ, Zhou XZ, Shen M and Lu KP (1999b) Function of WW domains as phosphoserine- or phosphothreonine-binding modules. Science 283: 1325–1328

Luan S, Li W, Rusnak F, Assmann SM and Schreiber SL (1993) Immunosuppressants implicate protein phosphatase regulation of K+ channels in guard cells. Proc Natl Acad Sci USA 90: 2202–2206

Luan S, Albers MW and Schreiber SL (1994a) Light-regulated, tissue-specific immunophilins in a higher plant. Proc Natl Acad Sci USA 91: 984–988

Luan S, Lane WS and Schreiber SL (1994b) pCyP B: A chloroplast-localized, heat shock-responsive cyclophilin from fava bean. Plant Cell 6: 885–892

Luan S, Kudla J, Gruissem W and Schreiber SL (1996) Molecular characterization of a FKBP-type immunophilin from higher plants. Proc Natl Acad Sci USA 93: 6964–6969

Lupas A (1996) Coiled coils: New structures and new functions. Trends Biochem Sci 21: 375–382

Maleszka R, Lupas A, Hanes SD and Miklos GL (1997) The dodo gene family encodes a novel protein involved in signal transduction and protein folding. Gene 203: 89–93

Marivet J, Margis-Pinheiro M, Frendo P and Burkard G (1994) Bean cyclophilin gene expression during plant development and stress conditions. Plant Mol Biol 26: 1181–1189

Marivet J, Frendo P and Burkard G (1995) DNA sequence analysis of a cyclophilin gene from maize: Developmental expression and regulation by salicylic acid. Mol Gen Genet 247: 222–228

Marks AR (1996) Cellular functions of immunophilins. Physiol Rev 76: 631–649

Marks AR (1997) Intracellular calcium-release channels: Regulators of cell life and death. Am J Physiol 272: H597–605

Matouschek A, Rospert S, Schmid K, Glick BS and Schatz G (1995) Cyclophilin catalyzes protein folding in yeast mitochondria. Proc Natl Acad Sci USA 92: 6319–6323

Mattoo AK (1998) Peptidylprolyl *cis-trans*-isomerases from plant organelles. Methods Enzymol 290: 84–100

Mi H, Kops O, Zimmermann E, Jaschke A and Tropschug M (1996) A nuclear RNA-binding cyclophilin in human T cells. FEBS Lett 398: 201–205

Montague JW, Hughes FM, Jr. and Cidlowski JA (1997) Native recombinant cyclophilins A, B, and C degrade DNA independently of peptidylprolyl *cis-trans*-isomerase activity. Potential roles of cyclophilins in apoptosis. J Biol Chem 272: 6677–6684

Morris DP, Phatnani HP and Greenleaf AL (1999) Phospho-carboxyl-terminal domain binding and the role of a prolyl isomerase in pre-mRNA 3′-End formation. J Biol Chem 274: 31583–31587

Nicolli A, Basso E, Petronilli V, Wenger RM and Bernardi P (1996) Interactions of cyclophilin with the mitochondrial inner membrane and regulation of the permeability transition pore, and cyclosporin A-sensitive channel. J Biol Chem 271: 2185–2192

Owens-Grillo JK, Stancato LF, Hoffmann K, Pratt WB and Krishna P (1996) Binding of immunophilins to the 90 kDa heat shock protein (hsp90) via a tetratricopeptide repeat domain is a conserved protein interaction in plants. Biochemistry 35: 15249–15255

Pahl A, Brune K and Bang H (1997) Fit for life? Evolution of chaperones and folding catalysts parallels the development of complex organisms. Cell Stress Chaperones 2: 78–86

Peltier JB, Friso G, Kalume DE, Roepstorff P, Nilsson F, Adamska I and van Wijk KJ (2000) Proteomics of the chloroplast: Systematic identification and targeting analysis of lumenal and peripheral thylakoid proteins. Plant Cell 12: 319–341

Pissavin C and Hugouvieux-Cotte-Pattat N (1997) Characterization of a periplasmic peptidyl-prolyl *cis-trans* isomerase in *Erwinia chrysanthemi*. FEMS Microbiol Lett 157: 59–65

Pratt WB (1998) The hsp90-based chaperone system: Involvement in signal transduction from a variety of hormone and growth factor receptors. Proc Soc Exp Biol Med 217: 420–434

Pratt WB, Gehring U and Toft DO (1996) Molecular chaperoning of steroid hormone receptors. Exs 77: 79–95

Rahfeld JU, Rucknagel KP, Schelbert B, Ludwig B, Hacker J, Mann K and Fischer G (1994) Confirmation of the existence of a third family among peptidyl-prolyl *cis/trans* isomerases. Amino acid sequence and recombinant production of parvulin. FEBS Lett 352: 180–184

Ranganathan R, Lu KP, Hunter T and Noel JP (1997) Structural and functional analysis of the mitotic rotamase Pin1 suggests substrate recognition is phosphorylation dependent. Cell 89: 875–886

Rassow J, Mohrs K, Koidl S, Barthelmess IB, Pfanner N and Tropschug M (1995) Cyclophilin 20 is involved in mitochondrial protein folding in cooperation with molecular chaperones Hsp70 and Hsp60. Mol Cell Biol 15: 2654–2662

Reddy RK, Kurek I, Silverstein AM, Chinkers M, Breiman A and Krishna P (1998) High-molecular-weight FK506-binding proteins are components of heat- shock protein 90 hetero-complexes in wheat germ lysate. Plant Physiol 118: 1395–1401

Rokka A, Aro E-M, Herrmann RG, Andersson B and Vener AV (1998) Heat-induced stimulation of Photosystem II dephosphorylation - possible involvement of TLP40. In: Garab G (ed) Photosynthesis: Mechanisms and Effects. Vol IV, pp 2453–2456, Kluwer Academic Publishers, Dordrecht

Rokka A, Aro E-M, Herrmann RG, Andersson B and Vener AV (2000) Dephosphorylation of Photosystem II reaction center proteins as an immediate heat-shock response in plant photosynthetic membrane. Plant Physiol 123: 1525–1535

Ruediger R, Hentz M, Fait J, Mumby M and Walter G (1994) Molecular model of the A subunit of protein phosphatase 2A: Interaction with other subunits and tumor antigens. J Virol 68: 123–129

Saito T, Niwa Y, Ashida H, Tanaka K, Kawamukai M, Matsuda H and Nakagawa T (1999a) Expression of a gene for cyclophilin which contains an amino-terminal endoplasmic reticulum-targeting signal. Plant Cell Physiol 40: 77–87

Saito T, Tadakuma K, Takahashi N, Ashida H, Tanaka K, Kawamukai M, Matsuda H and Nakagawa T (1999b) Two cytosolic cyclophilin genes of *Arabidopsis thaliana* differently regulated in temporal- and organ-specific expression. Biosci Biotechnol Biochem 63: 632–637

Schiene C, Reimer U, Schutkowski M and Fischer G (1998) Mapping the stereospecificity of peptidyl prolyl *cis/trans* isomerases. FEBS Lett 432: 202–206

Schmid FX and Baldwin RL (1978) Acid catalysis of the formation of the slow-folding species of RNase A: Evidence that the reaction is proline isomerization. Proc Natl Acad Sci USA 75: 4764–4768

Schneuwly S, Shortridge RD, Larrivee DC, Ono T, Ozaki M and Pak WL (1989) Drosophila ninaA gene encodes an eye-specific cyclophilin (cyclosporine A binding protein). Proc Natl Acad Sci USA 86: 5390–5394

Scholz C, Rahfeld J, Fischer G and Schmid FX (1997a) Catalysis of protein folding by parvulin. J Mol Biol 273: 752–762

Scholz C, Stoller G, Zarnt T, Fischer G and Schmid FX (1997b) Cooperation of enzymatic and chaperone functions of trigger factor in the catalysis of protein folding. EMBO J 16: 54–58

Scholz C, Scherer G, Mayr LM, Schindler T, Fischer G and Schmid FX (1998) Prolyl isomerases do not catalyze isomerization of non-prolyl peptide bonds. Biol Chem 379: 361–365

Schreiber SL (1991) Chemistry and biology of the immunophilins and their immunosuppressive ligands. Science 251: 283–287

Schreiber SL (1992) Immunophilin-sensitive protein phosphatase action in cell signaling pathways. Cell 70: 365–368

Schutkowski M, Bernhardt A, Zhou XZ, Shen M, Reimer U, Rahfeld JU, Lu KP and Fischer G (1998) Role of phosphorylation in determining the backbone dynamics of the serine/threonine-proline motif and Pin1 substrate recognition. Biochemistry 37: 5566–5575

Scorrano L, Nicolli A, Basso E, Petronilli V and Bernardi P (1997) Two modes of activation of the permeability transition pore: The role of mitochondrial cyclophilin. Mol Cell Biochem 174: 181–184

Sheldon PS and Venis MA (1996) Purification and characterization of cytosolic and microsomal cyclophilins from maize (*Zea mays*). Biochem J 315: 965–970

Shen M, Stukenberg PT, Kirschner MW and Lu KP (1998) The essential mitotic peptidyl-prolyl isomerase Pin1 binds and regulates mitosis-specific phosphoproteins. Genes Dev 12: 706–720

Shou W, Aghdasi B, Armstrong DL, Guo Q, Bao S, Charng MJ, Mathews LM, Schneider MD, Hamilton SL and Matzuk MM (1998) Cardiac defects and altered ryanodine receptor function in mice lacking FKBP12. Nature 391: 489–492

Stoller G, Rucknagel KP, Nierhaus KH, Schmid FX, Fischer G and Rahfeld JU (1995) A ribosome-associated peptidyl-prolyl *cis/trans* isomerase identified as the trigger factor. EMBO J 14: 4939–4948

van Wijk KJ, Andersson B and Aro E-M (1996) Kinetic resolution of the incorporation of the D1 protein into Photosystem II and localization of assembly intermediates in thylakoid membranes of spinach chloroplasts. J Biol Chem 271: 9627–9636

Vener AV, Van Kan PJM, Rich PR, Ohad I and Andersson B (1997) Plastoquinol at the quinol oxidation site of reduced cytochrome bf mediates signal transduction between light and protein phosphorylation: Thylakoid protein kinase deactivation by a single-turnover flash. Proc Natl Acad Sci USA 94: 1585–1590

Vener AV, Ohad I and Andersson B (1998) Protein phosphorylation and redox sensing in chloroplast thylakoids. Curr Opin Plant Biol 1: 217–223

Vener AV, Rokka A, Fulgosi H, Andersson B and Herrmann RG (1999) A Cyclophilin-Regulated PP2A-like Protein Phosphatase in Thylakoid Membranes of Plant Chloroplasts. Biochemistry 38: 14955–14965

Vittorioso P, Cowling R, Faure JD, Caboche M and Bellini C (1998) Mutation in the *Arabidopsis* PASTICCINO1 gene, which encodes a new FK506- binding protein-like protein, has a dramatic effect on plant development. Mol Cell Biol 18: 3034–3043

Vucich VA and Gasser CS (1996) Novel structure of a high molecular weight FK506 binding protein from *Arabidopsis thaliana*. Mol Gen Genet 252: 510–517

Walsh CT, Zydowsky LD and McKeon FD (1992) Cyclosporin A, the cyclophilin class of peptidylprolyl isomerases, and blockade of T cell signal transduction. J Biol Chem 267: 13115–13118

Wang G, Yang J and Huibregtse JM (1999) Functional domains of the Rsp5 ubiquitin-protein ligase. Mol Cell Biol 19: 342–352

Weis E and Berry JA (1988) Plants and high temperature stress. Symp Soc Exp Biol 42: 329–346

Xu Q, Liang S, Kudla J and Luan S (1998) Molecular characterization of a plant FKBP12 that does not mediate action of FK506 and rapamycin. Plant J 15: 511–519

Yaffe MB, Schutkowski M, Shen M, Zhou XZ, Stukenberg PT, Rahfeld JU, Xu J, Kuang J, Kirschner MW, Fischer G, Cantley LC and Lu KP (1997) Sequence-specific and phosphorylation-dependent proline isomerization: A potential mitotic regulatory mechanism. Science 278: 1957–1660

Yin L, Braaten D and Luban J (1998) Human immunodeficiency virus type 1 replication is modulated by host cyclophilin A expression levels. J Virol 72: 6430–6436

Zarnt T, Tradler T, Stoller G, Scholz C, Schmid FX and Fischer G (1997) Modular structure of the trigger factor required for high activity in protein folding. J Mol Biol 271: 827–837

Chapter 11

Role of the Plastid Envelope in the Biogenesis of Chloroplast Lipids

Maryse A. Block*, Eric Maréchal and Jacques Joyard
Laboratoire de Physiologie Cellulaire Végétale,
UMR 5019 CNRS/CEA/Université J. Fourier,DBMS/PCV, CEA-Grenoble,
38054 Grenoble CEDEX 9, France

Summary

The plastid envelope is the primary site of production of a diverse array of specific lipidic molecules that are essential to build up chloroplast membranes. A survey of the precise localization of galactolipids, MGDG and DGDG, sulfolipids, chlorophylls, carotenoids and prenylquinones in the thylakoids and chloroplast envelope membranes is presented. Analyses of membrane subfractions and of mutants deleted in particular lipids indicate some specific association of lipid and protein in the membranes. Molecular identification of major enzymes allows a better definition of the metabolic pathways that lead to the biosynthesis of these lipids. For MGDG synthesis and for DGDG synthesis recent results point to the fact that several pathways exist. Furthermore, whereas synthesis of prokaryotic glycerolipids (C18/C16) is entirely realized in the chloroplast envelope, synthesis of chloroplast eukaryotic glycerolipids (C18/C18) probably relies on the transfer of precursors between the ER and the outer envelope membrane. The molecular species to be transported and more generally

*Author for correspondence, email: mblock@cea.fr

Eva-Mari Aro and Bertil Andersson (eds): Regulation of Photosynthesis, pp. 195–218.

the mode of transport of lipids necessary to generate chloroplast membranes are mostly unknown and therefore briefly discussed. Recent localization of some enzymes involved in chlorophyll synthesis indicates a highly compartmentalized process interacting with the envelope. Several steps are catalyzed by membrane associated enzymes that have not yet been fully characterized and localized. Due to the molecular species they manipulate they could delineate important metabolic regulation steps. Recent investigations show that enzymes influencing plastid terpenoid chains such as CHL P providing phytol for both α-tocopherol and chlorophyll synthesis could also have significant regulatory roles. Eventually, two types of post-translational protein prenylation have been detected in chloroplasts, including a non-classical prenylation type confined to chloroplasts depending on plastid gene expression and not involving a thioether bond. They could provide an important way of metabolic regulation.

I. Introduction

The chloroplast membrane system is made up of a limiting two-membrane envelope and a sophisticated network of thylakoid membranes devoted to the photosynthetic process including light harvesting. Biogenesis of the chloroplast membranes requires biosynthesis of a whole set of specific lipids. These lipids are on one hand glycerolipids necessary to constitute the membrane bulk matrix structure. On the other hand, several specific lipophylic compounds such as chlorophylls, carotenoids and prenylquinones are essential for the functioning of the photosynthetic apparatus. An overview of the complex metabolic pathways that lead to the biosynthesis of these compounds emphasizes the striking importance of the envelope membranes (Douce et al., 1984).

Chloroplast membranes are essentially enriched in glycolipids (galactolipids and sulfolipids) in contrast to other plant cell membranes in which phospholipids are the major lipid constituents. Fatty acids synthesized in the plastid stroma are either directly metabolized into glycerolipids within the envelope or exported across the envelope to the endoplasmic reticulum and incorporated into phospholipids, especially into phosphatidylcholine. Part of the glycolipids necessary for chloroplast membranes structure seems to derive from these PC molecules and their synthesis requires trafficking of lipid moieties between the ER and the envelope. The final galactosylation of DAG is realized within the chloroplast envelope. Recent works indicate that multiple isoenzymes are involved in galactolipid synthesis (Dörmann et al., 1999; Miège et al., 1999).

Chlorophyll synthesis occurs as a compartmentalized process in chloroplast that involves some highly regulated multisubunit complexes. Several results clearly indicate that although devoid of chlorophyll, the plastid envelope is implicated in both chlorophyll synthesis and degradation (Pineau et al., 1986; Joyard et al., 1990; Matringe et al., 1991; Nakayama et al., 1998; Matile et al., 1999).

Chloroplast prenylquinones, plastoquinone-9 and phylloquinone, are essential compounds for photosynthesis as electron carriers whereas α-tocopherol is believed to assure protection of chloroplast membranes from oxidation. Their synthesis involves a series of enzymes probably located in the inner envelope membranes.

This chapter summarizes our current understanding of the role of the plastid envelope in chloroplast lipid biogenesis taking into account recent enzyme identification and mutant analyses. We will focus on where specific molecular species of lipids are needed in the chloroplasts and how their synthesis can be coupled to the transport to their final destination. Necessary interaction between lipids and proteins in membranes may be favored by covalent modification of proteins by lipids. Therefore we will review the new insights on lipid modification of proteins in chloroplasts.

Abbreviations: ABA – abscissic acid; ACP – acyl carrier protein; CHL P – geranylgeranyl reductase; Chl – chlorophyll; Chlide – chlorophyllide; DAG – diacylglycerol; DGD – DGDG synthase; DGDG – digalactosyldiacylglycerol; ER – endoplasmic reticulum; GC-MS – gas chromatography-mass spectrometry; GGGT – galactolipid-galactolipid galactosyltransferase; GGPP – geranygeranyl pyrophosphate; HHPDase – hydroxyphenylpyruvate dioxygenase; LysoPC – lysophosphatidylcholine; MGD – MGDG synthase; MGDG – monogalactosyldiacylglycerol; PA – phosphatidic acid; PC – phosphatidylcholine; PChlide – protochlorophyllide; PE – phosphatidylethanolamine; PG – phosphatidylglycerol; PhytylPP – phytyl pyrophosphate; PI – phosphatidylinositol; POR – protochlorophyllide oxidoreductase; SQD – enzyme involved in sulfolipid synthesis; SQDG – sulfoquinovosyldiacylglycerol; UDP-gal – uridine diphosphate galactose

II. Lipid Composition of Chloroplast Membranes

A. Glycerolipids

All plastid membranes are characterized by the presence of large amounts of galactolipids: monogalactosyldiacylglycerol (MGDG) and digalactosyldiacylglycerol (DGDG) containing one or two galactose molecules attached to the *sn*-3 position of the glycerol backbone respectively (Table 1). Galactolipids represent up to 80% of thylakoid membrane glycerolipids, out of which MGDG constitutes the main part (50%). The galactolipids have the same structure in both thylakoids and envelope membranes (Siebertz et al., 1979). MGDG contains a large proportion of polyunsaturated fatty acids; in some species up to 95% of the total fatty acid is linolenic acid (18:3). However, this value is an average and does not point out some local MGDG compositions. For instance, a highly saturated MGDG molecule has been found to be specifically associated with the PS II reaction center complex (Murata et al., 1990). The most abundant molecular species of MGDG have, like PC and most eukaryotic lipids, C18 fatty acids at both *sn*-1 and *sn*-2 positions of glycerol backbone (Heinz, 1977). This structure with C18 at *sn*-2 position is referred to as an eukaryotic structure. Plants such as pea and cucumber, having almost only C18:3 in MGDG are called '18:3 plants'. In '16:3 plants,' such as spinach or *Arabidopsis*, part

Table 1: Glycerolipid, pigment and prenylquinone composition of spinach chloroplast membranes. Reverences for original data are reported by Douce and Joyard (1996).

	outer envelope membrane	inner envelope membrane	total envelope membranes	thylakoids
Total polar lipids (mg /mg protein)	2.5–3	1	1.2–1.5	0.6–0.8
Polar lipids (% of total)				
MGDG	17	55	32	57
DGDG	29	29	30	27
SQDG	6	5	6	7
PC	32	0	20	0
PG	10	9	9	7
PI	5	1	4	1
PE	0	0	0	0
Total Chlorophylls (μg /mg protein)			0.1–0.3	160
Chlorophylls (% of total)				
Chl *a*			86	72
Chl *b*			14	28
Chlorophyll precursors (Protochlorophyllide + Chlorophyllide, μg /mg protein)			0.41	0–0.35
Total Carotenoids (μg /mg protein)	2.9	7.2	6–12	20
Carotenoids (% of total)				
β-Carotene	9	12	11	25
Violaxanthine	49	47	48	22
Luteine + Zeaxanthine	16	23	21	37
Antheraxanthine	–	5	6	–
Neoxanthine	26	13	13	16
Total Prenylquinones (μg /mg protein)	4–12	4–11	4–11	4–7
Prenylquinones (% of total)				
α-Tocopherol + α-Tocoquinone	81	67	69	24
Plastoquinone-9 + Plastoquinol	18	32	28	70
Phylloquinone K1	1	1	3	6

of the MGDG molecular species contain C16 fatty acids at the *sn*-2 position instead of C18 fatty acids (Heinz, 1977). This structure is referred to as a prokaryotic structure because it is characteristic for cyanobacterial glycerolipids. Among Angiosperms, dicots are either 16:3 or 18:3 plants whereas monocots are mostly 18:3 plants (Fig. 1) suggesting that loss of the prokaryotic pathway occurred several times independently and at different rates (Mongrand et al., 1998).

The ratio of eukaryotic to prokaryotic molecular species is not identical in all glycerolipids of a given plant. For instance, although half of the spinach MGDG has a prokaryotic structure, this holds true for only 10–15% of DGDG (Bishop et al., 1985). In addition, C16 fatty acids in DGDG are more saturated than in MGDG (Heemskerk et al., 1990). This discrepancy between DGDG molecular species composition and that of MGDG, from which DGDG is supposed to be derived, is puzzling. This suggests that some MGDG molecular species are selectively used for DGDG synthesis. In addition, in the outer envelope membrane DGDG is present in a higher proportion than MGDG while it is the opposite in the inner envelope or in the thylakoids stressing another point of regulation involved in the processes for constituting galactolipid composition. Indications of DGDG functions are found in the thylakoids where DGDG has been shown to be important for the structure of LHC II (Nussberger et al., 1993).

The most important sulfolipid found in higher plants is 1′,2′-diacyl-3′O-(6-deoxy-6-sulfo-α-D-glucopyranosyl)-*sn*-glycerol (sulfoquinovosyl-DAG or SQDG). The sulfonic residue carries a negative charge at physiological pH. SQDG is present in both the envelope and thylakoid membranes but in rather low proportions (7–8% of total lipids). Like galactolipids, this glycolipid is present in the outer envelope membrane since it is accessible to specific antibodies on the cytosolic face of isolated intact chloroplasts (Billecocq et al., 1972). Furthermore, SQDG was proposed to interact with an annexin in a Ca^{2+} dependant manner on the outer surface of chloroplast (Seigneurin-Berny et al., 2000). Lipid analysis of chlorophyll-protein complexes in *C. reinhardtii* have shown than SQDG is present in the PS II core complex and in LHCII but absent from PS I (Sato et al., 1995). In a '16:3 plant' such as spinach, a higher proportion of SQDG has a prokaryotic structure as compared to MGDG, whereas in wheat, a 18:3 plant, SQDG is exclusively eukaryotic (Bishop et al., 1985). In addition, high proportions of 16:0/16:0 molecular species have been reported in spinach. SQDG contains a higher level of C16 fatty acid than either MGDG or DGDG and this fatty acid is fully saturated.

The chloroplast membranes have a low phospholipid content. They are even devoid of phosphatidylethanolamine, a major component of extraplastidial membranes. The main phospholipid in the inner envelope membrane and in the thylakoids is phosphatidylglycerol, which is an anionic lipid like SQDG. Chloroplast PG is unique because it has a prokaryotic structure and moreover it contains a unique $16{:}1_{trans}$ fatty acid at the *sn*-2 position of the glycerol backbone (Dubacq and Trémolières, 1983). Because of this uniqueness, the specific localization and function of PG in chloroplasts have been investigated. Major results indicate that PG is directly involved in the formation of trimers in LHCII (Nussberger et al., 1993) and that desaturation of PG is important for the turn over of the D1 protein at low temperature (Moon et al., 1995, see below).

Phosphatidylcholine (PC) is only a minor constituent of plastid membranes. In chloroplasts, it is concentrated in the outer leaflet of the outer envelope membrane and is absent from the inner envelope membrane and the thylakoids (Dorne et al., 1990).

B. Chlorophylls

Chlorophylls are the most conspicuous chloroplast pigments. Within thylakoids, these Mg-tetrapyrroles are bound to proteins associated with light-harvesting and reaction center complexes. In higher plants, Chl *a* occurs in the reaction centers and core complexes of photosystems whereas Chl *a* and Chl *b* along with carotenoids are components of peripheral light harvesting antennae. Some Chl *a* derivatives such as pheophytin *a* (Mg-free Chl *a*) are also present in the reaction centers. In the PS II reaction center, P680 complex is considered to act as the primary electron donor while pheophytin *a* is the primary electron acceptor. The overall arrangement of two molecules of Chl *a* in P680 is apparently one of the reasons why P680 has a higher redox potential than other primary donors (Rhee et al., 1998). Similarly, in the PS I reaction center two Chl *a* are essential functional components of the electron donor part, P700, while a monomeric Chl *a* is the initial electron acceptor (Malkin, 1996). In contrast to the thylakoids,

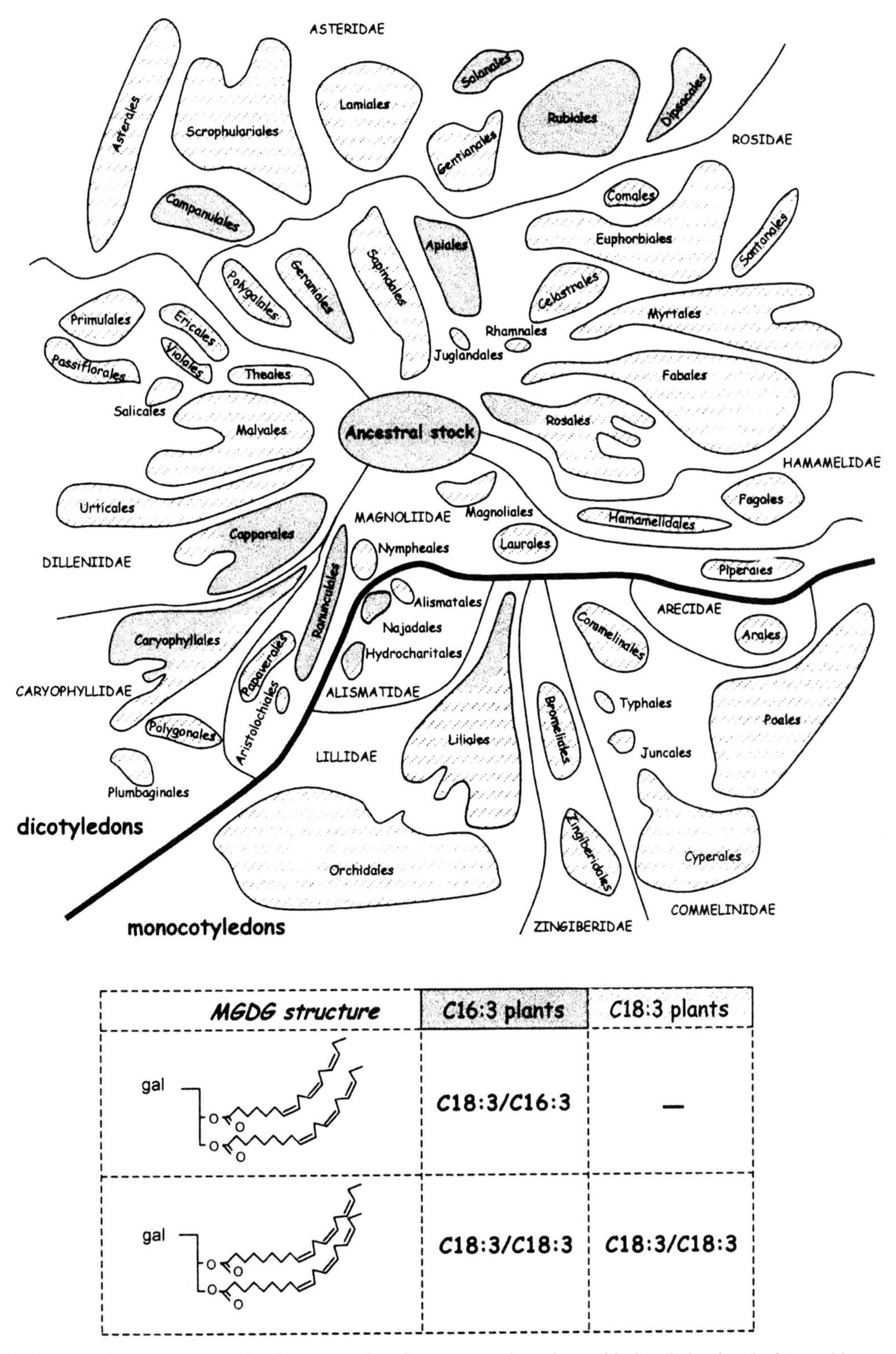

Fig. 1: Evolutionary diagram of most Angiosperm orders from ancestral stock considering their trienoic fatty acid composition in photosynthetic tissues according to Mongrand (1998). Surface is indicative of the amount of species in the order and distance from the ancestral stock is indicative of the evolution degree. 468 plant species were analyzed, using *cis* 7,10,13-hexadecatrienoic acid as a marker for the existence of the plastidial pathway. This figure is adapted from Mongrand (1998).

the envelope membranes are almost devoid of chlorophyll (Table 1). Only traces of Chl *a* and Chl *b* can be detected in envelope membrane preparations but the Chl *a*/Chl *b* ratio is higher than in the thylakoids indicating that these traces are not originating from contamination by small thylakoids fragments (Pineau et al., 1993). It was proposed that this small amount of Chl *a*/Chl *b* in the envelope represents a biosynthetic pool as envelope membranes contain also a significant concentration of chlorophyll precursors: protochlorophyllide and chlorophyllide (0.1 to 1.5 nmol/mg protein) (Pineau et al., 1986).

C. Carotenoids

Carotenoids are present in all plastid membranes and most likely restricted to plastids. In thylakoids, β-carotene is present in LHC and in the vicinity of the PS II reaction center whereas xanthophylls e.g. lutein, violaxanthin, neoxanthin, antheraxanthin and zeaxanthin are mainly associated with LHC. Lutein extends from the stromal to the lumenal surface of the LHCII monomer forming a cross brace between protein helices. These pigments contribute to the light harvesting and have important functions in dissipation of excited states and protection against photooxidation (for review see, Yamamoto and Bassi, 1996). Surprisingly, the envelope membranes of chloroplasts also contain carotenoids. In fact, carotenoids are present in both envelope membranes of plastid and violaxanthin is always the main carotenoid in the envelope (Joyard et al., 1990). The physiological significance of envelope carotenoids is not clear. Besides prevention of oxidative damage, a role in ABA synthesis is also considered. Relationship between carotenoid synthesis and ABA synthesis was demonstrated by analysis of a *Nicotiana plumbaginifolia* mutant. This mutant is deficient in ABA and is impaired in formation of violaxanthin and antheraxanthin due to a transposon insertion in a zeaxanthin epoxidase gene (Marin et al., 1996). The first enzyme specifically committed to ABA synthesis is a plastid 9-*cis*-epoxycarotenoid dioxygenase that cleaves violaxanthin or a 9-*cis* epoxycarotenoid to form xanthoxin (Cutler and Krochko, 1999). Finally, although various enzymes involved in carotenoid synthesis are present in the envelope, it is generally assumed that in the chloroplasts, the thylakoids play a central role in their own carotenoid synthesis (Britton, 1988).

D. Prenylquinones

Plants and especially photosynthetic tissues contain a wide variety and high amounts of plastid specific prenylquinones (Lichtenthaler, 1993), such as plastoquinone-9, α-tocopherol, α-tocoquinone and phylloquinone, as well as an inner mitochondrial membrane constituent, ubiquinone. In thylakoids, plastoquinone-9 is involved in the electron transfer between PS II reaction center and the cytochrome b_6f complex. Plastoquinone-9 is tightly bound to the D2 protein and constitutes the Q_A electron acceptor. Reduced Q_A subsequently transfers electron to Q_B, a mobile plastoquinone loosely bound to the D1 protein. Following double reduction and protonation of Q_B, the hydroquinone becomes free and is replaced in the Q_B pocket by a quinone from the membrane plastoquinone pool (for review see Wolfe and Hoober, 1996). Phylloquinone is associated with PS I and is involved in electron transfer to ferredoxin. Both envelope membranes contain plastoquinone-9 and phylloquinone. The role of these prenylquinones in the envelope membranes is unknown although they are presumed to act also as electron carriers (Jäger-Vottero et al., 1997). α-tocopherol is present in all chloroplast membranes but its concentration is higher in the envelope than in the thylakoids. Their best characterized function in biological membranes is to protect the membrane from free radicals generated by lipid peroxidation of polyunsaturated fatty acids (Grusak and DellaPenna, 1999). Plastoquinol also could be an antioxidant (Hundal et al., 1995). An antioxidative protection appears necessary in chloroplast membranes due to the presence of high level of unsaturated fatty acids. Importance of electron transfer reactions in plastid membranes, mostly in thylakoids but also in the envelope (Jäger-Vottero et al., 1997) and specific fatty acid peroxidation in the envelope (Blée and Joyard, 1996) may also explain the abundance of α-tocopherol in chloroplast membranes.

III. Biosynthesis of Chloroplast Glycerolipids

Plastid envelope membranes play a central role in the biosynthesis of glycerolipids since they are the site of assembly of fatty acids, glycerol and polar head groups. However, in vitro kinetics of acetate incorporation into chloroplast lipids have demonstrated that isolated chloroplasts can synthesize glycerolipids containing almost exclusively a C18/C16 diacylglycerol (DAG) backbone, but are apparently unable to catalyze the formation of phosphatidic acid and DAG containing only C18 fatty acids. Below we discuss that the building up of eukaryotic and prokaryotic structures proceeds from at least two distinct pathways, both involving envelope membranes.

A. The Fate of Fatty Acids Synthesized in the Plastid Stroma

The major site of synthesis of fatty acids in the plant cell is the plastid stroma. Fatty acids are produced predominantly as C18:1-ACP and C16:0-ACP. The critical importance of plastid fatty acids production for the overall structure is illustrated by the *fab2* mutation of *Arabidopsis*. The *fab2* mutant, characterized by a defective plastid 18:0-ACP desaturase, possesses a severe dwarf phenotype and many cell types in the mutant fail to expand (Lightner et al., 1994). Actually, recent investigations have shown that the plastid is not the unique site of fatty acid synthesis in plant cell. Plant mitochondria are able to synthesize C18 and C16 fatty acids although in a much lower rate than chloroplasts (Gueguen et al., 2000). Under normal conditions, fatty acids are either exported outside the plastids or locally incorporated into glycerolipids. An acyl-ACP thioesterase, localized in the inner envelope membrane, can hydrolyze acyl-ACP and release free fatty acids. On the outer membrane of the chloroplast envelope, an acyl-CoA synthetase assembles acyl-CoA thioesters (Joyard and Stumpf, 1981; Block et al., 1983) that are available outside the plastids to form glycerolipids within the ER. Several long chain acyl-ACP thioesterases have been identified in plants (Dörmann et al., 1995a). However, their subcellular localization and function are not yet clear. Concerning long chain acyl-CoA synthetases numerous sequences have been reported in *Arabidopsis*. Their potential subcellular localization and function within plant cells were discussed by Fulda et al. (1997).

B. Phosphatidic Acid Biosynthesis in the Envelope Membranes

As an alternative to export, fatty acid-ACPs remain in the plastid and are transferred to glycerol-3-phosphate or to lysophosphatidic acid to feed the plastid glycerolipid biosynthetic pathway (Fig. 2). The first enzyme of the pathway is a soluble glycerol-3-phosphate acyltransferase (E.C. 2.3.1.15), which produces lysophosphatidic acid. This enzyme is closely associated with the inner envelope membrane (Joyard and Douce, 1977) and catalyses the transfer of preferentially 18:1 from 18:1-ACP to exclusively the *sn*-1 position of glycerol (Frentzen et al., 1983). In fact, different isoforms of this enzyme have been reported in chloroplasts (Nishida et al., 1997). It has been suggested without any clear evidence that glycerol-3-phosphate acyltransferase isoforms have different specificity for acyl chains. The sequence for one of the *Arabidopsis* glycerol-3-phosphate acyltransferases was registered in Swiss-Prot under accession number Q 43307. LysoPA is further acylated to form PA by the action of a 1-acylglycerol-3-phosphate acyltransferase (E.C. 2.3.1.51) present in the envelope membranes of chloroplasts (Joyard and Douce, 1977) and non-green plastids (Alban et al., 1989). In spinach chloroplasts, both the outer and inner envelope membranes contain this acyltransferase activity (Block et al., 1983). Envelope fractions of spinach as well as pea direct almost exclusively 16:0 to the available *sn*-2 position (Frentzen et al., 1983). A plastidial form of this 16:0-specific enzyme has been cloned by functional complementation of an acyltransferase-deficient *E. coli* thermosensitive mutant (Genbank AAF73736) (Bourgis et al., 1999). This work points out that the enzyme precursor is imported into the chloroplasts where it is processed and that it is an integral membrane protein.

Finally, PA with 18:1 and 16:0 fatty acid at the *sn*-1 and *sn*-2 positions of glycerol respectively and a minor proportion of PA with 16:0 at both *sn* positions are synthesized in the chloroplasts. These structures are typical for prokaryotic glycerolipids. PA is further metabolized in the envelope into either PG or DAG.

C. Dual Origin of Diacylglycerol Molecular Species

A phosphatidate phosphatase located on the inner envelope membrane (Block et al., 1983) converts PA to DAG. The enzyme is clearly different from all other PA phosphatases since it is membrane bound, it exhibits a clear alkaline pH optimum and is inhibited by cations such as Mg^{2+} (Joyard and Douce, 1977, 1979). As Mg^{2+} concentration in chloroplasts varies from 1 to 6 mM (Leegood et al., 1985), the sensitivity of PA phosphatase to Mg^{2+} suggests a possible control of the enzyme. Furthermore, phosphatidate phos-

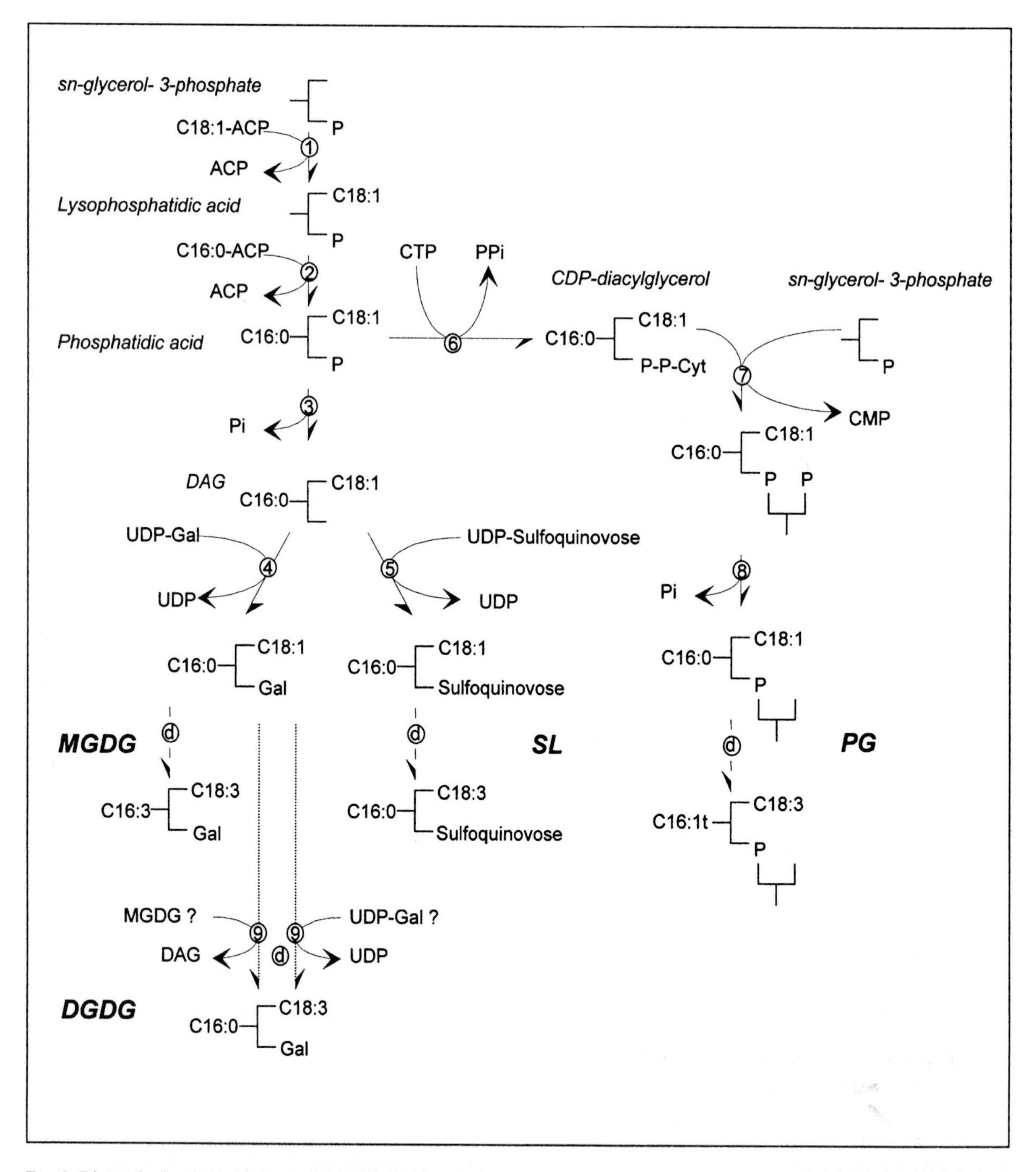

Fig. 2: Biosynthesis of plastid glycerolipids by plastid envelope membranes. The enzymes involved are (1) *sn*-glycerol-3-phosphate acyltransferase, (2) 1-acylglycerol-3-phosphate acyltransferase, (3) phosphatidic acid phosphatase, (4) MGDG synthase, (5) sulfolipide synthase, (6) CTP:phosphatidic acid cytosinetransferase, (7) *sn*-glycerol-3-phosphate:CDP-DAG phosphatidyltransferase, (8) phosphatidylglycerophosphatase and (d) desaturases. This biosynthetic pathway generates MGDG, SL and PG characterized by the presence of C18 and C16, in *sn*-1 and *sn*-2 positions of the glycerol backbone ('prokaryotic structure'), respectively. (9) Possible involvement of DGD is proposed either as a galactolipid:galactolipid galactosyltransferase or as a UDP-gal:galactolipid galactosyltransferase.

phatase activity is retro-inhibited by DAG (Malherbe et al., 1992). Feedback inhibition by DAG might lead to accumulation of PA and favor PG synthesis. In contrast to chloroplasts of 16:3 plants, those from 18:3 plants do not exhibit any phosphatidate phosphatase activity (Andrews et al., 1985). Therefore, PG is the only product of the prokaryotic pathway in chloroplasts of 18:3 plants.

The process described above does not account for synthesis of eukaryotic DAG. Indeed, the specificity of the envelope acyltransferases does not allow the formation of phosphatidic acid and DAG with C18 fatty acids at both *sn*-1 and *sn*-2 positions of glycerol (C18/C18), despite the fact that 18:1 is the main fatty acid produced by chloroplasts. In vivo kinetics of acetate and glycerol incorporation into chloroplast lipids have suggested that PC could provide the DAG backbone for plastid eukaryotic glycerolipids (Heinz, 1977; Slack et al., 1977). Furthermore desaturated PC molecular species are probably important in this process since analysis of the *Arabidopsis fad2* mutant lacking desaturation of PC 18:1 fatty acids has shown a decrease in eukaryotic MGDG molecular species with a parallel increase in prokaryotic species (Okuley et al., 1994). However, although PC is present in the outer envelope membrane, its de novo synthesis occurs in the ER. Hence a net import of some PC species from ER into chloroplasts is necessary. One hypothesis was that PC might be transported as a complete molecule via phospholipid transfer proteins (Kader, 1996). Re-examination of in vivo PC-MGDG relationship by Mongrand et al., (1997) suggests that LysoPC rather than PC could be transported. Positional distribution of labeled fatty acids was studied in PC and galactolipids during pulse chase experiments on 15 day old *Allium porrum* seedlings. The results show that labeled *sn*-2 fatty acid was not transferred from PC to galactolipids. Mongrand et al., (1997) proposed that transport of molecules between ER and chloroplasts could involve (a) a partial hydrolysis of PC in the ER, (b) a partition of the resulting LysoPC between the ER membrane, the cytosol aqueous phase and the envelope followed by (c) acylation of LysoPC by a plastidial LysoPC acyltransferase (Bessoule et al., 1995). Indeed, LysoPC is probably transiently formed in the ER for instance during the exchange of fatty acids between 18:1-CoA and unsaturated fatty acid esterified at the *sn*-2 position of PC (Stymne and Stobart, 1984). Furthermore, the presence in the chloroplast envelope of a specific LysoPC acyltransferase generating PC has been verified (Bessoule et al., 1995). However, a set of a specific (Lyso) PC phospholipase C and/or a specific C18 acyltransferase not yet detected in the envelope would be required to transform LysoPC or PC into eukaryotic MGDG. In addition, transport of LysoPC or PC between the ER and the chloroplast envelope remains unclear (see below). Recently, Williams et al. (2000) proposed that eukaryotic DAG could be synthesized in the cytosol using an acyl-CoA pool exchanging specifically with PC fatty acids.

D. MGDG Biosynthesis

The envelope of chloroplasts and non-green plastids is characterized by the presence of a 1,2-DAG 3-β-galactosyltransferase (E.C. 2.4.1.46) (or MGDG synthase) which transfers galactose from UDP-galactose to DAG (Douce et al., 1984). The biochemical properties of higher plant MGDG synthase were analyzed mostly in fractions derived from spinach chloroplast envelope membranes. Using mixed micelles containing DAG dispersed in CHAPS, PG and a partially purified delipidated envelope fraction, Maréchal et al., (1994a) showed that the MGDG synthase activity from spinach chloroplast envelope was able to use several DAG molecular species but with different affinity. The DAG species synthesized within the chloroplast i.e. 18:1/16:0 was a rather good substrate (K_m18:1/16:0 = 0.029-mol fraction), better than 16:0/18:1 (K_m16:0/18:1 = 0.042-mol fraction). The highest affinity was for dilinoleoylglycerol (K_m18:2/18:2 = 0.0089-mol fraction) as compared to any other species of DAG analyzed. The relevance of these results was confirmed further in envelope lipid vesicles (Maréchal et al., 1994b).

Ohta et al., (1995) reported the purification of a 47-kDa polypeptide associated with MGDG synthase activity from cucumber (*Cucumis sativus*, a '18:3 plant'). The corresponding cDNA was subsequently isolated and expressed as a recombinant protein that catalyzes the synthesis of MGDG (Shimojima et al., 1997). A characteristic UDP binding site can be recognized in the MGDG synthase sequence (Shimojima et al., 1997). A homologous cDNA from spinach (Miège et al., 1999) and three different homologous genes from *Arabidopsis* (Genbank AL031004, AJ000331, AC007187) have now been reported. From sequence analysis, MGDG synthase proteins can be classified in at least two families: MGD A, MGD B/C (Fig. 3). BLAST search does

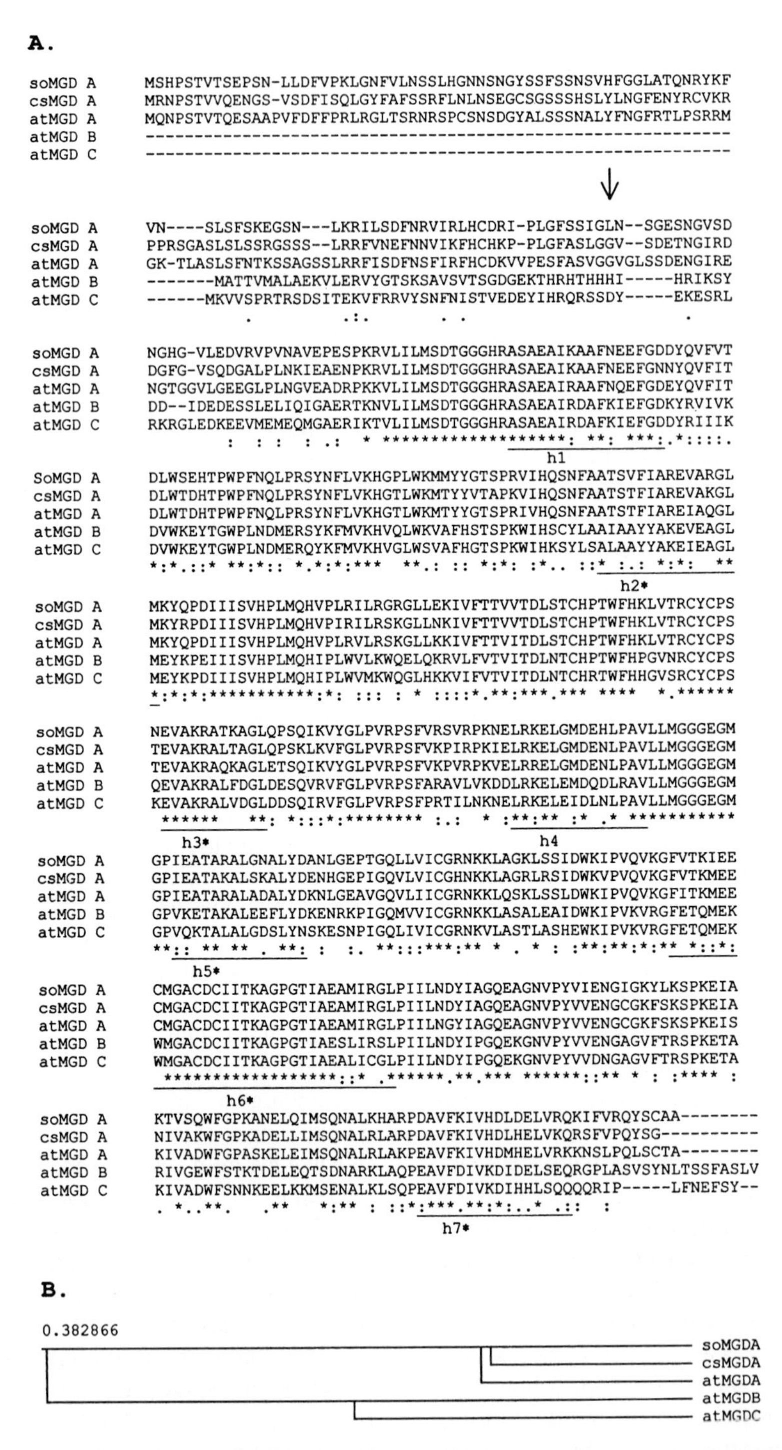

Fig. 3: Comparison of MGDG synthase sequences from spinach, cucumber and *Arabidopsis.* A. Primary MGDG synthase sequences from *Arabidopsis thaliana*, atMGD A (AL031004), atMGD B (AJ000331), at MGD C (AC007187), *Cucumis sativa* csMGD A (U62622), and *Spinacia oleracea* soMGD A (AJ249607) were compared with CLUSTAL W(1.5) multiple sequence alignment program. The black arrow indicates the transit peptide cleavage site for cucumber MGDG synthase, upstream of Gly104. The amino acid numbers starting from the initiation methionine are indicated on the right for each sequence. * and : are symbols for identical amino acids and conservative substitutions, respectively. Seven putative α-helices (h1 to h7) are detected in all sequences. The positions of these α-helices are indicated by a line under the corresponding aminoacid sequences. Putative amphipathic α-helices are h2*, h3* h5* h6* and h7*. B. Phylogenetic tree of the mature MGDG synthases. This figure is reproduced from Miège et al., (1999).

not allow the finding of any homologous MGDG synthase in cyanobacteria although these organisms possess galactolipids and are phylogenetically related to chloroplasts. This raises the question of the phylogenetic origin of MGDG synthase in present day chloroplasts (Jorasch and Heinz, 1999). A detailed analysis of recombinant spinach MGD A showed similar characteristics as purified chloroplast envelope MGDG synthase activity (Maréchal et al., 1995, Miège et al., 1999). Very interestingly, spinach MGD A catalyzes the synthesis of both prokaryotic and eukaryotic MGDG molecular species in vitro, with a selectivity for DAG similar to that of purified chloroplast envelope MGDG synthase activity that is to say a higher affinity for 18:2/18:2 DAG. Both spinach envelope MGDG synthase activity and spinach MGD A are sensitive to identical thiol reagents (dithiothreitol, N-ethyl-maleimide) and a hydrophobic chelating agent *(ortho*-phenanthroline). Together these results suggest that MGD A is a metalloprotein with no predictable canonical cation-binding site detected to date. Furthermore, in vitro import experiments carried out using isolated chloroplasts demonstrated that the spinach MGD A precursor is targeted to plastid membranes. Specific antibodies raised against the recombinant protein were used to demonstrate that spinach MGD A is a monotopic enzyme, associated with one leaflet of the inner membrane from the spinach chloroplasts envelope, a binding which may involve amphipatic α-helices. Indeed, five potential amphipathic α-helices are found in the sequence whereas no potential transmembrane domain was detected. Interaction with the membrane through amphipatic α-helices has already been described for another monotopic membrane protein, the prostaglandin H_2 synthase (Picot and Garavito, 1994). Otherwise amphipatic α-helices are sometimes involved in the formation of dimers. Interestingly, inactivation kinetics after γ-ray irradiation suggested that MGD A is likely to be active as a homodimer (Miège et al., 1999).

Spinach MGD A is most likely the enzyme catalyzing the synthesis of MGDG in the inner envelope of leaf chloroplasts. MGD A could be sufficient to catalyze MGDG from DAG originated from the plastid itself (prokaryotic structure) or from extraplastidial membranes (eukaryotic structure) provided that DAG transfer exists between the outer and the inner membranes of the envelope. The existence of other classes of MGDG synthase (MGD B and C in *Arabidopsis*) is therefore puzzling. It is possible that the distinct enzymes do not attribute to the biosynthesis of a given molecular species of MGDG in 16:3 or 18:3 plants, but could actually be (1) differentially present in the outer and inner envelope membrane and/or (2) differentially expressed within the plant tissues. Different locations of the enzymes in the envelope membranes is supported by the observation that MGDG synthase activity has also been found in the outer membrane fraction of pea chloroplasts (Cline and Keegstra, 1983). Furthermore MGD B and MGD C, in contrast to MGD A, lack a clearly recognizable chloroplast transit peptide. As several outer envelope membrane proteins have no transit peptide, this difference could be indicative of a location in the outer membrane. Future work will be conducted to analyze the expression in space and in time of the different genes and proteins. This should help in understanding the rationale for a multiplicity of isoenzymes.

E. DGDG Biosynthesis

The only DGDG-forming enzyme clearly described in plastids is the galactolipid:galactolipid galactosyltransferase. It catalyzes the enzymatic exchange of galactose between galactolipids with the production of DAG. Still, an alternative galactosylation of MGDG by transferring a galactose from a UDP-gal donor remains an attractive and simple hypothesis which, although lacking a direct successful assay, cannot be ruled out to date. It has been proposed that the GGGT is responsible for the synthesis of DGDG in vivo (Heemskerk and Wintermans, 1987). However, in vitro analysis of the enzyme activity led to unexpected results. In purified envelope fractions, the enzyme produces an unusually high concentration of DAG and unnatural additional galactolipids with polar heads containing more than two galactosyl residues. Under normal conditions, DAG is a transient lipid, poorly represented in cell membranes, and barely detectable (Miège and Maréchal, 1999). This suggests that, in vivo, the GGGT interacts with other lipid pathways that strongly regulate its activity. For instance, Sakaki et al., (1990) proposed that the enzyme is involved in the triacylglycerol biosynthesis pathway. Using ozone fumigated spinach leaves, they demonstrated that in vivo MGDG was converted into DAG by the GGGT and then to triacylglycerol owing to a DAG acyltransferase located in the envelope membranes (Martin and Wilson, 1984). Since the endogenous DGDG molecules have a different structure compared to MGDG, if we consider that the GGGT is indeed the DGDG-synthesizing enzyme, it should discriminate in vivo between the various available MGDG molecular species. However, in vitro the GGGT does not show any strong specificity for any MGDG molecular species. Spatial segregation of enzyme and substrates could lead to the same discrimination. Nothing is known about a local concentration of specific MGDG molecules in the plastid envelope but Dorne et al., (1982) have demonstrated that the GGGT is susceptible to proteolytic treatment of chloroplasts by thermolysin, a non-penetrating protease, and therefore located on the cytosolic side of the outer envelope membrane.

The analysis of an *Arabidopsis* mutant with a reduced DGDG content (*dgd1*) suggests the occurrence of multiple synthetic pathways for DGDG (Dörmann et al., 1995b). The *dgd1* mutant displays more than 90% reduction in the DGDG concentration as compared to the wild-type plant. DGDG labeling from ^{14}C-glucose was reduced in mutant seedlings to ~50% of the wild type levels while MGDG labeling was relatively stable. However, isolated chloroplasts from the *dgd1* mutant did not show any marked difference in galactose incorporation from UDP-[^{14}C]galactose into either MGDG or DGDG (Dörmann et al., 1995b). Together with the presence of residual DGDG in the *dgd1* mutant, these observations indicates that in the mutant (1) one pathway for the biosynthesis of DGDG is rendered inactive and (2) at least one other alternative pathway still exists. Indeed sequence analysis of the mutated locus lead to the identification of a protein called DGD 1 with weak similarities to bacterial and plant glycosyltransferases. When coexpressed in *E. coli* with cucumber MGD A that generates MGDG, the DGD 1 protein catalyzes the synthesis of DGDG, indicating that DGD 1 is able to convert MGDG into DGDG. This experiment does not inform whether DGD 1 uses MGDG or UDP-gal as a galactose donor. However, no motif characteristic for UDP-gal binding was found in the DGD 1 sequence. This leaves open the possibility that DGD 1 acts as a galactolipid: galactolipid galactosyltransferase. The predicted DGD 1 protein contains a N-terminal transit peptide typical for proteins imported into plastids. Interestingly, an ORF for a protein partially similar to DGD 1 and tentatively designated DGD 2 was located in the *Arabidopsis* genome (Dörmann et al., 1999). Expression study of the dgd2 gene and analysis of the protein might unravel major information concerning DGDG synthesis. Dörmann et al., (1999) proposed that DGD 1 is involved not only in the biosynthesis of DGDG but more generally in the assembly of thylakoid eukaryotic glycolipids. Considering that the *Arabidopsis act1* mutant is deficient in the synthesis of prokaryotic lipids (Kunst et al., 1988), a double mutant *act1,dgd1* was constructed in order to prevent compensation phenomena between eukaryotic and prokaryotic pathways. The double mutant showed a more extremely stunted phenotype than either parents. However, some eukaryotic MGDG and DGDG were still detected in further support for multiple biosynthetic pathways. Although the recent achievements to unravel DGDG synthesis are promising, key questions still remain. Is there more than one type of galactose donor (UDP-gal in addition to MGDG)? How many DGD isoforms do exist? When and where are these isoforms expressed? Why is the in vitro synthesis of galactolipids with more than two galactose residues not occurring in vivo in plants or in recombinant *E coli* expressing MGD A and DGD 1? A comprehensive study of the mechanism of DGDG synthesis will certainly be essential to analyze the function of DGDG.

F. Sulfolipid Biosynthesis

DAG molecules formed de novo in the inner envelope membrane constitute a pool of substrates shared by MGDG synthase and sulfolipid synthase (Joyard et al., 1986). A large array of evidence indicates that the sulfolipid is synthesized by a UDP-sulfoquinovose: 1,2-DAG 3-β-sulfoquinovosyltransferase, which transfers a sulfoquinovose from UDP-sulfoquinovose to DAG, as it occurs in photosynthetic bacteria (Benning, 1998). The enzyme activity is concentrated to the inner envelope membranes (Tietje and Heinz, 1998). Competition experiments between MGDG synthase and sulfolipid synthase were carried out in isolated envelope membranes supplied with UDP-gal and UDP-sulfoquinovose, loaded with 16:0/16:0 and/or 18:1/16:0 DAGs (Seifert and Heinz, 1992). Both DAG species could be used by MGDG synthase and sulfolipid synthase, but 16:0/16:0 was incorporated with a much higher efficiency into sulfolipid than into MGDG. This observation shows that the enzyme specificity for DAG molecular species may be responsible for the unique DAG structure of sulfolipid in the envelope membrane. Analyzing bacterial genes involved in sulfolipid synthesis and comparing their sequences with plant DNA databases, Benning and coworkers are presently working to identify the enzymes involved in plant sulfolipid synthesis. The *Arabidopsis* SQD 1 protein is believed to catalyze the transfer of SO_3^- to UDP-glucose to form UDP-sulfoquinovose essential for sulfolipid synthesis (Mulichak et al., 1999). Two likely candidate genes for photosynthetic bacteria and cyanobacteria UDP-sulfoquinovose: diacylglycerol sulfoquinovosyltransferases have been identified, sqd D and sqd X respectively. These genes were shown essential for sulfolipid synthesis. Both deduced proteins exhibit similarity to glycosyltransferases utilizing activated sugars (Güler et al., 2000). An equivalent SGDG synthase has not yet been identified in higher plants.

G. Phosphatidylglycerol Biosynthesis

Mudd and de Zacks (1981) first demonstrated the ability of intact chloroplast to synthesize PG. The complete pathway for PG synthesis is localized in the inner envelope membrane: phosphatidate cytidyltransferase (E.C. 2.7.7.41), CDP-DAG: glycerol-3-phosphate 3-phosphatidyltransferase (E.C. 2.7.8.5), and phosphatidylglycerol phosphatase (E.C. 3.1.3.27) (Andrews and Mudd, 1985). PG synthesis in the plastid envelope membranes differs from glycolipid synthesis by the fact that it does not utilize DAG. Plastid PG from '16:3' and '18:3 plants' has the same structure as prokaryotic PA synthesized by the envelope membrane. By similarity with bacterial genes some putative CDP-DAG: glycerol-3-phosphate phosphatidyltransferase genes have been reported in *Arabidopsis* genome (AAC28995 and CAB 82698).

H. Fatty Acid Desaturation

Palmitic acid (16:0) and oleic acid (*cis*-9-18:1) incorporated into glycerolipids are desaturated when they are esterified to polar lipids (for review see Joyard and Douce, 1987). In higher plant chloroplasts, desaturation results in incorporation of double bonds in a few preferential positions and always in the same order. In PG, a *trans* desaturation of palmitic acid occurs on C3 (as numbered from the fatty acid carboxyl group; resulting in a Δ3 desaturated fatty acid) whereas in glycolipids, palmitic acid is desaturated first on C7, then on C10 and finally on C13 positions with *cis* configuration. In all glycerolipids, oleic acid is desaturated first on C12 and then on C15 positions with *cis* configuration. Investigation of membrane desaturases by traditional biochemical approaches has been limited because their solubilization and purification have proven very difficult (Schmidt and Heinz, 1993). However, desaturation of MGDG was achieved by isolated chloroplasts (Heinz and Roughan, 1983) and pure envelope membranes have been used as a source of enzymes for desaturation of oleic acid to linoleic acid (Schmidt and Heinz, 1990). In addition, it has been indicated that desaturation was dependent on electron transport elements (Andrews et al., 1989) and preliminary evidence for the involvement of a ferredoxin:NADPH oxidoreductase has been presented (Schmidt and Heinz, 1990). In fact, envelope membranes very likely contain the enzymatic machinery necessary for fatty acid desaturation (Schmidt et al., 1994) including the electron carriers involved (Jäger-Vottero et al., 1997).

Our understanding of chloroplast desaturases has considerably benefited from the pioneering characterization of *Arabidopsis* mutants, each deficient in a specific desaturation step (Ohlrogge and Browse, 1995). Five loci, i.e. fad4, fad5, fad6, fad7 and fad8, affect chloroplast lipid desaturation. Two of these desaturases are highly substrate specific: the fad4 gene product is responsible for inserting a Δ3-*trans* double bond into the 16:0 esterified to *sn*-2 glycerol position of PG, and the fad5 gene product is responsible for the synthesis of Δ7 16:1 on MGDG and possibly on DGDG. The 16(18):1 desaturase is dependent of the fad6 gene, whereas two isozymes necessary for 16(18):2 desaturation are encoded by fad7 and fad8.

Specialized roles have been demonstrated for desaturated fatty acids including production of jasmonic acid, which is a key mediator of several processes as diverse as wound signaling (Blée and Joyard, 1996). Moreover, polyunsaturation is essential for maintaining cellular function and plant viability at low temperature (Ohlrogge and Browse, 1995). However with the exception of the *fab2* mutant of *Arabidopsis* (blocked in the desaturation of 18:0; see above), most of the single-gene desaturase mutants do not show any visible phenotype at normal growth temperatures (25 °C) or following several days of exposure to low temperatures (Somerville and Browse, 1996). In order to block possible compensation by alternative pathways, double and triple mutants were constructed (McConn and Browse, 1998). The *fad2-fad6* double mutant can not desaturate 18:1 and 16:1. Very interestingly, the double mutant is not capable of autotrophic growth but, on sucrose media, the plants develop and chloroplasts contain thylakoids although with a highly reduced chlorophyll content. This supports the idea that fatty acid desaturation is important for general plant cell metabolism and can be achieved by several parallel pathways that can be compensating for each other. Moreover, photosynthesis appears as a process critically affected by the absence of polyunsaturated lipids. Other data indicate a role for specific unsaturated lipids in thylakoid functioning. For instance, the rate of damage and repair of the D1 protein at low temperature was much slower in transgenic tobacco plants in which the proportion of saturated plastid PG was increased by expression of an isoform of glycerol-3-phosphate acyltransferase which prefers saturated fatty acyl substrates (Moon et al., 1995).

IV. Chlorophyll Biosynthesis

Although we refer the reader to Chapter 12 of this book for detailed presentation of chlorophyll synthesis, we wish to develop here the contribution of the chloroplast envelope to this biosynthetic pathway. Chlorophyll is made of two parts with distinct origins: the porphyrin (i.e. Chlide) and the prenyl chain (i.e. phytol). Envelope membranes are involved in the formation of the Chlide moieties while addition of the prenyl chain to Chlide, a reaction catalyzed by chlorophyll synthase, is specifically associated with thylakoids (Block et al., 1980). The first steps in Chlide biosynthesis, namely those from Δ-amino levulinate to protoporphyrinogen, occur in the soluble phase of the chloroplasts but all the subsequent steps are membrane-bound. Several enzymes of this second part of the pathway are associated with the envelope. Matringe et al., (1991) demonstrated that envelope membranes from mature spinach chloroplasts contain protoporphyrinogen oxidase. This enzyme that synthesizes protoporphyrin IX is the last common step to chlorophyll and heme biosynthesis. Actually, protoporphyrinogen oxidase is also present in thylakoids but, considering that it has the same location as ferrochelatase, thylakoid protoporphyrinogen oxidase is most likely involved in heme synthesis. In the chlorophyll pathway, the next enzyme after protoporphyrinogen oxidase is Mg-chelatase. It has been demonstrated by analysis of plant mutants (Koncz et al., 1990; Hudson et al., 1993) and by in vitro reconstruction of activity (Papenbrock et al., 1997) that plant Mg-chelatase is also composed of three subunits, Chl H, Chl D and Chl I., like in photosynthetic bacteria. The subunits are soluble proteins but Chl H binds to protoporphyrin IX synthesized in membranes (Hinchigeri et al., 1997) and moreover, in the presence of Mg^{2+}, Chl H was shown to associate with the envelope (Nakayama et al., 1998). As the concentration of Mg^{2+} in chloroplasts was estimated to vary from 1 to 6 mM during shift from darkness to light (Leegood et al., 1985), this suggests a Mg^{2+} dependent change in the distribution of Chl H between stroma and envelope. Further supporting that Mg-chelatase can be present in the envelope membranes, Fuesler et al. (1984) have shown that Mg-chelatase is accessible to molecules unable to cross the inner envelope membrane of intact chloroplasts. Recently, Grafe et al., (1999) proposed a model for the assembly of the interacting subunits required for the active Mg-chelatase complex. In a preactivation step, which is Mg^{2+} and ATP dependent, Chl I interacts with Chl D and the Chl D subunits subsequently multimerize. Finally, in the chelation step, a direct interaction between Chl H and Chl D occurs.

The next reactions in the chlorophyll biosynthesis convert Mg protoporphyrin into PChlide. These involve three enzymatic steps that are poorly understood in plants: methylation of ring 3 propionate side chain by s-adenosyl methionine, cyclization of this methylated chain and reduction of ring 4 vinyl side chain (for review see Suzuki et al., 1997). The plant methyltransferase has not been formally identified but sequencing of *Arabidopsis thaliana* (Genbank AL035523) and *Oryza sativa* (Genbank AP000529) genomes allows the detection of genes with a high homology with the *Synecchococus* methyltransferase previously characterized by Smith et al., (1996). The plant enzyme is known as a membrane bound enzyme in chloroplasts (Hinchigeri et al., 1981; Fuesler et al., 1982). No light induction of the enzyme was found. However, the substrate and product never accumulate and they are suspected to play a role in expression of some light dependent genes (Johanningmeier and Howell, 1984; Kropat et al., 1997). The cyclase utilizes O_2 as an oxygen donor and needs NADH or NADPH during a complex multistep process that very likely involves soluble and membranous enzymes (Wong and Castelfranco, 1984). Hydrogenation of the ring 4 vinyl group presumably occurs at different stages in the chlorophyll synthesis pathway. In addition, two parallel chlorophyll pathways might exist, one leading to divinyl chlorophyll and the other to monovinyl, monoethyl chlorophyll, making it difficult to know how many vinyl reductase enzymes exist and what their roles are (Rüdiger and Schoch, 1988).

PChlide oxidoreductase (or POR, E.C. 1.3.1.33) is massively present in etioplast prolamellar bodies (Shaw et al., 1985). It binds PChlide and NADPH and reduces PChlide into Chlide only when PChlide absorbs light. In chloroplasts, envelope membranes consistently contain small pools of PChlide and Chlide (Pineau et al., 1986, 1993). Incubation of chloroplast envelope membranes in low light in the presence of NADPH induces an enzymatic decrease in the level of PChlide (Pineau et al., 1986, Joyard et al., 1990). Chloroplast POR is found on the cytosolic side of the outer envelope membrane (Joyard et al., 1990). POR was also reported to associate with the thylakoid membranes of developing and mature plastids (Dahlin et al., 1999). Recent investigations demonstrated the existence of two differentially light regulated por genes in barley and *Arabidopsis* (for review see, Reinbothe et al., 1996). Corresponding proteins POR A and POR B exist in etiolated tissues. Import of POR B occurs constitutively whereas it is thought that POR A is imported into the plastid by a translocation mechanism involving the formation of a complex between the precursor of POR A and the envelope PChlide (Reinbothe et al., 1995). Therefore, the accumulation of POR A in the etioplast prolamellar bodies depends strictly on the plastid envelope PChlide content. Under illumination, PChlide is reduced by envelope POR; consequently cytosolic POR A precursor can no longer be imported into the plastid. POR A disappears from illuminated seedlings whereas POR B is still present. It is not yet known where POR B is located in mature chloroplasts. Reinbothe et al., (1999) found that in etiolated barley POR A and POR B form a stable LHPP (Light harvesting POR-PChlide) complex containing PChlide *b* and PChlide *a*. They proposed that PChlide *b* plays a special role in light harvesting necessary for the initial reduction of PChlide in etioplasts. However, the presence of PChlide *b* in etiolated tissues is a matter of debate (Scheumann et al., 1999) and questions the existence of the LHPP complex (Armstrong et al., 2000).

Finally, different situations should be considered concerning regulation of chlorophyll biosynthesis. Chlorophyll biosynthesis is necessary for chloroplast maintenance but also for generation of chloroplasts from other plastids; proplastids, etioplasts or chromoplasts. In etiolated leaves, chlorophyll synthesis is blocked at the level of light dependent POR. In tissues containing chromoplasts, the membrane bound enzymes necessary for the formation of the isocyclic ring are not present (Lützow and Kleinig, 1990). In photosynthetic leaves, little is known about regulation of chlorophyll biosynthesis. A future challenge will be to identify the numerous enzymes involved, to conclusively locate these enzymes and to identify the possible metabolite transfers to eventually address the specific regulatory processes involved.

V. Plastid Prenylquinone Biosynthesis

Chloroplast envelope membranes were shown to catalyze the synthesis of plastid prenylquinones (Soll et al., 1980, 1985). Homogentisic acid, the precursor for both α-tocopherol and plastoquinone-9, is synthesized from 4-hydroxyphenylpyruvate (a product of the plastid shikimate pathway) by a 4-hydroxyphenylpyruvate dioxygenase (E.C. 1.13.11.27) probably localized in the envelope membranes (Fielder et al., 1982). The cDNA encoding the HPPDase has been identified from carrot cells by Garcia et al., (1997) and the ortholog *Arabidopsis* gene identified by complementation of the *Arabidopsis pds1* mutation (Norris et al., 1998). Tocopherols are then synthesized in the inner envelope membrane by condensation of homogentisic acid and C20-prenyl pyrophosphate to form 2-methyl-6-prenylquinone (Soll et al., 1980, 1985). PhytylPP and GGPP are both accepted for in vitro prenylation. Reduction of the GGPP into phytylPP is catalyzed by CHL P, classically involved in reduction of geranylgeranyl in chlorophyll synthesis pathway. Therefore CHL P appears at a branch point between chlorophyll synthesis and tocopherol synthesis (Keller et al., 1998; Tanaka et al., 1999). In addition, recent analysis of mutants expressing antisens RNA for CHL P suggest that reduction of GGPP is necessary for tocopherol synthesis (Tanaka et al., 1999).

By a series of methylations and cyclizations, the 2-methyl-6-prenylquinol gives successively rise to 2,3-dimethyl-6-prenylquinol, γ-tocopherol or γ-tocotrienol in the inner envelope (Soll et al., 1980, 1985). The same pathway was demonstrated in pepper chromoplasts (Camara et al., 1982). The other major chloroplast prenylquinone, plastoquinone-9, is also synthesized in the inner envelope membrane by condensation of homogentisic acid and solanesyl-pyrophosphate to form 2-methyl-6-solanesylquinone, which is further methylated and oxidized to form successively plastoquinol-9 and plastoquinone-9 (Soll et al., 1985). To date, the biochemical identification of the proteins involved in prenylquinone biosynthesis have been hampered by the difficult purification procedure, mostly because of their presence in minute amounts in the cell, their low activity within the membranes as well as the need for detergents for their purification and assay. In contrast, molecular genetic approaches have proved to be successful.

Recently, Shintani and DellaPenna (1999) cloned the γ-tocopherol methyltransferase using a highly sophisticated and remarkably efficient genomic-based strategy including the following steps; (a) the investigation of *Synechocystis* PCC6803 to search for the gene ortholog to the *Arabidopsis* HPPDase (Norris et al., 1998), (b) the analysis of the ten-gene operon where the *Synechocystis* HPPDase gene was localized, (c) the search for methyltransferase features, including S-adenosyl-methionine-binding domains in the different *Synechocystis* ORFs, (d) analysis of the tocopherol content of a null *Synechocystis* mutant for one of the candidate ORF SLR0089, (e) the identification of the probable plant ortholog gene in the *Arabidopsis* EST database, and (f) the functional expression of this gene in *E. coli*. Finally, *Arabidopsis* plants overexpressing the γ-tocopherol methyltransferase were demonstrated to present a strongly increased level (nine-fold) of α-tocopherol in their seeds although the total level of tocopherol remained constant in the plants (Shintani and DellaPenna, 1999). These results demonstrate that it was possible to manipulate the level of this essential vitamin (vitamin E) to benefit the human health (Grusak and DellaPenna, 1999).

One question that remains to be addressed is where the corresponding gene product is localized. The fact that the *Arabidopsis* γ-tocopherol methyltransferase gene was cloned through its cyanobacterial ortholog strongly suggests that the higher plant enzyme is a plastid protein. However, despite previous analysis of the localization of tocopherol and plastoquinone biosynthesis in plastids (Soll et al., 1980, 1985) and of ubiquinones in mitochondria (Lütke-Brinkhaus et al., 1984), there is still a controversy. Indeed, such a view was challenged by Swiezewska et al., (1993) who found nonaprenyl-4-hydroxybenzoate and nonaprenyl-2-methylquinol transferase activities in ER and Golgi preparations. They therefore proposed that ubiquinone and plastoquinone-9 biosynthesis occurs in the ER-Golgi system, followed by a selective transfer of ubiquinone to the mitochondria and of plastoquinone-9 to the chloroplast. A partial answer to this question was provided by Avelange-Macherel and Joyard (1998) who constructed a stable mutant of *Saccharomyces cerevisiae* deleted for coq3 gene, encoding a 3,4-dihydroxy-5-hexaprenylbenzoate methyltransferase (which catalyses the fourth step in the biosynthesis of ubiquinone from *p*-hydroxy-benzoic acid). They transformed the mutant with an *Arabidopsis thaliana* cDNA library in order to isolate plant genes that could be involved in quinone synthesis. A series of evidence, and especially (a) the presence of an N-terminal additional sequence having features for targeting into mitochondria, (b) the immunolocalization of the protein in the inner membrane of purified mitochondria, demonstrated that the coq3 gene product was only present in mitochondria. A similar situation is likely to occur in plastids for the enzymes involved in plastid prenylquinone biosynthesis.

VI. Transport of Lipids From ER to Chloroplasts and Between Chloroplast Membranes

Final assembly of most of the thylakoid lipids is done in the envelope. Therefore a massive transport of the lipids from the envelope to the thylakoids should take place during development. The process involved in mature chloroplasts is probably different from that necessary in young plastid development. In developing plastids, transport may occur through the stroma via vesicles derived from the inner membrane (Wellburn, 1982; Morré et al., 1991). Since the inner envelope membrane and thylakoids do not differ in polar lipid composition, natural fusion of vesicles produced by inner membrane with growing thylakoids is possible. According to Rawyler et al., (1992), vesicle transfer does not occur in mature chloroplasts. Transfer would then be achieved either by transient fusion between the inner envelope and the thylakoids and subsequent lateral diffusion (Rawyler et al., 1995) or by specific lipid transfer proteins transporting monomers (Nishida and Yamada, 1986). We refer to Douce and Joyard (1996) for a detailed discussion of these possible transport mechanisms.

Some ER derived molecular species are also necessary to build up the thylakoids eukaryotic galactolipids indicating some transport between the ER and the envelope. The chloroplast exports fatty acids and PC synthesized in the ER from these fatty acids is most likely necessary to build up eukaryotic plastid glycerolipids (galactolipids and sulfolipid). One way for fatty acids and PC to cross the cytosol would be their transformation into an amphipathic form i.e. acyl-CoA and LysoPC that can partition between the cytosol and the membrane compartments. However, this transport mechanism would favor random partition of the molecules in any soluble or membrane compartment and would therefore require a specific and compartmentalized recruiting and regulation system. In addition, free acyl-CoA or LysoPC as detergents may destabilize membrane structure. Proteins, binding lipophilic molecules to be transported, could play three different roles: (1) to help to solubilize the lipid in the cytosol, (2) to protect membranes from destabilization and (3) to direct insertion of the molecules to the correct membrane site. The recent discovery in plants of cytosolic, soluble and membrane associated acyl-CoA binding proteins suggests that such proteins could be involved in fatty acid transport (Engeseth et al., 1996; Chye et al., 1999). Phospholipid transfer proteins are also present in plant cells. Although it is not clear why phospholipid transfer proteins can be secreted outside the cell, their involvement in PC transfer between ER and chloroplasts is still possible (Kader, 1996). It would be interesting to check whether a protein dependence of LysoPC transfer occurs as a prerequisite to verify the hypothetical transfer of LysoPC from the ER to the envelope (Bessoule et al., 1995). In vitro reconstitution with isolated compartments, i.e. the ER, the cytosol and the chloroplast envelope and in vivo experiments with mutants would ultimately help to understand how eukaryotic DAG backbone of glycolipids is built up.

Below we address the questions of what are the lipids transported between the two envelope membranes and how they are transported. MGDG and DGDG are synthesized from DAG in the envelope. To date we know that MGD A is present in the inner membrane of spinach chloroplast envelope. This could imply that the inner membrane synthesis of eukaryotic MGDG requires a transport of the eukaryotic DAG structure from the outer membrane to the inner membrane (Miège et al., 1999). From MGD B and MGD C sequence analysis, it is possible that one or more other MGDG synthases could be located in the outer membrane. Likewise, synthesis of DGDG occurs in the outer envelope because of the presence of a GGGT activity on the chloroplast surface. The transit peptide identified in DGD 1 sequence suggests that this DGDG synthase (DGD 1) is likely to be present in the inner envelope membrane. Although we lack a clear localization for these enzymes, except for MGD A present in the inner envelope membrane, one can speculate that both envelope membranes might be autonomous for MGDG and DGDG synthesis (Fig. 4). We can not know whether transfer of galactolipids between the two envelope membranes is necessary since transfer of DAG could be sufficient. Transient contact points exist between the two membranes and may facilitate transport or diffusion of lipids. Since the *dgd1* mutant has a lower ratio of eukaryotic MGDG as compared to the wild type and exhibits a moderate growth, Dörmann et al., (1999) proposed that DGD 1 is involved in the process necessary to transfer eukaryotic DAG backbone from the outer envelope membrane to the inner envelope membrane. Furthermore, lipid synthesis may interact with transfer since local increase in lipid concentration may favor disruption of the membrane bilayer structure and thus increase lipid diffusion. Concentration of MGDG, as a non bilayer-forming lipid, may be particularly critical (Gounaris and Barber, 1983).

VII. Lipid Modifications of Proteins in Chloroplasts

We have described that interactions between lipids and proteins are susceptible to occur at many different stages during chloroplast lipid biogenesis. Unfortunately, very little is known about these interactions. The main information concerns covalent modifications of proteins by lipids, a widespread process. Lipid modification of proteins facilitates the attachment of extrinsic proteins to biological membranes but also enables protein-protein interaction and in some cases shuttling of protein to a membrane or a compartment (Yalovsky et al., 1999). Furthermore, protein lipid modifications seem to be involved in the regulation of various cellular processes and intracellular signaling pathways (Yalovsky et al., 1999). At least three types of protein

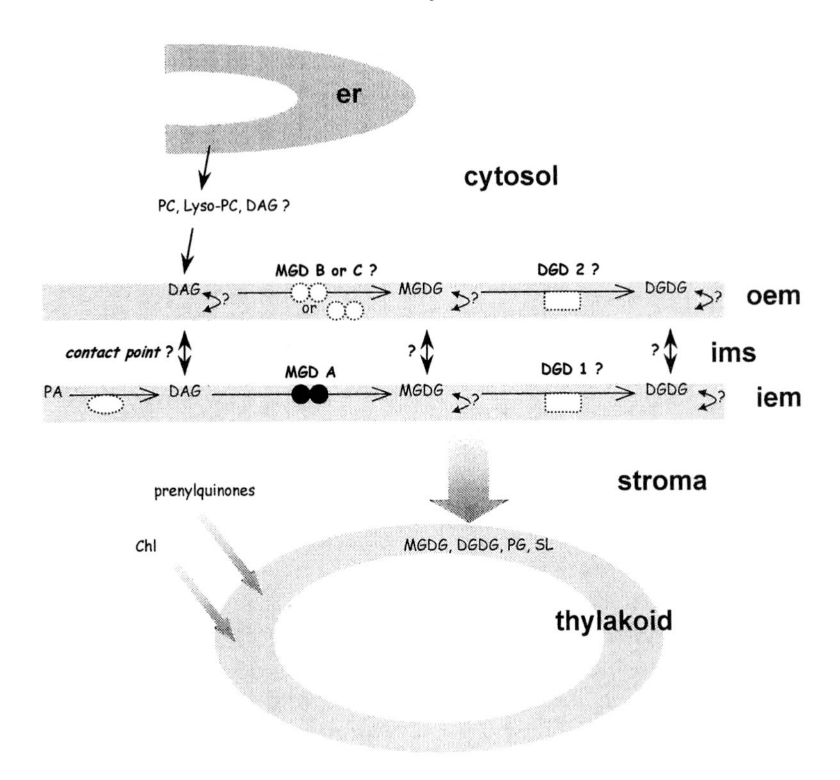

Fig. 4: Transport of lipids from the ER to the chloroplast and between chloroplast membranes. This scheme shows the hypothetical localization of galactolipid synthesizing enzymes according to their predicted transit sequences. The presence of MGD A in the outer leaflet of the inner membrane is the only data supported by experimental analysis (Miège et al., 1999). DGD 1 was proposed to be responsible for the GGGT activity in the outer membrane (Jorash and Heinz, 1999) however, in the absence of proper experimental demonstration, we hypothesize that it is in the inner membrane due to its predictable transit peptide. Likewise, MGD B, MGD C and DGD 2, lacking a long transit peptide, are supposed to be in the outer envelope membrane. In this scheme, it is possible that galactolipid synthesis occur independently in the outer and inner membranes. However, because MGD A can synthesize eukaryotic MGDG owing to its very high affinity for C18:2/C18:2 DAG, it is very likely that DAG originating from endomembrane PC/LysoPC can be transported to the inner envelope membrane. Vice versa, a transport from the inner envelope to the outer envelope is required to feed the outer envelope glycerolipid synthesis with prokaryotic DAG generated by the phosphatidate phosphatase localized within the inner envelope membrane. Lipid exchanges between outer and inner membrane could occur at the level of contact points. Eventually, transport to the thylakoids does not segregate the glycerolipid since the glycerolipid composition of the thylakoids and the inner envelope are similar. Transport of prenylquinones and chlorophyll precursors shows a different level of regulation.

lipid modifications involving the saturated fatty acids palmitic or myristic, or isoprenoid compounds have been reported. In the case of chloroplasts, Mattoo and Edelman (1987) reported that palmitic and myristic acids can be selectively attached to chloroplast proteins. For instance the D1 protein is palmitoylated after its maturation and insertion into the stromal lamellae. Thc palmitoylation occurs during the translocation of the protein to the stacked granal lamellae where the D1 protein interacts with PS II. Several other major chloroplast proteins such as Rubisco might also be modified by acylation.

When it is involved in membrane anchoring, protein acylation is often associated with protein prenylation. As plastids are an important site for terpenoid biosynthesis, prenylation of plastid proteins were presumed to exist, thereby providing an interesting way to coordinate several biogenesis pathways. Parmryd et al., (1997) detected protein prenylation in chloroplasts by in vivo ^{3}H-mevalonate labeling of spinach leaves. Some of these prenylations were detected at the cytoplasmic surface of the chloroplast envelope being sensitive to thermolysin treatment of intact chloroplasts whereas other protein prenylations were detected in the thylakoids. A farnesyl protein transferase was measured in the thylakoids. Furthermore, considering that the specific plastidic isoprenoid pathway does not start from mevalonate but from 1-deoxy-D-xylulose-5-phosphate (for review see Lichtenthaler, 1999), it is possible that the 1-deoxy-D-xylulose-5-phosphate pathway would favor other types of specific chloroplast prenylations. In fact, two different prenylation pathways could exist in chloroplasts. Parmryd et al., (1999) analyzed in vivo metabolic ^{3}H-mevalonate labeling in combination with the protein synthesis inhibitors cycloheximide or chloramphenicol. In spinach cotyledons they observed that this post-translational protein modification was divided into two categories. One represented the conventional prenylation involving farnesyl and geranylgeranyl groups and may be longer isoprenoid groups such as hexaprenoid (Parmryd and Dallner, 1999) bound to cysteine residues via thioether linkages. This category revealed a pattern of prenylated proteins similar to that observed in mammalian cells and depending on nuclear gene expression. Farnesyl pyrophosphate synthesized in the cytosol may be imported into the chloroplast as such or as a derivative to be used for protein modification. The other category was shown to represent a type of prenylation confined to chloroplasts, depending on plastid gene expression and not involving a thioether bond. The modifying isoprenoid could be released from the chloroplastic polypeptides by alkaline treatment and was identified as phytol upon GC-MS analysis. The phytol could readily be derived from all-*trans*-[^{3}H]farnesol, which, like all-*trans*-[^{3}H]geranylgeraniol, was taken up by the cotyledons, resulting in incorporation of radiolabel into proteins. Thai et al., (1999) also reported that farnesol as geranylgeraniol can be utilized by tobacco cells for protein prenylation. According to Parmryd et al., (1999) chloroplastic prenylated proteins seem to be of low abundance and are presumably long-lived. They also suggest that such proteins may be involved in the regulation of chloroplast functions, catabolism, or biosynthesis, as well as stress responses. Finally, protein prenylation in plastids might be an important way of metabolic regulation but is still a open field of research.

VIII. Conclusion

The plastid envelope is a major metabolic site for a wide variety of lipidic compounds such as fatty acids, glycerolipids, prenylquinones and chlorophylls. These lipids are required for the biogenesis of the complex chloroplast membranous system, including the thylakoids. In addition, fatty acids are used by the whole cell. Interestingly, the envelope is also an active border between the cytosol and the stroma, particularly for plastid protein precursors import. Coordination of lipid synthesis with protein import is therefore structurally favored.

In the last decade, genes and cDNA characterization have been particularly impressive and promising for glycerolipid and prenylquinone synthesizing enzymes. However, important activities involved in key steps of the glycerolipid metabolism still remain to be discovered. The cloning of an envelope PA phosphatase is an important challenge for evolution studies; possible pseudo-genes for envelope PA phosphatase in C18:3 plants should be very informative to understand the evolutionary loss of PA phosphatase in C16:3 ancestors. The molecular characterization of numerous membrane activities involved in chlorophyll metabolism is in progress.

Because of the very stable lipidic composition of biomembranes in general, and particularly in plastid membrane, a tight structural and functional coordination between different processes involved in lipid synthesis must occur. Our understanding of the mechanisms undergone to maintain a stable lipid composition is still very poor. Dynamic processes (enzymatic reactions, metabolite transport and transient enzyme association to membranes) are controlled in the planar membranes and at the interface with aqueous phases. On one hand, regulatory crossroads are well identified when an intermediate is used for divergent pathways, e.g. at the level of PA, DAG, GGPP, protoporphyrin IX. On the other hand, regulation of dynamic processes are important (1) to compensate for physical constraints, e.g. MGDG as a non-bilayer forming lipid is balanced with DGDG as a bilayer forming lipid, (2) to capture exogenous substrates, such as the DAG structure originated from the ER which is further used for plastid glycerolipid biosynthesis and (3) to transport products from its site of synthesis, the envelope, to its recipient, e.g. the thylakoids.

In the case of glycerolipid metabolism, redundant enzymatic activities were classically investigated in yeast or animal cells in the dual context of membrane biogenesis and second messengers production. For example, DAG can be quantitatively produced by a PA phosphatase to build up bilayer glycerolipids whereas it is also transiently generated by a PA phosphatase activity as a second messenger. The occurrence of redundant activities in chloroplast membranes (MGDG synthases, DGDG-producing enzymes) has not yet been analyzed out of the context of membrane biogenesis. The galactolipid:galactolipid galactosyltransferase activity is particularly puzzling since its activity in vivo is very constrained as compared to its activity in vitro. The GGGT activity producing DGDG and DAG has been associated to the DGD family of genes. Are all DGD gene products involved in membrane biogenesis? Could the transiently produced DAG be involved in an information transfer?

Lipid metabolism is therefore an interesting candidate to elucidate the dialogue between the plastid and the rest of the cell. Some reactions could be important sensors controlled by external signals, and lipidic metabolites such as PA, DAG, Mg-protoporphyrin and methyl-Mg-protoporphyrin could play a role as second messengers.

In the near future, the molecular characterization of proteins involved in envelope lipid metabolism will hopefully be completed. The construction of DNA chips using corresponding probes will be an important tool to address the control of metabolic pathways in the envelope and in other compartments and to investigate possible functions differing from membrane biogenesis, i.e. signal transduction, photosynthesis control, etc. Thus, sustained efforts on biochemical characterizations together with genomic and proteomic studies will help in understanding regulatory and evolutionary mechanisms and produce the first schemes to understand the complex compartmentalized metabolisms of lipophilic compounds.

Acknowledgment

We thank S. Mongrand for allowing us to reproduce Fig. 1 from his Ph.D. thesis.

References

Alban C, Joyard J and Douce R (1989) Comparison of glycolipid biosynthesis in non-green plastids from sycamore (*Acer pseudoplatanus*) cells and cauliflower (*Brassica oleracea*) buds. Biochem J 259: 775–783

Andrews J and Mudd JB (1985) Phosphatidylglycerol synthesis in pea chloroplasts. Pathways and localization. Plant Physiol 79: 259–265

Andrews J, Ohlrogge JB and Keegstra K (1985) Final step of phosphatidic acid synthesis in pea chloroplasts occurs in the inner envelope membrane. Plant Physiol 78: 459–465

Andrews J, Schmidt H and Heinz E (1989) Interference of electron inhibitors with desaturation of monogalactosyldiacylglycerol in intact chloroplasts. Arch Biochem Biophys 270: 611–622

Armstrong GA, Apel K and Rüdiger W (2000) Does a light-harvesting protochlorophyllide *a/b* binding protein complex exist? Trends Plant Sci 5: 41–44

Avelange-Macherel MH and Joyard J (1998) Cloning and functional expression of *Atcoq3,* the *Arabidopsis* homologue of the yeast *coq3* gene, encoding a methyltransferase from plant mitochondria involved in ubiquinone biosynthesis. Plant J 14: 203–213

Benning C (1998) Biosynthesis and function of the sulfolipid sulfoquinovosyldiacylglycerol. Annu Rev Plant Physiol Plant Mol Biol 49: 53–75

Bessoule JJ, Testet E and Cassagne C (1995) Synthesis of phosphatidylcholine in the chloroplast envelope after import of lysophosphatidylcholine from endoplasmic reticulum membranes. Eur J Biochem 228: 490–497

Billecocq A, Douce R and Faure M (1972) Structure des membranes biologiques: Localization des galactosyldiglycérides dans les chloroplastes au moyen des anticorps spécifiques. C R Acad Sci Paris 275: 1135–1137

Bishop DG, Sparace SA and Mudd JB (1985) Biosynthesis of sulfoquinovosyldiacylglycerol in higher plants: The origin of the diacylglycerol moiety. Arch Biochem Biophys 240: 851–858.

Blée E and Joyard J (1996) Envelope membranes from spinach chloroplasts are a site of metabolism of fatty acid hydroperoxides. Plant Physiol 110: 445–454

Block MA, Joyard J and Douce R (1980). Site of synthesis of geranylgeraniol derivatives in intact spinach chloroplasts. Biochim Biophys Acta 631: 210–219

Block MA, Dorne A-J, Joyard J and Douce R (1983) Preparation and characterization of membrane fractions enriched in outer and inner envelope membranes from spinach chloroplasts. II-Biochemical characterization. J Biol Chem 258: 13281–13286

Bourgis F, Kader JC, Barret P, Renard M, Robinson D, Robinson C, Delseny M and Roscoe T (1999) A plastidial lysophosphatidic acid acyltransferase from oilseed rape. Plant Physiol 120: 913–921

Britton G (1988) Biosynthesis of carotenoids. In: Goodwin TW (ed) Plant Pigments, pp 133–182. Academic Press, London

Camara B, Bardat F, Seye A, d'Harlingue A and Monéger R. (1982) Terpenoid metabolism in plastids. Localization of α-tocopherol synthesis in *Capsicum* chromoplasts. Plant Physiol 70: 1562–1563

Chye ML, Huang BQ and Zee SY (1999) Isolation of a gene encoding *Arabidopsis* membrane-associated acyl-CoA binding protein and immunolocalization of its gene product. Plant J 18: 205–214.

Cline K and Keegstra K (1983) Galactosyltransferases involved in galactolipid biosynthesis are located in the outer envelope membranes of pea chloroplasts. Plant Physiol 71: 366–372

Cutler AJ and Krochko JE (1999) Formation and breakdown of ABA. Trends Plant Sci 12: 472–478

Dahlin C, Aronsson H, Wilks HM, Lebedev N, Sundqvist C and Timko MP (1999) The role of protein surface charge in catalytic activity and chloroplast membrane association of the pea NADPH: Protochlorophyllide oxidoreductase (POR) as revealed by alanine scanning mutagenesis. Plant Mol Biol 39: 309–323

Dörmann P, Voelker TA and Ohlrogge JB (1995a) Cloning and expression in *Escherichia coli* of a novel thioesterase from *Arabidopsis thaliana* specific for long-chain acyl-acyl carrier proteins. Arch Biochem Biophys 316: 612–618

Dörmann P, Hoffmann-Benning S, Balbo I and Benning C (1995b) Isolation and characterization of an *Arabidopsis* mutant deficient in the thylakoid lipid digalactosyl DAG. Plant Cell 7: 1801–1810

Dörmann P, Balbo I and Benning C (1999) *Arabidopsis* galactolipid biosynthesis and lipid trafficking mediated by DGD1. Science 284: 2181–2184

Dorne A-J, Block MA, Joyard J and Douce R (1982) The galactolipid:galactolipid galactosyltransferase is located on the outer membrane of the chloroplast envelope. FEBS Lett 145: 30–34

Dorne A-J, Joyard J and Douce R (1990) Do thylakoids really contain phosphatidylcholine? Proc Natl Acad Sci USA 87: 71–74

Douce R and Joyard J (1996) Biosynthesis of thylakoid membrane lipids. In: Ort DR and Yocum CF (eds) Oxygenic Photosynthesis: The Light Reactions, pp 69–101. Kluwer Academic Publishers, Dordrecht

Douce R, Block MA, Dorne A-J and Joyard J (1984) The plastid envelope membranes: Their structure, composition, and role in chloroplast biogenesis. Subcell Biochem 10: 1–86

Dubacq JP and Trémolières A (1983) Occurrence and function of phosphatidyl glycerol containing Δ3-trans-hexadecenoic acid in photosynthetic lamellae. Physiol Veg 2: 293–312

Engeseth NJ, Pacovsky RS, Newman T and Ohlrogge JB (1996) Characterization of an acyl-CoA-binding protein from *Arabidopsis thaliana.* Arch Biochem Biophys 331: 55–62

Fiedler E, Soll J and Schultz G (1982). The formation of homogentisate in the biogenesis of tocopherol and plastoquinone in spinach chloroplasts. Planta 155: 511–515

Frentzen M, Heinz E, Mc Keon TA and Stumpf PK (1983) Specificities and selectivities of glycerol 3-phophate acyltransferase from pea and spinach chloroplasts. Eur J Biochem 129: 629–639

Fuesler TP, Hanamoto CM and Castelfranco PA (1982) Separation of Mg-Protoporphyrin IX and Mg-protoporphyrin IX monomethylester synthesized de novo by developing cucumber etioplasts. Plant Physiol 69: 421–423

Fuesler TP, Wong YS and Castelfranco PA (1984). Localization of Mg-chelatase and Mg-protoporphyrin IX mono methyl ester (oxidative) cyclase activities within isolated developing cucumber chloroplasts. Plant Physiol 75: 662–664

Fulda M, Heinz E and Wolter FP (1997) Brassica napus cDNAs encoding fatty acyl-CoA synthetase. Plant Mol Biol 33: 911–922

Garcia I, Rodgers M, Lenne C, Rolland A, Sailland A and Matringe M (1997) Subcellular localization and purification of a p-hydroxyphenylpyruvate dioxygenase from cultured carrot cells and characterization of the corresponding cDNA. Biochem J 325: 761–769

Gounaris K and Barber J (1983) Monogalactosyldiacylglycerol: The most abundant polar lipid in nature. Trends Biochem Sci 8: 378–381

Grafe S, Saluz HP, Grimm B and Hanel F (1999) Mg-chelatase of tobacco: The role of the subunit CHL D in the chelation step of protoporphyrin IX. Proc Natl Acad Sci USA 96: 1941–1946

Grusak MA and DellaPenna D (1999) Improving the nutrient composition of plants to enhance human nutrition and health. Annu Rev Plant Physiol Plant Mol Biol 50: 133–161

Gueguen V, Macherel D, Jaquinod M, Douce R and Bourguignon J (2000) Folic acid and lipoic acid biosynthesis in higher plant mitochondria. J Biol Chem 275: 5016–5025

Güler S, Essigmann B and Benning C (2000) A cyanobacterial gene, sqd X, required for biogenesis of the sulfolipid sulfoquinovosyldiacylglycerol. J Bact 182: 543–545

Heemskerk JWM and Wintermans JFGM (1987) The role of the chloroplast in the leaf acyl-lipid synthesis. Physiol Plant 70: 558–568

Heemskerk JWM, Storz T, Schmidt RR and Heinz E (1990) Biosynthesis of digalactosyldiacylglycerol in plastids from 16:3 and 18:3 plants. Plant Physiol 93: 1286–1294

Heinz E (1977) Enzymatic reactions in galactolipid biosynthesis. In: Tevini M and Lichtenthaler HK (eds) Lipids and Lipid Polymers, pp 102–120. Springer Verlag, Berlin

Heinz E and Roughan PG (1983) Similarities and differences in lipid metabolism of chloroplasts isolated from 18:3 and 16:3 plants. Plant Physiol 72: 273–279

Hinchigeri SB, Chan JC-S and Richards WR (1981) Purification of S-adenosyl-L-methionine: magnesium protoporphyrin methyltransferase by affinity chromatography. Photosynthetica 15: 351–359

Hinchigeri SB, Hundle B and Richards WR (1997) Demonstration that the BchH protein of *Rhodobacter capsulatus* activates S-adenosyl-L-methionine:magnesium proptoporphyrin IX methyltransferase. FEBS Lett 407: 337–342

Hudson A, Carpenter R, Doyle R and Coen ES (1993) Olive: A key gene required for chlorophyll biosynthesis in *Antirrinum majus.* EMBO J 12: 3711–3719

Hundal T, Forsmark-André P, Ernster L and Anderson B (1995) Antioxidant activity of reduced plastoquinone in chloroplast thylakoid membranes. Arch Biochem Biophys 324: 117– 122

Jäger-Vottero P, Dorne A-J, Jordanov J, Douce R and Joyard J (1997) Redox chains in chloroplast envelope membranes: Spectroscopic evidence for the presence of electron carriers, including iron-sulfur centers. Proc Natl Acad Sci USA 94: 1597–1602

Johanningmeier U and Howell SH (1984) Regulation of light-harvesting chlorophyll-binding protein mRNA accumulation in *Chlamydomonas reinhardii.* Possible involvement of chlorophyll synthesis precursors. J Biol Chem 259: 13541–13549

Jorasch P and Heinz E (1999) The enzymes for galactolipid biosynthesis are nearly all cloned: So what next? Trends Plant Sci 12: 469–471

Joyard J and Douce R (1977) Site of synthesis of phosphatidic acid and DAG in spinach chloroplasts. Biochim Biophys Acta 486: 273–285

Joyard J and Douce R (1979) Characterization of phosphatidate phosphohydrolase activity associated with chloroplast envelope membranes. FEBS Lett 102: 147–150

Joyard J and Douce R (1987) Galactolipid biosynthesis. In: PK Stumpf (ed) The Biochemistry of Plants, Vol 9, pp 215–274. Academic Press, New York

Joyard J and Stumpf PK (1981) Synthesis of long chain acyl-CoA in chloroplast envelope membranes. Plant Physiol 67: 250–256.

Joyard J, Blée E and Douce R (1986) Sulfolipid synthesis from $^{35}SO_4^{2-}$ and [1-^{14}C]-acetate in isolated intact spinach chloroplasts. Biochim Biophys Acta 879: 78–87

Joyard J, Block MA, Pineau B, Albrieux C and Douce R (1990) Envelope membranes from mature spinach chloroplasts contain a NADPH:protochlorophyllide reductase on the cytosolic side of the outer membrane. J Biol Chem 265: 21820–21827

Kader JC (1996) Lipid transfer protein in plants. Annu Rev Plant Physiol Plant Mol Biol 47: 627–655

Keller Y, Bouvier F, d'Harlingue A and Camara B (1998) Metabolic compartmentation of plastid prenyllipid biosynthesis-evidence for the involvement of a multifunctional geranylgeranyl reductase. Eur J Biochem 251: 413–417.

Koncz C, Mayerhofer R, Koncz-Kalman Z, Nawrath C, Reiss B, Redei GP and Schell J (1990) Isolation of a gene encoding a novel chloroplast protein by T-DNA tagging in *Arabidopsis thaliana*. EMBO J 9:1337–1346

Kropat J, Oster U, Rudiger W and Beck CF (1997) Chlorophyll precursors are signals of chloroplast origin involved in light induction of nuclear heat-shock genes. Proc Natl Acad Sci USA 94: 14168–14172

Kunst L, Browse J and Somerville CR (1988) Altered regulation of lipid biosynthesis in a mutant of *Arabidopsis* deficient in chloroplast glycerol-3-phosphate acyltransferase activity. Proc Natl Acad Sci USA 85: 4143–4147

Leegood RC, Walker DA and Foyer CH (1985) Regulation of the Benson Calvin cycle. In: Barber J and Barber NR (eds), Photosynthetic Mechanism and the Environment, pp 190–258. Elsevier Science Publishers, Amsterdam

Lichtenthaler HK (1993) The plant prenyllipids, including carotenoids, chlorophylls, and prenylquinones. In: Moore ST Jr (ed) Lipid Metabolism in Plants, pp 427–470. CRC press, Boca Raton

Lichtenthaler HK (1999) The 1-deoxy-D-xylulose-5-phosphate pathway of isoprenoid biosynthesis in plants. Annu Rev Plant Physiol Plant Mol Biol 50: 47–65

Lightner J, James DW Jr., Dooner HK and Browse J (1994) Altered body morphology is caused by increased stearate levels in a mutant of *Arabidopsis*. Plant J 6: 401–412

Lütke-Brinkhaus F, Liedvogel B and Kleinig H (1984) On the biosynthesis of ubiquinones in plant mitochondria. Eur J Biochem 14: 537–541

Lützow M and Kleinig H (1990) Chlorophyll-free chromoplasts from daffodil contain most of the enzymes for chlorophyll synthesis in a highly active form. Arch Biochem Biophys 277: 94–100

Malherbe A, Block MA, Joyard J and Douce R (1992) Feedback inhibition of phosphatidate phosphatase from spinach chloroplast envelope membrane by DAG. J Biol Chem 267: 23546–23553

Malkin R (1996) Photosystem I electron transfer reactions. Components and kinetics. In: Ort DR and Yocum CF (eds) Oxygenic Photosynthesis: The Light Reactions, pp 313–332. Kluwer Academic Publishers, Dordrecht

Maréchal E, Block MA, Joyard J and Douce R (1994a) Kinetic properties of MGDG synthase from spinach chloroplast envelope membranes. J Biol Chem 269: 5788–5798

Maréchal E, Block MA, Joyard J and Douce R (1994b) Comparison of the kinetic properties of MGDG synthase in mixed micelles and in envelope membranes from spinach chloroplast. FEBS Lett 352: 307–310

Maréchal E, Miège C, Block MA, Douce R and Joyard J (1995) The catalytic site of MGDG synthase from spinach chloroplast envelope membranes: Biochemical analysis of the structure and of the metal content. J Biol Chem 270: 5714–5722

Marin E, Nussaume L, Quesada A, Gonneau M, Sotta B, Hugueney P, Frey A and Marion-Poll A (1996) molecular identification of zeaxanthin epoxidase of *Nicotiana plumbaginifolia,* a gene involved in abscisic acid biosynthesis and corresponding to the ABA locus of *Arabidopsis thaliana*. EMBO J 15: 2331–2342

Martin BA and Wilson RF (1984) Subcellular localization of triacylglycerol biosynthesis in spinach leaves. Lipids 19: 117–121

Matile P, Hörtensteiner S and Thomas H (1999) Chlorophyll degradation. Annu Rev Plant Physiol Plant Mol Biol 50: 67–95

Matringe M, Camadro JM, Block MA, Joyard J, Scalla R, Labbe P and Douce R (1991) Localization within spinach chloroplasts of protoporphyrinogen oxidase, the target enzyme for diphenylether herbicides. J Biol Chem 267: 4646–4651

Mattoo AK and Edelman M (1987) Intramembrane translocation and posttranslational palmitoylation of the chloroplast 32-kDa herbicide-binding protein. Proc Natl Acad Sci USA 84: 1497–1501

McConn M and Browse J (1998) Polyunsaturated membranes are required for photosynthetic competence in a mutant of *Arabidopsis*. Plant J 15: 521–530

Miège C and Maréchal E (1999) 1,2-sn-diacylglycerol in plant cells: Product, substrate and regulator. Plant Physiol Biochem 37: 1–14

Miège C, Maréchal E, Shimojima M, Awai K, Block MA, Otha H, Takamiya K-I, Douce R and Joyard J (1999) Biochemical and topological properties of type A MGDG synthase, a spinach chloroplast envelope enzyme catalyzing the synthesis of both prokaryotic and eukaryotic MGDG. Eur J Biochem 265: 990–1001

Mongrand S (1998) De l'origine des lipides chloroplastiques eucaryotiques—étude in vivo—Distribution dans le règne végétal des lipides procaryotiques plastidiaux. Doctorat de l'université V Segalen Bordeaux 2, France.

Mongrand S, Bessoule JJ and Cassagne C (1997) A re-examination in vivo of the phosphatidylcholine-galactolipid metabolic relationship during plant lipid biosynthesis. Biochem J 327: 853–858

Mongrand S, Bessoule JJ, Cabantous F and Cassagne C (1998) The C16:3/C18:3 fatty acid balance in photosynthetic tissues from 468 plant species. Phytochemistry 49: 1049–1064

Moon BY, Higashi S, Gombos Z and Murata N (1995) Unsaturation of the membrane lipids of chloroplasts stabilizes the photosynthetic machinery against low-temperature photoinhibition in transgenic tobacco plants. Proc Natl Acad Sci USA 92: 6219–6223

Morré DJ, Morré JT, Morré SR, Sundqvist C and Sandelius AS (1991) Chloroplast biogenesis. Cell-free transfer of envelope monogalactosylglycerides to thylakoids. Biochim Biophys Acta 1070: 437–445

Mudd JB and de Zacks R (1981) Synthesis of phosphatidylglycerol by chloroplasts from leaves of *spinacia oleracea* L. Arch Biochem Biophys 209: 584–591

Mulichak AM, Theisen MJ, Essigmann B, Benning C and Garavito M (1999) Crystal structure of SQD 1, an enzyme involved in the biosynthesis of the plant sulfolipid headgroup donor UDP-sulfoquinovose. Proc Natl Acad Sci USA 96: 13097–13102

Murata N, Higashi S-I and Fujimura Y (1990) Glycerolipids in various preparations of Photosystem II from spinach chloroplasts. Biochim Biophys Acta 1019: 261–268

Nakayama M, Masuda T, Bando T, Yamagata H, Ohta H and Takamiya K (1998) Cloning and expression of the soybean chlH gene encoding a subunit of Mg-chelatase and localization of the Mg^{2+} concentration-dependent ChlH protein within the chloroplast. Plant Cell Physiol 39: 275–284

Nishida I and Yamada M (1986) Semisynthesis of a spin labeled monogalactosyldiacylglycerol and its application in the assay for galactolipid transfer activity in spinach leaves. Biochim Biophys Acta 813: 298–306

Nishida I, Frentzen M, Ishizaki O and Murata N (1997) Purification of isomeric forms of acyl-(acyl-carrier-protein): glycerol-3-phosphate acyltransferase from greening squash cotyledons. Plant Cell Physiol 28: 1071–1079

Norris SR, Shen X and DellaPenna D (1998) Complementation of the *Arabidopsis* pds1 mutation with the gene encoding p-hydroxyphenylpyruvate dioxygenase. Plant Physiol 117: 1317–1323

Nussberger S, Dorr K, Wang DN and Kühlbrandt W (1993) Lipid-protein interactions in crystals of plant light-harvesting complex. J Mol Biol 234: 347–356

Ohlrogge J and Browse J (1995) Lipid biosynthesis. Plant Cell 7: 957–970

Ohta H, Shimojima M, Arai T, Masuda T, Shioi Y and Takamiya K-I (1995) UDP-galactose:DAG galactosyltransferase in cucumber seedlings: Purification of the enzyme and its activation by phosphatidic acid. In: J-C Kader and P Mazliak (eds) Plant lipid Metabolism, pp. 152–155. Kluwer Academic Publishers, Dordrecht.

Okuley J, Lightner J, Feldmann K, Yadav N, Lark E and Browse J (1994) *Arabidopsis* FAD2 gene encodes the enzyme that is essential for polyunsaturated lipid synthesis. Plant Cell 6: 147–158

Papenbrock J, Grafe S, Kruse E, Hanel F and Grimm B (1997) Mg-chelatase of tobacco: identification of a Chl D cDNA sequence encoding a third subunit, analysis of the interaction of the three subunits with the yeast two-hybrid system, and reconstitution of the enzyme activity by co-expression of recombinant CHL D, CHL H and CHL I. Plant J 12: 981–990

Parmryd I and Dallner G (1999) In vivo prenylation of rat proteins: Modification of proteins with penta- and hexaprenyl groups. Arch Biochem Biophys 364: 153–160

Parmryd I, Shipton CA, Swiezewska E, Dallner G and Andersson B (1997) Chloroplastic prenylated proteins. FEBS Lett 414:527–531

Parmryd I, Andersson B and Dallner G (1999) Protein prenylation in spinach chloroplasts. Proc Natl Acad Sci USA 96: 10074–10079

Picot D and Garavito RM (1994) Prostaglandin H synthase: Implication for membrane structure. FEBS Lett 346: 21–25

Pineau B, Dubertret G, Joyard J and Douce R (1986) Fluorescence properties of the envelope membranes from spinach chloroplasts. Detection of protochlorophyllide. J Biol Chem 261: 9210–9215

Pineau B, Gérard-Hirne C, Douce R and Joyard J (1993) Identification of the main species of tetrapyrrolic pigments in envelope membranes from spinach chloroplasts. Plant Physiol 102: 821–828

Rawyler A, Meylan M and Siegenthaler PA (1992) Galactolipid export from envelope to thylakoid membranes in intact chloroplasts. I. Characterization and involvement in thylakoid lipid asymmetry. Biochim Biophys Acta 1104: 331–341

Rawyler A, Meylan-Bettex M and Siegenthaler PA (1995) (Galacto) lipid export from envelope to thylakoid membranes in intact chloroplasts. II. A general process with a key role for the envelope in the establishment of lipid asymmetry in thylakoid membranes. Biochim Biophys Acta 1233: 123–133

Reinbothe S, Runge S, Reinbothe C, van Cleve B and Apel K (1995) Substrate-dependent transport of the NADPH:protochlorophyllide oxidoreductase into isolated plastids. Plant Cell 7: 161–172

Reinbothe S, Reinbothe C, Lebedev N and Apel K (1996) POR A and POR B, two light-dependent protochlorophyllide-reducing enzymes of angiosperm chlorophyll biosynthesis. Plant Cell 8: 763–769

Reinbothe R, Lebedev N and Reinbothe S (1999) A protochlorophyllide light-harvesting complex involved in de-etiolation of higher plants. Nature 397: 80–84

Rhee KH, Morris EP, Barber J and Kuhlbrandt W (1998) Three-dimensional structure of the plant Photosystem II reaction centre at 8 Å resolution. Nature 396: 283–286

Rüdiger W and Schoch S (1988) Chlorophylls. In: Goodwin TW (ed) Plant Pigments, pp 1–59. Academic Press, London

Sakaki T, Kondo N and Yamada M (1990) Pathway for the synthesis of triacylglycerol from monogalactosyldiacylglycerols in ozone-fumigated spinach leaves. Plant Physiol 94: 773–780

Sato N, Sonoike K, Tsuzuki M and Kawaguchi A (1995) Impaired Photosystem II in a mutant of *Chlamydomonas reinhardtii* defective in sulfoquinovosyl diacylglycerol. Eur J Biochem 234: 16–23

Scheumann V, Klement H, Helfrich M, Oster U, Schoch S and Rudiger W (1999) Protochlorophyllide *b* does not occur in barley etioplasts. FEBS Lett 445: 445–448

Schmidt H and Heinz E (1990) Involvement of ferredoxin in desaturation of lipid-bound oleate in chloroplasts. Plant Physiol 94: 214–220

Schmidt H and Heinz E (1993) Direct desaturation of intact galactolipids by a desaturase solubilized from spinach (*Spinacia oleracea*) chloroplast envelopes. Biochem J 289: 777–782

Schmidt H, Dresselhaus T, Buck F and Heinz E (1994) Purification and PCR-based cDNA cloning of a plastidial n-6 desaturase. Plant Mol Biol 26: 631–642

Seifert U and Heinz E (1992) Enzymatic characteristics of UDP-sulfoquinovose: DAG sulfoquinovosyltransferase from chloroplast envelopes. Botanica Acta 105: 197–205

Seigneurin-Berny D, Rolland N, Dorne A-J and Joyard J (2000) Sulfolipid is a potential candidate for annexin binding to the outer surface of chloroplast. Biochem Biophys Res Commun 272: 519–524

Shaw P, Henwood J, Oliver R and Griffiths T (1985) Immunogold localization of protochlorophyllide oxidoreductase in barley etioplasts. Eur J Cell Biol 39: 50–55

Shimojima M, Ohta H, Iwamatsu A, Masuda T, Shioi Y, Takamiya K (1997) Cloning of the gene for monogalactosyldiacylglycerol synthase and its evolutionary origin. Proc Natl Acad Sci USA 94: 333–337

Shintani D and DellaPenna D (1999) Elevating the vitamin E content of plants through metabolic engineering. Science 282: 2098–2100

Siebertz HP, Heinz E, Linscheid M, Joyard J and Douce R (1979) Characterization of lipids from chloroplast envelopes. Eur J Biochem 101: 429–438

Slack CR, Roughan PG and Balasingham N (1977) Labelling studies in vivo on the metabolism of the acyl and glycerol moieties of the glycerolipids in the developing maize leaf. Biochem J 162: 289–296

Smith CA, Suzuki JY and Bauer CE (1996) Cloning and characterization of the chlorophyll biosynthesis gene chlM from *Synechocystis* PCC 6803 by complementation of a bacteriochlorophyll biosynthesis mutant of *Rhodobacter capsulatus*. Plant Mol Biol 30: 1307–1314

Soll J, Douce R and Schultz G (1980) Site of biosynthesis of α-tocopherol in spinach chloroplasts. FEBS Lett 112: 243–246

Soll J, Schultz G, Joyard J, Douce R and Block MA (1985) Localization and synthesis of prenylquinones in isolated outer and inner envelope membranes from spinach chloroplast. Arch Biochem Biophys 238: 290–299

Somerville C and Browse J (1996) Dissecting desaturation: Plants prove advantageous. Trends Cell Biol 6: 148–153

Stymne S and Stobart AK (1984) Evidence for the reversibility of the acyl-CoA:lysophosphatidylcholine acyltransferase in microsomal preparations from developing safflower (*Carthamus tinctorius L.*) cotyledons and rat liver. Biochem J 223: 305–314

Suzuki JY, Bollivar DW and Bauer CE (1997) Genetic analysis of chlorophyll biosynthesis. Annu Rev Genet 31: 61–89

Swiezewska E, Dallner G, Andersson B and Ernster L (1993) Biosynthesis of ubiquinone and plastoquinone in the endoplasmic reticulum-Golgi membranes of spinach leaves. J Biol Chem 268: 1494–1499

Tanaka R, Oster U, Kruse E, Rüdiger W and Grimm B (1999) Reduced activity of geranylgeranyl reductase leads to loss of chlorophyll and tocopherol and to partially geranylgeranylated chlorophyll in transgenic tobacco plants expressing antisense RNA for geranylgeranyl reductase. Plant Physiol 120: 695–704

Thai L, Rush JS, Maul JE, Devarenne T, Rodgers DL, Chappell J and Waechter CJ (1999) Farnesol is utilized for isoprenoid biosynthesis in plant cells via farnesyl pyrophosphate formed by successive monophosphorylation reactions. Proc Natl Acad Sci USA 96: 13080–13085

Tietje C and Heinz E (1998) Uridine-diphospho-sulfoquinovose: Diacylglycerol sulfoquinovosyltransferase activity is concentrated in the inner membrane of chloroplast envelopes. Planta 206: 72–78

Wellburn AR (1982) Bioenergetic and ultrastructural changes associated with chloroplast development. Int Rev Cytol 80: 133–191

Williams JP, Imperial V, Khan MU and Hodson JN (2000) The role of phosphatidylcholine in fatty acid exchange and desaturation in *Brassica napus* L. leaves. Biochem J 349: 127–133

Wolfe GR and Hoober JK (1996) Evolution of thylakoid structure. In: Ort DR and Yocum CF (eds) Oxygenic Photosynthesis: The Light Reactions, pp 31–40. Kluwer Academic Publishers, Dordrecht

Wong Y-S and Castelfranco PA (1984) Resolution and reconstitution of Mg-protoporphyrin IX monomethyl ester (oxidative) cyclase, the enzyme system responsible for the formation of the chlorophyll isocyclic ring. Plant Physiol 75: 658–661

Yalovsky S, Rodriguez-Concepcion M and Gruissem W (1999) Lipid modifications of proteins—slipping in and out of membranes. Trends Plant Sci 11: 439–445

Yamamoto HY and Bassi R (1996) Carotenoids: Localization and function. In: Ort DR and Yocum CF (eds) Oxygenic Photosynthesis: The Light Reactions, pp 539–563. Kluwer Academic Publishers, Dordrecht

Chapter 12

Pigment Assembly—Transport and Ligation

Harald Paulsen*
Institut für Allgemeine Botanik, Universität Mainz, Müllerweg 6, D-55099 Mainz, Germany

Summary

The ligation of pigments to proteins involved in photosynthesis appears to be strictly regulated and, in turn, to have an important regulatory impact on the biogenesis of the photosynthetic apparatus. Even so, the molecular mechanism of pigment-protein assembly is largely unknown. However, data are now accumulating on the co-translational transport of chlorophyll *a* proteins and the post-translational transport of chlorophyll *a/b* proteins into the thylakoid membrane. The molecular apparatus in the thylakoid membrane presumably occupied with protein insertion may also be involved in pigment ligation. Similarly, the last steps of pigment biosynthesis, whose location has not been fully established yet, will probably also provide a lead to the mechanism of pigment-protein assembly. Reconstitution studies with recombinant chlorophyll *a/b* proteins in vitro showed that the specificity of pigment binding varies—some, but not all, chlorophyll binding sites can be occupied with chlorophyll *a* or chlorophyll *b* almost equally well, and carotenoids can be structurally replaced with some other carotenoids, such as lutein with zeaxanthin. Finally, the possibility needs to be considered that pigments are assembled with proteins not only during the biogenesis of monomeric complexes but also during the assembly of multi-protein complexes of the photosynthetic apparatus.

*Email: Paulsen@mail.uni-mainz.de

Eva-Mari Aro and Bertil Andersson (eds): Regulation of Photosynthesis, pp. 219–233.

I. Introduction

The assembly of pigments of the photosynthetic apparatus with their apoproteins is a mass process that is very visible at the beginning of each warm season in temperate climate zones. 10^9 tons of chlorophyll (Chl) and carotenoids are estimated to be synthesized and assembled with pigment-binding proteins in photosynthesizing organisms every year. We know fairly well how these pigments are synthesized and how their biogenesis is regulated. On the other hand, our concept of how pigments are assembled with proteins is rather vague. This may be the reason why only a few recent reviews have focused on this topic (Nechushtai et al., 1995; Brand and Drews, 1997; Paulsen, 1997; Hoober et al., 1998; Hoober and Eggink 1999).

Our lack of information on the assembly of photosynthetic pigments with their apoproteins is astonishing, not only because this assembly occurs at such abundance but also because pigment-protein ligation appears to be an important regulatory step during the development of plants. The accumulation of Chls is tightly coupled with the assembly of Chl-protein complexes, and most Chl-binding proteins do not accumulate in a non-pigmented form. It may be this tight control of the assembly process that has made its elucidation so complicated. However, as outlined in this chapter, progress has been made recently in investigating the late steps of pigment biosynthesis, transport of pigment-binding proteins to the photosynthetic membranes as well as the molecular aspects of pigment binding using recombinant pigment-binding proteins in vitro. These are promising examples in the process of unraveling the mechanism of pigment-protein ligation.

Abbreviations: CAO – chlorophyllide-*a* oxygenase; Chl – chlorophyll; CP24, CP26, CP29 – minor light-harvesting chlorophyll *a/b* complexes of Photosystem II with apoproteins Lhcb6, Lhcb5, and Lhcb4, respectively; cpSRP – plastidic signal recognition particle; LHCI – light-harvesting chlorophyll *a/b* complex of Photosystem I; LHCI-680 – subunit of LHCI with apoproteins Lhca2 and Lhca3; LHCI-730 – subunit of LHCI with apoproteins Lhca1 and Lhca4; LHCII – light-harvesting chlorophyll *a/b* complex of Photosystem II with apoproteins Lhcb1-6; LHCIIb – major light-harvesting chlorophyll *a/b* complex of Photosystem II with apoproteins Lhcb1 and Lhcb2; PS I, PS II – Photosystems I and II, respectively

II. Assembly of Chlorophyll *a/b*-Protein Complexes

At least four different pathways exist for the insertion of proteins into the thylakoid membrane (Schnell, 1998). The Chl *a/b*–binding proteins of PS II, Lhcb1-6, enter the thylakoid membrane via one of these routes, the so-called cpSRP-dependent pathway (Li et al., 1995; Schuenemann et al., 1998). The components of this pathway known so far are (Fig. 1): cpSRP54 and cpSRP43, forming the chloroplast recognition particle, the chloroplast analog of FtsY, and GTP (Tu et al., 1999). The stromal loop of Lhcb1 appears to be involved in cpSRP binding, because cpSRP binds to a peptide, termed L18, in the stromal loop domain of Lhcb1 (Fig. 2) (DeLille et al., 2000). More specifically, L18 interacts with cpSRP43, whereas cpSRP54 seems to bind to (a) hydrophobic domain(s) in the protein (Tu et al., 2000). In the thylakoid membrane, ALB3 may be part of the translocase transporting Lhcb1 (Moore et al., 2000) (Fig. 1). ALB3 is an analog of Oxa1p, a component of a translocase in the inner mitochondrial membrane that inserts a number of inner membrane proteins. By analogy with the function of this mitochondrial translocase, it may be expected that ALB3 transports the luminal loop domain and the C-terminal domain of Lhcb1 across the membrane (Moore et al., 2000). This topography of Lhcb1 insertion into the thylakoid has also been suggested on the basis of in vitro experiments with Lhcb1 experiments carrying His_6 tags in various domains (Kosemund et al., 2000).

Mutants of *Arabidopsis thaliana* lacking cpSRP43 (Klimyuk et al., 1999) or containing reduced amounts of cpSRP54 (Pilgrim et al., 1998) both have a phenotype of reduced Chl *a/b* complexes. Interestingly, however, the phenotypes of these two mutants differ, suggesting that the subunits of cpSRP have some non-overlapping functions (Klimyuk et al., 1999). Both mutants do still contain Chl *a/b* proteins indicating that these proteins do not strictly depend on a complete and functional cpSRP for membrane insertion and assembly. Either the two subunits retain some activity in the absence of the other one, or there is some alternative pathway for the thylakoid insertion of Chl *a/b* proteins.

There are several indications that cpSRP is not only involved in the insertion of Chl *a/b* proteins into the thylakoid. The cpSRP54-less mutants have also reduced levels of Chl *a* proteins as (Klimyuk et al.,

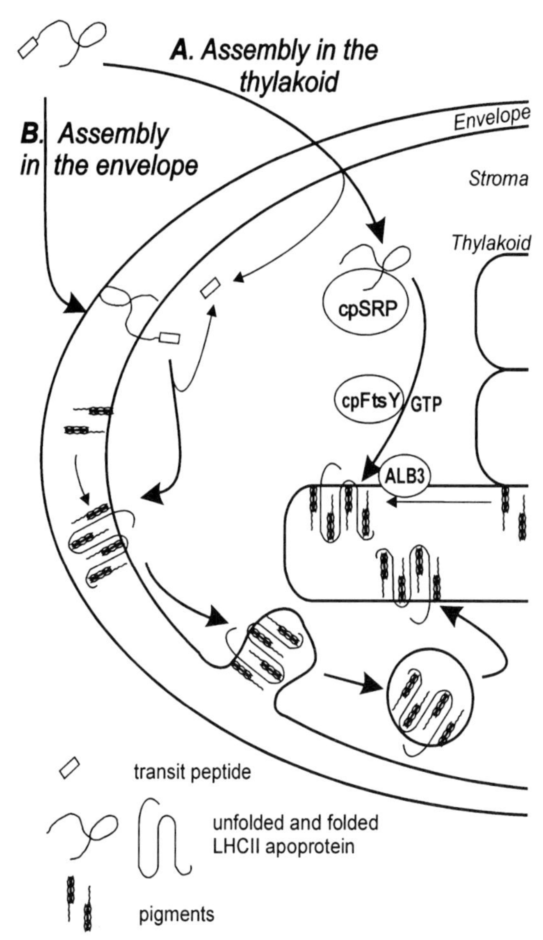

Fig. 1. Hypothetical pathways of LHCII apoprotein from its site of biosynthesis, the cytoplasm, towards the thylakoid membrane. In pathway A the protein crosses the stroma as a complex with the plastidic signal recognition particle (cpSRP), is transported to the thylakoid with the help of the plastid analog of FtsY, cpFtsY, and GTP, and finally becomes assembled with pigments in the thylakoid membrane. In pathway B the protein is ligated with pigments in the inner envelope membrane and then travels to the thylakoid in lipid vesicles budding from the inner envelope membrane. See text for references to the literature supporting either model.

1999). Crosslinking experiments showed that cpSRP54 selectively binds to a subset of thylakoid proteins or their precursors, including Lhcb1, cytochrome f and the Rieske protein (High et al., 1997). Another chloroplast-encoded protein besides cytochrome f that also has been shown to interact with cpSRP54 co-translationally is the D1 protein (Nilsson et al., 1999) (see below).

Nothing is known yet about any implication of pigments in the cpSRP pathway. As the pigments, before their ligation to proteins, most likely are membrane-localized, it is possible that the putative protein translocation apparatus in the thylakoid, including the ALB3 component, is also involved in ligating pigments to the translocated proteins.

An entirely different pathway of Lhcb1,2 insertion into the thylakoid membrane has been proposed (Fig. 1) based on observations with 'yellow-in-the-dark' mutants of *Chlamydomonas reinhardtii* (Hoober et al., 1998; Eggink and Hoober, 2000). These mutants allow the observation of very rapid greening when transferred from the dark into light, as they accumulate translatable mRNA for LHCII apoproteins already in the dark when grown at a raised temperature. Upon illumination, greening starts without a lag phase, as LHCII apoproteins immediately begin to enter the plastid and assemble with newly formed Chl. Interestingly, in parallel with this process, LHCII apoprotein also accumulates outside the plastid, in small soluble vacuoles (White et al., 1996). Even more striking is the observation that these LHCII apoproteins have the size of the mature protein, i.e. they presumably have lost their transit peptide responsible for their targeting to the plastid (White et al., 1996; Park and Hoober, 1997). If one assumes that the processing enzyme, converting LHCII apoproteins from their precursor to the mature form, is exclusively located in the plastid stroma, this implies that the N-terminal domain of these proteins has transiently crossed the thylakoid membrane. This observation, among others, led Hoober et al. (Hoober and Eggink, 1999) to propose that LHCII is assembled in the envelope rather than in the thylakoid. According to this model, the protein is (partially) inserted into the envelope, exposing the N-terminus to the stroma so that the transit sequence is cleaved off. If Chl is available in sufficient amounts, it binds to the amino acid sequence Glu-X-X-His/Asn-X-Arg, termed retention motif, which is found in the first and third trans-membrane domain of Lhcb1 (Fig. 2) and all other Chl *a/b*-binding proteins. In fact, Chl binding to a peptide maquette containing a retention motif could be demonstrated by fluorescence resonance energy transfer from a tryptophan in the peptide to Chl *a* (Eggink and Hoober, 2000). If Chl binds to the retention motif, the protein is assembled and then transported to the thylakoid, possibly via lipid vesicles that are thought to convey lipids from the envelope, the site of lipid biogenesis, to the thylakoid. The Chl

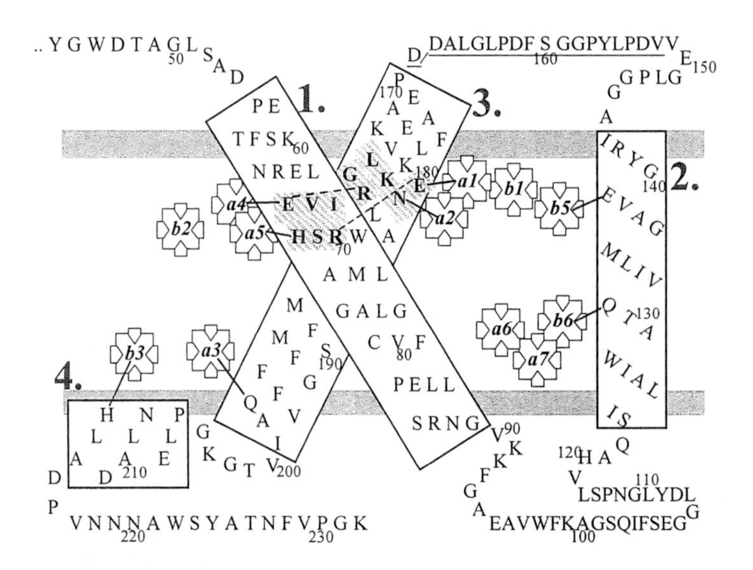

Fig. 2. Structural model of LHCIIb, according to Kühlbrandt et al. (1994). The α-helical protein domains (boxes) are numbered, starting from the N-proximal one. The surfaces of the thylakoid membrane are indicated by thick gray lines. The hatched sections of the protein sequence correspond to the so-called retention motifs (Eggink and Hoober, 2000). The underlined protein sequence represents the L18-peptide thought to interact with cpSRP (DeLille et al., 2000).

a/b proteins that fail to get assembled with pigments in the envelope are retracted into the cytoplasm and sequestered into small vacuoles where they are finally degraded (Park et al., 1999).

It is difficult to reconcile this model with the one of cpSRP-mediated transport of LHCIIb apoprotein into the thylakoid. If the site of LHCII assembly is the envelope, then it makes no sense to have a cpSRP machinery in the stroma, involved in transporting the apoprotein to the thylakoid. Moreover, the synthesis of chlorophyllide, the precursor of Chl, is thought to take place in the thylakoid (Armstrong et al., 1995; Holtorf et al., 1995) although the last enzyme in the Chl biosynthesis pathway, Chl synthetase, has not been unambiguously localized yet (see below, Section IV.A) and Chl has been found in the envelope (Douce and Joyard, 1996). Since it seems unlikely that pathways of Chl *a/b* proteins to the thylakoid are completely different between *C. reinhardtii* and higher plants, there is a possibility that both pathways exist in either organism. The presence of cpSRP in *C. reinhardtii* plastids has neither been shown nor excluded. Hoober et al. (1994) argue that the cpSRP-mediated assembly of Lhcb1,2 imported into isolated immature pea plastids might not properly reflect the situation in vivo, as the process is rather slow (Reed et al., 1990; Yuan et al., 1993), certainly slower than the import of lumen-targeted proteins (Cline and Henry, 1996). Also the observation that *A. thaliana* mutants deficient in cpSRP43 (Klimyuk et al., 1999) or cpSRP54 (Pilgrim et al., 1998) are still able to assemble Chl *a/b* proteins is consistent with the possible existence of an alternative pathway of Chl *a/b* proteins into the thylakoid (Fig. 1)

III. Assembly of Chlorophyll *a*-Protein Complexes

The plastid-encoded Chl *a*-binding proteins of the PS I and PS II core complexes are co-translationally inserted into the thylakoid. It is unknown whether pigments are bound to these proteins during membrane insertion or post-translationally. It is clear, however, that Chl *a* proteins, like the Chl *a/b* proteins, are stabilized by the binding of pigments (Eichacker et al., 1996a,b). However, pigments are not absolutely required for the accumulation, at least to a low extent, of Chl *a* proteins in the thylakoid: In etioplast membranes from barley a PS II pre-complex was found consisting of cytochrome b_{559}, the D2 protein and the precursor form of the D1 protein. Chl *a* synthesis then triggers the accumulation of monomers and, subsequently, dimers of the PS II reaction center (Müller and Eichacker, 1999).

Protein intermediates due to ribosome pausing have been observed upon biosynthesis of the D1 protein in etioplasts and chloroplasts (Kim et al., 1991; Kim et al., 1994); it has been suggested that ribosome pausing facilitates co-translational binding of Chl *a* to the protein. Efficient translation elongation and accumulation of the D1 protein in chloroplasts requires light (van Wijk and Eichacker, 1996; Edhofer et al., 1998; Mühlbauer and Eichacker, 1998) which again may be connected with a requirement for Chl *a*, newly synthesized in a light-dependent fashion.

The insertion of the plastid encoded D1 protein appears to involve cpSRP54 (Nilsson et al., 1999). Nascent chains of the D1 protein could be crosslinked to cpSRP54 as long as they were still attached to ribosomes. No interaction was detected with cpSRP43 or secA protein. Again, these observations give no information yet on the mechanism of pigment binding to the D1 protein but they may give a clue of where to look for a machinery involved in this process.

IV. How are Pigments Synthesized and Transported?

A. Chlorophylls

The rather simple and straight-forward question of where chlorophylls and carotenoids are localized immediately before they are assembled into Chl protein complexes of the photosynthetic apparatus cannot yet be answered. Particularly for the Chls, the pathway towards assembly is a complicated problem. Chls can become quite dangerous for their surroundings in the presence of oxygen and light. If Chl molecules, upon excitation by light, cannot get rid of their excitation energy fast enough, e.g. by transferring their energy to another Chl molecule or by using their excitation energy for electron transfer, they can adopt a triplet excited state. Triplet Chl itself is harmless but its lifetime is long enough that it can give rise to potentially very harmful singlet oxygen. Near-by carotenoids quench both triplet Chls and singlet oxygen. Therefore, it is reasonable to believe that Chls that have no efficient acceptor for their excitation energy will be bound in the vicinity of carotenoid molecules. Consequently, Chl biosynthesis must either take place very close to the site of Chl-carotenoid complex assembly, or there is some carotenoid-equipped Chl carrier involved in the translocation of Chls from one site to the other. This function has been proposed for a number of members of the extended Chl *a/b* protein family (see below).

In order to retrace the steps of Chl from the site of its biosynthesis to the site of its assembly with proteins, we need to know where its biosynthesis is finished. The last enzyme in the Chl *a* biosynthesis pathway is the Chl synthase, esterifying chlorophyllide with the long-chain alcohol phytol or geranylgeraniol. This enzyme from cyanobacteria has recently been expressed in *Escherichia coli* (Oster et al., 1997). Synthase activity has been found in the thylakoid membrane of chloroplasts (Soll et al., 1983) and in transforming prolamellar bodies of etioplasts (Lindsten et al., 1993). Now that antibodies of Chl synthase become accessible, it will be interesting to see where the protein resides in the plastid in order to facilitate the investigation of the Chl *a* pathway towards its target proteins in the photosynthetic apparatus.

The site of Chl *b* biosynthesis is also still unclear. In *C. reinhardtii*, six Chl *b*-less mutants were generated by insertional mutagenesis; the target gene was the same in all six mutants and showed homology to methyl monooxygenases (Tanaka et al., 1998). This suggests that Chl *b* is synthesized by action of a single monooxygenase on Chl *a* or one of its precursors, confirming the earlier observation that molecular oxygen is a substrate in Chl *b* biosynthesis (Porra et al., 1994). In fact, a chlorophyllide oxygenase (CAO) from *A. thaliana* capable of producing chlorophyllide *b* from chlorophyllide *a* has recently been overexpressed in *E. coli* (Oster et al., 2000). The recombinant activity does not oxidize Chl *a* or protochlorophyllide *a*, suggesting that Chl *b* synthesis takes place only via chlorophyllide *b*. A pathway for the formation of the 7-formyl group in Chl *b* via protochlorophyllide *b* has been proposed (Reinbothe et al., 1999), but questioned by a number of groups (Scheumann et al., 1999; Armstrong et al., 2000).The localization of CAO in the chloroplast may help to find out where the conversion of Chl *a* to Chl *b* actually takes place. It has been suggested that Chl *a* is oxidized in situ, that is when it is already bound to a Chl *a/b* protein (Porra et al., 1994; Plumley and Schmidt, 1995). Binding of chlorophyllide *a* to these proteins has not been shown yet but a transient interaction may be hard to detect. Moreover, it is possible that protein-bound but not unbound Chl *a* may be a substrate for CAO.

Chl *a* can also be produced by the reduction of

Chl *b* (Ito et al., 1996). This reaction initiates Chl *b* for degradation which can occur only via Chl *a* exclusively (Hörtensteiner et al., 1998). However, it is also possible that the resulting Chl *a* is recycled into Chl *a*-containing complexes, e.g. during the transition from low to high light when the antenna/reaction center ratio is reduced in most plants (Ito et al., 1994). Some enzymatic properties of Chl *b* reductase have been elucidated (Scheumann et al., 1998) but, again, the exact localization of the enzymatic apparatus involved in the reduction of Chl *b* is still unclear.

If there is a Chl carrier involved in the transport of Chl molecules from their site(s) of biogenesis to the site of Chl protein-complex assembly, it would be expected to be thylakoid-bound, contain carotenoids, and transiently bind Chls. As possible candidates for such a function, virtually all members of the Chl *a/b* protein family that are not assigned a light-harvesting function have been considered. These include the early light-induced protein (ELIP) (Adamska, 1997), the S unit of PS II (PsbS) (Funk et al., 1995a,b), and the small relatives containing only one-helix-spanning domain, the high light-induced protein (HLIP) in cyanobacteria (Miroshnichenko-Dolganov et al., 1995), the small Cab-like proteins (SCP) in cyanobacteria (Funk and Vermaas, 1999), and the one-helix protein (OHP) in *A. thaliana* (Jansson et al., 2000). Recently, stress-enhanced proteins (Sep) in *A. thaliana*, with two predicted trans-membrane domains, have been added to the family (Heddad and Adamska, 2000). Pigment binding has only been shown so far for PsbS (Funk et al., 1995a,b) and ELIP (Adamska et al., 1999). However, all of these proteins share sequence homology with the first and third trans-membrane helices of Chl *a/b* proteins and, therefore, may be expected to share at least some pigment binding sites as well. Several of the proteins in this group show stress-induced expression. This may point to an alternative (or additional) function of these proteins in scavenging Chls that are released during the stress-induced degradation of Chl-containing protein complexes (see Chapter 28, Adamska).

B. Carotenoids

The most common carotenoids in the photosynthetic apparatus of higher plants are β-carotene which occurs in Chl *a* complexes and some Chl *a/b* light-harvesting complexes, and xanthophylls (lutein, violaxanthin, antheraxanthin, zeaxanthin, and neoxanthin) assembled into Chl *a/b* complexes. There may be other carotenoids in the photosynthetic apparatus of some plants, such as lactucaxanthin replacing lutein in *Lactuca sativa* (Phillip and Young, 1995) or lutein-5,6-epoxide replacing neoxanthin in *Cuscuta reflexa* (Bungard et al., 1999). The xanthophylls are derived from α- and β-carotene: Hydroxylation in the C3 position of each ring in β-carotene and α-carotene produces zeaxanthin and lutein, respectively. Epoxidation and further modification of zeaxanthin then leads to antheraxanthin, violaxanthin, and neoxanthin. Violaxanthin is de-epoxidized to antheraxanthin and zeaxanthin in the regulatory xanthophyll cycle (Niyogi, 1999).

All carotenoids are synthesized in the plastids. According to a hypothesis by Cunningham and Gantt (1998) the terminal enzymes of the carotene biosynthesis pathway, leading to α- and β-carotene formation, are organized into a multi-subunit complex located in the thylakoid membrane, although it is difficult to exclude the possibility that these enzymes are actually bound to the envelope membrane. The hydroxylases generating lutein and zeaxanthin from α- and β-carotene are also thought to be bound to the thylakoid membrane (Cunningham and Gantt, 1998). The epoxidase converting zeaxanthin to antheraxanthin and then to violaxanthin is bound to the stromal surface of the thylakoid whereas the de-epoxidase, catalyzing the reverse conversions, has been found in the thylakoid lumen or, upon lumen acidification, bound to the lumenal side of the thylakoid membrane (Eskling et al., 1997). The site of the conversion of violaxanthin to neoxanthin, including the formation of an allenic group and a 9′-*cis* isomerization (Takaichi and Mimuro, 1998) is still unknown. Both xanthophylls are present in the thylakoid but are also the predominant carotenoids in the envelope membrane (Block et al., 1983).

If all of the carotenoids that are constituents of the photosynthetic apparatus are synthesized in the thylakoid membrane, they may just be partitioned into the lipid bilayer and become available for assembly into pigment-protein complexes, or they may be transported to the site of their assembly in some more specific way. Carotenoids such as lutein appear to be always protein-bound (Siefermann-Harms, 1985) but it is difficult to exclude the transient occurrence of uncomplexed pigment. If some carotenoids are synthesized in the envelope, they must be transported to the thylakoid, possibly via a route together with the envelope-synthesized thylakoid lipids.

V. How Do Pigments Find Their Correct Binding Site?

Pigment-binding proteins of the photosynthetic apparatus in plants always seem to bind their full complement of pigments. No substantial accumulation of non- or incompletely pigmented forms of these proteins has been shown, although low amounts of Chl *a*-binding proteins have been detected in etioplasts from barley (see above, Section III). This situation may be rationalized along the same line as the presumed existence of Chl carriers. A partially pigmented Chl-containing complex may leave a Chl molecule uncoupled to a proper excitation energy acceptor or to a triplet-quenching carotenoid and, therefore, pose the risk of producing toxic singlet oxygen. The full pigmentation of these proteins is most likely ensured by the stabilizing effect of the pigments: Proteins that lack some of their pigments are susceptible to thylakoid proteases and thus, rapidly degraded. The PsbS protein is the only potentially pigment-binding protein that has been reported to be stable also in the absence of Chls (but still containing carotenoids) as would be expected for a transient Chl carrier (Funk et al., 1995a,b). Clearly, Chls are required for the stable insertion of Chl *a/b* proteins into isolated thylakoids (Kuttkat et al., 1997; Bossmann et al., 1999). The stabilizing effect of Chl on Chl *a*-binding proteins (the PsaA and B proteins, the D1 and D2 proteins as well as CP43 and CP47) has also been shown (Eichacker et al., 1996a,b). Possibly, Chls are not only necessary in order to stabilize these proteins towards proteolysis but also for correct protein folding during membrane insertion (Paulsen et al., 1993; Paulsen, 1995).

Each pigment-protein complex of the photosynthetic apparatus is found to be pigmented at a constant pigment-protein ratio. Moreover, each complex has a specific pigment composition that seems quite independent of plant species, developmental state, and physiological condition (with a few exceptions like complexes containing the xanthophyll-cycle carotenoids). This is even more remarkable when the relatively small chemical difference between, e.g., Chl *a* and Chl *b* or lutein and zeaxanthin is considered. These observations suggest that pigment-binding sites in pigment-protein complexes are highly specific for a certain pigment species, and that the complexes are accumulated only when all pigment binding sites are filled. Specific pigment binding may be required for a well-defined architecture of each pigment-protein complex which in turn probably is a prerequisite for its correct functioning. But how is this binding specificity in pigment-binding proteins brought about?

A. Specificity of Chlorophyll Binding

The Chl *a/b* complexes are quite different in their Chl *b* requirement for assembly. In Chl *b*-deficient barley mutants, only three Chl *a/b* proteins are fully absent: Lhca4, Lhcb1, and Lhcb6. The accumulation of Lhca1, Lhca2, Lhca3, and Lhcb5 on the other hand is unaffected, whereas the remaining Chl *a/b* apoproteins are accumulated at reduced levels (Bossmann et al., 1997). It has not been established whether those Chl *a/b* proteins that do accumulate in the mutants are able to fill their Chl *b* binding sites with Chl *a* or can accumulate with these binding sites unoccupied.

A certain plasticity in the binding specificity of Chl *a* and *b* has been observed upon pigment reconstitution of over-expressed recombinant Chl *a/b*-protein complexes in vitro. When LHCIIb is reconstituted under stringent conditions, i.e. allowing only for rather stable recombinant complexes to be isolated, the resulting Chl *a/b* ratio stays at roughly 1.1 even when an excess amount of either Chl *a* or Chl *b* is used during reconstitution (Paulsen et al., 1990). However, if reconstitution and subsequent complex isolation is performed under milder conditions (e.g. with a less strongly denaturing detergent at a lower temperature), the Chl *a/b* ration in the recombinant complexes can vary within rather wide limits (Rogl et al., 1998, Kleima et al., 1999). Furthermore, a monomeric LHCIIb complex can be formed in the absence of Chl *a*; however, no stable pigment-protein complex is obtained in the absence of Chl *b* (Rogl et al., 1998). It should be noted however that the recombinant LHCIIb versions significantly deviating from the native Chl *a/b* ratio are considerably less stable than the native complex or recombinant complexes reconstituted to obtain a Chl *a/b* ratio of about 1.1.

Similar observations have been made with reconstituted complexes of the minor Chl *a/b* proteins, Lhcb4 (Giuffra et al., 1996, 1997), Lhcb5 (Ros et al., 1998), and Lhcb6 (Pagano et al., 1998). When recombinant Lhcb4 and Lhcb6 complexes are subjected to increasing amounts of Chl *a* relative to Chl *b*, the Chl *a/b* ratio within the reconstituted complex increases except for a plateau region where the ratio of protein-bound Chl *a* and Chl *b* does not change. This plateau is reached at a Chl *a/b* ratio of 1 for CP24 and at a ratio of 3 for CP29. These Chl *a/b* ratios are close to the values found in the corresponding native complexes, suggesting that this Chl composition presents an energy minimum for the recombinant complexes (Sandoná et al., 1998).

The variation of Chl *a/b* ratios in recombinant complexes suggests that some binding sites can readily be filled with the non-specific Chl molecule. A careful titration of recombinant Lhcb1 over a large range of Chl *a/b* ratios during reconstitution revealed that among the 12 Chl binding sites there are five Chl *b* positions that are highly specific. A maximum of two binding sites may have a high preference for Chl *a* but the titration curves can also be fitted assuming seven positions with only a low preference for Chl *a* (S. Hobe and H. Paulsen, unpublished). Consistently, mutation analyses of individual Chl binding sites in Chl *a/b* complexes indicated that some of these sites can be filled with either Chl *a* or Chl *b* (see next section). On the basis of these observations, it seems difficult to understand how a constant Chl *a/b* ratio is maintained in Chl *a/b* complexes in situ. The most likely explanation is that Chl binding into non-specific binding sites results in a significant loss of stability for the complex in the thylakoid membrane. In the case of LHCIIb, the selection for properly pigmented proteins possibly occurs on the level of trimeric complexes, as trimer formation in vitro seems to be more critical with respect to the proper Chl *a/b* ratio than the reconstitution of monomers (Rogl et al., 1998; Kleima et al., 1999).

B. Assignment of Chlorophyll Binding Sites

LHCIIb is the only Chl *a/b*-binding protein whose structure has been solved to the near-atomic level. Twelve Chl molecules are visible; however, the resolution is not sufficient to distinguish between Chl *a* and Chl *b*. Seven of the 12 Chl molecules have been assigned to Chl *a*, based on their proximity to the two carotenoids in the center of the complex, which are thought to be luteins (Kühlbrandt et al., 1994). Most Chls are seen to interact, via their central Mg atom, with amino acid side chains of the protein. The minor chlorophyll *a/b* proteins, Lhcb4-6, bind fewer Chls. Their binding sites have been assigned to Chl *a* and Chl *b* by analogy with LHCIIb, assuming that all these complexes share the same basic structure (Sandoná et al., 1998).

In order to test the assignment of Chl *a* and Chl *b* to individual binding sites in Lhcb4, Bassi et al. (1999) mutated single amino acids that are thought to coordinate Chl molecules. Biochemical analysis of the mutant pigment-protein complexes reconstituted in vitro indicated that four of the five Chls bound to the closely interacting protein helices 1 and 3 in the center of the complex are Chl *a* molecules. Mutation of the 5th binding site in the central domain as well as of the 4 more peripheral binding sites led to the loss of both Chl *a* and Chl *b*, suggesting that these are 'mixed' Chl positions. Dichroism spectra of mutant Lhcb4 compared to the wild-type complex were interpreted to obtain the orientation of each Chl within the complex (Simonetto et al., 1999).

Several groups performed a similar mutational analysis of some (Yang et al., 1999) or most of the Chl binding sites in LHCIIb (Remelli et al., 1999; Rogl and Kühlbrandt, 1999). Most of the original assignments (see Fig. 2) were confirmed with a few exceptions. Upon mutation of His212 in the amphiphilic helix binding Chl b3, Rogl und Kühlbrandt (1999) as well as Yang et al. (1999) saw a loss of predominantly Chl *a* suggesting that position b3 actually binds Chl *a*. Remelli et al. (2000) propose a mixture of Chl *a* and Chl *b* binding to position b3 as well as to position a3 and a6. Moreover, their data suggest that position a7 binds Chl *b* and position b1 binds Chl *a*.

LHCIIb contains approximately 12 Chl molecules; therefore, in order to distinguish between the loss of either one Chl *a* or one Chl *b* or a mixture of both upon mutating a Chl binding site requires an accuracy in measuring pigment stoichiometries better than 5%. This is not easy to achieve and becomes even more difficult when the exchange of a Chl-binding amino acid affects Chl molecules bound to other positions as well. Therefore it appears reasonable to consider at least some of the Chl assignments resulting from mutation analyses as preliminary.

C. Specificity of Carotenoid Binding

The plasticity of photosynthetic pigment-protein complexes with regard to their carotenoid composition appears to be even larger than that concerning the distinction between Chl *a* and Chl *b*. *A. thaliana* mutants lacking lutein show no obvious phenotype although lutein is a component of all Chl *a/b* light-harvesting complexes (Pogson et al., 1996). The carotenoid stoichiometries found in this mutant suggest that other carotenoids, in particular the xanthophyll cycle carotenoids violaxanthin, antheraxanthin and zeaxanthin can replace lutein in Chl *a/b* complexes. On the other hand, a *Scenedesmus obliquus* mutant deficient in β,ε-carotenoids has a significantly reduced content of Chl *a/b* complexes (Bishop, 1996).

A. thaliana mutants that accumulate zeaxanthin plus lutein or even zeaxanthin as their only xanthophyll are fully viable although they appear chlorotic at early stages of greening and, therefore, are retarded in their development (Pogson et al., 1998). Fully greened mutant plants exhibit photosynthesis rates close to those of the wild type but the rates of non-photochemical quenching are reduced. These observations show that, in spite of the remarkable plasticity of the assembly and light-harvesting function of the Chl *a/b* antenna with regard to their carotenoid composition, the native set of xanthophylls does appear to be required for normal development during greening and for the optimum response of the plants to light stress. The Chl *a/b* ratios in the various carotenoid mutants indicate that the amount of the Chl *a/b* light-harvesting complex is not reduced when the plants are deficient in lutein. By contrast, the lack of epoxidation products of zeaxanthin, that is neoxanthin, violaxanthin, and antheraxanthin or, even more so, the lack of these components plus lutein, leads to reduction in the amount of Chl *a/b* complexes. Apparently, neoxanthin (or another epoxidation product of zeaxanthin) is required for the accumulation of Chl *a/b* complexes to normal levels. Furthermore, the lack of lutein has no effect on the assembly of antenna complexes unless the epoxidation products of zeaxanthin are also missing. This may mean, as proposed by Pogson et al. (1998) that violaxanthin and/or antheraxanthin can more easily replace lutein than zeaxanthin, possibly because the epsilon ring of lutein is kinked out of plane in a similar way as the epoxy ring(s) of violaxanthin and antheraxanthin whereas zeaxanthin is a rather flat molecule. An alternative explanation is that the lack of both lutein and neoxanthin is more deleterious to Chl *a/b* complexes than the absence of either of these pigments. The observation in reconstitution experiments (see below) that zeaxanthin very easily replaces lutein in the Lhcb1 complex favors the second possible explanation, suggesting that lutein and neoxanthin somehow cooperate in stabilizing Chl *a/b* complexes.

The finding that lutein-deficient *A. thaliana* mutants accumulate Chl *a/b* complexes (see above) strongly suggests that the complex-stabilizing function ascribed to lutein can also be performed by other carotenoids, e.g. the xanthophyll-cycle carotenoids. On the other hand, *C. reinhardtii* mutants lacking the xanthophyll cycle pigments are still able to tolerate excess light conditions, probably because lutein also contributes to non-photochemical quenching (Niyogi et al., 1997). Now, if structural carotenoids can also dissipate energy and xanthophyll-cycle carotenoids can also stabilize the complex, the question is whether each xanthophyll can perform different functions in one and the same binding position. As an alternative scenario, different functions may be associated with different carotenoid binding sites in Chl *a/b* complexes, and xanthophylls may take on a different function by replacing another xanthophyll in its binding site.

Two carotenoids are visible in the LHCIIb structure (Kühlbrandt et al., 1994) that have been assigned to be luteins. The single molecule of neoxanthin and the substoichiometric amounts of violaxanthin (or its de-epoxidation products) associated with LHCIIb are not seen in the crystal structure. A localization of neoxanthin has been proposed on the basis of biochemical and spectroscopic data (Croce et al., 1999a). Mutations in the protein eliminating the binding site of Chl b5 (Fig. 2) also abolish neoxanthin binding, indicating that neoxanthin is associated with this Chl molecule in the vicinity of helix C. The orientation of neoxanthin with respect to the membrane plane was then determined by linear dichroism difference spectra of complexes plus/minus neoxanthin. The neoxanthin binding site was found to be highly specific whereas the two lutein binding sites also bound violaxanthin or zeaxanthin with lower affinity (Croce et al., 1999b).

In order to obtain information about the specificity of xanthophyll binding to LHCIIb, reconstitution experiments were performed in which two carotenoids competed for binding sites in the complex (Hobe et al., 2000). The reconstitution mixtures contained lutein and one other carotenoid so that the total amount of carotenoids stayed constant but the ratio between the two carotenoid species varied. Relative affinities (based on the comparison to lutein) of three binding sites were calculated as depicted in Fig. 3. Two binding sites were found to have a high preference of lutein over neoxanthin while the third one bound neoxanthin much more strongly than lutein. Violaxanthin had a lower affinity than lutein to the lutein-specific binding sites and a higher affinity than lutein to the neoxanthin-specific binding site but the specificity was not nearly as pronounced as compared to neoxanthin. Interestingly, the binding behavior of zeaxanthin was hardly distinguished from that of lutein, its affinities to all three binding sites being similar to those of lutein within a factor of 3 to 5. These observations would be consistent with a hypothetical scenario in which lutein and xanthophyll cycle carotenoids can replace each other in their binding sites when they take on different functions.

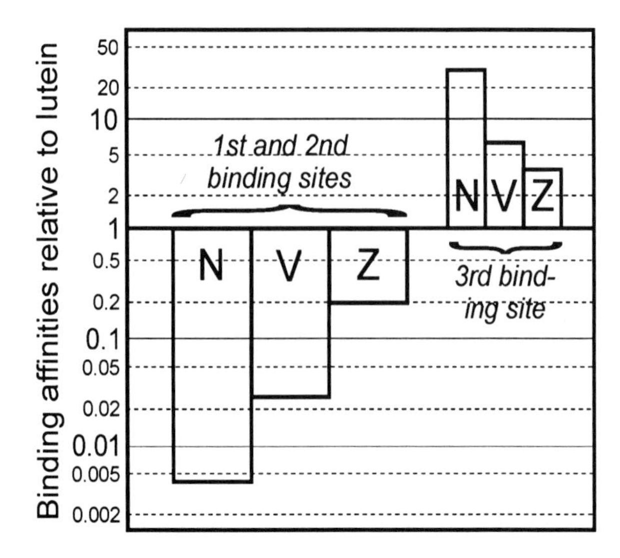

Fig. 3. Binding affinities of neoxanthin (N), violaxanthin (V), and zeaxanthin (Z), as compared to the affinity of lutein, to the 'lutein-specific' binding sites (1st and 2nd binding sites) and to the 'neoxanthin-specific' binding site (3rd binding site) of recombinant Lhcb1-pigment complex. Bars pointing upward and downward indicate higher affinities and lower affinities, respectively than those of lutein. Note the logarithmic scale of relative affinities.

The plasticity of the Chl *a/b* complexes with regard to different carotenoid structures can be extended even to structurally less related carotenoids. Two foreign carotenoids, capsaxanthin and capsorubin, are accumulated in tobacco plants that have been infected with a viral RNA vector carrying a gene for capsanthin-capsorubin synthase (Kumagai et al., 1998). This enzyme converts antheraxanthin and violaxanthin into capsaxanthin and capsorubin, respectively. These two xanthophylls are built into monomeric and trimeric LHCIIb but not into the PS I holocomplex or PS II core complex. Capsaxanthin and capsorubin seem to be able to replace lutein and violaxanthin.

A series of heterologous carotenoids has been tested replacing the native carotenoids lutein, neoxanthin and violaxanthin in the reconstitution of overexpressed Lhcb1 with pigments in vitro (D. Phillip, S. Hobe, A. Young, H. Paulsen, unpublished). Among the carotenoids supporting formation of the recombinant complex as the only carotenoid component were astaxanthin, okenone, and fucoxanthin which are all structurally quite different from the native carotenoids in this complex. A hydroxyl group in position 3 of at least one of the cyclohexane rings seems to be important for binding to and stabilizing the complex. Similarly, LHCIIb of *Chlorella fusca*, reconstituted from a heat-denatured native complex, binds prasinoxanthin but not peridinin when these pigments are added to the reconstitution mixture (Meyer and Wilhelm, 1993).

VI. Are Pigments Involved in the Assembly of Multi-protein Complexes of the Photosynthetic Apparatus?

The biogenesis of the photosynthetic apparatus is not complete with the assembly of apoproteins and pigments. These sub complexes then form the multi-subunit assemblies of the photosystems, including oligomeric states of antenna proteins. Protein phosphorylation has been shown to play a role in several assembly steps such as LHCIIb trimer and PS II dimer formation (see Chapters 23 (Rintamäki and Aro) and 24 (Ohad et al.)). Is the assembly of pigmented proteins into multi subunit complexes connected with pigment ligation, involving binding sites for example at the interface between several subunits? Such pigments might then stabilize complex-complex interaction in the holocomplex. Two of these processes of assembly into multiprotein complexes have been studied in some more detail: dimer formation of LHCI subunits and trimer formation of LHCIIb.

A. LHCI Dimer Formation

Reconstituted monomeric Lhca1 and Lhca4 have been reported to contain less Chl per apoprotein than the reconstituted or native heterodimer, LHCI-730 (Schmid et al., 1997). This observation may indicate that additional Chl molecules are bound during dimerization and then stabilize the dimer, in a similar way as lipids have been found to be essential for stabilizing LHCI-730 (Schmid et al., 1998). However, the different pigment composition of the LHCI-730 dimer and its subunits could also be explained by a lower stability of the monomeric vs. the dimeric form of the complex which in turn would lead to a preferential loss of Chls from the monomers during isolation. Whether additional pigments need to be added in order to form dimeric LHCI-680 from its subunits Lhca2 and Lhca3 remains to be established.

B. LHCIIb Trimer and Oligomer Formation

Whether trimeric LHCIIb has pigment binding sites that are not present in monomeric complexes is difficult to test experimentally. There is no report on different pigment stoichiometries being different between trimeric and monomeric LHCIIb, but the loss of one Chl molecule out of 12-15 upon monomerization of trimeric LHCIIb may be difficult to measure with any accuracy. The dissociation of isolated trimeric LHCIIb inevitably leads to the generation of some unbound pigment, probably due to an intrinsically lower stability of the monomeric compared to the trimeric complex. This may in turn mean that at least some of the pigments are more stably bound to the trimer than to the monomer, and hence these pigments may stabilize the trimer.

Among Lhcb1 mutants in which single Chl-binding amino acids have been exchanged, only those affecting the binding of Chl *a*5 and Chl *a*4 (Fig. 2) lose their ability to form trimers, both in a detergent environment (Rogl and Kühlbrandt, 1999) and upon insertion of the mutant proteins into isolated thylakoids (Yang et al., 1999). According to the structural model of LHCIIb, Chl *a*5 is thought to be located near the very center of the trimer, at the interface between monomers which may explain its potential impact on trimer stability. Chl *a*4 is neighboring *a*5 and could indirectly destabilize the trimer by destabilizing Chl *a*5. It may be interesting in this context that, as mentioned above (Section V.A), LHCIIb monomers can be reconstituted in vitro with Chl *b* as the only Chl, however for trimer formation, at least one Chl *a* per trimer is needed (Rogl et al., 1998; Kleima et al., 1999). It is therefore tempting to speculate that this would be a Chl *a* near the center of the trimeric complex like one in a Chl *a*5 position.

Whether carotenoids also are involved in LHCIIb trimer formation remains to be established. However, an important role of xanthophyll cycle carotenoids in the formation of LHCII aggregates has been reported. When zeaxanthin is added to detergent-dissolved isolated LHCII, it stimulates oligomerization and fluorescence quenching of these complexes. Violaxanthin added instead of zeaxanthin has the opposite effect by stabilizing a less aggregated state of LHCII in which fluorescence is not quenched. Horton and coworkers suggest this to be the molecular mechanism for how zeaxanthin and violaxanthin act in the xanthophyll cycle in which the de-epoxidation of violaxanthin forming antheraxanthin and then zeaxanthin, together with an acidification of the lumen, triggers energy dissipation in the Chl *a*/*b* antenna of PS II (Ruban et al., 1997). These carotenoids could either directly control the interaction between LHCII complexes or the reversible binding of one of these xanthophylls could trigger a conformational change in LHCII which then would influence oligomerization.

An alternative model of the xanthophyll cycle-mediated non-photochemical quenching has been proposed in which zeaxanthin or antheraxanthin directly accept excitation energy in the antenna complex and then dissipate it (Owens and Albrecht, 1992). Formation of zeaxanthin from violaxanthin in the xanthophyll cycle increases the number of conjugated double bonds from nine to eleven. Therefore, zeaxanthin (and antheraxanthin) are thought to have a lowered energy level of their first excited state that they can accept excitation energy from Chl *a* and, thus, quench Chl excitation. In order to distinguish between the two models of non-photochemical quenching by xanthophylls, Horton and co-workers tested whether the quenching effect of xanthophylls bound to detergent-solubilized isolated LHCII in fact correlates with their number of conjugated double bonds. They found that auroxanthin, a di-epoxy carotenoid with only seven conjugated double bonds mimicked the fluorescence quenching effect on isolated LHCII of zeaxanthin (eleven conjugated double bonds) although the energy level of its first excited state is thought to be even higher than that of violaxanthin (nine double bonds) (Ruban et al., 1998). Like zeaxanthin, auroxanthin has a much more planar shape than violaxanthin, so it may be this flat shape that enables zeaxanthin and auroxanthin to promote aggregation and fluorescence quenching of LHCII in vitro. This view is further supported by the recent finding of Frank et al. (2000) that zeaxanthin and violaxanthin have very similar energy levels of their first electronically exiled states.

One may speculate that pigments are also involved in the interaction between further subunits of the photosynthetic apparatus, in addition to factors that are known to control these interactions such as lipids (Nußberger et al., 1993; Kruse et al., 2000) or protein phosphorylation (Gal et al., 1997; Kruse et al., 1997). The improving structural information available from both PS I (Schubert et al., 1997; Klukas et al., 1999) and PS II (Rhee et al., 1997, 1998; Boekema et al., 1999a; Barber and Kühlbrandt, 1999) confirms that the photosystems have a well-defined architecture with each subunit in a fixed position, although there seems to be some flexibility in the arrangement of LHCIIb which may even have an impact on the regulation of photosynthesis (Boekema et al., 1999b). The possible role of pigments as a stabilizing factor in the photosystem holocomplexes will be an interesting topic to look into as soon as we know more about the molecular interactions that define and stabilize the architecture of multi-protein complexes in the photosynthetic apparatus.

Acknowledgments

Work in the author's lab has been funded by the *Deutsche Forschungsgemeinschaft*, *Stiftung Rheinland Pfalz für Innovation*, and *Fonds der Chemischen Industrie*.

References

Adamska I (1997) Elips: light induced stress proteins. Physiol Plant 100: 794–805

Adamska I, Roobol-Bóza M, Lindahl M and Andersson B (1999) Isolation of pigment-binding early light-inducible proteins from pea. Eur J Biochem 260: 453–460

Armstrong GA, Runge S, Frick G, Sperling U and Apel K (1995) Identification of NADPH:protochlorophyllide oxidoreductases A and B: A branched pathway for light-dependent chlorophyll biosynthesis in *Arabidopsis thaliana*. Plant Physiol 108: 1505–1517

Armstrong GA, Apel K and Rüdiger W (2000) Does a light-harvesting protochlorophyllide *a/b*-binding protein complex exist? Trends Plant Sci 5: 40–44

Barber J and Kühlbrandt V (1999) Photosystem II. Curr Opin Struct Biol 9: 469–475

Bassi R, Croce R, Cugini D and Sandonà D (1999) Mutational analysis of a higher plant antenna protein provides identification of chromophores bound into multiple sites. Proc Natl Acad Sci USA 96: 10056–10061

Bishop, NI (1996) The β,ε carotenoid, lutein, is specifically required for the formation of the oligomeric forms of the light harvesting complex in the green alga, *Scenedesmus obliquus*. J Photochem Photobiol B Biol 36: 279–283

Block MA, Dorne AJ, Joyard J and Douce R (1983) Preparation and characterization of membrane fractions enriched in outer and inner envelope membranes of spinach chloroplasts. J Biol Chem 258: 13281–86

Boekema EJ, van Roon H, van Breemen JFL and Dekker JP (1999a) Supramolecular organization of Photosystem II and its light-harvesting antenna in partially solubilized Photosystem II membranes. Eur J Biochem 266: 444–452

Boekema EJ, van Roon H, Calkoen F, Bassi R and Dekker JP (1999b) Multiple types of association of Photosystem II and its light-harvesting antenna in partially solubilized Photosystem II membranes. Biochemistry 38: 2233–2239

Bossmann B, Knötzel J and Jansson S (1997) Screening of chlorina mutants of barley (*Hordeum vulgare L.*) with antibodies against light harvesting proteins of PS I and PS II: Absence of specific antenna proteins. Photosynth Res 52: 127–136

Bossmann B, Grimme LH and Knötzel J (1999) Protease-stable integration of Lhcb1 into thylakoid membranes is dependent on chlorophyll *b* in allelic *chlorina-f2* mutants of barley (*Hordeum vulgare* L.). Planta 207: 551–558

Brand M and Drews G (1997) The role of pigments in the assembly of photosynthetic complexes in *Rhodobacter capsulatus*. J Basic Microbiol 37: 235–244

Bungard RA, Ruban AV, Hibberd JM, Press MC, Horton P and Scholes JD (1999) Unusual carotenoid composition and a new type of xanthophyll cycle in plants. Proc Natl Acad Sci USA 96: 1135–1139

Cline K and Henry R (1996) Import and routing of nucleus encoded chloroplast proteins. Ann Rev Cell Develop Biol 12: 1–26

Croce R, Remelli R, Varotto C, Breton J and Bassi R (1999a) The neoxanthin binding site of the major light harvesting complex (LHCII) from higher plants. FEBS Lett 456: 1–6

Croce R, Weiss S and Bassi R (1999b) Carotenoid-binding sites of the major light-harvesting complex II of higher plants. J Biol Chem 274: 29613–29623

Cunningham FX and Gantt E (1998) Genes and enzymes of carotenoid biosynthesis in plants. Annu Rev Plant Physiol 49: 557–583

DeLille J, Peterson EC, Johnson T, Moore M, Kight A and Henry R (2000) A novel precursor recognition element facilitates posttranslational binding to the signal recognition particle in chloroplasts. Proc Natl Acad Sci USA 97: 1926–1931

Douce R and Joyard J (1996) Biosynthesis of thylakoid membrane lipids. In: Ort DR and Yokum CF (eds) Oxygenic Photosynthesis: The Light Reactions, Vol. 4, pp 69–101. Kluwer Academic Publishers, Dordrecht

Edhofer I, Mühlbauer SK and Eichacker LA (1998) Light regulates the rate of translation elongation of chloroplast reaction center protein D1. Eur J Biochem 257: 78–84

Eggink LL and Hoober JK (2000) Chlorophyll binding to peptide maquettes containing a retention motif. J Biol Chem 275: 9087–9090

Eichacker LA, Helfrich M, Rüdiger W and Müller B (1996a) Stabilization of chlorophyll *a*-binding apoproteins P700, CP47, CP43, D2, and D1 by chlorophyll *a* or Zn-pheophytin a. J Biol Chem 271: 32174–32179

Eichacker LA, Müller B and Helfrich M (1996b) Stabilization of the chlorophyll-binding apoproteins, P700, CP47, CP43, D2, and D1, by synthesis of Zn-pheophytin *a* in intact etioplasts from barley. FEBS Lett 395: 251–256

Eskling M, Arvidsson PO and Akerlund HE (1997) The xanthophyll cycle, its regulation and components. Physiol Plant 100: 806–816

Frank HA, Bautista JA, Josua JS and Young AJ (2000) Mechanism of nonphotochemical quenching in green plants: Energies of the lowest excited singlet states of violaxanthin and zeaxanthin. Biochemistry 39: 2831–2837

Funk C, Adamska I, Green BR, Andersson B and Renger G (1995a) The nuclear-encoded chlorophyll-binding Photosystem II-S protein is stable in the absence of pigments. J Biol Chem 270: 30141–30147

Funk C, Schröder WP, Napiwotzki A, Tjus SE, Renger G and Andersson B (1995b) The PS II-S protein of higher plants: A new type of pigment-binding protein. Biochemistry 34: 11133–11141

Funk, C and Vermaas, W (1999) A cyanobacterial gene family coding for single-helix proteins resembling part of the light-harvesting proteins from higher plants. Biochemistry 38: 9397–9404

Gal A, Zer H and Ohad I (1997) Redox controlled thylakoid protein phosphorylation: News and views. Physiol Plant 100: 869–885

Giuffra E, Cugini D, Croce R and Bassi R (1996) Reconstitution and pigment-binding properties of recombinant CP29. Eur J Biochem 238: 112–120

Giuffra E, Zucchelli G, Sandoná D, Croce R, Cugini D, Garlaschi FM, Bassi R and Jennings RC (1997) Analysis of some optical properties of a native and reconstituted Photosystem II antenna complex, CP29: Pigment binding sites can be occupied by chlorophyll *a* or chlorophyll *b* and determine spectral forms. Biochemistry 36: 12984–12993

Heddad M and Adamska I (2000) Light stress-regulated two-helix proteins in *Arabidopsis thaliana* related to the chlorophyll *a/b*-binding gene family. Proc Natl Acad Sci USA 97: 3741–3746

High S, Henry R, Mould RM, Valent Q, Meacock S, Cline K, Gray JC and Luirink J (1997) Chloroplast SRP54 interacts with a specific subset of thylakoid precursor proteins. J Biol Chem 272: 11622–11628

Hobe S, Niemeier H, Bender A and Paulsen H (2000) Carotenoid binding sites in LHCIIb - Relative affinities towards major xanthophylls of higher plants. Eur J Biochem 267: 616–624

Holtorf H, Reinbothe S, Reinbothe C, Bereza B and Apel K (1995) Two routes of chlorophyllide synthesis that are differentially regulated by light in barley (*Hordeum vulgare* L.). Proc Natl Acad Sci USA 92: 3254–3258

Hoober JK and Eggink LL (1999) Assembly of light-harvesting complex II and biogenesis of thylakoid membranes in chloroplasts. Photosynth Res 61: 197–215

Hoober JK, White RA, Marks DB and Gabriel JL (1994) Biogenesis of thylakoid membranes with emphasis on the process in *Chlamydomonas*. Photosynth Res 39: 15–31

Hoober JK, Park H, Wolfe GR, Komine Y and Eggink LL (1998) Assembly of light-harvesting systems. In: Rochaix JD, Goldschmidt-Clermont M and Merchant S (eds) The Molecular Biology of Chloroplasts and Mitochondria in *Chlamydomonas*, pp 363–376. Kluwer Academic Publishers, Dordrecht

Hörtensteiner S, Wüthrich KL, Matile P, Ongania KH and Kräutler B (1998) The key step in chlorophyll breakdown in higher plants—Cleavage of pheophorbide *a* macrocycle by a monooxygenase. J Biol Chem 273: 15335–15339

Ito H, Takaichi S, Tsuji H and Tanaka A (1994) Properties of synthesis of chlorophyll *a* from chlorophyll *b* in cucumber etioplasts. J Biol Chem 269: 22034–22038

Ito H, Ohtsuka T and Tanaka A (1996) Conversion of chlorophyll *b* to chlorophyll *a* via 7-hydroxymethyl chlorophyll. J Biol Chem 271: 1475–1479

Jansson S, Andersson J, Jung Kim S and Jackowski G (2000) An *Arabidopsis thaliana* protein homologous to cyanobacterial high-light-inducible proteins. Plant Mol Biol 42: 345–351.

Kim J, Klein PG and Mullet JE (1991) Ribosomes pause at specific sites during synthesis of membrane-bound chloroplast reaction center protein D1. J Biol Chem 266: 14931–14938

Kim J, Klein PG and Mullet JE (1994) Synthesis and turnover of Photosystem II reaction center protein D1 - Ribosome pausing increases during chloroplast development. J Biol Chem 269: 17918–17923

Kleima FJ, Hobe S, Calkoen F, Urbanus ML, Peterman EJG, van Grondelle R, Paulsen H and van Amerongen H (1999) Decreasing the Chl *a/b* ratio in reconstituted LHCII: Structural and functional consequences. Biochemistry 38: 6587–6596

Klimyuk VI, Persello-Cartieaux F, Havaux M, Contard-David P, Schuenemann D, Meiherhoff K, Gouet P, Jones JDG, Hoffman NE and Nussaume L (1999) A chromodomain protein encoded by the *Arabidopsis CAO* gene is a plant-specific component of the chloroplast signal recognition particle pathway that is involved in LHCP targeting. Plant Cell 11: 87–99

Klukas O, Schubert WD, Jordan P, Krauss N, Fromme P, Witt HT and Saenger W (1999) Photosystem I, an improved model of the stromal subunits PsaC, PsaD, and PsaE. J Biol Chem 274: 7351–7360

Kosemund K, Geiger I and Paulsen H (2000) Insertion of light-harvesting chlorophyll *a/b* protein into the thylakoid—Topographical studies. Eur J Biochem 267: 1138–1145

Kruse O, Zheleva D and Barber J (1997) Stabilization of Photosystem II dimers by phosphorylation: Implication for the regulation of the turnover of D1 protein. FEBS Lett 408: 276–280

Kruse O, Hankamer B, Konczak C, Gerle C, Morris E, Radunz A, Schmid GH and Barber J (2000) Phosphatidylglycerol is involved in the dimerization of Photosystem II. J Biol Chem 275: 6509–6514

Kühlbrandt W, Wang DN and Fujiyoshi Y (1994) Atomic model of plant light-harvesting complex by electron crystallography. Nature 367: 614–621

Kumagai MH, Keller Y, Bouvier F, Clary D and Camara B (1998) Functional integration of non-native carotenoids into chloroplasts by viral-derived expression of capsanthin-capsorubin synthase in *Nicotiana benthamiana*. Plant J 14: 305–315

Kuttkat A, Edhofer I, Eichacker LA and Paulsen H (1997) Light harvesting chlorophyll *a/b* binding protein stably inserts into etioplast membranes supplemented with Zn pheophytin *a/b*. J Biol Chem 272: 20451–20455

Li X, Henry R, Yuan JG, Cline K and Hoffman NE (1995) A chloroplast homologue of the signal recognition particle subunit SRP54 is involved in the posttranslational integration of a protein into thylakoid membranes. Proc Natl Acad Sci USA 92: 3789–3793

Lindsten A, Wiktorsson B, Ryberg M and Sundqvist C (1993) Chlorophyll synthetase activity is relocated from transforming prolamellar bodies to developing thylakoids during irradiation of dark-grown wheat. Physiol Plant 88: 29–36

Meyer M and Wilhelm C (1993) Reconstitution of light-harvesting complexes from *Chlorella fusca* (*Chlorophyceae*) and *Mantoniella squamata* (*Prasinophyceae*). Z Naturforsch C 48: 461–473

Miroshnichenko-Dolganov NAM, Bhaya D and Grossman AR (1995) Cyanobacterial protein with similarity to the chlorophyll *a/b*-binding proteins of higher plants: Evolution and regulation. Proc Natl Acad Sci USA 92: 636–640

Moore M, Harrison MS, Peterson EC and Henry R (2000) Chloroplast Oxa1p homolog albino3 is required for post-translational integration off the light harvesting chlorophyll-binding protein into thylakoid membranes. J Biol Chem 275: 1529–1532

Mühlbauer SK and Eichacker LA (1998) Light-dependent formation of the photosynthetic proton gradient regulates translation elongation in chloroplasts. J Biol Chem 273: 20935–20940

Müller B and Eichacker LA (1999) Assembly of the D1 precursor in monomeric Photosystem II reaction center precomplexes precedes chlorophyll *a*-triggered accumulation of reaction center II in barley etioplasts. Plant Cell 11: 2365–2377

Nechushtai R, Cohen Y and Chitnis PR (1995) Assembly of the chlorophyll-protein complexes. Photosynth Res 44: 165–181

Nilsson R, Brunner J, Hoffman NE and van Wijk KJ (1999) Interactions of ribosome nascent chain complexes of the chloroplast-encoded D1 thylakoid membrane protein with cpSRP54. EMBO J 18: 733–742

Niyogi KK (1999) Photoprotection revisited: Genetic and molecular approaches. Annu. Rev. Plant Physiol. 50: 333–359

Niyogi KK, Björkman O and Grossman AR (1997) The roles of specific xanthophylls in photoprotection. Proc Natl Acad Sci USA 94: 14162–14167

Nußberger S, Dörr K, Wang DN and Kühlbrandt W (1993) Lipid-protein interactions in crystals of plant light-harvesting complex. J Mol Biol 234: 347–356

Oster U, Bauer CE and Rüdiger W (1997) Characterization of chlorophyll *a* and bacteriochlorophyll *a* synthases by heterologous expression in *Escherichia coli*. J Biol Chem 272: 9671–9676

Oster U, Tanaka R, Tanaka A and Rüdiger W (2000) Cloning and functional expression of the gene encoding the key enzyme for chlorophyll *b* biosynthesis (CAO) from *Arabidopsis thaliana*. Plant J 21: 305–310

Owens TG, Shreve AP and Albrecht AC (1992) Dynamics and mechanism of singlet energy transfer between carotenoids and chlorophylls: light harvesting and non-photochemical fluorescence quenching. In: Murata N (ed) Research in Photosynthesis, pp 179–186. Kluwer Academic Publishers, Dordrecht

Pagano A, Cinque G and Bassi R (1998) *In vitro* reconstitution of the recombinant Photosystem II light-harvesting complex CP24 and its spectroscopic characterization. J Biol Chem 273: 17154–17165

Park H and Hoober JK (1997) Chlorophyll synthesis modulates retention of apoproteins of light-harvesting complex II by the chloroplast in *Chlamydomonas reinhardtii*. Physiol Plant 101: 135–142

Park H, Eggink LL, Roberson RW and Hoober JK (1999) Transfer of proteins from the chloroplast to vacuoles in *Chlamydomonas reinhardtii* (Chlorophyta): A pathway for degradation. J Phycol 35: 528–538

Paulsen H (1995) Chlorophyll *a/b*-binding proteins. Photochem Photobiol 62: 367–382

Paulsen H (1997) Pigment ligation to proteins of the photosynthetic apparatus in higher plants. Physiol Plant 100: 760–768

Paulsen H, Rümler U and Rüdiger W (1990) Reconstitution of pigment-containing complexes from light-harvesting chlorophyll *a/b*-binding protein overexpressed in *E. coli*. Planta 181: 204–211

Paulsen H, Finkenzeller B and Kühlein N (1993) Pigments induce folding of light-harvesting chlorophyll *a/b*-binding protein. Eur J Biochem 215: 809–816

Phillip D and Young AJ (1995) Occurrence of the carotenoid lactucaxanthin in higher plant LHC II. Photosynth Res 43: 273–282

Pilgrim ML, vanWijk KJ, Parry DH, Sy DAC and Hoffman NE (1998) Expression of a dominant negative form of cpSRP54 inhibits chloroplast biogenesis in *Arabidopsis*. Plant J 13: 177–186

Plumley FG and Schmidt GW (1995) Light-harvesting chlorophyll *a/b* complexes: Interdependent pigment synthesis and protein assembly. Plant Cell 7: 689–704

Pogson B, McDonald KA, Truong M, Britton G and DellaPenna D (1996) *Arabidopsis* carotenoid mutants demonstrate that lutein is not essential for photosynthesis in higher plants. Plant Cell 8: 1627–1639

Pogson BJ, Niyogi KK, Björkman O and DellaPenna D (1998) Altered xanthophyll compositions adversely affect chlorophyll accumulation and nonphotochemical quenching in *Arabidopsis* mutants. Proc Natl Acad Sci USA 95: 13324–13329

Porra RJ, Schäfer W, Cmiel E, Katheder I and Scheer H (1994) The derivation of the formyl-group oxygen of chlorophyll *b* in higher plants from molecular oxygen—achievement of high enrichment of the 7-formyl-group oxygen from $^{18}O_2$ in greening maize leaves. Eur J Biochem 219: 671–679

Reed JE, Cline K, Stephens LC, Bacot KO and Viitanen PV (1990) Early events in the import assembly pathway of an integral thylakoid protein. Eur J Biochem 194: 33–42

Reinbothe C, Lebedev N and Reinbothe S (1999) A protochlorophyllide light-harvesting complex involved in de-etiolation of higher plants. Nature 397: 80–84

Remelli R, Varotto C, Sandonà D, Croce R and Bassi R (1999) Chlorophyll binding to monomeric light-harvesting complex - A mutation analysis of chromophore-binding residues. J Biol

Chem 274: 33510–33521
Rhee KH, Morris EP, Zheleva D, Hankamer B, Kühlbrandt W and Barber J (1997) Two dimensional structure of plant Photosystem II at 8 Å resolution. Nature 389: 522–526
Rhee KH, Morris EP, Barber J and Kühlbrandt W (1998) Three-dimensional structure of the plant Photosystem II reaction centre at 8 Å resolution. Nature 396: 283–286
Rogl H and Kühlbrandt W (1999) Mutant trimers of light-harvesting complex II exhibit altered pigment content and spectroscopic features. Biochemistry 38: 16214–16222
Rogl H, Lamborghini M and Kühlbrandt W (1998) Chlorophyll exchange on reconstituted LHCII: Chlorophyll *a* is essential for trimerisation. In: Garab G (ed) Photosynthesis: Mechanisms and Effects, Vol 1, pp 361–364. Kluwer Academic Publisher, Dordrecht
Ros F, Bassi R and Paulsen H (1998) Pigment-binding properties of the recombinant Photosystem II subunit CP26 reconstituted *in vitro*. Eur J Biochem 253: 653–658
Ruban AV, Phillip D, Young AJ and Horton P (1997) Carotenoid dependent oligomerization of the major chlorophyll *a/b* light harvesting complex of Photosystem II of plants. Biochemistry 36: 7855–7859
Ruban AV, Phillip D, Young AJ and Horton P (1998) Excited-state energy level does not determine the differential effect of violaxanthin and zeaxanthin on chlorophyll fluorescence quenching in the isolated light-harvesting complex of Photosystem II. Photochem Photobiol 68: 829–834
Sandoná D, Croce R, Pagano A, Crimi M and Bassi R (1998) Higher plants light harvesting proteins. Structure and function as revealed by mutation analysis of either protein or chromophore moieties. Biochim Biophys Acta 1365: 207–214
Scheumann V, Schoch S and Rüdiger W (1998) Chlorophyll *a* formation in the chlorophyll *b* reductase reaction requires reduced ferredoxin. J Biol Chem 273: 35102–35108
Scheumann V, Klement H, Helfrich M, Oster U, Schoch S and Rüdiger W (1999) Protochlorophyllide *b* does not occur in barley etioplasts. FEBS Lett 445: 445–448
Schmid VHR, Cammarata KV, Bruns BU and Schmidt GW (1997) In vitro reconstitution of the Photosystem I light-harvesting complex LHCI-730: Heterodimerization is required for antenna pigment organization. Proc Natl Acad Sci USA 94: 7667–7672
Schmid V, Beutelmann P, Schmidt GW and Paulsen H (1998) Ligand requirement for LHCI reconstitution. In: Garab G (ed) Photosynthesis: Mechanisms and Effects, Vol I, pp 425–428. Kluwer Academic Publishers, Dordrecht
Schnell DJ (1998) Protein targeting to the thylakoid membrane. Annu Rev Plant Physiol Plant Mol Biol 49: 97–126
Schubert WD, Klukas O, Krauss N, Saenger W, Fromme P and Witt HT (1997) Photosystem I of *Synechococcus elongatus* at 4 Å resolution: Comprehensive structure analysis. J Mol Biol 272: 741–769
Schuenemann D, Gupta S, PerselloCartieaux F, Klimyuk VI, Jones JDG, Nussaume L and Hoffman NE (1998) A novel signal recognition particle targets light-harvesting proteins to the thylakoid membranes. Proc Natl Acad Sci USA 95: 10312–10316
Siefermann-Harms D (1985) Carotenoids in photosynthesis. I. Location in photosynthetic membranes and light-harvesting function. Biochim Biophys Acta 811: 325–355
Simonetto R, Crimi M, Sandonà D, Croce R, Cinque G, Breton J and Bassi R (1999) Orientation of chlorophyll transition moments in the higher-plant light-harvesting complex CP29. Biochemistry 38: 12974–12983
Soll J, Schultz G, Rüdiger W and Benz J (1983) Hydrogenation of geranylgeraniol. Two pathways exist in spinach chloroplasts. Plant Physiol 71: 849–854
Takaichi S and Mimuro M (1998) Distribution and geometric isomerism of neoxanthin in oxygenic phototrophs: 9′-*cis*, a sole molecular form. Plant Cell Physiol 39: 968–977
Tanaka A, Ito H, Tanaka R, Tanaka NK, Yoshida K and Okada K (1998) Chlorophyll *a* oxygenase (CAO) is involved in chlorophyll *b* formation from chlorophyll *a*. Proc Natl Acad Sci USA 95: 12719–12723
Tu CJ, Schuenemann D and Hoffman NE (1999) Chloroplast FtsY, chloroplast signal recognition particle, and GTP are required to reconstitute the soluble phase of light-harvesting chlorophyll protein transport into thylakoid membranes. J Biol Chem 274: 27219–27224
Tu CJ, Peterson EC, Henry R and Hoffman NE (2000) The L18 domain of light-harvesting chlorophyll proteins binds to chloroplast signal recognition particle 43. J Biol Chem 275: 13187–13190
van Wijk KJ and Eichacker L (1996) Light is required for efficient translation elongation and subsequent integration of the D1-protein into Photosystem II. FEBS Lett 388: 89–93
White RA, Wolfe GR, Komine Y and Hoober JK (1996) Localization of light-harvesting complex apoproteins in the chloroplast and cytoplasm during greening of *Chlamydomonas reinhardtii* at 38 °C. Photosynth Res 47: 267–280
Yang CH, Kosemund K, Cornet C and Paulsen H (1999) Exchange of pigment-binding amino acids in light-harvesting chlorophyll *a/b* protein. Biochemistry 38: 16205–16213
Yuan JG, Henry R and Cline K (1993) Stromal factor plays an essential role in protein integration into thylakoids that cannot be replaced by unfolding or by heat shock protein Hsp70. Proc Natl Acad Sci USA 90: 8552–8556

Chapter 13

Chlorophyll Biosynthesis—Metabolism and Strategies of Higher Plants to Avoid Photooxidative Stress

Klaus Apel*
Swiss Federal Institute of Technology (ETH), Institute of Plant Sciences – Plant Genetics, Universitätstr. 2, CH 8092 Zürich, Switzerland

Summary

Photosynthetic organisms are prone to various forms of oxidative stress. Upon illumination excited porphyrin molecules such as chlorophyll (Chl) may transfer the excitation energy directly to oxygen, thus leading to the formation of highly reactive singlet oxygen. Most of the chlorophyll is bound to proteins and in this state may use various quenching mechanisms to dissipate absorbed light energy. Its biosynthetic precursors, however, occur mostly in a free form and are potentially much more destructive when illuminated. Thus, in order to avoid photooxidative damage plants had to evolve highly efficient strategies to prevent the accumulation of Chl intermediates. Two major rate-limiting steps, at which biosynthesis of Chl may be regulated, have been studied in greater details: the light-dependent reduction of protochlorophyllide (Pchlide) to chlorophyllide (Chlide) and the synthesis of 5-aminolevulinic acid (ALA), a precursor common to all tetrapyrroles. Both steps cooperate closely during the regulation of tetrapyrrole synthesis. In etiolated seedlings ALA synthesis is blocked once a critical level of Pchlide has been reached. Only after illumination, when Pchlide is photo-reduced and synthesis of ALA released from the feed-back inhibition, does the tetrapyrrole biosynthesis resume. This regulation of ALA synthesis has often been seen as being important for the protection of higher plants against photooxidative

*Email: klaus.apel@ipw.biol.ethz.ch

Eva-Mari Aro and Bertil Andersson (eds): Regulation of Photosynthesis, pp. 235–252.

damage. However, with the recent discovery of Chl intermediates acting as signals that control the nuclear plastid interaction and activate the defense against oxidative stress it seems likely that the feed-back control of ALA synthesis plays a more general role for the regulation of chloroplast development and the expression of some of the nuclear genes.

I. Introduction

Tetrapyrroles play an essential role in all living organisms. They are involved in various metabolic processes such as energy transfer, signal transduction and catalysis. In higher plants at least three structurally and functionally distinct classes of tetrapyrroles can be distinguished, Mg^{2+}-porphyrins (chlorophylls), $Fe^{2+/3+}$-porphyrins (hemes) and phycobilins (phytochromobilin). Chlorophylls (Chl) are the most abundant tetrapyrrole compounds. They play a fundamental role in the energy absorption and transduction activities of photosynthetic organisms. Hemes, the second class of tetrapyrroles, act as redox-active cofactors or prosthetic groups in many proteins. They are bound to various cytochromes of the plastidic and mitochondrial electron transport chains and to soluble enzymes such as catalases and peroxidases (Beale and Weinstein, 1990; von Wettstein et al., 1995; Reinbothe and Reinbothe, 1996). In contrast to Chls and hemes, the third group, phycobilins, are linear tetrapyrrole molecules. Their major constituent in higher plants, phytochromobilin, functions as a chromophore of the phytochromes, a small family of photoreceptors that regulate various aspects of light-controlled plant development (Fankhauser and Chory, 1997).

All of the tetrapyrroles described have extensive conjugated double-bond systems and thus strongly absorb light of visible as well as ultraviolet-A wavelengths. In the case of Chl and its derivatives, absorption of light quanta results in excitation of the molecules to the singlet state. If the energy of the absorbed quanta cannot be dissipated through photosynthetic electron transport or some other quenching mechanisms, the excited Chl changes from the singlet state into the much longer-living triplet (^{3}Chl) state. Triplet-state tetrapyrroles are highly active photosensitizers. Photodynamic damage can occur either directly by a reaction between ^{3}Chl and substrates such as fatty acids to produce a free radical cascade, or by reacting with O_2 to produce highly active singlet oxygen (Spikes and Bommer, 1991; Hideg et al., 1998). Singlet oxygen, one of the most potent reactive oxygen species, causes various types of cellular damage including lipid peroxidation, membrane lysis, DNA strand breakage and protein inactivation (Alscher et al., 1997).

Plants that accumulate free tetrapyrroles are particularly prone to the risk of photooxidative damage (Rebeiz et al., 1988). Most of the Chls, hemes and phycobilins are therefore bound to proteins and in this state may use various quenching mechanisms to dissipate absorbed light energy. Their biosynthetic precursors, however, occur mostly in a free form and are potentially much more destructive when illuminated. Thus, in order to avoid photooxidative damage, plants have had to evolve highly efficient strategies to prevent accumulation of intermediates during tetrapyrrole biosynthesis (Thomas, 1997). While in recent years enormous progress has been made in the identification and characterization of enzymes involved in tetrapyrrole biosynthesis (von Wettstein et al., 1995), up to now rather little is known about how this biosynthetic pathway is regulated. In this review emphasis will be given to aspects of Chl biosynthesis regulation that seem to be of special importance for protecting higher plants against photooxidative damage.

II. Tetrapyrrole Biosynthesis and Photooxidative Stress

A. Biosynthesis Pathways

In all living organisms, 5-aminolevulinic acid (ALA) is considered to be the first committed precursor of all tetrapyrroles. There are two distinct pathways by which ALA is synthesized. In animals and fungi the single mitochondrial enzyme, ALA-synthase, catalyzes the condensation of glycine with succinyl-coenzyme A to generate ALA; this route is called the Shemin pathway (Kikuchi et al., 1958). A different biosynthetic route, however, originating from

Abbreviations: ALA – 5-aminolevulinic acid; BChl – bacteriochlorophyll; cab protein – light-harvesting Chl *a/b* protein; Chl – chlorophyll; Chlide – chlorophyllide; LHPP – light harvesting POR-Pchlide complex; Pchlide – protochlorophyllide; POR – NADPH-protochlorophyllide oxidoreductase

glutamate and termed the C_5 pathway, operates in plants, cyanobacteria, most eubacteria and archaebacteria (Beale et al., 1975). In this pathway ALA is synthesized from glutamate in two sequential enzymatic steps (Kumar et al., 1996) (Fig. 1). In the first step, tRNA charged with glutamate is reduced to glutamate-1-semialdehyde in a reaction catalyzed by Glu-tRNA reductase. The Glu-tRNA is common to the biosynthesis of both porphyrins and proteins. Glu-1-semialdehyde aminotransferase then catalyzes the conversion of glutamate-1-semialdehyde to ALA (Kannangara et al., 1988). All of the carbon and nitrogen atoms in the porphyrin nucleus of tetrapyrroles are derived from ALA. Assembly of the tetrapyrrole skeleton begins by condensation of two groups of two ALA molecules to form porpho-

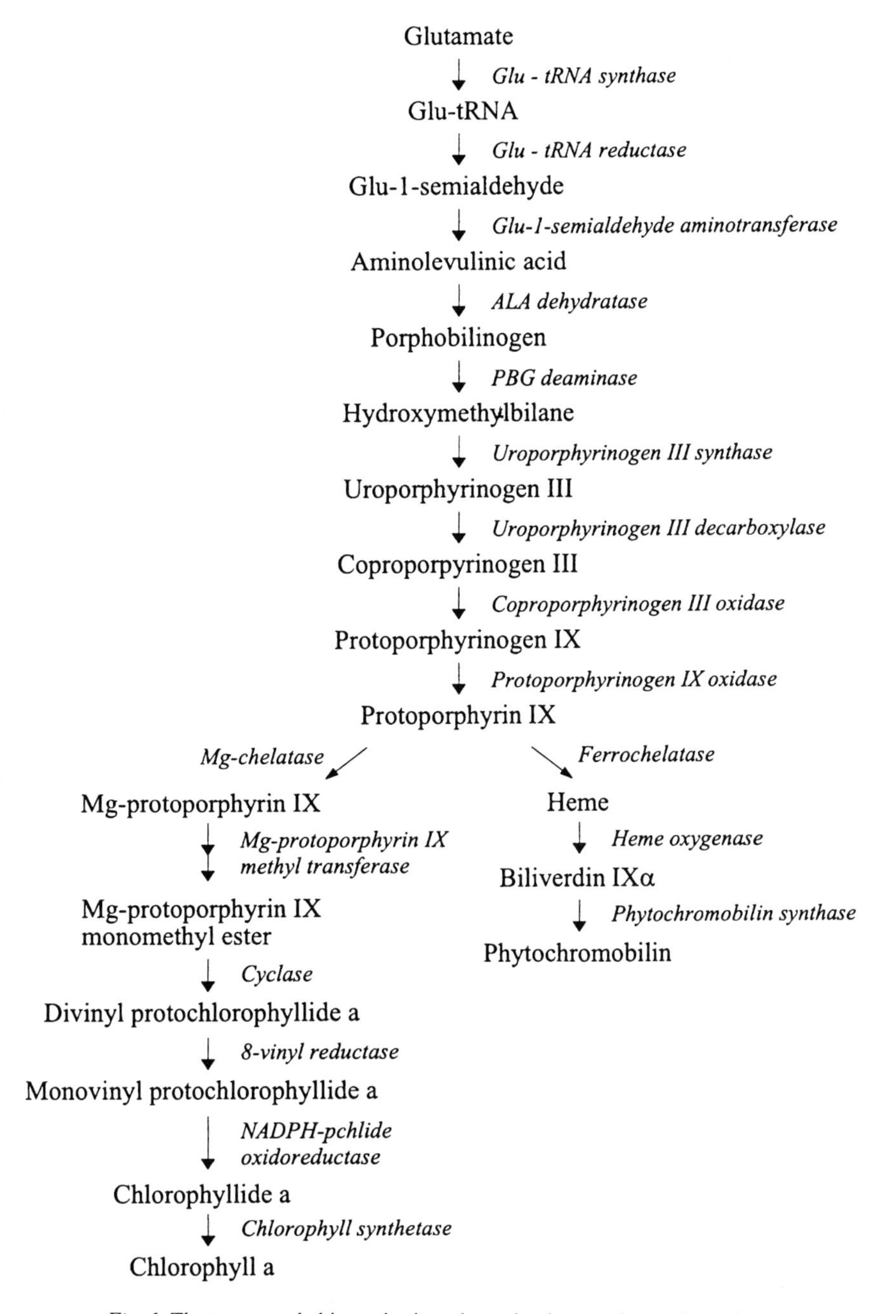

Fig. 1. The tetrapyrrole biosynthesis pathway in plants and cyanobacteria.

bilinogen units. Next, four of these units are linked together in a head-to-tail sequence to produce a linear tetrapyrrole, hydroxymethylbilane. This molecule is then enzymatically closed to form the first cyclic tetrapyrrole, uroporphyrinogen III, which is subsequently modified by three enzymes: uroporphyrinogen decarboxylase, coproporphyrinogen III oxidase and protoporphyrinogen IX oxidase. Conversion of uroporphyrinogen III to coproporphyrinogen III is accomplished by decarboxylation of the four acetic acid substituents, leaving methyl groups. Then, two of the four propionic acid moieties are oxidatively decarboxylated to vinyl groups forming protoporphyrinogen IX. Finally, protoporphyrinogen IX oxidase catalyzes a six-electron oxidation of protoporphyrinogen IX to protoporphyrin IX and establishes the system of conjugated double-bounds in this tetrapyrrole intermediate (Fig. 1).

Up until the level of protoporphyrin IX, the synthesis of heme, phycobilins and Chl shares a common pathway. However, at the point of metal ion insertion the pathway diverges, one route being directed to the synthesis of heme and phycobilins, and the other giving rise to the formation of Chl (Fig. 1). In heme and phycobilin synthesis a ferrous iron is inserted into protoporphyrin IX by the enzyme ferrochelatase. The first step in Chl formation begins with the insertion of a Mg^{2+} ion into protoporphyrin IX and is catalyzed by Mg-chelatase. The resulting Mg^{2+}-protoporphyrin IX is esterified to Mg^{2+}-protoporphyrin IX monomethylester and then the isocyclic ring of the macrocycle is formed. This reaction requires atmospheric O_2. The final product of the Mg^{2+} branch is protochlorophyllide *a* (Pchlide *a*), the immediate precursor of chlorophyllide (Chlide). Most oxygenic photosynthetic organisms have two options to synthesize Chlide *a* (see Section III.B.1). In a light-independent reaction Pchlide can be reduced to Chlide through a Pchlide-reducing enzyme that requires three different polypeptides, Ch L, Ch B and Ch N, that in plants are encoded by chloroplast DNA. In addition to this light-independent enzyme, they also contain a second light- and NADPH-dependent Pchlide-reducing enzyme, the NADPH-Pchlide oxidoreductase (POR) (Fig. 2). The last step in the synthesis of Chl *a* is catalyzed by Chl synthetase. This enzyme esterifies the propionic acid side chain of ring IV with either phytyl pyrophosphate or with geranyl geraniol pyrophosphate with the subsequent reduction of the geranyl geranyl group to phytyl.

B. Tetrapyrrole Intermediates and Photooxidative Stress

A direct correlation between the accumulation of cyclic tetrapyrrole intermediates and photooxidative leaf damage has been often demonstrated. Several steps of the tetrapyrrole biosynthetic pathway have been blocked experimentally either by expressing antisense RNA in transgenic plants or by treating plants with enzyme-specific inhibitors. Antisense expression technology has been used to reduce levels of glutamate-1-semialdehyde aminotransferase (Höfgen et al., 1994), uroporphyrinogen decarboxylase (Mock and Grimm, 1997), coproporphyrinogen III oxidase (Kruse et al., 1995) and protopor-

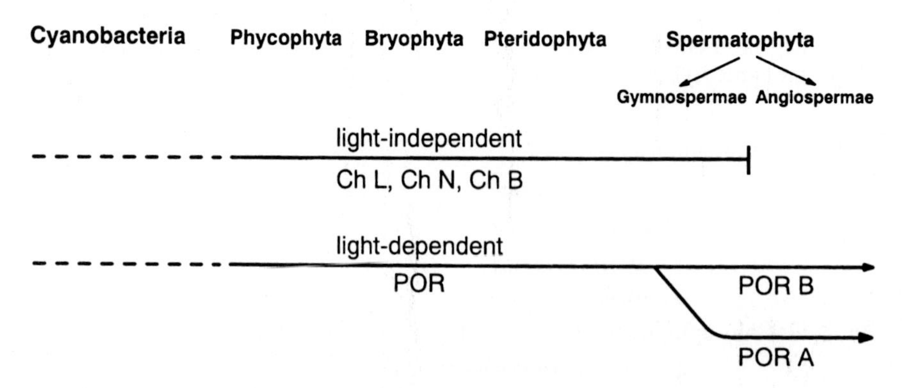

Fig. 2. Light-dependent and light-independent reductions of protochlorophyllide to chlorophyllide in oxygenic photosynthetic organisms. POR A = NADPH-protochlorophyllide oxidoreductase A; POR B = NADPH-protochlorophyllide oxidoreductase B; Ch L, Ch N, Ch B = Subunits of the light-independent protochlorophyllide reductase. For details, see text.

phyrinogen IX oxidase (Molina et al., 1999). In the latter three cases, the accumulation of cyclic tetrapyrrole intermediates induced photooxidative leaf damage. Similar deleterious effects of accumulated tetrapyrroles were also evident in plants treated with herbicides that inhibited protoporphyrinogen IX oxidase (Matringe et al., 1989) and in a mutant that carried a lesion in the gene encoding the uroporphyrinonogen decarboxylase (Hu et al., 1998). Furthermore, upon feeding ALA to seedlings of angiosperms during the dark period, large amounts of Pchlide accumulated. When such seedlings were exposed to the light the leaves suffered from severe photooxidative damage (Rebeiz et al., 1988). Most likely this bleaching of the seedlings was due to extensive destruction of the cellular integrity by reactive oxygen species. In some cases less severe phenotypes were seen. Instead of extensive bleaching, necrotic spots were developed in light-exposed leaves, these spots resembling those induced during a hypersensitive disease resistance response of plants to pathogens. One of the first changes in response to pathogen attack is a transient accumulation of reactive oxygen, the so-called oxidative burst, that is followed by a drastic increase in the expression of genes that encode various pathogen- and stress-defense proteins (Low and Merida, 1996). The expression of most of these defense reactions was also enhanced when plants that had accumulated minor amounts of tetrapyrrole intermediates were exposed to light (Hu et al., 1998; Mock et al., 1998; Molina et al, 1999).

The close correlation between the presence of photosensitizing tetrapyrrole intermediates and stress defense responses suggests that reactive oxygen species may not only act as toxins that directly cause deleterious effects within the cell but also act as second messengers controlling stress responses. This latter function is strongly supported by the recent finding of a systemic stress defense response that spreads rapidly within the plant after the photooxidative stress has been administered locally to a leaf (Karpinski et al., 1999). Reactive oxygen species have consequently been proposed to either diffuse into the neighboring leaf areas and induce directly subsequent defense reactions, or to activate, at the site of the applied stress stimulus, a signal transduction pathway that controls defense reactions in other leaf areas (Karpinski et al., 1999).

III. Regulatory Steps in Tetrapyrrole Biosynthesis

A. Mg-Chelatase

Severe photooxidative damage that occurs in plants suffering from a disturbance of their tetrapyrrole biosynthetic pathway clearly demonstrates the plant's need for a strict control of this pathway to avoid the accumulation of intermediates that may otherwise act as photosensitizers. Indirect evidence suggests that most enzymes in tetrapyrrole biosynthesis are non-rate-limiting, thus directing metabolites to only a few steps at which their flow may be tightly regulated according to the prevailing metabolic conditions. For instance, feeding ALA to etiolated seedlings of angiosperms leads only to the massive accumulation of Pchlide but not to accumulation of heme or other tetrapyrrole intermediates (Granick, 1959). The insertion of ions (either magnesium or iron) seems to be one of the key steps at which synthesis of tetrapyrroles is regulated. Up to protoporphyrin IX, heme and Chl share a common biosynthetic route (Fig. 1). The Chl branch of this pathway begins with the insertion of Mg^{2+}. Since Chl and heme are needed in various amounts according to the developmental stage and/or light conditions, the relative activities of Mg-chelatase and ferrochelatase must be regulated with respect to the plant's requirement for these two end products. During greening of etiolated seedlings, for instance, the ratio of extractable Mg^{2+}-porphyrins to $Fe^{2+/3+}$-porphyrins increases from 1.7 in dark-grown plants to almost 70 after 24 h in the light (Castelfranco and Jones, 1975; Stillman and Gassman, 1978). During this change there is only a marginal increase in heme content, so that the increase in Chl content accounts for most of the change in the ratio.

These drastic changes in the flow of protoporphyrin IX into the branched pathway towards Chl and heme require a sophisticated control mechanism. Mutants with a defect at the Mg-chelatase step have been used to characterize this control mechanism. Several of these mutants have been shown to carry lesions in one of the three subunits of the Mg-chelatase (Koncz et al., 1990; Hudson et al., 1993; Kannangara et al., 1997; Papenbrock et al., 1997). Surprisingly, these mutants do not exhibit the same severe photooxidative damage when exposed to light as seen in plants with lesions at earlier or later steps of Chl biosynthesis. The metabolic flow of tetrapyrrole biosynthesis has been shown to be regulated at the step of ALA synthesis (Granick, 1959). Angiosperms have evolved a feed-back inhibitory loop through which ALA synthesis will be stopped once a critical level of Pchlide has been reached in the dark. The existence of this regulatory circuit has been demonstrated by applying exogenous ALA to etiolated seedlings. Large amounts of Pchlide accumulate under these conditions that may lead to the apparent greening of the seedlings. Only after illumination when Pchlide has been photoreduced, the synthesis of ALA is released from the feed-back inhibition and tetrapyrrole biosynthesis will resume. Unexpectedly, Mg-chelatase-deficient mutants grown in the light do not accumulate tetrapyrrole intermediates. Only after they have been fed with exogenous ALA do these mutants contain larger amounts of protoporphyrin IX. Thus, a lesion of this enzymatic step seems to block synthesis of tetrapyrroles at a site prior to the synthesis of ALA (Mascia, 1978). An ATP-dependent activation of the Mg-chelatase has been proposed to be the key to understanding the enzyme's regulation (Walker and Willows, 1997). However, light-dependent fluctuations in the amounts of the three different subunits of the Mg-chelatase may also play an important role in determining the flux through the Chl and heme branches of the pathway. Thus far no coherent picture of the regulation of this enzymatic step has emerged.

B. Reduction of Protochlorophyllide to Chlorophyllide

1. Light-Independent and Light-Dependent Reactions

Photosynthetic organisms have evolved two genetically distinct strategies to synthesize Chl/bacteriochlorophyll (BChl) (Armstrong, 1998; Fujita et al., 1998) (Fig. 2). The more primitive strategy involves the light-independent reduction of Pchlide to Chlide, a precursor of Chl/BChl. In *Rhodobacter*, an anoxygenic photosynthetic bacterium, this reaction is catalyzed by an enzyme that requires three polypeptides encoded by *bchL*, *bchN* and *bchB* (Zsebo and Hearst, 1984; Yang and Bauer, 1990). These three polypeptides share a high sequence similarity with nitrogenase subunits encoded by the *nifH, D, K* operon of *Azotobacter vinelandii* (Ohyama et al., 1988). They are also closely related to three polypeptides that are implicated in light-independent Pchlide reduction in oxygenic photosynthetic organisms (Fujita et al., 1991,1992,1996; Burke et al., 1993a,b). In plants these polypeptides are encoded by the three chloroplastic genes *chlL*, *chlN* and *chlB* (Roitgrund and Mets, 1990; Choquet et al., 1992; Burke et al., 1993b; Li et al., 1993). Mutants with disruptions of these genes are no longer able to reduce Pchlide to Chlide in the dark. The presence of these genes has been established in cyanobacteria, green algae, bryophytes, pteridophytes and gymnosperms (Armstrong, 1998). However, in angiosperms these genes have been lost from the plastid genome and also the nuclear genome of these plants lacks related DNA sequences (Lindholm and Gustafsson, 1991; Suzuki and Bauer, 1992). In angiosperms the apparent absence of a light-independent Pchlide reductase complex is correlated with the inability of etiolated angiosperm seedlings to reduce Pchlide to Chlide in the dark (Apel et al., 1984). Even though angiosperms cannot synthesize Chl in the dark through this route, a light-independent Chl synthesis has been reported for some angiosperm species (Adamson et al., 1997). Intriguingly, such light-independent Chl synthesis does not occur in etiolated seedlings but only after seedlings had been pre-exposed to white light (Apel et. al, 1984; Packer and Adamson, 1986). According to these data a novel Pchlide-reducing enzyme has to be implicated with the chemical reduction of Pchlide in the absence of light which seems to be unique to angiosperms.

The second, light-dependent and highly regulated strategy of plants to complete Chl biosynthesis is mediated by the plastid-localized NADPH-Pchlide oxidoreductase (POR) (Griffiths, 1978; Apel et al., 1980) (Fig. 2). POR, which has no sequence similarity to any of the subunits of the light-independent enzyme and which is encoded by nuclear genes, catalyzes the photoreduction of Pchlide to Chlide in all oxygenic photosynthetic organisms including cyanobacteria, algae, liverworts, mosses, gymnosperms and angiosperms (Spano et al., 1992a; Forreiter and Apel, 1993; Suzuki and Bauer, 1995; Fujita, 1996; Li and Timko, 1996; Fujita et al., 1998; Takio et al., 1998; Skinner and Timko, 1999).

The occurrence of the light-independent and the light-dependent Pchlide reducing enzymes in most oxygenic photosynthetic organisms has led to a question of why two different enzymes are used to catalyze an identical reaction (Reinbothe et al., 1996a). Under anaerobic conditions photooxidative damage should not occur because of the lack of oxygen. Indeed, *Rhodobacter*, a facultative anaerobic purple bacterium, cannot synthesize BChl and perform photosynthesis in the presence of large amounts of oxygen. As mentioned above the three polypeptides required for light-independent reduction of Pchlide in *Rhodobacter* show significant similarities to three subunits of the nitrogenase of *Azotobacter* (Ohyama et al., 1988). This latter enzyme is known to be sensitive to oxygen. These data may suggest that the ancient form of a light-independent Pchlide reductase had been sensitive to oxygen. Under the anaerobic conditions of the Archaean Earth's atmosphere this enzyme might have been well suited for BChl/Chl synthesis. Only during the transition from the original anaerobic to an aerobic atmosphere might this enzyme have become inadequate. Since porphyrins may act as photosensitizers the increase in oxygen concentration may also enhance the risk of photooxidative damage through energy transfer from excited tetrapyrrole molecules to oxygen, leading to the formation of highly reactive singlet oxygen. If one accepts such a scenario, the appearance of the POR enzyme may have been due to a strong selective pressure towards evolution of a new Pchlide-reducing enzyme that was less sensitive to oxygen and was able to bind Pchlide such that upon illumination absorbed light energy could be used to drive the energy-requiring reduction of Pchlide to Chlide. This new quenching mechanism to dissipate absorbed light energy would provide an efficient protection against photooxidative damage. The universal presence of the light-dependent POR in all oxygenic photosynthetic organisms, ranging from cyanobacteria to angiosperms (Spano et al., 1992; Forreiter and Apel, 1993; Suzuki and Bauer, 1995; Fujita, 1996; Li and Timko, 1996; Fujita et al., 1998; Takio et al., 1998; Skinner and Timko, 1999), indicates that such a continuous need for a protection mechanism against photooxidative damage persisted throughout the evolution of plants. Experimental evidence that supports such a function of the POR enzyme has been obtained.

Among all enzymes POR is unique because it is a photoenzyme requiring light for its catalytic activity and using Pchlide itself as a photoreceptor (Lebedev and Timko, 1998) (Fig. 3). POR is encoded by the nucleus, translated as a precursor protein in the cytosol and ultimately transported into plastids (Reinbothe et al., 1996b). Although plants require freshly synthesized Chl molecules throughout their lifetimes to meet the demands of growth and pigment turnover, light exerts a rapid and dramatic negative regulation on POR-mediated Pchlide reduction at the levels of enzyme activity and protein accumulation (Mapleston and Griffiths, 1980; Santel and Apel, 1981). The drastic decline of POR activity and protein concentration following illumination of etiolated seedlings seems to be common among many angiosperms (Forreiter et al., 1990). When Chl accumulation in illuminated dark-grown seedlings has reached its maximum rate, POR activity has dropped beyond the limit of detection and only trace amounts of the enzyme protein are measurable. This apparent paradox was solved by the discovery of a second, closely-related POR protein (POR B) (Armstrong et al., 1995; Holtorf et al., 1995) (Fig. 2). This enzyme is also light-dependent, requires NADPH as a cosubstrate and shares an overall sequence identity of approximately 75–88 % with the previously described POR protein that accumulates transiently in etiolated seedlings. The first found enzyme was renamed POR A (Holtorf et al., 1995). POR A and POR B are encoded by two differentially regulated nuclear genes (Armstrong et al., 1995; Holtorf et al., 1995). Both are expressed in etiolated tissues. Accumulation of POR A mRNA is, however, negatively regulated by light and temporally

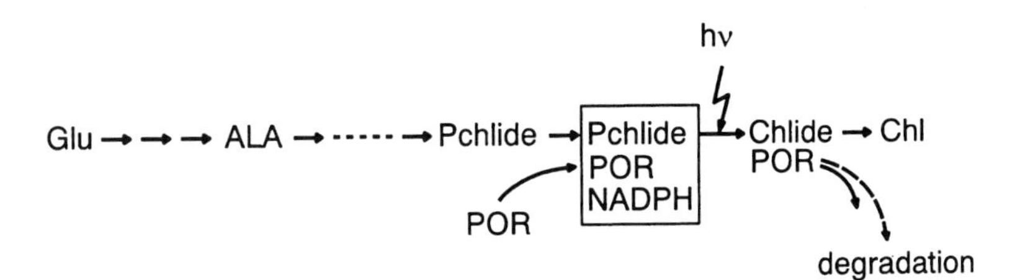

Fig. 3. Formation of the photoactive ternary POR-NADPH-Pchlide complex in higher plants in darkness. Light induces rapid synthesis of chlorophyll (Chl) and degradation of NADPH-Pchlide oxidoreductase (POR). Glu = glutamate; ALA = 5-aminolevulinic acid; Pchlide = protochlorophyllide; Chlide = chlorophyllide.

restricted to young seedlings (Armstrong et al., 1995). This negative effect of light on the mRNA is mediated by phytochrome. The continuous expression of POR B throughout the plant life cycle apparently compensates for the negative light regulation of POR A activity and protein levels in a manner that the steady state level of POR B is sufficient for Chl synthesis.

2. Functions of NADPH-Protochlorophyllide Oxidoreductases A and B

Angiosperm plants have lost the capability of light-independent reduction of Pchlide to Chlide during the night and are thus no longer able to avoid accumulation of Pchlide (Lindholm and Gustafsson, 1991; Suzuki and Bauer, 1992). However, also in some gymnosperms POR A- and POR B-like proteins have been described whose abundances are differentially regulated by light very similarly to the two enzymes in angiosperms (Spano et al, 1992a; Forreiter and Apel, 1993; Skinner and Timko, 1999) (Fig. 2). Thus, the first appearance of a second POR protein must have occurred during evolution prior to the loss of the light-independent enzyme. Since upon illumination Pchlide acts as a photosensitizer, angiosperms must have evolved new protection mechanisms to cope with the risk of photooxidative damage. The differences in expression patterns between POR A and POR B suggest that POR A may represent a specialized enzyme form that is being used as part of a new adaptive strategy for photoprotection during illumination of etiolated seedlings.

Ideally, one would want to test this proposed role of POR A by using POR-deficient mutants. In higher plants no such mutant is currently available, although such a mutant of *Arabidopsis thaliana* was reported to have been identified many years ago (Röbbelen, 1956). Subsequent attempts to isolate a completely POR-deficient mutant were, however, unsuccessful (Runge et al., 1995). The only photosynthetic eukaryote for which a POR mutant has been identified is *Chlamydomonas reinhardtii* and this green algae contains only a single POR gene (Li and Timko, 1996). Therefore, a transgenic approach was used to constitutively overexpress POR protein in *Arabidopsis* seedlings that contain little or no Chl and were severely depleted of endogenous POR. POR depletion was achieved either in wildtype seedlings grown in continuous far-red light which acts through the phytochrome photoreceptor to abolish POR A and strongly downregulates POR B mRNA accumulation (Runge et al., 1996; Sperling et al., 1997), or in the dark-grown *cop1* constitutive photomorphogenic mutant in which POR mRNA accumulation is drastically reduced even in the absence of light (Lebedev et al., 1995; Sperling et al., 1998). Such POR-depleted seedlings were characterized by the complete or nearly complete absence of photoactive Pchlide and the prolamellar body, the enrichment of nonphotoactive, free Pchlide and a drastically enhanced susceptibility to photooxidative damage.

The role of POR A was examined in POR-depleted seedlings that had been transformed with a POR transgene under the control of the CAMV 35S promoter. Overexpression of POR A in these seedlings restored the formation of both prolamellar bodies and photoactive Pchlide and at the same time offered substantial protection against photooxidative damage. These results indicate that indeed one of the functions of POR A, which accumulates in etiolated seedlings and the amount of which declines again soon after the beginning of illumination, is to protect angiosperm seedlings against photooxidative damage during the transition from dark to light. It is important to note that in the transgenic lines not only the overexpression of POR A but also that of POR B conferred protection against photooxidation. Thus, the proposed photoprotective function of POR A seems not to be due to an intrinsic difference between POR A and POR B proteins, but rather be caused by the different patterns of expression of the two POR genes at the beginning of light-induced chloroplast formation in wildtype seedlings (Sperling et al., 1997, 1998).

3. Does a Light-Harvesting NADPH-Protochlorophyllide Oxidoreductase-Protochlorophyllide Complex Exist in Vivo?

A very different view of the possible role of POR A and POR B during the initial phase of the greening process has been advanced by Reinbothe et al. (1999). According to their model POR A and POR B form part of a light-harvesting POR-Pchlide complex (LHPP) that is involved in photoprotection of dark-grown angiosperm seedlings. This complex has been proposed to consist of five POR A-Pchlide *b*-NADPH and one POR B-Pchlide *a*-NADPH ternary complexes. The POR A-Pchlide *b* complex is supposed to initially act as a light-harvesting complex by transferring excitation energy to Pchlide *a* for its photoreduction to Chlide *a*, a function catalyzed by POR B. In contrast to the Pchlide *a* that is reduced quickly, Pchlide *b* is not photoreduced rapidly during light-harvesting. Excess light energy that is not transferred from the POR A-Pchlide *b* complex to Pchlide *a* is proposed to be emitted as fluorescence. Thus, under high light intensity the Pchlide *b*-POR A complex would have a photoprotective function. Only at a later stage of greening when this complex is dissociated from the LHPP could absorbed light energy be used for the POR A-catalyzed photoreduction of Pchlide *b* to Chlide *b*. This hypothesis, however, relies solely on in vitro experiments and thus its in vivo relevance has been questioned (Scheumann et al., 1999; Armstrong et al., 2000).

Considering the above-mentioned in vitro data, the dark-grown angiosperm seedlings should contain predominantly Pchlide *b* rather than Pchlide *a* in their etioplast inner membranes. Furthermore, flash illumination of dark-grown seedlings should result in photoreduction of only ⅙ of the total Pchlide. No in vivo data exist to support these conclusions. On the contrary, HPLC analysis of the pigments in etiolated seedlings revealed the complete absence of Pchlide *b* (Scheumann et al., 1999). Flash illumination of isolated prolamellar bodies, where the aggregated photoactive Pchlide-POR complexes accumulate, results in immediate photoreduction of almost all of the Pchlide (Ryberg and Sundqvist, 1988). The proposed structure of LHPP also predicts a large excess of POR A to POR B in dark-grown seedlings. This is certainly not true in *Arabidopsis* (Armstrong et al., 1995) or in pea (Spano et al., 1992b). Furthermore this latter species contains only one POR gene (Spano et al., 1992b; M. Timko, personal communication). Overexpression of either POR A or POR B in the POR-deficient *Arabidopsis cop1* mutant reconstitutes etioplast inner membranes and photoreducible Pchlide-POR complexes comparable to those of the wildtype (Sperling et al., 1998). Thus, the published in vitro data are insufficient to conclude that the proposed LHPP exists in vivo.

4. Substrate-Dependent Import of NADPH-Protochlorophyllide Oxidoreductase A into Plastids

A unique property of POR A that distinguishes it from POR B and may relate to its proposed function as a photoprotectant, has been revealed during studies of its import into isolated plastids. While the barley POR B precursor protein is translocated into isolated etioplasts and chloroplasts with equal efficiency, the POR A precursor is imported only into isolated etioplasts but not into chloroplasts (Reinbothe et al., 1995a,b). Subsequent studies revealed that the import of the POR A precursor protein depends on the presence of the enzyme's substrate, Pchlide (Reinbothe et al., 1995a). Because the concentration of Pchlide in plastids rapidly declines during illumination of etiolated seedlings, this pigment soon becomes a limiting factor for the import reaction. The competence of chloroplasts to import POR A can be restored by feeding ALA to isolated barley chloroplasts in the dark, which leads to the rapid accumulation of Pchlide (Reinbothe et al., 1995a). Because part of the Pchlide has previously been shown to be synthesized in the plastid envelope (Pineau et al., 1993), the proposed interaction between Pchlide and the POR A precursor could occur already in the envelope, thereby triggering the actual translocation step. If this interaction takes place in close proximity of the import apparatus, it would make an efficient mechanism to prevent the release of free hazardous intermediates or products of pigment biosynthesis, such as Pchlide, that are known to cause photooxidative damage within the plastid compartment.

This protective mechanism would only operate efficiently if an excess of non-processed POR A precursor protein would be available outside of the plastid compartment. Indeed, POR A precursor proteins were shown to be bound to the outer surface of the envelope membrane of etiochloroplasts in a transport-competent conformation (Joyard et al., 1990; Reinbothe et al., 1996c). The translocation of this precursor across the outer and inner plastid envelope membranes was arrested because of the lack of Pchlide. When Pchlide was produced by feeding the plastids with ALA, the precursor of POR A was chased into the organelle and appeared there in the processed mature form. The substrate-dependent uptake of the POR A precursor seems to ensure the immediate assembly of the ternary photoactive POR-NADPH-Pchlide complex during or immediately after the passage of the precursor protein through the envelope membrane. This complexed POR A protein, in contrast to its free form, is very stable and well-protected against proteolytic attack (Reinbothe et al., 1995b, c). After photoreduction of Pchlide the resulting Chlide-POR-complex is highly susceptible to proteolytic attack and is rapidly degraded (Fig. 3). The POR A-degrading protease activity is not detectable in etioplasts but is induced during illumination (Reinbothe et al., 1995c).

Collectively, all of these data demonstrate that in etiolated seedlings of angiosperms POR A seems to play an important role as a photoprotectant. Its function is confined to the initial phase of light-induced chloroplast formation. While POR A is present abundantly in etioplasts it selectively disappears soon after the beginning of illumination. It is interesting to note that negative light regulation of POR A is mediated simultaneously at various levels and by different means: First, the concentration of POR A mRNA declines drastically during illumination of dark-grown seedlings (Apel, 1981; Holtorf et al., 1995). Second, the plastid's ability to import the precursor of POR A is reduced during the transition from etioplasts to chloroplasts. This effect is due to a rapid decline in the plastidic level of Pchlide, which is required for the translocation of the POR A precursor. Third, POR A becomes destabilized soon after the beginning of illumination and is degraded by a light-induced protease.

C. Feed-Back Inhibition of Aminolevulinic Acid Synthesis

In order to compensate for the loss of the light-independent Pchlide-reducing enzyme and to avoid the accumulation of Pchlide in the dark, angiosperms seem to have evolved a regulatory circuit that controls and restricts the synthesis of ALA. The existence of this feedback inhibitory loop was demonstrated by applying exogenous ALA to etiolated seedlings (Granick, 1959). Under these conditions large amounts of Pchlide accumulate and may lead to the apparent 'greening' of etiolated seedlings in the dark. Mutants that are defective in this regulation have been identified e.g. in barley and *Arabidopsis*. Etiolated seedlings of the *tigrina* mutants of barley accumulate excessive amounts of Pchlide (Nielsen, 1974; von Wettstein et al., 1995). When these mutant plants are exposed to light their leaves bleach and often develop necrotic areas. In mutants kept under continuous illumination, however, the Pchlide is immediately photoreduced and chloroplast development proceeds normally as in wildtype seedlings. We have isolated similar, so-called *flu* mutants of *Arabidopsis thaliana* (K. Apel and M. Nater, unpublished results) (Fig. 4). It is possible to maintain and propagate these mutants by growing them in continuous light. Whenever the mutant plants are transferred back to the dark they start immediately to accumulate Pchlide (Fig. 5). In young *flu* mutant seedlings that are either initially exposed or reexposed to light this accumulation of Pchlide has severe consequences. After a brief illumination they bleach and die. Light-adapted more mature mutant plants at the rosette leaf stage respond, however, differently. During a brief dark period they also accumulate detectable amounts of Pchlide, but upon illumination they do not bleach and their Chl content is not reduced. Instead, during the first 24 h of light they show typical stress symptoms. These results clearly indicate that the feed-back loop operates not only in young etiolated seedlings but is also needed in fully grown light-adapted plants to avoid the accumulation of free Pchlide during a night period.

The inhibition of ALA synthesis in dark-grown wildtype seedlings is overcome by exposure of plants to light. Light might affect the synthesis of ALA in two ways, either by inducing the synthesis of enzymes required, or by removal of an inhibitor affecting the activity of these enzymes (Beale and Weinstein, 1990). When etiolated seedlings are first illuminated the initial photoreduction of Pchlide to Chlide is not immediately followed by maximal Chl synthesis. This lag period in Chl accumulation could indicate

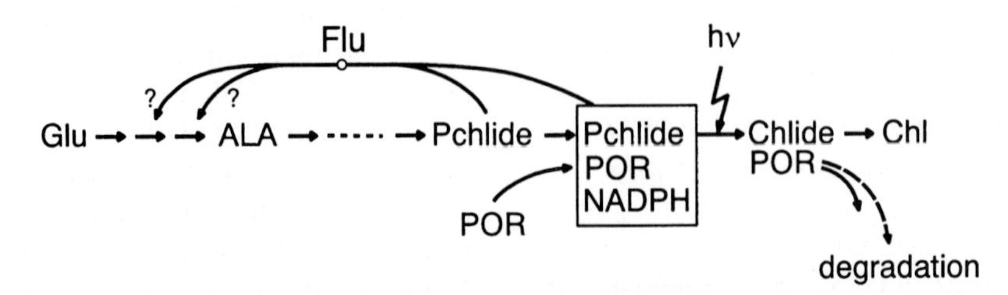

Fig. 4. A putative feed-back inhibitory loop in regulation of ALA synthesis in *Arabidopsis thaliana*. In addition to Pchlide, also the Flu gene product is necessary for repression of ALA synthesis in darkness. For abbreviations, see Fig. 3. For details, see text.

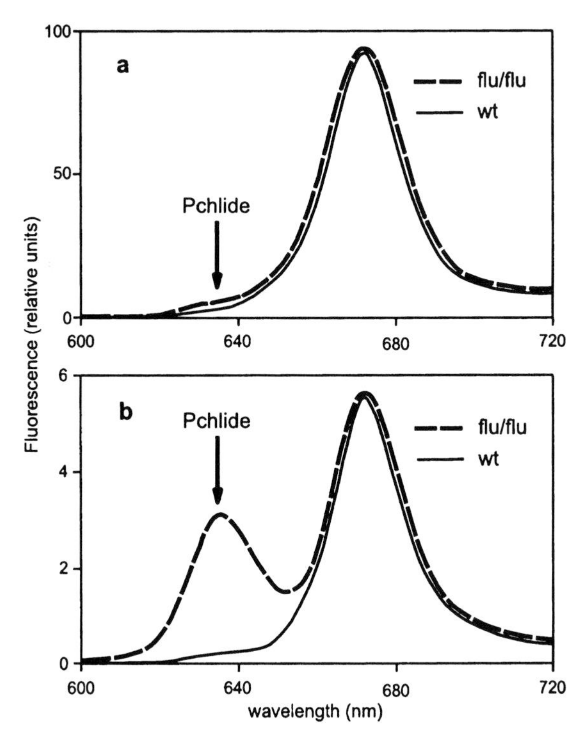

Fig. 5. Accumulation of Pchlide in the *flu* mutant, defective in regulation of ALA synthesis. Room temperature emission spectra of (a) total porphyrins and (b) of nonesterified porphyrins lacking phytol, extracted from the wildtype and *flu* mutant plants. Arabidopsis plants were grown for 3 weeks under continuous white light and then transferred to the dark for 8 h before the pigments were extracted. Fluorescence emission bands observed with an excitation wavelength of 440 nm include Pchlide and Chl(ide) with fluorescence emission maxima at 632 and 672 nm, respectively. Note that in the *flu* mutant Pchlide makes up only a very minor fraction of the total porphyrin extract. Pchlide = protochlorophyllide; Chlide = chlorophyllide; ALA = 5-amino-levulinic acid.

the need for a light-induced synthesis of enzymes responsible for ALA formation. Two enzymes are involved in the conversion of the initial metabolite, Glu-tRNA, into ALA: Glu-tRNA reductase and Glu-1- semialdehyde aminotransferase (Fig. 1). Changes in their mRNA concentrations were followed during the initial phase of the light-induced greening of etiolated *Arabidopsis* seedlings (Ilag et al., 1994). Whereas the level of mRNA for the Glu-tRNA reductase remained undetectable throughout the first 24 h of illumination, the level of Glu-1-semialdehyde aminotransferase mRNA gradually increased over the same period. However, no drastic changes in the amounts of these mRNA species were seen that might be expected if indeed the abolition of ALA synthesis inhibition in dark-grown seedlings would be due to the light-induced *de novo* synthesis of the enzyme system responsible for ALA formation. This result is in good agreement with the previous studies, which have shown that ALA synthesis activity in barley leaves increased only 3-fold during the first 12 h of illumination (Kannangara and Gough, 1978).

A different effect of light on the mRNAs for Glu-tRNA reductase and Glu-1-semialdehyde amino-transferase was seen when more mature light-adapted *Arabidopsis* plants were analyzed (Ilag et al., 1994). The two transcript species were present at high levels in plants grown in the light. When these plants were transferred to the dark the Glu-tRNA reductase mRNA completely disappeared whereas the Glu-1-seminaldehyde aminotransferase mRNA was expressed at a considerably lower level. Transcripts of both genes reached the original level when the plants were returned to the light (Ilag et al., 1994). These drastic changes in mRNA concentrations during light/dark cycles may form part of the regulatory loop that suppresses ALA synthesis during a night period in mature light-adapted *Arabidopsis* plants. In such a case, however, one might expect to find different levels of these transcripts in the *flu* mutants that are defective in this regulatory circuit and hence accumulate larger amounts of Pchlide in the dark. In fact, both transcript species showed the same fluctuations during a day/night cycle in the mutants as in the wildtype (R. Meskauskiene and K. Apel, unpublished). The drastic changes in mRNA abundance therefore seem to reflect the operation of a distinct and additional regulatory mechanism that is apparently used by plants to adjust tetrapyrrole synthesis during the day/night cycle.

A light-induced increase in the amount of Glu-tRNA reductase has also been described in barley. It was shown that both the ALA synthesizing capacity and the mRNA levels for Glu-tRNA reductase and Glu-1-semialdehyde aminotransferase were under developmental and circadian control (Kruse et al., 1997). The levels of Glu-tRNA reductase mRNA and its protein product cycled parallel to the changes in ALA-forming capacity. On the other hand, the levels of Glu-1-semialdehyde aminotransferase mRNA oscillated in an inverse phase with the tRNA reductase mRNA and the enzyme protein concentration remained constant under diurnal and circadian growth conditions. The very close correlation between the expression of Glu-tRNA reductase and ALA synthesis capacity suggests that the light-induced modulation of this enzyme might be a rate-limiting step during

ALA synthesis in light-adapted barley plants. The observed changes in ALA synthesis capacity followed closely the changes in the expression pattern of Chl-binding proteins. Therefore, the major function of the light-controlled modulation in the amount of Glu-tRNA reductase, essential for the production of ALA, might be to coordinate the assembly of pigment-binding proteins and to avoid the overproduction of free Chl that would enhance the risk of photooxidative damage (Kruse et al., 1997).

D. Protochlorophyllide and Heme as Putative Regulatory Factors

When greening plants are returned to darkness, synthesis of ALA ceases almost immediately. This inhibition is rapidly reversed upon illumination (Fluhr et al., 1975). Such a rapid on-and-off switching of synthesis during dark-to-light and light-to-dark transitions is more plausible in terms of direct metabolic control than in terms of enzyme synthesis and breakdown. Metabolic feed-back control of Chl synthesis could be induced by an intermediate of the pathway that accumulates in the dark and disappears in the light. One candidate for such a regulatory factor is Pchlide (Bogorad, 1976) (Fig. 6). ALA accumulation was found to be inversely proportional to the level of photoactive Pchlide, i. e. part of the ternary NADPH-POR-Pchlide complex in etiolated seedlings or leaves that had been pretreated with levulinic acid to block the ALA dehydratase (Ford and Kasemir, 1980; Stobart and Ameen-Bukhari, 1984). This result suggests that the level of phototransformable Pchlide could modulate the capacity to synthesize ALA.

Another intermediate of tetrapyrrole biosynthesis, heme, has also been implicated as a possible regulator (Fig. 6). Heme is a potent inhibitor of ALA formation in isolated plastids (Chereskin and Castelfranco, 1982). Exogenous heme administered to intact barley shoots (Hendry and Stobart, 1978) and to *Chlamydomonas* cultures (Hoober and Stegeman, 1973) inhibited the formation of Chl in vivo. It is also likely that the increased ALA accumulation observed in intact tissues treated with Fe chelators (Duggan and Gassman, 1974) was due to the interference with heme formation thereby decreasing concentration of this putative feedback inhibitor. The role of heme in

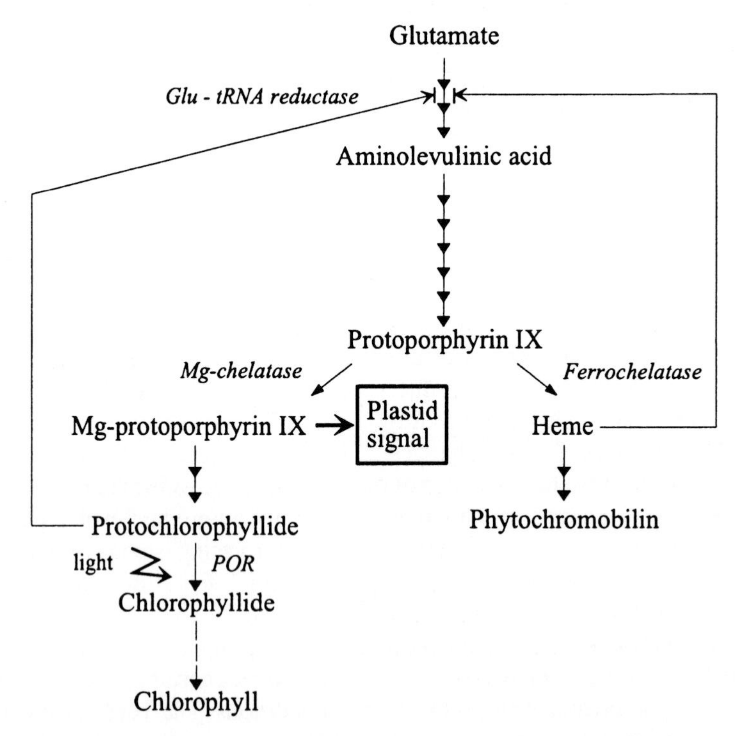

Fig. 6. Control of tetrapyrrole biosynthesis in higher plants. Enzymes that may be subject to a control and biosynthesis intermediates that may act as effector molecules are indicated. Mg^{2+}-protoporphyrin IX, an intermediate of tetrapyrrole biosynthesis, has been postulated to function as a plastid signal to regulate nuclear gene expression. POR = NADPH-Pchlide oxidoreductase.

regulating ALA formation in vivo is supported by genetic evidence. The *aurea* (*au*) and the *yellow-green-2* (*y-g-2*) mutants of tomato are unable to synthesize the linear tetrapyrrole phytochromobilin (Fig. 6). The *au* mutant is specifically deficient in the phytochromobilin synthase, whereas the *y-g-2* mutant is blocked in the preceding step of the pathway and cannot synthesize biliverdin IXa from heme (Fig. 1) (Terry and Kendrick, 1999). Etiolated seedlings of both tomato mutants have reduced levels of Pchlide due to an inhibition of its synthesis. Feeding experiments with ALA demonstrate that the pathway between ALA and Pchlide is intact in both mutants suggesting that the Pchlide-deficiency results from the inhibition of ALA synthesis. Since higher levels of heme accumulated in both mutants than in wildtype seedlings it seems likely that this tetrapyrrole intermediate inhibits ALA synthesis.

Various attempts have been made to study the mechanism by which heme and Pchlide may inhibit ALA synthesis. The three enzymes involved in the conversion of glutamate into ALA were isolated and the inhibitory effects of various tetrapyrrole derivatives on their activity in vitro were tested. Pchlide inhibition of Glu-tRNA synthase has been reported (Dörnemann et al., 1989); however, the physiological significance of this effect is questionable since an inhibition of this enzyme in vivo would not only block tetrapyrrole synthesis, but would also have a strong impact on plastid protein synthesis at the same time. The second of the three enzymes involved in ALA synthesis from glutamate, Glu-tRNA reductase, has been shown to be inhibited by heme (Pontoppidan and Kannangara, 1994). This enzyme is believed to catalyze the rate-limiting step in ALA synthesis and therefore it has been considered to be a likely target for inhibition by heme (Fig. 6). The feedback inhibitory loop that represses ALA synthesis in darkness in vivo has also been defined genetically (Nielsen, 1974, Meskauskiene and Apel, unpublished results). In *Arabidopsis*, mutations in the *Flú* gene seem to be sufficient to abolish this control mechanism. However, the position of the *Flú* gene on the genetic map of *Arabidopsis* is different from the genes encoding the three enzymes involved in ALA synthesis (R. Meskauskiene and K. Apel, unpublished results). Therefore, besides Pchlide and/or heme, at least one extra protein factor, encoded by the *Flu* gene, is required for the repression of ALA synthesis (Fig. 4).

Collectively, the data available indicate that with the loss of the light-independent Pchlide-reducing enzyme higher plants had to evolve a new strategy to control the accumulation of Pchlide in the dark and to avoid photooxidative damage during subsequent illumination. For instance in gymnosperms, which have conserved the ability to synthesize Chl in the dark, ALA is synthesized at similar rates in dark-grown and illuminated seedlings and thus does not limit the Chl synthesis in the dark. In angiosperms, however, ALA synthesis is rapidly switched off once the seedlings are transferred from light to dark, and thus ALA defines the major rate-limiting step in the tetrapyrrole biosynthetic pathway of these plants. Apparently two different mechanisms are being used by angiosperms to control the synthesis of ALA. First, there is a light-dependent change in the concentration of Glu-tRNA reductase and the Glu-1-semialdehyde aminotransferase, the enzymes involved in the formation of ALA from Glu-tRNA. Second, an inhibitory feed-back loop seems to exist through which tetrapyrrole intermediates block the function of Glu-tRNA reductase. Mainly two intermediates of tetrapyrrole biosynthesis, Pchlide and heme, have been implicated to act as effectors in this regulatory circuit. However, other intermediates such as protoporphyrin IX or Mg^{2+}-protoporphyrin IX cannot be excluded from this control. Pchlide in particular seems to be an attractive candidate for such an inhibitory molecule, because of its light-dependent reduction to Chlide that starts immediately after the beginning of illumination and closely correlates with the light-induced onset of ALA synthesis. It is not known whether heme and Pchlide act through different regulatory loops and if so, whether they interact with different target sites. Genetic evidence demonstrates that suppression of Pchlide accumulation in dark-grown seedlings requires, besides an effector molecule and a target enzyme, at least one additional factor, the Flu protein. Whether this protein interacts with Pchlide and/or heme directly is not known yet. One may expect that with the identification of the Flu protein the function of this key regulator and the operation of tetrapyrrole biosynthesis control in dark-grown angiosperms can be unraveled.

IV. Tetrapyrrole Derivatives as Plastid Signals

Thus far, several rate-limiting steps of tetrapyrrole

biosynthesis have been identified that may be used by plants to regulate the metabolic flow and to minimize the risk of photooxidative damage. The major resource allocation of tetrapyrrole biosynthesis in plants is directed towards the massive accumulation of Chl. Chl in its free form would cause extensive photooxidative damage if exposed to light. Thus, formation of Chl has to be closely coordinated and linked to the synthesis of other cellular constituents, such as Chl-binding proteins that are required for the assembly of photosynthetic membranes and that allow the quenching of absorbed light energy to occur. Since most Chl-binding proteins are not encoded by plastid genes but by the nucleus, it is not surprising that Chl accumulation during light-induced chloroplast development is based on an intimate interaction with the nuclear genetic system. The effect of light on the transcription of nuclear genes encoding plastid proteins, such as the apoproteins of several light-harvesting Chl *a/b*-containing complexes (*cab* proteins) has been well documented (Apel and Kloppstech, 1978; Silverthorne and Tobin, 1984; Mösinger et al, 1985). These studies, however, have also revealed the existence of a reverse type of control in which intact plastids release signals that affect the transcription of a small set of nuclear genes encoding proteins whose activities relate to the photosynthetic function of the chloroplast (Oelmüller, 1989; Taylor, 1989; Susek and Chory, 1992). Mutations or herbicides such as Norflurazon, which block the accumulation of colored carotenoids also result in extensive photooxidative damage within the chloroplast and in destruction of the chloroplast signal. For example, the expression of *cab* genes stops in plants that no longer contain intact plastids due to photooxidative damage (Mayfield and Taylor, 1984; Batschauer et al., 1986; Oelmüller and Mohr, 1986).

Various molecular candidates for the chloroplast signals have been discussed, including RNAs, proteins, products of photosynthesis, porphyrins or intermediates of other chloroplast metabolic pathways (Bradbeer et al., 1979; Johanningmeier and Howell, 1984; Batschauer et al., 1986; Johanningmeier, 1988; Sheen, 1990). Recently the first plastid factor has been identified (Kropat et al., 1997). Illumination of *Chlamydomonas* cells was shown to induce the expression of two nuclear heat shock genes HSP70 A and HSP70 B encoding cytosolic and plastid-localized heat shock proteins, respectively. Mutants defective in the synthesis of Mg^{2+}-protoporphyrin IX were no longer able to induce the expression of these two genes. Feeding of Mg^{2+}-protoporphyrin IX or its dimethylester to the wildtype or mutant cells in the dark resulted in the induction of these two genes. Other intermediates of Chl synthesis, such as protoporphyrin IX, Pchlide or Chlide, were unable to induce the expression of HSP70A and HSP70B genes. Thus, Mg^{2+}-protoporphyrin IX and its methylester are the first Chl intermediates that clearly qualify as plastidic signals (Fig. 6). These results raise the question of whether Mg^{2+}-protoporphyrin IX and its methylester account for all of the known physiological effects of plastid factors. There are reasons why this may not be the case. Thus far, the only genes, besides HSP70A and HSP70B in *Chlamydomonas* shown to be affected by Mg^{2+}-protoporphyrin IX and its methylester, are those encoding a cytosolic thioredoxin and one of the two small subunit isozymes of ribulose-1,5-bisphosphate carboxylase, but none of the *cab* genes (C. Beck, personal communication). Accumulation of Mg^{2+}-protoporphyrin IX within the cell causes photooxidative damage once the cells are illuminated. However, the signaling effect of this intermediate is not a consequence of photooxidative stress since such a signaling may also occur in darkness after addition of exogenous Mg^{2+}-protoporphyrin IX. The porphyrin-induced signaling in the dark could inform the cell of an impending dangerous stimulus and switch on an adaptive gene expression pattern. Chaperones like the heat shock proteins that are induced in the dark-grown *Chlamydomonas* cells by Mg^{2+}-protoporphyrin or its methylester might protect chloroplasts during the following light stress conditions. It has recently been shown that constitutive overexpression of the heat shock protein HSP70 reduces the extent of photoinhibition in transgenic *Chlamydomonas* cells when kept at higher light intensities (Schoda et al., 1999). It is not known whether the concentration of the endogenous Mg^{2+}-protoporphyrin and its methylester is sufficiently high to control this adaptive protection of the photosynthetic apparatus under normal light/dark cycles in nontransgenic cells.

Different plastid factors have been identified in higher plants. The redox status of the plastoquinone pool in chloroplasts functions as a photon-sensing system that controls the transcription of several plastidic and nuclear genes, including the *cab* genes, in response to light intensity (Escoubas et al., 1995, Pfannschmidt et al., 1999). It appears unlikely that the same signaling pathway acts also in etiolated seedlings. Besides intermediates of Chl biosynthesis and the plastoquinone redox state, additional types of plastid signals might be expected to exist in higher plants to coordinate the expression of nuclear and plastidic genes during the light-induced chloroplast development.

V. Outlook

Thus far studies of photooxidative stress in green plants have been focused mainly on describing the various forms of photooxidative damage and the protection mechanisms that are being used by plants to alleviate this damage. Through work on other organisms such as animals and yeast there has been increasing recognition in the past few years that the effects of oxidative stress are not only mediated through damage of cellular constituents but that reactive oxygen species can alter cell functions by acting as second messengers in signal transduction (Forman and Cadenas, 1997; Ryter and Tyrrell, 1998). Most plants synthesize and accumulate much larger amounts of tetrapyrroles than animals and their survival depends on the continuous utilization of light as a primary energy source. Thus, photooxidative stress is a condition that plants have to cope with throughout most of their life cycle. Only recently it has been realized that photooxidative stress in plants may also exert effects on signal transduction and that intermediates of tetrapyrrole biosynthesis, which intrinsically act as highly potent photosensitizers, and reactive oxygen species may be used directly as signals to control defense reactions against various stressors (Kropat et al., 1997; Karpinski et al., 1999).

Acknowledgments

The author is grateful to Dr. G. Armstrong for reading the manuscript, to Dr. D. Rubli for art work, to Mena Nater and Rasa Meskauskiene for providing unpublished data and to Renate Langjahr for typing the manuscript. Parts of the work were supported by the Swiss National Science Foundation.

References

Adamson HJ, Hiller RG and Walmsley J (1997) Protochlorophyllide reduction and greening in angiosperms: An evolutionary perspective. Photochem Photobiol 41: 201–221

Alscher RG, Donahue JL and Cramer CL (1997) Reactive oxygen species and antioxidants: Relationships in green cells. Physiol Plant 100: 224–233

Apel K (1981) The protochlorophyllide holochrome of barley (*Hordeum vulgare* L.). Phytochrome-induced decrease of translatable mRNA coding for the NADPH:protochlorophyllide oxidoreductase. Eur J Biochem 120: 89–93

Apel K and Kloppstech, K (1978) The plastid membranes of barley (*Hordeum vulgare*). Light-induced appearance of mRNA coding for the apoprotein of the light-harvesting chlorophyll *a/b* protein. Eur J Biochem 85: 581–588

Apel K, Santel HJ, Redlinger TE and Falk H (1980) The protochlorophyllide holochrome of barley (*Hordeum vulgare* L.). Isolation and characterization of the NADPH-protochlorophyllide oxidoreductase. Eur J Biochem 111: 251–258

Apel K, Motzkus M and Dehesh K (1984) The biosynthesis of chlorophyll in greening barley *(Hordeum vulgare)*. Is there a light-independent protochlorophyllide reductase? Planta 161: 550–554

Armstrong GA (1998) Greening in the dark: Light-independent chlorophyll biosynthesis from anoxygenic photosynthetic bacteria to gymnosperms. J Photochem Photobiol B: Biol 43: 87–100

Armstrong GA, Runge S, Frick G, Sperling U and Apel K (1995) Identification of NADPH: protochlorophyllide oxidoreductases A and B: A branched pathway for light-dependent chlorophyll biosynthesis in *Arabidopsis thaliana*. Plant Physiol 108: 1505–1517

Armstrong GA, Apel K and Rüdiger, W (2000) Does a light-harvesting protochlorophyllide *a/b*-binding protein complex exist? Trends Plant Sci 5: 40–44

Batschauer A, Mösinger E, Kreuz K, Dörr I and Apel K (1986) The implication of a plastid-derived factor in the transcriptional control of nuclear genes encoding the light-harvesting chlorophyll *a/b* protein. Eur J Biochem 154: 625–634

Beale SI and Weinstein JD (1990) Tetrapyrrole metabolism in photosynthetic organisms. In: Dailey HA (ed) Biosynthesis of Heme and Chlorophyll, pp 287–391. McGraw-Hill, New York

Beale SI, Gough SP and Granick S (1975) Biosynthesis of δ-aminolevulinic acid from the intact carbon skeleton of glutamic acid in greening barley. Proc Natl Acad Sci USA 72: 2719–2723

Bogorad L (1976) Chlorophyll biosynthesis. In: Goodwin TW (ed) Chemistry and Biochemistry of Plant Pigments, 2nd ed, Vol 1, pp 64–148. Academic Press, London, New York

Bradbeer JW, Atkinson JE, Börner T and Hagemann RC (1979) Cytoplasmic synthesis of plastid polypeptides may be controlled by plastid-synthesized RNA. Nature 279: 816–817

Burke DH, Alberti M and Hearst JE (1993a) The *Rhodobacter capsulatus* chlorin reductase-encoding locus, *bchA*, consists of the three genes *bchX*, *bchY*, and *bchZ*. J Bacteriol 175: 2407–2413

Burke DH, Alberti M and Hearst JE (1993b) *bchFNBH* bacteriochlorophyll synthesis genes of *Rhodobacter capsulatus* and identification of the third subunit of light-independent protochlorophyllide reductase in bacteria and plants. J Bacteriol 175: 2414–2422

Castelfranco PA and Jones OTG (1975) Protoheme turnover and chlorophyll synthesis in greening barley tissue. Plant Physiol 55: 485–490

Chereskin BA and Castelfranco PA (1982) Effects of iron and oxygen on chlorophyll biosynthesis. II. Observations on the biosynthetic pathway in isolated etiochloroplasts. Plant Physiol 69: 112–116

Choquet Y, Rahire M, Girard-Bascou J, Erickson J and Rochaix J-D (1992) A chloroplast gene is required for the light-independent accumulation of chlorophyll in *Chlamydomonas reinhardtii*. EMBO J 11: 1697–1704

Dörnemann D, Kotzabasis K, Richter P, Breu V and Senger H (1989) The regulation of chlorophyll biosynthesis by the action of protochlorophyllide on ^{Glu}tRNA-ligase. Bot Acta 102: 112–115

Duggan J and Gassman ML (1974) Induction of porphyrin synthesis in etiolated bean leaves by chelators of iron. Plant Physiol 53: 206–215

Escoubas JM, Lomas M, LaRoche J and Falkowski PG (1995) Light intensity regulation of *cab* gene transcription is signaled by the redox state of the plastoquinone pool. Proc Natl Acad Sci USA 92: 10237–10241

Fankhauser C and Chory J (1997) Light control of plant development. Annu Rev Cell Dev Biol 13: 203–229

Fluhr R, Harel E, Klein S and Meller E (1975) Control of aminolevulinic acid and chlorophyll accumulation in greening maize leaves upon light-dark transitions. Plant Physiol 56: 497–501

Ford MJ and Kasemir H (1980) Correlation between 5-aminolaevulinate accumulation and protochlorophyll photoconversion. Planta 50: 206–210

Forman HJ and Cadenas E (1997) Oxidative stress and signal transduction. Chapman and Hall, New York, London

Forreiter C and Apel K (1993) Light-independent and light-dependent protochlorophyllide-reducing activities and two distinct NADPH-protochlorophyllide oxidoreductase polypeptides in mountain pine (*Pinus mugo*). Planta 190: 536–545

Forreiter C, van Cleve B, Schmidt A and Apel K (1990) Evidence for a general light-dependent negative control of NADPH-protochlorophyllide oxidoreductase in angiosperms. Planta 183: 126–132

Fujita Y (1996) Protochlorophyllide reduction: a key step in the greening of plants. Plant Cell Physiol 37: 411–421

Fujita Y, Takahashi Y, Shonai F, Ogura Y and Matsubara H (1991) Cloning, nucleotide sequences and differential expression of the *nifH* and *nifH*-like (*frxC*) genes from the filamentous nitrogen-fixing cyanobacteria *Plectonema boryanum*. Plant Cell Physiol 32: 1093–1106

Fujita Y, Takahashi Y, Chuganji M and Matsubara H (1992) The *nifH*-like (*frxC*) gene is involved in the biosynthesis of chlorophyll in the filamentous cyanobacterium *Plectonema boryanum*. Plant Cell Physiol 33: 81–92

Fujita Y, Takagi H and Hase T (1996) Identification of the *chlB* gene and the gene product essential for the light-independent chlorophyll biosynthesis in the cyanobacterium *Plectonema boryanum*. Plant Cell Physiol 37: 313–323

Fujita Y, Takagi H and Hase T (1998) Cloning of the gene encoding a protochlorophyllide reductase: the physiological significance of the co-existence of light-dependent and -independent protochlorophyllide reduction systems in the cyanobacterium *Plectonema boryanum*. Plant Cell Physiol 39: 177–185

Granick S (1959) Magnesium porphyrin is formed in barley seedlings treated with delta-aminolevulinic acid. Plant Physiol Suppl 34: 18

Griffiths WT (1978) Reconstitution of chlorophyllide formation by isolated etioplast membranes. Biochem J 174: 681–692

Hendry GAF and Stobart AK (1978) The effect of haem on chlorophyll synthesis in barley leaves. Phytochemistry 17: 73–77

Hideg E, Kalai T, Hideg K and Vass I (1998) Photoinhibition of photosynthesis in vivo results in singlet oxygen production. Detection via nitroxide-induced fluorescence quenching in broad bean leaves. Biochemistry 37: 11405–11411

Höfgen R, Axelsen KB, Kannangara CG, Schüttke I, Pohlenz H-D, Willmitzer L, Grimm B and von Wettstein D (1994) A visible marker for antisense mRNA expression in plants: Inhibition of chlorophyll synthesis with a glutamate-1-semialdehyde aminotransferase antisense gene. Proc Natl Acad Sci USA 91: 1726–1730

Holtorf H, Reinbothe S, Reinbothe C, Bereza B and Apel K (1995) Two distinct NADPH-protochlorophyllide oxidoreductases are differentially expressed during the light-induced greening of dark-grown barley (*Hordeum vulgare* L.) seedlings. Proc Natl Acad Sci USA 92: 3254–3258

Hoober JK and Stegeman WJ (1973) Control of the synthesis of a major polypeptide of chloroplast membranes of *Chlamydomonas reinhardi*. J Cell Biol 56: 1–12

Hu G, Yalpani N, Briggs SP and Johal GS (1998) A porphyrin pathway impairment is responsible for the phenotype of a dominant disease lesion mimic mutant of maize. Plant Cell 10: 1095–1105

Hudson A, Carpenter R, Doyle S and Coen ES (1993) *Olive*: A key gene required for chlorophyll biosynthesis in *Antirrhinum majus*. EMBO J 12: 3711–3719

Ilag LL, Kumar AM and Söll D (1994) Light regulation of chlorophyll biosynthesis at the level of 5-aminolevulinate formation in *Arabidopsis*. Plant Cell 6: 265–275

Johanningmeier U (1988) Possible control of transcript levels by chlorophyll precursors in *Chlamydomonas*. Eur J Biochem 177: 417–424

Johanningmeier U and Howell S (1984) Regulation of light-harvesting chlorophyll-binding protein mRNA accumulation in *Chlamydomonas reinhardi*. J Biol Chem 265: 21820–21827

Joyard J, Block M, Pineau B, Albrieux C and Douce R (1990) Envelope membranes from mature spinach chloroplasts contain a NADPH:protochlorophyllide reductase on the cytosolic side of the outer membrane. J Biol Chem 265: 21820–21827

Kannangara CG and Gough SP (1978) Biosynthesis in greening barley leaves. II. Induction of enzyme synthesis by light. Carlsberg Res Comm 44: 1–20

Kannangara CG, Gough SP, Bruyant P, Hoober JK, Kahn A and von Wettstein D (1988) tRNA $^{Glu-1}$ as a cofactor in δ-aminolevulinate biosynthesis: steps that regulate chlorophyll synthesis. Trends Biochem Sci 13: 139–143

Kannangara CG, Vothknecht UC, Hansson M and von Wettstein D (1997) Magnesium chelatase: Association with ribosomes and mutant complementation studies identify barley subunit xantha –G as functional counterpart of *Rhodobacter* subunit BchD. Mol Gen Genet 254: 85–92

Karpinski S, Reynolds H, Karpinka B, Wingsle G, Creissen G and Mullineaux P (1999) Systemic signaling and acclimation in response to excess excitation energy in *Arabidopsis*. Science 284: 654–657

Kikuchi G, Kumar A, Talmage P and Shemin D (1958) The enzymatic synthesis of 5-aminolevulinic acid. J Biol Chem 233: 1214–1219

Koncz C, Mayerhofer R, Koncz-Kalman Z, Nawrath C, Reiss B, Redei GP and Schell J (1990) Isolation of a gene encoding a novel chloroplast protein by T-DNA tagging in *Arabidopsis thaliana*. EMBO J 9: 1137–1146

Kropat J, Oster U, Rüdiger W and Beck CF (1997) Chlorophyll precursors are signals of chloroplast origin involved in light induction of nuclear heat-shock genes. Proc Natl Acad Sci USA 94: 14168–14172

Kruse E, Mock H-P and Grimm B (1995) Reduction of coproporphyrinogen oxidase level by antisense RNA synthesis leads to deregulated gene expression of plastid proteins and affects the oxidative defense system. EMBO J 14: 3712–3720

Kruse E, Grimm B, Beator J and Kloppstech K (1997) Developmental and circadian control of the capacity for δ-aminolevulinic acid synthesis in green barley. Planta 202: 235–241

Kumar AM, Schaub U, Söll D and Ujwal ML (1996) Glutamyl-transfer RNA: At the crossroad between chlorophyll and protein synthesis. Trends Plant Sci 1: 371–376

Lebedev N and Timko MP (1998) Protochlorophyllide photoreduction. Photosynth Res 58: 5–23

Lebedev N, van Cleve B, Armstrong GA and Apel K (1995) Chlorophyll synthesis in a deetiolated (det 340) mutant of Arabidopsis without NADPH-protochlorophyllide (Pchlide) oxidoreductase (POR) A and photoactive Pchlide-F655. Plant Cell 7: 2081–2090

Li J and Timko MP (1996) The *pc-1* phenotype of *Chlamydomonas reinhardtii* results from a deletion in the nuclear gene for NADPH:protochlorophyllide oxidoreductase. Plant Mol Biol 30: 15–37

Li J, Goldschmidt-Clermont M and Timko MP (1993) Chloroplast-encoded *chlB* is required for light-independent proto-chlorophyllide reductase activity in *Chlamydomonas reinhardtii*. Plant Cell 5: 1817–1829

Lindholm J and Gustafsson P (1991) Homologues of the green algal *gidA* gene and the liverwort *frxC* gene are present in the chloroplast genomes of conifers. Plant Mol Biol 17: 787–798

Low PS and Merida JR (1996) The oxidative burst in plant defence—Function and signal transduction. Physiol Plant 96: 533–542

Mapleston ER and Griffiths WT (1980) Light modulation of the activity of the protochlorophyllide oxidoreductase. Biochem J 189: 125–133

Mascia P (1978) An analysis of precursors accumulated by several biosynthetic mutants of maize. Mol Gen Genet 161: 237–244

Matringe M, Camadro JM, Labbe P and Scalla R (1989) Protoporphyrinogen oxidase as a molecular target for diphenylether herbicides. Biochem J 260: 231–235

Mayfield SP and Taylor WC (1984) The appearance of photosynthetic proteins in developing maize leaves. Eur J Biochem 144: 79–84

Mock H-P and Grimm B (1997) Reduction of uroporphyrin decarboxylase by antisense RNA expression affects activities of other enzymes in tetrapyrrole biosynthesis and leads to light-dependent necrosis. Plant Physiol 113: 1101–1112

Mock H-P, Keetman U, Kruse E, Rank B and Grimm B (1998) Defense responses to tetrapyrrole-induced oxidative stress in transgenic plants with reduced uroporphyrinogen decarboxylase or coproporphyrinogen oxidase activity. Plant Physiol 116: 107–116

Molina A, Volrath S, Guyer D, Maleck K, Ryals J and Ward E (1999) Inhibition of protoporphyrinogen oxidase expression in *Arabidopsis* causes a lesion-mimic phenotype that induces systemic acquired resistance. Plant J 17: 667–678

Mösinger E, Batschauer A, Schäfer E and Apel K (1985) Phytochrome control of in vitro transcription of specific genes in isolated nuclei from barley (*Hordeum vulgare*). Eur J Biochem 147: 137–142

Nielsen OF (1974) Macromolecular physiology of plastids. XII. *Tigrina* mutants of barley. Hereditas 76: 269–304

Oelmüller R (1989) Photooxidative destruction of chloroplasts and its effect on nuclear gene expression and extraplastidic enzyme levels. Photochem Photobiol 49: 229–239

Oelmüller R and Mohr H (1986) Photooxidative destruction of chloroplasts and its consequences for expression of nuclear genes. Planta 167: 106–113

Ohyama K, Kohchi T, Sano T and Yamada Y (1988) Newly identified groups of genes in chloroplasts. Trends Biochem Sci 13: 19–22

Packer N and Adamson H (1986) Incorporation of 5-aminolevulinic acid into chlorophyll in darkness in barley. Physiol Plant 68: 222–230

Papenbrock J, Gräfe S, Kruse E, Hänel F and Grimm B (1997) Mg-chelatase of tobacco: identification of a *Chl D* cDNA sequence encoding a third subunit, analysis of the interaction of the three subunits with the yeast two-hybrid system, and reconstitution of the enzyme activity by co-expression of recombinant CHL D, CHL H and CHL I. Plant J 12: 981–990

Pfannschmidt T, Nilsson A and Allen JF (1999) Photosynthetic control of chloroplast gene expression. Nature 397: 625–628

Pineau B, Gerard-Hirne C, Douce R and Joyard J (1993) Identification of the main species of tetrapyrrolic pigments in envelope membranes from spinach chloroplasts. Plant Physiol 102: 821–828

Pontoppidan B and Kannangara CG (1994) Purification and partial characterization of barley glutamyl-tRNA^{Glu} reductase, the enzyme that directs glutamate to chlorophyll biosynthesis. Eur J Biochem 225: 529–537

Rebeiz CA, Montazer-Zouhoor A, Mayasich JM, Tripathy BC, Wu S-M and Rebeiz C (1988) Photodynamic herbicides. Recent developments and molecular basis of selectivity. CRC Critical Rev Plant Sci 6: 385–436

Reinbothe S and Reinbothe C (1996) The regulation of enzymes involved in chlorophyll biosynthesis. Eur J Biochem 237: 323–343

Reinbothe S, Runge S, Reinbothe C, van Cleve B and Apel K (1995a) Substrate-dependent transport of the NADPH: protochlorophyllide oxidoreductase into isolated plastids. Plant Cell 7: 161–172

Reinbothe S, Reinbothe C, Holtorf H and Apel K (1995b) Two NADPH:protochlorophyllide oxidoreductases in barley: Evidence for the selective disappearance of POR A during the light-induced greening of etiolated seedlings. Plant Cell 7: 1933–1940

Reinbothe C, Apel K and Reinbothe S (1995c) A light-induced protease from barley plastids degrades NADPH: proto-chlorophyllide oxidoreductase complexed with chlorophyllide. Mol Cell Biol 15: 6206–6212

Reinbothe S, Reinbothe C, Apel K and Lebedev N (1996a) Evolution of chlorophyll biosynthesis—The challenge to survive photooxidation. Cell 86: 703–705

Reinbothe S, Reinbothe C, Lebedev N and Apel K (1996b) POR A and POR B, two light-dependent protochlorophyllide-reducing enzymes of angiosperm chlorophyll biosynthesis. Plant Cell 8: 763–769

Reinbothe S, Reinbothe C, Neumann D and Apel K (1996c) A plastid enzyme arrested in the step of precursor translocation in vivo. Proc Natl Acad Sci USA 93:12026–12030

Reinbothe C, Lebedev N and Reinbothe S (1999) A protochlorophyllide light-harvesting complex involved in de-etiolation of higher plants. Nature 397: 80–84

Röbbelen G (1956) Ueber die Protochlorophyllidreduktion in einer Mutante von *Arabidopsis thaliana* (L.) Heynh. Planta 47: 532 546

Roitgrund C and Mets L (1990) Localization of two novel chloroplast functions: *Trans*-splicing of RNA and protochlorophyllide reduction. Curr Genet 17: 147–153

Runge S, van Cleve B, Lebedev N, Armstrong GA and Apel K (1995) Isolation and classification of chlorophyll-deficient xantha mutants of *Arabidopsis thaliana*. Planta 197: 490–500

Runge S, Sperling U, Frick G, Apel K and Armstrong GA (1996) Distinct roles for light-dependent NADPH-protochlorophyllide oxidoreductases (POR) A and B during greening in higher plants. Plant J 9: 513–523

Ryberg M and Sundqvist C (1988) The regular ultrastructure of isolated prolamellar bodies depends on the presence of membrane-bound NADPH-protochlorophyllide oxidoreductase. Physiol Plant 73: 218–226

Ryter SW and Tyrrell RM (1998) Singlet molecular oxygen ($^1O_2$): A possible effector of eukaryotic gene expression. Free Rad Biol Med 24: 1520–1534

Santel HJ and Apel K (1981) The protochlorophyllide holochrome of barley (*Hordeum vulgare* L.). The effect of light on the NADPH-protochlorophyllide oxidoreductase. Eur J Biochem 120: 95–103

Scheumann V, Klement H, Helfrich M, Oster U, Schoch S and Rüdiger W (1999) Protochlorophyllide *b* does not occur in barley etioplasts. FEBS Lett 445: 445–448

Schoda M, Vallon O, Wollmann FA and Beck C (1999) A chloroplast targeted heat shock protein 70 (HSP70) contributes to the photoprotection and repair of photosystem II during and after photoinhibition. Plant Cell 11: 1165–1178

Sheen JC (1990) Metabolic repression of transcription in higher plants. Plant Cell 2: 1027–1038

Silverthorne J and Tobin EM (1984) Demonstration of transcriptional regulation of specific genes by phytochrome action. Proc Natl Acad Sci USA 81: 1112–1116

Skinner JS and Timko MP (1999) Differential expression of genes encoding the light-dependent and light-independent enzymes for protochlorophyllide reduction during development in loblolly pine. Plant Mol Biol 39: 577–592

Spano AJ, He Z-H and Timko MP (1992a) NADPH:protochlorophyllide oxidoreductases in white pine (*Pinus strobus*) and loblolly pine (*P. taeda*). Mol Gen Genet 236: 86–95

Spano AJ, Zenghui H, Michel H, Hunt DF and Timko MP (1992b) Molecular cloning, nuclear gene structure, and developmental expression of NADPH-protochlorophyllide-oxidoreductase in pea (*Pisum sativum* L.). Plant Mol Biol 18: 967–972

Sperling U, van Cleve B, Frick G, Apel K and Armstrong GA (1997) Overexpression of light-dependent POR A or B in plants depleted of endogenous POR by far-red light enhances seedling survival in white light and protects against photooxidative damage. Plant J 12: 649–658

Sperling U, Franck F, van Cleve B, Frick G, Apel K and Armstrong GA (1998) Etioplast differention in *Arabidopsis*: Both POR A and POR B restore the prolamellar body and photoactive protochlorophyllide-F655 to the *cop1* photomorphogenic mutant. Plant Cell 10: 283–296

Spikes JD and Bommer JC (1991) Chlorophylls and related pigments as photosensitizers in biology and medicine. In: Scheer H (ed) Chlorophylls, pp 1181–1204. CRC Press, Boca Raton, Fl, USA

Stillman LC and Gassman ML (1978) Characterization of protoheme levels in etiolated and greening tissues. Plant Physiol 62: 182–184

Stobart AK and Ameen-Bukhari I (1984) Regulation of δ-aminolevulinic acid synthesis and protochlorophyllide regeneration in the leaves of dark-grown barley (*Hordeum vulgare*) seedlings. Biochem J 222: 419–426

Susek RE and Chory J (1992) A tale of two genomes: Role of a chloroplast signal in coordinating nuclear and plastid genome expression. Aust J Plant Physiol 19: 387–399

Suzuki JY and Bauer CE (1992) Light-independent chlorophyll biosynthesis: Involvement of the chloroplast gene *chlL* (frxC). Plant Cell 4: 929–940

Suzuki JY and Bauer CE (1995) A prokaryotic origin for light-dependent chlorophyll biosynthesis of plants. Proc Natl Acad Sci USA 92: 3749–3753

Takio S, Nakao N, Suzuki T, Tanaka K, Yamamoto I and Satoh T (1998) Light-dependent expression of protochlorophyllide oxidoreductase gene in the liverwort, *Marchantia paleacea* var. *diptera*. Plant Cell Physiol 39: 665–669

Taylor WC (1989) Regulatory interactions between nuclear and plastid genomes. Annu Rev Plant Physiol Plant Mol Biol 40: 211–233

Terry MJ and Kendrick RE (1999) Feedback inhibition of chlorophyll synthesis in the phytochrome-deficient *aurea* and *yellow-green-2* mutants of tomato. Plant Physiol 119: 143–152

Thomas H (1997) Chlorophyll: A symptom and a regulator of plastid development. New Phytol 136: 163–181

von Wettstein D, Gough S and Kannangara CG (1995) Chlorophyll biosynthesis. Plant Cell 7: 1039–1057

Walker CJ and Willows RD (1997) Mechanism and regulation of Mg-chelatase. Biochem J 327: 321–333

Yang Z and Bauer CE (1990) *Rhodobacter capsulatus* genes involved in early steps of bacteriochlorophyll biosynthetic pathway. J Bacteriol 172: 5001–5010

Zsebo KM and Hearst JE (1984) Genetic-physical mapping of a photosynthetic cluster from *R. capsulata*. Cell 37: 937–947

Chapter 14

Transport of Metals: A Key Process in Oxygenic Photosynthesis

Himadri Pakrasi[1]*, Teruo Ogawa[2] and Maitrayee Bhattacharrya-Pakrasi[1]
[1] *Department of Biology, Campus Box 1137, Washington University, St. Louis, MO 63130-4899, U.S.A.;* [2] *Bioscience Center, Nagoya University, Chikusa, Nagoya 464-8601, Japan*

Summary

Metals play important roles in all phases of oxygenic photosynthesis in cyanobacteria, eukaryotic algae and green plants. For the photosynthetic electron transport reactions in the thylakoid membranes, iron, copper, manganese and magnesium are essential cofactors in various proteins and pigment-protein complexes. Zinc, iron and magnesium also play critical roles during the carbon-fixation reactions. In addition, iron, copper and zinc are constituents of superoxide dismutase and other protective enzymes that are essential to maintain the integrity and function of the photosynthetic apparatus in its highly reactive environment. Inside any living cell, concentrations of various metals are maintained within specific ranges. If the concentration of any metal is below a lower threshold level, organisms suffer from this metal ion deficiency. On the other hand, excess amount of many metals can be toxic. Since metals are both essential and potentially toxic, they are under strict homeostatic control that requires a balance between their uptake and efflux. Usually, metals are transported across cell and organellar membranes via specific transporters. Many families of metal uptake and efflux

*Author for correspondence, email: Pakrasi@biology.wustl.edu.

Eva-Mari Aro and Bertil Andersson (eds): Regulation of Photosynthesis, pp. 253–264.

transporters have recently been described in both prokaryotes and eukaryotes. In 1996, the publication of the complete genome sequence of the cyanobacterium *Synechocystis* sp. PCC 6803, an oxygenic photosynthetic organism, provided valuable information about a number of potential metal transporters in this organism. During the past few years, genetic and biochemical dissection of some of these transporters has yielded important functional data about both the transport processes and their regulations, for a number of metals. Recent completion of the genome sequence of the *Arabidopsis thaliana* has opened additional exciting opportunities for functional genomic analysis of metal homeostasis and its influence on photosynthesis. In this chapter, we discuss transport of iron, copper, manganese, zinc and magnesium, primarily using the *Synechocystis* 6803 paradigm.

I. Introduction

Oxygenic photosynthesis, a principal bioenergetic process in plants, algae and cyanobacteria, is intimately dependent on the presence of a number of metals in various pigment-protein complexes and metabolic enzymes. The green color of these organisms originates from chlorophyll molecules, each of which contains a magnesium atom at the center of its tetrapyrrole ring. Numerous metals are present in the three supramolecular complexes that catalyze light-mediated charge separation and mediate subsequent electron transfer reactions in the thylakoid membranes (Buchanan et al., 2000). According to the currently accepted model, the Photosystem II (PS II) complex contains one non-heme iron, one or two *b*-type hemes, one *c*-type heme (in cyanobacterial and some algal PS II) and four manganese ions. The cytochrome (Cyt) b_6f complex has two *b*-type hemes, one *c*-type heme and one iron-sulfur cluster, whereas Photosystem I (PS I) has three iron-sulfur clusters. In addition, electron transfer between Cyt b_6f complex and PS I is mediated by plastocyanin, a copper-containing protein, or Cyt c_6, a *c*-type heme-containing protein. Finally, ferredoxin, an electron acceptor protein for PS I, contains an iron-sulfur complex. Other than the magnesium atoms in the chlorophyll molecules, there are as many as 28 iron, copper and manganese ions present in the equivalent of one linear electron transfer chain from two water molecules to one $NADP^+$ molecule.

Zinc is an essential cofactor in the enzyme carbonic anhydrase. The extent to which cyanobacterial photosynthetic processes are dependent upon the activities of the carbonic anhydrases is not well established. However, it is possible that the carbon dioxide requirement during photosynthetic carbon fixation is partially supplied by the activity of this enzyme. Magnesium is an important cofactor for at least three enzymes in the C3 Calvin-Benson cycle of carbon fixation. In addition, radical scavenging enzymes such as superoxide dismutases contain iron, copper, manganese and zinc. Under the highly redox active conditions generated by photosynthetic reactions, such enzymes are essential for the stability of the enzymes, proteins and other macromolecules in the photosynthetic apparatus.

The enormous importance of the above mentioned metals in virtually all phases of photosynthesis underscores the need for a carefully regulated supply of these metals in cyanobacterial cells and chloroplasts. Transport of proteins across biological membranes usually needs the presence of membrane-spanning transporters, most of which exhibit specificity for one or a small number of metals. This review will limit its discussion to the transporters for the following metals: iron, copper, manganese, zinc and magnesium. In particular, we will focus our discussion on such metal transporters present in the unicellular cyanobacterium *Synechocystis* 6803, a model oxygenic photosynthetic organism whose genome has been completely sequenced (Kaneko et al., 1996).

II. Different Classes of Transporters

In general there are several mechanisms of movement of metals across membranes that act to contain and partition the cell. These include: 1) active transport systems that require the hydrolysis of ATP or other alternative sources of energy, 2) transport mechanisms that depend on the electrochemical gradient present

Abbreviations: Cyt – cytochrome; EST – expressed sequence tag; Fur – Ferric uptake regulator; Mnt – manganese transporter; PS I – Photosystem I; PS II – Photosystem II; Znt – zinc transporter; Zur – zinc uptake regulator

across the membrane where the transporter complex is located and 3) classic diffusion without the requirements for energy (Saier, 2000; Williams et al., 2000).

Table 1 lists the metal transporters (some of which are still in the search for exact functions!) pertinent to this review, as deduced from a bioinformatics-based analysis of the complete genome sequence of *Synechocystis* 6803. Recently, Saier and colleagues have introduced a 'Transport Commission' (TC) nomenclature (comparable to the EC system for enzymes) for various classes of transporters (Saier, 1998; Saier, 2000). In Table 1, we have attempted to categorize various (putative) metal transporters in *Synechocystis* 6803 according the TC system. For a number of such transporters, the powerful forward and reverse genetic systems available for this organism have recently unraveled which metal is transported by which transporter. In addition, the directionality of such transport processes (uptake or egress) has also been elucidated.

A. ATP-Dependent Transporters

For the import of transition metals to the intracellular compartments from external sources, ATP linked processes are of central interest (Saier, 2000; Williams et al., 2000). These include the ABC (ATP binding cassette) transporter systems that operate to move metals from the periplasmic compartment to the cytoplasmic side of the plasma membrane and ATPase linked transporters (e.g., P-type ATPase). Both ABC transporters and ATPase linked plasma membrane localized transport systems have been identified in cyanobacteria.

Canonical ABC transporter complexes at the protein level consist of two or more membrane spanning domains linked either covalently or by association to two cytoplasmic ATP binding domains. In addition, there are usually periplasmic, solute binding, sometimes lipid anchored domains that facilitate metal binding and association with the membrane spanning components (Driessen et al., 2000). Analysis of the *Synechocystis* 6803 genome sequence has revealed that the family of ABC cassette containing transport proteins in this organism comprises of products of at least 32 genes and gene clusters. Homologs for most of these proteins can also be found in *Anabaena* 7120 and *Synechococcus* 7942. To date, a subset of these proteins in *Synechocystis* 6803 has been experimentally shown to transport iron, manganese and zinc to the intracellular compartment.

Membrane spanning ATPase translocators predominantly facilitate the transport of charged solutes by the hydrolysis of ATP. These translocators are usually linked to the electrochemical gradient across the membrane as well. The transport of a metal can occur in either direction across the plasma membrane and accompanies the co-transport of a cation. As indicated in Table 1, a number of *Synechocystis* 6803 genes encode members of the P-type ATPase superfamily. The catalytic subunits of such transporters are usually phosphorylated, and are hence termed the 'P-type' ATPase. CPx-ATPase proteins of this family have a cysteine-proline-X (where X is any amino acid) motif and can link the transport of a metal to either the symport or antiport transport of Na^+ ions (Driessen et al., 2000; Williams et al., 2000). Proteins similar to this class of the ATPase family exist in both *Synechocystis* 6803 and *Anabaena* 7120. In *Synechocystis* 6803, the CPx-ATPases are known to act in efflux mechanisms for zinc and cobalt, and uptake mechanisms for copper. Calcium/proton translocator ATPases link the transport activity to the pumping of calcium or hydrogen ions and have also been predicted to be involved in heavy metal transport in a number of organisms. The magnesium transporters MgtA and MgtB are members of this group of transport ATPases.

B. Secondary and Unclassified Transporters

As shown in Table 1, a number of metal transporters in *Synechocystis* 6803 cells are either secondary (not directly ATP hydrolysis driven) or unclassified transporters. Additionally, the Nramp family of transporter proteins has been found to exist in all organisms whose genomes have been analyzed to date. These are integral membrane proteins that were originally identified in the endosomal membrane of human macrophages. These proteins regulate the transport of divalent cations across membranes to limit their availability to engulfed bacteria to prevent bacterial growth that require transition elements. Nramp proteins have been found to be involved in the transport of manganese, copper, iron, zinc and cobalt. Analysis of the *Synechocystis* 6803 genome reveals a few Nramp homologs with relatively low degrees of similarity.

Table 1. Metal Transport Proteins in *Synechocystis* sp. PCC 6803

ATP-dependent

3.A.1 ATP-binding Cassette (ABC) Superfamily

	Proteins		Substrate (probable)
ATP-binding	*Membrane*	*Metal Binding*	
Sll1878	Slr0327	Slr0513/Slr1295	ferric iron
Slr1318	Slr1316/Slr1317	Slr1319	(ferric dicitrate)
–	Slr1488	Slr1491/Slr1492	(ferric dicitrate)
Sll1599	Sll1600	Sll1598	manganese
Slr2044	Slr2045	Slr2043	zinc
–	Sll0739	Sll0738	(molybdenum)

3.A.3 P-type ATPase Superfamily

Protein	Substrate (probable)
Slr0797	cobalt
Slr0798	zinc
Sll1920	copper
Slr1950	copper
Sll0672	(magnesium)
Slr0822	(magnesium)
Sll1614	(magnesium)
Sll1076	(cation)

Secondary Transporters

Protein	Substrate (probable)
2.A.1 Major Facilitator Superfamily (MFS)	
Slr0796	nickel, cobalt
2.A.6 Resistance-Nodulation-Cell Division (RND) Superfamily	
Slr0794	nickel

Unclassified

Protein	substrate (probable)
9.A.8 Ferrous Iron Uptake (FeoB) Family	
Slr1392	ferrous iron
9.A.10 Oxidase-dependent Fe^{2+} Transporter (OfeT) Family	
Slr0964	(ferrous iron)
9.A.19 MgtE family	
Slr1216	magnesium
9.A.17 MIT (metal ion transporter) family	
Sll0671	(magnesium)
Sll0571	(magnesium)

Bioinformatics based analysis of metal transporters encoded in the genome of *Synechocystis* 6803. The newly installed 'Transport Commission' (TC) nomenclature for various families and superfamilies of transporters has been used (Saier, 1998). Adapted and modified from http://www.biology.ucsd.edu/~msaier/transport/. The metals shown within parenthesis have not been experimentally demonstrated as substrates for the respective predicted transporters.

Few diffusion linked means of obtaining the necessary amounts of metals for intracellular processes are known. Transition metals are rarely expected to diffuse across membranes that are functionally intact. Diffusion coefficients for cationic species are at least in the range of 10^{-12} when compared with water. So it is highly improbable for the transition metals under discussion here to freely diffuse across the bacterial membrane. Recently, however, a family of metal transporters, called cation diffusion facilitators, has been described in plants and non plant eukaryotes. These proteins are membrane facilitating heavy metal transport proteins termed ZIP proteins (Guerinot, 2000). Whether there are ZIP proteins in the cyanobacteria to facilitate transport of heavy metals remains to be determined.

Finally, a small family of proteins involved in copper transport has been identified in *Arabidopsis* with homologs found in human expressed sequence tag (EST) databases. The ZiaA zinc exporter from *Synechocystis* 6803 shows limited homology to this novel class of transporters. It is possible that this class of transporters may be somewhat similar in structure to the CPx-ATPase family, while containing two different metal complexing strategies via separate binding motifs and different affinities for the metals.

III. Iron

A. Transport

Iron serves as components of heme and iron sulfur centers integrated into a variety of proteins essential for photosynthesis. Biological accessibility of this element is, however, very low since iron is present mostly as insoluble ferric iron (Fe^{3+}) compounds under oxygenic conditions. Ferric iron is accessed by photosynthetic organisms via two distinct strategies (Guerinot, 1994; Marschner and Römheld, 1994). Dicots and non-graminaceous monocots utilize a strategy that involves solubilization of Fe^{3+} by extracellular acidification, reducing Fe^{3+} to Fe^{2+} by plasma membrane localized redox systems, followed by the uptake of Fe^{2+} by a specific transporter. Graminaceous plants and cyanobacteria utilize a different strategy that involves the export of iron-specific chelators such as siderophores, followed by the uptake of the Fe^{3+}-chelator complex.

The complete genome sequence of the cyanobacterium *Synechocystis* 6803 has revealed many genes encoding proteins with similarity to those involved in iron-acquisition in non-photosynthetic prokaryotes or genes encoding subunit polypeptides of ABC transporters of unknown function (Kaneko et al., 1996). Analysis of mutants disrupted for these genes have led to the identification of a novel ABC-type Fe^{3+} transporter essential for iron acquisition. This transporter consists of the subunits in periplasmic space encoded by slr0513 (*futA2*) and slr1295 (*futA1*), that have redundant or overlapping Fe^{3+} binding function (Table 1), and the integral membrane subunit and nucleotide binding subunit encoded by slr0327 (*futB*) and sll1878 (*futC*), respectively (Katoh et al., 2000; Katoh et al., 2001). Both FutA1 and FutA2 are major polypeptide components in the periplasmic space of *Synechocystis* 6803 cells. All of the *fut* genes are expressed in cells grown in complete medium (20 μM Fe^{3+}) and expression of *futA1*, *fuA2* and *futC* is strongly enhanced in iron deprived cells. The K_m value of the *fut*-dependent Fe^{3+} transport in cells grown in iron-replete medium is 0.5 μM for Fe^{3+}. Both the maximal rate and K_m value are increased in iron starved cells. The mutants inactivated for *fut* gene(s) grow very poorly under Fe^{3+} concentrations below 1 μM. Interestingly, high rates of Fe^{2+} transport are induced in these mutants during growth in complete medium, whereas high rates of Fe^{2+} uptake are only observed in iron-deprived wild type cells. This iron starvation-induced Fe^{2+} transport activity has the K_m value of 2.9 μM for Fe^{2+} and is greatly reduced in a strain disrupted for slr1392 (Table 1). The activity of the slr1392–dependent Fe^{2+} transport is very high when ascorbate is added to the medium, but in the absence of ascorbate is only one third the Fe^{3+} transport activity in wild type. Thus, the Fe^{3+} reducing activity limits the uptake of Fe^{2+} under oxygenic conditions. The uptake of Fe^{3+} and Fe^{2+}, either in the presence or absence of ascorbate, proceeds in the dark. Light does not have a stimulatory effect on the accumulation of iron. Hence, respiration and other dark metabolic reactions generate a sufficient supply of ATP to energize both Fe^{3+} and Fe^{2+} transport.

B. Regulation

The regulation of bacterial genes in response to iron is to a great extent controlled by the ferric uptake regulatory (Fur) protein (Escolar et al., 1999). This small protein (approximately 15–20 kDa, 17 kDa in *E. coli*) controls the expression and repression of genes that are related to iron metabolism including siderophore production, various alternate sigma factors, and outer membrane proteins that mediate uptake of iron in Fe^{3+} form. In addition since a slight excess of iron tends to be toxic, the intracellular concentrations of iron must be tightly regulated. More often than not, this control is exhibited by turning uptake mechanisms on and off rather than initiating the efflux of iron and iron related compounds (Escolar et al., 1999). In *Synechocystis* 6803, the regulation of the *fut* family of genes (see above) appears to follow this paradigm, and is noteworthy in that these genes are not organized in an operon.

In the cyanobacteria *Synechococcus* 7942 and *Synechocystis* 6803, two genes have been identified that encode proteins that are homologs of the Fur protein. In *Synechocystis* 6803, these ORFs are sll0567 and sll1937. Sll0567 is the closest homolog of the Fur protein from *E. coli*, and despite repeated attempts (Katoh and Ogawa, unpublished; Bartsevich and Pakrasi, unpublished; Hageman, unpublished), it has not been possible to generate a completely segregated knockout mutant for this ORF. Even under low iron conditions that have allowed the *E. coli fur* mutants to segregate, the cyanobacteria have remained heteroallelic. Similar results were obtained when a *fur* homolog in the cyanobacterium *Synechococcus* 7942 was inactivated. The mutant did not segregate to homogeneity. However, this mutant did exhibit iron-stress symptoms when grown in iron replete medium (Ghassemian and Straus, 1996). These results indicate that regulation by Fur is an essential function in cyanobacteria.

Structural studies of the Fur protein in *E. coli* have shown that it exists in a zinc bound form in its active state (Althaus et al., 1999). The Fur binding site is generally conserved within many species of bacteria, and a consensus 'Fur-box' sequence has been determined (Escolar et al., 1999). This sequence is found 35 times in the promoter regions of *E. coli*. The same consensus sequence appears 22 times in possible promoter regions of *Synechocystis* 6803 (Bhattacharyya-Pakrasi, unpublished). Interestingly, one of these sites is in the promoter of the *mntCAB* operon (encoding a Mn-transporter, see section V), indicating possible control of transcription of this operon by iron availability. In *E. coli*, Fur mediates iron responsive regulation of Mn-Fe superoxide dismutases, and it is conceivable that regulation by Fur would be an important control element to optimize relative levels of transition metals within the cell.

The second Fur homolog in *Synechocystis* 6803 is 23% identical to Fur at the amino acid level, and is encoded by an ORF adjacent to the *zntCAB* operon that encodes an ABC transporter system for the high affinity uptake of zinc (see section VI).

IV. Copper

In the environment, copper is second only to iron in its available concentration to living cells. Copper in either of its +1 or +2 states cannot exist as a free metal within living systems, being highly toxic even at low concentrations. This transition metal can be linked to a number of metabolic processes within the cyanobacterial cell. An established requirement for copper exists in the transduction of electrons from the Cyt b_6f complex to PS I via the copper containing protein plastocyanin. Copper is also an essential cofactor in metabolic enzymes such as superoxide dismutases and in a number of small proteins that are thought to be involved in synthetic roles within the cell. The expression of at least two genes in cyanobacteria that encode plastocyanin and Cyt c_6 are regulated in response to copper levels that the cyanobacteria are exposed to (Zhang et al., 1992; Zhang et al., 1994).

A. Transport

The machinery for transport of copper to the intracellular compartment has been addressed extensively in eukaryotic cells and higher plants and bacteria. Within the cyanobacteria, several P-type ATPases have been identified in *Synechococcus* 7942 (Kanamaru et al., 1993). Such a transporter protein is localized to the thylakoid membrane of *Synechococcus*, and is the prototype for the PacS family of P-type ATPases found in prokaryotic cells (Kanamaru et al., 1994). Such heavy metal transporting ATPases have a common CPx motif embedded within one of the membrane spanning regions of the protein (Williams et al., 2000). Among the family of P-type ATPases that are identifiable from the genomic information of *Synechocystis* 6803, at least six possess this motif (Table 1), two of which have been shown by Robinson and colleagues to be involved in zinc and cobalt export (Robinson et al., 2000). Two of these open reading frames, slr1950 and sll1920 also possess the requirements to be included within the CPx family of P-type ATPases. Both are highly homologous to PacS, with 37% and 56% identity, respectively, at the amino acid level.

The results of genetic ablation of these two genes were recently made available in the 'Cyanomutants' database (http: //www.kazusa.or.jp/cyano/mutants/) by N. J. Robinson, indicating that these are indeed copper transporters in *Synechocystis* 6803. An slr1950-deficient mutant shows reduced cellular copper accumulation and has impaired switching to plastocyanin. In a cytochrome c_6 mutant background, the slr1950 deletion strain is sensitive to low (copper depleted) conditions generated by treatment with a copper-chelator. Sll1950 may be the primary copper uptake transporter in this cyanobacterium. The sll1920 deletion mutant is copper-sensitive. Since Sll1920 is most homologous to the *Synechococcus* 7942 PacS protein, which has been localized to the thylakoid membrane, an interesting possibility is that the Slr1950 protein is a plasma membrane localized transporter, while Sll1920 has a role in the provision of copper to the photosynthetic components within the thylakoid membrane system. It is also possible that sll1920 encodes a plasma membrane localized exporter for copper because of the copper sensitivity exhibited by the knockout mutant. In this regard, comparison to the *cop* operon of *Enterococcus hirae* suggests interesting possibilities for copper homeostasis in cyanobacteria. The *copYZAB* operon of the gram positive *Enterococcus hirae* is the primary locus to control copper homeostasis in this organism (Harrison et al., 2000). The *copA* and *copB* genes encode proteins for copper uptake and copper export, respectively, somewhat similar to the expected functions of Slr1950 and Sll1920 in *Synechocystis* 6803. The *copY* and *copZ* encoded proteins specify a repressor and an activator. In this process of regulation by copper in enterobacteria, the activator with bound copper interacts with a zinc-bound repressor to replace the zinc moiety in the repressor, resulting in the inactivation of the repressor and ensuing release of the DNA and the transcription of the *copAB* genes. Although such repressors are not found within an operon in cyanobacteria, it would be interesting to determine if the regulation of the copper transport machinery is dependent upon the presence or absence of zinc, or other transition metals, especially those involved in photosynthesis, within the cell. A very low degree of homology exists between the N-terminus of the Sll1950 protein and the CopZ activator protein.

V. Manganese

Manganese is an essential trace element in plant, microbial and animal cells, usually present at a level about 200-fold lower than that of iron (Supek et al., 1997). Mn is an essential cofactor in a growing list of enzymes from both prokaryotic and eukaryotic organisms (Dismukes, 1996). These include redox enzymes such as Mn-superoxide dismutase (Mn-SOD), Mn-catalase, ribonucleotide reductase, that strictly require Mn at their active sites. In cyanobacteria and plants, Mn is absolutely required for light-induced dissociation of water to molecular oxygen (Cheniae and Martin, 1970). During oxygenic photosynthesis, Photosystem II (PS II) mediates electron transfer from water to plastoquinone molecules (Debus, 1992; Pakrasi, 1995). The lumen-exposed domain of PS II contains four Mn atoms that play important redox functions during the oxidation of water. As an inevitable consequence of its normal function, the PS II complex undergoes a damage-repair cycle. During this process, the Mn-ensemble is dissociated, and it reassembles in a newly formed PS II complex. In the face of the rapid turnover of this metal cluster, transport of Mn is crucially important for the survival of oxygenic photosynthetic organisms, such as *Synechocystis* 6803.

A. Transport

In *Synechocystis* 6803, uptake of Mn is a crucially important process for the photoautotrophic life style of this organism. In this organism, a high affinity ABC transporter encoded by the *mntCAB* operon is expressed in cells grown under Mn-stress conditions, and functions effectively in the presence of very low nanomolar levels (~20 nM) of Mn (Bartsevich and Pakrasi, 1995). Higher Mn-levels during growth repress this system. Mn can also be transported by a second high affinity transport system in *Synechocystis* 6803 cells grown in the presence of *micromolar* concentrations of manganese (Bartsevich and Pakrasi, 1996). It is noteworthy that kinetic analysis of the initial rates of Mn uptake into cells suggested that in *Synechocystis* 6803 cells, Mn could also be accumulated by a third, low-affinity system. If the characteristics of manganese acquisition in this organism is used as a paradigm, one can extrapolate that the majority of cyanobacteria should have multiple Mn-uptake systems that operate under different external concentrations of this metal during growth.

The MntABC complex is the *first* specific Mn-transporter identified in any organism. The MntC protein is a periplasmic Mn-binding protein, whereas the MntA protein is the cytoplasmic ATPase subunit of this transporter. The MntB protein is an integral membrane protein with nine transmembrane domains (Bartsevich and Pakrasi, 1999). Interestingly, all three polypeptides in the MntABC transporter show significant sequence similarities to several ATP-binding proteins, integral membrane proteins and substrate-binding proteins, respectively, in a number of Gram-positive bacteria. Based on the sequence similarities with the MntABC proteins from *Synechocystis* 6803, we have earlier postulated that the biological function of such protein complexes in these Gram-positive bacteria is the transport of Mn or some other metal (Bartsevich and Pakrasi, 1995). During the past four years, a number of studies have demonstrated that these genes in Gram-positive bacteria indeed encode ABC-type metal permeases. Kolenbrander and coworkers (Kolenbrander et al., 1998) have provided genetic and biochemical evidence that the *scaA* operon in *Streptococcus gordonii* encodes a Mn-transporter complex. Also, Dintilhac and colleagues have shown that the PsaA protein, a homolog of MntC, is a component of an ABC-type Mn-transporter in *Streptococcus pneumoniae*. In addition, a homologous *adc* gene cluster in the same organism encodes a zinc transporter. As described in Section VI, in *Synechocystis* 6803, an ABC transporter encoded by the slr2043-slr2044-slr2045 (Table 1) gene cluster (closest homolog of *mntCAB*) is a Zn-transporter.

In an earlier publication, Tam and Saier (Tam and Saier, 1993) defined eight families or clusters of bacterial solute-binding proteins. According to our analysis, the MntC protein in *Synechocystis* 6803 and its homologs in the Gram-positive bacteria do not belong to any of these eight clusters. In a more recent manuscript, Saier has provided a more detailed classification of various ABC transporters (Saier, 1998). In this new scheme, MntABC is the prototypic member of a new class (cluster 15) of metal permeases.

We have examined the effects of several divalent cations on ^{54}Mn accumulation by *Synechocystis* 6803 cells (Bartsevich and Pakrasi, 1996). Cadmium, and the micronutrients cobalt and zinc competitively inhibit Mn accumulation via the MntABC transporter. Copper and iron do not inhibit Mn accumulation through either of the two high affinity transporter systems. For most bacterial species, cadmium is a highly specific competitive inhibitor of Mn-uptake and, perhaps, it is an alternative substrate for such high-affinity Mn-transport systems. Co and Zn may be other possible substrates for these systems. The macronutrient elements, magnesium and calcium, do not inhibit Mn uptake. Interestingly, addition of calcium significantly increased the rate of Mn-uptake, indicating that the activity of the MntABC transporter is enhanced by high concentrations of Ca.

It is noteworthy that the activity of the second high affinity Mn-transporter is not inhibited by Cd, Co, Zn, or any other metal tested so far. Thus this transporter appears to be specific for Mn alone. As yet, the identity of the second high affinity transporter for manganese in *Synechocystis* 6803 is not known.

VI. Zinc

A. Transport

Zinc is an abundant transition metal in the biosphere. Because of its Lewis acidity, flexible coordination geometry, rapid ligand exchange property and absence of redox activity, Zn is used as a metal of choice in many biological reactions (Lipscomb and Sträter, 1996). In addition, Zn-binding Zn-finger proteins form an important class of DNA-binding transcriptional regulators. During oxygenic photosynthesis in cyanobacteria and chloroplasts, carbonic anhydrase, a Zn-containing metalloenzyme, plays a critical role in the conversion of CO_2 to bicarbonate. Reactions of photosynthesis can generate reactive radicals that are potentially harmful. Several cyanobacterial species use Zn in Cu/Zn superoxide dismutase enzyme (Chadd et al., 1996) for protection against oxidative damage. Finally, Fur, a key transcriptional regulator in bacterial (including cyanobacteria) iron metabolism (Section III) requires Zn for its activity (Althaus et al., 1999).

During recent years, Eide, Guerinot and colleagues have identified a number of Zn transporter systems in yeast, human and plants (Eide, 1998; Guerinot and Eide, 1999). In a number of bacteria, Zn uptake and export transporters have also been characterized during the same period of time. Significant among these studies are two recent reports from Patzer and Hantke (Patzer and Hantke, 1998; Patzer and Hantke, 2000) who have identified an ABC-type Zn-permease in *E. coli*. They have also characterized Zur, a Zn-binding repressor protein similar to Fur, that regulates the transcription of the *znu* operon that encodes components of the Znu transporter in this organism. It is noteworthy that the Znu proteins are members of the broad class of ABC-type metal transporters discussed in Section V.

Robinson and coworkers have recently identified ZiaA, a Zn exporter protein, in the cyanobacterium *Synechocystis* 6803 (Thelwell et al., 1998; Robinson et al., 2000). As indicated in Table 1, the slr0798 ORF in the completely sequenced genome of *Synechocystis* 6803 was predicted to encode a metal-transporting CPx-type ATPase. A mutant strain of *Synechocystis* 6803 in which this ORF (*ziaA* gene) was inactivated was more sensitive to added Zn in its growth medium. This mutant also had reduced Zn content in its periplasm, suggesting that the amount of Zn transported from the cytoplasm to the periplasm was decreased. Moreover, expression of this gene, *ziaA*, enhanced Zn-tolerance of another freshwater cyanobacterium, *Synechococcus* 7942. The ZiaA transmembrane protein has one consensus metal binding motif in its N-terminal part, and it has been suggested that Zn-chaperones (as yet unidentified in cyanobacteria) may interact with this metal binding domain in ZiaA.

As mentioned in section V, the closest homolog of the *mntCAB* operon (encoding a Mn-permease) in the genome of *Synechocystis* 6803 is the slr2043-slr2044-slr2045 cluster of ORFs. Inactivation of each of these ORFs resulted in mutant strains that grow poorly in the absence of added Zn in the growth medium (M. Shibata, T. Ogawa, H. Pakrasi and M. Bhattacharyya-Pakrasi, unpublished). Moreover, such a mutant exhibited reduced ^{65}Zn-uptake activity. Analysis of these mutant strains have established that this operon encodes components of a Zn-uptake transporter (Znt) in *Synechocystis* 6803. Consistent with the nomenclature for the homologous Mn-transporter in the same organism, we have named this operon as the *zntCAB* operon. Thus, the *zntA* gene (slr2044) encodes the ATP-binding component, the *zntB* gene (slr2045) encodes the integral membrane protein component and the *zntC* gene (slr2043) encodes the periplasmic Zn-binding protein. As described in section VI, the Znt transporter is a member of the ABC metal permease family whose first identified member is the Mnt transporter in *Synechocystis* 6803. The steady state level of the *zntCAB* transcript is dramatically increased when wild-type *Synechocystis* 6803 cells are incubated in Zn-free growth medium.

B. Regulation

In the genome of *Synechocystis* 6803, immediately upstream of the *ziaA* gene is a divergently transcribed gene *ziaR* (sll0792) (Thelwell et al., 1998). The *ziaA* – *ziaR* divergon is part of a large gene cluster involved in metal (Ni, Co and Zn) homeostasis in *Synechocystis* 6803 (Garcia-Dominguez et al., 2000; Robinson et al., 2000). ZiaR, a Zn-binding protein acts as a repressor for the transcription of *ziaA*, by binding to the promoter region of the transporter gene. Site-directed mutagenesis experiments have identified two cysteine and one histidine residue in the ZiaR protein as possible ligands for Zn. ZiaR is a homolog of the SmtB protein, a well studied Zn-responsive repressor (Morby et al., 1993; Cook et al., 1998), that regulates the expression of a metallothionein gene in *Synechococcus* 7942, and is a member of the ArsR family of repressors.

Immediately upstream of the *zntCAB* operon in *Synechocystis* 6803 is a divergently transcribed ORF sll1937. The Sll1937 protein is a homolog of the Fur protein in this organism (Section III), and exhibits a high degree of sequence identity with the well-characterized Zn-responsive regulator Zur in *E. coli* (Patzer and Hantke, 2000). Moreover, in a sll1937 inactivation mutant strain, the *zntCAB* operon is constitutively expressed (M. Shibata, T. Ogawa, H. Pakrasi and M. Bhattacharyya-Pakrasi, unpublished). Based on these findings, we have called the sll1937 ORF the *zur* gene.

VII. Magnesium

Magnesium is the most abundant divalent cation in bacterial, plant and animal cells. Its concentration inside cyanobacterial cells and plant chloroplasts is in the mM range. Mg is the central atom in a chlorophyll molecule, and is hence, one of the most abundant metals in the photosynthetic apparatus (Buchanan et al., 2000). In addition, Mg is an important metal for the function of the ATP synthase complex, and at least three of the Calvin cycle enzymes, RuBP carboxylase, fructose-1,6-bisphosphatase and phosphoribulokinase. In addition, the absolute concentration of Mg in the stroma of chloroplasts is modulated by light, a phenomenon that has important regulatory effect on the function of numerous enzymes and proteins involved in photosynthesis. Little is, however, known about the process of Mg transport in and out of photosynthetic cells and organelles.

A. Transport

During the last decade, Maguire and colleagues have performed a comprehensive study on Mg transport in *Salmonella typhimurium* (Smith and Maguire, 1998; Moncrief and Maguire, 1999). There are three separate Mg transport systems, Cor, MgtA and MgtB, in *S. typhimurium*. Cells with any one of these systems can grow on micromolar concentrations of Mg. However, mutant strains with defects in all three of these transporters require 100 mM Mg for growth. Although Mg is the preferred substrate for all three systems, each of them has its own range of divalent cation specificity. Ni is transported through all of them, whereas Co is transported only by the Cor system. The Cor system is the primary Mg transport system (both uptake and efflux), with a high V_{max}, in *Salmonella* as well as in other gram-negative bacteria. The presence of this system is largely unaffected by external Mg concentrations for bacterial growth. The Cor transporter contains four proteins, CorA, CorB, CorC and CorD. Mutations in the genes for the last three proteins do not affect the K_m or V_{max} for Mg uptake, but result in lowered Mg efflux. CorA is the only protein involved in Mg uptake by this transporter. Elegant gene fusion studies have shown that the CorA protein has three membrane spanning domains. Interestingly, its periplasmic domain contains an unusually high concentration of both acidic and basic residues. The other two transporter systems, MgtA and MgtB, are inducibly regulated by Mg availability, and expressed only under Mg starvation conditions. They have relatively high affinity but lower V_{max}'s for Mg as compared to the Cor transporter. The MgtB protein has ten transmembrane spans. Both of these proteins are P-type ATPases, whereas the Cor system is in its own unique class of transporters.

In the *Synechocystis* 6803 genome, three ORFs, sll0672, slr0822 and sll1614, are homologs of *mgtA* and *mgtB* genes from *Salmonella*. Two other ORFs, sll0507 and sll0671, exhibit significant homology to the *corA* genes from *Salmonella*. We have termed them *corA1* and *corA2*, respectively. In addition, the slr1216 ORF is a homolog of the *mgtE* gene, encoding a putative Mg-transporter in *Bacillus firmus* OF4 (Smith and Maguire, 1998). Finally, the slr0014 ORF is a homolog of another putative Mg-transporter gene, *mgtC*. Inactivation of each of the genes mentioned above has indicated that the *mgtE* gene encodes the principal Mg-transporter in *Synechocystis* 6803 (M. Bhattacharyya-Pakrasi, C. Arrett, A. Ruch and H. Pakrasi, unpublished). Detailed molecular and physiological analysis of the *mgtE*-less mutant strain is currently under progress.

VIII. Concluding Remarks

Analysis of the forms and functions of various metal transporters is an important area of investigation in today's biology. Mechanisms of metal transport in photosynthetic organisms, a subject of limited interest in the second half of the twentieth century, is of topical interest at the current time. The recent publication of the genome sequence of *Arabidopsis thaliana* (Walbot, 2000), and the subsequent functional genomics analysis of numerous predicted metal transporters will be of tremendous value in the fields of plant nutrition and human health. From the point of view of organismic evolution, studies on various metal transporters in different cyanobacterial species will be of great interest. This class of organisms originally developed in anoxygenic environment where the bioavailable pools of various metals were vastly different from those present in today's 21% oxygen enriched air, a direct consequence of the photosynthetic activities of cyanobacteria and their relatives. The future research in this important area of regulation of photosynthesis holds exciting potentials.

Acknowledgments

The preparation of this article was supported in part by grants from NIH and USDA-NRICGP to HBP and MBP, and by a grant from 'Research for the Future' program (JSPS-RFTF97R16001) and by a grant from the Human Frontier Science Program (RG0051/1997M) to TO.

References

Althaus EW, Outten CE, Olson KE, Cao H and O'Halloran TV (1999) The ferric uptake regulation (Fur) repressor is a zinc metalloprotein. Biochemistry 38: 6559–6569

Bartsevich VV and Pakrasi HB (1995) Molecular identification of an ABC transporter complex for manganese: Analysis of a cyanobacterial mutant strain impaired in the photosynthetic oxygen evolution process. EMBO J 14: 1845–1853

Bartsevich VV and Pakrasi HB (1996) Manganese transport in the cyanobacterium *Synechocystis* sp. PCC 6803. J Biol Chem 271: 26057–26061

Bartsevich VV and Pakrasi HB (1999) Membrane topology of MntB, the transmembrane protein component of an ABC transporter system for manganese in the cyanobacterium *Synechocystis* sp. strain PCC 6803. J Bacteriol 181: 3591–3593

Buchanan BB, Gruissem W and Jones RL (2000) Biochemistry and Molecular Biology of Plants. American Society of Plant Physiologists, Rockville

Chadd HE, Newman J, Mann NH and Carr NG (1996) Identification of iron superoxide dismutase and a copper/zinc superoxide dismutase enzyme activity within the marine cyanobacterium *Synechococcus* sp. WH 7803. FEMS Microbiol Lett 138: 161–165

Cheniae GM and Martin IF (1970) Sites of function of manganese within Photosystem II. Roles in O_2 evolution and system II. Biochim Biophys Acta 197: 219–239

Cook WJ, Kar SR, Taylor KB and Hall LM (1998) Crystal structure of the cyanobacterial metallothionein repressor SmtB: a model for metalloregulatory proteins. J Mol Biol 275: 337–346

Debus RJ (1992) The manganese and calcium ions of photosynthetic oxygen evolution. Biochim Biophys Acta 1102: 269–352

Dismukes CJ (1996) Manganese enzymes with binuclear sites. Chem. Rev. 96: 2909–2926

Driessen AJ, Rosen BP and Konings WN (2000) Diversity of transport mechanisms: Common structural principles. Trends Biochem Sci 25: 397–401

Eide DJ (1998) The molecular biology of metal ion transport in *Saccharomyces cerevisiae*. Annu Rev Nutr 18: 441–469

Escolar L, Perez-Martin J and de Lorenzo V (1999) Opening the iron box: Transcriptional metalloregulation by the Fur protein. J Bacteriol 181: 6223–6229

Garcia-Dominguez M, Lopez-Maury L, Florencio FJ and Reyes JC (2000) A gene cluster involved in metal homeostasis in the cyanobacterium *Synechocystis* sp. strain PCC 6803. J Bacteriol 182: 1507–1514

Ghassemian M and Straus NA (1996) Fur regulates the expression of iron-stress genes in the cyanobacterium *Synechococcus* sp. strain PCC 7942. Microbiology 142: 1469–1476

Guerinot ML (1994) Microbial iron transport. Annu Rev Microbiol 48: 743–772

Guerinot ML (2000) The ZIP family of metal transporters. Biochim Biophys Acta 1465: 190–198

Guerinot ML and Eide D (1999) Zeroing in on zinc uptake in yeast and plants. Curr Opin Plant Biol 2: 244–249

Harrison MD, Jones CE, Solioz M and Dameron CT (2000) Intracellular copper routing: the role of copper chaperones. Trends Biochem Sci 25: 29–32

Kanamaru K, Kashiwagi S and Mizuno T (1993) The cyanobacterium, *Synechococcus* sp. PCC7942, possesses two distinct genes encoding cation-transporting P-type ATPases. FEBS Lett 330: 99–104

Kanamaru K, Kashiwagi S and Mizuno T (1994) A copper-transporting P-type ATPase found in the thylakoid membrane of the cyanobacterium *Synechococcus* species PCC7942. Mol Microbiol 13: 369–377

Kaneko T, Sato S, Kotani H, Tanaka A, Asamizu E, Nakamura Y, Miyajima N, Hirosawa M, Sugiura M, Sasamoto S, Kimura T, Hosouchi T, Matsuno A, Muraki A, Nakazaki N, Naruo K, Okumura S, Shimpo S, Takeuchi C, Wada T, Watanabe A, Yamada M, Yasuda M and Tabata S (1996) Sequence analysis of the genome of the unicellular cyanobacterium *Synechocystis* sp. strain PCC6803. II. Sequence determination of the entire genome and assignment of potential protein-coding regions. DNA Res 3: 109–136

Katoh H, Grossman AR, Hagino N and Ogawa T (2000) A gene of *Synechocystis* sp. strain PCC 6803 encoding a novel iron transporter. J Bacteriol 182: 6523–6524

Katoh H, Hagino N, Grossman AR and Ogawa T (2001) Genes essential to iron transport in the cyanobacterium *Synechocystis* sp. strain PCC6803. J Bacteriol 183: 2779–2784

Kolenbrander PE, Andersen RN, Baker RA and Jenkinson HF (1998) The adhesion-associated sca operon in *Streptococcus gordonii* encodes an inducible high-affinity ABC transporter for Mn^{2+} uptake. J Bacteriol 180: 290–295

Lipscomb WN and Sträter N (1996) Recent advances in zinc enzymology. Chem. Rev. 96: 2375–2433

Marschner H and Römheld V (1994) Strategies of plants for acquisition of iron. Plant Soil 165: 261–274

Moncrief MB and Maguire ME (1999) Magnesium transport in prokaryotes. J Biol Inorg Chem 4: 523–527

Morby AP, Turner JS, Huckle JW and Robinson NJ (1993) SmtB is a metal-dependent repressor of the cyanobacterial metallothionein gene smtA: Identification of a Zn inhibited DNA-protein complex. Nucleic Acids Res 21: 921–925

Pakrasi HB (1995) Genetic analysis of the form and function of Photosystem I and Photosystem II. Annu Rev Genet 29: 755–776

Patzer SI and Hantke K (1998) The ZnuABC high-affinity zinc uptake system and its regulator Zur in *Escherichia coli*. Mol Microbiol 28: 1199–1210

Patzer SI and Hantke K (2000) The zinc-responsive regulator Zur and its control of the znu gene cluster encoding the ZnuABC zinc uptake system in *Escherichia coli*. J Biol Chem 275: 24321–24332

Robinson NJ, Rutherford JC, Pocock MR and Cavet JS (2000) Metal metabolism and toxicity: Repetitive DNA. In: Whitton BA and Potts M (eds) The Ecology of Cyanobacteria, pp 443–463. Kluwer Academic Publishers, Dordrecht

Saier MH, Jr. (1998) Molecular phylogeny as a basis for the classification of transport proteins from bacteria, archaea and eukarya. Adv Microb Physiol 40: 81–136

Saier MH, Jr. (2000) A functional-phylogenetic classification system for transmembrane solute transporters. Microbiol Mol Biol Rev 64: 354–411

Smith RL and Maguire ME (1998) Microbial magnesium transport: unusual transporters searching for identity. Mol Microbiol 28: 217–226

Supek F, Supekova L, Nelson H and Nelson N (1997) Function of metal-ion homeostasis in the cell division cycle, mitochondrial protein processing, sensitivity to mycobacterial infection and brain function. J Exp Biol 200: 321–330

Tam R and Saier MH, Jr. (1993) Structural, functional, and evolutionary relationships among extracellular solute-binding receptors of bacteria. Microbiol Rev 57: 320–346

Thelwell C, Robinson NJ and Turner-Cavet JS (1998) An SmtB-like repressor from *Synechocystis* PCC 6803 regulates a zinc exporter. Proc Natl Acad Sci U S A 95: 10728–10733

Walbot V (2000) *Arabidopsis thaliana* genome. A green chapter in the book of life. Nature 408: 794–795

Williams LE, Pittman JK and Hall JL (2000) Emerging mechanisms for heavy metal transport in plants. Biochim Biophys Acta 1465: 104–126

Zhang L, McSpadden B, Pakrasi HB and Whitmarsh J (1992) Copper-mediated regulation of cytochrome *c*553 and plastocyanin in the cyanobacterium *Synechocystis* 6803. J Biol Chem 267: 19054–19059

Zhang L, Pakrasi HB and Whitmarsh J (1994) Photoautotrophic growth of the cyanobacterium *Synechocystis* sp. PCC 6803 in the absence of cytochrome *c*553 and plastocyanin. J Biol Chem 269: 5036–5042

Chapter 15

Chloroplast Proteases and Their Role in Photosynthesis Regulation

Zach Adam*

Department of Agricultural Botany, The Hebrew University of Jerusalem, Rehovot 76100, Israel

Summary

Decades of photosynthetic research have yielded a wealth of information on degradation of specific chloroplast proteins. These are found in the stroma and in the thylakoid membrane and lumen, and include unassembled and abnormal proteins, proteins lacking their prosthetic groups, and proteins damaged by exposure to unfavorable growth conditions, particularly high light intensities. The degradation processes characterized to date represent primarily housekeeping and repair mechanisms, whereby nonfunctional proteins can be recycled and their building blocks reused for protein synthesis. In some cases, like that of the D1 protein of Photosystem II's reaction center, removal of the photodamaged protein is an essential step in an elaborate process involving transcription and translation of the protein and its assembly into a functional complex. Although detailed characterizations of degradation processes have preceded the identification of the proteases involved, progress has been made on this front in recent years as well. Proteases and peptidases, all homologues of known enzymes in bacterial systems, have been cloned and characterized. Peptidases responsible for processing precursor and intermediate proteins have been identified in the stroma, associated with the thylakoid membrane, and in the lumen. ATP-dependent proteases, which are likely to carry out most proteolysis in the organelle, have been found in the stroma and thylakoid membrane. The two-component multimeric complex of Clp protease,

*Email: zach@agri.huji.ac.il

Eva-Mari Aro and Bertil Andersson (eds): Regulation of Photosynthesis, pp. 265–276.

composed of the serine proteolytic subunit ClpP and the regulatory ATPase subunit ClpC, is found in the stroma. The thylakoid membrane contains the metalloprotease FtsH, which has both its proteolytic and ATPase domains on the same protein. These functional domains are exposed to the stroma, suggesting a role in degrading both membrane and stromal substrates. On the lumenal side, thylakoid membranes have a tightly bound ATP-independent serine protease, DegP. Now that a number of chloroplast proteases have been identified, current challenges include linking the known substrates to the identified proteases, identifying other substrates—primarily short-lived proteins—that may play a regulatory role in chloroplast gene expression, and understanding substrate specificity and recognition.

I. Introduction

Proteolytic processes regulate a wide range of functions throughout the life cycle of the cell. Proteases are intimately involved in the control of the cell cycle, gene expression, differentiation, protein targeting and sorting, protein quality control and programmed cell death (for review, see Kumar, 1995; Coux et al., 1996; Gottesman, 1996; Suzuki et al., 1997; Hershko and Ciechanover, 1998). Thus, chloroplast proteases are expected to be involved in the regulation of photosynthesis at several different levels as well. However, the role of proteases at most of these levels is still unknown. The development of proplastids and etioplasts into chloroplasts involves not only the synthesis of a new set of proteins, but also massive degradation of the previous population. Gene expression within the chloroplast might require degradation of positive and negative regulators. Limited proteolytic processing is an essential step in the import and sorting of nuclear-encoded chloroplast proteins, and in the maturation of some chloroplast-encoded proteins. Proteases are likely to participate in the biogenetic processes of some protein complexes. Perhaps the most explored aspect of protein degradation in chloroplasts is its role in adapting to changing environmental conditions. Increasing light intensities lead to photodamage, the repair of which requires accelerated removal of damaged proteins. Adaptation to lower light intensities also involves selective proteolytic degradation. Similarly, heat-denatured proteins that fail to refold properly are probably degraded.

Research in chloroplast protein degradation has evolved primarily along two parallel tracks. Detailed characterizations of the degradation of specific proteins, especially in response to changing light intensities, have dominated the field for many years. Attempts have also been made to isolate the proteases responsible for these processes, with only limited success. In recent years, strategies for the identification of chloroplast proteases, independent of their substrates, have been employed. These have led to the identification and characterization of a number of chloroplast proteases. This chapter will survey protein degradation processes in chloroplasts, describe the proteases that have been characterized and cloned, try to link the degradation of specific substrates with known proteases, and evaluate how these processes may regulate photosynthesis.

Abbreviations: CPE – chloroplast processing enzyme; ELIP– early light-induced protein; LSU – large subunit of Rubisco; OEC 23 – the 23-kDa subunit of the oxygen-evolving complex; OEC 33 – the 33-kDa subunit of the oxygen-evolving complex; PC – plastocyanin; SSU – small subunit of Rubisco

II. Substrates for Proteolysis

Chloroplast substrates for proteolysis have been previously reviewed (Adam, 1996; Andersson and Aro, 1997). Degradation of specific proteins has been reported in different compartments of the organelle: stroma, thylakoid membrane and thylakoid lumen. In most cases, degradation is related to environmental stress, limited availability of cofactors, genetic mutations, and imbalances in the stoichiometry of multisubunit complexes. We will briefly categorize and describe known substrates in different chloroplast compartments.

A. Precursor and Intermediate Proteins

Most chloroplast proteins are encoded in the nucleus, synthesized in the cytosol as precursor proteins, and imported post-translationally into the organelle. During, or shortly after import, their amino-terminal transit peptide is proteolytically removed by a stromal peptidase, leaving the mature protein intact (for review, see Cline and Henry, 1996). Proteins destined for the thylakoid lumen are synthesized with a bipartite transit peptide. These precursor proteins are first cleaved to an intermediate protein upon translocation across the envelope, and then further cleaved to their mature size after translocation across the thylakoid membrane. This second cleavage event is carried out by a thylakoid processing peptidase (for review, see Robinson et al., 1998).

Some chloroplast-encoded proteins, such as the D1 protein of PS II's reaction center (Marder et al., 1984) and Cyt *f* (Kuras et al., 1995)—a component of the Cyt b_6f complex, are also synthesized in precursor form. The D1 protein is synthesized with a carboxy-terminal extension that is removed within the lumen prior to assembly into functional PS II (Diner et al., 1988). Cyt *f* is synthesized with an amino-terminal extension that is not translocated across the thylakoid membrane: this pre-sequence is removed on the stromal side of the thylakoid membrane. Whether the precursor protein is encoded by the plastid genome or the nuclear one, and whether the pre-sequence is on the amino- or carboxy-terminus of the protein, the removed extensions cannot be detected in chloroplasts, suggesting that they are further degraded by proteases (van't Hof and de Kruijff, 1995; Richter and Lamppa, 1999), probably down to the level of free amino acids which can be reutilized for protein synthesis within the organelle.

B. Stromal Substrates

Most known stromal proteins appear to be stable. However, under certain conditions, they may become unstable. Unbalanced stoichiometry of multisubunit complexes results in rapid degradation of the excess component. This was first demonstrated when inhibition of chloroplast protein synthesis, including that of the large subunit of Rubisco (LSU), led to the rapid degradation of newly imported small subunit (SSU) (Schmidt and Mishkind, 1983). Similarly, a point mutation in LSU that prevented its assembly to a holoenzyme resulted in failure of the unassembled LSU and SSU to accumulate (Avni et al., 1989). Proteins that are mis-sorted within the chloroplast can also get degraded in the stroma. When the transit peptide of the lumenal protein OEC 33 was modified to target it to the stroma, the protein was degraded after being imported into isolated chloroplasts, unlike the wild-type protein which was stable in the lumen (Halperin and Adam, 1996). These examples represent situations resulting from impaired control of gene expression, mutations, or defects in the protein-sorting mechanism. Such proteins never become functional and are therefore rapidly degraded.

Inherently stable functional proteins may also become unstable, primarily due to environmental insults such as high temperature, leading to heat denaturation, or high light intensities, leading to oxidative stress. Whereas heat denaturation of specific stromal proteins has never been documented, oxidative stress was shown to cause protein degradation in the stroma. When stress conditions were applied either in vivo or in vitro, Rubisco became unstable (Penarrubia and Moreno, 1990; Mehta et al., 1992; Roulin and Feller, 1997). Other stromal enzymes such as glutamine synthetase, phosphoribulokinase, nitrite reductase and NADPH:protochlorophyllide oxidoreductase also demonstrate increased instability upon exposure of intact chloroplasts to high light (Mitsuhashi and Feller, 1992; Reinbothe et al., 1995; Roulin and Feller, 1997).

C. Thylakoid-Membrane Substrates

Many of the thylakoid-membrane proteins are organized into four multisubunit complexes: PS II, Cyt b_6f, PS I and ATP synthase. Thus, imbalances in the stoichiometry of these complexes lead to degradation of the other subunits. In the absence of a single subunit, it is not uncommon for the other components of a complex to be unstable. When the gene encoding PsaC in *Chlamydomonas* was inactivated, other subunits of the PS I complex were synthesized, but failed to accumulate (Takahashi et al., 1991). Disruption of the genes encoding PsbK and PsbO led to destabilization of other PS II subunits (Takahashi et al., 1994). A *Lemna* mutant lacking the Rieske Fe-S protein failed to accumulate the other components of the Cyt b_6f complex (Bruce and Malkin, 1991). Thus, it appears that the full complement of a complex's components is pertinent to its stability: in the absence of one component, the others become extremely sensitive to proteolysis.

The absence of an entity smaller than a whole protein subunit can also lead to protein instability. Chlorophyll-binding proteins appear to be very unstable when the pigment is missing. Regardless of whether the lack of chlorophyll is due to inhibition of synthesis or mutation, such proteins are rapidly degraded (Apel and Kloppstech, 1980; Bennett, 1981; Mullet et al., 1990; Christopher and Mullet, 1994; Kim et al., 1994). Since the integration of a chlorophyll-binding protein into its respective complex requires several steps of assembly, one may envision that each step involves a conformational change in the protein. Failure to bind a cofactor, such as a pigment, may interfere with further steps of folding, and eventually, with assembly into a multisubunit complex. This partly folded protein may serve as a good substrate for proteolysis.

Whereas the aforementioned examples represent proteins that have never reached a functional stage, functioning proteins may also become unstable. Photoinhibition of photosynthesis, which is manifested primarily in the thylakoid membrane due to its high capacity for light-energy absorbance, results from photodamage to thylakoid proteins. These, in turn, become highly susceptible to proteolysis. The best-characterized protein in this respect is the D1 protein of PS II's reaction center (Andersson and Aro, 1997), but other components of PS II, such as D2 and CP43, are degraded as well (Schuster et al., 1988; Christopher and Mullet, 1994; Jansen et al., 1996). This degradation process is an essential step in a repair mechanism allowing the removal of damaged subunits and their replacement with newly synthesized ones. Plant adaptation to different light conditions is often accompanied by adjustments in the size of the photosynthetic antenna. Upon transition from low to high light intensity, the size of the PS II light-harvesting complex (LHC II) is reduced as a result of specific degradation of a subpopulation of chlorophyll *a/b*-binding proteins (Lindahl et al., 1995; Tziveleka and Argyroudi-Akoyunoglou, 1998; Yang et al., 1998). Under a converse change, the early light-inducible protein (ELIP), which is structurally related to LHC II, is degraded, allowing reincorporation of chlorophyll into LHC II (Adamska et al., 1993, 1996).

D. Lumenal Substrates

Specific degradation of lumenal proteins has also been demonstrated, plastocyanin (PC) being the best-characterized example. When *Chlamydomonas* cells are grown in a Cu^{2+}-deficient medium, apo-PC is synthesized, imported into the chloroplast and translocated across the thylakoid membrane, but it fails to accumulate due to rapid degradation (Merchant and Bogorad, 1986). A similar observation was made in an in vitro system. Apo-PC was sensitive to proteolytic digestion, unlike the native and reconstituted holoproteins, which were insensitive (Li and Merchant, 1995). This increased sensitivity of the apoprotein probably results from its lack of the characteristic secondary structure displayed by the holoprotein, as revealed by circular dichroism spectroscopy (Li and Merchant, 1995). In another study, it was demonstrated that truncated forms of OEC 23 (the 23-kDa subunit of the oxygen-evolving complex) were correctly targeted to the lumen where they were processed to their mature size, but then rapidly degraded (Roffey and Theg, 1996). These results suggest that the lumen compartment contains its own proteolytic machinery, capable of degrading unassembled or damaged soluble proteins. It also suggests that lumenal proteases participate in the degradation of thylakoid-membrane proteins exposed to the lumen.

III. Proteolytic Enzymes

A. Stromal Enzymes

1. Stromal Processing Peptidase

Newly imported proteins, whether they are to remain in the stroma or are further targeted to the thylakoid, are processed to remove their transit peptide. This reaction is catalyzed by a specialized endopeptidase. This stromal peptidase, known as chloroplast processing enzyme (CPE), is a ca. 140-kDa metalloprotease containing a zinc-binding motif, that is related to the mitochondrial processing peptidase and Protease III of *E. coli* (Oblong and Lamppa, 1992; Vandervere et al., 1995). In vitro assays with the recombinant enzyme demonstrated that it is active towards the precursor of LHCP (Richter and Lamppa, 1998). Its ability to release an intact transit peptide from the precursor of ferredoxin supports its identity as an endopeptidase that processes its substrate by a single cleavage event. CPE was also able to process a wide range of chloroplast precursors, suggesting that it functions as a general processing peptidase responsible for the removal of the stromal targeting domains from all imported proteins (Richter and Lamppa, 1998). This contention is supported by in vivo experiments where expression of an antisense construct of CPE resulted in major disruption of chloroplast biogenesis (Wan et al., 1998). Thus it seems that other stromal proteins cannot compensate for loss of the stromal processing peptidase. Nevertheless, a second processing enzyme that can cleave a number of precursor proteins was recently purified (Koussevitzky et al., 1998). It is a metallopeptidase of 75–80 kDa that is not recognized by an antibody against the 140-kDa CPE. However, although it processes a number of chloroplast precursors, it does not recognize LHCP precursors. Confirmation of the identity of this peptidase awaits its cloning.

2. Clp

Bacterial Clp protease is a barrel-like multimeric complex composed of two heptameric rings of the proteolytic subunit ClpP (Wang et al., 1997), capped by one hexameric ring on each side of a regulatory subunit, either ClpA or ClpX (Grimaud et al., 1998). This structure resembles that of the eukaryotic 26 S proteasome (Kessel et al. 1995). It functions as an ATP-dependent serine protease (for review, see Gottesman, 1996), but the regulatory ATPases can function independently of the proteolytic subunit as molecular chaperones (Wickner et al., 1994; Levchenko et al., 1995; Gottesman et al., 1997). Homologues of Clp protease subunits are found in chloroplasts. ClpP is encoded by the chloroplast genome (Gray et al., 1990; Maurizi et al., 1990) and is expressed constitutively (Shanklin et al., 1995; Ostersetzer et al., 1996). It is the only protease encoded in the plastid. To date, five other isomers of ClpP have been found in the nuclear genome of higher plants (Schaller and Ryan, 1995; Clarke, 1999; Nakabayashi et al. 1999). All ClpP isomers are in the range of 21–25 kDa and contain the typical catalytic triad of serine proteases composed of His-Asp-Ser. At least one of them is imported into the chloroplast post-translationally (Sokolenko et al., 1998), whereas another one is targeted to the mitochondria (T. Halperin et al., unpublished). The cellular location of the other three ClpPs is still unknown (Clarke, 1999). As for regulatory subunits, chloroplasts contain two different homologues of ClpA, designated ClpC and ClpD. The molecular mass of the mature proteins is 88 to 90 kDa, and they both contain two ATP-binding sites, each composed of Walker A and Walker B motifs. ClpC was first identified in tomato (Gottesman et al., 1990), then in pea (Moore and Keegstra, 1993). It was shown to be targeted to chloroplasts (Moore and Keegstra, 1993) and expressed constitutively (Shanklin et al., 1995; Ostersetzer and Adam, 1996). The related protein ClpD was identified in *Arabidopsis* and localized to the chloroplast, and its expression was shown to be induced by desiccation, high salt, dark and senescence (Kiyosue et al., 1993; Nakashima et al., 1997). ClpC and ClpP form a multimeric complex (Sokolenko et al., 1998) located in the stroma (Shanklin et al., 1995; Halperin and Adam, 1996; Ostersetzer et al., 1996), where it can degrade mistargeted proteins such as OEC 33 (Halperin and Adam, 1996), fully or partially assembled Cyt b_6f (Majeran et al., 2000) and probably other abnormal proteins. Similar to its bacterial homologues, ClpC can function in chloroplasts on its own. It has been found associated with the protein import machinery on the stromal side of the inner envelope membrane, without ClpP, where it is believed to participate in late stages of protein import into the chloroplast as a chaperone (Nielsen et al., 1997).

Although the plant genome contains a homologue of ClpX, the 50-kDa regulatory subunit with one ATP-binding site found associated with ClpP in bacteria (Gottesman et al., 1993; Wojtkowiak et al., 1993), it appears to be absent from chloroplasts. Instead, it is found in mitochondria, probably associated with one of the nuclear-encoded copies of ClpP (T. Halperin et al., manuscript submitted). It should be noted that *E. coli* contains another member of the Clp family, the complex of ClpY/ClpQ (also designated HslU/HslV) (Missiakas et al., 1996; Rohrwild et al., 1996). ClpY is very similar to ClpX, whereas ClpQ shows high homology to the β-subunit of the eukaryotic proteasome. However, attempts to find homologues of ClpY in chloroplasts, or elsewhere in the plant cell have failed (T. Halperin and Z. Adam, unpublished).

B. Thylakoid-Membrane Enzymes

1. Thylakoid Processing Peptidase

The peptidase that removes the thylakoid transfer domain from proteins translocated across the thylakoid membrane has recently been cloned and characterized (Chaal et al., 1998). It is a 30-kDa thylakoid-membrane protein, with a Ser-Lys catalytic dyad that can process intermediate forms of proteins targeted to the thylakoid lumen to their mature size. It shows similarity to Type I leader peptidases from *E. coli*, cyanobacteria and yeast mitochondria. This similarity makes sense from an evolutionary point of view since the translocation machinery of proteins across the thylakoid membrane resembles the secretion machinery of bacterial cells (Robinson et al., 1998); hence the enzyme cleaving the signal peptide is also conserved.

2. FtsH

In *E. coli*, FtsH is the only essential ATP-dependent protease (Herman et al., 1993). It is a metalloprotease bound to the plasma membrane, with its ATP-binding and proteolytic (zinc-binding) sites exposed to the cytoplasm (Tomoyasu et al., 1995). Yeast mitochondria contain three homologues of this protein, designated YTA10, 11 and 12, bound to the inner membrane (Arlt et al., 1996; Leonhard et al., 1996). Whereas YTA10 and 12 are exposed to the matrix, YTA11 faces the intermembrane space. *Synechocystis* PCC6803 contains four homologues of this protein. Current sequence data from *Arabidopsis thaliana* suggests the existence of five different FtsH proteins in the plant cells (Fig. 1). One of them most resembles YTA11, another one is related to YTA10 and 12, and the other three are more related to the cyanobacterial proteins (Fig. 1). Thus, these relationships suggest that three of the five plant FtsH proteins are located in the chloroplast. One of these proteins, AtFtsH-1 was positively identified and characterized in chloroplasts (Lindahl et al., 1996). It is an integral thylakoid-membrane protein, whose ATP-binding and catalytic sites are exposed to the stroma, and thus it can potentially digest both membrane and stromal proteins. Unassembled forms of the Rieske Fe-S protein, accumulating on the stromal surface of the thylakoid membrane after being imported into isolated chloroplasts in vitro, were indeed shown to be degraded by this protease (Ostersetzer and Adam, 1997). Interestingly, this process was stimulated by light, suggesting that conformational changes induced by the absorbance of light energy contribute to destabilization of the substrate.

The proteolytic enzymes involved in degrading the D1 protein of PS II's reaction center have long been an enigma. Recently, it was observed that following the initial GTP-stimulated cleavage of the protein to 23- and 10-kDa fragments, the further degradation of the 23-kDa fragment is stimulated by ATP and zinc (Spetea et al., 1999). Whereas the initial cleavage was observed in both photoinhibited thylakoid membranes and core complexes, degradation of the 23-kDa fragment was observed only in

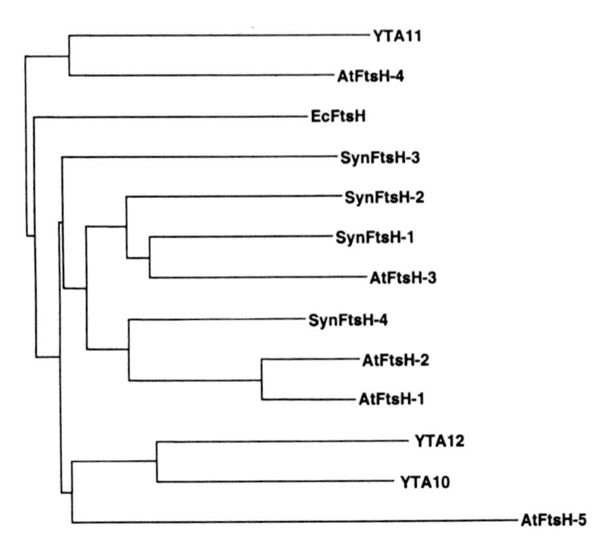

Fig. 1. Phylogenetic relationship between selected FtsH sequences. FtsH-like sequences from *E. coli* (Ec), yeast (YTA), *A. thaliana* (At) and *Synechocystis* sp. PCC 6803 (Syn) were aligned and graphically depicted using the program Clustal W v. 1.7. Analysis with the PAUP program gave similar results.

membranes. These observations led to the suggestion that FtsH, which is a zinc-stimulated, ATP-dependent, membrane-bound protease (Lindahl et al., 1996), is involved in complete degradation of the D1 protein after its initial cleavage. Indeed, when photodamaged PS II core complexes were incubated with purified recombinant FtsH, the 23-kDa fragment was degraded in an ATP-dependent manner (Lindahl et al., 2000). Moreover, mild trypsin treatment of thylakoid membranes resulted in removal of FtsH from the membrane. When these membranes, depleted of FtsH, were exposed to photoinhibitory treatment, disappearance of the 23-kDa fragment was inhibited, and it could be reconstituted with recombinant FtsH (Lindahl et al., 2000).

C. Lumenal Enzymes

1. DegP

E. coli and other bacteria contain a family of periplasmic serine proteases known as DegP, DegQ and DegS (or HtrA, HhoA and HhoB) (Skorko-Glonek et al., 1995, 1997; Kolmar et al., 1996). DegP is a heat-shock protein, essential for survival at elevated temperatures (Lipinska et al., 1990). Homologous genes are also found in cyanobacteria. A protein related to this family was recently cloned and characterized in chloroplasts (Itzhaki et al., 1998).

It is a 40-kDa protein, strongly associated with the lumenal side of the thylakoid membrane, similar to the bacterial homologue that is associated with the periplasmic side of the plasma membrane (Skorko-Glonek et al., 1997). It is present in chloroplasts under normal growth conditions, but its expression is stimulated upon exposure to higher temperatures. The deduced amino acid sequence of the cloned cDNA is consistent with the determined location of the protein, as the precursor protein contains the amino-terminal bipartite transit peptide typical of proteins targeted to the thylakoid lumen. Within the mature portion of the protein, the catalytic triad His-Asp-Ser is highly conserved. No trans-membrane helices are found in the sequence, in agreement with the peripheral nature of the protein found in the thylakoid fraction (Itzhaki et al., 1998). However, it is not yet clear why this protein is tightly associated with the membrane, nor are substrates of this protease known, but its location suggests that it might be involved in the degradation of both membrane and soluble lumenal proteins. Of particular interest is the possibility of its involvement in D1 protein degradation by cleaving its lumenal loops. However, this suggestion needs to be experimentally challenged.

2. Tsp

The bacterial tail-specific protease (Tsp) is a periplasmic protease, capable of either processing certain substrates by a single proteolytic cleavage, or total degradation of other substrates (Silber et al., 1992). The plant homologue of this protease, also known as CtpA, was found in the thylakoid lumen as the processing peptidase responsible for the carboxy-terminal cleavage of the precursor of the D1 protein to its mature form (Anbudurai et al., 1994; Shestakov et al., 1994; Inagaki et al., 1996; Oelmuller et al., 1996), an essential step in establishing functional PS II (Diner et al., 1988). Recently, a single base deletion in the *ctpA* gene in *Scenedesmus* was shown to result in a frame-shift mutation, and hence inability to process pre-D1 (Trost et al., 1997). A suppressor mutation with a single base insertion cured the original mutation, suggesting that Tsp (CtpA) is the only peptidase responsible for pre-D1 processing. However, whether this lumenal protease is exclusively dedicated to pre-D1 processing, or, like its bacterial homologue, is involved in total degradation of other substrates is not yet clear.

IV. Conclusions and Future Prospects

Degradation of specific proteins in the chloroplast has been documented for the past 15 years. Most of these are enzymes and proteins involved directly in different aspects of the photosynthetic process. As these can generally be considered stable proteins, their destabilization is primarily due to exposure to changing environmental conditions. Most cases are related to photoinhibition by increasing light intensities, where both membrane and soluble proteins can become oxidatively damaged (e.g., PS II D1 protein, Rubisco, etc.). Under these conditions, proteolytic processes play a role in a wider repair mechanism that allows replacement of damaged components by newly synthesized ones (Adam, 1996; Andersson and Aro, 1997). However, degradation in response to decreased light intensity, as in the case of the adjustment of the photosynthetic antenna's size by degradation of LHC II, has also been described (Lindahl et al., 1995; Yang et al., 1998).

The other large group of substrates is that of proteins or complexes lacking a structural element. The missing component can range from a single ion, a cofactor, or even a whole subunit in a protein complex (Adam, 1996). The reasons for the absence of one component may also vary: genetic mutations, unavailability of nutrients, environmental conditions, the common feature being the nonfunctional form of the protein or complex; the rapid degradation can be considered a recycling of building blocks for reuse. Regardless of whether the missing component directly affects function or not, its absence may be manifested in a structural change that leads to its increased sensitivity to proteolysis.

The role of structural changes in regulating protein stability should be considered in another context as well. An intensive search for induced proteases that are associated with certain environmental conditions has, so far, been unsuccessful. Moreover, most known chloroplast proteases appear to be expressed constitutively. Finally, under conditions that make one protein extremely sensitive to proteolysis, many other proteins are stable. Thus, it is difficult to envision protease availability as a major factor regulating protein degradation. Alternatively, the absence of components, or oxidative or thermal damage, may result in structural changes, however minor. Such changes may expose sites sensitive to proteolysis. Once a single cut is made, further structural changes occur, leading to further cuts in the protein. Thus, most degradation events of inherently stable proteins are likely to be regulated by the substrate's structure rather than by expression of a new protease.

There is another class of proteins, whose member's identities are mostly unknown, that is likely to provide substrates for proteolysis. In both eukaryotic and prokaryotic systems, many aspects of gene expression are controlled by short-lived, low-abundance, regulatory proteins, often described as 'timing proteins' (for review, see Gottesman, 1996). As studies on the regulation of chloroplast gene expression progress, many of these regulators are likely to be explored. It is expected that, as in other systems, the availability of some of these regulatory proteins will be found to play a major role in regulation. As more of these proteins are discovered, it is anticipated that their half-lives will be determined and the regulation of their degradation characterized.

All the proteases discovered in chloroplasts to date are homologues of bacterial proteases. They are distributed between the stroma, thylakoid membrane and thylakoid lumen (see Fig. 2). It should be noted that no protease was discovered yet in the chloroplast envelope, and there is no much information on specific envelope substrates yet. The homology between chloroplast and bacterial proteases is consistent with the evolutionary origin of chloroplasts. It also provides us with a framework for predicting the presence of other chloroplast proteases. A comparison between the sequence of *E. coli* proteases and *Arabidopsis* expressed sequence tags (ESTs), combined with genome sequencing data, should give us a clue as to other potential chloroplast proteases. As described above, in addition to the chloroplast-encoded copy of ClpP, other isomers are encoded in the nucleus as well. Whether nuclear and plastid-encoded ClpPs coexist in the same complex, and whether they have different specificities, either in association with their cognate regulatory subunits or with substrates, is totally unknown. It is believed that in *E. coli,* substrate specificity is determined primarily by the regulatory subunit, either ClpA or ClpX (Gottesman, 1996). A similar situation may exist in chloroplasts, where ClpC and ClpD coexist. However, it remains to be demonstrated whether ClpD is indeed capable of forming a complex with ClpP. Such an association cannot be taken for granted as *E. coli* possesses ClpB, which is highly homologous to ClpA, but cannot associate with ClpP (Woo et al., 1992). Instead, it functions exclusively as a molecular chaperone. A similar distribution of tasks could conceivably exist in chloroplasts between ClpC and ClpD. As for ClpX, it does exist in plant cells, but the only copy discovered so far is targeted to mitochondria, where it potentially forms a complex with one of the nuclear-encoded copies of ClpP. The related bacterial protein ClpY (HslU) is absent from currently available plant sequences. Similarly, its cognate proteolytic subunit ClpQ (HslV), although found as a proteasome subunit, is not found with a potential targeting sequence that

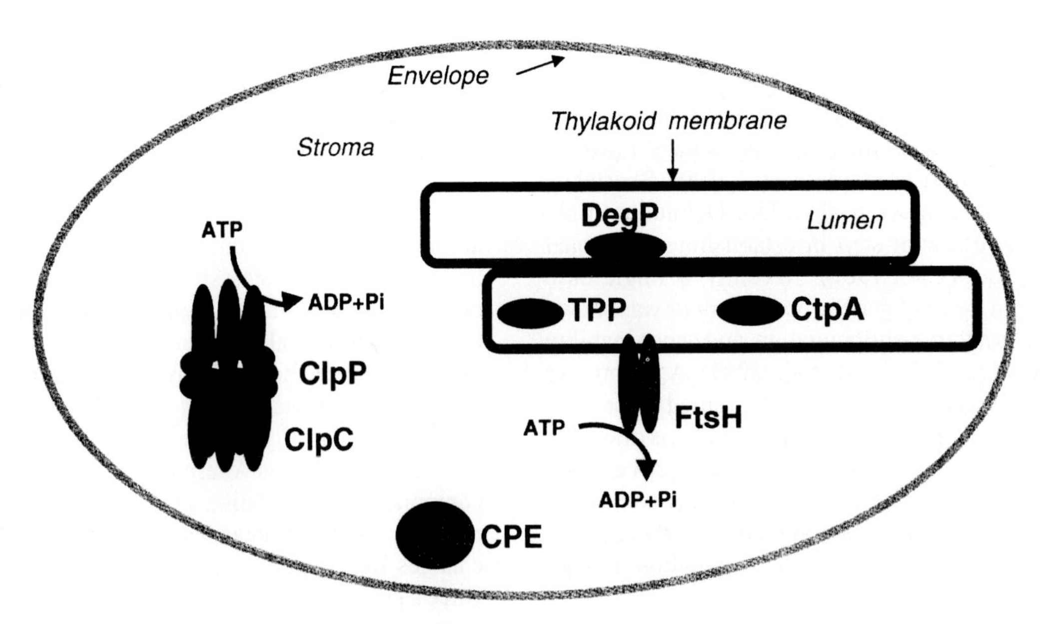

Fig. 2. Distribution of chloroplast proteases.

could direct it to an intracellular organelle. Thus, it appears that plant cells do not contain a homologue of this bacterial protease as a whole.

Whereas *E. coli* contains a single form of the essential protease FtsH, at least five different genes are found in plants. Of these, one is clearly targeted to chloroplasts, whereas another two more closely resemble mitochondrial homologues of FtsH (YTA proteins). Although the remaining two resemble the chloroplast-targeted one, their localization to chloroplasts needs to be demonstrated. It is still possible that they are targeted to the mitochondria, since yeast mitochondria contain three different homologues of FtsH associated with the inner membrane. Two of them form a complex and are oriented towards the matrix whereas the third one is oriented toward the intermembrane space. A similar situation may exist in plant cells.

Another bacterial ATP-dependent protease that has homologues in the plant cell is Lon. Two genes encoding Lon are found in both *Arabidopsis* and maize. Lon was detected in mitochondria, and its absence from the genome of *Synechocystis* suggests that it does not exist in chloroplasts. However, some preliminary results of immunological detection studies suggest the contrary (O. Ostersetzer and Z. Adam, unpublished).

Three periplasmic ATP-independent *E. coli* proteases/peptidases also have chloroplast homologues: Tsp, DegP, and the leader peptidase. The chloroplast homologue of Tsp is responsible for pre-D1 processing, whereas the homologue of the leader peptidase removes the thylakoid transfer domain of translocated proteins. The location of these proteins in the thylakoid lumen is consistent with the view of the lumen being the orientational homologue of the bacterial periplasm. This is the compartment into which protons are pumped, and the compartment into which proteins are translocated by the chloroplast homologue of the bacterial protein-secretion pathway. Thus, it is conceivable that the thylakoid lumen contains homologues of other periplasmic proteases. However, the two bacterial proteins related to DegP, DegQ and DegS do not have plant homologues in the database, and the lumen proteolytic inventory may include homologues of only several periplasmic proteases.

Can we expect to find more proteases in chloroplasts? The rapidly accumulating genomic data might give us a clue. If the contents of completely sequenced bacterial genomes, including cyanobacteria, can be considered a model for the proteolytic machinery of chloroplasts, the list is almost complete. Bacterial proteases and peptidases found also in cyanobacteria are likely to be found in the chloroplast. Confirmation of the cellular distribution, primarily between chloroplasts and mitochondria, of the products of identified homologous genes is likely to yield a complete list of chloroplast proteases in the near future. However, unknown proteases may still be discovered, and these may have homologues among bacterial sequences that currently have no assigned functions.

Although the identity of a number of chloroplast proteases has been revealed in recent years, the challenge of relating these to known substrates has, for the most part, not been met. This is especially true in the context of protein degradation in response to changing light intensities. Degradation of these substrates may be mediated by redundant activities, making it impossible to attribute degradation of a certain substrate to a single protease. However, attempts to identify or generate mutants in specific proteases are likely to enable a reevaluation of the degradation of specific substrates in these plants. This approach is likely to yield a link between specific substrates and specific proteases.

Acknowledgments

Work in the author's laboratory was supported by grants from the US-Israel Binational Science Foundation (BSF), the US-Israel Binational Agricultural Research & Development Fund (BARD), the European Union, the Israeli Ministry of Science, and the Israel Science Foundation.

References

Adam Z (1996) Protein stability and degradation in chloroplasts. Plant Mol Biol 32: 773–783

Adamska I, Kloppstech K and Ohad I (1993) Early light-inducible protein in pea is stable during light stress but is degraded during recovery at low light intensity. J Biol Chem 268: 5438–5444

Adamska I, Lindahl M, Roobol-Boza M and Andersson B (1996) Degradation of the light-stress protein is mediated by an ATP-independent, serine-type protease under low-light conditions. Eur J Biochem 236: 591–599

Anbudurai PR, Mor TS, Ohad I, Shestakov SV and Pakrasi HB (1994) The *ctpA* gene encodes the C-terminal processing protease for the D1 protein of the Photosystem II reaction center complex. Proc Natl Acad Sci USA 91: 8082–8086

Andersson B and Aro E-M (1997) Proteolytic activities and proteases of plant chloroplasts. Physiol Plant 100: 780–793

Apel K and Kloppstech K (1980) The effect of light in the biosynthesis of the light-harvesting chlorophyll *a/b* protein. Evidence for the stabilization of the apoprotein. Planta 150: 426–430

Arlt H, Tauer R, Feldmann H, Neupert W and Langer T (1996) The YTA10-12 complex, an AAA protease with chaperone-like activity in the inner membrane of mitochondria. Cell 85: 875–885

Avni A, Edelman M, Rachailovich I, Aviv D and Fluhr R (1989) A point mutation in the gene for the large subunit of ribulose 1,5-bisphosphate carboxylase/oxygenase affects holoenzyme assembly in *Nicotiana tabacum*. EMBO J 8: 1915–1918

Bennett J (1981) Biosynthesis of the light-harvesting chlorophyll *a/b* protein. Polypeptide turnover in darkness. Eur J Biochem 118: 61–70

Bruce BD and Malkin R (1991) Biosynthesis of the chloroplast cytochrome b_6f complex: Studies in a photosynthetic mutant of *Lemna*. Plant Cell 3: 203–212

Chaal BK, Mould RM, Barbrook AC, Gray JC and Howe CJ (1998) Characterization of a cDNA encoding the thylakoidal processing peptidase from *Arabidopsis thaliana*. Implications for the origin and catalytic mechanism of the enzyme. J Biol Chem 273: 689–692

Christopher DA and Mullet JE (1994) Separate photosensory pathways co-regulate blue light/ultraviolet-A-activated psbD-psbC transcription and light-induced D2 and CP43 degradation in barley (*Hordeum vulgare*) chloroplasts. Plant Physiol 104: 1119–1129

Clarke AK (1999) ATP-dependent Clp proteases in photosynthetic organisms—a cut above the rest! Ann Bot 83: 593–599

Cline K and Henry R (1996) Import and routing of nucleus-encoded chloroplast proteins. Annu Rev Cell Dev Biol 12: 1–26

Coux O, Tanaka K and Goldberg AL (1996) Structure and functions of the 20S and 26S proteasomes. Annu Rev Biochem 65: 801–847

Diner BA, Ries DF, Cohen BN and Metz JG (1988) COOH-terminal processing of polypeptide D1 of the Photosystem II reaction center of *Scenedesmus obliquus* is necessary for the assembly of the oxygen-evolving complex. J Biol Chem 263: 8972–8980

Gottesman S (1996) Proteases and their targets in *Escherichia coli*. Annu Rev Genet 30: 465–506

Gottesman S, Squires C, Pichersky E, Carrington M, Hobbs M, Mattick JS, Dalrymple B, Kuramitsu H, Shiroza T, Foster T, Clark WP, Ross B, Squires CL and Maurizi MR (1990) Conservation of the regulatory subunit for the Clp ATP-dependent protease in prokaryotes and eukaryotes. Proc Natl Acad Sci USA 87: 3513–3517

Gottesman S, Clark WP, de Crecy-Lagard V and Maurizi MR (1993) ClpX, an alternative subunit for the ATP-dependent Clp protease of *Escherichia coli*: Sequence and in vivo activities. J Biol Chem 268: 22618–22626

Gottesman S, Wickner S and Maurizi MR (1997) Protein quality control: Triage by chaperones and proteases. Genes Dev 11: 815–823

Gray JC, Hird SM and Dyer TA (1990) Nucleotide sequence of a wheat chloroplast gene encoding the proteolytic subunit of an ATP-dependent protease. Plant Mol Biol 15: 947–950

Grimaud R, Kessel M, Beuron F, Steven AC and Maurizi MR (1998) Enzymatic and structural similarities between the *Escherichia coli* ATP-dependent proteases, ClpXP and ClpAP. J Biol Chem 273: 12476–12481

Halperin T and Adam Z (1996) Degradation of mistargeted OEE33 in the chloroplast stroma. Plant Mol Biol 30: 925–933

Herman C, Ogura T, Tomoyasu T, Hiraga S, Akiyama Y, Ito K, Thomas R, D'Ari R and Bouloc P (1993) Cell growth and λ phage development controlled by the same essential *Escherichia coli* gene, *ftsH/hflB*. Proc Natl Acad Sci USA 90: 10861–10865

Hershko A and Ciechanover A (1998) The ubiquitin system. Annu Rev Biochem 67: 425–479

Inagaki N, Yamamoto Y, Mori H and Satoh K (1996) Carboxyl-terminal processing protease for the D1 precursor protein: Cloning and sequencing of the spinach cDNA. Plant Mol Biol 30: 39–50

Itzhaki H, Naveh L, Lindahl M, Cook M and Adam Z (1998) Identification and characterization of DegP, a serine protease associated with the luminal side of the thylakoid membrane. J Biol Chem 273: 7094–7098

Jansen MAK, Gaba V, Greenberg BM, Mattoo AK and Edelman M (1996) Low threshold levels of ultraviolet-B in a background of photosynthetically active radiation trigger rapid degradation of the D2 protein of Photosystem-II. Plant J 9: 693–699

Kessel M, Maurizi MR, Kim B, Kocsis E, Trus BL, Singh SK and Steven AC (1995) Homology in structural organization between *E. coli* ClpAP protease and the eukaryotic 26 S proteasome. J Mol Biol 250: 587–594

Kim J, Eichacker LA, Rudiger W and Mullet JE (1994) Chlorophyll regulates accumulation of the plastid-encoded chlorophyll proteins P700 and D1 by increasing apoprotein stability. Plant Physiol 104: 907–916

Kiyosue T, Yamaguchi-Shinozaki K and Shinozaki K (1993) Characterization of cDNA for a dehydration-inducible gene that encodes a Clp A, B-like protein in *Arabidopsis thaliana L.* Biochem Biophys Res Commun 196: 1214–1220

Kolmar H, Waller PRH and Sauer RT (1996) The DegP and DegQ periplasmic endoproteases of *Escherichia coli*: Specificity for cleavage sites and substrate conformation. J Bacteriol 178: 5925–5929

Koussevitzky S, Ne'eman E, Sommer A, Steffens JC and Harel E (1998) Purification and properties of a novel chloroplast stromal peptidase. Processing of polyphenol oxidase and other imported precursors. J Biol Chem 273: 27064–27069

Kumar S (1995) ICE-like proteases in apoptosis. Trends Biochem Sci 20: 198–202

Kuras R, Buschlen S and Wollman FA (1995) Maturation of pre-apocytochrome f in vivo—a site-directed mutagenesis study in *Chlamydomonas reinhardtii*. J Biol Chem 270: 27797–27803

Leonhard K, Herrmann JM, Stuart RA, Mannhaupt G, Neupert W and Langer T (1996) AAA proteases with catalytic sites on opposite membrane surfaces comprise a proteolytic system for the ATP-dependent degradation of inner membrane proteins in mitochondria. EMBO J 15: 4218–4229

Levchenko I, Luo L and Baker TA (1995) Disassembly of the *mu* transposase tetramer by the ClpX chaperone. Gene Develop 9: 2399–2408

Li HH and Merchant S (1995) Degradation of plastocyanin in copper-deficient *C. reinhardtii*—evidence for a protease-

susceptible conformation of the apoprotein and regulated proteolysis. J Biol Chem 270: 23504–23510
Lindahl M, Yang DH and Andersson B (1995) Regulatory proteolysis of the major light-harvesting chlorophyll *a/b* protein of Photosystem II by a light-induced membrane-associated enzymic system. Eur J Biochem 231: 503–509
Lindahl M, Tabak S, Cseke L, Pichersky E, Andersson B and Adam Z (1996) Identification, characterization, and molecular cloning of a homologue of the bacterial FtsH protease in chloroplasts of higher plants. J Biol Chem 271: 29329–29334
Lindahl M, Spetea C, Hundal T, Oppenheim AB, Adam Z and Andersson B (2000) The thylakoid FtsH protease plays a role in the light-induced turnover of the Photosystem II D1 protein. Plant Cell 12: 419–431
Lipinska B, Zylicz M and Georgopoulos C (1990) The HtrA (DegP) protein, essential for *Escherichia coli* survival at high temperatures, is an essential endopeptidase. J Bacteriol 172: 1791–1797
Majeran W, Wollman F-A and Vallon O (2000) Evidence for a role of ClpP in the degradation of the chloroplast cytochrome b_6f complex. Plant Cell 12: 137–149
Marder JB, Goloubinoff P and Edelman M (1984) Molecular architecture of the rapidly metabolized 32-kilodalton protein of Photosystem II: Indications for COOH-terminal processing of a chloroplast membrane polypeptide. J Biol Chem 259: 3900–3908
Maurizi MR, Clark WP, Kim SH and Gottesman S (1990) ClpP represents a unique family of serine proteases. J Biol Chem 265: 12546–12552
Mehta RA, Fawcett TW, Porath D and Mattoo AK (1992) Oxidative stress causes rapid membrane translocation and in vivo degradation of ribulose-1,5-bisphosphate carboxylase/oxygenase. J Biol Chem 267: 2810–2816
Merchant S and Bogorad L (1986) Rapid degradation of apoplastocyanin in Cu(II)-deficient cells of *Chlamydomonas reinhardii*. J Biol Chem 261: 15850–15853
Missiakas D, Schwager F, Betton JM, Georgopoulos C and Raina S (1996) Identification and characterization of HsIV-HsIU (ClpQ-ClpY) proteins involved in overall proteolysis of misfolded proteins in *Escherichia coli*. EMBO J 15: 6899–6909
Mitsuhashi W and Feller U (1992) Effects of light and external solutes on the catabolism of nuclear-encoded stromal proteins in intact chloroplasts isolated from pea leaves. Plant Physiol 100: 2100–2105
Moore T and Keegstra K (1993) Characterization of a cDNA clone encoding a chloroplast-targeted Clp homologue. Plant Mol Biol 21: 525–537
Mullet JE, Klein PG and Klein RR (1990) Chlorophyll regulates accumulation of the plastid-encoded chlorophyll apoprotein-CP43 and apoprotein-D1 by increasing apoprotein stability. Proc Natl Acad Sci USA 87: 4038–4042
Nakashima K, Kiyosue T, Yamaguchi-Shinozaki K and Shinozaki K (1997) A nuclear gene, *erd1*, encoding a chloroplast-targeted Clp protease regulatory subunit homolog is not only induced by water stress but also developmentally up-regulated during senescence in *Arabidopsis thaliana*. Plant J 12: 851–861
Nakabayashi K, Ito M, Kiosue T, Shinozaki K and Watanabe A (1999) Identification of *clp* genes expressed in senescing *Arabidopsis* leaves. Plant Cell Physiol 40: 504–514
Nielsen E, Akita M, Davilaaponte J and Keegstra K (1997) Stable association of chloroplastic precursors with protein translocation complexes that contain proteins from both envelope membranes and a stromal HSP 100 molecular chaperone. EMBO J 16: 935–946
Oblong JE and Lamppa GK (1992) Identification of two structurally related proteins involved in proteolytic processing of precursors targeted to the chloroplast. EMBO J 11: 4401–4409
Oelmuller R, Herrmann RG and Pakrasi HB (1996) Molecular studies of CtpA, the carboxyl-terminal processing protease for the D1 protein of the Photosystem II reaction center in higher plants. J Biol Chem 271: 21848–21852
Ostersetzer O and Adam Z (1996) Effects of light and temperature on expression of ClpC, the regulatory subunit of chloroplastic Clp protease, in pea seedlings. Plant Mol Biol 31: 673–676
Ostersetzer O and Adam Z (1997) Light-stimulated degradation of an unassembled Rieske FeS protein by a thylakoid-bound protease: The possible role of the FtsH protease. Plant Cell 9: 957–965
Ostersetzer O, Tabak S, Yarden O, Shapira R and Adam Z (1996) Immunological detection of proteins similar to bacterial proteases in higher plant chloroplasts. Eur J Biochem 236: 932–936
Penarrubia L and Moreno J (1990) Increased susceptibility of ribulose-1,5-bisphosphate carboxylase/oxygenase to proteolytic degradation caused by oxidative treatments. Arch Biochem Biophys 281: 319–323
Reinbothe C, Apel K and Reinbothe S (1995) A light-induced protease from barley plastids degrades NADPH:protochlorophyllide oxidoreductase complexed with chlorophyllide. Mol Cell Biol 15: 6206–6212
Richter S and Lamppa GK (1998) A chloroplast processing enzyme functions as the general stromal processing peptidase. Proc Natl Acad Sci USA 95: 7463–7468
Richter S and Lamppa GK (1999) Stromal processing peptidase binds transit peptides and initiates their ATP-dependent turnover in chloroplasts. J Cell Biol 147: 33–43
Robinson C, Hynds PJ, Robinson D and Mant A (1998) Multiple pathways for the targeting of thylakoid proteins in chloroplasts. Plant Mol Biol 38: 209–221
Roffey RA and Theg SM (1996) Analysis of the import of carboxyl-terminal truncations of the 23-kilodalton subunit of the oxygen-evolving complex suggests that its structure is an important determinant for thylakoid transport. Plant Physiol 111: 1329–1338
Rohrwild M, Coux O, Huang HC, Moerschell RP, Yoo SJ, Seol JH, Chung CH and Goldberg AL (1996) HslV-HslU: A novel ATP-dependent protease complex in *Escherichia coli* related to the eukaryotic proteasome. Proc Natl Acad Sci USA 93: 5808–5813
Roulin S and Feller U (1997) Light-induced proteolysis of stromal protein in pea chloroplasts: Requirement for intact organelles. Plant Science 128: 31–41
Schaller A and Ryan CA (1995) Cloning of a tomato cDNA encoding the proteolytic subunit of a Clp-like energy dependent protease. Plant Physiol 108: 1341
Schmidt GW and Mishkind ML (1983) Rapid degradation of unassembled ribulose 1,5-bisphosphate carboxylase small subunit in chloroplasts. Proc Natl Acad Sci USA 80: 2623–2636
Schuster G, Timberg R and Ohad I (1988) Turnover of thylakoid

Photosystem II proteins during photoinhibition of *Chlamydomonas reinhardtii*. Eur J Biochem 177: 403–410

Shanklin J, Dewitt ND and Flanagan JM (1995) The stroma of higher plant plastids contains ClpP and ClpC, functional homologs of *Escherichia coli* ClpP and ClpA: An archetypal two-component ATP-dependent protease. Plant Cell 7: 1713–1722

Shestakov SV, Anbudurai PR, Stanbekova GE, Gadzhiev A, Lind LK and Pakrasi HB (1994) Molecular cloning and characterization of the *ctpA* gene encoding a carboxyl-terminal processing protease—analysis of a spontaneous Photosystem II-deficient mutant strain of the cyanobacterium *Synechocystis* sp. PCC 6803. J Biol Chem 269: 19354–19359

Silber KR, Keiler KC and Sauer RT (1992) Tsp: A tail-specific protease that selectively degrades proteins with nonpolar C-termini. Proc Natl Acad Sci USA 89: 295–299

Skorko-Glonek J, Wawrzynow A, Krzewski K, Kurpierz K and Lipinska B (1995) Site-directed mutagenesis of the HtrA(DegP) serine protease, whose proteolytic activity is indispensable for *Escherichia coli* survival at elevated temperatures. Gene 163: 47–52

Skorko-Glonek J, Lipinska B, Krzewski K, Zolese G, Bertoli E and Tanfani F (1997) HtrA heat shock protease interacts with phospholipid membranes and undergoes conformational changes. J Biol Chem 272: 8974–8982

Sokolenko A, Lerbs-Mache S, Altschmied L and Herrmann RG (1998) Clp protease complexes and their diversity in chloroplasts. Planta 207: 286–295

Spetea C, Hundal T, Lohmann F and Andersson B (1999) GTP bound to the chloroplast thylakoid membranes is required for light-induced multi-enzyme degradation of the Photosystem II D1 protein. Proc Natl Acad Sci USA 96: 6547–6552

Suzuki C, Rep M, Vandijl JM, Suda K, Grivell LA and Schatz G (1997) ATP-dependent proteases that also chaperone protein biogenesis. Trends Biochem Sci 22: 118–123

Takahashi Y, Goldschmidt-Clermont M, Soen SY, Franzen LG and Rochaix JD (1991) Directed chloroplast transformation in *Chlamydomonas reinhardtii*: insertional inactivation of the *psaC* gene encoding the iron sulfur protein destabilizes Photosystem I. EMBO J 10: 2033–2040

Takahashi Y, Matsumoto H, Goldschmidt-Clermont M and Rochaix JD (1994) Directed disruption of the *Chlamydomonas* chloroplast *psbK* gene destabilizes the Photosystem II reaction center complex. Plant Mol Biol 24: 779–788

Tomoyasu T, Gamer J, Bukau B, Kanemori M, Mori H, Rutman AJ, Oppenheim AB, Yura T, Yamanaka K, Niki H, Hiraga S and Ogura T (1995) *Escherichia coli* FtsH is a membrane-bound, ATP-dependent protease which degrades the heat-shock transcription factor σ32. EMBO J 14: 2551–2560

Trost JT, Chisholm DA, Jordan DB and Diner BA (1997) The D1 C-terminal processing protease of Photosystem II from *Scenedesmus obliquus*. Protein purification and gene characterization in wild type and processing mutants. J Biol Chem 272: 20348–20356

Tziveleka LA and Argyroudi-Akoyunoglou JH (1998) Implications of a developmental-stage-dependent thylakoid-bound protease in the stabilization of the light-harvesting pigment-protein complex serving Photosystem II during thylakoid biogenesis in red kidney bean. Plant Physiol 117: 961–970

Vandervere PS, Bennett TM, Oblong JE and Lamppa GK (1995) A chloroplast processing enzyme involved in precursor maturation shares a zinc-binding motif with a recently recognized family of metalloendopeptidases. Proc Natl Acad Sci USA 92: 7177–7181

van't Hof R and de Kruijff B (1995) Characterization of the import process of a transit peptide into chloroplasts. J Biol Chem 270: 22368–22373

Wan J, Bringloe D and Lamppa GK (1998) Disruption of chloroplast biogenesis and plant development upon down-regulation of a chloroplast processing enzyme involved in the import pathway. Plant J 15: 459–468

Wang J, Hartling JA and Flanagan JM (1997) The structure of ClpP at 2.3 Å resolution suggests a model for ATP-dependent proteolysis. Cell 91: 447–456

Wickner S, Gottesman S, Skowyra D, Hoskins J, Mckenney K and Maurizi MR (1994) A molecular chaperone, ClpA, functions like DnaK and DnaJ. Proc Natl Acad Sci USA 91: 12218–12222

Wojtkowiak D, Georgopulos C and Zylicz M (1993) ClpX, a new specificity component of the ATP-dependent *Escherichia coli* Clp protease, is potentially involved in λ DNA replication. J Biol Chem 268: 22609–22617

Woo KM, Kim KI, Goldberg AL, Ha DB and Chung CH (1992) The heat-shock protein ClpB in *Escherichia coli* is a protein-activated ATPase. J Biol Chem 267: 20429–20434

Yang DH, Webster J, Adam Z, Lindahl M and Andersson B (1998) Induction of acclimative proteolysis of the light-harvesting chlorophyll *a/b* protein of Photosystem II in response to elevated light intensities. Plant Physiol 118: 827–834

Chapter 16

Senescence and Cell Death in Plant Development: Chloroplast Senescence and its Regulation

Philippe Matile*
Institute of Plant Biology, University of Zürich, Zollikerstrasse 107, CH-8008 Zürich, Switzerland

Summary

Senescence and death play prominent roles in plant development. Such vitally important processes as water transport in the xylem or the recycling of nutrients from one part of the plant to another are associated with cell differentiations leading to death. Leaf senescenceis a fascinating phenomenon and is even attractive from an esthetical point of view. The degreening of leaves is due to the transition of chloroplasts into gerontoplasts which is associated with extensive breakdown of photosynthetic pigments and remobilization of proteins. Whereas, in broad outline, the catabolic machinery responsible for the breakdown of chlorophyll is known, the orderly dismantling of the photosynthetic apparatus and digestion of proteins have largely remained enigmatic. As a matter of course, senescence is currently investigated in terms of differential gene expression. A considerable number of genes that are expressed senescence-specifically have been identified and many of them are waiting for being attributed with a function. Even cellular death is genetically programmed but its mechanism and underlying gene expression are not yet fully clarified.

*Email: phibus@botinst.unizh.ch

Eva-Mari Aro and Bertil Andersson (eds): Regulation of Photosynthesis, pp. 277–296.

I. Introduction

A. Terminological Consideration

Senescence has been defined as being 'the final phase in ontology of the organ in which a series of normally irreversible events is initiated that leads to cellular breakdown and death of the organ' (Sacher, 1973). In recent years the relevant terminology has been enriched by two terms, 'apoptosis' and 'programmed cell death.' Whereas 'apoptosis' and 'programmed cell death' are clearly synonymous, it seems necessary to examine whether they have an identical meaning as 'senescence.' Developmental programs that are terminated by cell death are quite different. Whereas, for example, cells that eventually constitute a lysigenous duct are first induced to produce a large amount of essential oil before they commit suicide, leaf cells run through a period of extensive degradation in the chloroplasts. In a pathological development of leaf cells such as in the hypersensitive response, programmed cell death is induced without preceding senescence. Hence, cellular death may be regarded as a common item of otherwise very different developmental programs. It is justified, therefore, to make a distinction between apoptosis as being the final and irreversible event of development and senescence as being a basically reversible process leading to cells that are predisposed for the last exit.

Incidentally, apoptosis is an ancient Greek word which etymologically (apo, away; ptosis, falling) refers to the shedding of petals from flowers and yellowed leaves from trees. Although the term apoptosis has originally been coined for designating a program of controlled cell deletion in animals, it has legitimately been adopted by plant biologists.

B. Significance of Senescence and Death

Senescence and death are attendant processes of normal plant development. Nothing could illustrate more convincingly the significance of cell death than the differentiation of procambium cells into the dead and particle-free capillaries of tracheids: such a vitally important function as water transport depends on programmed death of individual cells in the xylem.

Senescence and death occur at all ontogenetic stages and at different levels of organization. In leaves, the differentiation of tracheids and sklereids is terminated by apoptosis taking place during expansion and greening. Likewise, the final phase of development occurs tissue-specifically. The yellowing of leaves is primarily due to senescence in mesophyll cells while epidermal cells and the phloem retain their original functions. Even within the mesophyll, senescence does not comprise the entire set of compartments, but concerns rather selectively the chloroplasts. Hence, senescence of an organ, such as a leaf, represents a complex of programs for each cell phenotype and for each organelle and subcellular compartment. This complex is also referred to as 'senescence syndrome.'

In the case of leaves, senescence may be valued negatively because it is associated with declining rates of photosynthesis. And yet, it is highly significant for development because senescing leaves take a turn from sources of carbon to sources of nitrogen, phosphorus and other essential nutrient elements which, in contrast to the photosynthetically fixed carbon, are growth-limiting factors in most instances. Leaves are induced to senesce when the products of extensive degradation of cytoplasmic constituents in the mesophyll are required for development in other parts of the plant such as newly expanding leaves, seeds or subterranean storage organs. Nutrient recycling as associated with senescence is astonishingly efficient. This is demonstrated, for example, by nitrogen-starved oat plants which were able to complete a life cycle on the basis of the N-supply in the seeds and eventually produce a single viable grain (Mei and Thimann, 1984).

During normal development, apoptotic deletion occurs in cells that have temporary functions as is the case with short-lived root hairs, cells of the root cap, suspensor cells of embryos or cells of reserve tissues of seeds. It is also involved in the lobing of leaves and, in species with unisexual flowers, in the elimination of primordia of sex organs.

II. Leaf Senescence

A. Transition of Chloroplasts into Gerontoplasts

The bulk of protein present in leaf tissue is located in

Abbreviations: Chl – chlorophyll; Cpl – chloroplast; ctDNA – plastidic DNA; FCC – fluorescing chlorophyll catabolite; GL – galactolipid; Gpl – gerontoplast; NCC – nonfluorescing chlorophyll catabolite; nDNA – nuclear DNA; PaO – pheophorbide *a* oxygenase; pFCC – primary fluorescing chlorophyll catabolite; RCC – red chlorophyll catabolite; Rubisco – ribulose-bisphosphate-carboxylase/oxygenase

the chloroplasts (Cpls) and, hence, the withdrawal of N from senescing leaves is primarily due to remobilization of plastidic protein. In mesophyll cells, the population of Cpls appears to be stable until very late stages of senescence (Martinoia et al., 1983; Ford and Shibles, 1988). Cpls persist and run through a senescence-specific differentiation. The final product of this transition is termed 'gerontoplast' (Gpl, Sitte et al., 1980). Such terminological distinction is justified because, unlike other types of plastids, including chromoplasts, developing Gpls are characterized by an entirely catabolic metabolism. Thus, senescing leaves do not turn yellow because carotenoids are newly synthesized (as is the case in developing chromoplasts), but because the yellow pigments are partially retained and progressively unmasked as chlorophyll (Chl) is lost.

The transition of Cpls into Gpls is characterized by structural changes in the thylakoids such as loosening and disorganization of grana stacks, swelling, vesiculation and progressive loss of membranes (see references in Thomson and Whatley, 1980; Parthier, 1988; Matile, 1992; Thomas, 1997). As stroma components are also lost, volume and density decrease. A conspicuous feature of transition is the increase in number and size of plastoglobules (Fig. 1) in which undigested or undigestible lipophilic components of thylakoids accumulate. At the end of the senescence period, fully differentiated Gpls consist of a still intact envelope wrapping up a large number of plastoglobules.

Persistence and orderly development of plastids in senescent leaf tissue are also illustrated by the potentiality for regreening, in some species even of completely yellowed leaves (Mothes and Baudisch, 1958; Mothes, 1960; Greening et al., 1982; Venkatrayappa et al., 1984; Marek and Stewart, 1992; McLaughlin and Smith, 1995). Thus, developing Gpls seem to retain the internal organization that is required for reverse transition into Cpls.

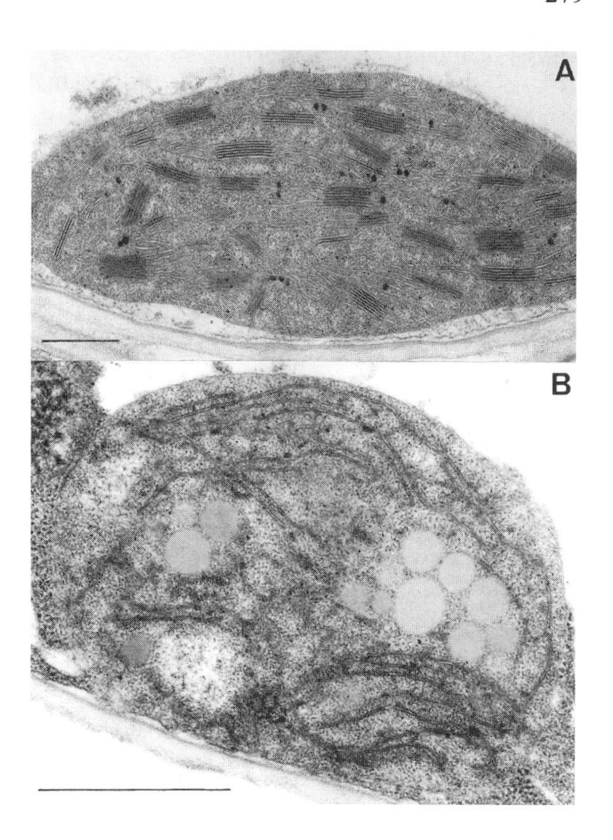

Fig. 1. Electron micrographs of thin-sectioned barley mesophyll cells showing a mature chloroplast (A) and a developing gerontoplast (B). Seedlings were allowed to develop for 7 days under photoperiod (A). Primary leaves were excised after 7 days and induced to senesce in permanent darkness for 5 days (B). Note the marked loss of plastid volume associated with gerontoplast development; grana have completely disappeared and few profiles of thylakoid membranes are retained. Intactness of Gpl envelope is as conspicuous as are the large plastoglobules. Bars indicate 1 μm. Courtesy of G. Wanner and V. Scheumann, Institute of Botany, University of Munich.

B. Decline of Photosynthesis

Progressive loss of Chl is symptomatic for the decline of photosynthesis during leaf senescence. However, the efficiency of photochemical energy conversion as based on contents of residual Chl remains high (Adams et al 1990; Lu and Zhang, 1998) suggesting that changes in the thylakoids take place in an orderly fashion. Even when isolated at a premortal stage of senescence, thylakoids had a surprisingly high photochemical activity (Schmidt, 1988). The primary event responsible for the decline of photosynthesis appears to be the preferential loss of activity and amount of ribulose bisphosphate carboxylase/ oxygenase (Rubisco). This abundant protein complex is a major source of nitrogen that is exported from senescing leaves (Peterson and Huffaker, 1975). As documented in a large number of studies (reviewed by Grover, 1993) Rubisco declines preferentially and concomitant with photosynthesis. Indeed, throughout the life span of leaves, activities and amounts of Rubisco appear to be directly correlated with carbon fixation (Mae et al., 1987; Jiang et al., 1993). As Rubisco is lost, the photochemical capacity in the thylakoids is adjusted to the declining rates of carbon fixation. The corresponding changes appear to take place in such a manner that the residual components still function and can satisfy the energy demand in the developing gerontoplasts. The protein complements of the components of Photosystem II and of the cytochrome *b/f* complex are particularly labile during senescence (Ben-David et al., 1983; Holloway et al., 1983; Woolhouse and Jenkins, 1983; Roberts et al., 1987; Jackowski et al., 1991).

C. Genetic Control

Leaf senescence and development of Gpls are controlled by the genome of the nucleus. In enucleated subprotoplasts of *Elodea* leaflets Cpls failed to senesce whereas in the presence of a nucleus, degreening took place (Yoshida, 1961). The requirement for cytoplasmic protein synthesis appears also from the effect of specific inhibitors. Protein synthesis in Cpls appears to be shut off as senescence is induced. Indeed, in rice leaves ctDNA has been demonstrated to be degraded even before the loss of Chl indicated the onset of senescence (Sodmergen et al., 1989, 1991). Genetic control is also documented by the existence of mutants that are unable to degrade Chl. The most thoroughly investigated case of such a stay-green genotype is Bf993 of *Festuca pratensis* in which the mutation in a nuclear gene is responsible for deficiency regarding a Chl catabolic enzyme (Thomas, 1987; Vicentini et al., 1995). In legumes, similar mutants were identified in cultivars that are characterized by the retention of Chl during seed maturation. They include several varieties of soybean (Guiamét et al., 1991; Guiamét and Giannibelli, 1994), French bean (Ronning et al., 1991; Bachmann et al., 1994) and pea cultivars such as used by Gregor Mendel (Thomas et al., 1996). Still another example of genetic control is the green-flesh mutation in tomato (Akhtar et al., 1999).

Perhaps the most convincing evidence for genetic control is derived from recent identifications of a large number of genes with senescence-associated changes of expression. These genes have been classified by Smart (1994) and Buchanan-Wollaston (1997); as illustrated in Fig. 2 some genes are exclusively expressed during senescence while the abundance of messages from others is merely enhanced or is also high at other stages of development. Whereas the senescence-specific expression of genes such as those encoding digestive enzymes or enzymes that play a role in nutrient

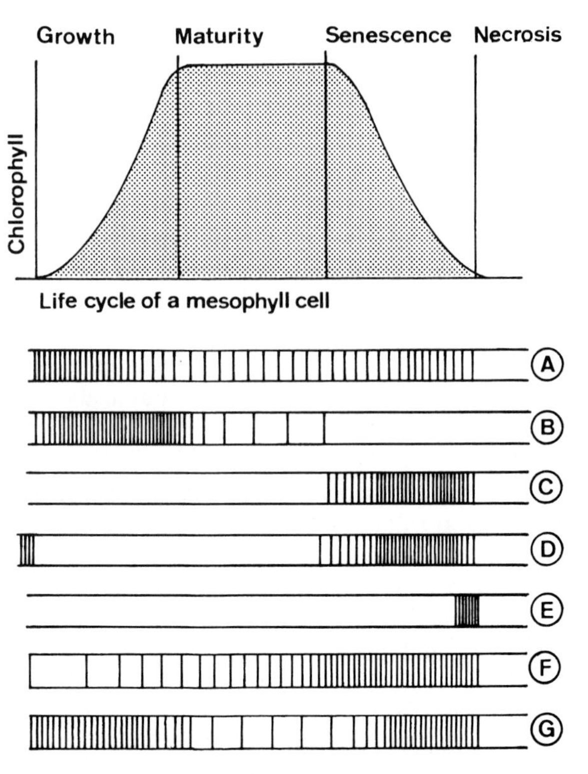

Fig. 2. Differential gene expression during leaf development. Classes of genes with reference to Smart (1994) and Buchanan-Wollaston (1997). A, genes related to house-keeping metabolism (e.g. respiration, protein synthesis); B, genes encoding components of chloroplasts (e.g. Rubisco, apoproteins of pigment protein complexes, enzymes of Chl biosynthesis); C, genes coding for catabolic enzymes (e.g. PaO, proteases); D, genes of enzymes having a function at other developmental stages such as seed germination (e.g. enzymes of the glyoxylic acid cycle); E, genes of putative proteins required for cell autolysis. F, genes with increased expression during senescence (e.g. cytosolic glutamine synthetase, metallothionein, ferritin. G, genes with high expression during senescence but also during leaf growth (e.g. enzymes of glycolysis as gylceraldehyde 3 phosphate dehydrogenase). For references see Buchanan-Wollaston (1997).

recycling or in gluconeogenesis (see also Fig. 5) is plausible, the expression of others such as genes of metallothioneins or cytochromes P_{450} (review: Buchanan-Wollaston, 1997) is unexpected and not easily understandable. The functions associated with a number of senescence-associated genes remains to be elucidated.

Regarding differential gene expression during leaf senescence it is worth mentioning that despite rapid decline of protein contents, proteins are synthesized at surprisingly high rates; in fact, rates of ^{35}S methionin incorporation into senescing oat leaves have been determined to be nearly as high as during leaf expansion and greening (Klerk et al., 1992). Hence, proteins that are newly synthesized during senescence appear to be turned over rapidly.

D. Regulation

Studies dealing with the regulation of leaf senescence are almost invariably based on Cpl-associated parameters such as contents of Chl or soluble protein (Rubisco). Hence, knowledge about the regulation of leaf senescence concerns the transition of Cpls into Gpls.

Among the various factors known to play a role in Cpl senescence, light is perhaps the most important one. In most cases light causes the retardation of degreening. For reasons of convenient experimentation, detached leaves allowed to senesce in permanent darkness have frequently been employed. In attached leaves kept under photoperiod, transition from photosynthetic to catabolic metabolism is gradual and the senescence process is slow, whereas in detached leaves, the abrupt change of developmental programs causes a comparatively rapid degreening. Light-dependent retardation of Cpl senescence is mediated by phytochrome (Biswal et al., 1983; Okada et al., 1992; Rousseaux et al., 1997). Effects of both, photoperiod (Kulkarni and Schwabe, 1985; Kar, 1986) and light dosage (Noodén et al., 1990) have been demonstrated. Whereas light normally causes marked delays of degreening, in some species the opposite effect has been observed (Maunders and Brown, 1983; Kar and Choudhuri, 1985).

In many annuals leaf senescence is induced upon flowering and seed development. Such monocarpic senescence can experimentally be delayed e.g. by the removal of developing pods in soybean (Wittenbach, 1982; Crafts-Brandner et al., 1984; Noodén and Guiamet, 1989) or by phloem interruption below the ear in a cereal (Fröhlich and Feller, 1991). The nature of the senescence-inducing signal which is sent out by the seeds is unknown, but otherwise several examples of hormonal control are well known. The most convincing example concerns the retarding effects of cytokinins. The roots represent the main source of cytokinins, suggesting that the rapid senescence observed in detached leaves may, in part, be due to the lacking supply of cytokinin via the transpiration stream. In explants of soybean consisting of an internode with a leaf and a pod, the supply of cytokinin via the stem largely overrides the senescence-inducing influence of the pod (Noodén and Letham, 1993). Tobacco is an annual plant characterized by sequential(progressive) leaf senescence; levels of endogenous cytokinins decrease markedly concomitant with leaf yellowing, and the application of exogenous cytokinin causes retardation (Singh et al., 1992). The effect of cytokinin on age-dependent senescence of tobacco leaves has also been demonstrated in plants transformed with a bacterial gene for cytokinin synthesis, isopentenyl transferase (Smart et al., 1991; Gan and Amasino, 1995; Wingler et al., 1998).

Of the five classical growth regulators, ethylene (Warman and Solomos, 1988; Knee, 1991; Philosoph-Hadas et al., 1994) and abscisic acid are known to accelerate leaf senescence, the latter especially in conjunction with light (Zhi-Yi et al., 1988; Rodoni et al., 1998). In some species the most potent antagonist of cytokinin is jasmonic acid methylester (Ueda et al., 1981) which is regarded to represent a stress hormone produced, for example, under conditions of drought (Sembdner and Parthier, 1993). In detached barley leaves exposed to light and methyljasmonate as much as half of Chl and Rubisco were degraded during the first 24 h (Weidhase et al., 1987). Jasmonate is known to induce a species- and tissue-specific pattern of newly synthesized proteins (Herrmann et al., 1989). It is thus to be expected that some of the genes turned on in leaves of barley (Lehmann et al., 1995) and probably other cereals as well, may play a decisive role in Cpl to Gpl transition.

It is interesting that the sequences of promoters analyzed so far contain no common *cis*-acting elements, suggesting that multiple regulatory pathways may be involved in the activation of the various genes (review: Gan and Amasino, 1997).

III. Biochemistry of Breakdown in Senescing Chloroplasts

A. Breakdown of Pigments

1. Detoxification of Chlorophyll

Most porphyrins are photodynamically active and hence potentially hazardous. Chl is a particularly interesting example of photodynamism because its function in photosynthesis is not only linked to exposure to light but is also associated with the production of oxygen. In the thylakoids, Chl is complexed together with carotenoids which prevent the occurrence of activated oxygen either by quenching triplet states of Chl or by scavenging singlet oxygen and thereby dissipating excessive energy. It is obvious that the dismantling of pigment-protein complexes during Gpl development is a delicate business because it implies abolition of protection from photodynamic damage. It is plausible, therefore, that the accumulation of unprotected Chl is avoided by virtue of a most efficient breakdown and photodynamic inactivation of Chl. Indeed, Chl breakdown in senescent leaves must be interpreted as a kind of detoxification.

In degreening leaves the porphyrin moiety of Chl is ultimately converted to colorless linear-tetrapyrrolic compounds in which the original conjugation among the pyrrole units is completely abolished. These 'nonfluorescing Chl catabolites' (NCCs; Fig. 3) accumulate in the vacuoles of senescing mesophyll cells, i.e. in the same cell compartment in which the potentially autotoxic secondary compounds such as phenolics are also detoxified (review: Matile et al., 1999).

The PaO pathway of Chl breakdown in senescent leaves is initiated by chlorophyllase which is responsible for dephytylation, and Mg dechelatase which exchanges the central Mg ion of chlorophyllide for two protons. The substrate of PaO, pheophorbide, is the product of these initial steps. Downstream of the ring-opening reaction, the ethyl side chain of pyrrole B is hydroxylated. As suggested by the structures of NCCs, this modification is common in all plant species examined so far. Other modifications and conjugations of FCCs occur species-specifically. Ultimately, the FCCs are transported across the tonoplast and tautomerized into NCCs under the acidic conditions of the vacuolar sap (review: Hörtensteiner, 1999).

The pathway by which Chl is catabolized into NCCs has turned out to be unexpectedly complicated (Fig. 3). The decisive step is represented by the oxidative opening of the porphyrin macrocycle at the methine bridge linking pyrroles A and B. It causes the loss of green color and is responsible for extensive photodynamic inactivation. The first identifiable product of oxygenolytic ring opening is a colorless blue-fluorescing derivative of pheophorbide *a* (Mühlecker et al., 1997). This primary fluorescing chlorophyll catabolite (pFCC) is produced by the joint action of two enzymes, pheophorbide *a* oxygenase (PaO), a monooxygenase located in the inner membrane of the Gpl envelope, and a stroma-located reductase which is responsible for the saturation of the C20/C1 double bond in the red colored intermediary product, RCC (Fig. 3; Rodoni et al., 1997). Both enzymes, PaO and RCC reductase require reduced ferredoxin for the driving of redox cycles (Hörtensteiner et al., 1995; Rodoni et al., 1997). Such requirement emphasizes the importance of intactness of Gpls with regard to the regeneration of reduced ferredoxin either photochemically at Photosystem I or, in the dark, through NADPH and its continuous reduction via pentose-phosphate cycle. It may also explain why a ferredoxin gene is expressed during leaf senescence (Smart et al., 1995).

A most remarkable property of PaO is its absolute specificity for pheophorbide *a* as substrate (Hörtensteiner et al., 1995). On one hand, this specificity explains the fact that all NCCs described so far are derived from Chl *a* while, on the other hand, it poses a problem regarding the fate of Chl *b*. Quantitative analyses of breakdown suggest that the NCCs accumulated account for both forms of Chl (Ginsburg and Matile, 1993; Curty and Engel, 1996). Hence, during degreening Chl *b* appears to be fed into the pool of Chl *a* forms. The corresponding reduction of the 7^1-formyl (via -hydroxyl) to -methyl group in pyrrole B is due to the action of two reductases, collectively termed 'Chl *b* reductase' (Scheumann et al., 1998 and references cited therein). Its role in the recycling of Chl *b* into Chl *a* has recently been demonstrated in senescing barley leaves (Scheumann et al., 1999).

The regulation of Chl breakdown is not yet entirely clear. The activity of PaO has been found to be present exclusively in senescent leaves and also in ripening berries of sweet pepper and tomato (review: Matile et al., 1999). In barley leaves the activity is positively correlated with rates of Chl breakdown and is modulated under the influence of hormones (Rodoni et al., 1998). The gene of PaO has not yet been cloned but it is probably identical with one of the various senescence-associated genes waiting for being attributed with a function.

In contrast to PaO all other enzymes of the pathway, including chlorophyllase, are constitutive. Hence, breakdown is very likely to be regulated not only at the level of PaO but in addition at a step upstream of the ring-opening reaction. In a PaO-deficient stay-green mutant of *Festuca pratensis* the dephytylated green catabolites, chlorophyllides *a* and *b* and pheophorbide *a* are accumulated upon the induction

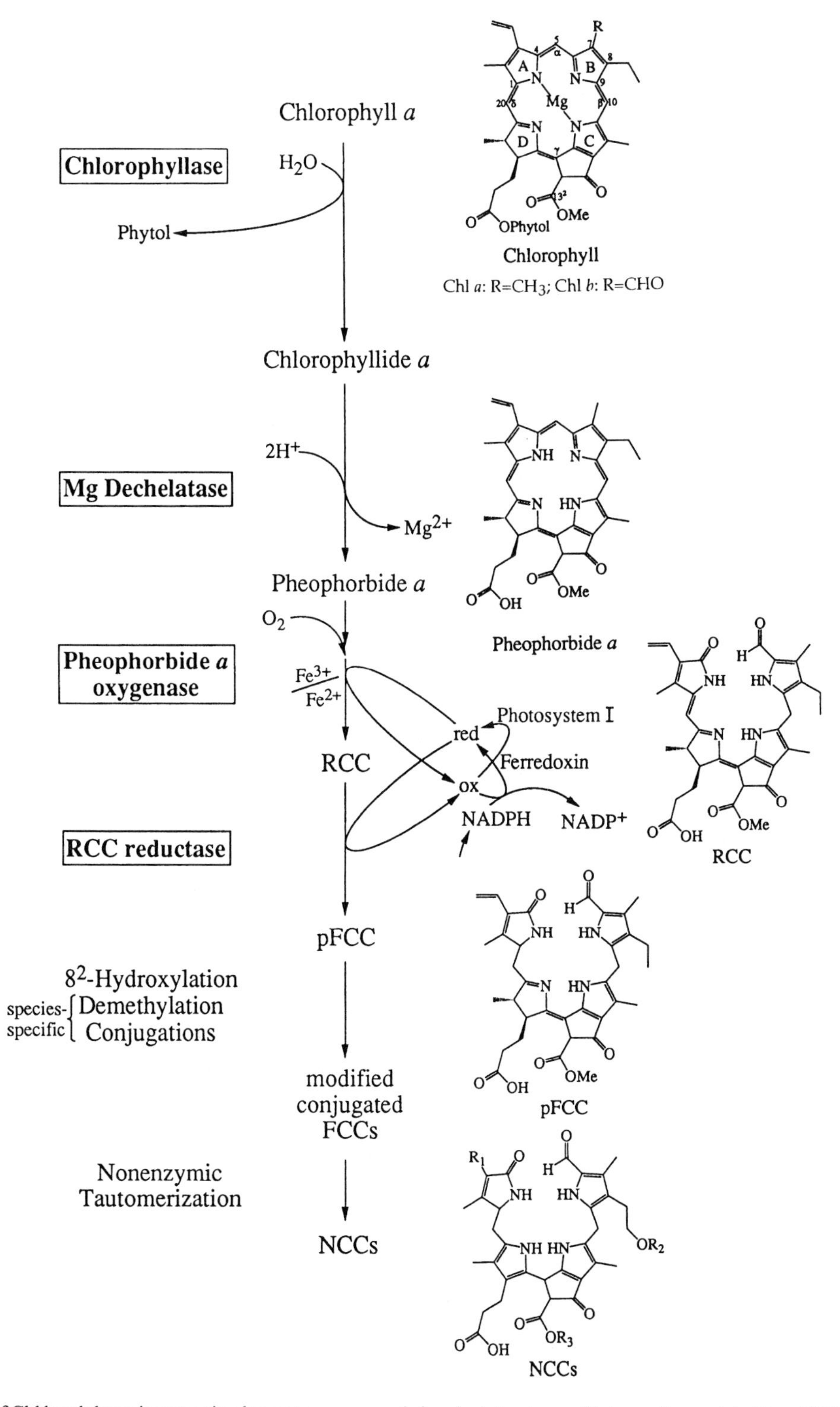

Fig. 3. Pathway of Chl breakdown in senescing leaves: enzymes and chemical structures of intermediary catabolites. All final catabolites (NCCs) identified so far are hydroxylated in the ethyl side-chain of pyrrole B. In two NCCs from canola, this hydroxyl group is either glucosylated or esterified with malonic acid. All NCCs from canola are demethylated (R_3=H). In NCCs from other species the methylester is preserved. In an NCC from barley the vinyl side chain of pyrrole A is hydroxylated (R_1= dihydroxyethyl). For references see Matile et al., (1999).

of leaf senescence (Vicentini et al., 1995); although chlorophyllase is constitutive, the dephytylation requires cytoplasmic protein synthesis (Thomas et al., 1989) suggesting that a decisive as yet unidentified regulatory mechanism is responsible for serving chlorophyllase with its substrate.

Chlorophyllase is a hydrophobic protein of Cpl membranes but in preparations of isolated membranes it has no contact with its substrates unless the membranes are solubilized in the presence of detergent (Amir-Shapira et al., 1986; Matile et al., 1997). The latency of activity appears to be due to the differential location of Chl in the thylakoids and chlorophyllase which has been localized in the Cpl envelope (Matile et al., 1997). Since PaO as well as presumably Mg dechelatase are also located in the envelope, the Chl complexed in the thylakoids appears to be properly separated from the catabolic machinery (Fig. 4). As transition of Cpls into Gpls is induced, chlorophyllase seems to be served with substrate but the mechanism by which the pigment-protein complexes of thylakoids are dismantled and Chl molecules transported to the site of chlorophyllase is unknown. The elucidation of this mechanism is very important for the understanding of Chl breakdown as a whole. Recent investigations of water-soluble Chl-protein complexes may be significant. A hypothetical carrier protein marked X in Fig. 4 could be identical with the apoprotein of the water-soluble Chl complex discovered in leaves of Brassicaceae (Kamimura et al., 1997 and references cited therein). This protein has some unique properties which may be relevant with regard to Chl degradation. Thus, the corresponding gene is induced under conditions that are likely to induce the catabolism of Chl (Ilami et al., 1997; Nishio and Satoh, 1997). Moreover, Satoh et al. (1998) have shown that in vitro the protein is able to remove Chl from thylakoids and, hence, to dismantle pigment-protein complexes. A contrasting possibility of Chl transfer from thylakoids across the stroma to the envelope is suggested by the presence of Chl in the plastoglobules (Young et al., 1991; Picher et al., 1993). Plastoglobules contain a specific set of proteins (Kessler et al., 1999) and should not be underestimated as products of passive accumulation of lipophilic material. Loaded with Chl they may represent the natural substrate of chlorophyllase and this would even explain why phytol is accumulated in the plastoglobules (Bortlik, 1990).

Chl breakdown is not only important in its own right but is also a prerequisite for the remobilization of about a third of total chloroplastic protein present

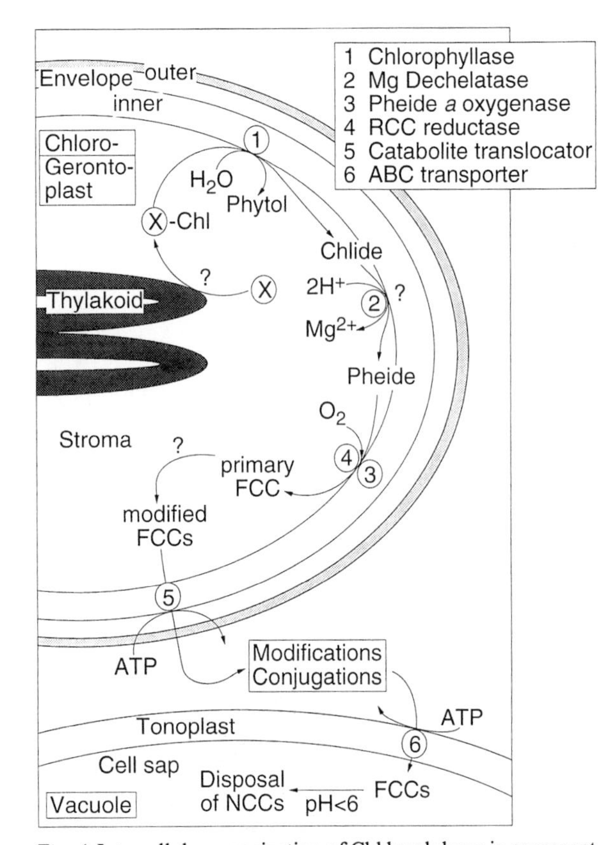

Fig. 4. Intracellular organization of Chl breakdown in senescent mesophyll cells. Abbreviations: Chlide, chlorophyllide; Pheide, pheophorbide. From Matile et al., (1999).

in pigment-protein complexes. In stay-green mutants the apoproteins of these complexes are retained (Bachmann et al., 1994 and references cited therein). Hence, such mutants pay a penalty due to incomplete N-recycling and this may explain why practically all of them are either diazotrophic legumes such as bean, soybean and pea, or cultivated species such as tomato or sweet pepper which normally don't suffer from N-deficiency.

2. Breakdown of Carotenoids

During Gpl development, carotenoids disappear concomitant with Chl, but usually at a slower rate (Gut et al., 1987). In many species, retention of carotenoids is pronounced such as to cause a bright yellow coloration of the autumn foliage. A most remarkable example is represented by the senescing leaves of *Ginkgo biloba* in which lutein is partially retained in esterified form (Matile et al., 1992). Such esters, not normally found in mature green leaf tissue, have also been detected in drought-stressed barley seedlings (Barry et al., 1992). Very little is known about the biochemical background of net loss of yellow pigments. In senescent barley leaves, the increased epoxidation of carotenoids of the xanthophyll cycle has been suggested to be connected with the increase of endogenous abscisic acid (Afitlhile et al., 1993), a growth regulator which is known to induce or hasten senescence. Whether or not observations of enzymic oxidative bleaching of carotenoids and carotene in vitro (Friend and Mayer, 1960; Matile and Martinoia, 1982) are relevant for catabolism in vivo is uncertain.

B. Utilization of Galactolipids

The galactolipids (GLs) represent the main components of the thylakoidal lipid bilayers. They are probably the most abundant lipids of the biosphere. During transition of Cpls into Gpls, GLs appear to serve as an energy source for the fueling of metabolism in senescent leaf cells.

Metabolism in senescent leaves is intensive and, hence, energy consuming. The demand for metabolic energy in senescing leaves is reflected by high respiratory activity as well as by the intactness of mitochondria throughout the senescence period (Romani, 1987). As photosynthesis declines, carbohydrates tend to be depleted and are replaced by other sources of metabolic energy. It has long been known that leaf senescence is associated with a drop of respiratory quotients to values below 1 (Yemm, 1935) but only recently it has been discovered that this phenomenon is due to the utilization of acyl residues of GLs for gluconeogenesis (Gut and Matile, 1988).

In senescent leaves, contents of GLs decline at about the same rate as Chl (Koiwai et al., 1981; Düggelin et al., 1988; Meir and Philosoph-Hadas, 1995) whereby the monogalactosyl species declines faster and earlier than the digalacto-form (Gut and Matile, 1989). The disappearance of GLs is not accompanied by the accumulation of free fatty acids (as one would expect if breakdown occurred by the action of a galactolipase) but rather by the transient increase of phosphatidylcholine having acylation patterns reminiscent of GLs (Gut and Matile, 1989). Radiolabeling of acyl residues in the GLs during de-etiolation of barley primary leaves and tracing of ^{14}C during subsequent dark-induced senescence confirmed that GLs are converted to phosphatidylcholine (Wanner et al., 1991), seemingly in a process reversing GL biosynthesis (Joyard and Douce, 1987).

Some important details of the pathway as, for example, the export of phosphatidylcholine from developing Gpls have not yet been studied but radiolabeling has revealed that the carbon of acyl residues originating from GLs is eventually utilized for gluconeogenesis (Fig. 5) and is ultimately respired (Wanner et al., 1991). Indeed, typical enzymes of the glyoxylic acid cycle are induced during leaf senescence (Gut and Matile, 1988; DeBellis and Nishimura, 1991; Pistelli et al., 1992; Pastori and DelRio, 1994; McLaughlin and Smith, 1995) and this is due to induced gene expression (Graham et al., 1992). Hence, leaf senescence is associated with the functional transition of peroxisomes from photo-

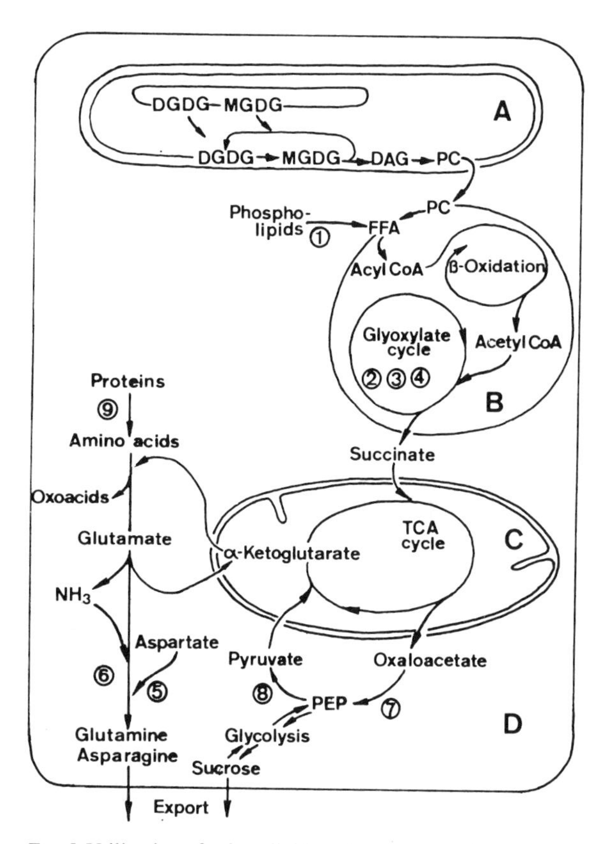

Fig. 5. Utilization of galactolipids and other metabolic processes taking place in senescing mesophyll cells. A, developing gerontoplast; B, peroxisome (gerontosome); C, mitochondrion; D, cytosol. The numbers refer to genes of enzymes that have been demonstrated to be expressed senescence-specifically. 1, phospholipase D; 2, isocitrate lyase; 3, malate synthase; 4, NAD malate dehydrogenase; 5, asparagine synthetase; 6, cytosolic glutamine synthetase; 7, PEP carboxykinase; 8, pyruvate orthophosphate dikinase; 9, various proteases. Redrawn with reference to Buchanan-Wollaston (1997). Abbreviations: MGDG, monogalactosyldiacylglycerol; DGDG, digalactosyldiacylglycerol; DAG, diacylglycerol; PC, phosphatidylcholine; FFA, free fatty acid; PEP, phosphoenolpyruvate. For references see Smart (1994) and Buchanan-Wollaston (1997).

respiratory to glyoxisomal metabolism (DeBellis et al., 1990; Landolt and Matile, 1990; Nishimura et al., 1993). The type of peroxisome which develops in senescing mesophyll cells is quite unique since the set of enzymes comprises residual components of the photorespiratory system along with senescence-induced glyoxisomal enzymes and, in addition, uricase which probably has a function in the catabolism of purines. For reasons of such uniqueness, a special term such as 'gerontosome' (Vicentini and Matile, 1993) may be justified for this senescence-specific form of multifunctional peroxisome.

The development of gerontosomes appears to depend on the supply of carbohydrates. Thus, the induction of glyoxisomal enzymes is weak under natural conditions as long as photosynthesis continues, and it is pronounced in excised leaves allowed to senesce in darkness (Pistelli et al., 1991; Vicentini and Matile, 1993). In fact, the expression of glyoxylate cycle genes appears to depend on the availability of sugars as demonstrated by the repression upon feeding sucrose to carbon-starved cultured cucumber cells (Graham et al., 1994).

C. Remobilization of Protein

The recycling of nitrogen from senescing leaves to other parts of the plant is associated with the remobilization of protein, primarily of Rubisco and of Chl-binding proteins, in developing Gpls. In view of the quantitative extent and of the significance of proteolysis in senescent leaves, it is disquieting that so little is known about this process.

It may be reasonable to assume that a proteolytic system is newly synthesized as senescence and wholesale breakdown of protein in Cpls are induced. Indeed, enhanced activities of various endopeptidases have been measured in several systems commonly employed in studies on leaf senescence (review: Huffaker, 1990). More convincing than changes of activity are data resulting from the screening of cDNA libraries derived from transcripts accumulated in senescent leaves. A number of senescence-associated genes have been identified to encode endopeptidases (Hensel et al., 1993; Lohman et al., 1994; Smart et al., 1995). As deduced from known localizations of homologous enzymes, these proteases are likely to be located in vacuoles and, hence, their role in the breakdown of plastidic proteins is questionable. A promising work on a senescence-induced serine protease located in thylakoids (Kawasaki and Takeuchi, 1989) has not been followed up, possibly because of experimental shortcomings due to contamination of isolated Cpls with a vacuolar protease. As judged by the senescence-associated decline of mRNA abundances of subunits of ATP-powered clp protease complexes as well as clp proteins, a function of this Cpl-located system in bulk degradation is rather unlikely (Crafts-Brandner et al., 1996; Humbeck and Krupinska, 1996).

The existence of a proteolytic system in Cpls is documented by a wealth of findings. Thus, proteins such as protochlorophyllide oxidoreductase, an abundant protein of etioplasts, is extensively broken down during greening of etiolated leaves (Reinbothe et al., 1995a). Likewise, a protein of the reaction center of Photosystem II, D1 (formerly Q_B) is turned over rapidly in the light (Mattoo et al., 1989). In both cases the action of endopeptidases that recognize specific cleavage sites appears from the occurrence of breakdown fragments (Reinbothe et al., 1995b; Barbato et al., 1995). Still another group of examples of proteolysis within Cpls concerns the rapid degradation of proteins which, for one reason or another, are not functionally integrated (Schmidt and Mishkind, 1983; Merchant and Bogorad, 1986; Kuwabara and Suzuki, 1994; Plumley and Schmidt, 1995; Halperin and Adam, 1996; Park and Hoober, 1997). The functional integration as, for example, the correct stoichiometry in pigment-protein complexes, appears to protect chloroplastic proteins from proteolytic cleavage by constitutively present endopeptidases.

Some phenomena observed in isolated Cpls demonstrate the presence of highly active endopeptidases. Convincing results have been achieved with isolated Cpls which were repurified after incubation under various conditions in order to rule out effects due to Cpl lysis during incubation and protein degradation by contaminating (vacuolar) proteases. Depending on the conditions of incubation, the immunological analysis of several stromal proteins revealed selective and surprisingly fast degradation (Mitsuhashi and Feller, 1992; Mitsuhashi et al., 1992). Glutamine synthetase, for example, had completely disappeared after only 2 h of incubation in the light (Thoenen and Feller, 1998). Interestingly, such rapid proteolysis of stromal proteins depends on the intactness of Cpls; when the organelles were lysed, the proteins were stable and proteolysis was not restored by the addition of a putative cofactor such as ATP (Roulin and Feller, 1997). Collectively, these findings suggest that breakdown in developing Gpls is not so much dependent on a proteolytic system that is newly synthesized during senescence, but rather on processes which render proteins susceptible to proteolytic attack. In the case of apoproteins of pigment-protein complexes,dismantling as associated with Chl breakdown may represent such a process. This view is corroborated by observations in stay-green genotypes which during senescence not only retain Chl but also the apoproteins (Thomas, 1982; Hilditch, 1986; Guiamét et al., 1991; Cheung et al., 1993; Bachmann et al., 1994). An interesting case is represented by the aforementioned D1 protein of Photosystem II reaction center; like other Chl binding proteins it is retained in senescent leaves of a stay-green mutant of meadow fescue but degradation during turnover in the light is normal (Hilditch et al., 1986). Hence, the proteolytic principle involved in the turnover of apoproteins during plastid assembly and in the mature Cpls seems to be different from the system responsible for degradation during senescence (Nock et al., 1992).

Susceptibility to proteolysis may also be determined by the availability of substrates, by the redox-status within the organelle, or by oxidative modifications of proteins (Mehta et al., 1992; Thomas and Feller, 1993; Desimone et al., 1998; Ishida et al., 1998; Kamber and Feller, 1998; Roulin and Feller, 1998). However, it is uncertain whether the phenomena described so far are relevant with regard to, for example, the preferential degradation of Rubisco during senescence.

It is intriguing that protein breakdown in Cpls has frequently been observed to be associated with the accumulation of distinct proteolytic fragments that can be immunologically attributed to a specific protein. This phenomenon points to the action of endopeptidases which recognize distinct cleavage sites, but also to the absence of further hydrolysis into small peptides and amino acids. It has not been considered, so far, that the primary cleavage products could be exported from developing Gpls and be digested in another subcellular compartment such as the vacuole. This possibility may be inferred from observations in a mutant of *Chlamydomonas reinhardtii* which is disturbed in the light-dependent regulation of the synthesis of Chl-binding proteins. In the dark, the synthesis of Photosystem II apoproteins of the light-harvesting complex occurs at nearly the same rate as in the light. As Chl is not available for the stabilization of pigment-protein complexes, the apoproteins (or fragments of them carrying epitopes) seem to be exported from the Cpls and are accumulated in vacuoles (Park and Hoober, 1997) i.e. in the lytic compartment. Hence, a role of vacuoles in the breakdown of Cpl proteins should not be excluded as long as senescence-associated mechanisms of protein breakdown have not been elucidated otherwise.

D. Breakdown of Nucleic Acids

Whereas ctDNA has been shown to be degraded during (Nii et al., 1988) or even before the onset of Cpl to Gpl transition (Sodmergen et al., 1989, 1991), levels of nuclear DNA remain constant and begin to decrease abruptly only at a late stage of senescence (Matile and Winkenbach, 1971; Lamattina et al., 1985). In rice Cpls, a Zn^{2+}-dependent nuclease may be responsible for the degradation of ctDNA (Sodmergen et al 1991). The loss of nDNA is probably a consequence of cell autolysis (see Section IV B). In senescent wheat leaves RNA contents declined at a similar rate as protein contents (Lamattina et al., 1985). Increased RNase activities and corresponding gene expression are symptomatic of leaf senescence (Blank and McKeon, 1991; Taylor et al., 1993; see also Laurière, 1983). Nothing is known about subcellular locations, and regulation of RNA degradation. Equally unknown is the metabolism of purines and pyrimidines that may be involved in nutrient recycling as associated with the breakdown of nucleic acids. It is imaginable that acid unspecific phosphatase, a prominent hydrolase of the lytic compartment, is responsible for the production of inorganic phosphate from nucleotides during autolysis.

IV. Programmed Cell Death

A. Avoidance of Death

Aerobic life depends on atmospheric oxygen and is, at the same time, exposed to hazards caused by activated species of oxygen. Normal metabolism is associated with the production of various forms of activated oxygen and cells are equipped with multifarious enzymatic tools and antioxidants which are responsible for the avoidance of oxidative damage by singlet oxygen and oxygen radicals (selected reviews: Elstner, 1987; Halliwell and Gutteridge,

1989; Dalton, 1995). With regard to senescence, oxidative damage has been, and still is, widely regarded as a decisive cause of decline of vital functions eventually leading to death. This theory assumes that the gradual loss of capacity to detoxify reactive species of oxygen results in progressive deteriorations, particularly in membranes, and finally in the collapse of subcellular order (selected review: Thompson et al., 1987). These conclusions are based on several lines of evidence concerning the decrease of membrane fluidity, phospholipid degradation, lipid peroxidation and accumulation of products of peroxidative degradation of polyunsaturated fatty acids, changes of activities of enzymes playing a role in detoxification, changes in membrane proteins, and progressive leakage of ions and small molecules. Collectively, these observations suggest that death is a more or less coincidental event comparable to the crashing of a machine that had been kept in repair insufficiently.

Although the evidence is seemingly convincing, the theory is unsatisfactory because it fails to explain the proper metabolic function in senescing cells. For example, the expression of genes throughout the entire senescence period (Smart, 1994; Buchanan-Wollaston, 1997) is hardly imaginable under conditions of declining homoeostasis in the cytoplasm. The house-keeping metabolism and particularly protein synthesis with its regulation depend on the maintenance of permeability properties of, and transport across the membranes. The subcellular compartmentation of ions and small molecules is indispensable for homoeostasis in the cytoplasm. For example, the escape of phenolics from vacuoles would cause fatal damage in the cytosol and, indeed, as demonstrated by the sudden browning of degreened leaf tissue, it occurs only at the very end of senescence. Likewise, lipid peroxidation is likely to represent a postmortal event which, mistakenly, has been interpreted as an attendant symptom of senescence because the analysis is normally performed with whole organs or tissues (Panavas and Rubinstein, 1998). Recent investigations on the occurrence of cell death within tissues have revealed conspicuous temporal and spatial patterns (Wang et al., 1996b; Inada et al., 1998). Hence, depending on developmental stages, membrane preparations and extracts from whole organs will inevitably yield data referring to changing proportions of viable and dead cells. It is true that senescence is associated with increasing proportions of solutes in the free space (Sacher, 1973). Progressive leakage is particularly impressive during senescence in short-lived petals (Hanson and Kende, 1975; Yamane et al., 1993). However, in attached primary leaves of barley the increase in leakage is modest during the period of degreening and is marked only after completion of senescence (Matile, unpublished data). It seems to be justified to distinguish between 'metabolic leakage' during senescence (export of solutes as associated with recycling of nutrients) and 'apoptotic leakage' due the collapse of solute compartmentation. In *Gladiolus* florets treated with cycloheximide, the increase of ion leakage is inhibited (Yamane and Ogata (1995) suggesting that even cell autolysis requires cytoplasmic protein synthesis. Hence, it seems that the program of senescence includes the final expression of a gene encoding a kind of suicide protein.

In developing Gpls, protection from oxidative damage is a prerequisite for membrane integrity as, for example, required for continuous reduction of ferredoxin during Chl breakdown. It is intriguing in this respect that α-tocopherol, a scavenger of free radicals and quencher of singlet oxygen (Fryer, 1992) is accumulated during leaf senescence (Rise et al., 1989) and its biosynthesis may even be an item of the genetic program as suggested by the observation of enhanced gene expression related to the biosynthesis of homogentisic acid in barley (Kleber-Janke and Krupinska, 1997).

B. Apoptosis: The Mechanism of Death

In animals, the genetically programmed deletion of individual cells is associated with characteristic events such as the condensation and fragmentation of the nucleus, blebbing of the plasmalemma and cleavage of the cell into apoptotic bodies (review: Martin et al., 1994). One of the hallmarks of apoptosis in animal cells is the hydrolysis of nDNA by a specific endonuclease whereby oligonucleosomal fragments of variable length are produced. In the TUNEL-assay the 3′OH ends of the fragments are converted into a fluorescence signal which allows the detection of apoptotic cells in situ (Gavrieli et al., 1992). The oligomeric series of DNA fragments manifests itself also upon electrophoresis; this is commonly refereed to as 'DNA laddering.'

Some of these phenomena associated with programmed cell death are also characteristic of apoptosis in plant cells. Above all, internucleosomal cleavage of DNA has been observed in several instances (Wang et al., 1996a; Orzaez and Granell, 1997; Katsuhara, 1997; McCabe et al., 1997; Young et al., 1997; Koch et al., 1998), including leaf senescence (Yen and Yang, 1998) and differentiation of tracheids (Mittler and Lam, 1995a).

For the study of cell death, the transdifferentiation of cultured mesophyll cells of *Zinnia elegans* into tracheary elements (Kohlenbach and Schmidt, 1975) offers unique opportunities of biochemical and molecular analysis (review: Fukuda, 1997). With regard to the apoptotic program, it is of particular interest that several proteases and nucleases are synthesized during the last stage of differentiation preceding autolysis (Thelen and Northcote, 1989; Ye and Varner, 1993; Minami and Fukuda, 1995; Beers and Freeman, 1997). These hydrolases are likely to be located in the vacuole. The intensification of the digestive potential in the vacuole is followed by the disruption of the tonoplast and subsequent rapid disappearance of all structured components of the cytoplasm (Fukuda, 1997). Hence, the decay of the tonoplast appears to mark the abrupt transition from life to death.

It has recently been demonstrated that Ca^{2+} plays a role in the signal transduction cascade associated with the induction of cell death (Grover and Jones, 1999; Navarre and Wolpert, 1999). In the case of tracheid differentiation, a critical role in the execution of cellular collapse has been attributed to the secretion of a serine protease which, during secondary wall formation, appears to play a role in the dissolution of the extracellular matrix (Groover and Jones, 1999). In suspension-cultured soybean cells apoptosis, as induced by exogenous hydrogen peroxide, has been shown to be associated with the activation of cysteine proteases; transformation of cells with a gene encoding a specific inhibitor protein of cysteine proteases caused a marked decrease of induced cell death (Solomon et al., 1999). Hence, proteases appear to play an instrumental role in the process of programmed cell death.

The accumulation of digestive enzymes in the vacuole followed by the disruption of the tonoplast may even turn out to be a general principle of apoptosis in plant cells. Thus, a number of genes associated with senescence or with hypersensitive response encode proteases and nucleases (Taylor et al., 1993; Lohman et al., 1994; Mittler and Lam, 1995b; Smart et al., 1995; Drake et al., 1996; Buchanan-Wollaston and Ainsworth, 1997; Xu and Chye, 1999). The same phenomenon has been observed with regard to the activities of hydrolytic enzymes (review: Matile, 1978) as well as to the disruption of the tonoplast (Matile and Winkenbach, 1971; Berjak and Lawton, 1973; Inada et al., 1998). It is also conceivable that cell wall lytic enzymes which are responsible for tissue disintegration, are first accumulated in vacuoles and released upon the collapse of lysosomal compartmentation. Examples of this kind of apoptotic program are represented by the short-lived petals of daylily (Panavas et al., 1998), by lysigenous aerenchyma formation in corn roots (He et al., 1994; Saab and Sachs, 1996), and by rapid decomposition of the fruiting body in *Coprinus* (Iten and Matile, 1970).

The expression of genes encoding apoptotic proteins is regulated by hormones such as ethylene (He et al., 1996; Navarre and Wolpert, 1999) gibberellin and abscisic acid (Wang et al., 1996b) or brassinosteroids (Yamamoto et al., 1997) depending on the system. In animal cells, the genetic dissection of apoptosis has lead to the identification of suppressor genes. A homolog of one of them, DAD-1, has recently been discovered in the genomes of rice (Tanaka et al., 1997) and apple (Dong et al., 1998). It is very intriguing that results from in situ hybridization suggest this gene being expressed mainly in the vascular bundles, most likely in the phloem which still has a function in nutrient recycling when senescence is completed in adjacent tissue and autolysis is induced.

C. Autolysis and Nutrient Recycling

Apoptotic development in fully senesced cells may be important for the extensive withdrawal of nutrients. The hydrolases which are released from disrupted vacuoles will do away quickly with all digestible remains of the cytoplasm. Neighboring cells that are still viable may take up solutes and store them in the vacuole until they autolyse likewise. A hypothesis of nutrient recycling which assumes that cell autolysis in the senesced mesophyll occurs asynchronously, is illustrated in Fig. 6 (Matile, 1997). The efficiency of solute withdrawal from a cell as taking place via autolysis, accumulation in the neighboring viable cells and transport to the phloem, may be very high. Nevertheless, in fully senesced leaves the residual protein accounts for some 10–40% of total protein in mature leaves. This is probably a consequence of differential senescence in the various cell phenotypes;

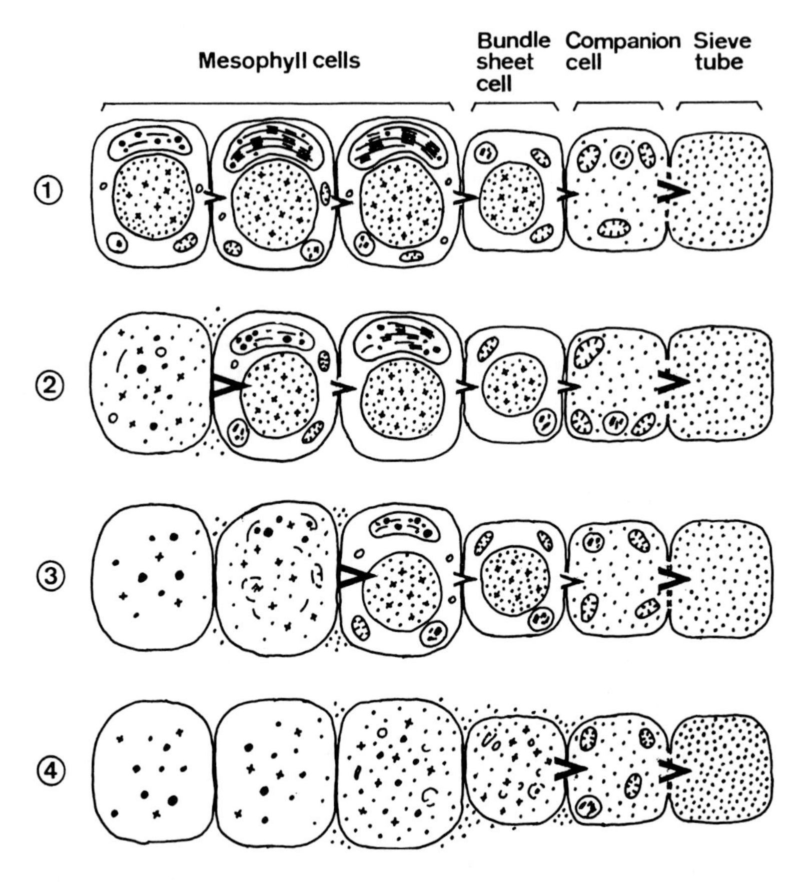

Fig. 6. Withdrawal of solutes from senescent mesophyll as associated with cell autolysis. It is assumed that solutes remain accumulated in senescing cell until autolysis takes place. Solutes released from autolysing cells will diffuse into the apoplast and may be absorbed by neighboring viable cells. Digestion of residual cytoplasmic components by the vacuolar hydrolases is another important consequence of the collapse of subcellular compartmentation. The resulting solutes will also be available for reabsorption and eventual loading into the sieve tubes. Solutes may also be directed to the phloem via the symplastic route. Rows 1 to 4 indicate time-dependent development in the leaf tissues, starting with incipient senescence in row 1 and ending with the final phase of the senescence period in row 4. Redrawn from Matile, 1997).

the epidermal cells may account for the bulk of residual protein.

V. Outlook

It is easy to predict that, in the near future, the understanding of senescence and death will be improved primarily through research on genes and gene expression. The number of genes identified as being expressed senescence-specifically is likely to increase considerably and certainly faster than the number of genes with clearly defined functions. Hence, plant biochemistry will probably also play an important role because the easy identification of gene functions by means of searching for homologous genes with known functions will not always produce useful results. The critical items of future research may be not so much the gene structures spat out by the sequencing machines but rather novel concepts that may lead to the discovery of novel mechanisms, as for example regarding regulation of senescence. It is amazing that so many proteases are being discovered, not only in the context of senescence and programmed death. Metabolism depends so much on the intactness of proteins and appears to take place in an apparently hostile world full of proteases. Undoubtedly, some additional knowledge about regulation of orderly proteolysis in chloroplasts is still required for the understanding of this as well as many other aspects of senescence.

References

Adams WW, Winter K, Schreiber U and Schramel (1990) Photosynthesis and chlorophyll fluorescence characteristics in relationship to changes in pigment and element composition of leaves of *Platanus occidentalis* L. during autumnal leaf senescence. Plant Physiol 92: 1184–1190

Afitlhile MM, Dent RM and Cowan AK (1993) Changes in carotenoid composition in senescing leaves of *Hordeum vulgare* L. cv. Dyan. J Plant Physiol 142:43–49

Akhtar MS, Goldschmidt EE, John I, Rodoni S, Matile P and Grieson D (1999) Altered patterns of senescence and ripening in *gf*, a stay-green mutant of tomato (*Lycopersicon esculentum* Mill.). J Exp Bot 50: 1115–1122

Amir-Shapira D, Goldschmidt EE and Altman A (1986) Autolysis of chlorophyll in aqueous and detergent suspensions of chloroplast fragments. Plant Sci 43: 201–206

Bachmann A, Fernandez-Lopez J, Ginsburg S, Thomas H, Bouwkamp JC, Solomos T and Matile P (1994) Stay-green genotypes of *Phaseolus vulgaris* L - chloroplast proteins and chlorophyll catabolites during foliar senescence. New Phytol 126: 593–600

Barbato R, LaRocca N and Rascio N (1995) Identification and characterization of D1 and D2 protein breakdown fragments in cotyledon thylakoids from *Ceratonia siliqua* L. J Plant Physiol 147: 168–174

Barry P, Evershed RP, Young A, Prescott MC and Britton G (1992) Characterization of carotenoid acyl esters produced in drought-stressed barley seedlings. Phytochemistry 31: 3163–3168

Beers EP and Freeman TB (1997) Proteinase activity during tracheary element differentiation in *Zinnia* mesophyll cultures. Plant Physiol 113: 873–880

Ben-David H, Nelson N and Gepstein S (1983) Differential changes in the amount of protein complexes in the chloroplast membrane during senescence of oat and bean leaves. Plant Physiol 73: 507–510

Berjak P and Lawton JR (1973) Prostelar autolysis: A further example of a programmed senescence. New Phytol 72: 625–637

Biswal UC, Bergfeld R and Kasemir H (1983) Phytochrome-mediated delay of plastid senescence in mustard cotyledons: Changes in pigment contents and ultrastructure. Planta 157: 85–90

Blank A and McKeon TA (1991) Expression of three RNase activities during natural and dark-induced senescence in wheat leaves. Plant Physiol 97: 1409–1413

Bortlik K (1990) Chlorophyllabbau: Charakterisierung von Kataboliten in seneszenten Gerstenblättern. Ph.D. Thesis, University of Zürich

Buchanan-Wollaston V (1997) The molecular biology of leaf senescence. J Exp Bot 48: 181–199

Buchanan-Wollaston V and Ainsworth C (1997) Leaf senescence in *Brassica napus*: Cloning of senescence related genes by subtractive hybridisation. Plant Mol Biol 33: 821–834

Cheung AY, McNellis T and Piekos B (1993) Maintenance of chloroplast components during chromoplast differentiation in the tomato mutant green flesh. Plant Physiol 101: 1223–1229

Crafts-Brandner SJ, Below FE, Harper JE and Hageman RH (1984) Effects of pod removal on metabolism and senescence of nodulating and nonnodulating soybean isolines. II. Enzymes and chlorophyll. Plant Physiol 75: 318–322

Crafts-Brandner SJ, Klein RR, Klein P, Hölzer R and Feller U (1996) Coordination of protein and mRNA abundances of stromal enzymes and mRNA abundance of the Clp protease subunits during senescence of *Phaseolus vulgaris* (L.) leaves. Planta 200: 312–318

Curty C and Engel N (1996) Detection, isolation and structure elucidation of a chlorophyll *a* catabolite from autumnal senescent leaves of *Cercidiphyllum japonicum*. Phytochemistry 42: 1531–1536

Dalton DA (1995) Antioxidant defense of plants and fungi. In: Ahmad S (ed) Oxidant-Induced Stress and Antioxidant Defenses in Biology, pp 298–355. Chapman and Hall, New York

DeBellis L and Nishimura M (1991) Development of enzymes of the glyoxylate cycle during senescence of pumpkin cotyledons. Plant Cell Physiol 32: 555–561

DeBellis L, Picciarelli P, Pistelli L and Alpi A (1990) Localization of glyoxylate-cycle marker enzymes in peroxisomes of senescent leaves and green cotyledons. Planta 180: 435–439

Desimone M, Wagner E and Johanningmeier U (1998) Degradation of active-oxygen-modified ribulose-1,5-bisphosphate carboxylase/oxygenase by chloroplastic proteases requires ATP-hydrolysis. Planta 205: 459–466

Dong Y-H, Zhan X-C, Kvarnheden A, Atkinson RG, Morris BA and Gardner RC (1998) Expression of a cDNA from apple encoding a homologue of DAD1, an inhibitor of programmed cell death. Plant Sci 139: 165–174

Drake R, John I, Farrell A, Cooper W, Schuch W and Grierson D (1996) Isolation and analysis of cDNAs encoding tomato cysteine proteases expressed during leaf senescence. Plant Mol Biol 30: 755–767

Düggelin T, Bortlik K, Gut H, Matile P and Thomas H (1988) Leaf senescence in a non-yellowing mutant of *Festuca pratensis*: accumulation of lipofuscin-like compounds. Physiol Plant 74: 131–136

Elstner EF (1987) Metabolism of activated oxygen species. In: Davies DD (ed) The Biochemistry of Plants, Vol 11, pp 253–315. Academic Press, New York

Ford DM and Schibles R (1988) Photosynthesis and other traits in relation to chloroplast number during soybean senescence. Plant Physiol 86: 108–111

Friend J and Mayer AM (1960) The enzymic destruction of carotenoids by isolated chloroplasts. Biochem Biophys Acta 41: 422–429

Fröhlich V and Feller U (1991) Effect of phloem interruption on senescence and protein remobilization in the flag leaf of field-grown wheat. Biochem Physiol Pflanzen 187: 139–147

Fryer MJ (1992) The antioxidant effects of thylakoid vitamin E (a-tocopherol). Plant Cell Environ 15: 381–392

Fukuda H (1996) Xylogenesis: Initiation, progression and cell death. Annu Rev Plant Physiol Plant Mol Biol 47: 299–325

Fukuda H (1997) Tracheary element differentiation. Plant Cell 9: 1147–1156

Gan S and Amasino RM (1995) Inhibition of leaf senescence by autoregulated production of cytokinin. Science 270: 1986–1988

Gan S and Amasino RM (1997) Making sense of senescence. Molecular genetic regulation and manipulation of leaf senescence. Plant Physiol 113: 313–319

Gavrieli Y, Sherman Y and Ben-Sasson SA (1992) Identification of programmed cell death in situ via specific labeling of nuclear DNA fragmentation. J Cell Biol 119: 493–501

Ginsburg S and Matile P (1993) Identification of catabolites of chlorophyll-porphyrin in senescent rape cotyledons. Plant Physiol 102: 521–527

Graham IA, Leaver CJ and Smith (1992) Induction of malate synthase gene expression in senescent and detached organs of cucumber. Plant Cell 4: 349–357

Graham IA, Denby KJ and Leaver CJ: (1994) Carbon catabolite repression regulates glyoxylate cycle gene expression in cucumber. Plant Cell 6: 761–772

Greening MT, Butterfield FJ and Harris N (1982) Chloroplast ultrastructure during senescence and regreening of flax cotyledons. New Phytol 92: 279–285

Groover A and Jones AM (1999) Tracheary element differentiation uses a novel mechanism coordinating programmed cell death and secondary wall synthesis. Plant Physiol 119: 375–384

Grover A (1993) How do senescing leaves lose photosynthetic activity. Curr Sci 64: 226–234

Guiamét JJ and Giannibelli MC (1994) Inhibition of the degradation of chloroplast membranes during senescence in nuclear 'stay-green' mutant of soybean. Physiol Plant 91: 395–402

Guiamét JJ, Schwartz E, Pichersky E and Noodén LD (1991) Characterization of cytoplasmic and nuclear mutations affecting chlorophyll and chlorophyll-binding proteins during senescence in soybean. Plant Physiol 96: 227–231

Gut H and Matile P (1988) Apparent induction of key enzymes of the glyoxylic acid cycle in senescent barley leaves. Planta 176: 548–550

Gut H and Matile P (1989) Breakdown of galactolipids in senescent barley leaves. Bot Acta 102: 31–36

Gut H, Rutz C, Matile P and Thomas H (1987) Leaf senescence in a non-yellowing mutant of *Festuca pratensis*: Degradation of carotenoids. Physiol Plant 70: 659–663

Halliwell B and Gutteridge JMC (1989) Free radicals in biology and medicine. 2nd ed. Clarendon Press, Oxford

Halperin T and Adam Z (1996) Degradation of mistargeted OEE33 in the chloroplast stroma. Plant Mol Biol 30: 925–933

Hanson AD and Kende H (1975) Ethylene-enhanced ion and sucrose efflux in morning glory flower tissue. Plant Physiol 55: 663–669

He C-J, Drew MC and Morgan PW (1994) Induction of enzymes associated with lysigenous aerenchyma formation in roots of *Zea mays* during hypoxia or nitrogen starvation. Plant Physiol 105: 861–865

He C-J, Morgan PW and Drew MC (1996) Transduction of an ethylene signal is required for cell death and lysis in the root cortex of maize during aerenchyma formation induced by hypoxia. Plant Physiol 112: 463–472

Hensel LL, Grbic V, Baumgarten DA and Bleeker AB (1993) Developmental and age-related processes that influence the longevity and senescence of photosynthetic tissues in *Arabidopsis*. Plant Cell 5: 553–564

Herrmann G, Lehmann J, Peterson A, Sembdner G, Weidhase RA and Parthier B (1989) Species and tissue specificity of jasmonate-induced abundant proteins. J Plant Physiol 134: 703–709

Hilditch P (1986) Immunological quantification of the chlorophyll *a/b* binding protein in senescing leaves of *Festuca pratensis* Huds. Plant Sci 45: 95–99

Hilditch P, Thomas H and Rogers LJ (1986) Two processes for the breakdown of the Q_B protein of chloroplasts. FEBS Lett 208: 313–316

Holloway PJ, Maclean DJ and Scott KJ (1983) Rate-limiting steps of electron transport in chloroplasts during ontogeny and senescence of barley. Plant Physiol 72: 795–801

Hörtensteiner S (1999) Chlorophyll breakdown in higher plants. Cell Mol Life Sci 56: 330–347

Hörtensteiner S, Vicentini F and Matile P (1995) Chlorophyll breakdown in senescent leaves: Enzymic cleavage of phaeophorbide *a* in vitro. New Phytol 129: 237–246

Huffaker RC (1990) Proteolytic activity during senescence of plants. New Phytol 116: 199–231

Humbeck K and Krupinska K (1996) Does the Clp protease play a role during senescence-associated protein degradation in barley leaves? J Photochem Photobiol B 36: 321–326

Ilami G, Nespoulous C, Huet J-C, Vartanian N and Pernollet J-C (1997) Characterization of BnD22, a drought-induced protein expressed in *Brassica napus* leaves. Phytochemistry 45: 1–8

Inada N, Sakai A, Kuriowa H and Kuriowa T (1998) Three-dimensional analysis of the senescence program in rice (*Oryza sativa* L.) coleoptiles. Planta 205: 153–164

Ishida H, Shimizu S, Makino A and Mae T (1998) Light-dependent fragmentation of the large subunit of ribulose-1,5-bisphosphate carboxylase/oxygenase in chloroplasts isolated from wheat leaves. Planta 204: 305–309

Iten W and Matile P (1970) Role of chitinase and other lysosomal enzymes of *Coprinus lagopus* in the autolysis of fruiting bodies. J Gen Microbiol 61: 301–309

Jackowski G, Li DS and Schneider J (1991) Immunoquantitation of the apoprotein of the light-harvesting chlorophyll-*a/b* protein complex of Photosystem II in senescing leaves as influenced by cytokinin. Acta Physiol Plant 13: 263–270

Jiang C-Z, Rodermel SR and Shibles RM (1993) Photosynthesis, Rubisco activity and amount, and their regulation by transcription in senescing soybean leaves. Plant Physiol 101: 105–112

Joyard J and Douce R (1987) Galactolipid synthesis. In: Stumpf PK (ed) The Biochemistry of Plants, 9: 215–274. Academic Press, New York

Kamber L and Feller U (1998) Influence of the activation status and of ATP on phosphoribulokinase degradation. J Exp Bot. 49: 139–144

Kamimura Y, Mori T, Yamasaki T and Katoh S (1997) Isolation, properties and a possible function of a water-soluble chlorophyll *a/b*-protein from Brussels sprouts. Plant Cell Physiol 38: 133–138

Kar M (1986) The effect of photoperiod on chlorophyll loss and lipid peroxidation in excised senescing rice leaves (1986) J Plant Physiol 123: 389–393

Kar RK and Choudhuri MA (1985) Senescence in *Hydrilla* leaves in light and darkness. Physiol Plant 63: 225–230

Katsuhara M (1997) Apoptosis-like cell death in barley roots under salt stress. Plant Cell Physiol 38: 1091–1093

Kawasaki S and Takeuchi J (1989) Senescence-induced, thylakoid-bound diisopropylfluorophosphate-binding protein in spinach. Plant Physiol 90: 338–344

Kessler F, Schnell D and Blobel G (1999) Isolation of plastoglobules from pea (*Pisum sativum* L.) chloroplasts and identification of associated proteins. Planta 208: 107–113

Klerk H, Tophof S and Van Loon LC (1992) Synthesis of proteins during the development of the first leaf of oat (*Avena sativa*). Physiol Plant 85: 595–605

Kleber-Janke T, and Krupinska K (1997) Isolation of cDNA clones for genes showing enhanced expression in barley leaves during dark-induced senescence as well as during senescence under field conditions. Planta 203: 332–340

Koch W, Wagner C and Seitz HU (1998) Elicitor-induced cell death and phytoalexin synthesis in *Daucus carota* L. Planta 206: 523–532

Kohlenbach HW and Schmidt B (1975) Cytodifferenzierung in Form einer direkten Umwandlung isolierter Mesophyllzellen zu Tracheiden. Z Pflanzenphysiol 75: 369–374

Knee M (1991) Role of ethylene in chlorophyll degradation in radish cotyledons. J Plant Growth Regul 10: 157–162

Koiwai A, Matsuzaki T, Suzuki F and Kawashima N (1981) Changes in total polar lipids and their fatty acid composition in tobacco leaves during growth and senescence. Plant Cell Physiol 22: 1059–1065

Kulkarni VJ and Schwabe WW (1985) Graft transmission of longday-induced leaf senescence in *Kleinia articulata*. J Exp Bot 36: 1620–1633

Kuwabara T and Suzuki K (1994) A prolyl endoproteinase that acts specifically on the extrinsic 18-kDa protein of Photosystem II: Purification and further characterization. Plant Cell Physiol 35: 665–675

Lamattina L, Pont-Lezica R and Conde RD (1985) Protein metabolism in senescing wheat leaves. Determination of synthesis and degradation rates and their effects on protein loss. Plant Physiol 77: 587–590

Landolt R and Matile P (1990) Glyoxisome-like microbodies in senescent spinach leaves. Plant Sci 72: 159–163

Laurière C (1983) Enzymes and leaf senescence. Physiol Vég 21: 1159–1177

Lehmann J, Atzorn R, Brücker C, Reinbothe S, Leopold J, Wasternak C and Parthier B (1995) Accumulation of jasmonate, abscisic acid, specific transcripts and proteins in osmotically stressed barley leaf segments. Planta 197: 156–162

Lohman KN, Gan S, John MC and Amasino RM (1994) Molecular analysis of natural leaf senescence in *Arabidopsis thaliana*. Physiol Plant 92: 322–328

Lu C and Zhang J (1998) Modifications in Photosystem II photochemistry in senescent leaves of maize plants. J Exp Bot 49: 1671–1679

Mae F, Makino A and Ohira K (1987) Carbon fixation changes with senescence in rice leaves. In: Thomson WW, Nothnagel EA and Huffaker RC (eds) Plant Senescence: Its Biochemistry and Physiology, pp 123–131. American Society of Plant Physiologists, Rockville

Marek LF and Stewart CR (1992) Photosynthesis and photorespiration in presenescent, senescent, and rejuvenated soybean cotyledons. Plant Physiol 98: 694–699

Martin SJ, Green DR and Cotter TG (1994) Dicing with death: dissecting the components of the apoptosis machinery. Trends Biolo Sci 19: 26–30

Martinoia E, Heck U, Dalling MJ and Matile P (1983) Changes in chloroplast number and chloroplast constituents in senescing barley leaves. Biochem Physiol Pflanzen 178: 147–155

Matile P (1978) Biochemistry and function of vacuoles. Annu Rev Plant Physiol 29: 193–213

Matile P (1992) Chloroplast senescence. In: Baker NR and Thomas H (eds) Crop Photosynthesis: Spatial and Temporal Determinants, pp 413–440. Elsevier, Amsterdam

Matile P (1997) The vacuole and cell senescence. Adv Bot Res 25: 87–112

Matile P and Winkenbach F (1971) Function of lysosomes and lysosomal enzymes in the senescing corolla of the morning glory (*Ipomoea purpurea*). J Exp Bot 22: 759–771

Matile P and Martinoia E (1982) Catabolism of carotenoids: involvement of peroxidase? Plant Cell Rep 1: 244–246

Matile P, Flach B and Eller BM (1992) Spectral optical properties, pigments and optical brighteners in autumn leaves of *Ginkgo biloba* L. Bot Acta 105: 13–17

Matile P, Schellenberg M and Vicentini F (1997) Localization of chlorophyllase in the chloroplast envelope. Planta 201: 96–99

Matile P, Hörtensteiner S, and Thomas H (1999) Chlorophyll degradation. Annu Rev Plant Physiol Plant Mol Biol 50: 67–95

Mattoo AK, Marder JB and Edelman M (1989) Dynamics of the Photosystem II reaction center. Cell 56: 241–246

Maunders MJ and Brown SB (1983) The effect of light on chlorophyll loss in senescing leaves of sycamore (*Acer pseudoplatanus* L.) Planta 158: 309–311

McCabe PF, Levine A, Meijer P-J, Tapon NA and Pennell RI (1997) A programmed cell death pathway activated in carrot cells cultured at low cell density. Plant J 12: 267–280

McLaughlin JC and Smith SM (1995) Glyoxylate cycle enzyme synthesis during the irreversible phase of senescence of cucumber cotyledons. J Plant Physiol 146: 133–138

Mehta RA, Fawett TW, Porath D and Mattoo AK (1992) Oxidative stress causes rapid membrane translocation and in vivo degradation of ribulose-1,5-bisphosphate carboxylase (oxygenase). J Biol Chem 267: 2810–2816

Mei HS and Thimann KV (1984) The relation between nitrogen deficiency and leaf senescence. Physiol Plant 62: 157–161

Meir S and Philosoph-Hadas S (1995) Metabolism of polar lipids during senescence of watermelon leaves. Plant Physiol Biochem 33: 241–249

Merchant S and Bogorad L (1986) Rapid degradation of apoplastocyanin in Cu(II)-deficient cells of *Chlamydomonas reinhardtii*. J Biol Chem 261: 15850–15853

Minami A and Fukuda H (1995) Transient and specific expression of a cysteine endopeptidase associated with autolysis during differentiation of *Zinnia* mesophyll cells into tracheary elements. Plant Cell Physiol 36: 1599–1606

Mitsuhashi W and Feller U (1992) Effects of light and external solutes on the catabolism of nuclear-encoded stromal proteins in intact chloroplasts isolated from pea leaves. Plant Physiol 100: 2100–2105

Mitsuhashi W, Crafts-Brandner SJ and Feller U (1992) Ribulose-1,5-bis-phosphate carboxylase/oxygenase degradation in isolated pea chloroplasts incubated in the light or in the dark. J Plant Physiol 139: 653–658

Mittler R and Lam E (1995a) In situ detection of nDNA fragmentation during the differentiation of tracheary elements in higher plants. Plant Physiol 108: 489–493

Mittler R and Lam E (1995b) Identification, characterization, and purification of a tobacco endonuclease activity induced upon hypersensitive response cell death. Plant Cell 7: 1951–1962

Mothes K (1960) Über das Altern der Blätter und die Möglichkeit ihrer Wiederverjüngung. Naturwissenschaften 47: 337–351

Mothes K and Baudisch W (1958) Untersuchungen über die Reversibilität der Ausbleichung grüner Blätter. Flora 146: 521–531

Mühlecker W, Ongania K-H, Kräutler B, Matile P and Hörtensteiner S (1997) Tracking down chlorophyll breakdown in plants: elucidation of the constitution of a fluorescent chlorophyll catabolite. Angew Chem Int Ed Engl 36: 401–404

Navarre DA and Wolpert TJ (1999) Victorin induction of an apoptotic/senescene-like response in oats. Plant Cell 11: 237–249

Nishimura M, Takeuchi Y, deBellis L and Hara-Nishimura I (1993) Leaf peroxisomes are directly transformed to glyoxisomes during senescence of pumpkin cotyledons. Protoplasma 175: 131–137

Nii N, Kawano S, Nakamura S and Kuroiwa T (1988) Changes in the fine structure of chloroplast and chloroplast DNA of peach leaves during senescence. J Japan Soc Hort Sci 57: 390–398

Nishio N and Satoh H (1997) A water-soluble chlorophyll protein in cauliflower may be identical to BnD22, a drought-induced, 22 kilodalton protein in rapeseed. Plant Physiol 115: 841–846

Nock LP, Rogers LJ and Thomas H (1992) Metabolism of protein and chlorophyll in leaf tissue of *Festuca pratensis* during chloroplast assembly and senescence. Phytochemistry 31: 1465–1470

Noodén LD and Guiamet JJ (1989) Regulation of assimilation and senescence by the fruit in monocarpic plants. Physiol Plant 77: 267–274

Noodén LD and Letham DS (1993) Cytokinin metabolism and signalling in the soybean plant. Aust J Plant Physiol 20: 639–653

Noodén LD, Hillsberg JW and Schneider MJ (1996) Induction of leaf senescence in *Arabidopsis thaliana* by long days through a light-dosage effect. Physiol Plant 96: 491–495

Okada K, Inoue Y, Satoh K and Katoh S (1992) Effects of light on degradation of chlorophyll and proteins during senescence of detached rice leaves. Plant Cell Physiol 33: 1183–1191

Orzaez D and Granell A (1997) DNA fragmentation is regulated by ethylene during carpel senescence in *Pisum sativum*. Plant J 11: 137–144

Panavas T and Rubinstein B (1998) Oxidative events during programmed cell death of daylily (*Hemerocallis hybrid*) petals. Plant Sci 133: 125–138

Panavas T, Reid PD and Rubinstein B (1998) Programmed cell death of daylily petals: Activities of wall-based enzymes and effects of heat shock. Plant Physiol Biochem 36: 379–388

Park H and Hoober JK (1997) Chlorophyll synthesis modulates retention of apoproteins of light-harvesting complex II by the chloroplasts in *Chlamydomonas reinhardtii*. Physiol Plant 101: 135–142

Parthier B (1988) Gerontoplasts—the yellow end in the ontogenesis of chloroplasts. Endocyt Cell Res 5: 163–190

Pastori GM and DelRio LA (1994) An activated-oxygen-mediated role for peroxisomes in the mechanism of senescence of *Pisum sativum* L. leaves. Planta 193: 385–391

Peterson LW and Huffaker RC (1975) Loss of ribulose 1,5-diphosphate carboxylase and increase in proteolytic activity during senescence of detached primary barley leaves. Plant Physiol 55: 1009–1015

Philosoph-Hadas S, Meir S and Aharoni N (1994) Role of ethylene in senescence of watercress leaves. Physiol Plant 90: 553–559

Picher M, Grenier G, Purcell M, Proteau L and Beaumont G (1993) Isolation and purification of intralamellar vesicles from *Lemna minor* L. chloroplasts. New Phytol 123: 657–663

Pistelli L, DeBellis L and Alpi A (1991) Peroxisomal enzymes activities in attached senescing leaves. Planta 184: 151–153

Pistelli L, Perata P and Alpi A (1992) Effect of leaf senescence on glyoxylate cycle enzyme activities. Aust J Plant Physiol 19: 723–729

Plumley FG and Schmidt GW (1995) Light-harvesting chlorophyll *a/b* complexes: Interdependent pigment synthesis. Plant Cell 7: 689–704

Reinbothe S, Reinbothe C, Holtorf H and Apel K (1995a) Two NADP:protochlorophyllide oxidoreductases in barley: evidence for the selective disappearance of POR A during light-induced greening of etiolated seedlings. Plant Cell 7: 1933–1940

Reinbothe C, Apel K and Reinbothe S (1995b) A light-induced protease from barley plastids degrades NADPH:protochlorophyllide oxidoreductase complexed with chlorophyllide. Mol Cell Biol 15: 6206–6212

Rise M, Cojocaru M, Gottlieb HE and Goldschmidt EE (1989) Accumulation of α-tocopherol in senescing organs as related to chlorophyll degradation. Plant Physiol 89: 1028–1030

Roberts DR, Thompson JE, Dumbroff EB, Gepstein S and Mattoo AK (1987) Differential changes in the synthesis and steady-state levels of thylakoid proteins during bean leaf senescence. Plant Mol Biol 9: 343–353

Rodoni S, Mühlecker W, Anderl M, Kräutler B, Moser D, Thomas H, Matile P and Hörtensteiner S (1997) Chlorophyll breakdown in senescent chloroplasts. Cleavage of pheophorbide *a* in two enzymic steps. Plant Physiol 115: 669–676

Rodoni S, Schellenberg M and Matile P (1998) Chlorophyll breakdown in senescing barley leaves as correlated with phaeophorbide *a* oxygenase activity. J Plant Physiol 152: 139–144

Romani RJ (1987) Mitochondrial activity during senescence. In: Thomson WW, Nothnagel EA and Huffaker RC (eds.) Plant Senescence: Its Biochemistry and Physiology, pp 81–88. The American Society of Plant Physiologists, Rockville

Ronning CM, Bowkamp JC and Solomos T (1991) Observations on the senescence of a mutant non-yellowing genotype of *Phaseolus vulgaris* L. J Exp Bot 42: 235–241

Roulin S and Feller U (1997) Light-induced proteolysis of stromal proteins in pea (*Pisum sativum* L.) chloroplasts: requirement for intact organelles. Plant Sci 128:31–41

Roulin S and Feller U (1998) Dithiothreitol triggers photooxidative stress and fragmentation of the large subunit of ribulose-1,5-bisphosphate carboxylase/oxygenase in intact pea chloroplasts. Plant Physiol Biochem 36: 849–856

Rousseaux MC, Ballaré CL, Jordan ET and Vierstra RD (1997) Directed overexpression of PHY A locally suppresses stem elongation and leaf senescence responses to far-red radiation. Plant Cell Environ 20: 1551–1558

Saab IN and Sachs MM (1996) A flooding-induced xyloglucan endotransglycosylase homologue in maize is responsive to ethylene and associated with aerenchyma. Plant Physiol 112: 385–391

Sacher JA (1973) Senescence and postharvest physiology. Annu Rev Plant Physiol 24: 197–224

Satoh H, Nakyama K and Okada M (1998) Molecular cloning and functional expression of a water-soluble chlorophyll-protein, a putative carrier of chlorophyll molecules in cauliflower. J Biol Chem 273: 30568–30575

Scheumann V, Schoch S and Rüdiger W (1998) Chlorophyll *a* formation in the chlorophyll *b* reductase reaction requires reduced ferredoxin. J Biol Chem 273: 35102–35108

Scheumann V, Schoch S and Rüdiger W (1999) Chlorophyll *b* reduction during senescence of barley seedlings. Planta 209: 364–370

Schmidt GW and Mishkind ML (1983) Rapid degradation of unassembled ribulose 1,5-bisphosphate carboxylase small subunits in chloroplasts. Proc Natl Acad Sci USA 80: 2632–2636

Schmidt HO (1988) The structure and function of grana-free thylakoid membranes in gerontoplasts of senescent leaves of *Vicia faba* L. Z Naturforsch 43c: 149–154

Sembdner G and Parthier B (1993) The biochemistry and the physiological and molecular actions of jasmonates. Annu Rev Plant Physiol Plant Mol Biol 44: 569–589

Singh S, Letham DS and Palni LMS (1992) Cytokinin biochemistry in relation to leaf senescence. VII. Endogenous cytokinin levels and exogenous applications of cytokinins in relation to sequential leaf senescence of tobacco. Physiol Plant 86: 388–397

Sitte P, Falk H and Liedvogel B (1980) Chromoplasts. In: Czygan F-C (ed.) Pigments in Plants, pp 117–148. Gustav Fischer, Stuttgart

Smart CM (1994) Gene expression during leaf senescence. New Phytol 126: 419–448

Smart CM, Scofield SR, Bevan MW and Dyer TA (1991) Delayed leaf senescence in tobacco plants transformed with tmr, a gene for cytokinin production in *Agrobacterium*. Plant Cell 3: 647–656

Smart CM, Hosken SE, Thomas H, Greaves JA, Blair BG and Schuch W (1995) The timing of maize leaf senescence and characterization of senescence-related cDNAs. Physiol Plant 93: 673–682

Sodmergen S, Kawano S, Tano S and Kuriowa T (1989) Preferential digestion of chloroplast nuclei (nucleoids) during senescence of the coleoptile of *Oryza sativa*. Protoplasma 152: 65–68

Sodmergen S, Kawano S, Tano S and Kuriowa T (1991) Degradation of chloroplast DNA in second leaves of rice (*Oryza sativa*) before leaf yellowing. Protoplasma 160: 89–98

Solomon M, Belenghi B, Delledonne M, Menachem E and Levine A (1999) The involvement of cysteine proteases and protease inhibitor genes in the regulation of programmed cell death in plants. Plant Cell 11: 431–443

Tanaka Y, Makishima T, Sasabe M, Ichinose Y, Shiraishi T, Nishimoto T and Yamada T (1997) dad-1, A putative programmed cell death suppressor gene in rice. Plant Cell Physiol 38: 379–383

Taylor CB, Bariola PA, Delcardayre SB, Raines RT and Green PJ (1993) RNS2: A senescence-associated RNase of Arabidopsis that diverged from s-RNases before speciation. Proc Natl Acad Sci USA 90: 5118–5122

Thelen MP and Northcote DH (1989) Identification and purification of a nuclease from *Zinnia elegans* L.: A potential molecular marker for xylogenesis. Planta 179: 181–195

Thoenen M and Feller U (1998) Degradation of glutamine synthetase in intact chloroplasts isolated from pea (*Pisum sativum*) leaves. Aust J Plant Physiol 25: 279–286

Thomas H (1982) Leaf senescence in a non-yellowing mutant of *Festuca pratensis*. I. Chloroplast membrane polypeptides. Planta 154: 212–218

Thomas H (1987) Sid: A Mendelian locus controlling thylakoid membrane disassembly in senescing leaves of *Festuca pratensis*. Theor Appl Genet 73: 551–555

Thomas H (1997) Chlorophyll, a symptom and regulator of plastid development. New Phytol 136: 163–181

Thomas H and Feller U (1993) Leaf development in *Lolium temulentum*: differential susceptibility of transaminase isoenzymes to proteolysis. J Plant Physiol 142: 37–42

Thomas H, Bortlik K, Rentsch D, Schellenberg M and Matile P (1989) Catabolism of chlorophyll in vivo: Significance of polar chlorophyll catabolites in a non-yellowing senescence mutant of *Festuca pratensis* Huds. New Phytol 111: 3–8

Thomas H, Schellenberg M, Vicentini F and Matile P (1996) Gregor Mendel's green and yellow pea seeds. Bot Acta 109: 3–4

Thompson JE, Legge RL and Barber RF (1987) The role of free radicals in senescence and wounding. New Phytol 105: 317–344

Thomson WW and Whatley JM (1980) Development of nongreen plastids. Annu Rev Plant Physiol 31: 375–394

Ueda J, Kato H, Yamane N and Takahashi N (1981) Inhibitory effect of methyl jasmonate and its related compounds on kinetin-induced retardation of oat leaf senescence. Physiol Plant 52: 305–309

Venkatrayappa T, Fletcher RA and Thompson JE (1984) Retardation and reversal of senescence in bean leaves by benzyladenine and decapitation. Plant Cell Physiol 25: 407–418

Vicentini F and Matile P (1993) Gerontosomes, a multifunctional type of peroxisome in senescent leaves. J Plant Physiol 142: 50–56

Vicentini F, Hörtensteiner S, Schellenberg M, Thomas H and Matile P (1995) Chlorophyll breakdown in senescent leaves: Identification of the lesion in a stay-green genotype of *Festuca pratensis*. New Phytol 129: 247–252

Wang H, Li J, Bostock RM and Gilchrist DG (1996a) Apoptosis: A functional paradigm for programmed cell death induced by host-selective phytotoxin and invoked during development. Plant Cell 8: 375–391

Wang M, Oppedijk BJ, Lu X, vanDuijn B and Schilperoort RA (1996b) Apoptosis in barley aleurone during germination and its inhibition by abscisic acid. Plant Mol Biol 32: 1125–1134

Wanner L, Keller F and Matile P (1991) Metabolism of radiolabelled galactolipids in senescent barley leaves. Plant Sci 78: 199–206

Warman TW and Solomos T (1988) Ethylene production and action during foliage senescence in *Hedera helix* L. J Exp Bot 39: 685–694

Weidhase RA, Lehmann J, Kramell H, Semdner G and Parthier B (1987) Degradation of ribulose-1,5-bisphosphate carboxylase and chlorophyll in senescing barley leaf segments triggered by jasmonic acid methylester and counteraction by cytokinin. Physiol Plant 69: 161–166

Wingler A, von Schaewen A, Leegood RC, Lea PJ and Quick WP (1998) Regulation of leaf senescence by cytokinin, sugars, and light. Effects on NADH-dependent hydroxypyruvate reductase. Plant Physiol 116: 329–335

Wittenbach VA (1982) Effect of pod removal on leaf senescence in soybean. Plant Physiol 70: 1544–1548

Woolhouse HW and Jenkins GI (1983) Physiological responses, metabolic changes and regulation during leaf senescence. In: Dale JE and Milthorpe FL (eds) The Growth and Functioning of Leaves, pp 449–487. Cambridge University Press

Xu F-X and Chye M-L (1999) Expression of cysteine proteinase during developmental events associated with programmed cell death in brinjal. Plant J 17; 321–327

Yamamoto R, Demura T and Fukuda H (1997) Brassinosteroids induce entry into the final stage of tracheary element differentiation in cultured *Zinnia* cells. Plant Cell Physiol 38: 980–983

Yamane K and Ogata R (1995) Effects cycloheximide on physiological parameters of *Gladiolus* florets during growth and senescence. J Jap Soc Hort Sci 64: 411–416

Yamane K, Abiru S, Fujishige N, Sakiyama R and Ogata R (1993) Export of soluble sugars and increase in membrane permeability of *Gladiolus* florets during senescence. J Japan Soc Hort Sci 62: 575–580

Ye Z-H and Varner JE (1993) Gene expression patterns associated with in vitro tracheary element formation from isolated mesophyll cells of *Zinnia elegans*. Plant Physiol 103: 805–813

Yemm EW (1935) Respiration of barley leaves. II. Carbohydrate concentration and CO_2 production in starving leaves. Proc R Soc London ser B 117: 504–525

Yen C-H and Yang C-H (1998) Evidence for programmed cell death during leaf senescence in plants. Plant Cell Physiol 39: 922–927

Yoshida Y (1961) Nuclear control of chloroplast activity in *Elodea* leaf cells. Protoplasma 54: 476–492

Young AJ, Wellings R and Britton G (1991) The fate of chloroplast pigments during senescence of primary leaves of *Hordeum vulgare* and *Avena sativum*. J. Plant Physiol 137: 701–705

Young TE, Gallie DR and DeMason DA (1997) Ethylene-mediated programmed cell death during maize endosperm development of wild-type and shrunken 2 genotypes. Plant Physiol 115: 737–751

Zhi-Yi T, Veierskov B, Park J and Thimann KV (1988) Multiple actions of abscisic acid in senescence of oat leaves. J Plant Growth Regul 7: 213–227

Chapter 17

Dynamics of Photosynthetic CO_2 Fixation: Control, Regulation and Productivity

Steven Gutteridge* and Douglas B. Jordan
DuPont Agricultural Products, Stine Haskell Research Center, PO Box 30, Newark, DE 19714 U.S.A.

Summary

Biotechnology is providing the tools to dissect complex interacting pathways in plants and also the means to alter plant traits based on understanding these interactions. There has been a concerted and systematic effort to apply these techniques to alter the levels of enzymes involved in the photosynthetic carbon reduction cycle. This has been informative in locating those reactions of the cycle that exert significant control over the flux of carbon through the cycle and the subsequent effects on partitioning of the photoassimilate in downstream processes. Biotechnology has also provided some understanding of the complexities in communication between the various compartments of an actively photosynthesizing plant cell in both C3 and C4 variants. This account attempts to provide some insight into the nature of this regulation in terms of the structural features of the enzymes involved and, based on this understanding, reveal where there may be opportunities to influence the productivity of photosynthesis.

*Author for correspondence, email: steven.gutteridge@usa.dupont.com

Eva-Mari Aro and Bertil Andersson (eds): Regulation of Photosynthesis, pp. 297–312.

I. Crop Yields, Land Use and Population Growth

There is little doubt we are experiencing the increasing momentum of another revolution in agriculture as biotechnology opens up access to both the nuclear and plastid genomes of plants as potential recipients of foreign genes from across phylogenetic boundaries. Whereas green revolution(s) have spawned an industry that could, until recently, guarantee steady annual increases in yield of about 2% through use of improved varieties, crop protection products, fertilizers and irrigation, the time is fast approaching (if not already with us) when these increases can no longer be sustained. Present data, in fact, suggest yields of rice, corn and wheat have about reached their maximum (Mann, 1999). With an ever-increasing world population, it is likely per capita production will initially remain constant and thereafter show a steady decline.

Arguably, the biotechnological revolution has arrived just in time with its potential to influence agriculture for decades to come—but can it promise to deliver as significant an impact on crop yields as its forerunners? Previous increases in productivity have been achieved with minimal expansion of agricultural land area. As the present revolution gathers momentum, an interesting competition is about to develop over land, between that used to produce crops for human and animal consumption and the area intended for specialized purposes, such as enhanced traits, biomaterials or industrial commodities. Superimposed on these potentially conflicting uses is the apparent trend of developing countries to shift from grain-based diets to one containing meat; coincidentally often occurring in countries with the largest or most rapidly increasing populations. Conversion of feed into animal bulk is particularly costly which, in turn, increases the demand for high-grade agricultural products. Most simply put, there has to be continued investment in biotechnology to improve both crop yields and traits, especially if present farming practices remain unchanged. We are just beginning to understand the myriad factors and complexities of the interacting systems that influence productivity and yield in plants, systems that are being dissected with the same technology driving the latest revolution.

Abbreviations: 2CA1P – 2′-carboxy arabinitol 1-phosphate; 2CABP – 2′-carboxy arabinitol 1,5-bisphosphate; 3P-glycerate – 3-phospho-D-glycerate; FBPase – fructose-1,6-bisphosphate phosphatase; GAP – glyceraldehyde 3-phosphate; MDH – malate dehydrogenase; NADP-MDH – NADP-malate dehydrogenase; PCR – photosynthetic carbon reduction; PEP – phosphoenolpyruvate; PPdK – pyruvate, Pi dikinase; PRK – phosphoribulokinase; ribulose-P_2 – ribulose 1,5-bisphosphate; Rubisco – ribulose-1,5-bisphosphate carboxylase/oxygenase

II. Photosynthesis—Light, Capture, Action

The photosynthetic carbon reduction (PCR) cycle links the almost instantaneous electronic events of the photochemical light reactions to the synthesis of the principal storage carbohydrates of the plant (Benson and Calvin, 1950). The cycle can be divided into three distinct sectors (Fig. 1). The first one is the complicated and energetically difficult reaction of CO_2 fixation catalyzed by ribulose-1,5-bisphosphate carboxylase/oxygenase (Rubisco) involving the carboxylation of ribulose 1,5-bisphosphate (ribulose-P_2) in the chloroplast stroma. There is then a series of reductive reactions of the 3C products through keto-aldol interchanges leading to 6C precursor of the storage carbohydrate, starch, and the principal transport sugar, sucrose. Finally, completing the cycle is the regeneration of the bisphosphate substrate of Rubisco. Somehow the PCR cycle must be able to maintain the link to the photochemistry, balance the flow of cofactors and intermediates under widely varying conditions (e.g., from extremes of high light and limiting CO_2 to limiting light and abundant CO_2) and not exhaust the plant of essential inorganic ions such as phosphate. The communication between these processes involves the thylakoid localized photochemical centers, the stromal components consuming the products of the light reactions to form carbohydrate, and the cytosol where carbohydrate is converted to transportable form and distributed throughout the plant. If the photorespiratory cycle is also included, then the extent of the network includes both peroxisomes and mitochondria (Tolbert, 1971). The enzyme that catalyses the primary step of CO_2 fixation and initiates photorespiration, Rubisco is replete with opportunities for controlling the flux and output of the PCR cycle.

A. Rubisco

1. Synthesis and Assembly

In higher plants, Rubisco assembly requires the coordinate expression of genes located in the nucleus

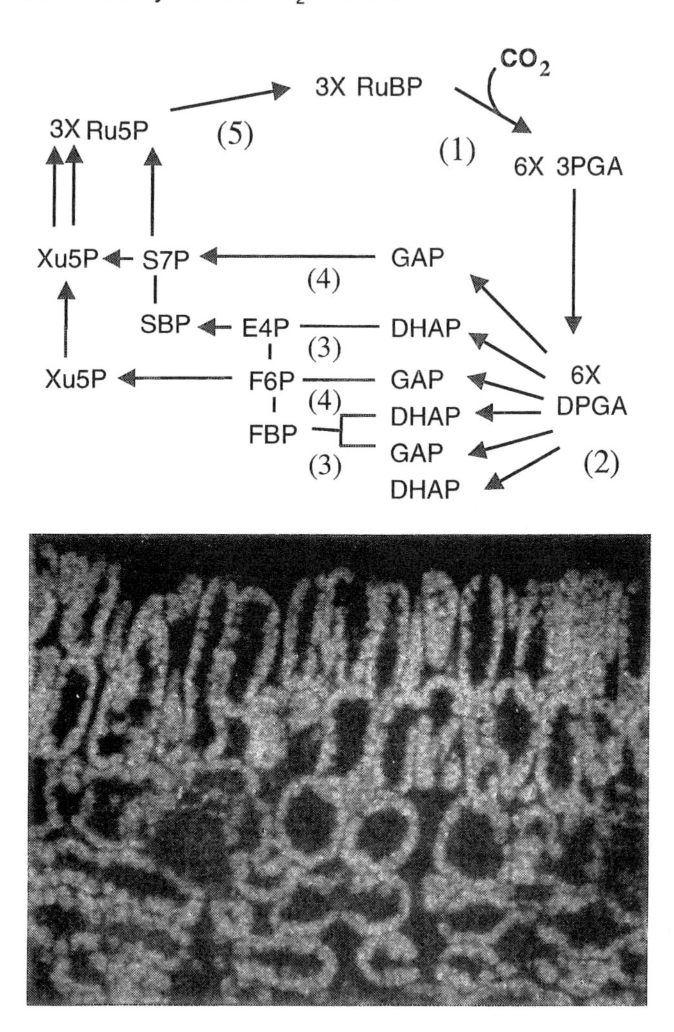

Fig. 1. The photosynthetic carbon reduction cycle. Rubisco (1) catalyses the carboxylation of ribulose-P_2 using atmospheric CO_2. The carboxylated intermediate is hydrolytically cleaved on the enzyme to produce two molecules of 3P-glycerate (3PGA). Following phosphorylation with ATP and reduction to triose phosphate (2), one molecule constitutes the product of the cycle and the rest go to regenerate ribulose-P_2. This one molecule is transferred to the cytosol by the Pi translocator where it can be converted to sucrose for transport throughout the plant. The regenerative reactions of the cycle involve a series of keto-aldol transfer and condensations involving aldolase (3) and transketolase (4) that lead to formation of ribulose 5-phosphate (Ru5P). Phosphorylation of ribulose 5-phosphate by phosphoribulokinase (5) completes the cycle back to ribulose-P_2 (RuBP). Fructose bisphosphatase and seduheptulose bisphosphatase release Pi from the 5- and 7-position of fructose bisphosphate (FBP) and seduheptulose bisphosphate (SBP), respectively, during regeneration. An epimerase converts the two molecules of xylulose 5-phosphate (Xu5P) to ribose 5-phosphate (R5P) and ribose 5-phosphate isomerase converts this to ribulose 5-phosphate. Fructose 6-phosphate (F6P) provides a means for plants to store the carbon fixed by photosynthesis as starch. DPGA is 1,3-diphosphoglycerate, GAP is glyceraldehyde 3-phosphate, DHAP is dihydroxyacetone phosphate and E4P, erythrose 4-phosphate. The cycle is shown above a transverse section of a C3 leaf, in this case from tobacco. The chloroplasts are visualised in this micrograph with an immunofluorescent antibody raised against the L subunit of Rubisco.

and the plastid (Gatenby, 1990; Roy and Andrews, 2000, for a review). Precursors of the S subunit, produced on cytosolic ribosomes, are imported into the plastid. Following processing to the mature protein by specific proteases, the S subunits assemble with L subunits that are organized as an octameric core. Formation of the core is not spontaneous because L subunits do not fold correctly after translation in the chloroplast. They instead require the assistance of specific chaperonins to ensure the majority, if not all the newly synthesized protein, folds appropriately for association (Gutteridge and Gatenby, 1995, and references cited therein). At least 70% of the soluble protein in the stroma of C3 plants is Rubisco.

The cyanobacterial enzyme is one of a few species of Rubisco with subunits that can reversibly dissociate to an octameric core and individual S subunits, in vitro. Both L and S subunits can be separately expressed in heterologous hosts and used to study how the smaller polypeptide affects enzyme activity. In the absence of S subunits, the core still has about 1% of the activity of the holoenzyme and retains the same relative reaction specificities. Saturation of the core with S subunits confers not only a hundred-fold increase in turnover and higher affinities for the individual substrates, but also a stabilizing effect on the continuity of the homo-octomer (Andrews, 1988, 1997; Gutteridge, 1991).

2. Activation of the Enzyme

A fascinating aspect of the reactions involving Rubisco that must precede any catalytic event, is activation. This involves the carbamylation of an active site lysine residue with a molecule of CO_2 to complete the co-ordination site of the essential Mg^{2+} ion (see Andrews and Lorimer, 1987, for a review). Until recently, the role of the active site carbamate and Mg^{2+} in catalysis was not understood. Spectroscopic work, combined with single turnover experiments, indicated that the enzyme required at least one active site residue to act as a base and initiate catalysis. Additionally, a second base is required to ensure that the correct stereochemical products of CO_2 fixation are generated (Lorimer et al., 1989). Crystallographic analysis of the activated enzyme with a transition intermediate analog bound and stabilizing the activated complex revealed the critical function of the two cofactors in substrate turnover (Newman and Gutteridge, 1993). Models of the quaternary complex of the *Synechococcus* enzyme show that one of the oxygen atoms of the carbamate is well-positioned to accept the C3 proton from ribulose-P_2 rendered more acidic by the polarizing effect of the metal. Subsequent events involving this enediol form of the bisphosphate, particularly proton movements associated with the appearance of the carboxylated intermediate formed between enediol and CO_2 substrate, must be orchestrated by the metal bound carbamate (Gutteridge and Gatenby, 1995). Cleavage of this six-carbon intermediate, 2′-carboxy, 3-keto arabinitol bisphosphate, to generate the first molecule of 3P-glycerate also involves the metal-bound carbamate. A different active-site Lys residue protonates the second molecule of 3P-glycerate in a stereochemically precise fashion (Cleland et al., 1998).

From the perspective of in vivo functionality and plant productivity, here lies a significant dilemma. If the carbamate is essential for Rubisco to fix CO_2, why then is the concentration of CO_2 required to achieve optimal activation of Rubisco so high relative to that expected to exist in the stroma of C3 plants during photosynthetic activity? Based on in vitro observations, at prevailing atmospheric concentrations of the gas, only one third of the enzyme should be activated in the stroma. If photosynthetic capacity is limited by the activity of Rubisco, two thirds of which might be inactive, then there is significant over investment of N in the enzyme. In other words, Rubisco is an exotic form of storage protein. Predictably, the true picture is somewhat more complex.

Studies of the enzyme in vitro in the presence of low molecular weight intermediates and products of photosynthesis that accumulate during fixation, influence the equilibria between CO_2, Mg^{2+} and the enzyme. Some favor the formation of the activated ternary complex, (enzyme. CO_2. Mg^{2+} or ECM complex) and others interfere in ECM formation (see Fig. 2). Therefore the assumption is, that during normal photosynthetic activity, the active site of the enzyme is rarely unoccupied by either cofactors or effectors which, in the absence of bisphosphate substrate, stabilize the activated state. In fact some of these effectors have a positive influence on the enzyme, such as when substrate displaces them from the site, the enzyme cycles at a small but significantly

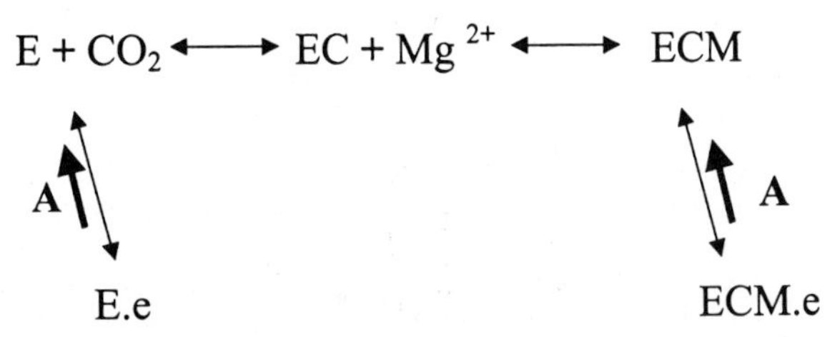

Fig. 2. Activation of Rubisco involves the reversible carbamylation of an active site Lys residue with a molecule of CO_2. Coordination of Mg^{2+} to the active site carbamate and adjacent acidic residues gives the active ternary complex ECM that is catalytically competent. The ability of the substrate ribulose-P_2 and similar effector (e) molecules to bind to either the non-activated form, such as xylulose-P_2 or active form, as with 2CA1P provides a means of regulating enzyme activation and turnover. In the plant these influences are modulated by the interaction of activase (A) with the enzyme binary (E.e) or quaternary effector (ECM.e) complex.

higher rate than the active ternary complex alone (Gutteridge et al., 1982), possibly through a mechanism of active site preorganization. It is not clear whether the general response to effectors can be extrapolated to stromal conditions, but there is evidence that in conditions of high irradiance at least 80% of the Rubisco must be catalytically active to achieve the rates of photosynthetic CO_2 fixation measured. Figure 2 illustrates the reversible equilibria between Rubisco, its inorganic activating cofactors CO_2 and Mg^{2+} and effectors with associated conformational embellishments. The basic catalytically competent form of the enzyme is the ternary complex, ECM. The effectors favoring this form bind in a way that overlaps the ribulose-P_2 binding pocket. Until recently, the order of association between the enzyme and cofactors as shown, was based on the rate of appearance of catalytic activity, i.e. CO_2 and slow carbamylation preceded the more rapid association of Mg^{2+} ions. However, given the co-ordination site of the metal is mainly acidic residues, it was still unclear whether Mg^{2+} binds before CO_2 to shield the proximal carboxylate sidechains that would prevent carbamylation of the active site Lys. However a three dimensional structure of the inactive enzyme (Taylor and Andersson, 1996) shows that the Lys involved in carbamylation is close to the sidechain of a His residue which most likely influences the pKa of the Lys amino group favoring carbamylation. Whatever the order of events is that leads to activation, the nature of the equilibrium with CO_2 and Mg^{2+} is such that in the stroma, in addition to active enzyme there are also various inactive forms, each one requiring different conditions to return to the active state.

3. Reaction Order and Mechanism

One early model of whole plant photosynthetic CO_2 fixation based on the kinetic parameters of Rubisco provided a good approximation for C3 plants before many details of activation mechanism were known. One of the assumptions of this model included a specific order of events in terms of substrate binding to the enzyme active site; namely, the bisphosphate binds first (Farquhar et al., 1980). Subsequently, spectroscopic data showed this was irrefutably the case for the enzyme from higher plants (Gutteridge et al., 1984), and later shown to be the case also for the *Rhodospirillum rubrum* enzyme (Pierce et al., 1986). The obvious implication is that no matter the origins of the enzyme, ribulose-P_2 must bind first.

Now that a convincing reaction mechanism has been constructed incorporating structural models, spectroscopic and solution chemistry data, it is clear that binding of the bisphosphate initiates all subsequent events. Details of the mechanism have been given elsewhere (Cleland et al., 1998) and will not be elaborated further; however, suffice to say that we know much about the carboxylase reaction. What we are less sure about structurally, is how a significant fraction of the bisphosphate becomes diverted into other reactions. Maybe it should not be surprising that, an enzyme which stabilizes at least four reactive intermediates during carboxylation, may produce more than just 3P-glycerate. Indeed there is a significant drain of ribulose-P_2 away from carboxylation due to the inability of the enzyme to effectively distinguish CO_2 from O_2 (Jordan and Ogren, 1981). In atmospheric concentrations of the gases, at least 1 out of every 4 cycles of the enzyme in a C3 plant results in formation of 2P-glycolate, a wasteful 2C product that requires ATP consumption during subsequent photorespiratory reactions to recycle the carbon back to the PCR cycle.

Oxygenation is not the only process that highlights the fidelity, or lack of it, of this enzyme. During turnover, at a small but significant rate, the enzyme also acts as an epimerase, where protons abstracted from ribulose-P_2 become misdirected and generate at least one epimer of the bisphosphate. The most notable is xylulose-P_2 which is a potent inhibitor of higher plant Rubisco, stabilizing the inactivated state (Zhu and Jensen, 1991).

Modulation of Rubisco through shifts in the various equilibria between active and inactive complexes, are features of the higher plant enzyme that confuses attempts to deconvolute all the factors influencing the efficiency of carboxylation and their physiological ramifications. The ability of various low molecular weight stromal components to occupy the site would provide a reliable theme, if they just all favored the activated enzyme-ternary complex. However, it complicates the picture to find that a number of them also have relatively high affinity for the inactivated form including, most perverse of all, the substrate itself (Fig. 2). The off-rate for ribulose-P_2 from the binary complex with the enzyme (E.R) is slow (minutes time scale) in vitro (Jordan and Chollet, 1983) and similar to that observed in leaves incubated in the dark (Cardon and Mott, 1989); reactivation of the binary complex involving xylulose-P_2 (E.X) is almost undetectable.

Three aspects of these particular associations are worth highlighting, not least their relevance to the conformational gymnastics that are a necessary component of Rubisco catalysis. First, the active site of the enzyme without bound bisphosphate is in an 'open' conformation between the N-terminal domain of one L subunit and C-terminal domain of a second L subunit. As part of the encounter between activated Rubisco and ribulose-P_2, at least 5 flexible segments or loops of both L subunits close over the active site, thereby preventing loss of the enediol and subsequent intermediates of the reaction chemistry dissociating from the site. The reason for tight binary associations between the non-active form of the enzyme and substrate or analogs is because the active site of Rubisco without activating cofactors can still adopt a 'closed' conformation, almost identical to the one that exists during carboxylation. The structure of the xylulose-P_2 binary complex with the *Synechococcus* enzyme has shown that the loops which normally stabilize the intermediate immediately following carboxylation, by closing off the active site from the surrounding solution, are also firmly closed with xylulose-P_2 in residence (Newman and Gutteridge, 1994). Indeed, the keto group of this pseudo-substrate is hydrated, in which form it can make extra H-bond contacts with the loop segments keeping them closed. This must mean the activating cofactors are not, in themselves, essential for this closed conformation to be adopted. Second, there is evidence that some bisphosphates may preferentially bind in an inverted orientation at the active site (Lundqvist and Schneider, 1989). Finally, this lethal combination would effectively shutdown all CO_2 fixation in plants unless there was some mechanism that intervened to rescue the enzyme from these multiply dead-end states.

The mechanism of recovery from dead binary complexes that is most widespread in plants is one involving a second protein, activase. Its initial discovery was in mutants of *Arabidopsis* which were severely challenged in their ability to survive unless grown in enriched CO_2 conditions (Portis, 1990 and refs cited). There is clear evidence that activase does not catalyze addition of CO_2 or Mg^{2+} to Rubisco directly. Rather, activase (A in Fig. 2) serves to increase the off-rate of effectors from non-activated and ECM forms of Rubisco. By increasing the dissociation rate of effectors from unproductive binary complexes, (as depicted by E.e in fig. 2) Rubisco has more opportunity to form the active ECM ternary complex (Andrews, 1997). The nature of the interaction between activase and Rubisco has yet to be fully explored due to its transitory nature. There is a nucleotide binding site for ATP which is hydrolyzed to ADP and Pi. It is speculated that hydrolysis of ATP is coincident with a change in the conformation of activase, one that recognizes bisphosphate containing inactive complexes of Rubisco. However, a physical protein-protein complex between activase and Rubisco has not been observed directly, rather the mechanism is surmised based on restoration of activity of the enzyme rescued from an E.R binary complex (Portis, 1990).

In the stroma it could therefore be envisioned that, as irradiance increases during the day and ATP levels rise, activase adopts a conformation which is compatible with recognition of some structural element of Rubisco. Once this element has associated with activase, the effect of this fleeting association is to concurrently relay changes to the active site of Rubisco which responds by releasing the bound antagonist, as ATP is consumed. A segment of Rubisco most likely to be involved is a surface loop of the L subunit (Portis, 1995). Species differences and similarities seen in this segment correlated with the specificity of activase depending on the plant source. Thus activases from *Solanaceae* are most effective reactivating Rubisco from the same genus (Wang et al., 1992).

Clearly, association of Rubisco and activase should not influence productive interactions, such as those involving bound intermediates during catalysis. For example, the very tight quaternary complex between activated Rubisco and the intermediate analog of carboxylation, 2′-carboxyarabinitol 1,5-bisphosphate (2CABP) is not affected appreciably by activase. If it was, then by extrapolation, the carboxylated reaction intermediate should be less tightly bound with adverse implications for CO_2 fixation. Possibly, once Rubisco has reached the point of forming a carboxyl intermediate, either the element on the surface recognized by activase may no longer be available for any external interaction, or once formed, the intermediate is too rapidly hydrolyzed to products to be recognized. The former possibility seems unlikely because the structures of unactivated and activated Rubisco liganded with 2CABP are quite similar (Newman and Gutteridge, 1993, 1994). Indeed, activase can influence the affinity of effectors bound to the activated form of the enzyme (see below). The second possibility also seems unlikely in that the turnover rate for Rubisco is similar to that of the

ATPase activity of activase (Robinson and Portis, 1989). Perhaps the over-riding consideration is the magnitude of the activation energy required to disrupt the association between activated Rubisco and its reaction intermediates. The machinery of activase may simply be insufficient for disruption of extremely tight complexes. To wit, activase can indeed influence the affinity of less tightly bound effectors bound to the activated form of the enzyme.

B. A Natural Inhibitor of Rubisco

A natural inhibitor of Rubisco, 2′-carboxyarabinitol 1-phosphate (2CA1P) has been identified which, like 2CABP resembles the reaction intermediate of carboxylation. 2CA1P is identical to 2CABP except for the absence of the phosphate at the 5- position (Gutteridge et al., 1986; Berry et al., 1987). Lack of the second phosphate adversely affects binding affinity by almost three orders of magnitude. Binding of 2CA1P is tight enough to stabilize the activated ECM state of Rubisco, but not tight enough to lock the enzyme irreversibly into an inactive quaternary complex as is the case with 2CABP. In plants that synthesize 2CA1P (Andralojc et al., 1996), this inhibitor accumulates during the night, trapping Rubisco in its activated state. As light intensity increases next day, the inhibitor is released. This can happen by simple dissociation, but plants that have reduced amounts of activase show slow rates of reactivation relative to plants with normal amounts of activase (see below).

Once 2CA1P is released from the active site of Rubisco by dissociation or the intervention of activase, the inhibitor is dephosphorylated by a specific phosphatase localized in the stroma (Gutteridge and Julien, 1989; Salvucci and Holbrook, 1989). The appearance and disappearance of 2CA1P and the activity of the phosphatase follows a diurnal cycle. The enzyme functions during periods of active photosynthesis but is inactive at night. These are characteristics of enzymes that sense the redox state of the stromal milieu through changes in the levels of reduced thioredoxin or glutathione. In this case, the purified phosphatase requires the presence of reduced thiol to retain activity. Most recently there is evidence that the activity of the enzyme has two distinct sets of thiol groups that are involved in its regulation (Heo and Holbrook, 1999).

III. Modulating Rubisco Activity and the Response of Photosynthetic CO_2-fixation in planta

A. Subunits of Rubisco

Historically, a question of much debate has been the extent to which the growth of C3 plants is limited by Rubisco. Models based on the kinetic parameters of the enzyme were convincing but were based on certain assumptions and could not take into account all the factors influencing other enzymes of the cycle. However, with the development of robust techniques to alter the expression of proteins in some plants, this question could be addressed more directly. Until recently, only cytosolic transcription processes could be modulated, whereas now, technology has developed to the point where plastid transformation can be contemplated. In the case of Rubisco composed of subunits originating from both cytosolic and plastid translation processes, this provides an ideal opportunity to understand better the importance of each of the subunits on photosynthesis. For example, if turnover is to be altered, this might be better achieved through changes to the S subunit gene, whereas relative specificities may best be achieved through L subunit mutations

1. S Subunits

Vector constructs of S subunit DNA cloned in an antisense orientation behind a 35S promoter suppressed the production of S subunits in tobacco (Rodermel et al., 1988; Quick et al., 1991). The amount of Rubisco was reduced by as much as 80% relative to control plants. Response of the plants to the absence of S subunits depended on the growth conditions in terms of light intensity, N supply and CO_2 availability (Hudson et al., 1992). In general, Rubisco of C3 plants exerts a large controlling effect on photosynthesis, with a control coefficient of 0.6 – 0.8 in high irradiance, enhanced further when N or CO_2 is also limiting. However, at low or more moderate levels of irradiance and intercellular CO_2, control shifts from Rubisco to the regeneration of ribulose-P_2 (Fig. 1). Transgenic plants with reduced levels of Rubisco through S subunit antisense expression, showed photosynthetic rates always limited by Rubisco over a wide range of growth conditions.

2. L subunits

Until recently, the ability to alter the amounts of the L subunit of Rubisco in chloroplasts has been beyond the reach of experimental inquiry. However, the development of robust methods to target the plastid genome for stable expression of foreign genes will see an increasing number of experiments investigating changes to the L subunit, at least in tobacco. The first experiments designed to investigate the feasibility of replacing the L-subunit of tobacco plants with a foreign L-subunit in the plastid genome, also provided an opportunity of assessing the effect of altering the kinetics of the enzyme on the efficiency of CO_2 fixation (Kanevski et al., 1998, 1999). Tobacco plants were transformed with plastid DNA containing the gene for the L subunit of sunflower Rubisco. The trans-plastid plants that were recovered produced an active Rubisco hybrid composed of sunflower L subunits and tobacco S subunits. It was estimated that only about 33% of the normal amounts of active enzyme were assembled in the stroma. Superimposed were the inferior kinetic parameters of a hybrid composed of a foreign L subunit. Compared to normal wild type, the transgenic plants suffered some 80% loss of CO_2 fixing ability and to reach maturity and set seed required grafting transgenic shoots onto wild-type root stock. Even then, regenerated plants clearly were unable to accumulate the normal levels of chlorophyll and the chloroplasts were poorly developed. Nevertheless this approach offers a means of, not only studying the effects of chimeric forms of Rubisco on plant development and growth, but also is a potential source of higher plant Rubisco mutants for detailed structural and kinetic analysis.

In another report describing homologous recombination of the L subunit gene in tobacco plastids, Whitney et al., (1999) introduced a site specific mutation at a Leu residue that is adjacent to an essential Lys in loop 6 of the L subunit. Replacement of Leu by Val at this position reduces CO_2 fixation and enhances the oxygenase activity of the enzyme. Consistent with these characteristics, plants with the mutation grew much slower than untransformed plants, although they did eventually flower and reproduce normally.

B. Activase

Tobacco plants transformed with vectors designed to alter the amounts of Rubisco activase (Mate et al., 1993), provided some useful insight into the influence of sugar phosphates and natural inhibitors on the activation of Rubisco in the plant and, by extension, how this alters CO_2 fixation. Plants deficient in activase are like the *Arabidopsis* mutants first used to identify the protein, requiring enriched CO_2 to maintain growth. At moderate levels of irradiance in plants lacking 30–50% activase, little or no effect was apparent. However, when irradiance is high activase accounts for as much as 25% of the flux control through relief of inhibition of Rubisco. More recently, Rubisco activity in leaves of plants with less than 5% activase has been compared to the activity of the enzyme immediately isolated from these plants, leading to further speculation about the function of activase in vivo (Andrews et al., 1998).

A second outcome of the antisense experiments showed that the release of 2CA1P from Rubisco active site is not simply dependent on dissociation of the inhibitor and processing by a specific phosphatase, rather its removal is enhanced by activase resulting in a much reduced affinity between 2CA1P and Rubisco. Unfortunately, the gene for 2CA1P phosphatase has proved elusive and so similar suppression experiments of this gene product in whole plants cannot be reported.

C. CO_2, Mg^{2+} and pH

CO_2 clearly has a dual effect on Rubisco activity. The first is as a substrate and the second is through its requirement as an activating cofactor. As cofactor, its effect is further influenced by the availability of Mg^{2+} ions which cycles through large changes in concentration, depending on the intensity of light radiation. In the light, the activation of Rubisco is directly influenced and favored by the elevated pH and Mg^{2+} concentrations in the stroma. During the dark, Mg^{2+} is exchanged for protons; as a consequence stromal pH drops and Rubisco is inactivated in response to both the change in pH and loss of Mg^{2+}.

Processes having most influence over CO_2 availability in C3 plants are those of stomatal conductance and the activity and disposition of carbonic anhydrase. However, plants with reduced amounts of Rubisco and thus elevated intercellular CO_2, rather surprisingly seem not to show altered stomatal conductance (Mate et al., 1993). Plants with carbonic anhydrase suppressed by as much as 50-fold through antisense techniques, had little reduced photosynthesis (at most 4%) despite a drop of 15 μbar in intracellular CO_2 (Price et al., 1994).

IV. Modulating Activities of Other Enzymes of the PCR Cycle

A more profound understanding of the regulation of the CO_2-fixation cycle and the influence of any of the enzymes is complicated, since plants have the ability to modulate the flux of the cycle by a number of different mechanisms. These include responses to photochemical processes through the accumulation of ATP and NADPH as co-substrates, and reducing equivalents in the form of thioredoxin. Reductive power can therefore influence flux through both its availability in enzyme reductive reactions, or altering enzyme conformation. ATP similarly has direct effects on the chemistry of reactions or alternatively will influence activity by structural changes to enzymes through phosphorylation.

Two enzymes of the cycle whose activities are influenced through structural changes by the state of the reduced stromal milieu and direct interaction with thioredoxin, are phospho-ribulose kinase (PRK) and fructose 1,6-bisphosphate phosphatase (FBPase). This form of control could not be considered fine-tuning of CO_2-fixation since in ambient conditions there is a surfeit of reducing power. Rather the depletion of reduced thioredoxin during periods of darkness serve to switch the enzymes to inactive forms and, in the case of FBPase, prevent futile cycling between the kinase and the phosphatase.

A. Phosphoribulokinase

Phosphoribulokinase (PRK) is the enzyme responsible for the production of ribulose-P_2 through phosphorylation of ribulose 5-P. It is responsive to the photochemical reactions in two ways. The sequence of the enzyme contains two Cys residues that are sensitive to changes in reduction potential of the stroma through their interaction with thioredoxin. When reductive power declines with daily loss of irradiance, these residues readily form a disulfide bond that shuts down the ability of the enzyme to produce ribulose-P_2. Increasing photon flux and accumulation of reducing power as the pool of thioredoxin builds the next day, reverses the process by reducing the disulfide bridge (see Buchanan, 1980 for review). The enzyme is reactivated via a transient mixed disulfide with thioredoxin (Brandes et al., 1996).

The thiol residues implicated in formation of the regulatory disulfide of spinach PRK are Cys 16 and Cys 55 (Hirasawa et al., 1998). As the ratio of reduced and oxidized thioredoxin shifts toward more oxidizing conditions, oxidized thioredoxin mediates the formation of an intrasubunit bridge between these residues through thiol/disulfide exchange. The residues are close to the active site of the enzyme, with Cys 55 moderately influencing catalysis and substrate binding. Once the thiol groups are tethered by the bridge, the enzyme is inactivated, showing less flexibility in those segments that form the active site and causing a drop in affinity for ATP by almost three orders of magnitude. Thus, loss of activity is, in part, due to non-availability of the thiol of Cys 55 and the restrictions on structural dynamics that the disulfide imparts on the enzyme. Site-specific mutants have revealed the impact of these conformational changes on the affinities for substrates and have shown that the enzyme can be converted to a permanently active form by replacing Cys 16 with Ser, removing the sensitivity to redox conditions (Hirasawa et al., 1998). The high resolution structure of the PRK enzyme from *Rhodobacter sphaeroides* has been solved and used as the model to design other site-directed mutations to determine the function of individual residues in catalysis and loop dynamics (Harrison et al., 1998a)

The second effect of the light reactions on PRK is through the availability of ATP as a necessary substrate and source of the phosphate that ultimately occupies the 1-position of ribulose-P_2. In this case, since the amount of ATP is most likely close to saturating concentrations in the stroma (Km ~0.1 mM), variation in the ATP/ADP ratio will have little influence on enzyme turnover. The co-substrate would need to drop a significant amount from those levels achieved during normal photosynthesis, before enzyme activity would be affected directly.

Transgenic plants have been produced expressing the antisense form of PRK. In one report it was concluded that a decrease in activity of as much as 85% did not have a dramatic effect on growth (Banks et al., 1999). However, another study reported that plants with a modest reduction in the amount of the kinase did suffer severe photoinhibition and stunting even when grown in low light conditions (Furbank and Taylor, 1995).

B. Fructose-1,6-Bisphosphatase

Site-specific mutagenesis has also been used to identify the Cys residues of chloroplast fructose bisphosphatases involved in redox regulation. The chloroplast enzymes have seven cysteines, with four located close to, or as part of a short segment of primary sequence not found in the cytosolic bisphosphatases. Five Cys were singled out for replacement with Ser (Jacquot et al., 1997) and only one change, C153S (pea enzyme) was found to completely alleviate the redox control of the enzyme activity. Other replacements either had no, or only partial effects on activity and the influence of thiol reductants. Clearly, another Cys residue must interact with Cys 153 to form the disulfide bridge and inactivate the bisphosphatase. Based on the response of the other Ser mutants and on the crystal structure of the spinach enzyme (Jacquot et al., 1995), it is possible that Cys 153 has some opportunity to form a bridge with either of two other residues, Cys 173 and Cys 178. The reason for three residues being involved is not yet clear, and may be associated with the nature of the rather specific interaction between the bisphosphatase and thioredoxin *f*.

Antisense expression of plastid FBPase has been successful using transgenic potato plants. These experiments produced a series of plants with different levels of enzyme activity. Plants with activity below 40% of wild-type levels had lower photosynthetic capacity, yet apparently produced normal tubers. However, in those plants with less than 15% of expected activity, there was growth impairment of plants and tuber development with increased partitioning of carbohydrate into soluble sugars (Kossmann et al., 1994).

C. Seduheptulose-1,7-Bisphosphatase

Antisense experiments have also been recently performed on seduheptulose bisphosphatase, another of the enzymes with activity sensitive to the redox conditions of the stroma. Tobacco plants transformed with antisense constructs of seduheptulose bisphosphatase cDNA behind the CaMV promoter produced primary transformants with a range of bisphosphatase activities and protein levels that correlated with lower amounts of bisphosphatase mRNA (Harrison et al., 1998b). Western blot analysis showed that no other enzymes of the PCR cycle were affected although plants with less than 20% bisphosphatase activity showed a range of phenotypes including growth reduction and chlorosis. Even plants with moderately reduced amounts of the bisphosphatase showed reduced carbon assimilation rates and PS II quantum efficiencies. Reduced carbon allocation as starch was evident, although sucrose levels were maintained in all but the most severely affected transformants.

D. Glyceraldehyde Phosphate Dehydrogenase

Glyceraldehyde phosphate (GAP) dehydrogenase is an enzyme essential for the regenerative part of the cycle leading to the formation of ribulose-P2. Antisense constructs of the tobacco enzyme targeting the A-subunit resulted in transgenic plants with activities ranging from 7% to wild-type amounts of activity (Price et al., 1995). The effect of the changes in expression of the enzyme were determined by assessing the response of the 3P-glycerate and ribulose-P_2 pools as well as the overall influence of reduced enzyme on photosynthetic capacity. Photosynthesis was unaffected unless only 20% of wild-type levels remained, however an interesting relationship between the ratio of the ribulose-P_2 pool and Rubisco active site concentration was discerned. Clearly, sub-optimal amounts of GAP dehydrogenase should lead to reduced ability to regenerate ribulose-P_2, however, CO_2 assimilation was not affected until ribulose-P_2 had dropped to less than 40% of normal levels and a ratio of ribulose-P_2 per Rubisco active site lower than 2. In normal ambient photosynthetic rates there was enough enzyme activity to maintain the apparently critical ratio of 4 ribulose-P_2 molecules per Rubisco, and this ratio only fell in conditions approaching maximum photosynthesis.

E. Aldolase

Regenerated potato transformants expressing full-length antisense transcripts of plastid aldolase were found to show a range of effects depending on the amount of enzyme activity remaining (Haake et al., 1998). Plants with less than 20% of wild-type activity showed severe growth impairment, however plants with as much as 40% activity still showed 40% limitation in growth. Such dramatic effects are surprising for an enzyme that catalyses a reversible reaction and has no known apparent regulatory function in metabolism or plant growth.

F. Pi Translocator

One of the major products of the PCR cycle and substrate for multiple biosynthetic pathways outside the chloroplast is triose phosphates. These 3-carbon photoassimilates, in the form of either 3P-glyceraldehyde or its interchangeable counterpart through the action of triose-P isomerase, dihydroxy acetone phosphate, provide a primary means of communication between the stroma and the cytosol. The protein that facilitates this communication is the Pi-translocator, a 29 kDa protein located in the chloroplast membrane that plays a pivotal role in balancing the movement of Pi into the stroma resulting from cytosolic assimilatory process and the flow of triose-P and 3P-glycerate into the cytosol. Therefore the ratio of Pi to 3P-glycerate provides a direct measure of photosynthetic efficiency. Regulation of the translocator should have direct effects on both Pi import, essential for photoassimilate production, and triose-P export, critical for assimilation. Altering the levels of this protein would provide some quantitative assessment of how changes in assimilate partitioning influence CO_2 fixation in vivo and the importance of stromal-cytosolic communication.

Transgenic potato plants expressing an antisense construct of the Pi translocator generated plants with both reduced message and protein (Reismeier et al., 1993). Plants with about 20–30% reduction in the ability to import Pi into chloroplasts showed a reduction in maximum photosynthesis of 40–60%. There was also a shift in C-partitioning to starch rather than sucrose and amino acids. In terms of growth, the plants showed a dwarf-like phenotype during early development, with fewer differences at later stages, as compared to control plants. However, the transgenic plants still showed a marked difference in assimilate partitioning into starch. Under ambient conditions, the rate of photosynthesis was unaffected by the reduced activity of the translocator, yet there was still a significant change in the allocation of fixed carbon.

V. C4 Metabolism

A. General Aspects

With increasing temperature the ratio of the two dissolved gases, O_2 and CO_2, increases due to differences in their water solubilities (Ku and Edwards, 1977). A 20° change, e.g., from 15 °C to 35 °C, alters this ratio by about 50%. Consequently, for plants that grow in high temperature regions this dramatic change in the relative concentration of O_2 to CO_2 on CO_2-fixation could be dramatic, given the sensitivity of Rubisco to CO_2 concentrations, both as substrate and activator, and O_2 as an alternate substrate. Plants that flourish in conditions of high temperatures and high light have evolved a means of ensuring that CO_2 is not limiting. Indeed the process is so successful in maintaining high intracellular concentrations, that not only does the plant need less Rubisco, but the particular variant of the enzyme isolated often shows less affinity for both ribulose-P_2 and CO_2 as substrates. These plants are described as practicing C4 carbon fixation (see Furbank et al., 2000 for a review) because they have the ability to store CO_2 in the form of a second carboxyl group in C4 dicarboxylic acids, oxalacetate, malate and aspartate.

The most evolved forms of C4 plants not only have genes specific for production of C4 acids but usually have a distinct cellular structure compared to that found in C3 plants. Figure 3 shows the reactions involved in the carboxylation and decarboxylation processes of a C4 plant that uses malate for CO_2 storage. Mesophyll cells generally form the outer ring of a concentric series of cells directed toward the vein tubes of C4 leaves. Within the chloroplasts of the mesophyll, little Rubisco is detectable and fixed CO_2 is primarily all in some form of C4 acid. Mesophyll chloroplasts also have the bulk of the PS II complex with water splitting and oxygen evolving capability.

The inner ring of cells, closest to developing veins, is the bundle sheath with chloroplasts that contain Rubisco and those enzymes which decarboxylate the C4 acid. The CO_2 released serves as the substrate and activator of Rubisco. Consequently the chloroplasts in these cells are the site of the bulk of carboxylation of ribulose-P_2 and the other reactions of the PCR cycle. This type of cell architecture is thus well developed to prevent as little loss of ribulose-P_2 into alternate reactions as possible by ensuring saturating concentrations of CO_2 at the site of carboxylation. Rubisco in this environment is thus optimally activated and photorespiration is essentially absent. Even oxygen from photochemical events is less likely to accumulate near Rubisco because of this cellular compartmentation.

Although both cell types have the genetic capability to produce enzymes of both C4 and PCR cycles in response to light induction, it appears that there is a light independent cell-specific mechanism that ensures differential expression; e.g., little Rubisco is detectable in mesophyll and little PEP carboxylase found in the bundle sheath.

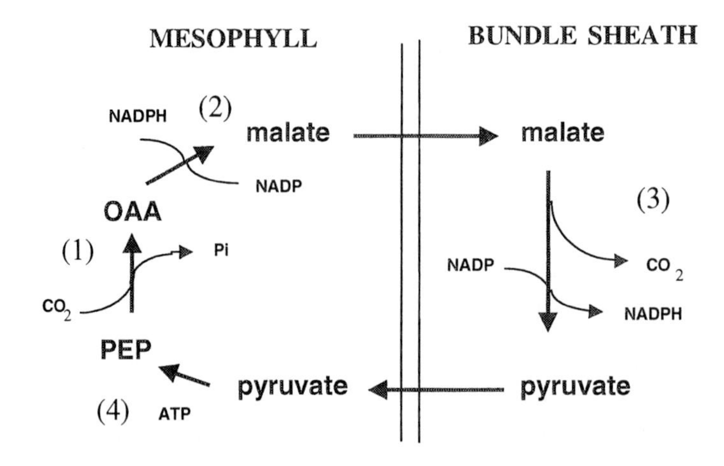

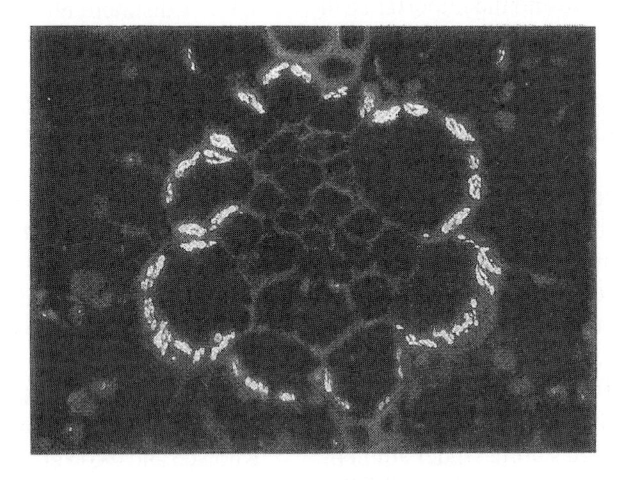

Fig. 3. The carboxylation and decarboxylation reactions of C4 photosynthesis. In this particular example of C4 cycle requiring NADP-malic enzyme (NADP-ME), phosphoenolpyruvate (PEP) is the acceptor of CO_2 in the form of HCO_3^- through the action of PEP carboxylase (1) in mesophyll cells. The oxaloacetate (OAA) that is formed is reduced to malate by NADP-malate dehydrogenase in mesophyll chloroplasts and the malate is transported to the chloroplast of the bundle sheath cells. The action of NADP-malic enzyme (3) releases CO_2 forming pyruvate. The CO_2 is subsequently fixed by Rubisco and all further steps of photosynthesis are the same as those depicted in Fig. 1 for C3 plants. Pyruvate returns to the mesophyll chloroplasts where it is converted to PEP by the action of pyruvate Pi dikinase (4). The NADP-ME form of C4 fixation shown is above a thin section micrograph of a maize leaf. Only the chloroplasts of the bundle sheath cells are readily apparent from the fluorescent immunolabelling of the Rubisco L subunit in their 'centrifugal' location relative to the vasculature.

B. Enzymes Involved in the C4 Acid Cycle

There are a number of variations of the C4 acid cycle that other authors have covered in some detail (see for example, Edwards and Walker, 1983). Only the simplest C4 system, which has been subjected to transgenic modifications of various enzymes is considered here. The plant most used to study the effects of changes in the amounts of the enzymes involved in C4 photosynthesis, is *Flaveria bidentis*. The plant is a C4 dicot, providing a relatively amenable system for introducing antisense constructs of genes involved in CO_2-fixation.

Transformation of *F. bidentis* with vectors expressing Rubisco S subunit antisense inserts, give progeny that show message levels linear with respect to the amount of protein produced (Furbank et al., 1997). Leaves of the transformants were found with 80% reduction in the amount of S subunit message and an 85% loss of Rubisco activity, but unaffected in the amounts of other enzymes of photosynthesis. The stunted phenotype of the transformants and the reduced rates of photosynthesis in ambient air at high light indicated that, like C3 plants, Rubisco is a major determinant of photosynthetic flux in C4 plants with a control coefficient of 0.7.

F. bidentis transformed with constructs expressing a gene coding for NADP- malate dehydrogenase (NADP-MDH) from a C4 monocot, sorghum, in attempts to increase this activity, produced plants that also showed a range of suppressed levels of the native dehydrogenase (Trevanion et al., 1997). The enzyme in both mesophyll and bundle sheath cells was co-suppressed. Plants with less than 10% of normal amounts of the dehydrogenase gave rates of photosynthesis unaffected in low light and low CO_2 and only limiting in rate when in high light and high CO_2. From an analysis of flux control in those transformants expressing a range of dehydrogenase activities indicates that NADP-MDH is not important in regulating C4 photosynthesis in *F. bidentis* in steady state conditions. The study also concluded that the enzyme in the bundle sheath cells is not limiting photosynthesis.

Irrespective of its passive regulatory role in assimilatory flux, NADP-MDH is itself regulated by sensing the redox conditions of the cell. In this case the Cys residues involved in thioredoxin interaction are located at the N- and C-termini of the enzyme (Issakidis et al., 1994). With the determination of the three-dimensional structure of the enzyme from *F. bidentis* (Carr et al., 1999) there is an explanation of the role of the C-terminal extension in determining the effects of co-enzyme on the rate of reductive activation.

Another enzyme that is covalently modified in response to the photosynthetic status of the cell is pyruvate Pi dikinase (PPdK). In this case, however, the enzyme is inactivated by phosphorylation at a Thr residue by a specific regulatory protein (Ashton et al., 1984). This protein is unusual in that it uses ADP as the source of the phosphate group for phosphorylation. The same protein reactivates PPdK, not through a phosphatase activity, but rather by phosphorylation of Pi generating PPi.

The amount of measurable PPdK activity in mesophyll cells is just enough to account for observed rates of photosynthesis. Thus plants reduced in PPdK to 10–20% of wild-type amounts can be only maintained in enriched CO_2 concentrations (1%) or on sucrose supplemented media. Those grown on sucrose suffer severe photoinhibition. Interestingly, no transformants were isolated that had intermediate levels of the enzyme and the regenerated plants fell into two distinct groups. The conclusion was drawn that transcript amounts are non-linear with respect to protein and once below a certain level, PPdK content drops dramatically (Furbank et al., 1997). The authors also reported a control coefficient of 0.2–0.3 for PPdK.

One other enzyme of C4 pathway that is also activated by light is PEP carboxylase located in the cytosol of mesophyll cells. This enzyme is responsible for the first step of C4 acid synthesis, carboxylating PEP to form oxaloacetate. Unlike PPdK, PEP carboxylase is activated by phosphorylation. The kinase is specific, phosphorylating a Ser residue, and the effect of the modification is characterized by a loss in affinity for malate that otherwise inhibits the enzyme.

In *Amaranthus* plants chemically treated with azide, mutants were regenerated with PEP carboxylase activities down to 5% of wild-type levels (Dever et al., 1997). Plants with about 50% of wild-type amounts showed reduced rates of maximum photosynthesis and carboxylation efficiency. In heterozygotes of these plants, the enzyme showed less malate sensitivity and increased phosphorylation. Interestingly, mutants with 40–50% PEP carboxylase activity had higher maximal photosynthesis in comparison to mutants with almost doubled PEP carboxylase activity. The postulate was that a compensatory activation mechanism is involved to deal with the loss of activities below 55% of wild type. It was also concluded that activation is more a binary switch rather than a means of fine control of enzyme activity.

VI. Concluding Remarks

There can be no question that we have a much better understanding of the factors influencing CO_2 fixation and the relative importance of the enzymes in regulating flux through this and connected cycles from antisense and suppression transformations of both C3 and C4 systems. As befits a technique that allows investigation by more direct means there have been some surprises; e.g., aldolase not expected to have the impact that it has. Antisense techniques only allow one part of the conundrum to be dissected, namely disrupting an integrated system through reducing the level of a particular enzyme. This has shown how CO_2 fixation responds to a drop in flow at any one point in the cycle, but can be misleading since regulation is often due to kinetic modulation of enzyme activity through substrate availability, allosteric effects, etc. There should be equally as much to learn by over-producting the enzymes in question, or alternatively expressing mutant forms designed to be insensitive to feedback inhibition. Given problems associated with co-suppression, the outcome of experiments designed to produce more enzyme is less predictable than antisense approaches. However, using genes from phylogenetically unrelated organisms may be one means of avoiding such problems.

Some of the general findings that these studies have highlighted include direct comparison of the influence of any change on multiple components of the system. Similar series of investigations should become routine, e.g., rates of fixation in vivo, determined in ambient as well as maximal conditions over short and long terms to assess effects of changes in assimilate partitioning on plant growth and development. Growth in limiting conditions of nitrogen and CO_2 also provides a more complete picture of control and regulation. A necessary extension of these types of studies must be to assess the effect of reduced levels of the enzyme of interest on the other enzymes of the system. If the activities of these are unchanged, then the resulting symptomology and physiology are a direct result of the modification to the primary target.

One general conclusion that can be drawn from these elegant series of investigations using altered plants is that Rubisco has significant control over the productivity of CO_2 fixation. This might not be such a surprise in C3 systems, where the plant often suffers CO_2 limitation, but is of some consequence for explaining the factors involved in regulating C4 systems, where effects of reduced amounts of Rubisco extend far beyond a direct reduction in carboxylation. It has often been hypothesized that C3 plants will respond positively if a more efficient Rubisco were discovered with increased affinity for CO_2 at the expense of O_2. However, it is also clear that it would potentially be just as important to alter a C4 recipient to find out if there are other advantages to expressing an improved Rubisco variant than suppression of oxygenation; e.g., nitrogen or water use.

Such experiments are now feasible and variants of Rubisco exist with at least 50% increased carboxylase efficiency compared with the higher plant enzyme (Read and Tabita, 1994; see also Sugawara et al., 1998). Differences in sequence of these marine algal species extend to L and S subunits, but since both nuclear and plastid genomes are now accessible to the molecular biologist, there is no technical impediment to producing transgenic plants expressing an algal Rubisco. Whether a functional enzyme would result, given the apparent necessity of specific chaperones for folding, cannot be predicted, nor whether once assembled, a productive interaction with activase would be possible. Answers to such questions await further investigation of these and other versions of Rubisco.

Currently we have basic findings which point quite clearly to mechanisms (and genes) where appropriate changes should provide enhancements in the efficiency of photosynthesis. Application of these findings to generate crops with improved photosynthesis would underscore the power of biotechnology in leading another agricultural revolution. It would mirror the achievements of earlier revolutions that applied advances in plant breeding and genetics to alleviate susceptibilities of crops to disease and environmental stress. Irrespective of the social and political controversy presently surrounding biotechnology and agriculture there has to be significant advantages from being able to develop tools specifically for plant related investigations, if nothing else than to understand the complex interactions of fundamental metabolic processes. Plants are being considered as production platforms for pharmaceuticals and industrial commodities, applications that will surely increase the pressure on land presently used for food production.

Acknowledgment

We would like to acknowledge the help of Dr. Todd Jones who produced the immunofluorographs used in the figures.

References

Andralojc PJ, Keys AJ, Martindale W, Dawson GW and Parry MAJ (1996) Conversion of D-hamamelose into 2-carboxy-D-arabinitol and 2-carboxy-D-arabinitol 1-phosphate in leaves of *Phaseolus vulgaris* L. J Biol Chem 271: 26803–26809

Andrews TJ (1988) Catalysis by cyanobacterial ribulose bisphosphate carboxylase large subunit in the complete absence of small subunit. J Biol Chem 263: 12213–12219

Andrews TJ (1997) The bait in the rubisco mousetrap. Nat Struct Biol 3: 3–7

Andrews TJ and Lorimer GH (1987) Rubisco: Structure, mechanisms, and prospects for improvement. In: Hatch MD and Boardman NK (eds) The Biochemistry of Plants, Vol 10, pp 131–218. Academic Press, San Diego

Andrews TJ, von Caemmerer S, He Z, Hudson GS and Whitney S (1998) Rubisco catalysis in vitro and in vivo. In: Garab G (ed) Photosynthesis: Mechanisms and Effects, Vol V, pp 3307–3313. Kluwer Academic Publishers, Dordrecht

Ashton AR, Burnell JN and Hatch MD (1984) Regulation of C4 photosynthesis: Inactivation of pyruvate Pi dikinase by ADP dependent phosphorylation and activation by phosphorylase. Arch Biochem Biophys 230: 492–503

Banks FM, Driscoll SP, Parry MA, Lawlor DW, Knight JS, Gray JC and Paul MJ (1999) Decrease in phosphoribulokinase activity by antisense RNA in transgenic tobacco. Relationship between photosynthesis, growth and allocation at different nitrogen levels. Plant Physiol 119: 1125–1136

Benson AA and Calvin M (1950) Carbon dioxide fixation by green plants. Ann Rev Plant Physiol 1: 25–40

Berry JA, Lorimer GH, Pierce J, Seemann JR, Meeks J and Freas S (1987) Isolation, identification, and synthesis of 2-carboxyarabinitol 1-phosphate, a diurnal regulator of ribulose bisphosphate carboxylase. Proc Natl Acad Sci USA 84: 734–738

Brandes HK, Larimer FW and Hartman FC (1996) The molecular pathway for the regulation of phosphoribulokinase by thioredoxin f. J Biol Chem 271: 3333–3335

Buchanan BB (1980) Role of light in the regulation of chloroplast enzymes. Ann Rev Plant Physiol 31: 341–374

Cardon ZG and Mott KA (1989) Evidence that ribulose 1,5-bisphosphate (rubp) binds to inactive sites of rubp carboxylase in vivo and an estimate of the rate constant for dissociation. Plant Physiol 89: 1253–1257

Carr PD, Verger D, Ashton AR and Ollis DL (1999) Chloroplast NADP-malate dehydrogenase: Structural basis of light-dependent regulation of activity by thiol oxidation and reduction. Structure 7: 461–475

Cleland WW, Andrews TJ, Gutteridge S, Hartman F and Lorimer GH (1998) Mechanism of Rubisco: The carbamate as general base. Chem Rev 98: 549–561

Dever L, Bailey KJ, Leegood RC and Lea PJ (1997) Control of photosynthesis in *Amaranthus edulis* mutants with reduced amounts of PEP carboxylase. Aust J Plant Physiol 24: 469–476

Edwards G and Walker D (1983) C3, C4: Mechanisms and cellular and environmental regulation of photosynthesis. Blackwell Scientific Publications. University of California Press, Los Angeles

Farquhar GD, von Caemmerer S and Berry JA (1980) A biochemical model of photosynthetic CO_2 fixation in leaves of C3 species. Planta 149: 78–90

Furbank RT and Taylor WC (1995) Regulation of photosynthesis in C3 and C4 plants: A molecular approach. Plant Cell 7: 797–807

Furbank RT, Chitty JA, Jenkins CLD, Taylor WC, Trevanion SJ, von Caemmerer S and Ashton AR (1997) Genetic manipulation of key photosynthetic enzymes in the C4 plant *Flaveria bidentis*. Aust J Plant Physiol 24: 477–485

Furbak RT, Hatch MD and Jenkins C (2000) C_4 photosynthesis: Mechanism and regulation. In: Leegood R, Sharkey T and von Caemmerer S (eds) Photosynthesis: Physiology and Metabolism, pp 435–457. Kluwer Academic Publishers, Dordrecht

Gatenby AA (1990) Chaperone function: The assembly of ribulose bisphosphate carboxylase-oxygenase. Annu Rev Cell Biol 6: 125–149

Gutteridge S (1991) The relative catalytic specificities of the large subunit core of *Synechococcus* ribulose bisphosphate carboxylase-oxygenase. J Biol Chem 266: 7359–7362

Gutteridge S and Gatenby AA (1995) Rubisco synthesis, assembly, mechanism and regulation. Plant Cell 7: 809–819

Gutteridge S and Julien B (1989) A phosphatase from chloroplast stroma of *Nicotiana tabacum* hydrolyses 2′-carboxyarabinitol 1-phosphate, the natural inhibitor of Rubisco to 2′-carboxyarabinitol. FEBS Lett 254: 225–230

Gutteridge S, Parry MAJ and Schmidt CNG (1982) The reactions between active and inactive forms of wheat ribulose bisphosphate carboxylase and effectors. Eur J Biochem 126: 597–602

Gutteridge S, Parry MA, Schmidt CNG and Feeney J (1984) An investigation of ribulose bisphosphate carboxylase activity by high-resolution proton NMR. FEBS Lett 170: 355–359

Gutteridge S, Parry MA, Keys AJ, Mudd A and Pierce J (1986) A nocturnal inhibitor of carboxylation in leaves. Nature 324: 274–276

Haake V, Zrenner R, Sonnewald U and Stitt M (1998) A moderate decrease of plastid aldolase activity inhibits photosynthesis, alters the levels of sugars and starch, and inhibits the growth of potato plants. Plant J 14: 147–157

Harrison DHT, Runquist JA, Holub A, and Miziorko HM (1998a) Crystal structure of phosphoribulokinase from *Rhodobacter sphaeroides* reveals a fold similar to that of adenylate kinase. Biochemistry 37: 5074–5085

Harrison EP, Willingham NM, Loyd JC and Raines CA (1998b) Reduced sedoheptulose-1,7-bisphosphatase levels in transgenic tobacco lead to decreased photosynthetic capacity and altered carbohydrate accumulation. Planta 204: 27–36

Heo J and Holbrook GP (1999) Regulation of 2′-carboxyarabinitol 1 phosphate phosphatase: Activation by glutathione and interaction with thiol reagents. Biochem J 338: 409–416

Hirasawa M, Brandes HK, Hartman FC and Knaff DB (1998) Oxidation-reduction properties of the regulatory site of spinach phosphoribulokinase. Arch Biochem Biophys 350: 127–131

Hudson GS, Evans JR, von Caemmerer S, Arvidsson YBC and Andrews TJ (1992) Reduction of ribulose bisphosphate carboxylase/oxygenase content by anitsense RNA reduces photosynthesis in transgenic tobacco plants. Plant Physiol 98: 294–302

Issakidis E, Saarinen M, Decottignies P, Jacquot J-P, Cretin C, Gadal P and Miginiac-Maslow M (1994) Identification and characterization of the second regulatory disulfide bridge of recombinant sorghum leaf NADP-malate dehydrogenase. J Biol Chem 269: 3511–3517

Jacquot JP, Lopez-Jaramilto J, Chueca A, Cherfils J, Lemaire S, Chedozeau B, Miginiac-Maslow M, Decottignies P, Wolosiuk RA and Lopez-Gorge J (1995) High-level expression of recombinant pea chloroplast fructose-1,6-bisphosphatase and mutagenesis of its regulatory site. Eur J Biochem 229: 675–681

Jacquot JP, Lopez-Jaramilto J, Miginiac-Maslow M, Lemaire S, Cherfils J, Chueca A and Lopez-Gorge J (1997) Cysteine-153 is required for redox regulation of pea chloroplast fructose-1,6-bisphosphatase. FEBS Lett 401: 143–147

Jordan DB and Chollet R (1983) Inhibition of ribulose bisphosphate carboxylase by substrate ribulose 1,5-bisphosphate. J Biol Chem 258: 13752–13758

Jordan DB and Ogren WL (1981) Species variation in the specificity of ribulose bisphosphate carboxylase/oxygenase. Nature 291: 513–515

Kanevski I, Maliga P, Rhoades DF and Gutteridge S (1998) Engineering Rubisco in higher plants. In: Garab G (ed) Photosynthesis: Mechanisms and Effects, Vol V, pp 3351–3354. Kluwer Academic Publishers, Dordrecht

Kanevski I, Maliga P, Rhoades DF and Gutteridge S (1999) Plastome engineering of ribulose bisphosphate carboxylase/oxygenase in tobacco to form a sunflower large subunit and tobacco small subunit hybrid. Plant Physiol 119: 133–141

Kossmann J, Sonnewald U and Wilmitzer L (1994) Reduction of the chloroplastic fructose-1,6-bisphosphatase in transgenic potato plants impairs photosynthesis and plant growth. Plant J 6: 637–650

Ku SB and Edwards GE (1977) Oxygen inhibition of Photosystem I. Temperature dependence and relation to O_2/CO_2 solubility ratio. Plant Physiol 59: 986–990

Lorimer GH, Gutteridge S and Reddy GS (1989) The orientation of substrate and reaction intermediates in the active site of ribulose-1,5-bisphosphate carboxylase. J Biol Chem 264: 9873–9879

Lundqvist T and Schneider G (1989) Crystal Structure of the complex of ribulose-1,5-bisphosphate carboxylase and a transition state analogue, 2-carboxy-D-arabinitol 1,5-bisphosphate. J Biol Chem 264: 7078–7083

Mann CM (1999) Crop scientists seek a new revolution. Science 283: 310–314

Mate CJ, Hudson GS, von Caemmerer S, Evans JR and Andrews TJ (1993) Reduction of ribulose bisphosphate carboxylase activase levels in tobacco by antisense RNA reduces ribulose bisphosphate carboxylase activation and impairs photosynthesis. Plant Physiol 102: 1119–1128

Newman JM and Gutteridge S (1993) The X-ray structure of *Synechococcus* ribulose-bisphosphate carboxylase/oxygenase-activated quaternary complex at 2.2 Å resolution. J Biol Chem 268: 25876–25886

Newman JM and Gutteridge S (1994) Structure of an effector-induced inactivated state of ribulose 1,5-bisphosphate carboxylase/oxygenase: The binary complex between enzyme. Structure 2: 495–502

Pierce J, Lorimer GH and Reddy GS (1986) Kinetic mechanism of ribulosebisphosphate carboxylase: Evidence for an ordered, sequential reaction. Biochemistry 25: 1636–1644

Portis A (1990) Rubisco activase. Biochim Biophys Acta 1015: 15–28

Portis A (1995) Regulation of rubisco by rubisco activase. J Exp Bot 46: 1285–1291

Price GD, von Caemmerer S, Evans JR, Yu JW, Lloyd J, Oja V, Kell P, Harrison K, Gallagher A and Badger MR (1994) Specific reduction of the chloroplast carbonic anhydrase activity by antisense RNA in transgenic tobacco plants has a minor effect on photosynthetic CO_2 assimilation. Planta 193: 331–340

Price GD, Evans JR, von Caemmerer S, Yu JW and Badger MR (1995) Specific reduction of glyceraldehyde-3-phosphate dehydrogenase activity by antisense RNA reduces CO_2 assimilation via a reduction in RUBP regeneration in transgenic tobacco plants. Planta 195: 369–378

Quick WP, Schurr U, Scheibe R, Schultze E-D, Rodermel SR, Bogorad L and Stitt M (1991) Decreased ribulose bisphosphate carboxylase-oxygenase in transgenic tobacco transformed with 'antisense' rbcS. I. Impact on photosynthesis in ambient growth conditions. Planta 183: 542–554

Read BA and Tabita FR (1994) High substrate specificity factor ribulose bisphosphate carboxylase/oxygenase from eukaryotic marine algae and properties of recombinant cyanobacterial rubisco containing 'algal' residue modifications. Arch Biochem Biophys 312: 210–218

Reismeier JW, Flugge UI, Schutz B, Heineke D, Heldt HW, Willmitzer L and Frommer WB (1993) Antisense repression of the chloroplast triose phosphate translocator affects carbon partitioning in transgenic potato plants. Proc Natl Acad Sci USA 90: 6160–6164

Robinson SP and Portis AR (1989) Adenosine triphosphate hydrolysis by purified rubisco activase. Arch Biochem Biophys 268: 93–99

Rodermel SR, Abbott MS and Bogorad L (1988) Nuclear organelle interactions: nuclear antisense gene inhibits ribulose bisphosphate carboxylase enzyme levels in transformed tobacco plants. Cell 55: 673–684

Roy H and Andrews J (2000) Rubisco: Assembly and mechanism. In: Leegood R, Sharkey T and von Caemmerer S (eds) Photosynthesis: Physiology and Metabolism, pp 53–83. Kluwer Academic Publishers

Salvucci ME and Holbrook GP (1989) Purification and properties of 2-carboxy-D-arabinitol 1-phosphatase. Plant Physiol 90: 679–685

Sugawara H, Yamamoto H, Inoue T, Miyake C, Yokata A and Kai Y (1998) Crystal structure of rubisco from a thermophilic red alga, *Galdiera partita*. In: Garab G (ed) Photosynthesis: Mechanisms and Effects, Vol V, pp 3339–3342. Kluwer Academic Publishers, Dordrecht

Taylor TC and Andersson I (1996) Structural transitions during activation and ligand binding in hexadecameric rubisco inferred from the crystal structure of the activated unliganded spinach enzyme. Nat Struct Biol 3: 95–101

Tolbert NE (1971) Microbodies—peroxisomes and glyoxysomes. Annu Rev Plant Physiol 22: 45–74

Trevanion SJ, Furbank RT and Ashton AR (1997) NADP-malate dehydrogenase in the C4 plant *Flaveria bidentis*. Plant Physiol 113: 1153—1165

Wang ZY, Snyder GW, Esau BD, Portis AR and Ogren WL (1992) Species-dependent variation in the interaction of substrate bound ribulose-1,5-bisphosphate carboxylase/oxygenase (rubisco) and rubisco activase. Plant Physiol 100: 1858–1862

Whitney SM, von Caemmerer S, Hudson GS and Andrews TJ (1999) Directed mutation of the Rubisco large subunit of tobacco influences photorespiration and growth. Plant Physiol 121: 579–588

Zhu G and Jensen RG (1991) Xylulose 1,5-bisphosphate synthesized by ribulose bisphosphate carboxylase/oxygenase during catalysis binds to decarbamylated enzyme. Plant Physiol 97: 1348–1353

Chapter 18

Chloroplastic Carbonic Anhydrases

Göran Samuelsson* and Jan Karlsson
Umeå Plant Science Center, Department of Plant Physiology, Umeå University, S-901 87 Umeå, Sweden

Summary

In biological systems the interconversion between CO_2 and $HCO_3^- + H^+$ is catalyzed by carbonic anhydrases (CAs). They belong to three evolutionary unrelated gene families designated α-, β- and γ-CA, with no significant sequence homologies between representatives of the different CA families. In the stroma of higher plant chloroplasts a β-CA is one of the most abundant proteins. Its specific function is unclear but it is generally assumed to facilitate diffusion of CO_2 to the active site of Rubisco. In the green algae *Chlamydomonas reinhardtii*, another CA, of α-type, was recently discovered. It is located inside the thylakoid lumen and is required for growth of *C. reinhardtii* under low CO_2 conditions (0.035% CO_2 in air). This CA is suggested to catalyze the production of CO_2 and H_2O inside the acidic lumenal compartment from HCO_3^- that is pumped into the lumen by an as yet unknown mechanism. The third and most recently identified chloroplast CA was identified in *Arabidopsis thaliana.* The function of this α-type CA is not known but it is located in the chloroplast stroma. No chloroplastic γ-CA has yet been reported.

* Author for correspondence, email: goran.samuelsson@ plantphys.umu.se

Eva-Mari Aro and Bertil Andersson (eds): Regulation of Photosynthesis, pp. 313–320.

I. Introduction

Carbonic anhydrase (CA) belongs to a small group of enzymes that catalyze inorganic reactions. Other enzymes in this group are superoxide dismutase and catalase (Newman and Raven, 1993), both with high turnover rates. Carbonic anhydrase is a ubiquitous zinc metalloenzyme that catalyzes the reversible interconversion of CO_2 and HCO_3^- with very high turnover rates, reaching $10^6\ s^{-1}$ (Khalifah, 1971). The enzyme was first discovered in red blood cells, after a long search for a catalytic factor that was theoretically predicted as necessary for the rapid transfer of the HCO_3^- from the erythrocyte to the pulmonary capillary. The early results were published by Meldrum and Roughton (1933) and by Stadie and O'Brien (1933). Since then, CAs has been found in most organisms studied, including animals, plants, archaebacteria and eubacteria. The CA catalyzed reactions may influence carboxylase and decarboxylase rates. Therefore pH fluctuations, ion regulation, ion exchange and inorganic carbon diffusion, processes that are fundamental to photosynthesis, respiration, renal tubular acidification, and bone resorption, are also influenced (Longmuir et al., 1966; Broun et al., 1970; Reed and Graham, 1981; Smith, 1988; Tashian, 1989; Raven, 1995).

Interestingly, several iso-enzymes of CA are usually present in organisms and for example in humans, at least ten have so far been identified.

The CAs are grouped into three independent gene families (Hewett-Emmett and Tashian, 1996), α-CA, β-CA, and γ-CA. These have no primary sequence similarities and seem to have evolved independently. It is noteworthy that some organisms have representatives of two (*C. reinhardtii*) or all three families of CAs (*A. thaliana*) (Hewett-Emmett and Tashian, 1996). However, in animals only the α-type has so far been found.

Despite their structural independency, the three CA isoforms share the same general catalytic mechanism (Lindskog, 1997). The metal atom in the active site binds an OH^- ion that attacks a CO_2 molecule to form a zinc-bound HCO_3^- that in a subsequent step is displaced by a water molecule.

Abbreviations: CA – carbonic anhydrase, carbonate dehydratase, carbonate hydro-lyase; C_i – inorganic carbon; PCR – photosynthetic carbon reduction; PS II – Photosystem II; Rubisco – ribulose 1,5-bisphosphate carboxylase/oxygenase

CO_2 dissolved in water is hydrated, forming H_2CO_3, which equilibrates with HCO_3^- and CO_3^{2-} as shown below:

$$CO_2 + H_2O \leftrightarrow H_2CO_3 \leftrightarrow HCO_3^- + H^+ \leftrightarrow$$

$$CO_3^{2-} + 2H^+$$

The uncatalyzed hydration-dehydration reactions are slow, while the dissociation reactions are considered instantaneous. The equilibrium is pH-dependent; at pH levels below the first dissociation constant ($pK_1 \approx 6.4$) CO_2 dominates; at pH between 6.4 and 10.3 (pK_2) HCO_3^- dominates; whereas above the pH of 10.3, CO_3^{2-} dominates. Temperature and/or the ionic strength affect these pK values.

II. Gene Families

All three CA gene families are coding for zinc containing enzymes that catalyze the same chemical reaction, the reversible hydration of CO_2. Despite this, there are no sequence homologies between the different types of CAs. Apparently, they have evolved at three different periods and may represent an example of convergent evolution of catalytic function (Hewett-Emmett and Tashian, 1996).

A. α-Carbonic anhydrase

The α-type is the most studied CA and it has been found in animals, plants, eubacteria and viruses. However, previously it was common to designate the α-CA as the animal CA but recent findings show that α-CA can be found both in cyanobacteria (Soltes-Rak et al., 1997) and in chloroplasts of the green algae *C. reinhardtii* (Karlsson et al., 1998). In addition, a novel α-type CA with a putative location in the chloroplast stroma of the higher plant *A. thaliana* was recently reported (accession # U73462) (S. Larsson, personal communication).

B. β-Carbonic anhydrase

The β-CAs, so far found only in plants and eubacteria, are less studied. Three distinct monophyletic groups are found within this family, based on their amino acid sequences. They can be divided into monocot, dicot and an eubacterial CA types (Hewett-Emmett and Tashian, 1996). The mitochondrial β-CA from

C. reinhardtii (Eriksson et al., 1996) and the recently found cytosolic algal β-CA from *Coccomyxa* sp. PA (Hiltonen et al., 1998) seem to fall into the eubacterial group. The similarity between the different dicot and monocot β-CAs is around 70%, whereas the similarity within each group is higher (80%).

C. γ-Carbonic anhydrase

The γ-CA was recently discovered in the archaebacterium *Methanosarcina thermophila* (Alber and Ferry, 1994). A homologous protein was discovered earlier in the cyanobacterium *Synechococcus* PCC 7942 (Price et al., 1993). Lately, also plant γ-CA representatives were revealed by the presence in the databases of partial *A. thaliana* Expressed Sequence Tags (ESTs) (Newman et al. 1994).

Recent analysis of the *A. thaliana* database reveals at least 14 genes potentially encoding carbonic anhydrases (6α-CA, 5β-CA, 3γ-CA) (Moroney et al., 2001).

III. Structure

As evident from the large difference in primary protein structures of the various CAs, their secondary and tertiary structures must be different. The three-dimensional structures have now been resolved for members of all CA families and despite the differences in primary amino acid sequences, they show a striking similarity in their metal-coordinating sites (Lindskog, 1997; Kimber and Pai, 2000).

A. α-Carbonic anhydrase

In general, α-CAs are monomeric with estimated molecular sizes of about 30 kDa. The molecular structure is dominated by antiparallel β-sheets forming a spherical molecule with two halves. The active site is located in a funnel shaped crater with the zinc atom located near the bottom. The zinc atom is coordinated to the nitrogen atoms of three conserved histidine residues (Lindskog, 1997). These three histidine residues are present in all known active α-CAs. It should be noted that there are several proteins with sequences homologous to the α-type CAs, but they are lacking one or several of the conserved active site forming residues (Hewett-Emmett and Tashian, 1996). One such example is the yam storage protein (dioscorin) (Conlan et al., 1995). An example of a heteromeric α-type enzyme is the glycosylated periplasmic CA of the green unicellular alga *C. reinhardtii* (Kamo et al., 1990). This isozyme is a heterotetramer consisting of two small (4 kDa) and two large (35–37 kDa) subunits linked with disulfide bonds.

B. β-Carbonic anhydrase

It was not until recently that the three-dimensional structure of a higher plant chloroplastic β-CA was resolved (Kimber and Pai, 2000). The structure of the pea β-CA was shown to be an octamer, a native form that had also been suggested earlier (Aliev et al., 1986; Björkbacka et al., 1997). Unlike the mammalian enzymes that exist as monomers in their native state, plant β-CAs are oligomers ranging in size from 42 to 220 kDa depending on the species. The plant β-CAs can further be divided into one high molecular mass group (140 to 220 kDa) containing CAs from dicots (Reed and Graham, 1981) and one low molecular mass group (lower than 100 kDa) containing CAs from monocots (Atkins et al., 1972; Atkins, 1974). Based on the results of Björkbacka et al.(1997) it seems reasonably to suggest that it is the dicots that have their stromal β-CA organized as octamers.

The β-CA found in *Escherichia coli* is also oligomeric (dimer or trimer) with a subunit of 24 kDa, each binding one zinc ion (Guilloton et al., 1992). A β-CA has also been found in the mitochondria of *C. reinhardtii* (Eriksson et al., 1996). Two nearly identical genes code for identical enzymes, which migrate as 22 kDa polypeptides on SDS-PAGE.

Higher plant β-CAs are dependent on a reduced environment for a fully active enzyme (Johansson and Forsman, 1993). The β-CA from the green algae *Coccomyxa* sp. PA, on the other hand, is fully active regardless of whether the reducing agents are present or not (Hiltonen et al., 1998). It has been shown that inactive oxidized β-CA from pea can be re-activated by the addition of SH-reducing agents and it is thought that two cysteine residues are responsible for the sensitivity of the enzyme to oxidation (Björkbacka et al., 1997). The β-CA from *Coccomyxa* sp. PA has a tryptophane and a methionine at the positions corresponding to the two cysteines of the pea β-CA; this may be the reason that *Coccomyxa* sp. PA β-CA is insensitive to oxidation. Moreover, in β-CAs from *E. coli* and *Synechococcus* sp. PCC 7942, the activities of which are also independent of a reducing environment, these two cysteines are missing (Guilloton et al., 1992; Price et al., 1992). Of the nine cysteine residues present in the primary sequence of *Coccomyxa* sp. PA β-CA, only cysteine 47 and cysteine 106 are thought to act as zinc ligands. These two residues are conserved in all β-CAs.

C. γ-Carbonic anhydrase

Recently, the γ-CA structure has been solved and it appears to be very different from the α-CAs. The γ-CA of *M. thermophila* is a trimer with three zinc-containing active sites located at the interface between two monomers (Kisker et al., 1996). Each subunit is dominated by a left-handed β-helix. Metal coordination is shared between two neighboring subunits. However, super-positioning of the metal centers of CA in *M. thermophila* and human α-CA II reveals a stunning similarity (Lindskog, 1997).

IV. Inhibitors

Inhibitors with specific binding to the enzyme's active site have been particularly useful in studies of the function and localization of CAs in cells and organelles. Although all three types of CAs can be rather specifically inhibited, detailed information about the CA-inhibitor interaction is available only for some of the α-CAs. The most important inhibitors belong to the class of sulfonamides. Azetazolamide and etoxyzolamide are examples of two compounds from this category that have been frequently used to inhibit CA activity in plant material (Reed and Graham, 1981). They inhibit CA activity by binding to the metal ion in the active site. The two compounds have very different rates of diffusion through lipid bilayers and this property has been used for selective inhibition of various isozymes of CA. Ethoxyzolamide permeates readily through biological membranes and therefore tends to inhibit CA activity within cells or organelles, while the poor diffusion rate of azetazolamide has been used when extracellular CAs are to be selectively inhibited. Most monovalent anions also inhibit CA activity, although with a large variety in dissociation constants. The anions interact with the metal ion of the active site, preventing the formation of the coordinated OH^- ion, which is essential for catalytic CO_2 hydration (Lindskog, 1997). Other inhibitors also exist, for example phenol, imidazol and azide. Also, a naturally occurring endogenous inhibitor of CA activity has been isolated from blood plasma (Hill, 1986).

V. Carbonic anhydrase catalysed functions in Chloroplasts

Carbonic anhydrase is apparently ubiquitous in plants. The fact that CAs catalyze equilibrium reactions has made it difficult to resolve their specific biological functions. CAs have been found in most organism groups and they often occur in several isoforms. In the chloroplast, only the α-CA and β-CA families have been identified. That the CA genes, corresponding to various isozymes, have been retained during evolution has given us the reason to believe that their biological function is in some way crucial. Despite this, no really obvious biological function has been proven for CAs in plant cells, except for the cytosolic β-CA of mesophyll cells in C_4 plants that is involved in the carboxylation of phosphoenolpyruvate (Hatch and Burnell, 1990). A typical example of the difficulties to define a function for stromal β-CA of higher plant chloroplasts comes from studies where the β-CA was down-regulated by antisense techniques in tobacco plants to levels less than 2% of the controls. Despite the low level of CA in these transformed plants there was no effect on photosynthesis per unit leaf area and only a small increase in stomatal conductance (Majeau et al., 1994; Price et al., 1994). It is possible that the chloroplast β-CA played a more critical role in photosynthesis during the early expansion of land plants in the Cretaceous era when CO_2 concentrations were lower than they are today. The increase of CO_2 in the atmosphere may have rendered the chloroplast β-CA expendable except during times when CO_2 is limiting. Another possibility is that the function of the stromal β-CA is only crucial under conditions of severe abiotic stress and therefore the effect was not seen in the above experiments.

A. Putative functions of α-Carbonic anhydrase

It has been known for many years that a portion of the cellular CA activity is often associated with a pelletable fraction both in algae (Semenenko et al., 1977) and higher plants (Komarova et al., 1982). Based on differential centrifugation techniques and studies of protein mobility, it was further concluded by the same authors that this activity was mainly associated with the thylakoid fraction. Vaklinova et al. (1982) further showed that the CA activity was enriched in isolated PS II particles (for review see Stemler, 1997).

As mentioned before, in plants including green algae, two chloroplast α-CAs have so far been identified. A complete coding region for an α-CA in *A. thaliana* was obtained by screening a cDNA library with a homologous EST-clone encoding part of a putative α-CA (accession number Z18493) (S. Larsson, unpublished). Western blot hybridization with specific antibodies against various cell fractions and Immuno-Gold localization experiments show a chloroplastic location of this CA (denoted as *Arabidopsis* Cah1). The full-length cDNA sequence does not contain a typical transit peptide, although it is obvious that it is located in the chloroplast. Northern and western blot hybridization experiments on plants exposed to various kinds of physiological stress, like cold exposure, high light and cadmium exposure, indicate that the expression of this CA is either up-regulated (cadmium) or down-regulated (cold, drought and anoxia). The function of this CA remains to be determined.

A recently isolated chloroplastic α-CA (Cah3) from the green alga *C. reinhardtii* was found to be located in the thylakoid lumen (Karlsson et al., 1995,1998). Thylakoid membranes and purified pyrenoids were isolated and used for western blot hybridization experiments with Rubisco- and α-CA-specific antibodies. The pyrenoid is a special proteinaceous body that contains most, if not all of the Rubisco (Kuchitsu et al., 1988). Immunological analyses showed that the α-CA antibody did not cross-react with the pyrenoid fraction while it gave a strong signal with the thylakoid membrane preparation (J. Karlsson and G. Samuelsson, unpublished). Contrarily, the antibody against Rubisco reacted strongly with the pyrenoid, but not with the thylakoid preparations. This is a further indication that Cah3 is located within the lumen, but not in the pyrenoid of *C. reinhardtii*. A mutant (*cia-3*), lacking the activity of Cah3, is unable to grow at low CO_2 conditions (0.035% CO_2). Complementation of this mutant with the wild-type cah3 gene restored the capacity for growth in air (Karlsson et al., 1998). The protein is expressed both under high (5%) and low (0.035%) CO_2 environment, with about a doubling of the corresponding mRNA levels in the latter condition. Cah3 does not seem to be directly linked to the inducible carbon concentrating mechanism found in many micro algae, but it is nevertheless vitally important for growth at low CO_2 conditions. In experiments where the cultures are grown with an organic supplement (acetate) in darkness, the CA is totally degraded within 4-5 days. Under this growth condition, the cells can still grow well, demonstrating that the lumenal CA in *C. reinhardtii* is not required during heterotrophic growth.

It should be stated that the specific function of the lumenal CA is not completely resolved. However, several different (partly conflicting) hypotheses about the function of a thylakoid bound CA have emerged from the time of the first publication on this subject by Warburg and Krippahl (1960). The two main principal ideas can be summarized as follows: CA is located at or close to the PS II complex and is in some unknown way required for an efficient oxidation of water to molecular oxygen, protons and electrons. This involvement may occur either directly by supplying substrate to the oxygen evolving centers or by supplying an inorganic carbon species (most likely HCO_3^-) as a ligand to stabilize the manganese complex known to be involved in this reaction. The first hypothesis assumes of course that oxygen is not formed directly by the splitting of water, but by the splitting of water bound in the bicarbonate ion (see Stemler, 1997). The other hypothesis that was put forward first by Pronina and Semenenko (1990) and later by Raven (1997) states that the lumenal CA is mainly responsible for the rapid formation of CO_2 from bicarbonate within the lumen. According to this model, bicarbonate is actively pumped into the lumen by some yet unknown mechanism (Fig. 1). The CO_2 formed by dehydration of bicarbonate in the lumen could thereafter readily diffuse out through the thylakoid membrane to the pyrenoid in the stroma where Rubisco is located.

In a recent paper, Park et al. (1999) aimed at resolving the physiological role and the specific localization of the lumenal CA in *C. reinhardtii*. Interestingly, experiments with western blot hybridization techniques indicated that the lumenal CA is associated with PS II particles but not with PS I particles. This strengthen the hypothesis that CA in some way is involved in the oxygen production. Using Chl *a* fluorescence techniques and simultaneous measurements of oxygen evolution they were able to confirm, by comparing the mutant cia-3 and WT cells, that the lumenal CA in *C. reinhardtii* is indeed supporting Calvin cycle activity by providing

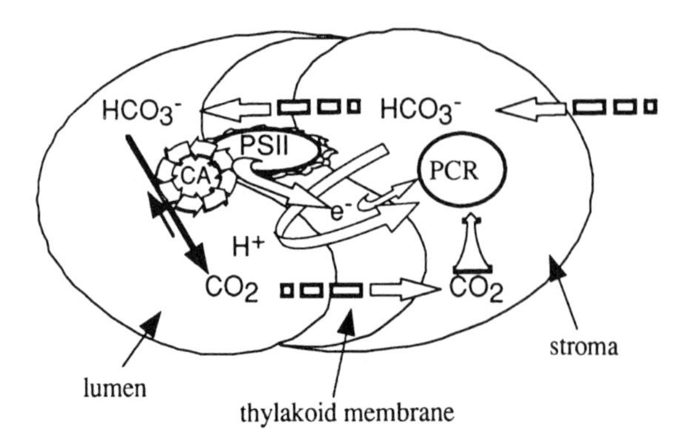

Fig. 1. A model for the thylakoid localized PS II-CA-driven supply of CO_2 for carbon fixation.

CO_2 to Rubisco. It was further proposed that the pH gradient formed over the membrane by light-driven PS II electron and proton transport is the driving force for transport of bicarbonate into the lumen. According to the model, in most microalgae, Rubisco is found in the pyrenoid that in *C. reinhardtii* is transversed by thylakoid membranes. CO_2 from the intra thylakoid space is then diffusing out to the pyrenoid that contains most, if not all of the Rubisco. This would lead to a local increase of CO_2 at the active site of Rubisco, providing that the pyrenoid itself does not contain any CA activity. A prediction of the structure of Cah3 indicates that it can occur in two different molecular forms (T. Shutova, personal communication). Assuming that the structure of Cah3, and thereby its binding properties, can change depending on the chemical conditions in the lumen, one might suggest that it exerts its function at or close to PS II, promoting electron donation to the PS II under conditions when electron transport is limiting photosynthesis. Under conditions of C_i limitations it may on the other hand support an efficient carbon fixation according to the model of Raven (1997).

Specific antibodies against the Cah3 enzyme cross-react with total leaf homogenates of many different higher plants. In plants like *A. thaliana*, *Populus tremula x tremuloides* and *Hordeum vulgare* one specific immunoreaction occurs involving a protein of a molecular size corresponding to the lumenal α-CA in *C. reinhardtii*, 30 kDa (G. Samuelsson, unpublished).

It is not easy to speculate about the putative function of the lumenal α-CA in C_3 plants, especially since these are assumed to lack a carbon concentrating mechanism and, contrary to green algae, have a soluble Rubisco distributed rather equally in the stroma. This fact alone would favor a different function for CA in higher plants than in algae. Furthermore, the function of this putative CA is difficult to resolve since a stromal β-CA isoform is an abundant protein of higher plant chloroplasts; and most membrane sub-fractions are therefore likely to be contaminated with small amounts of stromal β-CA.

Whether there is a functional α-CA in the thylakoid membranes of higher plant chloroplasts still remains to be proven. The genomic sequence of *A. thaliana*, published recently (The Arabidopsis Genome Initiative, 2000) reveals five putative α-CAs in addition to the chloroplastic Cah1. None of these five putative isozymes have a transit peptide that indicates that the protein is targeted to the thylakoid lumen. Therefore, the localization and function of these still remains to be elucidated.

B. Putative functions of β-Carbonic anhydrase

In principle, CAs might be of importance in all carboxylation/decarboxylation reactions. In the chloroplast, the dominating carboxylating reaction is that of Rubisco where CO_2 from the atmosphere is fixed forming 3-phosphoglycerate and simple sugars in the Calvin cycle. In the chloroplast stroma of C_3 plants, β-CA is an abundant protein making up somewhere between 0.5–2% of the total soluble leaf protein. The role of this chloroplastic CA is believed to be to facilitate C_i diffusion from the chloroplast envelope to Rubisco, where it is delivered (in the form of CO_2) to the enzyme's active site for fixation. One important tool to study the function of enzymes is the production of a knock-out mutant for a specific gene product. This is not easily achieved in plants, but by transforming plants with the gene of interest in an anti-sense direction a somewhat similar effect can be obtained. In an elegant experiment using anti-sense RNA techniques, the CA activity could be decreased in some of the transgenic tobacco plants by 98–99% (Majeau et al., 1994; Price et al., 1994). Despite this decrease, there was no effect on photosynthesis per unit leaf area at ambient levels of CO_2. A slight increase in stomatal conductance was obtained in the transformants, indicating that the plants can compensate for the reduction of CA activity by increasing intercellular CO_2 levels. Price et al. (1994) further showed that the ^{13}C isotope discrimination of leaf dry matter was reduced and it was calculated to give a 15-20 μbar decrease in chloroplast CO_2 concentration. These data support, although indirectly, the idea that β-CA in the chloroplast stroma is in some way involved in the facilitated diffusion of CO_2.

The conclusion is that although we believe that CAs in the chloroplast as in plants in general are of vital importance for biological functions, more work is required to resolve the specific function of each individual isozyme.

References

Albcr BE and Ferry JG (1994) A carbonic anhydrase from the archaeon *Methanosarcina thermophila.* Proc Natl Acad Sci USA 91: 6909–6913

Aliev DA, Guliev NM, Mamedov TG, and Tsuprun VL (1986) Physicochemical properties and quaternary structure of chick pea leaf carbonic anhydrase. Biokhimiya 51:1785–1794

Atkins CA (1974) Occurrence and some properties of carbonic anhydrase from legume root nodules. Phytochem 13: 93–98

Atkins CA, Patterson BD and Graham D (1972) Plant carbonic anhydrases. Plant Physiol 50: 218–223

Björkbacka H, Johansson IM, Skärfstad E and Forsman C (1997) The sulfhydryl groups of Cys 269 and Cys 272 are critical for the oligomeric state of chloroplast carbonic anhydrase from *Pisum sativum.* Biochemistry 36: 4287–4294

Broun G, Selegny E, Minh CT and Thomas D (1970) Facilitated transport of CO_2 across a membrane bearing carbonic anhydrase. FEBS Lett 7: 223–226

Conlan RS, Griffiths LA, Napier JA, Shewry PR, Mantell S, Ainsworth C (1995) Isolation and characterisation of cDNA clones representing the genes encoding the major tuber storage protein (dioscorin) of yam (*Dioscorea cayenensis* Lam.). Plant Mol Biol 28: 369–380

Eriksson M, Karlsson J, Ramazanov Z, Gardeström P and Samuelsson G (1996) Discovery of an algal mitochondrial carbonic anhydrase: Molecular cloning and characterization of a low-CO_2-induced polypeptide in *Chlamydomonas reinhardtii.* Proc Natl Acad Sci USA 93: 12031–12034

Guilloton MB, Korte JJ, Lamblin AF, Fuchs JA and Anderson PM (1992) Carbonic anhydrase in *Escherichia coli* A product of the *cyn* operon. J Biol Chem 267: 3731–3734

Hatch MD and Burnell JN (1990) Carbonic anhydrase activity in leaves and its role in the first step of C_4 photosynthesis. Plant Physiol 93: 825–828

Hewett-Emmett D and Tashian RE (1996) Functional diversity, conservation, and convergence in the evolution of the α-, β-, and γ-carbonic anhydrase gene families. Mol Phylogenet Evol 5: 50–77

Hill EP (1986) Inhibition of carbonic anhydrase by plasma of dogs and rabbits. J Appl Physiol 60:191–197

Hiltonen T, Björkbacka H, Forsman C, Clarke AK and Samuelsson G (1998) Intracellular β-carbonic anhydrase in the unicellular green alga *Coccomyxa.* Plant Physiol 117:1341–1349

Johansson IM and Forsman C (1993) Kinetic studies of pea carbonic anhydrase. Eur J Biochem 218: 439–446

Kamo T, Shimogawara K, Fukuzawa H, Muto S and Miyachi S (1990) Subunit constitution of carbonic anhydrase from *Chlamydomonas reinhardtii.* Eur J Biochem 192: 557–562

Karlsson J, Hiltonen,T, Husic HD, Ramazanov Z and Samuelsson G (1995) Intracellular carbonic anhydrase of *Chlamydomonas reinhardtii.* Plant Physiol 109: 533–539

Karlsson K, Clarke AK, Chen Z, Hugghins SY, Park Y-I, Husic D, Moroney JV and Samuelsson G (1998) A novel α-type carbonic anhydrase associated with the thylakoid membrane in Chlamydomonas reinhardtii is required for growth at ambient CO_2. EMBO J 17:1208–1216

Khalifah RG (1971) The carbon dioxide hydration activity of carbonic anhydrase. Stop-flow kinetic studies on the native human isoenzymes B and C. J Biol Chem 246: 2561–2573

Kimber MS and Pai EF (2000) The active site architecture of *Pisum sativum* β-carbonic anhydrase is a mirror image of that of α-carbonic anhydrases. EMBO J 19: 1407–1418

Kisker C, Schindelin H, Alber BE, Ferry JG and Rees DC (1996) A left-handed β-helix revealed by the crystal structure of a carbonic anhydrase from the archaeon *Methanosarcina thermophila.* EMBO J 15: 2323–2330

Komarova Y, Doman N and Shaposhnikov G (1982) Two forms of carbonic anhydrase from bean chloroplasts. Biokhimiya/ Biochemistry (English translation). 47: 1027–1034

Kuchitsu K, Tsuzuki M and Miyachi S (1988) Characterization of the pyrenoid isolated from the unicellular green alga *Chlamydomonas reinhardtii*: Particulate form of Rubisco protein. Protoplasma 144: 17–24

Lindskog S (1997) Structure and mechanism of carbonic anhydrase. Pharmacol Theor 74: 1–20

Longmuir IS, Forster RE and Woo C-Y (1966) Diffusion of carbon dioxide through thin layers of solution. Nature 209: 393–394

Majeau N, Arnoldo M and Coleman, JR (1994) Modification of carbonic anhydrase activity by antisense and over-expression constructs in transgenic tobacco. Plant Mol Biol 25: 377–385

Meldrum NU and Roughton FJW (1933) Carbonic anhydrase: Its preparation and properties. J Physiol 80: 113–142

Moroney JV, Bartlett SG, and Samuelsson G (2001) Carbonic anhydrases in plants and algae. Plant Cell Environ 24: 141–154

Newman JR and Raven JA (1993) Carbonic anhydrase in *Ranunculus penicillatus* spp. Pseudofluitans: Activity, location and implications for carbon assimilation. Plant Cell Environ 16: 491–500

Newman T, deBruijn FJ, Green P, Keegstra K, Kende H, McIntosh L, Ohlrogge J, Raikhel N, Somerville S, Thomashow M, Retzel E and Somerville C (1994) Genes galore: A summary of methods for accessing results from large-scale partial sequencing of anonymous *Arabidopsis* cDNA clones. Plant Physiol 106: 1241–1255

Park Y-I, Karlsson J, Rojdestvenski I, Pronina N, Klimov V, Öquist G and Samuelsson G. (1999) The role of the novel Photosystem II-associated carbonic anhydrase in carbon concentrating mechanism (CCM) in *Chlamydomonas reinhardtii.* FEBS Lett 444: 102–105

Price GD, Coleman JR and Badger MR (1992) Association of carbonic anhydrase activity with carboxysomes isolated from the cyanobacterium *Synechococcus.* PCC7942. Plant Physiol 100: 784–793

Price GD, Howitt SM, Harrison K and Badger MR (1993)

Analysis of genomic DNA region from the cyanobacterium *Synechococcus* sp strain PCC7942 involved in carboxysome assembly and function. J Bacteriol 175: 2871–2879

Price GD, Caemmerer Sv, Evans JR, Yu J-W, Lloyd J, Oja V, Kell P, Harrison K, Gallagher A and Badger MR (1994) Specific reduction of chloroplast carbonic anhydrase activity by antisense RNA in transgenic tobacco plants has a minor effect on photosynthetic CO_2 assimilation. Planta 193: 331–340

Pronina NA and Semenenko, VE (1990) Membrane-bound carbonic anhydrase takes part in CO_2 concentration in algae cells. In: Baltscheffsky M (ed) Current Research in Photosynthesis, Vol IV, pp 489–492. Kluwer Academic Publishers, Dordrecht

Raven JA (1995) Photosynthetic and non-photosynthetic roles of carbonic anhydrase in algae and cyanobacteria. Phycologia 34: 93–101

Raven JA (1997) CO_2 concentrating mechanisms: A direct role for thylakoid lumen acidification? Plant Cell Environ 20: 147–154.

Reed ML and Graham, D. (1981) Carbonic anhydrase in plants: distribution, properties and possible physiological roles. In: Reinholdt L, Harborne JB, and Swain T (eds) Progress in Phytochemistry, Vol 7, pp 47–94. Pergamon Press, Oxford

Semenenko VE, Avramova S, Georgiev D and Pronina NA (1997) Comparative study of activity and localisation of carbonic anhydrase in cells of Chlorella and Scenedesmus. Sov Plant Physiol 24: 852–856

Smith RG (1988) Inorganic carbon transport in biological systems. Comp Biochem Physiol 90B: 639–654

Soltes-Rak E, Mulligan ME and Coleman JR (1997) Identification of a gene encoding a vertebrate-type carbonic anhydrase in cyanobacteria. J Bacteriol 179: 769–774

Stadie WC and O'Brien H (1933) The catalysis of the hydration of carbon dioxide and dehydration of carbonic acid by the enzyme isolated from red blood cells. J Biochem 103: 521–529

Stemler AJ (1997) The case for chloroplast thylakoid carbonic anhydrase. Physiol Plant 99: 348–353

Tashian RE (1989) The carbonic anhydrases: Widening perspectives on their evolution, expression and function. Bioessays 10: 186–192

The Arabidopsis Genome Initiative (2000) Analysis of the genome sequence of the flowering plant *Arabidopsis thaliana* The Arabidopsis Genome Initiative. Nature 408: 796–815

Vaklinova SG, Goushtina LM and Lazova GN (1982) Carboanhydrase activity in chloroplasts and chloroplast fragments. C R Acad Bulg Sci 35: 1721–1724

Warburg O and Krippahl G (1960) Notwendigkeit der kohlensaure für die chinon und ferricyanid-reaktionen in grünen grana. Z Naturforsch 15: 367–369

Chapter 19

Thioredoxin and Glutaredoxin: General Aspects and Involvement in Redox Regulation

Arne Holmgren*

Medical Nobel Institute for Biochemistry, Department of Medical Biochemistry and Biophysics, Karolinska Institutet, S-171 77 Stockholm, Sweden

Summary

Redox-active thiols play a major role in the mechanisms of essential biosynthetic and repair enzymes, defense against oxidative stress or thiol redox control of transcription, translation and enzyme activity. Thioredoxins are strong reductants having a dithiol in the conserved active site sequence: Cys-Gly-Pro-Cys which is essential for catalytic activity in disulfide reduction. Thioredoxin reductase regenerates the dithiol form of thioredoxin with electrons from NADPH. Unique to photosynthesis, chloroplasts have ferredoxin-thioredoxin reductase, which reduces target specific thioredoxins *f* and *m* by electrons from light to control the activity of CO_2 assimilation enzymes. The other major disulfide reducing system is comprised of glutaredoxins, glutathione and glutathione reductase which together catalyze disulfide reductions by NADPH, either in dithiol reactions with functions overlapping that of thioredoxins, or unique in monothiol reactions. In this way, changes in the redox potential of the cellular redox buffer of glutathione and its disulfide can be transmitted for reversible regulation of protein function by thiol redox control. The thioredoxin superfamily of proteins comprise thioredoxins, glutaredoxins, glutathione peroxidases, thioredoxin peroxidases, glutathione transferases or protein disulfide isomerases all of which encompass the thioredoxin fold, a three-dimensional structure with a $\beta\alpha\beta\alpha\beta\beta\alpha$ unit and a unique *cis*-proline residue.

Recent studies on three-dimensional structures of ferredoxin-thioredoxin reductase and target enzymes, such as malate dehydrogenase and fructose bisphosphatase having unique extensions with regulatory disulfides,

*Email: Arne.Holmgren@mbb.ki.se

Eva-Mari Aro and Bertil Andersson (eds): Regulation of Photosynthesis, pp. 321–330.

have greatly advanced the knowledge about redox signaling systems unique to plants. Plants have unusually many thioredoxin and glutaredoxin genes with as yet incompletely known functions. Remarkably, plants have retained the low molecular weight type of thioredoxin reductase found in Archae, eubacteria and fungi. In contrast, *Caenorhabditis elegans* and mammalian thioredoxin reductases are selenocysteine-dependent enzymes of higher molecular weight with different mechanism and many additional functions. An exciting emerging field concerns redox regulation of transcription and translation via members of the thioredoxin superfamily.

I. Introduction

Thioredoxin (Trx), NADPH and thioredoxin reductase (TrxR), (the thioredoxin system) is a hydrogen donor for ribonucleotide reductase (RNR) the universal essential enzyme providing the deoxyribonucleotides for DNA synthesis in all living systems (Holmgren, 1985, 1989; Jordan and Reichard, 1998). Ribonucleotide reductases operate by a radical mechanism to reduce all four ribonucleotides (Jordan and Reichard, 1998) and each deoxyribonucleotide produced generates a disulfide in the enzyme, which is reduced by thioredoxin at the expense of NADPH (Fig. 1). Thioredoxins, which exists in all living species are general protein disulfide-oxidoreductases operating by a dithiol mechanism (Holmgren, 1985, 1989). The relatively strong reductive capacity of thioredoxin makes it the cellular equivalent of dithiothreitol (DTT). However, thioredoxin is a sophisticated specific catalyst with a large number of functions in different biological systems through specific protein-protein interactions. Redox regulation by thiol redox control was first described in plant photosynthesis in 1976 (Buchanan, 1991) and has received a great deal of attention also in many other systems. The chloroplast has a unique way of transmitting energy from light to disulfide reduction via the enzyme ferredoxin-thioredoxin reductase (FTR).

Glutathione (GSH) occurs in almost all organisms including plants in high (1–10 mM) concentrations and is a major redox buffer in cells with a wide range of functions (Meister, 1995). Glutaredoxin was discovered as an enzyme coupling the oxidation of glutathione to the synthesis of deoxyribonucleotides (Fig. 1) in E. coli mutants lacking thioredoxin (Holmgren, 1976). Glutaredoxins are not reduced by thioredoxin reductase and catalyze GSH-disulfide oxidoreductions either in monothiol or dithiol reactions and are coupled to reduction of GSSG by glutathione reductase utilizing NADPH (the glutaredoxin system) or de novo synthesis of GSH.

Thioredoxin and glutaredoxin have a common three-dimensional structure (the thioredoxin fold) also present in other proteins composed of multiple thioredoxin domains such as protein disulfide isomerases. The growing thioredoxin superfamily of proteins are important in redox reactions, thiol redox control, defense against oxidative stress by oxygen free radicals or detoxification. This chapter summarizes knowledge of the redoxin systems with particular emphasis on plants.

Abbreviations: DTNB – 5,5′-dithiobis (2-nitrobenzoic acid); DTT – dithiothreitol; FTR – ferredoxin-thioredoxin reductase; Grx – glutaredoxin; NMR – nuclear magnetic resonance; PAPS′ – 3′-phosphoadenyl sulfate; Trx – thioredoxin; TrxR – thioredoxin reductase

II. Thioredoxins

A. Functions

Thioredoxins have a large number of functions many of which are linked to their activity as a disulfide reductases (Table 1) coupling the reductive power of NADPH to enzyme-dependent substrate reduction. Unique to plants is the linking of reduced ferredoxin to cleavage of disulfide bonds in enzyme regulation (Buchanan et al; Dai et al., 2000). The thioredoxin are versatile enzymes able to catalyze not only reduction of disulfides, but also formation of such covalent cross links by its general protein dithiol-disulfide oxidoreductase activity (Holmgren, 1985). The many functions of thioredoxins can be classified in different categories. A first function is obviously to act as electron carriers for biosynthetic enzymes such as ribonucleotide reductases, methionine sulfoxide reductases and sulfate reductases (Table 1). A second function is to act as structural components of another enzyme by forming a complex. This is best known for T7 DNA polymerase, where Trx-$(SH)_2$ binds with high affinity (3 nM) to the DNA polymerase (Gene 5 protein) forming a 1:1 complex

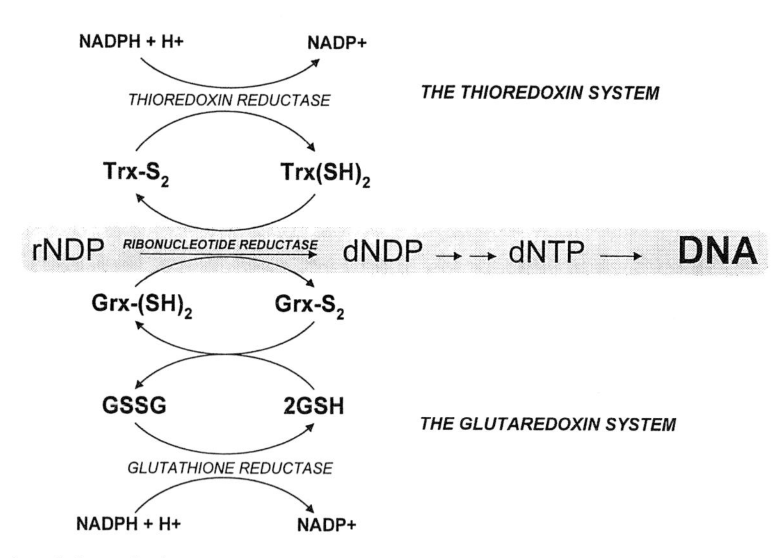

Fig. 1. Role of the thioredoxin and glutaredoxin systems as hydrogen donors for ribonucleotide reductase in reducing ribonucleonucleoside diphosphates (rNDP) to deoxyribonucleoside diphosphates (dNDP). Deoxyribonucleotides (dNTP) are substrates for DNA-synthesis.

Table 1. Major functions of thioredoxins in different biological systems

Function	Organisms	Selected references
Hydrogen donor of ribonucleotide reductase (DNA synthesis)	Bacteria, yeast, plants, mammals	Holmgren, 1989
Hydrogen donor for sulfate reduction (PAPS reductase) to sulfite (assimilation of sulfur)	Bacteria, yeast	Lillig et al., 1999
Hydrogen donor for methionine sulfoxide reductases (protein repair)	All organisms	Brot et al., 1986; Hassouni et al., 1999
Subunit of phage T 7 DNA polymerase. Processivity factor for DNA synthesis	*E. coli* / T 7 virus	Huber et al., 1987
Role in assembly of filamentous Trx a phages	*E. coli* / f 1, M 13 virus	Russel and Model, 1986
Regulation of photosynthetic enzymes in chloroplast. Enzyme regulation by light via ferredoxin	Plant, cyanobacteria?	Buchanan, 1991
General protein disulfide reductase many protein, e.g. insulin	All organisms	Holmgren, 1984
Redox regulation of transcription factor; NFkB, AP 1 etc.	Mammalian cells	Schenk et al., 1994
Defense against oxidative stress by acting via thioredoxin peroxidases (peroxiredoxins)	Many organisms	Kang et al., 1998
Regulation of apoptosis	Mammalian cells	Satoh et al., 1998

giving the enzyme high processivity (Huber et al, 1987). Remarkably, only Trx-$(SH)_2$ binds whereas Trx-S_2 is unable to bind despite their closely related structures (Holmgren, 1995). A similar role is assumed for phage assembly (Russel and Model, 1986). A striking example of the ability of thioredoxins to regulate a process is that of mammalian Trx-$(SH)_2$, which makes a complex with apoptosis signaling kinase 1 (ASK1) and prevents downstream signaling for apoptosis (Saitoh et al., 1998). Similar to the case for T7 DNA polymerase Trx-S_2 does not form a complex and leaves the ASK1 activity turned on giving a redox mechanism for apoptosis control. A third type of function is regulation of enzymes or transcription factors by thiol redox control (Table 1). A fourth type of function is to participate in defense mechanisms against oxidative stress (Table 1) by acting as electron donors to the ubiquitous family of thioredoxin peroxidases or peroxiredoxins (six members in mammalian cells) (Kang et al., 1998). Peroxiredoxins reduce H_2O_2 with a low K_m-value via cysteine residues forming a disulfide with a sulfenic acid as an intermediate. Oxidative stress in cells may lead to formation of artificial disulfides in proteins, which will be reduced by the thioredoxin system. It is also clear that the thioredoxin system has a major role in creating and keeping a reducing environment inside of cells (Prinz et al., 1997). Proteins in the cytosol have no or few disulfides in contrast to the cell surface and extracellular proteins which are stabilized by many disulfides acting as stable covalent crosslinks.

B. Structures and Mechanisms

Thioredoxins have the conserved active site sequence Cys-Gly-Pro-Cys (Fig. 2). The three-dimensional structure of oxidized *E. coli* thioredoxin was solved by X-ray crystallography (Holmgren, 1975). This structure, which now is known as the archetypical thioredoxin fold, showed a central core of five β-strands surrounded by four α-helices. The active site disulfide is located at the end of one β-strand and in the beginning of the long α-helix (Fig. 3). The complete structures for both oxidized and reduced *E. coli* thioredoxin were determined in solution by NMR and are very similar but with detailed local structure changes in the surface around the disulfide/dithiol (Jeng et al., 1994; Holmgren, 1995). Even though the disulfide of thioredoxin only links a short loop, the reduced form has more conformational substrates (Jeng et al., 1994). The result of the local conformational change is dramatically illustrated by the high affinity binding of only reduced thioredoxin to the gene 5 protein of T7 DNA polymerase; oxidized thioredoxin has at least five orders of magnitude lower binding affinity (Huber et al., 1987).

The structures of glutaredoxins, glutathione transferases and glutathione peroxidases and the *E. coli* B Dsb A have also demonstrated the

Fig. 2. Sequences of selected thioredoxins. The conserved active site CGPC and other conserved residues are dashed. Numbering and secondary structure elements are from *E. coli* Trx 1. (See Fig. 3).

thioredoxin fold (Martin, 1995). This has led to the insight that the thioredoxin fold is a widespread motif found in many enzymes in a growing superfamily of proteins.

The thioredoxin catalytic mechanism was formulated on the basis of the finding that the N-terminal active site Cys residue in *E. coli* thioredoxin has a low pKa-value (around 7) (Kallis and Holmgren, 1980) as also determined by NMR, and that the

C-terminal active site Cys-residue is shielded and not reactive with alkylating agents (Holmgren, 1985; Dyson et al., 1997). The mechanism (Fig. 4) also involves an initial noncovalent complex formation by docking a flat hydrophobic surface of thioredoxin surrounding the active site to a target protein. This may involve conformational rearrangements in the target protein to expose partly buried disulfides. The thiolate of Cys 32 (*E. coli* thioredoxin numbering) then attacks the target disulfide forming a transient covalent mixed disulfide intermediate. The speed of the thiol-disulfide interchange will be enhanced by a hydrophobic environment. The mixed disulfide intermediate is rearranged by attacking the Cys 35 thiolate to give a reduced protein substrate dithiol and the disulfide form of thioredoxin. The mechanism is reversible and the direction will be determined by the relative redox potentials of thioredoxin and the disulfide in the substrate protein. Buried disulfides will not react. Based on this a method based on specific reduction of disulfides has been developed (Holmgren, 1984), which enables quantitative selective reduction of disulfides by coupling to thioredoxin reductase recording NADPH oxidation. If the C-terminal Cys active site residue is mutated to a Ser the mutant thioredoxin will make stable mixed disulfides with a target protein peptide. Thus, the structure of a peptide from the transcription factor NF_kB bound to thioredoxin has been solved by NMR (Qin et al., 1995). The recent solution structure of

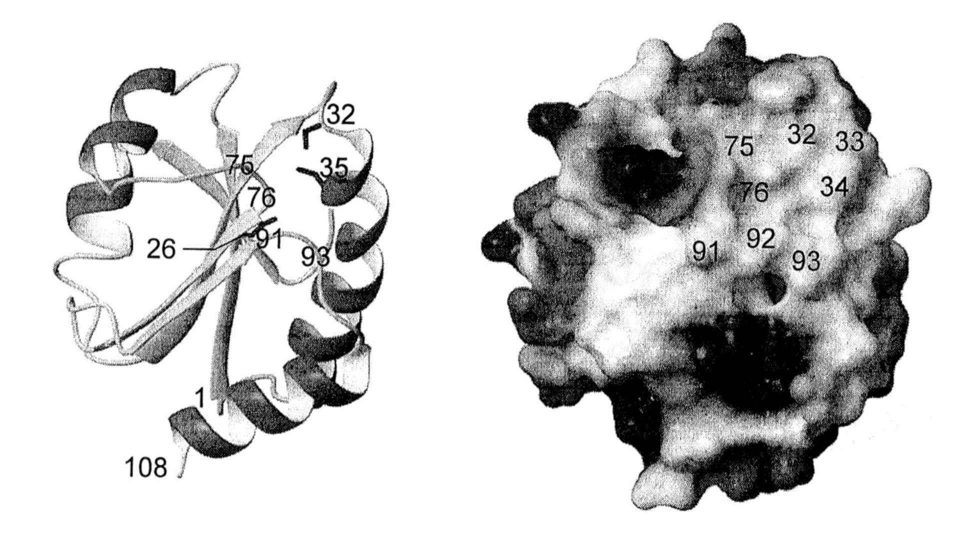

Fig. 3. Structure of *E. coli* thioredoxin (left) ribbon diagram with selected residues (see Fig. 2) and (right) surface for peptide binding (Jeng et al., 1994).

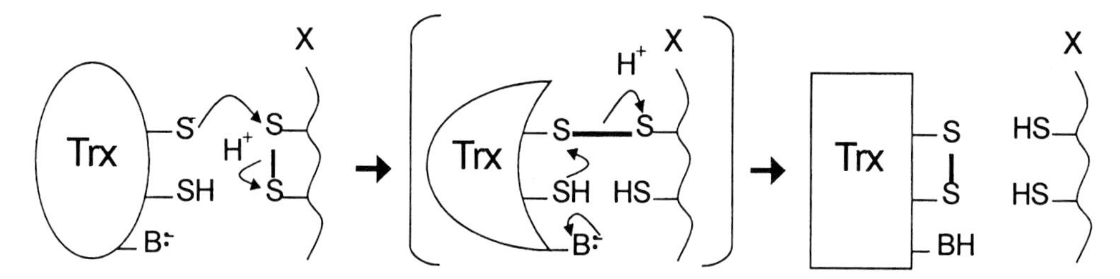

Fig. 4. Mechanism of thioredoxin (Trx) as a protein disulfide reductase with the transient covalent intermediate involving Cys 32 and the target thiol (Holmgren, 1985, 1995).

GSH in complex with glutaredoxin 3 from *E. coli* demonstrates a back-bone hydrogen bonding system involving the residue N-terminal to the conserved *cis*-Pro residue to align the GSH tripeptide as a β-strand to the protein. This ligand binding is conserved in other members of the thioredoxin superfamily (Nordstrand et al., 1999).

The redox potentials of thioredoxins will determine their reactions with disulfides in proteins (Åslund et al., 1997). *E. coli* thioredoxin has a redox potential of −270 mV determined from the equilibrium with NADPH at pH 7.0. Plant thioredoxin *f* has midpoint redox potentials of −290 mV for both the spinach and the pea enzymes and −300 mV for both *C. reinhardtii* and spinach thioredoxin m (Hirasawa et al., 1999). These potentials are slightly more oxidizing than that determined for the redox regulatory disulfide of spinach fructose-1,6-bisphosphatase (Hirasawa et al., 1999).

A major determinant of the redox potential in thioredoxins is found in the residues between the active the half-cystines as first observed with a mutant p34H of *E. coli* thioredoxin mimicking the active site of protein disulfide isomerase (Krause et al., 1991). Redox potentials for the whole thioredoxin/glutaredoxin family have been determined and it is clear that apart from the peptide sequence, also other factors in the protein folding will affect the redox potential (Åslund et al., 1997; Chivers et al., 1997). Clearly, protein disulfide isomerase which has a redox potential of −180 mV, makes it suitable for its function as a dithiol oxidant and particularly in its role as an isomerase of disulfides to reach the final native state (Lundström and Holmgren, 1993).

III. Thioredoxin Reductases

A. Plant Enzymes

1. Ferredoxin Thioredoxin Reductase

The thioredoxin-linked regulation of photosynthetic enzymes by light was discovered by Buchanan and coworkers and is covered in an extensive review in Chapter 20 of this volume (Schürmann and Buchanan, 2000). Ferredoxin thioredoxin reductase (FTR) uses two molecules of reduced ferredoxin to reduce the disulfide of thioredoxin *f* or *m* to generate the dithiol form (see review by Buchanan, 1991). The structure of *Synechocystis* FTR which is a heterodimer of 13 kDa subunits has recently been solved by X-ray crystallography (Dai et al., 2000). FTR is unique since it can reduce disulfides by an iron sulfur cluster, a property that is explained by the tight contact of its active site disulfide and the iron-sulfur center (Dai et al., 2000). The thin and flat shape of the FTR molecule makes the ferredoxin mediated two electron reduction possible by forming on one side a mixed disulfide with thioredoxin and on the opposite site providing access to ferredoxin for delivering the electrons (Dai et al., 2000). Another major recent advance is that the structure of the target protein malate hydrogenase (Johansson et al., 1999) has been solved by X-ray crystallography. Also the plant fructose-1,6-bisphosphatase structure was recently reported (Chiadmi et al., 1999). This progress sets the scene for a rather complete understanding of the unique thioredoxin-dependent chloroplast redox regulatory system.

2. NADPH-Thioredoxin Reductase

Thioredoxin reductase which is universally expressed in all living cells is also present in the cytosol of plant tissues. The enzyme is of the low molecular weight type Mr 70.000 with two 35 kDa subunits homologous to the *E. coli* thioredoxin reductase (Jacquot, J.-P. et al., 1994, Williams Jr, 1995). This enzyme system together with thioredoxin *h* in the cytosol is assumed to participate in the usual thioredoxin dependent reactions as in other organisms including peroxiredoxin-dependent H_2O_2 reduction (Verdoucq et al., 1999). The structures of *Arabidopsis thaliana* thioredoxin reductase has been solved to 2.5 Å by X-ray crystallography (Dai et al., 1996).

B. Other Biological Systems

Genome sequencing has demonstrated that thioredoxin reductase is present in all living cells including Archae, eubacteria, fungi and plants. However, it has long been known that the mammalian enzymes have very different properties compared with the well-characterized *E. coli* enzyme (Luthman and Holmgren, 1982; Holmgren and Björnstedt, 1995). Today we know that there is a major division between Archae, eubacteria, fungi and plants in that higher organisms and mammals have a high molecular weight thioredoxin reductase (M_r 114.000) with a completely different structure and mechanism (Arscott et al., 1997). Thus, mammalian thioredoxin reductase contains FAD and has two identical 55 kDa subunits homologous to glutathione reductase (Zhong et al., 1998). The enzymes contain an essential selenocysteine residue in a 16 residue extension with the conserved C-terminal sequence Gly-Cys-SeCys-Gly (Zhong and Holmgren, 2000; Zhong et al., 2000). The essential role of selenium in the mammalian enzymes make this a completely different system susceptible to inhibitors which specifically target the selenocysteine residue (Nordberg et al., 1998). The mammalian enzymes also directly reduces hydrogen peroxide and has a very wide range of substrates (Arnér et al., 1999) and functions in redox regulation and defense against oxido-like stress (Holmgren et al., 1998).

IV. Glutaredoxins

A. Functions

Glutaredoxins use the reductive power of the ubiquitous tripeptide glutathione, which is present in very high concentrations in essentially all cells to reduce disulfides (Holmgren, 1976, 1989). The functions of glutaredoxins are partly overlapping those of thioredoxins. Thus, glutaredoxins catalyze GSH-dependent electron transfer for enzymes such as ribonucleotide reductase, or PAPS reductase (Holmgren and Åslund, 1995; Lillig et al., 1999).

Glutaredoxins are able to deliver electrons to glutathione peroxidases (Björnstedt et al., 1994) and can be regarded as enzymes required to make the relatively unreactive glutathione operate as an electron donor for reductive processes. Several plant glutaredoxins have been studied (Minakuchi et al., 1994; Morrel et al., 1995; Szederkenyi et al., 1997) and more discovered in gene sequencing from plants (Meyer et al., 1999). A particular function of glutaredoxins in defense against oxidative stress is specific reduction of mixed disulfides with glutathione. Reactive oxygen species will raise the intracellular concentration of GSSG glutathione disulfide the oxidized form of glutathione and may

```
human      ---AQEFVNCKIQPGKVVVFIKP----TCPYCRRAQEILSQLPIKQG--LLEFVDITATN--H
pig        ---AQAFVNSKIQPGKVVVFIKP----TCPFCRKTQELLSQLPFKEG--LLEFVDITATS--D
yeast 2    MVSQETIKHVKDLIAENEIFVAS--KTYCPYCHAALNTLFEKLKVPRSKVLVLQLNDMKE---
T4 phage   ---------------MFKVYGYDSNIHKCVYCDNAKRLLTVKKQ-----PFEFINIMPEKGVF
E. coli 1  --------------MQTVIFGRS----GCPYCVRAKDLAEKLSNERDDFQYQYVDIRA-----
structure        α          β      10          α20         30   β     40

human      TNEIQDYLQQLTGAR-----TVPRVFIG-KDCIGGCSDLVSLQ-QSGELLTRLKQIGALQ 105
pig        TNEIQDYLQQLTGAR-----TVPRVFIG-KECIGGCTDLESMH-KRGELLTRLQQIGALK 105
yeast 2    GADIQAALYEINGQR-----TVPNIYI-NGKHIGGNDDLQELRETGELEELLEPILAN   108
T4 phage   DDEKIAELLTKLGRDTQIGLTMPQVFAPDGSHIGGFDQLREYFK                  87
E. coli 1  EGITKEDLQQKAGKPVE---TVPQIFV-DQQHIGGYTDFAAWVKENLDA             85
structure        α 50         60   β     β 70      α  80       α
```

Fig. 5. Sequences of glutaredoxins from different species. T4 is phage T4 glutaredoxin. The conserved active site residues CPYC are dashed, as well as residues involved in GSH binding. (See Fig. 6).

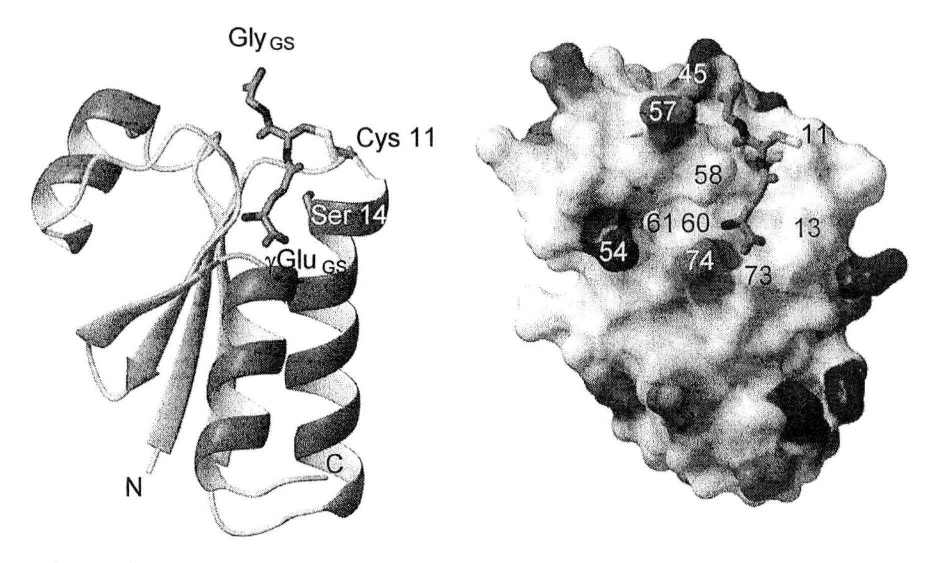

Fig. 6. Structure of glutaredoxin (Grx 1) in a covalent complex with GSH (left) ribbon diagram with selected protein residues and the bound GSH (γ-Glu-Cys-Gly) (right). Surface with GSH bound (Bushweller et al., 1994).

thereby modify specific susceptible proteins (Holmgren et al., 1998). Reversal of this process is catalyzed by glutaredoxins. This property of glutaredoxins requires only the N-terminal Cys residue (Bushweller et al., 1992) and a recently discovered new family of monothiol glutaredoxins have been described in yeast (Rodriguez-Manzaneque et al., 1999). The members of this family have a Cys to Ser substitution in the active site and appear to be particularly important for defense against oxidative stress. The Grx motif is incorporated in high molecular weight proteins and these new redoxins exist in all kingdoms including plants (Rodriguez-Manzaneque et al., 1999) .

E. coli contains three known glutaredoxins (Holmgren and Åslund, 1995) but the functions of glutaredoxin 2 and 3 are still largely unknown (Lillig et al., 1999), although it is clear that glutaredoxin 2 is a prime electron donor of arsenate reductase in *Escherichia coli* (Shi et al., 1999). Such a function has not been reported in plants.

B. Structures and Mechanisms

The structures of selected glutaredoxins are shown in Fig. 5. The three-dimensional structure of Grx1 from *E. coli* was solved by NMR and also a complex with GSH has been determined (Fig. 6). This was accomplished by solving the structure of an active site mutant with the C-terminal Cys active site residue mutated to Ser. The structure showed also the binding site for glutathione (Bushweller et al., 1994).

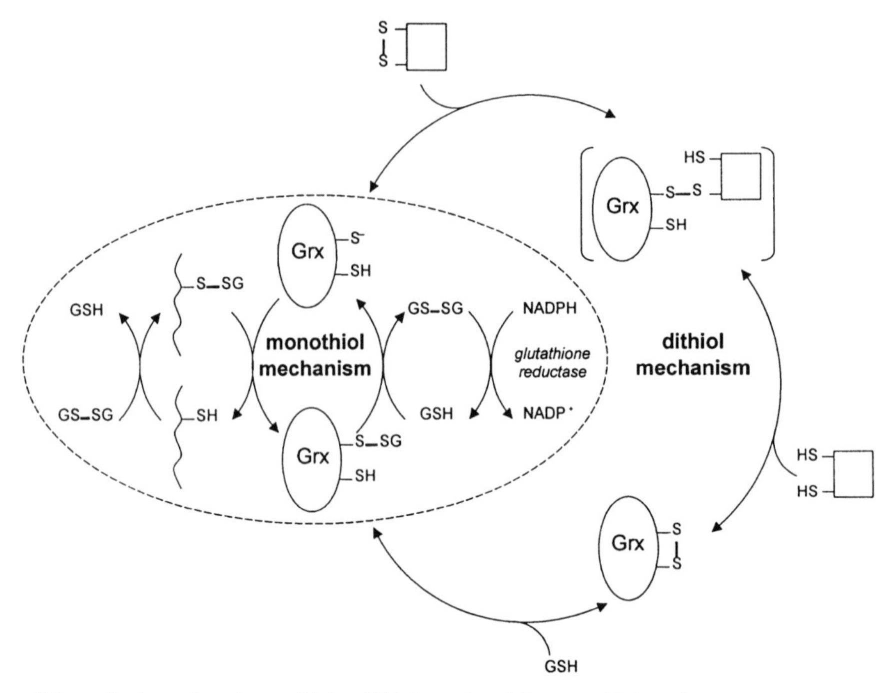

Fig. 7. Mechanism of glutaredoxin catalyzed monothiol or dithiol reactions. The monothiol mechanism involves formation and cleavage of GSH-mixed disulfides. The dithiol mechanism as in ribonucleotide reductase (Fig. 1) involves reduction of a disulfide.

The mechanism of glutaredoxin involves either monothiol reactions catalyzing formation and cleavage of mixed disulfides with proteins or dithiol reductions giving reduction of disulfides with the thioredoxin-like mechanism (Fig. 7).

Glutaredoxins and thioredoxins now appear to play major roles in cellular redox regulation. Thiol redox control (Holmgren 1985, 1989) involves different potential mechanisms as follows:

(i) A thiol group may form a disulfide with another thiol group in the protein. This mechanism of reversible conversion of a disulfide to a dithiol is catalyzed by both thioredoxins and glutaredoxins as well as protein-disulfide-isomerase all members of the thioredoxin superfamily of proteins. A prime example is the role of chloroplast thioredoxins in regulating photosynthetic enzymes as described above.

(ii) A thiol group may form a mixed disulfide with glutathione in a spontaneous reaction or catalyzed by glutaredoxins. The levels of GSH and GSSG will set the redox potential (E_0'). The redox pair of 2 GSH/GSSG is a monothiol system with a potential of –240 mV at pH 7.0 with 1 M of both GSH and GSSG (Åslund et al., 1997). This is equivalent to an E_0' of –240 mV for 10 mM GSH and 0.01 mM GSSG which is close to the measured in vivo concentration in e.g. mammalian liver (Gilbert, 1990). However, an increase in GSSG occurring after H_2O_2 reduction by a GSH-peroxidase or a decrease in intracellular GSH-following export, consumption by GSH-S-transferases or decreased synthesis by cystine limitation will change the ratio and increase the E_0' value of the redox buffer. Glutaredoxins have a redox potential of their active site disulfide/dithiol either 35 or 70 mV higher than thioredoxin as recently determined by protein-protein equilibria (Åslund et al., 1997). Human glutaredoxin may be expected to be similar to that of *E. coli* Grx1 with an E_0' value of –233 mV. Thus, under conditions of high GSSG the active site disulfide in Grx will not be reduced and its activity as a GSH-disulfide reductase will be shut off.

(iii) A thiol group may also be converted to a sulfenic acid by reaction with H_2O_2. This derivative may be stable or converted to a disulfide as recently shown for the oxyR system, where glutaredoxin 1 controls the activity of this transcription factor in *E. coli* (Zheng et al., 1998). For the control of the activity of protein-tyrosine phosphatase1B in A431 cells the catalytically active Cys residue is converted to a sulfenic acid by H_2O_2. Regeneration to a thiol can occur either by reduced thioredoxin or glutaredoxin (Lee et al., 1998).

Acknowledgments

Supported by grants from the Swedish Cancer Society, the Swedish Medical Research Council, the Knut and Alice Wallenberg Foundation and the Inga-Britt and Arne Lundberg Foundation.

References

Arnér ESJ, Zhong L and Holmgren A (1999) Preparation and assay of mammalian thioredoxin and thioredoxin reductase. Methods Enzymol 300: 226–239

Arscott LD, Gromer S, Schirmer RH, Becker K and Williams CH Jr (1997) The mechanism of thioredoxin reductase from human placenta is similar to the mechanisms of lipoamide dehydrogenase and glutathione reductase and is distinct from the mechanism of thioredoxin reductase from *Escherichia coli*. Proc Natl Acad Sci USA 94: 3621–3626

Åslund F, Berndt KD and Holmgren A (1997) Redox potentials of glutaredoxins and other thiol-disulfide oxidoreductases of the thioredoxin superfamily determined by direct protein-protein redox equilibria. J Biol Chem 272: 30780–30786

Björnstedt M, Xue J, Huang W, Åkesson B and Holmgren A (1994) The thioredoxin and glutaredoxin systems are efficient electron donors to human plasma glutathione peroxidase. J Biol Chem 269: 29382–29384

Brot N, Fliss H and Weissbach H (1986) The Biochemistry of Reduction of Methionine Sulfoxide Residues in Proteins. In: Holmgren A, Brändén C-I, Jörnvall H and Sjöberg B-M (eds) Thioredoxin and Glutaredoxin Systems: Structure and Function, pp 141–153, Raven Press, New York.

Buchanan BB (1991) Regulation of CO_2 assimilation in oxygenic photosynthesis: The ferredoxin/thioredoxin system. Perspective on its discovery, present status, and future development. Arch Biochem Biophys 288: 1–9

Buchanan BB, Schürmann P, Decottignies P and Lozano RM (1994) Thioredoxin: A multifunctional regulatory protein with a bright future in technology and medicine. Arch Biochem Biophys 314: 257–260

Bushweller JH, Åslund F, Wüthrich K and Holmgren A (1992) Structural and functional characterization of the mutant *Escherichia coli* glutaredoxin (C14→S) and its mixed disulfide with glutathione. Biochemistry 31: 9288–9293

Bushweller JH, Billeter M, Holmgren A and Wüthrich K (1994) The nuclear magnetic resonance solution structure of the mixed disulfide between *Escherichia coli* glutaredoxin(C14S) and glutathione. J Mol Biol 235: 1585–1597

Chiadmi M, Navaza A, Miginiac-Maslow M, Jacquot JP and Cherfils J (1999) Redox signalling in the chloroplast: Structure of oxidized pea fructose-1,6-bisphosphate phosphatase. EMBO J 18: 6809–6815

Chivers PT, Prehoda KE and Raines RT (1997) The CXXC motif: A rheostat in the active site. Biochemistry 36: 4061–4066

Dai S, Saarinen M, Ramaswamy S, Meyer Y, Jacquot JP and Eklund H (1996) Crystal structure of *Arabidopsis thaliana* NADPH dependent thioredoxin reductase at 2.5 Å resolution. J Mol Biol 264: 1044–1057

Dai S, Schwendtmayer C, Schürmann P, Ramaswamy S and Eklund H (2000) Redox signaling in chloroplasts: Cleavage of disulfides by an iron-sulfur cluster. Science 287: 655–658

Dyson HJ, Jeng MF, Tennant LL, Slaby I, Lindell M, Cui DS, Kuprin S and Holmgren A (1997) Effects of buried charged groups on cysteine thiol ionization and reactivity in *Escherichia coli* thioredoxin: Structural and functional characterization of mutants of Asp 26 and Lys 57. Biochemistry 36: 2622–2236

Gilbert HF (1990) Molecular and cellular aspects of thiol-disulfide exchange. Adv Enzymol Relat Areas Mol Biol 63: 69–172

Hassouni ME, Chambost JP, Expert D, Van Gijsegem F and Barras F (1999) The minimal gene set member msrA, encoding peptide methionine sulfoxide reductase, is a virulence determinant of the plant pathogen *Erwinia chrysanthemi*. Proc Natl Acad Sci USA 96: 887–892

Hirasawa M, Schürmann P, Jacquot JP, Manieri W, Jacquot P, Keryer E, Hartman FC and Knaff DB (1999) Oxidation-reduction properties of chloroplast thioredoxins, ferredoxin:thioredoxin reductase and thioredoxin *f*-regulated enzymes. Biochemistry 38: 5200–5205

Holmgren A (1976) Hydrogen donor system for *Escherichia coli* ribonucleoside-diphosphate reductase dependent upon glutathione. Proc Natl Acad Sci USA 73: 2275–2279

Holmgren A (1984) Enzymatic reduction-oxidation of protein disulfides by thioredoxin. Methods Enzymol 107: 295–300

Holmgren A (1985) Thioredoxin. Annu Rev Biochem 54: 237–271

Holmgren A (1989) Thioredoxin and glutaredoxin systems. J Biol Chem 24: 13963–13966

Holmgren A (1995) Thioredoxin structure and mechanism: conformational changes on oxidation of the active-site sulfhydryls to a disulfide. Structure 3: 239–243

Holmgren A and Åslund F (1995) Glutaredoxin. Methods Enzymol 252: 283–292

Holmgren A and Björnstedt M (1995) Thioredoxin and thioredoxin reductase. Methods Enzymol 252: 199–208

Holmgren A, Söderberg BO, Eklund H and Brändén CI (1975) Three-dimensional structure of *Escherichia coli* thioredoxin-S2 to 2.8 Å resolution. Proc Natl Acad Sci USA 72: 2305–2309

Holmgren A, Arnér ESJ, Åslund F, Björnstedt M, Zhong L, Ljung J, Nakamura H and Nikitovic D (1998) Redox regulation by the thioredoxin and glutaredoxin systems. In: Montagnier L, Olivier R, Pasquier C, (eds) Oxidative Stress, Cancer, AIDS and Neurodegenerative Diseases, pp 229–246. Marcel Dekker, Inc., New York,

Huber HE, Tabor S and Richardson CC (1987) *Escherichia coli* thioredoxin stabilizes complexes of bacteriophage T7 DNA polymerase and primed templates. J Biol Chem 262: 16224–16232

Jacquot JP, Rivera-Madrid R, Marinho P, Kollarova M, Le Marechal P, Miginiac-Maslow M and Meyer Y (1999) *Arabidopsis thaliana* NAPHP thioredoxin reductase. cDNA characterization and expression of the recombinant protein in *Escherichia coli*. J Mol Biol 235: 1357–1363

Jeng MF, Campbell AP, Begley T, Holmgren A, Case DA, Wright PE and Dyson HJ (1994) High-resolution solution structures of oxidized and reduced *Escherichia coli* thioredoxin.

Structure 2: 853–868

Johansson K, Ramaswamy S, Saarinen M, Lemaire-Chamley M, Issakidis-Bourguet E, Miginiac-Maslow M and Eklund H (1999) Structural basis for light activation of a chloroplast enzyme: The structure of sorghum NADP-malate dehydrogenase in its oxidized form. Biochemistry 38: 4319–4326

Jordan A and Reichard P (1998) Ribonucleotide reductases. Annu Rev Biochem 67: 71–98

Kallis GB and Holmgren A (1980) Differential reactivity of the functional sulfhydryl groups of cysteine-32 and cysteine-35 present in the reduced form of thioredoxin from *Escherichia coli*. J Biol. Chem 255: 10261–10265

Kang SW, Chae HZ, Seo MS, Kim K, Baines IC and Rhee SG (1998) Mammalian peroxiredoxin isoforms can reduce hydrogen peroxide generated in response to growth factors and tumor necrosis factor-alpha. J Biol Chem 273: 6297–6302

Krause G, Lundström J, Barea JL, Pueyo de la Cuesta C and Holmgren A (1991) Mimicking the active site of protein disulfide-isomerase by substitution of proline 34 in *Escherichia coli* thioredoxin. J Biol Chem 266: 9494–9500

Lee SR, Kwon KS, Kim SR and Rhee SG (1998) Reversible inactivation of protein-tyrosine phosphatase 1B in A431 cells stimulated with epidermal growth factor. J Biol Chem 273: 15366–15372

Lillig CH, Prior A, Schwenn JD, Åslund F, Ritz D, Vlamis-Gardikas A and Holmgren A (1999) New thioredoxins and glutaredoxins as electron donors of 3′-phosphoadenylylsulfate PAPS reductase. J Biol Chem 274: 7695–7698

Lundström J and Holmgren A (1993) Determination of the reduction-oxidation potential of the thioredoxin-like domains of protein disulfide-isomerase from the equilibrium with glutathione and thioredoxin. Biochemistry 32: 6649–6655

Luthman M and Holmgren A (1982) Rat liver thioredoxin and thioredoxin reductase: Purification and characterization. Biochemistry 21: 6628–6633

Martin JL (1995) Thioredoxin—a fold for all reasons. Structure 3: 245–250

Meister A (1995) Glutathione metabolism. Methods Enzymol 251: 3–7

Meyer, Y, Verdoucq, L and Vignols, F (1999) Plant thioredoxins and glutaredoxins: identity and putative roles. Trends Plant Sci 4: 388–394

Minakuchi K, Yabushita T, Masumura T, Ichihara K and Tanaka K (1994) Cloning and sequence analysis of a cDNA encoding rice glutaredoxin. FEBS Lett 337: 157–160

Morell S, Follmann H and Haberlein I (1995) Identification and localization of the first glutaredoxin in leaves of a higher plant. FEBS Lett 369: 149–152

Nordberg J, Zhong L, Holmgren A and Arnér ESJ (1998) Mammalian thioredoxin reductase is irreversibly inhibited by dinitrohalobenzenes by alkylation of both the redox active selenocysteine and its neighboring cysteine residue. J Biol Chem 273: 10835–10842

Nordstrand K, Åslund F, Holmgren A, Otting G and Berndt KD (1999) NMR structure of *Escherichia coli* glutaredoxin 3-glutathione mixed disulfide complex: Implications for the enzymatic mechanism. J Mol Biol 286: 541–552

Prinz WA, Åslund F, Holmgren A and Beckwith J (1997) The role of the thioredoxin and glutaredoxin pathways in reducing protein disulfide bonds in the *Escherichia coli* cytoplasm. J Biol Chem 272: 15661–15667

Qin J, Clore GM, Kennedy WM, Huth JR and Gronenborn AM (1995) Solution structure of human thioredoxin in a mixed disulfide intermediate complex with its target peptide from the transcription factor NF kappa B. Structure 3: 289–297

Rodriguez-Manzaneque MT, Ros J, Cabiscol E, Sorribas A and Herrero E (1999) Grx5 glutaredoxin plays a central role in protection against protein oxidative damage in *Saccharomyces cerevisiae*. Mol Cell Biol 12: 8180–8190

Russel M and Model P (1986) The role of thioredoxin in filamentous phage assembly. Construction, isolation, and characterization of mutant thioredoxins. J Biol Chem 261: 14997–5005

Saitoh M, Nishitoh H, Fujii M, Takeda K, Tobiume K, Sawada Y, Kawabata M, Miyazono K and Ichijo H (1998) Mammalian thioredoxin is a direct inhibitor of apoptosis signal-regulating kinase (ASK) 1. EMBO J 17: 2596–2606

Schenk H, Klein M, Erdbrugger W, Dröge W and Schulze-Osthoff K (1994) Distinct effects of thioredoxin and antioxidants on the activation of transcription factors NF-kappa B and AP-1. Proc Natl Acad Sci USA 91: 1672–1676

Shi J, Vlamis-Gardikas A, Åslund F, Holmgren A and Rosen BP (1999) Reactivity of glutaredoxins 1, 2, and 3 from *Escherichia coli* shows that glutaredoxin 2 is the primary hydrogen donor to ArsC-catalyzed arsenate reduction. J Biol Chem 274: 36039–36042

Szederkenyi J, Komor E and Schobert C (1997) Cloning of the cDNA for glutaredoxin, an abundant sieve-tube exudate protein from *Ricinus communis* L. and characterisation of the glutathione-dependent thiol-reduction system in sieve tubes. Planta 202: 349–356

Tabor S, Huber HE and Richardson CC (1987) Escherichia coli thioredoxin confers processivity on the DNA polymerase activity of the gene 5 protein of bacteriophage T7. J Biol Chem 262: 16212–16223

Verdoucq L, Vignols F, Jacquot JP, Chartier Y and Meyer Y (1999) In vivo characterization of a thioredoxin *h* target protein defines a new peroxiredoxin family. J Biol Chem 274: 19714–19722

Williams CH Jr (1995) Mechanism and structure of thioredoxin reductase from *Escherichia coli*. FASEB J 9: 1267–1276

Zheng M, Åslund F and Storz G (1998) Activation of the OxyR transcription factor by reversible disulfide bond formation. Science 279: 1718–1721

Zhong L and Holmgren A (2000) Essential role of selenium in the catalytic activities of mammalian thioredoxin reductase revealed by characterization of recombinant enzymes with selenocysteine mutations. J Biol Chem 275: 18121–18128

Zhong L, Arnér ESJ, Ljung J, Åslund F and Holmgren A (1998) Rat and calf thioredoxin reductase are homologous to glutathione reductase with a carboxyl-terminal elongation containing a conserved catalytically active penultimate selenocysteine residue. J Biol Chem 273: 8581–8591

Zhong L, Arnér ESJ and Holmgren A (2000) Structure and mechanism of mammalian thioredoxin reductase: The active site is a redox-active selenol thiol/selenenylsulfide formed from the conserved cysteine-selenocysteine sequence. Proc Natl Acad Sci USA 97: 5854–5859

Chapter 20

The Structure and Function of the Ferredoxin/Thioredoxin System in Photosynthesis

Peter Schürmann*
Laboratoire de Biochimie végétale, Université de Neuchâtel, Rue Emile-Argand 11, CH-2007 Neuchâtel, Switzerland

Bob B. Buchanan
Department of Microbial and Plant Biology, University of California, 111 Koshland Hall, Berkeley, CA 94720, U.S.A.

*Author for correspondence, email: peter.schurmann@bota.unine.ch

Eva-Mari Aro and Bertil Andersson (eds): Regulation of Photosynthesis, pp. 331–361.

Summary

The demonstration that thioredoxin could function in regulation grew out of CO_2 assimilation experiments initiated more than thirty years ago with isolated chloroplasts. This work ultimately led to the description of the ferredoxin/thioredoxin system, whereby the activity of enzymes, key to oxygenic photosynthetic processes, is linked to light. Electrons provided by excited chlorophyll are transferred to ferredoxin and then sequentially to the enzyme ferredoxin:thioredoxin reductase (FTR) and a thioredoxin (*f* or *m* in chloroplasts). FTR converts an electron signal to a sulfhydryl (SH) signal that can be recognized by specific enzymes with a complementary disulfide (S-S) site. Through a reversible reduction of regulatory disulfide elements, thioredoxin brings about a structural change that alters the catalytic properties of the target enzymes. By selectively activating and deactivating regulatory enzymes of opposing pathways, the ferredoxin/thioredoxin system makes it possible for the oxygenic photosynthetic system (chloroplast or prokaryote) to house biosynthetic and catabolic processes in a single compartment and separate their activities diurnally—i.e., by the presence or absence of light. In this way autotrophic (photosynthetic) and heterotrophic lifestyles can be accommodated and function in a single organism. We summarize below the recent developments on the structure, mechanism of action and function of the individual thiol components of the ferredoxin/thioredoxin system. As a result of impressive progress made during this decade, we now have a better understanding of the events underlying the regulatory role of thioredoxin in oxygenic photosynthesis and associated processes linked to light. This work will also assist in the understanding of the growing array of heterotrophic plant and animal processes found to be regulated by thioredoxin reduced enzymatically with NADPH.

I. Introduction

Light is not only a substrate but also an important regulatory factor in photosynthesis and related metabolic processes. It is now recognized that, in contrast to earlier views, the carbon reduction phase of photosynthesis is not independent of light, but needs light for the activation of key enzymes. One of the principal ways this activation is achieved is via a series of protein components that constitute the light-dependent ferredoxin/thioredoxin system. This system regulates the activity of key photosynthetic regulatory enzymes by a light-driven reduction of disulfide bonds. Electrons provided by the excitation of chlorophyll are transferred via ferredoxin to the enzyme ferredoxin:thioredoxin reductase (FTR). This enzyme, unique to oxygenic eukaryotic and prokaryotic photosynthetic cells, reduces thioredoxins, which in turn interact with selected target enzymes and covalently modify them through a reversible reduction of disulfide to sulfhydryl groups. These modifications bring about structural changes which alter the catalytic properties of target enzymes (Fig. 1). In the dark this process is reversed by oxygen (see Buchanan, 1980 for an early review). Since oxygen is produced in illuminated photosynthetic cells containing the ferredoxin/thioredoxin system, the redox equilibrium state of the target enzymes, and hence their activity, will depend on the 'electron pressure' built up by the photosystems. The subsequent fine tuning of the target enzymes activated by thioredoxin depends on other cellular factors such as pH and metabolite (effector) concentrations, both of which are linked to light (Buchanan, 1980; Scheibe, 1991).

The direction in which the activity of a regulated enzyme is changed depends on its function. Photosynthetic enzymes functional in assimilation are activated by reduction, whereas dissimilatory enzymes are reductively inactivated. Since these enzymes, and hence the associated opposing pathways, coexist in the same cell compartment, a regulatory system is indispensable in serving as a switch between light and dark metabolism. In fulfilling this function, thioredoxin, in a sense, acts as an 'eye' by which regulatory enzymes of biochemical processes associated with oxygenic photosynthesis distinguish light from dark. Light is recognized as the reduced (sulfhydryl) form of thioredoxin and dark as the oxidized (disulfide) form. With this form of light recognition, chloroplasts can

Abbreviations: DTT – dithiothreitol; FBPase – fructose 1,6-bisphosphatase; FTR – ferredoxin:thioredoxin reductase; GAPDH – glyceraldehyde 3-phosphate dehydrogenase; G6PDH – glucose 6-phosphate dehydrogenase; NADP-MDH – NADP-dependent malate dehydrogenase; PRK – phosphoribulokinase; Rubisco – ribulose 1,5-bisphosphate carboxylase/oxygenase; SBPase – sedoheptulose 1,7-bisphosphatase

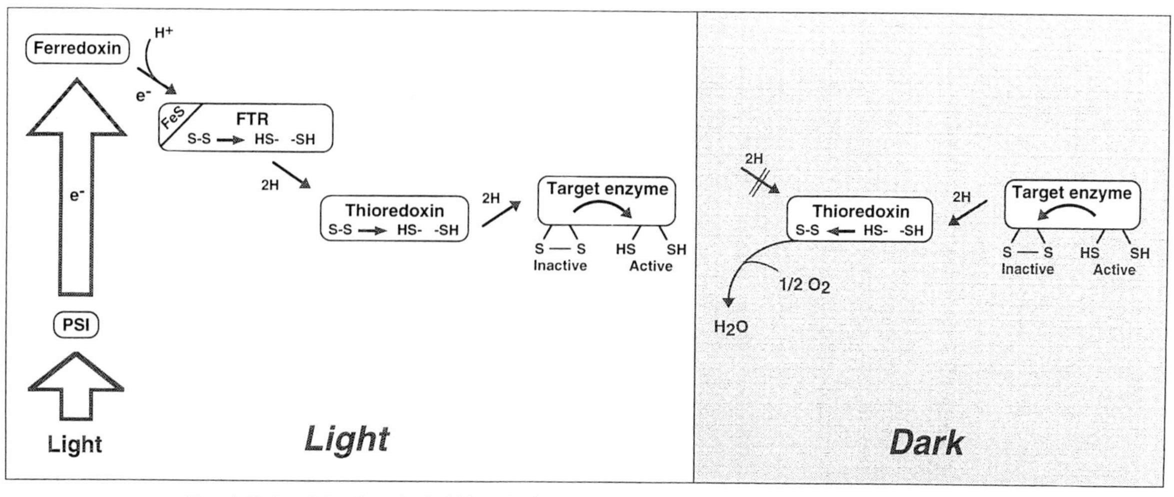

Fig. 1. Role of the ferredoxin/thioredoxin system in the activation of photosynthetic enzymes.

minimize futile cycles and maximize photosynthetic yield.

An assessment of the role and properties of the ferredoxin/thioredoxin system in oxygenic photosynthesis has been made in several contemporary reviews (Buchanan, 1991, 1992; Buchanan et al., 1994a,b; Scheibe, 1994; Follmann and Häberlein, 1996; Jacquot et al., 1997a; Meyer et al., 1999; Ruelland and Miginiac-Maslow, 1999; Schürmann and Jacquot, 2000). In this chapter we summarize recent developments, including new structural information, that provide a better understanding of the chemical events underlying the regulatory role of thioredoxin.

II. Biochemical Setting for Thioredoxin-Linked Regulation

A question of continued interest is how oxygenic photosynthetic systems (chloroplasts and cyanobacteria) are able to use thioredoxin for reversible redox regulation under homeostatic conditions. That is, how do chloroplasts maintain disulfide (S-S) bonds in the dark while these bonds appear to remain continuously reduced in compartments such as the cytosol? This question has recently been experimentally clarified by a study with a heterotrophic bacterium. *E. coli* mutants deficient in NADP:thioredoxin reductase were found to contain increased levels of protein disulfides relative to wild type cells (Derman et al., 1993). This and other observations prompted the conclusion that thioredoxin, reduced by NADPH, serves to maintain protein sulfur atoms in the reduced (sulfhydryl) state. It thus becomes clear that lack of an NADP:thioredoxin reductase in chloroplasts and cyanobacteria allows for the disulfide groups of thioredoxin-targeted enzymes to remain oxidized in the absence of photoreduced thioredoxin, i.e., in the dark. An active NADP-linked reductase would interfere with diurnal redox changes by keeping the targeted disulfides reduced by the NADPH generated in the dark. This situation does not apply to certain plant organs, such as the seeds of cereals, which are developmentally interrupted. In this case, the endosperm tissues become metabolically sluggish and their proteins oxidized during maturation and drying of the grain, thus allowing for specific thioredoxin effects during germination. Other cellular reductants, which are reduced in the dark—that is, glutathione and glutaredoxin—do not present a problem with thioredoxin-linked enzymes as they are not recognized, that is, they do not interact with the disulfide bonds of the target enzymes.

III. Thioredoxin Regulated Processes

Carbon dioxide assimilation in chloroplasts is the best understood thioredoxin-regulated process. Here, light functions in regulating selected biosynthetic reactions of the carbon assimilation pathways in C_3, C_4 and CAM plants (Buchanan, 1980). Four enzymes of the Calvin cycle have been known for many years to be activated by thioredoxin, i.e., fructose 1,6-bisphosphatase (FBPase), sedoheptulose 1,7-

bisphosphatase (SBPase), phosphoribulokinase (PRK) and glyceraldehyde 3-phosphate dehydrogenase (GAPDH) (Buchanan, 1980). Quite recently, a fifth member of the cycle, ribulose 1,5-bisphosphate carboxylase/oxygenase (Rubisco) has been added to this list (Zhang and Portis Jr, 1999). By contrast, the analysis of cyanobacterial FBPases suggests that in these organisms the Calvin cycle enzymes might not be regulated by light via the ferredoxin-thioredoxin system (Tamoi et al., 1998).

In C_4 plants the activity of NADP-dependent malate dehydrogenase (NADP-MDH), which is an essential catalyst of the carbon trapping and transport mechanism in C_4 mesophyll cells, was early found to be entirely light-dependent (Hatch, 1987, 1992) and then shown to require thioredoxin (Wolosiuk et al., 1977). C_3 plants contain a corresponding enzyme whose activity is also dependent on light for activation (Jacquot et al., 1976). However, in C_3 species this enzyme is not involved in enhancing CO_2 concentration but serves a different purpose in the export of reducing equivalents in the form of malate from chloroplast to cytosol. This export system has been named the 'malate valve' (Scheibe, 1987).

The activity of another important chloroplast enzyme, ATP synthase or coupling factor CF_1, was early recognized as being switched on upon illumination (Inoue et al., 1979; Mills and Hind, 1979). It was shown that reduced thioredoxin stimulates the appearance of ATPase activity in vitro (McKinney et al., 1978) and that the thioredoxin functional in activation could be photosynthetically reduced via the ferredoxin/thioredoxin system (Mills et al., 1980).

Glucose 6-phosphate dehydrogenase (G6PDH) is so far the only known enzyme for which light decreases its activity (Scheibe and Anderson, 1981). G6PDH is the first committed enzyme of the oxidative pentose phosphate pathway in chloroplasts which functions in supplying reduced NADP and pentoses in the dark. The observed effect of light is fully compatible with the function of G6PDH in the chloroplast since it allows the suppression of carbohydrate breakdown thereby preventing energy loss during active CO_2 assimilation.

Fatty acid synthesis in spinach chloroplasts is also light dependent. The formation of malonyl-CoA by acetyl-CoA carboxylase in intact chloroplasts is stimulated by illumination. Formerly this stimulation was considered to be due to changes of the ATP/ADP ratio, the pH and Mg^{++} concentration in the chloroplast stroma (Sauer and Heise, 1983). More recent experiments with the isolated chloroplast enzyme suggest, however, that redox regulation may also control fatty acid biosynthesis (Sasaki et al., 1997).

Reports of thioredoxin-stimulated enzymes of nitrogen assimilation, notably glutamine synthetase and ferredoxin-glutamate synthase or GOGAT, have appeared over the years. Recently, evidence was presented to extend the regulatory role of thioredoxin to the assimilation of nitrogen in higher plants. Chloroplast glutamate synthase from spinach and soybean was found to be activated by reduced thioredoxin (Lichter and Häberlein, 1998).

Translation is yet another process considered to be regulated by thioredoxin. The synthesis of certain chloroplast proteins is enhanced about 50- to 100-fold after illumination. This effect is due to an increase in the rate of translation. One of the protein products showing a light-dependent increase in translation is psbA which encodes the D1 protein of Photosystem II. The translation of psbA RNA by *Chlamydomonas reinhardtii* preparations was enhanced by the addition of dithiol compounds such as dithiothreitol (DTT), most effectively in the presence of thioredoxin (Danon and Mayfield, 1994). Reduction by thioredoxin increased the binding of translational activator proteins to psbA messenger RNA thereby accelerating the synthesis of the protein product. Thioredoxin from *E. coli* was used in the reported experiments, suggesting that chloroplast *m*-type species are active in vivo. No information is yet available on the structure of the putative regulatory disulfide of the activator protein responsible for this enhancement effect. The proposed regulatory mechanism provides a direct link between light and the translation of the D1 protein, which is known to undergo high turnover in the PS II reaction center during illumination. Regulation by the ferredoxin/thioredoxin system makes it possible to adjust the rate of replacing reaction centers destroyed by photooxidation.

IV. Structure and Function of the Proteins in the Regulatory Chain

During the past decade, dramatic progress has been made in the understanding of the ferredoxin/thioredoxin system, with respect to function and especially to structure. The crystal structure of each protein member is now known (Fig. 2). Advances made in recent years are highlighted below.

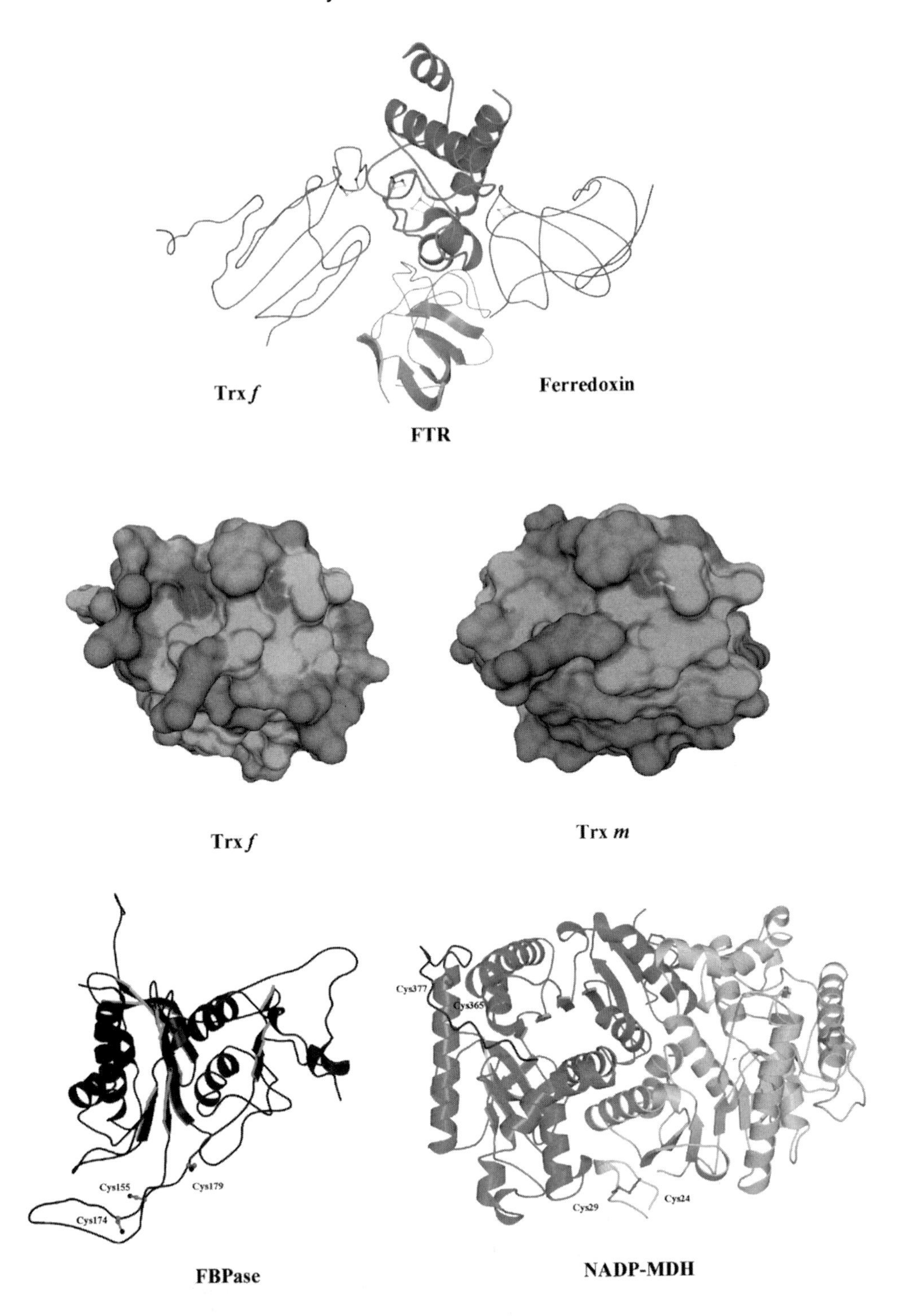

Fig. 2. Structure of proteins constituting the ferredoxin/thioredoxin system and two target enzymes. The top of the figure shows the FTR (*Synechocystis*) from the side with the variable subunit in green and the catalytic subunit in red. The concave disk shape of the heterodimer allows simultaneous docking of a thioredoxin (on the left) and of a ferredoxin molecule (on the right). Two spinach chloroplast thioredoxins are given in surface view in the middle . The colors (as shown in the color-plate section of this book) represent the following residue type: green—Cys; red—charged (+ or -); blue—polar; yellow—polar; gray—backbone. The bottom structures represent spinach FBPase (only one subunit), with the regulatory loop carrying the Cys, extending out of the core structure, and oxidized sorghum NADP-MDH dimer with the disulfide bonds implicated in regulation. Also see Color Plate 5.

A. Ferredoxin

As reviewed earlier in this series (Knaff, 1996), ferredoxin mediates electron transfer between chloroplast photosystem I and several ferredoxin-dependent enzymes, including ferredoxin:NADP reductase, nitrite reductase, glutamate synthase and FTR. Ferredoxin is negatively charged and displays a strong interaction with its electron acceptor proteins. The interaction is sufficient to be exploited for preparative purposes. It is a common practice to purify ferredoxin-dependent enzymes by affinity chromatography on ferredoxin-Sepharose columns as has been done, for example, with FTR (Droux et al., 1987a). The complexes formed between ferredoxin and its reaction partners are stabilized by electrostatic interactions where ferredoxin contributes the negative charges and the reaction partner the complementary positive charges. Although the amino acid sequence similarity among plant ferredoxins is generally very high, there can be differences in charge distribution which influence the stability of the interaction. The complementarity of the charges may lend specificity to the interaction between ferredoxin and the electron acceptor protein which will be strongest in a homologous system. Spinach ferredoxin has been shown to form a high affinity, electrostatically stabilized 1:1 complex with spinach FTR (Kd for ferredoxin $<1 \times 10^{-7}$ M; Table 1) (Hirasawa et al., 1988). Exact complementarity is not an absolute requirement since it is possible to reduce ferredoxin-dependent enzymes in heterologous systems. In such a case, the affinity between ferredoxin and enzyme is significantly reduced. Thus, spinach ferredoxin and *Synechocystis* FTR showed a Kd of 2×10^{-5} M (Schwendtmayer et al., 1998) suggesting that despite the high degree of sequence similarity the charge complementarity between the two heterologous proteins is not optimal.

Chemical modifications and site-directed mutagenesis experiments have been performed on ferredoxin to identify charged surface residues important in the interaction with FTR as compared to ferredoxin:NADP reductase. Ferredoxin features two distinct domains of strong negative surface potential on either side of the [2Fe-2S] cluster. Differential chemical modification of free and target bound ferredoxin indicates that binding to the ferredoxin:NADP reductase involves both of these negative surface domains. The binding site for FTR, however, includes mainly only one of the two electrostatic surface potential areas of ferredoxin. Residues Asp65, Glu92, Glu93, Glu94 and the carboxyl group of the carboxy-terminal Ala97 on one side of the cluster and Asp34 on the other were protected in the complex from chemical modification. This was taken as evidence that these residues interact closely with positive charges on the FTR surface (De Pascalis et al., 1994).

Table 1. Stability of ferredoxin-enzyme complexes of chloroplasts

Enzyme	Kd (Ferredoxin) μM
Ferredoxin:thioredoxin reductase[1] (in 30 mM Tris-Cl pH 8.0)	<0.1
Ferredoxin:NADP reductase[2] (in 50 mM Hepes pH 8.0)	<0.05
Ferredoxin:nitrite reductase[3] (in 10 mM K-phosphate pH 7.7)	0.63
Ferredoxin:glutamate synthase[3] (in 10 mM K-phosphate pH 7.5)	14.5

[1] (Hirasawa et al., 1988); [2] (Batie and Kamin, 1984); [3] (Hirasawa et al., 1986)

Supporting, yet somewhat different results were obtained by site-directed mutagenesis of *Chlamydomonas reinhardtii* ferredoxin (Jacquot et al., 1997b). Replacement of Glu91 (corresponding to Glu92 in the spinach protein) with glutamine or lysine completely abolished the capacity of the mutant ferredoxin to reduce *Chlamydomonas* FTR. This observation clearly confirms the results obtained by chemical modification. However, surprisingly, mutation of Glu92 (corresponding to Glu93 in spinach) which was protected from modification in the complex, had no effect on the reduction of FTR. In addition, a triple mutant involving Asp25, Glu28 and Glu29 (corresponding to Asp26, Glu29 and Glu30 in spinach FTR), was totally inactive. These three residues were not protected from modification in the complex and, therefore, are not considered to be involved in the interaction with FTR (De Pascalis et al., 1994). It may be that the reported differences are due to subtle structural discrepancies between the spinach and algal enzyme. The availability of a crystal structure of a spinach ferredoxin mutant (Binda et al., 1998) will allow an approach to these questions by modeling. No chemical modifications or mutations have yet been made to study the importance of the positive charges on the FTR surface in the interaction.

B. Ferredoxin:Thioredoxin Reductase

FTR, the central enzyme of the ferredoxin/thioredoxin system, converts an 'electron signal' received from ferredoxin to a 'thiol signal' which is transmitted to thioredoxin. FTR has been purified and characterized from different organisms: spinach (Schürmann, 1981; Droux et al., 1987a), maize and *Nostoc* (Droux et al., 1987a), soybean (P. Schürmann, unpublished) and *Chlamydomonas reinhardtii* (Huppe et al., 1990). The purified enzymes, all yellowish-brown, have very similar spectral properties, but vary with apparent molecular masses between 20 to 25 kDa, depending on the source. FTR is composed of two dissimilar subunits—designated the catalytic and variable subunit. The catalytic subunit of the enzyme from different organisms has a constant size of about 13 kDa, whereas the variable subunit ranges between 8 and 13 kDa.

1. Primary Structures

The primary structure based on amino acid or gene sequencing (or both) is known for FTR from several organisms (Fig. 3a,b). For the enzymes from spinach (Iwadate et al., 1994; Chow et al., 1995) and maize (Iwadate et al., 1996, 1998) the primary structures have been established by amino acid sequencing. Several other primary structures of FTR have been deduced from the nucleotide sequences. The gene for the variable subunit from *Anacystis nidulans*, which has been known for some time (Szekeres et al., 1991), codes for a protein with 77 residues. This gene is quite similar to the one from another cyanobacterium, *Synechocystis* sp. PCC6803.

Three cDNA clones have been isolated for the spinach FTR variable subunit, two are identical, corresponding to the protein sequence, and one differs slightly (Falkenstein et al., 1994; Gaymard and Schürmann, 1995). In addition a database search revealed a putative variable subunit from *Arabidopsis* and tomato for which a protein product has so far not been described. All higher plant genes contain the information for a typical chloroplast transit peptide thus confirming that the variable subunit is a nuclear encoded protein in these organisms. Sequence alignment of the mature proteins (Fig. 3a) shows a rather low conservation of the variable subunits, 46 to 60% identity within the eukaryotes and 33 to 40% between eukaryotes and prokaryotes. The most striking difference is the N-terminal extension present in the eukaryotic enzymes, which is rather long in spinach and probably also in *Arabidopsis* and tomato (their N-termini have been defined by similarity), and somewhat shorter in the maize FTR. The N-terminal extension of spinach FTR was found to be unstable, being degraded to discrete shorter peptides (Tsugita et al., 1991). The resulting truncated FTRs exhibit no differences in functional properties.

Nine complete nucleotide sequences from eukaryotes and one from a prokaryote are known for the catalytic subunit of FTR—spinach (2×), soybean, tomato, *Arabidopsis*, maize, *Porphyra purpurea*, *Cyanidium*, *Guillardia theta* and *Synechocystis*, respectively. Additional partial sequences can be retrieved from the EST database, two of them, for ice plant and *Chlamydomonas*, are rather complete. In the red algae *Porphyra* and *Cyanidium* and the cryptomonad *Guillardia* the catalytic subunit is plastid encoded and its gene therefore contains no transit peptide sequence. The genes of the higher plant proteins, by contrast, all include typical chloroplast transit peptides. The two cDNAs of the spinach FTR are different. One codes for a peptide which corresponds exactly to the isolated native protein (Gaymard and Schürmann, 1995). The other shows residue substitutions in the transit peptide as well as in the mature protein part (Falkenstein et al., 1994) where the two sequences have 81% identity. These observations suggest that there are at least two genes for each subunit in spinach.

In contrast to their variable counterparts the catalytic subunits have a highly conserved primary structure (Fig. 3b). Among the strictly conserved residues are seven Cys, six of them are organized in one CHC and two CPC motifs. These six are the functionally essential residues that constitute the redox active disulfide bridge and ligate the Fe-S cluster. The ligation of the cluster does not follow known consensus motifs, but shows a new arrangement with the following fingerprint: **C**P**C**X_{16}**C**P**C**X_8**C**H**C** (cluster ligands are in bold). In spinach FTR Cys54 and Cys84 form the active site disulfide. Cys54 is accessible to the solvent whereas Cys84 is protected. The four remaining cysteines, Cys52, Cys71, Cys73 and Cys82 are ligands to the iron center. This arrangement places the redox-active disulfide bridge adjacent to the cluster (Chow et al., 1995).

Interestingly, the sequencing of the complete genome of two archaebacteria revealed the presence of genes with striking resemblances to the catalytic

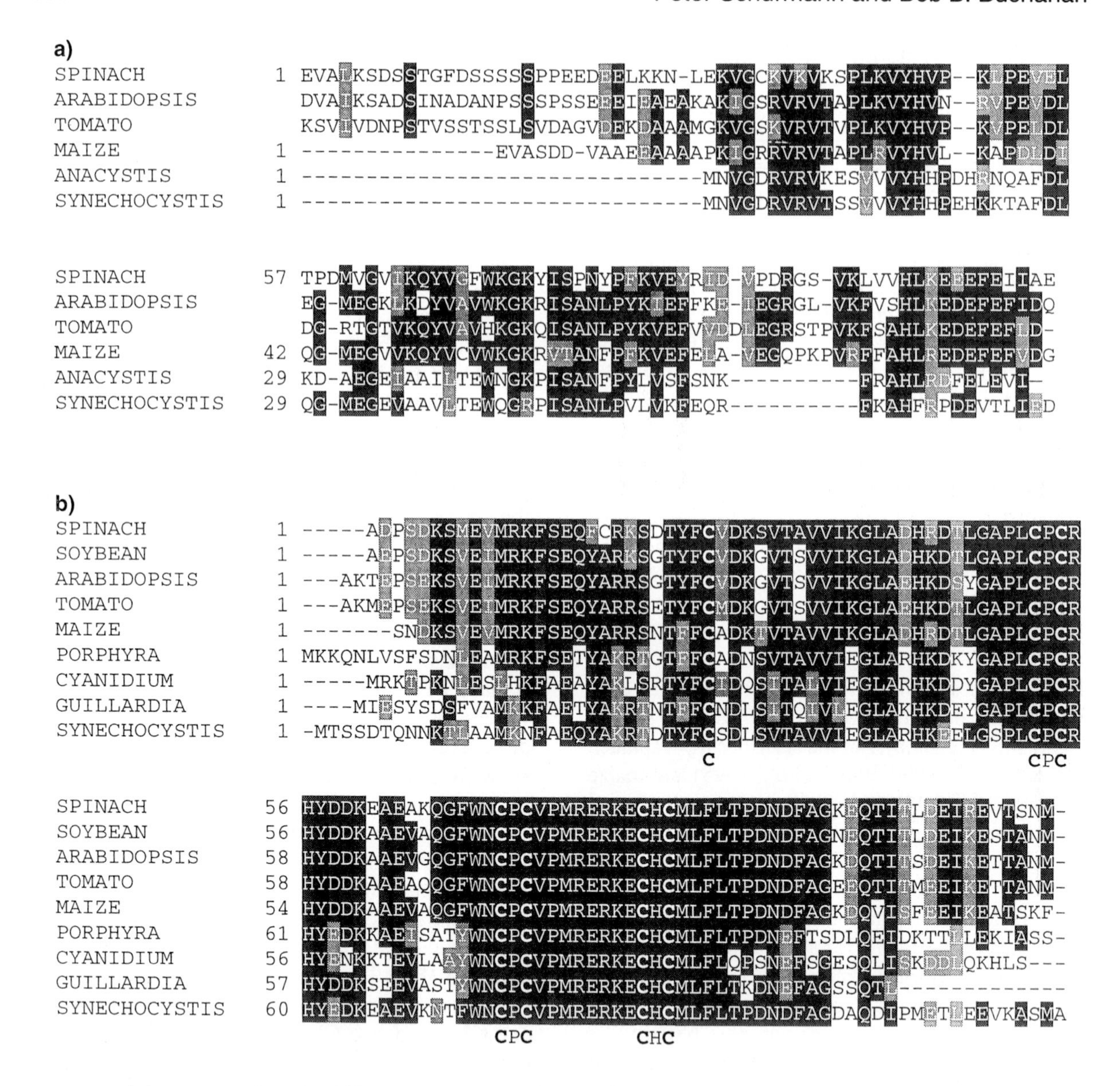

Fig. 3. Multiple sequence alignment of the variable (a) and catalytic (b) subunits of FTR from different organisms. The sequences were aligned using CLUSTALX (Thompson et al., 1997) and formatted with BOXSHADE (http://ulrec3.unil.ch/software/BOX_form.html). The N-termini of the eukaryotic subunits were determined by similarity. The numbering of the cyanobacterial and plastid encoded subunits starts after the initial excised Met residue. The sequences, retrieved from the Swiss Prot and the EMBL nucleotide data banks have the following accession numbers: variable subunit (a) spinach P38365; *Arabidopsis* AB018110; tomato AW092527; maize P80680;*Anacystis* P24018; *Synechocystis* Q55781; catalytic subunit (b) spinach P41348; soybean O49856; *Arabidopsis* AC006955 (AAD22336); tomato AW041151; maize P41347; *Porphyra* P51386; *Cyanidium* AF022186 (AAF13004); *Guillardia* O78461; *Synechocystis* Q55389.

subunit of FTR. *Archaeoglobus fulgidus*, a hyperthermophilic, sulfate-reducing archaeon (Klenk et al., 1997), contains a gene coding for a putative FTR catalytic subunit with 36% identity to the one from *Guillardia*. In the thermophilic archaeon *Methanobacterium thermoautotrophicum* (Smith et al., 1997) a gene for a hypothetical 17.4 kDa protein was sequenced. It has 25 to 27% sequence identity to the FTR from *Porphyra*, spinach and soybean. Although the overall identities between the photosynthetic FTR and the archaebacterial proteins are rather low, the residues found to be essential for the function of the

FTR, the CXC motifs, are found at conserved positions. There are, however, no functions known for those proteins in the archaea. Nonetheless, the striking structural similarities suggest that the catalytic subunit of photosynthetic FTR might be derived from an ancient precursor protein whose function has been adapted during evolution.

2. Expression

Two recombinant FTRs have been expressed in *E. coli*: spinach (Gaymard and Schürmann, 1995) and *Synechocystis* (Schwendtmayer et al., 1998). The same strategy has been adopted for the expression of both proteins. The genes coding for the two subunits were inserted in series in the expression vector, separated by a spacer region and a second ribosome binding site. These constructs yielded soluble, heterodimeric enzymes that contained the correctly inserted Fe-S cluster, demonstrated by spectral properties and enzyme activity. Probably owing to the prokaryotic nature of its genes, the *Synechocystis* protein is significantly better expressed than the spinach FTR. The recombinant cyanobacterial enzyme, which was also more stable than the spinach counterpart (Schwendtmayer et al., 1998) proved ideal for structural studies.

3. Crystal Structure

Recombinant *Synechocystis* FTR crystallized as dark brown crystals 10 to 100 μm in size. These crystals diffracted well and permitted structural resolution to 1.6 Å (Fig. 2) (Dai, 1998; Dai et al., 1998, 2000). The variable subunit has an open β-barrel structure whereas the catalytic subunit is mainly α-helical. Looking at the 'front' of the protein, the catalytic subunit is sitting on top of the heart-shaped variable subunit. From the side, the FTR heterodimer is revealed as an unusually thin molecule in the shape of a concave disc measuring only 10 Å across the center (Fig. 2). A cubane type [4Fe-4S] cluster accessible from one side of the molecule is present in the center with the active site disulfide bridge accessible from the other side. This arrangement allows simultaneous docking of a ferredoxin and a thioredoxin molecule—a prerequisite for the proposed reaction mechanism.

The docking sites on both sides of the molecule contain highly conserved residues. The ferredoxin docking area shows shape complementarity and contains four positive charges (Lys9, Lys23, Lys47, Arg80) which are thought to be responsible for the correct orientation of ferredoxin. In organisms in which the FTR is plastid encoded (*Porphyra* and *Guillardia*), the residue corresponding to Lys9 is replaced by a negatively charged Asp.

The thioredoxin docking area contains mainly hydrophobic residues and only one positive (Lys63) and one negative (Glu84) charged residue. The presence of mainly hydrophobic residues in the thioredoxin docking area makes it compatible for the interaction with different thioredoxins, since hydrophobic interactions are sterically not particularly specific.

While the function of the catalytic subunit is clear, the function of the variable subunit remains ill defined. The variable subunit may stabilize the interaction of the catalytic subunit with both ferredoxin and thioredoxin. It is interesting to note that the shape of the variable subunit shows striking similarities to PsaE, the ferredoxin binding protein of photosystem I, with the SH3 domain and with GroES although there are no sequence similarities with either protein (Dai, 1998; Dai et al., 2000).

C. Chloroplast Thioredoxins

Thioredoxins are messengers, which transmit the light signal, converted by FTR to a thiol signal, to target enzymes—some involved in carbon assimilation, others in processes that use photosynthetic energy to a different end. Higher plant chloroplasts contain two types of thioredoxins (Buchanan et al., 1978; Jacquot et al., 1978; Wolosiuk et al., 1979; Buchanan, 1980; Schürmann et al., 1981), a prokaryotic type, thioredoxin *m*, and a eukaryotic type, thioredoxin *f*, also present in algae (Huppe et al., 1990). These two types of thioredoxins can be clearly distinguished by their primary structures and specificity toward individual target enzymes.

Thioredoxin *f* specifically activates chloroplast FBPase, SBPase, PRK (Nishizawa and Buchanan, 1981), NADP-GAPDH (Buchanan et al., 1978), rubisco activase (Zhang and Portis Jr, 1999), chloroplast H^+-ATPase (Schwarz et al., 1997) and fatty acid metabolism (Sasaki et al., 1997), whereas reduced thioredoxin *m* is most effective for the activation of NADP-MDH (Jacquot et al., 1978), deactivation of G6PDH (Wenderoth et al., 1997) and translation (Danon and Mayfield, 1994; Levings, III and Siedow, 1995).

Both types of chloroplast thioredoxins appear to be reduced by FTR indiscriminately although quantitative data have not yet been reported to support this conclusion. Their interaction with the target proteins, however, displays specificity. This specificity necessitates recognition between thioredoxin and target enzyme, perhaps on the basis of charge and shape complementarity. Information has been gained on these points by site-directed mutagenesis.

1. Primary Structure

a. m*-Type Thioredoxins*

The *m*-type representatives from oxygenic prokaryotes, algae, monocots and dicots strongly resemble the thioredoxin from anoxygenic prokaryotes, both heterotrophic and photosynthetic. Thioredoxins of the *m*-type are, therefore, also known as bacterial-type thioredoxins. Owing to structural relatedness the bacterial and *m*-type thioredoxins are functionally similar and can be used interchangeably. A database search reveals 19 protein or gene sequences from 16 organisms, including both prokaryotes, eukaryotic algae and higher plants. While the different *m*-type thioredoxins are clearly related, they display less sequence similarity than their *f*-type counterparts sequenced to date (Fig. 4 and below).

b. f*-Type Thioredoxins*

Fewer sequences are known for thioredoxin *f* which is considered to be restricted to eukaryotic organisms. Seven complete primary structures of *f*-type thioredoxins are known, one for spinach, pea, ice plant, rape, rice and two for *Arabidopsis*, all from higher plants (Fig. 5). Two incomplete sequences, for soybean and watermelon respectively, can be found in the EST database. Thioredoxin *f* is a nuclear encoded protein and the deduced amino acid sequences show a typical chloroplast transit peptide. The amino acid sequence has been determined experimentally only for spinach thioredoxin *f*. The N-terminus of the protein was, however, blocked (Kamo et al., 1989). Indirect evidence strongly suggests that Met69 of the precursor is the N-terminal residue of the mature protein (Aguilar et al., 1992; del Val et al., 1999).

f-Type thioredoxins are highly conserved with 75 to 90% residue identity in the mature protein from dicots. Thioredoxin *f* from rice is the most divergent having only 60 to 67% residue identity with the dicot proteins. Owing to the additional amino acids at their N-terminus, the *f* thioredoxins are slightly longer than other types. Interestingly, the C-terminal part of the sequence resembles classical animal thioredoxin in containing a third, strictly conserved Cys (Cys73 in spinach).

2. Three-dimensional Structures

While extensive data are available for thioredoxins from nonphotosynthetic organisms, less structural information exists for photosynthetic counterparts. The crystal structure has been determined for an unusual thioredoxin from the cyanobacterium *Anabaena* (Saarinen et al., 1995). Although displaying some thioredoxin *f* activity in vitro, it cannot be considered a member of the thioredoxin *f* family on the basis of sequence similarity. Furthermore, its overall three-dimensional structure is similar to *E. coli* thioredoxin. Single amino acid substitutions around the protein interaction area may account for the unusual biochemical properties observed in vitro.

Based on its solution structure, extraplastidic (cytosolic) thioredoxin *h* from *Chlamydomonas reinhardtii* resembles human thioredoxin in accord with its phylogenetic history (Mittard et al., 1997). As expected this structure is quite different from both the *f*- and *m*-type chloroplast thioredoxins. An earlier NMR study showed that the secondary structure and the global protein folding of thioredoxin *m* from *Chlamydomonas reinhardtii* was similar to its *E. coli* counterpart (Lancelin et al., 1993).

a. Crystal Structure of Spinach Thioredoxin f

The structure has been solved for two forms of thioredoxin *f*, for the wt-recombinant or 'long form' and an N-terminus truncated, or 'short form' (Fig. 2). The long form of thioredoxin *f* has a slightly modified, three residue longer N-terminus than the native protein. Nonetheless, the two proteins are enzymically active and functionally indistinguishable from the native protein isolated from leaves (Aguilar et al., 1992). Both structures are essentially identical aside from the N-terminus, which contains an additional α-helix in the wt-recombinant, and with all probability also in the chloroplast protein.

The overall structure of thioredoxin *f* does not differ markedly from a typical thioredoxin, showing

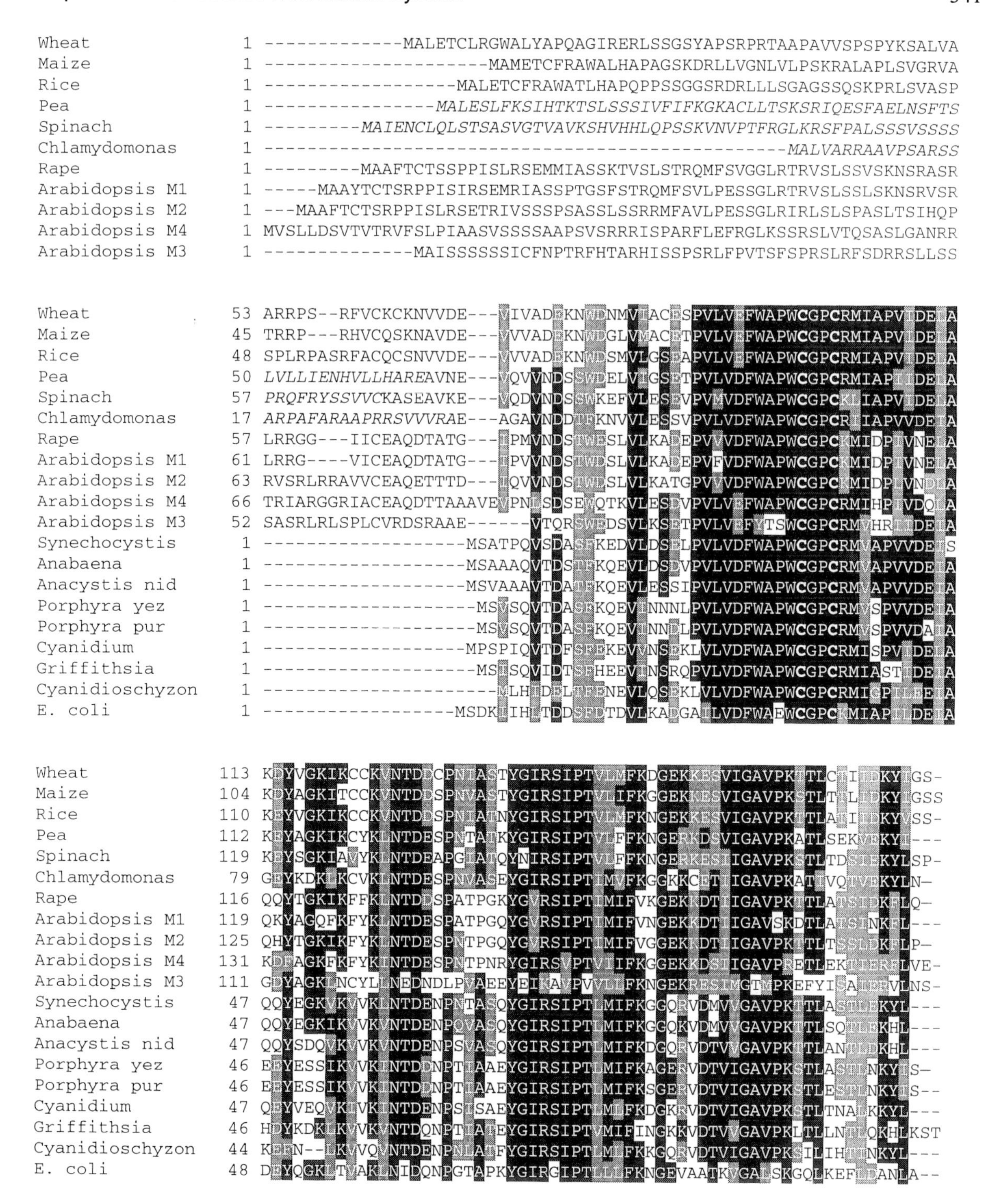

Fig. 4. Multiple sequence alignment of *m*-type thioredoxins from different organisms. The sequences were aligned using CLUSTALX (Thompson et al., 1997) and formatted with BOXSHADE (http://ulrec3.unil.ch/software/BOX_form.html). The transit peptides of the spinach, pea and *Chlamydomonas* thioredoxin, known from N-terminal sequencing, are printed in *italics*. The sequences, retrieved from the Swiss Prot and the EMBL nucleotide data banks have the following accession numbers: wheat AJ005840; maize Q41864; rice AJ005841; pea P48384; spinach P07591; *Chlamydomonas* P23400; rape O03043; *Arabidopsis* M1 AF095749, M2 AF095750, M3 AF095751, M4 AF095752; *Synechocystis* P72643; *Anabaena* P06544; *Anacystis* P12243; *Porphyra yezoensis* P50254; *Porphyra purpurea* P51225; *Cyanidium* P37395; *Griffithsia* P50338; *Cyanidioschyzon* O22022; *E.coli* P00274.

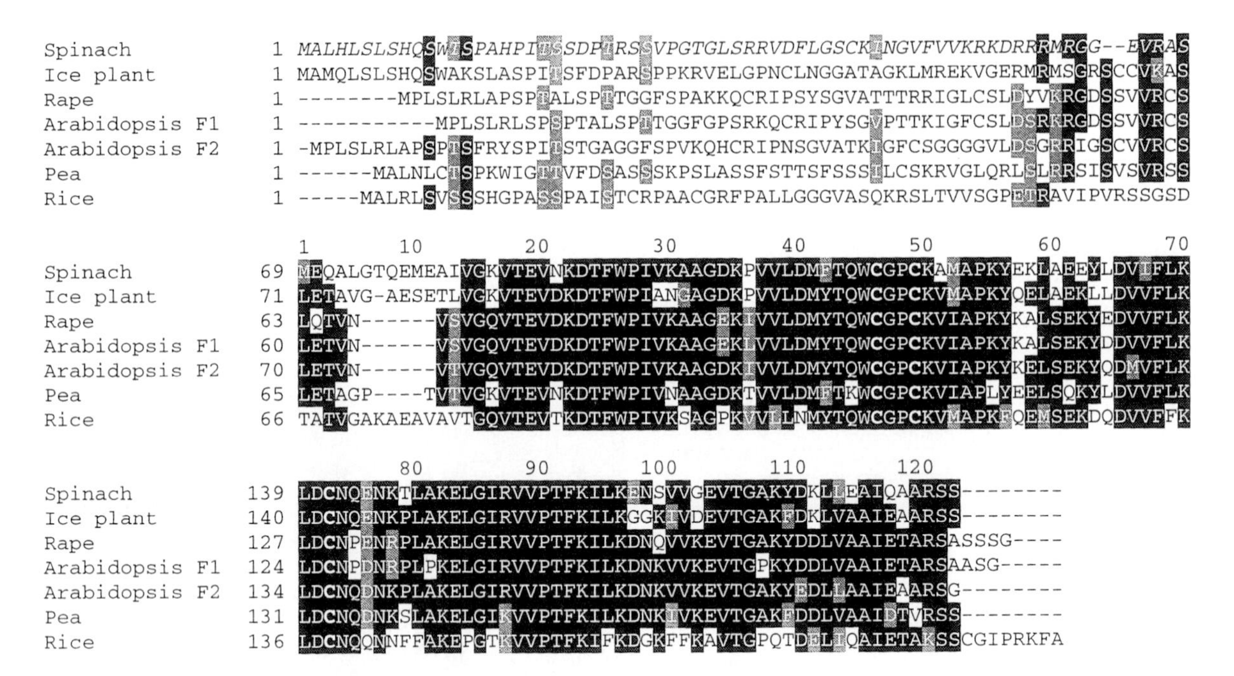

Fig. 5. Multiple sequence alignment of *f*-type chloroplast thioredoxins from different organisms. The sequences were aligned using CLUSTALX (Thompson et al., 1997) and formatted with BOXSHADE (http://ulrec3.unil.ch/software/BOX_form.html). The putative transit peptide of spinach thioredoxin *f* is printed in *italics*. The sequences, retrieved from the Swiss Prot and the EMBL nucleotide data banks have the following accession numbers: spinach P09856; ice plant O81332; rape O48897; *Arabidopsis* F1 AF144385, F2 AF144386; pea P29450; rice AF022741.

a central twisted β-sheet surrounded by α-helices (Capitani et al., 1998, 2000). However, its surface topography is distinct from that of other thioredoxins (Fig.2). A striking difference is the presence of the third Cys exposed on the surface (Cys73 in spinach), 9.7 Å away from the accessible Cys of the active site (Cys46). As already mentioned, this third Cys is conserved in all *f* thioredoxins. In addition to this difference, thioredoxin *f* has a substantially altered distribution of polar, charged and hydrophobic residues around the active site with respect to *E. coli* or *m*-type thioredoxin (Fig. 2). The active site of thioredoxin *f* is surrounded by a number of positive charges (Lys50, Lys78, Lys82, Arg87, and further away, Lys108) which may be instrumental in orienting thioredoxin *f* correctly with target proteins. The hydrophobic residues may be more important in a nonspecific interaction with FTR which reduces all thioredoxins tested. The structural model also confirms that the active site Cys with the lower sequence number is the attacking nucleophile in the reduction of the target disulfide (Brandes et al., 1993). This Cys is exposed whereas its active site partner is buried, a situation found in practically all proteins with redox active disulfides.

b. Crystal Structure of Spinach Thioredoxin m

The structure of thioredoxin *m* has been solved for the oxidized and reduced protein at 2.1 and 2.3 Å resolution, respectively (Capitani et al., 1998, 2000). The structure is very similar to that of *E. coli* (Katti et al., 1990) thioredoxin which corroborates the biochemical evidence showing the proteins are functionally interchangeable. An analysis reveals that the *m*-type and *E. coli* thioredoxins are nearly identical in secondary structure, whereas differences emerge with respect to thioredoxin *f*. The surface around the active site Cys is, as expected, quite different from that of thioredoxin *f* (Fig. 2) and resembles largely its *E. coli* counterpart. Superimposing the active sites of oxidized and reduced thioredoxin *m* reveals no large conformational change, thus confirming and extending observations reported for *E. coli* and human thioredoxins (Jeng et al., 1994; Weichsel et al., 1996). However, some slight structural changes in the main chain conformation of the active site renders the solvent-exposed Cys (Cys37) more accessible upon reduction.

V. Target Enzymes

The identification of an enzyme showing a light-dependent increase in catalytic activity has traditionally been made on the basis of biochemical evidence obtained with cell-free extracts. Here, a particular activity was shown to be enhanced specifically by a dithiol which can mimic the effect of light. Typically, activation by a dithiol such as DTT became dependent on thioredoxin once the enzyme was purified. Still further analysis revealed the presence of at least one reversibly reducible disulfide bond on the target enzyme and a coupling of activation to the ferredoxin/thioredoxin system.

An analysis of the primary structures of light regulated target enzymes from various oxygenic photosynthetic organisms provided evidence for the presence of conserved Cys residues absent from counterparts not regulated by thioredoxin. Although there can be strict specificity of a particular thioredoxin, there is no general consensus enzyme sequence for regulation. For certain enzymes, particularly those also occurring as cytosolic isoforms, the regulatory disulfide structures are located on extra loops or extensions, indicating their addition during adaptation to photosynthetic function. The analyses of the regulatory sites of the other enzymes suggest that the active cyst(e)ines need not be proximate and can be separated by many amino acids. The results suggest that the adaptation of enzymes to light-linked regulation arose multiple times during evolution (Table 2).

A. Fructose 1,6-bisphosphatase

Chloroplast FBPase is one of the classical light-regulated enzymes. It is a homotetramer of about 160 kDa. A total of seven primary structures of the chloroplast enzyme can be retrieved from the database, only one based also on protein sequence (Marcus and Harrsch, 1990). All sequences are quite homologous and, compared with the cytosolic isoenzyme, contain an insert in the middle of the primary structure, where the putative regulatory site is located (Marcus et al., 1988). This insert contains three Cys residues, two are separated by four hydrophobic residues (C173VVNVC178 in pea) and the third, Cys153, is twenty residues upstream toward the N-terminus (Table 2). These three Cys are candidates for the regulatory site and their involvement in a regulatory disulfide bridge has been probed by site-directed mutagenesis. While the replacement of Cys153 results in a permanently fully active enzyme, the replacement of either of the remaining Cys (C173, C178) results in a partly active enzyme that still requires reduction by thioredoxin for full activity. These results suggest that all three Cys are involved in regulation. While Cys153 is an obligatory part of the regulatory disulfide, the remaining two Cys components (C173, C178) act interchangeably in constituting its bonding partner (Jacquot et al., 1995, 1997c; Rodriguez-Suárez et al., 1997).

The structure of the spinach chloroplast FBPase has been analyzed by X-ray crystallography (Villeret et al., 1995), but unfortunately the results did not

Table 2. Regulatory disulfide sites of thioredoxin-linked target enzymes. The number of sequences compared is given in brackets. The variable residues in the regulatory motifs are indicated by X. Cysteines forming regulatory disulfides are in bold type.

Target enzyme		Regulatory site sequence
Fructose 1,6-bisphosphatase (7)		DE**C**X_n**C**XVXVCQPG (n = 14–19)
Sedoheptulose 1,7-bisphosphatase (4)		TAS**C**XGTX**C**VN
Phosphoribulokinase (8)		SG**C**GKX_{34}VX**C**LD
		X_{17} Synechocystis
		X_{38} Odontella (diatom)
ATP synthase (6)		GEX**C**DXXGX**C**VDA
NADP malate dehydrogenase (6)	N-terminal	XX**C**XGXF**C**XT
	C-terminal	EKX**C**XAHLTGEGXAX**C**DXP
Glucose 6-P-dehydrogenase (6)		LT**C**RIDXRXX**C**XX
Rubisco activase 46 kDa isoform (3)		XG**C**TDX$_{14}$GX**C**XYXX

provide evidence on the regulatory disulfide. Recently the X-ray structure of the oxidized pea FBPase and of a regulatory site mutant have been solved (Chiadmi et al., 1999). This model shows that the insert (C153–C173–C178) is on a loop extending out of the core structure of the enzyme, making it accessible to thioredoxin, and that there is a disulfide bridge between Cys153 and Cys173. In the oxidized enzyme the loop is stabilized by the disulfide bridge whereas in the C153S mutant, which probably resembles the reduced enzyme, it is more disordered. The oxidized conformation of the loop has an allosteric effect on the 20 Å distant active site where it disrupts the catalytic Mg^{2+} binding site by the displacement of Glu105. Through reduction of the regulatory disulfide, the active site becomes catalytically competent enabling the binding of Mg^{2+} and substrate. It is well known that the need for reductive activation can be partly bypassed in vitro by high pH and high Mg^{2+} concentrations. These non-physiological conditions must have a similar effect on the FBPase structure. Based on the structural analyses of the WT and the C153S mutant the third Cys (Cys178 in pea) of the regulatory loop does not seem to be normally involved in any disulfide bond. However, it might form an artifactual disulfide bridge in the mutant proteins, which could explain their partial thioredoxin dependency. There are four additional Cys residues conserved in the chloroplast FBPases that lie in well defined secondary structures. Their replacement with Ser had no significant effect on either activation or catalysis (Rodriguez-Suárez et al., 1997).

Several groups have carried out site-directed mutagenesis experiments on thioredoxin to understand its interaction with chloroplast FBPase (Lamotte-Guéry et al., 1991; Geck et al., 1996; Mora-García et al., 1998; del Val et al., 1999). The results confirm the importance of positive charges in thioredoxin *f* (Lys58, Lys108).

B. Sedoheptulose 1,7-bisphosphatase

A substrate-specific SBPase is found only in oxygenic photosynthetic eukaryotes and is unique to the Calvin cycle.

In higher plants SBPase is located in the chloroplast and is encoded in the nuclear genome. There is no cytosolic counterpart (Raines et al., 1999). Four deduced primary structures are known: spinach (Martin et al., 1996), wheat (Raines et al., 1992), *Arabidopsis* (Willingham et al., 1994) and *Chlamydomonas* (Hahn et al., 1998). Each codes for a protein of 387 to 393 residues which include a putative chloroplast transit peptide of about 70 residues.

Excluding putative transit peptide sequences these four proteins have a high degree of relatedness (up to 98% similarity). The mature SBPase is a homodimer of 76 kDa. Its sequence, and probable three-dimensional structure, show significant overall similarity with FBPase but lack the regulatory Cys insert. In autotrophic prokaryotes, SBPase activity appears to be contributed by a FBPase which recognizes both sugar bisphosphate substrates.

The SBPase enzyme from different sources contains a variable number of Cys residues. Four of the Cys residues are found at strictly conserved positions, in pairs, in the N-terminal part. The two most N-terminal Cys are arranged in the **C**XXXX**C** motif (**C**52GGTA**C**57 in wheat), typical of redox-active groups in other enzymes (Table 2). Although proposed as the regulatory Cys of SBPase for a number of years, these two residues have only recently been shown to be involved in redox regulation. Mutation of these two Cys to Ser resulted in an active, redox-insensitive SBPase (Dunford et al., 1998). Like FBPase, the regulatory Cys residues of SBPase are not part of the catalytic site. Based on the structural similarities between the two enzymes, the authors proposed that the regulatory disulfide is located on a flexible loop near the junction between the two subunits. Oxidation or reduction of the regulatory disulfide may alter the conformation of the dimer and thereby change the activity of the enzyme (Raines et al., 1999).

Like its FBPase counterparts, SBPase is specifically activated by thioredoxin *f* and, once reduced, is also regulated by stromal pH and Mg^{++} level as well as sugar bisphosphate substrate and by the products of the reaction (Wolosiuk et al., 1982; Schimkat et al., 1990). In contrast to FBPase, the SBPase shows an absolute requirement for a dithiol for activation—a property revealed when the enzyme was discovered (Breazeale et al., 1978; Nishizawa and Buchanan, 1981).

C. Phosphoribulokinase

Long recognized as a light-regulated enzyme, PRK was originally found to be activated most effectively by thioredoxin *f* (Wolosiuk and Buchanan, 1978a) and, in the case of spinach and *Chlamydomonas*, was shown to be nuclear encoded (Milanez and Mural,

1988; Roesler and Ogren, 1990). The deduced amino acid sequence of the spinach enzyme codes for a rather long transit peptide and a mature protein of 351 residues. In its native form PRK appears as a homodimer of 80 kDa. Each subunit contains four cysteine residues, two close to the N-terminus and two to the C-terminus. These cysteines are found at conserved positions in the eight primary structures from higher plants, algae and cyanobacteria that can be retrieved from the Swiss-Prot data bank. One exception is the *Odontella* enzyme in which the nonregulatory C-terminal pair (Cys244, Cys250) is missing. Two of the four Cys residues (Cys16 and Cys55, in spinach) were identified as being regulatory by chemical modification and protein sequencing (Table 2). These Cys residues form an intramolecular disulfide bond in the oxidized, inactive enzyme (Porter et al., 1988). Both regulatory Cys residues are located in the nucleotide binding domain of the active site and Cys55 was proposed to play a facilitative role in catalysis (Porter and Hartman, 1990; Porter et al., 1990). PRK is the only example of a thioredoxin-linked enzyme with a regulatory Cys as part of its active site.

The function of the different Cys residues in PRK was confirmed and extended by site-directed mutagenesis. The replacement of the two nonconserved cysteines, Cys244 and Cys250, was of no consequence to enzyme activity. Replacement of Cys55, by contrast, drastically reduced catalytic activity indicative that this residue is important in the active site, where it probably functions in the binding of the sugar phosphate substrate (Milanez et al., 1991). In addition, Cys55 forms the transient heterodisulfide with Cys46 of thioredoxin *f* during reductive activation (Brandes et al., 1996). Since one of the reduced Cys residues seems to be involved in substrate binding, the formation of a disulfide bridge very effectively blocks catalytic activity, a situation found also in NADP-MDH. The crystal structure of PRK from *Rhodobacter sphaeroides* is now available (Harrison et al., 1998). This enzyme is not redox regulated, but the structural data will allow homology modeling and provide useful information for further studies of the regulatory events taking place in the photosynthetic enzyme.

D. ATP Synthase (CF_1-ATPase)

Chloroplast ATP synthase or CF_0CF_1, a constituent of thylakoids, is composed of the integral membrane portion CF_0 and the hydrophilic CF_1, the latter consisting of five different subunits. Chloroplast CF_1 ATPase is a latent enzyme and, in contrast to its mitochondrial and bacterial counterparts, must be activated before showing ATPase activity. Activation is achieved in vivo by the transmembrane electrochemical proton potential difference which induces a conformational rearrangement (Ort and Oxborough, 1992). The potential gradient thus acts not only as driving force for phosphorylation but also as an activating factor for the reversible conversion of the complex to catalytic competence. In addition to the electrochemical activation, thioredoxin has been shown to reduce the enzyme from higher plants and thereby modulate its activity (McKinney et al., 1978; Ort and Oxborough, 1992).

The structural element allowing for thiol modulation is a sequence motif in the CF_1 gamma subunit which contains two Cys residues forming a disulfide bond in the oxidized enzyme (between Cys199 and Cys205 in the spinach enzyme; Table 2). This motif is present in the higher plant (Miki et al., 1988) and green algal enzyme (Yu and Selman, 1988) but not in the counterpart of cyanobacteria (Cozens and Walker, 1987; McCarn et al., 1988; Werner et al., 1990) or diatoms (Pancic and Strotmann, 1993). It appears that the regulatory disulfide bridge is inaccessible in the dark when the enzyme is inactive and becomes exposed upon activation by the transmembrane potential difference.

Different types of thioredoxins have been used for the reduction of the enzyme in vitro. Thioredoxin *f* was shown to activate thylakoid-bound ATPase in the light (Mills et al., 1980) more effectively than *E. coli*, *m*-type and human thioredoxins (Galmiche et al., 1990). A recent comparison of the efficiencies of the two spinach chloroplast and *E. coli* thioredoxins clearly showed that thioredoxin *f* is the most efficient activator. DTT alone showed less than 0.01% the effectiveness of thioredoxin *f* (Schwarz et al., 1997). Interestingly, thioredoxin *m* was less active than the *E. coli* protein. This difference was thought to be due to three charged surface residues present in thioredoxin *m* (R91, K92, E93), but absent from the other two proteins. However, based on the crystal structure of thioredoxin *m* (Capitani et al., 1998) these three residues are on the backside of the molecule and therefore apparently unimportant for the interaction. Some other structural difference must be responsible for the observed difference in affinity.

Reduction of CF_1 is very rapid, even in weak light.

This high rate may be due to the redox potential of the regulatory disulfide, which seems to be about equal to that of thioredoxin *f* or even slightly more positive (Table 3) (see below). The main purpose of reduction does not seem to be the modulation of enzyme activity, but rather to permit a higher rate of ATP formation at limiting electrochemical potential. Regulation linked to thioredoxin also allows the enzyme to be switched off in the dark so as to avoid wasteful hydrolysis of ATP. Analyses of the structural changes brought about by reduction of the regulatory disulfide should give new information as to how this gain of efficiency is achieved. It is noted that when the CF_1 serves a dual function in both oxidative and photophosphorylation, the enzyme should not be turned off in the dark. Thus CF_1 lacks the regulatory disulfide segment seen with chloroplast enzymes in organisms such as the cyanobacteria (Cozens and Walker, 1987; McCarn et al., 1988; Werner et al., 1990).

E. NADP-dependent Malate Dehydrogenase

Although having one of the most complex redox regulation mechanisms, NADP-MDH is the best understood light-activated enzyme of chloroplasts. Details of its regulation have been described in timely recent reviews (Miginiac-Maslow et al., 1997; Ruelland and Miginiac-Maslow, 1999). Primary structures from six different enzymes can be retrieved from the Swiss-Prot data bank. They have a high degree of homology (80 to 90% identical residues). The chloroplast enzyme, a homodimer of 85 kDa, differs from its NAD-dependent counterpart by the presence of N- and C-terminal extensions and eight Cys residues at strictly conserved positions. Two are located in the N-terminal extension in a CXXXXC motif, and another two, separated by eleven residues, are in the C-terminal region (Table 2). The four remaining Cys are located in the core part of the protein.

Table 3. Oxidation-reduction potentials of disulfide/sulfhydryl couples active in the ferredoxin/thioredoxin system. Potentials have been normalized at pH 7.0 assuming a slope of Em vs. pH of –59 mV/pH unit.

Protein		Redox potential E_m (pH = 7.0) in mV Error = ±10 mV	
Ferredoxin		–420	
Ferredoxin:thioredoxin reductase	spinach[1]	–320	
	Synechocystis[2]	–320	
Thioredoxin *f*	spinach	–290[1]	–270[3]
	pea[1]	–290	
Thioredoxin *m*	maize[4], spinach[1], *Chlamydomonas*[1]	–300	
Fructose 1,6-bisphosphatase	spinach	–305[5]	–330[1]
	pea[1]	–315	
	tomato[3]	–295	
Sedoheptulose 1,7-bisphosphatase	tomato[3]	–300	
Phosphoribulokinase	spinach[6]	–295	
	tomato[3]	–255	
ATP synthase	spinach[3]	–280	
NADP malate dehydrogenase	activation: maize[4], sorghum[7]	–330	
	N-terminal S24-S29: sorghum[7]	–280	
	C-terminal S365-S377: sorghum[7]	–330	
	transient S24-S207: sorghum[7]	–310	
Glucose 6-P-dehydrogenase	pea[8]	–300	
Glyceraldehyde 3-phosphate dehydrogenase	tomato[3]	–310	

[1] (Hirasawa et al., 1999); [2] (Schwendtmayer et al., 1998); [3] (Hutchison et al., 2000); [4] (Rebeille and Hatch, 1986); [5] (Y. Balmer, M. Hirasawa, D. B. Knaff and P. Schürmann, unpublished); [6] (Hirasawa et al., 1998); [7] (Hirasawa et al., 2000); [8] estimated by Kramer et al. (1990) based on results by Scheibe et al. (1989).

Systematic analysis of the functions of the different Cys residues in the chloroplast enzyme provided evidence for two regulatory disulfides (Jenkins et al., 1986), both in the terminal regions. One is formed by the two most N-terminal (Cys24, Cys29 in sorghum) (Decottignies et al., 1988; Scheibe et al., 1991) and a second by the two most C-terminal Cys residues (Cys365, Cys377 in sorghum) (Issakidis et al., 1992; Reng et al., 1993; Issakidis et al., 1994). Studies on the mechanism of activation revealed that these two regulatory sites are not equivalent but have different functions (Hatch and Agostino, 1992; Issakidis et al., 1996; Ruelland et al., 1998). Removal of the N-terminal disulfide by site-directed mutagenesis yielded an inactive enzyme, which still required activation by thioredoxin. However, its activation kinetics were dramatically different. The slow activation, seen with the native enzyme containing both regulatory disulfides, changed to almost instantaneous activation of the mutated enzyme upon addition of reduced thioredoxin. Removal of the C-terminal disulfide produced a mutant enzyme with activation properties very similar to the wild-type, but the activation was no longer inhibited by NADP. In addition, the Km for oxaloacetate was strongly increased. Removal of N- and C-terminal disulfides resulted in a thioredoxin-insensitive, permanently active enzyme (Issakidis et al., 1993; Reng et al., 1993; Issakidis et al., 1994).

A fifth internal Cys residue (Cys207 in sorghum) may also be involved in the regulatory modifications by forming a transient disulfide bridge, most probably with one of the N-terminal Cys (Ruelland et al., 1997). This internal Cys was recently shown to become accessible when there is no N-terminal disulfide and to be capable of forming a hetero-disulfide with thioredoxin (Goyer et al., 1999). A tentative activation scheme was developed based on the functional analyses of the modified proteins. The N-terminal disulfide bridge is responsible for the low rate of activation and brings about a slow conformational change of the active site toward higher catalytic activity. The reduction of the C-terminal disulfide opens access to the active site in a rapid reaction (Ruelland and Miginiac-Maslow, 1999).

The crystal structure has been solved for two chloroplast NADP-MDHs, both in the oxidized form: one from sorghum (Johansson et al., 1999) and the other, with bound NADP, from *Flaveria* (Carr et al., 1999). These studies, which provide a structural basis for redox modulation, confirm the presence of two surface-exposed, thioredoxin-accessible disulfide bonds. Perhaps the most striking feature is the C-terminal extension which, when the protein is oxidized, bends back over the surface and reaches with its tip into the active site, obstructing access of the C_4 acid substrate. Recent NMR observations have shown that the C-terminal peptide acquires increased mobility upon reduction, thus releasing the autoinhibition of the enzyme (Krimm et al., 1999). The tip of the C-terminus carries two negative charges which interact with NADP, but not with NADPH. These interactions slow down the release of the C-terminal extension and explain the observed inhibition of the reductive activation by NADP (Scheibe and Jacquot, 1983; Ashton and Hatch, 1983; Ruelland et al., 1998).

The structural changes taking place on reduction of the N-terminal disulfide are less clear. Two possible effects are proposed. The N-terminal extension is apparently flexible and its three-dimensional structure not well defined. The N-terminal disulfide might lock domains and thus maintain an unfavorable conformation of the active site (Johansson et al., 1999), or the N-terminal residues might reach over the surface of the molecule toward the adenosine end of the active site and limit substrate access (Carr et al., 1999). The mobility of the N-terminal extension would also allow for the formation of a transient disulfide bridge between a N-terminal and an internal Cys (Ruelland et al., 1997), probably between two subunits rather than within a single subunit (Carr et al., 1999).

An analysis of the redox activation of NADP-MDH clearly shows a complex picture. The strict regulation by redox equilibrium and the NADPH/NADP ratio is understandable in C_3 plants, where an export of NADPH via the malate/oxaloacetate shuttle should be allowed only in case of surplus reducing power. This export can be efficiently controlled by the proposed regulatory mechanism. C_4 plants have a different need for regulating the enzyme—that is, to prevent futile cycling in the dark. The strict dependency of the NADP-MDH by thioredoxin insures that the C_4 pathway of the mesophyll cells is inactive in the dark, thereby minimizing the wasteful consumption of ATP for the synthesis of phosphoenolpyrvuate by pyruvate, Pi dikinase. The requirement of pyruvate, Pi dikinase and phosphoenolpyruvate carboxylase for reversible light activation also helps to insure that the C_4 pathway is switched off in darkness.

NADP-MDH does not display a high degree of specificity for thioredoxin *m* since it can be activated by other thioredoxins. Under certain conditions *f*-type thioredoxins are equally or even more efficient (Hodges et al., 1994; Geck et al., 1996). There is no evidence that the enzyme forms a stable complex with an activating thioredoxin (Lunn et al., 1995), as can be observed with FBPase. It is possible that the thioredoxin interaction sites can change according to conditions in the chloroplast and that both types of thioredoxin can activate this enzyme in vivo.

F. Ribulose 1,5-bisphosphate Carboxylase/ Oxygenase (via Rubisco Activase)

Rubisco catalyzes the sole carboxylation step of the Calvin cycle. Early extensive attempts to link the regulation of Rubisco to the ferredoxin/thioredoxin system proved unsuccessful. The results were puzzling because the enzyme was known to be activated by light on the basis of several lines of evidence, including the enhancement of its activity by Rubisco activase. In the latter case, the enzyme is deinhibited by removal of bound ribulose 1,5-bisphosphate substrate in a reaction requiring the hydrolysis of ATP. A regulatory link to light via Rubisco activase has until recently existed only indirectly via change in the ADP/ATP ratio (Portis Jr, 1995).

Recent findings have added new insight to the problem (Zhang and Portis Jr, 1999). It has been found that the larger of the two forms of Rubisco activase from *Arabidopsis* is activated by reduced thioredoxin *f* at physiological ratios of ADP/ATP. The regulatory site on the enzyme was also identified ($C392X_{18}C411$ in *Arabidopsis*) (Table 2). This finding provides a direct link of Rubisco activity to light via the ferredoxin/thioredoxin system. A feature that remains to be clarified is the mode of light-induced regulation in those plants such as tobacco that appear to lack the large thioredoxin-linked (46 kDa) form of Rubisco activase. One means, already recognized, is via the inhibitor 2-carboxyarabinitol-1-phosphate, which in tobacco and certain other species accumulates at night and disappears during the day (Hammond et al., 1998; Parry et al., 1999).

G. Glucose 6-phosphate Dehydrogenase

G6PDH catalyzes the first step of the oxidative pentose phosphate cycle, the oxidation of glucose 6-phosphate with concomitant reduction of NADP. There are two major isoforms, one in the chloroplast and the other in the cytosol. Only the chloroplast enzyme is subject to redox regulation. So far G6PDH is the only chloroplast enzyme found to be reductively inactivated by thioredoxin under physiological conditions. G6PDH is therefore the opposite of biosynthetic enzymes in being inactive in the light (reduced state) and active in the dark (oxidized state). This regulation is the logical consequence of the regulatory role of thioredoxin in photosynthetic carbon metabolism—that is, to minimize simultaneous carbohydrate synthesis (via the Calvin cycle) and degradation (via the oxidative pentose phosphate cycle).

Redox modulation has been demonstrated for G6PDH from cyanobacteria (Cossar et al., 1984; Gleason, 1996) and higher plant chloroplasts (Johnson, 1972; Scheibe and Anderson, 1981). The chloroplast form of the enzyme contains four to six Cys residues at positions not conserved in the cytosolic isoform. Despite their similar response to light, the redox-modulated cyanobacterial and plastid G6PDH show considerable overall sequence differences. The cyanobacterial enzymes contain fewer Cys and the two conserved putative regulatory Cys occur at entirely different positions separated by 297 or 298 residues.

The possible function of the Cys residues in the plastid enzyme of potato has been investigated by individually replacing them with serines (Wenderoth et al., 1997). This enzyme contains six Cys residues, five at homologous positions relative to the counterpart from other plants. All six are located in the N-terminal half of the protein, within the NADP-binding domain. The recombinant plastid enzyme was clearly inhibited by reduction with DTT in a reaction markedly accelerated by thioredoxin *m*, but not by thioredoxin *f*. The mutants C149S and C157S were no longer inhibited by reduction whereas all other mutants behaved essentially like wild-type enzyme. The Km for G6P increased by about 30-fold for the inhibited mutants and the reduced wild-type enzyme. These results suggest that Cys149 and Cys157 are engaged in a regulatory disulfide bridge in the active enzyme, **C**149RIDKREN**C**157 (Table 2), that is specifically reduced by thioredoxin *m*. Homology modeling using the crystal coordinates of the *Leuconostoc* enzyme locates the proposed regulatory cysteines on an exposed loop, close enough to permit a disulfide bridge, and freely accessible for interaction with thioredoxin.

The recent findings on the thioredoxin specificity for G6PDH and enzymes of CO_2 assimilation (e.g., Rubisco activase and NADP-MDH) once again prompt the question on the basis for the presence of two distinct types of thioredoxin in chloroplasts. Based on the new work, it seems possible that, at least as far as carbohydrate metabolism is concerned, thioredoxin *f* functions primarily in enzyme activation (i.e., enhancing the rate of biosynthesis) whereas thioredoxin *m* acts mainly in enzyme deactivation (i.e., enhancing the rate of degradation). Future work is expected to add information relevant to this question.

H. Other Target Enzymes

1. Glyceraldehyde 3-phosphate Dehydrogenase

GAPDH was the first chloroplast enzyme with an extractable activity shown to be increased by illumination (Ziegler and Ziegler, 1965), but ironically both the importance and mechanism of its regulation remain ill defined. In catalyzing the reduction step of the carbon reduction cycle, GAPDH reduces 1,3 bisphosphoglycerate to glyceraldehyde 3-phosphate preferentially with NADPH as coenzyme. GAPDH is the only regulatory member of the Calvin cycle that catalyzes a freely reversible reaction. A number of factors have been reported to increase the activity of this enzyme, among them reduced thioredoxin (Wolosiuk and Buchanan, 1978b). The need for light activation is presumably to minimize loss of NADPH in the dark, because the NAD-linked activity associated with the enzyme is not affected by reduced thioredoxin or metabolite effectors (Wolosiuk and Buchanan, 1978b; Baalmann et al., 1995). The NADP-linked enzyme, a heterotetramer, is distinct from the cytosolic isoenzyme which is NAD-specific and a homotetramer.

GAPDH displays complex behavior in the chloroplast. It appears as a large aggregate of 600 kDa in the dark and dissociates into 150 kDa heterotetramers in the light. The tendency to disaggregate depends on illumination and parallels the state of activation. Recent reevaluation of the different activation states and the influence of effector molecules has given additional information on possible regulatory mechanisms (Baalmann et al., 1994, 1995). According to these results the transformation from aggregate to heterotetrameric enzyme, i.e., from inactive to fully active enzyme, is brought about by the combined action of a reductant and different effector molecules. Since the effect of DTT and thioredoxin can be largely achieved by low concentrations of GSH, it was concluded that either reductant may function in activation of this enzyme in vivo, at least under the conditions tested. The main effect of reduction seems to be to increase the sensitivity of the enzyme to the primary metabolic effector molecule, 1,3-bisphosphoglycerate (Wolosiuk et al., 1986; Baalmann et al., 1995).

Several Cys residues, including those of the active site are conserved in the chloroplast enzyme. There are, however, no structural data available to indicate an involvement in redox regulation. Two Cys residues are on the C-terminal extension of subunit B. This extension is found only in the light regulated enzyme and seems to be responsible for the reversible aggregation (Baalmann et al., 1996). Removal of the extension provides a recombinant homotetrameric enzyme composed of only subunit B which, although having a specific activity comparable to the enzyme isolated from chloroplasts, does not aggregate. Site-directed mutagenesis experiments are needed to show whether the two Cys in the extension, or other conserved Cys residues, are involved in regulation.

2. Acetyl CoA Carboxylase

Acetyl CoA carboxylase catalyzes the first committed step in fatty acid biosynthesis—the carboxylation of acetyl CoA to malonyl CoA. This enzyme plays a central role in regulating fatty acid synthesis which has long been known to increase in the light and decrease in the dark (Sauer and Heise, 1983). Two different forms of acetyl CoA carboxylase have been identified in plants, a prokaryotic form in plastids, and a eukaryotic form in the cytosol. The prokaryotic form isolated from pea chloroplasts is subject to redox regulation whereas the cytosolic counterpart from the same plant is not (Sasaki et al., 1997). The activity of the enzyme is significantly increased by the addition of millimolar concentrations of dithiols, such as lipoic acid or DTT, but not by monothiols. Comparable stimulation of activity is achieved at micromolar concentrations of reduced thioredoxin. The results suggest that thioredoxin is the natural reductant activator. Three different thioredoxins were tested in the activation of acetyl CoA carboxylase—*E. coli,* spinach *m* and *f*. Thioredoxin *f* was the most efficient, suggesting an in vivo role as specific activator. The prokaryotic acetyl CoA carboxylase is a complex enzyme consisting of four different polypeptides. Sequences of all of these polypeptides from different species are known to contain varying numbers of conserved Cys residues. There is as yet no indication as to which Cys are involved in regulation. Further experiments will be needed to clarify this point.

3. Glutamine Synthetase and Ferredoxin:Glutamate Synthase

It has been suggested that thioredoxin might function in the regulation of nitrogen assimilation in algae (Huppe and Turpin, 1994). This proposal is based on a number of reports that algal glutamine synthetase and ferredoxin:glutamate synthase, two enzymes incorporating nitrite-derived ammonia into amino acids, are dependent on thioredoxin (Schmidt, 1981; Tischner and Schmidt, 1982; Ip et al., 1984; Papen and Bothe, 1984; Marques et al., 1992; Florencio et al., 1993). Until recently neither of the two enzymes was shown to be thioredoxin dependent in higher plants.

A recent report changes the picture and provides evidence that spinach chloroplast ferredoxin:glutamate synthase (but not glutamine synthetase or nitrite reductase) is activated by thioredoxin (Lichter and Häberlein, 1998). Using a new enzyme assay system designed to avoid possible non-enzymatic deamination of the reaction products, the authors demonstrated that the activity of the ferredoxin:glutamate synthase is significantly stimulated by DTT, but not by GSH. Among the three thioredoxins tested (*E. coli*, spinach *f*- and *m*-types) thioredoxin *m* was the most efficient, indicating a degree of thioredoxin specificity. The main effect of reduction was an increase in reaction velocity. There was little effect on the affinity of the enzyme for its substrates. Although these results suggest that thioredoxin is an activator of ferredoxin:glutamate synthase, further experiments are needed to demonstrate the presence of a functional regulatory disulfide in this enzyme.

VI. Mechanism for Reduction of Thioredoxins and Target Enzymes

In a series of decisive experiments with isolated chloroplasts and purified protein components, it was shown about ten years ago that the light signal—represented by an electron from reduced ferredoxin—is converted by FTR to a thiol signal which is transmitted sequentially to a thioredoxin and target enzyme (Droux et al., 1987b; Crawford et al., 1989). Two electrons provided by ferredoxin, a highly reducing one electron carrier (E_m = –420 mV), are needed for the reduction of thioredoxin as well as the target enzyme. Since ferredoxin forms a 1:1 complex with FTR (Hirasawa et al., 1988; De Pascalis et al., 1994), the complete reduction of the active site disulfide has to proceed in two sequential one-electron transfer events. Although FTR contains a [4Fe-4S] cluster (Droux et al., 1987a; Chow et al., 1995; Staples et al., 1996; Dai et al., 2000) which could potentially serve as the intermediary electron acceptor, it has been shown that this cluster cannot be reduced either chemically or with illuminated thylakoids (De La Torre et al., 1982). The cluster can, however, be oxidized with ferricyanide and result in a S=1/2 $[4Fe\text{-}4S]^{3+}$ state (Staples et al., 1996).

Spectroscopic studies on native and active-site modified FTR have addressed the question of how this cluster, with a redox potential of +340 mV to +420 mV (De La Torre et al., 1982; Salamon et al., 1995; Staples et al., 1996), participates in the two-electron reduction of the disulfide active site with a redox potential of –320 mV (Hirasawa et al., 1999). The results of these studies (Staples et al., 1996; Staples et al., 1998), combined with information gained from the crystal analysis of FTR (Dai et al., 2000) are summarized in Fig. 6. The proposed reaction mechanism constitutes a new role for the cluster in stabilizing a one-electron reduced reaction intermediate. Such a reaction sequence seems compatible with the chemical and structural information which allow for the concomitant docking of a thioredoxin and a ferredoxin molecule to the FTR (see Fig.2).

Reduction by the first electron results in the cleavage of the active site disulfide of FTR producing a solvent accessible sulfhydryl (Cys57 in the *Synechocystis* enzyme) and a solvent-inaccessible cysteine-based thiyl radical (Cys87). This radical is stabilized by the covalent attachment of the cysteine to the cluster. Based on the crystal structure of FTR the most probable atom for this attachment is a cluster iron, which is at a distance of 3.2 Å from Cys87. The cluster becomes oxidized through this transient attachment of Cys87. The solvent accessible sulfhydryl of the active site (Cys57), in turn, cleaves the disulfide of a thioredoxin molecule by nucleophilic attack. This reaction anchors the thioredoxin molecule through a heterodisulfide linkage to the thioredoxin

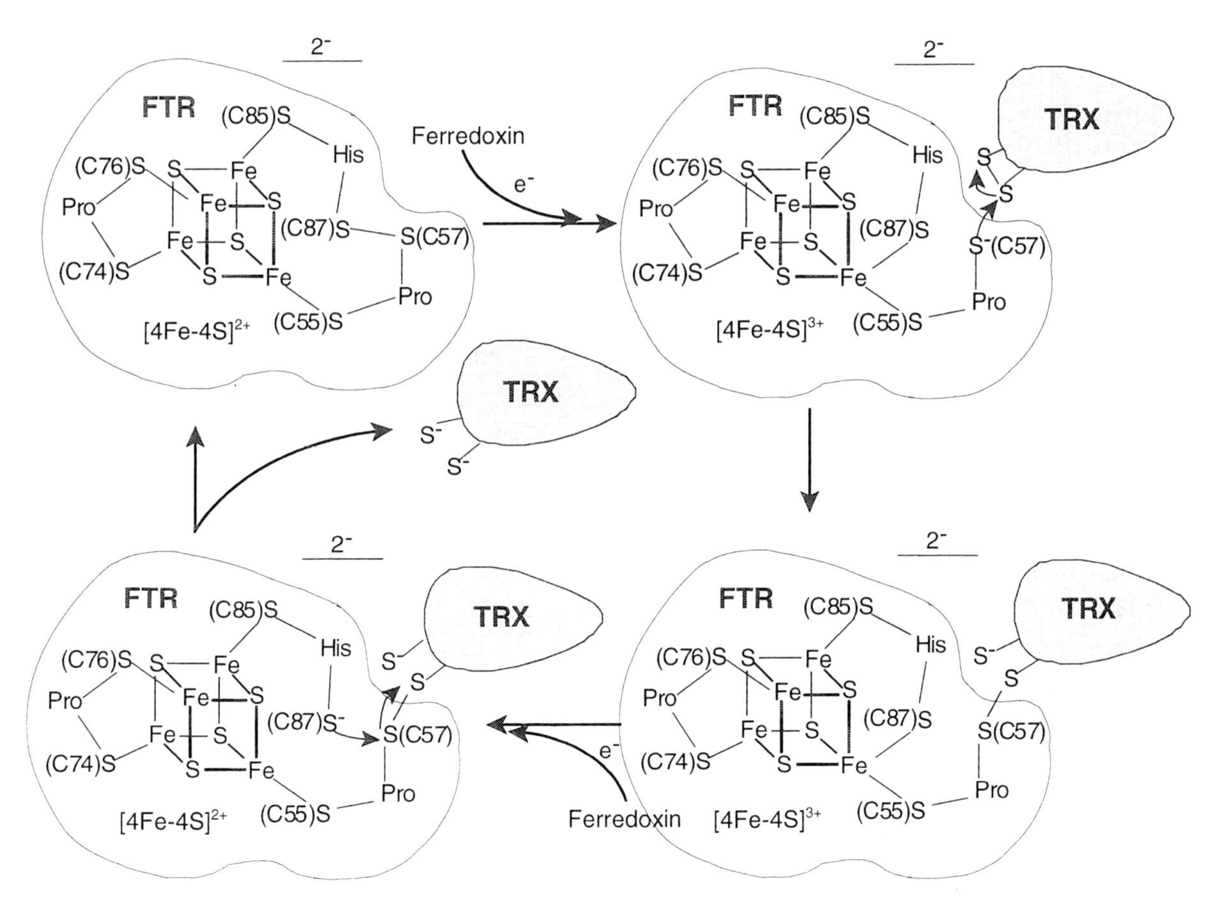

Fig. 6. Sequence of events taking place on FTR in the catalytic reduction of thioredoxin with reduced ferredoxin. Schematic drawing adapted from Staples et al. (1998) and Dai (1998)

docking area, which is on one side of the FTR molecule. The docking area, on the opposite side, stays free for a second interaction with a reduced ferredoxin. The second electron reduces the cluster-ligated Cys87 and reestablishes the original oxidation state of the cluster. The newly reduced internal Cys87 attacks and cleaves the heterodisulfide linkage between FTR and thioredoxin, thus liberating the reduced thioredoxin. A closing of the active site disulfide bridge completes the cycle. The crystal structure showing that the binding sites for ferredoxin and thioredoxin are on opposite sides of the protein supports such a two-step mechanism with a transient attachment of thioredoxin.

The reduction of the regulatory disulfides of the target enzymes by thioredoxin also proceeds with the formation of a transient heterodisulfide complex between the two reaction partners. The reactive Cys which is the solvent accessible one close to the N-terminus of thioredoxin, cleaves the target disulfide by nucleophilic attack, thereby forming a covalently linked mixed disulfide. In a fast second step, the second sulfhydryl, which is inaccessible to solvent, attacks the mixed disulfide to produce oxidized thioredoxin and reduced target enzyme.

Whereas structural evidence for a heterodisulfide has been demonstrated for human thioredoxin with its target peptide (Qin et al., 1995), the structure of such a complex has not yet been solved in the plant system. There is however biochemical evidence for a complex of this type between thioredoxin *m* and NADP-MDH (Goyer et al., 1999) and between thioredoxin *f* and PRK (Brandes et al., 1996) and FBPase (Y. Balmer and P. Schürmann, unpublished). Following this initial reaction, the regulatory disulfide of the target enzyme is reduced by the second active sulfhydryl of thioredoxin. In PRK, FBPase, SBPase, CF_1 and G6PDH, activation (or deactivation) is achieved in a single step—i.e., only one disulfide bond is reduced. For FBPase experiments with mutant enzyme suggested that this disulfide bond may involve interchangeable partners (see above). However, in the light of the three-dimensional structure these results may be due to an artifactual disulfide bond in the mutant proteins (Chiadmi et al., 1999). For NADP-MDH, on the other hand, the reduction of a single disulfide bond is insufficient and a second regulatory disulfide bond must be reduced by thioredoxin to achieve full activity (Fig. 7).

The oxidation-reduction properties of the different disulfides participating in the ferredoxin/thioredoxin system influence the activation state of the individual target enzymes. An understanding of the kinetics of activation (and deactivation) and the known differential effects of light intensity on the in situ activities of the target enzymes (Kramer et al., 1990) requires accurate knowledge of the oxidation-reduction midpoint potentials (E_m) of the disulfide/sulfhydryl couples involved. Determination of midpoint potentials during the past several years has provided this information for the members of the regulatory chain and several target enzymes. The initial values obtained with cyclic voltammetry (Salamon et al., 1995) were found to be overly positive and have, therefore, been superseded by the measurements made with activity or a fluorescent probe (Table 3).

There is reasonable agreement on the redox potentials of the disulfides active in the ferredoxin/thioredoxin system. The redox potential of the disulfide of FTR (–320 mV) is significantly more reducing than that of either of the two chloroplast thioredoxins, *f* and *m* (each approximately –300 mV), thereby insuring their ready reduction in the light. On the other hand, the potential of the two thioredoxins is either more oxidizing or equal to that of two of the

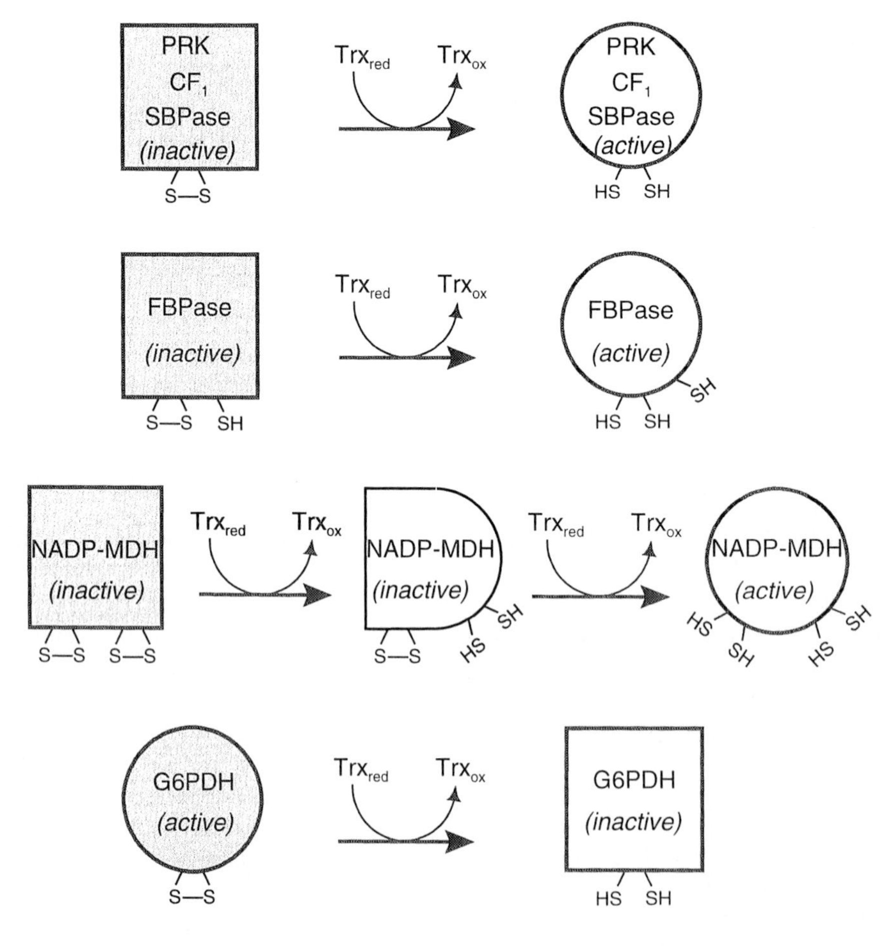

Fig. 7. Reduction steps required for the activation or deactivation of enzymes targeted by the ferredoxin/thioredoxin system.

target enzymes of CO_2 assimilation, FBPase (–305 mV in spinach) and NADP-MDH (–330 mV in sorghum). These differences in oxidation reduction potential insure that these enzymes will become active only in the light—that is, after a significant amount of thioredoxin has been reduced—to prevent futile cycles. The redox differences also provide an explanation for the delay in activating the enzymes of CO_2 assimilation after they are subjected to light. Another reason for this delay rests on their hysteretic nature—that is, the rate of activation of most thioredoxin-linked enzymes of CO_2 assimilation is slow relative to their rate of catalysis (Buchanan, 1980). It is noteworthy that CF_1 is an exception and is activated immediately after turning on the light, possibly as a result of its significantly more electropositive redox potential (Table 3).

The differences in redox potentials between thioredoxins and the various target enzymes suggest there is an order to activation of chloroplast enzymes in the light with the most critical processes given priority in the sequence of regulatory events (Table 4). These redox differences further suggest that the redox equilibria reached will result in different degrees of activation. These equilibria may also be modified by enzyme effectors (Faske et al., 1995). It seems possible, therefore, that effectors within the chloroplast may allow for fine-tuning of activity once the enzymes are switched on by light as was proposed in early enzyme regulation studies (Buchanan, 1980).

VII. Phylogenetic History of Thioredoxins and Photosynthetic Target Enzymes

There is evidence that the thioredoxins participating in enzyme regulation in oxygenic photosynthetic systems differ in their phylogenetic history. Gene and protein sequencing, together with biochemical activity measurements, indicate that thioredoxin *m* is of bacterial origin, whereas thioredoxin *f* is eukaryotic (Hartman et al., 1990; Sahrawy et al., 1996; Jacquot et al., 1997a; Meyer et al., 1999). According to this view, the *m*-type thioredoxins of chloroplasts were incorporated with the invading cyanobacteria and adapted to function in the regulation of oxygenic photosynthesis (via PRK) and the oxidative pentose phosphate pathway (via G6PDH). The photosynthetic thioredoxin *m* was retained when the plastid developed and continued to function in the regulation of G6PDH and a number of other enzymes. Thioredoxin *f*, on the other hand, was apparently derived from the nucleus of the eukaryotic host and adapted for transport to the chloroplast. With time, thioredoxin *f* became the principal regulator of photosynthetic enzymes, including the PRK incorporated with the cyanobacterium.

Table 4. Sequence of thioredoxin-linked events in oxygenic photosynthesis during sunrise based on redox potentials of target enzymes. No. 1 represents the first response to light.

Enzyme	Process	Effect
1. CF_1	→ Photophosphorylation	+
2. PRK	→ Calvin Cycle	+
3. G6PDH	→ Oxidative Pentose P Cycle	–
4. FBPase	→ Calvin Cycle	+
	→ Starch Synthesis	+
5. NADP-MDH	→ C4 Pathway	+
	→ C3 Shuttle	+

Recent chloroplast genome sequencing supports this sequence of events. The sequences for red algae (*Porphyra purpurea, P. yezoensis, Cyanidium, Griffithsia pacifica*) revealed a thioredoxin *m* gene in keeping with the conclusion that *m*-type thioredoxins in photosynthetic eukaryotes originated from an engulfed bacterial endosymbiont (Reynolds et al., 1994; Reith and Munholland, 1997). Similarly, a comparison of the protein and gene sequences, as well as the location of introns, indicates that thioredoxin *f* is most closely related to animal thioredoxin (Hartman et al., 1990; Sahrawy et al., 1996; Meyer et al., 1999). Thioredoxin *h*, an extraplastidic protein, also appears to have been of nuclear origin.

Phylogenetic analyses of thioredoxins have thus added a new dimension to our understanding of photosynthesis (Fig. 8). A recent review extends this picture and emphasizes the importance of endosymbiotic gene transfer in the evolution of the Calvin cycle and related processes of chloroplasts (Martin and Herrmann, 1998).

VIII. Concluding Remarks

The first evidence for the ferredoxin/thioredoxin system was obtained more than thirty years ago (Buchanan et al., 1967). Studies on CO_2 assimilation with isolated chloroplasts uncovered a role for ferredoxin in the activation of FBPase. Ensuing work

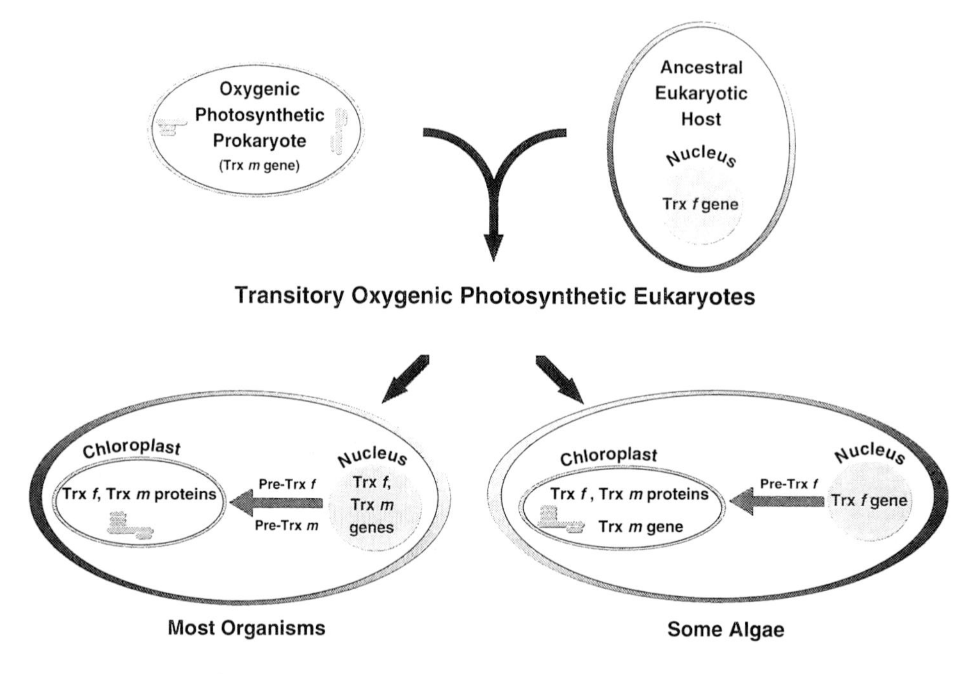

Fig. 8. Phylogenetic history of chloroplast thioredoxins.

revealed that ferredoxin did not act alone but required a 'protein factor' for activation (Buchanan et al., 1971). The 'protein factor' was later shown to consist of two components (Schürmann et al., 1976) that were identified as thioredoxin and FTR (Holmgren et al., 1977; Wolosiuk and Buchanan, 1977). Follow-up work demonstrated that the thioredoxin fraction was made up of two different species, showing specificity for target enzymes:thioredoxin *f* (specific for FBPase) and thioredoxin *m* (specific for NADP-MDH under the conditions tested) (Buchanan et al., 1978; Jacquot et al., 1978).

Starting in 1967, the first decade of research on the ferredoxin/thioredoxin system was primarily dedicated to its discovery and definition (Buchanan, 1980), the second to the determination of its function and distribution (Buchanan, 1991) and the third to elucidation of the structures and chemical properties of its component proteins (FTR, thioredoxins *f* and *m*) and target enzymes (FBPase, NADP-MDH, PRK) (Besse and Buchanan, 1997; Jacquot et al., 1997a; Ruelland and Miginiac-Maslow, 1999; Schürmann and Jacquot, 2000). The early biochemical studies on CO_2 assimilation in chloroplasts laid the foundation for later work that demonstrated a regulatory function of thioredoxin, first in photosynthesis, then in heterotrophic plant and animal processes (Fig. 9). To follow were experiments designed to apply the unique properties of thioredoxins, here reduced by NADPH, to solving a host of societal problems ranging from agriculture to food technology to medicine (Buchanan et al., 1994a; Besse and Buchanan, 1997).

The question arises as to what the future holds for the thioredoxin system of chloroplasts. While the remaining fundamental problems will likely be solved—e.g., mechanism for deactivating target enzymes, control of gene expression—the new century may primarily witness the application of the ferredoxin/thioredoxin system to practical problems, as is currently happening with its NADP-linked counterpart. While it is difficult to predict the direction the work may take, it seems likely that these applications will take advantage of the capability of chloroplasts and cyanobacteria to provide reduced thioredoxin in a sustained, inexpensive manner. In other words, the elegant mechanism nature designed for reducing thioredoxin to provide the fundamentals for life through photosynthesis may be used to improve the quality of life through new technologies.

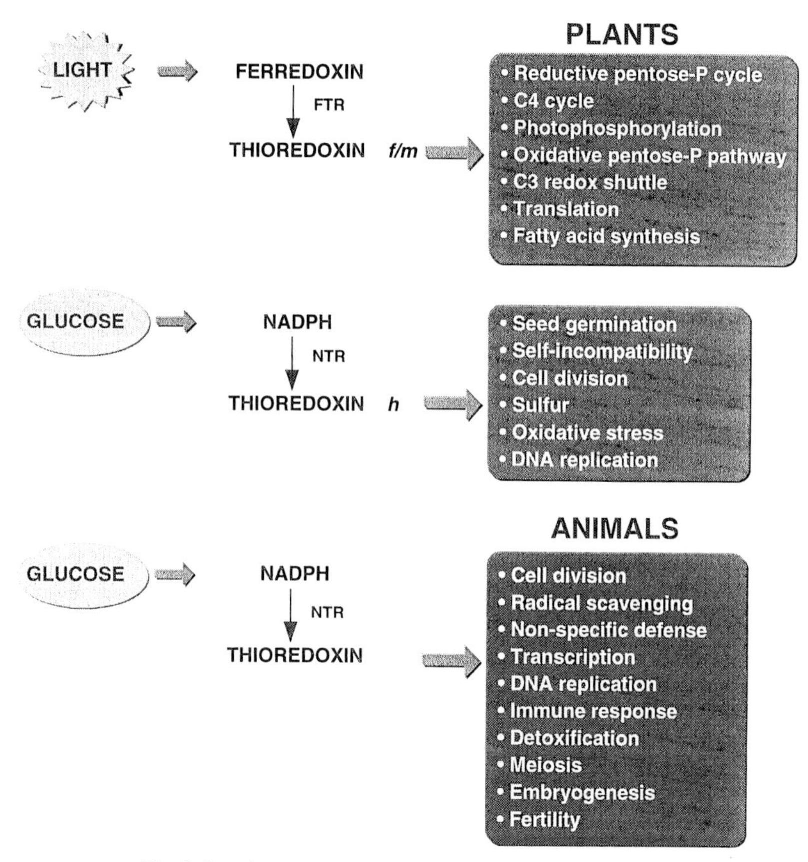

Fig. 9. Regulatory role of thioredoxin in plants and animals.

Acknowledgments

The authors thank Drs. Gregorio del Val, Guido Capitani and Shaodong Dai for providing figures used in this chapter and Drs. Hans Eklund, David B. Knaff, Myroslawa Miginiac-Maslow and Don Ort for access to material prior to publication. The authors also acknowledge helpful discussions with Drs. Capitani, Dai, Eklund, Knaff and Meyer.

References

Aguilar F, Brunner B, Gardet-Salvi L, Stutz E and Schürmann P (1992) Biosynthesis of active spinach-chloroplast thioredoxin *f* in transformed *E. coli*. Plant Mol Biol 20: 301–306

Ashton AR and Hatch MD (1983) Regulation of C_4 photosynthesis: Regulation of activation and inactivation of NADP-malate dehydrogenase by NADP and NADPH. Arch Biochem Biophys 227: 416–424

Baalmann E, Backhausen JE, Kitzmann C and Scheibe R (1994) Regulation of NADP-Dependent Glyceraldehyde 3-Phosphate Dehydrogenase Activity in Spinach Chloroplasts. Bot Acta 107: 313–320

Baalmann E, Backhausen JE, Rak C, Vetter S and Scheibe R (1995) Reductive modification and nonreductive activation of purified spinach chloroplast NADP-dependent glyceraldehyde-3-phosphate dehydrogenase. Arch Biochem Biophys 324: 201–208

Baalmann E, Scheibe R, Cerff R and Martin W (1996) Functional studies of chloroplast glyceraldehyde-3-phosphate dehydrogenase subunits A and B expressed in *Escherichia coli*: Formation of highly active A_4 and B_4 homotetramers and evidence that aggregation of the B_4 complex is mediated by the B subunit carboxy terminus. Plant Mol Biol 32: 505–513

Batie CJ and Kamin H (1984) Ferredoxin:NADP$^+$ oxidoreductase. Equilibria in binary and ternary complexes with NADP$^+$ and ferredoxin. J Biol Chem 259: 8832–8839

Besse I and Buchanan BB (1997) Thioredoxin-linked plant and animal processes: The new generation. Bot Bull Acad Sin (Taipei) 38: 1–11

Binda C, Coda A, Aliverti A, Zanetti G and Mattevi A (1998) Structure of Mutant E92K of [2Fe-2S] Ferredoxin I from *Spinacia oleracea* at 1.7 Å Resolution. Acta Cryst D54: 1353–1358

Brandes HK, Larimer FW, Geck MK, Stringer CD, Schürmann P and Hartman FC (1993) Direct Identification of the Primary

Nucleophile of Thioredoxin *f*. J Biol Chem 268: 18411–18414
Brandes HK, Larimer FW and Hartman FC (1996) The molecular pathway for the regulation of phosphoribulokinase by thioredoxin *f*. J Biol Chem 271: 3333–3335
Breazeale VD, Buchanan BB and Wolosiuk RA (1978) Chloroplast sedoheptulose 1,7-bisphosphatase: Evidence for regulation by the ferredoxin/thioredoxin system. Z Naturforsch 33c: 521–528
Buchanan BB (1980) Role of light in the regulation of chloroplast enzymes. Annu Rev Plant Physiol Plant Mol Biol 31: 341–374
Buchanan BB (1991) Regulation of CO_2 assimilation in oxygenic photosynthesis: The ferredoxin/thioredoxin system: Perspective on its discovery, present status, and future development. Arch Biochem Biophys 288: 1–9
Buchanan BB (1992) Carbon dioxide assimilation in oxygenic and anoxygenic photosynthesis. Photosynth Res 33: 147–162
Buchanan BB, Kalberer PP and Arnon DI (1967) Ferredoxin-activated fructose diphosphatase in isolated chloroplasts. Biochem Biophys Res Commun 29: 74–79
Buchanan BB, Schürmann P and Kalberer PP (1971) Ferredoxin-activated fructose diphosphatase of spinach chloroplasts. J Biol Chem 246: 5952–5959
Buchanan BB, Wolosiuk RA, Crawford NA and Yee BC (1978) Evidence for three thioredoxins in leaves. Plant Physiol. 61: 38S
Buchanan BB, Schürmann P, Decottignies P and Lozano RM (1994a) Thioredoxin: A multifunctional regulatory protein with a bright future in technology and medicine. Arch Biochem Biophys 314: 257–260
Buchanan BB, Schürmann P and Jacquot J-P (1994b) Thioredoxin and metabolic regulation. Seminars in Cell Biology 5: 285–293
Capitani G, Markovic-Housley Z, Jansonius JN, del Val G, Morris M and Schürmann P (1998) Crystal structures of thioredoxins *f* and *m* from spinach chloroplasts. In: Garab G (ed) Photosynthesis: Mechanisms and Effects, Vol 3, pp 1939–1942. Kluwer Academic Publishers, Dordrecht
Capitani G, Markovic-Housley Z, del Val G, Morris M, Jansonius JN and Schürmann P (2000) Crystal structures of two functionally different thioredoxins in spinach chloroplasts. J Mol Biol 302: 135–154
Carr PD, Verger D, Ashton AR and Ollis DL (1999) Chloroplast NADP-malate dehydrogenase: Structural basis of light-dependent regulation of activity by thiol oxidation and reduction. Structure 7: 461–475
Chiadmi M, Navaza A, Miginiac-Maslow M, Jacquot JP and Cherfils J (1999) Redox signalling in the chloroplast: Structure of oxidized pea fructose-1,6-bisphosphate phosphatase. EMBO J 18: 6809–6815
Chow L-P, Iwadate H, Yano K, Kamo M, Tsugita A, Gardet-Salvi L, Stritt-Etter A-L and Schürmann P (1995) Amino acid sequence of spinach ferredoxin:thioredoxin reductase catalytic subunit and identification of thiol groups constituting a redox active disulfide and a [4Fe-4S] cluster. Eur J Biochem 231: 149–156
Cossar JD, Rowell P and Stewart WDP (1984) Thioredoxin as a modulator of glucose-6-phosphate dehydrogenase in a N_2-fixing cyanobacterium. J Gen Microbiol 130: 991–998
Cozens AL and Walker JE (1987) The organization and sequence of the genes for ATP synthase subunits in the cyanobacterium *Synechococcus* 6301. Support for an endosymbiotic origin of chloroplasts. J Mol Biol 194: 359–383
Crawford NA, Droux M, Kosower NS and Buchanan BB (1989) Function of the ferredoxin/thioredoxin system in reductive activation of target enzymes of isolated intact chloroplasts. In: Briggs WR (ed) Plant Biology, pp 425–436. Alan R. Liss, Inc., New York
Dai S (1998) Structural and Functional Studies of NADPH and Ferredoxin Dependent Thioredoxin Reductases. PhD Thesis. Swedish University of Agricultural Sciences, Uppsala
Dai S, Schwendtmayer C, Ramaswamy S, Eklund H and Schürmann P (1998) Crystallization and crystallographic investigations of ferredoxin:thioredoxin reductase from *Synechocystis* sp. PCC6803. In: Garab G (ed) Photosynthesis: Mechanisms and Effects, Vol 3, pp 1931–1934. Kluwer Academic Publishers, Dordrecht
Dai S, Schwendtmayer C, Schürmann P, Ramaswamy S and Eklund H (2000) Redox Signaling in Chloroplasts: Cleavage of Disulfides by an Iron-Sulfur Cluster. Science 287: 655–658
Danon A and Mayfield SP (1994) Light-regulated translation of chloroplast messenger RNAs through redox potential. Science 266: 1717–1719
De La Torre A, Lara C, Yee BC, Malkin R and Buchanan BB (1982) Physiochemical properties of ferralterin, a regulatory iron-sulfur protein functional in oxygenic photosynthesis. Arch Biochem Biophys 213: 545–550
De Pascalis AR, Schürmann P and Bosshard HR (1994) Comparison of the binding sites of plant ferredoxin for two ferredoxin-dependent enzymes. FEBS Lett 337: 217–220
Decottignies P, Schmitter J-M, Miginiac-Maslow M, Le Marechal P, Jacquot J-P and Gadal P (1988) Primary structure of the light-dependent regulatory site of corn NADP-malate dehydrogenase. J Biol Chem 263: 11780–11785
del Val G, Maurer F, Stutz E and Schürmann P (1999) Modification of the reactivity of spinach chloroplast thioredoxin *f* by site-directed mutagenesis. Plant Sci 149: 183–190
Derman AI, Prinz WA, Belin D and Beckwith J (1993) Mutations that allow disulfide bond formation in the cytoplasm of Escherichia coli. Science 262: 1744–1747
Droux M, Jacquot J-P, Miginiac-Maslow M, Gadal P, Huet JC, Crawford NA, Yee BC and Buchanan BB (1987a) Ferredoxin-thioredoxin reductase, an iron-sulfur enzyme linking light to enzyme regulation in oxygenic photosynthesis: Purification and properties of the enzyme from C_3, C_4, and cyanobacterial species. Arch Biochem Biophys 252: 426–439
Droux M, Miginiac-Maslow M, Jacquot J-P, Gadal P, Crawford NA, Kosower NS and Buchanan BB (1987b) Ferredoxin-thioredoxin reductase: A catalytically active dithiol group links photoreduced ferredoxin to thioredoxin functional in photosynthetic enzyme regulation. Arch Biochem Biophys 256: 372–380
Dunford RP, Durrant MC, Catley MA and Dyer T (1998) Location of the redox-active cysteines in chloroplast sedoheptulose-1,7-bisphosphatase indicates that its allosteric regulation is similar but not identical to that of fructose-1,6-bisphosphatase. Photosynth Res 58: 221–230
Falkenstein E, von Schaewen A and Scheibe R (1994) Full-length cDNA sequences for both ferredoxin-thioredoxin reductase subunits from spinach (*Spinacia oleracea* L.). Biochim Biophys Acta 1185: 252–254
Faske M, Holtgrefe S, Ocheretina O, Meister M, Backhausen JE and Scheibe R (1995) Redox equilibria between the regulatory

thiols of light/dark-modulated chloroplast enzymes and dithiothreitol: Fine-tuning by metabolites. Biochim Biophys Acta 1247: 135–142

Florencio FJ, Gadal P and Buchanan BB (1993) Thioredoxin-linked activation of the chloroplast and cytosolic forms of *Chlamydomonas reinhardtii* glutamine synthetase. Plant Physiol Biochem 31: 649–655

Follmann H and Häberlein I (1996) Thioredoxins: Universal, yet specific thiol-disulfide redox cofactors. BioFactors 5: 147–156

Galmiche JM, Girault G, Berger G, Jacquot J-P, Miginiac-Maslow M and Wollman E (1990) Induction by different thioredoxins of ATPase activity in coupling factor 1 from spinach chloroplasts. Biochimie 72: 25–32

Gaymard E and Schürmann P (1995) Cloning and expression of cDNAs coding for the spinach ferredoxin:thioredoxin reductase. In: Mathis P (ed) Photosynthesis: From Light to Biosphere, Vol 2, pp 761–764. Kluwer Academic Publishers, Dordrecht

Geck MK, Larimer FW and Hartman FC (1996) Identification of residues of spinach thioredoxin *f* that influence interactions with target enzymes. J Biol Chem 271: 24736–24740

Gleason FK (1996) Glucose-6-phosphate dehydrogenase from the cyanobacterium, *Anabaena* sp PCC 7120: Purification and kinetics of redox modulation. Arch Biochem Biophys 334: 277–283

Goyer A, Decottignies P, Lemaire S, Ruelland E, Issakidis-Bourguet E, Jacquot J-P and Miginiac-Maslow M (1999) The internal Cys207 of sorghum leaf NADP-malate dehydrogenase can form mixed disulfides with thioredoxin. FEBS Lett 444: 165–169

Hahn D, Kaltenbach C and Kück U (1998) The Calvin cycle enzyme sedoheptulose-1,7-bisphosphatase is encoded by a light-regulated gene in *Chlamydomonas reinhardtii*. Plant Mol Biol 36: 929–934

Hammond ET, Andrews TJ and Woodrow IE (1998) Regulation of ribulose-1,5-bisphosphate carboxylase/oxygenase by carbamylation and 2-carboxyarabinitol 1-phosphate in tobacco: Insights from studies of antisense plants containing reduced amounts of rubisco activase. Plant Physiol 118: 1463–1471

Harrison DH, Runquist JA, Holub A and Miziorko HM (1998) The crystal structure of phosphoribulokinase from *Rhodobacter sphaeroides* reveals a fold similar to that of adenylate kinase. Biochemistry 37: 5074–5085

Hartman H, Syvanen M and Buchanan BB (1990) Contrasting evolutionary histories of chloroplast thioredoxins *f* and *m*. Mol Biol Evol 7: 247–254

Hatch MD (1987) C_4 photosynthesis: A unique blend of modified biochemistry, anatomy and ultrastructure. Biochim Biophys Acta 895: 81–106

Hatch MD (1992) C_4 photosynthesis: An unlikely process full of surprises. Plant Cell Physiol 33: 333–342

Hatch MD and Agostino A (1992) Bilevel disulfide group reduction in the activation of C_4 leaf nicotinamide adenine dinucleotide phosphate-malate dehydrogenase. Plant Physiol 100: 360–366

Hirasawa M, Boyer JM, Gray KA, Davis DJ and Knaff DB (1986) The interaction of ferredoxin with chloroplast ferredoxin-linked enzymes. Biochim Biophys Acta 851: 23–28

Hirasawa M, Droux M, Gray KA, Boyer JM, Davis DJ, Buchanan BB and Knaff DB (1988) Ferredoxin-thioredoxin reductase: Properties of its complex with ferredoxin. Biochim Biophys Acta 935: 1–8

Hirasawa M, Brandes HK, Hartman FC and Knaff DB (1998) Oxidation-reduction properties of the regulatory site of spinach phosphoribulokinase. Arch Biochem Biophys 350: 127–131

Hirasawa M, Schürmann P, Jacquot J-P, Manieri W, Jacquot P, Keryer E, Hartman FC and Knaff DB (1999) Oxidation-reduction properties of chloroplast thioredoxins, ferredoxin:thioredoxin reductase and thioredoxin *f*-regulated enzymes. Biochemistry 38: 5200–5205

Hirasawa M, Ruelland E, Schepens I, Issakidis-Bourguet E, Miginiac-Maslow M and Knaff DB (2000) Oxidation-reduction properties of the regulatory disulfides of sorghum chloroplast NADP-malate dehydrogenase. Biochemistry 39: 3344–3350

Hodges M, Miginiac-Maslow M, Decottignies P, Jacquot J-P, Stein M, Lepiniec L, Crétin C and Gadal P (1994) Purification and characterization of pea thioredoxin *f* expressed in *Escherichia coli*. Plant Mol Biol 26: 225–234

Holmgren A, Buchanan BB and Wolosiuk RA (1977) Photosynthetic regulatory protein from rabbit liver is identical with thioredoxin. FEBS Lett 82: 351–354

Huppe HC and Turpin DH (1994) Integration of carbon and nitrogen metabolism in plant and algal cells. Annu Rev Plant Physiol Plant Mol Biol 45: 577–607

Huppe HC, Lamotte-Guéry Fd, Jacquot J-P and Buchanan BB (1990) The ferredoxin-thioredoxin system of a green alga, *Chlamydomonas reinhardtii*. Identification and characterization of thioredoxins and ferredoxin-thioredoxin reductase components. Planta 180: 341–351

Hutchison RS, Groom Q and Ort DR (2000) Differential effects of low temperature induced oxidation on thioredoxin mediated activation of photosynthetic carbon reduction cycle enzymes. Biochemistry 39: 6679–6688

Inoue Y, Kobayashi Y, Shibata K and Heber U (1979) Synthesis and hydrolysis of ATP by intact chloroplasts under flash illumination and in darkness. Biochim Biophys Acta 504: 142–152

Ip SM, Rowell P, Aitken A and Stewart WDP (1984) Purification and characterization of thioredoxin from the N_2-fixing cyanobacterium *Anabaena cylindrica*. Eur J Biochem 141: 497–504

Issakidis E, Miginiac-Maslow M, Decottignies P, Jacquot J-P, Crétin C and Gadal P (1992) Site-directed mutagenesis reveals the involvement of an additional thioredoxin-dependent regulatory site in the activation of recombinant sorghum leaf NADP-malate dehydrogenase. J Biol Chem 267: 21577–21583

Issakidis E, Decottignies P and Miginiac-Maslow M (1993) A thioredoxin-independent fully active NADP-malate dehydrogenase obtained by site-directed mutagenesis. FEBS Lett 321: 55–58

Issakidis E, Saarinen M, Decottignies P, Jacquot J-P, Crétin C, Gadal P and Miginiac-Maslow M (1994) Identification and characterization of the second regulatory disulfide bridge of recombinant sorghum leaf NADP-malate dehydrogenase. J Biol Chem 269: 3511–3517

Issakidis E, Lemaire M, Decottignies P, Jacquot JP and Miginiac-Maslow M (1996) Direct evidence for the different roles of the N- and C-terminal regulatory disulfides of sorghum leaf NADP-malate dehydrogenase in its activation by reduced thioredoxin. FEBS Lett 392: 121–124

Iwadate H, Yano K, Kamo M, Gardet-Salvi L, Schürmann P and Tsugita A (1994) Amino acid sequence of spinach ferredoxin-thioredoxin reductase variable subunit. Eur J Biochem 223: 465–471

Iwadate H, Tsugita A, Chow L-P, Kizuki K, Stritt-Etter A-L, Li J and Schürmann P (1996) Amino acid sequence of the maize ferredoxin:thioredoxin reductase variable subunit. Eur J Biochem 241: 121–125

Iwadate H, Kizuki K, Stritt-Etter A-L, Li J, Schürmann P and Tsugita A (1998) Primary structure of maize ferredoxin:thioredoxin reductase catalytic subunit. Research Communication in Biochemistry and Cell & Molecular Biology 2: 105–112

Jacquot J-P, Vidal J and Gadal P (1976) Identification of a protein factor involved in dithiothreitol activation of NADP malate dehydrogenase from French bean leaves. FEBS Lett 71: 223–227

Jacquot J-P, Vidal J, Gadal P and Schürmann P (1978) Evidence for the existence of several enzyme-specific thioredoxins in plants. FEBS Lett 96: 243–246

Jacquot J-P, López-Jaramillo J, Chueca A, Cherfils J, Lemaire S, Chedozeau B, Miginiac-Maslow M, Decottignies P, Wolosiuk RA and Gorgé JL (1995) High-level expression of recombinant pea chloroplast fructose-1,6-bisphosphatase and mutagenesis of its regulatory site. Eur J Biochem 229: 675–681

Jacquot J-P, Lancelin J-M and Meyer Y (1997a) Thioredoxins: Structure and function in plant cells. New Phytol 136: 543–570

Jacquot JP, Stein M, Suzuki K, Liottet S, Sandoz G and Miginiac-Maslow M (1997b) Residue Glu-91 of *Chlamydomonas reinhardtii* ferredoxin is essential for electron transfer to ferredoxin-thioredoxin reductase. FEBS Lett 400: 293–296

Jacquot JP, López-Jaramillo J, Miginiac-Maslow M, Lemaire S, Cherfils J, Chueca A and Lopez-Gorge J (1997c) Cysteine-153 is required for redox regulation of pea chloroplast fructose-1,6-bisphosphatase. FEBS Lett 401: 143–147

Jeng M-F, Campbell AP, Begley T, Holmgren A, Case DA, Wright PE and Dyson HJ (1994) High-resolution solution structures of oxidized and reduced *Escherichia coli* thioredoxin. Structure 2: 853–868

Jenkins CLD, Anderson LE and Hatch MD (1986) NADP-malate dehydrogenase from *Zea Mays* Leaves: Amino acid composition and thiol content of active and inactive forms. Plant Sci 45: 1–7

Johansson K, Ramaswamy S, Saarinen M, Lemaire-Chamley M, Issakidis-Bourguet E, Miginiac-Maslow M and Eklund H (1999) Structural basis for light activation of a chloroplast enzyme. The structure of sorghum NADP-malate dehydrogenase in its oxidized form. Biochemistry 38: 4319–4326

Johnson HS (1972) Dithiothreitol: An inhibitor of glucose-6-phosphate dehydrogenase activity in leaf extracts and isolated chloroplasts. Planta 106: 273–277

Kamo M, Tsugita A, Wiessner Ch, Wedel N, Bartling D, Herrmann RG, Aguilar F, Gardet-Salvi L and Schürmann P (1989) Primary structure of spinach-chloroplast thioredoxin *f*. Protein sequencing and analysis of complete cDNA clones for spinach-chloroplast thioredoxin *f*. Eur J Biochem 182: 315–322

Katti SK, LeMaster DM and Eklund H (1990) Crystal structure of thioredoxin from *Escherichia coli* at 1.68 Å resolution. J Mol Biol 212: 167–184

Klenk HP, Clayton RA, Tomb JF, White O, Nelson KE, Ketchum KA, Dodson RJ, Gwinn M, Hickey EK, Peterson JD, Richardson DL, Kerlavage AR, Graham DE, Kyrpides NC, Fleischmann RD, Quackenbush J, Lee NH, Sutton GG, Gill S, Kirkness EF, Dougherty BA, McKenney K, Adams MD, Loftus B and Venter JC (1997) The complete genome sequence of the hyperthermophilic, sulphate-reducing archaeon *Archaeoglobus fulgidus*. Nature 390: 364–370

Knaff DB (1996) Ferredoxin and Ferredoxin-Dependent Enzymes. In: Ort DR and Yocum CF (eds) Oxygenic Photosynthesis: The Light Reactions, pp 333–361. Kluwer Academic Publishers, Dordrecht

Kramer DM, Wise RR, Frederick JR, Alm DM, Hesketh JD, Ort DR and Crofts AR (1990) Regulation of coupling factor in field-grown sunflower: A Redox model relating coupling factor activity to the activities of other thioredoxin-dependent chloroplast enzymes. Photosynth Res 26: 213–222

Krimm I, Goyer A, Issakidis-Bourguet E, Miginiac-Maslow M and Lancelin J-M (1999) Direct NMR observation of the thioredoxin-mediated reduction of the chloroplast NADP-malate dehydrogenase provides a structural basis for the relief of autoinhibition. J Biol Chem 274: 34539–34542

Lamotte-Guéry Fd, Miginiac-Maslow M, Decottignies P, Stein M, Minard P and Jacquot J-P (1991) Mutation of a negatively charged amino acid in thioredoxin modifies its reactivity with chloroplastic enzymes. Eur J Biochem 196: 287–294

Lancelin J-M, Stein M and Jacquot J-P (1993) Secondary structure and protein folding of recombinant chloroplastic thioredoxin Ch2 from the green alga *Chlamydomonas reinhardtii* as determined by ^{1}H NMR. J Biochem (Tokyo) 114: 421–431

Levings CS, III and Siedow JN (1995) Regulation by redox poise in chloroplasts. Science 268: 695–696

Lichter A and Häberlein I (1998) A light-dependent redox signal participates in the regulation of ammonia fixation in chloroplasts of higher plants—Ferredoxin:glutamate synthase is a thioredoxin-dependent enzyme. J Plant Physiol 153: 83–90

Lunn JE, Agostino A and Hatch MD (1995) Regulation of NADP-malate dehydrogenase in C_4 plants: Activity and properties of maize thioredoxin *m* and the significance of non-active site thiol groups. Aust J Plant Physiol 22: 577–584

Marcus F and Harrsch PB (1990) Amino acid sequence of spinach chloroplast fructose-1,6-bisphosphatase. Arch Biochem Biophys 279: 151–157

Marcus F, Moberly L and Latshaw SP (1988) Comparative amino acid sequence of fructose-1,6-bisphosphatases: Identification of a region unique to the light-regulated chloroplast enzyme. Proc Natl Acad Sci USA 85: 5379–5383

Marques S, Merida A, Candau P and Florencio FJ (1992) Light-mediated regulation of glutamine synthetase activity in the unicellular cyanobacterium *Synechococcus* sp. PCC 6301. Planta 187: 247–253

Martin W and Herrmann RG (1998) Gene transfer from organelles to the nucleus: How much, what happens, and why? Plant Physiology 118: 9–17

Martin W, Mustafa AZ, Henze K and Schnarrenberger C (1996) Higher-plant chloroplast and cytosolic fructose-1,6-bisphosphatase isoenzymes: Origins via duplication rather than prokaryote-eukaryote divergence. Plant Mol Biol 32: 485–491

McCarn DF, Whitaker RA, Alam J, Vrba J and Curtis SE (1988) Genes encoding the alpha, gamma, delta, and four F_0 subunits of ATP synthase constitute an operon in the cyanobacterium *Anabaena* sp. strain PCC 7120. J Bacteriol 170: 3448–3458

McKinney DW, Buchanan BB and Wolosiuk RA (1978) Activation of Chloroplast ATPase by Reduced Thioredoxin. Phytochemistry 17: 794–795

Meyer Y, Verdoucq L and Vignols F (1999) Plant thioredoxins and glutaredoxins: Identity and putative roles. Trends Plant Sci 4: 388–394

Miginiac-Maslow M, Issakidis E, Lemaire M, Ruelland E, Jacquot JP and Decottignies P (1997) Light-dependent activation of NADP-malate dehydrogenase: A complex process. Aust J Plant Physiol 24: 529–542

Miki J, Maeda M, Mukohata Y and Futai M (1988) The g-subunit of ATP synthase from spinach chloroplasts primary structure deduced from the cloned cDNA sequence. FEBS Lett 232: 221–226

Milanez S and Mural RJ (1988) Cloning and sequencing of cDNA encoding the mature form of phorphoribulokinase from spinach. Gene 66: 55–63

Milanez S, Mural RJ and Hartman FC (1991) Roles of cysteinyl residues of phosphoribulokinase as examined by site-directed mutagenesis. J Biol Chem 266: 10694–10699

Mills JD and Hind G (1979) Light-induced Mg^{2+} ATPase activity of coupling factor in intact chloroplasts. Biochim Biophys Acta 547: 455–462

Mills JD, Mitchell P and Schürmann P (1980) Modulation of coupling factor ATPase activity in intact chloroplasts. The role of the thioredoxin system. FEBS Lett 112: 173–177

Mittard V, Blackledge MJ, Stein M, Jacquot JP, Marion D and Lancelin JM (1997) NMR solution structure of an oxidised thioredoxin *h* from the eukaryotic green alga *Chlamydomonas reinhardtii*. Eur J Biochem 243: 374–383

Mora-García S, Rodriguez-Suárez RJ and Wolosiuk RA (1998) Role of electrostatic interactions on the affinity of thioredoxin for target proteins. Recognition of chloroplast fructose-1,6-bisphosphatase by mutant *Escherichia coli* thioredoxins. J Biol Chem 273: 16273–16280

Nishizawa AN and Buchanan BB (1981) Enzyme regulation in C_4 photosynthesis. Purification and properties of thioredoxin-linked fructose bisphosphatase and sedoheptulose bisphosphatase from corn leaves. J Biol Chem 256: 6119–6126

Ort DR and Oxborough K (1992) In situ regulation of chloroplast coupling factor activity. Annu Rev Plant Physiol Plant Mol Biol 43: 269–291

Pancic PG and Strotmann H (1993) Structure of the nuclear encoded g-subunit of CF_0CF_1 of the diatom *Odontella sinensis* including its presequence. FEBS Lett 320: 61–66

Papen H and Bothe H (1984) The activation of glutamine synthetase from the cyanobacterium *Anabaena cylindrica* by thioredoxin. FEMS Microbiol Lett 23: 41–46

Parry MA, Andralojc PJ, Lowe HM and Keys AJ (1999) The localisation of 2-carboxy-D-arabinitol 1-phosphate and inhibition of Rubisco in leaves of *Phaseolus vulgaris* L. FEBS Lett 444: 106–110

Porter MA and Hartman FC (1990) Exploration of the function of a regulatory sulfhydryl of phosphoribulokinase from spinach. Arch Biochem Biophys 281: 330–334

Porter MA, Stringer CD and Hartman FC (1988) Characterization of the regulatory thioredoxin site of phosphoribulokinase. J Biol Chem 263: 123–129

Porter MA, Potter MD and Hartman FC (1990) Affinity labeling of spinach phosphoribulokinase subsequent to *S*-methylation at Cys16. J Protein Chem 9: 445- 452

Portis Jr AR (1995) The regulation of Rubisco by Rubisco activase. J Exp Bot 46: 1285–1291

Qin J, Clore GM, Poindexter Kennedy WM, Huth JR and Gronenborn AM (1995) Solution structure of human thioredoxin in a mixed disulfide intermediate complex with its target peptide from the transcription factor NFkappaB. Structure 3: 289–297

Raines CA, Lloyd JC, Willingham NM, Potts S and Dyer TA (1992) cDNA and gene sequences of wheat chloroplast sedoheptulose-1,7-bisphosphatase reveal homology with fructose-1,6-bisphosphatases. Eur J Biochem 205: 1053–1059

Raines CA, Lloyd JC and Dyer TA (1999) New insights into the structure and function of sedoheptulose-1,7-bisphosphatase; an important but neglected Calvin cycle enzyme. J Exp Bot 50: 1–8

Rebeille F and Hatch MD (1986) Regulation of NADP-malate dehydrogenase in C_4 plants: effect of varying NADPH to NADP ratios and thioredoxin redox state on enzyme activity in reconstituted systems. Arch Biochem Biophys 249: 164–170

Reith M and Munholland J (1997) Complete Nucleotide Sequence of the *Porphyra purpurea* Chloroplast Genome. Plant Mol Biol Reptr 13: 333–335

Reng W, Riessland R, Scheibe R and Jaenicke R (1993) Cloning, site-specific mutagenesis, expression and characterization of full-length chloroplast NADP-malate dehydrogenase from *Pisum sativum*. Eur J Biochem 217: 189–197

Reynolds AE, Chesnick JM, Woolford J and Cattolico RA (1994) Chloroplast encoded thioredoxin genes in the red algae *Porphyra yezoensis* and *Griffithsia pacifica*: Evolutionary implications. Plant Mol Biol 25: 13–21

Rodriguez-Suárez RJ, Mora-García S and Wolosiuk RA (1997) Characterization of cysteine residues involved in the reductive activation and the structural stability of rapeseed (*Brassica napus*) chloroplast fructose-1,6-bisphosphatase. Biochem Biophys Res Commun 232: 388–393

Roesler KR and Ogren WL (1990) *Chlamydomonas reinhardtii* phosphoribulokinase. Sequence, purification, and kinetics. Plant Physiol 93: 188–193

Ruelland E and Miginiac-Maslow M (1999) Regulation of chloroplast enzyme activities by thioredoxins: Activation or relief from inhibition? Trends Plant Sci 4: 136–141

Ruelland E, Lemaire-Chamley M, Le Maréchal P, Issakidis-Bourguet E, Djukic N and Miginiac-Maslow M (1997) An internal cysteine is involved in the thioredoxin-dependent activation of sorghum leaf NADP-malate dehydrogenase. J Biol Chem 272: 19851–19857

Ruelland E, Johansson K, Decottignies P, Djukic N and Miginiac-Maslow M (1998) The autoinhibition of sorghum NADP malate dehydrogenase is mediated by a C-terminal negative charge. J Biol Chem 273: 33482–33488

Saarinen M, Gleason FK and Eklund H (1995) Crystal structure of thioredoxin-2 from *Anabaena*. Structure 3: 1097–1108

Sahrawy M, Hecht V, López-Jaramillo J, Chueca A, Chartier Y and Meyer Y (1996) Intron position as an evolutionary marker of thioredoxins and thioredoxin domains. J Mol Evol 42: 422–431

Salamon Z, Tollin G, Hirasawa M, Knaff DB and Schürmann P (1995) The oxidation-reduction properties of spinach thioredoxins *f* and *m* and of ferredoxin:thioredoxin reductase. Biochim Biophys Acta 1230: 114–118

Sasaki Y, Kozaki A and Hatano M (1997) Link between light and fatty acid synthesis: Thioredoxin-linked reductive activation of plastidic acetyl-CoA carboxylase. Proc Natl Acad Sci USA 94: 11096–11101

Sauer A and Heise K-P (1983) On the light dependence of fatty acid synthesis in spinach chloroplasts. Plant Physiol 73: 11–15

Scheibe R (1987) $NADP^+$-malate dehydrogenase in C_3-plants: Regulation and role of a light-activated enzyme. Physiol Plant 71: 393–400

Scheibe R (1991) Redox-modulation of chloroplast enzymes. A common principle for individual control. Plant Physiol 96: 1–3

Scheibe R (1994) Lichtregulation von Chloroplastenenzymen. Naturwissenschaften 81: 443–448

Scheibe R and Anderson LE (1981) Dark modulation of NADP-dependent malate dehydrogenase and glucose-6-phosphate dehydrogenase in the chloroplast. Biochim Biophys Acta 636: 58–64

Scheibe R and Jacquot J-P (1983) NADP regulates the light activation of NADP-dependent malate dehydrogenase. Planta 157: 548–553

Scheibe R, Geissler A and Fickenscher K (1989) Chloroplast glucose-6-phosphate dehydrogenase: Km shift upon light modulation and reduction. Arch Biochem Biophys 274: 290–297

Scheibe R, Kampfenkel K, Wessels R and Tripier D (1991) Primary structure and analysis of the location of the regulatory disulfide bond of pea chloroplast NADP-malate dehydrogenase. Biochim Biophys Acta 1076: 1–8

Schimkat D, Heineke D and Heldt HW (1990) Regulation of sedoheptulose-1,7-bisphosphatase by sedoheptulose-7-phosphate and glycerate, and of fructose-1,6-bisphosphatase by glycerate in spinach chloroplasts. Planta 181: 97–103

Schmidt A (1981) A thioredoxin activated glutamine synthetase in *Chlorella*. Z Naturforsch 36c: 396–399

Schürmann P (1981) The ferredoxin/thioredoxin system of spinach chloroplasts. Purification and characterization of its components. In: Akoyunoglou G (ed) Photosynthesis IV. Regulation of Carbon Metabolism, Vol 4, pp 273–280. Balaban Intern. Sci. Services, Philadelphia

Schürmann P and Jacquot J-P (2000) Plant thioredoxin systems revisited. Annu Rev Plant Physiol Plant Mol Biol 51: 371–400

Schürmann P, Wolosiuk RA, Breazeale VD and Buchanan BB (1976) Two proteins function in the regulation of photosynthetic CO_2 assimilation in chloroplasts. Nature 263: 257–258

Schürmann P, Maeda K and Tsugita A (1981) Isomers in thioredoxins of spinach chloroplasts. Eur J Biochem 116: 37–45

Schwarz O, Schürmann P and Strotmann H (1997) Kinetics and thioredoxin specificity of thiol modulation of the chloroplast H^+-ATPase. J Biol Chem 272: 16924–16927

Schwendtmayer C, Manieri W, Hirasawa M, Knaff DB and Schürmann P (1998) Cloning, expression and characterization of ferredoxin:thioredoxin reductase from *Synechocystis* sp. PCC6803. In: Garab G (ed) Photosynthesis: Mechanisms and Effects, Vol 3, pp 1927–1930. Kluwer Academic Publishers, Dordrecht

Smith DR, Doucette-Stamm LA, Deloughery C, Lee H, Dubois J, Aldredge T, Bashirzadeh R, Blakely D, Cook R, Gilbert K, Harrison D, Hoang L, Keagle P, Lumm W, Pothier B, Qiu D, Spadafora R, Vicaire R, Wang Y, Wierzbowski J, Gibson R, Jiwani N, Caruso A, Bush D and Reeve JN (1997) Complete genome sequence of *Methanobacterium thermoautotrophicum* deltaH: Functional analysis and comparative genomics. J Bacteriol 179: 7135–7155

Staples CR, Ameyibor E, Fu W, Gardet-Salvi L, Stritt-Etter A-L, Schürmann P, Knaff DB and Johnson MK (1996) The nature and properties of the iron-sulfur center in spinach ferredoxin:thioredoxin reductase: A new biological role for iron-sulfur clusters. Biochemistry 35: 11425–11434

Staples CR, Gaymard E, Stritt-Etter AL, Telser J, Hoffman BM, Schürmann P, Knaff DB and Johnson MK (1998) Role of the $[Fe_4 S_4]$ cluster in mediating disulfide reduction in spinach ferredoxin:thioredoxin reductase. Biochemistry 37: 4612–4620

Szekeres M, Droux M and Buchanan BB (1991) The ferredoxin-thioredoxin reductase variable subunit gene from *Anacystis nidulans*. J Bacteriol 173: 1821–1823

Tamoi M, Murakami A, Takeda T and Shigeoka S (1998) Acquisition of a new type of fructose-1,6-bisphosphatase with resistance to hydrogen peroxide in cyanobacteria: Molecular characterization of the enzyme from *Synechocystis* PCC 6803. Biochim Biophys Acta 1383: 232–244

Thompson JD, Plewniak F, Jeanmougin F and Higgins DG (1997) The ClustalX windows interface: Flexible strategies for multiple sequence alignment aided by quality analysis tools. Nucleic Acids Res 25: 4876–4882

Tischner R and Schmidt A (1982) A thioredoxin-mediated activation of glutamine synthetase and glutamate synthase in synchronous *Chlorella sorokiniana*. Plant Physiol 70: 113–116

Tsugita A, Yano K, Gardet-Salvi L and Schürmann P (1991) Characterization of spinach ferredoxin-thioredoxin reductase. Protein Seq Data Anal 4: 9–13

Villeret V, Huang S, Zhang Y, Xue Y and Lipscomb WN (1995) Crystal structure of spinach chloroplast fructose-1,6-bisphosphatase at 2.8 Å resolution. Biochemistry 34: 4299–4306

Weichsel A, Gasdaska JR, Powis G and Montfort WR (1996) Crystal structures of reduced, oxidized, and mutated human thioredoxins: Evidence for a regulatory homodimer. Structure 4: 735–751

Wenderoth I, Scheibe R and von Schaewen A (1997) Identification of the cysteine residues involved in redox modification of plant plastidic glucose-6-phosphate dehydrogenase. J Biol Chem 272: 26985–26990

Werner S, Schumann J and Strotmann H (1990) The primary structure of the g-subunit of the ATPase from *Synechocystis* 6803. FEBS Lett 261: 204–208

Willingham NM, Lloyd JC and Raines CA (1994) Molecular cloning of the *Arabidopsis thaliana* sedoheptulose-1,7-biphosphatase gene and expression studies in wheat and *Arabidopsis thaliana*. Plant Mol Biol 26: 1191–1200

Wolosiuk RA and Buchanan BB (1977) Thioredoxin and glutathione regulate photosynthesis in chloroplasts. Nature 266: 565–567

Wolosiuk RA and Buchanan BB (1978a) Regulation of chloroplast phosphoribulokinase by the ferredoxin/thioredoxin system. Arch Biochem Biophys 189: 97–101

Wolosiuk RA and Buchanan BB (1978b) Activation of Chloroplast NADP-linked Glyceraldehyde-3-Phosphate Dehydrogenase by the Ferredoxin/thioredoxin System. Plant Physiol 61: 669–671

Wolosiuk RA, Buchanan BB and Crawford NA (1977) Regulation of NADP-malate dehydrogenase by the light-actuated

ferredoxin/thioredoxin system of chloroplasts. FEBS Lett 81: 253–258
Wolosiuk RA, Crawford NA, Yee BC and Buchanan BB (1979) Isolation of three thioredoxins from spinach leaves. J Biol Chem 254: 1627–1632
Wolosiuk RA, Hertig CM, Nishizawa AN and Buchanan BB (1982) Enzyme regulation in C_4 photosynthesis. Role of Ca^+ in thioredoxin-linked activation of sedoheptulose bisphosphatase from corn leaves. FEBS Lett 140: 31–35
Wolosiuk RA, Hertig CM and Busconi L (1986) Activation of spinach chloroplast NADP-linked glyceraldehyde-3-phosphate dehydrogenase by concerted hysteresis. Arch Biochem Biophys 246: 1–8
Yu LM and Selman BR (1988) cDNA sequence and predicted primary structure of the gamma subunit from the ATP synthase from *Chlamydomonas reinhardtii*. J Biol Chem 263: 19342–19345
Zhang N and Portis Jr AR (1999) Mechanism of light regulation of Rubisco: A unique role for the larger Rubisco activase isoform involving reductive activation by thioredoxin-*f*. Proc Natl Acad Sci USA 96: 9438–9443
Ziegler H and Ziegler I (1965) The influence of light on the $NADP^+$-dependent glyceraldehyde-3-phosphate dehydrogenase. Planta 65: 369–380

Chapter 21

Reversible Phosphorylation in the Regulation of Photosynthetic Phosphoenolpyruvate Carboxylase in C_4 Plants

Jean Vidal*, Sylvie Coursol, Jean-Noël Pierre
Institut de Biotechnologie des Plantes, UMR CNRS 8618, Université de Paris-Sud, 91405 Orsay Cedex, France

Summary

C_4 species have a specific isoform of phosphoenolpyruvate carboxylase (PEPC) that catalyzes primary CO_2 fixation in the C_4 photosynthesis pathway. It has long been known that the enzyme in the cytosol of the mesophyll cells is subject to allosteric control by opposing photosynthesis-related metabolites. The discovery of a phosphorylation process acting on C_4 PEPC, via a complex light-signal transduction cascade, has revitalized interest in this enzyme and the ensuing wealth of data has highlighted one of a few signaling cascades known so far in the regulation of plant metabolism. The cascade depends upon a cross-talk between the two neighboring photosynthetic cell types, involves classical second messengers like pH, Inosital-1,4 5-trisphosphate ($Ins(1,4,5)P_3$) and calcium, and upregulates the activity of a Ca^{2+}-independent, C_4 PEPC-specific protein-serine/threonine kinase, which finally phosphorylates PEPC. The final activity of C_4 PEPC and the resulting carbon flux to bundle sheath cells are dependent on the mutual interaction between metabolite and covalent control mechanisms acting on this enzyme.

Author for correspondence, email: jean.vidal@ibp.u-psud.fr

Eva-Mari Aro and Bertil Andersson (eds): Regulation of Photosynthesis, pp. 363–375.

I. Introduction

Phosphoenolpyruvate carboxylase (EC 4.1.1.31, PEPC) is a widely distributed enzyme in plants, green algae and micro-organisms, but is absent in animals (Andreo et al., 1987). The enzyme was first characterized by Bandurski and Greiner (1953) in spinach leaves and for a long time it was considered as a subsidiary plant carboxylase together with the ribulose-1,5-bisphosphate carboxylase/oxygenase (Rubisco). However, the discovery of C_4 photosynthesis in the mid-1960s and the involvement of a specific isoform of PEPC in this pathway (which eventually led to the general photosynthetic grouping of C_3, C_4 or CAM plants) considerably boosted interest in this cytosolic enzyme (Hatch, 1992). More recently, the extensive development and use of molecular and biochemical techniques have generated an impressive wealth of data and significantly advanced our understanding of the complex nature of the regulatory processes concerning this enzyme. For a comprehensive overview of regulation of various enzymes in C_4 photosynthesis, see Furbank et al. (2000).

II. C_4 Phosphoenolpyruvate Carboxylase (PEPC) in the Physiological Context of C_4 Photosynthesis

C_4 plants exhibit specific anatomical and biochemical features. Their leaf architecture conforms, in most cases, to the classical 'Kranz' anatomy characterized by concentrically organized photosynthetic tissues, i.e. the outer mesophyll cells (MC) surrounding the inner bundle sheath cells (BSC). In terms of metabolic adaptation, there exist diverse types of C_4 plants, but the general metabolic scheme of division of labor is conserved; two cycles, C_4 and the Calvin cycle, working in concert to assimilate CO_2. In 'L-malate formers', the primary fixation of carbon dioxide (under its hydrated form) is carried out by PEPC in the MC cytoplasm to form oxaloacetate (OAA) that is reduced to L-malate in chloroplasts; exportation of L-malate to BSC and its decarboxylation by NADP-dependent malic enzyme in the chloroplast stroma, generate reducing power (NADPH) and CO_2 for operation of the Calvin cycle (Fig. 1). Because in some C_4 plants, like *Sorghum* and sugar cane, BSC chloroplasts are deficient in photosystem II (PS II) activity and energy production, 3-phosphoglyceric acid (3-PGA) formed in this cell compartment moves to MC chloroplasts to be transformed to triose phosphate (TP). The fate of TP is then partially to feed sucrose synthesis in the MC and to partially return to BSC where it re-enters the Calvin cycle. This intense metabolite trafficking between the photosynthetic cells is gradient-driven through a network of plasmodesmata in the cell walls. This C_4 adaptation largely prevents the wasteful production of CO_2 by photorespiration and loss of carbon from leaves. In arid environments, this adaptation confers on the C_4 plant a better water use efficiency and a higher productivity with respect to C_3 plants (Hatch, 1977,1987,1992). Such a complex and highly integrated pathway, involving different tissues and cell compartments, needs a high degree of coordination and enzymatic control. In this review we give an updated account for what we have learnt about the regulatory mechanisms which act on C_4 PEPC to adjust its activity to the carbon demand by the Calvin cycle in the C_4 photosynthesis pathway.

Abbreviations: 3-PGA – 3-phosphoglyceric acid ; BSC – Bundle sheath cell; CDPK – Calmodulin-like domain protein kinase ; CHX – Cycloheximide; DAG – 1,2-diacylglycerol; DCMU – 3-(3,4 dichlorophenyl)-1,1-dimethyl urea; Ins(1,4,5)P_3 – Inositol-1,4,5-trisphosphate; MC – Mesophyll cell; PEP– Phosphoenolpyruvate; PEPC – Phosphoenolpyruvate carboxylase ; PEPCk – Phosphoenolpyruvate carboxylase kinase; PI-PLC – Phosphoinositide-specific phospholipase C; S(P) – Phosphorylated serine 8; S8D– Serine 8 replaced by aspartate; TP – Triose phosphate; W-7– N-[6-aminohexyl]-5-chloro-1-naphthalenesulfonamide

III. Properties of C_4 PEPC

PEPC catalyzes the exergonic β-carboxylation of phosphoenolpyruvate (PEP) by HCO_3^- ($\Delta G = -7$ kcal mol^{-1}) in the presence of a divalent cation, generally Mg^{2+}. The reaction proceeds through a stepwise mechanism that involves a reversible, rate-limiting formation of carboxyphosphate and the enolate of pyruvate. Carboxyphosphate is split into inorganic phosphate and free CO_2 within the active site of PEPC, CO_2 then reacting with the enolate species to form oxaloacetate (Chollet et al., 1996). PEPC is thought to behave as a homotetramer with each subunit having an approximate mass of 110 kDa (Chollet et al., 1996). Active dimeric PEPC species have been detected in plant protein extracts; however

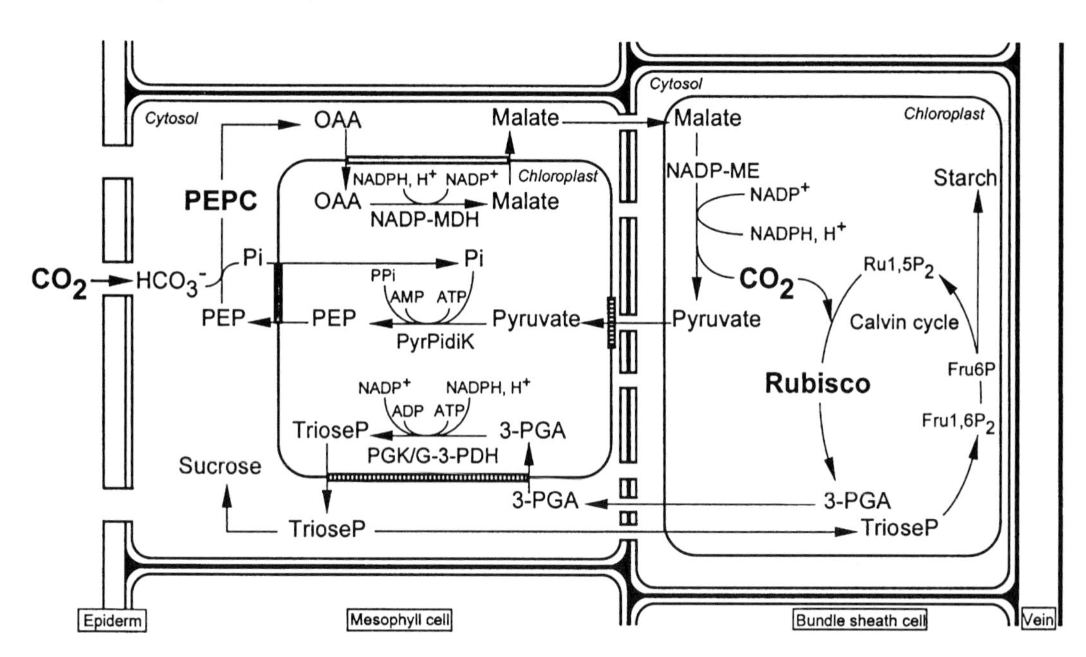

Fig. 1. Pathway of C_4 photosynthesis in 'L-malate former' C_4 plants. Enzymes catalyzing CO_2 fixation in the cytosol of mesophyll cells (PEPC) and in the chloroplasts of bundle sheath cells (Rubisco) are indicated in bold. Fru6P: fructose-6-phosphate; Fru1,6P_2: Fructose-1,6-bisphosphate; NADP-MDH: NADP-dependent malate dehydrogenase; NADP-ME: NADP-dependent malic enzyme; OAA: oxaloacetate; 3-PGA: 3-phosphoglyceric acid; PEP: phosphoenolpyruvate; PEPC: phosphoenolpyruvate carboxylase; PGK/G-3-PDH: phosphoglycerate kinase/glyceraldehyde-3-phospho dehydrogenase; PyrPidiK: pyruvate-inorganic phosphate dikinase; Ru1,5P_2: ribulose-1,5-bisphosphate; Rubisco: ribulose-1,5-bisphosphate carboxylase/oxygenase.

it is not yet known whether the dimer/tetramer equilibrium has any physiological role in vivo (McNaughton et al, 1989; Willeford and Wedding, 1992). Recently, X-ray crystallographic analysis has shed light on the three-dimensional structure of the *Escherichia coli* enzyme. The four subunits of the bacterial PEPC are organized in a 'dimer-of-dimers' form resulting in an overall square arrangement (Kai et al., 1999). The active site and the regulatory domains have been localized in the enzyme subunit.

Experiments with a recombinant C_4 PEPC (non-phosphorylated, see hereafter) from *Sorghum* have confirmed that the enzyme is subject to two opposing control mechanisms: feedback inhibition by the end-product L-malate (K_i: 0.17 mM) and allosteric activation by glucose-6P (K_a: 1.3 mM) when assayed at 1 mM PEP, pH 7.3 (Duff et al., 1995). Other sugar-P, such as dihydroxyacetone-P (Gao and Woo, 1996; Bakrim et al., 1998), or amino acids (e.g., aspartate, glutamate and glycine) are able to alter positively the enzyme activity. L-malate appears to act as a competitive inhibitor (Duff et al., 1995) whereas glucose-6P increases the apparent affinity of C_4 PEPC to PEP by lowering the K_m value for the substrate; thus this positive effector enhances the substrate's ability to compete with L-malate (Gao and Woo, 1996). This is particularly striking at pH 7, a value which is likely to be close to the physiological pH of the plant cytosol (Raghavendra et al., 1993). Because the primary structures of *E. coli* and plant PEPC are highly similar in the domains that are essential for protein folding, catalytic activity and allosteric regulation (with the notable exception of the N-terminal phosphorylation domain), the mechanism of the competitive inhibition by L-malate in plants can be inferred from the 3D-structure of the bacterial PEPC. In bacteria, L-malate interacts with arginine 587 in the highly conserved glycine-rich loop of PEPC, which contributes to the active site. Upon binding, the loop is removed from the catalytic site thus causing a loss of catalytic activity (Kai et al., 1999). This finding also explains why L-malate apparently behaves as a competitive inhibitor of PEP.

Thus, it appears that the activity of C_4 PEPC in vivo is fine-tuned by the balance of negative/positive effector metabolites, and this is further modulated by cytosolic pH (Echevarria et al., 1994).

IV. C_4 PEPC Activity is Reversibly Modulated in vivo by a Regulatory Phosphorylation Cycle

The phosphorylation/dephosphorylation-dependent regulation of PEPC activity was initially reported in the case of the photosynthetic isoform from the CAM plant *Bryophyllum* (Nimmo et al., 1984), and shortly afterwards in maize, a C_4 species (Budde and Chollet, 1986). Since then, a great deal of data on the enzyme's covalent control has been obtained, changing radically the dogma for C_4 PEPC regulation. Notably, the data accumulated has led to identification of components of the light-signaling cascade including the PEPC kinase (PEPCk), and to a working model for the signaling circuitry controlling the phosphorylation state of this enzyme in relation to C_4 photosynthesis.

A. C_4 PEPC as a Target for Phosphorylation

It soon became evident that phosphorylation of a specific serine residue on C_4 PEPC modulated the opposing effects of the aforementioned metabolites. Indeed, extensive phosphorylation (1 S(P)/enzyme subunit) in vitro of the *Sorghum* C_4 PEPC enzyme caused only a modest effect on the K_m for PEP but an approximately 2-fold increase in V_{max}, a 7-fold increase in the K_i for L-malate, and a 4.5-fold decrease in the K_a for glucose-6P (as measured at suboptimal pH and PEP concentration) (Duff et al., 1995).

Phosphorylation of C_4 PEPC in *Sorghum* was complete within 1–2 h upon induction by light (as estimated by the enzyme's sensitivity to L-malate), and the final ratio of phosphorylated/dephosphorylated enzyme was found to depend on the light fluence rate (Bakrim et al., 1992). Dephosphorylation, presumably by a type 2A protein phosphatase (Carter et al., 1990) followed a similar time course upon return of the plants to darkness. The phosphorylation domain (E/D-R/K-X-X-**S(P)**-I-D-A-Q-L/M-R) was found to be plant invariant and located very close to the N-terminus of the polypeptide containing a single serine residue as the target for phosphorylation (at the positions 8 and 15 in C_4 PEPC from *Sorghum* and maize, respectively) (Chollet et al., 1996; Vidal and Chollet 1997). The use of site-directed mutagenesis and recombinant protein technology clearly showed that phosphorylation could be functionally mimicked by the introduction of a negative charge (S8D) to the N-terminal domain of the protein (Lepiniec et al., 1994; Duff et al., 1995; Chollet et al., 1996). Based on the recently reported three-dimensional structure of the bacterial enzyme, it has been assumed that the negatively charged N-terminus extension (S(P) wild type or S8D mutant) of the plant C_4 PEPC impairs access of L-malate to the inhibitor site, thereby reducing the enzyme's sensitivity to the feedback inhibitor (Kai et al., 1999). Furthermore, investigation of the local structural requirements for phosphorylation of C_4 PEPC by its Ca^{2+}-independent protein-serine/threonine kinase (see below) suggested that a secondary site of interaction in C_4 PEPC was needed for the efficient in vitro phosphorylation of the target serine by PEPCk (Li et al., 1997).

B. Identification of the C_4 Phosphoenolpyruvate Carboxylase Kinase (PEPCk)

Both calcium-dependent and -independent protein kinases have been shown to phosphorylate the target C_4 PEPC in in vitro reconstitution assays. This raises the question of how to distinguish between physiologically relevant and other (gratuitous) phosphorylation events. Well established molecular and physiological characteristics of C_4 plant PEPC have provided evidence that the authentic PEPCk is a protein-serine/threonine kinase that, upon light-dependent specific phosphorylation of the target serine in the N-terminal domain of PEPC elicits the expected changes in the catalytic and regulatory properties of the enzyme (Chollet et al., 1996).

The first in vitro demonstration of the regulatory phosphorylation of C_4 PEPC was obtained using a soluble protein fraction partially purified from illuminated maize leaves (Jiao and Chollet, 1989). Two candidate PEPCk polypeptides, with molecular masses in the range of 30 to 39 kDa (Wang and Chollet, 1993; Li and Chollet, 1994), were subsequently identified. Assays carried out in vitro, at optimal pH values, revealed that the Ca^{2+}-independent protein kinase displayed apparent K_m values of 2.5 μM and 40 μM for C_4 PEPC (calculated on a subunit-basis) and total ATP, respectively. The optimum pH for function of the protein kinase was close to 8, and its catalytic activity decreased sharply with decreasing pH from 8 to 7. On the other hand, a Ca^{2+}-dependent PEPCk identified from maize was found to be inhibited by the calmodulin antagonist N-[6-aminohexyl]-5-chloro-1-naphthalenesulfonamide (W-7) as well as by the potent inhibitor of myosin light chain kinase KT5926 ($IC_{50,}$ 2.5 μM) (Izui et al.,

1995). The latter PEPCk was reminiscent of a plant calmodulin-like domain protein kinase (CDPK), a protein-serine/threonine kinase which possesses an intrinsic calcium-binding regulatory domain (with four typical EF-hand motifs, linked to an N-terminal catalytic domain (Roberts and Harmon, 1992). This particular CDPK has not been found in animals (Roberts and Harmon, 1992). Recently, a PEPC CDPK was partially purified from illuminated maize leaves by gel permeation and hydroxylapatite column chromatography and shown to be a dimer of 50 kDa subunits with K_m values for C_4 PEPC and ATP being approximately 0.3 μM and 30 μM, respectively (Ogawa et al., 1998).

To be physiologically relevant, the candidate protein kinase has to be subject to reversible light-activation in vivo (Bakrim et al., 1992; Chollet et al., 1996). Pretreatment of plants with cycloheximide (CHX), a cytosolic protein synthesis inhibitor, efficiently blocked both PEPCk upregulation and C_4 PEPC phosphorylation during the subsequent illumination period (Bakrim et al., 1992; Chollet et al., 1996). This result strengthened the view that the light-dependent increase in the phosphorylation state of C_4 PEPC reflected rapid de novo synthesis of PEPCk or, alternatively, synthesis of an unknown protein required for activation of PEPCk in vivo. Consistent with these results, the upregulation of PEPCk and the phosphorylation of C_4 PEPC were markedly blocked by CHX in isolated MC protoplasts from the C_4 species *Digitaria sanguinalis* (Giglioli-Guivarc'h et al., 1996) (see next section). Moreover, it has been reported recently that PEPCk from C_4, CAM and C_3 species was regulated at the level of translatable mRNA in response to light (C_4, C_3), or a circadian rhythm (CAM) (Hartwell et al., 1996).

An alternative strategy to identify the C_4 PEPCk was based on denaturing electrophoresis of soluble proteins from illuminated maize leaves followed by the in-situ renaturation and subsequent in-vitro assay for protein kinase activity (Li and Chollet, 1993). Three polypeptides with molecular masses of approximately 57, 37 and 30 kDa phosphorylated the target serine in C_4 PEPC. Interestingly, the 57-kDa protein kinase required calcium to catalyze phosphorylation whereas the two other polypeptides did not. It was established that the activity of the Ca^{2+}-independent PEPCk was strongly depressed in dark-adapted leaves and also in light-adapted leaves pretreated with CHX (Bakrim et al., 1992; Chollet et al., 1996). As a marked contrast, the Ca^{2+}-dependent 57-kDa protein kinase was virtually unaffected by these treatments in vivo. In addition, the Ca^{2+}-dependent PEPCk, partially purified from maize leaves, did not cause significant changes in the malate sensitivity of PEPC and was not specific to this protein target in vitro (Ogawa et al., 1998). Accordingly, since PEPCk catalyses phosphorylation of a synthetic peptide encompassing the N-terminal phosphorylation domain of C_4 PEPC, the authors have proposed that this protein kinase could play a role in the fine control of C_4 PEPC in response to unexpected and rapid changes in environmental conditions e.g., light intensity (Ogawa et al., 1998). Furthermore, recent results have established that the Ca^{2+}-independent protein kinase is the major PEPCk species in protein extracts from MC protoplasts (Nhiri et al., 1998). Taken together, this suggests that the Ca^{2+}-independent, 30- to 37 kDa polypeptides are the PEPCk phosphorylating C_4 PEPC in vivo. Whether both polypeptides are committed to this task is currently being examined in several laboratories.

The Ca^{2+}-independent PEPCk appears to be unique in that its activity is not modulated directly by second messengers (such as calcium or cyclic nucleotides) or by phosphorylation/dephosphorylation processes, but rather through rapid changes in its turnover rate (Bakrim et al., 1992; Chollet et al., 1996). In addition to this, reconstitution assays containing the C_4-leaf PEPCk or the mammalian type A protein kinase (PKA: previously shown to be able to phosphorylate C_4 PEPC in vitro) (Terada et al., 1990), revealed that phosphorylation of the purified C_4 PEPC was markedly inhibited by L-malate via a substrate effect (Wang and Chollet, 1993; Echevarria et al., 1994), whereas glucose-6P and PEP relieved this inhibition. The effect of the metabolites on C_4 PEPC phosphorylation was also suggested to occur in situ in MC protoplasts from *D. sanguinalis* (see hereafter and Bakrim et al., 1998). This indirect means of regulating protein phosphorylation might allow an individual, target-dependent control of a multi-substrate protein kinase. However, to date, all available evidence suggest that this highly regulated Ca^{2+}-independent PEPCk is specific for plant PEPC (Chollet et al., 1996).

C. The Light-Signal Transduction Cascade

As mentioned above, the enzymes of the photosynthetic pathway in C_4 leaves are distributed in specialized, concentrically organized photosynthetic cells, the outer MC, containing the enzymes for the C_4-cycle (with the exception of NADP-dependent malic enzyme), and the inner BSC containing enzymes for the Calvin cycle (Hatch, 1992).

Inhibitor-based experiments performed in plants (Bakrim et al., 1992; Chollet et al., 1996) revealed that the electron transport chain [inhibited by 3-(3,4 dichlorophenyl)-1,1-dimethyl urea (DCMU)] and ATP synthesis (inhibited by gramicidin) in MC chloroplasts as well as a functional Calvin cycle in BSC chloroplasts (inhibited by DL-glyceraldehyde) were required for upregulation of PEPCk and phosphorylation of C_4 PEPC in the MC cytosol. A working hypothesis was that light-transduction involved intercellular cross-talk, possibly mediated through changes in the level of a photosynthetic metabolite and/or energy charge. To address this question and to further identify the components of the light-transduction cascade, a cellular approach using isolated MC protoplasts from the C_4 grasses *D. sanguinalis* and *Sorghum* (Giglioli-Guivarc'h et al., 1996) was developed. Illuminated isolated protoplasts showed a marked decrease in L-malate sensitivity and an increase in catalytic activity of the endogenous C_4 PEPC when a weak base, like NH_4Cl or methylamine, was added to the suspension medium (Table 1). Notably, these changes were caused by a marked, light-induced stimulation of a Ca^{2+}-independent PEPCk (Bakrim et al., 1992) which was found to be sensitive to CHX in situ (Giglioli-Guivarc'h et al., 1996).

1. Alkalization of the Cytosol in C_4 Mesophyll Cells

The weak bases that triggered the in-situ phosphorylation of C_4 PEPC permeate into protoplasts in their neutral form and subsequently tend to increase cytosolic pH following protonation. The weak base-induced alkalization of the cytosol was experimentally documented by loading MC protoplasts with the fluorescent pH-probe, 2′,7′-bis-(2-carboxyethyl)-5-(and-6)carboxyfluorescein, acetoxymethyl ester (BCECF-AM) and performing in-situ fluorescence imaging by confocal microscopy. As an alternative approach, applying the 'null-point' method (Van der Veen et al., 1992) and monitoring induced changes in protoplast fluorescence by flow cytometry also provided estimates of the cytosolic pH values. Increase in cytosolic pH from about 6.4 to 7.3 with the concomitant increase in the activity of PEPCk and the apparent phosphorylation state of C_4 PEPC were found to be well correlated with the concentration of the exogenous weak-base inducer (Giglioli-Guivarc'h et al., 1996). Therefore, intracellular alkalization of MC protoplasts was implicated as an early signaling element in the C_4 PEPC phosphorylation circuitry. This observation provided an important clue as to the possible nature of the putative intercellular message. The most likely candidate was 3-PGA, the Calvin cycle intermediate that diffuses into MC chloroplasts for subsequent phosphorylation/ reduction. As transport by the chloroplast phosphate translocator proceeds only via the partially protonated 2^- form of 3-PGA, pumping of protons from the cytosol into the stroma would ensue, and a net alkalization of the cytosol should take place. Indeed, when 3-PGA was substituted for the weak base in the MC-protoplast suspension, this metabolite elicited the predicted changes in cytosolic pH (as judged by confocal microscopy). This resulted in upregulation of the Ca^{2+}-independent PEPCk and phosphorylation state of C_4 PEPC (Giglioli-Guivarc'h et al., 1996). Notably, both light and alkalization of MC cytosolic pH are needed for the induction of C_4 PEPCk activity. Consistent with these findings are the observations that this induction pathway was blocked by the photosynthesis inhibitors gramicidin and DCMU, while the increase in cytosolic pH was unaffected (Giglioli-Guivarc'h et al., 1996) (Table 1).

2. Phosphoinositide-Specific Phospholipase C and the Second Messenger Inositol-1,4,5-trisphosphate

In animal cells, activation of the phosphoinositide-specific phospholipase C (PI-PLC)(EC 3.1.4.11) plays a key role in transducing signals from both G-protein and tyrosine kinase-coupled receptors affecting growth, proliferation, secretion, contraction, sensory perception and over-all metabolism (reviewed by Berridge, 1993; Clapham, 1995). Briefly, activated PI-PLC catalyzes the splitting of phosphatidylinositol-4,5-bisphosphate (PtdIns(4,5)P_2) in the cell membrane into its hydrophobic and hydrophilic components, 1,2-diacylglycerol (DAG) and inositol-1,4,5-trisphosphate (Ins(1,4,5)P_3), respectively. Both of these two compounds are well established second messengers having major roles in cellular physiology. Specific receptors are localized on specialized Ca^{2+}-containing endomembrane compartments, and upon

Table 1. Identification of various signaling elements of the C_4 PEPC phosphorylation cascade by a pharmacological approach

treatment			reported effect	phosphorylation state of C_4 PEPC	putative component of the phosphorylation cascade
DARK				low	
LIGHT					
	NH_4Cl (20 mM)			high	cytosolic pH
	3-PGA (20 mM)			high	
		Neomycin (500 μM)	PI-PLC inhibitor	+	
		U-73122 (20 μM)	PI-PLC inhibitor	+	PI-PLC and Ins(1,4,5)P_3
		U-73343 (20 μM)	inactive analog of U-73122	–	
		EGTA + A23187 (5 mM) (1 μM)	Ca^{2+} chelator	+	
		TMB-8 (200 μM)	Ca^{2+} channel blocker (tonoplast)	+	
		Diltiazem (150 μM)	Ca^{2+} channel blocker (plasmalemma)	–	vacuolar Ca^{2+}
		Nifedipine (150 μM)	Ca^{2+} channel blocker (plasmalemma)	–	
		W-7 (200 μM)	Calmodulin antagonist	+	Ca^{2+}/CaM protein
		Compound 48/80 (5 μg/ml)	Calmodulin antagonist	+	
		DCMU (50 μM)	PS e^- transport inhibitor	+	
		Gramicidin (50 μM)	ATP synthesis inhibitor	+	NADPH/ATP synthesis
		CHX (5 μM)	protein synthesis inhibitor	+	PEPC kinase

A study at the cellular level was performed on the isolated MC protoplasts from the C_4 grasses *Digitaria sanguinalis* and *Sorghum*. Drugs blocking upregulation of PEPCk and phosphorylation of C_4 PEPC in situ are indicated as (+); drugs having no effect on these processes are indicated as (–). A23187: calcimycin; CHX: cycloheximide; DCMU: 3-(3,4 dichlorophenyl)-1,1-dimethyl urea; Ins(1,4,5)P_3: inositol-1,4,5-trisphosphate; PEPC: phosphoenolpyruvate carboxylase; 3-PGA: 3-phosphoglyceric acid; PI-PLC: phosphoinositide-specific phospholipase C; PS: photosystem; TMB-8: 8-[N,N-diethylamino]octyl-3,4,5-trimethoxybenzoate; U-73122: 1-(6-{[17β-3methoxyestra-1,3,5(10)-trien-17-xyl]amino}hexyl)-1H-pyrrole-2,5-dione; U-73343: 1-(6-{[17β-3methoxyestra-1,3,5(10)-trien-17-xyl]amino}hexyl)-1H-pyrrolidinedione; W-7: N-[6-aminohexyl]-5-chloro-1-naphthalenesulfonamide.

binding of Ins(1,4,5)P_3, Ca^{2+} is rapidly released from these intracellular stores. DAG, however, remains in the plasma membrane matrix, where it modulates the activity of type C protein kinase (PKC). Most components in the PI-PLC signaling system have structural or functional equivalents in plants, and evidence is emerging that they are involved in signaling (for reviews see Drøbak 1992; Coté and Crain, 1993; Munnik et al., 1998). A preliminary experiment has shown that preincubation of illuminated, weak-base-treated MC protoplasts with PI-PLC antagonists, neomycin and 1-(6-{[17β-3methoxyestra-1,3,5(10)-trien-17-xyl]amino}hexyl)-1H-pyrrole-2,5-dione (U-73122), inhibited phosphorylation of C_4 PEPC. In contrast, 1-(6-{[17β-3methoxyestra-1,3,5(10)-trien-17-xyl]amino}hexyl)-2,5-pyrrolidinedione (U-73343), an inactive analog of U-73122, had no inhibitory activity on this process (S. Coursol, unpublished) (Table 1). Furthermore, illuminated, weak-base-treated MC protoplasts

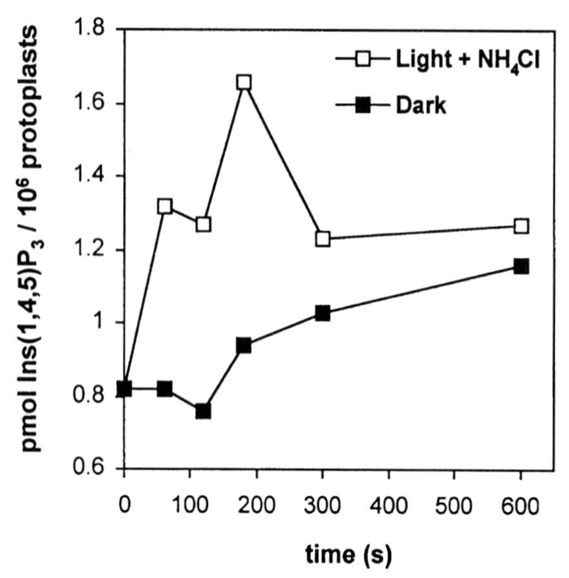

Fig. 2. Increase in the amount of Ins(1,4,5)P_3 in induced MC protoplasts from the C_4 plant *Digitaria sanguinalis*. Ins(1,4,5)P_3 levels in MC protoplasts were measured after addition of 40 mM NH_4Cl and illumination (500 $\mu E\ m^{-2}\ s^{-1}$) for 10 min. Control experiments were performed in the dark. Commercially available ^3H-Ins(1,4,5)P_3 assay kit (Amersham) was used to estimate the amount of Ins(1,4,5)P_3. Ins(1,4,5)P_3: inositol-1,4,5-trisphosphate.

showed a marked and transient increase in Ins(1,4,5)P_3 level (Fig. 2). This finding indicated that PI-PLC was potentially an upstream component of the C_4 PEPC phosphorylation chain in MC protoplasts. If this is the case, how might PI-PLC be activated and what is the role of Ins(1,4,5)P_3 in the induced MC protoplasts? Little is known about the precise mode of action of plant PI-PLCs, although it has been established that this enzyme is totally dependent on Ca^{2+} in the physiological range when assayed in vitro (Melin et al., 1992; Hirayama et al., 1995; Huang et al., 1995; Shi et al., 1995; Franklin-Tong et al., 1996; Kopka et al., 1998). Therefore, activation of PI-PLC by light and cytosolic alkalization are probably indirect effects. They may be mediated by Ca^{2+} in a process possibly involving Ca^{2+} influx across the plasma membrane, or perhaps by Ins(1,4,5)P_3-induced Ca^{2+} release suggesting multilevel feedback mechanisms. Information of the localization and concentration of the Ins(1,4,5)P_3-mobilizable Ca^{2+} stores would be crucial for understanding how PI-PLC is activated in the C_4-PEPC phosphorylation cascade. Currently, the vacuole is considered to be the major Ins(1,4,5)P_3-sensitive Ca^{2+} store in higher plants (Schumaker and Sze, 1987; Ranjeva et al., 1988; Brosnan and Sanders, 1993). However, there is evidence for Ins(1,4,5)P_3-mediated Ca^{2+} release from stores other than the vacuole in plants (Muir and Sanders, 1997).

3. Calcium and an Upstream Calcium-Dependent Protein Kinase(s)

It has been shown that a pretreatment of MC protoplasts with the calcium ionophore A23187 (calcimycin) and EGTA inhibited phosphorylation of C_4 PEPC. Specific recovery, however, was achieved if excess Ca^{2+} was reintroduced to protoplasts in the presence of A23187 (Pierre et al., 1992). The question about the origin of calcium and its mobilization into the cytosol of MC protoplasts was investigated by testing various pharmacological reagents. TMB-8, a tonoplast, Ins(1,4,5)P_3-gated Ca^{2+} channel blocker, severely inhibited the in-situ stimulation of PEPCk activity and C_4 PEPC phosphorylation by light together with cytosolic alkalization, whereas nifedipine and diltiazem (considered to act as plasma membrane Ca^{2+} channel inhibitors) did not have any effect on PEPCk activity or C_4 PEPC phosphorylation (Giglioli-Guivarc'h et al., 1996). Such findings support the view that Ca^{2+} is somehow involved in the light-transduction pathway as an upstream element. Given that the 30–37-kDa PEPCk are Ca^{2+}-independent enzymes (Li and Chollet, 1993; Wang and Chollet, 1993; Duff et al., 1996), a multicyclic protein kinase cascade, involving upstream Ca^{2+} elements had to be involved in transducing the light-signal. In good agreement with this idea, W-7 (an inhibitor of Ca^{2+}/calmodulin-regulated protein kinases or CDPK) was found to have a marked inhibitory effect on transducing the light signal in situ (Giglioli-Guivarc'h et al., 1996). Thus, these results suggest that the transduction chain might involve an upstream, Ca^{2+}-dependent protein-kinase(s) exerting an effect in the signaling pathway (Table 1).

A model for the spatio-temporal organization of the light-signal transduction chain controlling the activity of PEPCk and, thus the phosphorylation of C_4 PEPC is illustrated in Fig. 3.

V. Significance of the Regulatory Phosphorylation of C_4 PEPC

The uptake of the cytosolic, protein synthesis inhibitor CHX by an excised leaf of *Sorghum* or maize performing steady state photosynthesis, caused a

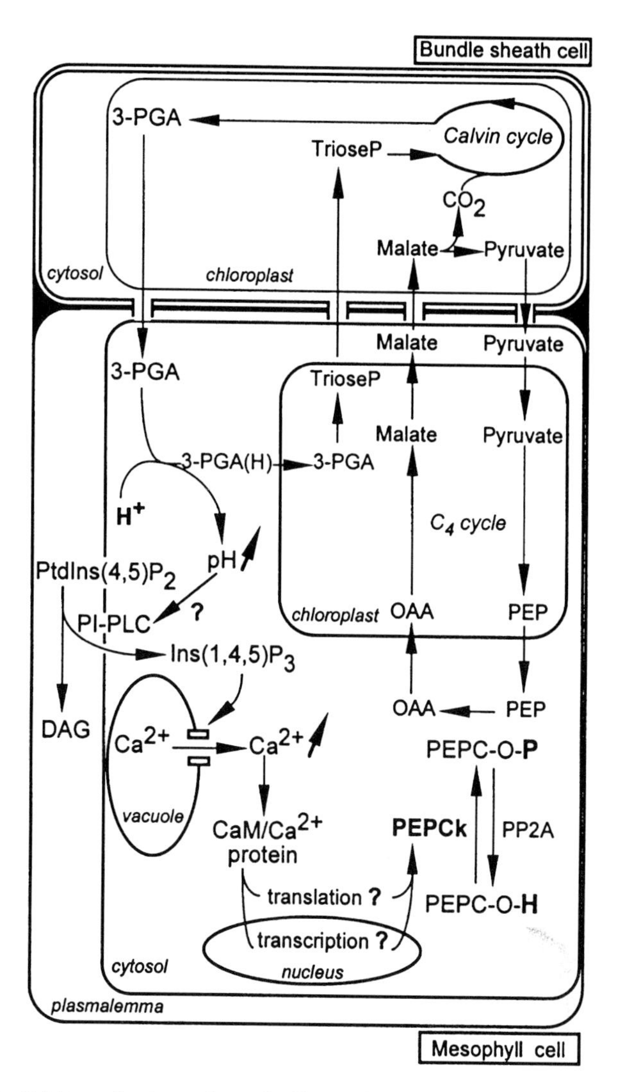

Fig. 3. Hypothetical model of the light-transduction pathway leading to the phosphorylation of C_4 PEPC in an illuminated C_4 leaf. For details, see text. CaM: calmodulin; DAG: 1,2-diacylglycerol; Ins(1,4,5)P_3: inositol-1,4,5-trisphosphate; OAA: oxaloacetate; PEP: phosphoenolpyruvate; PEPC-O-H: nonphosphorylated phosphoenolpyruvate carboxylase; PEPC-O-P: phosphorylated phosphoenolpyruvate carboxylase; PEPCk: phosphoenolpyruvate carboxylase kinase; 3-PGA: 3-phosphoglyceric acid; 3-PGA(H): 3-phosphoglyceric acid(protonated); PI-PLC: phosphoinositide-specific phospholipase C; PP2A: type 2A protein phosphatase; PtdIns(4,5)P_2: phosphatidylinositol-4,5-bisphosphate.

progressive and well correlated decrease in C_4 PEPC phosphorylation state, and in the CO_2 assimilation rate of the leaf, as monitored by Infra Red Gas Analysis (Bakrim et al., 1993). Clearly, C_4 PEPC phosphorylation has a crucial regulatory role in the overall functioning of C_4 photosynthesis.

A. Building of a Large Malate Pool in Mesophyll Cells: The C_4 PEPC Dilemma

In an illuminated C_4 leaf at high irradiance, C_4 PEPC is faced with millimolar concentrations of L-malate required for diffusive transport of this metabolite to the neighboring BSC. These levels of L-malate are sufficiently high (10 to 20 mM, as deduced from theoretical calculations) to severely impair the catalytic activity of the dephosphorylated enzyme (K_i for L-malate is about 0.2 mM). Reconstitution assays were performed in the presence of 4 mM glucose-6P, at pH 7.3, in order to simulate the physiological conditions likely to prevail in the mesophyll cytosol in light. It was observed that the fully phosphorylated form of C_4 PEPC had a markedly reduced sensitivity to L-malate (IC_{50} = 15 mM) (Echevarria et al., 1994). Certain amino acids (such as glycine and alanine), which are known to activate C_4 PEPC in monocot species, synergistically increased the K_i of the maize enzyme for L-malate (Gao and Woo, 1996; Bakrim et al., 1998). Therefore, the regulatory role of C_4 PEPC phosphorylation appeared to be to attenuate the inhibitory effect of L-malate on the enzyme by modulating its affinity for these opposing metabolite effectors. This would enable C_4 PEPC to continue to fix carbon when malate concentrations are high in the MC cytosol.

B. Fine-Tuning of C_4 PEPC Phosphorylation and its Coupling to the Operation of Calvin Cycle: A Role of Metabolites ?

The steady state phosphorylation of C_4 PEPC in vivo is reached after approximately 1h illumination of the C_4 leaf. Rapid fluctuations of environmental conditions in the minute-time-scale are therefore not directly related to the posttranslational phosphorylation process. As mentioned above (see Section IV.B), while phosphorylation of C_4 PEPC has an impact on the sensitivity of this enzyme to metabolite effectors, these metabolites, in turn, control the phosphorylation state of C_4 PEPC by modulating (L-malate inhibits while the positive effectors antagonize the effect of L-malate) the phosphorylation rate by PEPCk (Echevarria et al., 1994; Wang and Chollet, 1993). In contrast with the relatively slow upregulation of the PEPCk elicited by the light-dependent cascade, this effect is expected to rapidly modulate the phosphorylation state and catalytic properties of C_4 PEPC. Thus, phosphorylation appeared not only to protect C_4 PEPC against L-malate but also to adjust its catalytic activity according to the demand of the Calvin cycle for a C_4 acid-derived supply of CO_2. Any imbalance in the ratio of positive/negative effectors would result in a corresponding change in PEPCk activity and C_4 PEPC phosphorylation state. Along these lines, 3-PGA, which is assumed to trigger the cascade and the phosphorylation of PEPC, and to be the precursor of the positive effectors, TP and glucose-6P, act as a metabolite message connecting Calvin cycle in BSC with the activity of C_4 PEPC in MC (Fig. 4). This complex regulatory mechanism provides flexibility for adjusting carbon flow in C_4 plants to altered environmental conditions and ensures coordination of the two physically segregated metabolic cycles involved in C_4 photosynthesis.

C. C_4 PEPC, a Hysteretic Enzyme

Finally, as the phosphorylation-induced changes in the regulatory properties of C_4 PEPC are relatively slow, a hysteretic behavior of the enzyme is expected (Neet and Ainslie, 1980). A temporary imbalance between the light-dependent formation of PEP (MC chloroplast pyruvate-Pi dikinase is rapidly activated

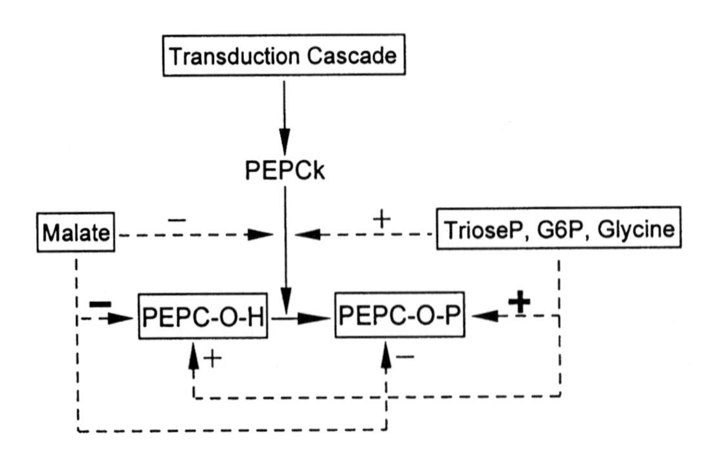

Fig. 4. Metabolite control of C_4 PEPC phosphorylation and modulation of the catalytic activity of C_4 PEPC in the cytosol of mesophyll cell. The effects [inhibition (–) or activation (+)] of C_4 photosynthesis-related metabolite are indicated with dotted lines. Symbols [(–) and (+)] in bold indicate a high sensitivity of C_4 PEPC to the metabolite effector. G6P: glucose-6-phosphate; PEPC-O-H: nonphosphorylated phosphoenolpyruvate carboxylase; PEPC-O-P: phosphorylated phosphoenolpyruvate carboxylase; PEPCk: phosphoenolpyruvate carboxylase kinase.

in the light) and its utilization by C_4 PEPC apparently causes a pool of this metabolite to build up during the induction phase of C_4 photosynthesis. Accumulation of millimolar concentrations of PEP (levels in the dark: 0.3 mM; levels in the light: 3 mM) in illuminated MC is needed for C_4 PEPC to gain significant catalytic activity as required by C_4 photosynthesis.

VI. Conclusions and Perspectives

The intense research performed during the last decade has led to the view that C_4 PEPC phosphorylation is a cardinal regulatory event in C_4 photosynthesis. The proposed transduction chain that links the light stimulus to upregulation of PEPCk involves several classical second messengers as found earlier in animal cells, including pH, Ins(1,4,5)P_3 and calcium. Our current research is focusing on the molecular characterization of the mesophyll PI-PLC and on the mechanism of its transient activation following the increase in cytosolic pH. Isolation and cloning of the Ca^{2+}-independent PEPCk and subsequent analysis of gene expression and functional/regulatory properties of this protein kinase are eagerly awaited. A good deal of structural data on *E. coli* PEPC has also become available and should accelerate the high-resolution X-ray crystallographic investigation of C_4 PEPC. Finally, the exciting possibility of complementing a C_4 PEPC-deficient mutant of *Amaranthus edulis* (Dever et al., 1995, 1997) with mutated C_4 PEPC will open the way to genetic manipulations aimed at understanding the role and impact of covalent regulation of this key C_4 enzyme in plants.

References

Andreo CS, Gonzalez DH and Iglesias AA (1987) Higher plant phosphoenolpyruvate carboxylase: Structure and regulation. FEBS Lett 213: 1–8

Bakrim N, Echevarria C, Crétin C, Arrio-Dupont M, Pierre JN, Vidal J, Chollet R and Gadal P (1992) Regulatory phosphorylation of *Sorghum* leaf phosphoenolpyruvate carboxylase: Identification of the protein-serine kinase and some elements of the signal-transduction cascade. Eur J Biochem 204: 821–830

Bakrim N, Prioul JL, Deleens E, Rocher JP, Arrio-dupont M, Vidal J, Gadal P and Chollet R (1993) Regulatory phosphorylation of C_4 phosphoenolpyruvate carboxylase: A cardinal event influencing the photosynthesis rate in *Sorghum* and maize. Plant Physiol 101: 891–897

Bakrim N, Nhiri M, Pierre JN and Vidal J (1998) Metabolite control of *Sorghum* C_4 phosphoenolpyruvate carboxylase catalytic activity and phosphorylation state. Photosynth Res 58: 153–162

Bandurski RS and Greiner CM (1953) The enzymatic synthesis of oxalacetate from phosphenolpyruvate and carbon dioxide. J Biol Chem 204: 781–786

Berridge MJ (1993) Inositol trisphosphate and calcium signalling. Nature 361: 315–325

Brosnan JM and Sanders D (1993) Identification and characterisation of high-affinity binding sites for inositol trisphosphate in red beet. Plant Cell 5: 931–940

Budde RJA and Chollet R (1986) In vitro phosphorylation of maize leaf phosphoenolpyruvate carboxylase. Plant Physiol 82: 1107–1114

Carter PJ, Nimmo HG, Fewson CA and Wilkins MB (1990) *Bryophyllum fedtschenkoi* protein phosphatase type 2A can dephosphorylate phosphoenolpyruvate carboxylase. FEBS Lett 263: 233–236

Chollet R, Vidal J and O'Leary MH (1996) Phosphoenolpyruvate carboxylase: A ubiquitous, highly regulated enzyme in plants. Annu Rev Plant Physiol Plant Mol Biol 47: 273–298

Clapham DE (1995) Calcium signalling. Cell 80: 259–268

Coté CG and Crain RC (1993) Biochemistry of phosphoinositides. Annu Rev Plant Physiol Plant Mol Biol 44: 333–3

Dever LV, Baron AC, Leegood RC and Lea PJ (1995) The characterisation of a mutant of *Amaranthus edulis* lacking phosphoenolpyruvate carboxylase. In: Mathis P (ed). Photosynthesis: From Light to Biosphere, Vol V, pp 277–280. Kluwer Academic Publishers, Dordrecht

Dever LV, Bailey KJ, Leegood RC and Lea PJ (1997) Control of photosynthesis in *Amaranthus edulis* mutants with reduced amounts of PEP carboxylase. Aust J Plant Physiol 24: 469–476

Drøbak BK (1992) The Plant phosphoinositide system. Biochem J 288: 697–712

Duff SMG, Andreo CS, Pacquit V, Lepiniec L, Sarath G, Condon SA, Vidal J, Gadal P and Chollet R (1995) Kinetic analysis of the non-phosphorylated, in vitro phosphorylated, and phosphorylation-site-mutant (Asp8) forms of intact recombinant C_4 phosphoenolpyruvate carboxylase from *Sorghum*. Eur J Biochem 228: 92–95

Duff SMG, Giglioli-Guivarc'h N, Pierre JN, Vidal J, Condon SA and Chollet R (1996) In-situ evidence for the involvement of calcium and bundle-sheath-derived photosynthetic metabolites in the C_4 phosphoenolpyruvate-carboxylase (PEPC) kinase signal-transduction chain. Planta 199: 467–474

Echevarria C, Pacquit V, Bakrim N, Osuna L, Delgado B, Arrio-Dupont M and Vidal J (1994) The effect of pH on the covalent and metabolic control of C_4 phosphoenolpyruvate carboxylase from *Sorghum* leaf. Arch Biochem Biophys 315: 425–430

Franklin-Tong VE, Drøbak BK, Allan AC, Watkins PAC and Trewavas AJ (1996) Growth of pollen tubes of *Papaver rhoeas* is regulated by a slow-moving calcium wave propagated by inositol 1,4,5-trisphosphate. Plant Cell 8: 1305–1321

Furbank RT, Hatch MD and Jenkins CLD (2000) C_4 photosynthesis: Mechanism and regulation. In: Leegood RC, Sharkey TD and von Caemmerer S (eds) Photosynthesis: Physiology and Metabolism, pp 435–457. Kluwer Academic Publishers, Dordrecht

Gao Y and Woo KC (1996) Regulation of phosphoenolpyruvate carboxylase in *Zea mays* by protein phosphorylation and metabolites and their roles in photosynthesis. Aust J Plant Physiol 23: 25–32

Giglioli-Guivarc'h N, Pierre JN, Brown S, Chollet R, Vidal J and Gadal P (1996) The light-dependent transduction pathway controlling the regulatory phosphorylation of C_4 phosphoenolpyruvate carboxylase in protoplasts from *Digitaria sanguinalis*. Plant Cell 8: 573–586

Hartwell J, Smith L, Wilkins MB, Jenkins GI and Nimmo HG (1996) Higher plant phosphoenolpyruvate carboxylase kinase is regulated at the level of translatable mRNA in response to light or a circadian rhythm. Plant J 10: 1071–1078

Hatch MD (1977) C_4 pathway photosynthesis: Mechanism and physiological function. Trends Biochem Sci 2: 199–201

Hatch MD (1987) C_4 photosynthesis: A unique blend of modified biochemistry, anatomy and ultrastructure. Biochim Biophys Acta 895: 81–106

Hatch MD (1992) C_4 photosynthesis: An unlikely process full of surprises. Plant Cell Physiol 33: 333–342

Hirayama T, Ohto C, Mizoguchi T and Shinozaki K (1995) A gene encoding a phosphatidylinositol-specific phospholipase C is induced by dehydration and salt stress in *Arabidopsis thaliana*. Proc Natl Acad Sci USA 92: 3903–3907

Huang CH, Tate BF, Crain RC and Coté GG (1995) Multiple phosphoinositide-specific phospholipases C in oat roots: Characterization and partial purification. Plant J 8: 257–26

Izui K, Yabuta N, Ogawa N, Ueno Y, Furumoto T, Saijo Y, Hata S and Sheen J (1995) Enzymological evidence for the involvement of a calcium-dependent protein kinase in regulatory phosphorylation of PEP carboxylase in maize. In: Mathis P (ed) Photosynthesis: from Light to Biosphere, Vol V, pp 163–166. Kluwer Academic Publishers, Dordrecht

Jiao JA and Chollet R (1989) Regulatory seryl-phosphorylation of C_4 phosphoenolpyruvate carboxylase by a soluble protein kinase from maize leaves. Arch Biochem Biophys 269: 526–535

Kai Y, Matsumura H, Inoue T, Terada K, Nagara Y, Yoshinaga T, Kihara A, Tsumura K and Izui K (1999) Three-dimensional structure of phosphoenolpyruvate carboxylase: A proposed mechanism for allosteric inhibition. Proc Natl Acad Sci USA 96: 823–828

Kopka J, Pical C, Gray JE and Müller-Röber B (1998) Molecular and enzymatic characterization of three phosphoinositide-specific phospholipase C isoforms from potato. Plant Physiol 116: 239–250

Lepiniec L, Vidal J, Chollet R, Gadal P and Cretin C (1994) Phosphoenolpyruvate carboxylase: structure, regulation and evolution. Plant Sci 99: 111–124

Li B and Chollet R (1993) Resolution and identification of C_4 phosphoenolpyruvate-carboxylase protein-kinase polypeptides and their reversible light activation in maize leaves. Arch Biochem Biophys 307: 416–419

Li B and Chollet R (1994) Salt induction and the partial purification/characterization of phosphoenolpyruvate carboxylase protein-serine kinase from an inducible Crassulacean-acid-metabolism (CAM) plant, *Mesembryanthemum crystallinum* L. Arch Biochem Biophys 314, 247–254

Li B, Pacquit V, Jiao J, Duff SMG, Maralihalli GB, Sarath G, Condon SA, Vidal J and Chollet R (1997) Structural requirements for phosphorylation of C_4-leaf phosphoenolpyruvate carboxylase by its highly regulated protein-serine kinase. A comparative study with synthetic-peptide substrates and native, mutant target proteins. Aust J Plant Physiol 24: 443–449

McNaughton GAL, Fewson CA, Wilkins SMB and Nimmo HG (1989) Purification, oligomerisation state and malate sensitivity of maize leaf phosphoenolpyruvate carboxylase. Biochem J 261: 349–355

Melin PM, Pical C, Jergil B and Sommarin M (1992) Phosphoinositide phospholipase C in wheat root plasma membranes. Partial purification and characterisation. Biochim Biophys Acta 1123: 163–169

Muir SR and Sanders D (1997) Inositol 1,4,5-trisphosphate-sensitive Ca^{2+} release across non-vacuolar membranes in cauliflower. Plant Physiol 114: 1511–1521

Munnik T, Irvine RF and Musgrave A (1998) Phospholipid signalling in plants. Biochim Biophys Acta 1389: 222–272

Neet KE and Ainslie GRJr (1980) Hysteretic enzymes: Historical background and development of the hysteretic concept of nonrapid enzyme transitions. In: Purish DL (ed) Methods in Enzymology, Vol 64, pp 192–226. Academic Press Inc

Nhiri M, Bakrim N, Pacquit V, El Hachimi-Messouak Z, Osuna L and Vidal J (1998) Calcium-dependent and -independent phosphoenolpyruvate carboxylase kinases in *Sorghum* leaves: further evidence for the involvement of the calcium-independent proteine kinase in the in situ regulatory phosphorylation of C_4 phosphoenolpyruvate carboxylase. Plant Cell Physiol 39: 241–246

Nimmo GA, Nimmo HG, Fewson CA and Wilkins MB (1984) Diurnal changes in the properties of phosphoenolpyruvate carboxylase in *Bryophyllum* leaves: A possible covalent modification. FEBS Lett 178: 199–203

Ogawa N, Yabuta N, Ueno Y and Izui K (1998) Characterization of a Maize Ca^{2+}-dependent protein kinase phosphorylating phosphoenolpyruvate carboxylase. Plant Cell Physiol 39: 1010–1019

Pierre JN, Pacquit V, Vidal J and Gadal P (1992) Regulatory phosphorylation of phosphoenolpyruvate carboxylase in protoplasts from *Sorghum* mesophyll cell and the role of pH and Ca^{2+} as possible components of the light transduction pathway. Eur J Biochem 210: 531–537

Raghavendra AS, Yin Z and Heber U (1993) Light-dependent pH changes in leaves of C_4 plants. Planta 189: 278–287

Ranjeva R, Carrasco A and Boudet AM (1988) Inositol trisphosphate stimulates the release of calcium from intact vacuoles isolated from *Acer* cells. FEBS Lett 230: 137–141

Roberts DM and Harmon AC (1992) Calcium-modulated proteins: targets of intracellular calcium in higher plants. Annu Rev Plant Physiol Plant Mol Biol 43: 375–414

Schumaker K and Sze H (1987) Inositol 1,4,5-trisphosphate releases Ca^{2+} from vacuolar membrane vesicles of Oat roots. J Biol Chem 262: 3944–3946

Shi J, Gonzales RA and Bhattacharyya MK (1995) Characterization of a plasma membrane-associated phosphoinositide-specific phospholipase C from soybean. Plant J 8: 381–390

Terada K, Kai T, Okuno S, Fujisawa H and Izui K (1990) Maize leaf phosphoenolpyruvate carboxylase: Phosphorylation of Ser15 with a mammalian cyclic AMP-dependent protein kinase diminishes sensitivity to inhibition by L-malate. FEBS Lett 259: 241–244

Van der Veen R, Heimovaara-Dijkstra S and Wang M (1992)

Cytosolic alkalinization mediated by abscisic acid is necessary, but not sufficient, for abscisic acid-induced gene expression in barley aleurone protoplasts. Plant Physiol 100: 699–705
Vidal J and Chollet R (1997) Regulatory phosphorylation of C_4 PEP carboxylase. Trends in Plant Science 2: 230–237
Wang YH and Chollet R (1993) Partial purification and characterization of phosphoenolpyruvate carboxylase protein-serine kinase from illuminated maize leaves. Arch Biochem Biophys 304: 496–502
Willeford KO and Wedding RT (1992) Oligomerization and regulation of higher plant phosphoenolpyruvate carboxylase. Plant Physiol 99: 755–758

Chapter 22

Photodamage and D1 Protein Turnover in Photosystem II

Bertil Andersson*
Division of Cell Biology, Linköping University, SE-581 85 Linköping, Sweden, and Department of Biochemistry and Biophysics, Arrhenius Laboratories for Natural Sciences, Stockholm University, SE-106 91 Stockholm, Sweden

Eva-Mari Aro
Department of Biology, University of Turku, FIN-20014 Turku, Finland

Summary

Photosystem II is frequently undergoing photoinduced damages targeted to its reaction center in a process normally referred to as photoinhibition. Photosynthesis is maintained through an intricate repair mechanism involving degradation of the damaged D1 reaction center protein and insertion of a new protein copy into the photosystem. The photoinhibition process is induced by inoptimal electron transfer at the acceptor or donor side of Photosystem II. Photoinhibition induced from the acceptor side is caused by a stepwise accumulation of stably reduced abnormal Q_A species that lead to formation of chlorophyll triplets and production of singlet oxygen, resulting in oxidative damage to the D1 protein. Photoinhibition induced from the donor side involves

*Author for correspondence, email: bertil.andersson@rek.liu.se

Eva-Mari Aro and Bertil Andersson (eds): Regulation of Photosynthesis, pp. 377–393.

formation of long-lived highly oxidizing $P680^+$ and finally D1 protein damage. A damaged D1 protein is triggered to be degraded via a multistep proteolytic reaction requiring GTP and ATP and catalyzed by chloroplast Deg P2 and FtsH, homologues to known bacterial proteases. Synthesis and assembly of the new D1 copy into PS II is also a multistep process. After targeting of the *psbA* mRNA ribosome complex to the thylakoid membrane, the elongating D1 protein is cotranslationally inserted into the thylakoid membrane and concomitantly assembled with the D2 protein. Both the cotranslational and post-translational assembly steps of the D1 protein into PS II are under strict redox control.

I. Introduction

Oxygenic photosynthesis provides the earth with nearly all the organic material to support present-day life. Oxidation of water, however, is not a trivial process, and plants have acquired unique structures and mechanisms to make them capable of utilizing light energy for water splitting. Nevertheless, light still remains an elusive substrate for plants and water splitting in Photosystem II (PS II) involves a dangerous series of reactions affecting both the functional and structural properties of the complex. It is thus logical that concomitantly with the evolution of the water splitting PS II complex, a number of protecting systems have evolved in the thylakoid membrane to dissipate excess light and to scavenge various radicals and active oxygen species created as side products in this process. Despite these protection mechanisms including the capability of plants to dissipate excess light energy as heat, there is always an inherent inhibition occurring in PS II with a very low quantum yield (Tyystjärvi and Aro, 1996; Anderson et al., 1997). Thus, even at low light irradiances oxidative damages are occurring in PS II and lead to an irreversible damage to the PS II reaction center D1 protein (Keren et al., 1997). At increasing irradiances such damages occur more and more frequently. Based on a grand design of PS II, the visible light-induced oxidative damage is directed mainly towards one specific target protein, the D1 protein in the reaction center of PS II (Mattoo et al., 1984, 1989; Prasil et al., 1992), whereas the other protein components of PS II remain largely unaffected. As long as plants can cope with the light-induced damage to the D1 protein, i.e. as long as they can rapidly replace the damaged D1 protein in the heart of PS II with a new D1 protein copy, no visible or measurable symptoms of photoinhibition of photosynthesis can be detected in vivo. Thus, this constant turnover of the D1 protein in PS II cannot be detected by monitoring chlorophyll fluorescence parameters or by measurements of the steady state oxygen evolution. However, as soon as the capacity for repair of damaged PS II centers is exceeded at increasing irradiances by the more frequently occurring damaging reactions, an irreversible inhibition of PS II can be recorded in vivo as a decrease in light-saturated PS II oxygen evolution capacity or in the chlorophyll fluorescence ratio Fv/Fm (Aro et al., 1993a). Such photoinhibition of PS II is not common in vivo but occurs more often when high irradiances prevail in combination with other stress factors like drought, low or high temperatures, CO_2 limitation or nutrient deprivation, just to mention a few occasions. As will be described by Hihara and Sonoike (Chapter 29), there are also photoinhibition effects targeted to PS I.

In this chapter we will review the turnover of PS II in regard to its light-induced vulnerability and consequent rapid turnover of its D1 protein. We will not participate in the dispute about the term 'photoinhibition' but simply define it here as an event finally leading to an irreversible damage in the D1 reaction center protein. Inactivation and damaging mechanisms, proteolytic degradation of the D1 protein and subsequent de novo synthesis and reassembly with PS II are discussed, including recent achievements in the laboratories of the authors.

Abbreviations: CP43 – 43 kDa chlorophyll *a*-binding protein of PS II; CP47 – 47 kDa chlorophyll *a*-binding protein of PS II; D1 protein – D1 reaction center protein of PS II; D2 protein – D2 reaction center protein of PS II; DBMIB – 2,5-dibromo-3-methyl-6-isopropyl-p-benzoquinone; DCMU – 3-(3,4-dichlorophenyl)-1,1-dimethylurea; DTT – dithiothreitol; EPR – electron paramagnetic resonance; Fm – maximal chlorophyll fluorescence; Fv – variable chlorophyll fluorescence; LHCII – light-harvesting chlorophyll *a/b* protein complex; NEM – N-ethylmaleimide; pD1 – precursor D1 protein; PS – Photosystem; Q_A and Q_B – primary and secondary quinone acceptor of PS II, respectively

II. Light-induced Inactivation and Damage to the Photosystem II Reaction Center

The molecular mechanisms behind photoinactivation of electron transport and turnover of the D1 protein involve several events that often can be described in sequential terms (Aro et al., 1993b; Andersson and Barber, 1996). Conditions that lead to imbalance between the various steps during linear electron transfer can lead to impairment of the photosynthetic process which in turn may lead to accumulation of toxic intermediates causing irreversible oxidative damage to the PS II reaction center, particularly to the D1 protein. Such photodamages are in turn inducing some kind of modification of the targeted protein thereby triggering it to be removed by proteolysis allowing the insertion of a new protein copy in order to restore the functionality of the reaction center. The events that give rise to such a photodamage-repair cycle can be initiated either from the acceptor or donor side of the PS II reaction center.

A. Acceptor Side Mechanism

Early studies on the mechanism of photoinhibition (Kyle et al., 1984; Ohad et al., 1984; Powles, 1984) were performed without the crucial information that the D1 protein was one of the two subunits forming the reaction center heterodimer of PS II (Trebst, 1986; Michel and Deisenhofer, 1988). This protein was at the time called 'the rapidly turning-over protein,' 'the shield protein' or most commonly 'the 32 kDa Q_B-binding protein.' At the time it was therefore reasonable to assume that the damaging mechanism during photoinhibition involved molecular oxygen interacting with a reactive form of plastoquinone bound to the Q_B site. Herbicides, such as 3-(3,4-dichlorophenyl)-1,1-dimethylurea (DCMU) and atrazine, both with a high affinity for this site and able to displace plastoquinone, prevented the rapid turnover of the protein (Kyle et al., 1984; Mattoo et al., 1984; Trebst et al., 1988; Jansen et al., 1993). Furthermore, no turnover of the 32 kDa protein was seen under anaerobic conditions (Arntz and Trebst, 1988).

When it became clear that the D1 protein indeed was a subunit of the reaction center of PS II involved not only in the binding of plastoquinone but also participating in the ligation of most redox-active reaction center components (Trebst, 1986; Michel and Deisenhofer, 1988), the phenomenological and conceptual questions concerning its rapid turnover became a vital issue within photosynthesis research. A new generation of experimental activities taking advantage of this new concept revealed that the site for inactivation of electron transport involved the Q_A site, the damaging reactions were associated with the reaction center chlorophylls while the Q_B site was found to be essential for triggering the damaged D1 protein for multistep proteolytic degradation (Prasil et al., 1992; Aro et al., 1993; Andersson and Barber, 1996).

1. Photoinactivation of Electron Transport

A better understanding of the mechanism for acceptor side induced photoinactivation was achieved when electron paramagnetic resonance (EPR) and fluorescence measurements were applied to isolated thylakoid membranes under anaerobic conditions (Vass et al., 1992). This set of experiments took advantage of the fact that in the absence of oxygen the early stages of photoinactivation are reversible and that D1 protein degradation does not occur (Hundal et al., 1990; Kirilovski and Etienne, 1991). Upon subjecting the thylakoid membranes to strong illumination, four kinetically distinct and sequential intermediates were resolved by fluorescence measurements (Vass et al., 1992). Mainly by using EPR spectroscopy these intermediates could be characterized. (i) The plastoquinone pool becomes fully reduced leaving the Q_B site unoccupied, which in turn makes the plastoquinone tightly bound to the Q_A site and to accumulate in an unusually long-lived singly reduced state (Q_A^-) (ii) This long-lived semiquinone becomes further stabilized via protonation producing $Q_A^- - H^+$. (iii) Since the primary charge separation remains functional such Q_A species receives a second electron, which leads to the formation of abnormal Q_AH_2. (iv) Finally, this double reduced Q_A detaches from its binding site in the reaction center. This fourth intermediate containing an empty Q_A site is not reversible while the first three intermediates are reversible in darkness via restoration of Q_A to Q_B electron transfer. As will be discussed below, this fourth intermediate is not very likely to be abundant under aerobic conditions since the earlier stages will give rise to reactions that will lead to damage and degradation of the D1 protein.

2. Irreversible Photodamage to the D1 Protein

The study described above (Vass et al., 1992) did not only provide evidence for the mechanism of PS II photoinactivation but also connected this inactivation with the photodamage of the reaction center and subsequent degradation of the D1 protein. It was shown by EPR that the slower phases of photo-inactivation (ii-iv) produced a chlorophyll triplet state in the light. The triplet arises from recombination of the primary radical pair $P680^+/Pheo^-$, a reaction promoted by impairment of the Q_A function. Chlorophyll triplets themselves are not toxic but can readily react with molecular oxygen to form singlet oxygen ($^1O_2$),

$$^3Chl + {}^3O_2 \rightarrow {}^1Chl + {}^1O_2$$

When thylakoids photoinactivated under anaerobic conditions were flushed with oxygen the chlorophyll triplet was quenched and D1 proteolysis was initiated. It was therefore concluded that singlet oxygen is the primary damaging species to the PS II reaction center during acceptor side induced photoinhibition (Vass et al., 1992). The production of singlet oxygen under photoinhibitory conditions has later been detected by several methods in isolated thylakoid membranes (Hideg et al., 1994) and various PS II particle preparations (Macpherson et al., 1993; Telfer et al., 1994). In fact, the presence of singlet oxygen has even been detected during photoinhibition in vivo (Hideg et al., 1998).

Normally chlorophyll triplets are quenched by carotenoids (Halliwell and Gutteridge, 1989), as is the case in chlorophyll antenna complexes. The question can therefore be asked why the carotenoids of the PS II reaction center are not efficient enough to protect from oxidative damage. It has been argued (Andersson and Barber, 1996) that PS II has an intrinsic limitation to be an efficient triplet quencher since $P680^+$ has a redox potential which is sufficiently high to oxidize any carotenoid molecule placed close enough.

The toxicity of singlet oxygen lies in the fact that it is highly reactive, able to oxidize nearby pigments, redox components and amino acids. It has been shown that one target for singlet oxygen is the P680 chlorophylls (Telfer et al., 1991). More recent studies involving mass spectroscopy techniques have revealed the presence of oxidized amino acid residues in the PS II reaction center, particularly in the D1 protein but also in the D2 protein (Sharma et al., 1997). The proportion of oxidized amino acids was enhanced in light-treated material. Furthermore, the oxidized amino acids were derived from domains of the D1 protein containing binding sites for redox active cofactors.

Although the majority of experimental evidence points towards singlet oxygen as the main damaging species of PS II, other toxic oxygen and hydroxyl radicals may be formed under certain conditions and contribute to the photodamage of the reaction center. Interestingly, superoxide and hydrogen peroxide formed in PS I has recently been shown to be able to affect also PS II and induce degradation of the D1 protein (Tjus et al., 2001).

The photodamaging events initiated from the acceptor side of PS II outlined above are normally associated with rather high light intensities. However, an acceptor side induced photodamage has been demonstrated to occur at light intensities that limit electron transport (Keren et al., 1997). A conclusive experiment was carried out using single turnover flashes spaced with different dark intervals whereby the occupancy of the Q_B-binding site by semiquinone and quinone can be controlled. When long periods between the flashes were used to permit recombination of the charges between the Q_B/S_2, S_3 states, an enhanced D1 protein degradation was observed. Actually, with a train of flashes it was possible to induce oscillation of the degradation related to the level of Q_B^-/Q_B^{2-} induced by the flash sequence. The efficiency of this low light-induced degradation of the D1 protein, thought to occur via triplet formation and singlet oxygen production, considerably exceeded that seen at high light. These experiments (Keren et al., 1997) may provide explanation for the D1 protein turnover seen under low light conditions in vivo.

B. Donor Side Mechanism

There are compelling experimental evidence, at least in vitro, to suggest that there is also an oxygen independent route for photodamage and degradation of the D1 protein which is initiated from the donor side of PS II (Aro et al., 1993; Andersson and Barber, 1996). This takes place when the rate of electron donation to the PS II reaction center does not keep up with the rate of electron removal to the acceptor side. Under such conditions the lifetime of the $P680^+$ and Tyr_Z^+, the oxidized form of the redox active tyrosine of PS II, will increase. In particular, $P680^+$, which has an oxidizing potential greater than 1.0 V, can extract electrons from its surrounding resulting in an irreversible oxidative damage. It is generally agreed that the donor side inactivation occurs between the manganese cluster and P680 (Blubaugh and Cheniae, 1990; Jegerschöld et al., 1990; Eckert et al., 1991) and that the primary site of damage is between Tyr_Z and P680 (Eckert et al., 1991). The experimental evidence for this type of photoinactivation comes from in vitro experiments where the donor side has been purposely manipulated prior to the photo-inhibitory illumination. Such experiments involve release of the manganese cluster by hydroxylamine (Callahan et al., 1986) or Tris (Wang et al., 1992a), chloride depletion (Critchley et al., 1984; Jegerschöld et al., 1990) or exposure to low temperatures (Wang et al., 1992b). Also, photosynthetic mutants with lesion on the PS II donor side such as the LF-1 mutant of *Scenedesmus obliquus* (Gong and Ohad, 1991) and the PsbO-less mutant of *Synechocystis* 6803 (Mayes et al., 1991; Philbrick et al., 1991) readily undergo a donor side type of photoinactivation.

As in the case of acceptor side photoinhibition the donor side photoinhibition leads to a selective degradation of the D1 protein, which however can occur in the absence of oxygen.

C. UV-Induced Damages

Although UV-B light is only a minor component of sunlight reaching the earth, its effects on the photosynthetic apparatus have long been recognized. As with visible light, one of the primary targets of UV-B-induced damage in plants and cyanobacteria is the PS II complex (for reviews see Bornman, 1987; Vass, 1996; Jansen et al., 1998). The mechanisms, however, by which these two spectral regions of light, visible and UV-B, seem to exert their damaging effects on PS II are likely to be quite different. The most obvious support for this claim comes from experimental consensus indicating that the UV-B-induced inactivation of PS II leads to damage and degradation of both PS II reaction center proteins, the D1 and D2 proteins (Greenberg et al., 1989; Melis et al., 1992; Friso et al., 1994), whereas visible light targets only the D1 protein for degradation. Experiments resolving the mechanisms by which UV-B light inactivates and damages the PS II complex, on the other hand, have induced much more controversial results.

A vast number of experiments have been conducted both in vitro and in vivo, and using both plants and cyanobacteria as experimental material, to resolve the mechanism(s) of UV-B-induced inactivation of PS II. Data obtained from studies with PS II preparations strongly point to a donor side type of inactivation mechanism, most likely at the Mn cluster of the water splitting system and/or at the Tyr_D and Tyr_Z donors (Renger et al., 1989; Vass et al., 1996). Compelling evidence, however, has also accumulated on the detrimental effects of UV-B radiation on the function of the acceptor side of PS II (Jansen et al., 1996, 1998). These apparently controversial results were recently subjected to thorough reinvestigations using cyanobacterial cells as experimental material, and measuring the relaxation of flash-induced variable chlorophyll fluorescence on cells exposed to UV-B radiation (Vass et al., 1999). This method allows simultaneous monitoring of both the acceptor and donor side events in PS II. It was concluded that in intact systems UV-B light impairs electron transfer from the Mn cluster of water oxidation to Tyr_Z^+ and $P680^+$, similar to that earlier reported for isolated systems. UV-B light was, however, observed to induce also acceptor side modifications, particularly in the Q_B pocket of the D1 protein.

UV-A radiation, whose intensity in the natural sunlight is at least ten times higher than that of UV-B, was recently shown to exert damaging effects on PS II with mechanisms similar to those induced by UV-B radiation (Turcsányi and Vass, 2000).

As mentioned above, UV-B radiation induces a degradation of both the D1 and D2 proteins. It is still highly controversial whether the initial degradation of these proteins takes place via direct photocleavage or whether specific proteases are involved. Also, the mechanisms for repair of PS II after UV-B damage, including simultaneous synthesis and assembly of both the D1 and D2 reaction center proteins, are still mostly unresolved.

III. Proteolysis of the Damaged D1 Protein

The photoinactivation and photodamages described above are determined by intrinsic properties of the PS II reaction center and its light-induced chemical reactions. Much of the photodamage is targeted to the D1 protein and restoration of photosynthetic function calls upon replacement of the damaged protein copy. This process requires a proteolytic enzyme system able to discriminate between functional and photodamaged reaction centers. This means that the irreversible photodamage should induce some kind of 'triggering' of the reaction center and its D1 protein making it a substrate for proteolysis. The latter process will be determined by the proteolytic system and the cleavage will occur as a function of the specificity and location of the enzyme catalytic site. This means that the site of damage and primary proteolytic cleavage does not necessarily need to be the same.

A simple but informative experiment addressing the distinction between photodamage and protein degradation was designed by Aro et al. (1990). In this experiment isolated thylakoid membranes were subjected to strong illumination at 2 °C for 40 min followed by transfer to complete darkness and room temperature (Fig. 1). During the first experimental phase there was a pronounced photoinactivation of electron transfer but no degradation of the D1 protein. During the second phase there was no further inactivation of electron transport but notably there was an onset of D1 protein degradation. These results produced the following pieces of information, (i) it is possible to make a distinct experimental discrimination between the photoinactivation and degradation phases, (ii) evidence for a triggering mechanism of the photodamaged D1 protein to permit its subsequent degradation upon the raise of the temperature in darkness, (iii) excludes the concept of a photocleavage of the D1 protein in favor of a proteolytic degradation, (iv) provided an explanation for increased photoinhibitory damage at low temperatures in terms of an inhibited proteolysis obstructing a new D1 protein to be inserted into PS II.

Below we will discuss the triggering for D1 protein degradation and report on progress when it comes to our understanding of the proteolytic process with special emphasis on the enzymes and nucleotides involved.

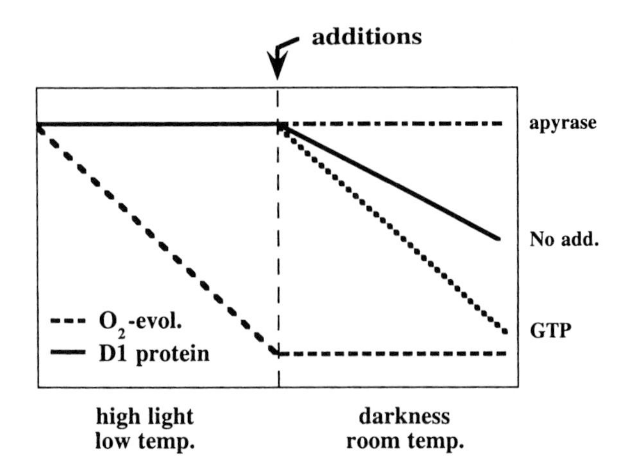

Fig. 1. Schematic illustration of an experiment designed to discriminate between the phase of photoinactivation of electron transport and D1 protein degradation during the photoinhibitory process. Photoinhibitory illumination of thylakoid membranes was performed at low temperatures followed by transfer of the sample to darkness and room temperature. The effects of added GTP and the release of tightly bound GTP (apyrase) on D1 protein degradation are also illustrated.

A. Triggering for Proteolysis

The 'light/cold–dark/warm' experiment (Aro et al., 1990) strongly suggests a triggering mechanism for the damaged D1 protein allowing it to be recognized by the responsible protease. There are several pieces of evidence to suggest that such a triggering involves a conformational change of the D1 protein. This possibility is supported by the well-characterized protective effect of urea and triazine type herbicides on the D1 protein degradation (Kyle et al., 1984; Mattoo et al., 1984; Trebst et al., 1988; Janssen et al., 1993). These herbicides bind tightly in the Q_B site of the D1 protein and may thereby restrict a conformational change triggering the protein for proteolysis. This possibility is supported by the addition of DCMU to thylakoids in the type of photoinhibition experiments outlined above (Virgin et al., 1992). When DCMU is added at the start of the experiment or immediately prior to warming up the sample, the normal protective effect of the herbicide can be seen. In contrast, if the DCMU is given after transferring the sample to a higher temperature it can no longer reduce the extent of D1 protein degradation. A plausible conclusion is that upon increasing the temperature of the photoinactivated thylakoids there is a conformational change around the Q_B site, which lowers its affinity for DCMU, and that the new conformation is essential for making the D1 protein a substrate for proteolysis. A similar conclusion was made by Zer and Ohad (1995) using an in vivo approach on D1 protein degradation in a *Chlamydomonas reinhardtii* mutant. The concept of a conformational triggering around the Q_B site is also supported by experiments using the PS II inhibitor PNO8 (Nakajima et al., 1995). This plastoquinone analogue can induce D1 protein degradation in thylakoid membranes in darkness without any preceding photoinhibitory illumination. It was concluded that PNO8 binds to the Q_B site inducing a conformational change similar to that occurring during the triggering process following photodamage to the D1 protein.

The fact that the D1 protein undergoes transient phosphorylation has suggested some role for this post-translational modification in the triggering process. However, as discussed by Rintamäki and Aro (Chapter 23), the phospho-D1 protein is a poor substrate for degradation and its phosphorylation appears to have more of a regulatory function during the turnover process.

B. The Proteolytic Process

The D1 protein is an integral membrane protein with five transmembrane helices (designated A-E) with its N-terminus and C-terminus exposed at the stromal and lumenal thylakoid surfaces, respectively (Hankamer et al., 1997). It is forming a heterodimer with the homologous D2 protein, which is surrounded by approximately 25 other subunits forming the PS II supercomplex. The proteolytic degradation of such a transmembrane spanning protein in a giant multisubunit complex is apparently not a trivial process and requires, as will be discussed below, a special molecular arrangement.

1. Fragmentation Patterns

In order to understand the complicated proteolytic process much experimental efforts have been made to identify the D1 protein degradation fragments. Certainly, the complete degradation of the D1 protein will produce a large number of small fragments but in particular one has tried to identify the primary fragments in order to understand where the protein initially is cleaved. This primary proteolytic cleavage has been located to the stromal DE-loop that encompasses the Q_B site of the D1 protein (Greenberg at al., 1987). Thus, it was possible in vivo to identify an N-terminal 23.5 kDa fragment. Later a C-terminal 10 kDa fragment has been identified after photo-inhibitory treatment in vivo (Canovas and Barber, 1993). The identification of degradation products in vivo is not easy due to the fact that the primary fragment rapidly will be subjected to secondary proteolytic cleavages. Consequently most data available stems from in vitro photoinhibitory experiments on isolated thylakoids or purified PS II preparations. Typically, an N-terminal 23 kDa fragment and a C-terminal 10 kDa fragment are identified in vitro (Aro et al., 1993; Andersson and Barber, 1996) corroborating the primary cleavage in the stromal DE-loop. However, additional D1 fragments have been detected, in particular in the 16 kDa molecular weight region suggesting an additional cleavage site located in the lumenal loop between helices C and D (Salter et al., 1992a; Andersson and Barber, 1996).

The fragmentation pattern described above is typically seen under conditions of acceptor side induced photoinhibition. Surprisingly, it was shown that if PS II particles were subjected to donor side photoinhibition, the primary D1 degradation fragments were identified as an N-terminal 9 kDa fragment and a C-terminal 24 kDa fragment (Barbato et al., 1991; De Las Rivas et al., 1992; Andersson and Barber, 1996). This observation suggests that the primary cleavage under donor side conditions is not the stromal DE-loop but rather the lumenal loop between helices A and B. It remains to be established whether the different fragmentation patterns associated with the two photoinhibitory mechanisms are due to triggering effects at the substrate level or the enzymatic action of different proteases.

2. Specific Proteases and Their Nucleotide Requirement

The identification of the protease or proteases responsible for the D1 protein degradation following photoinhibitory damage has been a central issue in the research field (Aro et al., 1993; Andersson and Barber, 1996) It was clear from early in vitro experiments that the proteolytic system is associated with the thylakoid membrane. (Ohad et al., 1985; Aro et al., 1990). Several biochemical experiments even suggest that the proteolytic activity is closely associated with the PS II complex since at least a slow degradation could be induced in various isolated PS II particles (Shipton and Barber, 1991; De Las Rivas et al., 1992; Salter et al., 1992b; Aro et al., 1993; Andersson and Barber, 1996). Using diagnostic inhibitors the protease was found to be of the serine type (De Las Rivas et al., 1992,1993). Despite intense efforts for more than a decade to identify and isolate a D1 specific protease, biochemical approaches were not successful. Possibly the responsible enzymes are not very abundant and today we know from the *Arabidopsis thaliana* genome that the chloroplast contains as much as 38 putative proteolytic enzymes (The Arabidopsis Genome Initiative, 2000). The key to the progress on protease identification that has been made in recent years (Lindahl et al., 2000; Haussühl et al., 2001) came from combined genomic and biochemical analysis of bacterial proteases and adopting this knowledge to the chloroplast con-

Fig. 2. Model for a multistep proteolysis of the D1 protein following photoinhibitory damage. The enzyme and nucleotide requirements for the two proteolytic steps are indicated.

sidering the prokaryotic origin of this organelle (Lindahl et al., 2000; Haussühl et al., 2001; Chapter 15, Adam).

Another unsolved question concerned the nucleotide requirement for the degradation of the D1 protein considering that proteolysis of membrane proteins normally is dependent on ATP (Goldberg, 1992). However, photoinhibition experiments performed in vitro have not revealed any stimulation of ATP on the D1 degradation process (Ohad et al., 1985; Aro et al., 1990; Salter et al., 1992b). Surprisingly, it was recently discovered that the primary D1 proteolytic step is specifically requiring GTP while the subsequent proteolytic steps are ATP- dependent (Spetea et al., 1999). Below we will describe these novel pieces of information and present a current picture on the enzyme and nucleotide requirement for the primary and secondary steps during D1 protein degradation (Fig. 2).

The chloroplast contains several enzymes that are homologous to bacterial proteases (Chapter 15, Adam). Recently a homologue of the cyanobacterial protease designated DegP was found in chloroplasts of *Arabidopsis* and pea (Itzhaki et al., 1998). This protease is located in the lumenal space of the thylakoid membrane system. Genomic analysis has revealed the presence in *Arabidopsis* of a highly conserved homologue to this DegP protease which has been designated DegP2 (Haussühl et al., 2001). It is encoded as a precursor protein of 66.8 kDa containing a transit peptide typical for chloroplast targeted proteins. The mature protein is of 60 kDa and has no predicted membrane span but a short hydrophobic domain at its C-terminal region. It contains the typical catalytical triad (serine, histidine and aspartic acid) of serine proteases. Membrane topological experiments suggest the protein to be located at the outer thylakoid surface of the stroma-exposed thylakoid regions.

Overexpression of the DegP2 protein in *Escherichia coli* followed by His-tag affinity purification was performed in order to experimentally test the potential role of DegP2 in D1 protein degradation. When added to photoinhibited thylakoid membranes where the endogenous enzyme had been removed by salt washing, DegP2 could induce a significant degradation of the D1 protein with a concomitant production of an N-terminal 23 kDa fragment. Under the conditions used no other proteins were affected by the purified protease. DegP2 could also degrade the D1 protein in thylakoids subjected to elevated temperatures. On the other hand the overexpressed enzyme itself appeared to be more active at lower temperatures, which is an apparent contradiction to the previous in situ experiments demonstrating only residual D1 protein degradation at low temperatures (Aro et al., 1990). Still, the DegP2 protease appears currently as the main candidate for the elusive primary D1 protease, a view that is supported by its requirement for GTP (see below).

As described above, ATP does not promote light-induced D1 protein degradation. However, if ATP is replaced by GTP there is a pronounced enhancement of the proteolytic activity (Spetea et al., 1999) (Fig. 1). This stimulation is associated with GTP hydrolysis and the degradation can be completely abolished upon addition of non-hydrolysable GTP analogues or by removing tightly membrane-bound nucleotides by apyrase. These results provide evidence that indeed proteolysis of the D1 protein is dependent on nucleotide hydrolysis and that the active nucleotide species is GTP (Fig. 2). Furthermore, these results suggest that earlier demonstrations of D1 protein degradation in vitro must have relied on GTP tightly bound to the isolated thylakoid membranes which in turn points to the existence of thylakoid GTP-binding proteins.

The proteolytic removal of the N-terminal 23 kDa primary fragment was however not requiring GTP but ATP (Fig. 2). The most efficient degradation of this fragment was obtained when the ATP was supplemented with zinc ions. These latter observations made a logical connection to the existence of another bacterial type of protease, FtsH, earlier proven to be located in the stroma-exposed regions of plant thylakoid membranes (Lindahl et al., 1996). FtsH is a 78 kDa integral membrane protein with two predicted membrane-spanning regions. Its C-terminus is comprised of a bulky hydrophilic region exposed to the chloroplast stroma, which contains the catalytical and nucleotide binding sites. Its enzymatic activity is dependent on ATP and Zn^{2+}. Overexpressed and affinity purified *Arabidopsis* FtsH was shown to specifically degrade the accumulated N-terminal 23 kDa fragment both in isolated thylakoids as well as in PS II core complexes (Lindahl et al., 2000). In experiments involving thylakoids the catalytic domain of the endogenous FtsH was first removed by trypsin, a treatment that resulted in the accumulation of the 23 kDa fragment upon illumination. Based upon these experiments, it was concluded (Lindahl et al., 2000) that the secondary proteolysis during D1 protein degradation is catalysed by FtsH in a process requiring ATP and Zn^{2+}.

These novel results suggest that the D1 protein is degraded in a multistep process (Fig. 2), which is not unexpected considering its complicated folding. Probably even more proteases may be involved considering the need for cleavages also in the loops exposed at the inner thylakoid surface. A possible candidate for such a protease would be the lumenal DegP (now termed DegP1) (Itzhaki et al., 1998).

It also remains to be determined if the above mentioned proteolytic enzymes are involved only in response to acceptor side induced photoinactivation or if they also are responsible for D1 protein degradation under donor side conditions. Possibly the different fragment patterns seen under the two photoinhibitory conditions reflect a different triggering of the D1 substrate rather than a difference in enzymology.

3. Events Associated with the Degradation of Photodamaged D1 Protein

Considering the fact that the D1 protein is associated with the PS II supercomplex comprising at least 25 different subunits and involved in the ligation of several redox components of the reaction center, it is not surprising that its high turnover will induce secondary effects. There are several lines of evidence that the triggering and proteolysis of the D1 protein lead to monomerization and partial disassembly of the PS II complex including release of manganese (Prasil et al., 1992; Aro et al., 1993; Hankamer et al., 1997). Such a partial disassembly may be a necessity to facilitate the insertion of a new D1 protein copy but could also impose a difficulty with respect to the stabilization of the PS II complex. Ideally only the D1 protein should be degraded and replaced during the repair of photodamaged PS II centers, however, this is not always the case (Ohad et al., 2000). The D2 protein can show a significant rate and extent of degradation although normally to a lesser degree as compared to the D1 protein. Also the chlorophyll *a*-binding proteins of PS II, CP47 and CP43, can show a limited degradation under prolonged illumination. Furthermore, the nuclear-encoded low molecular weight subunit PsbW shows considerable degradation that can be comparable to that of the D1 protein (Hagman et al., 1997).

Is the degradation of subunits, in addition to the D1 protein, a reflection that these subunits have suffered photooxidative damage and need to be replaced or simply that subunits of a partially disassembled complex becomes destabilized and proned for secondary proteolysis? In the case of CP43 and CP47, the latter possibility appears plausible. When it comes to the D2 protein the situation is not clear particularly since mass spectrometric analysis has revealed the presence of oxidized amino acids in this protein (Sharma et al., 1997). The PsbW protein has been suggested (Shi et al., 2000) to be required for the dimer formation of PS II complexes (Hankamer et al., 1997). Since monomerization of the PS II complex seems to be a part of the photoinhibition repair cycle it could be speculated that the PsbW protein is degraded to allow this event to happen.

IV. Location of Photosystem II Damage and Repair in the Thylakoid Membrane

Functional PS II centers are mainly located in the appressed grana membranes, where also the photoinactivation and primary photodamage to PS II are occurring (Cleland et al., 1986, Mäenpää et al., 1987, Adir et al., 1990). It is, however, well established that the repair of PS II and insertion of the newly synthesized D1 copy into the thylakoid membrane take place in the stroma-exposed thylakoid domains (Mattoo and Edelman, 1987). Moreover, the new D1 copy is directly inserted into existing PS II after depletion of the damaged D1 protein (Adir et al., 1990; Zhang et al., 1999). In which composition, and by which mechanisms the damaged PS II is migrating from the grana to the stroma-exposed membranes for repair, is not completely understood. It is clear, however, that the degradation of the D1 protein is rapidly followed by an insertion of a new D1 copy into PS II. Many lines of evidence suggest that only a rather small and fixed number of PS II centers can undergo repair in stroma-exposed thylakoids at a certain time point (Kettunen et al., 1997). Indeed, it has turned out that under conditions that induce severe inactivation of PS II, most of the damaged PS II centers remain in the grana, with their D1 protein in a phosphorylated state (Chapter 23, Rintamäki and Aro). Phosphorylation of PS II core proteins in the grana appressions is likely to prevent a premature disassembly of photodamaged PS II complexes and probably also to prevent D1 protein degradation under conditions where the synthesis of the new copy is not possible. Only after the migration of the damaged PS II complex to stroma-exposed thylakoid domains and dephosphorylation of the PS II core proteins, processes that both require light, the damaged D1 protein becomes susceptible to proteolytic degradation. PS II in cyanobacteria, which lack both the grana structures and core protein phosphorylation, have their reaction centers constantly exposed to the cytosol. In these organisms the inactivation of PS II and subsequent photodamage to the D1 protein are more directly followed by proteolytic degradation of the D1 protein than in higher plant chloroplasts with spatial segregation of the damaging and repair processes to grana and stroma thylakoids, respectively. The main features of PS II repair in stroma thylakoids are depicted in Fig. 3.

V. Biogenesis and Assembly of the New D1 Copy into Photosystem II

A. PsbA *Transcript Levels and Translation Initiation in Cyanobacteria and Chloroplasts*

The PS II reaction center protein D1 is encoded by the *psbA* gene (Zurawski et al., 1982). Cyanobacteria generally have a small *psbA* gene family with three to four gene copies (Curtis and Haselkorn, 1984; Jansson et al., 1987). In the plastome, however, only a single *psbA* gene encodes the D1 protein, although the number of plastomes per chloroplast may be hundreds. Not only the number of genes but also the regulation of cyanobacterial *psbA* genes differs from that in chloroplasts. In the cyanobacterium *Synechocystis* 6803, the *psbA1* gene is silent but the *psbA2* and *psbA3* genes are up-regulated in light (Mohamed and Jansson, 1989, Tyystjärvi et al. 1996) and the transcripts produced by these genes have a half-life of 10 to 20 min (Mulo et al., 1998; Herranen et al., 2001). During the dark phase, the *psbA* gene transcription is repressed and the transcripts are completely degraded with a half-time significantly longer than that in the light (Mohamed and Jansson, 1991; Herranen et al., 2001). In another well-studied cyanobacterium, *Synechococcus* 7942, the *psbA* genes are differentially transcribed as a response to environmental cues; the *psbAI* gene is active under normal growth conditions whereas the *psbAII* and *psbAIII* genes are up-regulated under stress conditions with concomitant down-regulation of the *psbAI* gene (Golden et al., 1986; Golden, 1994; Campbell et al., 1995; Sippola and Aro, 1999).

In chloroplasts, on the contrary, the *psbA* transcripts are very stable with the half-life ranging between 10 and 40 h (Mullet and Klein, 1987; Klaff and Gruissem, 1991). Although some up-regulation in *psbA* gene transcription as well as in the steady-state level of *psbA* transcripts takes place at high irradiances, the diurnal fluctuations in *psbA* transcripts are only minor as compared to cyanobacteria. Indeed, in cyanobacteria the main regulation of *psbA* gene expression resides at the level of transcription initiation (Golden et al., 1986; Mohamed and Jansson, 1989) while in higher plants the corresponding main regulatory level resides on translation initiation of *psbA* mRNAs. Mayfield and co-workers (see Chapter 8 and references therein, Somanchi and Mayfield) have elucidated a complicated light-driven activation of *psbA* mRNA translation initiation occurring by binding of a nuclear-encoded protein complex to the 5′ untranslated region of the *psbA* mRNA. This allows a proper association of ribosomes and the scanning of *psbA* mRNA to be initiated. Such highly regulated mechanism for translation initiation is apparently missing in cyanobacteria, and in *Synechocystis* 6803 the *psbA* transcripts are probably always attached to ribosomes (Tyystjärvi et al., 2001). Interestingly, in *Synechocystis* 6803 the cellular content of *psbA* transcripts is closely related to the rate of D1 protein turnover in visible light, although the detailed mechanisms for such causal relationship are yet to be elucidated. Moreover, the *psbA3* gene in the same species was demonstrated to be specifically

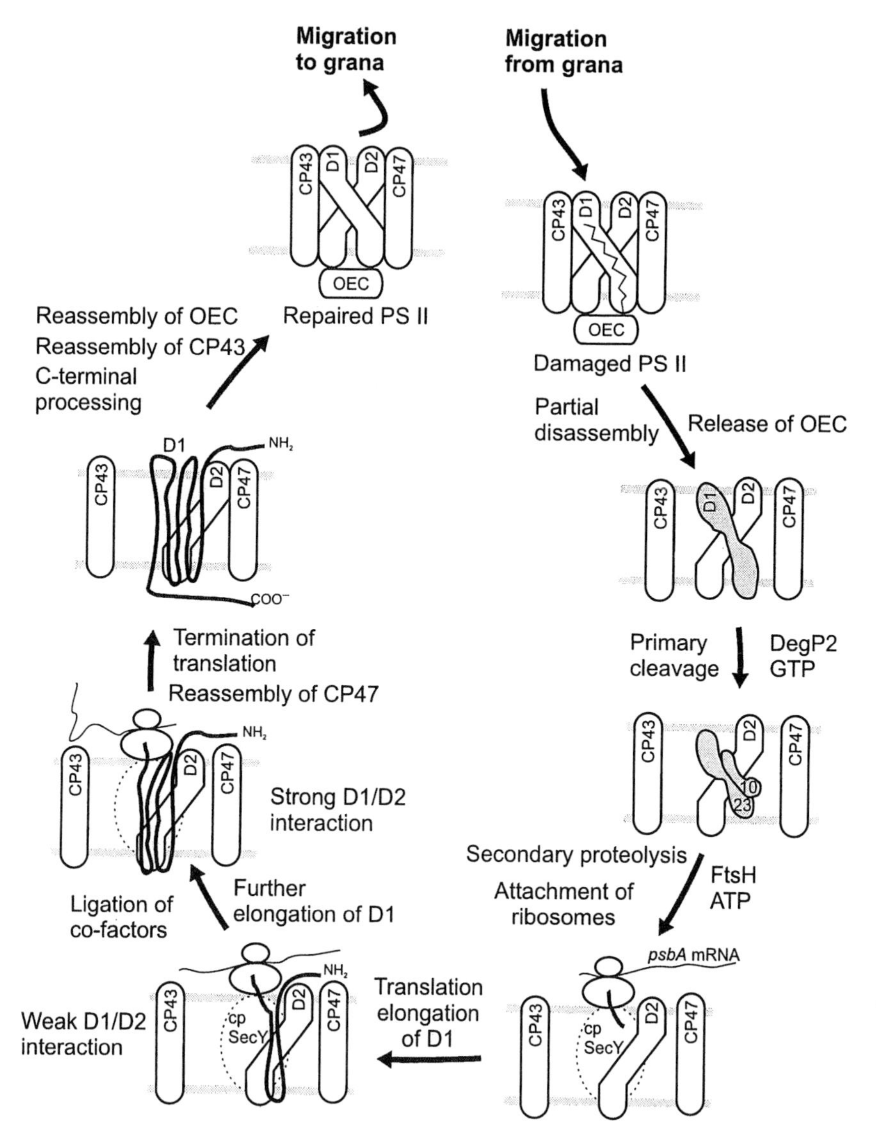

Fig. 3. Repair of PS II centers via D1 protein turnover in stroma-exposed thylakoid membranes. Different phases of D1 protein proteolysis, resynthesis and assembly with existing PS II components are depicted. OEC, oxygen evolving complex.

up-regulated by UV-B light (Màtè et al., 1998).

B. Regulation of psbA *mRNA Translation Elongation*

Although only low light intensities are required to saturate the initiation of *psbA* mRNA translation in chloroplasts, and in cyanobacteria all the *psbA* transcripts seem to be associated with ribosomes irrespective of light conditions, the D1 protein is, however, not synthesized in uncontrolled amounts neither in cyanobacteria nor in the chloroplasts. We have recently shown with *Synechocystis* 6803 that the targeting of the ribosome nascent D1 chain complexes to the thylakoid membrane is probably one important component in the regulation of D1 elongation in cyanobacterial cells (Tyystjärvi et al., 2001). In chloroplasts, the *psbA* mRNA ribosome complexes are efficiently targeted to the thylakoid membrane in light, possibly via binding of cpSRP54 (Nilsson et al., 1999), a chloroplast homologue of the signal recognition particle (SRP) that targets ribosome nascent chain complexes to endoplasmic reticulum for cotranslational secretion in eukaryotic systems. Despite efficient association of the *psbA* mRNA ribosome complexes with the thylakoid membrane in light, they have, however, turned out to be under elongation arrest if the membranes, or rather the PS II complexes, are not in the need for new D1 copies (Kettunen et al., 1997; Zhang et al., 1999). It is conceivable that only after cleavage and at least partial degradation of the damaged D1 protein, the synthesis of a new D1 copy can be accomplished (Fig. 3).

Concomitantly with the elongation process, the D1 protein is incorporated and assembled into the PS II complex (Zhang et al., 1999). How is this occurring? Already ten years ago Mullet and co-workers (Kim et al., 1991) reported an intriguing pausing of ribosomes at specific sites on the *psbA* mRNA. It was postulated that these pausing sites correspond to the occasions when the transmembrane helices are inserted into the thylakoid membrane and/or the cofactors are ligated to the D1 protein. More recently, experimental evidence has accumulated indicating that translation elongation of the D1 protein plays a key regulatory role also in the assembly of the new D1 copy with other PS II components (Zhang et al., 1999, 2000).

In vitro translation in intact chloroplasts and subsequent 'protective' fractionation of the thylakoid membrane has provided an excellent method for studies of the regulation of D1 protein elongation and assembly into PS II (Zhang et al., 1999, 2000). Applying very short pulses of only a couple of minutes to incorporate enough radiolabeled methionine into elongating D1 chains, it has been possible to chase the label into precursor D1 protein via different D1 elongation intermediates and finally to follow the maturation of the D1 protein through C-terminal processing. It has become clear that already during the elongation phase the D1 protein starts interacting with other PS II proteins. Thus, besides translation initiation, the regulation of translation elongation of the *psbA* mRNA has also been under intense research during the past few years.

A crucial regulatory role of translation elongation in the biogenesis of the D1 protein has been recognized for a long time (Kim et al. 1991, 1994; Taniguchi et al., 1993; Kuroda et al., 1996; Edhofer et al., 1998). Labeling studies with intact chloroplasts have made it possible to get new insights into the regulation of D1 translation elongation without interference from complicated regulation of translation initiation (Chapter 8, Somanchi and Mayfield). *PsbA* mRNA translation elongation was shown to be dependent on electron transport, probably by providing the system with essential reducing compounds via PS I electron flow (Kuroda et al., 1996; Zhang et al., 1999), and also by maintaining transmembrane proton gradient (Mühlbauer and Eichacker, 1998, Zhang et al., 2000). The former claim is supported by the ability of either the PS I electron transfer or a reduced thiol reactant dithiotreitol (DTT), to restore D1 elongation under conditions where PS II electron transfer was blocked with DCMU. Dissipation of the proton gradient with nigericin or other uncouplers completely blocks the elongation process (Mühlbauer and Eichaker, 1998; Zhang et al., 2000). These observations are in accordance with several earlier experiments (Aro et al., 1994) indicating that the synthesis of the new D1 copy and repair of PS II centers are mostly prohibited in darkness.

Light, although obligatory for PS II repair, is required only in very low fluence rates to saturate the synthesis of the D1 protein and to allow maximal recovery of plants from photoinhibition. Besides for the initiation and elongation of the *psbA* mRNA translation, light is probably also essential for migration of the damaged PS II complexes from the grana membranes to the stroma-exposed membranes as well as for subsequent dephosphorylation and degradation of the D1 protein (Chapter 23, Rintamäki and Aro).

C. Cotranslational Steps in the Assembly of the New D1 Protein into Photosystem II

Most of the cofactors in PS II are ligated to the reaction center proteins D1 and D2. Ligation of cofactors to the D1 protein during its turnover cycle has been suggested to occur upon pausing of ribosomes at specific sites on the *psbA* mRNA (Kim et al., 1991). This has been considered also to stabilize the nascent D1 chains against proteolytic degradation. Further information about the assembly of the D1 protein into PS II was obtained by searching possible interactions between the ribosome D1 nascent chain complexes and the other PS II core proteins and/or yet unknown proteins possibly involved in D1 protein biogenesis (Zhang et al., 1999, 2000). To this end, the ribosome nascent chain complexes were isolated from intact chloroplasts after a short pulse labeling. Subsequently, an association of labeled nascent D1 chains with other interacting proteins was studied by immunoprecipitation, co-immunoprecipitation and cross-linking experiments. An intriguing discovery (L. Zhang, M. Suorsa, V. Paakkarinen and EM Aro, unpublished) was a close interaction of ribosome D1 nascent chain complexes with cpSecY, a chloroplast homologue of the bacterial translocation channel protein SecY (Roy and Barkan, 1998). This strongly suggests that the cpSecY translocon in thylakoid membranes not only functions in post-translational translocation and insertion of nuclear-encoded proteins (Schnell, 1998; Keegstra and Cline, 1999) but also guides the translocation and insertion of chloroplast-encoded membrane proteins (Fig. 3). Strong interaction of ribosomes with cpSecY suggests that the ribosomes anchor at this translocon channel, which then starts guiding the membrane insertion of the elongating D1 protein.

Protein-protein interactions during D1 protein elongation were not restricted to cpSecY. It was revealed that the 17 kDa D1 intermediate, composed of two transmembrane α-helices, with the third helix still in the ribosome tunnel, is already interacting with the D2 protein (Zhang et al., 1999). This association, however, is rather weak and easily disrupted e.g. by detergents. When four transmembrane helices of D1 have been translated and inserted into the thylakoid membrane, a strong interaction between elongating D1 and the D2 protein was evident. Ligation of most cofactors is probably accomplished at this stage (Kim et al., 1991), but also a formation of a transient disulfide bridge between the nascent D1 chain and the D2 protein is likely to strengthen this interaction (Zhang et al., 2000). Indeed the elongation of the nascent D1 chain and its interaction with the D2 protein are strongly hampered if the thiol groups are chemically reduced by DTT or alkylated by N-ethylmaleimide (NEM) (Zhang et al., 2000).

Further elongation and finally the termination of *psbA* mRNA translation result in a release of a precursor D1 protein (pD1) from ribosomes. This is followed by fast insertion of the fifth transmembrane helix into the membrane with concomitant traversal of the C-terminus with its cleavable extension into the thylakoid lumen.

D. Post-translational Assembly Steps in Photosystem II

C-terminal processing of the precursor D1 protein is conducted by a lumenal protease (Anbudurai et al., 1994) soon after termination of translation. This step in the repair of PS II is particularly temperature-dependent and completely inhibited at low temperatures (Kanervo et al., 1997). As described above, the newly synthesized D1 is cotranslationally associated with a pre-existing D2 protein during the repair of damaged PS II centers. This repair step is followed by an attachment of a pre-existing CP47 protein to the PS II subcomplex (Fig. 3). The possibility of CP47 not dissociating from PS II during the repair process cannot, however, be completely excluded, as the biochemical isolation procedures employed include detergents, which may destabilize the PS II subcomplexes undergoing repair (van Wijk et al., 1996, 1997). Finally, the CP43 protein is attached to the PS II complex. Opposite to the uncertainty in the case of CP47, reversible dissociation and reassociation of CP43 with PS II during the repair process is indisputable. Labeling in intact chloroplasts show that either a nonlabeled and thus pre-existing CP43 or a newly synthesized CP43 can associate with PS II after replacement of the damaged D1 protein (Zhang et al., 2000). Moreover, an antibody against CP47 but not the one against the CP43 protein is capable in co-immunoprecipitation of D1 elongation intermediates. All the PS II assembly steps discussed above require oxidizing conditions in the thylakoid membrane. Moreover, it is likely that the association of CP43 with PS II involves a formation of a disulfide bond (Zhang et al., 2000).

Chase of the newly synthesized D1 protein into PS II assemblies revealed that either a precursor or a mature form of the D1 protein could be present in the assemblies also including D2 and CP47. On the contrary, the CP43 protein was stably assembled only in those PS II complexes where the D1 protein had undergone maturation via C-terminal processing (Zhang et al., 2000). Whether the processing of the C-terminal extension of the pD1 protein is a prerequisite for association of CP43 with PS II or vice versa, the association of CP43 induces proper conformation and rapid processing of the C-terminal extension of the D1 protein, cannot be distinguished by our experimental approaches. Coordination of the processing of the C-terminal extension of the D1 protein and the assembly of CP43 might be important for 'safe' ligation of the Mn cluster and subsequent photoactivation of the oxygen evolving complex. These processes have been shown to be extremely vulnerable to light-inactivation (Rova et al., 1998). Removal of the C-terminal extension has long been known to be a prerequisite for reassociation of the oxygen evolving complex and restoration of the water splitting activity of PS II (Diner et al., 1988; Bowyer et al., 1992).

All the assembly steps discussed above take place on stroma-exposed thylakoid domains where the PS II monomers are assembled. Repair of PS II by such assembly steps is not directly dependent on de novo synthesis of chlorophyll or carotenoid pigments, evidenced by efficient synthesis of D1 and its normal assembly into PS II despite the presence of inhibitors of chlorophyll or carotenoid biosynthesis (Zhang et al., 2000). PS II monomers, properly assembled in stroma thylakoids, then migrate to the grana where peripheral LHCII complexes are attached and where the repaired PS II complexes can be found in their dimer form.

Acknowledgments

Research in the authors laboratories has been supported by the Academy of Finland, EU INCO-Copernicus (IC15-CT98-0126), Nordiskt Kontaktorgan för Jordbruksforskning and the Swedish Natural Research Council. The contribution of Drs Torill Hundal and Cornelia Spetea in preparing this chapter is highly appreciated.

References

Adir N, Shochat S and Ohad I (1990) Light-dependent D1 protein synthesis and translocation is regulated by reaction center II. Reaction center II serves as an acceptor for the D1 precursor. J Biol Chem 265: 12563–12568

Anbudurai PR, Mor TS, Ohad I, Shestakov SV and Pakrasi HB (1994) The *ctpA* gene encodes the C-terminal processing protease for the D1 protein of the Photosystem II reaction center complex. Proc Natl Acad Sci USA 91: 8082–8086

Anderson JM, Park YI and Chow WS (1997) Photoinactivation and photoprotection of Photosystem II in nature. Physiol Plant 100: 214–223

Andersson B and Barber J (1996) Mechanisms of photodamage and protein degradation during photoinhibition of Photosystem II. In: Baker NR (ed) Photosynthesis and the Environment, pp 101–121. Kluwer Academic Publishers, Dordrecht

Arntz B and Trebst A (1986) On the role of the Q_B protein of Photosystem II in photoinhibition. FEBS Lett 194: 43–49

Aro EM, Hundal T, Carlberg I and Andersson B (1990) In vitro studies on light-induced inhibition of Photosystem II and D1-protein degradation at low temperatures. Biochim Biophys Acta 1019: 269–275

Aro EM, McCaffery S and Anderson JM (1993a) Photoinhibition and D1 protein degradation in peas acclimated to different growth irradiances. Plant Physiol 103: 835–843

Aro EM, Virgin I and Andersson B (1993b) Photoinhibition of Photosystem II. Inactivation, protein damage and turnover. Biochim Biophys Acta 1143: 113–134

Aro EM, McCaffery S and Anderson JM (1994) Recovery from photoinhibition in peas (*Pisum sativum L.*) acclimated to varying growth conditions. Plant Physiol 104: 1033–1041

Barbato R, Shipton CA, Giacometti, GM and Barber J (1991) New evidence suggests that the initial photoinduced cleavage of the D1 protein may not occur near the PEST sequence. FEBS Lett 290: 162–166

Blubaugh DJ and Cheniae GM (1990) Kinetics of photoinhibition in hydroxylamine-extracted Photosystem II membranes: Relevance to photoactivation and sites of electron donation. Biochemistry 29: 5109–5118

Bornman JF (1989) Target sites of UV-B radiation in photosynthesis of higher plants. J Photochem Photobiol B4: 145–158

Bowyer JR, Packer JCL, McCormack BA, Whitelegge JP, Bobinson C and Taylor M (1992) Carboxyl-terminal processing of the D1 protein and photoactivation of water splitting in Photosystem II. J Biol Chem 267: 5424–5433

Callahan FE, Becker DW and Cheniae GM (1986) Studies on the photoactivation of the water-oxidizing enzyme. II Characterization of weak light photoinhibition of PS II and its light-induced recovery. Plant Physiol 82: 261–268

Campbell D, Zhou G, Gustafsson P, Öquist G and Clarke AK (1995) Electron transport regulates exchange of two forms of Photosystem II D1 protein in the cyanobacterium *Synechococcus*. EMBO J 14: 5457–5466

Cánovas PM and Barber J (1993) Detection of a 10 kDa breakdown product containing the C-terminal of the D1 protein in photoinhibited wheat leaves suggests an acceptor side mechanism. FEBS Lett 324: 341–344

Cleland RE, Melis A and Neale JP (1986) Mechanism of photoinhibition: Photochemical reaction center inactivation in system II of chloroplasts. Photosynth Res 9: 79–88

Critchley C, Andersson B, Ryrie IJ and Anderson JM (1984) Studies on oxygen evolution of inside-out thylakoid vesicles from mangroves: Chloride requirement, pH dependence and polypeptide composition. Biochim Biophys Acta 767: 532–539

Curtis SE and Haselkorn R (1984) Isolation, sequence and expression of the two members of the 32 kd thylakoid membrane protein gene family from the cyanobacterium *Anabaena* 7120. Plant Mol Biol 3: 249–258

De Las Rivas J, Andersson B and Barber J (1992) Two sites of primary degradation of the D1 protein induced by acceptor or donor side photoinhibition in Photosystem II core complexes. FEBS Lett 30: 246–252

De Las Rivas J, Shipton CA, Ponticos M and Barber J (1993) Acceptor side mechanism of photo-induced proteolysis of the D1 protein in Photosystem II reaction centers. Biochemistry 32: 6944–6950

Diner BA, Ries DF, Cohen BN and Metz JG (1988) COOH-terminal processing of polypeptide D1 of the Photosystem II reaction center of *Scenedesmus obliquus* is necessary for the assembly of oxygen evolving complex. J Biol Chem 263: 8972–8980

Douglas SE (1994) Chloroplast origin and evolution. In: Bryant DA (ed) The Molecular Biology of Cyanobacteria, pp 91–118. Kluwer Academic Publishers, Dordrecht

Eckert H J, Geiken B, Bernarding J, Napiwotski A, Eichler H J and Renger G (1991) Two sites of photoinhibition of the electron transfer in oxygen evolving and Tris-treated PS II membrane fragments from spinach. Photosynth Res 27: 97–108

Edhofer I, Mühlbauer SK and Eichacker LA (1998) Light regulates the rate of translation elongation of chloroplast reaction center protein D1. Eur J Biochem 257: 78–84

Friso G, Spetea C, Giacometti GM, Vass I and Barbato R (1994) Degradation of Photosystem II reaction center D1 protein induced by UV-B radiation in isolated thylakoids. Identification and characterization of C-and N-terminal breakdown products. Biochim Biophys Acta 1184: 78–84

Goldberg AL (1992) The mechanism and functions of ATP-dependent proteases in bacterial and animal cells. Eur J Biochem 203: 9–23

Golden SS (1994) Light-responsive gene expression and the biochemistry of Photosystem II reaction center. In: Bryant DA (ed) The Molecular Biology of Cyanobacteria, pp 693–714. Kluwer Academic Publishers, Dordrecht

Golden SS, Brusslan J and Haselkorn R (1986) Expression of a family of *psbA* genes encoding a Photosystem II polypeptide in the cyanobacterium *Anacystis nidulans* R2. EMBO J 5: 2789–2798

Gong H and Ohad I (1991) The PQ/PQH$_2$ ratio and the occupancy of Photosystem II QB site by plastoquinone control the degradation of D1 protein during photoinhibition in vivo. J Biol Chem 226: 21293–21299

Greenberg BM, Gaba V, Mattoo AK and Edelman M (1987) Identification of a primary in vivo degradation product of the rapidly-turning over 32 kD protein of Photosystem II. EMBO J 6: 2865–2869

Greenberg BM, Gaba V, Canaani O, Malkin S, Mattoo AK and Edelman M (1989) Separate photosensitizers mediate degradation of the 32-kDa Photosystem II reaction center protein in the visible and UV spectral region. Proc Natl Acad Sci USA 86: 6617–6620

Hagman Å, Shi L-X, Rintamäki E, Andersson B and Schröder WP (1997) The nuclear-encoded PsbW protein subunit of Photosystem II undergoes light-induced proteolysis. Biochemistry 36: 2666–2671.

Halliwell B and Gutteridge JM (1989) Free radicals in biology and medicine. Clarendon Press, Oxford

Hankamer B, Boekema EJ and Barber J (1997) Structure and membrane organization of Photosystem II in green plants. Annu Rev Plant Physiol Plant Mol Biol 48: 641–671

Haussühl K, Andersson B and Adamska I (2001) A chloroplast DegP2 protease performs the primary cleavage of the photodamaged D1 protein in plant Photosystem II. EMBO J 20: 1–10

Herranen M, Aro EM and Tyystjärvi T (2001) Two distinct mechanisms regulate the transcription of Photosystem II genes in *Synechocystis* sp. PCC 6803. Physiol Plant, in press

Hideg É, Spetea C and Vass I (1994) Singlet oxygen and free radical production during acceptor-and donor side-induced photoinhibition. Studies with spin trapping EPR spectroscopy. Biochim Biophys Acta 1186: 143–152

Hideg É, Kálai T, Hideg K and Vass I (1998) Photoinhibition of photosynthesis in vivo results in singlet oxygen production. Detection via nitroxide-induced fluorescence quenching in broad bean leaves. Biochemistry 37: 11405–11411

Hundal T, Aro EM, Carlberg I and Andersson B (1990) Restoration of light-induced Photosystem II inhibition without de novo protein synthesis. FEBS Lett 267: 203–206

Itzhaki H, Naveh L, Lindahl M, Cook M and Adam Z (1998) Identification and characterization of DegP, a serine protease associated with the lumenal side of the thylakoid membrane. J Biol Chem 273: 7094–7098

Jansen MAK, Depka B, Trebst A and Edelman, M (1993) Engagement of specific sites in plastoquinone niche regulates degradation of D1 protein in Photosystem II. J Biol Chem 268: 21246–21252

Jansen MAK, Gaba V, Greenberg BM, Mattoo A and Edelman M (1996) Low threshold levels of ultraviolet-B in a background of photosynthetically active radiation trigger rapid degradation of the D2 protein of photosystem-II. Plant J 9: 693–699

Jansen MAK, Gaba V and Greenberg BM (1998) Higher plants and UV-B radiation: Balancing damage, repair and acclimation. Trends Plant Sci 3: 131–135

Jansson C, Debus RJ, Osiewacz HD, Gurevitz M and McIntosh L (1987) Construction of an obligate photoheterotrophic mutant of the cyanobacterium *Synechocystis* 6803. Inactivation of the psbA family. Plant Physiol 85: 1021–1025

Jegerschöld C, Virgin I and Styring S (1990) Light-dependent degradation of the D1 protein in Photosystem II is accelerated after inhibition of the water splitting reaction. Biochemistry 29: 6179–6186

Kanervo E, Murata N and Aro EM (1997) Membrane lipid unsaturation modulates processing of the Photosystem II reaction-center protein D1 at low temperatures. Plant Physiol 114: 841–849

Keegstra K and Cline K (1999) Protein import and routing system of chloroplasts. Plant Cell 11: 557–570

Keren N, Berg A, van Kan PJM, Levanon H and Ohad I (1997) Mechanism of Photosystem II inactivation and D1 protein degradation at low light: The role of back electron flow. Proc Natl Acad Sci USA 94: 1579–1584

Kettunen R, Pursiheimo S, Rintamäki E, van Wijk KJ and Aro EM (1997) Transcriptional and translational adjustments of *psbA* gene expression in mature chloroplasts during photoinhibition and subsequent repair of Photosystem II. Eur J Biochem 247: 441–448

Kim J, Klein PG and Mullet JE (1991) Ribosomes pause at specific sites during synthesis of membrane-bound chloroplast reaction center protein D1. J Biol Chem 266: 14931–14938

Kim J, Eichaker LA, Rudiger W and Mullet JE (1994) Chlorophyll regulates accumulation of the plastid-encoded chlorophyll-proteins P700 and D1 by increasing apoprotein stability. Plant Physiol 104: 907–916

Kirilovsky, D and Etienne, A L (1991) Protection of reaction center II from photodamage by low temperature and anaerobiosis in spinach chloroplasts. FEBS Lett 279: 201–204

Klaff P and Gruissem W (1991) Changes in chloroplast mRNA stability during leaf development. Plant Cell 3: 517–529

Kulkarni RD and Golden SS (1994) Adaptation to high light intensity in *Synechococcus* sp. Strain PCC 7942: regulation of three *psbA* genes and two forms of the D1 protein. J Bacteriol 176: 959–965

Kuroda M, Kobashi K, Kaseyama H and Satoh K (1996) Possible involvement of a low redox potential component(s) downstream of Photosystem I in the translational regulation of the D1 subunit of the Photosystem II reaction center in isolated pea chloroplasts. Plant Cell Physiol 37: 754–761

Kyle DJ, Ohad I and Arntzen CJ (1984) Membrane protein

damage and repair; selective loss of a quinone protein function in chloroplast membranes. Proc Natl Acad Sci USA 81: 4070–4074

Lindahl, M., Tabak, S., Cseke, L., Pickersky, E., Andersson, B. and Adam, Z. (1996) Identification, characterization and molecular cloning of a homologue to the bacterial FtsH protease in chloroplasts of higher plants. J Biol Chem 271: 29329–29334.

Lindahl M, Spetea C, Hundal T, Oppenheim AB, Adam Z and Andersson B (2000) The thylakoid FtsH plays a role in the light-induced turn-over of the Photosystem II D1 protein. Plant Cell 12: 419–432

Macpherson AN, Telfer A, Barber J and Truscott TG (1993) Direct detection of singlet oxygen from isolated PS II reaction centers. Biochim Biophys Acta 1143: 301–309

Mátè Z, Sass L, Szekeres M, Vass I and Nagy F (1998) UV-B-induced differential transcription of *psbA* genes encoding the D1 protein of Photosystem II in the cyanobacterium *Synechocystis* 6803. J Biol Chem 273: 17439–17444

Mattoo AK and Edelman M (1987) Intramembrane translocation and posttranslational palmitoylation of the chloroplast 32-kDa herbicide-binding protein. Proc Natl Acad Sci USA 84: 1497–1501

Mattoo AK, Hoffman-Falk H, Marder JB and Edelman M (1984) Regulation of protein metabolism: coupling of photosynthetic electron transport to in vivo degradation of the rapidly metabolized 32-kilodalton protein of the chloroplast membranes. Proc Natl Acad Sci USA 81: 1380–1384

Mattoo AK, Marder JB and Edelman M (1989) Dynamics of the Photosystem II reaction center. Cell 56: 241–246

Mayes SR, Cook KM, Self SJ, Zhang Z-H and Barber J (1991) Deletion of the gene encoding the Photosystem II 33 kDa protein from *Synechocystis* sp PCC 6803 does not inactivate water splitting but increases vulnerability to photoinhibition. Biochem Biophys Acta 1060: 1–12

Melis A, Nemson JA and Harrison MA (1992) Damage to functional components and partial degradation of Photosystem II reaction center proteins upon exposure to ultraviolet-B radiation. Biochim Biophys Acta 1100: 312–320

Michel H and Deisenhofer J (1988) Relevance of the photosynthetic reaction center of purple bacteria to the structure of Photosystem II. Biochemistry 27: 1–7

Mohamed A and Jansson C (1989) Influence of light on accumulation of photosynthesis-related transcripts in the cyanobacterium *Synechocystis* 6803. Plant Mol Biol 13: 693–700

Mohamed A and Jansson C (1991) Photosynthetic electron transport controls degradation but not production of *psbA* transcripts in the cyanobacterium *Synechocystis* 6803. Plant Mol Biol 16: 891–897

Mohamed A, Eriksson J, Osiewacz HD and Jansson C (1993) Differential expression of the *psbA* genes in the cyanobacterium *Synechocystis* 6803. Mol Gen Genet 238: 161–168

Mori H, Summer EJ, Ma X and Cline K (1999) Component specificity for the thylakoid Sec and ΔpH-dependent protein transport pathways. J Biol Chem 146: 34–56

Mühlbauer SK and Eichaker LA (1998) Light-dependent formation of the photosynthetic proton gradient regulates translation elongation in chloroplasts. J Biol Chem 273: 20935–20940

Mullet JE and Klein RR (1987) Transcription and mRNA stability are important determinants of higher plants chloroplast RNA levels. EMBO J 6: 1571–1579

Mulo P, Eloranta T, Aro EM and Mäenpää P (1998) Disruption of a spe-like open reading frame alters polyamine content and *psbA2* mRNA stability in the cyanobacterium *Synechocystis* sp. PCC 6803. Bot Acta 111: 71–76

Mäenpää P, Andersson B and Sundby C (1987) Difference in sensitivity to photoinhibition between Photosystem II in the appressed and non-appressed thylakoid region. FEBS Lett 215: 31–36

Nakajima Y, Yoshida S, Inoue Y, Yoneyama K and Ono T (1995) Selective and specific degradation of the D1 protein induced by binding of novel Photosystem II inhibitor to the Q_B site. Biochim Biophys Acta 1230: 38–44

Nilsson R, Brunner J, Hoffman NE and van Wijk KJ (1999) Interactions of ribosome nascent chain complexes of the chloroplast-encoded D1 thylakoid membrane protein with cpSRP54. EMBO J 18: 733–742

Ohad I, Kyle DJ and Arntzen CJ (1984) Membrane protein damage and repair, removal and replacement of inactivated 32kDa polypeptide in chloroplast membranes. J Cell Biol 99: 481–485

Ohad I, Kyle DJ and Hirschberg J (1985) Light-dependent degradation of the Q_B protein in isolated pea thylakoids. EMBO J 4: 1655–1659

Ohad I, Sonoike K and Andersson B (2000) Photoinactivation of the two photosystems in oxygenic photosynthesis: Mechanisms and regulation. In: Yunus M, Pathre U and Mohanty P (eds) Probing Photosynthesis, pp 293–309. Taylor and Francis, London

Philbrick JB, Diner BA and Zilinskas BA (1991) Construction and characterization of cyanobacterial mutants lacking the manganese-stabilising polypeptide of Photosystem II. J Biol Chem 266: 13370–13376

Powles SB (1984) Photoinhibition of photosynthesis induced by visible light. Annu Rev Plant Physiol 35: 15–44

Prasil O, Adir N and Ohad I (1992) Dynamics of Photosystem II: Mechanisms of photoinhibition and recovery process. In: Barber J (ed) The Photosystems: Structure, Function and Molecular Biology, Vol 11, pp 295–348. Elsevier Science Publishers, Amsterdam.

Renger GM, Völker HJ, Eckert R, Fromme S, Hohm-Veit S and Graber P (1989) On the mechanism of Photosystem II deterioration by UV-B-irradiation. Photochem Photobiol 49: 97–105

Rova M, Mamedov F, Magnuson A, Fredriksson PO and Styring S (1998) Coupled activation of the donor and the acceptor side of Photosystem II during photoactivation of the oxygen evolving cluster. Biochemistry 37: 11039–11045

Roy M and Barkan A (1998) A SecY homologue is required for the elaboration of the chloroplast thylakoid membrane and for normal chloroplast gene expression. J Biol Chem 141: 385–395

Salter AH, De Las Rivas J, Barber J and Andersson B (1992a) On the molecular mechanisms of light-induced D1-protein degradation. In: Murata N (ed) Research in Photosynthesis, Vol IV, pp 395–402. Kluwer Academic Publishers, Dordrecht

Salter AH, Virgin I, Hagman Å and Andersson B (1992b) On the molecular mechanism of light-induced D1 protein degradation in Photosystem II core particles. Biochemistry 31: 3990–3998

Schnell DJ (1998) Protein targeting to the thylakoid membrane

(1998) Annu Rev Plant Physiol Plant Mol Biol 49: 97–106
Sharma J, Panico M, Shipton CA, Nilsson F, Morris HR and Barber J (1997) Primary structure characterization of the Photosystem II D1 and D2 subunits. J Biol Chem 272: 33158–33166
Shi L-X, Lorkovic ZJ, Oelmüller R and Schröder WP (2000) The low molecular mass PsbW protein is involved in the stabilization of the dimeric Photosystem II complex in *Arabidopsis thaliana*. J Biol Chem 275: 37945–37950
Shipton CA and Barber J (1991) Photoinduced degradation of the D1 polypeptide in isolated reaction centers of Photosystem II: Evidence for an autoproteolytic process triggered by the oxidizing side. Proc Natl Acad Sci USA 88: 6691–6695
Sippola K and Aro EM (1999) Thiol redox state regulates expression of *psbA* genes in *Synechococcus* sp. PCC 7942. Plant Mol Biol 41: 425–433
Spetea, C., Hundal, T., Lohmann, F. and Andersson, B. (1999) GTP bound to chloroplast thylakoid membranes is required for light-induced multi-enzyme degradation of the Photosystem II D1 protein. Proc Natl Acad Sci USA 96: 6547–6552.
Taniguchi F, Yamamoto Y and Satoh K (1995) Recognition of the structure around the site of cleavage by the carboxyl-terminal processing protease for D1 precursor protein of the Photosystem II reaction center. J Biol Chem 270: 10711–10716
Taniguchi M, Kuroda H and Satoh K (1993) ATP-dependent protein synthesis in isolated pea chloroplasts. FEBS Lett 317: 57–61
The Arabidopsis Genome Initiative (2000) Analysis of the genome sequence of the flowering plant *Arabidopsis thaliana*. Nature 408: 796–815
Telfer A, De Las Rivas J and Barber J (1991) Beta-carotene within the isolated Photosystem II reaction center; photo-oxidation and irreversible bleaching of this chromophore by oxidized P680. Biochim Biophys Acta 1060: 106–114
Telfer A, Bishop S M, Phillips D and Barber J (1994) The isolated photosynthetic reaction of PS II as a sensitiser for the formation of singlet oxygen; detection and quantum yield determination using a chemical trapping technique. J Biol Chem 269: 13244–13253
Tjus SE, Scheller HV, Andersson B and Lindberg-Møller B (2001) Active oxygen produced during selective excitation of Photosystem I is damaging not only to Photosystem I, but also to Photosystem II. Plant Physiol 125: 2007–2015
Trebst A (1986) The topology of the plastoquinone and herbicide binding peptides of Photosystem II in the thylakoid membrane. Z Naturforsch 41c: 240–245
Trebst A, Depka B, Kraft B and Johanningmeier U (1988) The Q_B site modulates the conformation of the Photosystem II reaction center polypeptide. Photosynth Res 18: 163–177
Turcsányi E and Vass I (2000) Inhibition of photosynthetic electron transport by UV-A radiation targets the Photosystem II complex. Photochem Photobiol 72: 513–520
Tyystjärvi E and Aro EM (1996) The rate constant of photoinhibition, measured in lincomycin-treated leaves, is directly proportional to light intensity. Proc Natl Acad Sci USA 93: 2213–2218
Tyystjärvi T, Mulo P, Mäenpää P and Aro EM (1996) D1 polypeptide degradation may regulate *psbA* gene expression at transcriptional and translational levels in *Synechocystis* sp. PCC 6803. Photosynth Res 47: 111–120
Tyystjärvi T, Herranen M, Aro EM (2001) regulation of translation elongation in cyanobacteria: Membrane targeting of the ribosome nascent-chain complexes controls the synthesis of D1 protein. Mol Microbiol 40: 476–484
van Wijk KJ, Andersson B and Aro EM (1996) Kinetic resolution of the incorporation of the D1 protein into Photosystem II and localization of assembly intermediates in thylakoid membranes of spinach chloroplasts. J Biol Chem 271: 9627–9636
van Wijk KJ, Roobol-Boza M, Kettunen R, Andersson B and Aro EM (1997) Synthesis and assembly of the D1 protein into Photosystem II: Processing of C-terminus and identification of the initial assembly partners and complexes during Photosystem II repair. Biochemistry 36: 6178–6186
Vass I (1996) Adverse effects of UV-B light on the structure and function of the photosynthetic apparatus. In: Pessarakli M (ed) Handbook of Photosynthesis, pp 931–950. Marcel Dekker Inc., New York
Vass I, Styring S, Hundal T, Koivuniemi A, Aro EM and Andersson B (1992) Reversible and irreversible intermediates during photoinhibition of Photosystem II—stable reduced Q_A species promote chlorophyll triplet formation. Proc Natl Acad Sci USA 89: 1408–1412
Vass I, Sass L, Spetea C, Bakou A, Ghanotakis F and Petrouleas V (1996) UV-B-induced inhibition of Photosystem II electron transport studied by EPR and chlorophyll fluorescence. Impairment of donor and acceptor side components. Biochemistry 35: 8964–8973
Vass I, Kirilovsky D and Etienne AL (1999) UV-B radiation-induced donor- and acceptor side modifications of Photosystem II in the cyanobacterium *Synechocystis* sp. PCC 6803. Biochemistry 38: 12786–12794
Vink M, Zer H, Herrmann RG, Andersson B and Ohad I (2000) Regulation of Photosystem II core proteins phosphorylation at the substrate level: Light induces exposure of the CP43 chlorophyll *a* protein complex to thylakoid protein kinase(s). Photosynth Res 64: 209–219
Virgin I, Salter AH, Hagman Å, Vass I, Styring S and Andersson B (1992) Molecular mechanisms behind light-induced inhibition of Photosystem II electron transport and degradation of reaction centre polypeptides. Biochim Biophys Acta 1101: 139–142
Wang WQ, Chapman D and Barber J (1992) Inhibition of water splitting increases the susceptibility of photosystem-II to photoinhibition. Plant Physiol 99: 16–20
Zhang LX, Paakkarinen V, van Wijk KJ and Aro EM (1999) Cotranslational assembly of the D1 protein into Photosystem II. J Biol Chem 274: 16062–16067
Zhang L, Paakkarinen V, van Wijk KJ and Aro EM (2000) Biogenesis of the chloroplast-encoded D1 protein: Regulation of translational elongation, insertion, and assembly into Photosystem II. Plant Cell 12: 1769–1781
Zurawski G, Bohnert HJ, Whitfeld PR and Bottomley W (1982) Nucleotide sequence of the gene for M_r 32,000 thylakoid membrane protein from *Spinacia oleracea* and *Nicotiana debneyi* predicts a totally conserved primary translation product of M_r 38,950. Proc Natl Acad Sci USA 79: 7699–7703

Chapter23

Phosphorylation of Photosystem II Proteins

Eevi Rintamäki* and Eva-Mari Aro
Department of Biology, University of Turku, FIN-20014 Turku, Finland

Summary

Light induces phosphorylation of a number of Photosystem II-related proteins in the thylakoid membrane. Four proteins of Photosystem II (PS II) complex, the D1 and D2 reaction center proteins, the 43-kDa chlorophyll *a*-binding protein and the *psbH* gene product are reversibly phosphorylated. Three proteins of the PS II antenna, Lhcb1 and Lhcb2 (designated as LHCII), as well as Lhcb4 also undergo light-dependent phosphorylation. Several studies support the existence of two distinct kinases in the thylakoid membrane, one for phosphorylation of the PS II core proteins and another for the LHCII proteins. Reduction of plastoquinone activates phosphorylation of the PS II core proteins, while both the reduction of a plastoquinone and binding of plastoquinol to the cytochrome b_6f complex are required to induce LHCII phosphorylation. Moreover, LHCII phosphorylation is down-regulated under high-light conditions in vivo, and this inactivation is likely to be mediated via the

* Author for correspondence, email: evirin@utu.fi

Eva-Mari Aro and Bertil Andersson (eds): Regulation of Photosynthesis, pp. 395–418.

ferredoxin-thioredoxin system in chloroplasts. Both light-dependent and light-independent dephosphorylation of the PS II proteins have been reported. Physiological implications of thylakoid protein phosphorylation include (i) the control of the location and timing of the proteolytic degradation of photodamaged D1 protein by phosphorylation of the PS II core proteins, (ii) the induction of state transitions by reversible LHCII phosphorylation, and (iii) a putative role of the PS II and Lhcb phosphoproteins, or the corresponding kinases and phosphatases, in the relay of signals within chloroplast and from chloroplast to the nucleus to initiate the acclimatization of plants to prevailing environmental conditions.

I. Introduction

More than forty years ago it was recognized that cellular proteins can undergo enzymic reversible phosphorylation. Such covalent modifications of proteins were first demonstrated to control enzyme activities (Krebst, 1994). Over the years, however, the phosphoproteins and their protein kinases and phosphatases, which themselves can be regulated by reversible phosphorylation, have been shown to participate in diverse cellular functions, with growing body of interest being currently paid on their key role in the signal transduction pathways.

Chloroplast phosphoproteins were first found in thylakoid membranes by Bennett (1977, 1980). Later, reversible phosphorylation of proteins has been shown to occur also in the soluble stroma and envelope membranes (Foyer, 1985; Bhalla and Bennett, 1987; Soll and Bennett, 1988; Soll et al., 1988; Tiller and Link, 1993; Danon and Mayfield, 1994). Phosphorylation of thylakoid proteins is regulated by light via the redox state of electron transfer components in the thylakoid membrane (Bennett, 1977; Allen and Bennett, 1981; Allen et al., 1981). Of the 13 thylakoid phosphoproteins labeled by ^{32}P-ATP, 11 proteins are phosphorylated under reducing conditions and two under oxidizing conditions (Silverstein et al., 1993a).

Abbreviations: CP43 protein – 43-kDa chlorophyll *a*-binding protein of PS II; D1 protein – D1 reaction center protein of PS II; D2 protein – D2 reaction center protein of PS II; DBMIB – *2,5*-dibromo-3-methyl-6-isopropyl-*p*-benzoquinone; DCMU – 3-(3′4-dichlorophenyl)-1,1-dimethylurea; DTT – dithiothreitol; *Lhcb* gene – gene encoding Lhcb protein; Lhcb – light-harvesting chlorophyll *a/b* proteins associated with PS II; LHCII – Lhcb1 and Lhcb2 light-harvesting chlorophyll *a/b* proteins; P-proteins – phosphorylated form of thylakoid proteins; PFD – photon flux density; PS – Photosystem; *psaAB* genes – genes encoding A and B subunits of Photosystem I; *psbA* gene – gene encoding D1 protein; PsbH protein – *psbH* gene product of PS II; Q_A and Q_B – primary and secondary quinone acceptor of PS II, respectively; Q_o – quinol oxidation site of the cytochrome b_6f complex; SDS-PAGE – sodium dodecyl sulfate polyacrylamide gel electrophoresis; Ser – serine residue; TAKs – thylakoid-associated kinases; Thr – threonine residue

The identity of the redox sensor has been under extensive study, plastoquinone and cytochrome b_6f complex being the most promising candidates to regulate protein phosphorylation in the thylakoid membrane (Bennett, 1991; Allen, 1992; Gal et al., 1997). Dephosphorylation of thylakoid phosphoproteins, on the other hand, occurs in darkness (Bennett, 1980), but both light-dependent and light-stimulated dephosphorylation of thylakoid proteins have also been reported (Elich et al., 1993; Ebbert and Godde, 1994; Rintamäki et al., 1996a). So far, only membrane-bound kinases have been reported to phosphorylate thylakoid proteins, while both soluble and membrane-bound protein phosphatases are active in the dephosphorylation processes.

Thorough reviews have been periodically published dealing with the identity of the kinases and phosphatases, regulation mechanisms and the physiological implications of thylakoid protein phosphorylation (Bennett, 1991; Allen, 1992, 1995; Gal et al., 1997; Vener et al., 1998). In this chapter we will focus on the taxonomic diversity, regulation and physiological significance of the phosphoproteins associated with Photosystem II (PS II) and its light-harvesting chlorophyll *a/b* protein complexes. Special attention will be paid to processes occurring in vivo.

II. Thylakoid Phosphoproteins

A. Characterized Thylakoid Phosphoproteins

Thylakoid proteins that are most intensively labeled with ^{32}P, belong to the PS II and its antenna complexes. Four proteins of the multisubunit PS II complex of higher plants, the reaction center proteins D1 (D1 protein) and D2 (D2 protein), 43-kDa chlorophyll *a*-binding protein (CP43 protein) and the *psbH* gene product (PsbH protein) undergo reversible phosphorylation (Ikeuchi et al., 1987; Michel and Bennett, 1987; Michel et al., 1988).

The PS II antenna consists of six different

polypeptides from which three have been found to become reversibly phosphorylated. The antenna polypeptides are denoted Lhcb1 through Lhcb6, and they are encoded by nuclear genes *Lhcb1* to *Lhcb6* (for nomenclature see Jansson et al., 1992). The most abundant antenna proteins, Lhcb1 and Lhcb2, generally denoted LHCII, accumulate the highest amounts of ^{32}P-label in the thylakoid membrane (Bennett, 1991). These proteins form, together with Lhcb3, the major trimeric light-harvesting complexes (Jansson, 1994; Simpson and Knoetzel, 1996; Boekema et al., 1999), which can be divided into two subclasses. The inner trimeric complexes are tightly associated with PS II, while the outer trimeric complexes can reversibly detach from the PS II antenna upon phosphorylation of the Lhcb1 and Lhcb2 proteins (Larsson et al., 1987b). Lhcb4 (also named CP29), which belongs to the minor chlorophyll *a/b* antenna proteins located in the close vicinity of the PS II core (Jansson, 1994; Bassi et al., 1997), is also phosphorylated in thylakoid membranes (Hayden et al., 1988; Bergantino et al., 1995).

Moreover, there is a 12-kDa protein in the thylakoid membrane that is moderately labeled in the presence of ^{32}P (Bennett, 1991). It has been reported to be released from the membrane with a mild detergent treatment indicating that it is not an integral thylakoid protein (Bhalla and Bennett, 1987; Lindahl et al., 1995b). However, the identity and function of this protein are still unknown. New phosphoproteins are still found in the thylakoid membrane. Thylakoid associated kinases (TAKs) itself were recently reported to undergo phosphorylation (Snyders and Kohorn, 1999). A most recently identified thylakoid phosphoprotein is a novel 15,2-kDa polypeptide of cytochrome b_6f complex, also called subunit V (Hamel et al., 2000).

B. Phosphorylation Site of Photosystem II and Lhcb Phosphoproteins

All the PS II and Lhcb phosphoproteins identified so far are phosphorylated on the threonine residue (Thr) at the N-terminus of the protein locating at the stromal surface of the thylakoid membrane (Mullet, 1983; Michel and Bennett, 1987, 1989; Michel et al., 1988; Testi et al., 1996). The N-terminus of the D1 and D2 proteins undergoes three types of post-translational modifications (Michel et al., 1988). The initiating N-formylmethionine residue is first removed followed by α-N-acetylation of the second amino acid residue, Thr-2. The acetylated Thr residue is then prone to reversible phosphorylation.

The initiating N-formylmethionine is also removed from the N-terminus of the PsbH protein (Michel and Bennett, 1987). Contrary to the D1 and D2 proteins, the N-terminus of PsbH can be sequenced by Edman degradation, indicating that it is not acetylated (Michel and Bennett, 1987). The first Thr residue in the sequence, Thr-3, undergoes reversible phosphorylation (Michel and Bennett, 1987). Recently, however, double phosphorylation of Thr-3 and Thr-5 has been reported in thylakoid membranes of *Arabidopsis* (Vener et al., 2001).

Also CP43 undergoes three types of post-translational modifications. In addition to the removal of the first N-formylmethionine, the next 13 amino acids are removed followed by α- N-acetylation of the Thr-15 residue, which then can undergo reversible phosphorylation (Michel et al., 1988). Instead of the Thr-15, the Thr-3 residue has been occasionally pointed as a phosphorylation site of CP43 in the literature. However, the Thr-3 is not a conserved residue in higher plants: it is missing from many monocot species and also from a liverwort *Marchantia* (Fig. 1). Moreover, alternative translation start codons for the gene encoding CP43 protein (Bricker, 1990) results in a variation in the length of the N-terminus. For example, the N-terminus of the CP43 protein in *Chlamydomonas reinhardtii*, *Chlamydomonas eugametos* (Fig. 1) and *Synechococcus PCC7942* (Golden and Stearns, 1988) is 12 amino acid residues shorter, and that of the red alga *Porphyra purpurea* is 14 amino acids longer than the N-terminus of CP43 in higher plants (Fig. 1). If this variation in the length of the CP43 N-terminus is taken into consideration, the phosphorylation site (corresponding to the Thr-15 residue in higher plant sequence) and its flanking region are very conserved in oxygenic photosynthetic organisms (Fig. 1).

The phosphorylation site of the Lhcb1 protein is also close to the N-terminus (Mullet, 1983), but the exact residue is not as well defined as for the PS II phosphoproteins. Pre-LHCII protein is first processed at a conserved methionyl-arginyl bond followed by acetylation of the N-terminal arginine residue (Michel et al., 1991). In the synthetic peptide analogue of the N-terminus of pea Lhcb1 protein, Thr-5 was the most intensively phosphorylated amino acid (Michel and Bennett, 1989), when the numbering begins from the arginine residue of the processed protein. If this Thr residue was removed from the synthetic

```
D1 PROTEIN:                                                   D2 PROTEIN:

                                ◆                                 ◆
Spinacia oleracea          a)M T A I L E R R E S   E S L W G...   b)M T I A V G K F T K   D E K D L...
Secale cereale             c)- - - - - - - - - -   T - - - -...   d)- - - - L - R I P -   E - N - -...
Pinus thunbergii           e)- - - - I - - - - -   A N - - S...   f)- - - - L - - S S -   E - - T -...
Marchantia polymorpha      g)- - - T - - - - - -   A - I - -...   g)- - - - I - - S S -   E P - G -...
Chlamydomonas reinhardtii  h)- - - - - - - - - N   S - - - A...   i)- - - - I - T * Y Q   E K R T W...
Porphyra purpurea          j)- - - T - Q - - - -   A - - - E...   j)- - - - I - Q E K T   R G G F D...
Synechocystis PCC6803      k)- - T T - Q Q - - -   A - - - E...   k)- - - - - - R * A P   V - R G W...


CP43 PROTEIN:
                                                                                       ◆
Spinacia oleracea                                          b)M K T L Y S L R R F   Y P V E T L F N G T...
Secale cereae                                              d)- - I - - - - - - -   - H - - - - - - - -...
Pinus thunbergii                                           f)- - - - - - - - - S   - - - - - - - - - -...
Marchantia polymorpha                                      g)- - I - - - Q - - -   - - - - - - - - - -...
Chlamydomonas reinhardtii                                                             l)M - - - - - - -...
Chlamydomonas eugametos                                                               m)M - - - - - - -...
Chlorella vulgaris                                         n)- - N - - - - - - -   - H - - - - - - - S...
Porphyra purpurea          j) M K V F V L G W L L K I N L    - - - - - - Q - - -   - H - - - P - - T N...
Synechocystis PCC6803                                      k)- - - - S - - - - -   S - - V - - S - T S...
```

Fig. 1. The aligned N-terminal sequence of the D1, D2 and CP43 proteins, containing the phosphorylated Thr residue (◆), from taxonomically divergent photosynthetic oxygenic organisms. The sequences presented in the figure are from a dicot (*Spinacia oleracea*) and a monocot plant (*Secale cereale*), from a conifer (*Pinus thunbergii*), from a liverwort (*Marchantia polymorpha*), from green algae (*Chlamydomonas reinhardtii*, *Chlamydomonas eugametos* and *Chlorella vulgaris*), from a red alga (*Porphyra purpurea*), and from a cyanobacterium (*Synechocystis* PCC 6803). Identical residues are represented by –, and the gaps from the alignment analysis (Svensson et al., 1991) are marked by *. [a)] Zurawski et al., 1982, [b)] Holschuh et al., 1984, [c)] Kolosov et al., 1989, [d)] Bukharov et al., 1989, [e)] Tsudzuki et al., 1992, [f)] Wakasugi et al., 1994, [g)] Ohyama et al., 1986, [h)] Erickson et al., 1984, [i)] Erickson et al., 1986, [j)] Reith and Munholland, 1995, [k)] Kaneko et al., 1996, [l)] Rochaix et al., 1989, [m)] Turmel et al., 1993, [n)] Wakasugi et al., 1997.

peptide, other Thr or serine (Ser) residues present in the sequence were phosphorylated (Michel and Bennett, 1989). However, the Thr-5 residue is not a conserved amino acid in the Lhcb1 or Lhcb2 proteins (Jansson, 1999), indicating that also other Thr and Ser residues at the N-terminus of these proteins are candidates for phosphorylation sites (Dilly-Hatrwig et al., 1998). Michel et al. (1991) isolated two Lhcb1 and one Lhcb2 N-terminal peptides with phosphorylated Thr-3 residues, as well as one Lhcb1 N-terminal peptide with a phosphorylated Ser-3 residue from thylakoid membranes of spinach.

Phosphorylation site of the Lhcb4 protein differs from that of the other Lhcb and PS II proteins being not localized to the immediate N-terminus of the protein. In the thylakoid membranes of maize, the Thr-83 residue was identified as a phosphorylation site upon cold treatment of leaves (Testi et al., 1996). The N-terminal, stroma-exposed domain flanking the Thr-83 residue in the Lhcb4 protein, is not homologous to the other members of the light-harvesting protein family but it is conserved in the Lhcb4 proteins of maize, barley and *Arabidopsis thaliana* (Bergantino et al., 1998; Jansson, 1999).

III. Reversible Phosphorylation of Thylakoid Proteins

A. Redox-Dependent Phosphorylation of Thylakoid Proteins

Application of electron transport inhibitors in the phosphorylation studies with isolated thylakoid membranes has indicated that light activates the protein kinase via electron transport rather than via direct excitation of a kinase or via production of ATP in photosynthesis (Bennett, 1979; Allen and Bennett,

1981). A number of studies of the redox regulation of PS II and LHCII protein phosphorylation, have pointed plastoquinone as a key regulator of thylakoid protein kinase(s) (see the reviews of Bennett, 1991; Allen, 1992; Gal et al., 1997; Chapter 24, Ohad et al.). The inhibitors blocking electron transport between the secondary quinone acceptor Q_B of PS II and the plastoquinone pool completely abolish thylakoid protein phosphorylation in light (Fig. 2A) (Allen and Bennett, 1981; Allen et al., 1981), while incubation of isolated thylakoid membranes with reducing agents, such as duroquinole (Schuster et al., 1986) or NADPH and ferredoxin (Larsson et al., 1987b), induces PS II and LHCII protein phosphorylation even in darkness. Blocking of electron transport between the plastoquinone pool and the cytochrome b_6f complex, on the other hand, has a differential effect on phosphorylation of LHCII and PS II core proteins. Such inhibition prevents the light-induced phosphorylation of LHCII but not that of the PS II core proteins (Fig. 2A) (Farchaus et al., 1985; Bennett et al., 1988; Gal et al., 1990; Frid et al., 1992), indicating that the redox activation of LHCII protein phosphorylation is mediated via reduction of the cytochrome b_6f complex. Taken together, the reduction of plastoquinone in the thylakoid membrane seems to fulfill the requirements for inducing PS II core protein phosphorylation, while both the reduction of plastoquinone and subsequent binding of plastoquinol to the cytochrome b_6f complex are required to activate LHCII phosphorylation.

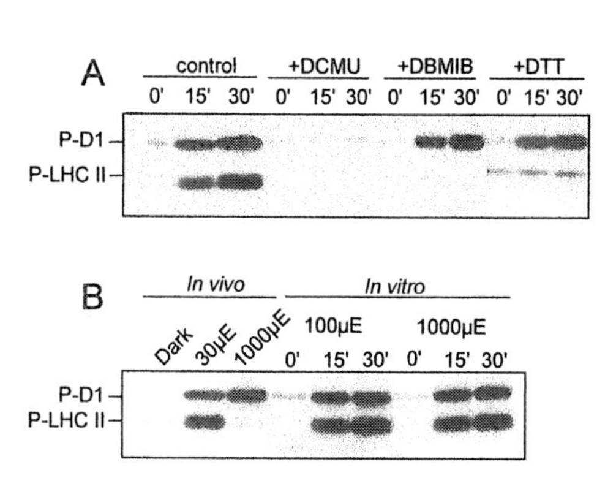

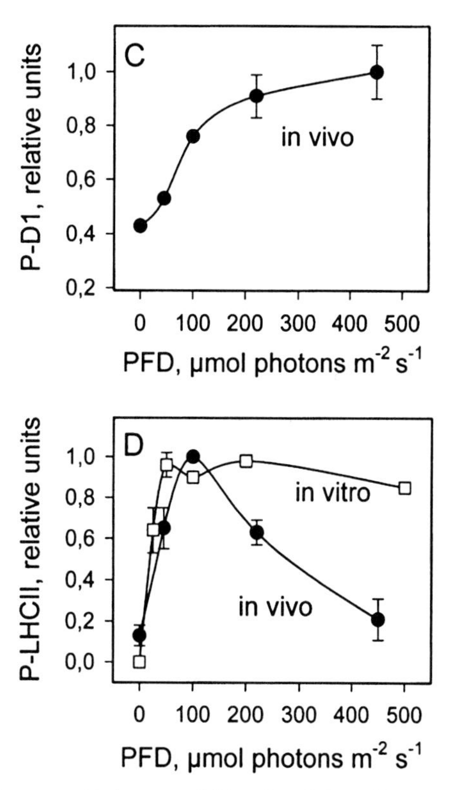

Fig. 2. Phosphorylation of PS II and LHCII proteins in isolated thylakoid membranes and in leaves. (A) Phosphorylation of the PS II core proteins is induced by reduction of plastoquinone pool, while LHCII protein phosphorylation additionally requires the reduction of cytochrome b_6f complex and is inhibited by thiol reagents. Thylakoid membranes were isolated from dark-adapted leaves and incubated at a chlorophyll concentration of 0.4 mg ml^{-1} with 20 μM 3-(3′4-dichlorophenyl)-1,1-dimethylurea (DCMU), 10 μM *2,5*-dibromo-3-methyl-6-isopropyl-*p*-benzoquinone (DBMIB) or 2 mM dithiothreitol (DTT) for 10 min. Thylakoid proteins were then phosphorylated at 100 μmol photons m^{-2} s^{-1} (μE) in the presence of 0.4 mM ATP for indicated times. (B) Inactivation of LHCII phosphorylation occurs in vivo but not in isolated thylakoids when illuminated at high light intensities. The leaf discs were illuminated under indicated irradiances for two hours before isolation of thylakoid membranes (in vivo) or thylakoids were isolated from dark-adapted leaves and phosphorylated in vitro at indicated light intensities (in vitro) as described in A. C and D) Phosphorylation level of spinach D1 protein in vivo (C) and LHCII proteins both in vivo and in vitro (D) under different light intensities. Illumination in vivo and in vitro at indicated photon flux densities (PFD) was carried out as described in A and B. The phosphorylation levels of PS II and LHCII proteins were detected by immunoblotting using a phosphothreonine antibody.

B. How Many Kinases?

Phosphorylation of the threonine residue both in PS II core and LHCII proteins indicates that the kinase(s) involved in thylakoid protein phosphorylation belongs to the group of serine-threonine-specific kinases. However, these thylakoid protein kinases are not inhibited by a conventional kinase inhibitor, staurosporine (Rintamäki et al., 1998). Considering the number of kinases involved in the phosphorylation of PS II and LHCII proteins, both a single kinase (Silverstein et al., 1993a; Vener et al., 1995) or two distinct kinases (Bennett, 1991; Gal et al., 1997) have been suggested. Support for the idea of a single kinase comes from studies on the redox titration of the phosphorylation of 13 pea thylakoid proteins (Silverstein et al., 1993a) and on the induction of phosphorylation of the spinach thylakoid proteins by a pH shift treatment (Vener et al., 1995). On the other hand, the existence of two different kinases, one for LHCII and another for PS II core proteins with distinct redox regulation systems, is supported by several observations made both in vivo and in vitro: i) application of electron transport inhibitors (Fig. 2A) (Bennett et al., 1988) and thiol reagents (Millner et al., 1982; Carlberg et al., 1999, Rintamäki et al., 2000) modulates differentially the phosphorylation of the LHCII and PS II core proteins, ii) LHCII phosphorylation, but not PS II core protein phosphorylation, is inhibited in the mutants deficient in the cytochrome b_6f complex (Gal et al., 1987; Bennett et al., 1988; Vener et al., 1995), iii) PS II and LHCII protein phosphorylation is differentially regulated by ambient light conditions in vivo (Schuster et al., 1986; Ebbert and Godde, 1994; Rintamäki et al., 1997).

Despite extensive research, none of the thylakoid-associated kinases have yet been specifically characterized (see the review of Gal et al., 1997). Kinase activity has been shown to copurify both with the cytochrome b_6f complex (Gal et al., 1990; Frid et al., 1992) and with the PS II core complex (Race and Hind, 1996). Recently, however, progress has been made in the isolation of thylakoid associated kinases (Weber et al., 1998; Vink et al., 1998; Snyders and Kohorn, 1999). Four apparent protein kinases with molecular masses of 85, 64, 55 and 30 kDa were isolated from spinach thylakoids by combination of several techniques, such as electroelution, perfusion chromatography, sodium dodecyl sulfate polyacrylamide gel electrophoresis (SDS-PAGE) and isoelectric focusing (Weber et al., 1998; Vink et al., 1998). The kinases of 64 and 55 kDa were shown to phosphorylate isolated LHCII proteins and CP43 in isolated PS II particles (Vink et al., 1998). The 55 kDa protein has a moderate homology score to the potential histidine kinase in *Synechocystis*, but no homology was found for the 64 kDa protein in data bases (Weber et al., 1998).

The yeast assay system that allows the selection of cDNAs encoding activities for a specific target, was applied for the isolation of thylakoid protein kinases (Smith and Kohorn, 1991, Snyders and Kohorn, 1999). Using the amino terminus of the LHCII protein containing the Thr phosphorylation site as a target, resulted in cloning of a family of proteins called thylakoid-associated kinases (Snyders and Kohorn, 1999). The predicted amino acid sequences of these three kinases with molecular masses of 50 to 56 kDa, showed 90% identity. These kinases were copurified with the cytochrome b_6f complex and such preparations appeared to phosphorylate LHCII proteins, but not PS II core proteins. One of the kinases cross-reacted with the antibodies raised against phosphothreonine or phosphoserine peptides, indicating that this kinase itself may be regulated by phosphorylation.

Phosphorylation of serine (Garcia Vescovi and Lucero, 1990) and tyrosine residues (Tullberg et al., 1998) of the thylakoid membrane proteins has also been reported. However, these phosphoproteins remain only poorly characterized and the kinases involved are unknown.

The enormous amount of data that have accumulated since the discovery of thylakoid protein phosphorylation, speaks for the existence of several protein kinases in the thylakoid membrane. The idea of distinct threonine kinases for PS II core and LHCII proteins is supported by several experimental approaches. The current list of putative kinases involved in phosphorylation of thylakoid proteins is, to our knowledge, the following: i) 30, 55, 64 and 80 kDa kinases, one of which has homology to a histidine kinase (Vink et al., 1998; Weber et al., 1998), ii) the PS II core associated kinase (Race and Hind, 1996), iii) a family of TAK kinases which are the best characterized thylakoid kinases so far (Snyders and Kohorn, 1999), iv) a putative tyrosine kinase (Tullberg et al., 1998), v) a putative D1-specific kinase (A.K. Mattoo, personal communication). Only isolation and sequencing of the genes encoding these enzyme proteins will finally make possible to reveal the exact localization of the enzymes in different thylakoid compartments and to determine their substrate specificity and regulation in chloroplasts.

C. What Proportions of PS II Cores and Lhcb Proteins are Phosphorylated in Leaves?

1. Methodological Aspects

Determination of the pool sizes of phosphoproteins under different developmental, physiological and environmental conditions provides information about the function and regulation of the phosphoproteins in cell metabolism. Methodologically, protein phosphorylation in the thylakoid membrane has been primarily investigated by labeling the substrate proteins with ^{32}P-orthophosphate or ^{32}P-ATP. This method, however, gives only relative changes in the pool sizes of phosphoproteins under the conditions studied, but does not reveal e.g. the number of PS II complexes containing phosphoproteins in their phosphorylated form. Neither does it detect the phosphorylated proteins present before the addition of the radioactive reagent, which may result in an underestimation of the phosphorylated pool size of a protein. These problems are partially overcome by the recent development of immunodetection methods of phosphoproteins (Rintamäki et al., 1997). This analytical tool also enables extensive studies of protein phosphorylation in vivo in intact plant systems (Pursiheimo et al., 1998) thereby excluding the problems with transport and uneven labeling of the cells faced in the experiments with radioactive compounds.

A clear separation of the phosphorylated and nonphosphorylated forms of a protein is required for simultaneous analysis of the two forms with protein-specific antibodies. Moreover, it has to be tested that the protein-specific antibodies crossreact equally with both forms of the protein (Rintamäki et al., 1995). Callahan et al. (1990) first reported that the phosphorylated form of the D1 protein can be separated from the non-phosphorylated one by slower mobility in SDS-PAGE. Also the phosphorylated D2 and CP43 proteins displayed similar retarded migration during electrophoresis as observed for the phosphorylated D1 protein (deVitry et al., 1991; Rintamäki et al., 1997).

Use of the phosphothreonine (Giardi et al., 1995; Rintamäki et al., 1997) and phosphotyrosine (Tullberg et al., 1998) antibodies give, like the radioactive labeling, only relative proportions of the phosphoproteins in the thylakoid membrane. However, the immunological approach has two advantages as compared to the radioactive labeling assay. First, antibodies detect the endogenous phosphorylation level of the protein prevailing in the beginning of the experiment, thus providing a fundamental control for studies of thylakoid protein phosphorylation. Dark incubation of plants and isolated thylakoids is routinely used to induce complete dephosphorylation of the thylakoid phosphoproteins before phosphorylation experiments. This is not always a sufficient treatment, since variation of the phosphorylation level, especially of the D1, CP43 and LHCII proteins is observed in plants incubated in darkness (Fig. 3A) (Rintamäki et al., 1997; Pursiheimo et al., 1998). Secondly, the phosphorylation level of proteins can be analyzed directly from the leaf samples collected e.g. from the field experiments and rapidly frozen in liquid nitrogen, without any pretreatment period of leaves with labeling chemicals.

All the threonine, serine and tyrosine antibodies tested so far, crossreact with the thylakoid proteins, particularly with the LHCII proteins. Only the phosphothreonine antibodies detect the light- and redox-dependent phosphorylation of PS II and LHCII proteins (Fig. 3A). Several commercial antibodies raised against phosphotyrosine (Fig. 3B) and phosphoserine (E. Rintamäki and E.-M. Aro, unpublished) crossreact with several thylakoid proteins, without any response, however, to the redox state of the chloroplast. Therefore, the redox-regulated reversible phosphorylation is likely to occur on Thr residues of thylakoid phosphoproteins. The role and specificity of the crossreaction of serine and tyrosine antibodies with thylakoid proteins must be challenged until some reversible changes can be detected in the phosphorylation level of Ser or Tyr residues of these phosphoproteins under varying physiological conditions.

Analysis of phosphoproteins based on mass spectrometry (Michel et al., 1988) was recently launched for quantification of thylakoid phosphoproteins (Vener et al., 2001). This method, however, is rather laborious for routine analysis of thylakoid phosphoproteins.

2. The Amount of Phosphorylated PS II Complexes in Vivo

Based on the present knowledge of the structure of

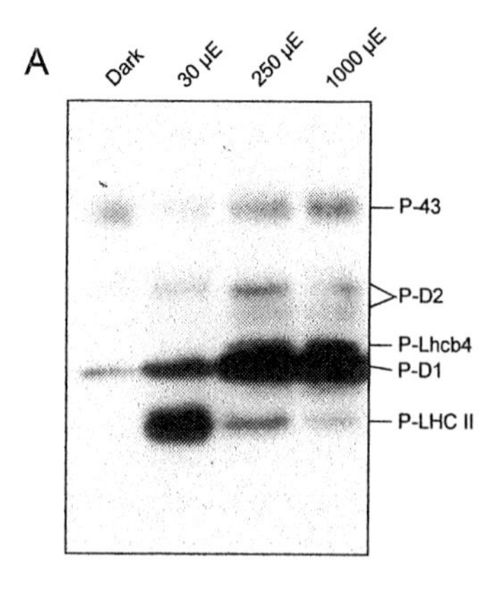

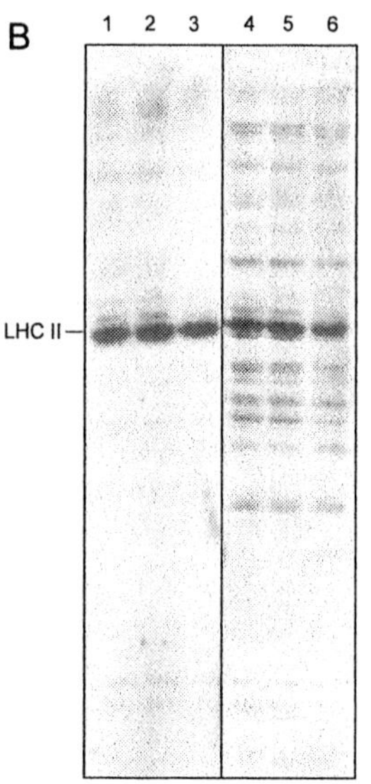

Fig. 3. (A) Steady-state phosphorylation level of PS II and LHCII proteins in maize leaves at different light intensities. The leaf discs were illuminated for two hours under irradiances indicated above each lane, after which the thylakoid membranes were isolated. The phosphorylation level of PS II and Lhcb proteins was detected by immunoblotting using a polyclonal phospho-threonine antibody (Rintamäki et al., 1997). (B) Thylakoid proteins recognized by a polyclonal phosphotyrosine antibody (lanes 1 to 3) or stained with Coomassie Brilliant Blue (lanes 4 to 6). Pumpkin leaves were incubated in darkness (lanes 1 and 4) or illuminated at 30 μmol photons m^{-2} s^{-1} (μE) (lanes 2 and 5) and at 1000 μmol photons m^{-2} s^{-1} (lanes 3 and 6) for two hours before isolation of the thylakoid membranes. After electrophoretic separation, the proteins were transferred to polyvinylidene fluoride membranes and probed with a polyclonal phosphotyrosine antibody (PY99, Santa Cruz Biotechnology) or stained with Coomassie Brilliant Blue.

the PS II complex (Rhee et al,. 1998; Zouni et al., 2001) and the phosphorylation sites of PS II core proteins (Michel and Bennett, 1987; Michel et al., 1988; Vener et al., 2001), there is only one phosphoprotein of each type in the PS II core monomer and only one phosphate group can be bound to each phosphoprotein.

Separation of the phosphorylated and nonphosphorylated forms of the D1 protein in SDS-PAGE has been used to analyze by immunodetection the proportion of the two forms of D1 under different experimental and environmental conditions (Aro et al., 1992; Elich et al., 1992, 1993; Koivuniemi et al., 1995; Ebbert and Godde, 1996; Rintamäki et al., 1996a, 1997). Accumulation of the phosphorylated form of the D1 protein increased with increasing irradiances up to the growth light level in higher plant leaves (Fig. 2B and C) (Rintamäki et al., 1996a, 1997). This maximal phosphorylation level, corresponding to 70–80% of the PS II centers with the D1 protein in its phosphorylated form, was maintained also at higher irradiance (Rintamäki et al., 1997).

The most important factor regulating the proportion of the phosphorylated D1 protein in thylakoids is not the light intensity per se, but rather the redox state of electron transfer components between PS II and PS I (Salonen et al., 1998). Not only the light intensity, but also all the factors affecting the energy supply and consumption in photosynthesis, modulate the redox state of the plastoquinone pool (Huner et al., 1996). The redox state of chloroplast in illuminated leaves can be estimated as a proportion of the primary electron acceptor Q_A of PS II in the reduced state (Dietz et al., 1985; Huner et al., 1996), which has been proposed to reflect also the redox state of the electron transfer components between PS II and PS I (Huner et al., 1996). Maximal number of PS II centers with phosphorylated D1 protein (80%) is observed already at a relatively low redox state of electron transfer components in thylakoids, with less than 20% of Q_A in a reduced state (Salonen et al., 1998).

The surface structure and thickness of the leaf influence the penetration and reflection of radiation into and inside the leaves (Cui et al., 1991), thereby determining the excitation state of PS II in chloroplasts. The metabolic state of the plant also affects the actual redox state of plastoquinone in the thylakoid membrane (Huner et al., 1996; Chapter 30, Nixon and Mullineaux). These differences in the structure and metabolism of leaves explain the extensive variation in the accumulation of the phosphorylated D1 protein observed in distinct plant species and under various environmental conditions

(Elich et al., 1992; Rintamäki et al., 1996a, 1997; Pursiheimo et al., 1998; Baena Gonzales et al., 1999; Vener et al., 2001). For example, differential accumulation of phosphorylated D1 protein was observed in spinach and pumpkin leaves illuminated at similar ambient light intensities (Rintamäki et al., 1997). Furthermore, abiotic and biotic stresses have been shown to induce high phosphorylation level of the D1 protein independent of light intensity (Ebbert and Godde, 1994, 1996; Hollinderbäumer et al., 1997; Salonen et al., 1998). This may be due to a high redox state of thylakoid components, caused by an imbalance between energy supply and consumption in photosynthesis under stress conditions.

Extensive studies on the proportions of other PS II core phosphoproteins than D1 in the thylakoid membrane have not been carried out. However, light response of the phosphorylation level of D2 and CP43 proteins follows exactly the curve determined for the phosphorylation of the D1 protein, with a large pool of phosphorylated proteins as compared to the pool of nonphosphorylated ones in high light (Rintamäki et al., 1997). Changes in the pool size of the phosphorylated PsbH protein under varying conditions have not been determined.

3. Phosphorylation of Lhcb Proteins in Vivo

One phosphate group has been found to be present in the N-terminus of spinach Lhcb1 and Lhcb2 proteins (Michel et al., 1991), but the number of LHCII phosphoproteins associated with one PS II center varies depending on the plant species and growth conditions (Staehelin, 1986; Larsson et al., 1987a; Lindahl et al., 1995a; Boekema et al., 1999). Moreover, several phosphorylation sites for the Lhcb1 and Lhcb2 proteins have recently been postulated (Dilly-Hartwig et al., 1998). Electrophoretic separation of the phosphorylated and non-phosphorylated forms of the Lhcb1 and Lhcb2 proteins has not been reported. The proportion of phosphorylated Lhcb1 and Lhcb2 proteins has only been determined by quantitative SDS-PAGE and scintillation counting after phosphorylation of isolated thylakoid membranes in the presence of ^{32}P-ATP (Islam, 1987). 14 to 22% and 15 to 25% of Lhcb1 and Lhcb2 proteins, respectively, were phosphorylated under in vitro conditions (Islam,1987).

Determination of the redox-dependence of LHCII phosphorylation in vivo has indicated that the maximal phosphorylation of LHCII proteins occurs only at a low steady-state level of the reduced electron transfer components in thylakoid membranes, less than 10% of Q_A in its reduced form (Rintamäki et al., 1997; Salonen et al., 1998). Reduction of plastoquinone induced by a pH shift treatment of dark-adapted thylakoids indicated that the low reduction state of the plastoquinone pool (a low ratio of plastoquinol to plastoquinone) is sufficient to activate LHCII protein phosphorylation also in vitro (Vener et al., 1995). Strong reduction of the plastoquinone pool, on the other hand, does not inhibit LHCII phosphorylation. Indeed, illumination with PSII light keeps the LHCII kinase fully active (Pursiheimo et al., 2001).

Contrary to the PS II core protein phosphorylation, a strong down regulation of LHCII phosphorylation is observed in plants and in intact chloroplasts with increasing light intensities (Fig. 2D) (Ebbert and Godde, 1994; Rintamäki et al., 1997; Pursiheimo et al., 1998; Baena-Gonzalez et al., 1999). The lack of phosphorylated LHCII proteins in high light is due to an inhibition of LHCII phosphorylation, not to any rapid dephosphorylation of phosphorylated LHCII proteins (Rintamäki et al., 1997). Interestingly, such inactivation of LHCII protein phosphorylation does not occur under illumination of isolated thylakoids (Fig. 2B and D) (Rintamäki et al., 1997). This observation suggests that besides the cytochrome b_6f complex-dependent activation of the LHCII kinase, phosphorylation of LHCII polypeptides in vivo is additionally regulated by a mechanism not present in the thylakoid membrane (Rintamäki et al., 1997). The mechanism is likely to be under a control of the thiol redox state of the chloroplast. This hypothesis is based on experiments indicating that chloroplast thioredoxins and other sulfydryl reagents inhibit specifically the phosphorylation of the LHCII proteins (Fig. 2A) (Millner et al., 1982; Carlberg et al., 1999; Rintamäki et al., 2000). Further support for the hypothesis comes from treatment of leaves with reagents oxidizing thiols in proteins. Such treatments distinctively displaces the irradiance-dependent inactivation of LHCII phosphorylation in spinach leaves to higher light intensities (Carlberg et al., 1999).

We have recently shown that the two regulatory systems involved in the regulation of LHCII phosphorylation, activation via reduction of the cytochrome b_6f complex and inactivation via thiol-redox state of chloroplast, are not independent but function cooperatively (Fig. 4) (Rintamäki et al.,

2000). This further suggests that the LHCII kinase rather than its substrate, the LHCII protein, is a target component of the thiol regulation of LHCII phosphorylation. Activation of the LHCII kinase via reduction of cytochrome b_6f complex makes LHCII phosphorylation insensitive to thiol mediators and this active state prevails in chloroplast in vivo under low-light conditions where LHCII phosphorylation is maximal (Fig. 4). Upon high-light illumination of plant leaves the target thiols of LHCII phosphorylation become exposed and accessible to reduction by thiols produced in chloroplast, resulting in stable inactivation of the LHCII kinase (Fig. 4). How this exposure of the regulatory thiol site is induced in high light condition, is not currently known. Complex interactions between the LHCII kinase and cytochrome b_6f complex (Chapter 24, Ohad et al.) may be involved not only in the activation of the kinase (Vener et al., 1997, 1998) but also at high light leading to inactivation of the kinase. The synergetic function of the two regulation mechanisms of the LHCII kinase strictly couples the phosphorylation of LHCII proteins to the ambient redox state of chloroplast in vivo.

Phosphorylation of the Lhcb4 antenna protein that belongs to the minor antenna systems of PS II has been reported to occur under low temperature (Bergantino et al., 1995, 1998; Pursiheimo et al., 1998) and under moderate/high light intensities (Fig. 3A). The two forms of Lhcb4 are well separated from each other in Tris/sulfate/urea PAGE and in a high-molarity Tris buffer system without urea (Bergantino et al., 1998). The size of the phosphorylated protein pool varies depending on the plant species reaching 50% in published studies (Bergantino et al., 1995, 1998).

In summary, the phosphorylation of both PS II core and LHCII proteins is strictly under the redox control in chloroplasts. Imbalance between the energy supply and consumption in photosynthesis, recorded as an increase in the steady-state reduction level of electron transport components between PS II and PS I, induces a high steady-state phosphorylation level of the PS II core proteins, while mere electron flow through the plastoquinone and the cytochrome b_6f complex seems to be enough to activate and maintain the phosphorylation of Lhcb1 and Lhcb2 proteins. Furthermore, LHCII phosphorylation is additionally regulated by the thiol redox state of chloroplast resulting in a strong down regulation of phosphorylation at high light intensities.

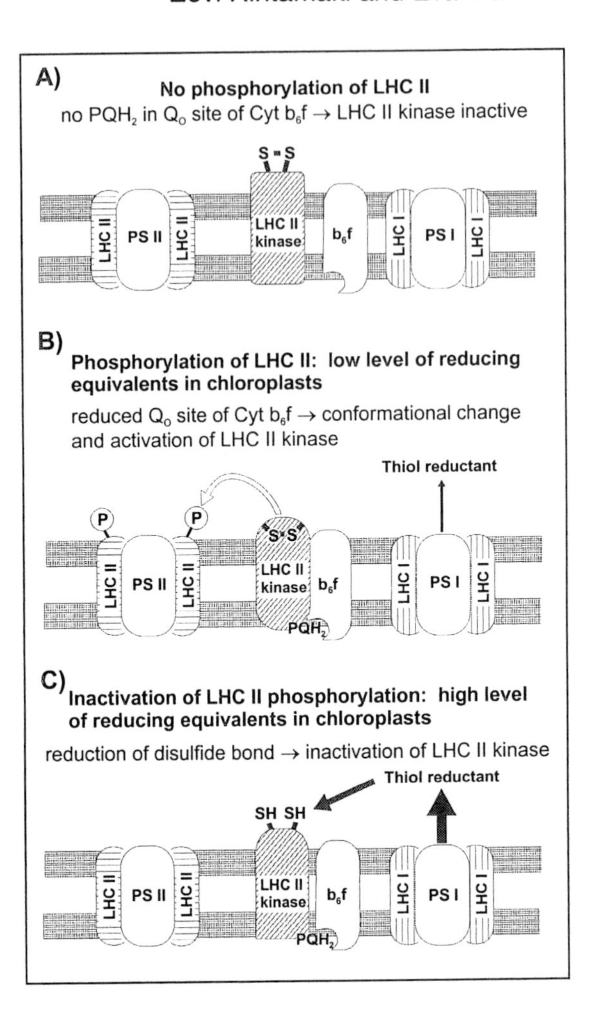

Fig. 4. Redox regulation of LHCII protein phosphorylation in vivo. The model depicts three hypothetical states of the LHCII kinase present under different redox conditions of chloroplasts. (A) LHCII kinase is inactive when the Q_o-site of cytochrome b_6f complex is not occupied by reduced plastoquinone (PQH_2). This state of the enzyme prevails in chloroplast when plastoquinone pool is oxidized. (B) Light activates the LHCII kinase via reduction of plastoquinone and subsequent occupation of the Q_o-site in cytochrome b_6f complex with plastoquinol. This activation induces a conformational change in the LHCII kinase with concomitant hiding of the thiol-regulatory site of the enzyme. Only low reduction level of plastoquinone pool, occurring under low-light illumination of leaves, is enough for activation of the kinase. (C) Increase in redox equivalents on the reducing side of PS I, e.g. under high-light illumination, induces a second conformational change in the LHCII kinase thus exposing the thiol-regulatory site of the LHCII kinase. The disulfide bridge in the thiol-regulatory site is directly reduced by thiol reductants produced in chloroplasts under such conditions, resulting in an inactivation of the LHCII kinase. The model is based on experiments presented in Rintamäki et al. (2000).

D. Regulation of Thylakoid Protein Dephosphorylation

Reversibility of thylakoid protein phosphorylation can be demonstrated by transferring illuminated thylakoid membranes, chloroplasts or leaves to darkness, by application of inhibitors deactivating thylakoid protein kinases and by chasing the radioactive ^{32}P-phosphate in the presence of non-labeled substrate. Considerable variation can be found in the published data concerning dephosphorylation rates of different thylakoid phosphoproteins, effects of activators and inhibitors on protein phosphatases, as well as the light-dependence of thylakoid protein dephosphorylation. Such discrepancies in the literature mainly result from differences in the adopted assay systems and experimental conditions. It is conceivable that this variation reflects a highly regulated dephosphorylation of thylakoid phosphoproteins, as is the case with the phosphorylation reactions. Such complex regulation is apparently related to the physiological function of phosphoproteins.

Several serine/threonine phosphatases, either membrane-integrated, membrane surface-associated or located in the stroma, have been partially purified from chloroplasts (Sun et al., 1989; Hammer et al., 1995a,b, 1997; Hast and Follmann, 1996; Vener et al., 1999). Nevertheless, neither the genes nor the protein sequences are available yet. Chloroplast protein phosphatases have mainly been characterized by using endogenous phosphorylated LHCII, N-terminal LHCII peptides released by chymotrypsin treatment, and synthetic phosphopeptide analogue of the N-terminal sequence of LHCII as a substrate. The specificity of these partially purified phosphatases towards PS II core phosphoproteins, however, remains largely unknown. A broad substrate specificity, at least for some chloroplast protein phosphatases, is suggested by inhibition of the dephosphorylation of both the PS II core and the LHCII phosphoproteins in thylakoid membranes by addition of a short phosphopeptide corresponding to the N-terminal sequence of LHCII (Cheng et al., 1994). On the other hand, differential regulation of LHCII and PS II core protein dephosphorylation (Ebbert and Godde, 1994) and a purified phosphatase preparation with higher specificity to PS II core proteins than to LHCII proteins (Vener et al., 1999) suggest distinct substrate specificity of particular chloroplast phosphatases.

Thylakoid protein dephosphorylation is generally inhibited by NaF and stimulated by millimolar concentrations of magnesium and dithiothreitol (DTT) (Bennett, 1980; Sun et al., 1989; Hammer et al., 1995b; Koivuniemi et al., 1995; Carlberg and Andersson, 1996). Magnesium-independent activity of thylakoid-associated phosphatases is, however, also reported (Hast and Follmann, 1996). Thylakoid associated phosphatases appear to have a neutral (Sun et al., 1989; Hammer et al., 1995b) or slightly acidic (Hast and Follmann, 1996) pH optimum. The 29-kDa stromal phosphatase isolated from pea chloroplasts has been shown to differ from the thylakoid membrane bound phosphatase in its slightly alkaline pH optimum, in its insensitivity to molybdate ions, and in the higher modulation of its activity by divalent cations (Hammer et al., 1995b).

Insensitivity of the thylakoid associated (Sun and Markwell, 1992; Cheng et al., 1994; Hammer et al., 1995b; Hast and Follmann, 1996) and stromal (Hammer et al., 1995b) phosphatases to the mammalian type PP1 and PP2A phosphatase inhibitors, okadaic acid and microcystin, suggests that chloroplast phosphatases can not be classified as conventional groups of serine/threonine phosphatases. However, it was recently shown that a purified thylakoid associated protein phosphatase was inhibited by okadaic acid and tautomycin, which indicates that this particular phosphatase belongs to the family of classical eukaryotic protein phosphatases (Vener et al., 1999). These differences in the sensitivity of chloroplast phosphatases to inhibitors may be due to the presence of several phosphatases in the thylakoid membrane. Alternatively, the membrane-bound phosphatase may have transmembrane helices with regulatory domains on the lumenal side of the thylakoid membrane (Vener et al., 1998), thus being unable to interact with inhibitors in intact membranes. Indeed, an immunophilin TLP40 that regulates dephosphorylation of the LHCII and PS II core phosphoproteins, has recently been isolated from the thylakoid membrane (Fulgosi et al., 1998; Vener et al., 1998) and localized to the chloroplast lumen (Fulgosi et al., 1998; Kieselbach et al., 1998).

Dephosphorylation of proteins in isolated thylakoid preparations does not display similar redox dependence typical for the thylakoid kinase reactions (Silverstein et al., 1993b). However, stimulation of LHCII dephosphorylation in light has been observed in isolated chloroplasts (Ebbert and Godde, 1994) and in leaves (Rintamäki et al., 1997). DTT has been shown to increase the activity of LHCII dephosphorylation in isolated thylakoid membranes in darkness (Sun et al., 1989; Hammer et al., 1995b; Carlberg and Andersson, 1996). Thus the stimulation of LHCII dephosphorylation by light in vivo may be due to a modulation of the phosphatase activity by the ferredoxin-thioredoxin system (Hammer et al., 1995b; Carlberg and Andersson, 1996).

Both the light dependence (Elich et al., 1993; Rintamäki et al., 1996a) and light inactivation (Ebbert and Godde, 1994) of the PS II core protein dephosphorylation in vivo have been reported. Modification of substrate by the experimental conditions adopted in a study may alter the regulation of its dephosphorylation. Phosphorylated D1 protein is a good example of the modifications occurring at the substrate level (Rintamäki et al., 1996a). In the functional PS II center, the phosphorylated D1 protein can be dephosphorylated in vivo both in darkness and in light. However, dephosphorylation of the D1 protein becomes strictly light-dependent upon irreversible photodamage to the D1 protein (Section V.A.2.b).

Evidence from dephosphorylation experiments of thylakoid phosphoproteins favors the existence of multiple protein phosphatases. Differential light regulation and sensitivity of dephosphorylation to various regulatory compounds imply that phosphatases, by regulating the reversibility of thylakoid protein phosphorylation, have specific physiological roles in the chloroplasts. Isolation of the genes and characterization of the enzymic properties is required to clarify the exact function of phosphatases in chloroplasts.

Finally, reversible phosphorylation of thylakoid proteins is a dynamic process. Based on the published data it can be concluded that both reactions, phosphorylation and dephosphorylation, are regulated, at least to some extent, by light via electron transport in the thylakoid membrane or by the electron acceptors in the stroma. However, the rate of the phosphate group turnover in PS II core and LHCII proteins in vivo is still fairly unknown due to technical problems. No specific kinase inhibitors of PS II core or LHCII protein phosphorylation are available without exerting concomitant effects on the redox level of the chloroplast.

IV. Photosystem II and Light-Harvesting Complex II Protein Phosphorylation in Taxonomically Divergent Oxygenic Photosynthetic Organisms

Since the discovery of light-dependent phosphorylation of PS II and LHCII proteins (Bennett, 1977), the phenomenon has been well established and found to be very similar among higher plants. In lower plant thylakoids and cyanobacteria, however, different patterns of thylakoid phosphoproteins have been reported.

Organisms performing oxygenic photosynthesis can be roughly divided into three distinct groups with respect to the reversible phosphorylation of thylakoid proteins (Pursiheimo et al., 1998). One group consists of cyanobacteria and red algae with the phycobilisome antenna system. The PS II core proteins typically phosphorylated in higher plants, do not undergo reversible phosphorylation within this group (Allen, 1992; Mann, 1994). Three phosphoproteins with molecular mass less than 23 kDa were detected in thylakoid membranes of *Synechococcus* after labeling of the cells with ^{32}P-orthophosphate (Sanders et al., 1986; Allen, 1992). Two proteins were detected in thylakoids from *Synechocystis* (24-kDa and 31-kDa proteins) and from a red alga (22-kDa and 31-kDa proteins) respectively by an immunoassay with phosphothreonine antibodies (Pursiheimo et al., 1998). None of these phosphorylated proteins was recognized with specific antibodies for the D1, D2 or CP43 proteins.

Lower plants (green algae, mosses, liverworts and ferns) form the second group with respect to the phosphorylation of thylakoid proteins. Lhcb1 and Lhcb2 proteins as well as D2, and CP43 proteins undergo reversible phosphorylation, while the phosphorylated form of the D1 protein has never been detected in these plant species (Wollman and Delepelaire, 1984; de Vitry et al., 1991; Rintamäki et al., 1995; Andronis et al., 1998; Pursiheimo et al., 1998).

Phosphorylation of the D1 protein occurs only in seed plants, which comprise the third group of species. Light-intensity dependent regulation of LHCII protein phosphorylation occurs in the plant species of the second and third groups with the maximal phosphorylation of LHCII proteins at low light and a remarkably lower phosphorylation level at higher irradiances (Fig. 3A) (Pursiheimo et al., 1998).

Phosphorylation of the D1 protein seems to be the latest event in the evolution of PS II protein phosphorylation. The lack of D1 protein phosphorylation in the species of the first and second groups cannot be explained by the amino acid sequence of the phosphorylation site. The Thr-2 residue is present in all the sequenced D1 proteins except in *Euglena* (Svensson et al., 1991), and the sequences flanking the putative phosphorylation site in the species of the first and second groups are very homologous to the sequences of seed plants (Fig. 1). Like in the D1 protein, the phosphorylated Thr residue and the two flanking amino acid residues in the D2 protein are very conserved, even in cyanobacteria and red algae which do not phosphorylate PS II proteins (Fig. 1). Conserved amino acid sequences flanking the phosphorylation site are also found in the CP43 protein (Fig. 1). Therefore, evolution of the kinase and/or its structural orientation with respect to the substrate, might be more crucial for the capability to phosphorylate PS II proteins, than the sequence of the putative phosphorylation site of the protein.

V. Physiological Role of Thylakoid Protein Phosphorylation

A. Reversible Phosphorylation of the PS II Core Proteins

A great number of physiological implications have been suggested for reversible phosphorylation of PS II core proteins: i) regulation of electron transport activity (Horton and Lee, 1984; Giardi et al., 1992), ii) effect on the sensitivity of PS II to photoinhibition (Horton and Lee, 1985; Harrison and Allen, 1991; Allen, 1992; Giardi, 1993), iii) regulation of the D1 protein turnover (Aro et al., 1992; Elich et al., 1993; Ebbert and Godde, 1994; Rintamäki et al., 1996a; Baena-Gonzalez et al., 1999), iv) general stabilization of PS II complexes or modulation of the oligomeric structure of PS II (Ebbert and Godde, 1996; Kruse et al., 1997; Salonen et al., 1998; Baena-Gonzalez et al., 1999).

1. Implications on PS II Electron Transfer and the Oligomeric Structure of PS II

D1 and D2 reaction center proteins bind all the redox active components of PS II. Thus the regulation of PS II electron transfer capacity via phosphorylation of these proteins sounds attractive. This hypothesis has been extensively studied during the last decades, with partially contradictory results (Hodges et al., 1987; Packham, 1987; Habash and Baker, 1990; Harrison and Allen, 1991; Giardi et al., 1992; Giardi, 1993). The discrepancies may be due to differences in plant material and in experimental approaches. Furthermore, the lack of systematic analysis of the phosphorylation state of PS II and LHCII proteins in many studies make it difficult to combine the observations with the phosphorylation of the target protein. So far, no direct coupling of the reversible phosphorylation of the PS II core proteins and the capacity of PS II electron transport has been reported. Recently, however, PS II membranes from low (5%) to high (65%) phosphorylation level of PS II core proteins were analyzed by fluorescence and low temperature EPR measurements (F. Mamedov, E. Rintamäki, J. Ström, E.-M. Aro, S. Styring and B. Andersson, unpublished). The phosphorylation level of the PS II core proteins (D1, D2 and CP43) did not affect the oxygen evolution activity, the electron transfer from Q_A to Q_B, or the EPR signals from the redox components of PS II. The absence of PS II core protein phosphorylation in the cyanobacteria and red algae, and the lack of D1 phosphorylation in the green algae, mosses and ferns (deVitry et al., 1991; Rintamäki et al., 1995; Pursiheimo et al., 1998) also dispute the importance of phosphorylation in the basic regulation of PS II electron transport. Furthermore, elimination of the phosphorylation site in the D2 protein (Andronis et al., 1998; Fleischmann and Rochaix, 1999) or in the PsbH protein (O'Connor et al., 1998) of *Chlamydomonas* did not impair the assembly or function of PS II. These findings speak against the idea of basic regulation of PS II electron transport via PS II core protein phosphorylation and favor a more specific function in plants.

The PS II complex can exist either as a monomeric or as a dimeric complex in the thylakoid membrane. The dimers are enriched in the grana region of thylakoid membranes and are active in electron transport (review of Rögner et al., 1996). The stromal PS II complexes exist exclusively as monomers (Rögner et al., 1996) and only a few Lhcb proteins are associated with these inactive PS II centers (Mäenpää and Andersson, 1989; Peter and Thornber, 1991, Jansson, 1994). Phosphorylation of the PS II core proteins has been proposed to stabilize the dimeric PS II complex in isolated systems, while dephosphorylation was reported to induce monomerization of the PS II complex (Kruse et al., 1997). This conclusion, however, is not supported by the experiments conducted with intact leaves. Clearly, the dimeric PS II complexes in the grana appressions can be either phosphorylated or nonphosphorylated, depending on light conditions (Baena-Gonzalez et al., 1999).

2. PS II Photoinhibition and Reversible Phosphorylation of PS II Core Proteins

Photosystem II is the main target for light-induced inactivation of photosynthesis, called photoinhibition. Photoinhibition of PS II is associated with both the induction of photoprotecting mechanisms as well as light-mediated PS II inactivation followed by oxidative damage to the D1 protein (reviewed by Krause and Weis, 1991; Prasil et al., 1992; Aro et al., 1993b). This protein damage is related to irreversible PS II photoinhibition which can be overcome only by de novo protein synthesis and repair of the photodamaged PS II centers. The repair process involves a partial disassembly and migration of the inactive PS II complex from the appressed thylakoid regions to the stroma exposed membranes, with subsequent degradation and replacement of the D1 protein via de-novo protein synthesis (Chapter 22, Andersson and Aro; Adir et al., 1990; Guenther and Melis, 1990; Barbato et al., 1992; Prasil et al., 1992; Aro et al., 1993b).

a. Does the Phosphorylation of the PS II Core Proteins Play a Role in the Susceptibility of PS II to Photoinhibition?

Phosphorylation of the PS II core proteins has been suggested both to expose PS II to photoinhibition (Giardi, 1993) and to protect it against photoinhibition (Horton and Lee, 1985). Particularly, the PsbH phosphoprotein has been proposed to be involved in the sensitivity of PS II to photoinhibition. It was reported (Giardi, 1993) that phosphorylation of the D1 and D2 proteins induces a detachment of the PsbH protein from the isolated PS II particles, which results in an enhanced susceptibility of PS II to photoinhibition. Furthermore, prevention of PsbH protein phosphorylation has been suggested to protect PS II against photoinhibition (Sundby et al., 1989). These experimental observations support the protective role of the non-phosphorylated PsbH protein in PS II photoinhibition. On the other hand, it has also been shown that an induction of a high phosphorylation level of the PS II core proteins in the thylakoid membranes did not affect the rate of inactivation of PS II oxygen evolution at high light (Koivuniemi et al., 1995; Rintamäki et al., 1996b), which speaks against a direct involvement of PS II core protein phosphorylation in determining the susceptibility of PS II to photoinactivation and damage.

b. Regulation of the Photoinhibition Repair Cycle by PS II Core Protein Phosphorylation

Photodamage to PS II clearly increases with increasing light intensity (Keren et al., 1995; Tyystjärvi and Aro, 1996) and concomitantly accelerates the degradation and synthesis of the D1 protein (Aro et al., 1993a). These two latter processes are saturated at very similar light intensities (Kettunen et al., 1997), suggesting that the reactions are controlled by parallel signal systems. Furthermore, despite the high turnover rate of the D1 protein (Schuster et al., 1988; Aro et al., 1993a), no significant accumulation of the PS II centers depleted of the D1 protein is observed in the leaves of higher plants under physiological light intensities (Kettunen et al., 1991, 1997; Schnettger et al., 1994; Rintamäki et al., 1996a). This coordination of degradation and synthesis of the D1 protein reduces the life-time of the D1-depleted PS II centers, thus preventing the collapse of the multisubunit complex and subsequent degradation of the other PS II subunits as well. The latter phenomenon can be demonstrated by a treatment of leaves with chloroplast protein synthesis inhibitors that prevent the synthesis of the new D1 protein (Rintamäki et al., 1996b). A number of studies on the function of the PS II repair cycle have indicated that the degradation of the photodamaged D1 protein is a key regulatory step in the cycle (Aro et al., 1992; Prasil et al., 1992; Kim et al., 1993; Bracht and Trebst, 1994; Ebbert and Godde, 1994, 1996; Zer et al., 1994; Komenda and Barber, 1995; Rintamäki et al., 1995, 1996a; Zer and Ohad, 1995). The observation that phosphorylation of the D1 protein may control its own degradation (Callahan et al., 1990; Kettunen et al., 1991; Aro et al., 1992; Elich et al., 1992) opens a novel view for the regulation of the PS II repair cycle by reversible phosphorylation of PS II core proteins. Since this original discovery, both in vivo and in vitro studies have shown that only the non-phosphorylated form of the D1 protein is prone to proteolysis (Koivuniemi et al., 1995; Ebbert and Godde, 1996; Rintamäki et al., 1996a).

Figure 5 describes the current model of the PS II repair cycle regulated by the reversible phosphorylation of PS II core proteins, combining the results from our own and other laboratories (Guenther and Melis, 1990; Adir et al., 1991; Barbato et al., 1992; Elich et al., 1992, 1993; Prasil et al., 1992; Aro et al., 1993b; Rintamäki et al., 1996a; Andersson and Aro, 1997). The model is based on two main principles that have been experimentally verified. First, phosphorylation of the D1, D2 and CP43 proteins is not coupled to the photodamage to PS II; the phosphorylation rate per se does not correlate either with the photodamage to PS II or with the degradation rate of the D1 protein (Aro et al., 1992; Ebbert and Godde, 1996; Rintamäki et al., 1996a; Salonen et al., 1998). Second, reversible phosphorylation of the D1 protein controls the repair cycle by retarding the proteolysis of the damaged D1 protein and by limiting the degradation process to the stroma exposed thylakoid membranes (Koivuniemi et al., 1995; Ebbert and Godde, 1996; Rintamäki et al., 1996a; Baena-Gonzalez et al., 1999).

Exposure of the leaves to light induces phosphorylation of the PS II core proteins, which occurs exclusively in the grana thylakoids (Step 1 in Fig. 5) (Callahan et al., 1990; Stefansson et al., 1995). The number of the PS II centers containing phosphorylated D1, D2 and CP43 proteins depends on the physiological stage of the leaves and on the environmental conditions (Section III). Under high light irradiances most of the PS II centers in the grana have these proteins in their phosphorylated form (Rintamäki et al., 1996a, 1997; Baena-Gonzalez et al., 1999). Therefore it is very likely that in high light the damage to D1 occurs in the phosphorylated PS II dimers (Step 2 in Fig. 5). Furthermore, a strict coordination of the degradation and synthesis of the D1 protein is particularly important at high irradiances

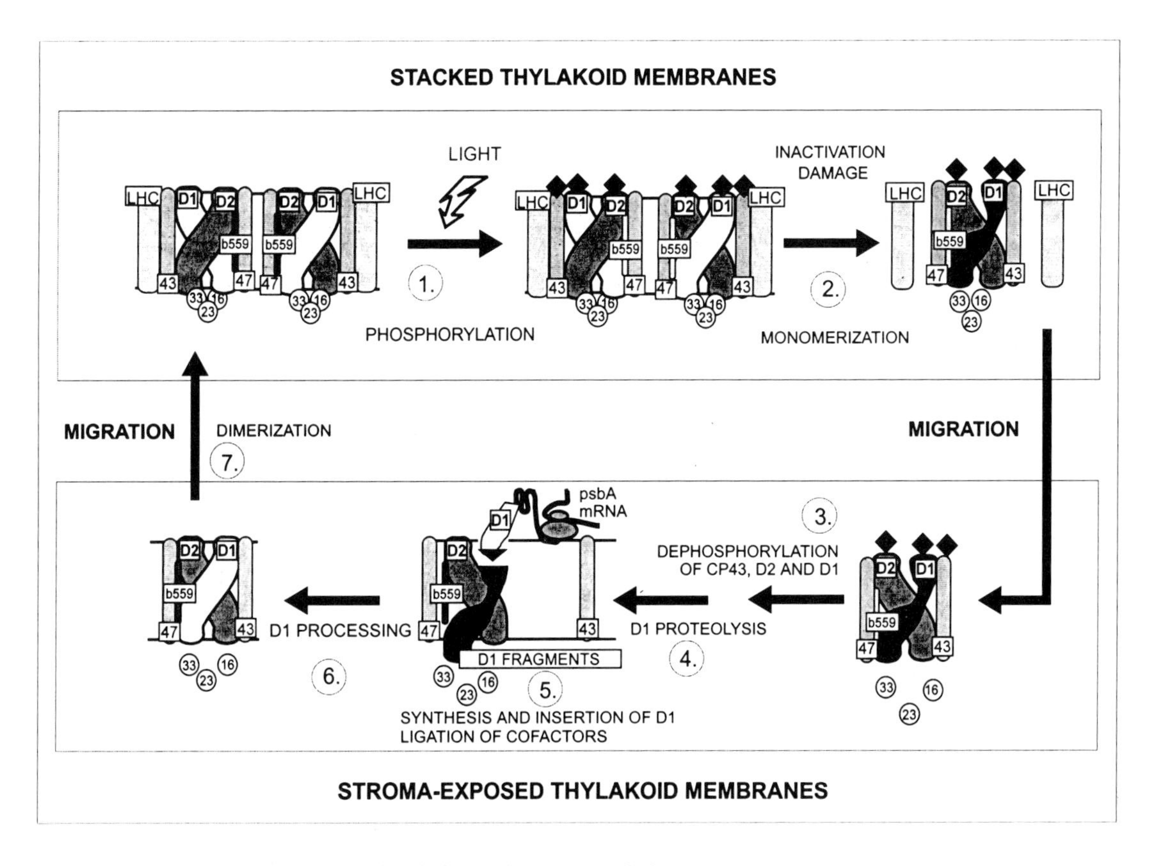

Fig. 5. Phosphorylation of thylakoid proteins during various stages of Photosystem II photodamage-repair cycle in higher plants. Phosphorylation and migration of the photodamaged PS II from grana region of thylakoids to stroma-exposed membranes for repair are presented. LHCII phosphorylation occurs mainly at low light and is not shown in the figure. The various steps of the cycle, marked from 1 to 7, are described in the text. See also Color Plate 6.

due to limited capacity of chloroplasts to repair photodamaged PS II complexes, which thus accumulate in the thylakoid membranes under high light condition (Greer et al., 1986; Tyystjärvi et al., 1992; Aro et al., 1993a, 1994; Rintamäki et al., 1996a).

Although the photodamage to PS II occurs exclusively in the grana lamellae (Guenther and Melis, 1990), the cotranslational incorporation of a new D1 protein into the partially disassembled PS II subcomplexes (Wijk et al., 1997; Zhang et al., 1999) is spatially possible only in stroma-exposed membranes. Accumulation of the D1 degradation fragments preferentially in the stroma lamellae also indicates that proteolysis of the D1 protein occurs in the stroma-exposed membranes (Barbato et al., 1991). Thus the photodamaged PS II centers in the grana dissociate from the antenna complex and migrate to the stroma-exposed thylakoid membranes. Barbato et al. (1992) showed that monomerization of the PS II complexes precedes the repair process (Step 2 in Fig. 5). Upon migration of PS II monomers to the stroma-exposed membranes, the D1, D2 and CP43 proteins remain phosphorylated (Baena-Gonzalez et al., 1999). This prevents the degradation of the damaged D1 protein in the place and at the time when a new D1 copy cannot be inserted. After dephosphorylation of these phosphoproteins in the stroma lamellae and dissociation of CP43 from the PS II subcomplex (Barbato et al., 1992; Baena-Gonzalez et al., 1999) (Step 3 in Fig. 5), proteolysis of the photodamaged D1 protein can take place (Step 4 in Fig. 5).

Dephosphorylation of the damaged D1 protein is an essential step in the PS II repair cycle (Rintamäki et al., 1996a). Moreover, this reaction is light-dependent and does not occur in darkness (Section III). This is either due to the light regulation of the migration of the damaged PS II complex from the grana region to the stroma-exposed membranes and/or to the light regulation of a specific protein phosphatase. The putative light-dependent phosphatase can be a soluble enzyme or a protein associated with the stroma-exposed membranes (Vener et al., 1999). Elich et al. (1993) have earlier reported that the light-induced dephosphorylation of the D1 protein in vivo is dependent on the excitation of PS I. Thus the light-dependent D1-phosphatase may be regulated by electron transport through PS I.

Dephosphorylation of the phosphorylated and damaged D1 is an excellent candidate for the regulatory step that controls the rate of D1 protein turnover: the activity of this reaction is adjusted to the capacity of D1 synthesis under various environmental conditions. Although an inactivation in the phosphate turnover of the D1 protein has been reported in illuminated chloroplasts (Ebbert and Godde, 1994, 1996), it is difficult to apply this argument to the in vivo observations, which clearly demonstrate that the active phosphorylation and dephosphorylation of the D1 protein occurs in light (Aro et al., 1993a, 1994, Elich et al., 1993; Rintamäki et al., 1996a). Therefore it seems reasonable to assume that isolated and illuminated chloroplasts are not able to sustain such an active phosphorylation/dephosphorylation cycle of the D1 protein detected in vivo.

Not only dephosphorylation of the damaged D1 protein but also the protease involved in the primary cleavage of the D1 protein has been suggested to control the turnover rate of the D1 protein (Kim et al., 1993; Bracht and Trebst, 1994; Trebst and Soll-Bracht, 1996). Photodamaged PS II complexes accumulate during high light illumination of green algae (Gong and Ohad, 1991; Kim et al., 1993; Zer et al., 1994; Zer and Ohad, 1995) or a moss (Rintamäki et al., 1994), the species, in which the regulation of D1 protein turnover via phosphorylation is missing (de Vitry et al., 1991; Pursiheimo et al., 1998). This delay in the degradation of the D1 protein in high light is either due to the modification of the substrate (Zer et al., 1994; Zer and Ohad, 1995) or to the regulation of the protease activity (Kim et al., 1993; Vasilikiotis and Melis, 1995) e.g. via reversible phosphorylation of the protease (Bracht and Trebst, 1994). Aggregation of the photodamaged PS II centers in a green alga, *Dunaliella,* has also been suggested to control D1 protein turnover by reducing the degradation of D1 in high light (Kim et al., 1993). Very little is known about regulation of the activity of the D1 protease in higher plants in vivo, and indeed the photodamaged D1 protein seems to undergo proteolytic cleavage immediately after dephosphorylation. This, however, does not rule out the possibility that the protease activity would be regulated also in higher plant chloroplasts. The D1 protease is of serine type (Salter et al., 1992) and recently it has been shown that the primary cleavage of the D1 protein in spinach thylakoids is a GTP-dependent process (Spetea et al., 1999). It will be interesting to find out, how the GTP-dependent regulation of the D1 protease is associated with the light-intensity-dependent turnover of the D1 protein.

Partial disassembly of the PS II complex in the stroma lamellae facilitates the degradation and subsequent cotranslational insertion of a new D1 copy into the complex (Step 5 in Fig. 5). Synthesis intermediates of the D1 protein have been shown to interact with the D2 protein, indicating co-translational assembly of the nascent D1 protein (Zhang et al., 1999). A very low turnover rate of the other PS II core proteins, D2, CP43 and CP47 (Kettunen et al., 1997; Wijk et al., 1997; Zhang et al., 2000), implies that these proteins are reused to build up the functional PS II. The precursor D1 protein is processed at the C-terminus (Diner et al., 1988) upon reassembly of the CP43 protein into the complex (Zhang et al., 1999) (Step 6 in Fig. 5) and before the remigration of the repaired PS II complex to the grana appressions and dimerization of the PS II complex (Step 7 in Fig. 5) (Chapter 22, Andersson and Aro).

As discussed in Section IV, reversible phosphorylation is not an universal mechanism to regulate the D1 protein turnover in organisms performing oxygenic photosynthesis. This raises the question, why has the phosphorylation of the D1 protein evolved in the seed plants? The need for a strict regulation of the repair of PS II may be reflected in the structural organization of thylakoid membranes built up into tightly appressed grana regions containing the functional PS II centers and into stroma-exposed membranes with the facilities for cotranslational protein insertion into PS II. Cyanobacteria and red algae have a unilamellar thylakoid organization with a dynamic association of soluble antenna complexes to the reaction centers (Rögner et al., 1996) thus providing a direct contact of the PS II complexes with the soluble compartment. On the other hand, grana organization in green algae and bryophytes is less compact than in higher plants (Aro, 1982; Staehelin, 1986). Furthermore, high-light induced thylakoid swelling has been reported in *Chlamydomonas* (Prasil et al., 1992; Topf et al., 1992) that may facilitate the repair process of PS II centers without any 'long-distance' migration of the complexes (Rintamäki et al., 1995). Moreover, a net loss of thc D1 protein in a green alga (Schuster et al., 1988) and in a moss (Rintamäki et al., 1994) under physiological light conditions indicates a lack of tight coordination between the proteolysis and synthesis of the D1 protein as observed in the seed plants.

B. Reversible Phosphorylation of Lhcb Proteins

The induction of state transition is a classical function attributed to the phosphorylation of the Lhcb1 and Lhcb2 proteins (Allen, 1992). The state transition is considered as a short-term acclimation mechanism of plants to changes in light climate by balancing the distribution of excitation energy between PS II and PS I. High light intensities, however, inactivate the phosphorylation of Lhcb1 and Lhcb2 proteins in vivo (Section III.C.3), and, on the other hand, induce a phosphorylation of the minor chlorophyll *a/b* protein, Lhcb4 (Fig. 3) (Bergantino et al., 1995). Phosphorylation of Lhcb4 has been proposed to protect PS II against photoinhibition (Bassi et al., 1997). This argument was based on parallel increase in the resistance of maize cultivars to cold-stress induced photoinhibition and in the phosphorylation level of Lhcb4 protein (Bergantino et al., 1995). Such protection may be acquired via different partition of excitation energy between the PS II center and its antenna after phosphorylation of Lhcb4 (Mauro et al., 1997).

VI. Future Perspective: Is Thylakoid Protein Phosphorylation Involved in the Relay of Signals for Acclimation Processes?

Phosphorylation-induced reduction of the light-harvesting capacity of PS II would, in theory, be a good mechanism to protect PS II against high-light-induced inactivation. Experiments on reversible phosphorylation and migration of LHCII complexes between PS II and PS I, conducted under nonphysiological light conditions (red and far red light to excite PS II and PS I, respectively) and with isolated systems, provided strong support for such a theory. Recently, however, it was shown that mere electron flow through the plastoquinone pool and the cytochrome b_6f complex, occurring in vivo at low light conditions, is enough to activate the LHCII kinase both in vivo and in vitro (Fig. 4) (Vener et al., 1995, 1997; Rintamäki et al., 1997; Salonen et al., 1998; Pursiheimo et al., 2001). Moreover, the phosphorylation of LHCII was shown to become inactivated in vivo under moderate and high light conditions (Rintamäki et al., 1997). Thus, the organizational rearrangements of the PS II antenna in vivo, based on the phosphorylation of Lhcb1 and Lhcb2 proteins, cannot provide any protection against high-light-induced photoinhibition of PS II. Indeed, the highest phosphorylation level of the Lhcb1 and Lhcb2 proteins in vivo occurs at light intensities well below the growth light level (Fig. 2) (Rintamäki et al., 1997). It is conceivable that under those suboptimal light conditions favoring the phosphorylation of LHCII, increased generation of ATP is required as compared to that of reducing power (Allen, 1992). Moreover, under low light conditions specific acclimation processes are initiated that result in an increase in the light-harvesting capacity of PS II. Accordingly, it can be asked, weather the reversible LHCII phosphorylation or the activation state of the kinase involved in LHCII phosphorylation, participates in the relay of information from chloroplasts to the nuclear genome in order to enhance the expression of e.g. *Lhcb* genes (Pursiheimo et al., 2001).

As discussed in Section III.C.3, the LHCII kinase is regulated via a complicated network (Fig. 4) involving redox control both via plastoquinol occupying the Q_0 site in the cytochrome b_6f complex (Gal et al., 1990; Vener et al., 1995,1997; Zito et al., 1999; Finazzi et al., 2001) and via thiol redox state of chloroplast (Rintamäki et al., 2000). Studies on LHCII phosphorylation in rye plants grown under combinations of various light intensities and temperatures have revealed LHCII phosphorylation to act as a sensitive sensor for changes in the ambient environment (Pursiheimo et al., 2001). Therefore, it can be speculated that the LHCII kinase is involved in signaling enhancing the expression of *Lhcb* genes, and probably also of other nuclear genes under conditions where the light harvesting efficiency and consequent electron transfer in PS II do not meet the requirements of carbon fixation reactions. Antisense plants of TAK kinases (Section III.B) have recently been constructed (S. Snyders and B. D. Kohorn, personal communication). Interestingly, these plants showed not only a loss of state transitions but also anomalies in plant development, suggesting a pivotal, still unknown role for TAK kinases.

Phosphorylation of the Lhcb4 protein has also been suggested to be involved in signal transduction pathways controlling the nuclear or chloroplastic gene expression (Testi et al., 1996; Bassi et al., 1997). The phosphorylation site of Lhcb4 shows homology to the motif in proteins phosphorylated by the casein kinase II (Testi et al., 1996), a known component of cellular signaling networks (Inagaki et al., 1994), making Lhcb4 a putative candidate for the chloroplast redox sensor involved in environmental acclimation processes (Bassi et al., 1997)

Although phosphorylation of PS II core proteins seems to play a crucial role in the regulation of PS II photoinhibition-repair cycle (Fig. 5), the reversible phosphorylation of D1 (or PS II core proteins in general) may also be involved, directly or indirectly, in signaling pathways regulating gene expression. Regulation of the chloroplast gene expression via the redox state of the plastoquinone pool has recently been reported (Pfannschmidt et al., 1999). These studies were based on growth of plants in light favoring PS II or PS I excitation and on a subsequent short-term transfer of plants into opposite light regimes. In these studies, the light favoring PS I excitation enhanced transcription of the chloroplastic *psbA* gene encoding the D1 protein, while the light favoring PS II excitation reduced the transcription rate of this gene (Pfannschmidt et al., 1999). Similar light quality experiments showed a converse effect on the transcription activity of the *psaAB* genes encoding reaction center proteins of PS I. Accordingly, such variations in light quality lead to optimization of the ratio of the two photosystems in thylakoid membranes. Similar changes in the photosystem stoichiometry can be observed by growing plants under different light intensities (Anderson et al., 1988; Anderson and Aro, 1994). On the other hand, both the quality and the quantity of light modulate the redox state of chloroplast, and accordingly also regulate the activation state of PS II kinases and/or phosphatases. It is conceivable that such induced reversible phosphorylations are in turn involved in signaling cascade regulating chloroplast gene expression.

Acknowledgments

The authors wish to thank Dr. Eira Kanervo for critical reading of the manuscript and Mr. Kurt Ståhle for preparing the figures for this chapter. This work was supported by the Academy of Finland and by EU INCO-Copernicus (IC15-CT98-0126).

References

Adir N, Shochat S and Ohad I (1990) Light-dependent D1 protein synthesis and translocation is regulated by reaction center II. J Biol Chem 265: 12563–12568

Allen JF (1992) Protein phosphorylation in regulation of photosynthesis. Biochim Biophys Acta 1098: 275–335

Allen JF (1995) Thylakoid protein phosphorylation, state-1–state-2 transitions, and photosystem stoichiometry adjustment: redox control at multiple levels of gene expression. Physiol Plant 93: 196–205

Allen JF and Bennett J (1981) Photosynthetic protein phosphorylation in intact chloroplasts: Inhibition by DCMU and by the onset of CO_2 fixation. FEBS Lett 123: 67–70

Allen JF, Bennett J, Steinback KE and Arntzen CJ (1981) Chloroplast protein phosphorylation couples plastoquinone redox state to distribution of excitation energy between photosystems. Nature 291: 21–25

Anderson JM and Aro EM (1994) Grana stacking and protection of Photosystem II in thylakoid membranes of higher plant leaves under sustained high irradiance: An hypothesis. Photosynth Res 41: 315–326

Anderson JM, Chow WS and Goodchild DJ (1988) Thylakoid membrane organisation in sun/shade acclimation. Aust J Plant Physiol 15: 11–26

Andersson B and Aro EM (1997) Proteolytic activities and proteases of plant chloroplasts. Physiol Plant 100: 780–793

Andronis C, Kruse O, Deak Z, Vass I, Diner BA and Nixon PJ (1998) Mutation of the residue threonine-2 of the D2 polypeptide and its effect on Photosystem II function in *Chlamydomonas reinhardtii*. Plant Physiol 117: 515–524

Aro EM (1982) A comparison of the chlorophyll-protein composition and chloroplast ultrastructure in two bryophytes and two higher plants. Z Pflanzenphysiol 108: 97–105

Aro EM, Kettunen R and Tyystjärvi E (1992) ATP and light regulate D1 protein modification and degradation. Role of D1* in photoinhibition. FEBS Lett 297: 29–33

Aro EM, McCaffery S and Anderson J (1993a) Photoinhibition and D1 protein degradation in peas acclimated to different growth irradiances. Plant Physiol 103: 835–843

Aro EM, Virgin I and Andersson B (1993b) Photoinhibition of Photosystem II. Inactivation, protein damage and turnover. Biochim Biophys Acta 1143: 113–134.

Aro EM, McCaffery S and Anderson JM (1994) Recovery from photoinhibition in peas (*Pisum sativum* L.) acclimated to varying growth irradiances. Plant Physiol 104: 1033–1041

Baena-Gonzalez E, Barbato R and Aro EM (1999) Role of phosphorylation in the repair cycle and oligomeric structure of Photosystem II. Planta 208: 196–204

Barbato R, Friso G, Giardi MT, Rigoni F and Giacometti GM (1991) Breakdown of the Photosystem II reaction center D1 protein under photoinhibitory conditions: Identification and localization of the C-terminal degradation products. Biochemistry 30: 10220–10226

Barbato R, Friso G, Rigoni F, Dalla Veccia F and Giacometti GM (1992) Structural changes and lateral redistribution of Photosystem II during donor side photoinhibition of thylakoids. J Cell Biol 119: 325–335

Bassi R, Sandona D and Croce R (1997) Novel aspect of chlorophyll *a/b*-binding proteins. Physiol Plant 100: 769–779

Bennett J (1977) Phosphorylation of chloroplast membrane polypeptides. Nature 269: 344 –346

Bennett J (1979) Chloroplast phosphoproteins. The protein kinase of thylakoid membranes is light-dependent. FEBS Lett 103: 342–344

Bennett J (1980) Chloroplast phosphoproteins. Evidence for a thylakoid-bound phosphoprotein phosphatase. Eur J Biochem 104: 85–89

Bennett J (1991) Protein phosphorylation in green plant chloroplasts. Annu Rev Plant Physiol Plant Mol Biol 42: 281–311

Bennett J, Shaw EK and Michel H (1988) Cytochrome b_6f complex is required for phosphorylation of light-harvesting chlorophyll *a/b* complex II in chloroplast photosynthetic membranes. Eur J Biochem 171: 95–100

Bergantino E, Dainese P, Cerovic Z, Sechi S and Bassi R (1995) A post-translational modification of the Photosystem II subunit CP29 protects maize from cold stress. J Biol Chem 270: 8474–8481

Bergantino E, Sandona D, Cugini D and Bassi R (1998) The Photosystem II subunit CP29 can be phosphorylated in both C3 and C4 plants as suggested by sequence analysis. Plant Mol Biol 36: 11–22

Bhalla P and Bennett J (1987) Chloroplast phosphoproteins: Phosphorylation of a 12-kDa stromal protein by the redox-controlled kinase of thylakoid membranes. Arch Biochem Biophys 252: 97–104

Boekema EJ, van Roon H, Calkoen F, Bassi R and Dekker JP (1999) Multiple types of association of Photosystem II and its light-harvesting antenna in partially solubilized Photosystem II membranes. Biochemistry 38: 2233–2239

Bracht E and Trebst A (1994) Hypothesis on the control of D1 protein turnover by nuclear coded proteins in *Chlamydomonas reinhardtii*. Z Naturforsch 49c: 439–446

Bricker TM (1990) The structure and function of CPa-1 and CPa-2 in Photosystem II. Photosynth Res 24: 1–13

Bukharov AA, Kolosov VL, Klezovich ON and Zolotarev AS (1989) Nucleotide sequence of rye chloroplast DNA fragment, comprising *psbD*, *psbC* and *trnS* genes. Nucl Acids Res 17: 798

Callahan FE, Ghirardi ML, Sopory SK, Mehta AM, Edelman M and Mattoo AK (1990) A novel metabolic form of the 32 kDa-D1 protein in grana-localized reaction center of Photosystem II. J Biol Chem 265: 15357–15360

Carlberg I and Andersson B (1996) Phosphatase activities in spinach thylakoid membranes—effectors, regulation and location. Photosynth Res 47: 145–156

Carlberg I, Rintamäki E, Aro EM and Andersson B (1999) Thylakoid protein phosphorylation and the thiol redox state. Biochemistry 38: 3197–3204

Cheng L, Spangfort MD and Allen JF (1994) Substrate specificity and kinetics of thylakoid phosphoproteins phosphatase reactions. Biochim Biophys Acta 1188: 151–157

Cui M, Vogelmann TC and Smith WK (1991) Chlorophyll and light gradients in sun and shade leaves of *Spinacia oleracea*. Plant Cell Environ 14: 493–500

Danon A and Mayfield SP (1994) ADP-dependent phosphorylation regulates RNA-binding in vitro: Implications in light-modulated translation. EMBO J 13: 2227–2235

de Vitry C, Diner BA, Popot JL (1991) Photosystem II particles from *Chlamydomonas reinhardtii*. Purification, molecular weight, small subunit composition, and protein phosphorylation. J Biol Chem 266: 16614–16621

Dietz KJ, Schreiber U and Heber U (1985) The relationship between the redox state of Q_A and photosynthesis in leaves at various carbon-dioxide, oxygen and light regimes. Planta 166: 219–226

Dilly-Hatrwig H, Allen JF, Paulsen H and Race HL (1998) Truncated recombinant light harvesting complex II proteins are substrates for a protein kinase associated with Photosystem II core complexes. FEBS Lett 435: 101–104

Diner BA, Ries DF, Cohen BN and Metz JG (1988) COOH-terminal processing of polypeptide D1 of the Photosystem II reaction center of *Scenedesmus obliquus* is necessary for the assembly of the oxygen-evolving complex. J Biol Chem 263: 8972–8980

Ebbert V and Godde D (1994) Regulation of thylakoid protein phosphorylation in intact chloroplasts by the activity of kinases and phosphatases. Biochim Biophys Acta 1187: 335–346

Ebbert V and Godde D (1996) Phosphorylation of PS II polypeptides inhibits protein-degradation and increases PS II stability. Photosynth Res 50: 257–269

Elich TD, Edelman M and Mattoo AK (1992) Identification, characterization, and resolution of the in vivo phosphorylated form of the D1 Photosystem II reaction center protein. J Biol Chem 267: 3523–3529

Elich TD, Edelman M and Mattoo AK (1993) Dephosphorylation of Photosystem II core proteins is light-regulated in vivo. EMBO J 12: 4857–4862

Erickson JM, Rahire M and Rochaix JD (1984) *Chlamydomonas reinhardtii* gene for the 32000 mol. wt. protein of Photosystem II contains four large introns and is located entirely within the chloroplast inverted repeat. EMBO J 3: 2753–2762

Erickson JM, Rahire M, Malnoe P, Girard-Bascou J, Pierre Y, Bennoun P and Rochaix JD (1986) Lack of the D2 protein in a *Chlamydomonas reinhardtii psbD* mutant affects Photosystem II stability and D1 expression. EMBO J 5: 1745–1754

Farchaus J, Dilley RA and Cramer WA (1985) Selective inhibition of the spinach thylakoid LHC II protein kinase. Biochim Biophys Acta 809: 17–26

Finazzi G, Zito F, Barbagallo RP and Wollman FA (2001) Contrasted effects of inhibitors of cytochrome b_6f complex on state transitions in *C. reinhardtii*: the role of Q_o site occupancy in LHCII-kinase activation. J Biol Chem 276: 9770–9774

Fleischmann MM and Rochaix JD (1999) Characterization of mutants with alteration of the phosphorylation site in the D2 Photosystem II polypeptide of *Chlamydomonas reinhardtii*. Plant Physiol 119: 1557–1566

Foyer CH (1985) Stromal protein phosphorylation in spinach (*Spinacia oleracea)* chloroplasts. Biochem J 231: 97–103

Frid D, Gal A, Oettmeier W, Hauska G, Berger S and Ohad I (1992) The redox-controlled light-harvesting chlorophyll *a/b* protein kinase. Deactivation by substituted quinones. J Biol Chem 267: 25908–25915

Fulgosi H, Vener AV, Altschmied L, Herrmann RG and Andersson B (1998) A novel multi-functional chloroplast protein: Identification of a 40 kDa immunophilin-like protein located in the thylakoid lumen. EMBO J 17: 1577–1587

Gal A, Shahak Y, Schuster G and Ohad I (1987) Specific loss of LHCII phosphorylation in *Lemna* mutant 1073 lacking the cytochrome b_6/f complex. FEBS Lett 221: 205–210

Gal A, Hauska G, Herrmann RG and Ohad I (1990) Interaction between light harvesting chlorophyll-*a/b* protein (LHCII) kinase and cytochrome b_6/f complex. In vitro control of kinase activity. J Biol Chem 265: 19742–19749

Gal A, Zer H and Ohad I (1997) Redox-controlled thylakoid protein phosphorylation. News and views. Physiol Plant 100: 863–868

Garcia Vescovi E and Lucero HA (1990) Phosphorylation of serine residues in endogenous proteins of thylakoids and subthylakoid particles in the dark under nonreducing conditions. Biochim Biophys Acta 1018: 23–28

Giardi MT (1993) Phosphorylation and disassembly of the Photosystem II core as an early stage of photoinhibition. Planta 190: 107–113

Giardi MT, Rigoni F and Barbato R (1992) Photosystem II core phosphorylation heterogeneity, differential herbicide binding, and regulation of electron transfer in Photosystem II preparations from spinach. Plant Physiol 100: 1948–1954

Giardi MT, Kucera T, Briantais JM and Hodges M (1995) Decreased Photosystem II core phosphorylation in a yellow-green mutant of wheat showing monophasic fluorescence induction curve. Plant Physiol 109: 1059–1068

Golden SS and Stearns GW (1988) Nucleotide sequence and transcript analysis of the three Photosystem II genes from the cyanobacterium *Synechococcus* sp. PCC7942. Gene 67: 86–96

Gong H and Ohad I (1991) The PQ/PQH_2 ratio and occupancy of Photosystem II-Q_B site by plastoquinone control the degradation of D1 protein during photoinhibition in vivo. J Biol Chem 286: 21293–21299

Greer DH, Berry JA and Björkman O (1986) Photoinhibition of photosynthesis in intact bean leaves: Role of light and temperature, and requirement for chloroplast-protein synthesis during recovery. Planta 168: 253–260

Guenther JE and Melis A (1990) The physiological significance of Photosystem II heterogeneity in chloroplasts. Photosynth Res 23: 105–109

Habash DZ and Baker NR (1990) Demonstration of two sites of inhibition of electron transport by protein phosphorylation in wheat thylakoids. J Exp Bot 41: 761–767

Hamel P, Olive J, Pierre Y, Wollman FA and de Vitry C (2000) A new subunit of cytochrome b6f complex undergoes reversible phosphorylation upon state transition. J Biol Chem 275: 17072–17079

Hammer MF, Sarath G and Markwell J (1995a) Dephosphorylation of the thylakoid membrane light-harvesting complex-II by stromal protein phosphatase. Photosynth Res 45: 195–201

Hammer MF, Sarath G, Osterman JC and Markwell J (1995b) Assessing modulation of stromal and thylakoid light-harvesting complex-II phosphatase activities with phosphopeptide substrates. Photosynth Res 44: 107–115

Hammer MF, Markwell J and Sarath G (1997) Purification of a protein phosphatase from chloroplast stroma capable of dephosphorylating the light-harvesting complex-II. Plant Physiol 113: 227–233

Harrison MA and Allen JF (1991) Light-dependent phosphorylation of Photosystem II polypeptides maintains electron transport at high light intensity: Separation from effects of phosphorylation of LHC-II. Biochim Biophys Acta 1058: 289–296

Hast T and Follmann H (1996) Identification of two thylakoid-associated phosphatases with protein phosphatase activity in chloroplasts of the soybean (*Glycine max)*. J Photochem Photobiol B 36: 313–319

Hayden DB, Covello PS and Baker NR (1988) Characterization of a 31 kDa polypeptide that accumulates in the light-harvesting apparatus of maize leaves during chilling. Photosynth Res 15: 257–270

Hodges M, Boussac A and Briantais JM (1987) Thylakoid

membrane protein phosphorylation modifies the equilibrium between Photosystem II quinone electron acceptors. Biochim Biophys Acta 894: 138–145

Hollinderbäumer R, Ebbert V and Godde D (1997) Inhibition of CO_2-fixation and its effect on the activity of Photosystem II, on D1-protein synthesis and phosphorylation. Photosynth Res 52: 105–116

Holschuh K, Bottomley W and Whitfeld PR (1984) Structure of the spinach chloroplast genes for the D2 and 44 kd reaction-centre proteins of Photosystem II and for tRNASer (UGA). Nucl Acids Res 12: 8819–8834

Horton P and Lee P (1984) Phosphorylation of chloroplast thylakoids decreases the maximum capacity of Photosystem 2 electron transfer. Biochim Biophys Acta 767: 563–567

Horton P and Lee P (1985) Phosphorylation of chloroplast membrane proteins partially protects against photoinhibition. Planta 165: 37–42

Huner NPA, Maxwell DP, Gray GR, Savitch LV, Krol M, Ivanov AG and Falk S (1996) Sensing environmental temperature change through imbalances between energy supply and energy consumption: Redox state of Photosystem II. Physiol Plant 98: 358–364

Ikeuchi M, Plumley FG, Inoue Y and Schmidt GW (1987) Phosphorylation of Photosystem II components, CP 43 apoprotein, D1, D2, and 10 to 11 kilodalton protein in chloroplast thylakoid of higher plants. Plant Physiol 85: 638–642

Inagaki N, Ito M, Nakano T and Inagaki M (1994) Spatiotemporal distribution of protein kinase and phosphatase activities. Trends Biochem Sci 19: 448–452

Islam K (1987) The rate and extent of phosphorylation of the two light-harvesting chlorophyll *a/b* binding protein complex (LHC-II) polypeptides in isolated spinach thylakoids. Biochim Biophys Acta 893: 333–341

Jansson S (1994) The light-harvesting chlorophyll *a/b*-binding proteins. Biochim Biophys Acta 1184: 1–19

Jansson S (1999) A guide to the *Lhc* genes and their relatives in *Arabidopsis*. Trends Plant Sci 4: 236–240

Jansson S, Pichersky E, Bassi R, Green BR, Ikeuchi M, Melis A, Simpson DJ, Spangfort M, Staehelin LA and Thornber JP (1992) A nomenclature for the genes encoding the chlorophyll *a/b* binding proteins of higher plants. Plant Mol Biol Rep 10: 242–253

Kaneko T, Sato S, Kotani H, Tanaka A, Asamizu E, Nakamura Y, Miyajima N, Hirosawa M, Sugiura M, Sasamoto S, Kimura T, Hosouchi T, Matsuno A, Muraki A, Nakazaki N, Naruo K, Okumura S, Shimpo S, Takeuchi C, Wada T, Watanabe A, Yamada M, Yasuda M and Tabata S (1996) Sequence analysis of the genome of the unicellular cyanobacterium *Synechocystis* sp. strain PCC6803. II. Sequence determination of the entire genome and assignment of potential protein-coding regions. DNA Res 3: 109–136

Keren N, Gong H and Ohad I (1995) Oscillation of the reaction center II-D1 protein degradation in vivo induced by repetitive light flashes. Correlation between the level of RCII-Q_B^- and protein degradation in low light. J Biol Chem 270: 806–814

Kettunen R, Tyystjärvi E and Aro EM (1991) D1 protein degradation during photoinhibition of intact leaves. A modification of the D1 protein precedes degradation. FEBS Lett 290: 153–156

Kettunen R, Pursiheimo S, Rintamäki E, Wijk KJ and Aro EM (1997) Transcriptional and translational adjustment of *psbA* gene expression in mature chloroplasts during photoinhibition and subsequent repair of Photosystem II. Eur J Biochem 247: 441–448

Kieselbach T, Hagman Å, Andersson B and Schröder WP (1998) The thylakoid lumen of chloroplasts. Isolation and characterization. J Biol Chem 273: 6710–6716

Kim JH, Nemson JA and Melis A (1993) Photosystem II reaction center damage and repair in *Dunaliella salina* (green alga). Analysis under physiological and irradiance-stress conditions. Plant Physiol 103: 181–189

Koivuniemi A, Aro EM and Andersson B (1995) Degradation of D1- and D2-proteins of Photosystem II in higher plants is regulated by reversible phosphorylation. Biochemistry 34: 16022–16029

Kolosov VL, Bukharov AA and Zolotarev AS (1989) Nucleotide sequence of the rye chloroplast *psbA* gene, encoding D1 protein of Photosystem II. Nucl Acids Res 17: 1759

Komenda J and Barber J (1995) Comparison of *psbO* and *psbH* mutants of *Synechocystis* PCC 6803 indicates that degradation of D1 protein is regulated by the Q_B site and dependent on protein synthesis. Biochemistry 34: 9625–9631

Krause GH and Weis E (1991) Chlorophyll fluorescence and photosynthesis: The basics. Annu Rev Plant Physiol Plant Mol Biol 42: 313–349

Krebs EG (1994) The growth of research on protein phosphorylation. Trends Biochem Sci 227: 439–518

Kruse O, Zheleva D and Barber J (1997) Stabilization of photosystem two dimers by phosphorylation: Implication for the regulation of the turnover of D1 protein. FEBS Lett 408: 276–280

Larsson UK, Anderson JM and Andersson B (1987a) Variation in the relative content of the peripheral and tightly bound LHC II subpopulations during thylakoid light adaptation and development. Biochim Biophys Acta 894: 69–75

Larsson UK, Sundby C and Andersson B (1987b) Characterization of two different subpopulations of spinach light-harvesting chlorophyll *a/b*-protein complex (LHCII): Polypeptide composition, phosphorylation pattern and association with Photosystem II. Biochim Biophys Acta 894: 59–68

Lindahl M, Yang D-H Andersson B (1995a) Regulatory proteolysis of the major light-harvesting chlorophyll *a/b* protein of Photosystem II by a light-induced membrane-associated enzymic system. Eur J Biochem 231: 503–509

Lindahl M, Carlberg I, Schröder WP and Andersson B (1995b) Characterization of a 12 kDa phosphoprotein from spinach thylakoid. In: Mathis P (ed) Photosynthesis: From Light to Biosphere, Vol III, pp 321–324. Kluwer Academic Publishers, Dordrecht

Mann NH (1994) Protein phosphorylation in cyanobacteria. Microbiology 140: 3207–3215

Mauro S, Dainese P, Lannoye R and Bassi R (1997) Cold-resistant and cold-sensitive maize lines differ in the phosphorylation of the Photosystem II subunit, CP29. Plant Physiol 115: 171–180

Michel HP and Bennett J (1987) Identification of the phosphorylation site of an 8.3 kDa protein from Photosystem II of spinach. FEBS Lett 212: 103–108

Michel H and Bennett J (1989) Use of synthetic peptides to study the substrate specificity of a thylakoid protein kinase. FEBS Lett 254: 165–170

Michel H, Hunt DF, Shabanowitz J and Bennett J (1988) Tandem mass spectrometry reveals that three Photosystem II proteins of spinach chloroplasts contain N-acetyl-O-phosphothreonine at their NH_2 termini. J Biol Chem 263: 1123–1130

Michel H, Griffin PR, Shabanowitz J, Hunt DF and Bennett J (1991) Tandem mass spectrometry identifies sites of three post-translational modifications of spinach light-harvesting chlorophyll protein II. J Biol Chem 266: 17584–17591

Millner PA, Widger WR, Abbott MS, Cramer WA and Dilley RA (1982) The effect of adenine nucleotides on inhibition of the thylakoid protein kinase by sulfhydryl-directed reagents. J Biol Chem 257: 1736–1742

Mullet JE (1983) The amino acid sequence of the polypeptide segment which regulates membrane adhesion (grana stacking) in chloroplasts. J Biol Chem 258: 9941–9948

Mäenpää P and Andersson B (1989) Photosystem II heterogeneity and long-term acclimation of light-harvesting. Z Naturforsch 44C: 403-406

O´Connor HE, Ruffle SV, Cain AJ, Deak Z, Vass I, Nugent JHA and Purton S (1998) The 9-kDa phosphoprotein of Photosystem II. Generation and characterisation of *Chlamydomonas* mutants lacking PS II-H and a site-directed mutant lacking the phosphorylation site. Biochim Biophys Acta 1364: 63–72

Ohyama K, Fukuzawa H, Kohchi T, Shirai H, Sano T, Sano S, Umesono K, Shiki Y, Takeuchi M, Chang Z, Aota S, Inokuchi H and Ozeki H (1986) Chloroplast gene organization deduced from complete sequence of liverwort *Marchantia polymorpha* chloroplast DNA. Nature 322: 572–574

Packham NK (1987) Phosphorylation of the 9 kDa Photosystem II-associated protein and the inhibition of photosynthetic electron transport. Biochim Biophys Acta 893: 259–266

Peter GF and Thornber JP (1991) Biochemical composition and organization of higher plant Photosystem II light-harvesting pigment-protein. J Biol Chem 266: 16745–16754

Pfansschmidt T, Nilsson A and Allen JF (1999) Photosynthetic control of chloroplast gene expression. Nature 397: 625–628

Prasil O, Adir N and Ohad I (1992) Dynamics of Photosystem II: Mechanism of photoinhibition and recovery processes. In: Barber J (ed) Topics in Photosynthesis, Vol. 11, pp 293–348. Elsevier, Amsterdam

Pursiheimo S, Rintamäki E, Baena-Gonzalez E and Aro EM (1998) Thylakoid protein phosphorylation in evolutionally divergent species with oxygenic photosynthesis. FEBS Lett 423: 178–182

Pursiheimo S, Mulo P, Rintamäki E and Aro EM (2001) Coregulation of light-harvesting complex II phosphorylation and *lhcb* mRNA accumulation in winter rye. Plant J 26: in press

Race HL and Hind G (1996) A protein kinase in the core of Photosystem II. Biochemistry 35: 13006–13010

Reith M and Munholland J (1995) Complete nucleotide sequence of the *Porhyra purpurea* chloroplast genome. Plant Mol Biol Rep 13: 333–335

Rhee KH, Morris EP, Barber J and Kuhlbrandt W (1998) Three-dimensional structure of the plant Photosystem II reaction centre at 8 Å resolution. Nature 396: 283–286

Rintamäki E, Salo R and Aro EM (1994) Rapid turnover of the D1 reaction-center protein of Photosystem II as a protection mechanism against photoinhibition in a moss, *Ceratodon purpureus* (Hedw.) Brid. Planta 193: 520–529

Rintamäki E, Salo R, Lehtonen E and Aro EM (1995) Regulation of D1 protein degradation during photoinhibition of Photosystem II in vivo: Phosphorylation of the D1 protein in various plant groups. Planta 195: 379–386

Rintamäki E, Kettunen R and Aro EM (1996a) Differential D1 dephosphorylation in functional and photodamaged Photosystem II centres. Dephosphorylation is a prerequisite for degradation of damaged D1*. J Biol Chem 271: 14870–14875

Rintamäki E, Salo R, Koivuniemi A and Aro EM (1996b) Protein phosphorylation and magnesium status regulate the degradation of D1 reaction centre protein of Photosystem II. Plant Science 115: 175–182

Rintamäki E, Salonen M, Suoranta UM, Carlberg I, Andersson B and Aro EM (1997) Phosphorylation of light-harvesting complex II and Photosystem II core proteins shows different irradiance-dependent regulation in vivo. Application of phosphothreonine antibodies to analysis of thylakoid phosphoproteins. J Biol Chem 272: 30476–30482

Rintamäki E, Carlberg I, Andersson B and Aro EM (1998) Irradiance-dependent regulation of thylakoid protein phosphorylation in vivo—The role of the thiol redox state. In: Garab G (ed) Photosynthesis: Mechanism and Effects, Vol III, pp 1899–1902. Kluwer Academic Publishers, Dordrecht

Rintamäki E, Martinsuo P, Pursiheimo S and Aro EM (2000) Cooperative regulation of light-harvesting complex II phosphorylation via plastoquinol and ferredoxin-thioredoxin system in chloroplast. Proc Natl Acad Sci USA 97: 11644–11649.

Rochaix JD, Kuchka M, Mayfield S, Schirmer-Rahire M, Girard-Bascou J and Bennoun P (1989) Nuclear and chloroplast mutations affect the synthesis or stability of the chloroplast *psbC* gene product in *Chlamydomonas reinhardtii*. EMBO J 8: 1013–1021

Rögner M, Boekema EJ and Barber J (1996) How does Photosystem 2 split water? The structural basis of efficient energy conversion. Trends Biochem Sci 21: 44–49

Salonen M, Aro EM and Rintamäki E (1998) eversible phosphorylation and turnover of the D1 protein under various redox states of Photosystem II induced by low temperature photoinhibition. Photosynth Res 58: 143–151

Salter AH, Virgin I, Hagman Å and Andersson B (1992) On the molecular mechanism of light-induced D1 protein degradation in Photosystem II core particles. Biochemistry 31: 3990–3998

Sanders CE, Holmes NG and Allen JF (1986) Membrane protein phosphorylation in the cyanobacterium *Synechococcus* 6301. Biochem Soc Trans 14: 66–67

Schnettger B, Critchley C, Santore UJ, Graf M and Krause GH (1994) Relationship between photoinhibition of photosynthesis, D1 protein turnover and chloroplast structure. Effects of protein synthesis inhibitors. Plant Cell Environ 17: 55–64

Schuster G, Dewit M, Staehelin LA and Ohad I (1986) Transient inactivation of the thylakoid Photosystem II light-harvesting protein kinase system and concomitant changes in intramembrane particle size during photoinhibition of *Chlamydomonas reinhardtii*. J Cell Biol 103: 71–80

Schuster G, Timberg R and Ohad I (1988) Turnover of thylakoid Photosystem II proteins during photoinhibition of *Chlamydomonas reinhardtii*. Eur J Biochem 177: 403–410.

Silverstein T, Cheng L and Allen JF (1993a) Redox titration of multiple protein phosphorylations in pea chloroplast thylakoids. Biochim Biophys Acta 1183: 215–220

Silverstein T, Cheng L and Allen JF (1993b) Chloroplast thylakoid protein phosphatase reactions are redox-independent and kinetically heterogeneous. FEBS Lett 332: 101–105

Simpson DJ and Knoetzel J (1996) Light-harvesting of plants and algae: Introduction, survey and nomenclature. In: Ort DR and Yocum CF (eds) Oxygenic Photosynthesis: The Light Reactions, pp 493–506. Kluwer Academic Publishers, Dordrecht

Smith TA and Kohorn BD (1991) Direct selection for sequences encoding proteases of known specificity. Proc Natl Acad Sci USA 88: 5159–5162

Snyders S and Kohorn BD (1999) TAKs, thylakoid membrane protein kinases associated with energy transduction. J Biol Chem 274: 9137–9140

Soll J and Bennett J (1988) Localization of a 64-kDa phosphoprotein in the lumen between the outer and inner envelopes of pea chloroplasts. Eur J Biochem 175: 301–307

Soll J, Fisher I and Keegstra K (1988) A guanosine 5′-triphosphate-dependent protein kinase is localized in the outer envelope membrane of pea chloroplasts. Planta 176: 488–496

Spetea C, Hundal T, Lohmann F and Andersson B (1999) GTP bound to chloroplast thylakoid membranes is required for light-induced multi-enzyme degradation of the Photosystem II D1 protein. Proc Natl Acad Sci USA 96: 6547–6552

Staehelin LA (1986) Chloroplast structure and supramolecular organization of photosynthetic membranes. Encyc Plant Physiol 19: 1–84

Stefansson H, Wollenberger L, Yu SG and Albertsson PÅ (1995) Phosphorylation of thylakoids and isolated subthylakoid vesicles derived from structural domains of the thylakoid membrane from spinach chloroplast. Biochim Biophys Acta 1231: 323–332

Sun G and Markwell J (1992) Lack of types 1 and 2A protein serine(P)/threonine(P) phosphatase activities in chloroplasts. Plant Physiol 100: 620–624

Sun G, Bailey D, Jones MW and Markwell J (1989) Chloroplast thylakoid protein phosphatase is a membrane surface-associated activity. Plant Physiol 89: 238–243

Sundby C, Larsson UK and Henrysson T (1989) Effects of bicarbonate on thylakoid protein phosphorylation. Biochim Biophys Acta 975: 277–282

Svensson B, Vass I and Styring S (1991) Sequence analysis of the D1 and D2 reaction center proteins of Photosystem II. Z Naturforsch 46c: 765–776

Testi MG, Croce R, Polverino-De Laureto P and Bassi R (1996) A CK2 site is reversibly phosphorylated in the Photosystem II subunit CP29. FEBS Lett 399: 245–250

Tiller K and Link G (1993) Phosphorylation and dephosphorylation affect functional characteristics of chloroplast and etioplast transcription systems from mustard (*Sinapis alba* L). EMBO J 12: 1745–1753

Topf J, Gong H, Timberg R, Mets L and Ohad I (1992) Thylakoid membrane energization and swelling in photoinhibited *Chlamydomonas* cells is prevented in mutants unable to perform cyclic electron flow. Photosynth Res 32: 59–69

Trebst A and Soll-Bracht E (1996) Cycloheximide retards high light driven D1 protein degradation in *Chlamydomonas reinhardtii*. Plant Sci 115: 191–197

Tsudzuki J, Nakashima K, Tsudzuki T, Hiratsuka J, Shibata M, Wakasugi T and Sugiura M (1992) Chloroplast DNA of black pine retains a residual inverted repeat lacking rRNA genes: Nucleotide sequences of *trnQ*, *trnK*, *psbA*, *trnI*, and *trnH* and the absence of *rps16*. Mol Gen Genet 232: 206–214

Tullber A, Håkansson G and Race HL (1998) A protein tyrosine kinase of chloroplast thylakoid membranes phosphorylates light harvesting complex II proteins. Biochem Biophys Res Commun 250: 617–622

Turmel M, Mercier JP and Cote MJ (1993) Group I introns interrupt the chloroplast *psaB* and *psbC* and the mitochondrial *rrnL* gene in *Chlamydomonas*. Nucl Acids Res 21: 5242–5250

Tyystjärvi E and Aro EM (1996) The rate constant of photoinhibition, measured in lincomycin-treated leaves, is directly proportional to light intensity. Proc Natl Acad Sci USA 93: 2213–2218

Tyystjärvi E, Ali-Yrkkö K, Kettunen R and Aro EM (1992) Slow degradation of the D1 protein is related to the susceptibility of low-light-grown pumpkin plants to photoinhibition. Plant Physiol 100: 1310–1317

Vasilikiotis C and Melis A (1995) The role of chloroplast-encoded protein biosynthesis on the rate of D1 protein degradation in *Dunaliella salina*. Photosynth Res 45: 147–155

Vener AV, van Kan PJM, Gal A, Andersson B and Ohad I (1995) Activation/deactivation cycle of redox-controlled thylakoid protein phosphorylation. Role of plastoquinone bound to the reduced cytochrome *bf* complex. J Biol Chem 270: 25225–25232

Vener AV, van Kan PJM, Rich PR, Ohad I and Andersson B (1997) Plastoquinol at the quinol oxidation site of the reduced cytochrome *bf* mediates signal transduction between light and protein phosphorylation: Thylakoid protein kinase deactivation by a single turnover flash. Proc Natl Acad Sci USA 94: 1585–1590

Vener AV, Ohad I and Andersson B (1998) Protein phosphorylation and redox sensing in chloroplast thylakoids. Current Opinion Plant Biol 1: 217–223

Vener AV, Rokka A, Fulgosi H, Andersson B and Herrmann RG (1999) A cyclophilin-regulated PP2A-like protein phosphatase in thylakoid membranes of plant chloroplasts. Biochemistry 38: 14955-14965

Vener AV, Harms A, Sussman MR and Vierstra RD (2001) Mass spectrometric resolution of reversible protein phosphorylation in photosynthetic membranes of *Arabidopsis thaliana*. J Biol Chem 276: 6959–6966

Vink M, Zer H, Herrmann RG, Ohad I and Andersson B (1998) Purification and identification of thylakoid protein kinase(s) that phosphorylates isolated LHCII and CP43 of Photosystem II. In: Garab G (ed) Photosynthesis: Mechanism and Effects, Vol III, pp 1887–1890. Kluwer Academic Publishers, Dordrecht

Wakasugi T, Tsudzuki J, Ito S, Nakashima K, Tsudzuki T and Sugiura M (1994) Loss of all ndh genes as determined by sequencing the entire chloroplast genome of the black pine *Pinus thunbergii*. Proc Natl Acad Sci USA 91: 9794–9798

Wakasugi T, Nagai T, Kapoor M, Sugita M, Ito M, Ito S, Tsudzuki J, Nakashima K, Tsudzuki T, Suzuki Y, Hamada A, Ohta T, Inamura A, Yoshinaga K and Sugiura M (1997) Complete nucleotide sequence of the chloroplast genome from the green alga *Chlorella vulgaris*: The existence of genes possibly involved in chloroplast division. Proc Natl Acad Sci USA 94: 5967–5972

van Wijk KJ, Roobol-Boza M, Kettunen R, Andersson B and Aro EM (1997) Synthesis and assembly of the D1 protein into Photosystem II: Processing of the C-terminus and identification of the initial assembly partners and complexes during

Photosystem II repair. Biochemistry 36: 6178–6186

Weber P, Fulgosi H, Sokolenko A, Karnauchov I, Andersson B, Ohad I and Herrmann RG (1998) Evidence for four thylakoid-located protein kinases. In: Garab G (ed) Photosynthesis: Mechanism and Effects, Vol III, pp 1883–1886. Kluwer Academic Publishers, Dordrecht

Wollman FA and Delepelaire P (1984) Correlation between changes in light energy distribution and changes in thylakoid membrane polypeptide phosphorylation in *Chlamydomonas reinhardtii*. J Cell Biol 98: 1–7

Zer H and Ohad I (1995) Photoinactivation of Photosystem II induces changes in the photochemical reaction center II abolishing the regulatory role of the Q_B site in the D1 protein degradation. Eur J Biochem 231: 448–453

Zer H, Prasil O and Ohad I (1994) Role of plastoquinol oxidoreduction in regulation of photochemical reaction center II D1 protein turnover in vivo. J Biol Chem 269: 17670–17676

Zhang L, Paakkarinen V, van Wijk KJ and Aro EM (1999) Co-translational assembly of the D1 protein into Photosystem II. J Biol Chem 274: 16062–16067

Zhang L, Paakkarinen V, van Wijk KJ and Aro EM (2000) Biogenesis of the chloroplast-encoded D1 protein: Regulation of translation elongation, insertion, and assembly into Photosystem II. Plant Cell 12: 1769–1781

Zito F, Finazzi G, Delosme R, Nitschke W, Picot D and Wollman FA (1999) The Qo site of cytochrome b_6f complexes controls the activation of LHCII kinase. EMBO J 18: 2961–2969.

Zouni A, Witt HT, Kern J, Fromme P, Krauss N, Saenger W and Orth P (2001) Crystal structure of Photosystem II from *Synechococcus elongatus* at 3.8 Ångstrom resolution. Nature 409: 739–743

Zurawski GR, Bohnert HJ, Whitfeld PR and Bottomley W (1982) Nucleotide sequence of the gene for the Mr 32,000 thylakoid membrane protein from *Spinacia oleracea* and *Nicotiana debneyi* predicts a totally conserved primary translation products of Mr 38950. Proc Natl Acad Sci USA 79: 7699–7703

Chapter 24

Novel Aspects on the Regulation of Thylakoid Protein Phosphorylation

Itzhak Ohad[1]*, Martin Vink[2], Hagit Zer[1], Reinhold G. Herrmann[3] and Bertil Andersson[2,4]

[1] Department of Biological Chemistry, The Hebrew University of Jerusalem, Jerusalem, 91904, Israel, [2] Department of Biochemistry and Biophysics, Arrhenius Laboratories for Natural Sciences, Stockholm University, SE-106 91 Stockholm, Sweden, [3] Botanisches Institute der Ludwig-Maximilians-Universität München, D-80638 Münich, Germany and [4] Division of Cell Biology, Linköping University, SE-581 85 Linköping, Sweden

Summary

Thylakoid membrane proteins are phosphorylated by different enzymes, which are subject to different control mechanisms. Activation of the light harvesting complex (LHCII) kinase is signaled by the redox state of plastoquinone and the cytochrome *b/f* complex and modulated by the thiol reduction state. Phosphorylation of Photosystem II (PS II) proteins may involve kinase(s) associated with the PS II core complex that do not involve the cytochrome *b/f* complex. Exposure of the phosphoprotein phosphorylation site(s) to protein kinases is regulated by light-induced conformational changes. Thus, thylakoid protein phosphorylation is regulated at both the enzyme and substrate levels. Thylakoid protein dephosphorylation is also under regulatory control, involving interaction between an immunophilin and a membrane-bound phosphatase. The physiological significance of thylakoid protein phosphorylation is not fully understood. Phosphorylation of LHCII is suggested to have a dual role: i) regulation of the LHCII/PS II/PS I interaction, underlying the mechanism of energy transfer balance and ii) prevention of the light-induced aggregation of LHCII or LHCII–PS II complexes. The formation of such macrodomains may affect the dynamics of the thylakoid membrane, which requires unhindered lateral diffusion of integral protein complexes. Phosphorylation of PS II subunits appear to be essential for the repair of photodamage to its reaction center occurring during light stress conditions.

*Author for correspondence, email: ohad@huji.ac.il

Eva-Mari Aro and Bertil Andersson (eds): Regulation of Photosynthesis, pp. 419–432.

I. Introduction

In their natural habitats, plants are exposed to both short and long term changes in light intensity and quality. This may lead to an imbalance in the excitation of PS II and PS I. As a result, the variations in the photosynthetic efficiency of the linear versus cyclic electron flow and light-induced damage to both PS II (Prasil et al., 1992; Aro et al., 1993) as well as to PS I (Sonoike, 1995) may occur to different degrees. Energy transfer between the light harvesting antennae and PS II or PS I in oxygen evolving organisms harboring the Chl *a/b* protein complex LHCII, can be modulated to counteract this imbalance, a phenomenon termed 'state transition' (Allen et al., 1981; Staehelin and Arntzen, 1983; Fork and Satoh, 1986) (see below). Soon it became clear that also the Chl *a* protein complex CP43 acting as a PS II core antennae and the D1/D2 proteins of the photochemical reaction center II (Millner et al., 1986; Ikeuchi et al., 1987; Marder et al., 1988; Michel et al., 1988) are phosphorylated by membrane-bound protein kinase(s). The discovery of the light-induced reversible phosphorylation of LHCII (Bennett, 1977) as well as of other PS II associated proteins led to intense research of the physiological significance and mechanism behind these phenomena. Imbalance between the photosystems excitation rates affects the redox state of the plastoquinone pool. It was found that reduction of plastoquinone by light-driven electron flow from PS II as well as by artificial reductants in the dark activates the LHCII phosphorylation process while oxidation of the plastoquinol pool is accompanied by protein kinase inactivation (Bennett, 1979; Allen and Horton, 1981; Allen et al., 1981; Vener et al., 1995).

Recent results suggest that thylakoid protein phosphorylation does not only respond to the above type of electron transport-dependent redox control but also involves redox control by thiol groups as well as light activation at the phospho-substrate level. Furthermore, regulatory events also appear to occur at the stage of phosphoprotein dephosphorylation. These regulatory aspects will be dealt with mainly from the in vitro perspective including their relation to the state transition phenomenon.

II. Redox Control of Thylakoid Protein Phosphorylation

Following the discovery of the light-induced phosphorylation of LHCII it was proposed that the redox level of plastoquinone (PQ) regulates this process. For example, mutants of the green alga *Chlamydomonas reinhardtii* impaired in plastocyanin or PS I activity, and thus unable to oxidize the plastoquinol pool in the light, could phosphorylate LHCII (Delosme et al., 1995). Thus, the thylakoid protein phosphorylation phenomenon represents a redox-controlled system in which plastoquinol (PQH_2) acts as the activator. Early redox titrations of the kinase activation process in darkness indicated that the slope of the titration curve is compatible with a two electron donor (Horton et al., 1981; Millner et al., 1982). Frid et al. (1992) reported inconclusive results suggesting that $n = 2$, while Silverstein et al. (1993) reported a value close to $n = 1$. The possibility that the redox component activating the kinase is a one-electron donor was already questioned by Bennett et al. (1988). Thus, the involvement of PQH_2 as a direct activator of the kinase could not be settled.

However, due to concerted efforts in several laboratories the basic aspects of the light-driven signal transduction system connecting the plastoquinone redox state with LHCII protein kinase(s) activation has recently been clarified. The first progress was achieved upon the finding that the cytochrome *b/f* complex is involved in the activation mechanism by interconnecting the plastoquinone redox state and the protein kinase(s) activation. This involvement was first indicated by the effect of specific cytochrome *b/f* inhibitors on the LHCII kinase activity (Bennett et al., 1987; Gal et al., 1988), and more specifically by the use of cytochrome *b/f* deficient mutants (Lemaire et al., 1986; Gal et al., 1987; Bennett et al., 1988; Coughlan, 1988; Wollman and Lemaire, 1988b). These studies demonstrated specific loss of LHCII phosphorylation in the cytochrome

Abbreviations: Chl – chlorophyll; CP43 – the PsbC chlorophyll *a*-protein complex; D1, D2 – reaction center II PsbA and PsbD proteins, respectively; DBMIB – 2,5-dibromo-3-methyl-6-isopropyl-*p*-benzoquinone; DCMU – 3-(3,4-dichlorophenyl)-1,1-dimethyl-urea; DTT – dithioerythreitol; EPR – electron paramagnetic resonance; NEM – N-ethylmaleimide; PPM – serine/threonine Protein Phosphatase, Mg^{2+} dependent; PPP – serine/threonine Protein Phosphatase; PQ, PQH_2 – plastoquinone and plastoquinol, respectively; PS I, PS II – Photosystem I and II, respectively; Qo site – quinol oxidation site of the cytochrome *b/f* complex; Qr site – quinol reduction site of the cytochrome *b/f* complex; TAK – Thylakoid Associated Kinase

deficient mutants while the phosphorylation of the other PS II phosphoproteins was only partially or not at all affected.

Further progress in the study of the cytochrome *b/f* role in the signal transduction chain was hampered by the fact that activation of the kinase(s) could be elicited only by light-dependent PS II-electron flow reducing the PQ pool or by addition of excess plastoquinol reductants in darkness. In both cases the high potential path components of the cytochrome *b/f* complex (the Rieske Fe-S center and cytochrome *f*) as well as plastocyanin, are fully reduced and thus, a 'redox buffer' is present throughout the activation experiment. A correlation between the reduction/oxidation state of the electron carrier components of the cytochrome *b/f* complex and that of kinase activation/deactivation could serve to identify the redox component(s) responsible for the process. As an example one can note that in *Acetabularia mediterranea* thylakoids, LHCII phosphorylation activity persists in darkness for hours despite the fact that under such conditions, the plastoquinone pool is mostly oxidized. Yet the kinase(s) could be deactivated by inhibitors of the cytochrome *b/f* complex that bind at its quinol oxidation (Qo) site (Gal et al., 1988; Frid et al., 1992).

However, such studies have not identified the electron carrier whose redox state is essential for the kinase activation/deactivation and therefore a new approach was needed. Significant progress in this direction followed the use of a new experimental system developed by Vener et al. (1995). Transient lowering of the pH of a thylakoid suspension in darkness to pH 4.3 for about 1–2 min was found to activate the protein phosphorylation. This activation was ascribed to formation of PQH_2 by the shift in the equilibrium PQ/PQH_2. This was sufficient to reduce in darkness the high potential path components of the cytochrome *b/f* complex, cytochrome *f* and cytochrome b_6 as well as plastocyanin, concomitant with the activation of the thylakoid phosphorylation process. While most of the plastoquinol pool generated by this procedure was rapidly re-oxidized by the ambient oxygen (<2 min), the high potential path components of the cytochrome *b/f* complex remained reduced in darkness and the kinase activity persisted since electron flow from plastocyanin via PS I is not active in darkness. The $t_{1/2}$ of the kinase-activated state was estimated to be approximately 4 min. The kinetics of the kinase deactivation in darkness paralleled the loss of the electron paramagnetic resonance (EPR) g_z signal (g = 2.3) (Vener et al., 1995). This signal is considered to reflect a change in the interaction of a reduced quinol molecule with the Qo site of the reduced cytochrome *b/f* complex (Malkin, 1982, 1986).

From these measurements it follows that the kinase is activated and maintained in its active state as long as a quinol is bound at, or can interact with, the Qo site without being immediately oxidized. If this was the case, one would predict that oxidation of this residual plastoquinol should inactivate the kinase. Indeed, oxidation of this plastoquinol at the Qo site of the 'kinase activating mode' of the cytochrome *b/f* complex following a single turnover flash to PS I was found to promptly inactivate the protein kinase (Vener et al., 1997). This was explained by flash activation of a cascade of electron flow from PQH_2 at the Qo site of the cytochrome *b/f* complex via the high potential path carriers including the Rieske Fe-S center and cytochrome *f* to plastocyanin which in turn is oxidized by PS I. The second electron removed from PQH2 via the cytochrome b_6 path reduces a plastoquinone molecule at the quinol reduction site (Qr) of the complex to a semiquinone. This scheme (Fig. 1) is in agreement with data indicating a value of n = 2 for the electron donor involved in the kinase activation process. Furthermore, displacement of plastoquinol from the Qo site by PQ-analogues such as DBMIB (Vener et al., 1997) as well as by various other substituted quinol analogues, inhibitors of the cytochrome *b/f* complex (Frid et al., 1992), deactivated the kinase demonstrating that the occupancy of the Qo site by plastoquinol is directly related to the kinase activation/deactivation processes (Vener et al., 1997). The concept of a plastoquinol bound to the Qo site as the activator of the kinase does not necessarily imply a tight binding but merely a transient interaction or occupancy. So far, the presence of a tightly associated plastoquinone molecule in the cytochrome *b/f* complex has not been unequivocally demonstrated. In mitochondria however, the presence of a bound ubiquinone that acts as an intermediate between the reduced ubiquinone pool and the cytochrome bc_1, homologous to the thylakoid cytochrome *b/f* complex, has been reported (Ding et al., 1995).

Further insight in understanding the mechanism whereby the cytochrome *b/f* may act in kinase activation was made possible following the structural determination of the respiratory chain cytochrome

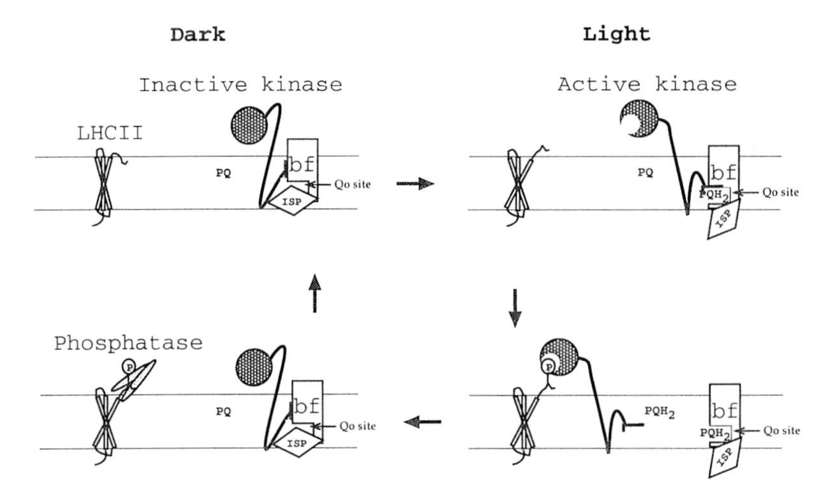

Fig. 1. Schematic representation of the protein–kinase and cytochrome *b/f* interactions during the enzyme activation/deactivation process. The cartoon shows the proposed mechanism for the redox control of the LHCII kinase, which is activated upon binding of plastoquinol (PQH_2) to the Qo site of the cytochrome *b/f* complex (bf). This binding changes conformation of the Rieske Iron Sulfur protein (ISP) that leads to the kinase activation and dissociation from the cytochrome *b/f* complex. The light-induced conformational changes of LHCII modulates the protein phosphorylation at the substrate level, which is demonstrated in the cartoon as well.

bc_1 complex (Yu et al., 1996, Iwata et al., 1998) as well as the resolution of the lumen-exposed cytochrome *f* subunit of the cytochrome *b/f complex* (Cramer et al., 1996; Carrell et al., 1999). Both the cytochrome bc_1 complex, and the cytochrome *b/f* complex posses many similarities. Both complexes assume a dimeric form in their active state (Cramer et al., 1994, 1996; Huang et al., 1994; Chain and Malkin, 1995; Pierre et al., 1995; Breyton et al., 1997). It was suggested that the dimeric form might be related to the ability of cytochrome *b/f* complex to interact with the kinase (Cramer et al., 1994, 1996; Hauska et al., 1996).

Besides the major subunits, the Rieske Fe-S center, cytochrome b_6 and *f* and the subunit IV that are homologues to the major components of cytochrome bc_1 complex, cytochrome *b/f* complex contains at least three low molecular mass hydrophobic subunits, the PetG protein (4.8 kDa in spinach and 4.1 kDa in *Chlamydomonas*), a putative nuclear gene product PetX (3.7 kDa in spinach and 3.8 kDa in *Chlamydomonas*) and a 3.4 kDa protein, PetL (Schmidt and Malkin, 1993; Cramer et al., 1996; Hauska et al., 1996). Recently, the presence in *Chlamydomonas* of a new component of the cytochrome *b/f* complex designated subunit V, has also been reported. This protein undergoes reversible phosphorylation upon state transition (Hamel et al., 2000) which may be relevant for the lateral migration of the cytochrome *b/f* complex (Vallon et al., 1991). Furthermore, it is now clearly demonstrated that cytochrome *b/f* also contains a tightly bound chlorophyll *a* molecule (Pierre et al., 1997). The minor components of the complex described above do not seem to be directly involved in the electron transfer process and may rather excert a regulatory function.

The recent demonstration of a significant transient structural change at the Qo site during the ubiquinol oxidation process in the mitochondrial cytochrome bc_1 complex (Crofts and Berry, 1998; Iwata et al., 1998; Zhang et al., 1998) pointed out the possibility that such a change may occur in the cytochrome *b/f* complex as well, and thus, may be involved in the process of the kinase redox activation. Briefly, these authors postulated that the domain of the Rieske Fe-S protein exposed at the intermembrane space, could oscillate between two positions during the turnover of ubiquinol/ubiquinone at the quinol oxidation site of the cytochrome bc_1 complex (Fig.1). In one position, termed the proximal position, the Rieske protein lies close to the membrane in the vicinity of cytochrome b_1, while the protein in its other position, the distal position, extends further into the intermembrane space interacting with the cytochrome c_1. Such a hinge movement of the lumenal domain of the Rieske Fe-S center in thylakoids is a good candidate for the structural link interconnecting the occupancy of the Qo site by plastoquinol and its oxidation, with the reversible kinase activation process (Vener et al., 1998; Zito et al., 1999). Thus, it was suggested that the Rieske protein, in its proximal position but not in its distal position, inhibits the LHCII kinase via its interaction with a putative transmembrane helix of the kinase on the lumenal side of the membrane (Fig. 1). The similarity of the operative redox potentials and electron donors of the cytochrome bc_1 and *b/f* complexes as well as the structural similarity of cytochrome *f* with the corresponding cytochrome c_1 (Hauska et al., 1996; Bron et al., 1999) as discussed above support these considerations.

Experiments directed towards obtaining further evidence in support of this model came from mutations of the PEWY sequence to PWYE in the subunit IV of the cytochrome *b/f* complex in *Chlamydomonas* (Zito et al., 1999). This sequence, located in the so-called EF loop of the protein, is lining the Qo pocket and is required for a functional binding of the plastoquinol at this site. Phosphorylation studies carried out with this mutant showed that, indeed, the radioactive phosphate labeling of the LHCII complex was completely abolished. Thus, these results by Zito et al. (1999) add the molecular aspect and give strong support to the notion that the conformational changes induced in the cytochrome *b/f* complex by the binding of plastoquinol at the Qo site mediates the LHCII kinase activation.

What then is the relation between the activation of the kinase and its association with the cytochrome *b/f* complex? In discussing this aspect of the redox control of the LHCII phosphorylation one should keep in mind that the kinase concentration is extremely low relative to the concentration of cytochrome *b/f* as well as that of the phosphorylation substrates (Gal et al., 1997). The thylakoid protein phosphorylation may require alternating interactions of the kinase with a cytochrome *b/f* complex and the phosphoprotein substrate. This interaction should therefore be followed within a limited time, compatible with the $t_{1/2}$ of the kinase active state (Vener et al., 1995) to that with an open phosphorylation site of a PS II-LHCII complex. Thus, lateral mobility of the above complexes constitutes an important rate limiting factor in the LHCII phosphorylation process (Bennett, 1991; Allen, 1992; Drepper et al., 1993; Gal et al., 1997).

Presently, there is no evidence indicating that only a specific sub-population of cytochrome *b/f* may interact with the thylakoid protein kinase(s). Thus, all cytochrome *b/f* complexes may assume the 'kinase activating mode,' that is, its high potential path electron carriers are reduced and plastoquinol may bind to the Qo site without being oxidized. Statistically the fraction of the cytochrome *b/f* population in this mode will rise with the increase in the reduction of the plastoquinone pool. In principal several different scenarios can be considered in terms of the nature of the kinase-cytochrome *b/f* interaction. Below we will outline one hypothetical possibility as a working model. Under non-reducing conditions the deactivated kinase is normally bound to a cytochrome *b/f* complex. Under reducing conditions this cytochrome complex will be turned into its activation mode and the kinase may be released as a consequence of the Rieske Fe-S domain going from its distal to its proximal position. Upon release from the cytochrome *b/f* complex the kinase becomes active and able to phosphorylate available LHCII substrates. The kinase may remain active as long as it does not interact or bind to a cytochrome complex that is not in its 'activating mode.' The simplicity in this model lies in that a free kinase is active while a kinase bound to the cytochrome *b/f* complex is inactive.

In this respect it is noteworthy that upon cytochrome *b/f* purification, the protein kinase activity comigrates with this complex (Gal et al., 1990b; Hamel et al., 2000). Furthermore, solubilized preparations, highly enriched in protein kinase activity following perfusion chromatography, still contain subunits of the cytochrome *b/f* complex (Gal et al., 1995; Vink et al., 1998). However, the protein kinase activity of such preparations has lost its redox control and is not enhanced by typical reductants of cytochrome *b/f* such as duroquinol, which in thylakoid membranes activates the kinase in darkness (Allen and Horton, 1981). This suggests that most if not all of the deactivated kinase in situ is bound to the cytochrome complex. Release of the protein kinase from the complex occurs during isolation by use of detergents and the kinase in such preparations is basically in its activated state.

Also the kinase-substrate interactions can be regarded in terms of a dynamic interactive system. The free active kinase may diffuse within the grana membrane and interact with dephosphorylated free LHCII or PS II-bound LHCII. Typically, under non-reducing conditions, most of the LHCII is assumed to be dephosphorylated and thus bound to PS II. Under more reducing conditions LHCII is phosphorylated, released from PS II and free to interact with PS I. However, although controversial (Elich et al., 1997), the protein phosphatase seems to be active probably under both reducing and non-reducing conditions (see below). Thus, part of the phosphorylated free or PS I-bound LHCII may be dephosphorylated and be a substrate for the free active kinase prior to its reassociation with the PS II complex. Thus, the players in the phosphorylation game are: oxidized and reduced cytochrome complexes, bound (inactive) and free (active) protein kinase, dephosphorylated LHCII bound to or free from PS II or PS I, and free phosphorylated LHCII. In such a dynamic model the phosphorylation activity is limited by the time span of the active state of the kinase dictated by the probability of its encountering/binding to a cytochrome complex and the availability of LHCII having its phosphorylation site open and exposed. The protein kinase activation system discussed above does not require, nor does it exclude, conformational changes in the kinase structure upon its binding to or release from the cytochrome *b/f* complex.

The mechanistic model discussed above predicts that at low electron flow rates, the ratio deactivated/active kinase and dephosphorylated/phosphorylated LHCII are highest. With increasing electron flow rate, the situation inverts and the ratio active kinase/closed phosphorylation site(s) increases reaching a maximum at high light intensities. This is also the case under in vitro experimental conditions. However, as will be discussed below, this is not the case under in vivo conditions (Chapter 23, Rintamäki and Aro) Under high light intensities and thus, reducing conditions, LHCII is mostly dephosphorylated indicating that additional factors are involved in the control of the LHCII phosphorylation process (Rintamäki et al., 1997, 2000; Zer et al., 1999).

The scheme illustrated in Fig. 1 may account for the redox control of LHCII phosphorylation by an LHCII-specific kinase. However, CP43 and the D2 polypeptide of PS II as well as some of the LHCII polypeptides in *Chlamydomonas* are phosphorylated under redox control irrespectively of the presence of or state of the cytochrome *b/f* complex (Delepelaire and Wollman, 1985; Lemaire et al., 1986; Wollman and Lemaire, 1988b; Gal et al., 1987, 1990a; Bennett et al., 1988; Coughlan, 1988; Wollman and Lemaire, 1988b). This raises the question whether additional membrane-bound protein kinases, different from the LHCII kinase, may be activated by interaction directly with plastoquinol (Bennett, 1991; Allen, 1992). Thus, it is likely that more than one kinase is involved in the process of phosphorylation of thylakoid membrane proteins (Weber et al., 2001; Chapter 23, Rintamäki and Aro) and the related state transition phenomenon. Participation of more than one protein kinase in the redox-dependent regulation of thylakoid membrane protein phosphorylation in *Chlamydomonas* was also implied before and a model of a kinase cascade was suggested, where one kinase is activated by redox control, and in turn activates other phosphorylation processes by phosphotransferase activity (Wollman and Lemaire, 1988a).

III. Role of Thiol Redox State in Kinase Activation/Deactivation Process in Isolated Thylakoids

In addition to the redox events involving plastoquinol and the cytochrome *b/f* complex discussed above it has become evident that the thylakoid protein-thiol redox state influences the protein phosphorylation process. The activity of an over-expressed thylakoid protein kinase of the TAK family is enhanced by dithiothreitol and inactivated by thiol group inhibitors (Snyders and Kohorn, 1999). An influence on the kinase activity by reduced dithiol was also reported for a protein kinase present in a crude thylakoid protein extract enriched in cytochrome *b/f* (Gal et al., 1990a). Recent work has demonstrated that the redox state of dithiols plays an important role in the thylakoid protein phosphorylation also in situ (Carlberg et al., 1999). By poising the ratio reduced/oxidized DTT in an in vitro thylakoid phosphorylation system it was demonstrated that thiol-reducing conditions stimulate the phophorylation of the D1/D2 proteins of PS II while lowering the phosphorylation of LHCII. On the other hand, thiol-oxidizing conditions had an opposite effect. These stimulation/down-regulation effects obtained in vitro are however in the range of two to eight-fold relative to control conditions as compared to the all or none effect of the LHCII-kinase activation by the PQH_2/cytochrome *b/f* redox system. Measurements of the effect of added *Escherichia coli* thioredoxin and reduced DTT on the level of LHCII phosphorylation could indicate that thioredoxin may serve as an intermediate in the kinase regulation via dithiol reducing agents (Carlberg et al., 1999). This regulation could be mediated via reduction/oxidation of the phospho-substrate´s thiol groups, the kinase and/or effecting the conformation of surrounding membrane proteins. Treatment with N-ethylmaleimide, a sulfhydryl alkylating agent, inactivates protein phosphorylation in thylakoids (Millner et al., 1982; Rintamäki et al., 2000) as well as that of isolated PS II core complex protein kinase(s) (Vink et al., 2000).

IV. Light-Induced Modulation of Thylakoid Protein Phosphorylation at the Substrate Level

So far, the control of the thylakoid protein phosphorylation has mainly been discussed at the enzyme level, but there are also recent indications of a regulation at the substrate level. In an attempt to study in more detail the interaction between the protein kinase and the LHCII substrate, an in vitro experimental system was recently developed (Zer et al., 1999) consisting of native isolated or recombinant reconstituted LHCII supplied with a solubilized thylakoid protein kinase preparation enriched in a 64 kDa enzyme (Vink et al., 1998). Using this system it was observed that illumination of the LHCII substrates prior to the addition of the kinase fraction, stimulates considerably the phosphorylation of native LHCII. No stimulation was observed for the recombinant apo-protein lacking carotenoids and chlorophyll. This light activation phenomenon was demonstrated to be due to an increased exposure of the LHCII N-terminal domain to the protein kinase. The light effect is saturated at relatively low light intensities (approximately 200 μmol of photons $m^{-2} s^{-1}$ for 25 μg Chl of LHCII ml^{-1}). The exposure of the N-terminal domain by illumination could be demonstrated also by an increase in its accessibility to trypsin cleavage (Zer et al., 1999). Notably, the light-induced conformational changes of the isolated LHCII were found to be reversible in darkness.

The effect of illumination on the exposure of the LHCII N-terminal domain was also studied in intact thylakoids (H. Zer, unpublished). To discriminate experimentally between the light-induced exposure of the phosphorylation sites from the light-dependent activation of the protein kinase, the thylakoids were incubated at relatively low light intensities in the absence of ATP and with the addition of DCMU to prevent the reduction of the PQ pool. The thylakoids were subsequently transferred to darkness, ATP was added and the protein kinase was activated by addition of duroquinol. Short preillumination at low irradiance of the thylakoid membranes increased the level of LHCII phosphorylation in darkness as compared to the dark-control activated by duroquinol alone (Zer et al., 1999). However, prolonged illumination of thylakoids or increased light intensity, (20 min, 400 μmol of photons $m^{-2}s^{-1}$) resulted in a drastic lowering of the phosphorylation level of LHCII assayed in darkness subsequent to the preillumination as above (Fig. 2). This latter effect was not the result of protein kinase inactivation since thylakoids exposed to the high light conditions were capable to phosphorylate exogenously added LHCII.

As in the case of the isolated LHCII, the short light treatment of the thylakoids stimulating LHCII phosphorylation also enhanced the exposure of the N-terminal domain to tryptic cleavage. Furthermore, the occlusion of the LHCII phosphorylation site in

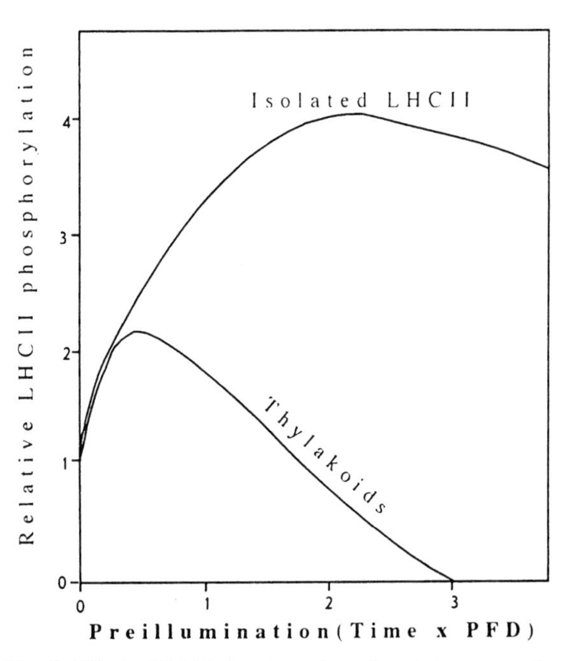

Fig. 2. Effect of light-induced conformational changes on the phosphorylation of isolated or membrane-bound LHCII. Preillumination of isolated LHCII or thylakoid membranes under conditions preventing phosphorylation activity exposes the N-terminal domain of the LHCII complex to subsequent protein kinase activity in darkness for both substrates. However, increased excitation of the thylakoid membrane causes a significant lowering of the membrane-bound LHCII phosphorylation. The isolated LHCII was phosphorylated by addition (in darkness) of a solubilized, protein kinase-enriched preparation and ATP; the thylakoids were preilluminated in presence of DCMU and without addition of ATP. In this case phosphorylation was carried out in darkness following addition of ATP and duroquinol to activate the membrane-bound protein kinase. Experimental conditions were as described in (Zer et al., 1999); the maximal light intensity (PFD) and illumination times were 800 μmol of photons $m^{-2} s^{-1}$ and 20 min.

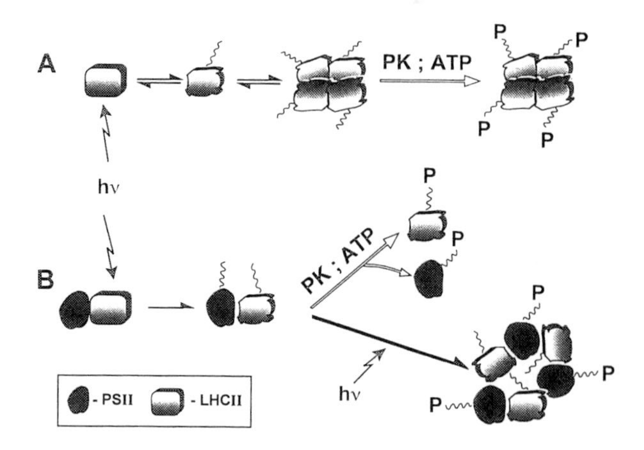

Fig. 3. Schematic representation of the light-induced conformational changes affecting the exposure of the N-terminal domain of LHCII to the protein kinase in isolated and membrane-bound LHCII. The black and shaded figures represents the PS II core and LHCII complex, respectively. A, isolated LHCII; B, PS II.LHCII complex in situ; PK, protein kinase either added to the isolated LHCII complex or the endogenous, membrane-bound, enzyme activated by duroquinol. Pre-illumination induces the exposure of the N-terminal domain of LHCII in both the isolated and membrane-bound LHCII as well as that of the phosphorylation site(s) of the PS II core complex phosphoproteins (squiggled lines). In the thylakoid membranes, the light-induced lateral aggregation of the non-phosphorylated LHCII/PS II may hinder the accessibility of the LHCII phosphorylation site to the membrane-bound LHCII kinase. However, the phosphorylation sites of the PS II core complex phosphoproteins remain accessible to the PS II core complex associated protein kinase (Zer et al., 1999, Vink et al., 2000).

situ following prolonged preillumination also inhibited the tryptic cleavage of this domain. Preliminary results indicate that a similar effect of substrate activation is obtained also for thylakoid protein phosphorylation in intact leaves (H. Zer, unpublished).

These unexpected results could be interpreted in terms of: a) at moderate illumination reversible light-induced exposure of the N-terminal domain of LHCII to the kinase occurring for both the isolated as well as membrane integrated complex and b), at stronger illumination light-induced changes in the overall organization of the protein complexes in the grana domains lowering the accessibility of the protein kinase to LHCII. Such a lowered accessibility of LHCII to the protein kinase in thylakoids subjected to stronger illumination while the phosphorylation is reduced may be due to light-induced aggregation of LHCII/PS II complexes (Fig. 3), and possibly to the lateral exclusion of the protein kinase and/or kinase-cytochrome *b/f* complex from the grana domains. In this context it is relevant to consider several earlier reports showing that illumination induces changes in the chlorophyll-protein organization within the grana domains (Garab et al., 1991; Horton et al., 1991; Jennings et al., 1991; Barzda et al., 1996; Garab and Mustardy, 1999) as well as in isolated LHCII or LHCII integrated into lipid bilayers.

Furthermore, it is noteworthy that exposure of *Chlamydomonas* cells to high light was reported to result in a low degree of thylakoid phosphoproteins (Schuster et al., 1986). The process was accompanied by changes in the intramembrane particle distribution as revealed by the freeze fracturing technique. However, the protein kinase appeared not inactivated in the photoinhibited thylakoids since brief sonication or freezing and thawing restored the phosphorylation activity thus suggesting that the inaccessibility of LHCII to the kinase was due to reversible changes in the organization of the chlorophyll-protein complexes (Schuster et al., 1986).

The question arises whether the light-induced conformational changes causing exposure of phosphorylation site(s) at low light intensities are restricted to LHCII or occur for the PS II core phosphoproteins CP43 and the D1/D2 proteins as well. Experimental results obtained using isolated PS II core complexes indicate that indeed this phenomenon can be demonstrated. Illumination of PS II core complexes in which their endogenous protein kinase (Race and Hind, 1996; Vink et al., 2000) has been inactivated by NEM, induces a significant increase in the exposure of the CP43 phosphorylation site(s) as judged by its increased phosphorylation when supplied with a partially purified protein kinase fraction (Vink et al., 2000). Illumination of PS II core complexes also enhances the exposure of CP43 to tryptic cleavage. No effects were obtained with respect to phosphorylation for the D1/D2 proteins.

In conclusion these novel findings demonstrate that light plays a dual role in the process of regulation of the thylakoid protein phosphorylation, that of enzyme activation via the redox signal transduction systems as well as modulating the exposure of the phosphorylation site(s) of the chlorophyll-protein substrate to the protein kinase.

V. Thylakoid Protein Dephosphorylation and its Regulation

The dephosphorylation of thylakoid phosphoproteins has been shown to be catalyzed predominantly by enzymes integral to the thylakoid membrane (Bennett, 1980, 1991). Serine/threonine phosphatases are classified according to their substrate specificity, requirement for divalent cations, and susceptibility to inhibitors (Cohen, 1989; Hubbard and Cohen, 1993; Barford, 1996; Cohen, 1997). Using these criteria two major categories of phosphatases have been defined, the Mg^{2+}-dependent PPM family and the Mg^{2+} -independent PPP family. The latter can be divided into subfamilies such as PP1, PP2A and PP2B. Many of these phosphatases do not only consist of a catalytic subunit but also regulatory and targeting subunits (Hubbard and Cohen, 1993; Fruman et al., 1994; Barford, 1996; Cohen, 1997).

The protein phosphatases of thylakoid membranes have previously been considered to be unrelated to the PPP family, mainly because of insensitivity to typical inhibitors of in membrano dephophorylation experiments (MacKintosh et al., 1991; Sun and Markwell, 1992; Carlberg and Andersson, 1996). However, recently a hydrophobic thylakoid protein of approximately 39 kDa in molecular weight and present in very low abundance could be identified (Vener et al., 1999). Its enrichment in detergent-solubilized fractions obtained from thylakoid membranes followed by a multi-step chromatographic procedure correlated with an increased capacity for dephosporylation of a mixture of small thylakoid phospho-peptide substrates. Notably, for each chromatographic step, the sensitivity of the phosphatase activity to okadaic acid and tautomycin increased and in the highly purified form the IC_{50} for okadaic acid and tautomycin was 0.4 nM and 25 nM, respectively (Vener et al., 1999). Such inhibition characteristics are indicative of the PP2A and the recently discovered PP2A-related protein phospahatase subfamilies (PP4–PP6) (Cohen, 1989; Barford, 1996; Cohen, 1997).

A unique feature of this thylakoid phosphatase is its interaction with an immunophilin located in the thylakoid lumen (Fulgosi et al., 1998; Vener et al., 1999). It is suggested that the PP2A-immunophilin interaction is essential for regulation of thylakoid protein dephosphorylation (Chapter 10, Vener). Although normally immunophilins are associated with the PP2B-type of phosphatases (calcineurins) (Liu et al., 1991; Walsh et al., 1992; Fruman, 1994; Marks, 1996).

As in the case of thylakoid kinases the presence of several phosphatases is likely (Allen, 1992; Hammer et al., 1995, 1997). The purified PP2A showed the highest specificity for the D1- and D2-phosphoproteins of the PS II reaction center although it was to some extent able to dephosphorylate most thylakoid phosphoproteins including LHCII (Vener et al., 1999). Furthermore, analysis of the dephosphorylation kinetics in vitro of the endogenous thylakoid-bound phosphatase(s) detected two distinct kinetic patterns (Carlberg and Andersson, 1996). The LHCII phosphoproteins and an uncharacterized 12 kDa protein show a rapid dephosphorylation behavior while the dephosphorylation of the PS II phosphoproteins is very slow. DDT stimulated the dephosphorylation of the former group of thylakoid phosphoproteins at variance with the PS II phosphoproteins.

VI. Thylakoid Protein Phosphorylation and State Transition: Open Questions

The process of state transition is defined as phosphorylation-induced dissociation of the mobile LHCII antennae from the grana located PS II thus lowering the energy transfer to the PS II reaction center. Lateral migration of free phospho-LHCII may allow its interaction and energy transfer to PS I localized in the stroma-exposed regions of the thylakoid membranes (Allen, 1992; Andersson and Barber, 1994). Dephosphorylation of phospho-LHCII should lead to re-association of LHCII with PS II. Thus, the reversible phosphorylation of LHCII is

PSII + LHCII ⟹ PSII.LHCII

Scheme a

PSII.LHCII ⟺ (Light) PSII + ***LHCII*** ⟺ (Activated Kinase) PSII + Phospho – **LHCII**

Scheme b

PSII.LHCII ⟹ (DQH2 – Activated Kinase) PSII – Phospho – **LHCII** ⟺ PSII + Phospho – **LHCII**

Scheme c

considered to be the underlying mechanism regulating energy distribution between the two photosystems (Gal et al., 1997). While the above mechanism is compatible with the observed experimental results in general terms, several questions concerning the details of this process remain so far unanswered.

Among these unanswered questions are the mechanism of dissociation of LHCII from PS II and its association with PS I. It is not clear whether in vivo light-induced conformational changes of LHCII can facilitate exposure of the N-terminal domain of LHCII allowing its dissociation from PS II prior to its phosphorylation. Binding of LHCII to PS II is a reversible reaction with an equilibrium constant that favors by far the association state as indicated in Scheme (a).

Illumination-induced conformational changes of both LHCII and PS II (Zer et al., 1999; Vink et al., 2000) may lower the association constant thereby transiently increasing the concentration of free unphosphorylated LHCII and thus exposing it to the protein kinase as indicated in Scheme (b).

In the scheme the size of the arrow head and variations in font indicate changes in the binding constants and the light-induced conformation changes exposing the LHCII N-terminal domain to the active kinase, respectively. This model does not exclude the possibility that the activated protein kinase may phosphorylate, albeit at a lower rate, also the PS II-bound LHCII. Such a reaction is demonstrated by the phosphorylation of LHCII following activation of the protein kinase by addition of duroquinol to thylakoid membranes in darkness (Gal et al., 1988) as indicated in Scheme (c).

It should be stressed that the difference between the processes described in Schemes (b) and (c) is of a kinetic nature and possibly could be detected by accurate measurements of the initial rates of LHCII phosphorylation under the above two different conditions.

The lateral migration of phospho-LHCII is considered to be a random process (Drepper et al., 1993), the phosphorylated antennae interacting either with the PS I located in the margins of the grana region (Stefansson et al., 1995) and therefore covering a short distance (estimated to be less than 50 nm from the grana centers) or further migration into the stroma membranes domains thus covering a perimeter of 100–150 nm (Stefansson et al., 1995). However, so far no direct experimental evidence was provided in favor of a random as opposed to directed migration of phospho-LHCII away from PS II or phospho-PS II core complexes. Mechanistically, by the addition of at least three negative charges/LHCII trimer (25 kDa, lhcb2) and at least an additional three to the phosphorylated PS II core that still contains the immobile phosphorylated component (27 kDa, lhcb1), as well as phospho-CP43 and -D1/D2 proteins, one could assume that electrostatic repulsion between the phosphorylated complexes may be a 'pseudo-electrophoretic' process directing the migration of LHCII away from PS II.

Finally, the association of phospho-LHCII with PS I should be considered. So far there is no proof of a tight binding of phospho-LHCII to PS I neither directly nor via the LHCI complex. There are experimental indications that also the LHCI proteins are phosphorylated, however, the turnover of their phosphate groups seems to be extremely low and thus these antennae proteins are practically phosphorylated under all light conditions (Knoetzel et al., 1995). Thus, one would expect that the interaction of the phosphorylated LHCII with LHCI may be hindered and thus occur at a different docking site. It is generally assumed that energy transfer by the Foerster mechanism may occur without physical contact between the donor and acceptor components across distances up to approximately 3 nm and thus there is no necessity for a tight binding between phospho-LHCII and PS I.

Recent results have however demonstrated that deletion of the PsaH subunit of PS I in *Arabidopsis thaliana*, inhibits energy transfer from LHCII to PS I under state 2 illumination indicating that this subunit of PS I may participate in the binding of phospho-LHCII to PS I (Lunde et al., 2000). However, while the phosphorylation of LHCII was not impaired in this mutant, energy transfer from phospho-LHCII to PS II was not lowered. These results are interpreted as evidence for two novel concepts: i), the energy transfer from phospho-LHCII to PS I is mediated by a specific PS I subunit implying protein-protein interaction and thus it is not a simple energy transfer process across an un-bridged distance; ii), phospho-LHCII may interact with PS II and transfer energy to this complex even in its phosphorylated state. This concept is in agreement with the scheme presented above describing the interactions of LHCII with PS II (Scheme (c)), and implies that the binding constant of phospho-LHCII to PS I is significantly higher than that of phospho-LHCII binding to PS II. This would suggest that mutants impaired in the PS I complex activity may phosphorylate LHCII and the energy transfer to PS II should not be affected in such mutants (Fleischmann et al., 1999).

These novel results and the underlying concepts open the way to further investigation of the mechanism of interactions between, and binding constants of, the various actors having a role in this play: PS II, Phospho-PS II, LHCII, Phospho-LHCII, PS I, LHCI, phospho-LHCI. While until now one could be satisfied mostly by common sense reasoning regarding the above interactions, the recent findings and data accumulated during the last few years require the reopening of the investigation of the phosphorylation of the photosystems and their antennae components and mechanisms of interactions.

Acknowledgments

The authors acknowledge the financial support of the Israeli National Science Foundation administered by the Israeli Academy of Sciences and the German Israeli Science foundation (GIF), the Swedish Natural Research Council, the Swedish Research Council for Agriculture and Forestry, the German Research Foundation (DFG-SFB 184), the Fonds der Chemischen Industrie and the Human Frontier Science Program.

References

Allen JF (1992) Protein phosphorylation in regulation of photosynthesis. Biochim Biophys Acta 1098: 275–335

Allen J and Horton P (1981) Chloroplast protein phosphorylation and chlorophyll fluorescence quenching activation by tetramethyl-p-hydroquinone, an electron donor to plastoquinone. Biochim Biophys Acta 638: 290–295

Allen JF, Bennett J, Steinback KE and Arntzen CJ (1981) Chloroplast protein phosphorylation couples plastoquinone redox state to distribution of excitation energy between photosystems. Nature 291: 25–29

Andersson B and Barber J (1994) Composition, organization and dynamics of thylakoid membranes. In: Bittar EE (ed) Advances in Molecular and Cell Biology. Processes of Photosynthesis, Vol 10, pp 1–53. Jai Press Inc, Greenwich

Barford D (1996) Molecular mechanisms of the protein serine/threonine phosphatases. Trends Biochem Sci 21: 407–412

Barzda V, Istokovics A, Simidjiev I and Garab G (1996) Structural flexibility of chiral macroaggregates of light-harvesting chlorophyll *a/b* pigment-protein complexes. Light- induced reversible structural changes associated with energy dissipation. Biochemistry 35: 8981–8985

Bennett J (1977) Phosphorylation of chloroplast membrane proteins. Nature 269: 344–346

Bennett J (1979) The protein kinase of the thylakoid membrane is light-dependent. FEBS Lett 103: 342–344

Bennett J (1980) Chloroplast phosphoproteins. Evidence for a thylakoid-bound phosphoprotein phosphatase. Eur J Biochem 104: 85–89

Bennett J (1991) Protein phosphorylation in green plant chloroplasts. Ann Rev Plant Physiol Plant Mol Biol 42: 281–311

Bennett J, Shaw EK and Baker S (1987) Phosphorylation of thylakoid proteins and synthetic peptide analogs: Differential sensitivity to inhibition by a plastoquinone antagonist. FEBS Lett 210: 22–26

Bennett J, Shaw EK and Michel H (1988) Cytochrome b6f

complex is required for phosphorylation of light-harvesting chlorophyll *a/b* complex II in chloroplast photosynthetic membranes. Eur J Biochem 171: 95–100

Breyton C, Tribet C, Olive J, Dubacq JP and Popot JL (1997) Dimer to monomer conversion of the cytochrome b_6f complex. Causes and consequences. J Biol Chem 272: 21892–21900

Bron P, Lacapere JJ, Breyton C and Mosser G (1999) The 9 Å projection structure of cytochrome b_6f complex determined by electron crystallography. J Mol Biol 287: 117–126

Carlberg I and Andersson B (1996) Phosphatase activities in spinach thylakoid membranes-effectors, regulation and location. Photosynth Res 47: 145–156

Carlberg I, Rintamäki E, Aro E-M and Andersson B (1999) Thylakoid protein phosphorylation and the thiol redox state. Biochemistry 38: 3197–3204

Carrell CJ, Schlarb BG, Bendall DS, Howe CJ, Cramer WA and Smith JL (1999) Structure of the soluble domain of cytochrome *f* from the cyanobacterium *Phormidium laminosum*. Biochemistry 38: 9590–9599

Chain RK and Malkin R (1995) Functional activities of monomeric and dimeric forms of the chloroplast cytochrome b_6f complex. Photosynth Res 46: 419–426

Cohen P (1989) The structure and regulation of protein phosphatases. Annu Rev Biochem 58: 453–508

Cohen PTW (1997) Novel protein serine/threonine phosphatases: Variety is the spice of life. Trends Biochem Sci 22: 245–251

Coughlan SJ (1988) Chloroplast thylakoid protein phosphorylation is influenced by mutations in the cytochrome *bf* complex. Biochim Biophys Acta 933: 413–422

Cramer WA, Martinez SE, Furbacher PN, Huang D and Smith JL (1994) The cytochrome b_6f complex. Curr Opin Struct Biol 4: 536–544

Cramer WA, Soriano GM, Ponomarev M, Huang D, Zhang H, Martinez SE and Smith, JL (1996) Some new structural aspects and old controversies concerning the cytochrome b_6f complex of oxygenic photosynthesis. Annu Rev Plant Physiol Plant Mol Biol 47: 477–508

Crofts AR and Berry EA (1998) Structure and function of the cytochrome bc_1 complex of mitochondria and photosynthetic bacteria. Curr Opin Struct Biol 8: 501–509

Delepelaire P and Wollman F-A (1985) Correlation between fluorescence and phosphorylation changes in thylakoid membranes of *Chlamydomonas reinhardtii* in vivo: A kinetic analysis. Biochim Biophys Acta 809: 277–283

Delosme R, Olive J and Wollman F-A (1995) Changes in light energy distribution upon state transitions: An in vivo photoacoustic study of the wild type and photosynthesis mutants from *Chlamydomonas reinhardtii*. Biochim Biophys Acta 1273: 150–158

Ding H, Moser CC, Robertson DE, Tokito MK, Daldal F and Dutton PL (1995) Ubiquinone pair in the Q_o site central to the primary energy conversion reactions of cytochrome bc_1 complex. Biochemistry 34: 15979–15996

Drepper F, Carlberg I, Andersson B and Haehnel W (1993) Lateral diffusion of an integral membrane protein: Monte Carlo analysis of the migration of phosphorylated light-harvesting complex II in the thylakoid membrane. Biochemistry 32: 11915–11921

Elich TD, Edelman M and Mattoo AK (1997) Evidence for light-dependent and light independent protein dephosphorylation in chloroplast. FEBS Lett 411: 236–238

Fleischmann MM, Ravanel S, Delosme R, Olive J, Zito F, Wollman F-A and Rochaix JD (1999) Isolation and characterization of photoautotrophic mutants of *Chlamydomonas reinhardtii* deficient in state transition. J Biol Chem 274: 30987–30994

Fork DC and Satoh K (1986) The control by state transitions of the distribution of excitation energy in photosynthesis. Ann Rev Plant Physiol Plant Mol Biol 37: 335–361

Frid D, Gal A, Oettmeier W, Hauska, G, Berger S and Ohad I (1992) The redox-controlled light-harvesting chlorophyll *a/b* protein kinase. J Biol Chem 267: 25908–25915

Fruman DA, Burakoff SJ and Bierer BE (1994) Immunophilins in protein folding and immunosuppression. FASEB J 8: 391–400

Gal A, Shahak Y, Schuster G and Ohad I (1987) Specific loss of LHCII phosphorylation in *Lemna* mutant 1073 lacking the cytochrome b_6/f complex. FEBS Lett 221: 205–210

Gal A, Schuster G, Frid D, Canaani O, Schwieger HG and Ohad I (1988) Role of the cytochrome b_6f complex in the redox-controlled activity of *Acetabularia* thylakoid protein kinase. J Biol Chem 263: 7785–779

Gal A, Hauska G, Herrmann RG and Ohad I (1990a) Interaction between light harvesting chlorophyll *a/b* protein (LHCII) kinase and cytochrome b_6/f complex. J Biol Chem 265: 19742–19749

Gal A, Mor T S, Hauska G, Herrmann RG and Ohad I (1990b) LHCII kinase activity associated with isolated cytochrome b_6/f complex. In: Baltscheffsky M (ed) Current Research in Photosynthesis, Vol II, pp 787–789. Kluwer Academic Publishers, Dordrecht

Gal A, Zer H, Roobol-Boza M, Fulgosi H, Herrmann R G, Ohad I and Andersson B (1995) Use of perfusion chromatography for the rapid isolation of thylakoid kinase enriched preparations. In: Mathis P (ed) Photosynthesis: From Light to Biosphere. Vol III, pp 341–344. Kluwer Academic Publishers, Dordrecht

Gal A, Zer H and Ohad I (1997) Redox controlled thylakoid protein kinase(s): News and views. Physiol Plant 100: 869–885

Garab G, and Mustardy L (1999) Role of LHCII-containing macrodomains in the structure, function and dynamics of grana. Aust J Plant Physiol 26: 649–658

Garab G, Kieleczawa J, Sutherland JC, Bustamante C and Hind G (1991) Organisation of pigment-protein complexes into macrodomains in the thylakoid membranes of wild-type and chlorophyll *b*-less mutant of barley as revealed by circular dichroism. Photochem Photobiol 54: 273–281

Hamel P, Olive J, Pierre Y, Wollman F-A and de Vitry C (2000) A new subunit of cytochrome b_6f complex undergoes reversible phosphorylation upon state transition. J Biol Chem 275: 17072–17079

Hammer MF, Sarath G, Osterman J C and Markwell J (1995) Assessing modulation of stromal and thylakoid light-harvesting complex-II phosphatase activities with phosphopeptide substrates. Photosynth Res 44: 107–115

Hammer MF, Markwell J and Sarath G (1997) Purification of a protein phosphatase from chloroplast stroma capable of dephosphorylating the light-harvesting complex-II. Plant Physiol 133: 227–233

Hauska G, Schütz M and Büttner M (1996) The cytochrome b_6f complex—composition, structure and function. In: Ort DR and Yocum CF (eds) Oxygenic Photosynthesis: The Light Reactions, Vol 4, pp 377–398. Kluwer Academic Publishers, Dordrecht

Horton P, Allen JF, Black MT and Bennett J (1981) Regulation of phosphorylation of chloroplast membrane polypeptides by the redox state of plastoquinone. FEBS Lett 125: 193–196

Horton P, Ruban AV, Rees D, Pascal AA, Noctor G and Young AJ (1991) Control of the light-harvesting function of chloroplast membranes by aggregation of the LHCII chlorophyll-protein complex. FEBS Lett 292: 1–4

Huang D, Everly RM, Cheng RH, Heymann JB, Schägger H, Sled V, Ohnishi T, Baker TS and Cramer WA (1994) Characterization of the chloroplast cytochrome b_6f complex as a structural and functional dimer. Biochemistry 33: 4401–4409

Hubbard MJ and Cohen P (1993) On target with a new mechanism for the regulation of protein phosphorylation. Trends Biochem Sci 18: 172–177

Ikeuchi M, Plumley FG, Inoue Y and Schmidt GW (1987) Phosphorylation of Photosystem II components, CP43 apoprotein, D1, D2, and 10 to 11 kilodalton protein in chloroplast thylakoids of higher plants. Plant Physiol 85: 638–642

Iwata S, Lee JW, Okada K, Lee JK, Iwata M, Rasmussen B, Link TA, Ramaswamy S and Jap BK (1998) Complete structure of the 11-subunit bovine mitochondrial cytochrome bc_1 complex. Science 281: 64–71

Jennings RC, Garlashi FM and Zucchelli G (1991) Light-induced fluorescence quenching in the light harvesting chlorophyll *a/b* protein complex. Photosynth Res 27: 57–64

Knoetzel J, Meyer DU and Grimme LH (1995) Phosphorylated Photosystem I antennae proteins in barley. In: Mathis P (ed) Photosynthesis: From Light to Biosphere, Vol I, pp 131–134. Kluwer Academic Publishers, Dordrecht

Lemaire C, Girard-Bascou J and Wollman F-A (1986) Characterization of the b_6/f complex subunits and studies on the LHC-Kinase in *Chlamydomonas reinhardtii* using mutant strains altered in the b_6/f complex. In: Biggins J (ed) Progress in Photosynthetic Research, Vol IV, pp 655–658 Nijhoff, Dordrecht

Liu J, Farmer JD, Lane WS, Friedman J, Weissman I and Schreiber SL (1991) Calcineurin is a common target of cyclophilin-cyclosporin A and FKBP-FK506 complexes. Cell 66: 807–815

Lunde C, Jensen PE, Haldrup A, Knoetzel J and Scheller HV (2000) The PS I-H subunit of Photosystem I is essential for state transitions in plant photosynthesis. Nature 408: 613–615

MacKintosh C, Coggins J and Cohen P (1991) Plant protein phosphatases. Biochem J 273: 733–738

Malkin R (1982) Interaction of photosynthetic electron transport inhibitors and the Rieske iron-sulfur center in chloroplasts and the cytochrome b_6/f Complex. FEBS Lett. 21: 2945–2950

Malkin R (1986) Interaction of stigmatellin and DNP-INT with the Rieske iron-sulfur center of the chloroplast cytochrome b_6f complex. FEBS Lett 208: 317–320

Marder JB, Telfer A and Barber J (1988) The D1 polypeptide subunit of the Photosystem II reaction centre has a phosphorylation site at its amino terminus. Biochim Biophys Acta 932: 362–365

Marks AR (1996) Cellular functions of immunophilins. Physiol Rev 76: 631–649

Michel HP, Hunt DF, Shabarkowitz J and Bennett J (1988) Tandem mass spectroscopy reveals that three Photosystem II proteins of spinach chloroplasts contain *N*-acetyl-*O*-phosphothreonine at their NH_2 termini. J Biol Chem 263: 1123–1130

Millner PA, Widger WR, Abbott MS, Cramer WA and Dilley RA (1982) The effect of adenine nucleotides on inhibition of the thylakoid protein kinase by sulfhydril-directed reagents. J Biol Chem 257: 1736–1742

Millner PA, Marder JB, Gounaris K and Barber J (1986) Localization and identification of phosphoproteins within the Photosystem II core of higher plant thylakoid membranes. Biochim Biophys Acta 852: 30–37

Pierre Y, Breyton C, Kramer D and Popot JL (1995) Purification and characterization of the cytochrome b_6f complex from *Chlamydomonas reinhardtii.* J Biol Chem 270: 29342–29349

Pierre Y, Breyton C, Lemoine Y, Robert B, Vernotte C and Popot JL (1997) On the presence and role of a molecule of chlorophyll a in the cytochrome b_6f complex. J Biol Chem 272: 21901–21908

Prasil O, Adir N and Ohad I (1992) Dynamics of Photosystem II: Mechanism of photoinhibition and recovery processes. In: Barber J (ed) The Photosystems: Structure, Function and Molecular Biology, pp 295–348. Elsevier, Amsterdam

Race HL and Hind G (1996) A protein kinase in the core of Photosystem II. Biochemistry 35: 13006–13010

Rintamäki E, Salonen M, Souranta U-M, Carlberg I, Andersson B and Aro E-M (1997) Phosphorylation of light-harvesting complex II and Photosystem II core proteins shows different irradiance-dependent regulation in vivo. J Biol Chem 272: 30476–30482

Rintamäki E, Martinsuo P, Pursiheimo S and Aro E-M (2000) Cooperative regulation of light-harvesting complex II phosphorylation via the plastoquinol and ferredoxin-thioredoxin system in chloroplasts. Proc Natl Acad Sci USA 97: 11644–11649

Schmidt CL and Malkin R (1993) Low molecular weight subunits associated with the cytochrome b_6f complex from spinach and *Chlamydomonas reinhardtii.* Photosynth Res 38: 73–81

Schuster G, Dewit M, Staehelin A and Ohad I (1986) Transient inactivation of the thylakoid Photosystem II light-harvesting protein kinase system and concomitant changes in intramembrane particle size during photoinhibition of *Chlamydomonas reinhardtii.* J Cell Biol 103: 71–80

Snyders S and Kohorn BD (1999) TAKs, thylakoid membrane protein kinases associated with energy transduction. J Biol Chem 274: 9137–9140

Sonoike K (1995) Selective photoinhibition of Photosystem I in isolated thylakoid membranes from cucumber and spinach. Plant Cell Physiol 36: 825–830

Staehelin LA and Arntzen CJ (1983) Regulation of chloroplasts membrane function: Protein phosphorylation changes the spatial organization of membrane components. J Cell Biol 97: 1327–1337

Stefansson H, Wollenberger L, Yu S-G and Albertsson P-Å (1995) Phosphorylation of thylakoids and isolated subthylakoid vesicles derived from different structural domains of the thylakoid membrane from spinach chloroplast. Biochim Biophys Acta 1231: 323–334

Sun G and Markwell J (1992) Lack of types 1 and 2A protein serine(P)/threonine (P) phosphatase activities in chloroplasts. Plant Physiol 100: 620–624

Vallon O, Bulté L, Dainese P, Olive J, Bassi R and Wollman F-A (1991) Lateral redistribution of cytochrome b6/f complexes along thylakoids membranes upon state transitions. Proc Natl Acad Sci USA 88: 8262–8266

Vener AV, van-Kan PJ M., Gal A, Andersson B and Ohad I (1995) Activation/deactivation cycle of redox-controlled thylakoid protein phosphorylation. J Biol Chem 270: 25225–25232

Vener A, van Kan PJ, Rich R, Ohad I and Andersson B (1997) Plastoquinol at the Qo-site of reduced cytochrome *b/f* mediates signal transduction between light and thylakoid phosphorylation. Thylakoid protein kinase deactivation by a single turnover flash. Proc Natl Acad Sci USA 94: 1585–1590

Vener A, Ohad I and Andersson B (1998) Protein phosphorylation and redox sensing in chloroplast thylakoids. Curr Opin Plant Biol 1: 217–223

Vener AV, Rokka A, Fulgosi H, Andersson B and Herrmann R G (1999) A cyclophilin-regulated PP2A-like protein phosphatase in thylakoid membranes of plant chloroplasts. Biochemistry 38: 14955–14965

Vink M, Zer H, Herrmann RG, Andersson B and Ohad I (2000) Regulation of Photosystem II core proteins phosphorylation at the substrate level: Light induces exposure of the CP43 chlorophyll a protein complex to thylakoid protein kinase(s). Photosynth Res, 64: 209–219

Walsh CT, Zydowsky LD and McKeon FD (1992) Cyclosporin A, the cyclophilin class of peptidylpropyl isomerases, and blockade of T cell signal transduction. J Biol Chem 267: 13115–13118

Weber P, Sokolenko A, Fulgosi H, Vener AV, Andersson B, Ohad I and Herrmann RG (2001) Elements of signal transduction involved in thylakoid membrane dynamics. In: Sopory SK, Oelmüller R and Maheshwari SC (eds) Signal Transduction in Plants: Current Advances. Kluwer Academic Publishers/Plenum Press, in press

Wollman F-A and Lemaire C (1988a) Phosphorylation processes interacting in vivo in the thylakoid membranes from *C. reinhardtii*. In: Hall DO and Grassi G (eds) Photocatalytic Production of Energy Rich Compounds, pp 210–214. Elsevier, Amsterdam

Wollman F-A and Lemaire C (1988b) Studies on kinase-controlled state transitions in Photosystem II and b_6f mutants from *Chlamydomonas reinhardtii* which lack quinone-binding proteins. Biochim Biophys Acta 933: 85–94

Yu CA, Xia JZ, Kachurin AM, Yu L, Xia D, Kim H and Deisenhofer J (1996) Crystallization and preliminary structure of beef heart mitochondrial cytochrome-bc_1 complex. Biochim Biophys Acta 1275: 47–53

Zer H, Vink M, Keren N, Dilly-Hartwig HG, Paulsen H, Herrmann RG, Andersson B and Ohad I (1999) Regulation of thylakoid protein phosphorylation at the substrate level: Reversible light-induced conformational changes expose the phosphorylation site of the light-harvesting complex II. Proc Natl Acad Sci USA 96: 8277–8282

Zhang Z, Huang L, Shulmeister VM, Chi Y-I, Kim KK, Hung W, Crofts AR, Berry EA and Kim S-H (1998) Electron transfer by domain movement in cytochrome bc_1. Nature 392: 677–684

Zito, F., Finazzi, G, Delosme, R, Nitschke, W, Picot, D and Wollmann F-A (1999) The Qo site of cytochrome b_6f complexes controls the activation of the LHCII kinase. EMBO J 18: 2961–2969

Chapter 25

Enzymes and Mechanisms for Violaxanthin-Zeaxanthin Conversion

Marie Eskling, Anna Emanuelsson and Hans-Erik Åkerlund*
Department of Plant Biochemistry, Lund University, POB 117, S-221 00 Lund, Sweden

Summary

The xanthophyll cycle is of great importance in relation to light stress. Particularly, interest has been focused on the possible photoprotective role of zeaxanthin. In higher plants under light stress, zeaxanthin is formed from violaxanthin in a reaction catalyzed by violaxanthin de-epoxidase (VDE). The reverse reaction is catalyzed by zeaxanthin epoxidase (ZE) under low light or in darkness. VDE has been purified from spinach and lettuce as a 43-kDa protein. The gene has been cloned and sequenced from several species, and a few mutants have been isolated. The gene is nuclear encoded and the transit peptide is characteristic for targeting to the thylakoid lumen. The activity of VDE is affected by factors such as a pH-dependent binding to the thylakoid membrane, concentration of ascorbic acid, temperature and availability of violaxanthin in relation to amount, type and distribution of pigment-protein complexes in the membrane. The information about ZE is more limited. The

*Author for correspondence, email: Hans-Erik.Akerlund@plantbio.lu.se

Eva-Mari Aro and Bertil Andersson (eds): Regulation of Photosynthesis, pp. 433–452.

enzyme has not yet been isolated but its gene has been cloned and sequenced and a number of mutants have been isolated. The role of the xanthophyll cycle in the dissipation of excess light energy will be discussed particularly in relation to the recent progress in studies on various mutants. The possible role of the xanthophyll cycle in other processes, such as protection against oxidative stress of lipids, regulation of membrane fluidity, participation in blue light responses, and regulation of abscisic acid synthesis will also be presented.

I. Introduction

Growth and development of plants are strongly determined by the efficiency of photosynthesis. Since plants usually cannot move, it is important that the photosynthetic machinery can acclimate dynamically to stress, imposed by changes in the external conditions. Factors like irradiance, water supply and temperature play important roles. At low irradiance i.e. cloudy days, as much light as possible must be collected. However, at high irradiance, there is a great risk for photodamage if the plant cannot acclimate to the surrounding light conditions.

An increasing interest has focused on the importance of carotenoids in the process of light capture, handling of excessive light energy and protection against harmful radicals. Of special importance is a group of carotenoids that participate in the so-called xanthophyll cycle. The xanthophyll cycle is a light dependent and reversible conversion of violaxanthin to zeaxanthin. Results from several research groups in recent years have shown that the xanthophyll cycle directly or indirectly is of importance for regulation of light harvesting, protection against overexcitation and recovery from photoinhibitory conditions. Especially zeaxanthin has been correlated to the dissipation of excess energy, thereby reducing the risk of photodamage. For other recent reviews see Demmig-Adams et al., 1997; Eskling et al., 1997; Gilmore, 1997; Havaux, 1998.

II. The Xanthophyll Cycle, Enzymes and Pigments

The xanthophyll cycle is widespread in nature and is present in thylakoid membranes of all higher plants, ferns, mosses and several algal groups. There are three variants, the violaxanthin cycle, which is more common and found in higher plants, the diadinoxanthin cycle found in some algal groups (see Table 3 in Adamska, 1997) and recently a third variant, involving a lutein to lutein-5,6-epoxide cycle, was found in the parasitic angiosperm *Cuscuta reflexa* (Bungard et al., 1999). Virtually all organisms performing aerobic photosynthesis have the capacity to form zeaxanthin or the analogue diatoxanthin (Hager, 1975). Cyanobacteria have neither the diadinoxanthin cycle nor the violaxanthin cycle but at high irradiances they can form zeaxanthin from β-carotene via the xanthophyll biosynthesis pathway (Demmig-Adams, 1990). While the xanthophyll cycle is confined to photosynthetic organisms, the presence of zeaxanthin is not. Zeaxanthin and, in lesser quantities, lutein are also present in the human eye (Demmig-Adams et al., 1997). Medical research has suggested a link between the presence of these carotenoids in the human diet and the prevention of age-related macular degeneration that can cause blindness (Seddon et al., 1994).

The light dependent conversion of xanthophylls was discovered by Sapozhnikov et al. (1957) and the main features of the xanthophyll cycle were established in the 1960s and 1970s through the pioneering studies of the groups of Yamamoto and Hager (Yamamoto et al., 1962; Hager, 1969; Yamamoto, 1979). The xanthophyll cycle involves two successive de-epoxidations of violaxanthin. During high light conditions (overexcitation of PS II) violaxanthin is converted into zeaxanthin via the intermediate antheraxanthin, a reaction reversed in darkness and at low irradiances that do not saturate photosynthesis (Fig. 1). The reactions are catalyzed by two different enzymes, violaxanthin de-epoxidase (VDE) for the de-epoxidation reaction and zeaxanthin epoxidase (ZE) for the epoxidation reaction.

A. Violaxanthin De-Epoxidase

The enzyme responsible for de-epoxidation, VDE is located in the thylakoid lumen. The activity of the enzyme is strongly pH dependent. At high pH in the

Abbreviations: ABA – Abscisic acid; CP29, CP26, CP24 – Minor chlorophyll protein complexes; DCCD – Dicyclohexylcarbodiimide; DGDG – Digalactosyldiacylglycerol; DTT – Dithiotreitol; ELIP – Early light induced protein; LHC – Light harvesting complex; MGDG – Monogalactosyldiacylglycerol; NPQ – Non-photochemical quenching; PS I, PS II – Photosystem I, II; VAZ – Violaxanthin+antheraxanthin+zeaxanthin; VDE – Violaxanthin de-epoxidase; ZE – Zeaxanthin epoxidase

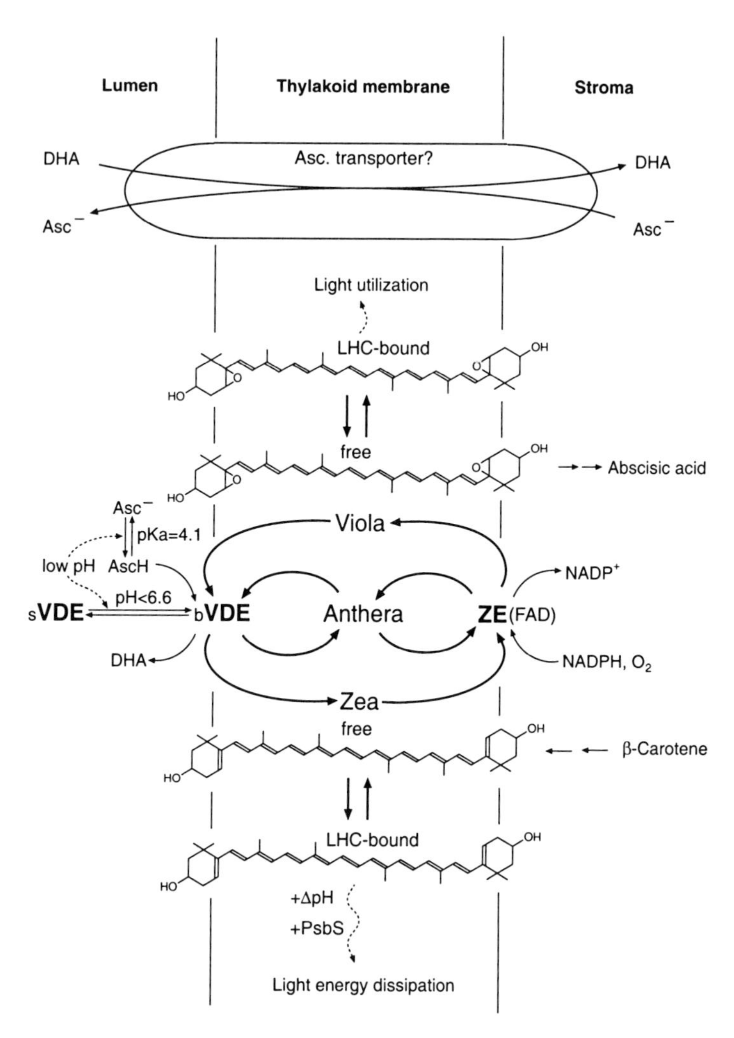

Fig. 1. The xanthophyll cycle of higher plants. DHA, dehydroascorbate; Asc⁻, ascorbate; AscH, ascorbic acid; s and bVDE, soluble and bound VDE.

thylakoid lumen, VDE is a soluble enzyme, but upon irradiation electron transport generates a proton gradient across the thylakoid membrane, the pH of the lumen drops and the enzyme becomes fully attached to the inner thylakoid membrane surface. VDE was found to be specific for xanthophylls that have the 3-hydroxy-5,6-epoxy group in a 3*S*, 5*R*, 6*S* configuration and are all *trans* in the polyene chain (Yamamoto and Higashi, 1978; Grotz et al., 1999). The conversion of violaxanthin requires ascorbate as well as the major thylakoid lipid monogalactosyldiacylglycerol (MGDG). A direct interaction between lipids and VDE was demonstrated by Rockholm and Yamamoto (1996). They found MGDG to be four times more efficient in precipitating VDE compared to the second most common lipid in thylakoids, digalactosyldiacylglycerol (DGDG) and up to 38 times more efficient than other lipids.

1. Isolation of VDE

VDE has been purified from spinach (Åkerlund et al., 1995; Arvidsson et al., 1996; Havir et al., 1997; Kuwabara et al., 1999) and from lettuce (Rockholm and Yamamoto, 1996). The isolated protein showed a molecular weight of 43 kDa on SDS-gel electrophoresis and moved as a monomer in gel filtration. The amount present in the thylakoid was estimated to be lower than 1 VDE per 20–100 electron-transport chains and a more than 10000-fold purification was required to reach homogeneity. Isoelectric focusing established the acid nature of VDE with a pI of 5.0–5.4. The sulfhydryl reagent DTT is a potent inhibitor of VDE (Yamamoto and Kamite, 1972). VDE has been found to change its mobility on gel electrophoresis, between 40–43 kDa (Kuwabara et al., 1999), and it was suggested that fully reduced VDE had the lowest electrophoretic mobility and depending on number of oxidized disulfide bridges faster mobility. Furthermore, the active site of VDE was suggested to contain a reactive aspartic acid residue, as the enzyme is sensitive to pepstatin A.

2. Cloning of VDE

Bugos and Yamamoto (1996) reported a cDNA clone from romaine lettuce (*Lactuca sativa*), which encoded a functionally active VDE when expressed in *Escherichia coli* (GenBank U31462). The recombinant protein showed properties in common with VDE isolated from higher plants; e.g. it was inhibited by DTT (Yamamoto and Kamite, 1972). The enzyme showed to be nuclear-encoded in higher plants and the VDE gene was found in a single copy.

Since then, two more VDE genes have been sequenced by Bugos and Yamamoto, from *Nicotiana tabacum* and *Arabidopsis thaliana* (GenBank U34817 and U44133, respectively). The VDE gene from *Spinacia oleracea* was recently sequenced by Emanuelsson et al. (GenBank AJ250433). Alignment of known amino acid sequences of VDE is shown in Fig. 2. These sequences contain the signal peptides, directing the polypeptides to the thylakoid lumen, preceding the mature protein sequence. The cleavage sites are, when investigated by a neural networks program (Nielsen et al., 1997), predicted to be located between the amino acids A and VDALK resulting in a 125, 134, 113 and 124 residues long signal sequence in *L. sativa*, *N. tabacum*, *A. thaliana* and *S. oleracea*, respectively, in agreement with the N-terminal sequence determined for the mature VDE from spinach (Arvidsson et al., 1996). The sequences of the signal peptides are divergent but share the common feature of a bipartite signal peptide, consisting of a long, charged N-terminal part targeting the protein for the chloroplast stroma and a short, hydrophobic, C-terminal part for thylakoid import (Brink et al., 1998; Bugos et al., 1998). There are few identical amino acids between the four signal peptides, some being present at the outermost N- and C-terminal ends. The (–3, –1) rule (von Heijne, 1983, 1985) of small neutral residues being present at positions –1 and –3 from the cleavage site for correct cleavage, is valid in all four cases.

The mature proteins show much greater identity than the signal peptides, with a pairwise identity ranging from 83% (*A. Thaliana–L. sativa*) to 78% (*S. Oleracea–L. sativa*). Comparing all four polypeptides, they show an overall identity of 68%. The amino acid composition reveals a high abundance of charged and polar residues, which is typical for water-soluble hydrophilic proteins. The sequence of VDE is unique, showing no apparent sequence homology with any non-VDE sequence in the databases.

Three interesting domains appear within the mature protein (Fig. 2), (1) a cysteine-rich region containing 11 of the 13 cysteines of VDE, located in the first 72 amino acids from the N-terminus, (2) a lipocalin signature (Bugos et al., 1998), which is a possible binding site for violaxanthin and MGDG, and (3) a highly charged region close to the C-terminus (approximately amino acids 255–295), which may be involved in binding of VDE to the thylakoid membrane (Jahns and Heyde, 1999). The cysteines in the N-terminal region probably form more than one disulfide linkage because partial inhibition of VDE activity with DTT results in an accumulation of antheraxanthin (Gilmore and Yamamoto, 1993; Havir et al., 1997). Additional evidence for the N-terminal region to be involved in the active site of VDE is the *Arabidopsis* mutant *npq1*, which have a substitution in the last cysteine in the N-terminal region (amino acid 72) and as a result of this becomes non-functional (Niyogi et al., 1998). The active site of VDE was suggested to contain a reactive aspartic acid residue, as the enzyme is sensitive to pepstatin A (Kuwabara et al., 1999).

3. Binding of VDE to the Thylakoid Membrane

An important factor in the regulation of VDE activity is its binding to the membrane. Hager and Holocher (1994) showed, by repeated freeze-thaw cycles, that VDE is tightly bound to the membrane at a pH below 6.5 and mobile in the lumen at pH above 7.0. Bratt et al. (1995) found the pH dependent release of VDE from the thylakoid membrane to occur in a narrow pH range, with half of the enzyme being released (inflexion point) at pH 6.6 and a cooperativity of 4 with respect to protons. The results showed that at pH values below 6, VDE was fully bound to the membrane and at pH values above 7, it was fully released. This is in accordance with the results of Pfündel and Dilley (1993), who found a cooperativity of 5.3 with respect to protons, where 1 unit should be considered

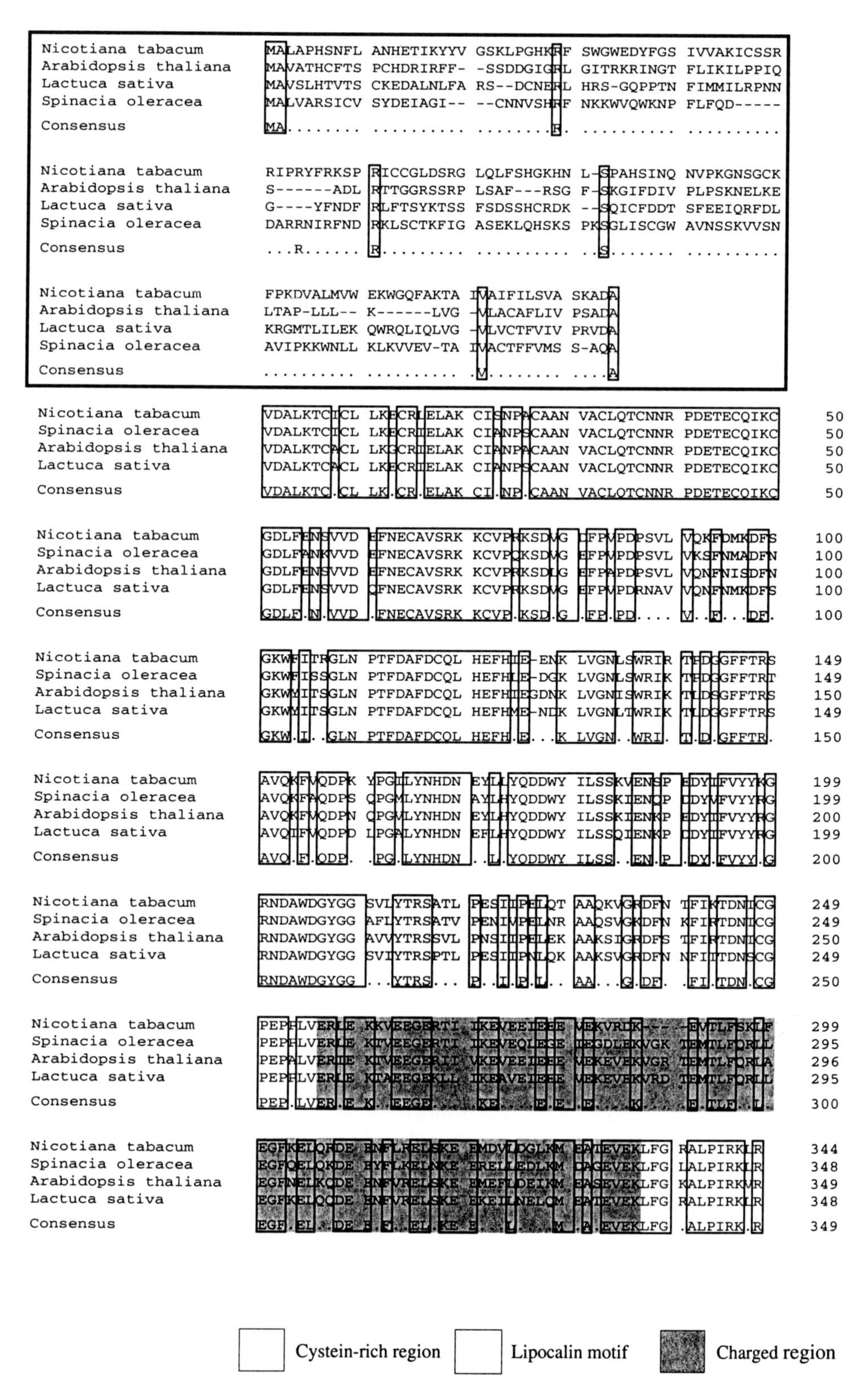

Fig. 2. Alignment of amino acid sequences of VDE, obtained from translated PCR products. The signal peptide for import into the chloroplast and thylakoid is boxed. Three specific regions within the protein is marked in grey scale: (1) a cysteine-rich region, (2) a lipocaline signature and (3) a highly charged region.

for protonation of ascorbate and 4.3 units for the binding. Modification of VDE by DCCD (dicyclohexylcarbodiimide), which binds to carboxylic acids, caused a shift in pH dependence of about 0.3 pH-units to more alkaline values and also caused a less pronounced cooperativity for the pH dependence of de-epoxidation (Jahns and Heyde, 1999). This indicates that the C-terminal region, rich in acidic amino acids, is involved in the binding of VDE to the thylakoid membrane. VDE clearly falls into the group of amphitropic proteins, that is, proteins interconverted between a soluble inactive form and a membrane bound active form. Several amphitropic proteins are known and the signal they respond to may vary (Cornell and Northwood, 2000).

The inhibition of VDE by DTT, that brakes disulfide bridges in the cysteine-rich N-terminal region of the enzyme, is a useful tool when studying the function of the xanthophyll cycle, but it has to be remembered that the inhibition is reversible (Yamamoto and Kamite, 1972). Irreversible inhibition of VDE has been achieved with a combination of DTT and iodoacetamide at pH 5.2 and 5.7 (Arvidsson et al., 1997). No inhibition was found at pH 7.2, indicating that the N-terminal disulfide bridges were not exposed at high pH. This suggests that VDE undergoes a conformational change upon binding to the thylakoid membrane and that the exposed disulfide bridges may be involved in the environment at the active site. When the endogenous VDE (VDE in the thylakoid lumen) was irreversibly inhibited by this method, conversion of violaxanthin to zeaxanthin could be restored by adding partly purified VDE to the outside of the thylakoids. This indicates that treatment of thylakoids with DTT and iodoacetamide acts on VDE and not on any other component required for de-epoxidation. The indicated conformational change of VDE has recently been strengthened by the pH dependent inhibition of VDE by pepstatin and also the demonstration, by ion exchange chromatography, of two interconvertible forms of VDE (Kawano and Kuwabara, 2000).

4. Regulation of VDE Activity by Ascorbate

Ascorbate (vitamin C), an abundant soluble antioxidant in plant chloroplasts, functions as a substrate for VDE in the xanthophyll cycle. Early Hager (1969) proposed that a light-driven acidification of the lumen regulated the VDE activity. This was based on the pH optimum for VDE, that was 4.8 for chloroplasts and 5.2 for the partly purified enzyme. Thus, under high light conditions the electron-transport chain creates an acidic lumen and at metabolic restrictions (e.g. shortage of ADP at the ATP synthase) the lumen would be acidified even further. Pfündel and Dilley (1993) showed that in isolated pea chloroplasts zeaxanthin formation was strongly upregulated within a narrow range from pH 5.8 to 6.3, with maximum activity below 5.8.

A detailed study of the activity as a function of both pH and ascorbate concentration was made by Bratt et al. (1995). The rise in activity and the pH optimum was found to be dependent on the ascorbate concentration, as both the pH optimum and the inflexion points were shifted to lower values when reducing the ascorbate concentration. These results lead to the conclusion that the acid form of ascorbate (pKa = 4.1) is the substrate for VDE. The ascorbate concentration in leaves and chloroplast stroma has been estimated to be in the range of 10–50 mM (Gillham and Dodge, 1986; Schöner and Krause, 1990; Foyer 1993; Eskling and Åkerlund, 1998). The way of transport to the lumen and the concentration in the lumen is not known. The uncharged acid form of ascorbate would be the only form able to penetrate the membrane by simple diffusion, and therefore the concentration of ascorbic acid in the lumen would be determined by the stroma pH, e.g. at 50 mM ascorbate and pH 8, the concentration of the acid form would be 8 μM, well below the K_m of 0.1 mM (Bratt et al., 1995). This argues for the presence of a transport system for ascorbate in the thylakoid membrane. The ascorbate availability could in principle regulate the xanthophyll cycle, as VDE can be inhibited by competitive consumption of ascorbate by ascorbate peroxidase when adding hydrogen peroxide (Neubauer and Yamamoto, 1994). In the case that the concentration of the acid form of ascorbate would be determined by the light-driven acidification of the thylakoid lumen, an elevated ascorbate concentration could promote zeaxanthin formation at higher pH in stress situations, e.g. a concentration of 0.1 mM ascorbic acid would be reached at a total ascorbate concentration of 10 mM, 25 mM and 50 mM at approximately pH 6.2, pH 6.6 and pH 6.9, respectively. In accordance with this, Eskling and Åkerlund (1998) found that an increase in the ascorbate concentration and the rate of violaxanthin de-epoxidation in spinach plants shifted from low to high light. Additionally to VDE, the radical scavenging system uses ascorbate, which involves the enzyme ascorbate peroxidase (Asada, 1992). Conklin et al. (1996) found an *Arabidopsis* mutant (*soz*-1) that contained 30% of the normal ascorbic acid concentration. It had elevated lipid peroxidation and was hypersensitive to SO_2 and UV-B. This mutant and the more recently identified ascorbic acid deficient *Arabidopsis* mutants (*vtc1*, *vtc2*, Conklin et al., 2000) could be valuable for further investigation on the role of ascorbic acid in regulation of the xanthophyll cycle.

B. Zeaxanthin Epoxidase (ZE)

The enzyme ZE catalyses the conversion of zeaxanthin to violaxanthin through antheraxanthin and has an activity optimum between pH 7 and pH 7.5, close to the pH of the stroma in darkness, and no activity at low pH (Siefermann and Yamamoto, 1975a). ZE is a monooxygenase using molecular oxygen and NADPH as substrates and FAD as a cofactor (Büch et al., 1995). The epoxidation reaction occurs in darkness but is stimulated by weak light (Hager, 1966), when O_2 and NADPH are formed through photosynthetic electron transport (Siefermann and Yamamoto, 1975b). Free fatty acids are potent inhibitors (Siefermann and Yamamoto, 1975a). Based on the pH optimum, NADPH dependence and washing experiments, the ZE has been concluded to be bound to the thylakoid membrane and exposed to the stroma (Siefermann and Yamamoto, 1975b). The enzyme has not yet been isolated but both LHCII in PS II (Strasser and Butler, 1976; Gruszecki and Krupa, 1993) and LHCII-depleted bundle sheath cells of *Sorghum bicolor* and stroma lamella fractions of isolated spinach thylakoids (Färber and Jahns, 1998) have been claimed to catalyze the epoxidation of zeaxanthin.

It has been observed that the rate of epoxidation depends on irradiation and the content of LHCII, CP29, CP26 and CP24, with a slower epoxidation rate in plants with only CP26 present (pea plants grown in intermittent light). Under prolonged exposure to excess light there was a decrease in the rate of epoxidation for normal and intermittent-light grown plants (Jahns, 1995). Härtel et al. (1996) have also shown that epoxidation rates decrease with a reduction in LHC proteins when investigating wild-type barley plants and the chlorophyll *b*-less mutant *chlorina* under continuous and intermittent light. However, results recently presented by Färber and Jahns (1998) show reduced epoxidation rates with increasing amounts of antenna proteins under continuous irradiation of pea plants grown in intermittent light. Even if the results for ZE activity presented by different authors cannot be correlated to antenna protein content, Färber and Jahns (1998) state that antenna proteins are required for maximum epoxidation rate. ZE may also become inactivated under physiological conditions allowing sustained high levels of zeaxanthin (Adams et al., 1995) possibly due to protein phosphorylation (Xu et al., 1999).

1. Cloning of Zeaxanthin Epoxidase

When Marin et al. (1996) analyzed *Nicotiana plumbaginifolia* mutants deficient in abscisic acid (ABA) biosynthesis they found an accumulation of zeaxanthin in a mutant, isolated by transposon tagging. The gene corresponded to the *aba* locus of *Arabidopsis thaliana*, and they identified the gene product as ZE. The *aba2* cDNA encoded a chloroplast-imported protein of 72.5 kDa, sharing similarities with different monooxygenases and oxidases of bacterial origin. It had an ADP-binding fold containing a glycine-rich motif common in cofactor-binding enzymes with binding properties of NAD(P) or FAD and a FAD-binding domain. Transcription-translation chloroplast import studies in vitro showed cleavage of a 50 amino acid signal peptide resulting in a mature protein of 67 kDa. The signal peptide showed characteristics ascribed to known chloroplast-targeted proteins that are nuclear encoded. However, the experiment did not show whether the mature protein was located in the soluble chloroplast stroma, or associated with, or integrated into, the thylakoid membrane. As the ABA2 protein, expressed in *Escherichia coli*, only exhibited ZE activity in vitro when incubated together with a stroma protein extract, the authors suggested that an unidentified factor, absent in *E. coli*, was necessary for activity. The stroma extract by itself showed no activity. Studies on the cloned and expressed ZE from pepper, showed that the reducing power from NADPH could be transferred to ZE via reduced ferredoxin (Bouvier et al., 1996). The enzyme was moderately hydrophobic and carried three possible PEST domains usually observed in rapidly degraded proteins, as well as a FAD-binding domain. The pepper ZE showed 88% identity to the amino acid sequence of the proposed ZE from *N. Plumbaginifolia*. Also, in addition to VDE, ZE was found to have a lipocalin motif (Bugos et al., 1998).

Expression studies done on the ZE gene in *Nicotiana plumbaginifolia* showed that ABA2 mRNA accumulated in all plant organs, but that transcript levels were higher in aerial parts (stem and leaves) than in roots and seeds (Audran et al., 1998). ABA2 mRNA accumulation displayed a day/night cycle in leaves, but not in roots. However, the ABA2 protein level remained constant. The authors propose that regulation of ABA2 transcript levels in leaves are more related to regulation of photosynthetic genes and that levels in non photosynthetic tissues are coordinately regulated with changes in ABA levels. These results agree with results from the tomato gene ZE where mRNA was shown to oscillate with a phase similar to LHCII mRNA (Thompson et al, 2000). The exact location of these ZE genes remains to be investigated.

C. Pigment Distribution in the Thylakoid Membrane

The pigments found in the thylakoid membrane are not evenly distributed between the grana and stroma lamellae. The stroma lamellae membranes are enriched in chlorophyll *a* and *β*-carotene, which correlates with a high content of PS I, whereas the grana membranes have a high content of PS II and LHCII and an enrichment in chlorophyll *b*, lutein and neoxanthin (Arvidsson et al., 1997). The procedure used to enrich thylakoid sub-fractions was mechanical fragmentation of spinach thylakoids in combination with aqueous two-phase partition. This technique was used to prevent the loss of pigments unavoidable with detergent extraction techniques. The VAZ pigments are unique in the sense that they are the only pigments that are almost uniformly distributed in the thylakoid membrane, especially under high light conditions (Juhler et al., 1993; Arvidsson et al., 1997). Already in 1976, Siefermann and Yamamoto found by a cruder differential centrifugation technique violaxanthin to be present in both photosystems and consequently evenly distributed.

The distribution of pigments between protein complexes has been studied more extensively in PS II than in PS I. The general theme in pigment-protein isolation has involved solubilization of PS II particles by mild detergent treatment and purification by either isoelectric focusing, sucrose gradient centrifugation or deriphat-PAGE techniques. Pigments are non-covalently bound to protein complexes but as the distribution is heterogeneous there must be preferred interaction sites for each pigment in the photosynthetic apparatus. A major proportion of the pigments is believed to be constituents of the antenna complexes (Siefermann-Harms, 1985; Peter and Thornber, 1991; Ruban et al., 1994) although some may only be loosely bound to the periphery of the complexes. Violaxanthin has been found associated with most pigment proteins (Bassi et al., 1993, 1997; Phillip and Young, 1995). The data reported from different laboratories are often contradictory, but despite this, all authors agree that the minor complexes CP24, CP26 and CP29 have a considerably higher VAZ/chlorophyll *a* ratio than the bulk LHCII. Yamamoto and Bassi (1996) also found a significant loss of pigment during pigment-protein purification through detergent solubilization, especially after light treatment. This could be one explanation for the large discrepancies present in the literature.

Violaxanthin is also present in membranes enriched in PS I (Siefermann and Yamamoto, 1976; Siefermann-Harms, 1985; Juhler et al., 1993; Arvidsson et al., 1997). Thayer and Björkman (1992) suggested that 30% of the VAZ pigments are associated with the LHCI and PS I core while Lee and Thornber (1995) found VAZ predominantly associated with LHCI.

1. Availability of Violaxanthin

The amount of violaxanthin available for conversion at high excitation pressures differs greatly in between publications. The maximum violaxanthin conversion, at saturating light conditions, is in the range of 50–80% for most plants. The reason for the upper limitation is not yet clear. Notably, the degree of conversion is higher in species with low photosynthetic capacity than in those with high photosynthetic capacity (Thayer and Björkman, 1990; Adams and Demmig-Adams, 1992; Demmig-Adams and Adams, 1992a).

Many studies have reported high degrees of conversion in plants deficient in LHC compared to the wild type (Leverenz et al., 1992; Falbel et al., 1994; Jahns and Krause, 1994; Lokstein et al., 1994; Andrews et al., 1995; Färber and Jahns, 1998). The only case where almost 100% de-epoxidation in high light has been reported is for pea plants grown under intermittent light with CP26 as the only PS II antenna pigment-protein present (Jahns, 1995). Also the chlorophyll *b*-less *chlorina* mutant of barley, grown under intermittent light conditions and lacking CP26

(Król et al., 1995) had a de-epoxidation of up to 98% (Härtel et al., 1996). These results are in apparent contradiction to the product inhibition of VDE by zeaxanthin found by Havir et al. (1997).

Taken together, it appears to be a correlation between a decreasing amount of pigment-binding proteins and an increasing convertibility of violaxanthin into zeaxanthin. This suggests that the organization of the pigment-protein complexes may control the degree to which violaxanthin is converted to zeaxanthin. In plants with a decreased amount of antenna proteins the VAZ pigments can either be free in the lipid matrix or be enriched in the few remaining LHC proteins, or even be associated with other proteins e.g. early light induced proteins, ELIPs (Król et al., 1995).

2. Location of Violaxanthin

The question of the site for conversion of violaxanthin is controversial and follows the disagreements on pigment location. In the results described below de-epoxidation had been induced prior to pigment-protein purification. Lee and Thornber (1995) found that the xanthophyll cycle occurs exclusively within the light-harvesting antenna of both photosystems. Thayer and Björkman (1992) found similar results in the case of PS II but for PS I the conversion of violaxanthin to zeaxanthin was found to be 40% in both LHCI and PS I core. Many groups working with PS II have found 50% of the light-induced zeaxanthin located in bulk LHCII and only lower amounts in the minor CP antenna (Peter and Thornber, 1991; Lee and Thornber, 1995; Phillip and Young, 1995). Ruban et al. (1994) found, upon light treatment of spinach, an increase in VAZ pigments in LHCII and a decrease in minor CPs, suggesting a xanthophyll cycle-induced reorganization of pigments. As these studies have been done on detergent-solubilized particles, loss of pigments is a complicating factor. Due to the preferential loss of xanthophylls it has been suggested that violaxanthin is weakly bound to the periphery of the chlorophyll antenna (Peter and Thornber, 1991, see also review by Ruban et al., 1994; Lee and Thornber, 1995; Horton et al., 1996). In light-treated barley leaves a low amount of zeaxanthin was found in LHCII prepared after de-epoxidation, despite the presence of violaxanthin before de-epoxidation (Tardy and Havaux, 1997). The released zeaxanthin from LHCII did not rebind to the minor CP complexes and was concluded to be in the lipid matrix after conversion. Dicyclohexylcarbodiimide (DCCD), an inhibitor of NPQ in vitro, retards the kinetics of de-epoxidation of violaxanthin to zeaxanthin, with the largest effect on antheraxanthin to zeaxanthin conversion. This effect of DCCD is restricted to the PS II antenna-bound xanthophylls and indicates binding of DCCD to glutamate side chains of specific LHC polypeptides (Heyde and Jahns, 1998).

In Arvidsson et al. (1997) sub-thylakoid membrane domains were isolated, without the use of detergents, to obtain vesicles with different amounts of PS II and PS I. In combination with added VDE and ascorbate at low pH they found almost the same degree of conversion of violaxanthin to zeaxanthin (73–78%) for different domains of the membrane. Apparently, the restriction for de-epoxidation is not dependent on whether violaxanthin is located in the grana or stroma lamellae. Arvidsson et al. (1997) found, using DTT plus iodoacetamide as a novel irreversible method to inhibit endogenous VDE, that violaxanthin could be converted into zeaxanthin from both sides of the thylakoid membrane, provided VDE was added exogenously. As the pigment-proteins have different faces at the two sides of the thylakoid membrane it was concluded that the conversion takes place in the lipid matrix rather than at the protein complexes. Hager and Holocher (1994) also suggested that VDE converts violaxanthin in the lipid matrix. Furthermore, an *aba1* mutant of *Arabidopsis thaliana*, that lacks ZE and accumulates zeaxanthin, had a substantial increase of pigments free in the membrane (mainly zeaxanthin and chlorophyll *b*) (Tardy and Havaux, 1996). In order for violaxanthin to be accessible from both sides of the membrane, the VAZ pigments are likely to be oriented in a vertical manner in the membrane. The model of vertical orientation (reviewed by Gruszecki, 1995) involves interaction of the polar groups of violaxanthin with the hydrophilic head groups of the lipids in the bilayer and is supported by the fact that the thickness of the hydrophobic core of the lipid bilayer and the length of violaxanthin are similar (30 Å).

III. The Conversion of Violaxanthin Depends on Temperature and Light

A. Temperature-Dependent Conversion of Violaxanthin

It is not only the amount of pigment-binding proteins that influence the de-epoxidation level but also temperature. Temperature has a strong influence on the rate of violaxanthin to zeaxanthin conversion and the degree of maximal conversion (Koroleva et al., 1995; Arvidsson et al., 1997). At low temperatures the zeaxanthin formation was found to be strongly suppressed (Demmig-Adams et al., 1989; Bilger and Björkman, 1991), which in turn was suggested to result in an increased susceptibility to photoinhibition (Krause, 1994). Moderately elevated temperature has also been found to increase the accessibility of violaxanthin in the membrane to VDE (Havaux and Tardy, 1996). A recent theory is that a limited conversion may be influenced by possible *cis-trans* isomerizations of violaxanthin (Gruszecki et al., 1997), with only *trans*-violaxanthin acting as substrate for de-epoxidation.

B. Acclimation of Plants to High Light Involves Parameters in the Xanthophyll Cycle

The photosystem is often subjected to extreme variations in photon flux densities and can be damaged by overexcitation. An excess of light can arise when there is an increase in photon flux density or a decrease in photosynthetic activity at constant photon flux density, such as might occur under chilling conditions or in response to water stress (Demmig-Adams and Adams, 1992b). A decrease in photosynthetic activity due to light stress is known as photoinhibition and often results in photoinactivation or photodamage, particularly of PS II (reviewed by Aro et al., 1993).

In long-term acclimation, light-acclimated plants have a larger pool of xanthophylls and an increased maximal conversion to zeaxanthin (Thayer and Björkman, 1990; Logan et al., 1996; Demmig-Adams et al., 1997). The light environment can also regulate the ascorbate concentration (Logan et al., 1996). Plants grown at high irradiances and low temperature contain higher levels of ascorbate and higher levels of oxygen radical scavenging enzymes than plants grown under low light (Gillham and Dodge, 1987; Schöner and Krause, 1990). In spinach plants the ascorbate concentration may rise to as much as 50 mM (Schöner and Krause, 1990).

Bugos et al. (1999) have reported an inverse nonlinear relation between changes in VDE and VAZ pigments in tobacco plants during leaf development. Also, acclimation of spinach leaves to high light resulted in an increase in the amount of VAZ, the rate of violaxanthin to zeaxanthin conversion and the amount of ascorbate but a decrease in the amount of VDE (Eskling and Åkerlund, 1998). The increased rate of conversion of violaxanthin to zeaxanthin found could only partly be explained by the increase in VAZ and ascorbate, therefor the most probable explanation for the faster conversion is an increased accessibility to violaxanthin in the membrane (Eskling and Åkerlund, 1998). An increase of ascorbate content made artificially in detached maize leaves de-epoxidize the xanthophyll-cycle pigments faster upon exposure to high irradiance, but to the same maximal level as in non-incubated leaves (Leipner et al., 2000). An increased ratio of VAZ to pigment-protein complexes after prolonged light stress has been reported and it was suggested that the excess VAZ might be present in the lipid matrix (Schäfer et al., 1994; Schmid and Schäfer, 1994). However, photoconvertible VAZ pigments have also been suggested to bind to ELIPs that replace normal LHC proteins under conditions of light stress (Król et al., 1995). Isolation of ELIPs in a native form and analysis of pigments bound to these proteins revealed that ELIPs could bind chlorophyll *a* and lutein (Adamska, 1997). These data indicate that ELIPs might represent chlorophyll-binding proteins, that have a transient function(s) during light stress.

During prolonged irradiation with strong light, an increase in the VAZ pigments can be caused by stimulated de novo synthesis directly from β-carotene to zeaxanthin (Schäfer et al., 1994; Schindler and Lichtenthaler, 1996). This has not been studied much but recently the gene for the enzyme β-carotene hydroxylase was cloned (Schumann et al., 1996; Sun et al., 1996).

IV. The Role of the Xanthophyll Cycle

The role of zeaxanthin and the xanthophyll cycle in the protection against photodestruction is almost exclusively based on fluorescence measurements. In many studies a good correlation between the fluorescence parameter, non-photochemical quenching and the amount of zeaxanthin (and antheraxanthin) has been found. Still a number of other functions for the xanthophyll cycle have been proposed, such as protection against oxidative stress of lipids, regulation of membrane fluidity, participation in a blue light response and regulation of abscisic acid synthesis.

A. The Role of the Xanthophyll Cycle in Energy Dissipation

Very little progress was made on the role of the xanthophyll cycle until the finding by Demmig et al. (1987) of a possible connection between zeaxanthin formation and dissipation of excess light energy, seen as fluorescence quenching. The main part of the fluorescence at room temperature comes from PS II and to our knowledge all functional studies have focused on PS II. However, the effects seen on PS II may well hold also for PS I, but have not yet been studied because of the lack of a convenient analysis method. When light energy is absorbed by PS II three main competitive routes occur. The energy is (1) re-emitted as light in the form of fluorescence, (2) emitted as heat or (3) used in photosynthesis. Thus the chlorophyll fluorescence gives indirect information on the activity of photosynthesis and is at room temperature thought to be predominantly emitted by the antenna of PS II. When measuring chlorophyll fluorescence, dark-adapted plant material may first be exposed to a modulated red light and then excited with a brief white saturating pulse of high irradiance. First a minimal fluorescence, F_0, is achieved corresponding to a situation when all Q_A are oxidized (open reaction centers), followed by the maximum level of fluorescence, F_M, when all Q_A are reduced (closed reaction centers)(Dau, 1994). The variable fluorescence, F_V, is $F_M - F_0$. The decrease from the maximum level of fluorescence is called quenching. There can be a decrease in fluorescence as a result of light energy being converted into chemical energy, so-called photochemical quenching (qP, PQ) (Gilmore 1997). Fluorescence quenching, which does not take part in the process of photochemistry, is denoted non-photochemical quenching (qN, NPQ). Three major mechanisms of NPQ have been described; (1) the pH gradient-dependent or energy-dependent mechanism (qE) (Briantais et al., 1979; Krause and Behrend, 1986) is characterized by rapid induction and relaxation kinetics, (2) the redistribution of energy from PS II to PS I (state transitions, qT) (Staehelin and Arntzen, 1983; Horton and Lee, 1985; Allen, 1992) and (3) photoinhibition (qI) (Demmig et al., 1987).

A correlation between an increase in NPQ and zeaxanthin formation under high light conditions suggests that zeaxanthin has a photoprotective function, dissipating excess energy by quenching (Demmig et al., 1987; Krause, 1988; Demmig-Adams, 1990; Krause and Weis, 1991; Demmig-Adams and Adams, 1992b; Pfündel and Bilger, 1994; Demmig-Adams et al., 1996). An even better correlation was obtained when the sum of antheraxanthin and zeaxanthin was correlated to NPQ (Gilmore and Yamamoto, 1993; Adams et al., 1995). There are exceptions though, and it has become apparent that ΔpH in addition to zeaxanthin or antheraxanthin is required for efficient energy dissipation (Thiele and Krause, 1994; Demmig-Adams et al., 1996). A correlation has also been found between turnover of D1 protein and photoinhibition, measured as a decrease in the ratio of variable to maximal fluorescence, under both low and high light (Tyystjärvi and Aro, 1996). Photoinhibition kinetics showed an agreement between loss of PS II activity and D1 protein degradation. A fast recovery after inhibition has been reported to be closely correlated with the conversion of zeaxanthin to violaxanthin and a slow recovery phase correlated to the resynthesis of the D1 protein (Thiele et al., 1996).

How this quenching occurs is still obscure. Two models have been proposed, the indirect model and the direct model. The indirect model includes proton-induced structural changes that allow antheraxanthin and zeaxanthin to form dissipating centers in specific antenna complexes, while violaxanthin inhibits such structural changes and causes elevated fluorescence (Horton et al., 1991; Walters et al., 1996; Ruban et al., 1997; Wentworth et al., 2000). In the direct model (Frank et al., 1994) violaxanthin was suggested to transfer energy to chlorophyll while chlorophyll would transfer energy to zeaxanthin. This model requires that the first excitation level of violaxanthin (S_1) is above that of chlorophyll Qy and that S_1 of zeaxanthin is below Qy. However, recent studies by Polívka et al. (1999) and Frank et al. (2000) showed that S_1 of both zeaxanthin and violaxanthin is below or close to chlorophyll Qy in energy in vitro, indicating that the direct model may not hold.

The site of quenching in the antenna of PS II has been under thorough investigation and one approach to solve the mystery has been the studies of mutants with reduced antenna. In chlorophyll *b*-deficient mutants of barley and wheat, with reduced LHCII and an increased ratio of VAZ to chlorophyll, a significant portion of xanthophyll cycle-dependent NPQ is virtually the same in the mutant and in the wild type, i.e. independent of the PS II antenna size (Lokstein et al., 1994; Andrews et al., 1995; Gilmore et al., 1996). Since LHCII is not required to facilitate quenching associated with zeaxanthin, Gilmore et al. (1996) suggested that in PS II the minor CP proteins might be the site of a rapid relaxing NPQ. This is in agreement with results by Ruban et al. (1996) who suggested that the minor CP complexes play a role in energy dissipation. Bassi et al. (1999) recently identified two xanthophyll-binding sites in CP29, one binding lutein and the other one being able to bind either violaxanthin or neoxanthin, indicating a possible exchange of xanthophylls upon shifting irradiation. Most authors agree that significant NPQ can arise even in the absence of LHCII, but Härtel et al. (1996) found that the capacity to develop NPQ decreased stepwise with a reduction in antenna size in wild-type barley and in a *Chlorina* mutant grown in continuous and intermittent light. Pea plants grown at intermittent light have also shown drastically reduced levels of NPQ (Jahns and Schweig, 1995).

The approach of screening knock out plant mutants for changes in fluorescence and xanthophyll cycle composition has proven to be prosperous in the search for components important for NPQ. These mutants provide new tools for solving questions about function and importance also for the xanthophyll cycle. A summary of all mutants mentioned below is presented in Table 1. Li et al. (2000) found an *Arabidopsis* mutant (*npq4-1*) that lacked an intrinsic chlorophyll-binding protein of PS II, the PsbS (also known as CP22)(Funk et al., 1995), to have normal pigment composition and normal xanthophyll cycle function. Despite this, the mutant is unable to dissipate excess absorbed light energy. The *npq4-1* mutant lacks the ΔpH and the zeaxanthin-dependent conformational change in the thylakoid membrane that is necessary for qE and the results suggest that PsbS is required for efficient NPQ in addition to ΔpH and zeaxanthin. PsbS was suggested to be the actual site for quenching (Li et al., 2000). An *Arabidopsis* T-DNA knockout mutant with reduced NPQ capacity possessing normal capacity for xanthophyll cycle activity and normal pigment composition have also been found by Peterson and Havir (2000). Their results suggested that the mutant is defective in sensing the trans thylakoid ΔpH associated with qE. It was proven to be a single nuclear mutation mapped to chromosome 1, where also *psbS* is located, and the authors speculate that these mutations are allelic, even though the reduction in quenching was not as dramatic. The use of the VDE inhibitor DTT indicated that a zeaxanthin dependent component of NPQ was specifically reduced in the mutant.

The photoprotective function of zeaxanthin has been seriously questioned. Several studies have been made on *aba* mutants of both *Arabidopsis thaliana* and *Chlamydomonas reinhardtii* (Tardy and Havaux, 1996; Hurry et al., 1997; Niyogi et al., 1997a). These *aba* mutants are ZE mutants that cannot convert zeaxanthin to antheraxanthin or violaxanthin. The wild type and the *aba* mutant could not be distinguished on the basis of their photosynthetic performance, and hence the action of the xanthophyll cycle did not seem to be required for survival in excessive light. When dark-adapted leaves were exposed to red light the strong non-photochemical quenching of fluorescence was virtually identical in both types of leaves despite the fact that the xanthophyll pool was all in the zeaxanthin form in the *aba* mutant and almost exclusively in the form of violaxanthin in the wild type. Tardy and Havaux (1996) concluded that zeaxanthin does not play any specific role in direct (thermal) energy dissipation in PS II. They suggested that only a few of the zeaxanthin molecules together with a proton gradient might be involved in NPQ. The future will show how these results correlate to the role of the PsbS protein in NPQ.

The first VDE mutant (*npq1*, Fig. 3) was identified in the green alga *Chlamydomonas* by video imaging of chlorophyll fluorescence quenching (Niyogi et al., 1997a). The mutant has a Cys to Tyr substitution at position 72, was unable to convert violaxanthin and showed an increased carotenoid/chlorophyll ratio. The mutant was not impaired in growth relative to the wild type under high light, demonstrating that de-epoxidation of violaxanthin is not absolutely required for survival of *Chlamydomonas* in high light. The *Arabidopsis npq1* mutant is also unable to convert violaxanthin to antheraxanthin and zeaxanthin. The *npq2* mutant, on the other hand, accumulates zeaxanthin and is hence a new allele of *aba1* (Niyogi et al., 1998). The D1 protein of the *npq2* mutant of *Chlamydomonas* was recently found to be more resistant to photoinhibitory degradation than D1 of wild type and the *npq1* mutant (Jahns et al., 2000). This gives further support for a protective role of zeaxanthin.

Arabidopsis mutants *lut1* and *lut2*, deficient in lutein synthesis, show that lutein is not absolutely required for photosynthesis in higher plants (Pogson

Table 1. Mutants related to the xanthophyll cycle

Mutant	Organism	Method of mutagenesis	Mutation	Mutant phenotype	Reference
aba1	*A. thaliana*	Ethyl methanesulfonate	ZE mutant	Z accumulation	Koornneef et al., 1982 Tardy and Havaux, 1996 Hurry et al., 1997
aba2	*N. plumbaginifolia*	Transposon tagging	ZE mutant	Z accomulation Reduced abscisic acid	Marin et al., 1996
lor1	*C. reinhardtii*	UV induction	ε-cyclase mutant	Reduced NPQ No lutein	Niyogi et al., 1997b
lsr1	*A. thaliana*	T-DNA knockout	Mutation in chromosome 1	Reduced NPQ	Peterson et al., 2000
lut1	*A. thaliana*	Ethyl methanesulfonate	ε-hydroxylase mutant	Zeinoxanthin accum. No lutein	Pogson et al., 1996 Pogson et al., 1998
lut2	*A. thaliana*	Ethyl methanesulfonate	ε-cyclase mutant	Reduced NPQ No lutein	Pogson et al., 1996 Pogson et al., 1998
lut1aba1	*A. thaliana*	–	ε-hydroxylase and ZE mutant	Delayed greening Low NPQ	Pogson et al., 1998
lut2aba1	*A. thaliana*	–	ε-cyclase and ZE mutant	Delayed greening Low NPQ	Pogson et al., 1998
npq1	*A. thaliana*	Ethyl methanesulfonate /fast-neutron bombardment	VDE mutant Cys-72-Tyr	Low NPQ	Niyogi et al., 1998
npq1	*C. reinhardtii*	Linearised plasmid with ARG7	VDE mutant	Reduced NPQ Increased car/chl	Niyogi et al., 1997a
npq2	*A. thaliana*	Ethyl methanesulfonate /fast-neutron bombardment	ZE mutant allelic to *aba1*	High NPQ Z accumulation	Niyogi et al., 1998
npq2	*C. reinhardtii*	Linearised plasmid with ARG7	ZE mutant	Z accumulation	Niyogi et al., 1997a
npq4	*A. thaliana*	Ethyl methanesulfonate /fast-neutron bombardment	PsbS mutant	No NPQ	Li et al., 2000
npq1lor1	*C. reinhardtii*	–	ε-cyclase and VDE mutant	Almost no NPQ Photobleaching No lutein	Niyogi et al., 1997b
npq4npq1	*A. thaliana*	–	VDE and PsbS mutant	High lipid peroxidation No NPQ	Havaux, 1999 Li et al., 2000

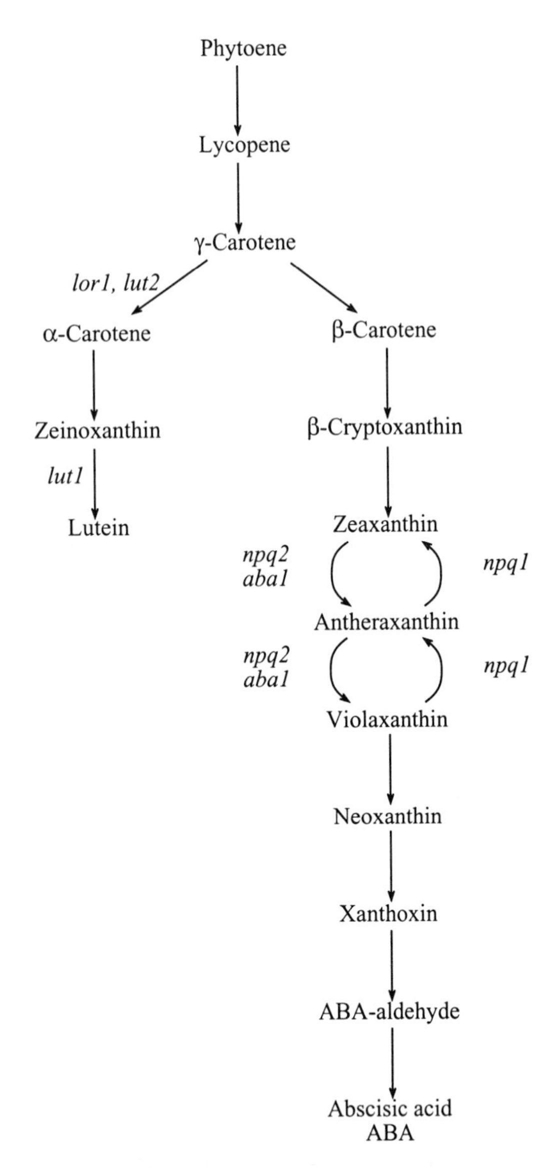

Fig. 3. Biosynthetic pathway for carotenoids with identified mutants marked.

et al., 1996). Other carotenoids are able to compensate functionally and structurally for lutein in the photosystems in the same way as zeaxanthin can replace missing violaxanthin and neoxanthin. The double mutant *lut2aba1*, defective in synthesis of both α-carotene and antheraxanthin (β-carotene branch) showed delayed greening which was often lethal when zeaxanthin was the only xanthophyll present (Pogson et al., 1998). These results indicate that both zeaxanthin and lutein contribute to non-photochemical quenching and plants with deficiency in either pathway still are viable under normal growth conditions. The *lor1* mutant from *Chlamydomonas* lack all xanthophylls in the α-carotene branch, and further indicates a role for lutein and/or zeaxanthin in NPQ, since *npq1lor1* double mutant lacked almost all NPQ and was very susceptible to photooxidative bleaching in high light (Niyogi et al., 1997b).

B. Other Possible Functions of the Xanthophyll Cycle

1. Protection Against Oxidative Stress of Lipids

Carotenoids, including the xanthophylls have essential roles as inhibitors of lipid peroxidation. Especially zeaxanthin has been suggested to protect thylakoid lipids against photodegradation. Zeaxanthin may react non-enzymatically as quencher of triplet chlorophyll (^{3}Chl) and various highly reactive oxygen species, e.g. singlet oxygen ($^1O_2$), which are formed in chloroplasts under prolonged high light conditions (Asada and Takahashi, 1987; Asada, 1992; Schindler and Lichtenthaler, 1996).

To investigate oxidative stress, pea leaves were subjected to high light (Havaux et al., 1991). The amount of unsaturated fatty acids decreased in favor of saturated fatty acids, an effect strongly enhanced when zeaxanthin production was inhibited with DTT. The authors suggested that the xanthophyll cycle is an adaptive mechanism providing violaxanthin as an accessory pigment in low light, and zeaxanthin as an efficient photoprotector in high light (reviewed by Havaux, 1998). Sarry et al. (1994) found that by using high light treatment of potato leaves at low temperature, lipid peroxidation (estimated as ethane production) was prevented by zeaxanthin. At low temperature or by treatment with DTT, a lower amount of zeaxanthin was formed followed by a higher ethane production. Havaux and Niyogi (1999) showed that lipid peroxidation is higher in *npq1* mutants, deficient in VDE, than in wild-type *Arabidopsis*. The *npq4* mutant, which has a normal xanthophyll cycle but no NPQ, was more tolerant to lipid peroxidation than the *npq1* mutant. The double mutant *npq4npq1* had a high susceptibility to photooxidative damage through lipid peroxidation, which indicates a role for the xanthophyll cycle in lipid protection.

2. Regulation of Membrane Fluidity

The synthesis of zeaxanthin has been associated with a modulation of the physical properties of the thylakoid membrane. Thus zeaxanthin was found to affect membrane fluidity in the peripheral region of the hydrophobic core (Gruszecki and Strzalka, 1991), cause a decrease in membrane fluidity proportional to the amount of zeaxanthin present in the membrane (Tardy and Havaux, 1997) and to decrease the rate of plastoquinone reoxidation at chilling temperatures (Havaux and Gruszecki, 1993). Zeaxanthin is more hydrophobic and has more conjugated double bonds (11) than violaxanthin (9), and could act like cholesterol and α-tocopherol that strongly interact with membrane lipids and decrease membrane fluidity (Fryer, 1992). Zeaxanthin formation in strong light has also been associated with increased thermostability of PS II and ionic permeability properties of the membrane (Havaux et al., 1996; Tardy and Havaux, 1997). For a more detailed discussion see Havaux (1998).

3. Participation in Blue Light Responses

A putative role of zeaxanthin has been suggested in blue light photoreception of *Zea mays* coleoptiles (Quiñones and Zeiger, 1994). Coleoptile tips converted violaxanthin into zeaxanthin in the light and manipulation of the zeaxanthin content by red light, red light plus darkness, or incubation with the VDE inhibitor DTT, resulted in blue light-induced bending proportional to zeaxanthin content. Phenomena induced by blue light include e.g. phototropism, solar tracking of leaves and the opening of stomata. DTT treatment of epidermal peels of *Vicia faba* also inhibited blue light-stimulated stomatal opening in a concentration dependent manner (Srivastava and Zeiger, 1995). Contradictory results suggest that zeaxanthin is not the photoreceptor for phototropism in *Zea mays* coleoptiles (Palmer et al., 1996). Seedlings that were carotenoid deficient either through genetic lesion in the gene encoding phytoene desaturase or chemically treated with norflurazon, an inhibitor of phytoene desaturase, showed the same light-dependent phototropism as those seedlings containing normal levels of carotenoids. However, the lack of a specific blue light response for stomata opening from the zeaxanthin-less *Arabidopsis npq1* mutant provides genetic evidence for the role of zeaxanthin as a blue light photoreceptor in guard cells (Frechilla et al., 1999).

4. Regulation of Abscisic Acid Synthesis

Abscisic acid (ABA) is a plant hormone involved in seed maturation and stress responses in plants, e.g. stomatal closure. Strong arguments for zeaxanthin as an intermediate in the synthesis of ABA in the carotenoid pathway have been presented by Parry et al. (1990) and Marin et al. (1996). The possible role of the xanthophyll cycle in the regulation of ABA synthesis is therefore compelling. Thus, the action of VDE under overexcitation conditions in leaves would be expected to reduce the synthesis of ABA. However, in high light irradiated, and ozone-exposed leaves of tobacco a degradation of violaxanthin was observed, which was not compensated for by transformation to antheraxanthin or zeaxanthin (Ederli et al., 1997). The authors measured a higher level of ABA and hypothesized that ABA was synthesized in response to strong oxidative stress from violaxanthin in the leaves concomitantly with a normal de-epoxidation of violaxanthin to zeaxanthin. It is well known that soil drying causes an increase in ABA concentration in the xylem and also reduced leaf conductance, i.e. stomatal closure (Liang et al., 1997). Also, moderate water deficit at constant light levels leads to increases in zeaxanthin, with further conversion of violaxanthin to zeaxanthin at severe deficits (Iturbe-Ormaetxe et al., 1998). Thus, in drought-stressed plants violaxanthin can be converted into both ABA and zeaxanthin. An up regulation of ZE mRNA during drought stress has been found in roots, but not in leaves (Thompson et al., 2000). The question remains whether there are separate pools of violaxanthin, that are converted to ABA in the root and zeaxanthin in the leaf, if the processes are connected and how this affects the photosynthetic activity of the mesophyll cell, e.g. inhibition of leaf photosynthesis under drought (Iturbe-Ormaetxe et al., 1998).

Acknowledgments

The authors wish to acknowledge the Swedish Natural Science Research Council and Carl Trygger Foundation for financial support.

References

Adams III WW and Demmig-Adams B (1992) Operation of the xanthophyll cycle in higher plants in response to diurnal

changes in incident sunlight. Planta 186: 390–398
Adams III WW, Demmig-Adams B, Verhoeven AS and Barker DH (1995) 'Photoinhibition' during winter stress: Involvement of sustained xanthophyll cycle-dependent energy dissipation. Aust J Plant Physiol 22: 261–276
Adamska I (1997) ELIPs – Light-induced stress proteins. Physiol Plant 100: 794–805
Åkerlund H-E, Arvidsson P-O, Bratt C and Carlsson M (1995) Partial purification of the violaxanthin de-epoxidase. In: Mathis P (ed) Photosynthesis – from Light to Biosphere, Vol IV, pp 103–106. Kluwer Academic Publishers, Dordrecht
Allen JF (1992) Protein phosphorylation in regulation of photosynthesis. Biochim Biophys Acta 1098: 275–335
Andrews JR, Fryer MJ and Baker NR (1995) Consequences of LHCII deficiency for photosynthetic regulation in *chlorina* mutants of barley. Photosynth Res 44: 81–91
Aro E-M, Virgin I and Andersson B (1993) Photoinhibition of Photosystem II. Inactivation, protein damage and turnover. Biochim Biophys Acta 1143: 113–134
Arvidsson P-O, Bratt CE, Carlsson M and Åkerlund H-E (1996) Purification and identification of the violaxanthin deepoxidase as a 43 kDa protein. Photosynth Res 49: 119–129
Arvidsson P-O, Carlsson M, Stefánsson H, Albertsson P-Å and Åkerlund H-E (1997) Violaxanthin accessibility and temperature dependency for de-epoxidation in spinach thylakoid membranes. Photosynth Res 52: 39–48
Asada K (1992) Ascorbate peroxidase—a hydrogen peroxide-scavenging enzyme in plants. Physiol Plant 85: 235–241
Asada K and Takahashi M (1987) Production and scavenging of active oxygen in photosynthesis. In: Kyle DJ, Osmond CB, Arntzen SJ (eds) Photoinhibition, pp 227–287. Elsevier Science Publishers, Amsterdam
Audran C, Borel C, Frey A, Sotta B, Meyer C, Simonneau T and Marion-Poll A (1998) Expression studies of the zeaxanthin epoxidase gene in *Nicotiana plumbaginifolia*. Plant Physiol 118: 1021–1028
Bassi R, Pineau B, Dainese P and Marquardt J (1993) Carotenoid-binding proteins of Photosystem II. Eur J Biochem 212: 297–303
Bassi R, Sandonà D and Croce R (1997) Novel aspects of chlorophyll *a/b*-binding proteins. Physiol Plant 100: 769–779
Bassi R, Croce R, Cugini D and Sandona D (1999) Mutational analysis of a higher plant antenna protein provides identification of chromophores bound into multiple sites. Proc Natl Acad Sci USA 96: 10056–10061
Bilger W and Björkman O (1991) Temperature dependence of violaxanthin de-epoxidation and non-photochemical fluorescence quenching in intact leaves of *Gossypium hirsutum* L. and *Malva parviflora* L. Planta 184: 226–234
Bouvier F, d'Harlingue A, Hugueney P, Marin E, Marion-Poll A and Camara B (1996) Xanthophyll biosynthesis. Cloning, expression, functional reconstitution, and regulation of beta-cyclohexenyl carotenoid epoxidase from pepper (*Capsicum annuum*). J Biol Chem 271: 28861–28867
Bratt CE, Arvidsson P-O, Carlsson M and Åkerlund H-E (1995) Regulation of violaxanthin de-epoxidase activity by pH and ascorbate concentrations. Photosynth Res 45: 169–175
Briantais JM, Vernotte C, Picaud M and Krause GH (1979) A quantitative study of the slow decline of chlorophyll a fluorescence in isolated chloroplasts. Biochim Biophys Acta 548: 128–138
Brink S, Bogsch EG, Edwards WR, Hynds PJ and Robinson C (1998) Targeting of thylakoid proteins by the ΔpH-driven twin-arginine translocation pathway requires a specific signal in the hydrophobic domain in conjunction with the twin-arginine motif. FEBS Lett 434: 425–430
Bugos RC and Yamamoto HY (1996) Molecular cloning of violaxanthin de-epoxidase from romaine lettuce and expression in *Escherichia coli*. Proc Natl Acad Sci USA 93: 6320–6325
Bugos RB, Hieber AD and Yamamoto HY (1998) Xanthophyll cycle enzymes are members of the lipocalin family, the first identified from plants. J Biol Chem 273: 15321–15324
Bugos RC, Chang SH and Yamamoto HY (1999) Developmental expression of violaxanthin de-epoxidase in leaves of tobacco growing under high and low light. Plant Physiol 121: 207–214
Bungard RA, Ruban AV, Hibberd JM, Press MC, Horton P and Scholes JD (1999) Unusual carotenoid composition and a new type of xanthophyll cycle in plants. Proc Natl Acad Sci USA 96: 1135–1139
Büch K, Stransky H and Hager A (1995) FAD is a further essential cofactor of the NAD(P)H and O_2-dependent zeaxanthin-epoxidase. FEBS Lett 376: 45–48
Conklin PL, Williams EH and Last RL (1996) Environmental stress sensitivity of an ascorbic acid-deficient *Arabidopsis* mutant. Proc Natl Acad Sci USA 93: 9970–9974
Conklin PL, Saracco SA and Norris SR (2000) Identification of ascorbic acid-deficient *Arabidopsis thaliana* mutants. Genetics 154: 847–856
Cornell RB and Northwood IC (2000) Regulation of CTP:phosphocholine cytidylyltransferase by amphitropism and relocalization. Trends Biol Sci 25:441–447
Dau H (1994) Short-term adaptation of plants to changing light intensities and its relation to Photosystem II photochemistry and fluorescence emission. J Photochem Photobiol B 26: 3–27
Demmig B, Winter K, Krüger A and Czygan F-C (1987) Photoinhibition and zeaxanthin formation in intact leaves. Plant Physiol 84: 218–224
Demmig-Adams B (1990) Carotenoids and photoprotection in plants: A role for the xanthophyll zeaxanthin. Biochim Biophys Acta 1020: 1–24
Demmig-Adams B and Adams III WW (1992a) Carotenoid composition in sun and shade leaves of plants with different life forms. Plant Cell Environ 15: 411–419
Demmig-Adams B and Adams III WW (1992b) Photoprotection and other responses of plants to high light stress. Annu Rev Plant Physiol Plant Mol Biol 43: 599–626
Demmig-Adams B, Winter K, Krüger A and Czygan F-C (1989) Zeaxanthin synthesis, energy dissipation, and photoprotection of Photosystem II at chilling temperatures. Plant Physiol 90: 894–898
Demmig-Adams B, Gilmore AM and Adams III WW (1996) In vivo functions of carotenoids in higher plants. FASEB J 10: 403–412
Demmig-Adams B, Adams III WW and Grace SC (1997) Physiology of light tolerance in plants. Horticultural Rev 18: 215–246
Ederli L, Pasqualini S, Batini P and Antonielli M (1997) Photoinhibition and oxidative stress: Effects on xanthophyll cycle, scavenger enzymes and abscisic acid content in tobacco plants. J Plant Physiol 151: 422–428
Eskling M and Åkerlund H-E (1998) Changes in the quantities of violaxanthin de-epoxidase, xanthophylls and ascorbate in

spinach upon shift from low to high light. Photosynth Res 57: 41–50

Eskling M, Arvidsson P-O and Åkerlund H-E (1997) The xanthophyll cycle, its regulation and components. Physiol Plant 100: 806–816

Falbel T, Staehelin L and Adams III WW (1994) Analysis of xanthophyll cycle carotenoids and chlorophyll fluorescence in light intensity-dependent chlorophyll-deficient mutants of wheat and barley. Photosynth Res 42: 191–202

Färber A and Jahns P (1998) The xanthophyll cycle of higher plants: Influence of antenna size and membrane organization. Biochim Biophys Acta 1363: 47–58

Foyer CH (1993) Ascorbic acid. In: Alscher RG, Hess JL (eds) Antioxidants in Higher Plants, pp 31–58. CRC Press, Boca Raton

Frank HA, Cua A, Chynwat V, Young A, Gosztola D and Wasielewski MR (1994) Photophysics of the carotenoids associated with the xanthophyll cycle in photosynthesis. Photosynth Res 41: 389–395

Frank HA, Bautista JA, Josue JS and Young AJ (2000) Mechanism of nonphotochemical quenching in green plants: Energies of the lowest excited singlet states of violaxanthin and zeaxanthin. Biochemistry 39: 2831–2837

Frechilla S, Zhu J, Talbott LD and Zeiger E (1999) Stomata from *npq1*, a zeaxanthin-less *Arabidopsis* mutant, lack a specific response to blue light. Plant Cell Physiol 40: 949–954

Fryer MJ (1992) The antioxidant effects of thylakoid Vitamin E (α-tocopherol). Plant Cell Environ 15: 381–392

Funk C, Schröder WP, Napiwotzki A, Tjus SE, Renger G and Andersson B (1995) The PS II-S protein of higher plants: A new type of pigment-binding protein. Biochemistry 34: 11133–11141

Gillham DJ and Dodge AD (1986) Hydrogen-peroxide-scavenging systems within pea chloroplasts. Planta 167: 246–251

Gillham DJ and Dodge AD (1987) Chloroplast superoxide and hydrogen peroxide scavenging systems from pea leaves: Seasonal variations. Plant Sci 50: 105–109

Gilmore AM (1997) Mechanistic aspects of xanthophyll cycle-dependent photoprotection in higher plant chloroplasts and leaves. Physiol Plant 99: 197–209

Gilmore AM and Yamamoto HY (1993) Linear models relating xanthophylls and lumen acidity to non-photochemical fluorescence quenching. Evidence that antheraxanthin explains zeaxanthin-independent quenching. Photosynth Res 35: 67–78

Gilmore AM, Hazlett TL, Debrunner PG and Govindjee (1996) Photosystem II chlorophyll *a* fluorescence lifetimes and intensity are independent of the antenna size differences between barley wild-type and *chlorina* mutants: Photochemical quenching and xanthophyll cycle-dependent nonphotochemical quenching of fluorescence. Photosynth Res 48: 171–187

Grotz B, Molnar P, Stransky H and Hager A (1999) Substrate specificity and functional aspects of violaxanthin de-epoxidase, an enzyme of the xanthophyll cycle. J Plant Physiol 154: 437–446

Gruszecki WI (1995) Different aspects of protective activity of the xanthophyll cycle under stress conditions. Acta Physiol Plant 17: 145–152

Gruszecki WI and Krupa Z (1993) LHCII, the major light-harvesting pigment-protein complex is a zeaxanthin epoxidase. Biochim Biophys Acta 1144: 97–101

Gruszecki WI and Strzalka K (1991) Does the xanthophyll cycle take part in the regulation of fluidity of the membrane? Biochim Biophys Acta 1060: 310–314

Gruszecki WI, Matula M, Ko-chi N, Koyama Y and Krupa Z (1997) *Cis-trans*-isomerization of violaxanthin in LHCII: Violaxanthin isomerization cycle within the violaxanthin cycle. Biochim Biophys Acta 1319: 267–274

Hager A (1966) Die Zusammenhänge zwischen lichtinduzierten Xanthophyll-Umwandlungen und Hill-Reaktion. Ber Dtsch Bot Ges 79: 94–107

Hager A (1969) Lichtbedingte pH-Erniedrigung in einem Chloroplasten-Kompartiment als Ursache der enzymatischen Violaxanthin-Zeaxanthin-Umwandlung; Beziehungen zur Photophosphorylierung. Planta 89: 224–243

Hager A (1975) Die reversiblen, lichtabhängigen Xanthophyllumwandlungen im Chloroplasten. Ber Dtsch Bot Ges 88: 27–44

Hager A and Holocher K (1994) Localisation of the xanthophyll-cycle enzyme violaxanthin de-epoxidase within the thylakoid lumen and abolition of its mobility by a (light-dependent) pH decrease. Planta 192: 581–589

Härtel H, Lokstein H, Grimm B and Rank B (1996) Kinetic studies on the xanthophyll cycle in barley leaves: Influence of antenna size and relations to nonphotochemical chlorophyll fluorescence quenching. Plant Physiol 110: 471–482

Havaux M (1998) Carotenoids as membrane stabilizers in chloroplasts. Trends Plant Sci 3: 147–151

Havaux M and Gruszecki WI (1993) Heat- and light-induced chlorophyll and fluorescence changes in potato leaves containing high or low levels of the carotenoid zeaxanthin: Indications of a regulatory effect of zeaxanthin on thylakoid membrane fluidity. Photochem Photobiol 58: 607–614

Havaux M and Niyogi KK (1999) The violaxanthin cycle protects plants from photooxidative damage by more than one mechanism. Proc Natl Acad Sci USA 96: 8762–8767

Havaux M and Tardy F (1996) Temperature-dependent adjustment of the thermal stability of Photosystem II in vivo: Possible involvement of xanthophyll-cycle pigments. Planta 198: 324–333

Havaux M, Gruszecki WI, Dupont I and Leblanc RM (1991) Increased heat emission and its relationship to the xanthophyll cycle in pea leaves exposed to strong light stress. J Photochem Photobiol 8: 361–370

Havaux M, Tardy F, Ravenel J, Chanu D and Parot P (1996) Thylakoid membrane stability to heat stress studied by flash spectroscopic measurements of the electrochromic shift in intact potato leaves: influence of the xanthophyll content. Plant Cell Environ 19: 1359–1368

Havir EA, Tausta SL and Peterson RB (1997) Purification and properties of violaxanthin de-epoxidase from spinach. Plant Science 123: 57–66

Heyde S and Jahns P (1998) The kinetics of zeaxanthin formation is retarded by dicyclohexylcarbodiimide. Plant Physiol 117: 659–665

Horton P and Lee P (1985) Phosphorylation of chloroplast membrane proteins partially protects against photoinhibition. Planta 165: 37–42

Horton P, Ruban AV, Rees D, Pascal AA, Noctor G and Young AJ (1991) Control of the light-harvesting function of chloroplast membranes by aggregation of the LHCII chlorophyll-protein complex. FEBS Lett 292: 1–4

Horton P, Ruban AV and Walters R (1996) Regulation of light harvesting in green plants. Annu Rev Plant Physiol Plant Mol Biol 47: 655–684

Hurry V, Anderson JM, Chow WS and Osmond CB (1997) Accumulation of zeaxanthin in abscisic acid-deficient mutants of *Arabidopsis* does not affect chlorophyll fluorescence quenching or sensitivity to photoinhibition in vivo. Plant Physiol 113: 639–648

Iturbe-Ormaetxe I, Escuredo P, Arrese-Igor C and Becana M (1998) Oxidative damage in pea plants exposed to water deficit or paraquat. Plant Physiol 116: 173–181

Jahns P (1995) The xanthophyll cycle in intermittent light-grown pea plants: Possible functions of chlorophyll *a/b*-binding proteins. Plant Physiol 108: 149–156

Jahns P and Heyde S (1999) Dicyclohexylcarbodiimide alters the pH dependence of violaxanthin de-epoxidation. Planta 207: 393–400

Jahns P and Krause GH (1994) Xanthophyll cycle and energy-dependent fluorescence quenching in leaves from pea plants grown under intermittent light. Planta 192: 176–182

Jahns P and Schweig S (1995) Energy-dependent fluorescence quenching in thylakoids from intermittent light grown pea plants: Evidence for an interaction of zeaxanthin and the chlorophyll *a/b* binding protein CP26. Plant Physiol Biochem 33: 683–687

Jahns P, Depka P and Trebst A (2000) Xanthophyll cycle mutants from *Chlamydomonas reinhardtii* indicate a role of zeaxanthin in D1 protein turnover. Plant Physiol Biochem 38: 371–376

Juhler R, Andreasson E, Yu S-G and Albertsson P-Å (1993) Composition of photosynthetic pigments in thylakoid membrane vesicles from spinach. Photosynth Res 35: 171–178

Kawano M and Kuwabara T (2000) pH-dependent reversible inhibition of violaxanthin de-epoxidase by pepstatin related to protonation-induced structural change of the enzyme. FEBS Lett 481: 101–104

Koornneef M, Jorna ML, Brinkenhorst-van der Swan DLC and Karssen CM (1982) The isolation of abscisic acid (ABA) deficient mutants by selection of induced revertants in non-germinating gibberellin sensitive lines of *Arabidopsis thaliana* (L.) Heynth. Theor Appl Genet 61: 385–393

Koroleva OY, Thiele A and Krause GH (1995) Increased xanthophyll cycle activity as an important factor in acclimation of the photosynthetic apparatus to high-light stress at low temperatures. In: Mathis P (eds) Photosynthesis: From Light to Biosphere, Vol IV, pp 425–428. Kluwer Academic Publishers, Dordrecht

Krause GH (1988) Photoinhibition of photosynthesis. An evaluation of damaging and protective mechanisms. Physiol Plant 74: 566–574

Krause GH (1994) Photoinhibition induced by low temperatures. In: Baker NR, Bowyer JR (eds) Environmental Plant Biology Series: Photoinhibition of Photosynthesis: From Molecular Mechanisms to the Field, pp 331-348. BIOS Scientific Publishers, Oxford

Krause GH and Behrend U (1986) pH-dependent chlorophyll fluorescence quenching indicating a mechanism of protection against photoinhibition of chloroplasts. FEBS Lett 200: 298–302

Krause GH and Weis E (1991) Chlorophyll fluorescence and photosynthesis: The basics. Annu Rev Plant Physiol Plant Mol Biol 42: 313–349

Król M, Spangfort MD, Huner NPA, Öquist G, Gustafsson P and Jansson S (1995) Chlorophyll *a/b*-binding proteins, pigment conversions, and early light-induced proteins in a chlorophyll *b*-less barley mutant. Plant Physiol 107: 873–883

Kuwabara T, Hasegawa M, Kawano M and Takaichi S (1999) Characterization of violaxanthin de-epoxidase purified in the presence of Tween 20: effects of dithiothreitol and pepstatin A. Plant Cell Physiol 40: 1119–1126

Lee A-C and Thornber JP (1995) Analysis of the pigment stoichiometry of pigment-protein complexes from barley (*Hordeum vulgare*). Plant Physiol 107: 565–574

Leipner J, Stamp P and Fracheboud Y (2000) Artificially increased ascorbate content affects zeaxanthin formation but not thermal energy dissipation or degradation of antioxidants during cold-induced photooxidative stress in maize leaves. Planta 210: 964–969

Leverenz JW, Öquist G and Wingsle G (1992) Photosynthesis and photoinhibition in leaves of chlorophyll *b*-less barley in relation to absorbed light. Physiol Plant 85: 495–502

Li X-P, Björkman O, Shih C, Grossman AR, Rosenquist M, Jansson S and Niyogi KK (2000) A pigment-binding protein essential for regulation of photosynthetic light harvesting. Nature 126: 213–222

Liang J, Zhang J and Wong MH (1997) Can stomatal closure caused by xylem ABA explain the inhibition of leaf photosynthesis under soil drying. Photosynth Res 51: 149–159

Logan BA, Barker DH, Demmig-Adams B and Adams III WW (1996) Acclimation of leaf carotenoid composition and ascorbate levels to gradients in the light environment within an Australian rainforest. Plant Cell Environ 19: 1083–1090

Lokstein H, Härtel H, Hoffmann P, Woitke P and Renger G (1994) The role of light-harvesting complex II in excess excitation energy dissipation: An in-vivo fluorescence study on the origin of high-energy quenching. J Photochem Photobiol 26: 175–184

Marin E, Nussaume L, Quesada A, Gonneau M, Sotta B, Hugueney P, Frey A and Marion-Poll A (1996) Molecular identification of zeaxanthin epoxidase of *Nicotiana plumbaginifolia*, a gene involved in abscisic acid biosynthesis and corresponding to the *ABA* locus of *Arabidopsis thaliana*. EMBO J 15: 2331–2342

Neubauer C and Yamamoto HY (1994) Membrane barriers and Mehler-peroxidase reaction limit the ascorbate available for violaxanthin de-epoxidase activity in intact chloroplasts. Photosynth Res 39: 137–147

Nielsen H, Engelbrecht J, Brunak S and von Heijne G (1997) Identification of procaryotic and eukaryotic signal peptides and prediction of their cleavage sites. Protein Engineering 10: 1–6

Niyogi KK, Björkman O and Grossman AR (1997a) *Chlamydomonas* xanthophyll cycle mutants identified by video imaging of chlorophyll fluorescence quenching. Plant Cell 9: 1369–1380

Niyogi KK, Björkman O and Grossman AR (1997b) The roles of specific xanthophylls in photoprotection. Proc Natl Acad Sci USA 94: 14162–14167

Niyogi KK, Grossman AR and Björkman O (1998) *Arabidopsis* mutants define a central role for the xanthophyll cycle in the regulation of photosynthetic energy conversion. Plant Cell 10: 1121–1134

Palmer JM, Warpeha KMF and Briggs WR (1996) Evidence that zeaxanthin is not the photoreceptor for phototropism in maize coleoptiles. Plant Physiol 110: 1323–1328

Parry AD, Babiano MJ and Horgan R (1990) The role of *cis*-carotenoids in abscisic acid biosynthesis. Planta 182: 118–128

Peter GF and Thornber JP (1991) Biochemical composition and organization of higher plant Photosystem II light-harvesting pigment-proteins. J Biol Chem 25: 16745–16754

Peterson RB and Havir EA (2000) A nonphotochemical-quenching-deficient mutant of *Arabidopsis thaliana* possessing normal pigment composition and xanthophyll-cycle activity. Planta 210: 205–214

Pfündel EE and Bilger W (1994) Regulation and possible function of the violaxanthin cycle. Photosynth Res 42: 89–109

Pfündel EE and Dilley RA (1993) The pH dependence of violaxanthin deepoxidation in isolated pea chloroplasts. Plant Physiol 101: 65–71

Phillip Y and Young AJ (1995) Occurrence of the carotenoid lactucaxanthin in higher plant LHCII. Photosynth Res 43: 273–282

Pogson B, McDonald KA, Truong M, Britton G, DellaPenna D (1996) *Arabidopsis* carotenoid mutants demonstrate that lutein is not essential for photosynthesis in higher plants. Plant Cell 8: 1627–1639

Pogson BJ, Niyogi KK, Björkman O and DellaPenna D (1998) Altered xanthophyll compositions adversely affect chlorophyll accumulation and nonphotochemical quenching in *Arabidopsis* mutants. Proc Natl Acad Sci USA 95: 13324–13329

Polivka T, Herek JL, Zigmantas D, Åkerlund H-E and Sundström V (1999) Direct observation of the (forbidden) S_1 state in carotenoids. Proc Natl Acad Sci USA 96: 4914–4917

Quiñones MA and Zeiger E (1994) A putative role of the xanthophyll, zeaxanthin, in blue light photoreception of corn coleoptiles. Science 264: 558–561

Rockholm DC and Yamamoto HY (1996) Violaxanthin de-epoxidase. Purification of a 43-kilodalton lumenal protein from lettuce by lipid-affinity precipitation with monogalactosyldiacylglyceride. Plant Physiol 110: 697–703

Ruban AV, Young AJ, Pascal AA and Horton P (1994) The effects of illumination on the xanthophyll composition of the Photosystem II light-harvesting complexes of spinach thylakoid membranes. Plant Physiol 104: 227–234

Ruban AV, Young AJ and Horton P (1996) Dynamic properties of the minor chlorophyll *a/b* binding proteins of Photosystem II, an in vitro model for photoprotective energy dissipation in the photosynthetic membrane of green plants. Biochemistry 35: 674–678

Ruban AV, Phillip D, Young AJ and Horton P (1997) Carotenoid-dependent oligomerization of the major chlorophyll *a/b* light harvesting complex of Photosystem II of plants. Biochemistry 36: 7855–7859

Sapozhnikov DI, Krasovskaya TA and Maevskaya AN (1957) Change in the interrelationship of the basic carotenoids of the plastids of green leaves under the action of light. Dokl Akad Nauk USSR 113: 465–467

Sarry J-E, Montillet J-L, Sauvaire Y and Havaux M (1994) The protective function of the xanthophyll cycle in photosynthesis. FEBS Lett 353: 147–150

Schäfer C, Schmid V and Roos M (1994) Characterization of high-light-induced increases in xanthophyll cycle pigment and lutein contents in photoautotrophic cell cultures. J Photochem Photobiol 22: 67–75

Schindler C and Lichtenthaler HK (1996) Photosynthetic CO_2-assimilation, chlorophyll fluorescence and zeaxanthin accumulation in field grown maple trees in the course of a sunny and a cloudy day. J Plant Physiol 148: 399–412

Schmid V and Schäfer C (1994) Alterations of the chlorophyll-protein pattern in chronically photoinhibited *Chenopodium rubrum* cells. Planta 192: 473–479

Schöner S and Krause GH (1990) Protective systems against active oxygen species in spinach: Response to cold acclimation in excess light. Planta 180: 383–389

Schumann G, Nurnberger H, Sandmann G and Krugel H (1996) Activation and analysis of cryptic crt genes for carotenoid biosynthesis from *Streptomyces griseus*. Mol Gen Genet 252: 658–666

Seddon JM, Ajani UA, Sperduto RD, Hiller R, Blair N, Burton TC, Farber MD, Gragoudas ES, Haller J and Miller DT (1994) Dietary carotenoids, vitamins A, C, and E, and advanced age-related macular degeneration. JAMA 172: 1413–1420

Siefermann D and Yamamoto HY (1975a) NADPH and oxygen-dependent epoxidation of zeaxanthin in isolated chloroplasts. Biochem Biophys Res Commun 62: 456–461

Siefermann D and Yamamoto HY (1975b) Properties of NADPH and oxygen-dependent zeaxanthin epoxidation in isolated chloroplasts. A transmembrane model for the violaxanthin cycle. Arch Biochem Biophys 171: 70–77

Siefermann D and Yamamoto H (1976) Light-induced de-epoxidation in lettuce chloroplasts. Plant Physiol 57: 939–940

Siefermann-Harms D (1985) Carotenoids in photosynthesis. I. Location in photosynthetic membranes and light-harvesting function. Biochim Biophys Acta 811: 325–355

Srivastava A and Zeiger E (1995) Guard cell zeaxanthin tracks photosynthetically active radiation and stomatal apertures in *Vicia faba* leaves. Plant Cell Environ 18: 813–817

Staehelin LA and Arntzen CJ (1983) Regulation of chloroplast membrane function: Protein phosphorylation changes the spatial organization of membrane components. J Cell Biol 97: 1327–1337

Strasser RJ and Butler WL (1976) Correlation of absorbance changes and thylakoid fusion with the induction of oxygen evolution in bean leaves greened by brief flashes. Plant Physiol 58: 371–376

Sun Z, Gantt E and Cunningham Jr FX (1996) Cloning and functional analysis of the β-carotene hydroxylase of *Arabidopsis thaliana*. J Biol Chem 271: 24349–24352

Tardy F and Havaux M (1996) Photosynthesis, chlorophyll fluorescence, light-harvesting system and photoinhibition resistance of a zeaxanthin-accumulating mutant of *Arabidopsis thaliana*. J Photochem Photobiol 34: 87–94

Tardy F and Havaux M (1997) Thylakoid membrane fluidity and thermostability during the operation of the xanthophyll cycle in higher-plant chloroplasts. Biochim Biophys Acta 1330: 179–193

Thayer SS and Björkman O (1990) Leaf xanthophyll content and composition in sun and shade determined by HPLC. Photosynth Res 23: 331–343

Thayer SS and Björkman O (1992) Carotenoid distribution and deepoxidation in thylakoid pigment-protein complexes from cotton leaves and bundle-sheath cells of maize. Photosynth Res 33: 213–225

Thiele A and Krause GH (1994) Xanthophyll cycle and thermal

energy dissipation in Photosystem II: Relationship between zeaxanthin formation, energy-dependent fluorescence quenching and photoinhibition. J Plant Physiol 144: 324–332

Thiele A, Schirwitz K, Winter K and Krause GH (1996) Increased xanthophyll cycle activity and reduced D1 protein inactivation related to photoinhibition in two plant systems acclimated to excess light. Plant Science 115: 237–250

Thompson AJ, Jackson AC, Parker RA, Morpeth, DR, Burbidge A and Taylor IB (2000) Abscisic acid biosynthesis in tomato: regulation of zeaxanthin epoxidase and 9-*cis*-epoxycarotenoid dioxygenase mRNAs by light/dark cycles, water stress and abscisic acid. Plant Mol Biol 42: 833–845

Tyystjärvi E and Aro E-M (1996) The rate constant of photoinhibition, measured in lincomycin-treated leaves, is directly proportional to light intensity. Proc Natl Acad Sci USA 93: 2213–2218

von Heijne G (1983) Patterns of amino acids near signal-sequence cleavage sites. Eur J Biochem. 133: 17–21

von Heijne G (1985) Signal sequences. The limits of variation. J Mol Biol 184: 99–105

Walters RG, Ruban AV and Horton P (1996) Identification of proton-active residues in a higher plant light- harvesting complex. Proc Natl Acad Sci USA 93: 14204–14209

Wentworth M, Ruban AV and Horton P (2000) Chlorophyll fluorescence quenching in isolated light harvesting complexes induced by zeaxanthin. FEBS Lett 471: 71-74

Xu CC, Jeon YA, Hwang HJ and Lee CH (1999) Suppression of zeaxanthin epoxidation by chloroplast phosphatase inhibitors in rice leaves. Plant Science 146: 27–34

Yamamoto HY (1979) Biochemistry of the violaxanthin cycle in higher plants. Pure Appl Chem 51: 639–648

Yamamoto HY and Bassi R (1996) Carotenoids: Localisation and function. In: Ort RR, Yocum CF, (eds) Oxygenic photosynthesis: The light reactions, pp 539–563. Kluwer Academic Publishers, Dordrecht

Yamamoto HY and Higashi RM (1978) Violaxanthin de-epoxidase. Lipid composition and substrate specificity. Arch Biochem Biophys 190: 514–522

Yamamoto HY and Kamite L (1972) The effects of dithiothreitol on violaxanthin de-epoxidation and absorbance changes in the 500-nm region. Biochim Biophys Acta 267: 538–543

Yamamoto HY, Nakayama TOM and Chichester CO (1962) Studies on the light and dark interconversions of leaf xanthophylls. Arch Biochem Biophys 97: 168–173

Chapter 26

The PsbS Protein: A Cab-Protein with a Function of Its Own

Christiane Funk*
Arrhenius Laboratories of Natural Sciences, Department of Biochemistry and Biophysics, Stockholm University, S-106 91 Stockholm, Sweden

Summary

Chlorophyll *a/b* binding proteins (Cab proteins) are the most abundant membrane proteins on earth. The intrinsic PsbS protein of Photosystem II is very peculiar among the family of the Cab proteins. It differs from the conventional light harvesting proteins by an additional putative fourth transmembrane helix. PsbS is able to bind chlorophyll *a* and *b*, but unlike other chlorophyll-binding proteins it does not take part in the process of light harvesting. It is present in etiolated plants and seems to be stable also in the absence of pigments. Therefore, it was suggested to have a function in transient pigment binding and act as a chlorophyll carrier protein, a role that is also postulated for its relatives, the early light induced proteins (ELIPs). Recently the PsbS protein received broad attention when it was shown, that an *Arabidopsis thaliana* mutant, which is not able to perform non-photochemical quenching, is deficient in the *psb*S gene. This chapter provides an overview of the data obtained for the PsbS protein so far, emphasizing its similarities and differences to the Cab-antenna proteins and ELIPs and discusses its possible function.

*Email: Christiane.Funk@dbb.su.se

Eva-Mari Aro and Bertil Andersson (eds): Regulation of Photosynthesis, pp. 453–467.

I. Introduction

The process of photosynthetic cleavage of water into dioxygen and hydrogen takes place in the multisubunit thylakoid protein complex, Photosystem II (PS II) (for recent reviews see Barber and Kühlbrandt, 1999; Rhee, 2001). PS II of green algae and higher plants contains almost 30 different polypeptides with molecular masses ranging from 3 to about 50 kDa. Several of these are known to bind pigments, such as chlorophylls and carotenoids. At least two types of chlorophyll-binding proteins can be distinguished: the chlorophyll *a* binding proteins and the chlorophyll *a/b* binding proteins (von Wettstein et al., 1995; Green and Durnford, 1996). The first group is represented by proteins forming the PS II reaction center and core complex, i.e., the gene products of *psb*A (D1), *psb*D (D2), *psb*B (CP47) and *psb*C (CP43). The redox active groups that catalyze a stable charge separation are located in a heterodimer consisting of the polypeptides D1 and D2 that are closely associated with three smaller polypeptides, the two subunits of Cyt *b*559 (*psb*E and *psb*F) and the *psb*I gene product, forming the reaction center of PS II. The D1/D2 heterodimer binds six chlorophyll *a* molecules, two pheophytins, and one to two carotenes (for reviews see Diner and Babcock, 1996; Renger, 1997). CP47 and CP43, each containing 15–22 chlorophyll *a* molecules (for review see Bricker, 1990), are assumed to act as the inner core antenna.

The second major type of chlorophyll-binding proteins in Photosystem II is the family of the chlorophyll *a/b* binding proteins (Cab proteins) forming the peripheral antenna system, which plays a key role in light-harvesting. These proteins are also involved in acclimation of the photosynthetic apparatus to different light environments (Jansson, 1994; von Wettstein et al., 1995; Pichersky and Jansson, 1996; Bassi et al., 1997; Paulsen, 1997; Jansson, 1999). The dominant complex is the trimeric LHCII that forms the major part of the antenna system and accounts for about 50% of all chlorophylls in the thylakoid membrane. The less abundant chlorophyll *a/b* antenna proteins are designated CP29, CP26 and CP24 according to size (Jansson, 1994; Pichersky and Jansson, 1996). The determination of the molecular structure of LHCII to 3.4 Å resolution (Kühlbrandt et al., 1994) helped understanding many features of the Cab proteins. Eighty percent of the polypeptide, 12 chlorophylls and two luteins have been fitted into the structure and eight to nine chlorophyll ligands have been identified. LHCII consists of three transmembrane helices and an additional short fourth α-helix at the C-terminus. This fourth helix consists of only three turns and is localized at the interface between the hydrophobic interior and the thylakoid lumen. The seven chlorophylls closest located to the carotenoids have been assigned as chlorophyll *a*, whereas the five chlorophylls adjacent to another chlorophyll were assigned as chlorophyll *b* molecules. Only two of the ligands are histidines (His 68 binding Chl *a*5 and His 212 binding Chl *b*3), three side-chain ligands are amides (Gln 131 binding Chl *b*6, Gln 197 binding Chl *a*3 and Asn 183 binding Chl *a*2). Another three appear to be charge-compensated glutamates, forming ion pairs with arginines either in the same helix (Glu 139-Arg 142 binding Chl *b*5), or in another helix (Glu 65-Arg 185 binding Chl *a*4, Glu 180-Arg 70 binding Chl *a*1). Chl *a*6 appears to be liganded by the peptide carbonyl of Gly 78. Probably a water molecule is involved in the binding. The remaining three chlorophylls (*b*1, *b*2, *a*7) are likely to be bound in a similar way by peptide carbonyls in the loop regions. The most striking feature of the 3.4 Å model is the two-fold symmetry relating the first and third transmembrane helices. This supported the suggestion, that these two helices were related as the result of an ancient gene duplication (Hoffman et al., 1987) (Fig. 1). Helix I and III cross each other and are held together by a symmetry-related pair of reciprocal ionic bonds between glutamate on one helix and arginine on the other, that serve to cross-link the two helices. The two molecules of lutein appear to play an important structural role in linking the flanking regions at the opposite ends of the helices I and III (Kühlbrandt et al., 1994).

Also the other chlorophyll *a/b* binding proteins of PS II consist most likely of three transmembrane helices. The amount of chlorophylls bound to isolated CP29, CP26 and CP24 varies between 5–12 for CP29, 8–14 for CP26 and 20–40 for CP24, depending on the isolation method (Bassi and Dainese, 1989; Henrysson et al., 1989; Dainese et al., 1990; Dainese and Bassi, 1991; Peter and Thornber, 1991). However, after in vitro pigment reconstitution, 8 chlorophylls could be bound to CP29 (Giuffra et al., 1996), 9

Abbreviations: Cab – chlorophyll *a,b* binding; DCMU – 3-(3,4-dichlorophenyl)-1,1-dimethylurea; ELIP – early light induced protein; LHC – light harvesting complex; PS I, PS II – Photosystem I and II, respectively

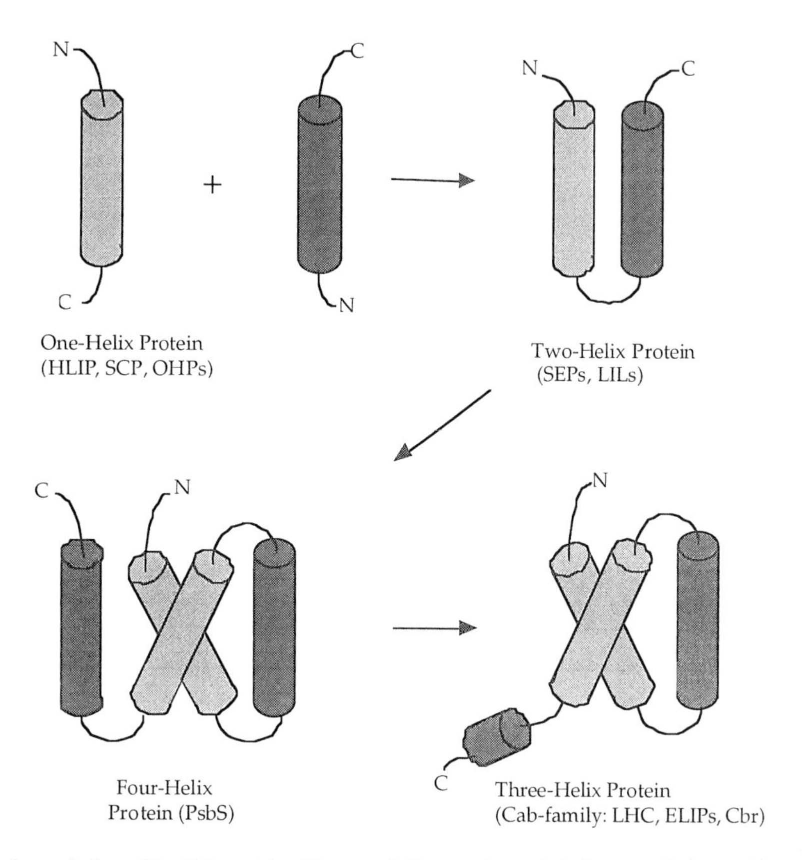

Fig. 1. Hypothesis for the evolution of the Cab proteins. Two one-helix proteins coded after gene fusion and internal duplication coded for a four helix protein, the ancestor of the PsbS protein. Shortage or loss of the fourth helix by deletion could have given rise to the ancestor the eukaryotic Cab proteins.

chlorophylls to CP26 (Ros et al., 1998) and 10 chlorophylls to CP24 (Pagano et al., 1998). Whereas reconstituted LHCII shows selectivity for chromophore binding (Paulsen et al., 1990), the less abundant chlorophyll *a/b* binding proteins of PS II can be stable reconstituted with various chlorophyll *a/b* ratios, but they are specific for the carotenoids they bind. The xanthophyll lutein is the only carotenoid necessary for reconstitution of CP29, CP26 can be reconstituted with violaxanthin and either lutein or violaxanthin are coordinated by CP24. Recently also in reconstituted LHCII most chlorophyll *a* molecules could be replaced with chlorophyll *b*, but not the other way around (Kleima et al., 1999).

Cab proteins are not only found in Photosystem II, but also function as light harvesting antenna of Photosystem I (Jansson, 1994; Pichersky and Jansson, 1996). The LHCI is composed of four Cab proteins (Lhca 1–4), that all show high similarity to the LHCII proteins. LHCI binds approximately 100–120 chlorophyll molecules (Haworth et al., 1983; Bassi and Simpson, 1987). Lhca1 and Lhca4 build a complex called LHCI-730 with a Chl *a/b* ratio of 2.3, and Lhca2 and Lhca3 together form LHCI-680, for which a Chl *a/b* ratio of 1.4 has been reported (Jansson, 1994). More distant relatives to the Cab proteins are the early light induced proteins (ELIPs) (Chapter 28, Adamska). This class of proteins also shows high sequence homology to the light harvesting proteins and possess three putative transmembrane domains. ELIPs from pea have been isolated binding chlorophyll *a* and lutein (Adamska et al., 1999). In contrast to other Cab family members ELIPs are induced only transiently during greening of etiolated seedlings or during light stress (Adamska, 1997) and their amounts are substochiometric. Recently, ELIP-related proteins with two or one predicted transmembrane helix have been identified in *Arabidopsis thaliana* (Jansson, 1999; Heddad and Adamska, 2000) as well as in cyanobacteria (Dolganov et al., 1995; Funk and Vermaas, 1999). They show a particularly high sequence homology to the first and third helix of the Cab-proteins and give additional support to the evolutionary origin of the family of Cab-proteins (Green and Pichersky, 1994) (Fig. 1).

A PS II subunit of 22 kDa, encoded by the nuclear *psb*S gene, exhibits unique properties, but has so far escaped unambiguous assignment of its functional role. The sequence of the *psb*S gene indicated homology to the Cab gene family (Kim et al., 1992; Wedel et al., 1992), ELIPs (early light induced proteins) (Adamska et al., 1993) and FCPs (fucoxanthin-chlorophyll *a/c* antenna proteins) (Green and Pichersky, 1994). However, in contrast to the Cab proteins and ELIPs, which have three transmembrane helices, the PsbS protein is inferred to have four (Kim et al., 1992; Wedel et al., 1992; Wallbraun et al., 1994). The protein was found to bind pigments, but also was stable in the absence of pigments (Funk et al., 1995a,b). Based on its molecular mass this protein was first called the 22kDa-protein, then CP22 due to its chlorophyll binding properties and in accordance with the terminology used for the minor light harvesting complex of Photosystem II. However, after sequencing its gene, CP22 is now referred to as the PsbS protein.

Recent evidence suggests that the PsbS protein is involved in the crucial photoprotective mechanism used by higher plants called non-photochemical quenching (NPQ) (Li et al., 2000). NPQ is thought to reflect a range of mechanisms associated with both the PS II reaction center and the light-harvesting antenna, which act to dissipate excess excitation energy safely as heat (Gilmore, 1997).

The purpose of this chapter is to present the information obtained on the PsbS protein, to classify it among the Cab-gene family and to discuss its possible function.

II. Early History of the PsbS Protein: A Mysterious Protein in Photosystem II

The PsbS protein was originally identified and isolated by Ljungberg et al. (1984, 1986) as a 22 kDa subunit of Photosystem II. Solubilization with detergents followed by co-immunoprecipitation with antibodies directed against the extrinsic PsbO and PsbQ proteins isolated not only the antigenic proteins, but also polypeptides of 24 kDa, 22 kDa and 10 kDa (Ljungberg, 1984). Following crosslinking studies showed that there is a near neighborhood relationship between the extrinsic PS II proteins and the 22 kDa protein (Bowlby and Frasch, 1986) and that the 22 kDa protein is one of the subunits in 'highly active' oxygen evolving PS II-core complexes (Tang and Satoh, 1985; Ghanotakis and Yocum, 1986). Ljungberg and coworkers (1986) claimed that the presence of the 22 kDa subunit was needed for reconstitution of the extrinsic PsbP protein. Therefore the PsbS protein was thought to be part of, or at least closely associated with photosynthetic water splitting. However, Merrit et al. (1987) were able to rebind the PsbP and PsbQ protein to PS II-core complexes depleted of the PsbS protein. Also later Mishra and Ghanotakis (1993) used the detergent octyl-thio-glucopyranoside to isolate the PsbS and PsbR proteins without depleting PS II of the extrinsic proteins. After removing the PsbS protein the PS II complexes showed a higher sensitivity to DCMU and in these complexes Cyt *b*559 was found to have totally converted to its low potential form (Mishra and Ghanotakis, 1993). The observation that depleting PS II of the PsbS protein alters the DCMU binding site (Q_B-pocket) had been made earlier (Henrysson et al., 1986; Ghanotakis et al., 1987b), but the use of sodium cholate for the extraction process might have been responsible for this effect (Bowlby and Yocum, 1993). Due to its close location to D1 and D2, PsbS was even discussed to have a function similar to the bacterial H subunit, which supports the two reaction center proteins L and M in purple bacteria (Andersson and Styring, 1991).

Depending on the isolation protocol used, the PsbS protein can be one of the last proteins which is depleted from the minimal PS II-core complex consisting of D1, D2, Cyt *b*559, *psb*I gene product, CP47, CP43 and the PsbO protein (Ghanotakis et al., 1987a). Different detergents seem to access/release the PsbS protein with various efficiency. Whereas digitonin and octyl-glycopyranoside do not remove PsbS from PS II particles, β-dodecyl-maltoside and sodium cholate are very effective (Harrer et al., 1998; Nield et al., 2000b).

CP29 (Ikeuchi et al., 1985; Ghanotakis et al., 1987b) or CP24 are often co-isolated with PsbS (Ljungberg et al., 1984; Nield et al., 2000b), suggesting a close connection between the PsbS protein and the minor antenna proteins. Selective extraction of the PsbS protein even leads to efficient separation of LHCII from the PS II reaction center core (Kim et al., 1994). Digestion studies with trypsin revealed that the LHCII might shield the N-terminus of PsbS against proteolysis (Kim et al., 1994). Independent of the detergent used the PsbS protein is easily co-isolated with the PsbR protein, which also indicates a close structural and maybe even functional role to this protein (Ljungberg et al., 1984, 1986; Ghanotakis et al., 1987a,b; Mishra and Ghanotakis, 1993; Kim et al., 1994). However, anti-sense potato plant mutants of the PsbR show no influence on PsbS transcription (Stockhaus et al., 1990).

In conclusion, the PsbS protein seems to be located close to the PS II-core complex proteins, probably connecting the core antenna proteins and the minor light harvesting proteins. A stoichiometry of one PsbS protein per PsbO protein has been established by immunological assays in plants grown at normal growth conditions (Funk et al., 1995a). However, the PsbS protein seems to be absent in the LHCII-PS II supercomplex for which a 3D structure has been obtained at 24Å resolution (Nield et al., 2000a; b). Therefore more detailed localization studies are required that are not only based on biochemical findings, but address the macromolecular systems recently identified by Boekema et al. (1999) in which LHCII-PS II supercomplexes are interconnected by additional LHCII trimers and Cab proteins.

III. The *psbS* Gene and Gene Product

Neither chromatographic purification nor isolation after different PS II treatments indicated in the early studies that the 22kDa-protein might bind pigments or any other cofactors (Ljungberg et al., 1986; Bowlby and Yocum, 1993; Mishra and Ghanotakis, 1993). Therefore it was very surprising, when the *psbS* gene was identified and showed high homology to the chlorophyll *a*/*b* binding protein family (Kim et al., 1992; Wedel et al., 1992), the early light induced proteins (ELIPs) (Adamska et al., 1993), and FCPs (fucoxanthin-chlorophyll *a*/*c* antenna proteins) (Green and Pichersky, 1994). The degree of similarity of the PsbS protein to the Cab-antenna proteins is comparable to that between Cab proteins and ELIPs (Kim et al., 1992). The PsbS protein in tomato has 44.4% similarity (35% identity) to the *Lhcb3* gene product and around 40% similarity (32.8% identity) to the various ELIPs (Table 1). Moreover, tertiary structure analyses of the Cab-antenna proteins and the PsbS protein showed high similarity: the protein folding remains conserved even when there is no detectable amino acid sequence homology (Kim et al., 1992). As mentioned earlier, one important difference based on protein sequences and hence on the predicted folding patterns is, that the PsbS protein is inferred to have four transmembrane helices (Kim et al., 1992; Wedel et al., 1992; Wallbraun et al., 1994), while only three are predicted for the Cab proteins and ELIPs (Grimm et al., 1989). Within the structural pattern derived from hydropathy analysis, the first and third helix of the PsbS protein are homologous to the corresponding transmembrane helices of the Cab proteins, while the second helix of the Cab proteins is related to the other two predicted helices of the PsbS protein (Kim et al., 1992; Wedel et al., 1992) (Fig. 1). The similarity with the first and third helices of the Cab-antenna proteins was noted several years ago (Hoffman et al., 1987). However, the two halves of the PsbS protein (i. e. helices I/II and III/IV) are more related to each other than the comparable parts of the Cab-proteins (i. e. helices I and III), and they have higher degree of identity with each other than with the conserved regions of any of the Cab-proteins (see Green and Pichersky, 1994). Therefore it was suggested that the PsbS protein must be the result of an internal gene duplication of an ancestor protein containing two transmembrane segments (Kim et al., 1992; Wedel et al., 1992; Green and Pichersky, 1994). Gene duplications and subsequent divergence have been crucial in the phylogeny of organisms. Indeed, various thylakoid structures have evolved by duplications of primordial DNA segments (Büttner et al., 1992). Usually this has resulted in two individual proteins (Wedel et al.,

Table 1. Similarity and identity of the PsbS protein to other proteins of the Cab family determined with two different programs

	best fit		gap	
	% similarity	% identity	% similarity	% identity
LHCII type III	44.4	35.18	33.5	26.0
CP29	55.9	52.94	31.9	23.8
ELIP *pea*	35.2	31.8	32.5	26.4
ELIP *Ara. th.**	40.0	32.8	33.0	24.0

**Ara. th.*: *Arabidopsis thaliana*

1992). In the case of the ancestral PsbS protein however, no divergence has occurred (Fig. 1). Furthermore shortage or loss of the fourth helix by deletion could have given rise to the ancestor or ancestors of the eukaryotic antenna LHCII and LHCI proteins and ELIPs (Green and Kühlbrandt, 1995). The two-helix ancestor may have been itself derived by the fusion of a one-helix transmembrane protein. Another possibility would be that the one-helix ancestral gene product was made as a precursor, which inserted into the membrane as a two-helix hairpin and was subsequently processed (Green and Kühlbrandt, 1995). One-helix as well as two-helix relatives to the Cab-proteins can still be found in prokaryotes and eukaryotes (Dolganov et al., 1995; Funk et al., 1999; Jansson, 1999; Chapter 28, Adamska; Heddad and Adamska, 2000) (Fig. 1).

From consideration of the LHCII crystal structure (Kühlbrandt et al., 1994) it is expected that also in the PsbS protein the first and third helix cross each other at an angle of about 56° and project above the plane of the membrane, while the second and fourth membrane are shorter and almost perpendicular to the membrane plane. Whereas the N-terminus contains many basic amino acid residues, the C-terminus is acidic. Both, N- and C-terminus are located at the stromal side of the thylakoid membrane (Kim et al., 1994). The theoretical isoelectric point of PsbS is higher than 6.5, but after isoelectric focusing it was found to migrate to a pH of 6.1 (Funk et al., 1995a). Due to its four transmembrane helices the PsbS protein has an extreme hydrophobic character (Ljungberg et al., 1986; Funk et al., 1995a).

The *psbS* gene has been identified and sequenced in several C3 and C4 plants (Kim et al., 1992; Wedel et al., 1992; Wallbraun et al., 1994; Iwasaki et al., 1997; Wyrich et al., 1998; Jansson, 1999). The protein has also been identified by polyclonal antibodies in the fern *Nephrolepis exaltata* and the green algae *Chlamydomonas reinhardtii*. However, the gene has not been detected in chromophyte algae and antibodies failed to detect PsbS in the red algae *Phorphyridium purpureun* and the cyanobacterium *Phormidium laminosum* (Ljungberg et al., 1986). Contrary to earlier reports (Nilsson et al., 1990), it is not present in the cyanobacterium *Synechocystis* sp. PCC 6803 (Kaneko et al., 1996). However, the existence of five small Cab-like proteins in *Synechocystis* 6803, which are able to form stable oligomers (Funk and Vermaas, 1999) might explain the crossreaction seen with the PsbS antibody by Nilsson et al. (1990). The *psbS* gene is very conserved in the different organisms, which suggests an essential function. In tomato it is 86% identical (92% with conservative substitutions) to spinach (Wallbraun et al., 1994), while in the monocotyl plant rice it is 84.4% homologous to tomato and 80.6% to spinach (Iwasaki et al., 1997).

IV. Pigment Binding

The sequence similarities between *psbS* and the Cab genes suggested that the PsbS protein could be pigment binding. Although only two amino acids of the predicted chlorophyll binding sites in the PsbS are identical to the ones of LHCII (Green and Pichersky, 1994; Kühlbrandt et al., 1994; Funk et al., 1995a), various other chlorophyll binding sites are clearly possible. Like LHCII, the PsbS protein could bind one chlorophyll molecule on amino acid Glu-180 (chlorophyll *a*1), one chlorophyll *a* molecule on Glu-65 (chlorophyll *a*4), and possibly another chlorophyll *a* molecule on Gly-78 (chlorophyll *a*6). However, the potentially Chl-ligating His and Asn (binding chlorophyll *a*5 and *a*2) in the first and third helix are replaced by Val, and the Gln/Glu and terminal Arg in the second transmembrane helix are missing (Fig. 2).

Experimental evidence that PsbS can bind pigments was first obtained by mild SDS-PAGE analysis (Funk et al., 1994). PsbS extracted from PS II complexes after treatment with octyl-thioglucopyranoside, resulted in a clear green band on partly denaturing SDS-gels. By comparing the electrophoretic migration of this green band with that of other chlorophyll-protein bands it was found to co-migrate with LHCII and was therefore probably overlooked in earlier studies. In the SDS-gel the PsbS protein was observed to bind chlorophyll *a*, chlorophyll *b* and carotenoids. Also transfer from chlorophyll *b* to chlorophyll *a* was detected spectroscopically. Therefore the PsbS protein was suggested to be another antenna protein and by analogy to the other Cab minor antenna proteins was renamed to CP22 (Funk et al., 1994).

Using isoelectric focusing and a mild nonionic detergent, PsbS was purified in a more native, pigment-binding state (Funk et al., 1995a). It was shown to bind only four to six chlorophyll molecules and one carotenoid per protein and to exhibit a chlorophyll *a*/*b* ratio of 6. Thus, the overall pigment content, and in particular the amount of chlorophyll *b* is lower for the PsbS protein as compared to the Cab proteins. As is also the case for these proteins, the isolated PsbS protein contained substoichiometric amounts of several carotenoid molecules (neoxanthin, violaxanthin, lutein, β-carotene) (Funk et al., 1995a). This could be explained by the possibility that some of these pigments may be bound in between subunits of an oligomeric complex in analogy to what was found for other Cab proteins (Bassi et al., 1991, 1993; Dainese and Bassi, 1991). The authors noticed a heterogencity with respect to bound chlorophylls per protein, which might have been due to the preparation method. Whereas some of the data supported the fact that PsbS could be an additional antenna protein, at the same time 30% of the chlorophyll molecules were more loosely attached to the protein than in other light harvesting proteins. There was also an extremely weak excitation coupling between the chlorophylls (Funk et al., 1995a). Even more astonishing was the fact that PsbS, contrary to other chlorophyll binding proteins, could be integrated into the membrane in a stable form in the absence of pigments (Funk et al., 1995b). PsbS transcript and translation products were identified in etiolated plants and it could be shown that neither carotenoid nor chlorophyll precursors stabilize the protein under these conditions. PsbS is the only known chlorophyll binding protein, which is stable in the absence of pigments. During greening it seems to be the first protein that binds chlorophyll (Funk et al., 1995b). Despite the fact that the chlorophyll *a* binding proteins are known to have a higher pigment affinity then the chlorophyll *a*/*b* binding proteins (see Paulsen, 1997), already after 4 h of greening the PsbS protein could be isolated with pigments bound to it. At the same time chlorophyll *a* binding proteins were detected, but chlorophyll *a*/*b* binding proteins were absent (Funk et al., 1995b). Furthermore the PsbS protein could be detected in chlorophyll-deficient mutants lacking the conventional Cab-proteins, e.g. the barley *chlorina-f*2 mutant grown under intermittent light, or the pigment free tobacco plastome mutant SR1V35 (Funk et al., 1995b). As discussed later, these results led to the suggestion that the PsbS protein might be involved in pigment storage/transport or possibly taking part in the PS II protective mechanism involving nonphotochemical quenching.

V. The PsbS Protein: An Early Ligh Induced Protein or Light Harvesting Protein or a Protein of Its Own?

A. Primary and Secondary Structure Comparison

Comparison of the amino acid sequence (Fig. 2) of PsbS, the Cab-antenna proteins and the ELIP family clearly shows differences despite the common overall features (Green et al., 1991; Adamska and Kloppstech, 1994; Green and Pichersky, 1994; Green and Kühlbrandt, 1995; Green and Durnford, 1996; Adamska, 1997, 2000). Conserved amino acid residues of the Cab-antenna proteins, the ELIPs and the PsbS protein involve in helix I two Glu (except in CP24), one Arg (except in CP24), Ala, Met and two Gly (except in ELIP1 of *Arabidopsis*). In helix II there is a conserved Leu while in helix III one Glu, Gly (except CP24), Arg, Asn and Ala are conserved. Pro and Leu, which can be found in helix IV of the PsbS protein are also found in the short C-terminal α-helix of the Cab-antenna proteins. This amphipathic α-helix is totally missing in the ELIPs. Differences in the primary structure are seen in the length of the hydrophilic amino acid regions, that connect the transmembrane helices, that are very short for ELIPs (Green and Kühlbrandt, 1995; Chapter 28, Adamska) (Fig. 1) and slightly expanded in the PsbS protein. In the Cab-antenna proteins however, these loops are more extended. Also typical conserved amino acid motives are missing in the PsbS protein. The Cab-antenna proteins have a motive called 'crown' (PGDYGWDTAGL and PGGSFDPLGL) in front of the first and third transmembrane helices (Green and Kühlbrandt, 1995), whereas in ELIPs two motives, the KOST and AFGHAN motives, are found in front of the first helix (Chapter 28, Adamska). Based on these sequence comparisons, the PsbS protein seems to be neither a Cab-antenna protein nor an ELIP, although a few homologous amino acids exist that could group the PsbS protein to one or the other group (Fig 1). Nevertheless, according to structural investigations the Cab-antenna proteins, ELIPs and the PsbS protein seem to have, besides the already described structural homology of the α-helices, also β-turns on similar positions in front of helix I and III. Therefore the folding pattern of the secondary structure seems to be conserved, even if the sequence similarity is no longer recognizable (Green et al., 1991).

```
LHCII-I      MATSTMALSSSTFAGK........AVKLSPSSSEITGNGGRVTMRKTATKAKPASSGSP........WYGPDRVKYLGPFSGES.
CP29         MAAATSLYVSEMLGSPVKFSGVARPAAPSPSSSATFKTVALFKKKAAAAPAKAKAAAVSPADDELAKWYGPDRRIFLPEGLLDRSEI
LHCI         MASNTLMSCGIPAVCPSFL...........SSTKSKF........AAAMPVSVGATNSMSRFSMSADWMPGQPR
CP24         MATTSAAVLNGLSSSFLTGGNKSQALLAAPLAARVGGAAPKRFTVLAAAAKKS..............WIPAVRGGGNLVD

PsbS         MAQTMLLTANAKVDLRSKESLVERLKPKPLSSLFLPSLPLRFSSSSTNASSSKFTSTTVALFKSKAKAPPKKVAPPKEKQKVED

ELIP  pea   MAVSSCQSIMSNSMTNI.............SSRSRVNQFTNIPSVYIPTLRRNVSLKVRSMAEGEPKEQSKVAVDPTTP
ELIP1 ara   MATASFNMQSVFAGGLTTRKINTNKLFSAGSFPNLKRNYPVGVRCMAEGGPTNEDSSPAPSTSAA
ELIP9 hor   MATMMSMSSFAGAAVVPR.........SSASSFGARSLPALGRRALVV


                                      HELIX I
                                  *  *  *       *
LHCII-I      .PSYLTGEFPGDYGWDTAGLSADPETFSKNRELEVIHSRWAMLGALGCVFPELLSRNGVKFG.EAVWFKAG
CP29         .PEYLNGEVPGDYGYDPFGLSKKPEDFAKYQAYELIHARWAMLGAAGFIIPEAFNKFGANCGPEAVWFKTG
LHCI         .PPWLDGSLPGDFGFDPLGLASDPESLRWNQQAELVHCRWAMLGAAGIFIPELLTKIGILNTP..SWYTAG
CP24         .PEWLDGSLPGDYGFDPLGLGKDPAFLKWYREAELIHg WAM..AAVLGIFVGQAWSGIP......WFEAG

PsbS         AKAPKKVEKPKLKVED..GLFGTSGGIGFTKENELFVgRVAMIGfAASLLGEGITgKGILSQ.........

ELIP  pea   .TASTPTPQPAYTRPpkmstKFSdLMafsgpapERIngRlAMIGfvAaMGvEIAKgQG............
ELIP1 ara   .QPLPKSPSPPPPMKpkmstKFSdLLafsgpapERIngRlAMVGfvAaLAvELSKgEN............
ELIP9 hor   .RAQTEGPSAPPPNKpkmstSIWdEMafsgpapERIngRlAMVGfvTaLAvEAGRgDG............


                                   HELIX II
                                  *         *  *
LHCII-I      SQIFSEGGLDYLGNPSLVHAQSILAIWATQVILMGAVEGYRIAGGPLGEVVDPL...................
CP29         ...ALLLDGNTLNYFGKNIPINLILAVVAEVVLVGGAEYYRIINGLDLEDKL.....................
LHCI         ...............EQEYFTDTTTLFIVELVLIGWAEGRRWADIIKPGCVNTDPIFPNNKLTGTDVG......
CP24         ............ADPGAIAPFSFGTLLGTQLILMGWVESKRWVDFFDPDSQSVEWATPWSKTAENFANFTGEQG

PsbS         ...........LNLETg.IPIYEAEPLlLFFILFTLL.gAIGALGDRGR.FVDEPTTGLEKAV...........

ELIP  pea   .......LSEqL.sGGgVAwFLGtSVLlSLasLIpFFQgVsV...............................
ELIP1 ara   .......VLAq.IsDGgVSwFLGtTAIlTLasLVpLFKgIsV...............................
ELIP9 hor   .......LLSqLGsGTgQAwFAYtVAVlSMasLVpLLQgEsA...............................
```

```
                                HELIX III
                                  * *        *
LHCII-I     .YPGGS.FDPLGLAD.DPEAFAELKVKELKNGRLAMFSM.FGFFVQAIVTG
CP29        .HPGGP.FDPLGLAK.DPDQAAILKVKEIKNGRLAMFSM.LGFFIQAYVTG
LHCI        .YPGGLWFDPLGWGSGSPAKIKELRTKEIKNGRLAMLAV.MGAWFQHIYTG
CP24        .YPGGKFFDPLALA..DTEKLERLKVaEIKHARLAMLAM.LIFYFeA.GQG

PsbS        IPPGKDVRSALGLKTK.GPLFGFTKSNELFVGRLAQLgFAFSLIGeII.TG

ELIP  pea  ...............eSKSKSimSSDaEFwNGRIAMLglvalaFteF.VKG
ELIP1 ara  ...............eSKSKGimTSDaELwNGRFAMLglvalaFteF.VKG
ELIP9 hor  ...............eGRAGAimNANaELwNGRFAMLglvalaAteII.TG


                       HELIX IV

LHCII-I     KG...........PLENLADHLADPVNNNAWSYATNFVPGK
CP29        QG...........PVENLAAHLSDPFGNNLLTVIGGASERVPTL
LHCI        TG...........PIDNLFAHLADPGHATIFAAFSPK
CP24        KT...........PLGALGL

PsbS        KGALAQLNIETGVPI.NEIEPLVLLNVVFFFIAAINPGTGKFITDDEEED

ELIP  pea TSLV
ELIP1 ara GTLV
ELIP9 hor APFINV
```

Fig. 2. Sequence alignment of the Cab-antenna proteins LHCII (Lhcb1), CP29, CP24 and LHCI, the PsbS protein from tomato and ELIPs from pea (*Pisum sativum*), *Arabidopsis thaliana*, and *Hordeum vulgare*. The predicted transmembrane helices are marked. Amino acids belonging to typical Cab-antenna motives are marked in bold, whereas ELIP motives are shown in small letters. The common Cab-family amino acids or motives are outlined. * indicates the predicted chlorophyll binding sites in Lhcb1 (as in Kühlbrandt et al., 1994).

B. Expression Pattern

A very pronounced difference between ELIPs and Cab-antenna proteins is their expression pattern (Chapter 28, Adamska). The light harvesting proteins are very abundant (Jansson, 1994; 1999). As part of the long term adaptation of plants, their expression varies dependent on light conditions (Jansson, 1994; Pichersky and Jansson, 1996; Simpson and Knoetzel, 1996). ELIPs are only transiently expressed under stress conditions, mainly light stress (Chapter 28, Adamska) and therefore their abundance is low. Under normal conditions ELIPs can not be detected in plants (Adamska et al., 1992a,b). Expression of the PsbS protein, however, is very stable. The protein can be detected in etiolated plants, independent of the age of the seedling and in various mutants that are deficient in other Cab-proteins (Funk et al., 1995b). During light-induced inhibition of Photosystem II and degradation of the D1 protein (Aro et al., 1990), as well as during acclimative reduction of the PS II-antenna in response to increased light (Lindahl et al., 1997), the level of the PsbS protein stays constant in spinach. However, it should be noted that in *Arabidopsis thaliana* the *psb*S mRNA was found to increase several fold in response to high light (Li et al., 2000).

Expression of the *psb*S gene has distinct differences compared to the other members of the *cab*-gene family. Accumulation of the Cab-antenna proteins is strictly light-dependent, even though transcript accumulation can be observed in dark grown plants. The accumulation of PsbS in etioplast membranes of spinach seedlings (Funk et al., 1995a,b) indicates that the *psb*S gene undergoes an endogenous, light independent regulation, which controls not only its transcript level, but also influences the level of the protein. But the accumulation of this protein and its transcript is also positively regulated by external stimuli like light. In etiolated plants the phytochrome receptor is involved in a positive regulation of the *psb*S gene expression, whereas red light negatively influences the post-transcriptional events and thus leads to a reduction of the *psb*S transcript abundance (Adamska et al., 1996). Contrary to ELIPs and the Cab-antenna proteins (Marrs and Kaufman, 1989; Adamska et al., 1992a, b) in spinach the *psb*S transcription seems to be influenced only by phytochrome and not by the blue light low fluence system (Adamska et al., 1996; Green and Salter, 1996; Iwasaki et al., 1997). With respect to the fluence rate, the *psb*S transcript is stimulated by a single pulse of red light (very low fluence response), similar to its relatives the Cab-antenna proteins and the ELIPs (Apel, 1979; Karlin-Neumann et al., 1988; Adamska, 1995; White et al., 1995).

Based on the RNA and protein expression pattern the PsbS protein resembles neither the Cab-antenna protein nor the ELIPs. Even though light stimulates the PsbS regulation, similar to the other cab-gene family members, the transcriptional and translational regulation of PsbS is mainly light-independent.

C. Targeting, Insertion and Localization in the Thylakoid Membrane

The targeting and localization within the thylakoid membrane differs much between the PsbS protein and its Cab-relatives, the antenna proteins and the ELIPs. Despite the high sequence homology between the different Cab-family members, the transit peptides are very different between PsbS, ELIPs and the Cab-antenna proteins and also differ between PsbS from tomato, spinach or rice (Iwasaki et al., 1997). Contrary to the gene-doubling hypothesis, the transit peptides could have arisen by exon shuffling or they must have had only very little selective constraint during evolution (Pichersky and Jansson, 1996). The first 69 amino acid residues of the PsbS precursor possess all attributes of a typical stroma-targeting chloroplast transit sequence. This sequence starts with a MAQAM motif, which is also found in the transit sequence of the extrinsic, lumenal PsbQ polypeptide (Wedel et al., 1992).

At least four different, protein-specific insertion pathways are known to import nuclear proteins into the thylakoid membrane (Robinson and Klösgen, 1994; Klösgen, 1997; Robinson et al., 1998). The Sec-dependent pathway requires ATP and a SecA protein in the stroma. The ΔpH-dependent route is exclusively dependent on the transthylakoid proton gradient, whereas the SRP (signal recognition particle)-dependent pathway requires a transthylakoid ΔpH, GTP and a chloroplast homologue to the 54 kDa subunit of the endoplasmic reticulum SRP, called cpFtsY. Finally the spontaneous protein import seems not to depend on any known factors. Thylakoid import studies performed for the Cab-antenna proteins LHCII (Lhcb1) (Li et al., 1995), CP26 (Lhcb5) and LHCI (Lhca1) (Kim et al., 1999) have been shown to require SRP together with GTP for insertion into the membrane. Recently it was also reported that a chloroplast Oxa1p homologue, Albino3, is required for the integration of Lhcb1 (Moore et al., 2000). It was shown that the PsbS protein as well as Elip2 are able to use the assisted SRP pathway (Kim et al., 1999). In contrast to this assisted pathway the PsbS protein as well as Elip2 of *Arabidopsis thaliana* seem also to be able to follow the spontaneous or non-assisted pathway. This pathway was found to operate for small proteins like the CF_0, PsbW or PsbX proteins. These mature proteins contain a single transmembrane span and are synthesized in the cytosol with a bipartite presequence in which a typical envelope transit signal is followed by a cleavable signal peptide. It has been suggested that the signal peptides probably serve an unusual function by simply providing an additional hydrophobic region that, together with the transmembrane segment in the mature protein, is able to drive the translocation of the hydrophilic intervening region into the thylakoid lumen (Thompson et al., 1998). PsbS and Elip2 were the first multi-spanning proteins inserted in vitro in the absence of SecA, SRP and dNTPs. The import pathway of the PsbS protein is in accordance with the suggested evolutionary origin of the PsbS protein that a one-helix ancestral gene product was made as a precursor, which inserted into the membrane as a two-helix hairpin and was subsequently processed (Green and Kühlbrandt, 1995). This sequence of events is homologues to that observed with the PsbW and PsbX precursor proteins. During evolution the processing site of PsbS might have been lost and even after gene duplication the same import pathway would have been retained. However, if the Cab-antenna proteins and ELIPs really arose by shortening or deleting the fourth helix from this ancestral four helix protein, it is hard to explain why the ELIPs still seem to use the same import pathway as PsbS, whereas the Cab-antenna proteins are dependent on SRP for thylakoid insertion.

Most obvious are the differences in the lateral distribution of LHCII, the ELIPs and the PsbS protein within the thylakoid membrane system. LHCII is mainly present in the appressed region of the thylakoid, but a subpopulation of LHCII is mobile and can relocate into the stroma region of the thylakoid, depending on the light conditions (state1-state2 transition) (Jansson, 1994; Pichersky and Jansson, 1996). ELIPs are found in the non-appressed regions of the thylakoid membranes (Adamska, 1997; Lindahl et al., 1997). The PsbS protein, however, is almost exclusively confined to the grana partition regions and exists only in trace amount in the stroma exposed thylakoid regions (Ljungberg et al., 1986; Funk et al., 1995a). It remains in the grana partitions even after in vitro induced photoinhibitory damage of the PS II reaction center, when other PS II subunits migrate to the stroma exposed thylakoid regions (Hundal et al., 1990). The pronounced lateral heterogeneity of the PsbS protein is reflected as well by its prominent mesophyll-specific presence in the C4 plant *Sorghum bicolor* (Wyrich et al., 1998).

Based on the data obtained for targeting and insertion into the thylakoid membrane the PsbS protein differs from the Cab-antenna proteins, but resembles that of the ELIPs. However, the localization of PsbS along the plane of the thylakoid membrane is opposite to that of the ELIPs. Therefore, based on these distinct differences, the PsbS protein should be classified neither as a Cab-antenna protein nor as an ELIP.

VI. The Function of the PsbS Protein

Several functions have been discussed for the PsbS protein. Its weak binding of pigments, the relative low number of chlorophyll molecules and the poor excitonic coupling between the chromophores suggested that it may act as a 'ligand chaperone', serving as a transient chlorophyll storage protein. In this way newly synthesized chlorophyll molecules would not be free in the membrane, but kept in association with the xanthophylls bound by the PsbS protein.

The PsbS protein binds violaxanthin (Funk et al., 1995a), which is known to play an essential role in mediating nonphotochemical quenching. This primary photoprotective mechanism is responsible for dissipating excess levels of excitation energy as heat in the light-harvesting antenna of Photosystem II (Horton et al., 1996; Gilmore, 1997). The formation of zeaxanthin has been particularly well correlated with nonphotochemical quenching of chlorophyll fluorescence in plants and some algae. It is generally accepted that a synergistic effect exists on quenching between the conversion of violaxanthin into zeaxanthin via the xanthophyll cycle and the formation of the transthylakoid pH gradient. The major fraction of nonphotochemical quenching, which depends on the transmembrane ΔpH is called energy-dependent quenching (qE). It forms within minutes of exposure to light and relaxes as fast in darkness (Horton et al., 1996). A second, smaller, component (qT) arises from the fluorescence decrease associated with a state transition and a third type (qI) is associated with stronger degrees of light stress when quenching increasingly becomes irreversible. As the PsbS protein binds xanthophylls and is also present in chlorophyll-deficient mutants like the barley *chlorina-f2* mutant (Funk et al., 1995b), that lack all conventional Cab-proteins when grown under intermittent light (Krol et al., 1995), it was suggested to be involved in non-photochemical quenching (Funk et al., 1995a,b). Recent data obtained in *Arabidopsis thaliana* (Li et al., 2000) support and expand this idea: a mutant deficient of nonphotochemical quenching was shown to have a single nuclear mutation in the *psb*S gene. This npq4 mutant is defective in the qE component of nonphotochemical quenching and lacks the transmembrane ΔpH and zeaxanthin dependent conformational change in the thylakoid membrane. However, it is not affected in pigment composition, light-harvesting or photosynthesis. Therefore, it was concluded that PsbS contributes to photoprotective energy dissipation rather than photosynthetic light harvesting. The exact role of PsbS in qE remains to be determined, but the lack of both qE and the protonation-induced conformational change in the PsbS-deficient mutant indicates that binding of one or more protons by PsbS may be a necessary feature of the qE mechanism (Li et al., 2000). Upon protonation of the PsbS protein, a conformational change could induce quenching of a single excited chlorophyll by direct energy transfer from chlorophyll to zeaxanthin bound to PsbS. Alternatively, protonation of PsbS might induce conformational changes that actually occur in the LHC proteins. It might also be possible that PsbS is necessary to maintain the LHC proteins in a proper supermolecular organization that allows qE to occur. In pine trees it was observed that during winter time an increase in the amount of PsbS is coupled to a major reorganization of the light-harvesting antenna proteins (Ottander et al., 1995). The structural data of PS II indicate that PsbS is located in the vicinity of PS II (Nield et al., 2000b), so it might be part of the more flexible LHCII pool and keep the environment stable for LHCII trimers and/or monomers that build up the supercomplex.

VII. Conclusion

Although significant progress has been made toward understanding the role of PsbS in Photosystem II, the function of this 22 kDa protein is still not unambiguously clarified. Due to its sequence similarity it belongs to the family of the chlorophyll *a*/*b* binding proteins. Indeed its ability to bind pigments has now been well established. Like the antenna proteins it binds chlorophyll *a*, chlorophyll *b* and carotenoids. However, the apparent weak binding of pigments, the relative low number of chlorophylls bound and the poor excitonic coupling between the pigments as well as the fourth membrane spanning helix make the PsbS protein distinct from all the other known Cab-proteins.

Several features suggest that the PsbS protein is more closely related to the ELIPs than to the Cab-antenna proteins, the most prominent one being the pigment binding. Recently an ELIP has been purified from light stressed pea leaves, that binds chlorophyll *a* and lutein (Adamska et al., 1999). Like the PsbS protein this ELIP exhibits a very weak excitonic coupling between the bound chlorophyll molecules, which suggests that these chlorophylls are not involved in energy transfer. Furthermore, it is still not experimentally confirmed that other members of the ELIPs strongly bind chlorophylls. They might—like the PsbS protein—bind pigments, but also be stable in the absence of pigments. Based on these findings a function as chlorophyll-carrier or pigment-exchanger has been discussed for PsbS as well as for the ELIPs. However, the PsbS can not be classified as an ELIP protein. Sequence comparison shows that the typical ELIP motifs (Chapter 28, Adamska) are missing in the PsbS amino acid sequence. Also the location of these proteins in the thylakoid membrane is opposite: Whereas the PsbS protein is found exclusively in the grana region of the thylakoid, ELIPs seem to be predominantly localized in the non-stacked region. Finally the PsbS protein is always expressed in plants, independent on the growth condition and or plant species.

No mutants for the various ELIPs have been found so far, but the identification of the npq4 mutant with the mutation in the *psb*S gene gives strong evidence that PsbS is involved in nonphotochemical quenching. Further characterization concerning the pigment binding ability under different growth conditions as well as in vitro pigment reconstitution studies will be needed to elucidate the exact function of the PsbS protein.

Acknowledgments

The author would like to thank Prof. B. Andersson, Prof. J. Barber and Dr. W. P. Schröder for valuable discussions on the manuscript. Financial support from the Swedish Research Foundation for Natural Sciences (NFR) is greatly acknowledged.

References

Adamska I (1995) Regulation of Early light-inducible protein gene expression by blue and red light in etiolated seedlings involves nuclear and plastid factors. Plant Physiol 107: 1167–1175

Adamska I (1997) Elips—light-induced stress proteins. Physiol Plant 100: 794–805

Adamska I and Kloppstech K (1994) The role of early light-induced proteins (ELIPs) during light stress. In: Baker NR and Bowyer JR (eds) Photoinhibition of Photosynthesis: From Molecular Mechanisms to the Field, pp 205–219. Bios Scientific Publishers, Oxford

Adamska I, Ohad I and Kloppstech K (1992a) Synthesis of the early light-inducible protein is controlled by blue light and related to light stress. Proc Natl Acad Sci USA 89: 2610–2613

Adamska I, Kloppstech K and Ohad I (1992b) UV light stress induces the synthesis of the early light-inducible protein and prevents its degradation. J Biol Chem 267: 24732–24737

Adamska I, Kloppstech K and Ohad I (1993) Early light-inducible protein in pea is stable during light stress but is degraded during recovery at low light intensity. J Biol Chem 268: 5438–5444

Adamska I, Funk C, Renger G and Andersson B (1996) Developmental regulation of the PsbS gene expression in spinach seedlings: The role of phytochrome. Plant Mol Biol 31: 793–802

Adamska I, Roobol-Boza M, Lindahl M and Andersson B (1999) Isolation of pigment-binding early light-inducible proteins from pea. Eur J Biochem 260: 453–460

Andersson B and Styring S (1991) Photosystem II: Molecular organization, function, and acclimation. Curr Topics Bioenerg 16: 1–69

Apel K (1979) Phytochrome-induced appearance of mRNA activity for the apoprotein of the light-harvesting chlorophyll *a/b* protein of barley (*Hordeum vulgare*). Eur J Biochem 97: 183–188

Aro E-M, Hundal T, Carlberg I and Andersson B (1990) In vitro studies on light-induced inhibition of Photosystem II and D1-protein degradation at low temperatures. Biochim Biophys Acta 1019: 269–275

Barber J and Kühlbrandt W (1999) Photosystem II. Curr Opin Struct Biol 9: 469–475

Bassi R and Dainese P (1989) The role of light harvesting complex II and of the minor chlorophyll *a/b* proteins in the organisation of the Photosystem II antenna system. Progress Photosynth Res 2: 209–216

Bassi R and Simpson D (1987) Chlorophyll-protein complexes of barley Photosystem I. Eur J Biochem 163: 221–230

Bassi R, Silvestri M, Dainese P, Moya I and Giacometti GM (1991) Effects of non-ionic detergent on the spectral properties and aggregation state of the light-harvesting chlorophyll *a/b* protein complex (LHCII). J Photochem Photobiol 9: 335–354

Bassi R, Pineau B, Dainese P and Marquardt J (1993) Carotenoid-binding proteins of Photosystem II. Eur J Biochem 212: 297–303

Bassi R, Sandona D and Croce R (1997) Novel aspects of chlorophyll *a/b*-binding proteins. Physiol Plant 100: 769–779

Boekema EJ, van Roon H, van Breemen JFL and Dekker JP (1999) Supramolecular organization of Photosystem II and its light-harvesting antenna in partially solubilized Photosystem II membranes. Eur J Biochem 266: 444–452

Bowlby NR and Frasch WD (1986) Isolation of a Manganese-Containing Protein complex from Photosystem II Preparations of Spinach. Biochemistry 25: 1402–1407

Bowlby NR and Yocum CF (1993) Effects of cholate on Photosystem II: Selective extraction of a 22 kDa polypeptide and modification of QB-site activity. Biochim Biophys Acta 1144: 271–277

Bricker TM (1990) The structure and function of CPa-1 and CPa-2 in Photosystem II. Photosynth Res 24: 1–13

Büttner M, Xie DL, Nelson H, Pinther W, Hauska G and Nelson N (1992), Photosynthetic reaction center genes in green sulfur bacteria and in Photosystem 1 are related. Proc Natl Acad Sci USA 89: 8135–8139

Dainese P and Bassi R (1991) Subunit stoichiometry of the chloroplast Photosystem II antenna system and aggregation state of the component chlorophyll *a/b* binding proteins. J Biol Chem 266: 8136–8142

Dainese P, Hoyer-Hansen G and Bassi R (1990) The resolution of chlorophyll *a/b* binding proteins by a preparative method based on flat bed isoelectric focusing. J Photochem Photobiol 51: 693–703

Diner BA and Babcock GT (1996) Primary electron transfer: Z-Q_A. In: Ort, DR and Yocum CF (eds) Oxygenic Photosynthesis: The Light Reactions, pp 213–247. Kluwer Academic Publishers, Dordrecht

Dolganov NA, Bhaya D and Grossman AR (1995) Cyanobacterial protein with similarity to the chlorophyll *a/b* binding proteins of higher plants: Evolution and regulation. Proc Natl Acad Sci USA 92: 636–640

Funk C and Vermaas W (1999) A cyanobacterial gene family coding for single-helix proteins resembling part of the light-harvesting proteins from higher plants. Biochemistry 34: 11133–11141

Funk C, Schröder WP, Green BR, Renger G and Andersson B (1994) The intrinsic 22 kDa protein is a chlorophyll-binding subunit of Photosystem II. FEBS Lett 342: 261–266

Funk C, Schröder WP, Napiwotzki A, Tjus SE, Renger G and Andersson B (1995a) The PS II-S protein of higher plants: A new type of pigment-binding protein. Biochemistry 34: 11133–11141

Funk C, Adamska I, Green BR, Andersson B and Renger G (1995b) The nuclear-encoded chlorophyll-binding Photosystem II-S protein is stable in the absence of pigments. J Biol Chem 270: 30141–30147

Ghanotakis DF and Yocum CF (1986) Purification and properties of an oxygen-evolving reaction center complex from Photosystem II membranes. FEBS Lett 197: 244–248

Ghanotakis DF, Demetriou DM and Yocum CF (1987a) Isolation and characterization of an oxygen-evolving Photosystem II reaction center core preparation and a 28 kDa Chl-*a*-binding protein. Biochim Biophys Acta 891: 15–21

Ghanotakis DF, Waggoner CM, Bowlby NR, Demetrious DM, Babcock GT and Yocum CF (1987b) Comparative structural and catalytic properties of oxygen-evolving Photosystem II preparations. Photosynth Res 14: 191–199

Gilmore AM (1997) Mechanistic aspects of xanthophyll cycle-dependent photoprotection in higher plant chloroplasts and leaves. Phys Plant 99: 197–209

Giuffra E, Cugini D, Croce R and Bassi R (1996) Reconstitution and pigment-binding properties of recombinant CP29. Eur J Biochem 238: 112–120

Green BR and Durnford DG (1996) The chlorophyll-carotenoid proteins of oxygenic photosynthesis. Annu Rev Plant Physiol Plant Mol Biol. 47: 685–714

Green BR and Kühlbrandt W (1995) Sequence conservation of light-harvesting and stress-response proteins in relation to the three-dimensional molecular structure of LHCII. Photosynth Res 44: 139–148

Green BR and Pichersky E (1994) Hypothesis for the evolution of three-helix Chl *a/b* and Chl *a/c* light-harvesting antenna proteins from two-helix and four-helix ancestors. Photosynth Res 39: 149–162

Green BR and Salter AH (1996) Light regulation of nuclear-encoded thylakoid proteins. In: Andersson, B, Salter AH, and Barber, J (eds) Molecular Genetics of Photosynthesis, pp 75–103. Oxford University Press, New York

Green BR, Pichersky E and Kloppstech K (1991) Chlorophyll *a/b*-binding proteins: An extended family. Trends Plant Sci 16: 181–186

Grimm B, Kruse E and Kloppstech K (1989) Transiently expressed early light-inducible thylakoid proteins share transmembrane domains with light-harvesting chlorophyll binding proteins. Plant Mol Biol 13: 583–593

Harrer R, Bassi R, Testi MG and Schäfer C (1998) Nearest-neighbor analysis of a Photosystem II complex from *Marchantia polymorpha* L. (liverwort), which contains reaction center and antenna proteins. Eur J Biochem 255: 196–205

Haworth P, Watson JL and Arntzen CJ (1983) The detection, isolation and characterization of a light-harvesting complex with is specifically associated with Photosystem I. Biochim Biophys Acta 724: 151–158

Heddad M and Adamska I (2000) Light stress-regulated two-helix proteins in *Arabidopsis thaliana* related to the chlorophyll *a/b*-binding gene family. Proc Natl Acad Sci USA 97: 3741–3746

Henrysson T, Ljungberg U, Franzen LG, Andersson B and Åkerlund HE (1987) Low molecular weight polypeptides in Photosystem II and protein dependent acceptor requirement for Photosystem II. In: Biggins, J (ed) Progress in Photosynthesis Research, Vol II, pp 125–128. Martinus Nijhoff Publishers, Dordrecht

Henrysson T, Schröder WP, Spangfort M and Åkerlund H-E (1989) Isolation and characterization of the chlorophyll *a/b* protein complex CP29 from spinach. Biochim Biophys Acta 977: 301–308

Hoffman NE, Pichersky E, Malik VS, Castresana C, Ko K, Darr SC and Cashmore AR (1987) A cDNA clone encoding a Photosystem I protein with homology to Photosystem II chlorophyll *a/b*-binding polypeptides. Proc Natl Acad Sci USA 84: 8844–8848

Horton P, Ruban AV and Walters RG (1996) Regulation of light harvesting in green plants. Annu Rev Plant Physiol Plant Mol Biol 47: 655–684

Hundal T, Virgin I, Styring S and Andersson B (1990) Changes in the organization of Photosystem II following light-induced D1-protein degradation. Biochim Biophys Acta 1017: 235–241

Ikeuchi M, Yuasa M and Inoue Y (1985) Simple and discrete isolation of an O_2-evolving PS II reaction center complex retaining Mn and the extrinsic 33 kDa protein. FEBS Lett 185: 316–322

Iwasaki T, Saito Y, Harada E, Kasai M, Shoji K, Miyao M and Yamamoto N (1997) Cloning of cDNA encoding the rice 22 kDA protein of Photosystem II (PS II-S) and analysis of light-induced expression of the gene. Gene 185: 223–229.

Jansson S (1994) The light-harvesting chlorophyll *a/b*-binding proteins. Biochim Biophys Acta 1184: 1–19

Jansson S (1999) A guide to the identification of the Lhc genes and their relatives in *Arabidopsis*. Trends Plant Sci 4: 236–240

Kaneko T, Sato S, Kotani H, Tanaka A, Asamizu E, Nakamura Y, Miyajima N, Hirosawa M, Sugiura M, Sasamoto S, Kimura T, Hosouchi T, Matsuno A, Muraki A, Nakazaki N, Naruo K, Okumura S, Shimpo S, Takeuchi C, Wada T, Watanabe A, Yamada M, Yasuda M and Tabata S (1996) Sequence analysis of the genome of the unicellular cyanobacterium *Synechocystis* sp. strain PCC6803. II. Sequence determination of the entire genome and assignment of potential protein-coding regions. DNA Res 3: 109–136

Karlin-Neuman GA, Sun L and Tobin EM (1988) Expression of light-harvesting Chl *a/b* protein genes is phytochrome-regulated in etiolated *Arabidopsis thaliana* seedlings. Plant Physiol 88: 1323–1331

Kim S, Sandusky P, Bowlby NR, Aebersold R, Green BR, Vlahakis S, Yocum CF and Pichersky E (1992) Characterization of a spinach *psb*S cDNA encoding the 22 kDa protein of Photosystem II. FEBS Lett 314: 67–71

Kim S, Pichersky E and Yocum CF (1994) Topological studies of spinach 22 kDa protein of Photosystem II. Biochim Biophys Acta 1188: 339–348

Kim SJ, Jansson S, Hoffman NE, Robinson C and Mant A (1999) Distinct 'assisted' and 'spontaneous' mechanisms for the insertion of polytopic chlorophyll-binding proteins into the thylakoid membrane. J Biol Chem 274: 4715–4721

Kleima FJ, Hobe S, Calkoen F, Urbanus ML, Peterman EJG, van Grondelle R, Paulsen H and van Amerongen H (1999) Decreasing the chlorophyll *a/b* ratio in reconstituted LHCII: Structural and functional consequences. Biochemistry 38: 6587–6596

Klösgen RB (1997) Protein transport into and across the thylakoid membrane. J Photochem Photobiol 38: 1–9

Krol M, Spangfort MD, Huner NPA, Öquist G, Gustafsson P and Jansson S (1995) Chlorophyll *a/b*-binding proteins, pigment conversions, and early light-induced proteins in a chlorophyll *b*-less barley mutant. Plant Physiol 70: 1242–1248

Kühlbrandt W, Wang DN and Fujiyoshi Y (1994) Atomic model of plant light-harvesting complex by electron crystallography.

Nature 367: 614–621

Li X, Henry R, Yuan J, Cline K and Hoffman NE (1995) A chloroplast homologue of the signal recognition particle subunit SPR54 is involved in the posttranslational integration of a protein into thylakoid membranes. Proc Natl Acad Sci USA 92: 3789–3793

Li XP, Björkman O, Shih C, Grossman AR, Rosenqvist M, Jansson S and Niyogi K (2000) A pigment-binding protein essential for regulation of photosynthetic light harvesting. Nature 403: 391–395

Lindahl M, Funk C, Webster J, Bingsmark S, Adamska I and Andersson B (1997) Expression of ELIPs and PsbS protein in spinach during acclimative reduction of the Photosystem II antenna in response to increased light intensities. Photosynth Res 54: 227–236

Ljungberg U, Åkerlund H-E, Larsson C and Andersson B (1984) Identification of polypeptides associated with the 23 and 33 kDa proteins of photosynthetic oxygen evolution. Biochim Biophys Acta 767: 145–152

Ljungberg U, Åkerlund H-E and Andersson B (1986) Isolation and characterization of the 10-kDa and 22-kDa polypeptides of higher plant Photosystem 2. Eur J Biochem 158: 477–482

Marrs KA and Kaufman LS (1989) Blue light regulation of transcription for nuclear genes in pea. Proc Natl Acad Sci USA 86: 4492–4495

Merrit S, Ernfors P, Ghanotakis DF and Yocum CF (1987) Binding of the 17 and 23 kDa water-soluble polypeptides to a highly-resolved PS II reaction center complex. In: Biggins, J (ed) Progress in Photosynthesis Research, Vol I, pp 689–692. Martinus Nijhoff Publishers, Dordrecht

Mishra RK and Ghanotakis DF (1993) Selective extraction of 22 kDa and 10 kDa polypeptides from Photosystem II without removal of 23 kDa and 17 kDa extrinsic proteins. Photosynth Res 36: 11–16

Moore M, Harrison MS, Peterson EC and Henry R (2000) Chloroplast Oxa1p homolog Albino3 is required for post-translational integration of the light harvesting chlorophyll-binding protein into thylakoid membranes. J Biol Chem 275: 1529–1532

Nield J, Orlova EV, Morris EP, Gowen B, van Heel M and Barber J (2000a) 3D map of the plant Photosystem two supercomplex obtained by cryoelectron microscopy and single particle analysis in higher plant PS II structure and membrane organisation. Nature Struct Biol 7: 44–47

Nield J, Funk C and Barber J (2000b) Supermolecular Structure of Photosystem Two and Localization of the PsbS Protein. Phil Trans R Soc Lond B 355: 1337–1344

Nilsson F, Andersson B and Jansson C (1990) Photosystem II characteristics of a constructed *Synechocystis* 6803 mutant lacking synthesis of the D1 polypeptide. Plant Mol Biol 14: 1051–1054

Ottander C, Campbell D and Öquist G (1995) Seasonal changes in Photosystem II organisation and pigment composition in *Pinus sylvestris.* Planta 197: 176–183

Pagano A, Cinque G and Bassi R (1998) In vitro reconstitution of the recombinant Photosystem II light harvesting complex CP24 (Lhcb6) and its spectroscopic characterization. J Biol Chem 273: 17154–17165

Paulsen H (1997) Pigment ligation to proteins of the photosynthetic apparatus in higher plants. Physiol Plant 100: 760–768

Paulsen H, Rümler U and Rüdiger W (1990) Reconstitution of pigment-containing complexes from light-harvesting chlorophyll *a/b*-binding protein overexpressed in *Escherichia coli.* Planta 181: 204–211

Peter GF and Thornber JP (1991) Biochemical composition and organization of higher plant Photosystem II light-harvesting pigment-proteins. J Biol Chem 266: 16745–16754

Pichersky E and Jansson S (1996) The light-harvesting chlorophyll *a/b*-binding polypeptides and their genes in angiosperm and gymnosperm species. In: Ort, DR and Yocum CF (eds) Oxygenic Photosynthesis: The Light Reactions, pp 507–521. Kluwer Academic Publishers, Dordrecht

Renger G (1997) Mechanistic and structural aspects of photosynthetic water oxidation. Physiol Plant 100: 828–841

Rhee KH (2001) Photosystem II: The solid structural E+a. Ann Rev Biophys Biomol Struct 30: 307–328

Robinson C and Klögen RB (1994) Targeting of proteins into and across the thylakoid membrane—a multitude of mechanisms. Plant Mol Biol 26: 15–24

Robinson C, Hynds PJ, Robinson D and Mant A (1998) Multiple pathways for the targeting of thylakoid proteins in chloroplasts. Plant Mol Biol 38: 209–221

Ros F, Bassi R and Paulsen H (1998) Pigment binding properties of the recombinant Photosystem II subunit CP26 reconstituted in vitro. Eur J Biochem 253: 653–658

Simpson DJ and Koetzel J (1996) Light-harvesting complexes of plants and algae: Introduction, survey and nomenclature. In: Ort, DR and Yocum, CF (eds) Oxygenic Photosynthesis: The Light Reactions, pp 493–506. Kluwer Academic Publishers, Dordrecht

Stockhaus J, Hofer M, Renger G, Westhoff P, Wydrzynski T and Willmitzer L (1990) Anti-sense RNA efficiently inhibits formation of the 10 kd polypeptide of Photosystem II in transgenic potato plants: Analysis of the role of the 10 kd protein. EMBO J 9: 3013–3021

Tang X-S and Satoh K (1985) The oxygen-evolving Photosystem II core complex FEBS Lett 179: 60–64

Thompson SJ, Kim SJ and Robinson C (1998) Sec-independent insertion of thylakoid membrane proteins: analysis of insertion forces and identification of a loop intermediate involving the signal peptide. J Biol Chem 273: 18979–18983

von Wettstein D, Gough S and Kannangara CG (1995) Chlorophyll biosynthesis. Plant Cell 7: 1039–1057

Wallbraun M, Kim S, Green BR, Piechulla B and Pichersky E (1994) Nucleotide sequence of a tomato psbS gene. Plant Physiol 106: 1703–1704

Wedel N, Klein R, Ljungberg U, Andersson B and Herrmann RG (1992) The single-copy gene psbS codes for a phylogenetically intriguing 22 kDa polypeptide of Photosystem II. FEBS Lett 314: 61–66

White MJ, Kaufman LS, Horwitz BA, Briggs WR and Thompson WF (1995) Individual members of the Cab gene family differ widely in fluence response. Plant Physiol 107: 161–165

Wyrich R, Dressen U, Brockmann S, Streubel M, Chang C, Qiang D, Paterson AH and Westhoff P (1998) The molecular basis of C4 photosynthesis in sorghum: isolation, characterization and RFLP mapping of mesophyll- and bundle-sheath-specific cDNAs obtained by differential screening. Plant Mol Biol 37: 319–335

Chapter 27

Redox Sensing of Photooxidative Stress and Acclimatory Mechanisms in Plants

Stanislaw Karpinski[1]*, Gunnar Wingsle[2], Barbara Karpinska[1] and Jan-Erik Hällgren[2]
[1]Department of Botany, Stockholm University, Lilla Frescativ. 5, Frescati, 106 91 Stockholm Sweden. [2]Department of Forest Genetics and Plant Physiology, Swedish University of Agricultural Sciences, 901 83 Umeå, Sweden

Summary

Unfavorable changes in light and temperature conditions can damage plant health and limit crop production. However, sensing of oxidative stress and subsequent regulation of acclimatory mechanisms in plants are poorly understood. This chapter discusses the pivotal role of reactive oxygen species (ROS) metabolism in the acclimatory mechanisms to photooxidative stress. Also, this chapter emphasizes the role of different molecules and photosynthetic electron transport in the regulation of acclimatory processes in plants. We demonstrate the effects of hydrogen peroxide and glutathione on acclimatory responses controlled by redox events in the proximity of the Photosystem II. By considering regulation of the expression of genes encoding enzymatic ROS-scavenging systems, we shall be able to make an evaluation of the role of ROS metabolism on the possible amelioration of acclimatory mechanisms in higher plants.

*Author for correspondence, email: Stanislaw.Karpinski@ botan.su.se

Eva-Mari Aro and Bertil Andersson (eds): Regulation of Photosynthesis, pp. 469–486.

I. Introduction

Land plants are sessile and have developed sophisticated mechanisms, which allow for both immediate and acclimatory responses to changing environments. In natural habitats plants can be exposed to rapid fluctuation in environmental conditions, for example, changes in light intensity and temperature, limitations in carbon dioxide (CO_2) availability, nutritional status and water supply. Large increases in light intensity for a short period can be beneficial for photosynthetic yields in low-light (LL) adapted plants (Long et al., 1994). However, if stress conditions persist, an imbalance can be created such that the energy absorbed through the light harvesting complex is in excess of that which can be dissipated or transduced by Photosystems II and I (PS II and PS I) or utilized by CO_2 assimilation (Karpinski et al., 1999). Under such an imbalance, excess excitation energy (EEE) can be generated by excess light (EL) alone or by chilling alone or both, that can be strongly enhanced by combination with many other abiotic and biotic factors.

Dissipation of EEE is an immediate response, which occurs through heat irradiation (Demmig-Adams and Adams, 1994; Krause, 1994). However, prolonged exposure to the conditions which cause EEE can lead to an increase in the generation of reactive oxygen species (ROS) such as singlet oxygen ($^1O_2$), superoxide anion ($O_2^{\cdot-}$), Haber-Weiss reaction (H_2O_2) and hydroxyl radical $OH^\cdot$ (Karpinski et al., 1997, 1999; Asada, 1999; Karpinska et al., 2000, Mullineaux et al., 2000). If accumulation of ROS under conditions of EEE exceeds the capacity of antioxidant systems to remove them, this can lead to irreversible photooxidative damage to the chloroplast and the cell. Thus over-production of ROS under EEE conditions can ultimately lead to permanent photodamage of leaf tissues, however the precise mechanisms are not understood (Fig. 1F).

Abbreviations: $^1O_2$ – singlet oxygen; APX – ascorbate peroxidase; AsA – ascorbate; CAT – catalase; cDNA – DNA complementary to mRNA; CDPK – Ca^{2+}-dependent protein kinase; Chl – chlorophyll; dAsA – dehydroascorbate; DHAR – dehydroascorbate reductase; EEE – excess excitation energy; EL – excess light; GOR – glutathione reductase; GPX – glutathione peroxidase; GSH – reduced glutathione; GSSG – oxidized glutathione; H_2O_2 – hydrogen peroxide; LL – low light; LT – low temperature; MAPKKK – mitogen-activated protein kinase kinase kinase; MDAR – monodehydroascorbate reductase; mdAsA – monodehydroascorbate; mRNA – messenger RNA; NO – nitric oxide; O_2 – dioxygen; $O_2^{\cdot-}$ – superoxide anion; $OH^\cdot$ – hydroxyl radical; POX – peroxidase; PQ – plastoquinone; PS I and II – Photosystem I and II; Q_A – primary quinone electron acceptor of PS II; Q_B – secondary quinone electron acceptor of PS II; RNA – ribonucleic acid; ROS – reactive oxygen species; SAA – systemic acquired acclimation; SOD – superoxide dismutase; V_{max} – maximal velocity

II. Stress and Acclimation

Plants can adjust to diurnal and seasonal changes in the light and temperature. Adaptation is regarded as a stable genotypic long-term response to seasonal changes in environment, while acclimation is seen as an environmentally induced short-term physiological response. Both processes cause phenotypic alterations with underlying physiological, biochemical and molecular changes. Not only higher plants but also mosses, lichens, algae and cyanobacteria have the ability to acclimate to EEE.

Many important crops like rice and maize, usually species of tropical or subtropical origin, exhibit poor tolerance to photooxidative stress generated by chilling and high light. They may fail to grow or are damaged by exposure to such conditions. It has been demonstrated recently, that redox changes in the proximity of primary or secondary quinone electron acceptor (Q_A and Q_B respectively) of PS II and plastoquinone (PQ), due to EEE, might be an environmental sensor, which control acclimatory mechanisms to photooxidative stress in plants (Escoubas et al., 1995, Maxwell et al., 1995; Karpinski et al., 1997, 1999; Huner et al., 1998; Pfannschmidt et al., 1999; Karpinska at al., 2000). Moreover, such mechanism controls a cellular and systemic induction of the genes encoding ROS-scavenging enzymes (Karpinski et al., 1997, 1999; Karpinska at al., 2000; Mullineaux et al., 2000).

A. Reactive Oxygen Species

Generation of ROS is an indispensable process for all aerobic organisms. Every living aerobic cell, under optimal conditions, has proper dynamic balance between ROS production and ROS utilization. However, in changing environmental conditions, for example during EEE, ROS production can be in excess and consequent oxidative stress appears.

The best known site for univalent reduction of O_2 to $O_2^{\cdot-}$ in a plant is the thylakoid-membrane-bound primary electron acceptor of PS I complex and of the peripheral reduced ferredoxin (Asada et al., 1974,

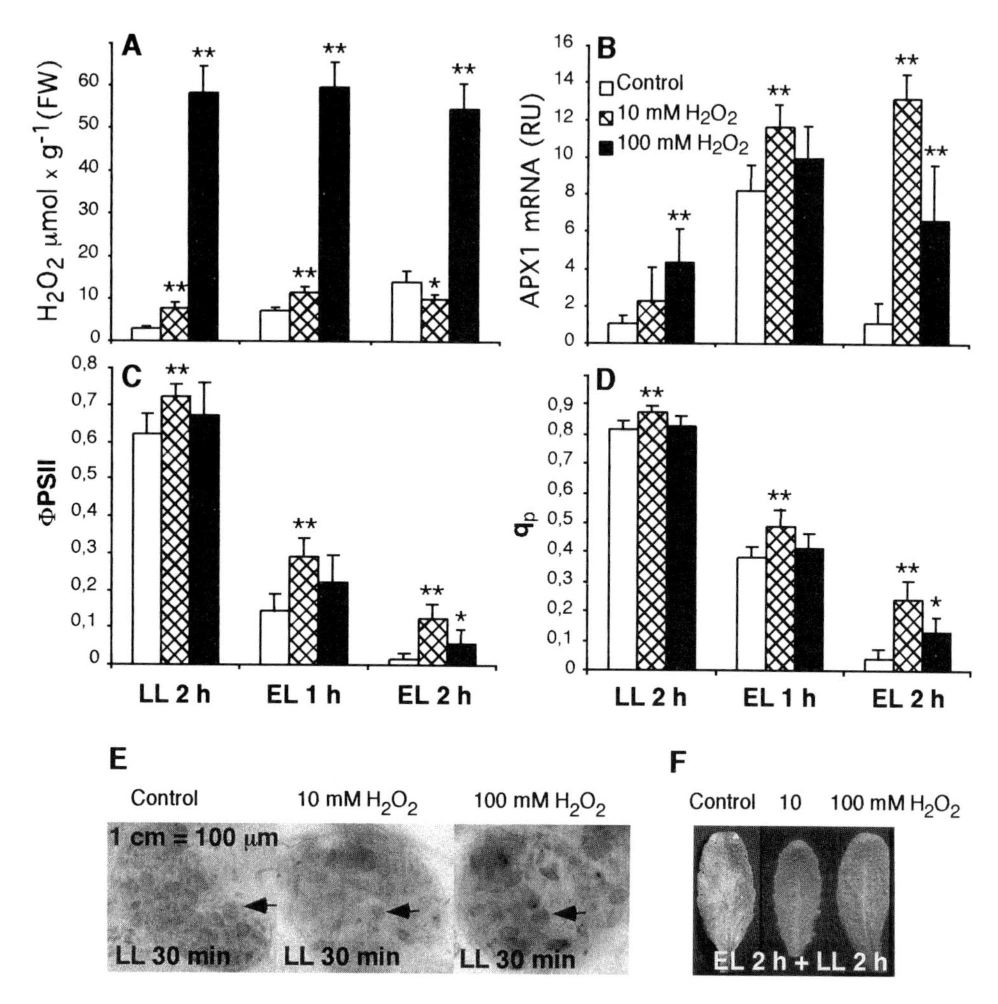

Fig. 1. The role of hydrogen peroxide (H_2O_2) in acclimation to excess excitation energy (EEE) in *Arabidopsis thaliana* leaf tissue (after Karpinska et al., 2000). (A) H_2O_2 levels in detached leaves treated with water (control), 10 mM H_2O_2 and 100 mM H_2O_2 for 2 h in low light (LL) and then exposed to excess light (EL), for 1 or 2 h. Parameters were measured in five different leaves obtained from two independent experiments, $n = 10 \pm$ SD. (B) APX1 mRNA levels – relative units (RU) in leaves treated as above. The values were obtained after scanning autoradiograms from slot blot hybridization experiments as described (Karpinski et al., 1997). Parameters were measured in three different pooled samples of leaves obtained from three independent experiments, $n = 3 \pm$ SD. (C and D) Photosynthetic electron transport efficiency in Photosystem II (ΦPS II) and photochemical quenching (q_p) in detached leaves treated as above. Parameters were measured in two different leaves obtained from each of three independent experiments, $n = 6 \pm$ SD. (E) 3,3-diaminobenzidine (DAB-HCl) staining of the *Arabidopsis* leaf tissue treated as above for 30 min in LL. Arrows indicate chloroplasts with strong DAB-HCl staining due to high H_2O_2 levels. Pictures shown are representative for three independent experiments. (F) Permanent photodamage and protection against photodamage in leaves treated as above and re-exposed for 2 h to LL. Picture shown is representative for five independent experiments ($n = 25$). Levels of significance were calculated from the ANOVA data (*$P < 0.05$, **$P < 0.01$). LL = 200 µmol of photons $m^{-2}\,s^{-1}$, EL = 2600 µmol of photons $m^{-2}\,s^{-1}$.

Asada, 1999). Under neutral pH range the rate of $O_2^{\cdot-}$ production is approximately 20 µmol mg $Chl^{-1}\,h^{-1}$ (Asada, 1999). It is not clear whether $O_2^{\cdot-}$ can also be produced at the PS II side (Chen et al., 1995). However, Cleland and Grace (1999) suggest that $O_2^{\cdot-}$ can be produced at PS II side. Most of the $O_2^{\cdot-}$ produced in the thylakoid membrane is converted to O_2 in O_2-mediated cyclic electron flow and by the non-catalytic dismutation to H_2O_2, before $O_2^{\cdot-}$ reaches the stroma space (Asada, 1999).

H_2O_2 is not a radical, by definition, since the two outer orbitals of the molecule are completely filled. It is the most stable of the ROS and can act as a reductant or as an oxidant (Salin 1987). The electron-transfer chains of chloroplasts, mitochondria and peroxisomes are well-documented sources of H_2O_2

(Cadenas 1989; Del Rio et al., 1992, Asada, 1999), and EEE has been observed to induce increased levels of the compound in LL acclimated *Arabidopsis thaliana* leaves (Karpinski et al., 1997, 1999). H_2O_2 has been reported to react with SH-groups. For instance, it has been shown to inhibit SH-enzymes in the Calvin cycle in isolated chloroplasts at relatively low (10 μM) concentrations (Kaiser 1979). It has been suggested that 10 to 12 mM H_2O_2 can induce cell death mechanisms in plant cell suspension culture (Levine et al., 1994; Lamb and Dixon 1997).

Contrary to the results above, we have found that treatment of intact *Arabidopsis* leaves with 10 or 100 mM H_2O_2 before EEE protects chloroplasts and the cell against EEE-induced photooxidative damage (Fig. 1A-F). Leaves treated with 100 mM H_2O_2 have higher concentrations of H_2O_2 in their cells and chloroplasts (Fig. 1A and E) than the concentration suggested to inhibit SH-enzymes in the Calvin cycle and induce cell death mechanisms in plants (Kaiser 1979; Levine et al., 1994; Lamb and Dixon 1997). Large increases in ascorbate peroxidase (*APX1*) mRNA levels were also induced by such treatments (Fig. 1B). Furthermore, leaves given H_2O_2 treatment were found to be more resistant to subsequent EL exposure. Photoinhibition of the photosynthetic electron transport developed more slowly in treated leaves (Figure 1C and D), and they showed no visual photodamage in their tissues (Fig. 1F). In contrast, exposing LL-acclimated control leaves for up to two hours to EL, caused faster decline of photosynthetic parameters and permanent photodamage (65% of leaves never recover from stress) of leaf tissue (Fig. 1C, D and F). H_2O_2 concentrations in such photo-damaged, control leaves were much lower than in 100 mM H_2O_2-treated leaves (Fig. 1A). Thus, it seems that higher H_2O_2 levels do not trigger EEE-induced permanent photodamage, and factors other than increased H_2O_2 production cause such damage (for example $^1O_2$, $O_2^{\cdot-}$, $OH^{\cdot}$). However, treatment of *Arabidopsis* leaves with 0.5 to 1 M H_2O_2 and subsequent exposure to EL accelerated permanent photodamage of leaf tissue. The lowest H_2O_2 concentration in our experimental system that trigger acclimation to EEE and caused significant changes in the chlorophyll *a* fluorescence parameters in LL was 5 mM (data not shown).

These results indicate that active chloroplasts and photosynthetic tissue from mature intact *Arabidopsis* leaves cultivated in short day have antioxidant capacities that are several orders of magnitude higher than those from in vitro cell cultures. Thus, in vitro experiments with isolated chloroplasts (Kaiser, 1979) and plant cell culture experimental systems, which have very low photosynthetic activity (Levine et al., 1994, Lamb and Dixon, 1997), have limited relevance to whole leaf stress physiology.

It is a well-known fact that H_2O_2 and $O_2^{\cdot-}$ can react together in biochemical systems to form $OH^{\cdot}$. Pure H_2O_2 and $O_2^{\cdot-}$ do not react at significant rates in vitro unless traces of iron or copper salts are present (Halliwell and Gutteridge, 1989, 1993). $O_2^{\cdot-}$ acts as a reductant of Fe(III) or Cu(II) and is a source of H_2O_2 (Halliwell and Gutteridge, 1989, 1993). Also other reductants such as semiquinone radicals, thiols and ascorbate (AsA) are able to reduce these metals. In addition, there are other metal-catalyzed reactions involving H_2O_2 that produce $OH^{\cdot}$. Thus, the cellular localization of metals and reductants such as thiols and AsA, and the site of production of both $O_2^{\cdot-}$ and H_2O_2, will determine the significance of the $OH^{\cdot}$ toxicity. No specific direct-scavengers of $OH^{\cdot}$ radicals have been found so far in any organism (Asada, 1999).

Other forms of ROS are the singlet species. Singlet chlorophyll (^{1}Chl*) is generated by light excitation. Carotenoid pigments appear to have a dual protective role quenching both ^{1}Chl* and $^1O_2$. The chloroplast membranes are particularly susceptible to $^1O_2$-induced lipid peroxidation since approximately 90% of the fatty acids of the thylakoid glycolipids, phospholipids and sulfolipids consist of the unsaturated fatty acid *[alpha]*-linolenate (Knox and Dodge, 1985).

Generally, EEE causes an increase in ROS levels and induces oxidative stress in plants. However, the precise mechanisms remain to be established. The roles of many other free radicals such as the AsA-, monodehydroascorbate (mdAsA)-, reduced glutathione (GSH)-, thiyl-radical and other organic radicals and peroxides, in relation to damage and acclimation to EEE are still not understood.

B. Antioxidants

During optimal conditions, a small portion of the total O_2 is converted to ROS (Asada, 1999). Plants have developed different types of protection mechanisms against oxidative stress. Protective mechanisms can be divided into two separate categories, those involved in removing ROS and those involved in reducing ROS production. The relative roles of these two processes are not known today and they may differ between plant species and cultivates. For example, plants can use O_2 'safely,' without releasing any ROS (e.g. as a substrate in respiration and the oxygenase reaction). Generally, defense systems against ROS in plant cells are a net result of suppression mechanisms, scavenging and repair systems. These systems interact to protect plant cell molecules and compartments against oxidation. They are active at different physiological stages of the cell and under both aerobic and anaerobic conditions. However, if oxidative stresses approach the maximum capacity of the defense system, ROS can initiate a cascade of cellular damage leading to cell death.

Higher plants contain numerous enzymatic and non-enzymatic ROS-scavengers and antioxidants, both water- and lipid-soluble, localized in different cellular compartments (Larsson 1988; Dalton 1995; Wise 1995, Noctor and Foyer, 1998; Asada, 1999). Non-enzymatic antioxidants include pigments, GSH, AsA, vitamin E and many others not discussed in this chapter. The membrane-localized antioxidants mainly scavenge $^1O_2$, while water-soluble antioxidants scavenge $O_2^{\cdot-}$ and H_2O_2. Dalton (1995), Noctor and Foyer (1998) and Asada (1999) have recently reviewed the interactions between different ROS-scavengers and antioxidants.

1. Non-enzymatic Antioxidants

One of the most acknowledged antioxidants in biological systems is α-tocopherol (Larsson 1988; Hess, 1993; Polle and Rennenberg, 1994) with main location within the chloroplast. α-Tocopherol is the most abundant tocopherol of the four forms found in plants (α-,β-,γ-, δ-tocopherol). This lipid-soluble vitamin functions as a ROS scavenger and plays an important role in protecting and maintaining the integrity of cell membranes, especially in the chloroplasts (Tappel, 1972). During scavenging of ROS α-tocopherol is oxidized to α-chromanoxyl radicals that can be regenerated by ascorbate or glutathione.

Glutathione is the most abundant thiol in higher plants (Foyer and Halliwell, 1976; Foyer, 1997; Mullineaux and Creissen, 1997; Noctor and Foyer, 1998; Asada, 1999; Wingsle et al., 1999). Plants have normally low oxidized glutathione (GSSG) level, for example, in Scots pine it is approximately 20-fold lower than the GSH content (Wingsle et al., 1989). Many environmental factors have been shown to change the ratio or redox status of glutathione [GSH/(GSSG +GSH)] in different organisms (Huerta and Murphym, 1989; Gilbert, 1990; O'Kane et al., 1996; Foyer, 1997; Karpinski et al., 1997; Noctor and Foyer, 1998; Wingsle et al., 1999), and an accumulation of GSSG can be an indicator of an oxidative stress (Smith et al., 1990).

The pivotal roles of AsA and dehydroascorbate (dAsA) in several physiological processes in plants and mammalian cells have been thoroughly reviewed (Arrigoni, 1994; Dalton, 1995; Polle, 1997; Noctor and Foyer, 1998; Asada, 1999; Wingsle et al., 1999). The roles of AsA as an ROS scavenger in the aqueous phase of cells and as a co-factor in structural protein organization are well known. The redox status of AsA in different tissues has been shown to be an indicator of oxidative stress in plants (Sgherri et al., 1994; Luwe and Heber, 1995). Another antioxidant, chlorogenic acid, has also been suggested to play an important role in the scavenging of ROS during EEE (Grace et al., 1998).

2. Enzymatic Antioxidants

The enzymatic ROS-scavenging system in the chloroplast consists of several enzymes like superoxide dismutase (SOD), APX, monodehydro ascorbate reductase (MDAR), dehydroascorbare reductase (DHAR), glutathione peroxidase (GPX) and glutathione reductase (GOR) (Foyer, 1997; Mullineaux and Creissen, 1997, Mullineaux et al., 1998; Noctor and Foyer, 1998; Asada, 1999). Similarly, the defense system against ROS in the cytosol is proposed to consist of APX, SOD, CAT, GPX, GOR, MDAR and DHAR. ROS enzymatic scavenging system plays several major physiological roles. In the chloroplast, it participates in the generation of proton gradient across the thylakoid membrane thus allowing for an appreciable flux of the linear electron flow. It keeps the glutathione and ascorbate at a reduced state in the chloroplast, thus scavenging the excess of electrons. It minimizes $O_2^{\cdot-}$ and H_2O_2 levels thus preventing a metal-catalyzed Haber-Weiss reaction, which forms $OH^{\cdot}$ radicals, and thus protects macromolecular complexes in the chloroplast. It is suggested that 10 to 30% of all generated electrons in the chloroplast lumen flows through $O_2^{\cdot -}$ and H_2O_2 and are captured again within the water molecule by the ROS-enzymatic scavenging system in the chloroplast stroma (Asada, 1999). However, other results suggest that Mehler reaction play a minor role in dissipating of EEE, and photorespiration is far more important component in this process (Badger et al., 2000).

SOD has evolved to be one of the enzymes with the fastest function ($V_{max} = 2 \times 10^9\ M^{-1}\ s^{-1}$) with an optimum close to the diffusion rate of $O_2^{\cdot -}$. SOD converts $O_2^{\cdot -}$ to H_2O_2 and constitutes the first link in the enzymatic scavenging system of ROS (McCord and Fridovich, 1969; Asada, 1999). Thirty years of research on this enzyme has shown that SOD plays an essential role in numerous physiological, biochemical and molecular processes in aerobic and anaerobic organisms. In humans and other animals, the involvement of SOD in the programming of neurone cell death has already been determined (Deng et al., 1993; Kane et al., 1993; Raff et al., 1993). Different SOD isoforms in plants are differentially expressed and also localized in different compartments within and outside the cell (Perl-Treves and Galun, 1991; Tsang et al., 1991; Wingsle et al., 1991; Bowler et al., 1992; Karpinski et al., 1992b,a, 1993; Streller and Wingsle, 1994; Wingsle and Karpinski, 1996; Schinkel et al., 1998). For example, different SOD isoforms are differentially expressed during the recovery from winter stress. A comparison of chloroplastic and cytosolic CuZn-*SOD* mRNA levels showed a four-fold higher transcript level for the chloroplastic form (Karpinski et al., 1993). This higher transcript level was accompanied with higher chloroplastic CuZn-SOD activity. Transcript levels for both chloroplastic and cytosolic CuZn-*SODs* were reduced after the repair process of the photosynthetic apparatus was completed and photosynthetic capacity had fully recovered from winter stress (Karpinski et al., 1993, 1994). In the same experiment (Karpinski et al., 1993), however, GOR enzyme activity was induced but the transcript levels of chloroplastic *GOR* gene were not altered. Later on it was demonstrated that GOR activity in Scots pine needles could be upregulated by redox intraconversion of the enzyme without any change in its mRNA and protein levels (Wingsle and Karpinski, 1996).

Are the levels of other members of the enzymatic ROS-scavenging system critical for the defense against EEE-induced oxidative stress? The general answer is yes. In the light, the key enzyme involved in H_2O_2 scavenging is APX, which catalyzes the reaction: $2\ AsA + H_2O_2 \rightarrow 2\ mdAsA + 2H_2O$. CAT can also scavenge H_2O_2, but it is a peroxisomal enzyme and does not require a reducing substrate. The role of CAT in the photorespiratory process, during EEE stress, was recently demonstrated and discussed (Willekens et al., 1998). The authors suggested that CAT is an important sink for H_2O_2, although their data did not indicate higher H_2O_2 levels in the *CAT*-antisense transgenic *Nicotiana* leaves, with reduced CAT activity however, higher glutathione levels were detected. A possible explanation of this result is that inhibition of the peroxisomal CAT might cause an inhibition or partial inactivation of the photorespiratory processes.

Chloroplasts photoregenerate AsA from mdAsA or dAsA. Reduced ferredoxin or NAD(P)H with MDAR converts mdAsA to AsA. DHAR is thought to regenerate AsA using GSH as an electron donor (Noctor and Foyer, 1998). In plants and animals, GPX has generated much attention as an important enzyme in scavenging H_2O_2 and the products of lipid peroxidation. Recently a plastidic GPX has been identified (Mullineaux et al., 1998). This indicates that enzymatic oxidation of GSH to GSSG in the chloroplast can also occur without an involvement of DHAR, and that glutathione and ascorbate cycles can also function independently in the chloroplast.

Another pertinent question is whether the expression of genes encoding different enzymes of the ROS-scavenging system is co-regulated during EEE-induced photooxidative stress? There is no clear answer to this question. Different plant species and different plant organs and tissues may differ in the regulation of expression of genes encoding these enzymes. Moreover, expression of genes encoding different isoforms of the same ROS-scavenging enzyme (e.g. SOD) may be regulated differently in response, for example, to low-temperature- and EL-induced photooxidative stress (Bowler et al., 1992; Karpinski et al., 1993). Surprisingly, only *APX1* and *APX2* genes, which encode cytosolic isoforms of this enzyme (but not SOD), were induced during EEE stress in *Arabidopsis* leaves (Karpinski et al., 1997, 1999; Karpinska et al., 2000). Moreover, transcript levels for stromal *APX4* and thylakoid-bound *APX5* do not change during EEE (S. Karpinski, unpublished). These results may suggest that in different plant species different acclimatory responses have evolved.

C. Photosynthesis during Photooxidative Stress

It is well known that photosynthesis generates ROS. There are several excellent reviews of ROS in photosynthesis, covering the role of oxygen in photoinhibition (Krause, 1994), antioxidant metabolism (Foyer, 1997) and chilling stress (Baker, 1994; Wise, 1995). Asada, (1999) has described the production and the role of ROS in the water-water cycle in the chloroplasts. Horton and co-workers (1996) have comprehensively described a biophysical mechanism for the regulation of light harvesting and energy dissipation in PS II, where the Δ-pH across thylakoid membranes controls the energy dissipation.

Low temperature and high light induce photooxidative stress, which is manifested as O_2-dependent bleaching of photosynthetic pigments. Chlorophyll bleaching in conifers is much greater in sun-exposed than shaded habitats (Karpinski et al., 1994). In Scots pine the chlorophyll concentration is lower during winter, and the carotenoid levels remain equal, or slightly increase. At the end of winter, when the quantum flux density is relatively high, the pigment levels are lowest (Linder, 1972; Karpinski et al., 1994). This coincides with a very low PS II efficiency (Strand and Öquist, 1985; Lundmark et al., 1988; Karpinski et al., 1994; Ottander et al., 1995). It has been suggested that the reorganization of the photosynthetic apparatus during fall and winter allows Scots pine to maintain a large fraction of chlorophyll in a quenched, photo-protected state (Öquist et al., 1992), allowing rapid recovery of photosynthesis in the spring (Lundmark et al., 1988, Karpinski et al., 1993, 1994; Ottander et al., 1995).

The PS II reaction center is generally described as the most sensitive part of the photosynthetic apparatus when plants are subjected to EEE-induced photooxidative stress. So-called 'acceptor- and donor-side photoinhibition' which is associated with proteolytic degradation of the D1 protein and induction of photooxidative stress can ultimately lead to photodamage of leaf tissue (Fig. 1) (Aro et al., 1993; Barber 1995, Van Wijk et al., 1997). Photoinactivation of Photosystem I (PS I) has been reported to be more severe than that of PS II during chilling stress (Sonoike, 1996). Chilling and EL lead to overproduction of ROS through EEE. In the chloroplast, light reactions will continue, while the energy consuming biochemical reactions are more limited in LT.

Chloroplasts subjected to EEE have a number of mechanisms to dissipate excess energy and thereby reduce the generation of ROS. Leaf and chloroplast movements, increased energy dissipation by various mechanisms and an increased use of reducing equivalents in other chloroplast processes are examples of such mechanisms. Increased energy dissipation can be achieved by number of mechanisms: 1) increasing the photorespiratory activity (Kozaki and Takeba, 1996), 2) by the Mehler-peroxidase reaction (Schreiber and Neubauer, 1990; Asada, 1999; Badger et al., 2000), 3) by an increased conversion of absorbed light into heat. The thermal dissipation process occurs within the antenna and the violaxanthin (V) + antheraxantin (A) + zeaxanthin (Z) (VAZ) cycle is suggested to play a major role (Demmig-Adams and Adams III, 1994; Horton et al., 1996). An increased use of anabolic pathways in the stroma also limits ROS production (Wise, 1995).

Two different strategies seem to have evolved in relation to adaptation of photosynthesis to photooxidative stress in over-wintering plants. One of them is to maintain photosynthetic capacity throughout the winter by different adjustments in the photosynthetic apparatus, and the other is to photosynthesize during warm periods and down-regulate photosynthesis during winter. In over-wintering cereals, positive correlation between photosynthetic capacity at low temperature and freezing tolerance results from photosynthesis providing energy for cellular metabolism (Huner et al., 1998). Cold acclimation, which induces significant freezing resistance of winter cereals and spinach, also increases resistance to photoinhibition. In contrast, cold acclimation of spring cereals lacking the ability to frost harden, does not induce increased resistance to photoinhibition (Somersalo and Krause, 1989; Hurry and Huner, 1992; Öquist et al., 1992). The temperature-induced resistance to photoinhibition in winter rye reflects adjustments of PS II to EEE (Gray et al., 1996).

In Scots pine cold acclimation does not affect the susceptibility of photosynthesis to photoinhibition, but there is a distinct increase in resistance to photoinhibition at the level of PS II reaction centers, limiting photodamage despite suppression of the capacity for photosynthesis (Krivosheeva et al., 1996). This is demonstrated in Fig. 2, where the photochemical quenching (q_p) is plotted as a function of photoinhibition of photosynthesis in non-acclimated and cold-acclimated Scots pine needles. This

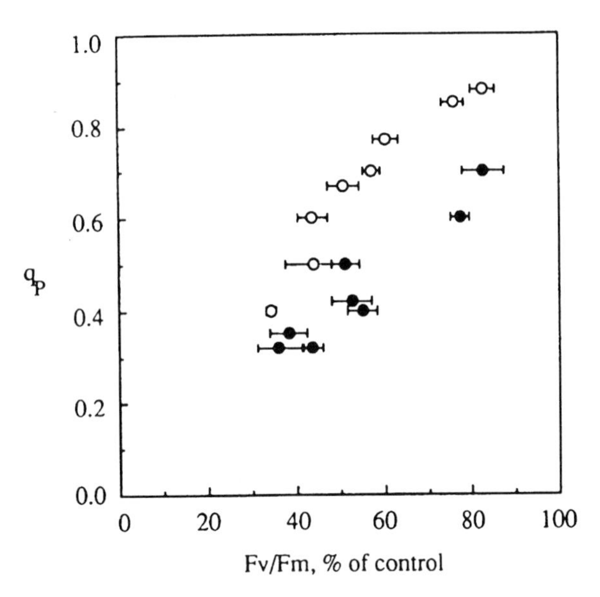

Fig. 2. Photochemical quenching (q_P), as the function of photoinhibition of photosynthesis (decrease in Fv/Fm) in non-acclimated (open symbols) and cold-acclimated (closed symbols) Scots pine needles. Measurements were done after 8 h exposure of the needles to 400, 600, 800, 1000 and 1200 μmol of photons m^{-2} s^{-1} at 20 °C. Each value represents a mean ±SE, $n = 6$ (after Krivosheeva et al., 1996).

experiment was designed to generate photoinhibition under the same PS II excitation pressures, for example at similar values of $1 - q_p$. Clearly, under the same excitation pressures of PS II, needles of cold-acclimated Scots pine were much more resistant to photoinhibition than needles of non-hardened pine. Unlike the winter varieties of rye and wheat, which respond to cold acclimation by increased capacities for photosynthesis, seedlings of Scots pine respond to cold acclimation by a 25% inhibition of photosynthesis over the studied range of absorbed photon flux density. Increased activities and levels of several enzymes and metabolites of the enzymatic ROS-scavenging system (Krivosheeva et al., 1996) accompanied this.

Two major oxygen-consuming reactions are associated with photosynthesis, the rubisco oxygenase reaction leading to photorespiration and donation of electrons to O_2 to form $O_2^{\cdot -}$ in a pseudocyclic electron flow in PS I proximity. This last reaction leads to formation of H_2O_2 in dismutation reaction with proton (Mehler reaction). Several studies have emphasized the significance of electron transfer to O_2 (Osmond and Grace, 1995, Asada, 1999). Biehler and Fock (1996) showed that photorespiration increased in wheat during drought stress when the availability of CO_2 was limited. The reaction of AsA with H_2O_2 is efficient and there is accumulating evidence that the Mehler reaction also serves as an important sink for excess electrons (Asada, 1999), although the relative roles of these two processes in dissipating EEE are still not well understood. Krivosheeva and co-workers (1996) hypothesize that the H_2O_2-scavenging system has two roles in protection of cold acclimated needles from photoinhibition: 1) protection from ROS formed upon excessive excitation in general, 2) allows O_2 to function as an electron acceptor, thus opening a fraction of Photosystem II reaction centers and consuming electrons in excess of the requirements of CO_2-fixation.

Protection against EEE via the photorespiratory pathway in LT is theoretically possible. However, to our knowledge, no data can be found in the literature to support the hypothesis that increased photorespiration during LT stress would protect the photosynthetic machinery in plant leaves. The argument is that LT generally slows down the enzymatic reactions more than the photosynthetic electron transport rate, and hence photorespiration would not serve as an effective protective mechanism. Krause (1994) and Kozaki and Takeba (1996) have comprehensively described the role of oxygen in photoinhibition of photosynthesis and the possible protective role of photorespiration during stress. This protective role cannot be explained simply in quantitative terms by energy dissipation. Heber and co-workers (Wu et al., 1991) argued that the limited rate of coupled electron flow facilitated by photorespiration protects the photosynthetic machinery in two ways: firstly, by maintaining the Q_A, in a partly oxidized state and secondly, by building up a high proton gradient over the thylakoid membrane. This highly energized state of the thylakoid membrane may be dependent on cyclic electron flow involving PS I (Wu et al., 1991, Asada, 1999). Acidification of the thylakoid lumen is supposed to be a control mechanism in PS II. It will lead to an increased dissipation of excitation energy via chlorophyll-fluorescence and this energy-dependent quenching mechanism is known to protect against photoinhibition (Krause and Weis, 1991; Horton et al., 1996).

Relative roles of avoidance mechanisms and antioxidant systems remain to be established for plants subjected to unfavorable temperatures and light. On the basis of data presented above, it can be concluded that the ability of plants to adjust their defense systems against EEE- and LT-induced photooxidative stress depends on a number of factors that should be considered in further studies on possible future amelioration of plant photooxidative stress tolerance.

D. Signaling and Regulation of Acclimation to Photooxidative Stress

Many compounds have been assigned agents involved in signaling both in biotic and abiotic stress responses. These include salicylic acid, H_2O_2 (Chen et al., 1993; Levine et al., 1994; Prasad et al., 1994; Bi et al., 1995; Neuenschwander et al., 1995; Alvarez et al., 1998; Karpinski et al., 1999), NO (Delledonne et al., 1998), $O_2^{\cdot-}$ (Doke, 1983; Tsang et al., 1991, Jabs et al., 1996); GSH and GSSG (Wingate et al., 1988; Hérouart et al., 1993; Wingsle and Karpinski, 1996; Karpinski et al., 1997), Ca^{2+} (Price et al., 1994; Monroy and Dhindsa, 1995; Knight et al., 1991, 1996), photoreceptors with Ca^{2+} (Neuhaus et al., 1993; Millar et al., 1995); abscisic acid (Giraudat et al., 1994; Giraudat, 1995) and recently the redox status of Q_B and/or PQ pool (Escoubas et al., 1995; Maxwell et al., 1995; Karpinski et al., 1997, 1999; Huner et al., 1998; Karpinski et al., 1999; Pfannschmidt et al., 1999; Karpinska at al., 2000). However, very little is known about the signaling cascades, initiated by these molecules. ROS are known to be involved in the regulation of such diverse processes as the hypersensitive response and systemic acquired resistance (Lamb and Dixon, 1997); acclimation to chilling temperatures (Prasad et al., 1994); cross tolerance to different abiotic stresses (Bowler et al., 1992); regulation of the photosynthetic electron transport and systemic acquired acclimation (SAA) (Karpinski et al., 1999; Karpinska at al., 2000; Mullineaux et al., 2000). Specific transcription factors, which regulate gene expression in response to oxidative stress can be involved in regulation of stress responses and acclimatory mechanisms in plants (Lu et al., 1996; Cao et al., 1997; Dietrich et al., 1998; Kranz et al., 1998; Tamagnone et al., 1998).

Generally, Ca^{2+} is considered to function as a secondary messenger in plant oxidative stress response (Neuhaus et al., 1993; Price et al., 1994; Millar et al., 1995; Monroy and Dhindsa, 1995; Knight et al., 1996). Ca^{2+} channels have been suggested to function as temperature-sensors during cold acclimation (Ding and Pickard, 1993). Cytosolic Ca^{2+} levels increased transiently as a result of oxidative stress induced by cold shock (Knight et al., 1991, 1996). It has also been suggested that cold damage in plants may be due to lack of Ca^{2+} homeostasis and subsequent Ca^{2+} toxicity (Minorsky, 1985). However, more recent data suggest that changes in Ca^{2+} levels are a necessary step in sensing and signaling low temperatures in plants.

Two cDNAs for genes encoding calcium dependent protein kinases (CDPK) have been isolated from *Arabidopsis* (Urao et al., 1994, 1998). The gene for CDPK from pea is induced by cold stress. In animals phospholipase generates two secondary messengers, inositol 3-phosphate and diacylglycerol. Inositol 3-phosphate induces the release of Ca^{2+} in the cytoplasm, and it has been suggested that a similar mechanism could exist in plants and inositol 3-phosphate and Ca^{2+} could function as secondary messengers during drought and cold conditions. Monroy and Dhindsa (1995) demonstrated that elevated levels of Ca^{2+} are needed for initiation of the induction of genes involved in the cold acclimation process in alfalfa but were not enough to sustain this induction for a longer time. These results suggest that other signals are needed to complete cold acclimation. Later it was demonstrated that changes in the levels of Ca^{2+} can be detected within seconds of cold shock treatment (Knight et al., 1996). This immediate increase in the cytosolic free Ca^{2+} concentration was detected in both chilling-sensitive tobacco and chilling-tolerant *Arabidopsis*. Price and co-workers (1994) demonstrated that Ca^{2+} can regulate some enzymatic ROS-scavengers (e.g. SOD) due to oxidative stress.

Protein phosphorylation and dephosphorylation processes, catalyzed by protein kinases and phosphatases, respectively, play a key role in regulating many aspects of plant growth, such as development, metabolism and stress responses. Protein phosphorylation has been suggested to be involved in a signal transduction pathway regulating the response to oxidative stress. It has been shown that changes in the pattern of protein phosphorylation occur during cold acclimation of alfalfa cell suspension cultures (Monroy et al., 1993). Changes induced by low temperature stress occur in phosphorylations of existing proteins and were inhibited by Ca^{2+} channel blockers and by an antagonist of calmodulin and CDPK (Monroy et al., 1993). Several genes involved in the mitogen-activated protein kinase

kinase kinase (MAPKKK) cascade are induced by LT stress (Mizoguchi et al., 1996). The MAPKKK and adenosine-triphosphate kinase 19 (ATPK19) are induced in low temperature, suggesting that they might be involved in signal transduction pathways under low temperature stress (Mizoguchi et al., 1996). Other genes encoding factors involved in signal transduction, such as transcription factors (Kusano et al., 1995, 1998) and phospholipase C, are also induced by cold and drought and have been discussed by Shinozaki and Yamaguchi-Shinozaki (1996). It has been shown by Kusano et al., (1995) that the gene encoding a βZIP DNA-binding factor is induced by LT. This protein binds to the histone motif ACGTCA and to the promoter region of a cold inducible gene, which suggests that it may control some cold-inducible genes in chilling-sensitive plants. Our data suggest that induction of the *APX1* and *APX2* genes in *Arabidopsis*, due to EEE, depends on calcium and therefore could involve a kinase-mediated cascade (S. Karpinski, unpublished).

A regulatory role for H_2O_2 as a signaling molecule in different secondary messenger systems in humans and animals is well documented (Ramasarma, 1982; Meyer et al., 1993; Ginnpease and Whisler, 1996). In plants the ability to control H_2O_2, $O_2^{\cdot-}$ and GSH levels is an important factor in biotic and abiotic stress responses (Jabs et al., 1996; Karpinski et al., 1997, 1999; Creissen et al., 1999; Karpinska et al., 2000). Exogenous application of H_2O_2 or, an $O_2^{\cdot-}$-generating compound menadione, to maize seedlings, suggested that mild oxidative stress at 27 °C might induce chilling acclimation (Prasad et al., 1994). Both of these compounds caused an increase in *CAT3* and peroxidase (*POX*) transcript levels and protein activities but SOD activity remained constant in these experiments. However, the exogenously applied and endogenously accumulated H_2O_2 failed to increase the activity of these enzymes at 4 °C. Hence, higher levels of H_2O_2 that could cause cell death at a lower temperature induced cold-acclimation at a higher temperature. The authors suggest that mechanism by which H_2O_2 or menadione, induce *CAT3* and *POX* gene expression in maize is sensitive to chilling. Moreover, it was suggested that in non-acclimated seedlings, chilling injury may be partly due to the excess of ROS, which promoted oxidation of proteins and lipids. In chilling-acclimated seedlings, the enhanced enzymatic ROS-scavenging system prevented the accumulation of ROS and therefore prevented damage to lipids and proteins at 4 °C (Prasad et al., 1994; Prasad, 1996). In tobacco, increased mRNA levels for Fe-*SOD* were triggered by $O_2^{\cdot-}$ generated in the proximity of PS I due to paraquat treatment (Tsang et al., 1991). It was also shown that CuZn-*SOD4* and CuZn-*SOD4A* transcript level in maize increase in response to H_2O_2 treatment (Kernodle and Scandalios, 1996).

The most relevant functions of GSH in the context of oxidative stress are those where GSH participates in redox reactions and therefore GSSG is generated (Foyer and Halliwell, 1976). In plants, high concentrations of GSH, but not GSSG, enhanced the expression of genes encoding enzymes involved in phytoalexin and lignin biosynthesis and suggesting thus a general role for GSH in signaling systems in biological stress (Wingate et al., 1988). Different thiols, such as GSH, cysteine and dithiothreitol, increased the transcript level of a reporter gene under the control of the cytosolic CuZn-*SOD* promoter in transgenic tobacco protoplasts (Hérouart et al., 1993).

The regulatory impact of glutathione on the transcript levels of CuZn-*SOD* and *GOR*, genes has been described (Wingsle and Karpinski, 1996). Our results, that GSH reduced the cytosolic CuZn-*SOD* transcript level, are in agreement with findings for human CuZn-*SOD* and Mn-*SOD* genes, which were found to be down-regulated by thiols (Suzuki et al., 1993). However, they are in contrast to those reported by Hérouart et al., (1993). The difference in results may be due to the fact that different systems were used. It has been suggested that the levels of GSH and GSSG, or the redox state of the glutathione pool, play an important role in the in vivo regulation of the expression of genes encoding the enzymatic ROS-scavenging enzymes in plants. We concluded that the mechanisms regulating the expression of *SOD* and *GOR* genes respond differently to the altered levels of GSH and GSSG in Scots pine needles (Wingsle and Karpinski, 1996). The activity of GOR increased per se (but not the *GOR* transcript level) in response to increasing levels of GSSG, suggesting that the enzyme itself undergoes redox intraconversion in vivo. However, the transcript levels of cytosolic and chloroplastic CuZn-SOD were strongly lowered by GSH.

Recently, we have demonstrated that in *Arabidopsis* leaves treated with 10 mM GSH or GSSG, photosynthetic electron transport and *APX1* and *APX2* gene expression are inhibited (Karpinski et al., 1997; Karpinska et al., 2000) (Fig. 3A-D). Leaves treated with 10mM GSH or GSSG and incubated in LL for

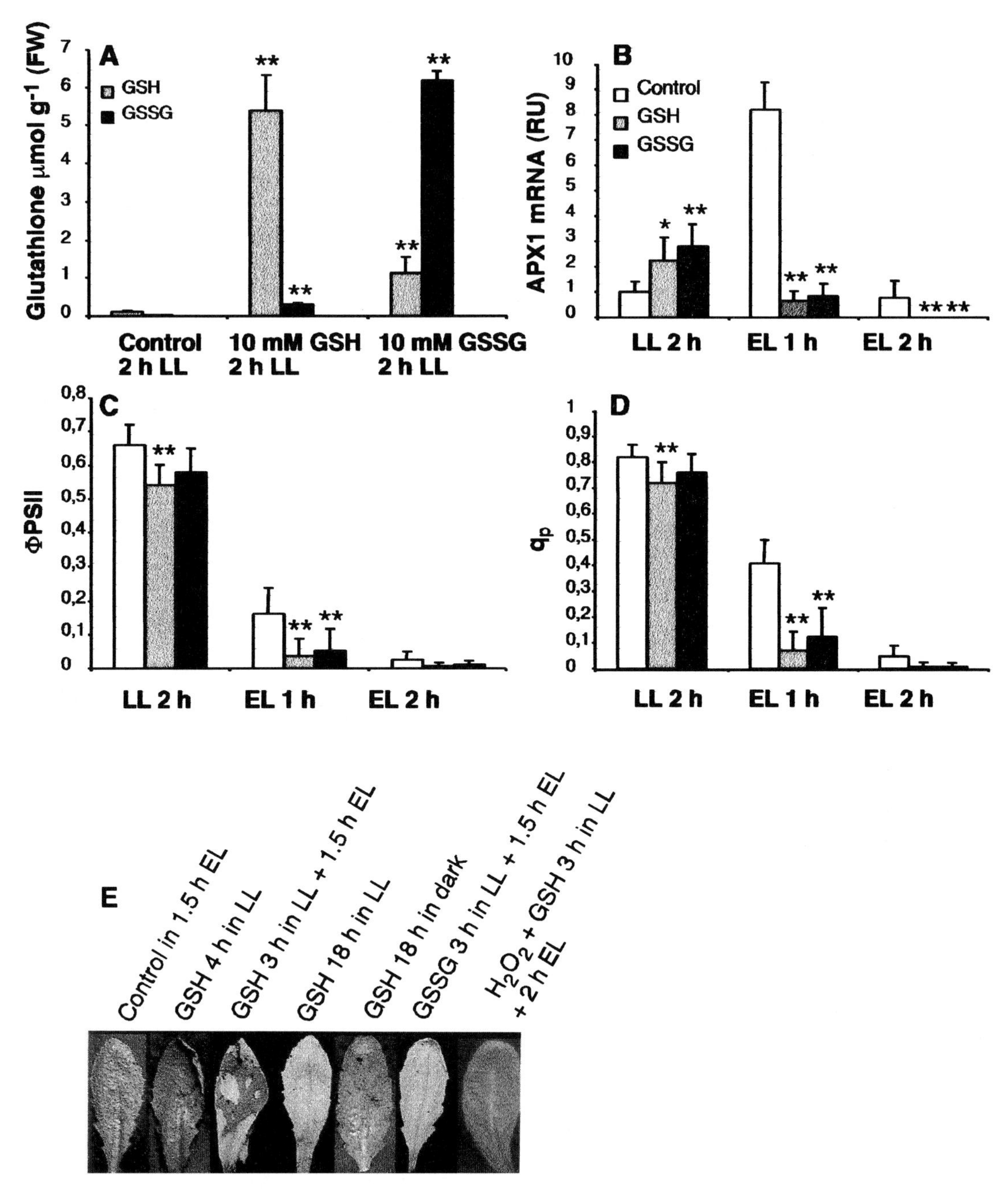

Fig. 3. The roles of reduced (GSH) and oxidized (GSSG) glutathione in acclimation to excess excitation energy (EEE) in *Arabidopsis thaliana* leaf tissue (after Karpinska et al., 2000). (A) GSH and GSSG levels in detached leaves treated with water (control), 10 mM GSH or 10 mM GSSG for 2 h (h) in LL. Parameters were measured in pooled leaf samples obtained from three independent experiments, $n = 6 \pm$ SD. (B) APX1 mRNA levels—relative units (RU)—in leaves treated as above and then exposed to EL for 1 or 2 h. The values were obtained after scanning autoradiograms from slot blot hybridization experiments as previously described (Karpinski et al., 1997). Parameters were measured in three different pooled samples of leaves obtained from three independent experiments, $n = 3 \pm$ SD). (C and D) Photosynthetic electron transport efficiency in Photosystem II (ΦPS II) and photochemical quenching (q_p) in detached leaves treated as above. Parameters were measured in six different leaves obtained from three independent experiments, $n = 6 \pm$ SD. (E) Permanent photodamage and protection against photodamage in leaves treated as described. In the combined 100 mM H_2O_2 and 10 mM GSH treatment, leaves were first treated with H_2O_2 for 1 h and then with GSH for 2 h, before being exposed to EL for 2 h. Picture shown is representative for five independent experiments. Levels of significance were calculated from the ANOVA data ($*P < 0.05$, $**P < 0.01$). LL = 200 μmol of photons m^{-2} s^{-1}, EL = 2600 μmol of photons m^{-2} s^{-1}.

18 h showed extensive photodamage (Fig. 3E). This damage could be accelerated after only 2 h incubation in GSH or GSSG followed by a 1.5 h EL treatment. (Fig. 3E). These effects were markedly reduced in combined treatments with 100 mM H_2O_2 followed by 10 mM GSH. Note that only small visual damage occurred in leaves incubated in the dark for 18 h with 10mM GSH (Fig. 3E).

The similar results of the GSH and GSSG treatments can be explained by the increase in GSH levels induced by both treatments (ten-fold in the GSSG treatment and 50-fold in GSH treatment, Fig. 3A). Inhibition of *APX1* and *APX2* (Karpinski et al., 1997) gene expression and photosynthetic electron transport in *Arabidopsis* (Fig. 3C and D) can be explained by highly reduced environment in the leaf cells, caused by excess of GSH. This leads to inhibition of the photosynthetic electron transport and redox signaling systems, and consequently accelerated photooxidative damage—bleaching of the chlorophyll (Karpinski et al., 1997; Åslund and Beckwith, 1999, Åslund et al., 1999; Creissen et al., 1999). Similar, negative regulatory effects and acceleration of permanent photodamage of leaf tissue after exposure to EL were observed in leaves treated with 25 mM AsA (Galley et al., 1996; S. Karpinski, unpublished)

1. Redox Control of Acclimation to Photooxidative Stress

We propose the following explanation of the observed phenomena (Figs. 1 and 3). The extent of reduction of the Q_A-Q_B-PQ pools, in light conditions, depends on many factors, including trans-thylakoid ΔpH and the electron transport efficiency in PS II (ΦPS II, Horton et al., 1996). ΦPS II reflects the functioning and turnover of the D1 protein. Photochemical quenching (q_p) indicates the extent of reduction of the primary electron acceptor, quinone A (Q_A), which is associated with the PS II complex. This parameter can also reflect, in specific conditions (e.g. following treatment with 2,5 dibromo-3-methyl-6-isopropyl-p-benzoquinone (DBMIB)), the extent of reduction of the Q_B and PQ pools. GSH/GSSG treatment increases the degree of Q_A reduction and decreases electron transport efficiency in PS II in LL (i.e. q_p and ΦPS II values decrease, Fig. 3D). This reduction is more pronounced in EL conditions, which increase the excitation pressure on PS II (Fig. 3A-D). In contrast, H_2O_2 treatment increases the oxidation of Q_A and electron transport efficiency in PS II in LL (i.e. q_p and ΦPS II values increase, Fig. 1 D). Moreover, the extent of Q_A reduction in EL conditions is lower than in control (H_2O-treated) leaves (Fig. 1D).

These results suggest that excess glutathione decreases, and excess H_2O_2 increases, the trans-thylakoid ΔpH gradient, thus H_2O_2 energizes and GSH de-energizes thylakoid membranes. Therefore, H_2O_2 and GSH may antagonistically regulate the redox status of the Q_A-Q_B-PQ pools and subsequent acclimatory responses to EEE (Karpinski et al., 1997, 1999, Karpinska et al., 2000) (Figs. 1 and 3). This can also explain the chlorophyll bleaching observed in glutathione treated leaves. This suggestion is supported by the results of treating leaves with excess GSH in the dark (Fig. 3E). Such leaves do not develop visual symptoms of chlorophyll bleaching, as do the leaves treated with GSH in LL. Furthermore, leaves treated with 100 mM H_2O_2, and then with 10 mM GSH, are also more resistant to EL than either control leaves or those treated solely with glutathione (Fig 3E). However, we have provided no evidence that GSH or GSSG was delivered to chloroplasts or leaf cells. Nevertheless, whether the effect of GSH is direct or indirect, the observations remain that excess of GSH has marked inhibitory effects on photosynthetic electron transport and acclimation to EEE (Fig. 3). Such inhibitory effects (developed in longer time) were also observed in leaves treated with 1 mM GSH (S. Karpinski, unpublished data). Concentration of the GSH in the active chloroplasts is estimated for 5 to 25 mM (Noctor and Foyer, 1998, Asada, 1999). We have also provided evidence that the H_2O_2-treatments easily penetrated to the chloroplast (Fig. 1E), and the H_2O_2 persisted in the leaf for the duration of the experiments (Fig. 1A), as previously demonstrated in other experimental systems (Prasad et al., 1994, Willekens et al., 1998).

2. Systemic Acquired Acclimation

Our recent work (Karpinski et al., 1997, 1999; Karpinska at al., 2000; Mullineaux et al., 2000) allows a unified and new view of acclimatory responses to any different fluctuating environmental conditions, which elicit EEE and subsequent redox changes of Q_A-Q_B or PQ pools. When a leaf experiences a stress condition like EL, one of the many cellular responses is an induction of antioxidant defenses, controlled at least in part by the redox status of the Q_A-Q_B or PQ pool or both (Karpinski et

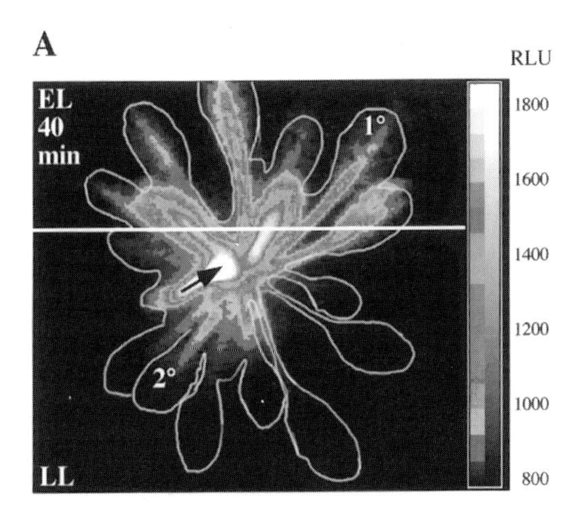

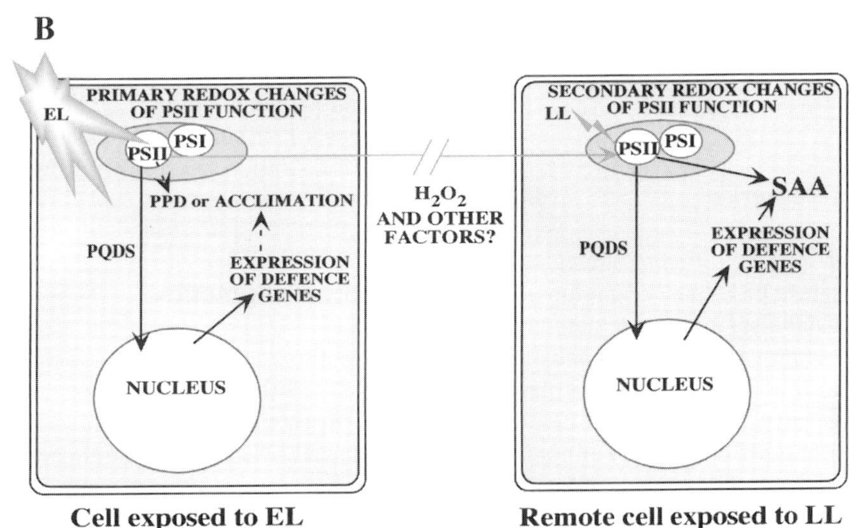

Fig. 4. Systemic induction of *APX2-LUC* expression in transgenic *Arabidopsis* leaf tissue and the scheme illustrating the systemic acquired acclimation (SAA) mechanism. (A) Image of luciferase activity in relative light units (RLU). A part of the whole rosette (as shown) exposed to EL for 40 min, the arrow indicates the apical region of the rosette. A typical primary (1°) EL-exposed leaf and secondary (2°) LL-exposed leaf are shown (after Karpinski et al., 1999). (B) EL-induced permanent photodamage (PPD) and the SAA mechanism in plants. Induction of expression of the defense genes in the cell is controlled by plastoquinone dependent signaling (PQDS). SAA = systemic acquired acclimation at 200 μmol of photons m^{-2} s^{-1}, EL = 2500 μmol of photons m^{-2} s^{-1}, LL = 200 μmol of photons m^{-2} s^{-1}. (See also Color Plate 7.)

al., 1997, 1999; Karpinska et al., 2000) (Fig. 4). However, cells suffering this stress also produce a systemic signal, a component of which is H_2O_2, which sets up an acclimatory response to EEE and consequent photooxidative stress in unstressed regions of the plant (Karpinski et al., 1999). Furthermore, given that changes in the photosynthetic parameters (e.g. F_v/F_m and q_p) have been observed in leaves not directly exposed to EL (2° leaves) (Karpinski et al., 1999), we suggest that a systemic signal, putatively mediated by H_2O_2 or GSSG (S. Karpinski, unpublished results) or both can promote redox changes in the proximity of PS II in non-stressed chloroplasts, thus inducing protective mechanisms in remote chloroplasts and cells. The summary of the SAA is presented (Fig. 4).

Recently it has also been demonstrated that the redox status of the Q_A-Q_B or PQ pool or both controls the transcriptional activity of the chloroplast-encoded genes and consequently regulates stoichiometric balance between the PS II and PS I in different light conditions (Pfannschmidt et al., 1999). It has also been reported that this redox status controls the transcriptional activity of nuclear genes encoding LHCB proteins (Escoubas et al., 1995; Maxwell et al 1995). All these results together strongly suggest that redox status of the Q_A-Q_B or PQ pool or both is a key redox sensing mechanism controlling plant acclimatory responses to EEE and consequent photooxidative stress.

The network of signaling pathways regulating expression of genes encoding the enzymatic ROS-scavenging system in plant cells is complex. One gene can be regulated by more than one signaling pathway. Interactions between different signaling pathways and the role of redox status of the glutathione or the ascorbate or both in stress sensing mechanisms are not understood. However, the redox sensing mechanisms of the oxidative stress response seem to play a primary role in regulation of the acclimatory responses in plants.

III. Concluding Remarks

In different plant species, different strategies have evolved to acclimate to EEE-induced photooxidative stress. An understanding of these strategies may pave the way to create improved stress tolerance in some plants. The signal transduction pathways that link day length, light intensity or temperature perception to increased or decreased contents of antioxidants, phytohormones and subsequent gene activation or deactivation mechanisms, remain to be elucidated. To increase the understanding of the photooxidative-stress responses induced by EEE in plants, we have also to pinpoint the subcellular compartments and processes, which initiate the specific signaling cascades regulating acclimation to EEE and photooxidative stress.

At present, it is difficult to understand how H_2O_2 can induce such diverse processes in plants as the hypersensitive response, systemic acquired resistance and SAA. Also it is difficult to understand how excess of glutathione or ascorbate can generate negative regulatory effects on acclimation mechanisms to EEE-induced photooxidative stress (Fig. 4). During the last decade, many researchers have tried to improve oxidative stress tolerance in plants by increasing their antioxidant capacity. Ironically, some transgenic plants with higher than normal GSH synthesis capacity, successfully generated in these trials, were impaired in development and were more susceptible to EEE and photooxidative stress than the wild-type parental strains (Creissen et al., 1999). Our and others results strongly suggest that transiently higher H_2O_2 levels in the chloroplast or cytosol prior stress may prove to be much more important for the amelioration of photooxidative stress tolerance in plants than simply permanently higher levels of antioxidants.

We have also to identify and isolate regulatory genes and the redox-activated transcription factors involved in acclimatory processes. Isolation of such regulatory genes controlling biotic stress responses (Cao et al., 1997; Dietrich et al., 1997; Shirasu et al., 1999) provided a great potential for manipulation of systemic acquired resistance in plants. Similarly, regulatory genes for SAA should be detected and isolated (Karpinski et al., 1999). Our recent unpublished work on isolation of *Arabidopsis* mutants impaired in regulation of *APX2* due to EEE strongly suggests that such genes exist (Mullineaux et al., 2000; L. Ball, S. Karpinski, P. Mullineaux, unpublished).

Acknowledgments

The authors acknowledge support from grants to SU and SUAS from the Swedish Councils: for Forestry and Agricultural Research (SJFR), for Natural Sciences (NFR) and for Strategic Research (SFF).

References

Alvarez ME, Pennell RI, Meijer PJ, Ishikawa A, Dixon RA and Lamb C (1998) Reactive oxygen intermediates mediate a systemic signal network in the establishment of plant immunity. Cell 92: 773–784

Aro EM, Virgin I and Andersson B (1993) Photoinhibition of Photosystem II. Inactivation, protein damage and turnover. Biochim Biophys Acta 1143: 113–134

Arrigoni O (1994) Ascorbate system in plant development. J Bioenerg Biomembr 26: 407–419

Asada K (1999) The water-water cycle in chloroplasts: Scavenging of active oxygens and dissipation of excess photons. Annu Rev Plant Physiol Plant Mol Biol 50: 601—639

Asada K, Kiso K and Yoshikawa K (1974) Univalent reduction of molecular oxygen by spinach chloroplasts on illumination. J Biol Chem 249: 2175–2179

Åslund F and Beckwith J (1999) Bridge over troubled water: Sensing stress by disulfite bond formation. Cell 96: 751–753

Åslund F, Zheng M, Beckwith J and Storz G (1999) Regulation of OxyR transcription factor by hydrogen peroxide and the cellular thiol-disulfite status. Proc Natl Acad Sci USA 96:

6161–6165

Badger MR, von Caemmerer S, Ruuska S and Nakano H (2000) Electron flow to oxygen in higher plants and algae: Rates and control of direct photoreduction (Mehler reaction) and rubisco oxygenase. Phil Trans R Soc Lond B 355: 1433–1446

Baker N (1994) Chilling stress and photosynthesis. In: Foyer CH and Mullineaux PM (eds) Causes of Photooxidative Stress and Amelioration of Defence Systems in Plants, pp 127–154. CRC Press, Boca Raton

Barber J (1995) Molecular basis of the vulnerability of Photosystem II to damage by light. Aust J Plant Physiol 22: 201–208

Bi YM, Kenton P, Mur L, Darby R and Draper J (1995) Hydrogen peroxide does not function downstream of salicylic acid in the induction of PR protein expression. Plant J 8: 235–245

Biehler K and Fock H (1996) Evidence for the contribution of the Mehler-peroxidase reaction in dissipating excess electrons in drought-stressed wheat. Plant Physiol 112: 265–272

Bowler C, Van Montagu M and Inzé D (1992) Superoxide dismutase and stress tolerance. Ann Rev Plant Physiol Plant Mol Biol 43: 83–116

Cadenas E (1989) Biochemistry of oxygen toxicity. Annu Rev Biochem 58: 79–110

Cao H, Glazebrook J, Clarke JD, Volko S, and Dong XN (1997) The *Arabidopsis NPR1* gene that controls systemic acquired resistance encodes a novel protein containing ankyrin repeats. Cell 88: 57—63

Chen GX, Blubaugh DJ, Hormann PH, Golbeck JH and Cheniae GM (1995) Superoxide contributes to the rapid inactivation of specific secondary donors of the Photosystem II reaction centre during photodamage of manganese-depleted Photosystem II membranes. Biochemistry 34: 2317–2332

Chen Z, Silva H and Klessig DF (1993) Active oxygen species in the induction of plant systematic acquired resistance by salicylic acid. Science 262: 1883–1885

Cleland RE and Grace SG (1999). Voltametric detection of superoxide production by Photosystem II. FEBS Letts 475: 384–352

Creissen G, Firmin J, Fryer M, Kular B, Leyland N, Reynolds H, Pastori G, Wellburn F, Baker N, Wellburn A and Mullineaux P (1999) Elevated glutathione biosynthetic capacity in the chloroplasts of transgenic tobacco plants paradoxically causes increased oxidative stress. Plant Cell 11: 1277–1291

Dalton DA (1995) Antioxidant defences of plants and fungi. In: Ahman S (ed) Oxidant-induced Stress and Antioxidant Defences in Biology, pp 298—355. Chapman and Hall, New York

Delledonne M, Xia Y, Dixon RA and Lamb C (1998) Nitric oxide functions as a signal in plant disease resistance. Nature 394: 585–587

Del Rio LA, Sandalio LM, Palma JM, Bueno P and Corpas FJ (1992) Metabolism of oxygen radicals in peroxisomes and cellular implications. Free Rad Biol Med 13: 557–580

Demmig-Adams B and Adams III WW (1994) Light stress and photoprotection related to the xanthophyll cycle. In: Foyer CH and Mullineaux PM (eds) Causes of Photooxidative Stress and Amelioration of Defence Systems in Plants, pp 105–126. CRC Press, Boca Raton FL

Deng HX, Hentati A, Tainer JA, Iqbal Z, Cayabyab A, Hung WY, Getzoff ED, Hu P, Herzfeldt B, Roos RP, Warner C, Deng G, Soriano E, Smyth C, Parge HE, Ahmed A, Roses AD, Hallewell RA, Pericak-Vance MA and Siddique T (1993) Amyotrophic lateral sclerosis and structural defects in Cu,Zn superoxide dismutase. Science 261: 1047–1051

Dietrich RA, Richberg MH Schmidt R Dean C and Dangl JL (1997) A novel zinc finger protein is encoded by the *Arabidopsis* LSD1 gene and functions as a negative regulator of plant cell death. Cell 88: 685–694

Ding JP and Pickard BG (1993) Modulation of mechanosensitive calcium-selective channels by temperature. Plant J 3: 713–720

Doke N (1983) Involvement of superoxide anion generation in hypersensitive response of potato tuber tissues to infection with an incompatible race of *Phytophthora infestans*. Physiol Plant Pathol 23: 345–347

Escoubas JM, Lomas M, Laroche J And Falkowski PG (1995) Light-intensity regulation of cab gene-transcription is signaled by the redox state of the plastoquinone pool. Proc Natl Acad Sci USA 92: 10237–10241

Foyer CH (1997) Oxygen metabolism and electron transport in photosynthesis. In: Scandalios JD (ed) Oxidative Stress and the Molecular Biology of Antioxidant Defenses, pp 587–622. Cold Spring Harbour Laboratory Press.

Foyer CH and Halliwell B (1976) The presence of glutathione and glutathione reductase in chloroplasts: A proposed role in ascorbic acid metabolism. Planta 133: 21–25

Galley HF, Davies MJ and Webster NR (1996) Ascorbyl radical formation in patients with sepsis: Effect of ascorbate loading. Free Rad Biol Med 1: 139–143

Gilbert HF (1990) Molecular and cellular aspects of thiol-disulphide exchange. Adv Enzymol 63: 69–172

Ginnpease ME and Whisler RL (1996) Optimal NF-κB mediated transcriptional responses in Jurkat T-cells exposed to oxidative stress are dependent on intracellular glutathione and co-stimulatory signals. Biochem Biophys Res Com 226: 695–702

Giraudat J (1995) Abscisic acid signalling. Curr Opin Cell Biol 7: 232–238

Giraudat J, Parcy F, Bertauche N, Gosti F, Leung J, Morris PC, Bouvierdurand M and Vartanian N (1994) Current advances in abscisic acid action and signalling. Plant Mol Biol 26: 1557–1577

Grace SC, Logan BA and Adams WW III (1998) Seasonal differences in foliar content of chlorogenic acid, a phenyl-propanoid antioxidant, in *Mahonia repens*. Plant Cell Environ 21: 513–521

Gray GR, Savitch LV, Ivanov AG and Huner NPA (1996) Photosystem II excitation pressure and development of resistance to photoinhibition. II. Adjustment to photosynthetic capacity in winter wheat and winter rye. Plant Physiol 110: 61–71

Halliwell B and Gutteridge JMC (1989) Free radicals in biology and medicine. Oxford University Press, Oxford

Halliwell B and Gutteridge JMC (1993) Biologically relevant metal ion-dependent hydroxyl radical generation. FEBS Let 307: 108–112

Hérouart D, Van Montagu M, and Inzé D (1993) Redox-activated expression of the cytosolic copper/zinc superoxide dismutase gene in *Nicotiana*. Proc Natl Acad Sci USA 90: 3108–3112

Hess JL (1993) Vitamin E, α-tocopherol. In: Alscher RG and Hess JL (eds), Antioxidants in Higher Plants, pp 11–134. CRC Press, Boca Raton

Horton P, Ruban AV and Walters RG (1996) Regulation of light harvesting in green plants. Annu Rev Plant Physiol Plant Mol

Biol 47: 656–684

Huerta AJ and Murphy TM (1989) Control of intracellular glutathione and its effect on ultraviolet radiation-induced K^+ efflux in cultured rose cells. Plant Cell Environ 12: 825–830

Huner NPA, Öquist G and Sarhan F (1998) Energy balance and acclimation to light and cold. Trends Plant Sci 3: 224–230

Hurry VM and Huner NPA (1992) Effects on cold hardening on sensitivity of winter and spring wheat leaves to short term photoinhibition and recovery of photosynthesis. Plant Physiol 100: 1283–1290.

Jabs T, Dietrich RA and Dangl JL (1996) Initiation of runaway cell death in an *Arabidopsis* mutant by extracellular superoxide. Science 273: 1853–1856

Kaiser WM (1979) Reversible inhibition of the Calvin cycle and activation of oxidative pentose phosphate cycle in isolated intact chloroplasts by hydrogen peroxide. Planta 145: 377–382.

Kane DJ Srafian TA, Hahn RAH, Gralla EB, Valentine JS, Örd T and Bredesen DE (1993) Bcl-2 inhibition of neural death: Decreased generation of reactive oxygen species. Science 262: 1274–1277

Karpinska B, Wingsle G and Karpinski S (2000). Antagonistic effects of hydrogen peroxide and glutathione on acclimation to excess excitation energy in *Arabidopsis*. IUBMB Life 50: 21–26

Karpinski S, Wingsle G, Karpinska B and Hällgren J-E (1992a) Differential expression of CuZn-superoxide dismutases in *Pinus sylvestris* (L.) needles exposed to SO_2 and NO_2. Physiol Plant 85: 689–696

Karpinski S, Wingsle G, Olsson O and Hällgren J-E (1992b) Characterization of cDNAs encoding CuZn-superoxide dismutases in Scots pine. Plant Mol Biol 18: 545–555

Karpinski S, Wingsle G, Karpinska B and Hällgren J-E (1993) Molecular responses to photooxidative stress in *Pinus sylvestris* (L.) II. Differential expression of CuZn-superoxide dismutases and glutathione reductase. Plant Physiol 103: 1385–1391

Karpinski S, Karpinska B, Wingsle G and Hällgren J-E (1994) Molecular responses to photooxidative stress in *Pinus sylvestris*. I. Differential expression of nuclear and plastid genes in relation to recovery from winter stress. Physiol Plant 90: 358–366

Karpinski S, Escobar C, Karpinska B, Creissen G and Mullineaux P (1997) Photosynthetic electron transport regulates the expression of cytosolic ascorbate peroxidase genes in *Arabidopsis* during excess light stress. Plant Cell 9: 627–642.

Karpinski S, Reynolds H, Karpinska B, Wingsle G, Creissen G and Mullineaux P (1999) Systemic signaling and acclimation in response to excess excitation energy in *Arabidopsis*. Science 284: 654–657.

Kernodle SP and Scandalios JG (1996) A comparison of the structure and function of the highly homologous maize antioxidant Cu/Zn superoxide dismutase genes, *Sod4* and *Sod4A*. Genetics 143: 317–328

Knight MR, Campbell AK, Smith SM and Trewavas AJ (1991) Transgenic plant aquorin reports the effects of touch and cold-shock and elicitors on cytoplasmatic calcium. Nature 352: 524–526

Knight H, Trewavas AJ and Knight M. (1996) Cold calcium signaling in *Arabidopsis* involves two cellular pools and change in calcium signature after acclimation. Plant Cell 8: 489–503

Knox PJ and Dodge AD (1985) Singlet oxygen and plants. Phytochem 24: 889–896

Kozaki A and Takeba G (1996) Photorespiration protects C_3 plants from photoinhibition. Nature 384: 557–560

Kranz HD, Denekamp M, Greco R, Jin H, Leyva A, Meissner RC, Petroni K, Urzainqui A, Bevan M, Martin C, Smeekens S, Tonelli C, PazAres J and Weisshaar B (1998) Towards functional characterization of the members of the R2R3-MYB gene family from *Arabidopsis thaliana*. Plant J 16: 263–276

Krause GH (1994) The role of oxygen in photoinhibition of photosynthesis. In: Foyer CH and Mullineaux PM (eds) Causes of Photooxidative Stress and Amelioration of Defense Systems in Plants, pp 43–76. CRC Press, Boca Raton

Krause GH and Weis E (1991) Chlorophyll fluorescence and photosynthesis: The basics. Annu Rev Plant Physiol Plant Mol Biol 42: 313–349

Krivosheeva A, Tao DL, Ottander C, Öquist G and Wingsle G (1996) Cold acclimation and photoinhibition of photosynthesis in Scots pine. Planta 200: 296–305.

Kusano T, Berberich T, Harada N, Suzuki N and Sugawara K (1995) A maize DNA-binding factor with a βZIP motif is induced by low temperature. Mol Gen Genet 248: 507–517

Kusano T, Sugawara K, Harada M and Berberich T (1998) Molecular cloning and partial characterization of a tobacco cDNA encoding a small βZIP protein. Biochim Biophys Acta 2: 171–175

Lamb C and Dixon RA (1997) The oxidative burst in plant disease resistance. Annu Rev Plant Phys Plant Mol Biol 48: 251–275

Larsson RA (1988) The antioxidants of higher plants. Phytochem 27: 969–978

Levine A, Tenhaken R, Dixon R and Lamb C (1994) H_2O_2 from the oxidative burst orchestrates the plant hypersensitive disease resistance response. Cell 79: 583–593

Linder S (1972) Seasonal variation of pigments in needles: A study of Scots pine and Norway spruce seedlings grown under different nursery conditions. Stud For Suec 100: 1–27

Long SP, Humphries S and Falkowski PG. (1994) Photoinhibition of photosynthesis in nature. Annu Rev Plant Physiol Plant Mol Biol 45: 633–662

Lu G, Paul AL, McCarty DR, Ferl RJ (1996) Transcription factor veracity: Is GBF3 responsible for ABA-regulated expression of *Arabidopsis* Adh? Plant Cell 8: 847–857

Lundmark T, Hällgren J-E and Hedén J (1988) Recovery from winter depression of photosynthesis in pine and spruce. Trees 2: 110–114

Luwe M and Heber U (1995) Ozone detoxification in the apoplasm and symplasm of spinach, broad bean and beech leaves at ambient and elevated concentrations of ozone in air. Planta 197: 448–455

Maxwell DP, Laudenbach DE and Huner NPA (1995) Redox regulation of light-harvesting complex-II and CAB messenger-RNA abundance in *Dunaliella-salina*. Plant Physiol 109: 787–794

McCord JM and Fridovich I (1969) Superoxide dismutase. An enzymatic function for erythrocuprein (*Hemocuprein*). J Biol Chem 244: 6049–6055

Meyer M, Schreck R and Baeuerle PA (1993) H_2O_2 and antioxidants have opposite effects on activation of NF-κB and AP-1 in intact cells: AP-1 as secondary antioxidant-responsive factor EMBO J 12: 2005–2015

Millar AJ, Straume M, Chory J. Chua NH and Kay SA (1995)

The regulation of circadian period by phototransduction pathways in *Arabidopsis*. Science 267: 1163–1166

Minorsky PV (1985) An heuristic hypothesis of chilling injury in plants: A role for calcium as the primary physiological transducer of injury. Plant Cell Environ 8: 75–94

Mizoguchi T, Irie K, Hirayama T, Hayashida N, Yamaguchi-Shinozaki K, Matsumoto K and Shinozaki K (1996) A gene encoding a MAP kinase kinase kinase is induced simultaneously with genes for a MAP kinase and an S6 kinase by touch, cold and water stress in *Arabidopsis thaliana*. Proc Natl Acad Sci USA 93: 765–769

Monroy AF and Dhindsa RS (1995) Low-temperature signal transduction: Induction of cold acclimation-specific genes of alfalfa by calcium at 25 °C. Plant Cell 7: 321–331

Monroy AF, Sarhan F and Dhindsa RS (1993) Cold-induced changes in freezing tolerance, protein phosphorylation, and gene expression: Evidence for role of calcium. Plant Physiol 102: 1227–1235

Mullineaux PM and Creissen GP (1997) Glutathione reductase: Regulation and role in oxidative stress. In: Scandalios JG (ed), Oxidative Stress and the Molecular Biology of Antioxidant Defenses, pp 667–714. Cold Spring Harbour Laboratory Press

Mullineaux PM, Karpinski S, Jimenéz A, Cleary SP, Robinson C and Creissen G (1998) Identification of cDNAs encoding plastid-targeted glutathione peroxidase. Plant J 13: 375–379

Mullineaux P, Ball L, Escobar C, Karpinska B, Creissen G, Karpinski S (2000). Are diverse signaling pathways integrated in the regulation of *Arabidopsis* antioxidant defence gene expression in response to excess excitation energy. Phil Trans R Soc Lond B 355: 1531–1540

Neuenschwander U, Vernooij B, Friedrich L, Uknes S, Kessmann H and Ryals J (1995) Is hydrogen peroxide a second messenger of salicylic acid in systemic acquired resistance? Plant J 8: 227–233.

Neuhaus G, Bowler C, Kern R and Chua NH (1993) Calcíum/calmodulin-dependent and calcium/calmodulin-independent phytochrome signal-transduction pathways. Cell 73: 937–952

Noctor G and Foyer CH (1998) Ascorbate and glutathione: Keeping active oxygen under control. Annu Rev Plant Physiol Plant Mol Biol 49: 249–279

O'Kane D, Gill V, Boyd P and Burdon B (1996) Chilling, oxidative stress and antioxidant responses in *Arabidopsis thaliana* callus. Planta 198: 371–377

Öquist G, Chow WS and Andersson JM (1992) Photoinhibition of photosynthesis represents a mechanism for the long term regulation of Photosystem II. Planta 186: 450–460

Osmond CB and Grace SC (1995) Perspective of photoinhibition and photorespiration in the field: Quintessential inefficiencies of the light and dark reactions in the terrestrial oxygenic photosynthesis? J Exp Bot 46: 1351–1362.

Ottander C, Campbell D and Öquist G (1995) Seasonal changes in Photosystem II: Organisation and pigment composition in *Pinus sylvestris*. Planta 197: 176–183

Perl-Treves R and Galun E (1991) The tomato Cu,Zn superoxide dismutase genes are developmentally regulated and respond to light and stress. Plant Mol Biol 17: 745–760

Pfannschmidt T, Nilsson A and Allen JF (1999) Photosynthetic control of chloroplast gene expression. Nature 397: 625–628

Polle A (1997) Defense against photooxidative damage in plants. In: Scandalios JG (ed), Oxidative Stress and the Molecular Biology of Antioxidant Defenses, pp 623–666. Cold Spring Harbour Laboratory Press

Polle A and Renenberg H (1994) Field studies on Norway spruce trees at high altitudes. II Defence systems against oxidative stress in needles. New Phytol 121: 635–642

Prasad TK (1996) Mechanisms of chilling-induced oxidative stress injury and tolerance in developing maize seedlings: Changes in antioxidant system, oxidation of proteins and lipids, and protease activities. Plant J 10: 1017–1026

Prasad TK, Anderson MD, Martin BA and Stewart CR (1994) Evidence for chilling-induced oxidative stress in maize seedlings and regulatory role of hydrogen peroxide. Plant Cell 6: 65–74

Price AH, Taylor A, Ripley SJ, Cuin T, Tomos D and Ashenden T (1994) Oxidative signals in tobacco increase cytosolic calcium. Plant Cell 6: 1301–1310

Raff MC, Barres BA, Burne JF, Coles HS, Ishizaki Y and Jacobson MD. (1993) Programmed cell death and the control of cell survival: Lessons from the nervous system. Science 262: 695–700

Ramasarma T (1982) Generation of H_2O_2 in biomembranes. Biochem Biophys Acta 694: 69–93

Salin ML (1987) Toxic oxygen species and protective systems of the chloroplast. Physiol Plant 72: 681–689

Schinkel H, Streller S and Wingsle G (1998) Multiple forms of extracellular superoxide dismutase in needles, stem tissues and seeds of Scots pine. J Exp Bot 49: 931–936

Schreiber U and Neubauer C (1990) O_2-dependent electron flow, membrane energization and the mechanism of non-photochemical quenching of chlorophyll fluorescence. Photosynth Res 25: 279–293

Sgherri CLM, Loggini B and Puliga S (1994) Antioxidant system in *Sporobolus stapfianus*: Changes in response to desiccation and rehydration. Phytochem 3: 561–565

Shinozaki K and Yamaguchi-Shinozaki K (1996) Molecular responses to drought and cold stress. Curr Opin Biotech 7: 161–167

Shirasu K, lahaye T, Tan M-W, Zhou F, Azevedo C and Schulze-Lefert P (1999) A novel class of eukaryotic zinc-biding proteins is required for disease resistance signalingin barley and development in *C. elegans*. Cell 99: 355–366

Smith I, Polle A and Rennenberg H (1990) Glutathione. In: Alscher RG and Cumming J (eds), Stress Responses in Plants: Adaptation and Acclimation Mechanisms, pp 201–217. Wiley-Liss, New York

Somersalo S and Krause GH (1989) Photoinhibition at chilling temperature. Fluorescence characteristics of unhardened and cold acclimated spinach leaves. Planta 177: 409–416

Sonoike K (1996) Photoinhibition of Photosystem I; Its physiological significance in the chilling sensitivity of plants. Plant Cell Physiol. 37: 239–247

Strand M and Öquist G (1985) Inhibition of photosynthesis by freezing temperatures and high light levels in cold-acclimated seedlings of Scots pine (Pinus sylvestris): I: Effect on the light limited and light saturated rates of CO_2 assimilation. Physiol Plant 64: 425–43

Streller S and Wingsle G (1994) *Pinus sylvestris* (L.) needles contain extracellular CuZn superoxide dismutase. Planta 192: 195–201

Suzuki H, Matsumori A, Matoba Y, Kyu B, Tanaka A, Fujita J and Sasayama S (1993) Enhanced expression of superoxide dismutase messenger RNA in viral myocarditis: An SH-dependent reduction of its expression and myocardial injury. J Clin Invest 6: 2727–2733

Tamagnone L, Merida A, Stacey N, Plaskitt K, Parr A, Chang CF, Lynn, D, Dow JM, Roberts K, Martin C (1998) Inhibition of phenolic acid metabolism results in precocious cell death and altered cell morphology in leaves of transgenic tobacco plants. Plant Cell 10: 1801–1816

Tappel AL (1972) Vitamin E and free radical peroxidation of lipids. Ann New York Acad Sci 203: 12–28

Tsang EWT, Bowler C, Herouart D, Van Camp W, Villarroel R. Genetello C, Van Montagu M and Inzé D (1991) Differential regulation of superoxide dismutases in plants exposed to environmental stress. Plant Cell 3: 783–792

Urao T, Katagiri T, Mizoguchi T, Yamaguchi-Shinozaki K, Hayashida N and Shinozaki K (1994) Two genes that encode Ca^{2+}-dependent protein kinases are induced by drought and high salt stresses in *Arabidopsis thaliana*. Mol Gen Genet 224: 331–340

Urao T, Yakubov B, Yamaguchi-Shinozaki K and Shinozaki K (1998) Stress-responsive expression of genes for two-component response regulator-like proteins in *Arabidopsis thaliana*. FEBS Lett 427: 175–178

Van Wijk KJ, Roobol-Boza M, Kettunen R, Andersson B and Aro EM (1997) Synthesis and assembly of the D1 protein into Photosystem II: Processing of the C-terminus and identification of the initial assembly partners and complexes during Photosystem II repair. Biochem 36: 6178–6186

Willekens H, Chamnongpol S, Davey M, Schraudner M, Langebartels C, VanMontagu M, Inzé D and VanCamp W (1998) Catalase is a sink for H_2O_2 and is indispensable for stress defence in C-3 plants. EMBO J 16: 4806–4816

Wingate VPM, Lawton MA and Lamb CJ (1988) Glutathione causes a massive and selective induction of plant defence genes. Plant Physiol 87: 206–210

Wingsle G and Karpinski S (1996) Differential redox regulation by glutathione of glutathione reductase and CuZn superoxide dismutase genes expression in *Pinus sylvestris* (L.) needles. Planta 198: 151–157

Wingsle G, Sandberg G and Hällgren J-E (1989) Determination of glutathione in Scots pine needles by high-performance liquid chromatography as its monobromobimane derivative. J Chromatogr 479: 335–344

Wingsle G, Gardeström P, Hällgren J-E and Karpinski S (1991) Isolation, purification, and subcellular localization of isozymes of superoxide dismutase from Scots pine (*Pinus sylvestris* L.) needles. Plant Physiol 95: 21–28

Wingsle G, Karpinski S and Hällgren JE (1999) Low temperature, high light stress and antioxidant defence mechanisms in higher plants. Phyton 39: 253–268

Wise RR (1995) Chilling-enhanced photooxidation: The production, action and study of reactive oxygen species produced during chilling in the light. Photosynth Res 45: 79–97

Wu J, Neimanis S and Heber U (1991) Photorespiration is more effective than the Mehler reaction in protecting the photosynthetic apparatus against photoinhibition. Bot Acta 104: 283–291

Chapter 28

The Elip Family of Stress Proteins in the Thylakoid Membranes of Pro- and Eukaryota

Iwona Adamska*
Department of Biochemistry, Arrhenius Laboratories of Natural Sciences, Stockholm University, S-10691 Stockholm, Sweden

Summary

Early light-induced protein (Elip) family consists of low-molecular-mass stress proteins localized in the thylakoid membranes of pro- and eukaryota. Various physiological conditions, such as light stress, dehydrative processes or morphogenesis, have been reported to trigger transient induction of these proteins in different plant species. According to the deduced amino acid sequences, all members of Elip family are related to the Cab (chlorophyll *a/b*-binding) proteins of Photosystem I and II. In terms of predicted protein structure, the 27 members of the Elip family are divided into three groups: (a) three-helix Elips and related proteins, (b) two-helix Seps (stress-enhanced proteins) called also Lils (light-harvesting-like), and (c) one-helix Hlips (high light-induced proteins), Scps (small Cab-like proteins) and Ohps (one-helix proteins). Transmembrane α-helices I and III of Elip family members are very conserved and share, in addition to the potential chlorophyll-ligands, Elip consensus motifs. Recently, a binding of chlorophyll was experimentally confirmed for the pea Elip, which might imply that also other proteins of this family bind pigments. Despite binding of chlorophyll, non-light-harvesting functions have been proposed for Elip family members. It is believed that these proteins fulfill a protective role within the thylakoids under stress conditions either by transient binding of free chlorophyll molecules and preventing the formation of free radicals and/or by acting as sinks for excitation energy. This chapter provides an overview of Elip family members, shows their similarities and differences with Cab proteins and discusses their possible function(s).

*Email: iwona@biokemi.su.se

Eva-Mari Aro and Bertil Andersson (eds): Regulation of Photosynthesis, pp. 487–505.

I. Introduction

The chlorophyll-binding proteins of pro- and eukaryota are located either in the core complexes of Photosystem I (PS I) and Photosystem II (PS II) or form light-harvesting antenna systems. The eukaryotic core complexes consist of several chlorophyll *a*-binding proteins represented by the PS I and PS II reaction center heterodimers, such as the PS I-A/B and the D1/D2, respectively, and by the CP43 and CP47 proteins of the PS II core (reviewed in von Wettstein et al., 1995; Green and Durnford, 1996). All these proteins are encoded by the chloroplast genome (Sugiura, 1995).

The chlorophyll-binding antenna complexes of higher plants, green algae (*Chlorophyta*) and *Chromophyta* consist of nuclear-encoded chlorophyll *a/b*-binding (Cab) or fucoxanthin-chlorophyll *a/c*-binding (Fcp) proteins associated with both photosystems (reviewed in Jansson, 1994, 1999; Green and Kühlbrandt, 1995; Paulsen 1995, 1997; von Wettstein et al., 1995; Green and Durnford, 1996; Horton et al., 1996; Bassi et al., 1997). According to the deduced amino acid sequences these proteins are related to each other and are assumed to share a common evolutionary origin (Cavalier-Smith, 1992; Wolfe et al., 1994; Green and Durnford, 1996). The Cab proteins of photosynthetic eukaryotes are encoded by multigene families consisting of at least 21 different genes, as was reported for *Arabidopsis thaliana* (Jansson, 1999). All known Cab and Fcp family members are polytopic membrane proteins with three transmembrane segments (Kühlbrandt et al., 1994; Green and Kühlbrandt, 1995; Green and Durnford, 1996). An exception here is the PS II-S protein of higher plants, a distant relative of the Cab family, which has four predicted α-helices (Kim et al., 1992; Wedel et al., 1992).

Abbreviations: ABA – abscisic acid; Cab – chlorophyll *a/b*-binding protein; Cbr – carotene biosynthesis-related protein; Dsp – desiccation stress protein; Elip – early light-induced protein; Fcp – fucoxanthin-chlorophyll *a/c*-binding protein; Hlip – high-light-induced protein; HV58 – low-molecular-mass Elip of *Hordeum vulgare*; HV60, HV90 – high-molecular-mass Elips of *Hordeum vulgare*; Isi – iron stress-induced; Lhca, Lhcb – light-harvesting chlorophyll *a/b*-binding proteins of PS I or PS II, respectively; Lil – light-harvesting-like protein; Ode – osmotic differentially expressed protein; Ohp – one-helix protein; PS I, PS II – Photosystem I and II, respectively; Scp – small Cab-like protein; Sdi – sunflower drought-induced protein; Sep – stress-enhanced protein; SRP – signal recognition particle; Wcr – wheat cold-regulated protein

Red algae (*Rhodophyta*) and cyanobacteria have a chlorophyll *a*-containing antenna complex functionally associated with PS I and phycobilisomes, which serve as a light-harvesting antenna in PS II (Glazer, 1989; Grossman et al., 1993; Golbeck, 1994; Wolfe et al., 1994). Chlorophyll *b* is missing from both of these systematic groups. There are significant differences in organization of the PS I antenna complexes between red algae and cyanobacteria. While in cyanobacteria chlorophylls are bound directly to the A and B subunits of the PS I reaction center (Golbeck, 1994), in red algae some additional antenna proteins are involved in binding of chlorophylls. These chlorophyll-binding polypeptides are related to those of higher plants, even though they bind only chlorophyll *a* (Wolfe et al., 1994). An exception are cyanobacteria-like *Prochlorophyta*, where phycobilisomes are missing and the light-harvesting antenna consists of both, chlorophyll *a* and chlorophyll *b* (Palenik and Haselkorn, 1992; Matthijs et al., 1994; Post and Bullerjahn, 1994). The chlorophyll *a/b* antenna proteins of *Prochlorophyta* fundamentally differ from those of Cab or Fcp families since they are closely related to the *IsiA* (iron stress-induced) gene from cyanobacterium *Synechococcus* sp. PCC 7942 (La Roche et al., 1996). A cyanobacterial *IsiA* gene encodes a chlorophyll *a*-binding protein related to CP43 of higher plants (referred to as CP43′), which is induced in response to iron starvation (Straus, 1994).

In the past few years, several potential chlorophyll-binding proteins were reported to be transiently expressed in pro- and eukaryota under various stress conditions. These proteins, called Elips (early light-induced proteins), were related to the Cab and Fcp family members as based on their sequence similarities. However, this group of proteins also contained several distinct characteristics. The purpose of this chapter is to provide an overview of such stress-induced chlorophyll-binding proteins and to address their relationships with Cab proteins.

II. What is an Elip? Past and Present Definitions

The initial definition of Elips was based mainly on their unusual expression pattern during greening of etiolated seedlings and a sequence similarity to the light-harvesting chlorophyll *a/b*-binding proteins of PS I (Lhca) and PS II (Lhcb). According to this definition Elips are nuclear-encoded, thylakoid membrane proteins with a high sequence similarity to the *Cab* gene family members but with a transient expression restricted to the early stages of seedling development.

Kloppstech and coworkers (Meyer and Kloppstech, 1984; Grimm and Kloppstech, 1987) described the first members of Elip family to be induced during greening of etiolated pea and barley seedlings. The Elip transcripts accumulated very rapidly after a transition from dark to light and preceded the accumulation of transcripts for other light-regulated genes. Based on their expression kinetics these proteins were named the early light-induced proteins. Since Elips were not detected in mature green plants grown under ambient light conditions it was assumed that the induction of *Elip* genes was restricted to the early stages of seedling development, where the corresponding proteins performed their physiological functions. Sequencing of *Elip* genes from pea (Scharnhorst et al., 1985; Kolanus et al., 1987) and barley (Grimm et al., 1989) and comparison of the deduced amino acid sequences revealed that these proteins share a high similarity in their primary and secondary structures with all Cab proteins (Grimm et al., 1989; Green et al., 1991). Both groups of proteins, the Elips and the Cabs, contain three predicted α-helices with a very distinctive pattern of conserved amino acid residues located in the related first and third transmembrane segments (Grimm et al., 1989; Green et al., 1991; Green and Pichersky, 1994; Green and Kühlbrandt, 1995)

During the following years several Elips and related proteins were described from different plant species. Considerable progress in understanding the structure and the regulation of these proteins has been made (see reviews Adamska and Kloppstech, 1994b; Adamska, 1997). Based on this information the initial definition of Elips needs to be revised.

First, not all Elip family members are nuclear-encoded proteins. The nuclear localization of *Elip* genes was reported for higher plants (Grimm et al., 1987; Green et al., 1991), ferns (Raghavan and Kamalay, 1993) and the green alga *Dunaliella bardawil* (Lers et al., 1991). Sequencing of the plant organelle genomes has revealed that in the red algae *Porphyra purpurea* (Reith and Munholland, 1995) and *Cyanidium caldarium* (Kessler et al., 1992) as well as in the Cryptophyta *Guillardia theta* (Douglas and Penny, 1999) *Elip* genes have remained in the chloroplast genome. In the Glaucocystophycae species, such as *Cyanophora paradoxa*, an *Elip*-like gene was found in the cyanelle genome (V.L. Stirewalt, C.B. Michalowski, W. Löffelhardt, H.J. Bohnert and D.A. Bryant, 1995, direct GenBank submission). Cyanelles are plastid-like organelles, which resemble cyanobacteria in morphology, in biochemical organization of their photosynthetic apparatus and in the presence of the peptidoglycan wall (Löffelhardt and Bohnert, 1994). In the cyanobacteria *Synechococcus* and *Synechocystis* several Elip-related sequences are encoded by a nucleoid DNA (Brahamsha and Haselkorn, 1991; Dolganov et al., 1995; Funk and Vermaas, 1999).

The second reason for a revision of the Elip definition is the predicted secondary structure of these proteins, which uncovered that some members of Elip family contain only one or two transmembrane α-helices, instead of three as initially reported (Grimm et al., 1989; Green et al., 1991).

Finally the expression of Elip family members is not restricted to young developing seedlings as it was originally assumed (Meyer and Kloppstech, 1984; Grimm and Kloppstech, 1987). The Elip family members have been reported to be induced also in mature green plants exposed to variety of physiological stresses. In addition to light stress (Lers et al., 1991; Adamska et al., 1992a,b; Pötter and Kloppstech, 1993; Dolganov et al., 1995; Moscovici-Kadouri and Chamovitz, 1997; Heddad and Adamska, 1998, 2000; Shimosaka et al., 1999), a nutrient deprivation (Lers et al., 1991) or dehydrative processes, such as desiccation (Bartels et al., 1992; Ouvrard et al., 1996; Shimosaka et al., 1999), cold stress (Adamska and Kloppstech, 1994a; Montané et al., 1996, 1999; Król et al., 1997; Shimosaka et al., 1999), salt stress (Shimosaka et al., 1999) or osmotic stress (Y.M. Verastegui-Pena and N. Ochoa-Alejo, 1999, direct GenBank submission) have also led to induction of *Elip* genes. Some of these stress conditions required light as a supplementary factor to effect the expression of Elip genes.

In view of the existing information a new definition of Elips is proposed: the Elip family consists of low-molecular-mass stress proteins located in the thylakoid membranes of pro- and eukaryota and related to the light-harvesting antenna proteins of higher plants and green algae. Despite the binding of chlorophyll, a non-light-harvesting function has been proposed for this group of proteins (see below).

III. Division of Elip Family Based on Predicted Protein Structure

Based on a predicted secondary structure, the members of the Elip family can be divided into three groups: (i) three-helix Elips and related proteins, (ii) two-helix Seps (stress-enhanced proteins), called also Lils (light-harvesting-like), and (iii) one-helix Hlips (high light-induced proteins), Scps (small Cab-like proteins) and Ohps (one-helix proteins). The helices I and III of the three-helix Elips and related proteins and the helix I of Seps, Hlips, Scps or Ohps are very conserved in their amino acid composition and contain an Elip consensus motif. The helix II is highly polymorphic and differs between Elip family members (Fig. 1).

Three-helix Elips and related proteins as well as one-helix Hlips, Scps and Ohps are predicted to be type II membrane proteins, with the N-terminus on the stromal and the C-terminus on the lumenal sides of the membrane. The Seps represent most likely type III membrane proteins, with both termini directed to the stromal compartment (Fig. 1). These predictions, however, still require an experimental proof.

Elips share a significant homology with Cab proteins and presumably have a similar three-dimensional structure. The three-dimensional structure of one Cab family member, the Lhcb1, has been determined at 3.4 Å resolution by electron crystallography of two-dimensional crystals (Kühlbrandt et al., 1994). It was shown that two of the three membrane-spanning α-helices of Lhcb1 are held together by ion pairs formed by charged residues. Therefore, it is expected that also helices I and III of Elips and their relatives are held together by the reciprocal ionic bonds between an arginine (R) on one helix and a glutamic acid (E) on the other that crosslink these two helices (Green and Pichersky, 1994; Kühlbrandt et al., 1994; Green and Kühlbrandt, 1995). In two-helix Seps the conserved arginine and glutamic acid residues are present only in the first conserved transmembrane helix, thus in order to form a similar two-fold symmetry structure in vivo, these proteins have to form homodimers (Heddad and Adamska, 2000). The homodimer structure was proposed in the past for cyanobacterial one-helix Hlips (Dolganov et al., 1995) and Scps (Funk and Vermaas, 1999), and it is expected also for higher plant Ohps.

A. Three-Helix Elips and Related Proteins

The occurrence of three-helix Elips seems to be restricted to higher plants and green algae as summarized in Table 1. All known three-helix Elips are nuclear-encoded, synthesized by cytoplasmic ribosomes as precursor proteins and posttranslationally imported into the chloroplasts. Prior to insertion into the thylakoid membranes the precursor proteins are processed to their mature forms by a stromal peptidase, which cleaves the N-terminal leader sequence.

Comparison of deduced amino acid sequences of Elips and related proteins from higher plants and the green alga *Dunaliella bardawil* is shown in Fig. 2. The three-helix Elips show a high degree of amino acid conservation, especially in the transmembrane helices I and III. The most diverse from the whole

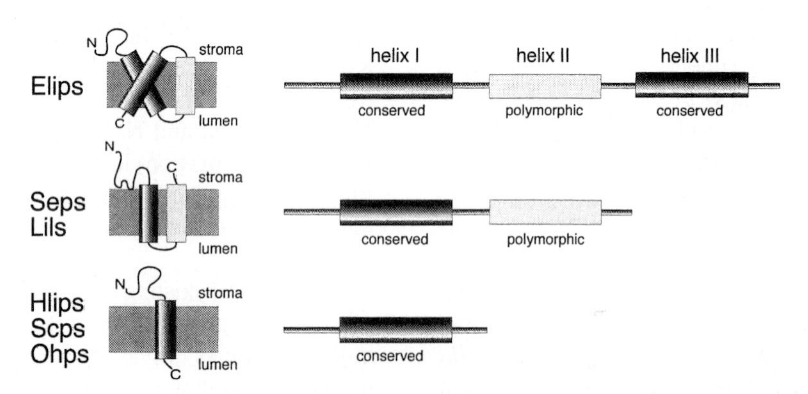

Fig. 1. Predicted protein structure and thylakoid membrane topology of Elip family members. Based on a predicted secondary structure the Elip family can be divided into three-helix Elips and related proteins, two-helix Seps (called also Lils), and one-helix Hlips, Scps and Ohps. While the transmembrane helices I and III are highly conserved for all members of the Elip family, helix II is polymorphic and differs between these proteins. The structure and topology of three-helix Elips is based on the model proposed for the Lhcb1 (Kühlbrandt et al., 1994).

Table 1. Characterization of the three-helix Elips and related proteins from higher plants and green algae

Plant species	Gene name	Gene size (bp)	Transcript size (bp)	Precursor protein (aa)	Mature protein (aa)	Transit peptide (aa)	GenBank accession	References
Higher plants								
Monocotyledones								
Glycine max	*Elip*	n.d.	796	192	149	43	U82810	(1)
Hordeum vulgare	*HV58*	n.d.	860	231	200	31	X15693	(2, 3)
	HV90	n.d.	660	172	144	38	X15692	(2, 3)
	HV60	n.d.	764	167	134	33	X15691	(2, 3)
Oryza sativa	*Elip*	876	783	202	160	42	AC007789	(4)
Triticum aestivum	*Wcr12*	n.d.	818	174	134	40	AB019617	(5)
Dicotyledones								
Arabidopsis thaliana	*Elip1*	1078	848	195	151	44	U89014	(6, 7, 8)
	Elip2	913	750	193	152	41	Z97336	(7, 8, 9)
Craterostigma plantagineum	*Dsp22*	n.d.	716	199	151	48	X66598	(10)
Helianthus annuus	*Sdi1*	n.d.	713	174	141	33	X92646	(11)
Pisum sativum	*Elip*	2922	783	196	150	46	X05979	(12)
Green algae								
Dunaliella bardawil	*Cbr*	1325	799	172	134	38	L32871	(13)

Following abbreviations are used: Cbr, carotene biosynthesis-related protein; Dsp, desiccation stress protein; Elip, early light-induced protein; HV58, high-molecular-mass Elip of *Hordeum vulgare*; HV90, HV60, low-molecular-mass Elips of *Hordeum vulgare*; Sdi, sunflower drought-induced protein; Wcr, wheat cold-regulated protein. N.d., no data available. References: (1) Yamagata and Bowler, 1996; (2) Grimm and Kloppstech, 1987; (3) Grimm et al., 1989; (4) M.I. Benito, 1999, direct GenBank submission; (5) Shimosaka et al., 1999; (6) Moscovici-Kadouri and Chamovitz, 1997; (7) Heddad and Adamska, 1998; (8) Heddad and Adamska, 1999; (9) Jansson, 1999; (10) Bartels et al., 1992; (11) Ouvrard et al., 1996; (12) Kolanus et al., 1987; (13) Lers et al., 1991.

group are the Dsp22 (desiccation stress protein) from *Craterostigma plantagineum* (Bartels et al., 1992) and the Cbr (carotene biosynthesis-related) protein from the green alga *Dunaliella* (Lers et al., 1991).

The transmembrane helices I and III of three-helix Elips are related to each other (Green et al., 1991; Green and Pichersky, 1994; Green and Kühlbrandt, 1995) containing Elip consensus motifs ERINGRLAMIGFVAALAVE and ELWNGRFAMLGLVALAFTE, respectively. Within these motifs, there are several universally conserved amino acids for all known Elip family members, such as two glutamic acids, one arginine and one methionine (M) in the first and third helices, as well as two alanines (A), two glycines (G) and one asparagine (N) in the third helix. The conserved pattern of small residues in Lhcb1, consisting of a glycine, an alanine and a methionine, which are present also in all Elip family members, has been reported to be essential for close packing of helices I and III in Lhcb1 (Kühlbrandt et al., 1994; Green and Kühlbrandt, 1995). Thus, these amino acid residues might have a similar function in Elips. Two conserved glutamic acids flanking the consensus motifs of Elips might act as a stop-transfer signal at the ends of the helix I and III, as suggested for the Cab proteins (Green and Pichersky, 1994).

The transmembrane helix II is polymorphic and shows a wide diversity between the Elip family members (Fig. 2). However, there is a significant conservation in this region of Elips within monocot or dicot plant species. Monocotyledon Elips and their relatives showed 76% similarity and 30% identity in the amino acid composition of helix II, and within dicotyledon species this conservation was even higher reaching 81% similarity and 38% identity.

In addition to the features common for all members of Elip family, three-helix Elips and related proteins possess several distinct characteristics. There are some conserved glycine, alanine and glutamine (Q) residues flanking Elip consensus sequences in the first and the third helices (Fig. 2). Two conserved amino acid motifs, KXST and AFSGPAP, are present in front of helix I, the functional significance of which is not yet understood. Interestingly, these motifs are missing in the Dsp22 and Cbr proteins (Fig. 2). Furthermore, the three-helix Elips and related proteins contain a hydrophobic, proline-rich region of 14–24 amino acids, which is located at the N-terminus

Fig. 2. Sequence alignment of the three-helix Elips and Elip-related protein precursors from higher plants and green algae. Identical amino acids are shown on a gray background, solid lines mark the positions of predicted transmembrane helices, dashed lines mark two conserved amino acid motifs and a dotted line shows the proline-rich region. The arrowheads indicate the positions of conserved amino acids discussed in text. Sequences were aligned using the ClustalW (Thompson et al., 1994) program with manual correction of gaps. For origin of sequences see Table 1.

close to helix I of the mature proteins. This region contains, depending on the plant species, from 5 to 12 proline (P) residues, two of them at the highly conserved positions. In Elips in *Arabidopsis,* rice and barley (a high-molecular-mass Elip), this region is also serine-rich and contains, in addition to prolines, several serine (S) residues. An exception here is the Dsp22, which contains only 2 prolines and a stretch of seven glutamine residues in this region.

B. Two-Helix Seps (Lils)

Recently, Elip-related proteins with two predicted transmembrane helices have been identified and described from *Arabidopsis,* these proteins being designated either as Seps (Heddad and Adamska, 2000) or as Lils (Jansson, 1999). In contrast to other members of the Elip family investigated so far, low amounts of Sep transcripts were present in leaves of *Arabidopsis* under ambient light conditions (Heddad and Adamska, 2000). Exposure of leaves to light stress resulted in a four- to ten-fold increase in Sep transcript level (Heddad and Adamska, 2000).

The Seps are nuclear-encoded proteins composed of 146–262 amino acids as precursors, which are processed to their mature forms of 103–223 amino acids (Table 2). The differences in the sizes of various Seps result mainly from the different lengths of their hydrophilic N- and C-terminal regions.

Comparison of amino acid sequences (Fig. 3) uncovered that the overall sequence similarity between *Arabidopsis* Seps is less than 25%. High similarity exists only in the helix I of Seps, which is conserved and contains an Elip consensus motif.

Table 2. Characterization of the two-helix Seps from *Arabidopsis thaliana*

Gene name	Gene size (bp)	Transcript size (bp)	Precursor protein (aa)	Mature protein (aa)	Transit peptide (aa)	GenBank accession	References
Sep1	1138	691	146	103	43	AF133716	(1)
Sep2	912	860	202	181	21	AF133717	(1)
Sep3	1087	937	262	223	39	Z97343	(2)

Following abbreviation is used: Sep, stress-enhanced protein. References: (1) Heddad and Adamska, 2000; (2) Jansson, 1999.

```
Arabidopsis-Sep1    1  MALSQ---VSASLAFS--LPNS--------------GALKLATI---TNPTSTCRVH-----------V
Arabidopsis-Sep2    1  MAMAT---RAIRYQLP--SPRF--------------RAPRCES----SEPIKQIQIQQ----------R
Arabidopsis-Sep3    1  MALFSPPISSSSLQNPNFIPKFSFSLLSSNRFSLLSVTRASSDSGSTSPTAAVSVEAPEPVEVIVKEP

Arabidopsis-Sep1   37  PQLA--------GIRST-FASGSP------LLPLK------LSMTRR----GGNRAASVSIRS----E-
Arabidopsis-Sep2   37  PRGG--------DLAENGKIVLQPR-----LCTLRSY-GSDMVIAKKDGGDGGGGGSDVELASPFF-ET
Arabidopsis-Sep3   69  PQSTPAVKKEEATAKNVAVEGEEMKTTESVVKFQDARWINGTWDKQFEKDGKTDWDSVIVAEAKRRKW

Arabidopsis-Sep1   76  -----QSTEGSSGL-----------------------DIWLGRGAMVGFAVAITVEISTGKGLLENFGVA
Arabidopsis-Sep2   91  LTDYIESSKKSQDF-----------------------ETISGRLAMIVFAVTVTEEIVTGNSLFKKLDVE
Arabidopsis-Sep3  137  LEENPETTSNDEPVLFDTSIIPWWAWIKRYHLPEAELLNGRAAMIGFFMAYFVDSLTGVGLVDQMGNF

Arabidopsis-Sep1  118  SPLPTVALAVTALVGVLAAVFIFQSSSKN-----------------------------------
Arabidopsis-Sep2  138  -GLSEAIGAGLAAMGCAAMFAWLTISRNRVGRIFTVSCNSFIDSLVDQIVDGLFYDTKPSDWSDDL
Arabidopsis-Sep3  205  -FCKTLLFVAVAGVLFIRKNEDVDKLKNLFDETTLYDKQWQAAWKNDDDESLGSKKK
```

Fig. 3. Sequence alignment of the two-helix Sep precursors from *Arabidopsis thaliana*. Identical amino acids are shown on a gray background, solid lines mark the positions of predicted transmembrane helices and arrowheads indicate the positions of conserved amino acids discussed in text. Sequences were aligned using the ClustalW (Thompson et al., 1994) program with manual correction of gaps. For origin of sequences see Table 2.

Helix II and the N-terminus of Seps are polymorphic and differ in their amino acid composition between all known Seps (Fig. 3). The positions of glycine, arginine, alanine, methionine and phenylalanine (F) residues in helix I are identical for all known Seps. One of the two conserved glutamic acid residues, which are universal for all Elip family members, is replaced by an aspartic acid (D) in *Arabidopsis* Sep1 and Sep3. In addition, conserved glycine, threonine (T) and leucine (L) residues are present in helix I or flank the Elip consensus motif (Fig. 3).

C. One-Helix Hlips, Scps and Ohps

To date, several one-helix proteins related to the three-helix Elips of higher plants have been described from various cyanobacteria, the red alga *Porphyra purpurea*, the Cryptophyta *Guillardia theta* and the Glaucocystophyceae *Cyanophora paradoxa* (Table 3). Analysis of data obtained from the genome sequencing projects <http://www.tigr.org/index.html> revealed that Elip-related proteins with one predicted transmembrane domain are present also in higher plants. The Ohps have been found in *Oryza sativa* (M.C. Lee, C.S. Kim and M.Y. Eun, 1997, direct GenBank submission) and *Arabidopsis thaliana* (Jansson, 1999). In addition, two EST clones (accessions: AI823025 and AI822673) with a high degree of similarity to *Arabidopsis* Ohp are now available from *Mesembryanthemum crystallinum* (J.C. Cushman, 1999, direct GenBank submission). Thus, the occurrence of one-helix Elips is not restricted to algae and prokaryota as was suggested before (Green and Kühlbrandt, 1995; Green and Durnford, 1996).

There are significant differences in the localization of genes encoding one-helix Elips. The higher plant Ohps are nuclear-encoded, synthesized as precursor proteins of 110-157 amino acids. After import into *Arabidopsis* or rice chloroplasts these proteins are processed to their mature forms of 69 and 118 amino acids, respectively. The one-helix proteins of the Elip family in various algal species or cyanobacteria are encoded by the chloroplast, cyanelle or nucleoid genomes and thus synthesized without an organelle targeting sequence. The size of these proteins is between 48 and 72 amino acids (Table 3).

Comparisons of deduced amino acid sequences of all known one-helix Elips are shown in Fig. 4. The transmembrane helix of Hlips, Scps and Ohps is conserved between all members of this group and contains an Elip consensus motif ERINGRLAMIGFVAALAVE. A conserved alanine and two glycine residues flank the Elip consensus motif. An exception

Table 3. Characterization of the one-helix Hlips, Scps and Ohps from prokaryota and eukaryota

Plant species	Gene name	Gene size (bp)	Transcript size (bp)	Precursor protein (aa)	Mature protein (aa)	Transit peptide (aa)	GenBank accession	References
Higher plants								
Arabidopsis thaliana	*Ohp*	n.d.	427	110	69	41	AF054617	(1)
Oryza sativa	*Ohp*	n.d.	818	157	118	39	AF017356	(2)
Red alga								
Porphyra purpurea	*ycf17*	n.d.	n.d.	chloroplast	48	-	AAC08241	(3)
Cryptophyta								
Guillardia theta	*ycf17*	162	162	chloroplast	53	-	AAC35610	(4)
Glaucocystophyta								
Cyanophora paradoxa	*ycf17*	n.d.	n.d.	cyanelle	49	-	P48367	(5)
Cyanobacteria								
Anabaena sp.	*orf2*	n.d.	n.d.	nucleoid	59	-	AAA22044	(6)
Synechococcus sp.	*HliA*	800	333	nucleoid	72	-	A55916	(7)
Synechocystis sp.	*ScpB*	213 bp	213	nucleoid	70	-	BAA17603	(8)
	ScpC	213 bp	213	nucleoid	70	-	BAA17209	(8)
	ScpD	213 bp	213	nucleoid	70	-	BAA17469	(8)
	ScpE	174 bp	174	nucleoid	57	-	BAA16949	(8)

The following abbreviations are used: Hli, high light-induced; Ohp, one-helix protein; Scp, small Cab-like protein. N.d. no data available. References: (1) Jansson, 1999; (2) M.C. Lee, C.S. Kim and M.Y. Eun, 1997, direct GenBank submission; (3) Reith and Munholland, 1995; (4) Douglas and Penny, 1999; (5) V.L. Stirewalt, C.B. Michalowski, W. Löffelhardt, H.J. Bohnert and D.A. Bryant, 1996, direct GenBank submission; (6) Brahamsha and Haselkorn, 1991; (7) Dolganov et al., 1995; (8) Funk and Vermaas, 1999.

Fig. 4. Sequence alignment of the one-helix Hlips, Scps and Ohps from pro- and eukaryota. Identical amino acids are shown on a gray background, a solid line marks the position of a predicted transmembrane helix and arrowheads indicate the positions of conserved amino acids discussed in text. Sequences were aligned using the ClustalW (Thompson et al., 1994) program with manual correction of gaps. For origin of sequences see Table 3.

here is the Ohp from rice, in which the conserved alanine is missing and one of the conserved glycines is replaced by an aspartic acid residue. A similar replacement of one of the conserved glycines by an asparagine residue has occurred in one-helix Elip from *Guillardia theta* (Fig. 4). In addition, the Ohp from rice differs from the other one-helix Elips by the presence of AFNGPAP motif, which is characteristic for three-helix Elips and their relatives from higher plants (see Section III.A). Interestingly, the Ohp from rice contains a relatively long C-terminal tail, with a stretch of prolines and several positively charged arginine and lysine (K) residues, whose the physiological relevance is still not known.

IV. Genomic Organization of Elip Family in *Arabidopsis thaliana*

Recently, genomic sequences of Elips and Seps were localized in the *Arabidopsis* chromosomes. The *Elip1* gene was localized in chromosome III (accession: AB02223), the *Elip2*, *Sep1* and *Sep3* in chromosome IV (accessions: Z97336, AL035521, Z97343, respectively) and the *Sep2* in chromosome II (accession: AC007019). The genomic localization of Ohp from *Arabidopsis* is still not known.

The coding regions of *Elip1* and *Elip2* genes have a similar genomic structure, comprising three exons and two introns. The positioning of the introns is conserved between the two *Elip* genes and is shown schematically in Fig. 5. More diverse genomic organization was reported for *Sep* genes (Heddad and Adamska, 2000). The number of introns varied here from one for *Sep2* to two and three for *Sep3* and *Sep1* genes, respectively. In contrast to Elips the positions of the introns in *Sep* genes were not conserved (Fig. 5). The scattered distribution of *Elip* and *Sep* genes in the *Arabidopsis* genome indicated that the translocation of these genes from chloroplast to nuclear genome has occurred individually. The identical positions of the introns in the *Elip1* and *Elip2* genes and their overall sequence homology suggest that these two genes might have evolved through a duplication event.

V. Similarities and Differences Between the *Elip* and *Cab* Gene Families

Based on the sequence similarities between Elips and Cab proteins, the Elips were counted among an extended *Cab* gene family (Grimm et al., 1989; Green et al., 1991; Green and Pichersky, 1994; Green and Kühlbrandt, 1995). The *Cab* gene family in *Arabidopsis* consists of at least 21 different members, which share the same characteristics (Jansson, 1999). During the last few years, a sufficient amount of information has been accumulated on Elips and related proteins (reviewed in Green et al., 1991; Adamska and Kloppstech, 1994b; Adamska, 1997) as well as on Cab family members (reviewed in Green and Pichersky 1994; Jansson, 1994, 1999; Green and Kühlbrand, 1995; Paulsen, 1995, 1997; Green and Durnford, 1996; Horton et al., 1996; Sandona et al., 1998) enabling a broad and detailed comparison between these two families of proteins.

Despite several similar features shared by Elips and Cabs there are also very pronounced differences between these two families of proteins. These differences concern the primary and secondary structure of proteins, gene expression patterns, the targeting, insertion and thylakoid location of proteins, the quantity and quality of bound pigments and their binding characteristics, and finally, as a consequence of these differences, their distinct physiological functions. The differences between Elips and Cab proteins are discussed below.

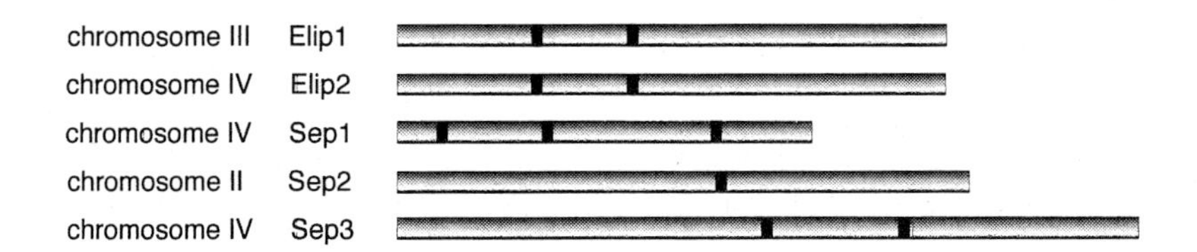

Fig. 5. Locations of *Elip* and *Sep* genes in the *Arabidopsis* chromosomes. Coding regions of *Elip* and *Sep* genes are indicated with the schematic positions of introns (black bars).

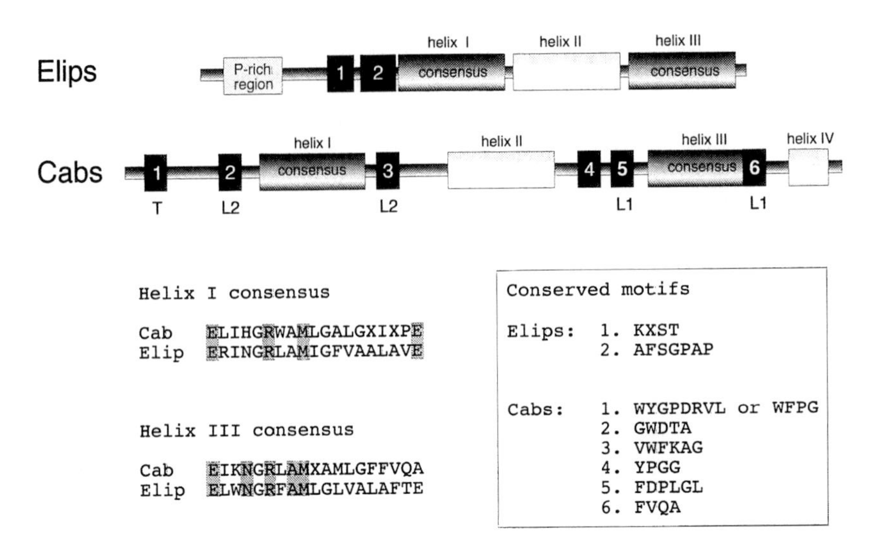

Fig. 6. Schematic comparison of the predicted primary and secondary structures of higher plant Elips and Cab proteins. Consensus sequences in helices I and III have several conserved amino acid residues common for both Elip and Cap protein families (indicated in grey background). Conserved motifs specific for Elips and Cabs sequences are indicated in a separate box. L1 and L2 mark lutein-binding sites and T indicates a site involved in trimerization of Lhcb (according to Bassi et al. 1997).

A. Primary and Secondary Structure of Elips and Cabs

The similarities and differences in the primary and secondary structure of three-helix Elips and Cab proteins are shown schematically in Fig. 6. Since several reviews published in the past covered this topic extensively (Green et al., 1991; Adamska and Kloppstech, 1994b; Green and Pichersky, 1994; Green and Kühlbrandt; 1995; Green and Durnford, 1996; Adamska, 1997), this aspect will be discussed here only very briefly.

Several conserved amino acid residues, located in helices I and III of Elips and related proteins and in helix I of Seps, Hlip, Scps or Ohps, are shared between the Elip and Cab families. These amino acid residues involve two glutamic acids, an arginine and a methionine in the helix I and a glutamic acid, an asparagine, an arginine, an alanine and a methionine in helix III. There are also several amino acids, which are similar in the three-helix Elips and Cab proteins. Helices II of Elips and Cab proteins are very variable and do not contain any identical or similar amino acids.

The major differences between the Elip and Cab primary and secondary structures are: (i) a fourth short amphipathic α-helix, which is present at the C-terminus of Cab proteins but missing in Elips; (ii) the connecting loops between transmembrane helices of Elips that are much shorter than those of Cab proteins; (iii) the presence of conserved amino acid motifs, which differ in both families of proteins; and (iv) the N-terminus of three-helix Elips and related proteins that contains a proline-rich domain, which is missing from Cab proteins. However, the two latter features are present only in three-helix Elips and related proteins. In two-helix Seps and one-helix Hlips, Scps or Ohps these N-terminal conserved motifs and the proline-rich domain are missing.

B. Expression Pattern of Elips and Cabs

The most pronounced difference between Elip and Cab families is that Cab proteins are structural components of the photosynthetic membranes, whereas Elips are induced only transiently under certain physiological conditions. In contrast to Cabs, which are the most abundant membrane proteins on earth (Jansson, 1999), Elips are less abundant and accumulate in the membranes in substoichiometric amounts. Based on in vivo labeling studies it has been calculated that one Elip molecule accumulated per 10–20 PS II reaction centers, when mature green pea leaves are exposed to light stress (Adamska, 1997).

Generally, three different types of conditions trigger the expression of Elip family members: light stress, dehydrative processes, and morphogenesis. In addition to the regulation of *Elip* gene expression by external factors, there is also an endogenous mechanism, the circadian rhythms, controlling the expression of *Elip* genes (Kloppstech, 1985; Adamska et al., 1991; Beator and Kloppstech, 1993).

Detailed studies on the expression of *Elip* genes performed on pea (M. Heddad and I. Adamska, unpublished) and *Arabidopsis* (Heddad and Adamska, 1998, 2000; M. Heddad, K. Haußühl and I. Adamska, unpublished) have revealed that Elip induction is strictly controlled by light stress in both plant species. Other physiological conditions, such as cold stress, heat shock, desiccation, wounding, salt stress, oxidative stress, hormone treatment or senescence have not been found to affect the expression of these genes. Similarly to Elips, transcription of the *Arabidopsis* Seps, for which low levels of transcripts were present under ambient light conditions, was also significantly enhanced by light-stress treatment but not by other abiotic stress conditions tested (Heddad and Adamska, 2000). Furthermore, the induction of Elips and Seps in *Arabidopsis* was triggered specifically by blue and UV-A light (M. Heddad and I. Adamska, unpublished). A similar regulation of *Elip* gene expression by blue and UV-A light was reported in the past for Elip in pea (Adamska et al., 1992a,b, 1993) or Hlip in *Synechococcus* (Dolganov et al., 1995). Thus, it is expected that also the expression of other members of the Elip family might be ascribed to the activation of a cryptochrome-like receptor, which senses the light stress signal and transduces this stimuli to *Elip* genes.

Light stress triggers the induction of Elips in barley (Pötter and Kloppstech, 1993) and spinach (Lindahl et al., 1997) as well as Elip-related proteins in wheat (Shimosaka et al., 1999) and the green alga *Dunaliella* (Lers et al., 1991). In some plant species these genes are induced not only by light stress but also by other physiological stress conditions, which are summarized in Table 4.

Several stress conditions, such as desiccation, osmotic stress, cold stress and salt stress, all of them reported to lead to dehydrative processes in plants, have been shown to induce the expression of Elip family members (Table 4). Since it is widely accepted that abscisic acid (ABA) mediates general adaptive responses to drought (Zeevaart and Creelman, 1988), the induction of Elips during a treatment of plants with this hormone is also interpreted as a dehydrative response (Table 4).

The induction of *Elip* and related genes during dehydrative responses is triggered either by the dehydration process alone or an additional light factor is required. With respect to the light intensity, needed for Elip induction, the reports have been very contradictory (Table 4). For example, light was not necessary for induction of the Sdi1 (sunflower drought-induced) protein in the common sunflower during desiccation (Ouvrard et al., 1996), the Wcr12 (wheat cold-regulated) protein in wheat during a cold-stress treatment (Shimosaka et al., 1999) or the Ode1 (osmotic differentially expressed) protein in *Capsicum* during an exposure to osmotic stress (Y.M. Verastegui-Pena and N. Ochoa-Alejo, 1999, direct GenBank submission). In contrast, low-intensity light was crucial for induction of the Dsp22 in *Craterostigma* (Bartels et al., 1992) and the Wcr12 protein in wheat (Shimosaka et al., 1999) during desiccation or treatment with ABA. Also the induction of the low-molecular-mass Elips in barley (Montané et al., 1996, 1997) and the Cbr protein in *Dunaliella* (Król et al., 1997) during cold stress was found to be dependent on low-intensity light. High-intensity light was obligatory for Elip induction in pea exposed to cold stress, otherwise cold stress did not cause any effect on *Elip* gene expression (Adamska and Kloppstech, 1994a).

Considering these reports the question arises, what is the exact role of photon fluxes in the induction of Elip genes? It was proposed that it is not the light intensity *per se*, but the reduced state of the electron-transport chain that might control Elip induction in barley (Montané et al., 1996, 1997) or *Dunaliella* (Król et al., 1997) during exposure to cold-stress treatment. However, it is also possible that the signal transduction pathways acting under different stress conditions might have some steps in common and as a result, affect the induction of Elip genes.

All Elip family members investigated so far have been shown to be induced transiently during photomorphogenesis of etiolated seedlings exposed to light (Table 4). However, it is difficult to distinguish whether the induction of Elips during greening is directly related to the synthesis and assembly of the photosynthetic membranes or to the other factors, e.g. light stress. The pigment content of developing seedlings is very low so that the ambient light intensities could be sufficient to cause a light-stress syndrome.

A heat-shock treatment of etiolated seedlings has been reported to lead to morphogenic changes that resemble those obtained after illumination with far red light (Kloppstech et al., 1991; Beator et al., 1992; Otto et al., 1992). The cyclic heat-shock treatment

Table 4. Expression of Elip family members in different plant species under varying developmental and environmental conditions

Induction	Gene name	References
Light		
Light stress	*Arabidopsis-Elip1, Arabidopsis-Elip2, Arabidopsis-Sep1, Arabidopsis-Sep2, Arabidopsis-Sep3, Dunaliella-Cbr, Hordeum-HV58, Hordeum-HV60, Hordeum-HV90, Pisum-Elip, Spinacia-Elip19, Spinacia-Elip17, Spinacia-Elip16, Spinacia-Elip15, Synechococcus-HliA, Synechocystis-ScpB, Synechocystis-ScpC, Synechocystis-ScpD, Synechocystis-ScpE, Triticum-Wcr12*	(1-11)
Acclimation to light stress	*Spinacia-Elip19, Spinacia-Elip17, Spinacia-Elip16, Spinacia-Elip15*	(11)
Dehydrative processes		
Desiccation in the dark	*Helianthus-Sdi1*	(12)
Desiccation in connection with low light	*Craterostigma-Dsp22, Triticum-Wcr12*	(10, 13)
Abscisic acid in connection with low light	*Craterostigma-Dsp22, Triticum-Wcr12*	(10, 13)
Cold stress in the darkness	*Triticum-Wcr12*	(10)
Cold stress in connection with low light	*Dunaliella-Cbr, Hordeum-HV60*	(14, 15)
Cold stress in connection with light stress	*Pisum-Elip*	(16)
Osmotic stress	*Capsicum-Ode1*	(17)
Salt stress	*Triticum-Wcr12*	(10)
Morphogenesis		
Greening of etiolated seedlings	*Hordeum-HV58, Hordeum-HV60, Hordeum-HV90, Pisum-Elip*	(18-20)
Cyclic heat shock	*Hordeum-HV60, Hordeum-HV90*	(21, 22)
Other conditions		
Sulfate deprivation	*Dunaliella-Cbr*	(7)
Absence of PS I	*Synechocystis-ScpB, Synechocystis-ScpC, Synechocystis-ScpD, Synechocystis-ScpE*	(4)
Circadian rhythmicity	*Hordeum-HV58, Hordeum-HV60, Hordeum-HV90, Pisum-Elip*	(23, 24)
No expression data available	*Anabaena-orf2, Arabidopsis-Ohp, Cyanophora-ycf17, Guillardia-ycf17, Oryza-Elip, Oryza-Ohp, Porphyra- ycf17, Glycine-Elip*	

For abbreviations see Tables 1-3. (1) Adamska et al., 1992a; (2) Adamska et al., 1993; (3) Dolganov et al., 1995; (4) Funk and Vermaas, 1999; (5) Heddad and Adamska, 1998; (6) Heddad and Adamska, 2000; (7) Lers et al., 1991; (8) Moscovici-Kadouri and Chamovitz, 1997; (9) Pötter and Kloppstech, 1993; (10) Shimosaka et al., 1999; (11) Lindahl et al., 1997; (12) Ouvrard et al., 1996; (13) Bartels et al., 1992; (14) Montané et al., 1996; (15) Król et al., 1997; (16) Adamska and Kloppstech, 1994a; (17) Y.M. Verastegui-Pena and N. Ochoa-Alejo, 1999, direct GenBank submission; (18) Meyer and Kloppstech, 1984; (19) Grimm and Kloppstech, 1987; (20) Grimm et al., 1989; (21) Beator et al., 1992; (22) Beator and Kloppstech, 1993; (23) Kloppstech, 1985; (24) Adamska et al., 1991.

applied in the dark resulted in the accumulation of transcripts for two low-molecular-mass Elips in barley (Beator et al., 1992). Interestingly, a similar treatment had no effect on expression of the *Elip* gene in pea (Kloppstech et al., 1991). Although these conditions allowed separation of the effects of light and morphogenesis, a clear assignment of Elip induction to one of these stimuli failed due to the fact that the heat-shock treatment induced a circadian rhythmicity in plants (Hoober et al., 1984; Nagy et al., 1988; Kloppstech et al., 1991). Endogenous regulation of gene expression by circadian rhythms has been experimentally proven for pea and barley Elips (Kloppstech, 1985; Adamska et al., 1991).

There are also reports indicating that the sulfate deprivation connected to carotenogenesis induces expression of the *Cbr* gene in *Dunaliella* (Lers et al., 1991). In *Synechocystis* sp. PCC6803 four Scps were expressed in a mutant deficient in PS I (Funk and Vermaas, 1999). Originally it was thought that the expression of *Scp* genes was restricted only to PS I-less strains since no significant accumulation of Scp transcripts and proteins was noticed in wild type cells exposed to moderately high-intensity light. More recent observations indicated, however, that two of the *Scp* genes, the *ScpC* and *ScpD*, were induced in wild type *Synechocystis* cells grown under light-stress conditions for 12 hours (Funk and Vermaas, 1999). For the other two *Scp* genes, the *ScpB* and *ScpE*, such observations have not been reported.

It can be concluded that the conditions for the expression of Elip family members vary dramatically depending on plant species. It is not clear whether these differences in the expression are due to the species-specific responses against environmental stimuli or due to the different experimental setups used by research groups. More systematic studies are necessary to clarify this point.

C. Targeting, Insertion and Localization of Elips and Cabs in the Thylakoid Membrane

There are significant differences in the targeting and in the thylakoid membrane insertion of eukaryotic Elips and Cab proteins, as well as in their intrathylakoid localization. Unfortunately, such information is not yet available for eukaryotic Seps or prokaryotic Hlips, Scps or Ohps.

At least four different, protein-specific insertion pathways into the thylakoid membranes of higher plant chloroplasts have been reported (Robinson and Klösgen, 1994; Klösgen, 1997; Robinson et al., 1998). The Sec-dependent pathway requires nucleoside triphosphates and a SecA protein in the stroma, the ΔpH-dependent route is exclusively dependent on the trans-thylakoid proton gradient, the SRP (signal recognition particle)-dependent pathway requires a trans-thylakoid ΔpH, GTP and a chloroplast homologue of the 54 kDa subunit of the SRP from endoplasmic reticulum and finally, a spontaneous protein transport mechanism also functions in the thylakoid membranes.

A spontaneous insertion mechanism into the thylakoid membrane has been reported for Elip2 in *Arabidopsis* (Kim et al., 1999). A similar spontaneous insertion has been suggested to occur for the low-molecular-mass Elips in barley (Kruse and Kloppstech, 1992). The latter proteins were shown to be inserted into the thylakoid membrane without a requirement for membrane surface proteins or stromal factors. However, N-terminal processing of the precursor Elips by a stromal processing peptidase has to precede the insertion of Elips into the thylakoid membrane (Kruse and Kloppstech, 1992).

The thylakoid insertion studies performed for three Cab gene family members, the Lhcb1 (Li et al., 1995), Lhcb5 and Lhca1 (Kim et al., 1999) have indicated the involvement of the SRP-dependent pathway. A requirement of stromal and membrane protein transport machinery (Kim et al., 1999) but no N-terminal processing (Cline et al., 1989) have been reported for the insertion of Cab proteins into the thylakoid membrane.

There is only very limited information available on the intrathylakoid localization of Elip family members, this information being restricted to higher plant and green algae Elips and related proteins. Localization studies carried out on pea (Cronshagen and Herzfeld, 1990; Adamska and Kloppstech, 1991) and barley (Montané et al., 1999) Elips have shown that these proteins are located in the non-appressed regions of the thylakoid membranes. These regions consist of stroma lamellae, outer parts and margins of the grana stacks. Crosslinking studies performed on pea Elip revealed that this protein is located in the vicinity of the D1-protein of the PS II reaction center (Adamska and Kloppstech, 1991). However, there is also a contradictory report indicating that Elips in the developing pea seedlings are associated with both PS I and PS II (Cronshagen and Herzfeld, 1990).

Recently, it was shown that the low-molecular-mass Elips of barley can form a high-molecular-mass complex of over 100 kDa with unknown components of the thylakoid membrane under combined light- and cold-stress conditions (Montané et al., 1999). Similar high-molecular-mass Elip complexes composed of several unidentified polypeptides of 24–26 kDa were found in pea leaves exposed to light stress (Adamska et al., 1999). The formation of a large protein complex with the minor light-harvesting antenna proteins was proposed for the Cbr protein from *Dunaliella* (Lers et al., 1991; Levy et al., 1992, 1993). Although the existence of

Elip complexes is experimentally well established for many plant species, their composition and molecular organization still remain to be elucidated.

VI. Are Elips Chlorophyll-Binding Proteins?

According to the high resolution structure determined for Lhcb1 (Kühlbrandt et al., 1994), seven chlorophyll *a*, five chlorophyll *b* and two lutein molecules, in addition to non-stoichiometric amounts of violaxanthin and neoxanthin (Kühlbrandt et al., 1994; Bassi et al., 1997) are bound to this protein. Two lutein binding sites, L1 and L2, have been localized by electron crystallography in helices I and III of Lhcb1 (Kühlbrandt et al., 1994). Recently, it was shown that neoxanthin was bound to a distinct N1 site and an additional peripheral violaxanthin-binding site (V1) was postulated (Croce et al., 1999). It was proposed that the V1 binding site might be stabilized by Lhcb1 trimerization or occur only in a subset of Lhcb1-3, whose expression is enhanced under light-stress conditions (Croce et al., 1999).

Recent purification of Elip from light-stressed pea leaves has shown that this protein binds chlorophyll *a* and lutein (Adamska et al., 1999). The chlorophyll *b* and other carotenoids were not detected in Elip preparations. The quality and quantity of pigments bound to Elip, as well as their binding characteristics, differed significantly from those reported for Cab proteins (Kühlbrandt et al., 1994). First, a very unusual chlorophyll : lutein ratio of 0.5 was calculated for purified Elip. This value markedly deviated from the values reported for other chlorophyll-binding proteins, such as PSII-S (chlorophyll : carotenoid = 4.9), Lhcb4 (chlorophyll : carotenoid = 3.1), Lhcb5 (chlorophyll : carotenoid = 3.9) or Lhcb6 (chlorophyll: carotenoid = 3.0) (Jansson, 1994; Kühlbrandt et al., 1994; Funk et al., 1995; Paulsen, 1995, 1997; Green and Durnford, 1996; Bassi et al., 1997). In this respect Elips were considered to be chlorophyll-binding proteins with an extremely high xanthophyll content (Adamska et al., 1999). Furthermore, a very weak excitonic coupling between bound chlorophyll molecules was measured (Adamska et al., 1999). This strongly suggests that these chlorophylls are not involved in energy transfer but fulfill completely different functions (see below).

It is not yet experimentally confirmed whether other members of the Elip family, like Seps, Hlips, Scps and Ohps, bind chlorophylls. Analysis of the conserved amino acid residues present in helix I of Seps, Hlips, Scps and Ohps have revealed that these proteins contain conserved asparagine, glutamic and aspartic acid residues capable of binding chlorophylls, as shown in studies of the purple bacterial reaction center (Coleman and Youvan, 1990) and of Lhcb1 from higher plants (Kühlbrandt et al., 1994). Four such putative chlorophyll-binding residues are present in helices I and III of all three-helix Elips and their relatives, one in helix I of Seps and two in helix I of Hlips, Scps and Ohps (Figs. 2, 3 and 4). However, to provide a fifth ligand to the Mg atom of chlorophyll, the negative charges of aspartic or glutamic acids have to be neutralized by an ionic bridge to an arginine side chains, as was reported for Lhcb1 (Kühlbrandt et al., 1994). In this respect all three-helix Elips and their relatives as well as the homodimers of Seps, Hlips, Scps and Ohps fulfill the theoretical requirements for binding of chlorophyll.

In addition, the stromal N-terminal part of Sep1 and Sep2 contains a few predicted myristoylation sites, such as GGNRAA located between 63–68 amino acids in Sep1 and GGDLAE and GGDGGG located between 39–44 amino acids and 71–76 amino acids in Sep2, respectively. Thus, it is possible that the covalent addition of myristate (a C14-saturated fatty acid) might promote the attachment of the N-terminal part of Seps to the membranes via lipids and thus provide additional ligands for binding of chlorophyll.

VII. Possible Physiological Functions of Elip Family Members

It is proposed that Elip family members might fulfill protective functions within the thylakoids either by binding free chlorophyll molecules and preventing the formation of free radicals (Adamska et al., 1992a,b) and/or by binding of the xanthophyll cycle pigments and participating in thermal dissipation of light energy (Król et al., 1995, 1997, 1999; Braun et al., 1996). These functions do not exclude each other and can be performed in parallel.

Transient pigment-carrier or chlorophyll-exchange functions have been postulated for Elips in higher plants and the green alga *Dunaliella* (Lers et al., 1991; Adamska et al., 1992a,b). Under light-stress conditions, an increased turnover of chlorophyll-binding proteins has been reported (reviewed in Prasil et al., 1992; Aro et al., 1993). A similar high turnover of pigment-proteins is expected to accompany other physiological conditions, such as dehydrative processes or morphogenesis. Degradation of damaged pigment-proteins is expected to be connected with the release of free chlorophylls, which in turn can be easily sensitized by light absorption and form triplet chlorophylls. These excited chlorophyll species are able to react with molecular oxygen and generate very reactive intermediates, like singlet oxygen, which can induce photooxidative damage within the thylakoid membrane. Binding of liberated chlorophylls by Elips and related proteins and preventing the formation of toxic oxygen species would consequently be very significant for the protection of thylakoid membrane components against photooxidative damage. A low excitonic coupling between chlorophyll molecules reported for isolated Elips (Adamska et al., 1999), as well as their localization in the non-appressed regions of the thylakoid membranes (Cronshagen and Herzfeld, 1990; Adamska and Kloppstech, 1991), known to be the site of assembly/disassembly for pigment-protein complexes, would speak in favor of such pigment-carrier or pigment-exchange functions.

Induction of Cbr in correlation with carotenogenesis in *Dunaliella* led to the conclusion that the role of this protein could be to deposit newly formed carotenoids in the thylakoid membranes in a form available for later integration into the light-harvesting and reaction center complexes (Lers et al., 1991). Another related function of Cbr would be to bind xanthophyll zeaxanthin and provide photoprotection of the assembling antenna and reaction center complexes (Lers et al., 1991; Braun et al., 1996; Król et al., 1997).

Under light stress conditions the xanthophyll zeaxanthin is formed through the light-regulated deepoxidation of violaxanthin with the antheraxanthin as an intermediate (Yamamoto, 1979). This conversion is referred to as the xanthophyll cycle (reviewed in Demmig-Adams and Adams, 1992; Pfündel and Bilger, 1994; Eskling et al., 1997). Evidence for the physical association of Cbr and zeaxanthin in *Dunaliella* was provided by non-denaturating gel electrophoresis (Levy et al., 1993). It was proposed that Cbr protein binds zeaxanthin to form photoprotective complexes within the minor light-harvesting antennae of PS II (Levy et al., 1993). Correlations were observed between zeaxanthin accumulation and the non-photochemical quenching of chlorophyll fluorescence under light intensities exceeding the photosynthetic capacity (Demmig-Adams and Adams, 1992; Pfündel and Bilger, 1994). Such excessive fluxes were generated either by high intensity illumination or by lower light intensities in connection with other stress conditions, such as cold stress or dehydration (Demmig-Adams and Adams, 1992). Thus, putative zeaxanthin-Cbr complexes were proposed to play a role in non-photochemical quenching of chlorophyll fluorescence and preventing photooxidative damage of the photosynthetic machinery (Braun et al., 1996).

However, there are some discrepancies between the experimental data and proposed Elip function(s). First, comparisons of the induction kinetics indicated that in light-stressed *Dunaliella* cells zeaxanthin accumulated more rapidly than Cbr protein (Braun et al., 1996). Also in higher plants the rates of Elip induction preceded that of the zeaxanthin accumulation (Adamska et al., 1992b; Demmig-Adams and Adams, 1992; Demmig-Adams et al., 1996). To explain this discordance it was proposed that two types of reversible non-photochemical quenching mechanisms exist in plants: the initial trans-thylakoid ΔpH-dependent mechanism and the other one requiring zeaxanthin-Cbr complexes (Braun et al., 1996).

Second, the purified Elip does not contain bound zeaxanthin (Adamska et al., 1999). The absence of the xanthophyll cycle pigments in isolated pea Elip could be easily explained by dissociation of part of the pigments from the protein and their loss during purification. It was reported recently (Croce et al., 1999) that the L1 and L2 carotenoid-binding sites in Lhcb1 have the highest affinity to lutein, but might also bind violaxanthin and zeaxanthin with lower affinity. If similar carotenoid binding sites, as in Lhcb1, exist in Elips it would not be surprising that the purified protein lacks the latter pigments.

More difficult to explain are controversies in the localization of Elips and xanthophyll cycle pigments in the thylakoid membrane. It has been reported that higher plant Elips accumulate in non-appressed regions of the thylakoid membranes (Cronshagen and Herzfeld, 1990; Adamska and Kloppstech, 1991), whereas the xanthophyll cycle pigments are found in both appressed and non-appressed regions of the thylakoids (reviewed in Eskling et al., 1997).

An extremely high carotenoid content of Elips in higher plants would support the function both in the scavenging of free chlorophylls and in dissipation of excessive energy. On one hand, the presence of carotenoids would allow the binding of free chlorophylls and protect the Elips against destruction (Demmig-Adams et al., 1996; Yamamoto and Bassi, 1996), on the other hand a high carotenoid content would increase the capability for non-photochemical quenching (Horton et al., 1996).

On the basis of available information no definite statement about Elip function(s) can be made and additional experiments are necessary to prove the present working hypothesis. One should mention here that the deletion of *Hli*A gene from cyanobacterium *Synechococcus* did not lead to any changes in the phenotype or growth of the cells (Dolganov et al., 1995). Single mutants lacking *ScpB*, *ScpC*, *ScpD* or *ScpE* genes in *Synechocystis* had a relatively normal phenotype (Funk and Vermaas, 1999). Also the Elip-antisense tobacco plants did not show any impairments, even when grown under light-stress conditions (Montané et al., 1999). Thus, the function(s) of Elips and related proteins seem to be more complex and will require more sophisticated genetic and/or biochemical approaches in order to be elucidated.

VIII. Evolutionary Aspects

It is assumed that the light-harvesting proteins of all photosynthetic eukaryotes have a common origin and this supports the idea that chloroplasts have a common ancestor (Wolfe et al., 1994; Tomitani et al., 1999). It has previously been proposed (Green and Pichersky, 1994; Green and Kühlbrandt, 1995) that the three-helix members of the Cabs, Fcps and Elips have originated from a four-helix protein. This hypothesis was based on the discovery of the Cab-related PSII-S protein with four transmembrane helices, where helices I and III as well as II and IV were clearly related (Kim et al., 1992; Wedel et al., 1992). The deletion of the fourth helix might give rise to the three-helix ancestors of the eukaryotic antenna proteins and Elips. Furthermore, it was suggested that the PSII-S protein arose from a two-helix protein as the result of internal gene duplication and fusion. It was speculated that the two-helix ancestor of the PSII-S protein might originate from a fusion of a one-helix *Hlip*-like gene with a gene for another one-helix transmembrane protein of unknown origin. On the basis of the high degree of homology between a cyanobacterial Hlip, the first helix of Seps and helices I and III of Cabs in *Arabidopsis,* the two-helix Seps are proposed to represent an evolutionary missing link between one- and three-helix antenna proteins of pro- and eukaryota (Heddad and Adamska, 2000).

IX. Concluding Remarks

This chapter summarizes our current knowledge about transiently expressed pigment-binding proteins related to the *Cab* gene family. Although significant progress has been made towards understanding of the regulation and function(s) of this group of proteins, there are still many open questions that have to be answered. The most basic questions are: How many different Elips are expressed in one plant species? Are all of them expressed with the same kinetics under similar stress conditions?

Very little is known about the location of Elips and related proteins within the thylakoid membrane and their interactions with other proteins. No information exists as to whether the different Elip species can form one functional multisubunit complex. Finally, the exact function(s) of these proteins is not yet clear.

The progress in sequencing plant genomes offers a good opportunity for identification and isolation of novel Elip family members. These proteins might escape our current detection using classical molecular biology or biochemistry methods due to the fact that their expression is related to very specific physiological conditions or their transcripts are expressed at a very low level. The information about the number of Elips in plants would be very useful for the functional analysis of these proteins. Since it is expected that the Elip family members can substitute for each other in their functions, double or even multiple Elip mutants will be needed to find out their role in plants.

Acknowledgments

This work was supported by research grants from the Swedish Natural Science Research Council, the Swedish Strategic Foundation and the Carl Tryggers Foundation. I thank Dr. Patrick Dessi for comments on the manuscript.

References

Adamska I (1997) Elips-light-induced stress proteins. Physiol Plant 100: 794–805

Adamska I and Kloppstech K (1991) Evidence for an association of the early light-inducible protein (Elip) of pea with Photosystem II. Plant Mol Biol 16: 209–223

Adamska I and Kloppstech K (1994a) Low temperature increases the abundance of early light-inducible transcript under light stress conditions. J Biol Chem 269: 30221–30226

Adamska I and Kloppstech K (1994b) The role of early light-induced proteins (Elips) during light stress. In: Baker NR and Bowyer JR (eds) Photoinhibition of Photosynthesis: From Molecular Mechanisms to the Field, pp 205–219. Bios Scientific Publishers, Oxford

Adamska I, Scheel B and Kloppstech K (1991) Circadian oscillations of nuclear-encoded chloroplast proteins in pea (*Pisum sativum*). Plant Mol Biol 14: 1055–1065

Adamska I, Kloppstech K and Ohad I (1992a) UV light stress induces the synthesis of the early light-inducible protein and prevents its degradation. J Biol Chem 267: 24732–24737

Adamska I, Ohad I and Kloppstech K (1992b) Synthesis of the early light-inducible protein is controlled by blue light and related to light stress. Proc Natl Acad Sci USA 89: 2610–2613

Adamska I, Kloppstech K and Ohad I (1993) Early light-inducible protein in pea is stable during light stress but is degraded during recovery at low light intensity. J Biol Chem 268: 5438–5444

Adamska I, Roobol-Bóza M, Lindahl M and Andersson B (1999) Isolation of pigment-binding early light-inducible proteins from pea. Eur J Biochem 260: 453–460

Aro EM, Virgin I and Andersson B (1993) Photoinhibition of Photosystem II. Inactivation, protein damage and turnover. Biochim Biophys Acta 1143: 113–134

Bartels D, Hanke C, Schneider K, Michel D and Salamini F (1992) A desiccation-related *Elip*-like gene from the resurrection plant *Craterostigma plantagineum* is regulated by light and ABA. EMBO J 11: 2771–2778

Bassi R, Sandona D and Croce R (1997) Novel aspects of chlorophyll *a/b*-binding proteins. Physiol Plant 100: 769–779

Beator J and Kloppstech K (1993) The circadian oscillator coordinates the synthesis of apoproteins and their pigments during chloroplast development. Plant Physiol 103: 191–196

Beator J, Pötter E and Kloppstech K (1992) The effect of heat shock on morphogenesis in barley. Plant Physiol 100: 1780–1786

Brahamsha B and Haselkorn R (1991) Isolation and characterization of the gene encoding the principal sigma factor of the vegetative cell RNA polymerase from the cyanobacterium *Anabaena* sp. strain PCC 7120. J Bacteriol 173: 2442–2450

Braun P, Banet G, Tal T, Malkin S and Zamir A (1996) Possible role of Cbr, an algal early light-induced protein, in nonphotochemical quenching of chlorophyll fluorescence. Plant Physiol 110: 1405–1411

Cavalier-Smith T (1992) The number of symbiotic origins of organelles. BioSystems 28: 91–106

Cline K, Fulsom DR and Viitanen PV (1989) An imported thylakoid protein accumulates in the stroma when insertion into thylakoids is inhibited. J Biol Chem 264: 14225–14232

Coleman WJ and Youvan DC (1990) Spectroscopic analysis of genetically modified photosynthetic reaction centers. Annu Rev Biophys Chem 19: 333–367

Croce R, Weiss S and Bassi R (1999) Carotenoid-binding sites of the major light-harvesting complex II of higher plants. J Biol Chem 274: 29613–29623

Cronshagen U and Herzfeld F (1990) Distribution of early light inducible proteins in the thylakoids of developing pea chloroplasts. Eur J Biochem 193: 361–366

Demmig-Adams B and Adams III WW (1992) Photoprotection and other responses of plants to high light stress. Annu Rev Plant Physiol Plant Mol Biol 43: 599–626

Demmig-Adams B, Gilmore AM and Adams III WW (1996) In vivo function of carotenoids in higher plants. FASEB J 10: 403–412

Dolganov NA, Bhaya D and Grossman AR (1995) Cyanobacterial protein with similarity to the chlorophyll *a/b* binding proteins of higher plants: evolution and regulation. Proc Natl Acad Sci USA 92: 636–640

Douglas SE and Penny SL (1999) The plastid genome of the cryptophyte alga, *Guillardia theta*: Complete sequence and conserved synteny groups confirm its common ancestry with red algae. J Mol Evol 48: 236–244

Eskling M, Arvidsson PO and Åkerlund HE (1997) The xanthophyll cycle: Its regulation and components. Physiol Plantarum 100: 806–816

Funk C and Vermaas W (1999) A cyanobacterial gene family coding for single-helix proteins resembling part of the light-harvesting proteins from higher plants. Biochemistry 38: 9397–9404

Funk C, Schröder W, Napiwotzki A, Tjus S, Renger G and Andersson B (1995) PSII-S protein of higher plants: A new type of pigment-binding protein. Biochemistry 34: 11133–11141

Glazer AN (1989) Light guides. Directional energy transfer in a photosynthetic antenna. J Biol Chem 264: 1–4

Golbeck JH (1994) Photosystem I in cyanobacteria. In: Bryant DA (ed) The Molecular Biology of Cyanobacteria, pp 319–360. Kluwer Academic Publishers, Dordrecht

Green BR and Durnford DG (1996) The chlorophyll-carotenoid proteins of oxygenic photosynthesis. Annu Rev Plant Physiol Plant Mol Biol 47: 685–714

Green BR and Kühlbrandt W (1995) Sequence conservation of light-harvesting and stress-response proteins in relation to the three-dimensional molecular structure of LHCII. Photosynth Res 44: 139–148

Green BR and Pichersky E (1994) Hypothesis for the evolution of three-helix chlorophyll *a/b* and chlorophyll *a/c* light-harvesting antenna proteins from two-helix and four-helix ancestors. Photosynth Res 39: 149–162

Green BR, Pichersky E and Kloppstech K (1991) Chlorophyll *a/b*-binding proteins: An extended family. Trends Biochem Sci 16: 181–186

Grimm B and Kloppstech K (1987) The early light-inducible proteins of barley: Characterization of two families of 2-h-specific nuclear-coded chloroplast proteins. Eur J Biochem 167: 493–499

Grimm B, Kruse E and Kloppstech K (1989) Transiently expressed early light-inducible thylakoid proteins share transmembrane domains with light-harvesting chlorophyll-binding proteins. Plant Mol Biol 13: 583–593

Grossman AR, Schäfer MR, Chiang GG and Collier JL (1993)

The phycobilisome, a light-harvesting complex responsive to environmental conditions. Microbiol Rev 57: 725–749

Heddad M and Adamska I (1998) Non-structural putative chlorophyll-binding proteins of *Arabidopsis thaliana*. In: Garab G (ed) Photosynthesis: Mechanisms and Effects, pp 389–392. Kluwer Academic Publishers, Dordrecht

Heddad M and Adamska I (2000) Light stress-regulated two-helix proteins in *Arabidopsis thaliana* related to the chlorophyll *a/b*-binding gene family. Proc Natl Acad Sci USA 97:3741–3746

Hoober JK, Marks DB, Keller BJ and Margulies MM (1984) Regulation of accumulation of the major thylakoid polypeptides in *Chlamydomonas reinhardtii*. J Cell Biol 95: 552–558

Horton P, Ruban AV and Walters RG (1996) Regulation of light harvesting in green plants. Annu Rev Plant Physiol Plant Mol Biol 47: 655–684

Jansson S (1994) The light-harvesting chlorophyll *a/b*-binding proteins. Biochim Biophys Acta 1184: 1–19

Jansson S (1999) A guide to the identification of the *Lhc* genes and their relatives in *Arabidopsis*. Trends Plant Sci 4: 236–240

Kessler U, Maid U and Zetsche K (1992) An equivalent to bacterial *omp'R* genes is encoded on the plastid genome of red algae. Plant Mol Biol 18: 777–780

Kim SJ, Sandusky P, Bowlby NR, Aebersold R, Green BR, Vlahakis S, Yokum CF and Pichersky E (1992) Characterization of a spinach *psbS* cDNA encoding the 22 kDa protein of Photosystem II. FEBS Lett 314: 67–71

Kim SJ, Jansson S, Hoffman NE, Robinson C and Mant A (1999) Distinct 'assisted' and 'spontaneous' mechanisms for the insertion of polytopic chlorophyll-binding proteins into the thylakoid membranes. J Biol Chem 274: 4715–4721

Kloppstech K (1985) Diurnal and circadian rhythmicity in the expression of light-induced plant nuclear messenger RNAs. Planta 165: 502–506

Kloppstech K, Otto B and Sierralta W (1991) Cyclic temperature treatments of dark-grown pea seedlings induce a rise in specific transcript levels of light-regulated genes related to photomorphogenesis. Mol Gen Gent 225: 468–473

Kolanus W, Scharnhorst C, Kühne U and Herzfeld F (1987) The structure and light-dependent transient expression of a nuclear-encoded chloroplast protein gene from pea (*Pisum sativum* L.). Mol Gen Genet 209: 234–239

Klösgen RB (1997) Protein transport into and across the thylakoid membrane. J Photochem Photobiol 38: 1–9

Król M, Spangfort MD, Huner NPA, Öquist G, Gustafsson P and Jansson S (1995) Chlorophyll *a/b*-binding proteins, pigment conversions, and early light-induced proteins in a chlorophyll *b*-less barley mutant. Plant Physiol 107: 873–883

Król M, Maxwell DP and Huner NPA (1997) Exposure of *Dunaliella salina* to low temperature mimics the high light-induced accumulation of carotenoids and the carotenoid-binding protein (Cbr). Plant Cell Physiol 38: 213–216

Król M, Ivanov AG, Jansson, S, Kloppstech K and Huner NP (1999) Greening under high light and cold temperature affects the level of xanthophyll-cycle pigments, early light-inducible proteins, and light-harvesting polypeptides in wild-type barley and the *chlorina f2* mutant. Plant Physiol 120: 193–204

Kruse E and Kloppstech K (1992) Integration of early light-inducible proteins into isolated thylakoid membranes. Eur J Biochem 208: 195–202

Kühlbrandt W, Wang DN and Fujiyoshi Y (1994) Atomic model of plant light-harvesting complex by electron crystallography. Nature 367: 614–621

La Roche J, van der Staay GWM, Partensky F, Ducret A, Aebersold R, Li R, Golden SS, Hiller RG, Wrench PM, Larkum AWD and Green BR (1996) Independent evolution of the prochlorophyte and green plant chlorophyll *a/b* light-harvesting proteins. Proc Natl Acad Sci USA 93: 15244–15248

Lers A, Levy H and Zamir A (1991) Co-regulation of a gene homologous to early light-induced genes in higher plants and *β*-carotene biosynthesis in the alga *Dunaliella bardawil*. J Biol Chem 266: 13698–13705

Levy H, Gokhman, I and Zamir A (1992) Regulation and light-harvesting complex II association of a *Dunaliella* protein homologous to early light-induced proteins in higher plants. J Biol Chem 267: 18831–18836

Levy H, Tal T, Shaish A and Zamir A (1993) Cbr, an algal homolog of plant early light-induced proteins is a putative zeaxanthin binding protein. J Biol Chem 268: 20892–20896

Li X, Henry R, Yuan J, Cline K and Hoffman NE (1995) A chloroplast homologue of the signal recognition particle subunit SRP54 is involved in the posttranslational integration of a protein into thylakoid membranes. Proc Natl Acad Sci USA 92: 3789–3793

Lindahl M, Funk C, Webster J, Bingsmark S, Adamska I and Andersson B (1997) Expression of Elips and PSII-S protein in spinach during acclimative reduction of the Photosystem II antenna in response to increased light intensities. Photosynth Res 54: 227–236

Löffelhardt W and Bohnert H (1994) Molecular biology of cyanelle. In: Bryant DA (ed) The Molecular Biology of Cyanobacteria, pp 65–89. Kluwer Academic Publishers, Dordrecht

Matthijs HCP, van der Staay GWM and Mur LR (1994) *Prochlorophytes*: the 'other' cyanobacteria? In: Bryant DA (ed) The Molecular Biology of Cyanobacteria, pp 49–64. Kluwer Academic Publishers, Dordrecht

Meyer G and Kloppstech K (1984) A rapidly light-induced chloroplast protein with a high turnover coded for by pea nuclear DNA. Eur J Biochem 138: 201–207

Montané MH, Dreyer S and Kloppstech K (1996) Post-translational stabilization of Elips and regulation of other light stress genes under prolonged light- and cold-stress in barley. In: Grilo S and Leone A (eds) Physical Stresses in Plants: Genes and their Products for Tolerance, pp 210–222. Springer, Berlin

Montané MH, Dreyer S, Triantaphylides C and Kloppstech K (1997) Early light-inducible protein during long-term acclimation of barley to photooxidative stress caused by light and cold: High level of accumulation by posttranscriptional regulation. Planta 202: 293–302

Montané MH, Petzold B and Kloppstech K (1999) Formation of early light-inducible protein complexes and status of xanthophyll levels under high light and cold stress in barley (*Hordeum vulgare* L.). Planta 208: 519–527

Moscovici-Kadouri S and Chamovitz DA (1997) Characterization of a cDNA encoding the early light-inducible protein (Elip) from *Arabidopsis*. Plant Physiol 115: 1287–1290

Nagy F, Kay SA and Chua NH (1988) A circadian clock regulates transcription of the wheat *cab-1* gene. Genes Devel 2: 376–382

Otto B, Ohad I and Kloppstech K (1992) Temperature treatments of dark-grown pea seedlings cause an accelerated greening in the light at different levels of gene expression. Plant Mol Biol 18: 887–896

Ouvrard O, Cellier F, Ferrare K, Tousch D, Lamaze T, Dupuis JM and Casse-Delbart F (1996) Identification and expression of water stress- and abscisic acid-regulated genes in a drought-tolerant sunflower genotype. Plant Mol Biol 31: 819–829

Palenik B and Haselkorn R (1992) Multiple evolutionary origins of *Prochlorophytes*, the chlorophyll *b*-containing prokaryotes. Nature 355: 265–267

Paulsen H (1995) Chlorophyll *a/b*-binding proteins. Photochem Photobiol 62: 367–382

Paulsen H (1997) Pigment ligation to proteins of the photosynthetic apparatus in higher plants. Physiol Plant 100: 760–768

Pfündel E and Bilger W (1994) Regulation and possible function of the violaxanthin cycle. Photosynth Res 42: 89–109

Post AF and Bullerjahn GS (1994) The photosynthetic machinery in *Prochlorophytes*: structural properties and ecological significance. FEMS Microbiol Rev 13: 393–413

Pötter E and Kloppstech K (1993) Effects of light stress on the expression of early light-inducible proteins in barley. Eur J Biochem 214: 779–786

Prasil O, Adir N and Ohad I (1992) Dynamics of Photosystem II: Mechanism of photoinhibition and recovery process. In: Barber J (ed) The Photosystems: Structure, Function and Molecular Biology Topics Photosynthesis, Vol 11, pp 295–348. Elsevier, Amsterdam

Raghavan V and Kamalay JC (1993) Expression of two cloned mRNA sequences during development and germination of spores of the sensitive fern, *Onoclea sensibilis* L. Planta 189: 1–9

Reith ME and Munholland J (1995) Complete nucleotide sequence of the *Porphyra purpurea* chloroplast genome. Plant Mol Biol Rep 13: 333–335

Robinson C and Klösgen RB (1994) Targeting of proteins into and across the thylakoid membrane-a multitude of mechanisms. Plant Mol Biol 26: 15–24

Robinson C, Hynds PJ, Robinson D and Mant A (1998) Multiple pathways for the targeting of thylakoid proteins in chloroplasts. Plant Mol Biol 38: 209–221

Sandonà D, Croce R, Pagano A, Crimi M and Bassi R (1998) Higher plants light harvesting proteins. Structure abd function by mutation analysis of either proteins or chromophore moieties. Biochim Biophys Acta 1365: 207–214

Scharnhorst C, Heinze H, Meyer G, Kolanus W, Bartsch K, Heinrichs S, Gudschun T, Möller M and Herzfeld F (1985) Molecular cloning of a pea mRNA encoding an early light induced, nuclear coded chloroplast protein. Plant Mol Biol 4: 241–245

Shimosaka E, Sasanuma T and Handa H (1999) A wheat cold-regulated cDNA encoding an early light-inducible protein (Elip): Its structure, expression and chromosomal localization. Plant Cell Physiol 40: 319–325

Straus NA (1994) Iron deprivation: physiology and gene regulation. In: Bryant DA (ed) The Molecular Biology of Cyanobacteria, pp 731–750. Kluwer Academic Publishers, Dordrecht

Sugiura M (1995) The chloroplast genome. Essays Biochem 30: 49–57

Thompson JD, Higgins DG and Gibson TJ (1994) Improving the sensitivity of progressive multiple sequence alignment through sequence weighting, position specific gap penalties and weight matrix choice. Nucleic Acid Res 22: 4673–4680

Tomitani A, Okada K, Miyashita H, Matthijs HCP, Ohno T and Tanaka A (1999) Chlorophyll *b* and phycobilins in the common ancestor of cyanobacteria and chloroplasts. Nature 400: 159–162

Von Wettstein D, Gough S and Kannangara CG (1995) Chlorophyll biosynthesis. Plant Cell 7: 1039–1057

Wedel N, Klein R, Ljungberg U, Andersson B and Herrmann RG (1992) The single copy gene *psbS* codes for a phylogenetically intriguing 22 kDa polypeptide of Photosystem II. FEBS Lett 314: 61–66

Wolfe RG, Cunningham FX, Durnford D, Green BR and Gantt E (1994) Evidence for a common origin of chloroplasts with light-harvesting complexes of different pigmentation. Nature 367: 566–568

Yamagata H and Bowler C (1996) Molecular cloning and characterization of a cDNA encoding early light-inducible protein from soybean (*Glycine max* L.). Biosci Biotechnol Biochem 61, 2143–2144

Yamamoto HY (1979) Biochemistry of the violaxanthin cycle in higher plants. Pure Appl Chem 51: 639–648

Yamamoto HY and Bassi R (1996) Carotenoids: Localization and Functions. In: Ort DR and Yocum CF (eds) Oxygenic Photosynthesis: The Light Reactions, pp 539–563. Kluwer Academic Publishers, Dordrecht

Zeevaart JAD and Creelman RA (1988) Metabolism and physiology of abscisic acid. Annu Rev Plant Physiol 39: 439–473

Chapter 29

Regulation, Inhibition and Protection of Photosystem I

Yukako Hihara
Department of Biochemistry and Molecular Biology, Faculty of Science, Saitama University, Shimo-okubo 255, Urawa-shi, Saitama 338-8570, Japan

Kintake Sonoike*
Department of Integrated Biosciences, Graduate School of Frontier Sciences, University of Tokyo, Hongo 7-3-1, Bunkyo-ku, Tokyo 113-0033, Japan

*Author for correspondence, email: sonoike@k.u-tokyo.ac.jp

Eva-Mari Aro and Bertil Andersson (eds): Regulation of Photosynthesis, pp. 507–531.

Summary

The dynamic nature of Photosystem I (PS I) reaction center complex is described with emphasis on regulatory aspects such as synthesis, assembly, degradation, inhibition and protection. Although the dynamic and regulatory aspects of PS I have been somewhat neglected, there have been several reports on the assembly of this photosystem during greening and its degradation during senescence. The PS I/PS II ratio is known to change in response to the surrounding environment. It is also established that not only PS II but also PS I can be photoinhibited. The dynamic aspects of PS I can now be viewed based on the well-defined structural image of the complex, and some recent findings in that respect will be described. The signal transduction pathway involved in the regulatory process will be also discussed.

I. Introduction

Photosystem (PS) I is a plastocyanin:ferredoxin oxidoreductase driven by light energy (for recent reviews, see Brettel (1997) on biophysical aspects, and Scheller et al., (1997) on biochemical aspects). PS I receives electrons from PS II through plastoquinone, cytochrome b_6f and plastocyanin in a linear electron flow, and it finally reduces ferredoxin and NADP. This electron transfer is coupled with the formation of ATP. PS I is also involved in cyclic electron transfer which produces only ATP without the participation of PS II. PS I exists in thylakoid membranes as a supramolecular complex composed of more than ten different subunits, about 100 chlorophyll (Chl) *a* molecules, 10–15 β-carotene molecules, two phylloquinone molecules and three iron-sulfur centers. Since the sequence of all the subunits are known and the 3-D structure of the PS I reaction center has been determined by X-ray crystallography to 4 Å resolution (Krauß et al., 1996), we have now a rough structural image of the PS I complex. However, we have very little information about the dynamic nature of PS I as compared to the vast information on the dynamics of the other photosystem, PS II.

The D1 protein, one of the hetero-dimer subunits of PS II reaction center, was first noticed as a protein with a very high turnover rate (Ellis 1981). PS II was supposed to be more sensitive than PS I to almost all kinds of stresses, in particular to light. Thus, the dynamic nature of PS II such as degradation and repair during photoinhibition has been a central issue in the field of plant physiology (Aro et al., 1993), while PS I has been much less studied in that respect. Furthermore, PS I is not a rate-limiting step in electron transfer under normal conditions. Regulation of the PS I content, however, is an intriguing problem considering the need for a balanced excitation of the two photosystems and adjustment of the ATP/NADPH ratio. Moreover, PS I must change dynamically for assembly during the greening, degradation during the inhibitory process or senescence, and during acclimatory processes caused by the change of environment.

Abbreviations: A_0 – primary acceptor chlorophylls; A_1 – the secondary electron acceptor phylloquinone; Chl – chlorophyll; DAD – 2,3,5,6-tetramethyl-*p*-phenylenediamine; DBMIB – 2,5-dibromo-3-methyl-6-isopropyl-*p*-benzoquinone; DCIP – dichlorophenol indophenol; DCMU – 3-(3,4-dichlorophenyl)-1,1-dimethylurea; EPR – electron paramagnetic resonance; FNR – ferredoxin:NADP$^+$ oxidoreductase; F_X, F_A, and F_B – the terminal electron acceptor iron-sulfur centers; P-700 – primary electron donor; PS I – Photosystem I; PS II – Photosystem II; SOD – superoxide dismutase

As will be discussed in Section II, when the environmental conditions change, the PS I/PS II ratio in the photosynthetic organisms must be adjusted to achieve an optimal excitation balance between the two photosystems and a proper ATP/NADPH ratio. The assembly and degradation of PS I during greening and senescence are also discussed in the same section. The activity of PS I is also known to be affected by the environmental factors even under the conditions where the quantity of PS I was not much affected as described in Section III. Regulation of the expression of the various PS I subunits and their assembly is discussed in Section IV. Although PS I had long been believed to be resistant to photoinhibition, it was recently found that PS I is quite sensitive to light especially at chilling temperatures. The inhibition of PS I and its mechanism will be addressed in Section V. Section VI is devoted to protective mechanisms for light stress of PS I.

II. Regulation of the Quantity of PS I

The organisms flexibly modulate their photosynthetic apparatus in order to acclimate to the changes in light or other environmental factors. Photosystem stoichiometry (PS I/PS II ratio) is modulated in various organisms under different environmental conditions, responding to the changes in light intensity, light quality as well as salt and CO_2 concentration (Table 1). In cyanobacteria, extensive studies have provided evidence that the synthesis of PS I, and not of PS II is responsible for the variations of photosystem stoichiometry (Fujita, 1997). Also in

Table 1. Regulation of photosystem stoichiometry

Species	Conditions	PS I[a]	PS II[a]	PS I/PS II	References
Response to Light Quality					
Pisum sativum	PS II → PS I	2.2 → 1.7	4.0 → 4.9	0.55 → 0.35	Melis and Harvey, 1981
Pisum sativum	PS II → PS I	1.7 → 1.0	1.8 → 2.0	0.99 → 0.51	Glick et al., 1986
Pisum sativum	PS II → PS I	1.7 → 1.0	2.0 → 2.7	0.88 → 0.37	Chow et al., 1990
Hordeum vulgare	PS II → PS I	2.0 → 1.4	2.9 → 3.7	0.71 → 0.37	Kim et al., 1993
Arabidopsis thaliana	PS II → PS I	1.4 → 1.0	2.2 → 2.7	0.62 → 0.37	Walters and Horton, 1994
Tradescantia albiflora	PS II → PS I	1.0 → 0.78	2.3 → 2.1	0.45 → 0.37	Liu et al., 1993
Spinacia oleracea	PS II → PS I	1.8 → 1.3	2.8 → 2.9	0.64 → 0.45	Chow et al., 1991a
Chlamydomonas reinhardtii	PS II → PS I	3.8 → 1.4[b]	4.3 → 3.4[b]	0.88 → 0.40	Melis et al., 1996
Porphyridium cruentum	PS II → PS I	11 → 7.5[b]	2.9 → 9.1[b]	3.8 → 0.83	Cunningham et al., 1990
Porphyridium cruentum	PS II → PS I	5.3 → 4.1	2.3 → 3.2	2.3 → 1.3	Fujita et al., 1985
Anabaena variabilis	PS II → PS I	6.9 → 6.2	2.0 → 3.5	3.5 → 1.8	Fujita et al., 1985
Anacystis nidulans	PS II → PS I	5.6 → 5.0	2.2 → 5.0	2.5 → 1.0	Fujita et al., 1985
Phormidium percicinum	PS II → PS I	9.8 → 9.1	0.94 → 1.8	10 → 5.0	Fujita et al., 1985
Synechococcus PCC 6301	PS II → PS I	6.4 → 6.1	1.7 → 4.3	3.7 → 1.4	Manodori and Melis, 1986
Response to Light Intensity					
Sinapis alba	Low → High	2.5 → 2.5	1.7 → 2.5	1.5 → 1.0	Wild et al., 1986
Tradescantia albiflora	Low → High	1.6 → 1.7	2.2 → 2.2	0.75 → 0.78	Chow et al., 1991b
Chlamydomonas reinhardtii	Low → High	1.1 → 1.0	1.5 → 2.6	0.74 → 0.38	Neale and Melis, 1986
Skeletonema costatum	Low → High	0.49 → 0.69[b]	1.1 → 0.73[b]	0.44 → 0.95	Falkowski et al., 1981
Cylindrotheca fusiformis	Low → High	0.59 → 1.0	2.3 → 1.3	0.26 → 0.77	Smith and Melis, 1988
Prochlorothrix hollandica	Low → High	2.6 → 4.1	1.2 → 3.0	2.3 → 1.4	Burger-Wiersma and Post, 1989
Anabaena variabilis	Low → High	1.5 → 0.55[b]	1.0 → 0.80[b]	1.5 → 0.7	Kawamura et al., 1979
Synechocystis PCC 6714	Low → High	0.11 → 0.039[b]	0.037 → 0.029[b]	2.9 → 1.3	Murakami and Fujita, 1991b
Synechocystis PCC 6803	Low → High	0.42 → 0.17[b]	0.20 → 0.15[b]	2.1 → 1.2	Hihara et al., 1998
Response to CO_2					
Anacystis nidulans	High → Low	5.2 → 5.1	2.1 → 1.3	2.5 → 4.2	Manodori and Melis, 1984
Synechocystis PCC 6714	High → Low	6.3 → 8.2	3.9 → 3.5	1.6 → 2.4	Murakami et al., 1997a
Response to Salt					
Synechocystis PCC 6714	Low → High	6.7 → 8.6	4.6 → 4.5	1.5 → 1.9	Murakami et al., 1997a
Synechocystis PCC 6803	Low → High	6.8 → 8.4	1.6 → 1.4	4.2 → 5.9	Jeanjean et al., 1993

[a] mmol/mol Chl unless otherwise stated; [b] mol/10^{18} Cells

the case of the green alga, *Chlamydomonas reinhardtii*, the variations of the photosystem stoichiometry seems to be induced mainly by changes in the synthesis of PS I (Murakami et al., 1997b). Moreover, as shown in Section IV, there have been some reports proposing that PS I synthesis is regulated at various processes such as transcription, translation and assembly. In conclusion, this regulation of PS I seems to be responsible for the acclimatory processes in the adjustment of photosystem stoichiometry.

A. Adjustment of PS I/PS II Ratio during Acclimation

1. Regulation of Photosystem Stoichiometry in Response to Light Quality—Its Importance for the Improvement of the Photosynthetic Efficiency

Why do photosynthetic organisms modulate photosystem stoichiometry upon the various environmental change? In general, the two photosystems in oxygenic photosynthetic organisms utilize different wavelengths of light with different efficiency, since they have different light-harvesting pigments. For example, in cyanobacteria and red algae, PS II which has phycobilisome antenna preferentially absorbs light in the 550-620 nm region, whereas in higher plant chloroplasts, PS II strongly absorbs light at 475 and 650 nm by its Chl *b*-containing antenna. On the other hand, in both plants and cyanobacteria, light at 435 and 680 nm is strongly absorbed by PS I whose main antenna pigment is Chl *a*. Thus, under light-limiting conditions, unbalanced excitation of the two photosystems could occur depending on the spectral qualities of light. Under such conditions, changes in photosystem stoichiometry have been observed widely among photosynthetic organisms (Table 1). The relative PS I/PS II ratio is high under PS II-excited conditions (PS II light) and low under PS I-excited conditions (PS I light) in chloroplasts of higher plants (Melis and Harvey, 1981; Glick et al., 1986; Chow et al., 1990, 1991a; Kim et al., 1993; Liu et al., 1993; Walters and Horton, 1994), in green algae (Melis et al., 1996) and in phycobilisome-containing organisms (Fujita et al., 1985; Manodori and Melis, 1986; Melis et al; 1989; Cunningham et al., 1990).

This type of response is important to these organisms, since the adjustment of photosystem stoichiometry improves the quantum efficiency of photosynthesis under any given light quality conditions (Murakami and Fujita, 1988; Melis et al., 1989; Chow et al., 1990; Walters and Horton, 1995; Melis et al., 1996). Analyzing flash-induced oxidation-reduction reactions of cytochrome *f* and P-700, Murakami and Fujita (1991a) investigated the redox state of the electron pool between the two photosystems before and after adjustment of photosystem stoichiometry. When cells with a high PS I/PS II ratio were shifted to PS I light, the electron pool between the two photosystems became extremely oxidized and most of the PS I reaction centers were closed. On the other hand, in cells with a low PS I/PS II ratio, the intermediate pool became extremely reduced upon the shift to PS II light and most of the PS II reaction centers were closed. Such temporal reduction of the intermediate pool may cause photoinhibition of PS II even under a low light-intensity condition (Fujita, 1999). Only after the adjustment of photosystem stoichiometry, the redox state was released from such a biased state. Therefore, by the regulation of the photosystem stoichiometry, both reaction centers seem to be kept in a relatively open state to maximize photosynthetic efficiency.

2. Regulation of Photosystem Stoichiometry in Response to Light Intensity—Its Importance for Avoidance of Photodamage

a. Physiological Significance

The photosystem stoichiometry also changes along with the variation in light intensity of constant quality (Table 1). It is widely accepted that cyanobacteria grown under low-intensity white light have a higher PS I/PS II ratio than those grown under high-intensity light due to a change in PS I content (for a review, see Hihara, 1999). In chloroplasts of higher plants (Leong and Anderson, 1984, 1986; Wild et al., 1986; Chow and Anderson, 1987; Evans, 1987; Chow and Hope, 1988, De la Torre and Burkey, 1990) and green alga (Neale and Melis, 1986), a similar change in PS I/PS II ratio was observed, although there are a few exceptions (Chow et al., 1991b). Interestingly, in diatoms, which have a unique antenna system composed of a Chl *a*-Chl *c*-fucoxanthin light harvesting complex, showed a higher, not lower, PS I/PS II ratio upon the shift to high-intensity light (Falkowski et al., 1981; Smith and Melis, 1988).

The change of photosystem stoichiometry in response to light intensity has been often explained as optimization of the excitation of two photosystems:

the antenna size of PS II decreases under high-intensity light compared with that of PS I (Section II.D.), and thus it is necessary to change the photosystem stoichiometry to compensate for the change of antenna size. However, it is not so easy to explain why photosystem stoichiometry should be optimized under high-intensity light conditions where the light energy is no longer limiting photosynthesis. A more attractive explanation is that an imbalance of excitation between the two photosystems may cause more damage under high-intensity light than would a balanced excitation.

To elucidate the physiological significance of the photosystem stoichiometry and to examine which hypothesis described above is more plausible, Hihara et al. (1998) characterized a mutant of *Synechocystis* sp. PCC 6803 (*pmgA* mutant) which is unable to regulate its photosystem stoichiometry under high-intensity light. In the wild type cells of this cyanobacterial species, the amount of PS I per cell basis decreased significantly when the cyanobacteria were shifted to a high-intensity light condition while the change in the amount of PS II is relatively small. Thus, the photosystem stoichiometry was regulated through the content of PS I in this case. This suppression of PS I accumulation at high-intensity light was largely abolished in the mutant strain of *pmgA* resulting in a constant PS I/PS II ratio. Many cyanobacterial mutants with varied PS I/PS II ratio have been reported so far, but the *pmgA* mutant is unique in the point that the PS I/PS II ratio under a low-intensity light condition is the same as that in wild type cells and that only the acclimatory process is affected by the mutation.

Unexpectedly, the *pmgA*-deleted strain grew better than the wild-type strain under photoautotrophic conditions (Hihara and Ikeuchi, 1997). In fact, a point mutant of *pmgA* that had appeared spontaneously in a wild-type strain completely replaced the wild-type culture for a year or so under photoautotrophic conditions. The higher PS I content seemed to enhance the photosynthetic rate and growth during a short-term exposure (ca 24 h) to high-intensity light. Thus, the adjustment of PS I/PS II ratio under high-intensity light is not for the efficiency of photosynthesis.

b. Avoidance of Photodamage

In *pmgA* mutants of *Synechocystis*, the selective reduction of PS I content under high-intensity light is not operational. The resulting higher abundance of PS I in the *pmgA* mutants extracted more electrons from the acceptor side of PS II even though the rate-limiting step in electron transport is oxidation of plastoquinone. As a result, photoinhibition of PS II was mitigated for a short term, and the whole electron flow was enhanced in spite of the constant PS II content. In fact, mitigation of the PS II photoinhibition was clearly observed in the *pmgA* mutants as a variable fluorescence was more resistant to a brief, high-intensity light treatment (>1000 μmol $m^{-2}s^{-1}$) than in the wild-type strain, when the cells were grown under high-intensity light (300 μmol $m^{-2}s^{-1}$) for 24 h (Sonoike et al., 1999).

Then, what is the nature of the disadvantage of the mutant cells with a non-regulated high PS I content when they are grown under extended high light conditions? More PS I together with mitigation of PS II photoinhibition maintains a high electron-transport activity and eventually makes the acceptor side of PS I more reduced. Supposedly, excessive generation of reactive species of oxygen at the PS I reducing side may exceed the capacity of the scavenging system in the *pmgA* mutants (Asada, 1999). This may cause oxidative damage to cellular components and finally lead to the loss of viability of the mutants under the prolonged high-intensity light.

3. Regulation of Photosystem Stoichiometry in Response to Other Environmental Changes—Requirement for ATP

Photosystem stoichiometry can also be modulated according to factors other than light conditions (Table 1). For example, cyanobacteria have more PS I when they are grown under a low CO_2 condition (Manodori and Melis, 1984; Murakami et al., 1997a) or in a high salinity medium such as a high NaCl concentration (Schubert and Hagemann, 1990; Jeanjean et al., 1993; Hibino et al., 1996; Murakami et al., 1997a). Under low CO_2 conditions, cells need more ATP for active transport of inorganic carbon. Similarly, energy demand increases under high salt conditions since additional ATP is required for the synthesis of compatible solutes and for extrusion of Na^+ ions (Joset et al., 1996). Thus, the relative amount of PS I in low CO_2-grown cells or cells in a high salinity medium increased in order to facilitate the generation of extra ATP via cyclic photophosphorylation. The relative PS I amount increases concomitantly with the up-regulation of the cyclic electron flow around PS I (Jeanjean et al., 1993; Hibino et al., 1996).

B. Accumulation of PS I during Greening

There are many studies on the development of photochemical activities during greening. It has been established that PS I activity can be detected at an earlier stage of greening than that of PS II (Egnéus et al., 1972; Henningsen and Boardman, 1973; Plesnicar and Bendall, 1973; Boardman and Anderson, 1978; Wellburn and Hampp, 1979; Kyle and Zalik, 1982; Ohashi et al., 1989), although the time of the appearance of these two activities varies greatly with the plant material, growth conditions, and light conditions. The most comprehensive study on the development of the photosystems during greening was done by Tsuji's group (Tanaka and Tsuji, 1985; Ohashi et al., 1989; Ohashi et al., 1992). By an improved method of SDS-polyacrylamide gel electrophoresis, they observed the appearance of labile Chl-protein complexes at the earliest stages of greening of the etiolated barley seedlings (Tanaka and Tsuji, 1985). Moreover, they investigated the development of the photosynthetic electron transport system during greening and related it to the time course of appearance of Chl-protein complexes (Ohashi et al., 1989). They showed that the activity of PS I and PS II was detectable 1 h and 1.5 h after the onset of illumination, respectively. After formation of the two photosystems, electron transport components between them and the PS I acceptor side developed in coordination with each other. As for the development of PS I after 1 h, it continued as follows: some components may be associated with the reaction center, and the reaction center began to reduce ferredoxin after 2 h. At 4 h, ferredoxin:$NADP^+$ oxido-reductase (FNR) became attached to the PS I core allowing reduction of NADP. The whole electron transport from water to NADP was operational after 4 h of development. Ohashi et al., (1992) examined the amount of several components of the photo-synthetic electron-transport system during greening on a per plastid basis. They estimated the amount of PS II and PS I by the assay of Q_A and P-700, respectively. Once PS I and PS II have appeared, the number of both photosystems increased at the same rates during the greening process, with the ratio remaining constant at around unity. After 6 h, the increase of both photosystems was accelerated, probably because the electron-transport activity driven by the two photosystems was sufficient to further support the development of plastids (Ohashi et al., 1989). As for the antenna of PS I, it was proposed that LHCI apoproteins are first assembled into monomeric pigmented complexes after 2 h of illumination and then assembly into a trimer before attaching to the pre-existing core complex to form a complete PS I holocomplex which in total required more than 4 h of greening (Dreyfuss and Thornber, 1994).

C. Degradation of PS I during Senescence

Photosynthetic activity and Chl content of a leaf decline after full expansion of the leaf (senescence). Similar decreases are observed in a shorter time range (several days) when plants are transferred to darkness. Although some reported a faster decline of PS I than that of PS II (Bricker and Neuman, 1982; Shinohara and Murakami, 1996), the site of the impairment of electron transfer was not determined conclusively (for a review, see Gepstein, 1988). Not so much is known about the degradation of PS I during senescence. Dark incubation of rice seedlings for 4 days resulted in the disappearance of P-700, but the application of very weak light (5 μmol m^{-2} s^{-1}) during the treatment completely suppressed this decrease (Okada and Katoh, 1998). Together with the fact that the antagonistic effect of red and far red light was observed on the loss of Chl in detached rice leaves (Okada et al., 1992), the degradation of PS I during senescence seems to be under the control of phytochrome.

D. Regulation of PS I Antenna

The antenna size of PS I has been assumed to be only slightly affected by changes in the light environment. In the case of a red or green alga, decrease of polypeptides in the PS I antenna upon the shift to high-intensity light was reported (Tan et al., 1995; Stefánsson et al., 1997). LHCI may dissociate from CP1 complex under natural shade environment where far red light preferentially exciting PS I is enriched (Burkey and Wells, 1996). It was also reported that there is heterogeneity in PS I with respect to the antenna size both in green algae and in spinach (Andreasson and Albertsson, 1993; Stefánsson et al., 1997). The physiological significance of the heterogeneity of the PS I antenna size is not clear at present.

III. Regulation of the Activity of PS I

Electron transfer rate of PS I in thylakoid membranes from higher plants was reported to increase upon heat treatment (e.g. Armond et al., 1978; Thomas et al., 1986) or high light treatment (Tjus and Andersson, 1993). However, the effect of heat seems to be ascribed to the release of superoxide dismutase (SOD) from thylakoid membranes (Lajkó et al., 1991; Boucher and Carpentier, 1993). In the absence of SOD, oxygen consumption is known to be overestimated since molecular oxygen is not regenerated by disproportionation of superoxide (Izawa, 1980). On the other hand, the increased PS I activity after high-light treatments was ascribed to the loss of the trans-thylakoid proton gradient (Tjus and Andersson, 1993). Enhanced electron transfer through PS I in cyanobacterial cells was also caused by heat (Babu et al., 1992) or high-intensity light (Hihara and Ikeuchi, 1998). The enhancement is mostly due to increased permeability of cells for the entry of acceptors and donors upon the treatments. Thus, these enhancements of electron transfer though PS I by high light or heat cannot be regarded as 'regulation of PS I.'

PS I activity has also been reported to increase upon winter acclimation. The rate of electron transfer through PS I increased 1.6-fold when winter rye was grown at cold-hardening temperatures (Huner, 1985). A similar increase of PS I activity was observed in the overwintering of periwinkle (*Vinca minor* L.) without any detectable changes in the relative Chl contents of the Chl-protein complexes (Huner et al., 1988). The stimulation was observed when DCIP was used as an electron carrier and when cations such as Na^+ or Mg^{2+} were present in thylakoid-membrane preparations during the measurements (Huner and Reynolds, 1989). A similar increase of PS I activity was observed in a shorter term upon the shift of plants to a low temperature, but the increase in this case was transient and decayed to below the initial activity after 96 to 120 h at the new temperature (Huner et al., 1993). The physiological significance of this phenomena is presently unknown.

IV. Regulation of PS I Expression

As described in Section II, photosynthetic organisms must modulate the accumulation of PS I to cope with variations in light environment. During such acclimation, organisms need sensors for monitoring the changing light regime, signal transduction systems and regulatory mechanisms for the controlled accumulation of each PS I subunit (see Fig. 1). The sensing mechanism of light conditions and subsequent signal transduction are poorly understood. However, when it comes to the greening process, the involvement of a phytochrome dependent signal-transduction pathway has been gradually clarified as will be discussed in Section IV.A.1. In the case of regulation of the photosystem stoichiometry, the redox state of the electron transport chain seems to play an important role as shown in Section IV.A.2. We have considerable information about regulation of individual PS I subunits although the signal mechanisms are still largely unknown. A regulated accumulation of each subunit is very crucial since the PS I complex requires subunits in stoichiometric amounts for the correct assembly. Moreover, in most photosynthetic eukaryotes, PS I comprises subunits of mixed cellular origin, organellar-encoded (PsaA, B, C, I, J) and nuclear-encoded (PsaD, E, F, G, H, K, L, N). Cooperation between the chloroplast and nucleus is indispensable in these organisms. Thus, regulatory mechanisms for organization of the PS I complex have been developed at various levels: transcriptional, post-transcriptional, translational, and post-translational levels. Although the knowledge we have of these mechanisms is not systematic at present, we will overview the current information in Section IV.B-D and discuss its significance for acclimation.

A. Signaling Mechanisms for the Regulation of PS I Accumulation

1. Signaling Mechanisms for Greening

There have been some studies indicating that phytochrome is involved in signaling of PS I gene expression (Zhu et al., 1985; Brunner et al., 1991; Lotan et al., 1993). However, while these observations were not verified and the signal transduction pathways linking phytochrome with gene expression were basically unknown, Neuhaus et al., (1993) solved these issues using a single-cell microinjection assay. They microinjected putative signaling compounds into hypocotyl cells of a PhyA-deficient mutant of tomato. The hypocotyl cells of this mutant could not develop chloroplasts and could not synthesize anthocyanins in response to light. However, by microinjecting an exogenous PhyA, they restored

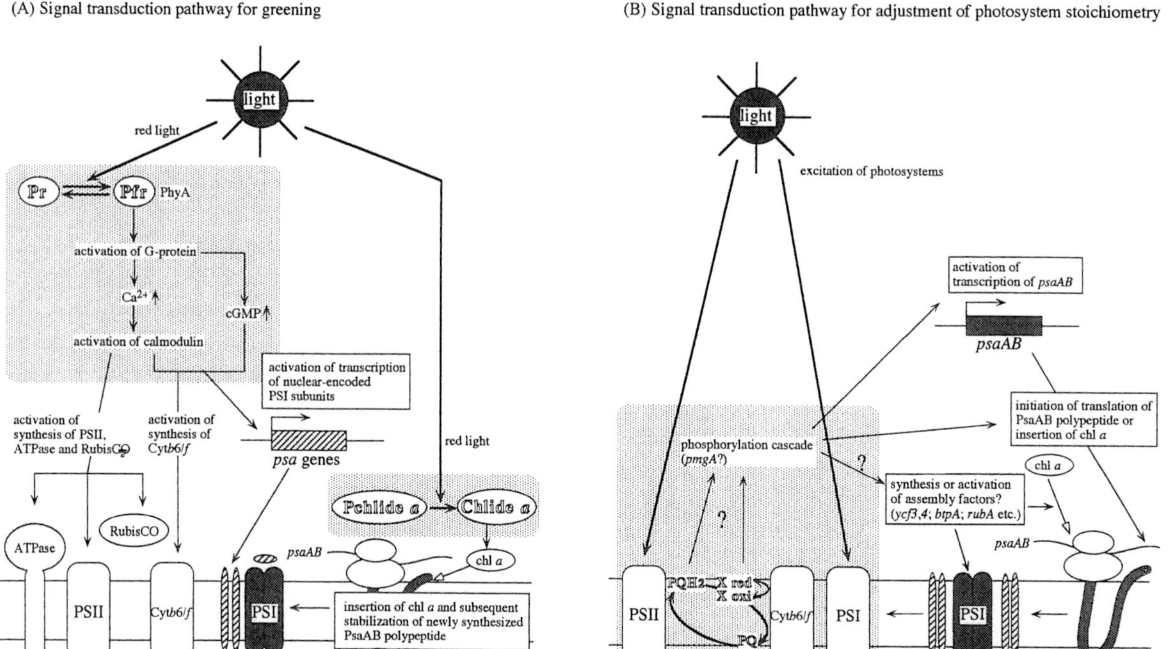

Fig. 1. Schematic representation of mechanism of sensing of light condition, signal transduction, and regulation of PS I accumulation in the case of (A) greening and (B) regulation of photosystem stoichiometry. Putative primary sensors are indicated with outline type. Mechanisms of sensing and transduction of light signals are indicated in hatched area. Please note that the mechanisms only observed in higher plants or in cyanobacteria is also shown (see text).

the chloroplast development and anthocyanin biosynthesis. Furthermore, G protein, calcium and calmodulin were found to be working further downstream in the PhyA-mediated pathway. Interestingly, PhyA and G protein could stimulate the synthesis and assembly of all the photosynthetic complexes in the plastids of injected cells, while calcium and calmodulin were ineffective in mediating the synthesis of the PS I and cytochrome b_6f complex. Subsequently, the induction of PS I and cytochrome b_6f was shown to require both cyclic GMP and calcium, in contrast to that of PS II which needed only calcium (Bowler et al., 1994). These results indicated clearly that synthesis of PS I and PS II are controlled by different phytochrome signaling intermediates. Moreover, using reporter genes for these different pathways, they demonstrated that cyclic GMP and calcium acted primarily by modulating gene expression. Since they used the promoter regions only from the ferredoxin:NADP$^+$ oxidoreductase gene (*fnr*) in order to assay the PS I gene expression, additional experiments are needed to further verify the transcriptional control of PS I genes by cyclic GMP and calcium.

2. Signaling Mechanisms for the Regulation of Photosystem Stoichiometry

Unlike in the greening process, several studies have suggested that phytochrome is not involved in the regulation of photosystem stoichiometry. Chow et al., (1990) showed that brief supplementary far red irradiance to pea plants, which caused marked morphological changes attributable to phytochrome regulation, had little or no effect on the amounts of the major chloroplast components. Smith et al., (1993) also argued against the involvement of phytochrome based upon experiments with a PhyA deficient mutant of tomato. Then what is the signaling mechanism for regulation of photosystem stoichiometry? Some information is available from studies on cyano-bacteria. Murakami and Fujita (1991a) suggested that the regulation of photosystem stoichiometry may occur in response to the redox state of the electron pool between two photosystems, possibly at the cytochrome b_6f complex. This notion seems plausible, since in some mutants that have a severe defect in the excitation of one particular photosystem, photosystem stoichiometry changes to correct the imbalance in light absorption and rate of electron transport between two photosystems. For example a phycobilisome-less mutant of *Synechococcus* sp. PCC 7002 maintained a low PS I/PS II ratio irrespective of given light irradiance (Kim et al., 1993). A similar behavior is seen in Chl *b*-less mutants of higher plants (Terao et al., 1996). In a mutant of *Synechococcus* sp. PCC 7942 lacking activity in one of its two iron superoxide dismutases, PS I selectively suffered photo-oxidative damage under high-intensity light. Under such conditions, the PS I/PS II ratio did not decrease since the PS I content was up-regulated to compensate for the low PS I efficiency (Samson et al., 1994). Furthermore, from the observation that 2-*n*-heptyl-4-hydroxyquinoline *N*-oxide (HQNO), an inhibitor of cytochrome b_6 oxidation, impaired the regulatory response with respect to photosystem stoichiometry, Murakami and Fujita (1993) proposed that the redox state of cytochrome b_6 in the cytochrome b_6f complex, worked as a sensor for the regulation of the photosystem stoichiometry. They hypothesized that an unknown component that was reduced by cytochrome b_6 may be a signal to regulate the synthesis of PS I (Fujita, 1997).

Redox regulation is observed not only in adjustment of the photosystem stoichiometry mentioned above but also in many regulatory processes of photosynthetic components. For example, transcription of the nuclear *cab* gene, encoding light harvesting complex II, was proposed to be coupled to light-intensity via the redox state of plastoquinone pool in *Dunaliella tertiolecta* (Escoubas et al., 1995). Translation of *psbA* mRNA for PS II reaction center D1 protein in *Chlamydomonas reinhardtii* was also shown to be regulated by light via a redox mechanism. The binding of putative translational factors to the *psbA* mRNA may depend on the redox state of thioredoxin and on the ADP levels of the cell (Danon and Mayfield, 1991, 1994a,b; Kim and Mayfield, 1997). Moreover, thylakoid protein phosphorylation was found to be activated by binding of a plastoquinol molecule to the quinol oxidation site (Qo site) of cytochrome b_6f complex (Vener et al., 1995, 1997). Many redox sensing mechanisms seem to work at various steps of regulatory processes in photosynthetic organisms.

Although we can imagine the outline of the upstream (signal sensing) and the downstream (regulation of PS I accumulation) of such a signal transduction chain, the intermediate signaling steps remain totally unknown. The *pmgA* gene, which is so far the only component identified as a regulatory factor of the photosystem stoichiometry in response to changes in light intensity (Hihara and Ikeuchi, 1997; Hihara et al, 1998), may be a key for the elucidation of the process. It codes for a 23 kDa polypeptide of 204 amino acid residues with limited homology to RsbW, which is a switch kinase of the partner-switching module of the stress-responsive signal transduction pathway in *Bacillus subtilis*, and to putative nucleotide-binding motifs of a sensor kinase of the two-component regulatory system. These characteristics indicate that *pmgA* may participate in a phosphorylation cascade. By searching for a factor or factors that interact with the gene product of *pmgA*, it may be possible to elucidate the signal sensing and transduction mechanisms for regulation of photosystem stoichiometry.

B. Regulatory Mechanism for Accumulation of Large Reaction Center Subunits

1. Transcriptional and Post-Transcriptional Regulation

The *psaA* and *psaB* genes, which encode the reaction center subunits of PS I, exist tandemly and are co-transcribed in the chloroplast genome of vascular plants (Fish et al., 1985; Kirsch et al., 1986) and of *Euglena* (Cushman et al., 1988; Stevenson and Hallick, 1994), as well as in the cyanobacterial genome (Cantrell and Bryant, 1987; Kaneko et al., 1996). The greening process was shown to be under the control of PhyA as described in Section IV.A.1, and there are several observations of light-induced expression (Rodermel and Bogorad, 1985; Schrubar et al., 1990) or phytochrome-dependent accumulation (Zhu et al., 1985; Grover et al., 1999) of *psaAB* transcripts in chloroplasts. However, it is widely accepted that etioplasts already contained high levels of *psaAB* mRNAs similar to the levels found in chloroplasts (Klein and Mullet, 1986, 1987; Kreuz et al, 1986; Laing et al., 1988). In the case of greening, *psaAB* seems to be under translational and/or post-translational controls as mentioned in Section IV.B.2 rather than under transcriptional control.

However, transcriptional and/or post-transcriptional controls can be observed in the process of acclimation to changing light environment. Glick et al., (1986) showed that in pea plants grown in PS I light, the ratio of a PS II gene (*psbB*) transcript to a PS I gene (*psaA*) transcript was 2.6 times greater than that in plants grown in PS II light. This change was largely due to significant changes of *psaA* mRNA levels under PS I or PS II light. Deng et al., (1989) performed a run-on transcription assay on spinach plastid genes and showed that the relative transcription activity of only the *psaA* gene was significantly influenced by the light quality. Recently, Pfannschmidt et al., (1999) examined the transcriptional activity of *psaAB* and *psbA* of *Sinapis alba* seedlings grown under different light qualities. The transcriptional rate of *psbA* was highest when plants were transferred from PS II light to PS I light and PS II became rate-limiting, whereas transcription of *psaAB* was highest upon the shift from PS I to PS II light. Taking into account the effect of DCMU and DBMIB, inhibitors of PS II and cytochrome b_6f complex respectively, on the rate of *psaAB* transcription, they concluded that *psaAB* transcription was promoted when plastoquinone was reduced, and retarded when it was oxidized. Though there is some discrepancy in the above three reports (Glick et al., 1986; Deng et al., 1989; Pfannschmidt, 1999), they suggest that the transcriptional control of *psaAB* in response to light quality may be common in higher plants.

In the case of cyanobacteria, Aizawa and Fujita (1997) reported that the *psaAB* mRNA level was not altered by the light quality. As for the acclimation to light intensity, however, cyanobacteria seem to modulate the expression level of *psaAB*: the amount of transcript decreased at photon flux densities above the growth level, while *psbA* mRNA increased (Herranen et al, 1998; Y. Hihara, unpublished). These observations are consistent with the fact that the PS I/PS II ratio decreases upon the shift to a high-intensity light condition. Unlike the acclimation to light quality, cyanobacteria may regulate photosystem stoichiometry at a transcriptional level in this case.

Chen et al., (1993) showed by competitive DNA-binding assays the presence of DNA-binding proteins that interact specifically with two regions located downstream and upstream of the transcription initiation site of the plastid *psaA-psaB-rps14* operon. Purification experiments using spinach chloroplasts identified two relevant proteins. One was a dimeric protein composed of 31 kDa subunits which specifically recognized the 5′-untranslated region of *psaA* (Cheng et al., 1997) and the other was a dimeric protein composed of 34 kDa subunits which bound to the promoter sequence (Wu et al., 1999).

2. Translational and Post-Translational Regulation

In the process of greening in higher plants, accumulation of PS I apoprotein is mainly regulated at the post-translational level. *psaAB* mRNA accumulated in plastids of dark-grown plants (Klein and Mullet, 1986, 1987; Kreuz et al., 1986; Laing et al., 1988) in association with membrane-bound polysomes (Kreuz et al., 1986; Klein et al., 1988a; Laing et al., 1988), while PS I apoproteins were not observed. Illumination of the dark-grown seedlings triggered the accumulation of Chl *a* and then PS I apoproteins (Vierling and Alberte, 1983; Klein and Mullet, 1986). Phytochrome involvement in controlling PS I apoprotein synthesis was tested by using red/far-red light illumination (Laing et al., 1988). Such treatment showed no far-red reversibility of red-induced apoprotein synthesis, indicating that the light induction of PS I apoprotein was not under the control of phytochrome. Instead, some studies showed that presence of Chl *a* was the primary determinant of accumulation of the PsaA/PsaB proteins. For example, the time course for the light-induced increase of the PsaA/PsaB proteins was similar to the time course of photoconversion of protochlorophyllide to chlorophyllide with the subsequent formation of Chl *a* (Vierling and Alberte, 1983; Klein and Mullet, 1986). Furthermore, the light-induced conversion of protochlorophyllide to Chl *a* was shown to be necessary for the accumulation of PsaA/PsaB, using a Chl *a*-deficient barley mutant (Klein et al., 1988b). De novo synthesis of Chl *a* in isolated barley etioplasts was necessary and sufficient to trigger PsaA/PsaB accumulation(Eichacker et al., 1990). Since the initiation and elongation of translation in etioplasts did not depend on the presence or absence of Chl *a*, Kim et al., (1994) concluded that light-induced Chl *a* biosynthesis triggered the increase of the PsaA/PsaB proteins by enhancement of apoprotein stability.

In cyanobacteria, the variable component responsible for the change of photosystem stoichiometry under different light qualities is PS I as discussed above (Fujita and Murakami, 1987; Aizawa et al., 1992). In *Synechocystis* sp. PCC 6714, the inhibition pattern of the synthesis of the PsaA/B proteins by chloramphenicol indicated that the regulation of PS I synthesis occurred at a step(s) other than peptide elongation (Aizawa and Fujita, 1997). It was assumed that the regulatory point might be at the initiation of translation or at the insertion of Chl *a* for the stabilization of the polypeptide. The insertion of Chl *a* into PS I apoproteins may be the regulatory step like the case for greening of etioplasts. Fujita et al., (1990) showed that a selective suppression in the accumulation of the PS I complex in cyanobacteria was induced by the inhibitors of Chl *a* synthesis. Though protochlorophyllide accumulated markedly in cells grown under PS I light which induced a low PS I/PS II ratio, the protochlorophyllide photoreduction step did not limit the synthesis of PS I (Fujita et al., 1995; Aizawa and Fujita, 1997). It was concluded that the accumulation of protochlorophyllide under PS I light resulted from the reduced level of PS I apoprotein. More studies will be required to clarify the regulatory point of PS I accumulation.

There is an interesting report about a nuclear locus specifically required for the initiation of *psaB* mRNA translation in *Chlamydomonas reinhardtii* (Stampacchia et al., 1997). A mutation at the TAB1 locus abolished the translation of *psaB*, leading to the concomitant loss also of the PsaA polypeptide. By suppressor mutation within a putative base-paired region near the *psaB* initiation codon, the synthesis of both PsaB and PsaA polypeptides was restored in the presence of the *tab1* mutation. This indicates that TAB1 is involved in the activation of translation of the *psaB* mRNA and that the accumulation of PsaA polypeptide strongly depends on PsaB synthesis.

C. Regulatory Mechanism for Accumulation of Nuclear-Encoded Subunits

Many of the low molecular weight subunits of PS I are nuclear-encoded. Recently, the promoters of the genes encoding seven of these PS I subunits have been determined, and the characterization of these promoter sequences revealed that none of them contained a TATA-box (J. Obokata, personal communication). Considering that more than 90% of promoter sequences of other plant nuclear-encoded genes characterized at the same time contained the TATA-box, the expression of PS I genes seems to be differently regulated compared to other genes. Some of the recent results on the regulation of the expression of PS I nuclear-encoded low molecular weight subunits are reviewed below.

1. Polymorphism of the Nuclear-Encoded Subunits

In contrast to the chloroplast genes which mainly have single-copy organization in the organelle genome, many nuclear genes including those of PS I constitute multi-gene families and produce sets of isoproteins. For example, *psaD*, *psaE*, *psaF*, *psaH* and *psaL* are all present in isoforms in *Nicotiana* spp. (Obokata et al., 1993, 1994; Yamamoto et al., 1993; Nakamura and Obokata, 1994). There are two main reasons for the polymorphism of PS I subunits; one is alloploidy, which is found in many cultivated plant species (Obokata et al., 1990) and the other is multigene families. Alternative integration of the isoproteins may result in molecular heterogeneity of the PS I complex. However, little is known about the physiological meaning of the polymorphism of PS I subunits. Yamamoto et al., (1993) showed that the two *psaD* genes of *Nicotiana sylvestris* were subjected to different regulations during leaf development. It is intriguing to examine whether the PS I polymorphism is involved in the fine tuning of PS I function.

2. Coordinated Accumulation

The light-regulated accumulation of the nuclear-encoded subunits of PS I seems to be highly coordinated. Their mRNA are already present at a low level in etiolated seedlings (Brunner et al., 1991; Lotan et al., 1993; Yamamoto et al., 1995a). After exposure to continuous white light, levels of all these PS I mRNAs showed a prominent increase, and then varied in a coordinated manner (Brunner et al., 1991; Yamamoto et al., 1995a). Brunner et al., (1991) showed that red light treatment (2 min) had the same effect on the PS I gene expression as continuous white light. The far-red illumination following the red-light treatment inhibited the effect of red light suggesting the involvement of phytochrome on the gene expression of the nuclear-encoded subunits of PS I. At the protein level, some authors proposed that the nuclear-encoded subunits of PS I accumulated in a sequential manner during the greening (Nechushtai and Nelson, 1985; Herrmann et al., 1991). However, by taking the existence of isoproteins and titers of the antibodies for individual subunits into account, Yamamoto et al., (1995a) claimed that all the subunits accumulated in a synchronous fashion in *Nicotiana sylvestris*.

In order to elucidate the mechanism of coordinated

expression of nuclear-encoded PS I genes, the *cis*-regulatory elements of PS I genes were analyzed. However, it was found that promoter sequences as well as the design and location of regulatory elements are strikingly different between individual PS I genes (Flieger et al., 1994; Lübberstedt et al., 1994). For example, the promoter and leader of spinach *psaF* fused to the GUS reporter gene showed a positive light response whereas the equivalent *psaD* regions conferred a negative light regulation to the GUS gene in the cotyledons of transgenic tobacco seedlings (Flieger et al., 1994). On the other hand, Nakamura and Obokata (1995) found that an octamer motif, CATGTATC, was commonly present in *psaDb*, *psaEb* and *psaHa* of *Nicotiana sylvestris*. In gel shift experiments with nuclear extracts and a labeled probe containing the octamer motif, three retarded bands were discerned. The sensitivity of the band to proteinase K and shift of mobility by treatment with alkaline phosphatase indicated binding of phosphoproteins to the octamer motif. Identification of these phosphoproteins will lead to the great progress in elucidation of coordinated expression mechanism of the nuclear-encoded PS I genes.

3. Regulatory Domain of the psaD *gene*

Among the nuclear-encoded PS I genes, PsaD has been the most extensively investigated from the viewpoint of its transcriptional and translational control. Lotan et al., (1993) examined the light-regulated accumulation of mRNA and polypeptide of PsaD in spinach. Upon exposure to continuous light, the mRNA, detected only at low levels in etiolated seedlings, began to accumulate. These authors made the same observation as Brunner et al., (1991) that phytochrome controlled *psaD* gene expression. *psaD* mRNA was present in the polysomal fraction only when seedlings were exposed to light, suggesting that the light-regulated step initiated the translation and that phytochrome was not involved in the regulation of translation. In *Nicotiana sylvestris*, one of the PsaD isoforms appeared earlier than other nuclear-encoded PS I subunits in response to illumination (Yamamoto et al., 1995a). This coincided with the early response of its mRNA, *psaDb*, during the initial greening phase. The organization of the light-responsive *cis*-elements of *psaDb* was investigated in transgenic tobacco by examining light-responsiveness of *psaDb* chimeric constructs using a GUS dependent assay and primer extension analysis (Yamamoto et al., 1997). Two light-responsive elements were found: one was located upstream of (−170 to +24) and the other was within the transcribed region (+1 to +861). The internal light-responsive element was utilized in etiolated seedlings but not in green leaves. Flieger et al. (1994) also pointed out the existence of *cis*-elements for the positive light response located within the coding region and/or even further downstream of the *psaD* gene in spinach. By comparison of the light-responsiveness of chimeric construct using genomic *psaD* sequence and *psaD* cDNA, it was revealed that the intron sequence contributed to the light-dependent expression of the spinach *psaD* gene (Bolle et al., 1996). An internal light-responsive element located downstream of the start codon was also observed for the ferredoxin gene from pea (Dicky et al., 1992, 1994; Gallo-Meagher et al., 1992) and *Arabidopsis thaliana* (Bovy et al., 1995).

There is another unusual feature in the regulatory domain of *psaD* gene. The 5′-leader of the *psaDb* gene (+1 to +23) of *Nicotiana sylvestris* was identified to contain a translational enhancer element (Yamamoto et al., 1995b), which was not involved in the light-response (Yamamoto et al., 1997). This *psaDb* leader shared two common motifs with the leader of an *Arabidopsis* ferredoxin gene, *Fed-A*. The *Fed-A* leader was also reported to act as an enhancer though it was not known if it influenced at the transcriptional or post-transcriptional levels (Casper and Quail, 1993). Since the PsaD subunit is the binding site of ferredoxin on the surface of PS I (Golbeck, 1992), it is reasonable that the genes encoding the two subunits are co-regulated at the transcriptional and translational levels.

D. Assembly of the PS I Complex

The assembly of the PS I complex may be the critical step for the regulation of photosystem stoichiometry. In fact, several genes have been documented to be specifically involved in the accumulation of PS I complexes but not with PS II complexes(Wilde et al., 1995; Bartsevich and Pakrasi, 1997; Boudreau et al., 1997; Ruf et al., 1997; Shen et al., 1998). Disruption of the chloroplast *ycf3* gene led to a complete loss of PS I complexes in both tobacco (Ruf et al., 1997) and *Chlamydomonas* (Boudreau et al., 1997). By disruption of the *ycf4* gene, the content of PS I relative to PS II decreased to about ⅓ compared to the wild type in *Synechocystis* sp. PCC 6803 (Wilde et al., 1995), whereas there was practically no accumulation of the PS I complex in the *Chlamy-*

domonas mutant (Boudreau et al., 1997). Another gene, *btpA*, seems to regulate a post-transcriptional process affecting biogenesis of the PS I complex in *Synechocystis* sp. PCC 6803 (Bartsevich and Pakrasi, 1997). A *btpA* disruption mutant had only 10–15% of PS I reaction center proteins compared to the wild type, while the PS II content remained unaffected. The BtpA protein is an extrinsic membrane protein which is exposed to the cytoplasmic face of the thylakoid membrane (Zak et al., 1999). Shen et al., (1998) reported cloning of a novel genc, *rubA*, that encodes a putative membrane-associated rubredoxin-like protein from *Synechococcus* sp. PCC 7002. Inactivation of *rubA* resulted in loss of three Fe-S clusters of PS I (F_x, F_A and F_B) and three of its peripheral subunits (PsaC, PsaD and PsaE)on the cytoplasmic side. The *rubA* gene product may consequently play an important role in the assembly or protection of the Fe-S clusters in PS I. Recently, it was found that disruption of *slr0228* in *Synechocystis* sp. PCC 6803 caused 60% reduction in the abundance of functional PS I with no effect on the content of PS II or phycobilisomes (Mann et al., 2000). *slr0228* is one of the four ORFs in *Synechocystis* sp. PCC 6803 encoding putative homologues to the FtsH protease. FtsH is known to be involved in a number of processes in *Escherichia coli*, including protein assembly in and through the cytoplasmic membrane. The product of *slr0228* may have a chaperone-like function in the assembly of PS I complex. There are some reports of mutants of barley (Nielsen et al., 1996) and maize (Heck et al., 1999) which showed a defect in PS I activity presumably due to a defect in translation or assembly of PS I subunits. In the latter case, the effect of the mutation was localized to the stromal side (PsaC, PsaD, and PsaE subunits) of PS I (Heck et al., 1999).

V. Inhibition of PS I by Environmental Factors

A. Photoinhibition of PS I

The term 'photoinhibition' was first defined by Kok (1956) as the decrease of photosynthesis by excess light illumination. Both PS I and PS II were reported as the site of inhibition in those early days (e.g. Satoh, 1970a,b,c). However, the term photoinhibition became gradually a synonym for 'photoinhibition of PS II,' since there was virtually no report about photoinhibition of PS I in vivo and only a few reports from in vitro studies. A few cases which reported the photoinhibition of PS I in vitro (Satoh and Fork, 1982b, Inoue et al., 1986, 1989) tended to be considered as a non-relevant artificial phenomenon. It was widely believed that the site of photoinhibition was PS II and that PS I was quite stable compared with PS II (e.g. Powles, 1984).

In 1994, the first evidence for a selective photoinhibition of PS I in vivo was reported (Terashima et al., 1994). Treatment of cucumber leaves at 4° C for 5 h at photon flux density (PFD) of less than 100 μmol m^{-2} s^{-1} resulted in a loss of PS I activity down to 20% while more than 80% of the PS II activity remained intact. Normally, PS I is not a rate-limiting step in photosynthetic electron transport, but after photoinhibition at chilling temperatures, whole chain electron transport was limited by the PS I activity. Initially, the photoinhibition of PS I was only observed in chilling sensitive plants such as cucumber (Terashima et al., 1994), common bean (Sonoike et al., 1995a) or pumpkin (Barth and Krause, 1998). Subsequent experiments showed that PS I can also be photoinhibited in chilling tolerant plants such as potato (Havaux and Davaud, 1994), winter rye (Ivanov et al., 1998) and barley (Tjus et al., 1998a, 1999) although the extent of the inhibition is smaller than in the chilling sensitive species. The damage could also be observed in barley plants in the field during chilling conditions verifying this phenomenon to be physiologically relevant (Tjus et al., 1998b). It was demonstrated that the photoinhibition of PS I could also be induced at room temperature in studies involving *Scenedesmus obliquus* (Harvey and Bishop, 1978), a marine diatom, *Amphora* sp. (Gerber and Burris, 1981), *Chlamydomonas reinhardtii* (Martin et al., 1997) and cultured cells of *Marchantia polymorpha* (Herrmann et al., 1997). In cyanobacteria, there has so far been no reports on the photoinhibition of PS I except for the mutant lacking detectable iron superoxide dismutase activity (Herbert et al., 1992).

1. Site of Inhibition

It was demonstrated that, under low photon flux densities, there was no decrease of P-700, the primary donor of PS I, when determined chemically. However, there was a substantial inhibition of the P-700 activity when determined photochemically (Sonoike and Terashima, 1994). This result suggests that some component(s) on the acceptor side of PS I is the site of inhibition. In fact, the destruction of iron-sulfur centers, F_A, F_B and F_X, was experimentally demonstrated by flash induced absorption change of P-700 in the sub-millisecond range combined with EPR spectroscopy (Sonoike et al., 1995b). It seems therefore that the iron-sulfur centers are the primary target during photoinhibition of PS I. When photon flux densities during the photoinhibitory treatment were increased, P-700 itself was also destroyed (Terashima et al., 1994).

The specific inhibition of the iron-sulfur centers leads to difficulties in analyzing the photoinhibition of PS I. Obviously, chemically determined P-700 is not influenced by the photoinhibition of PS I at low photon flux density. Moreover, oxygen consumption in the presence of methyl viologen, a conventional method to detect electron transfer activity through PS I, cannot be used to detect the photoinhibition, since methyl viologen can accept electrons from the electron acceptors, A_1 or A_0, in the absence of iron-sulfur centers (Fujii et al., 1990; Sonoike and Terashima, 1994). Furthermore, when in vivo determination of P-700 by a pulse amplified modulation system was used to detect PS I photoinhibition after light/chilling treatment, there was a difficulty to oxidize P-700 completely by a conventional far red light source, presumably due to enhanced cyclic electron transfer after chilling (Herbert et al., 1997; Sonoike 1998b). Thus, the content of P-700 would be underestimated by this method, unless very strong far red light is applied to oxidize P-700. The most reliable method to detect the photoinhibition of PS I is therefore to determine $NADP^+$ reduction in the presence of ferredoxin/FNR and an artificial electron donor such as DAD. Absorption changes due to oxidation of P-700 by continuous illumination in the presence of a low concentration ($<10\ \mu M$) of methyl viologen, or flash-induced absorption changes of P-700 at millisecond range can also be used to detect the photoinhibition of P-700 (Sonoike and Terashima, 1994). The difficulty to accurately determine the activity of PS I may be one of the reasons why the photoinhibition of this photosystem has been overlooked until recently.

2. Protein Degradation

It is well known that the D1 protein, one of the hetero-dimer subunits of the PS II reaction center, undergoes a rapid turnover during photoinhibitory conditions. Also during photoinhibition of PS I, degradation of one of the Chl-binding reaction center subunit was observed (Sonoike and Terashima, 1994). The subunit was identified as the PsaB protein using specific antisera (Sonoike, 1996b). There have been observed two types of peptide cleavages in connection with this degradation, one produced a fragment of apparent molecular mass of 51 kDa, and the other resulted in fragments of 45 kDa and 18 kDa. The site of the latter cleavage was determined as between Ala^{500} and Val^{501} of PsaB on the loop exposed to the lumenal side, between helices 7 and 8 (Sonoike et al., 1997). The site of the cleavage to produce the 51 kDa fragment could not be precisely determined, but the apparent molecular mass of the fragment suggested that the site could be on the loop exposed to the stromal side, between helices 8 and 9, where the ligands for the iron-sulfur center, F_X, is located.

The amount of the 51 kDa fragment was reduced in the presence of DCMU, DBMIB or methyl viologen when destruction of iron-slur centers was not observed. Thus, the formation of the 51 kDa fragment and destruction of the iron-sulfur centers may be coupled. On the other hand, the formation of the 45 kDa fragment and the 18 kDa fragment was enhanced by the addition of methyl viologen, suggesting that the formation of these fragments was not directly related to the destruction of iron-sulfur centers but rather related to the general production of superoxide (Sonoike et al., 1997). The formation of the 51 kDa protein seems to be suppressed in the presence of inhibitors for serine type proteases (Sonoike et al., 1997) and the involvement of the protease(s) in the process of PsaB degradation should be further examined.

Degradation of subunits other than PsaB has also been reported. Photoinhibition of isolated PS I reaction center complexes induced by very high-intensity light was claimed to cause a nonspecific degradation of PS I subunits (Baba et al., 1995). This degradation was suppressed by the presence of histidine, a scavenger of singlet oxygen. Very recently, Tjus et al., (1999) found that degradation of PsaA as well as PsaB was induced during in vivo photoinhibition of PS I in barley. They also reported that small subunits of PS I, especially the extrinsic proteins exposed to the stromal side, were also degraded in response to the photoinhibition. These results all indicate that the first site of damage is at the stromal side of PS I.

3. Mechanism of the Inhibition

As in most cases for PS II, photoinhibition of PS I is not observed in the absence of oxygen (Havaux and Davaud, 1994; Terashima et al., 1994; Sonoike, 1995). The addition of DCMU, an inhibitor of PS II, also suppressed the photoinhibition of PS I (Havaux and Davaud, 1994, Sonoike, 1995). These results imply that reduction of oxygen and formation of reactive oxygen species at the PS I reducing side by electrons from PS II are the cause of the inhibition. In fact, addition of *n*-propyl gallate, a scavenger of hydroxyl radicals and alchoxyl radicals, partially protected PS I from photoinhibition (Sonoike, 1996b). It is known that the hydroxyl radical is formed by the reaction between hydrogen peroxide and a reduced metal ion, the so-called Fenton reaction. Addition of hydrogen peroxide to thylakoid membranes in the darkness did not affect PS I (Sonoike et al., 1997; Forti et al., 1999). However, illumination of thylakoid membranes caused photoinhibition of PS I and the residual PS I activity was almost completely gone by the addition of hydrogen peroxide in the light (Sonoike et al., 1997). These results demonstrate that hydroxyl radicals, formed by the reaction between photoreduced iron in the iron-sulfur centers of PS I and hydrogen peroxide formed from photoreduced oxygen, is the cause of the inhibition. In vivo experiments also supported this mechanism of inhibition. The concentration of hydrogen peroxide in vivo was found to temporally increase after light/chilling treatment of cucumber leaves (Terashima et al., 1998). It should be noted that the addition of methyl viologen totally suppressed the photoinhibition of PS I in spite of the increased formation of superoxide and subsequent formation of reactive oxygen species (Sonoike, 1995). This phenomenon can be explained as follows: methyl viologen suppresses the formation of hydroxyl radical through the oxidation of iron-sulfur centers (i.e., decreased concentration of reduced metal ion) although the production of hydrogen peroxide should be enhanced in the presence of methyl viologen.

B. Role of Chilling Temperature

As mentioned above, photoinhibition of PS I in higher plants has been observed mainly in combination with chilling stress. When isolated thylakoid membranes from higher plants were subjected to weak light treatment, however, the selective inhibition of PS I was observed both in chilling sensitive and chilling tolerant plants regardless of the temperature during treatment (Sonoike, 1995). This observation revealed several important points. First of all, PS I, in itself, must be sensitive to photoinhibition. This sensitivity to photoinhibition may be ascribed to the low-redox characteristics of the PS I electron acceptors which can reduce oxygen and produce reactive oxygen species (Asada, 1999). The inhibition of PS I has probably been overlooked only because of the difficulty to detect the inhibition at the acceptor side as discussed in the Section V.A.1. Secondly, there must be some protective mechanism for PS I in vivo since the inhibition of PS I illuminated at room temperature in vivo is much smaller as compared to the inhibition in vitro. Thirdly, that protection must involve a temperature sensitive component in chilling sensitive plants. At present, we do not know the nature of the protection mechanism and how temperature affects it. Some of the putative protective mechanisms are discussed in Section VI below.

It is well known that photoinhibition of photosynthesis in general is aggravated at chilling temperatures (Baker et al., 1988; Krause, 1994; Sonoike, 1998a). It has been proposed that photoinhibition of PS II is enhanced at chilling temperatures both in chilling tolerant plants (Huner et al., 1996) and a chilling sensitive plant (Sonoike, 1999) through increased excitation pressure, i.e., more reduced condition at the acceptor side of PS II. In the case of photoinhibition of PS I, however, the correlation with excitation pressure is weak in a chilling sensitive plant (Sonoike, 1999). Photoinhibition of PS I seems to be induced at chilling temperatures by a different mechanism from that in the case of photoinhibition of PS II.

VI. Protection of Photosystem I from Photoinhibition

In the case of PS II photoinhibition, the damaged reaction center protein, D1, can be rapidly replaced in the range of hours due to its fast turnover rate (for a review, see Aro et al., 1993). On the other hand, the recovery of PS I from photoinhibition is not complete even after one week (H. Kudoh and K. Sonoike, unpublished). Thus, the photoinhibition of PS I is far more dangerous to the organism than that of PS II, if once induced. This may explain why in vivo photoinhibition of PS I is observed only under certain

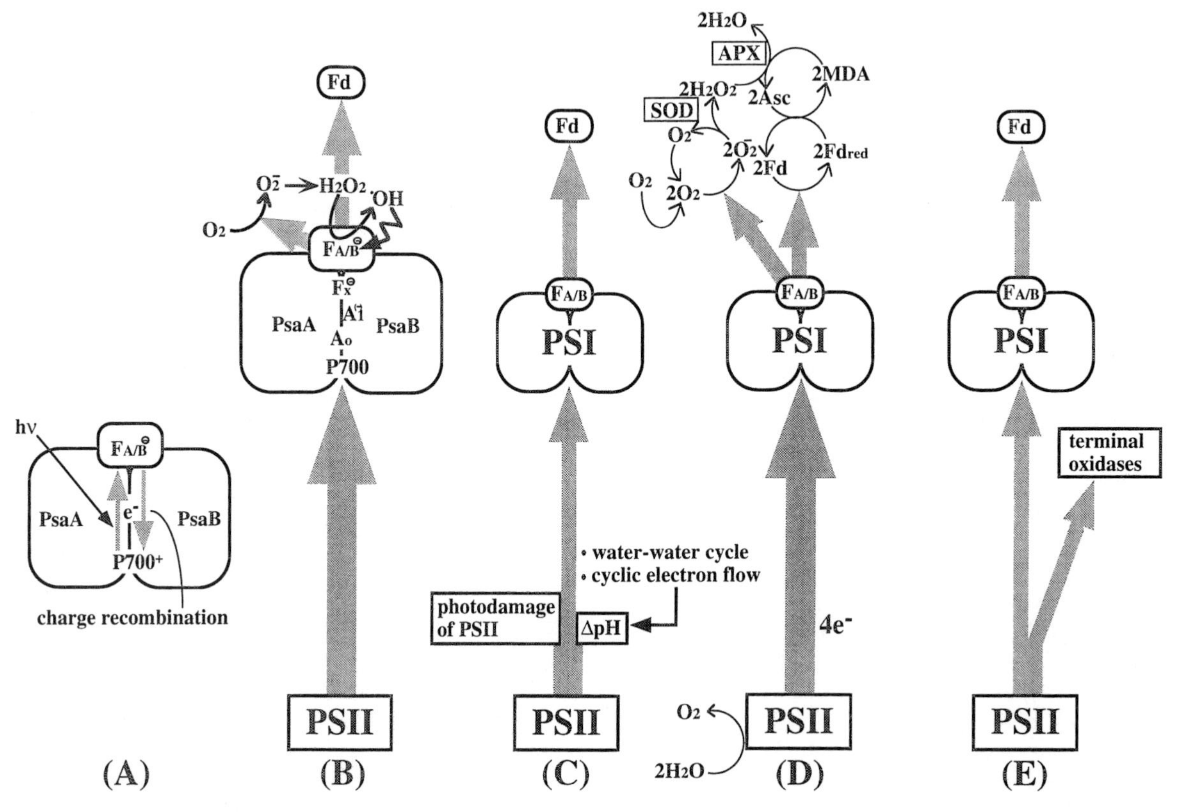

Fig. 2. Putative protection mechanisms from PS I photoinhibition. In the absence of reducing pressure from PS II, charge recombination protects PS I from photoinhibition (A). If any mechanism does not protect PS I, the iron sulfur centers are destroyed (B). Down regulation of electron transfer due to photodamage of PS II or proton gradient produced by water-water cycle or cyclic electron flow (C), oxygen scavenging system (water-water cycle) (D) and respiratory electron transfer (E) may play an important role to protect PS I from photoinhibition. (D) and (E) assumed PS I of higher plants and cyanobacteria, respectively.

conditions. As will be described below, there must be efficient protective mechanisms in vivo (see Fig. 2) to circumvent photoinhibitory damage to PS I.

A. Contribution from PS II

There have been several reports that photoinhibition of PS I is suppressed by the addition of the inhibitors of electron transfer such as DCMU or DBMIB (Havaux and Davaud, 1994; Sonoike, 1996b; Sonoike et al., 1997). These observations imply that any kind of decrease in PS II activity should protect PS I from photoinhibition. Thus, the photoinhibition of PS II itself can be regarded as the protective mechanism for PS I (Sonoike, 1996a).

The activity of PS II is down regulated when a pH gradient is built up across thylakoid membranes. Interestingly, treatment of cucumber leaves at a chilling temperature under low-intensity light, the condition which induces photoinhibition of PS I, was reported to cause uncoupling of thylakoid membranes isolated from the chilled leaves (Terashima et al., 1991a,b). A smaller extent of the proton gradient and of the down regulation of PS II under such a condition might promote the photoinhibition of PS I at a chilling temperatures.

Although there is no direct evidence to support a protection of PS I by photoinhibition of PS II so far, lack of PS II down-regulation seems to result in lowered viability in some green algae and cyanobacteria. For example, a cyanobacterial mutant (*pmgA* mutant) which cannot reduce its PS I content upon a shift to high-intensity light showed decreased sensitivity to photoinhibition of PS II. However, it led to lowered viability under prolonged high-intensity light conditions, indicating that lack of PS II down-regulation is harmful under this condition (Hihara and Ikeuchi, 1998; Sonoike et al., 1999). Also, in *Chlamydomonas reinhardtii* the efficiency of PS II electron transfer was reduced during sulfur deprivation, while SacI mutant did not show such an acclimatory process (Wykoff et al., 1998). This mutant became light-sensitive during sulfur deprivation because it cannot down-regulate photosynthetic electron transport. These results indicate that photoinhibition of PS II under a stress condition could be important for the survival of photosynthetic organisms.

B. Oxidized P-700

The light energy absorbed by PS I induces charge separation, i.e. the oxidation of P-700 and reduction of iron-sulfur centers which may be dangerous to PS I irrespective of the PS II activity. However, photoinhibition of PS I is not observed in the absence of PS II electron transport as described above. This may be explained in two ways. First, oxidized P-700 is very stable and does not cause any oxidative damage to its surroundings as is the case for oxidized P-680, since the redox potential of P-700 is much lower than that of P-680. In a low-redox photosystem like PS I, the reduced electron acceptors, not the oxidized electron donors, are potentially dangerous. Secondly, the charge recombination between oxidized P-700 and reduced iron-sulfur centers may produce triplet P-700, but not singlet oxygen even under photoinhibitory condition (Hideg and Vass, 1995). Thus, in the absence of reducing pressure from PS II, i.e. in the presence of oxidized P-700, the negative charge on the iron sulfur centers of PS I can be safely scavenged through recombination.

C. Scavenging Systems for Reactive Oxygen Species

As discussed in Section V.A.3, PS I under a light chilling condition is inhibited by hydroxyl radicals produced by the Fenton reaction which takes place between hydrogen peroxide and reduced iron sulfur centers. PS I, with its low redox potential, is able to reduce molecular oxygen. However, the electron transfer to molecular oxygen is a double-edged sword. Providing that the reactive oxygen species are safely scavenged, the reduction of molecular oxygen can be a sink for excess reducing power (Asada, 1999). However, when the reactive oxygen species cannot be scavenged, the low redox potential of the electron carriers of PS I together with the presence of iron on the acceptor side of PS I leads to the production of hydroxyl radicals which can destroy biological components with a diffusion limit. To avoid such a situation, the scavenging enzymes, SOD and ascorbate peroxidase, are localized near PS I (Miyake and Asada, 1992; Ogawa et al., 1995). Inhibition of PS I in intact chloroplasts from *Bryopsis* has been found to be enhanced under anaerobic conditions (Satoh and Fork, 1982a,b). On the other hand, the experiments using broken chloroplasts usually have resulted in less inhibition under anaerobic condition (Inoue et al., 1986; Sonoike, 1995,1996b). The difference between these results may be ascribed to the presence of the scavenging system for reactive oxygen species. In the intact chloroplast with an intact enzymatic scavenging system for reactive oxygen, the presence of oxygen works as a sink for electron transfer whereas formation of reactive oxygen species induces photoinhibition in broken chloroplasts lacking such enzymes. The fact that a mutant of a cyanobacterium with little SOD activity was susceptible to photoinhibition (Herbert et al., 1992) or to chilling (Thomas et al., 1999) also supports the importance of this scavenging system for avoidance of PS I photoinhibition. More recently, PS I was found to be susceptible to photoinhibition in an antisense mutant of tobacco which showed a decreased level of CuZn SOD (K. Asada, personal communication). In contrast to the case of SOD, there has been no evidence on importance of ascorbate peroxidase to protect PS I from photoinhibition. Terashima et al., (1998) reported that the activity of ascorbate peroxidase was not affected by a light chilling treatment in cucumber leaves.

D. Other Protective Mechanisms

In addition to the mechanisms discussed in the previous sections, there are several other reactions which might be involved in the protection of PS I from photoinhibition. Although direct evidence has not been obtained so far, some of those additional mechanisms may be functioning in cooperation with the more well established protection mechanisms mentioned above.

1. Cyclic Electron Transfer

Cyclic electron transfer around PS I is assumed to provide protection against photoinhibition. Since such a cyclic electron flow contributes to the formation of a trans thylakoid proton gradient, it leads to down regulation of PS II which in turn should protect PS I from photoinhibition. In addition, the cyclic electron transfer also stimulates ATP formation which enhances CO_2 fixation in the Calvin cycle and increases the sink activity for NADPH. Characterization of a *ndhB* disruption mutant of *Nicotiana tabacum* which showed reduced cyclic electron transfer revealed the protective role of cyclic electron transfer through the change of ATP/NADPH ratio at least for the photoinhibition of PS II (Endo et al., 1999). As for the photoinhibition of PS I, Chow and Hope (1998) reported that the P-700 absorption change of tobacco leaf at 820 nm decreased after photoinhibitory treatment when the leaf had been poisoned with Antimycin A, an inhibitor of cyclic electron flow. On the other hand, cyclic electron transfer could not contribute to the protection against photoinhibition of PS I in the light/chilling treatment of leaves of chilling sensitive species, where the uncoupling of thylakoid H^+-ATPase is induced (Terashima et al., 1991a,b). Under these circumstances, cyclic electron flow does not form any proton gradient, and light/chilling treatment seems to enhance cyclic electron flow (Sonoike, 1998b) rather than to reduce it. The relationship between cyclic electron flow and photoinhibition of PS I may therefore be totally different depending on the extent of coupling of thylakoid membranes.

2. Respiratory Electron Transfer

Although many reports on the photoinhibition of PS I have recently accumulated for many plants and algal species, there are so far virtually no reports for cyanobacteria. This may appear strange since the photosynthetic machinery is known to be very similar in plants and cyanobacteria. One big difference is that, in the case of cyanobacteria, the plastoquinone pool is shared between photosynthetic and respiratory electron transport (Aoki and Katoh, 1982). As discussed in Section V.A.3, accumulation of reducing power on the reducing side of PS I is the prerequisite for photoinhibition of PS I. The reducing power is supplied by electrons derived from PS II, but those electrons can also be transferred via the respiratory chain to a terminal oxidase, instead of PS I, in the case of cyanobacteria. Thus, the presence of such an alternative electron sink in cyanobacteria may protect PS I from photoinhibition.

3. Xanthophyll Cycle

The xanthophyll cycle, reversible conversion of violaxanthine, anthraxanthine and zeaxanthine through epoxidation and de-epoxidation, promotes energy dissipation under a high-intensity light condition and is a part of the down regulation of PS II (Pfündel and Bilger, 1994). The light harvesting complex of PS I also contains xanthophyll cycle pigments, which can perform interconversion through epoxidation and de-epoxidation (Thayer and Björkman, 1992; Lee and Thornber, 1995). Actually the xanthophyll-cycle pigments bound to PS I account for as much as 30–50% of the total xanthophyll-cycle pool (Färber et al., 1997). The addition of dithiothreitol (DTT), an inhibitor of violaxanthin de-epoxidase (VDE), did not induce PS I photoinhibition, but this may be due to the reducing action of DTT (Barth and Krause, 1998). Since fluorescence intensity from PS I antenna is not so much affected by the redox state of PS I reaction center, it is not yet clear whether the xanthophyll cycle in PS I contributes to protection against photodamage.

4. Unknown Factors

The susceptibility to the photoinhibition of PS I is affected by the growth photon flux densities in an acclimative manner: the higher the photon flux density is during growth, the more resistant PS I is to photoinhibition (Sonoike et al., 1995a). Furthermore, this acquired resistance to photoinhibition is specific to PS I (Ivanov et al., 1998). There may therefore be some unknown mechanism to protect PS I from photoinhibition, which can be induced by higher photon flux density during growth.

VII. Concluding Remarks

The study of the regulation of PS I activity and of the turnover of PS I is in an early stage. PS I may play an important role in the ability of the plant to respond to an ever fluctuating environment. Among more than eleven identified subunits of PS I, we know so far the physiological function of only a few. Many of the deletion mutants of PS I subunits in cyanobacteria as well as the cosuppression mutants of PS I subunits in higher plants showed normal growth under optimal conditions. This fact may indicate that such PS I subunits are involved in the assembly of the PS I complex or in the response or acclimation to a changing environment. The dynamic nature of PS I will without doubt be an important target for photosynthesis research.

Acknowledgments

We are very grateful to Drs. Masahiko Ikeuchi, Ayumi Tanaka, Junichi Obokata and Kouki Hikosaka for critical reading of an early version of the manuscript and for helpful discussions and suggestions. This work was supported in part by the Human Frontier Science Program and the Japan Society for the Promotion of Science.

References

Aizawa K and Fujita Y (1997) Regulation of synthesis of PS I in the cyanophytes *Synechocystis* PCC6714 and *Plectonema boryanum* during the acclimation of the photosystem stoichiometry to the light quality. Plant Cell Physiol 38: 319–326

Aizawa K, Shimizu T, Hiyama T, Satoh K, Nakamura Y and Fujita Y (1992) Changes in composition of membrane proteins accompanying the regulation of PS I/PS II stoichiometry observed with *Synechocystis* PCC 6803. Photosynth Res 32: 131–138

Andreasson E and Albertsson P-Å (1993) Heterogeneity in Photosystem I—the larger antenna of Photosystem Ia is due to functional connection to a special pool of LHCII. Biochim Biophys Acta 1141: 175–182

Aoki M and Katoh S (1982) Oxidation and reduction of plastoquinone by photosynthetic and respiratory electron transport in a cyanobacterium *Synechococcus* sp. Biochim Biophys Acta 682: 307–314

Armond PA, Schreiber U and Björkman O (1978) Photosynthetic acclimation to temperature in the desert shrub *Larrea divaricata*. II. Light harvesting efficiency and electron transport. Plant Physiol 61: 411–415

Aro EM, Virgin I and Andersson B (1993) Photoinhibition of Photosystem II. Inactivation, protein damage and turnover. Biochim Biophys Acta 1143: 113–134

Asada K (1999) The water-water cycle in chloroplasts: Scavenging of active oxygens and dissipation of excess photons. Annu Rev Plant Physiol Mol Biol 50: 601–639

Baba K, Itoh S and Hoshina S (1995) Degradation of Photosystem I reaction center proteins during photoinhibition in vitro. In: Mathis P (ed) Photosynthesis: From Light to Biosphere, Vol II, pp 179–182. Kluwer Academic Publishers, Dordrecht

Babu TS, Sabat SC and Mohanty P (1992) Heat induced alterations in the photosynthetic electron transport and emission properties of the cyanobacterium *Spirulina platensis*. J Photochem Photobiol 12: 161–171

Baker NR, Long SP and Ort DR (1988) Photosynthesis and temperature, with particular reference to effects on quantum yield. In: Long SP and Woodward FI (eds) Plants and Temperature, pp 347–375. Company of Biologists, Cambridge

Barth C and Krause GH (1998) Effects of light stress on Photosystem I in chilling-sensitive plants. In: Garab G (ed) Photosynthesis: Mechanisms and Effects, Vol IV, pp 2533–2536. Kluwer Academic Publishers, Dordrecht

Bartsevich VV and Pakrasi HB (1997) Molecular identification of a novel protein that regulates biogenesis of Photosystem I, a membrane protein complex. J Biol Chem 272: 6382–6387

Boardman NK and Anderson JM (1978) Composition, structure and photochemical activity of developing and mature chloroplasts. In: Akoyunoglou G and Argyroudi-Akoyunoglou JH (eds) Chloroplast Development, pp 1–14. Elsevier Science Publisher, Amsterdam

Bolle C, Herrmann RG and Oelmüller R (1996) Intron sequences are involved in the plastid- and light- dependent expression of the spinach *PsaD* gene. Plant J 10: 919–924

Boucher N and Carpentier R (1993) Heat-stress stimulation of oxygen uptake by Photosystem I involves the reduction of superoxide radicals by specific electron donors. Photosynth Res 35: 213–218

Boudreau E, Takahashi Y, Lemieux C, Turmel M and Rochaix JD (1997) The chloroplast *ycf3* and *ycf4* open reading frames of *Chlamydomonas reinhardtii* are required for the accumulation of the Photosystem I complex. EMBO J 16: 6095–6104

Bovy A, Van Den Berg C, De Vrieze G, Thompson WF, Weisbeek P and Smeekens S (1995) Light-regulated expression of the *Arabidopsis thaliana* ferredoxin gene requires sequences upstream and downstream of the transcription initiation site. Plant Mol Biol 27: 27–39

Bowler C, Neuhaus G, Yamagata H and Chua NH (1994) Cyclic GMP and calcium mediate phytochrome phototransduction. Cell 77: 73–81

Brettel K (1997) Electron transfer and arrangement of the redox cofactors in Photosystem I. Biochim Biophys Acta 1318: 322–373

Bricker TM and Newman DW (1982) Changes in the chlorophyll-proteins and electron transport activities of soybean (*Glycine max* L., Wayne) cotyledon chloroplasts during senescence. Photosynthetica 16: 239–244

Brunner H, Thümmler F, Song G and Rüdiger W (1991) Phytochrome-dependent mRNA accumulation for nuclear coded Photosystem I subunits in spinach seedlings. J Photochem Photobiol 11: 129–138

Burger-Wiersma T and Post AF (1989) Functional analysis of the photosynthetic apparatus of *Prochlorothrix hollandica* (Prochlorales), a chlorophyll *b* containing prokaryote. Plant Physiol 91: 770–774

Burkey KO and Wells R (1996) Effects of natural shade on soybean thylakoid membrane composition. Photosynth Res 50: 149–158

Cantrell A and Bryant DA (1987) Molecular cloning and nucleotide sequence of the *psaA* and *psaB* genes of the cyanobacterium *Synechococcus* sp. PCC7002. Plant Mol Biol 9: 453–468

Caspar T and Quail PH (1993) Promoter and leader regions involved in the expression of the *Arabidopsis* ferredoxin A gene. Plant J 3: 161–174

Chen MC, Cheng MC and Chen SCG (1993) Characterization of

the promoter of rice plastid *psaA-psaB-rps14* operon and the DNA-specific binding proteins. Plant Cell Physiol 34: 577–584

Cheng MC, Wu SP, Chen LFO and Chen SCG (1997) Identification and purification of a spinach chloroplast DNA-binding protein that interacts specifically with the plastid *psaA-psaB-rps14* promoter region. Planta 203: 373–380

Chow WS and Anderson JM (1987) Photosynthetic responses of *Pisum sativum* to an increase in growth irradiance II. Thylakoid membrane components. Aust J Plant Physiol 14: 9–20

Chow WS and Hope AB (1988) The stoichiometries of supramolecular complexes in thylakoid membranes from spinach chloroplasts. Aust J Plant Physiol 14: 21–28

Chow WS and Hope AB (1998) Redox reactions in photoinhibited tobacco leaf segments. In: Garab G (ed) Photosynthesis: Mechanisms and Effects, Vol III, pp 2123–2126. Kluwer Academic Publishers, Dordrecht

Chow WS, Melis A and Anderson JM (1990) Adjustments of photosystem stoichiometry in chloroplast improve the quantum efficiency of photosynthesis. Proc Natl Acad Sci USA 87: 7502–7506

Chow WS, Miller C and Anderson JM (1991a) Surface charges, the heterogeneous lateral distribution of the two photosystems, and thylakoid stacking. Biochim Biophys Acta 1057: 69–77

Chow WS, Adamson HY and Anderson JM (1991b) Photosynthetic acclimation of *Tradescantia albiflora* to growth irradiance: lack of adjustment of light-harvesting components and its consequences. Physiol Plant 81: 175–182

Cunningham FX, Dennenberg RJ, Jursinic PA and Gantt E (1990) Growth under red light enhances Photosystem II relative to Photosystem I and phycobilisomes in the red alga *Porphyridium crunetum*. Plant Physiol 93: 888–895

Cushman JC, Hallick RB and Price CA (1988) The two genes for the P-700 chlorophyll apoproteins on the *Euglena gracilis* chloroplast genome contain multiple introns. Curr Genet 13: 159–171

Danon A and Mayfield SP (1991) Light regulated translational activators: Identification of chloroplast gene specific mRNA binding proteins. EMBO J 10: 3993–4001

Danon A and Mayfield SP (1994a) ADP-dependent phosphorylation regulates RNA-binding in vitro: Implications in light modulated translation. EMBO J 13: 2227–2235

Danon A and Mayfield SP (1994b) Light-regulated translation of chloroplast messenger RNAs through redox control. Science 226: 1717–1719

De la Torre WR and Burkey KO (1990) Acclimation of barley to changes in light intensity: Chlorophyll organization. Photosynth Res 24: 117–125

Deng XW, Tonkyn JC, Peter GF, Thornber JP and Gruissem W (1989) Post-transcriptional control of plastid mRNA accumulation during adaptation of chloroplasts to different light quality environments. Plant Cell 1: 645–654

Dickey LF, Gallo-Meagher M and Thompson WF (1992) Light regulatory sequences are located within the 5′ portion of the *Fed-1* message sequence. EMBO J 11: 2311–2317

Dickey LF, Nguyen TT, Allen GC and Thompson WF (1994) Light modulation of ferredoxin mRNA abundance requires an open reading frame. Plant Cell 6: 1171–1176

Dreyfuss BW and Thornber JP (1994) Organization of the light-harvesting complex of Photosystem I and its assembly during plastid development. Plant Physiol 106: 841–848

Egnéus H, Reftel S and Selldén G (1972) The appearance and development of photosynthetic activity in etiolated barley leaves and isolated etio-chloroplasts. Physiol Plant 27: 48–55

Eichacker L, Soll J, Lauterbach P, Rüdiger W, Klein RR and Mullet JE (1990) In vitro synthesis of chlorophyll *a* in the dark triggers accumulation of chlorophyll *a* apoproteins in barley etioplasts. J Biol Chem 265: 13566–13571

Ellis RJ (1981) Chloroplast proteins: Synthesis, transport, and assembly. Annu Rev Plant Physiol 32: 111–137

Endo T, Shikanai T, Takabayashi A, Asada K and Sato F (1999) The role of chloroplastic NAD(P)H dehydrogenase in photoprotection. FEBS Lett 457: 5–8

Escoubas JM, Lomas M, LaRoche J and Falkowski PG (1995) Light intensity regulation of *cab* gene transcription is signaled by the redox state of the plastoquinone pool. Proc Natl Acad Sci USA 92: 10237–10241

Evans JR (1987) The relationship between electron transport components and photosynthetic capacity in pea leaves grown at different irradiances. Aust J Plant Physiol 14: 157–170

Färber A, Young AJ, Ruban AV, Horton P and Jahns P (1997) Dynamics of xanthophyll-cycle activity in different antenna subcomplexes in the photosynthetic membranes of higher plants. Plant Physiol 115: 1609–1618

Falkowski PG, Owens TG, Ley AC and Mauzerall DC (1981) Effects of growth irradiance levels on the ratio of reaction centers in two species of marine phytoplankton. Plant Physiol 68: 969–973

Fish LE, Kück U and Bogorad K (1985) Two partially homologous adjacent light inducible maize chloroplast genes encoding polypeptides of the P700 chlorophyll *a*-protein complex of Photosystem I. J Biol Chem 260: 1413–1421

Flieger K, Wicke A, Herrmann RG and Oelmüller R (1994) Promoter and leader sequences of the spinach *PsaD* and *PsaF* genes direct an opposite light response in tobacco cotyledons: *PsaD* sequences downstream of the ATG codon are required for a positive light response. Plant J 6: 359–368

Forti G, Barbagallo RP and Inversini B (1999) The role of ascorbate in the protection of thylakoids against photo-inactivation. Photosynth Res 59: 215-222

Fujii T, Yokoyama E, Inoue K and Sakurai H (1990) The sites of electron donation of Photosystem I to methyl viologen. Biochim Biophys Acta 1015: 41–48

Fujita Y (1997) A study on the dynamic features of photosystem stoichiometry: Accomplishments and problems for future studies. Photosynth Res 53: 83–93

Fujita Y (1999) An evidence indicating that a weak orange light absorbed by phycobilisomes causes inactivation of PS II in cells of the red alga *Porphyridium cruentum* grown under a weak red light preferentially exciting Chl *a*. Plant Cell Physiol 40: 924–932

Fujita Y and Murakami A (1987) Regulation of electron transport composition in cyanobacterial photosynthetic system: Stoichiometry among Photosystem I and II complexes and their light-harvesting antennae and cytochrome b_6/f complex. Plant Cell Physiol 28: 1547–1553

Fujita Y, Ohki K and Murakami A (1985) Chromatic regulation of photosystem composition in the photosynthetic system of red and blue-green algae. Plant Cell Physiol 26: 1541–1548

Fujita Y, Murakami A and Ohki K (1990) Regulation of the stoichiometry of thylakoid components in the photosynthetic system of cyanophytes: Model experiments showing that control of the synthesis or supply of Chl *a* can change the stoichiometric relationship between the two photosystems. Plant Cell Physiol 31: 145–153

Fujita Y, Murakami A and Aizawa K (1995) The accumulation of protochlorophyllide in cells of *Synechocystis* PCC 6714 with a low PS I/PS II stoichiometry. Plant Cell Physiol 36: 575–582

Gallo-Meagher M, Sowinski DA, Elliott RC and Thompson WF (1992) Both internal and external regulatory elements control expression of the pea *Fed*-1 gene in transgenic tobacco. Plant Cell 4: 389–395

Gepstein S (1988) Photosynthesis. In: Noodén LD and Leopold AC (eds) Senescence and Aging in Plants, pp 85–109. Academic Press, New York

Gerber DW and Burris JE (1981) Photoinhibition and P-700 in the marine diatom *Amphora* sp. Plant Physiol. 68: 699–702

Glick RE, McCauley SW, Gruissem W and Melis A (1986) Light quality regulates expression of chloroplast genes and assembly of photosynthetic membrane complexes. Proc Natl Acad Sci USA 83: 4287–4291

Golbeck JH (1992) Structure and function of Photosystem I. Annu Rev Plant Physiol Plant Mol Biol 43: 293–324

Grover M, Dhingra A, Sharma AK, Maheshwari SC and Tyagi AK (1999) Involvement of phytochrome(s), Ca^{2+} and phosphorylation in light-dependent control of transcript levels for plastid genes (*psbA*, *psaA* and *rbcL*) in rice (*Oryza sativa*). Physiol Plant 105: 701–707

Harvey GW and Bishop NI (1978) Photolability of photosynthesis in two separate mutants of Scenedesmus obliquus. Plant Physiol 62: 330–336

Havaux M and Davaud A (1994) Photoinhibition of photosynthesis in chilled potato leaves is not correlated with a loss of Photosystem-II activity—Preferential inactivation of Photosystem I. Photosynth Res 40: 75–92

Heck DA, Miles D and Chitnis PR (1999) Characterization of two photosynthetic mutants of maize. Plant Physiol 120: 1129–1136

Henningsen KW and Boardman NH (1973) Development of photochemical activities and the appearance of the high potential form of cytochrome b_{559} in greening barley seedlings. Plant Physiol 51: 1117–1126

Herbert SK, Samson G, Fork DC and Laudenbach DE (1992) Characterization of damage to Photosystem I and II in a cyanobacterium lacking detectable iron superoxide dismutase activity. Proc Natl Acad Sci USA 89: 8716–8720

Herbert SK, Sacksteder C and Kramer DM (1997) Accelerated cyclic electron transport is an early effect of chilling in chilling-sensitive plants. Plant Physiol 114: 219

Herranen M, Aro EM and Tyystjärvi T (1998) Expression of PS II and PS I genes in *Synechocystis* 6803. In: Garab G (ed) Photosynthesis: Mechanisms and Effects, Vol IV, pp 2913–2916. Kluwer Academic Publishers, Dordrecht

Herrmann B, Kilian R, Peter S and Schäfer C (1997) Light-stress-related changes in the properties of Photosystem I. Planta 201: 456–462

Herrmann RG, Oelmüller R, Bichler J, Schneiderbauer A, Steppuhn J, Wedel N, Tyagi AK and Westhoff P (1991) The thylakoid membrane of higher plants: Genes, their expression and interaction. In: Herrmann RG and Larkins B (eds) Plant Molecular Biology 2, pp 411–427. Plenum Press, New York

Hibino T, Lee BH, Rai AK, Ishikawa H, Kojima H, Tawada M, Shimoyama H and Takabe T (1996) Salt enhances Photosystem I content and cyclic electron flow via NAD(P)H dehydrogenase in the halotolerant cyanobacterium *Aphanothece halophytica*. Aust J Plant Physiol 23: 321–330

Hideg E and Vass I (1995) Singlet oxygen is not produced in Photosystem I under photoinhibitory conditions. Photochem Photobiol 62: 949–952

Hihara Y (1999) The molecular mechanism for acclimation to high light in cyanobacteria. Curr Topics Plant Biol 1: 37–50

Hihara Y and Ikeuchi M (1997) Mutation in a novel gene required for photomixotrophic growth leads to enhanced photoautotrophic growth of *Synechocystis* sp. PCC 6803. Photosynth Res 53: 243–252

Hihara Y and Ikeuchi M (1998) Toward the elucidation of physiological significance of *pmgA*-mediated high-light acclimation to adjust photosystem stoichiometry: Effects of the prolonged high-light treatment on *pmgA* mutants. In: Garab G (ed) Photosynthesis: Mechanisms and Effects, Vol IV, pp 2929–2932. Kluwer Academic Publishers, Dordrecht

Hihara Y, Sonoike K and Ikeuchi M (1998) A novel gene, *pmgA*, specifically regulates photosystem stoichiometry in the cyanobacterium *Synechocystis* species PCC 6803 in response to high light. Plant Physiol 117: 1205–1216

Huner NPA (1985) Acclimation of winter rye to cold-hardening temperatures results in an increased capacity for photosynthetic electron transport. Can J Bot 63: 506–511

Huner NPA and Reynolds TL (1989) Low growth temperature-induced increase in light saturated PS I electron transport is cation dependent. Plant Physiol 91: 1308–1316

Huner NPA, Krol M, Williams JP and Maissan E (1988) Overwintering periwinkle (*Vinca minor* L.) exhibits increased Photosystem I activity. Plant Physiol 87: 721–726

Huner NPA, Öquist G, Hurry VM, Krol M, Falk S and Griffith M (1993) Photosynthesis, photoinhibition and low temperature acclimation in cold tolerant plants. Photosynth Res 37: 19–39

Huner NPA, Maxwell DP, Gray GR, Savitch LV, Krol M, Ivanov AG and Falk S (1996) Sensing environmental temperature change through imbalances between energy supply and energy consumption: redox state of Photosystem II. Physiol Plant 98: 358–364

Inoue K, Sakurai H and Hiyama T (1986) Photoinactivation sites of Photosystem I in isolated chloroplasts. Plant Cell Physiol 27: 961–968

Inoue K, Fujii T, Yokoyama E, Matsuura K, Hiyama T and Sakurai H (1989) The photoinhibition site of Photosystem I in isolated chloroplasts under extremely reducing conditions. Plant Cell Physiol 30: 65–71

Ivanov AG, Morgan RM, Gray GR, Velitchkova MY and Huner NPA (1998) Temperature/light dependent development of selective resistance to photoinhibition of Photosystem I. FEBS Lett 430: 288–292

Izawa S (1980) Acceptors and donors for chloroplast electron transport. Methods Enzymol 69: 419–433

Jeanjean R, Matthijs HCP, Onana B, Havaux M and Joset F (1993) Exposure of the cyanobacterium *Synechocystis* PCC6803 to salt stress induces concerted changes in respiration and photosynthesis. Plant Cell Physiol 34: 1073–1079

Joset F, Jeanjean R and Hagemann M (1996) Dynamics of the response of cyanobacteria to salt stress: Deciphering the molecular events. Physiol Plant 96: 738–744

Kaneko T, Sato S, Kotani H, Tanaka A, Asamizu E, Nakamura, Miyajima N, Hirosawa M, Sugiura M, Sasamoto S, Kimura T,

Hosouchi T, Matsuno A, Muraki A, Nakazaki N, Naruo K, Okumura S, Shimpo S, Takeuchi C, Wada T, Watanabe A, Yamada M, Yasuda M and Tabata S (1996) Sequence analysis of the genome of the unicellular cyanobacterium *Synechocystis* sp. strain PCC 6803. II. Sequence determination of the entire genome and assignment of potential protein-coding regions. DNA Res 3: 109–136

Kawamura M, Mimuro M and Fujita Y (1979) Quantitative relationship between two reaction centers in the photosynthetic system of blue-green algae. Plant Cell Physiol 20: 697–705

Kim J and Mayfield SP (1997) Protein disulfide isomerase as a regulator of chloroplast translational activation. Science 278: 1954–1957

Kim J, Eichacker LA, Rudiger W and Mullet JE (1994) Chlorophyll regulates accumulation of the plastid-encoded chlorophyll proteins P-700 and D1 by increasing apoprotein stability. Plant Physiol 104: 907–916

Kim JH, Glick RE and Melis A (1993) Dynamics of photosystem stoichiometry adjustment by light quality in chloroplasts. Plant Physiol 102: 181–190

Kirsch W, Seyer P and Herrmann RG (1986) Nucleotide sequence of the clustered genes for two P700 chlorophyll *a* apoproteins of the Photosystem I reaction center and the ribosomal protein S14 of the spinach plastid chromosome. Curr Genet 10: 843–855

Klein RR and Mullet JE (1986) Regulation of chloroplast-encoded chlorophyll-binding protein translation during higher plant chloroplast biogenesis. J Biol Chem 261: 11138–11145

Klein RR and Mullet JE (1987) Control of gene expression during higher plant chloroplast biogenesis: Protein synthesis and transcript levels of psbA, psaA-psaB, and rbcL in dark-grown and illuminated barley seedlings. J Biol Chem 262: 4341–4348

Klein RR, Mason HS and Mullet JE (1988a) Light-regulated translation of chloroplast proteins. I. Transcripts of PsaA-PsaB, PsbA, and RbcL are associated with polysomes in dark-grown and illuminated barley seedlings. J Cell Biol 106: 289–301.

Klein RR, Gamble PE and Mullet JE (1988b) Light-dependent accumulation of radiolabeled plastid-encoded chlorophyll *a*-apoproteins requires chlorophyll *a*. Plant Physiol 88: 1246–1256

Kok B (1956) On the inhibition of photosynthesis by intense light. Biochim Biophys Acta 21: 234–244

Krause GH (1994) Photoinhibition induced by low temperatures. In: Baker NR and Bowyer JR (eds) Photoinhibition of Photosynthesis. From Molecular Mechanisms to the Field. pp 331–348. BIOS Scientific Publishers, Oxford

Krauß N, Schubert WD, Klukas O, Fromme P, Witt HT and Saenger W (1996) Photosystem I at 4 Å resolution represents the first structural mode of a joint photosynthetic reaction centre and core antenna system. Nat Struct Biol 3: 965–973

Kreuz K, Dehesh K and Apel K (1986) The light-dependent accumulation of the P-700 chlorophyll *a* protein of the Photosystem I reaction center in barley: Evidence for translational control. Eur J Biochem 159: 459–467

Kyle DJ and Zalik S (1982) Development of photochemical activity in relation to pigment and membrane protein accumulation in chloroplasts of barley and its virescens mutant. Plant Physiol 69: 1392–1400

Laing W, Kreuz K and Apel K (1988) Light-dependent, but phytochrome-independent, translational control of the accumulation of the P-700 chlorophyll-a protein of Photosystem I in barley (*Hordeum vulgare* L.). Planta 176: 269–276

Lajkó F, Kadioglu A and Garab G (1991) Involvement of superoxide dismutase in heat-induced stimulation of Photosystem I-mediated oxygen uptake. Biochem Biophys Res Com 174: 696–700

Lee AI-C and Thornber JP (1995) Analysis of the pigment stoichiometry of pigment-protein complexes from barley (*Hordeum vulgare*). The xanthophyll cycle intermediates occur mainly in the light-harvesting complexes of Photosystem I and Photosystem II. Plant Physiol 107: 565–574

Leong TY and Anderson JM (1984) Adaptation of the thylakoid membranes of pea chloroplasts to light intensities. II. Regulation of electron transport capacities, electron carriers, coupling factor (CF1) activity and rates of photosynthesis. Photosynth Res 5: 117–128

Leong TY and Anderson JM (1986) Light-quality and irradiance adaptation of the composition and function of pea-thylakoid membranes. Biochim Biophys Acta 850: 57–63

Liu LX, Chow WS and Anderson JM (1993) Light quality during growth of *Tradescantia albiflora* regulates photosystem stoichiometry, photosynthetic function and susceptibility to photoinhibition. Physiol Plant 89: 854–860

Lotan O, Cohen Y, Michaeli D and Nechushtai R (1993) High levels of Photosystem I subunit II (PsaD) mRNA result in the accumulation of the PsaD polypeptide only in the presence of light. J Biol Chem 268: 16185–16189

Lübberstedt T, Bolle CEH, Sopory S, Flieger K, Herrmann RG and Oelmüller R (1994) Promoters from genes for plastid proteins possess regions with different sensitivities toward red and blue light. Plant Physiol 104: 997–1006

Mann NH, Novac N, Mullineaux CW, Newman J, Bailey S and Robinson C (2000) Involvement of an FtsH homologue in the assembly of functional Photosystem I in the cyanobacterium *Synechocystis* sp. PCC 6803. FEBS Lett 479: 72–77

Manodori A and Melis A (1984) Photochemical apparatus organization in *Anacystis nidulans* (Cyanophyceae). Effect of CO_2 concentration during cell growth. Plant Physiol 74: 67–71

Manodori A and Melis A (1986) Cyanobacterial acclimation to Photosystem I or Photosystem II light. Plant Physiol 82: 185–189

Martin RE, Thomas DJ, Tucker DE and Herbert SK (1997) The effects of photooxidative stress on Photosystem I measured in vivo in *Chlamydomonas*. Plant Cell Environ 20: 1451–1461

Melis A and Harvey GW (1981) Regulation of photosystem stoichiometry, chlorophyll *a* and chlorophyll *b* content and relation to chloroplast ultrastructure. Biochim Biophys Acta 637: 138–145

Melis A, Mullineaux CW and Allen JF (1989) Acclimation of the photosynthetic apparatus to Photosystem I or Photosystem II light: Evidence from quantum yield measurement and fluorescence spectroscopy of cyanobacterial cells. Z Naturforsch 44c: 109–118

Melis A, Murakami A, Nemson JA, Aizawa K, Ohki K and Fujita Y (1996) Chromatic regulation of photosystem stoichiometry in *Chlamydomonas reinhardtii* alters thylakoid membrane structure-function and improves the quantum efficiency of photosynthesis. Photosynth Res 47: 263–265

Miyake C and Asada K (1992) Thylakoid-bound ascorbate peroxidase in spinach chloroplasts and photoreduction of its

primary oxidation product monodehydroascorbate radicals in thylakoids. Plant Cell Physiol 33: 541–553
Murakami A and Fujita Y (1988) Steady state of photosynthesis in cyanobacterial photosynthetic systems before and after regulation of electron transport composition: Overall rate of photosynthesis and PS I/PS II composition. Plant Cell Physiol 29: 305–311
Murakami A and Fujita Y (1991a) Steady state of photosynthetic electron transport in cells of the cyanophyte *Synechocystis* PCC 6714 having different stoichiometry between PS I and PS II: Analysis of flash-induced oxidation-reduction of cytochrome *f* and P-700 under steady state of photosynthesis. Plant Cell Physiol 32: 213–222
Murakami A, and Fujita Y (1991b) Regulation of photosystem stoichiometry in the photosynthetic system of the cyanophyte *Synechocystis* PCC 6714 in response to light-intensity. Plant Cell Physiol 32: 223–230
Murakami A and Fujita Y (1993) Regulation of stoichiometry between PS I and PS II in response to light regime for photosynthesis observed with *Synechocystis* PCC 6714: Relationship between redox state of cyt *b6-f* complex and regulation of PS I formation. Plant Cell Physiol 34: 1175–1180
Murakami A, Kim SJ and Fujita Y (1997a) Changes in photosystem stoichiometry in response to environmental conditions for cell growth observed with the cyanophyte *Synechocystis* PCC 6714. Plant Cell Physiol 38: 392–397
Murakami A, Fujita Y, Nemson JA and Melis A (1997b) Chromatic regulation in *Chlamydomonas reinhardtii*: Time course of photosystem stoichiometry adjustment following a shift in growth light quality. Plant Cell Physiol 38: 188–193
Nakamura M and Obokata J (1994) Organization of the *psaH* gene family of Photosystem I in *Nicotiana sylvestris*. Plant Cell Physiol 35: 297–302
Nakamura M and Obokata J (1995) Tobacco nuclear genes for Photosystem I subunits, *psaD*, *psaE* and *psaH* share an octamer motif bound with nuclear proteins. Plant Cell Physiol 36: 1393–1397
Neale PJ and Melis A (1986) Algal photosynthetic membrane complexes and the photosynthesis-irradiance curve: A comparison of light-adaptation responses in *Chlamydomonas reinhardtii* (Chlorophyta). J Phycol 22: 531–538
Nechushtai R and Nelson N (1985) Biogenesis of Photosystem I reaction center during greening of oat, bean and spinach leaves. Plant Mol Biol 4: 377–384
Neuhaus G, Bowler C, Kern R and Chua NH (1993) Calcium/calmodulin-dependent and -independent phytochrome signal transduction pathways. Cell 73: 937–952
Nielsen VS, Scheller HV and Møller BL (1996) The Photosystem I mutant *viridis-zb*^{63} of barley (*Hordeum vulgare*) contains low amounts of active but unstable Photosystem I. Physiol Plant 98: 637–644
Obokata J, Mikami K, Hayashida N and Sugiura M (1990) Polymorphism of a Photosystem I subunit caused by alloploidy in *Nicotiana*. Plant Physiol 92: 273–275
Obokata J, Mikami K, Hayashida N, Nakamura M and Sugiura M (1993) Molecular heterogeneity of Photosystem I. *psaD*, *psaE*, *psaF*, *psaH*, and *psaL* are all present in isoforms in *Nicotiana* spp. Plant Physiol 102: 1259–1267
Obokata J, Mikami K, Yamamoto Y and Hayashida N (1994) Microheterogeneity of PS I-E subunit of Photosystem I in *Nicotiana sylvestris*. Plant Cell Physiol 35: 203–209
Ogawa K, Kanematsu S, Kakabe K and Asada K (1995) Attachment of CuZn-superoxide dismutase to thylakoid membranes at the site of superoxide generation (PS I) in spinach chloroplasts: detection by immuno-gold labeling after rapid freezing and substitution method. Plant Cell Physiol 36: 565–573
Ohashi K, Tanaka A and Tsuji H (1989) Formation of the photosynthetic electron transport system during the early phase of greening in barley leaves. Plant Physiol 91: 409–414
Ohashi K, Murakami A, Tanaka A, Tsuji H and Fujita Y (1992) Developmental changes in amounts of thylakoid components in plastids of barley leaves. Plant Cell Physiol 33: 371–377
Okada K and Katoh S (1998) Two long-term effects of light that control the stability of proteins related to photosynthesis during senescence of rice leaves. Plant Cell Physiol 39: 394–404
Okada K, Inoue Y, Satoh K and Katoh S (1992) Effects of light on degradation of chlorophyll and proteins during senescence of detached rice leaves. Plant Cell Physiol 33: 1183–1191
Pfannschmidt T, Nilsson A and Allen JF (1999) Photosynthetic control of chloroplast gene expression. Nature 397: 625–628
Pfündel E and Bilger W (1994) Regulation and possible function of the violaxanthin cycle. Photosynth Res 42: 89–109
Plesnicar M and Bendall DS (1973) The photochemical activities and electron carriers of developing barley leaves. Biochem J 136: 803–812
Powles SB (1984) Photoinhibition of photosynthesis induced by visible light. Annu Rev Plant Physiol 35: 15–44
Rodermel SR and Bogorad L (1985) Maize plastid photogenes: Mapping and photoregulation of transcript levels during light-induced development. J Cell Biol 100: 463–476
Ruf S, Kössel H and Bock R (1997) Targeted inactivation of a tobacco intron-containing open reading frame reveals a novel chloroplast-encoded Photosystem I-related gene. J Cell Biol 139: 95–102
Samson G, Herbert SK, Fork DC and Laudenbach DE (1994) Acclimation of the photosynthetic apparatus to growth irradiance in a mutant strain of *Synechococcus* lacking iron superoxide dismutase. Plant Physiol 105: 287–294
Satoh Ka and Fork DC (1982a) The light induced decline of chlorophyll fluorescence as an indication of photoinhibition in intact *Bryopsis corticularis* chloroplasts illuminated under anaerobic conditions. Photobiochem Photobiophys 4: 153–162
Satoh Ka and Fork DC (1982b) Photoinhibition of reaction centers of Photosystems I and II in intact *Bryopsis* chloroplasts under anaerobic conditions. Plant Physiol 70: 1004–1008
Satoh Ki (1970a) Mechanism of photoinactivation in photosynthetic systems. I. The dark reaction in photoinactivation. Plant Cell Physiol 11: 15–27
Satoh Ki (1970b) Mechanism of photoinactivation in photosynthetic systems. II. The occurrence and properties of two different types of photoinactivation. Plant Cell Physiol 11: 29–38
Satoh Ki (1970c) Mechanism of photoinactivation in photosynthetic systems. III. Site and mode of photoinactivation in Photosystem I. Plant Cell Physiol 11: 187–197
Scheller HV, Naver H and Møller BL (1997) Molecular aspects of Photosystem I. Physiol Plant 100: 842–851
Schrubar H, Wanner G and Westhoff P (1990) Transcriptional control of plastid gene expression in greening Sorghum seedlings. Planta 183: 101–111

Schubert H and Hagemann M (1990) Salt effects on 77K fluorescence and photosynthesis in the cyanobacterium *Synechocystis* sp. PCC 6803. FEMS Microbiol Lett 71: 169–172

Shen G, Antonkine ML, Vassiliev IR, Golbeck JH and Bryant DA (1998) A rubredoxin-like protein plays an essential role in assembly of the F_A, F_B & F_X iron-sulfur clusters in Photosystem I. In: Garab G (ed) Photosynthesis: Mechanisms and Effects, Vol IV, pp 3147–3150. Kluwer Academic Publishers, Dordrecht

Shinohara K and Murakami A (1996) Changes in levels of thylakoid components in chloroplasts of pine needles of different ages. Plant Cell Physiol 37: 1102–1107

Smith BM and Melis A (1988) Photochemical apparatus organization in the diatom *Cylindrotheca fusiformis*: Photosystem stoichiometry and excitation distribution in cells grown under high and low irradiance. Plant Cell Physiol 29: 761–769

Smith H, Samson G and Fork DC (1993) Photosynthetic acclimation to shade: Probing the role of phytochromes using photomorphogenic mutants of tomato. Plant Cell Environ 16: 929–937

Sonoike K (1995) Selective photoinhibition of Photosystem I in isolated thylakoid membranes from cucumber and spinach. Plant Cell Physiol 36: 825–830

Sonoike K (1996a) Photoinhibition of Photosystem I: Its physiological significance in the chilling sensitivity of plants. Plant Cell Physiol 37: 239–247

Sonoike K (1996b) Degradation of *psaB* gene product, the reaction center subunit of Photosystem I, is caused during photoinhibition of Photosystem I: Possible involvement of active oxygen species. Plant Sci 115: 157–164

Sonoike K (1998a) Various aspects of inhibition of photosynthesis under light/chilling stress: 'Photoinhibition at chilling temperatures' versus 'Chilling damage in the light.' J Plant Res 111: 121–129

Sonoike K (1998b) Photoinhibition of Photosystem I in chilling sensitive plants determined in vivo and in vitro. In: Garab G (ed) Photosynthesis: Mechanisms and Effects, Vol III, pp 2217–2220. Kluwer Academic Publishers, Dordrecht

Sonoike K (1999) The different roles of chilling temperatures in the photoinhibition of Photosystem I and Photosystem II. J Photochem Photobiol 48: 136–141

Sonoike K and Terashima I (1994) Mechanism of the Photosystem I photoinhibition in leaves of *Cucumis sativus* L. Planta 194: 287–293

Sonoike K, Ishibashi M, Watanabe A (1995a) Chilling sensitive steps in leaves of *Phaseolus vulgaris* L. Examination of the effects of growth irradiances on PS I photoinhibition. In: Mathis P (ed) Photosynthesis: From Light to Biosphere, Vol IV, pp 853–856. Kluwer Academic Publishers, Dordrecht

Sonoike K, Terashima I, Iwaki M and Itoh S (1995b) Destruction of Photosystem I iron-sulfur centers in leaves of *Cucumis sativus* L. by weak illumination at chilling temperatures. FEBS Lett 362: 235–238

Sonoike K, Kamo M, Hihara Y, Hiyama T and Enami I (1997) The mechanism of the degradation of *psa*B gene product, one of the photosynthetic reaction center subunits of Photosystem I, upon photoinhibition. Photosynth Res 53: 55–63

Sonoike K, Hihara Y and Ikeuchi M (1999) Physiological significance of photosystem stoichiometry in a cyanobacterium upon light acclimation. Plant Cell Physiol 40: S43

Stampacchia O, Girard-Bascou J, Zanasco J-L, Zerges W, Bennoun P and Rochaix J-D (1997) A nuclear-encoded function essential for translation of the chloroplast *psaB* mRNA in *Chlamydomonas*. Plant Cell 9: 773–782

Stefánsson H, Andreasson E, Weibull C and Albertsson P-Å (1997) Fractionation of the thylakoid membrane from *Dunaliella salina*—heterogeneity is found in Photosystem I over a broad range of growth irradiance. Biochim Biophys Acta 1320: 235–246

Stevenson JK and Hallick RB (1994) The psaA operon pre-mRNA of the *Euglena gracilis* chloroplast is processed into Photosystem I and II mRNAs that accumulate differentially depending on the conditions of cell growth. Plant J 5: 247–260

Tan S, Wolfe GR, Cunningham FX and Gantt E (1995) Decrease of polypeptides in the PS I antenna complex with increasing growth irradiance in the red alga *Porphyridium cruentum*. Photosynth Res 45: 1–10

Tanaka A and Tsuji H (1985) Appearance of chlorophyll-protein complexes in greening barley seedlings. Plant Cell Physiol 26: 893–902

Terao T, Sonoike K, Yamazaki J-Y, Kamimura Y and Katoh S (1996) Stoichiometries of Photosystem I and Photosystem II in rice mutants differently deficient in chlorophyll *b*. Plant Cell Physiol 37: 299–306

Terashima I, Kashino Y and Katoh S (1991a) Exposure of leaves of *Cucumis sativus* L. to low temperature in the light causes uncoupling of thylakoids. I. Studies with isolated thylakoids. Plant Cell Physiol 32: 1267–1274

Terashima I, Sonoike K, Kawazu T and Katoh S (1991b) Exposure of leaves of *Cucumis sativus* L. to low temperature in the light causes uncoupling of thylakoids. II. Non-destructive measurements with intact leaves. Plant Cell Physiol 32: 1275–1283

Terashima I, Funayama S and Sonoike K (1994) The site of photoinhibition in leaves of *Cucumis sativus* L. at low temperatures is Photosystem I, not Photosystem II. Planta 193: 300–306

Terashima I, Noguchi K, Itoh-Nemoto T, Park Y-M, Kubo A and Tanaka K (1998) The cause of PS I photoinhibition at low temperatures in leaves of *Cucumis sativus*, a chilling-sensitive plant. Physiol Plant 103: 295–303

Thayer SS and Björkman O (1992) Carotenoid distribution and deepoxidation in thylakoid pigment-protein complexes from cotton leaves and bundle-sheath cells of maize. Photosynth Res 33: 213–225

Thomas DJ, Thomas JB, Prier SD, Nasso NE and Herbert SK (1999) Iron superoxide dismutase protects against chilling damage in the cyanobacterium *Synechococcus* PCC7942. Plant Physiol 120: 275–282

Thomas PG, Quinn PJ and Williams WP (1986) The origin of photosystem-I-mediated electron transport stimulation in heat-stressed chloroplasts. Planta 167: 133–139

Tjus SE and Andersson B (1993) Loss of the trans-thylakoid proton gradient is an early event during photoinhibitory illumination of chloroplast preparations. Biochim Biophys Acta 1183: 315–322

Tjus SE, Møller BL and Scheller HV (1998a) Photosystem I is an early target of photoinhibition in barley illuminated at chilling temperatures. Plant Physiol 116: 755–764

Tjus SE, Teicher HB, Møller BL and Scheller HV (1998b)

Photoinhibitory damage of barley Photosystem I at chilling temperatures induced under controlled illumination is also identified in the field. In: Garab G (ed) Photosynthesis: Mechanisms and Effects, Vol III, pp 2221–2224. Kluwer Academic Publishers, Dordrecht

Tjus SE, Møller BL and Scheller HV (1999) Photoinhibition of Photosystem I damages both reaction center proteins PS I-A and PS I-B and acceptor-side located small Photosystem I polypeptide. Photosynth Res 60: 75–86

Vener AV, Van Kan PJ, Gal A, Andersson B and Ohad I (1995) Activation/deactivation cycle of redox-controlled thylakoid protein phosphorylation. Role of plastoquinol bound to the reduced cytochrome *bf* complex. J Biol Chem 270: 25225–25232

Vener AV, Van Kan PJM, Rich PR, Ohad I and Andersson B (1997) Plastoquinol at the quinol oxidation site of reduced cytochrome *bf* mediates signal transduction between light and protein phosphorylation: Thylakoid protein kinase deactivation by a single-turnover flash. Proc Natl Acad Sci USA 94: 1585–1590

Vierling E and Alberte RS (1983) Regulation of synthesis of the Photosystem I reaction center. J Cell Biol 97: 1806–1814

Walters RG and Horton P (1994) Acclimation of *Arabidopsis thaliana* to the light environment: Changes in composition of the photosynthetic apparatus. Planta 195: 248–256

Walters RG and Horton P (1995) Acclimation of *Arabidopsis thaliana* to the light environment: Changes in photosynthetic function. Planta 197: 306–312

Wellburn AR and Hampp R (1979) Appearance of photochemical function in prothylakoids during plastid development. Biochem Biophys Acta 547: 380–397

Wild A, Höpfner M, Rühle W and Richter M (1986) Changes in the stoichiometry of Photosystem II components as an adaptive response to high-light and low-light conditions during growth. Z Naturforsch 41c: 597–603

Wilde A, Härtel H, Hübschmann T, Hoffmann P, Shestakov SV and Börner T (1995) Inactivation of a *Synechocystis* sp. strain PCC 6803 gene with homology to conserved chloroplast open reading frame 184 increases the Photosystem II-to-Photosystem I ratio. Plant Cell 7: 649–658

Wu SP, Cheng MC and Chen SCG (1999) Characterization of a spinach chloroplast sequence-specific DNA-binding factor for Photosystem I *psaA* operon promoter. Physiol Plant 106: 98–104

Wykoff DD, Davies JP, Melis A and Grossman AR (1998) The regulation of photosynthetic electron transport during nutrient deprivation in *Chlamydomonas reinhardtii*. Plant Physiol 117: 129–139

Yamamoto Y, Tsuji H and Obokata J (1993) Structure and expression of a nuclear gene for the PS I-D subunit of Photosystem I in *Nicotiana sylvestris*. Plant Mol Biol 22: 985–994

Yamamoto YY, Nakamura M, Kondo Y, Tsuji H and Obokata J (1995a) Early light-response of *psaD*, *psaE* and *psaH* gene families of Photosystem I in *Nicotiana sylvestris*: PS I-D has an isoform of very quick response. Plant Cell Physiol 36: 727–732

Yamamoto YY, Tsuji H and Obokata J (1995b) 5′-Leader of a Photosystem I gene in *Nicotiana sylvestris*, *psaDb*, contains a translational enhancer. J Biol Chem 270: 12466–12470

Yamamoto YY, Kondo Y, Kato A, Tsuji H, Obokata J (1997) Light-responsive elements of the tobacco PS I-D gene are located both upstream and within the transcribed region. Plant J 12: 255–265

Zak E, Norling B, Andersson B, Pakrasi HB (1999) Subcellular localization of the BtpA protein in the cyanobacterium *Synechocystis* sp. PCC 6803. Eur J Biochem 261: 311–316

Zhu YS, Kung SD and Bogorad L (1985) Phytochrome control of levels of mRNA complementary to plastid and nuclear genes of maize. Plant Physiol 79: 371–376

Chapter 30

Regulation of Photosynthetic Electron Transport

Peter J. Nixon*
Department of Biochemistry, Imperial College of Science, Technology and Medicine, London SW7 2AY, U.K.

Conrad W. Mullineaux
Department of Biology, University College London, Darwin Building, Gower St, London WC1E 6BT, U.K.

*Author for correspondence, email: p.nixon@ic.ac.uk

Eva-Mari Aro and Bertil Andersson (eds): Regulation of Photosynthesis, pp. 533–555.

Summary

Green plants and cyanobacteria possess a complex and diverse set of mechanisms to regulate their photosynthetic electron transport. These include both rapid mechanisms, and long-term responses involving changes in gene expression. Together they are believed to play a number of physiological roles, including the maintenance of a relatively constant flow of electrons in a changing light-environment (homeostasis), the adjustment of the mode of photosynthetic electron transport in response to the changing metabolic needs of the cell (adaptation) and the minimization of the destructive side-effects of light and redox reactions (protection). This Chapter concentrates on the rapid responses (those not involving changes in gene expression). There has been exciting progress in our understanding of these mechanisms in recent years through a multi-disciplinary approach combining biophysics, biochemistry and molecular genetics. We describe the major mechanisms that have been characterized to date, concentrating on recent progress towards understanding the mechanism at the molecular level and highlighting the questions that remain to be answered.

I. Introduction

Oxygenic photosynthetic organisms show a number of alternative modes of photosynthetic electron transport (Fig. 1). Linear photosynthetic electron transport involves the Photosystem two (PS II) and Photosystem one (PS I) reaction centers acting in series to transfer electrons from water to ferredoxin. To maintain this mode of electron transport efficiently, the turnover of the reaction centers must be regulated so that the turnover of PS II is roughly the same as that of PS I. In addition, there are alternative modes of electron transport. Various forms of cyclic electron transport around PS I have been proposed. Respiratory electron transfer involving NADH and succinate dehydrogenases (Complexes I and II, respectively), plastoquinone and terminal oxidases also occurs in the thylakoid membrane of cyanobacteria (reviewed by Schmetterer, 1994). There is now increasing evidence for similar activities in chloroplasts (Endo et al., 1997; Burrows et al., 1998). This raises the possibility that the redox state of the plastoquinone pool, which is important for the regulation of photosynthetic electron flow, is dependent on both photosynthetic and respiratory processes. The co-existence of respiratory and photosynthetic electron transfer chains in the same membrane also means that electrons could be extracted from water and passed back to oxygen via a terminal oxidase. Such pathways are readily observed in cyanobacterial mutants lacking one or other of the reaction centers (Vermaas, 1994) but they may also be important in wild-type cells.

A photosynthetic organism must be able to regulate the various possible modes of photosynthetic electron transport. Regulation is necessary for several reasons: First, it is important to maintain photosynthetic electron transport in a varying light environment (homeostasis). Photosynthetic organisms may be subject to huge and rapid changes in the intensity and spectral quality of illumination. They use regulatory mechanisms to buffer the effects of these changes, for example, by activating mechanisms for energy dissipation in strong illumination, or by adjusting the relative turnover of the two reaction centers to compensate for changes in the spectral quality of illumination. Second, regulation is necessary for adjusting the mode of electron transport in response to the metabolic needs of the cell. For example, cyclic electron transport generates a proton gradient (and hence ATP) but no reducing equivalents. Linear electron transport generates both ATP and reducing equivalents. It may therefore be necessary to regulate the extent of linear versus cyclic electron flow according to the requirement for ATP and reducing equivalents.

Third, regulation is needed to protect the organism from the destructive side-effects of light and redox

Abbreviations: FNR – ferredoxin:NADP$^+$ reductase; FQR – ferredoxin:quinone reductase; FTIR – Fourier transform infrared; IEC – intersystem electron transfer chain; LHC – light-harvesting complex; MDA – monodehydroascorbate; Ndh – NADH dehydrogenase; ORF – open reading frame; PET – photosynthetic electron transport; PS I – Photosystem I; PS II – Photosystem II; Q_A – primary plastoquinone electron acceptor of Photosystem two; Q_B – secondary plastoquinone electron acceptor of Photosystem two; qE – energy-dependent component of non-photochemical quenching; qI – non-photochemical quenching induced by photoinactivation; RC – reaction center; WT – wild type

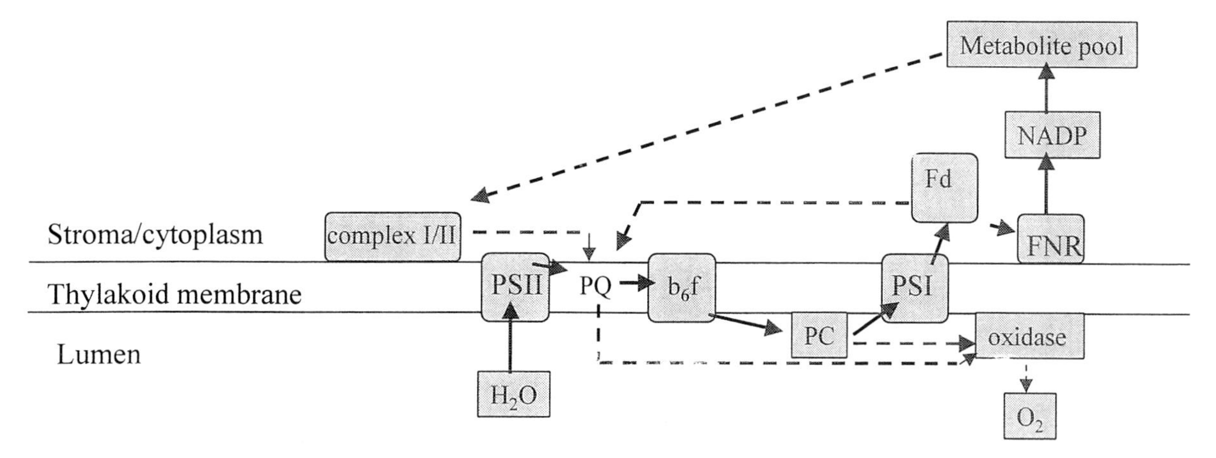

Fig. 1. Pathways of electron transport in chloroplasts and/or cyanobacteria. Solid lines: electron transfer steps involved in linear photosynthetic electron transport. Dashed lines: other routes of electron transfer. PC, plastocyanin (or cytochrome *c* in some bacteria); Fd, ferredoxin; FNR, ferredoxin:NADP$^+$ reductase; b_6f, cytochrome b_6f complex; PQ, plastoquinone.

reactions. Photosynthetic electron transport is a dangerous activity, since there is always a risk of generating reactive by-products such as singlet oxygen and free radicals, which can cause widespread damage in the cell. Mechanisms exist to down-regulate photosynthetic electron transport, particularly when the supply of the products exceeds the metabolic requirements of the cell.

Regulation often occurs via feedback control mechanisms. Many mechanisms can be effective for more than one of the functions listed above. For example, a mechanism that adjusts the relative turnover of the two photosystems could be employed both for maintaining similar rates of turnover in varying light environment (homeostasis) and for switching between different modes of electron transport (adjustment).

Numerous regulatory mechanisms have been proposed in plants and cyanobacteria. Regulation can occur on different timescales. Long-term mechanisms, generally involving changes in gene expression, are complemented by rapid mechanisms occurring on timescales of seconds to minutes. Regulation of gene expression is discussed elsewhere in this volume (see Part II: Gene Expression and Signal Transduction). This Chapter will therefore discuss the rapid mechanisms, concentrating on those mechanisms where there has been significant progress towards understanding the molecular basis of the mechanism.

II. Background Concepts

A. Photosynthetic Electron Transport and the Z Scheme

According to the Z scheme model of photosynthesis, PS I and PS II act in tandem to transfer electrons from water to ferredoxin and then to NADP$^+$ (Fig. 1). Accordingly, a minimum of eight photons are required (four for each photosystem) to produce one molecule of oxygen from the oxidation of water at PS II, and to generate the four reducing equivalents needed to fix one molecule of CO_2 in the Benson-Calvin cycle. The Z scheme therefore predicts that the maximum quantum yield for both oxygen evolution and CO_2 fixation is 0.125. Measured quantum yields in C3 plants are near this maximum value suggesting the operation of the Z scheme in vivo (discussed in Genty and Harbinson, 1996). A recent challenge to the ubiquity of the Z scheme, based on the analysis of PS I deficient mutants of the green alga *Chlamydomonas reinhardtii* (Greenbaum et al., 1995), has since proved to be unfounded (Redding et al., 1999). Although from a thermodynamic perspective it is feasible that PS II can reduce ferredoxin or NADP$^+$, recent experiments with well-characterized engineered PS I-deficient mutants of *C. reinhardtii* appear to exclude these as significant reactions in vivo (Redding et al., 1999). Despite this, the uncoupling of PS II from PS I under certain physiological conditions cannot yet be fully discounted.

B. Coordination of Electron Transport and Metabolism

The products of photosynthetic electron transport, ATP and reductant in the form of reduced ferredoxin, are used for a variety of purposes including not only the fixation of CO_2 into carbohydrate and the process of photorespiration but also key biosynthetic steps such as sulfate and nitrite assimilation. These pathways can be thought of as electron or energy sinks for photosynthetic reductant.

In classic light saturation curves for photosynthesis, two regions can be defined. At low light intensities, photosynthesis is light-limited. At higher light intensities, the rate of photosynthesis becomes saturated because of limitations in the metabolic demand for reductant. In the absence of metabolic electron sinks, there is a danger that stromal components and the photosynthetic electron transfer chain become over-reduced. This can then lead, by a variety of mechanisms, to the production of reactive oxygen species throughout the electron transport chain, with the ultimate result that there is an increased photodestruction of protein and pigment especially within PS II (Barber and Andersson, 1992; Heber and Walker, 1992). This potential problem means that metabolic demand for reductant and the photosynthetic electron transfer rate must be closely coordinated. In practice, damage to the photosynthetic electron transport chain is alleviated by (i) feed-forward mechanisms to activate stromal metabolism, (ii) feed-back mechanisms to slow photosynthetic electron transport when metabolic demand reduces, (iii) the use of alternate 'emergency' electron sinks that dissipate photosynthetic electron flow relatively harmlessly (Asada, 1999), (iv) the presence of detoxification systems to remove reactive oxygen species (Niyogi, 1999), and (v) rapid replacement of damaged protein, such as the D1 subunit of PS II, through highly coordinated repair processes (Barber and Andersson, 1992).

C. Feed-forward Activation of the Calvin Cycle

Upon illumination of chloroplasts, there is an increase of both pH and the Mg^{2+} concentration in the stroma. These ionic conditions activate enzymes within the Calvin cycle. Superimposed on this is the specific activation by reduced thioredoxin of several key enzymes, such as fructose-1, 6-bisphosphatase and sedoheptulose-1, 7-bisphosphatase (Buchanan, 1994). The mechanism involves reduction of disulphide bridges in the target enzyme by thioredoxin which in the chloroplast is reduced using a ferredoxin-dependent thioredoxin reductase and is thus controlled by photosynthetic electron flow (Buchanan, 1994). Thioredoxin also activates the chloroplast ATP-synthase (Stumpp et al., 1999).

D. The Use of Alternate Electron Sinks

Several electron sinks within the chloroplast have been viewed as 'safety valves' to prevent over-reduction of the photosynthetic electron transport and consequent damage (reviewed by Niyogi, 1999). These sinks act to dissipate the energy of photons that is in excess of that required for useful biosynthetic processes and as such play an important photo-protective role.

Molecular oxygen is an important physiological electron acceptor for photosynthetic electron flow. In the Mehler reaction (Mehler, 1951), oxygen is reduced by PS I to produce superoxide (O_2^-). This species is hazardous so there is an efficient scavenging system for its removal. It is first converted to hydrogen peroxide and oxygen by superoxide dismutase, then hydrogen peroxide is scavenged through a number of routes in the stroma and on the thylakoid membrane (Asada, 1999). In the thylakoid-associated pathway, hydrogen peroxide is removed by the ascorbate peroxidase catalysed conversion of ascorbate to monodehydroascorbate (Asada, 1999). The regeneration of ascorbate by reduction of monodehydroascorbate (MDA) requires photosynthetic reductant in the form of reduced ferredoxin (Asada, 1999) and so MDA itself acts as an electron sink. Overall the four electrons derived from the oxidation of two molecules of water to one molecule of oxygen are accounted for by the reduction of two molecules of oxygen to superoxide and by the reduction of two molecules of MDA to ascorbate. This process has been termed the water-water cycle (Asada, 1999) because electrons derived from the oxidation of water ultimately lead to the reduction of oxygen to water without the net production of oxygen. These steps form the basis of pseudo-cyclic electron flow in which linear electron flow is coupled to ATP synthesis without the net production of reductant. Oxygen also plays a major photoprotective role in C3 plants through its involvement in photorespiration (Osmond, 1981).

Excess NADPH can be removed from the stroma

by reducing oxaloacetate to malate in the 'malate valve' (Scheibe, 1987). This reaction is catalysed by malate dehydrogenase and is driven by export of malate from the chloroplast.

III. Feedback Control of the Photosynthetic Electron Transport Chain

A. Role of the Lumenal pH in Slowing the Rate of Electron Transfer

Because several of the electron transfer reactions of the photosynthetic electron transfer chain are coupled to deposition of protons in the lumen, a decrease in lumenal pH automatically slows the flow of electrons. This has been clearly demonstrated for the oxidation of water at PS II (Bowes and Crofts, 1981) and the oxidation of plastoquinol at the Q_o site of the Cyt b_6f complex (Kramer and Crofts, 1993). In addition the mid-point redox potential of plastocyanin is dependent on pH so that as the lumenal pH drops, reduction of $P700^+$ is less favored (discussed by Kramer and Crofts, 1996). As ATP synthesis at the ATP synthase is coupled to the dissipation of the proton gradient, a high ratio of stromal [ATP]/[ADP][P_i], which is indicative of a slow down in metabolic activity, will tend to maintain a low lumenal pH.

B. Role of Lumenal pH in Quenching Excitation Energy in PS II

'Non-photochemical quenching' (otherwise known as 'non-radiative quenching') represents a collection of mechanisms that regulate the conversion of light energy. In general these mechanisms involve the conversion of excitation energy to heat, thereby reducing the turnover of the photosystems and down-regulating photosynthetic electron transport. Non-photochemical quenching has been subdivided into components including energy-dependent quenching (qE) and photoinhibitory quenching (qI). State transitions (Section III.C) are sometimes considered to contribute to non-photochemical quenching. For the qE component, the triggering signal involves changes in the pH of the thylakoid lumen (Krause et al., 1983, Horton et al., 1996). When the pH in the lumen is low, quenching mechanisms are activated, thereby down-regulating electron transport. The reduction in electron flow will decrease the rate of

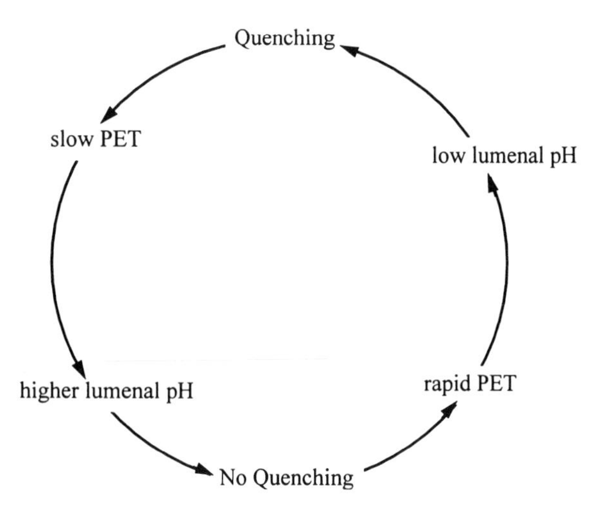

Fig. 2. Homeostatic feedback loop for non-photochemical quenching (NPQ). Because NPQ inhibits photosynthetic electron transport (PET) and is triggered by low lumenal pH, it buffers the lumenal pH, preventing the lumen from becoming excessively acidic under illumination.

proton pumping across the thylakoid membrane, and this will in turn allow the lumenal pH to increase. Thus qE should act as a homeostatic mechanism to maintain a relatively constant lumenal pH and rate of electron flux during variable illumination (Fig. 2). It is generally assumed that the major physiological role of qE is to protect the photosystems (particularly PS II) from photodamage during conditions of excess illumination. However, it is also possible that an excessively low pH could damage lumenally-exposed protein complexes. It is now becoming clear that the biochemistry of the lumen is much more complex than was thought (Fulgosi et al., 1998)

qE has been observed in numerous green plant species, although the extent of the quenching and the rate of induction vary widely according to species and habitat (Demmig-Adams, 1990; Johnson et al., 1993). There is less evidence for such a process in the phycobilin-containing organisms (cyanobacteria and red algae). Quenching mechanisms have been postulated in cyanobacteria (Campbell et al., 1996) and red algae (Delphin et al., 1996) but it is likely that the mechanisms are different from those operating in green plants.

qE in green plants is influenced by the levels of xanthophyll carotenoids in the light harvesting antenna (Demmig-Adams, 1990). In particular, there is a correlation between the level of zeaxanthin and the extent and rapidity of non-photochemical quenching (Demmig-Adams, 1990; Phillip et al.,

1996. This has led to the proposal that the conversion of violaxanthin to zeaxanthin, catalyzed by violaxanthin de-epoxidase, plays a crucial role in qE (Demmig-Adams, 1990).

There have been a number of proposals for the mechanism of qE in green plants, including: (i) energy dissipation within the Photosystem II reaction center, either by charge recombination (Krieger et al., 1992) or through the accumulation of a chlorophyll cation designated Chl_Z^+ (Schweitzer and Brudvig, 1997) bound to the D1 polypeptide, (ii) energy dissipation by direct transfer of singlet excitons from chlorophylls to xanthophyll carotenoids (Frank et al., 1994) and (iii) energy dissipation by the formation of quenching aggregates of chlorophylls in the light-harvesting antenna (Horton et al., 1996).

The extent to which these different processes contribute to qE in vivo remains controversial. The most complete mechanistic model that has been developed is that of Horton and co-workers (Horton et al., 1996; Gilmore et al., 1998). These authors propose that the conversion of excitation energy to heat occurs in pairs or aggregates of chlorophylls formed as a result of conformational changes in chlorophyll *a/b* binding proteins of the light-harvesting antenna. The conformational changes are triggered by the protonation of specific amino acids exposed to the lumen, or forming part of a channel that conducts protons away from the water-oxidizing complex of PS II. Specific residues on the minor light-harvesting complex (LHC) components, CP29 and CP26, have been implicated (Walters et al., 1996). It is further proposed that the extent of quenching can be modulated by the xanthophyll pigments associated with the light-harvesting complexes through the action of zeaxanthin as an allosteric activator of the quenching state (Phillip et al., 1996). This contrasts with the view of other authors who propose that zeaxanthin acts as a direct quencher of singlet excitons (Frank et al., 1994).

Mutants are now being used to clarify the role of the xanthophyll cycle pigments and to identify proteins involved in qE. Mutants affected in xanthophyll metabolism generally show altered qE, confirming the role of the xanthophyll cycle pigments in this mechanism (Niyogi, 1999). However, mutants affected in qE have also been isolated that show normal pigment composition and xanthophyll interconversions (Niyogi, 1999). One such mutant lacks the *psbS* gene (Li et al., 2000) which codes for a minor chlorophyll-binding subunit associated with the PS II complex (Funk et al., 1995). PsbS could be the major site for the conversion of excitation energy into heat, or it could play a crucial role in sensing a low lumenal pH.

For PS I, no such specialized quenching mechanism exists in the antenna system. The difference between PS I and PS II probably lies in the thermodynamic constraints placed on the oxidized primary electron donors. For PS I, the mid-point redox potential of the $P700^+/P700$ couple is ~0.5 V which means that its presence does not lead to damaging oxidative side reactions. Hence $P700^+$ can act as a quencher of excitation (Butler et al., 1979). In contrast, P680 in PS II has a much higher mid-point redox potential (~1.1V) consistent with its role in water oxidation. Unfortunately the use of $P680^+$ as a physiological quencher of excess excitation would lead to deleterious oxidative reactions within PS II. To avoid this, PS II has thus developed to be a shallow trap with the exciton more delocalized in the antenna system.

C. Regulation of Electron Flow through State Transitions

State 1-state 2 transitions (state transitions for short) are a rapid mechanism that adjusts the function of the light-harvesting apparatus in response to signals from the photosynthetic electron transport chain (Allen, 1992). Reduction of electron carriers between PS II and PS I triggers a switch to state 2, in which more excitation energy is transferred to PS I. Oxidation of the electron carriers triggers a switch to state 1, in which more energy is transferred to PS II. Thus the mechanism functions as a homeostatic control to maintain a moderate level of reduction of the intersystem electron carriers (Fig. 3).

State transitions are observed both in green plants and cyanobacteria (Allen, 1992). The phenomena observed in Chl-*b* and phycobilin-containing organisms show many features in common, although it is clear that some features of the mechanism must differ in organisms with fundamentally different light-harvesting complexes.

1. The Triggering Mechanism for State Transitions

State transitions are triggered by changes in the redox state of electron carriers between PS II and PS I, both in green plants (Allen, 1992) and in

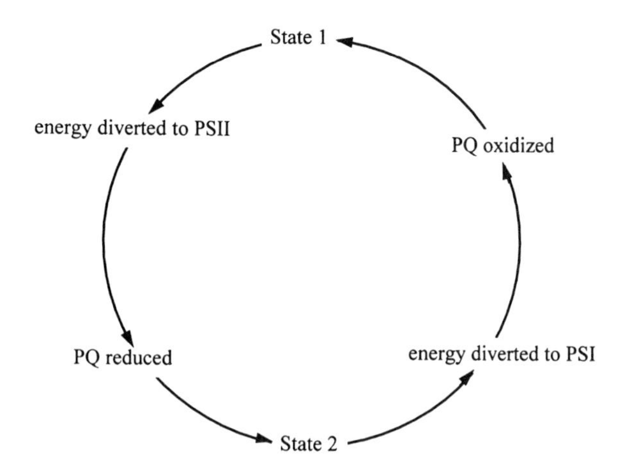

Fig. 3. Homeostatic feedback loop for state transitions. State transitions are triggered by changes in the redox state of intersystem electron carriers. They therefore balance the turnover of Photosystem II (PS II) and Photosystem I (PS I) and maintain a moderate level of plastoquinone reduction under illumination.

cyanobacteria (Mullineaux and Allen, 1990). In the case of green plants, mutagenesis studies have indicated that the cytochrome b_6f complex plays an essential role in state transitions (Wollman and Lemaire, 1988). Flash photolysis studies have shown the involvement of a specific quinol binding site (Vener et al., 1997). It is probable that cyanobacterial state transitions are triggered the same way, although the evidence is less conclusive because no mutants lacking the cytochrome b_6f complex are available.

Triggering by changes in the redox state of intersystem electron carriers provides a mechanism for homeostatic regulation of electron transport in a varying light-environment (Fig. 3). In cyanobacteria, where the respiratory and photosynthetic electron transport chains intersect (Fig. 1, Scherer et al., 1988), state transitions may be important in regulating the interaction between photosynthetic and respiratory electron transport (Schreiber et al., 1995). State transitions may also be involved in regulating the extent of linear versus cyclic electron transport (Section V.B). Some effects cannot straightforwardly be explained in terms of triggering by the redox state of intersystem electron carriers. It is therefore likely that other triggers are involved (i.e. a number of different sensing mechanisms trigger the same response). In green plants, it has recently been suggested that a direct light-induced conformational change may play a role in facilitating LHCII phosphorylation (Zer et al., 1999). It is also possible that the ADP/ATP ratio is a triggering signal (Bulté et al., 1990), but as yet we have no information on the molecular mechanisms that may be involved.

2. Alterations in Function of the Light-harvesting Complexes

State transitions change the association of the light-harvesting complexes with the reaction centers. In green plants, LHCII is predominantly associated with PS II in state 1. In state 2, a proportion of LHCII detaches from PS II and appears to associate with PS I instead (Allen, 1992). In cyanobacteria the phycobilisomes appear to be a mobile light-harvesting antenna (Mullineaux et al., 1997) that can associate with and transfer energy both to PS II and to PS I (Mullineaux, 1994). State transitions change the relative coupling of phycobilisomes to PS II and PS I (Mullineaux, 1992) although it is likely that other effects are also involved (Olive et al., 1997; Emlyn-Jones et al., 1999).

In green plants, the changes in LHCII association are brought about by phosphorylation of the LHCII complexes on threonine residues near the N-terminus (Allen, 1992). This seems to bring about a conformational change in LHCII (Nilsson et al., 1997) that causes its dissociation from PS II and reassociation with PS I. The biochemical mechanism in cyanobacteria is not known. Allen and co-workers correlated changes in the phosphorylation state of two polypeptides with state transitions in the cyanobacterium *Synechococcus* 6301 (Allen et al., 1985), but it now seems increasingly unlikely that there is a causal relationship (Emlyn-Jones et al., 1999). It is very probable that some form of covalent modification is involved, with the phycobilisome core the most probable site of modification.

There have been many attempts to isolate the protein kinase responsible for LHCII phosphorylation in green plants. The kinase must be activated as a result of the triggering redox signal, resulting in net phosphorylation of LHCII. The process is reversed by a phosphatase that seems to be constitutively active (Allen, 1992). A number of kinases have been isolated from thylakoid membranes (Coughlan and Hind, 1987; Gal et al., 1992; Snyders and Kohorn, 1999) but we do not know for sure which, if any, of these kinases are involved in state transitions in vivo. Furthermore, we know nothing of the signal transduction pathway that links the triggering redox signal to activation/inactivation of the kinase.

3. The Genetic Approach to State Transitions

Mutants specifically deficient in state transitions have recently been identified in the cyanobacterium *Synechocystis* 6803 (Emlyn-Jones et al., 1999) and the green alga *C. reinhardtii* (Kruse et al., 1999; Fleischmann et al., 1999). In the cyanobacterial case, the mutation results from insertional inactivation of a specific gene that codes for a protein of previously unknown function, showing no homology to any previously-characterized gene product (Emlyn-Jones et al., 1999). The genetic approach shows great promise as a method of identifying the proteins involved in the signal transduction pathway of state transitions. Furthermore, mutants specifically deficient in state transitions are a powerful tool for establishing the physiological role(s) of the mechanism. Cyanobacterial state transition mutants grow slower than the wild-type under very weak yellow illumination, suggesting that the primary role of state transitions in cyanobacteria is to maximize the efficiency of utilization of light absorbed by the phycobilisomes (Emlyn-Jones et al., 1999).

4. Higher Levels of Control

In green plants the level of LHCII phosphorylation is maintained by the balance between the activities of a protein kinase that is activated by a redox signal, and a constitutively-active phosphatase (Allen, 1992). Such a mechanism has a clear metabolic cost. The maintenance of any level of LHCII phosphorylation will require continual hydrolysis of ATP. Higher levels of kinase and/or phosphatase activity would result in more rapid state transitions and therefore a swifter response to rapidly changing illumination conditions. However, this would come at the expense of more rapid ATP hydrolysis. It is therefore possible that there are mechanisms to up-regulate or down-regulate the enzymes involved, depending on growth conditions. It could be imagined that the genes coding for the enzymes would be strongly expressed under low or variable illumination, but repressed under strong, constant illumination. The subject needs further investigation. In green plants, the phosphorylation of LHCII is inactivated in high light (Rintamäki et al., 1997). In the cyanobacterium *Synechocystis* 6803, state transitions are only observed in low-light grown cells (C. W. Mullineaux, unpublished).

IV. Regulation of Photosystem II Activity—The Bicarbonate Effect

It has long been known that carbon dioxide itself, in the form of bicarbonate, is able to modulate the activity of the photosynthetic electron transport chain (recently reviewed by van Rensen et al., 1999). More specifically depletion of bicarbonate leads to the downregulation of PS II activity which can be restored by subsequent addition of carbon dioxide/bicarbonate. Current models indicate two distinct effects of bicarbonate within PS II: an acceptor side effect at the iron quinone complex and a donor side effect at the manganese cluster.

A. The Acceptor-side Effect

Type II photosynthetic reaction centers (RCs) such as PS II and those found in purple non-sulfur photosynthetic bacteria contain on the acceptor side a non-haem iron located between two quinone molecules termed Q_A and Q_B (reviewed by Diner et al., 1991a). Q_A is tightly bound to the RC and acts as a one-electron carrier (Q_A/Q_A^-) which reduces Q_B first to the semiquinone anion, Q_B^-, then after a second charge separation reaction within the RC to the quinol (probably in the state Q_BH^-). The quinol is further protonated and released from the RC in the neutral form, QH_2. Because depletion of bicarbonate primarily slows electron transfer between Q_A and Q_B after formation of Q_B^-, it is likely that bicarbonate participates in the protonation reactions involved in formation of Q_BH_2 (Blubaugh and Govindjee, 1988). Despite the similarity in redox factors on the acceptor in type II RCs, only PS II shows this bicarbonate effect.

Fourier transform infrared (FTIR) experiments (Hinerwadel and Berthomieu, 1995) indicate that bicarbonate acts as a bidentate ligand to the non-haem iron in PS II. In bacterial RCs a glutamate residue, at position 232 in the M subunit of *Rhodopseudomonas viridis*, fulfils this role. Bicarbonate binding is reversible so that a wide range of anions including formate, glycolate, glyoxylate, and oxalate compete with bicarbonate for binding to the non-haem iron and slow electron transfer between Q_A and Q_B (Petrouleas et al., 1994). The presence of bicarbonate close to the Q_B site is thus consistent with a role in the protonation of Q_B^{2-} either directly or indirectly. Mutation of a number of potentially positively charged amino-acid residues within the D2 polypeptide (Diner et al., 1991b; Cao et al., 1991), modeled to be in the vicinity of the non-haem iron, reduce the affinity of PS II for bicarbonate. Based on the isolation of D1 mutants that are resistant to formate inhibition (Xiong et al., 1997, 1998), it has also been suggested that there is an additional bicarbonate bound within the Q_B niche which is responsible for protonation of Q_B (van Rensen et al., 1999).

Whether the bicarbonate effects observed for PS II in vitro also occur in vivo is a matter of debate. K_m values for bicarbonate are estimated to be 40-80 μM (Snel and van Rensen, 1984). The concentration of bicarbonate in the stroma is highly dependent on pH and temperature but has been estimated to be about 9.7 μM at pH 6.3 and 432 μM at pH 8.0 (van Rensen et al., 1999). Thus under optimal rates of photosynthesis, when the stromal pH reaches pH 8, the levels of bicarbonate do not appear to be controlling PS II activity. Only when levels of carbon dioxide, hence bicarbonate, decrease, such as upon the closure of stomata through for example drought or high temperature, would PS II activity be downregulated. This downregulation of PS II would also act to reduce the level of oxygen within the cell which otherwise would lead to enhanced photorespiration in C3 plants. Binding of bicarbonate to PS II has also been linked to enhanced resistance to photoinhibition (Sundby, 1990; Klimov et al., 1997).

B. The Donor-side Effect

A variety of indirect evidence has also hinted to the possibility that bicarbonate may have a role on the donor side of PS II (reviewed by Stemler, 1999; van Rensen et al., 1999; Klimov and Baranov, 2001). Recently Klimov and co-workers have interpreted FTIR difference spectra obtained for the donor side of PS II in terms of actual ligation of bicarbonate to the Mn cluster (Yruela et al., 1998). In the absence of bicarbonate assembly of the Mn cluster is also less efficient (Allakhverdiev et al., 1997). Further work is required to clarify these donor side effects particularly as it is plausible that binding of bicarbonate on the acceptor side of PS II causes structural effects that extend across the membrane to the donor side.

V. Cyclic Electron Flow

Historically, thylakoid membranes have been considered to perform two types of photophosphorylation: cyclic and non-cyclic. In cyclic electron flow, ATP synthesis is coupled to light-induced electron flow in a closed system around PS I without the net production of NADPH whereas in non-cyclic mode, or linear electron flow, both ATP and NADPH are synthesized. Cyclic photophosphorylation is therefore an ATP generating process that serves to increase the ratio of ATP/NADPH produced in photosynthetic electron flow (reviewed in detail by Fork and Herbert, 1993; Bendall and Manasse, 1995). In principle the metabolic needs of the cell can be met through appropriate modulation of the ratio of linear to cyclic electron flow. Although the water-water cycle also increases the ATP/NADPH ratio, cyclic electron flow is a more energy efficient way of generating ATP.

In general, cyclic electron flow involves the reduction of the intersystem electron transfer chain (IEC) linking PS II and PS I (Biggins, 1974) (Fig. 1). The terms 'PS II-independent electron flow' or 'non-photochemical reduction' are sometimes used to describe the various non-PS II electron-transfer pathways able to feed electrons into the IEC. In addition the term cyclic is often used loosely to describe all PS II-independent pathways for the reduction of PS I, including those leading to reduction of the IEC through dehydrogenases (Complex I and II) (Fig. 1).

According to the Z scheme, the linear transport of an electron from water to $NADP^+$ is thought to translocate one proton at PS II and two protons at Cyt b_6f, if a Q cycle operates constitutively (Rich, 1988). In total three protons are translocated per electron transported. If the synthesis of one molecule of ATP requires the influx of four protons at the ATP synthase (van Walraven et al., 1996), then the ratio of ATP/NADPH produced by linear electron flow is about 1.5. This value will depend on the amount of passive proton leakage across the thylakoid membrane at high pH, which would lower the effective number of protons pumped by the electron transport chain (Berry and Rumberg, 1999), and the precise number of protons required to synthesize ATP (3, 4 or indeed a non-integral value).

For C3 plants in ambient air, the ratio of NADPH/ATP needed to fix CO_2 and perform photorespiration is about 1.56 (discussed by Foyer and Harbinson, 1997). Hence ATP production through cyclic electron flow is not required for C3 plants during steady-state photosynthesis and indeed there is little evidence for its existence under these conditions (Herbert et al., 1990). It has been suggested that the main role for cyclic electron flow in C3 plants is to regulate electron flow through acidification of the lumen under conditions where electron sinks for photosynthetic electron transport are depleted (Heber and Walker, 1992; Katona et al., 1992; Heber et al., 1995). For C4 photosynthesis, there is a requirement of 2.5–3 ATP/NADPH, much greater than that can be generated in linear mode. The extra ATP is considered to be generated by cyclic electron flow around PS I. For algae and cyanobacteria, which need to adapt to large fluctuations in environmental conditions through energy dependent processes, cyclic electron flow appears to be an important component of photosynthetic electron flow (reviewed by Bendall and Manasse, 1995).

The precise rates of cyclic electron transfer in vivo are, however, difficult to assess because of the technical problems deconvoluting linear and cyclic electron flow, although a variety of techniques have been applied including photoacoustic spectroscopy (Herbert et al., 1990) and the rate of re-reduction of $P700^+$ using optical spectroscopy (Fork and Herbert, 1993; Bendall and Manasse, 1995).

A. Pathways of Cyclic Electron Transfer

As yet the precise pathways of cyclic electron transfer remain unclear although the consensus favors multiple routes (Hosler and Yocum, 1985; Ravenel et al., 1994). Currently two cycles are thought to dominate. One mediated by the Ndh complex involves the pyridine nucleotide pool and the other consists of ferredoxin-mediated reduction of the IEC. From studies on cyanobacteria (Yu et al., 1993) and algae (Ravenel et al., 1994) in vivo, the rates of cyclic electron flow have been estimated to be less than 10% of the maximum linear rates. Of this, recent estimates using isolated barley thylakoids suggest that the rates of Ndh-mediated cyclic electron flow is about 3% of the linear rates and the rate of ferredoxin-dependent cyclic electron flow is about 5% (Teicher and Scheller, 1998). Under stress conditions such as photoinhibition and nutrient deprivation, the contribution of cyclic electron flow may, however, become more important (discussed by Fork and Herbert, 1993).

1. Ferredoxin Cycles

Historically reduced ferredoxin was considered the stromal reductant of the membrane bound intersystem chain of electron carriers in chloroplasts (Arnon, 1991). This assignment stemmed from experiments in vitro in which ferredoxin was added back to thylakoid membranes in the absence of other stromal components (Tagawa et al., 1963). Whether this or some other cycle actually operated in vivo could not be assessed.

For many years the mechanism by which reduced ferredoxin reduced the intersystem chain was thought to involve the direct reduction of cytochrome b_6f. This hypothesis was based on the mistaken assumption that the binding site of antimycin A, a known inhibitor of ferredoxin-mediated cyclic electron flow, was the Cyt b_6f complex (discussed in Bendall and Manasse, 1995). The actual binding site for antimycin is still uncertain although there is some preliminary evidence to indicate an association with PS I (Davies and Bendall, 1987) and also that there may be an additional weaker binding site in the Ndh complex (Endo et al., 1998).

The inhibition of ferredoxin-mediated cyclic electron flow by inhibitors of the Q_o site of the Cyt b_6f complex, together with other data (Bendall and Manasse, 1995), has led to models in which cyclic electron flow involved reduction of the plastoquinone pool followed by its oxidation by Cyt b_6f. Reduction of the plastoquinone pool by reduced ferredoxin was therefore hypothesized to be catalyzed by a ferredoxin-plastoquinone reductase (FQR) (Moss and Bendall, 1984). The molecular identity of FQR still remains unknown. Recent evidence indicates that FQR may include a cytochrome, termed Cyt b-559(Fd) (Miyake et al., 1995). Whether this cytochrome is truly part of FQR or is reduced through redox equilibration with the plastoquinol pool is uncertain.

Given that there may be an antimycin-binding site close to PS I, attention has focused on whether PS I contains FQR activity (Bendall and Manasse, 1995). Mutation of the *psaE* gene, which encodes a stromally exposed subunit of PS I (Klukas et al., 1999), in *Synechococcus* (Yu et al., 1993) impairs cyclic electron flow when assayed by the dark reduction kinetics of $P700^+$. However photoacoustic spectroscopy has failed to detect changes in cyclic electron flow in this mutant (Charlebois and Mauzerall, 1999). Because PsaE is thought to be involved in the binding of ferredoxin (Barth et al., 1998) and ferredoxin:$NADP^+$ oxidoreductase (FNR) to PS I (van Thor et al., 1999a), the role of PsaE in cyclic electron flow may be to stabilize the binding of these components close enough to a quinone binding site within the PS I core complex to allow plastoquinone reduction.

2. NAD(P)H Cycles

a. Identification of the NADH Dehydrogenase (Ndh) Complex

Following the sequencing of the tobacco (Shinozaki et al., 1986) and liverwort (Ohyama et al., 1986) chloroplast genomes, several open reading frames (ORFs) were identified with significant sequence similarities to mitochondrial and eubacterial genes encoding subunits of the respiratory NADH:ubiquinone oxidoreductase (also known as Complex I or type I NADH dehydrogenase) (Fearnley and Walker, 1992). These ORFs were consequently designated *ndh* genes (NADH dehydrogenase). *ndh* genes have now been identified in cyanobacteria and in the chloroplast genomes of some but not all green algae (Turmel et al., 1999). The absence of *ndh* genes from various plastomes may reflect transfer to the nucleus or deletion from the organism. In the case of black pine, of the 11 *ndh* genes usually found in land plants, four have been lost from the plastome and seven remain as pseudogenes (Wakasugi et al., 1994).

Despite little evidence at the time to suggest a role for the 11 *ndh* genes in encoding a chloroplast analogue of Complex I (Fearnley and Walker, 1992), recent biochemical (Sazanov et al., 1998) and genetic studies (Burrows et al., 1998; Kofer et al., 1998; Shikanai et al., 1998) have provided strong evidence for this assignment (also reviewed by Nixon, 2000). The low abundance of the Ndh complex in the chloroplast, estimated to be at 1.5% the levels of PS II in tobacco (Burrows et al., 1998), helps to explain why it had escaped detection for so long. For chloroplasts, the Ndh complex is located in the stromal lamellae (Nixon et al., 1989; Sazanov et al., 1998) whereas in cyanobacteria there is still uncertainty about its location although it would appear to be in the thylakoid (Norling et al., 1998) and possibly cytoplasmic (Berger et al., 1991) membranes of *Synechocystis* 6803. Analysis of the cyanobacterial Ndh complex is also complicated by the presence of multigene families which may reflect structural and functional heterogeneity (Ohkawa et al., 1998; Klughammer et al., 1999).

Biochemical analysis of the Ndh complex remains incomplete. A 550 kDa Ndh complex with an associated NADH:ferricyanide oxidoreductase activity has been isolated from pea thylakoids (Sazanov et al., 1998). Of the estimated 16 subunits, five have been so far assigned to plastid Ndh proteins. For cyanobacteria, a 376 kDa hydrophilic subcomplex of the Ndh complex, which displays an NADPH:ferricyanide oxidoreductase activity, has been isolated from *Synechocystis* PCC 6803 (Matsuo et al., 1998). Of the nine subunits present, only NdhH could be assigned.

b. Substrate Specificity of the Ndh Complex in vitro and in vivo

The enzyme activity of the Ndh complex has been assayed mainly through the use of pyridine nucleotides as electron donors and artificial electron acceptors such as water-soluble quinones and ferricyanide, which probably accept electrons from the iron-sulfur clusters within the complex rather than from the natural quinone binding site (Sazanov et al., 1998).

Biochemical characterization of the Ndh complex has proved difficult because of its low abundance in chloroplast thylakoid membranes, its lability, the lack of a convincing inhibitor, possible mitochondrial contamination in chloroplast samples and the presence of multiple NAD(P)H dehydrogenase activities, including a high level of NADPH:ferricyanide oxidoreductase activity due to ferrodoxin NADP reductase (FNR). This latter difficulty may explain why FNR has been suggested to be a component of the Ndh complex (Guedeney et al., 1996). Consequently it is extremely difficult to assign enzyme activities within the thylakoid membrane to the Ndh complex without additional purification or verification. The instability of the Ndh complex also raises the possibility that the activity of the Ndh complex and its possible role in cyclic photophosphorylation may have been overlooked in early experiments.

Attempts to study the Ndh complex in tobacco thylakoid membranes, for which there are engineered *ndh* knock-out mutants available to help assign the activities of the complex, have proved difficult because of the apparent instability of the complex (Endo et al., 1998). Such behavior may explain why reduction of the plastoquinone pool in isolated thylakoid membranes using NAD(P)H is so low (Rich et al., 1998). In membranes the reduction of the plastoquinone pool by NADPH is known to be stimulated by ferredoxin (Mills et al., 1979; Rich et al., 1998). Rather than an Ndh-catalyzed reaction, the possible pathway may involve the membrane-bound FNR-mediated reduction of ferredoxin by NADPH followed by reduction of the plastoquinone pool using FQR.

The nature of the stromal reductant oxidized by the *isolated* detergent-solubilized plastid Ndh complex, of different degrees of purification, is still under debate with there being evidence for a specificity for NADH (Sazanov et al., 1998; Elortza et al., 1999), NADPH (Guedeney et al., 1996) or both NADH and NADPH (Quiles and Cuello, 1998; Funk et al., 1999). For an isolated sub-complex of the *Synechocystis* 6803 Ndh complex, a specificity for NADPH has been reported (Matsuo et al., 1998) whilst for the complex in thylakoid membranes, NADPH, NADH and reduced ferredoxin have all been implicated (Mi et al., 1995).

It is possible that in vivo the Ndh complex may be rather promiscuous interacting with a number of different reductants depending on plant species and physiological state. The resulting holocomplex(es) containing the plastid *ndh* gene products plus the as yet uncharacterized electron input module(s) (Friedrich et al., 1995) may be rather unstable. The identity of the subunits that constitute the sub-complex involved in NAD(P)H oxidation is particularly intriguing. As yet no obvious homologues of the NADH-binding subunits found in other Complex I species have been identified (Friedrich et al., 1995).

Although there has been controversy over whether plants can tolerate the inactivation of plastid *ndh* genes (Maliga and Nixon, 1998; Koop et al., 1998) the consensus now appears that under normal growth conditions loss of the Ndh complex does not result in an obvious growth defect (Burrows et al., 1998; Shikanai et al., 1998). Plastid *ndh* mutants do however appear to be more sensitive to the effects of high light damage, for which the reason is not yet clear (Endo et al., 1999). For cyanobacteria, the photoautotrophic growth of several, but not all, *ndh* mutants shows a requirement for enhanced CO_2 levels (Ohkawa et al., 1998; Klughammer et al., 1999). This observation is consistent with a role for Ndh subunits in generating the ATP needed for the active transport of dissolved inorganic carbon into the cell.

c. Role of the Ndh Complex

Despite the lack of a consensus concerning the precise substrate specificity of the Ndh complex, a role as an NAD(P)H:plastoquinone oxidoreductase has led to speculation that it may have a dual function in photosynthetic systems (Burrows et al., 1998). In the light one role for the Ndh complex may be to catalyze cyclic electron flow around PS I (either directly or through appropriate redox poising of the IEC), whereas in the dark the Ndh complex may function in respiration. Figure 4 summarizes the possible electron pathways involving an NADH-specific Ndh complex of the chloroplast. By analogy to the situation in purple bacteria, it is also possible that the Ndh complex may catalyze the reverse reaction in which the plastoquinol-mediated reduction of $NAD(P)^+$ to NAD(P)H is coupled to the proton motive force.

Experimental data to support a role of the Ndh complex in cyclic electron flow first came from studies on cyanobacterial *ndh* mutants (Mi et al., 1992; Yu et al., 1993) and more lately from analysis of chloroplast *ndh* mutants for which: (i) the post-illumination reduction of the plastoquinone pool is absent or strongly attenuated (Burrows et al., 1998; Kofer et al., 1998; Shikanai et al., 1998), (ii) the rate of re-reduction of $P700^+$ is slower in the dark after far-red induced oxidation of P700 (Burrows et al., 1998) and (iii) $P700^+$ accumulates more rapidly under far-red illumination, after a period of white light illumination to reduce the pool of stromal reductant (Shikanai et al., 1998).

Interestingly under conditions of water stress, when the levels of electron acceptors for linear flow are limited, the *ndh* mutants show a reduced ability to quench chlorophyll fluorescence non-photochemically during the early stages of the induction of photosynthesis in dark-adapted plants (Burrows et al., 1998). Such behavior is consistent with a role for the Ndh complex under conditions when the Calvin cycle enzymes are not fully activated and there is a build up of reductant in the stroma. Cyclic electron flow under these conditions would contribute to the pH gradient across the thylakoid enhancing qE, thus downregulating PS II, as well as removing stromal reductant which could lead to the generation of reactive oxygen species.

Further indirect evidence to support a role for the Ndh complex in cyclic electron flow has come from the finding that Ndh proteins are located in the stromal lamellae close to PS I (Nixon et al., 1989) and show elevated levels in the bundle sheath cells of C4 plants which lack PS II and carry out high levels of cyclic photophosphorylation (Kubicki et al., 1996). In the latter case, it is possible that a significant

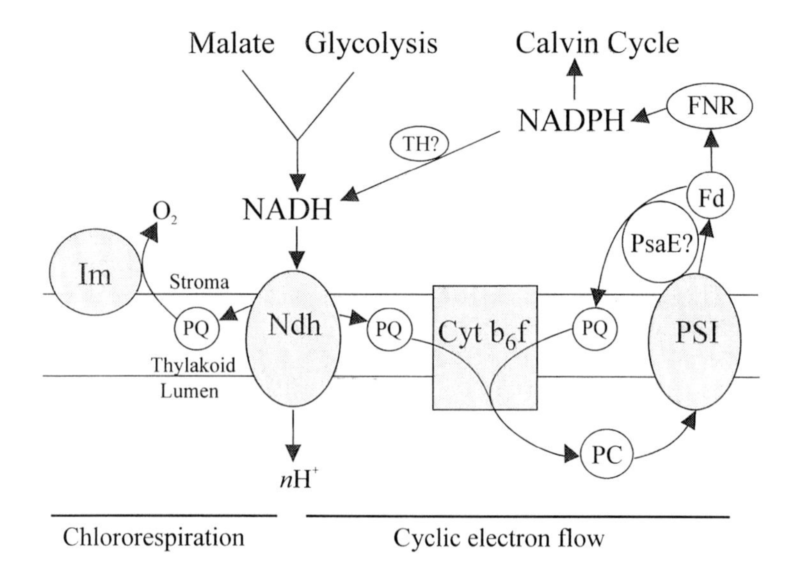

Fig. 4. Possible role of the chloroplast Ndh complex (adapted from Nixon, 2000). Scheme shows the Ndh complex as a component both of cyclic electron flow around PS I and of a putative chlororespiratory pathway involving Immutans (Im). Fd, ferredoxin; PQ, plastoquinone; FNR, ferredoxin:NADP$^+$ reductase; nH^+, an unknown number of protons pumped. Reduction of NAD$^+$ by NADPH may be catalyzed either directly through a putative transhydrogenase (TH) or indirectly through substrate cycles. An independent cyclic pathway catalyzed by PsaE of PS I is also indicated

electron transfer pathway involves the oxidation by the Ndh complex of NAD(P)H derived from malate dehydrogenation rather than the oxidation of NAD(P)H produced by PS I activity. A role for the Ndh complex outside photosynthesis is also suggested from its detection in non-photosynthetic tissue in higher plants (Berger et al., 1993).

B. Regulation of Cyclic Electron Flow

Although many of the components involved in cyclic electron flow have yet to be clarified, it is clear that for optimum rates of cyclic electron transport, there must be appropriate redox poising of the system so that it is neither too oxidizing nor too reducing. This is probably achieved in vivo by concurrent cyclic and linear electron flow (Arnon and Chain, 1975). A dynamic picture of cyclic electron flow would then be of cycle(s) in which electrons that leak out are replaced through both photochemical (PS II) and respiratory processes (e.g. oxidation of NADH by the Ndh complex).

The mechanism whereby cyclic flow is switched on at the expense of linear electron flow is unknown. Under normal steady state linear electron flow, when the Calvin cycle is fully activated, the cycling of electrons from NAD(P)H or reduced ferredoxin to the IEC is unlikely to compete with PS II-dependent reduction of the plastoquinone pool. In any event, the ferredoxin and NADP pools are relatively oxidized under these conditions because of forward electron transfer. Cyclic electron flow would however be promoted if the levels of NADPH and reduced ferredoxin increased because of a reduction in the rate of CO_2 fixation. Such physiological situations include (i) the early stages in the induction of photosynthesis in dark-adapted plants (Takahama et al., 1981), (ii) a reduction in the levels of the electron acceptors, CO_2 and O_2, through for example the water-stress induced closure of stomata (Katona et al., 1992), and (iii) the inhibition of the Calvin cycle enzymes through for instance heat stress (Weis, 1981; Havaux, 1996).

An increase in the activity of PS I compared with PS II may also act to enhance cyclic electron flow by poising the plastoquinone pool in a more oxidized state. This may occur physiologically when (i) PS II is downregulated through means of a state-transition, (ii) PS II has been inactivated through light damage (Canaani et al., 1989), and (iii) there are low levels of PS II complex such as in the bundle sheath cells of C4 plants.

How is the ratio of cyclic to linear electron flow controlled by the metabolic demands of the cell? In principle, changes in the level of ATP/ADP may regulate cyclic electron flow through allosteric modulation of FQR or the Ndh complex. However, neither ATP nor ADP modulate FQR activity (Cleland and Bendall, 1992). Effects of ATP on Ndh activity have not yet been tested although preliminary work has indicated that the thylakoid NAD(P)H dehydrogenase activity, assumed to be due to the Ndh complex, is activated by light and incubation in low-ionic-strength buffer (Teicher and Scheller, 1998).

There is good evidence, particularly in green algae, that the metabolic control of cyclic over linear electron flow appears to be exerted through state transitions (Section III.C). For *C. reinhardtii* state transitions are more dramatic than in higher plants (Delosme et al., 1996). State 2 is linked with cyclic electron flow whereas State 1 promotes linear electron flow (Finazzi et al., 1999). In state 2 there is little PS II-dependent oxygen evolution (Finazzi et al., 1999) which reflects not only a reallocation of excitation energy to PS I but also the migration of Cyt b_6f away from PS II to PS I (Vallon et al., 1991) so that linear electron flow is further diminished in favor of cyclic electron flow.

Thus when ATP demand is high, state 2 would be favored. Such a case occurs when nitrogen-limited cells of the green alga *Selenastrum minutum* are transferred into an NH_4^+ replete medium (Turpin and Bruce, 1990). NH_4^+ assimilation requires a high ATP/NADPH ratio with the extra ATP provided by cyclic electron flow. Chlorophyll fluorescence measurements confirm that without changing light quality the cells go into state 2 to enhance PS I activity. Concomitantly dark respiration is stimulated and CO_2 fixation is decreased probably because of the need to transfer carbon skeletons from the Calvin cycle (in the chloroplast) to the TCA cycle (in the mitochondrion) in order to synthesize amino acids (Elrifi and Turpin, 1986). Consequently the availability of electron acceptors in the chloroplast declines to produce redox conditions that would further favor cyclic electron transport and cyclic photophosphorylation. Transition to state 2 also occurs in response to hyperosmotic stress in *Chlamydomonas* (Endo et al., 1995). The role of state transitions in regulating cyclic electron flow can now be tested more directly using recently isolated state transition mutants (Kruse et al., 1999).

How cellular ATP levels are sensed within the chloroplast and converted into a state transition is uncertain. In principle metabolite translocators within the inner envelope membrane of chloroplasts allow changes in metabolite levels outside the chloroplast to be detected within the stroma. This signal can then be transduced into an appropriate response within the thylakoid membrane to regulate photosynthetic electron flow. For instance inhibition of mitochondrial respiration in *C. reinhardtii* in the dark, to prevent ATP synthesis, leads to a more reduced thylakoid plastoquinone pool (Gans and Rebeille, 1990; Bennoun, 1994) and a transition to state 2 (Gans and Rebeille, 1990) probably via a redox-controlled LHCII kinase. Cyclic electron flow would then be favored upon re-illumination. The mechanism of plastoquinone reduction has been suggested (Gans and Rebeille, 1990; Bulté et al., 1990) to be (i) a reduction in cellular ATP levels which is sensed in the chloroplast by a drop in ATP, (ii) an increase in the rates of starch degradation and chloroplast glycolysis to synthesize chloroplastic ATP, (iii) an increase in chloroplast NAD(P)H pools through glycolysis and the oxidative pentose phosphate pathway, and (iv) reduction of the plastoquinone pool using an NAD(P)H:PQ oxidoreductase. Whether it is the ATP level (Bulté et al., 1990) or the redox state of the plastoquinone pool which exerts the major control on state transitions in *C. reinhardtii* in vivo is a matter of debate (Finazzi et al., 1999).

It is possible that the need for enhanced cyclic electron flow results in the synthesis of new electron transfer components or the enhanced accumulation of pre-existing proteins. Under salt stress conditions, *Synechocystis* 6803 produces a flavodoxin which can substitute for ferredoxin and which promotes cyclic over linear electron flow (Hagemann et al., 1999). Under photooxidative stress conditions, when PS II activity declines, expression of the NdhA subunit of the Ndh complex increases (Martin et al., 1996).

Given that PS II and PS I in chloroplasts are spatially segregated (Andersson and Anderson, 1980), it is possible that cyclic electron flow is restricted to the stromal lamellae and linear flow to the grana. Upon state transitions, a further redistribution of electron transfer complexes within the membrane may occur, such as that observed for Cyt b_6f in *C. reinhardtii* and maize thylakoids (Vallon et al., 1991), to promote cyclic over linear electron flow.

FNR has also been widely speculated as having a direct role in cyclic electron flow but as yet there is no definitive evidence (Bendall and Manasse, 1995). In cyanobacteria, there are two populations of FNR, one directly associated with the thylakoid membrane and one attached to the phycobilisomes (van Thor et al, 1999b). It is possible that the two populations of FNR have different electron transport roles.

As a cyanobacterial *psaE/ndhF* double mutant shows little cyclic electron flow (Yu et al., 1993), it would appear that cyclic electron flow in this particular case is dominated by the Ndh and PsaE pathways. But what regulates the type of cycle? It can be hypothesized that the relative reduction levels of NAD(P)H and ferredoxin is an important determinant in the type of cycle, whether ferredoxin- or Ndh-mediated, used around PS I. Because the reduction of $NADP^+$ by reduced ferredoxin (via FNR) is favored, in the event that there is an accumulation of reductant on the acceptor side of PS I, it would be anticipated that the ratio of $NADPH/NADP^+$ would increase before that of reduced ferredoxin/oxidized ferredoxin. Such a simplistic analysis would suggest that the Ndh-mediated cycle is more likely to occur in vivo than ferredoxin cycles. Analysis of tobacco *ndh* mutants support this view (Burrows, 1998). Under moderate illumination the post-illumination reduction of the plastoquinone pool is absent in *ndh* mutants compared with WT (Burrows et al., 1998; Kofer et al., 1998; Shikanai et al., 1998), suggesting a block in the cycling of electrons into the pool. Under more extreme illumination conditions some post-illumination reduction of the pool now occurs, perhaps mediated by reduced ferredoxin that could not accumulate under the lower intensity light (Burrows, 1998).

Where the Ndh complex is specific for NADH rather than NADPH, then cyclic electron flow could involve a transhydrogenase reaction to produce NADH from NADPH (possibly via FNR) or the use of NAD- and NADP-dependent dehydrogenases in a substrate cycle (Sazanov et al., 1998). Recently chloroplasts have been shown to contain an NAD-linked malate dehydrogenase (Berkemeyer et al., 1998) in addition to the light-activated NADP-dependent malate dehydrogenase.

VI. Chlororespiration and Cyanobacterial Respiration

In 1982, Pierre Bennoun coined the term 'chlororespiration' to describe respiration within the chloroplast as distinct from mitorespiration which occurs in the mitochondrion (Bennoun, 1982). A chloroplast respiratory chain was invoked to explain a number of observations concerning the redox state of the plastoquinone pool in green algae in darkness. It was proposed that the plastoquinone pool could be reduced through the action of an NAD(P)H dehydrogenase and on the basis of inhibitor studies that it could be oxidized by oxygen at an oxidase. At the same time an electrochemical gradient could be generated across the thylakoid. While there is now ample evidence for non-photochemical reduction of the plastoquinone pool in a range of plants (Groom et al., 1993) the original conclusions concerning the presence of thylakoid oxidases are less convincing (Bennoun, 1998). It is now clear that the inhibitors of the putative chloroplast oxidase were in fact inhibiting the mitochondrial oxidase which then caused the indirect reduction of the chloroplast plastoquinone pool (Bennoun, 1994). Because of mitochondrial/chloroplast interactions (Section VII), mitochondrial oxidases are in principle able to drive the oxidation of the plastoquinone pool (Bennoun, 1998) through reverse electron flow from plastoquinol to $NAD(P)^+$ at a proton pumping NAD(P)H dehydrogenase, with the reaction driven by the proton electrochemical gradient across the membrane.

Despite the fact that the original data (and possibly later results from higher plants) can no longer be interpreted unambiguously, there are now biochemical and genetic data to support the concept of chlororespiration. First, the Ndh complex found in higher plant chloroplasts, but not yet in *C. reinhardtii*, appears to fulfill the role of the NAD(P)H dehydrogenase (see above). Second, a protein, designated Immutans, with strong sequence similarities to the alternative oxidases of plant mitochondria has been detected in chloroplasts of *Arabidopsis thaliana* (Wu et al., 1999; Carol et al., 1999). It is plausible that these two components together with plastoquinone may form a chloroplast respiratory chain (Fig. 4). Additional or alternative plastoquinone reductase and oxidase activities may be present depending on the species and physiological state. For instance there appears to be multiple thylakoid NADH dehydrogenase activities in tobacco thylakoids (Cornac et al., 1998). A plastoquinol peroxidase has also been suggested to be a component of chlororespiration (Casano et al., 2000).

The rate of chlororespiratory electron transfer in the dark is, however, quite small with rates in sunflower estimated to be at only about 0.3% of the light-saturated photosynthetic electron flow (Feild et al., 1998). The role of chlororespiration is uncertain although speculation has focused on metabolism in the dark such as the maintenance of a proton motive force (for ATP synthesis and activation of the ATP synthase) and the regulation of pyridine nucleotide levels for efficient starch mobilization. Given the low rates of electron transfer in chlororespiration, it is unlikely that it makes a significant contribution to ATP synthesis in the light in mature chloroplasts. In immature or non-photosynthetic plastids, the contribution may become important. However the Ndh component of the chain may have a role in cyclic electron flow in the light as described above, possibly directly or by poising the IEC in an appropriate state for cyclic electron transfer via ferredoxin-mediated pathways.

The *immutans* mutant lacking the putative chloroplast alternative oxidase shows a variegated phenotype with leaves consisting of white and green sectors. The white sectors develop because insufficient carotenoid is made to protect from photooxidative stress (Carol et al., 1999; Wu et al., 1999). Lack of Immutans may inhibit phytoene desaturation because the plastoquinone pool is too reduced.

For cyanobacteria there are a number of potential thylakoid plastoquinone reductases and oxidases (early work reviewed by Schmetterer, 1994). In *Synechocystis* 6803, there are three open reading frames encoding potential type 2 NADH dehydrogenases, designated *ndbA-C* (Howitt et al., 1999). In contrast to the type 1 dehydrogenases, type 2 dehydrogenases consist of single subunits, lack iron-sulfur centers and do not pump protons across the membrane. Because the activity of the *Synechocystis ndb* gene products appears to be low under laboratory growth conditions it has been suggested that they may act as redox sensors and serve a regulatory function (Howitt et al., 1999). The activity of the Ndb proteins under stress conditions, such as iron-depletion when the Ndh complex may be less active, awaits examination as does confirmation that the Ndb subunits are located in the thylakoid rather than cytoplasmic membrane.

The genome of *Synechocystis* 6803 also contains three sets of genes for terminal respiratory oxidases: a cytochrome aa_3-type cytochrome *c* oxidase (CtaI), a possible cytochrome *bo*-type quinol oxidase (CtaII) and a putative cytochrome *bd* quinol oxidase (Cyd). Recent conclusions based on the analysis of engineered mutants are that CtaI is the major oxidase in thylakoid membranes, Cyd is mainly located in the cytoplasmic membrane and that CtaII plays only a minor role in cellular respiration under the conditions tested (Howitt and Vermaas, 1998). Interestingly there is no obvious cyanobacterial homologue of Immutans.

VII. Interaction between Chloroplasts and Mitochondria

Chloroplasts, like mitochondria, possess a number of protein transporters within the inner envelope membrane which enable the selective passage of metabolites into and out of the organelle (Flügge, 1998). Although there is a chloroplast ADP/ATP antiport system, there is no translocator of NAD(P)H. Instead reducing equivalents are transported indirectly (Heineke et al., 1991). For instance the oxaloacetate/malate antiporter allows the transport of malate which can be oxidized by NAD^+- and $NADP^+$-dependent malate dehydrogenases in a number of cell compartments to yield NADH and NADPH respectively. The chloroplast is thus in metabolic communication with the rest of the cell and potentially (on a slower time scale) with the rest of the plant.

The potential importance of mitochondrial-chloroplast interactions in photosynthesis has been highlighted by the isolation of a photoautotrophic suppressor of the chloroplast mutant FUD50 of *C. reinhardtii* which lacks the chloroplast ATP synthase (Lemaire et al., 1988). The suppressor strain still lacks the chloroplast ATP synthase and has therefore had to develop an alternative route to accumulate the necessary ATP in the chloroplast for the Calvin cycle. A possible pathway involves the export from the chloroplast to the cytosol of reducing equivalents in the form of dihydroxyacetone phosphate which is first converted to glyceraldehyde 3-phosphate and then dehydrogenated to produce cytosolic NADPH. NADPH can be then oxidized at the external surface of the inner mitochondrial membrane and ATP synthesized through oxidative phosphorylation. ATP is then exported from the mitochondrial matrix to the chloroplast stroma using ADP/ATP translocators in both organelles. This pathway is energetically more costly than ATP synthesis by the chloroplast ATP synthase and does not seem to operate in the WT.

The role of the mitochondrial electron transport chain in photosynthetic electron flow has also been implicated from studies of PS I and Cyt b_6f-deficient strains of *C. reinhardtii*. Using mass spectrometry measurements to differentiate between oxygen uptake and oxygen release, a permanent electron flow could be established between water oxidation at PS II and oxygen reduction at mitochondrial terminal oxidases (Peltier and Thibault, 1988). In the absence of PS I, the reductant NAD(P)H was hypothesized to be produced by an NAD(P)H dehydrogenase driven in reverse by the proton motive force (Peltier and Thibault, 1988). As yet the identity of this complex is unknown although it would appear not to be the Ndh complex (Cornac et al., 1998). Transport of the reductant to the mitochondrion for oxidation would occur through metabolite shuttles as described above.

Evidence that the mitochondrial ATP synthesis may be important for high rates of photosynthesis in WT has come from studies on barley leaf protoplasts (Krömer et al., 1988). Selective inhibition of mitochondrial respiration using oligomycin resulted in an approximate 40% inhibition in oxygen evolution.

The role of the mitochondrion in photosynthetic electron flow may be important at two levels. First, from energetic considerations it has been suggested that ATP production through the mitochondrial oxidation of reductant produced by photosynthetic electron transport would be more efficient than that through cyclic electron flow in the chloroplast (Krömer et al., 1988). Thus when the ATP/NADPH ratio needs to be increased, reductant may be oxidized within the mitochondrion. Indeed the mitochondrial oxidation of photosynthetic reductant is considered the basis of the phenomenon of 'light-enhanced dark respiration' (Xue et al., 1996). Second, mitochondrial activity acts as an additional electron sink for photosynthetic electron transport and so may have a role in protecting the photosynthetic electron transport chain from damage (through the malate valve).

VIII. Questions for the Future

It is likely that the power of the genetic approaches now being applied in cyanobacteria, green algae and higher plants will lead to the identification of the genes and gene products required for all the known photosynthetic regulatory mechanisms within the next few years. This in turn should lead to the determination of the basic biochemistry of the mechanisms. Some other questions are then likely to come to the fore.

A. Are there other Regulatory Mechanisms still to be Characterized?

Even unicellular photosynthetic organisms are exceptionally complex systems, capable of responding to their environment in numerous and subtle ways. This is nicely illustrated by the complete sequencing of the genome of the cyanobacterium *Synechocystis* 6803 (Kaneko et al., 1996). In addition to a very large number of genes of completely unknown function, there are more than 80 open reading-frames coding for components of putative two-component signal transduction systems (Mizuno et al., 1996). We still have no idea of the function of the majority of these genes. It seems almost certain that any photosynthetic organism will respond to any environmental change with a whole suite of responses. We may have very little hope of dissecting out all these responses using conventional physiological and biophysical techniques. Specific mutants may offer the best way forward. Mutants can be used to address the question in two ways: First, the inactivation of specific genes may lead to the characterization of new regulatory mechanisms. For example, the inactivation of a putative DNA-binding response regulator in *Synechocystis* produces a phenotype suggesting that the response regulator is involved in a long-term mechanism for regulating the interaction of phycobilisomes with reaction centers, a mechanism that had not previously been suspected (Ashby and Mullineaux, 1999). Second, if a specific regulatory mechanism is inactivated by a mutation, what other responses can be seen in the mutant? The inactivation of one or more regulatory pathways can be a good way to reveal other control mechanisms that were masked in the wild-type. For example, a mutational study shows that 'state transitions' in cyanobacteria in fact consist of at least two independent responses (Emlyn-Jones et al., 1999).

B. What are the Physiological Roles of the Adaptation Mechanisms?

Our current thinking on these questions is based largely on plausible guesswork, combined with comparative studies of the occurrence of the regulatory mechanisms in organisms in different environments. The isolation of mutants with highly specific phenotypes will allow a much more direct approach. If we take a mutant specifically deficient in a particular regulatory mechanism, and place it in a particular environment, how will it fare in comparison to the wild type? This approach has been used, for example, to show that state transitions in cyanobacteria are important for regulating the efficiency of utilization of phycobilisome-absorbed light (Emlyn-Jones et al, 1999).

C. Are there Cell- and Tissue-specific Responses in Multicellular Photosynthetic Organisms?

Mechanisms for regulating photosynthetic electron transport have mainly been characterized in unicellular organisms, or in isolated chloroplasts. It seems almost certain that, in intact plants, there will be different responses in different cells and different tissues. Fluorescence video imaging can be used to show this at the macroscopic level (Scholes and Rolfe, 1996). This could be extended to microscopic scales by confocal fluorescence microscopy and microvolume spectroscopy. Tissue-specific gene expression studies should also provide a powerful approach to this question.

References

Allakhverdiev SI, Yruela I, Picorel R and Klimov VV (1997) Bicarbonate is an essential constituent of the water-oxidizing complex of Photosystem II. Proc Natl Acad Sci USA 94: 5050–5054

Allen JF (1992) Protein phosphorylation in regulation of photosynthesis. Biochim Biophys Acta 1098: 275–335

Allen JF , Sanders CE and Holmes NG (1985) Correlation of membrane protein phosphorylation with excitation energy distribution in the cyanobacterium *Synechococcus* 6301. FEBS Lett 193: 271–275

Andersson B and Anderson JM (1980) Lateral heterogeneity in the distribution of chlorophyll-protein complexes of the thylakoid membranes of spinach chloroplasts. Biochim Biophys Acta 593: 427–440

Arnon DI (1991) Photosynthetic electron transport: Emergence of a concept, 1949-59. Photosynth Res 29: 117–131

Arnon DI and Chain RK (1975) Regulation of ferredoxin-catalysed photosynthetic phosphorylations. Proc Natl Acad Sci USA 72: 4961–4965

Asada K (1999) The water-water cycle in chloroplasts: Scavenging of active oxygens and dissipation of excess photons. Annu Rev Plant Physiol Plant Mol Biol 50: 601–639

Ashby MK and Mullineaux CW (1999) Cyanobacterial ycf27 gene products regulate energy transfer from phycobilisomes to Photosystems I and II. FEMS Microbiology Lett 181: 253–260

Barber J and Andersson B (1992) Too much of a good thing: Light can be bad for photosynthesis. Trends Biochem Sci 17: 61–66

Barth P, Lagoutte B and Sétif P (1998) Ferredoxin reduction by Photosystem I from *Synechocystis* sp. PCC 6803: Toward an understanding of the respective roles of subunits PsaD and PsaE in ferredoxin binding. Biochemistry 37: 16233–16241

Bendall DS and Manasse RS (1995) Cyclic photophosphorylation and electron transport. Biochim Biophys Acta 1229: 23–38

Bennoun P (1982) Evidence for a respiratory chain in the chloroplast. Proc Natl Acad Sci USA 79: 4352–4356

Bennoun P (1994) Chlororespiration revisited: Mitochondrial-plastid interactions in *Chlamydomonas*. Biochim Biophys Acta 1186: 59–66

Bennoun P (1998) Chlororespiration, sixteen years later. In: Rochaix J-D, Goldschmidt-Clermont M and Merchant S (eds) The Molecular Biology of Chloroplasts and Mitochondria in Chlamydomonas, pp 675–683. Kluwer Academic Publishers, Dordrecht

Berger S, Ellersiek U and Steinmüller K (1991) Cyanobacteria contain a mitochondrial complex I-homologous NADH dehydrogenase. FEBS Lett 286: 129–132

Berger S, Ellersiek U, Westhoff P and Steinmüller P (1993) Studies on the expression of NDH-H, a subunit of the NAD(P)H-plastoquinone oxidoreductase of higher-plant chloroplasts. Planta 190: 25–31

Berkemeyer M, Scheibe R and Ocheretina O (1998) A novel, non-redox-regulated NAD-dependent malate dehydrogenase from chloroplasts of *Arabidopsis thaliana* L. J Biol Chem 273: 27927-27933

Berry S and Rumberg B (1999) Proton to electron stoichiometry in electron transport of spinach thylakoids. Biochim Biophys Acta 1410: 248–261

Biggins J (1974) The role of plastoquinone in the in vivo photosynthetic cyclic electron transport pathway in algae. FEBS Lett 38: 311–314

Blubaugh DJ and Govindjee (1988) The molecular mechanism of the bicarbonate effect at the plastoquinone reductase site of photosynthesis. Photosynth Res 19: 85–128

Bowes JM and Crofts AR (1981) The role of pH and membrane potential in the reactions of Photosystem II as measured by effects on delayed fluorescence. Biochim Biophys Acta 637: 464–472

Buchanan BB (1994) The ferredoxin-thioredoxin system: update on its role in the regulation of oxygenic photosynthesis. Adv Mol Cell Biol 10: 337–354

Bulté L, Gans P, Rebéillé F and Wollman F-A (1990) ATP control on state transitions in vivo in *Chlamydomonas reinhardtii*. Biochim Biophys Acta 1020: 72–80

Burrows PA (1998) Functional characterization of the plastid *ndh* gene products. PhD Thesis, University of London

Burrows PA, Sazanov LA, Svab Z, Maliga P and Nixon PJ (1998) Identification of a functional respiratory complex in chloroplasts through analysis of tobacco mutants containing disrupted plastid *ndh* genes. EMBO J 17: 868–876

Butler WL, Tredwell CJ, Malkin R and Barber J (1979) The relationship between the lifetime and yield of the 735 nm fluorescence of chloroplasts at low temperatures. Biochim Biophys Acta 545: 309–315

Campbell D, Bruce D, Carpenter C, Gustafsson P and Öquist G (1996) Two forms of the Photosystem II D1 protein alter energy dissipation and state transitions in the cyanobacterium *Synechococcus* sp. PCC7942. Photosynth Res 47: 131–144

Canaani O, Schuster G and Ohad I (1989) Photoinhibition in *Chlamydomonas reinhardtii*: Effect on state transition, intersystem energy distribution and Photosystem I cyclic electron flow. Photosynth Res 20: 129–146

Cao J, Vermaas WFJ and Govindjee (1991) Arginine residues in the D2 polypeptide may stabilize bicarbonate binding in Photosystem II of *Synechocystis* sp. PCC 6803. Biochim Biophys Acta 1059: 171–180

Carol P, Stevenson D, Bisanz C, Breitenbach J, Sandmann G, Mache R, Coupland G and Kuntz M (1999) Mutations in the *Arabidopsis* gene *IMMUTANS* cause a variegated phenotype by inactivating a chloroplast terminal oxidase associated with phytoene desaturation. Plant Cell 11: 57–68

Casano LM, Zapata JM, Martín M and Sabater B (2000) Chlororespiration and poising of cyclic electron transport. Plastoquinone as electron transporter between thylakoid NADH dehydrogenase and peroxidase. J Biol Chem 275: 942–948

Charlebois D and Mauzerall D (1999) Energy storage and optical cross-section of PS I in the cyanobacterium *Synechococcus* PCC 7002 and a *psaE*$^-$ mutant. Photosynth Res 59: 27–38

Cleland RE and Bendall DS (1992) Photosystem I cyclic electron transport: Measurement of ferredoxin-plastoquinone reductase activity. Photosynth Res 34: 409–418

Cornac L, Guedeney G, Joët T, Rumeau D, Latouche G, Cerovic Z, Redding K, Horvath E, Medgyesy P and Peltier G (1998) Non-photochemical reduction of intersystem electron carriers in chloroplasts of higher plants and algae. In: Garab G (ed) Photosynthesis: Mechanisms and Effects, Vol III, pp 1877–1882. Kluwer Academic Publishers, Dordrecht

Coughlan SJ and Hind G (1987) A protein kinase that phosphorylates light-harvesting complex is autophosphorylated and is associated with Photosystem II. Biochemistry 26: 6515–6521

Davies EC and Bendall DS (1987) The antimycin-binding site of thylakoid membranes from chloroplasts. In Biggins J (ed) Progress in Photosynthesis Research, Vol II, pp 485–488. Martinus Nijhoff, Dordrecht

Delosme R, Olive J and Wollman F-A (1996) Changes in the light energy distribution upon state transitions: An in vivo photoacoustic study of the wild type and photosynthesis mutants from *Chlamydomonas reinhardtii*. Biochim Biophys Acta 1273: 150–158

Delphin E, Duval J-C, Etienne A-L and Kirilovsky D (1996) State transitions or ΔpH-dependent quenching of Photosystem II fluorescence in red algae. Biochemistry 35: 9435–9445

Demmig-Adams B (1990) Carotenoids and photoprotection in plants: A role for the xanthophyll zeaxanthin. Biochim Biophys Acta 1020: 1–24

Diner BA, Petrouleas V and Wendoloski JJ (1991a) The iron-quinone electron-acceptor complex of Photosystem II. Physiol Plant 81: 423–436

Diner BA, Nixon PJ and Farchaus JW (1991b) Site-directed mutagenesis of photosynthetic reaction centers. Curr Opin Struct Biol 1: 546–554

Elortza F, Asturias JA and Arizmendi JM (1999) Chloroplast NADH dehydrogenase from *Pisum sativum*: Characterization of its activity and cloning of *ndhK* gene. Plant Cell Physiol 40: 149–154

Elrifi IR and Turpin DH (1986) Nitrate and ammonium induced photosynthetic suppression in N-limited *Selenastrum minutum*. Plant Physiol 81: 273–279

Emlyn-Jones D, Ashby MK and Mullineaux CW (1999) A gene required for the regulation of photosynthetic light-harvesting in the cyanobacterium *Synechocystis* 6803. Mol Microbiol 33: 1050–1058

Endo T, Schreiber U and Asada K (1995) Suppression of quantum yield of Photosystem II by hyperosmotic stress in *Chlamydomonas reinhardtii*. Plant Cell Physiol 36: 1253–1258

Endo T, Mi H, Shikanai T and Asada K (1997) Donation of electrons to plastoquinone by NAD(P)H dehydrogenase and ferredoxin-quinone reductase in spinach chloroplasts. Plant Cell Physiol 38: 1272–1277

Endo T, Shikanai T, Sato F and Asada K (1998) NAD(P)H dehydrogenase-dependent, antimycin A-sensitive electron donation to plastoquinone in tobacco chloroplasts. Plant Cell Physiol 39: 1226–1231

Endo T, Shikanai T, Takabayashi A, Asada K and Sato F (1999) The role of chloroplastic NAD(P)H dehydrogenase in photoprotection. FEBS Lett 457: 5–8

Fearnley IM and Walker JE (1992) Conservation of sequences of subunits of mitochondrial complex I and their relationships with other proteins. Biochim Biophys Acta 1140: 105–134

Feild TS, Nedbal L and Ort DR (1998) Non-photochemical reduction of the plastoquinone pool in sunflower leaves originates from chlororespiration. Plant Physiol 116: 1209–1218

Finazzi G, Furia A, Barbagallo RP and Forti G (1999) State transitions, cyclic and linear electron transport and photophosphorylation in *Chlamydomonas reinhardtii*. Biochim Biophys Acta 1413: 117–129

Fleischmann MM, Ravanel S, Delosme R, Olive J, Zito F, Wollman F-A and Rochaix J-D (1999) Isolation and characterization of photoautotrophic mutants of *Chlamydomonas reinhardtii* deficient in state transition. J Biol Chem 274: 30987–30994

Flügge U-I (1998) Metabolite transporters in plastids. Curr Opin Plant Biol 1: 201–206

Fork DC and Herbert SK (1993) Electron transport and photophosphorylation by Photosystem I in vivo in plants and cyanobacteria. Photosynth Res 36: 149–169

Foyer CH and Harbinson J (1997) The photosynthetic electron transport system: Efficiency and control. In: Foyer CH and Quick WP (eds) A molecular approach to primary metabolism in higher plants, pp 3-39. Taylor and Francis Ltd, London

Frank HA, Cua A, Chynwat V, Young A, Gosztola D and Wasielewski MR (1994) Photophysics of the carotenoids associated with the xanthophyll cycle in photosynthesis. Photosynth Res 41: 389–395

Friedrich T, Steinmüller K and Weiss H (1995) The proton-pumping respiratory complex I of bacteria and mitochondria and its homologue in chloroplasts. FEBS Lett 367: 107–111

Fulgosi H, Vener AV, Altschmied L, Herrmann RG and Andersson B (1998) A novel multi-functional chloroplast protein: identification of a 40 kDa immunophilin-like protein located in the thylakoid lumen. EMBO J 17: 1577–1587

Funk C, Schröder WP, Napiwotzki A, Tjus SE, Renger G and Andersson B (1995) The PS II-S protein of higher plants: A new type of pigment-binding protein. Biochemistry 34: 11133–11141

Funk E, Schäfer E and Steinmüller K (1999) Characterization of the Complex I-homologous NAD(P)H-Plastoquinone-Oxidoreductase (NDH-complex) of Maize Chloroplasts. J Plant Physiol 154: 16–23

Gal A, Herrmann RG, Lottspeich F and Ohad I (1992) Phosphorylation of cytochrome b_6 by the LHCII kinase associated with the cytochrome complex. FEBS Lett 298: 33–35

Gans P and Rebeille F (1990) Control in the dark of the plastoquinone redox state by mitochondrial activity in *Chlamydomonas reinhardtii*. Biochim Biophys Acta 1015: 150–155

Genty B and Harbinson J (1996) Regulation of light utilization for photosynthetic electron transport. In: Baker NR (ed) Photosynthesis and the Environment, pp 67–99. Kluwer Academic Publishers, Dordrecht

Gilmore A, Shinkarev VP, Hazlett TL and Govindjee (1998) Quantitative analysis of the effects of intrathylakoid pH and the xanthophyll cycle pigments on chlorophyll *a* fluorescence lifetime distributions and intensity in thylakoids. Biochemistry 37: 13582–13593

Greenbaum E, Lee JW, Tevault CV, Blankinship SL and Mets LJ (1995) CO_2 fixation and photoevolution of H_2 and O_2 in a mutant of *Chlamydomonas* lacking Photosystem I. Nature 376: 438–441

Groom QJ, Kramer DM, Crofts AR and Ort DR (1993) The non-photochemical reduction of plastoquinone in leaves. Photosynth Res 36: 205–215

Guedeney G, Corneille S, Cuiné S and Peltier G (1996) Evidence for an association of *ndhB*, *ndhJ* gene products and ferredoxin-NADP-reductase as components of a chloroplastic NAD(P)H dehydrogenase complex. FEBS Lett 378: 277–280

Hagemann M, Jeanjean R, Fulda S, Havaux M, Joset F and Erdmann N (1999) Flavodoxin accumulation contributes to enhanced cyclic electron flow around Photosystem I in salt-stressed cells of *Synechocystis* sp. strain PCC 6803. Physiol Plant 105: 670–678

Havaux M (1996) Short-term responses of Photosystem I to heat stress. Induction of a PS II-independent electron transport through PS I fed by stromal components. Photosynth Res 47: 85–97

Heber U and Walker D (1992) Concerning a dual function of coupled cyclic electron transport in leaves. Plant Physiol 100: 1621–1626

Heber U, Gerst U, Krieger A, Neimanis S and Kobayashi Y (1995) Coupled cyclic electron transport in intact chloroplasts and leaves of C3 plants: Does it exist ? If so, what is its function? Photosynth Res 46: 269–275

Heineke D, Riens B, Grosse H, Hoferichter P, Peter U, Flügge U-I and Heldt HW (1991) Redox transfer across the inner chloroplast envelope membrane. Plant Physiol 95: 1131–1137

Herbert SK, Fork DC and Malkin S (1990) Photoacoustic measurements in vivo of energy storage by cyclic electron flow in algae and higher plants. Plant Physiol 94: 926–934

Hinerwadel R and Berthomieu C (1995) Bicarbonate binding to the non-haem iron of Photosystem II investigated by Fourier transform infrared difference spectroscopy and ^{13}C-labeled bicarbonate. Biochemistry 34: 16288–16297

Horton P, Ruban AV and Walters RG (1996) Regulation of light-harvesting in green plants. Annu Rev Plant Physiol Plant Mol Biol 47: 655–684

Hosler JP and Yocum CF (1985) Evidence for two cyclic photophosphorylation reactions concurrent with ferredoxin-catalyzed non-cyclic electron transport. Biochim Biophys Acta 808: 21–31

Howitt CA and Vermaas WFJ (1998) Quinol and cytochrome oxidases in the cyanobacterium *Synechocystis* sp. PCC 6803. Biochemistry 37: 17944–17951

Howitt CA, Udall PK and Vermaas WFJ (1999) Type 2 NADH dehydrogenases in the cyanobacterium *Synechocystis* sp. strain PCC 6803 are involved in regulation rather than respiration. J Bacteriol 181: 3994–4003

Johnson GN, Scholes JD, Horton P and Young AJ (1993) The dissipation of excitation energy in British plant species. Plant, Cell and Environment 16: 673–679

Kaneko T, Sato S, Kotani H, Tanaka A, Asamizu E, Nakamura Y, Miyajima N, Hirosawa M, Sugiura M, Sasamoto S, Kimura T, Hosouchi T, Matsuno A, Muraki A, Nakazaki N, Naruo K, Okumura S, Shimpo S, Takeuchi C, Wada T, Watanabe A, Yamada M, Yasuda M and Tabata S (1996) Sequence analysis of the genome of the unicellular cyanobacterium *Synechocystis* sp. strain PCC6803. 2. Sequence determination of the entire genome and assignment of potential protein-coding regions. DNA Res 3: 109–136

Katona E, Neimanis S, Schönknecht G and Heber U (1992) Photosystem I-dependent cyclic electron transport is important in controlling Photosystem II activity in leaves under conditions of water stress. Photosynth Res 34: 449–464

Klimov VV and Baranov SV (2001) Bicarbonate requirement for the water-oxidizing complex of Photosystem II. Biochim Biophys Acta 1503: 187–196

Klimov VV, Baranov SV and Allakhverdiev SI (1997) Bicarbonate protects the donor side of Photosystem II against photoinhibition and thermoinactivation. FEBS Lett 418: 243–246

Klughammer B, Sultemeyer D, Badger MR and Price GD (1999) The involvement of NAD(P)H dehydrogenase subunits, NdhD3 and NdhF3, in high-affinity CO_2 uptake in *Synechococcus* sp. PCC7002 gives evidence for multiple NDH-1 complexes with specific roles in cyanobacteria. Mol Microbiol 32: 1305–1315

Klukas O, Schubert W-D, Jordan P, Krauß N, Fromme P, Witt HT and Saenger W (1999) Photosystem I, an improved model of the stromal subunits PsaC, PsaD and PsaE. J Biol Chem 274: 7351–7360

Kofer W, Koop H-U, Wanner G and Steinmüller K (1998) Mutagenesis of the genes encoding subunits A, C, H, I, J and K of the plastid NAD(P)H-plastoquinone-oxidoreductase in tobacco by polyethylene glycol-mediated plastome transformation. Mol Gen Genet 258: 166–173

Koop H-U, Kofer W and Steinmüller K (1998) Judging the homoplastomic state of plastid transformants—a reply. Trends Plant Sci 3: 377

Kramer DM and Crofts AR (1993) The concerted reduction of the low and high potential chains of the *bf* complex by plastoquinol. Biochim Biophys Acta 1183: 72–84

Kramer DM and Crofts AR (1996) Control and measurement of photosynthetic electron transport in vivo. In: Baker NR (ed) Photosynthesis and the Environment, pp 25–66. Kluwer Academic Publishers, Dordrecht

Krause GH, Briantais J-M and Vernotte C (1983) Characterization of chlorophyll fluorescence quenching in chloroplasts by fluorescence spectroscopy at 77 K. Biochim Biophys Acta 723: 169–175

Krieger A, Moya I and Weis E (1992) Energy-dependent

quenching of chlorophyll a fluorescence: Effect of pH on stationary fluorescence and picosecond-relaxation kinetics in thylakoid membranes and Photosystem II preparations. Biochim Biophys Acta 1102: 167–176

Krömer S, Stitt M and Heldt HW (1988) Mitochondrial oxidative phosphorylation participating in photosynthetic metabolism of a leaf cell. FEBS Lett 226: 352–356

Kruse O, Nixon PJ, Schmid GH and Mullineaux CW (1999) Isolation of state transition mutants of *Chlamydomonas reinhardtii* by fluorescence video imaging. Photosynth Res 61: 43–51

Kubicki A, Funk E, Westhoff P and Steinmüller K (1996) Differential expression of plastome-encoded *ndh* genes in mesophyll and bundle-sheath chloroplasts of the C4 plant *Sorghum bicolor* indicates that the complex I-homologous NAD(P)H-plastoquinone oxidoreductase is involved in cyclic electron transport. Planta 199: 276–281

Lemaire C, Wollman F-A and Bennoun P (1988) Restoration of photoautotrophic growth in a mutant of *Chlamydomonas reinhardtii* in which the chloroplastic *atp*B gene of the ATP synthase has a deletion: An example of mitochondrial dependent photosynthesis. Proc Natl Acad Sci USA 85: 1344–1348

Li X-P, Björkman O, Shih C, Grossman AR, Rosenquist M, Jansson S and Niyogi KK (2000) A pigment-binding protein essential for regulation of photosynthetic light harvesting. Nature 403: 391–395

Maliga P and Nixon PJ (1998) Judging the homoplastomic state of plastid transformants. Trends Plant Sci 3: 376–377

Martín M, Casano LM and Sabater B (1996) Identification of the product of *ndhA* gene as a thylakoid protein synthesised in response to photooxidative treatment. Plant Cell Physiol 37: 293–298

Matsuo M, Endo T and Asada K (1998) Properties of the respiratory NAD(P)H dehydrogenase isolated from the cyanobacterium *Synechocystis* PCC6803. Plant Cell Physiol 39: 263–267

Mehler AH (1951) Studies on the reaction of illuminated chloroplasts. I. Mechanism of the reduction of oxygen and other Hill reagents. Arch Biochem Biophys 33: 65–77

Mi H, Endo T, Schreiber U, Ogawa T and Asada K (1992) Electron donation from cyclic and respiratory flows to the photosynthetic intersystem chain is mediated by pyridine nucleotide dehydrogenase in the cyanobacterium *Synechocystis* sp. PCC 6803. Plant Cell Physiol 33: 1233–1237

Mi H, Endo T, Ogawa T and Asada K (1995) Thylakoid membrane-bound, NADPH-specific pyridine nucleotide dehydrogenase complex mediates cyclic electron transport in the cyanobacterium *Synechocystis* sp. PCC 6803. Plant Cell Physiol 36: 661–668

Mills JD, Crowther D, Slovacek RE, Hind G and McCarty RE (1979) Electron transport pathways in spinach chloroplasts. Reduction of the primary acceptor of Photosystem II by reduced nicotinamide adenine dinucleotide phosphate in the dark. Biochim Biophys Acta 547: 127–137

Miyake C, Schreiber U and Asada K (1995) Ferredoxin-dependent and antimycin A-sensitive reduction of cytochrome *b*-559 by far-red light in maize thylakoids: participation of a menadiol-reducible cytochrome *b*-559 in cyclic electron flow. Plant Cell Physiol 36: 743–748

Mizuno T, Kaneko T and Tabata S (1996) Compilation of all genes encoding two-component phosphotransfer signal transducers in the genome of the cyanobacterium *Synechocystis* sp. strain PCC6803. DNA Res 3: 407–414

Moss DA and Bendall DS (1984) Cyclic electron transport in chloroplasts: The Q cycle and the site of action of antimycin. Biochim Biophys Acta 767: 389–395

Mullineaux CW (1992) Excitation energy transfer from phycobilisomes to Photosystem I in a cyanobacterium. Biochim Biophys Acta 1100: 285–292

Mullineaux CW (1994) Excitation energy transfer from phycobilisomes to Photosystem I in a cyanobacterial mutant lacking Photosystem II. Biochim Biophys Acta 1184: 71–77

Mullineaux CW and Allen JF (1990) State 1-state 2 transitions in the cyanobacterium *Synechococcus* 6301 are controlled by the redox state of electron carriers between Photosystems I and II. Photosynth Res 22: 157–166

Mullineaux CW, Tobin, MJ and Jones GR (1997) Mobility of photosynthetic complexes in thylakoid membranes. Nature 390: 421–424

Nilsson A, Stys D, Drakenberg T, Spangfort MD, Forsen S and Allen JF (1997) Phosphorylation controls the three-dimensional structure of plant light harvesting complex II. J Biol Chem 272: 18350–18357

Nixon PJ (2000) Chlororespiration. Phil Trans R Soc Lond B 355: 1541–1547

Nixon PJ, Gounaris K, Coomber SA, Hunter CN, Dyer TA and Barber J (1989) *psbG* is not a Photosystem two gene but may be an *ndh* gene. J Biol Chem 264: 14129–14135

Niyogi KK (1999) Photoprotection revisited: Genetic and molecular approaches. Annu Rev Plant Physiol Plant Mol Biol 50: 333–359

Norling B, Zak E, Andersson B and Pakrasi H (1998) 2D-isolation of pure plasma and thylakoid membranes from the cyanobacterium *Synechocystis* sp. PCC 6803. FEBS Lett 436: 189–192

Ohkawa H, Sonoda M, Katoh H and Ogawa T (1998) The use of mutants in the analysis of the CO_2-concentrating mechanism in cyanobacteria. Can J Bot 76: 1035–1042

Ohyama K, Fukuzawa H, Kohchi T, Shirai H, Sano T, Sano S, Umesono K, Shiki Y, Takeuchi M, Chang Z, Aota S, Inokuchi H and Ozeki H (1986) Chloroplast gene organization deduced from complete sequence of liverwort *Marchantia polymorpha* chloroplast DNA. Nature 322: 572–574

Olive J, Ajlani G, Astier C, Recouvreur M and Vernotte C (1997) Ultrastructure and light adaptation of phycobilisome mutants of *Synechocystis* PCC6803. Biochim Biophys Acta 1319: 275–282

Osmond CB (1981) Photorespiration and photoinhibition: some implications for the energetics of photosynthesis. Biochim Biophys Acta 639: 77–98

Peltier G and Thibault P (1988) Oxygen-exchange studies in *Chlamydomonas* mutants deficient in photosynthetic electron transport: evidence for a Photosystem II-dependent oxygen uptake in vivo. Biochim Biophys Acta 936: 319–324

Petrouleas V, Deligiannakis Y and Diner BA (1994) Binding of carboxylate anions at the non-haem Fe(II) of PS II. 2. Competition with bicarbonate and effects on the Q_A/Q_B electron transfer rate. Biochim Biophys Acta 1188: 271–277

Phillip D, Ruban AV, Horton P, Asato A and Young AJ (1996) Quenching of chlorophyll fluorescence in the major light-harvesting complex of Photosystem II: A systematic study of the effect of carotenoid structure. Proc Natl Acad Sci USA 93: 1492–1497

Quiles MJ and Cuello J (1998) Association of ferredoxin-NADP oxidoreductase with the chloroplastic pyridine nucleotide dehydrogenase complex in barley leaves. Plant Physiol 117: 235–244

Ravenel J, Peltier G and Havaux M (1994) The cyclic electron pathways around Photosystem I in *Chlamydomonas reinhardtii* as determined in vivo by photoacoustic measurements of energy storage. Planta 193: 251–259

Redding K, Cournac L, Vassiliev IR, Golbeck JH, Peltier G and Rochaix J-D (1999) Photosystem I is indispensable for photoautotrophic growth, CO_2 fixation, and H_2 photoproduction in *Chlamydomonas reinhardtii*. J Biol Chem 274: 10466–10473

Rich PR (1988) A critical examination of the supposed variable proton stoichiometry at the chloroplast cytochrome *b/f* complex. Biochim Biophys Acta 932: 33–42

Rich PR, Hoefnagel MHN and Wiskich JT (1998) Possible chlororespiratory reactions of thylakoid membranes. In: Møller IM, Gardeström P, Glimelius K and Glaser E (eds) Plant Mitochondria: From Gene to Function, pp 17–23. Backhuys Publishers, Leiden

Rintamäki E, Salonen M, Suoranta U-M, Carlberg I, Andersson B and Aro E-M (1997) Phosphorylation of light-harvesting complex II and Photosystem II core proteins shows different irradiance-dependent regulation in vivo. J Biol Chem 272: 30476–30482

Sazanov LA, Burrows PA and Nixon PJ (1998) The plastid *ndh* genes code for an NADH-specific dehydrogenase: Isolation of a complex I analogue from pea thylakoid membranes. Proc Natl Acad Sci USA 95: 1319–1324

Scheibe R (1987) $NADP^+$-malate dehydrogenase in C3-plants: Regulation and role of a light-activated enzyme. Physiol Plant 71: 393–400

Scherer S, Almon H and Böger P (1988) Interaction of photosynthesis, respiration and nitrogen fixation in cyanobacteria. Photosynth Res 15: 95–114

Schmetterer G (1994) Cyanobacterial respiration. In: Bryant DA (ed) The Molecular Biology of Cyanobacteria, pp 409—435. Kluwer Academic Publishers, Dordrecht

Scholes JD and Rolfe SA (1996) Photosynthesis in localised regions of oat leaves infected with crown rust (*Puccinia coronata*): Quantitative imaging of chlorophyll fluorescence. Planta 199: 573–582

Schreiber U, Endo T, Mi H and Asada K (1995) Quenching analysis of chlorophyll fluorescence by the saturation pulse method: particular aspects relating to the study of eukaryotic algae and cyanobacteria. Plant Cell Physiol 36: 873–882

Schweitzer RH and Brudvig GW (1997) Fluorescence quenching by chlorophyll cations in Photosystem II. Biochemistry 36: 11351–11359

Shikanai T, Endo T, Hashimoto T, Yamada Y, Asada K and Yokota A. (1998) Directed disruption of the tobacco ndhB gene impairs cyclic electron flow around Photosystem I. Proc Natl Acad Sci USA 95: 9705–9709

Shinozaki K, Ohme M, Tanaka M, Wakasugi T, Hayashida N, Matsubayashi T, Zaita N, Chunwongse J, Obokata J, Yamaguchi-Shinozaki K, Ohto C, Torazawa K, Meng BY, Sugita M, Deno H, Kamogashira T, Yamada K, Kusuda J, Takaiwa F, Kato A, Tohdoh N, Shimada H and Sugiura M (1986) The complete nucleotide sequence of the tobacco chloroplast genome: its gene organization and expression. EMBO J 5: 2043–2049

Snel JFH and van Rensen JJS (1984) Reevaluation of the role of bicarbonate and formate in the regulation of photosynthetic electron flow in broken chloroplasts. Plant Physiol 75: 146–150

Snyders S and Kohorn BD (1999) TAKS, thylakoid membrane protein kinases associated with energy transduction. J Biol Chem 274: 9137–9140

Stemler AJ (1998) Bicarbonate and photosynthetic oxygen evolution: An unwelcome legacy of Otto Warburg. Indian J Exp Biol 36: 841–848

Stumpp MT, Motohashi K and Hisabori T (1999) Chloroplast thioredoxin mutants without active-site cysteines facilitate the reduction of the regulatory disulphide bridge on the γ-subunit of chloroplast ATP synthase. Biochem J 341: 157–163

Sundby C (1990) Bicarbonate effects on photoinhibition. Including an explanation for the sensitivity to photo-inhibition under anaerobic conditions. FEBS Lett 274: 77–81

Tagawa K, Tsujimoto HY and Arnon DI (1963) Role of chloroplast ferredoxin in the energy conversion process of photosynthesis. Proc Natl Acad Sci USA 49: 567–572

Takahama U, Shimidzu-Takahama M and Heber U (1981) The redox state of NADP in illuminated chloroplasts. Biochim Biophys Acta 637: 530–539

Teicher HB and Scheller HV (1998) The NAD(P)H dehydrogenase in barley thylakoids is photoactivatable and uses NADPH as well as NADH. Plant Physiol 117: 525–532

Turmel C, Otis C and Lemieux C (1999) The complete chloroplast DNA sequence of the alga *Nephroselmis olivacea*: Insights into the architecture of ancestral chloroplast genomes. Proc Natl Acad Sci USA 96: 10248–10253

Turpin DH and Bruce D (1990) Regulation of photosynthetic light-harvesting by nitrogen assimilation in the green alga *Selenastrum minutum*. FEBS Lett 263: 99–103

Vallon O, Bulté L, Dainese P, Olive J, Bassi R and Wollman F-A (1991) Lateral redistribution of cytochrome b_6f complexes along thylakoid membranes upon state transitions. Proc Natl Acad Sci USA 88: 8262–8266

van Rensen JJS, Xu C and Govindjee (1999) Role of bicarbonate in Photosystem II, the water-plastoquinone oxido-reductase of plant photosynthesis. Physiol Plant 105: 585–592

van Thor JJ, Geerlings TH, Matthijs HCP and Hellingwerf KJ (1999a) Kinetic evidence for the PsaE-dependent transient ternary complex Photosystem I/ferredoxin/ferredoxin:$NADP^+$ reductase in a cyanobacterium. Biochemistry 38: 12735–12746

van Thor JJ, Gruters OWM, Matthijs HCP and Hellingwerf KJ (1999b) Localization and function of ferredoxin:$NADP^+$ reductase bound to the phycobilisomes of *Synechocystis*. EMBO J 18: 4128–4136

van Walraven HS, Strotmann H, Schwarz O and Rumberg B (1996) The H^+/ATP coupling ratio of the ATP synthase from thiol-modulated chloroplasts and two cyanobacterial strains is four. FEBS Lett 379: 309–313

Vener AV, van Kan PJM, Rich PR, Ohad I and Andersson B (1997) Plastoquinol at the quinol oxidation site of reduced cytochrome bf mediates signal transduction between light and protein phosphorylation: Thylakoid protein kinase deactivation by a single-turnover flash. Proc Natl Acad Sci USA 94: 1585–1590

Vermaas W (1994) Molecular genetic approaches to study photosynthetic and respiratory electron transport in thylakoids from cyanobacteria. Biochim Biophys Acta 1187: 181–186

Wakasugi T, Tsudzuki J, Ito S, Nakashima K, Tsudzuki T and Sugiura M (1994) Loss of all *ndh* genes as determined by sequencing the entire chloroplast genome of the black pine *Pinus thunbergii*. Proc Natl Acad Sci USA 91: 9794–9798

Walters RG, Ruban AV and Horton P (1996) Identification of proton-active residues in a higher plant light-harvesting complex. Proc Natl Acad Sci USA 93: 14204–14209

Weis E (1981) Reversible heat-inactivation of the Calvin cycle: A possible mechanism of the temperature regulation of photosynthesis. Planta 151: 33–39

Wollman F-A and Lemaire C (1988) Studies on kinase-controlled state transitions in Photosystem II and b_6f mutants from *Chlamydomonas reinhardtii* which lack quinone-binding proteins. Biochim Biophys Acta 933: 85–94

Wu D, Wright DA, Wetzel C, Voytas DF and Rodermel S (1999) The IMMUTANS variegation locus of *Arabidopsis* defines a mitochondrial alternative oxidase homolog that functions during early chloroplast biogenesis. Plant Cell 11: 43–55

Xiong J, Hutchinson RS, Sayre RT and Govindjee (1997) Modification of the Photosystem II acceptor side function in a D1 mutant (arginine-269-glycine) of *Chlamydomonas reinhardtii*. Biochim Biophys Acta 1322: 60–76

Xiong J, Minagawa J, Crofts A and Govindjee (1998) Loss of inhibition by formate in newly constructed Photosystem II D1 mutants, D1-R257E and D1-R257M, of *Chlamydomonas reinhardtii*. Biochim Biophys Acta 1365: 473–491

Xue X, Gauthier DA, Turpin DH and Weger HG (1996) Interactions between photosynthesis and respiration in the green alga *Chlamydomonas reinhardtii*. Plant Physiol. 112: 1005–1014

Yruela I, Allakhverdiev SI, Ibarra JV and Klimov VV (1998) Bicarbonate binding to the water-oxidizing complex in the Photosystem II. A Fourier transform infrared spectroscopy study. FEBS Lett 425: 396–400

Yu L, Zhao J, Mühlenhoff U, Bryant DA and Golbeck JH (1993) PsaE is required for in vivo cyclic electron flow around Photosystem I in the cyanobacterium *Synechococcus* sp. PCC 7002. Plant Physiol 103: 171–180

Zer H, Vink M, Keren N, Dilly-Hartwig HG, Paulsen H, Herrmann RG, Andersson B and Ohad I (1999) Regulation of thylakoid protein phosphorylation at the substrate level: Reversible light-induced conformational changes expose the phosphorylation site of the light-harvesting complex II. Proc Natl Acad Sci USA 96: 8277—8282

Chapter 31

Functional Genomics in *Synechocystis* sp. PCC6803: Resources for Comprehensive Studies of Gene Function and Regulation

Takakazu Kaneko and Satoshi Tabata*
Kazusa DNA Research Institute, 1532-3 Yana, Kisarazu, Chiba 292-0812, Japan

Summary

The nucleotide sequence of the entire genome of the cyanobacterium *Synechocystis* sp. PCC6803 has been determined. The primary structures and the positions of all the genes in this genome can be obtained from the genome information database called CyanoBase. The information in the CyanoBase allows a variety of experimental methods to be utilized for the elucidation of the function of each gene and for the unraveling of the regulatory network controlling gene expression. Systematic gene disruption projects employing the natural transformation ability of this organism are in progress and the results are compiled in the database CyanoMutants, which is closely linked to CyanoBase. Further, CyanoBase permits the use of proteome and transcriptome technologies for the study of *Synechocystis* gene expression and regulation at the genome-wide level. These technologies can significantly accelerate screening of the transcription and translation of genes under various conditions. The combination of all these resources should allow us to rapidly and comprehensively identify the functions of all the *Synechocystis* genes, thus permitting the level of understanding of the functional and regulatory mechanisms of the *Synechocystis* genome that was not previously possible.

I. Introduction

Progress in DNA sequencing technologies has allowed us to rapidly accumulate much information about genomes. Since the publication of the entire nucleotide sequence of the genome of *Haemophilus influenzae* Rd in 1995 (Fleischmann et al., 1995), complete genome sequences of a number of organisms, including archae, eubacteria and lower eukaryotes, have been reported (Pallen, 1999). Moreover, sequencing of the genomes of fifty or more bacterial species as well as those of higher eukaryotes is currently in progress. The information on structures, organization, and composition of genes in various organisms is contained in the genome databases and can be easily accessed via internet (Genome research links; http://www.kazusa.or.jp/en/plant/database.html). A disadvantage of this approach, however, is that the knowledge obtained in this way tends to be limited to the structural features

*Author for correspondence, email: tabata@kazusa.or.jp

Eva-Mari Aro and Bertil Andersson (eds): Regulation of Photosynthesis, pp. 557–561.

of genes and genomes. Gene function can only be deduced by structural similarity to genes with known functions or by computer prediction. Thus, despite the large amounts of structural information available, conventional genetic and biochemical methods are still necessary for the determination of gene function. One such method is the overproduction of proteins in *Escherichia coli* cells followed by in vitro characterization (Fernandez-Gonzalez, 1997; Hughes et al., 1997; Okada et al., 1997; Masamoto et al., 1998). More recently, however, the sequencing of entire genomes has permitted development of more comprehensive approaches such as a large scale gene disruption, transcriptome and proteome analyses. These methods target a large number of genes at a time and demonstrate the useful combination of conventional methods with genome information (Dujon, 1998).

Synechocystis sp. PCC6803 is an unicellular cyanobacterium which has been used as a model organism for the study of photosynthetic processes because of its natural transformation characteristics, and because it is capable of photoheterotrophic growth in the presence of an appropriate carbon source (Williams, 1988). The complete genome structure of *Synechocystis* sp. PCC6803 was determined in 1996 (Kaneko et al., 1996). The genome is circular and 3,573,470 bp long, and 3,168 putative protein-coding genes could be identified by similarity searches and computer prediction using a GeneMark program (Borodovsky and McIninch, 1993). Of the putative genes, 1,404 (45%) could be assigned with known biological functions on the basis of sequence similarity, including 138 genes for photosynthesis and respiration. However, the functions of the remaining 1764 genes (55%) remain to be ascertained. Furthermore, the regulatory network controlling the genome is still not understood. In this chapter, we will describe databases and tools developed for the comprehensive analysis of gene function and regulation in the *Synechocystis* genome. These resources used in combination should allow the rapid and comprehensive identification of the functions of all the *Synechocystis* genes, thus permitting a level of understanding of the functional and regulatory mechanisms of the *Synechocystis* genome that was not previously possible.

Abbreviations: bp – base pairs; HDF – high density gene filter; WWW – the World Wide Web

II. CyanoBase and CyanoMutants—Genome Information Databases

CyanoBase (http://www.kazusa.or.jp/cyano) is a WWW database containing information on the genes and genome of *Synechocystis* sp. PCC6803 (Nakamura et al., 1998). CyanoBase provides an interface for access to annotation pages for each of the 3,167 putative protein-coding genes via three different menus: graphical physical maps of the genome, a gene category list, and a keyword search menu. All menus are accessible from the main page (Fig. 1). The circular and enlarged linearized physical maps show the relative positions of each gene in the genome. The gene category list and the classification table of functional categories are linked to the annotation page of each gene, which is useful for finding genes with related functions. From the gene annotation pages, the following information for each gene assigned with a function can be obtained: 1) the nucleotide sequence of the gene as well as the sequences of the upstream and downstream regions, 2) a translated amino acid sequence of the putative gene product, and 3) the result of the similarity search against the nr peptide sequence database. Two types of similarity search engines, BLASTN for the nucleotide sequences of all the putative protein-coding genes as well as the intergenic regions, and BLASTP for the translated amino acid sequences, are also provided in the database to allow browsing of the genome. CyanoBase, therefore, can be used as a tool to extract and utilize as much information as possible from the *Synechocystis* genome for the study of the genetic system.

The natural transformation ability of *Synechocystis* has permitted a variety of mutants, mostly of photosynthetic genes, to be generated by insertion of drug resistant cassettes into the target genes. The availability of the nucleotide sequences of all the gene candidates has made this process more feasible, and gene disruption projects on a genome-wide scale are on their way. The accumulated information on the mutants has been compiled in a web database named CyanoMutants (http://www.kazusa.or.jp/cyano/mutants/; Nakamura et al., 1999). The information on each individual mutant, specifically the name of the target gene, the method of mutant production, the phenotype of the mutant if observed, the proposed gene name, and the address of correspondence, is posted on the Mutant List page, and each mutant page is linked to the corresponding

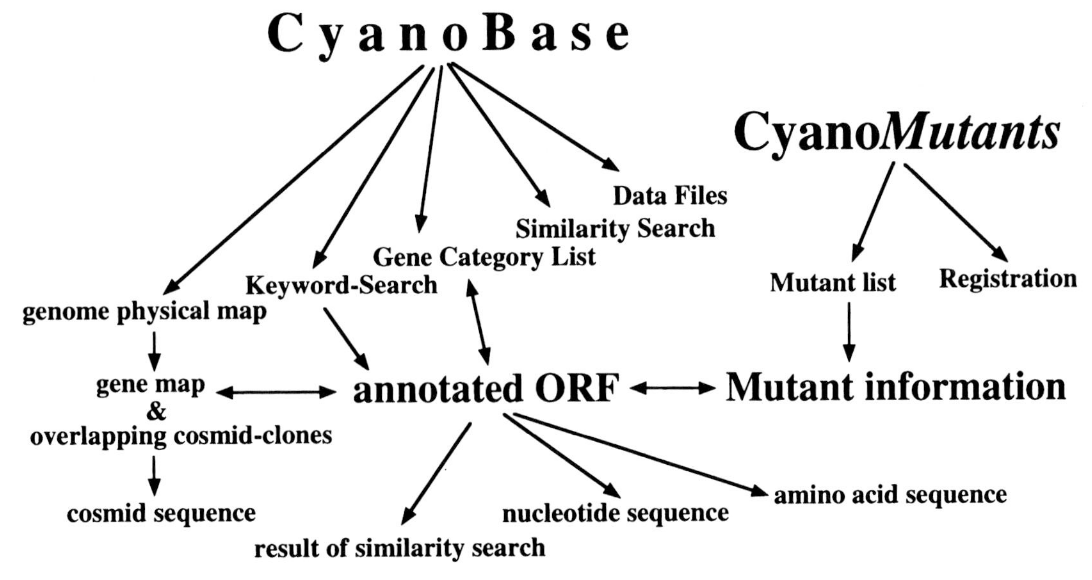

Fig. 1. Menus and linkage in CyanoBase and CyanoMutants

gene page in CyanoBase. By 1 October 1999, 240 mutants on 196 genes had been registered in the database. Thus, CyanoMutants in combination with CyanoBase provides an essential tool for the elucidation of the functions of all the genes in the *Synechocystis* genome.

III. Genome-wide Monitoring of Gene Expression by Proteome and Transcriptome Analyses

One of the best clues for determination of the biological role(s) of microbial genes of unknown function is to know, when the genes are expressed and where the gene products are required. To this end, two approaches, namely proteome and transcriptome analysis, can be applied if, as is the case for the *Synechocystis* genome, the structures of all the gene candidates are known.

A. Proteome Analysis

A conventional way to examine composition of soluble cellular proteins is to separate the total cellular protein extract by two-dimensional (2-D) gel electrophoresis on the basis of both isoelectric point and molecular mass. If the nucleotide sequence of the genome is known, each protein spot on the gel can be directly connected to the corresponding gene in the genome by N-terminal sequencing of the protein recovered from the spot, followed by comparison with the translated amino acid sequences in the database.

In *Synechocystis*, 227 major protein spots representing 143 independent protein species from the total cellular extract and thylakoid membrane fraction were partially sequenced and the corresponding genes were identified (Sazuka et al., 1999). Of these proteins, 109 appeared as single spots and 34 produced multiple spots due to post-translational modification. Among these proteins, there were 7 subunits of photosystems, 7 phycobilisome-related polypeptides, 4 ribosomal proteins, 3 chaperonins, and 55 proteins of unknown function. The 2-D gel images and the gene map showing locations of the identified genes are provided in Cyano2Dbase (http://www.kazusa.or.jp/cyano/cyano2D/index.html). Such proteome analysis can thus accelerate not only the identification of the genes induced under specific conditions(Sazuka and Ohara, 1997) but also permit identification of genes in the regulatory network composed of direct and indirect gene interaction.

B. Transcriptome Analysis

Knowing the nucleotide sequences of all putative genes in a genome allows the level of transcription of these genes under various conditions to be determined by transcriptome analysis. Here, DNA segments representing all of the putative genes in the genome

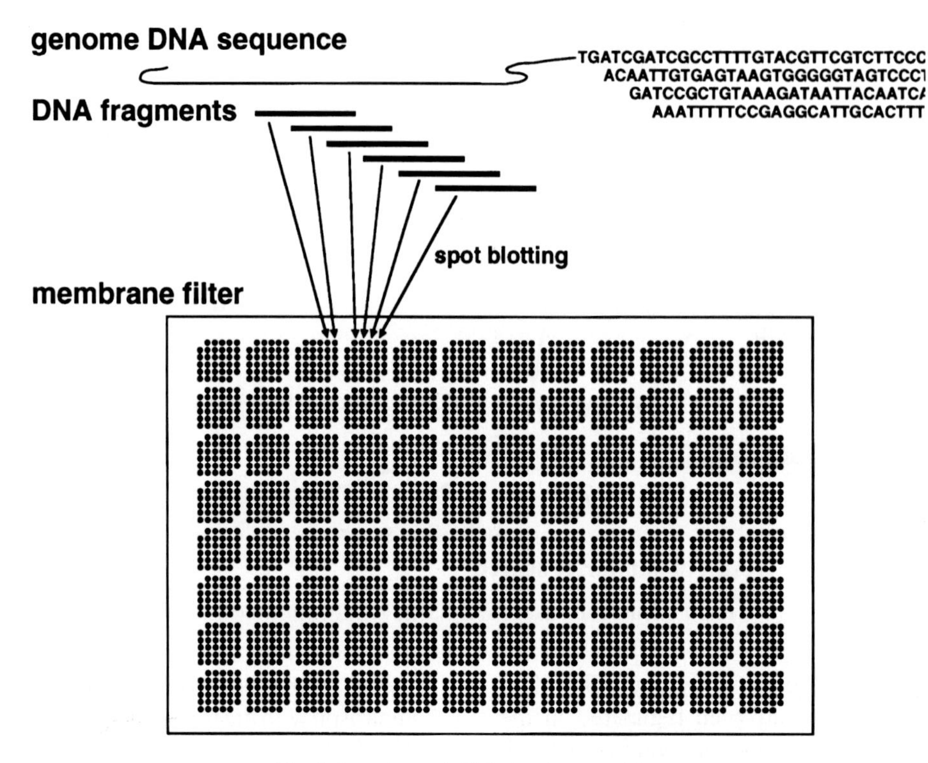

Fig. 2. Synechocystis high density gene filters

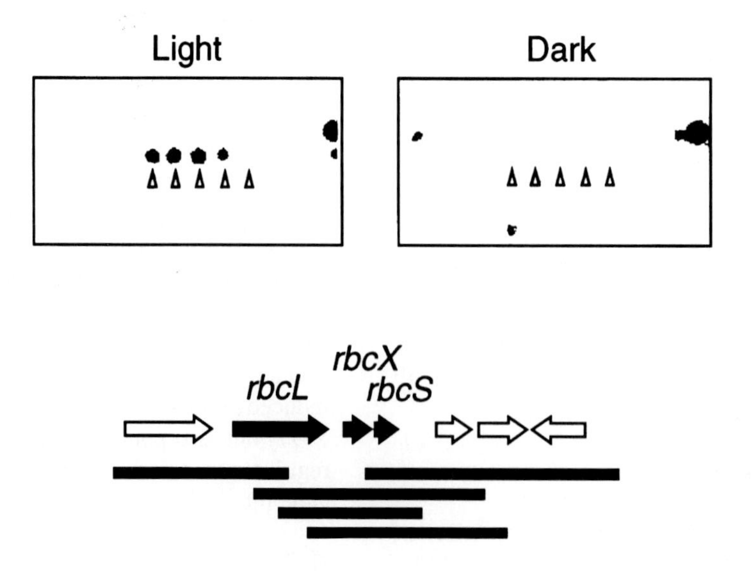

Fig. 3. Gene transcription of *Synechocystis* under different light conditions. HDFs were hybridized with the radiolabeled probes derived from RNAs from cells under light and dark conditions. Arrow heads indicate the spots corresponding to *rbcL-rbcX-rbcS* genes. A gene map of this region is also shown. Black arrows indicate *rbc*-genes. Gray horizontal bars indicate the positions of the DNA fragments spotted on the membrane.

are fixed onto either solid supports in high densities (microarray) (Brown and Botstein, 1999, Kehoe et al., 1999), or onto glass plates or nylon membranes in lower densities (macroarray). Fluorescent-labeled or radioisotopically labeled probes derived from RNAs extracted from the cells under various conditions or with mutations are then hybridized to the probe arrays.

We adapted this approach to our cyanobacterial research and developed *Synechocystis* high density gene filters (HDF) (T. Kaneko, unpublished). The DNA probes covering the entire *Synechocystis* genome consisted of 5,177 DNA fragments of 1-1.5 kb long which were amplified by PCR from either genomic clones or directly from genomic DNA with minimal overlaps (Fig. 2). The probes were spotted onto two sheets of nylon membranes of microtiter dish size in such a way that the alignment reflects the order in the genome. The validity of this tool was tested in a pilot experiment with labeled RNAs prepared from cells grown in two different light conditions, in the presence of light and after transition to dark. Figure 3 shows the *rbcL-rbcX-rbcS* gene region on the two sets of membranes. A significant decrease at the level of transcription of these genes after transition to dark can be seen. This observation has been previously reported using a conventional method (Mohamed and Jansson, 1991), thus our approach is validated. This system, particularly in combination with gene disruption experiments, has the potential to rapidly identify genes participating in gene regulation cascades.

References

Borodovsky M and McIninch JD (1993) GENEMARK: Parallel gene recognition for both DNA strands. Comput Chem 17: 123–133

Brown PO and Botstein D (1999) Exploring the new world of the genome with DNA microarrays. Nat Genet 21: 33–37

Dujon B (1998) European Functional Analysis Network (EUROFAN) and the functional analysis of the *Saccharomyces cerevisiae* genome. Electrophoresis 19: 617–624

Fernandez-Gonzalez B, Sandmann G and Vioque A (1997) A new type of asymmetrically acting beta-carotene ketolase is required for the synthesis of echinenone in the cyanobacterium *Synechocystis* sp. PCC 6803. J Biol Chem 272: 9728–9733

Fleischmann RD, Adams MD, White O, Clayton RA, Kirkness EF, Kerlavage AR, Bult CJ, Tomb JF, Dougherty BA, Merrick JM, McKenney K, Sutton G, FitzHugh W, Fields C, Gocayne JD, Scott J, Shirley R, Liu L, Glodek A, Kelly JM, Weidman JF, Phillips CA, Spriggs T, Hedblom E, Cotton MD, Utterback TR, Hanna MC, Nguyen DT, Saudek DM, Brandon RC, Fine LD, Fuhrmann JL, Geoghagen NSM, Gnehm CL, McDonald LA, Small KV, Frazer CM, Smith HO and Venter JC (1995) Whole-genome random sequencing and assembly of *Haemophilus influenzae* Rd. Science 269: 496–512

Hughes J, Lamparter T, Mittmann F, Hartmann E, Gärtner W, Wilde A and Börner T (1997) A prokaryotic phytochrome Nature 386: 663

Kaneko T, Sato S, Kotani H, Tanaka A, Asamizu E, Nakamura Y, Miyajima N, Hirosawa M, Sugiura M, Sasamoto S, Kimura T, Hosouchi T, Matsuno A, Muraki A, Nakazaki N, Naruo K, Okumura S, Shimpo S, Takeuchi C, Wada T, Watanabe A, Yamada M, Yasuda M and Tabata S (1996) Sequence analysis of the genome of the unicellular cyanobacterium *Synechocystis* sp. strain PCC6803. II. Sequence determination of the entire genome and assignment of potential protein-coding regions. DNA Res 3: 109–136

Kehoe DM, Villand P and Sommerville S (1999) DNA microarrays for studies of higher plants and other photosynthetic organisms. Trends in Plant Science 4: 38–41

Masamoto K, Misawa N, Kaneko T, Kikuno R and Toh H (1998) Beta-carotene hydroxylase gene from the cyanobacterium *Synechocystis* sp. PCC6803. Plant Cell Physiol 39: 560–564

Mohamed A and Jansson C (1991) Photosynthetic electron transport controls degradation but not production of *psbA* transcripts in the cyanobacterium *Synechocystis* 6803. Plant Mol Biol 16: 891–897

Nakamura Y, Kaneko T, Hirosawa M, Miyajima N and Tabata S (1998) CyanoBase, a www database containing the complete nucleotide sequence of the genome of *Synechocystis* sp. strain PCC6803. Nucleic Acids Res 26: 63–67

Nakamura Y, Kaneko T, Miyajima N and Tabata S (1999) Extension of CyanoBase. CyanoMutants: Repository of mutant information on *Synechocystis* sp. strain PCC6803. Nucleic Acids Res 27: 66–68

Okada K, Minehira M, Zhu X, Suzuki K, Nakagawa T, Matsuda H and Kawamukai M (1997) The *ispB* gene encoding octaprenyl diphosphate synthase is essential for growth of *Escherichia coli*. J Bacteriol 179: 3058–3060

Pallen MJ (1999) Microbial genomes. Mol Microbiol 32: 907–912

Sazuka T and Ohara O (1997) Towards a proteome project of cyanobacterium *Synechocystis* sp. strain PCC6803: Linking 130 protein spots with their respective genes. Electrophoresis 18: 1252–1528

Sazuka T, Yamaguchi M and Ohara O (1999) Cyano2Dbase updated: Linkage of 234 protein spots to corresponding genes through N-terminal microsequencing Electrophoresis 20: 2160–2171

Williams JGK (1988) Construction of specific mutations in Photosystem II photosynthetic reaction center by genetic engineering methods in *Synechocystis* 6803. Methods Enzymol 167: 766–778

Chapter 32

Arabidopsis Genetics and Functional Genomics in the Post-Genome Era

Wolf-Rüdiger Scheible*, Todd A. Richmond, Iain W. Wilson and Chris R. Somerville
Carnegie Institution of Washington, Department of Plant Biology, 260 Panama Street, Stanford, CA 94305-1297, U.S.A.

Summary

Expressed-sequence tag and genome sequencing projects for *Arabidopsis* and other plants have yielded unprecedented amounts of new information about plant genes. The pending completion of the genome sequences for *Arabidopsis* and rice will provide comprehensive knowledge of the coding capacity of higher plant genomes. It is now possible to envision assigning the specific biological functions to all genes in an angiosperm. A variety of new experimental approaches, which are enabled by large amounts of sequence information, will be used to this effect. In particular, positional cloning of *Arabidopsis* genes corresponding to mutations has become a standard technique. In addition, highly efficient transformation protocols for *Arabidopsis*

*Author for correspondence, email: sheible@andrew2.stanford.edu

Eva-Mari Aro and Bertil Andersson (eds): Regulation of Photosynthesis, pp. 563–592.

make it possible to generate very large transposon and T-DNA insertion collections. Mutants for any gene can be found in these collections with minimal effort using powerful polymerase chian reaction (PCR) screening procedures or by database searches of sequenced insertion sites. New or improved methods for targeted gene disruption and large-scale gene-silencing will also be useful, particularly for understanding the function of redundant genes. Large-scale surveys of transcript, protein and metabolite levels, and collection of the results in public databases will greatly extend our knowledge of gene-expression, gene-function and regulatory networks, and will facilitate the formulation of an integrated view of the information content and processing that regulates all aspects of plant growth and development.

I. Introduction

The determination of the complete nucleotide sequence of a free-living organism, the bacterium *Haemophilus influenza*, (Fleischmann et al., 1995) signaled the beginning of the genome era in 1995. Since then, the nucleotide sequences of at least twenty additional archaic, bacterial and eucaryotic genomes have been finished and published (Table 1, #1), including the photosynthetic cyanobacterium *Synechocystis* PCC 6803. Dozens more, including the photosynthetic bacterium *Rhodobacter capsulatum*, are currently being sequenced (Saier, 1998). At the end of 1998 the 97 Mb genome of the nematode *Caenorhabditis elegans* became available (The *C. elegans* Sequencing Consortium, 1998). In September 1999 Celera Genomics announced that the sequencing phase for the ~165 Mb genome of the fruit fly, *Drosophila melanogaster* has been finished (Table 1, #2), and it is expected that the next completed genome of comparable size will be the one of the flowering plant, *Arabidopsis thaliana*.

During the past decade, *Arabidopsis thaliana*, a member of the mustard family closely related to agricultural crops like canola, cabbage, turnip and cauliflower, has emerged as one of the most widely used model organisms for the study of the biology of higher plants (Meyerowitz and Somerville, 1994; Meinke et al, 1998). The technical advantages of *Arabidopsis* include a short generation time (about two months), the ability to grow large populations in limited space, the large number of progeny, the ability to self-fertilize or outcross, and the more recently developed techniques for tissue culture, mutagenesis, and transformation. In addition, a large, steadily growing number of isolated and characterized mutants, and a large collection of established molecular markers and genetic maps have become available (Meinke et al., 1998). *Arabidopsis* is also a useful model-organism for physiological and biochemical studies, as indicated by the growing number of publications in this area.

Arabidopsis was chosen for genome sequencing primarily for its highly compact genome (Pruitt and Meyerowitz, 1986). Early *Arabidopsis* sequencing efforts concentrated on expressed sequence tags

Abbreviations: AGI – *Arabidopsis* genome initiative; BAC – bacterial artificial chromosome; CAPS – cleaved amplified polymorphic sequence; EST – expressed sequence tag; FRET – fluorescence resonance energy transfer; Mb – megabase, million base pairs; MS – mass spectrometry; NCBI – National Center for Biotechnology Information; PCR – polymerase chain reaction; SNP – single nucleotide polymorphism; SSLP – single sequence length polymorphism; YAC – yeast artificial chromosome

Table 1. Cited Internet addresses

1. http://www.ncbi.nlm.nih.gov/Entrez/Genome/org.html
2. http://www.celera.com
3. http://www.tigr.org/tdb/agi
4. http://www.tigr.org/tdb/ogi/index.html
5. http://www.ncbi.nlm.nih.gov/dbEST/dbEST_summary.html
6. http://macgrant.agron.iastate.edu/soybeanest.html
7. http://www.arabidopsis.org/agi.html
8. http://www.arabidopsis.org
9. http://www.arabidopsis.org/cgi-bin/maps/Schrom
10. http://websvr.mips.biochem.mpg.de
11. http://www.nsf.gov/bio/pubs/awards/genome99.htm
12. ftp://ftp-igbmc.u-strasbg.fr/pub/ClustalX/
13. http://taxonomy.zoology.gla.ac.uk/rod/treeview.html
14. http://genome-www3.stanford.edu/cgi-bin/AtDB/Seqtable
15. http://www.tigr.org/tdb/at/atgenome/Ler.html
16. http://aims.cps.msu.edu/aims/
17. http://nasc.nott.ac.uk/
18. http://genetics.nature.com/chips_interstitial.html
19. http://www.incyte.com/science/index.html
20. http://www.monsanto.com/Arabidopsis
21. http://afgc.stanford.edu/
22. http://www.ciphergen.com
23. http://www.expasy.ch/ch2d/2d-index.html
24. http://www.biotech.wisc.edu/Arabidopsis
25. http://www.jic.bbsrc.ac.uk/Sainsbury-lab

(ESTs), generating over 45,000 sequences to date. In the past two years, efforts have been largely directed at genomic sequencing. One of the most important results of the large-scale EST- and genomic sequencing efforts for *Arabidopsis* and other plants is the identification of thousands of previously unknown genes. Database searches can usually suggest a putative biochemical function (e.g. protein-kinase, cytochrome P450) or cellular function (e.g. transporter, receptor) for about half of these genes, based on nucleotide or amino acid sequence similarity and common protein motifs. The elucidation of the biological functions of these new or insufficiently characterized genes—determining when, where and under which conditions these genes and their gene products are important for the plant—will represent a major step towards an integrated understanding of plant growth and development.

In this chapter we first outline the status, important issues and results of sequencing efforts for *Arabidopsis* and, where appropriate, for other plants. Then we describe how the new wealth of sequence information is currently revolutionizing classical genetics, for example by accelerating map-based cloning, as well as enabling the creation and use of the new tools of functional genomics and reverse genetics to rapidly determine the biological functions of genes.

II. *Arabidopsis* Expressed Sequence Tags and Genome Sequencing Projects

A. Expressed Sequence Tags

Expressed sequence tags (ESTs) often are single pass partial 5′-sequences of clones randomly chosen from cDNA libraries (Höfte et al., 1993; Newman et al., 1994). ESTs are typically 300–500 bp long and often contain sequence ambiguities, which can confound sequence comparisons. However, high-throughput random sequencing of anonymous cDNA clones is an attractive, rapid and cost-efficient way to obtain a working knowledge of the repertoire of expressed genes from an organism. The potential reading frames of each EST are compared to sequences in public databases, resulting in assignment of putative function to a significant proportion (45%) of the ESTs (Rounsley et al., 1996). ESTs without assigned function may represent previously unknown genes or the untranslated regions of mRNA, which are usually more divergent among different species than protein coding regions.

Because clones are randomly sampled, highly expressed transcripts (e.g. the small subunits of ribulose 1,5-bisphosphate carboxylase/oxygenase or light harvesting complex subunits) are sequenced multiple times in the course of a large-scale EST sequencing project, resulting in considerable redundancy and bias toward certain genes. This problem can partially be overcome by using uniform abundance, normalized cDNA libraries (Sankhavaram et al., 1991; Kohchi et al., 1995) or subtractive hybridization procedures (Fedoroff et al., 1999). The number of represented genes can also be increased (and the redundancy simultaneously decreased) by generating mixed-tissue libraries. For example, the EST project from Michigan State University, which produced most of the publicly available *Arabidopsis* ESTs in Genbank, used a cDNA library prepared from equal portions of poly-A mRNA from etiolated seedlings, roots, leaves and flowering inflorescences (Newman et al., 1994). Large-scale *Arabidopsis* EST sequencing of tissue-specific or developmental-specific libraries, as was practiced by a French consortium (Höfte et al., 1993), offers the advantage that the origin of the cDNA is known and also gives a relative expression pattern for abundant mRNAs in a given specific library. However, the significance of this information is likely to be supplanted by the use of DNA microarrays in the near future.

Once EST sequence information has been collected redundancy in the data set can be reduced computationally (Rounsley et al. 1996, Cooke et al., 1997). By using a computer algorithm (Sutton et al.,1995; Table 1, #3 for details) Rounsley et al. (1996) were able to group overlapping ESTs and known mRNAs representing the same gene into virtual transcripts (Table 1, #3). The comparison of redundant overlapping sequences can resolve many sequence ambiguities inherent in the ESTs. In addition, these virtual transcripts are usually longer than single ESTs, resulting in an increase in information content. However a potential danger of this approach is that ESTs from very similar but distinct genes can be assembled together, resulting in a loss of information and an underestimate of gene number. Double-stranded EST sequencing can help to differentiate similar ESTs, but many of the plant EST sequencing projects sequence only one end of random clones. Currently all singleton and assembled *Arabidopsis* ESTs, and mRNAs together form almost 16,000 virtual transcripts (Table 1, #3), representing 33–56% of the estimated 20,000–25,000 *Arabidopsis* genes (Rounsley et al., 1996; Bevan et al., 1998; Lin et al., 1999; Mayer et al., 1999). One of the major disadvantages of EST sequencing is the inability to complete the full expression profile of an organism. Even when normalized libraries from multiple tissues, developmental stages and experimental treatments are utilized, rare mRNAs are often not represented, and others are difficult or impossible to clone. It has also been observed that ESTs from areas of reduced genetic recombination in the genome, like pericentromeric regions, are rare (Parnell et al., 1999). Hence, some degree of incomplete gene coverage will remain when using large-scale EST-sequencing for the discovery of genes.

Since large-scale EST sequencing is fast and cost-effective more projects have been initiated for a variety of additional plant species, including rice (Yamamoto and Sasaki, 1997; Table 1, #4), tomato, soybean and maize. Currently more than 250,000 plant EST sequences are represented in the public NCBI (the National Center for Biotechnology Information) EST database (dbEST, Table 1, #5, Table 2) and the number and variety of represented plant species is expected to grow rapidly during the next years. The soybean genome project alone plans to sequence 300,000 ESTs in the next three years (Table 1, #6).

We expect that because of the high efficiency afforded by the new capillary fluorescence sequencers, large numbers of EST sequences will be available for all higher plants of economic significance within the next few years. Because orthologous genes from higher plants usually exhibit a significant degree of sequence similarity, the availability of large numbers of EST sequences will greatly facilitate the annotation of *Arabidopsis* and other plant genome sequences. In addition, once the *Arabidopsis* genome sequence is complete, it will become possible to unambiguously determine the extent to which other angiosperms contain genes (indicated by ESTs) that are not found in *Arabidopsis* or which show abnormally high degrees of sequence divergence. This should be helpful in understanding the genetic basis for biological diversity in higher plants.

B. Sequencing of the Arabidopsis *Nuclear Genome*

Sequencing a complete plant nuclear genome is orders of magnitude more cost- and labor-intensive than large-scale EST sequencing and therefore only plant genomes of relatively small size will be completely sequenced in the foreseeable future. Indeed, rice is the only plant other than *Arabidopsis* for which there is currently a public commitment to obtain a complete genome sequence. In addition, several private companies have recently announced that they will attempt to complete a full genome sequence for rice sometime in the year 2001.

In comparison to EST sequencing, whole-genome sequencing offers various important advantages. It gives complete information regarding the structure, order and position of genes on the chromosomes, which is important for activities like positional cloning (Section III.A). Unlike large-scale EST sequencing, whole-genome sequencing also reveals the complete set of genes of an organism along with their promoter regions and all the intergenic sequence. This resource is profoundly changing biological research in plants.

1. The Arabidopsis *Genome Initiative*

Arabidopsis thaliana was chosen for genome sequencing primarily due to its highly compact genome (early estimates were 70–100 Mb) and the availability of extensive physical maps for its five chromosomes based on YACs, BACs and P1 clones (Choi et al., 1995; Goodman et al., 1995; Liu et al. 1995a; Schmidt et al., 1995; Zachgo et al., 1996, Kotani et al., 1997; Schmidt et al., 1997; Camilleri et al., 1998; Mozo et al., 1998, 1999; Sato et al., 1998). The small amount of repetitive intergenic DNA further facilitates sequencing and cloning of genes (Meyerowitz, 1992). The Arabidopsis Genome Initiative (AGI) (Table 1, #7; Bevan et al., 1997), founded in 1996, is a coordinated collaboration of six research consortia in Japan, Europe and the United States. The various AGI labs have agreed to systematically sequence different defined regions of the *Arabidopsis* genome on the basis of minimal overlapping bacterial artificial chromosomes (BACs) or bacteriophage P1 clones, resulting in large stretches of continuous sequence (Bevan, 1997; also Kaneko et al., 1999 and references therein). This task includes establishing tiling paths by end-sequencing and/or fingerprinting BAC clones, double-stranded sequencing of these clones by a 'shotgun' approach, assembly of the individual sequences into a single contig, and the annotation of the sequence features. The final annotated sequence is then deposited into public

Table 2. Top fifteen plant species represented in the GenBank EST-subdivision (dbEST) as of 12/17/1999

Species	number of ESTs
Lycopersicon esculentum (tomato)	50,303
Oryza sativa (rice)	47,449
Arabidopsis thaliana (thale cress)	45,757
Zea mays (maize)	44,842
Glycine max (soybean)	33,277
Gossypium hirsutum (upland cotton)	9,369
Pinus taeda (loblolly pine)	8,405
Populus tremula x *Populus tremuloides*	4,809
Medicago truncatula (barrel medic)	3,613
Mesembryanthemum crystallinum (common ice plant)	3,377
Brassica napus (oilseed rape)	1,693
Citrus unshiu (mandarin orange)	1,251
Sorghum halepense (Johnson grass)	1,210
Brassica campestris (field mustard)	963
Brassica rapa subsp. *pekinensis* (Chinese cabbage)	934

databases like GenBank, EMBL or the DNA Database of Japan (DDBJ). However, access to the data is not limited to the final, annotated sequence. Some of the AGI labs release preliminary data from the shotgun-sequencing phase into the HTG subdivision of GenBank. Preliminary and final data, updated daily, are available as downloadable files, or in searchable form via the public BLAST servers of the AGI labs (Table 1, #7), NCBI and the Arabidopsis Information Resource (TAIR, Table 1, #8) at the Carnegie Institution Department of Plant Biology at Stanford University. The continuous data deposition currently progresses at a rate of several megabases per month. As of December 1999, the AGI has sequenced nearly 85% of the roughly 130 Mb *Arabidopsis* (ecotype Columbia) genome (Table 1, #9). Two of the five chromosomes (II and IV) are almost fully sequenced (Lin et al., 1999; Mayer et al., 1999), and the whole genome is now expected to be complete by the end of the year 2000 (Meinke et al., 1998).

2. Sequence Annotation

The annotation of DNA sequence, is one of the most problematic steps in knowledge acquisition for an anonymous sequence, especially for plants (Rouzé et al., 1999). Structural annotation includes determining sites and regions involved in gene and genome functionality (splice sites, introns, exons, untranslated regions, transcriptional and translational start and control regions, promoters, matrix attachment regions, etc.). In contrast, functional annotation assigns a basic cellular or biochemical function to a DNA sequence or its predicted protein product, based on DNA or protein similarity, multiple alignments, pattern and domain searches. Clearly, the quality of structural annotation will greatly affect the quality of functional annotation; proper annotation of genome data is therefore crucial for genomics.

Presently, annotation programs for the *Arabidopsis* genome are not ideal, and therefore each of the different AGI-labs uses a variety of programs (e.g. NetPlantGene, GENSCAN or GRAIL) for their annotation work (Rouzé et al., 1999). Structural annotation of genomic sequence can be greatly aided by the existence of a large set of EST sequences. The correct identification of splice sites in plant genes has always been difficult, since the exact splicing signals remain obscure. By comparison of the genomic and corresponding EST sequence, the precise location of splice-sites, the size of exons and the transcription initiation and termination sites can be determined (Rounsley et al., 1998). In this respect EST and genomic sequencing are complementary. The completion of the *Arabidopsis* genome should facilitate the annotation of other plant genomes, as coding regions tend to be very similar across different species even if promoters, terminators and other noncoding regions diverge. The comparison of orthologous sequences from various plant species can be as useful for defining coding regions and splice sites as ESTs. Thus, the rapid accumulation of EST sequences from many plant species will facilitate the annotation of plant genome sequences in general.

3. Features of the *Arabidopsis* Genome

In the first paper of its kind, Bevan et al. (1998) described features of a 1.9 Mb continuous stretch from *Arabidopsis* chromosome IV. More recently, two other landmark papers were published (Lin et al., 1999; Mayer et al., 1999). They report the sequences and analyses of two (II and IV) of the five *Arabidopsis* chromosomes.

Chromosome II contains 4,037 genes in just over 19.6 Mb and 3,744 protein-coding genes were detected on chromosome IV in 17.4 Mb. The total of 37 Mb is more than a quarter of the estimated size of the *Arabidopsis* genome. Since the gene-rich part of the genome is thought to be around 120 Mb, the entire genome should code for up to 25,000 proteins (Meyerowitz, 1999), which agrees well with previous estimates (Gibson and Somerville, 1993; Meyerowitz, 1994; Bevan et al., 1998). Based on the results from chromosomes II and IV, the average gene-density in the *Arabidopsis* nuclear genome is one gene per 4.5 kb. This is high compared to the estimated gene-density of one gene every 15, 50–100 or 500 kb in the nuclear genomes of rice, maize or wheat, respectively (Goff, 1999) or one gene every 6.4 kb in the *Arabidopsis* 367 kb mitochondrial genome (Klein et al., 1994; Unseld et al., 1997; Sato S, Nakamura Y et al., direct submission to GenBank, 9/9/1999). However the local gene-density in the *Arabidopsis* nuclear genome can vary enormously along a chromosome, being as low as one gene per 150 kb in the pericentromeric heterochromatin of chromosome IV. The average gene has about 2.1 kb coding region and an average number of five small (250–300 bp) exons. The GC content of exons (~44%) is considerably higher than in the, on average, even smaller (~180 bp) introns (~33%) or intergenic DNA (~32%), resulting in overall CG-content of 36% in the genome. About 20% of the genes consist of a single exon, and one of the largest genes identified (T20F21.17) contains 52 predicted exons. Genes don't appear to overlap on the same strand, but putative promoter regions are often smaller than 200 bp, indicating that regulatory sequences may be frequently in the coding regions of adjacent genes.

Database comparisons show that an actual or potential cellular role can be predicted for approximately 52–60% of the identified genes on chromosomes II and IV, based on their significant similarity to known genes from *Arabidopsis* and other organisms. Another ~25% share homology to predicted proteins of unknown function in other organisms. The residual fraction may represent plant-specific genes and spurious gene predictions. It is interesting to note that only a third of the predicted genes on chromosomes II and IV match at least one of the available expressed sequence tags with >90% similarity. Surprisingly, genes have been identified that were not believed to exist in *Arabidopsis*, such as genes with high similarity to hydroxynitrile lyase or tropinone reductase. Similarly, genomic sequencing has revealed homologs of genes not known to exist in plants, including genes similar to those encoding mammalian neurotransmitter receptors or genes related to insect virus inhibitors of apoptosis (ESSA, 1999). Hence, genomic sequencing can lead to the discovery of unknown metabolic or signal transduction pathways and allows prediction of the processes an organism has the ability to carry out.

Based on functional catalogs established for *Escherichia coli* and yeast, *Arabidopsis* genes with predicted or known function can be classified into 15 classes of putative cellular roles (for an updated version of the *Arabidopsis* functional catalog Table 1 #10). The largest class (~15–20%) contains genes involved in primary and secondary metabolism, which reflects the complex photoautotrophic metabolism of plants. Two other large classes, genes involved in transcription (~10%) and signal transduction (~8%) highlight the importance of information processing in complex multicellular organisms. Notably, only a small percentage of these latter classes of genes are represented as ESTs, reflecting their low average level of expression.

Other interesting features of the *Arabidopsis* genome include gene and chromosomal duplications. On chromosome II more than 60% (2,542 out of 4,037) of the predicted proteins have a significant match (FASTA, score: $p < 10^{-10}$) (Pearson and Lipman, 1988) to another *Arabidopsis thaliana* protein and more than 50% (2,138) match at least one other protein encoded on the same chromosome. Tandem gene duplications or even clusters of multigene family members are frequently observed (e.g. receptor-kinase-like proteins, cytochromes P-450, glutathion S-transferases, glutaredoxins and LRR-disease-resistance genes), indicating that simple duplication and subsequent divergence are a common mechanism for expanding gene families in *Arabidopsis*. Besides these intrachromosomal gene duplications, several interchromosomal duplications were detected. For example, a ~700 kb segment from the bottom arm of chromosome II, containing 170 genes, is duplicated on chromosome I with an inversion in the middle, and an even larger duplication (4.6 Mb), containing several rearrangements, is found between chromosome II and IV. The insertion of a ~270 kb fragment with near identity to the mitochondrial genome in the centromere of chromosome II reveals an unexpectedly large and recent organellar-to-nuclear gene-transfer event. These examples illustrate the dynamic nature and constant motion of the *Arabidopsis* genome.

Intergenic repetitive DNA-elements, like transposons, retroelements or ribosomal DNA, constitute only a relatively small part (15–20%) of the *Arabidopsis* genome. In the gene-rich arms of the two sequenced *Arabidopsis* chromosomes transposons and retroelements are relatively rare and occur usually just as single or double repeats. However they are much more frequent and highly multimeric close to the centromeres of the *Arabidopsis* chromosomes (Meyerowitz, 1992; Thompson et al., 1996a,b; Round et al., 1997; Heslop Harrison et al., 1999; Lin et al., 1999). Almost megabase-sized clusters of ribosomal DNA-repeats on the other hand are found at the tip of the short arms and comprise the nucleolar organizer regions of chromosomes II and IV. This situation is radically different in cereals (Section II.C), where repetitive elements are spread all over and account for large majorities of the genomes.

C. The Rice Genome and Comparative Genetics

The *Arabidopsis* genome will not be the only sequenced plant genome in the future. Genome sequencing for rice, as a representative of the cereals, has already begun (Sasaki, 1998). The experience gained during the sequencing of the *Arabidopsis* genome, new automated capillary DNA sequencers and whole genome shotgun-cloning approaches will reduce the labor, cost and time required for the successful execution. Choosing rice as model for cereal genomics is obvious for at least two reasons. First, about 50% of the world's population derives a significant proportion of their caloric intake from rice consumption and the demand for rice is expected to rise with the predicted doubling of the world-population over the next fifty years (Sasaki, 1998; Goff, 1999). Second, rice has the most compact genome (~450 Mb) among the cereals, containing about 3.5 times as much DNA as *Arabidopsis*, but only about 10–15% and 3% as much DNA as maize and wheat, respectively. The difference in genome size between the cereal genomes is mostly due to ploidy and amplification of interspersed repetitive sequences (Flavell et al., 1974; Hake and Walbot, 1980; San Miguel et al., 1996) but hardly reflected in chromosome number or gene content. The proposed number of genes (25,000–30,000) in the cereal genomes (Gale and Devos, 1998b; Goff 1999) is not much higher than in *Arabidopsis*, and may prove to be essentially identical, because it is unlikely that there has been adequate time for the evolution of many entirely novel genes since the ancestors of *Arabidopsis* and the cereals diverged less than 150 million years ago.

A striking feature of cereal genomes (e.g. rice, wheat, maize, sorghum and others) is their high degree of synteny or colinearity (Ahn et al., 1993; Bennetzen and Freeling, 1997; Gale and Devos, 1998a,b). Synteny among grass genomes makes them ideal candidates for evolutionary (Kellogg, 1998) and comparative genetics, and will add much value to the rice genome analysis. Cereal genome organization in terms of gene order, relative position of molecular markers or matrix attachment regions (Avramova et al., 1998) is very conserved, with only a few major chromosomal rearrangements; therefore they can be viewed as one genetic system (Bennetzen and Freeling, 1993). The National Science Foundation (NSF) has recently funded research to assess gene content, colinearity and evolution in rice, barley, maize, sorghum and wheat, which will proceed by examining large stretches of evolutionarily-related regions from each of these five plant species (Table 1, #11). Another more focused project is directed at the comparison of the maize genome with that of the much more compact sorghum genome.

Knowledge of gene order and organization in rice, as it will arise from genome sequencing, will be generally useful to pinpoint and positionally clone candidate orthologous genes of interest from other cereal species. For instance, if a genetic locus encoding a useful trait is mapped between a pair of closely linked markers in a cereal with a large genome, it should be possible to identify candidate genes for the rice ortholog by analyzing the rice genomic sequence located between the corresponding rice markers. This information will then be sufficient to go back and genetically target orthologs in cereals with less available genome information. In addition, the ongoing explosion of EST data in rice and maize (Table 2) and their localization on genetic or physical maps combined with the rapidly expanding gene sequence databases will make a powerful gene-mining tool for cereals.

Significant synteny also exists within dicot families such as *Brassicaceae* (Lagercrantz et al., 1996, 1998; Cavell et al., 1998; Schmidt et al., 1999) and *Solanaceae* (Tanksley et al., 1988), and to a lesser extent between monocots and dicots (Gale and Devos, 1998a). A large-scale systematic effort will be required to determine the extent to which synteny between *Arabidopsis* or rice and distant relatives is useful in predicting the identity of adjacent genes. A first step toward testing this possibility would be to develop markers for other plants that are based on *Arabidopsis* or rice sequences. Sequencing of the known polymorphic markers for other species may allow their placement on the *Arabidopsis* (and eventually) rice genome sequences if they contain at least some coding sequence that is likely to be conserved between species.

Synteny seems to exists also at the fine scale or gene sequence level (microsynteny; Chen et al., 1997,1998). The relative orientation of orthologous genes and their exon-sequences are highly conserved. This is less obvious however for non-coding intron sequences and completely different for the species-specific intergenic regions, and this might give an explanation in the future for e.g. the different strategies for photosynthetic carbon-acquisition (C3-type in wheat and C4-type in maize) and therewith connected morphological changes.

Figure 1 shows an excellent example of the utility of EST sequencing and comparative genomics. EST sequencing revealed a small family of cellulose synthase-like genes (*CslA*) in *Arabidopsis* (Cutler and Somerville, 1997). Genomic sequencing has revealed a number of new family members, based on amino acid similarity. EST sequencing from other plant species has identified orthologs of these genes in rice, cotton, maize, soybean and tomato. An examination of the predicted protein sequences indicates that the various genes are as similar among species as within species (Fig. 1A). Once the exact function of these genes is established in *Arabidopsis*, their putative function in other plant species can be predicted. These genes also provide an example of how the complete genomic sequence of *Arabidopsis* can help to annotate genes in other higher plants (cp. Section II.B 2). Figure 1B shows a comparison of the

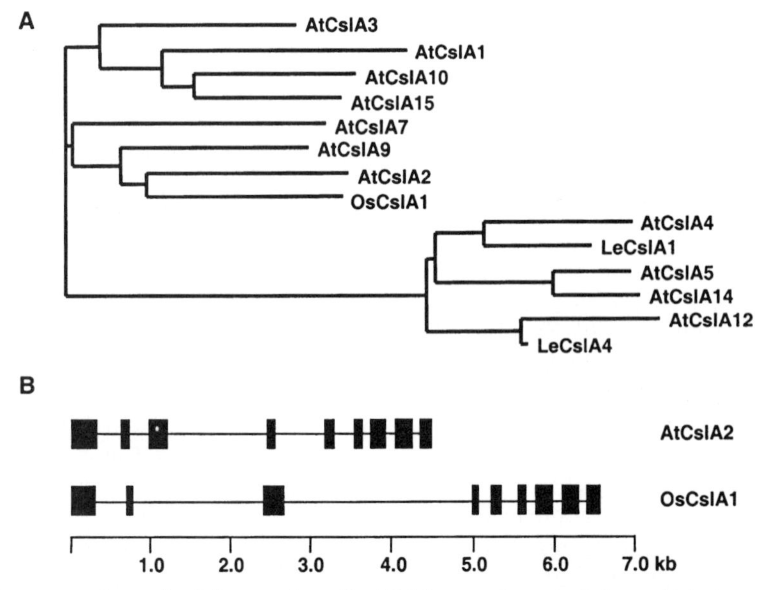

Fig. 1. (A) Unrooted, bootstrapped tree of cellulose synthase-like (*CslA*) genes, from *Arabidopsis* (At), rice (Os) and tomato (Le). Full length and near full length protein sequences were aligned and bootstrapped (n=5000) using ClustalX (Table 1, #12). The cladogram was generated using Treeview (Table 1, #13). (B) Gene structure of AtCslA2 (*Arabidopsis*) and OsCslA1 (rice). Exon sequences are depicted as black boxes and introns as thin lines.

gene structure of an *Arabidopsis* and a rice *CslA* gene. The exon sequences are highly conserved (73% nucleotide identity; 80% amino acid identity), while the intron sequences are very divergent. The presence of large introns in genes can cause difficulty with annotation and indeed, the first three exons of the rice gene are missing from the sequence annotation in the public database. Yet comparison to the *Arabidopsis* gene clearly demonstrates the presence of these exons and allows for the precise identification of splice sites and overall gene structure.

III. Classical *Arabidopsis* Genetics in the Post-Genome Era

Classical genetics links a phenotype to a genotype, and usually begins with the identification and isolation of mutants in the course of a screening-experiment, followed by their characterization, and ends in the cloning and determination of a biochemical or cellular function of the corresponding gene(s). The availability of extensive sequence information from the *Arabidopsis* genome is currently revolutionizing classical genetics in *Arabidopsis* by shortening the traditionally labor- and time-intensive cloning phase (Fig. 2).

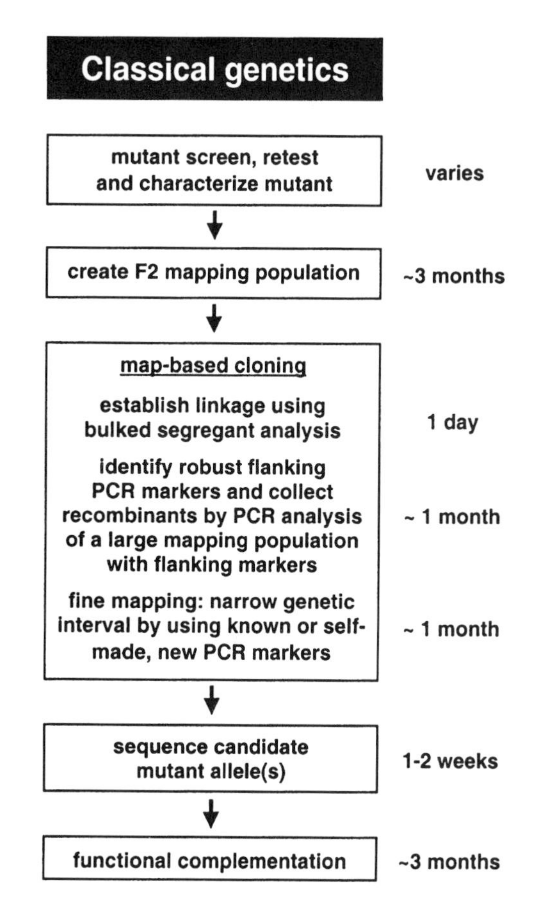

Fig. 2. Flowchart for classical genetics, depicting the steps and expected timeframes necessary for finding and linking a monogenic mutant phenotype to the corresponding gene by map-based cloning.

A. Progress in Positional Cloning

Chemical and physical mutagens [e.g. ethyl methanesulfonate (EMS) and fast neutron radiation] lead to frequent point mutations and small deletions in a genome. Like DNA insertion mutagens (Section III.B) they can lead to loss of gene function (e.g. by introduction of early stop codons, frameshifts, unacceptable substitutions or elimination of splice-sites). Chemically induced point mutations in the coding region of a gene also yield single amino acid substitutions in the gene product, which can give rise to functionally altered protein isoforms and additional phenotypes, attributes that are highly appreciated by biochemists and geneticists. Unlike DNA insertion mutations, however, point mutations or small deletions require positional (map-based) cloning of the corresponding gene. Traditionally this has been the major difficulty with EMS-mutants. Until recently, map-based cloning of *Arabidopsis* genes has been very time-consuming (1 year and more) and dependent on several critical factors. These include the ability to generate and screen large mapping populations for recombination events in the target region, the possession of an extensive collection of mapped molecular markers, the availability of genomic libraries, and the transformability of the plant species. In all of these, *Arabidopsis* excels in comparison to other plants, though until recently the number of publicly known molecular markers for genotyping recombinants and genetic fine mapping was still a limiting factor.

With the availability of over 80% of the complete genomic sequence for the Columbia (Col) accession of *Arabidopsis*, this bottleneck is disappearing. When a gene maps to a fully sequenced region of the genome, discovery of new co-dominant, PCR based molecular markers, like simple sequence length polymorphism markers (SSLP markers, Tautz, 1989; Bell and Ecker, 1994), can now be reduced to a simple database search for putative polymorphic sites in a given target region. SSLPs are usually long mono- and dinucleotide repeats or even more complex repetitive sequence stretches. The Arabidopsis database (AtDB) project has facilitated the process of finding SSLPs by constructing a table that includes long nucleotide repeats from all sequenced BAC and P1 clones (Table 1, #14). PCR with a primer pair designed to give a product that includes the repetitive sequence, will frequently result in a PCR-product that differs in size when amplified from genomic DNA from the Columbia and another *Arabidopsis* accession (Fig. 3). The genotype of the recombinants at that position in the genome can then be scored based on the size of the PCR product. Positional mapping of *Arabidopsis* genes with SSLP markers is further facilitated by their compatibility with high-throughout DNA preparations (Klimyuk et al., 1993) that enable an individual to prepare DNA from 1000 plants in a single day. Combined with instrumentation for working in 96-well plates, a researcher can genotype thousands of plants in less than a week. In addition, SSLP markers can be designed for simultaneous amplification of multiple markers in one PCR-plate with a standardized thermocycler program (Lukowitz et al., 2000) or for multiplex PCR (Ponce et al., 1999). These improvements make the determination of a rough map position for a mutation and the initial phase of fine mapping very straightforward (compare Fig. 2).

Sequencing a fragment from another accession and comparison with the corresponding Columbia sequence from the database will also often yield information like single nucleotide polymorphisms (SNPs), that can be used to detect and design other co-dominant, PCR-based molecular markers. Two related types of these markers are the cleaved amplified polymorphic sequence (CAPS) marker (Fig. 3; Konieczny and Ausubel, 1993) and derived

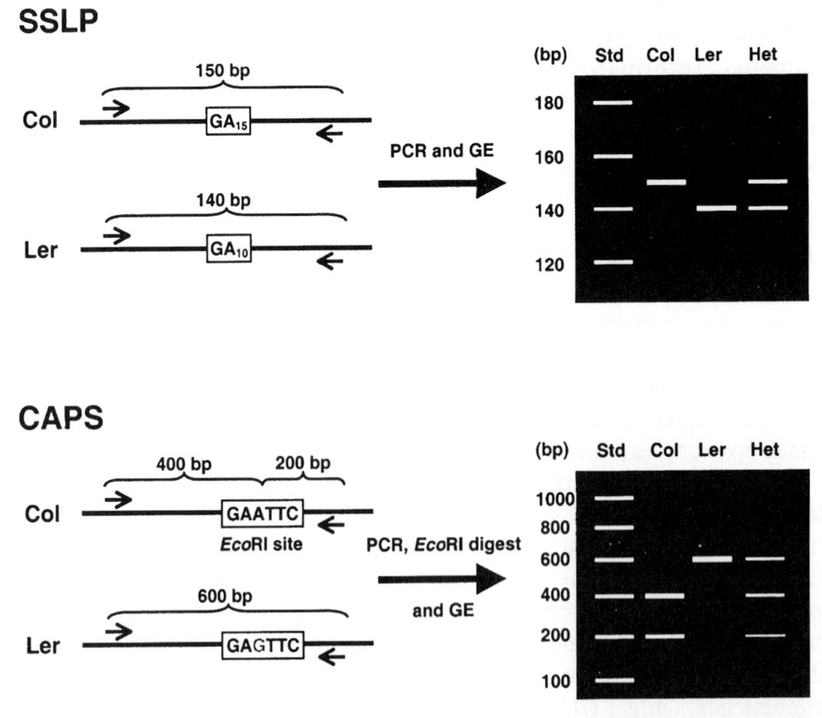

Fig. 3. Examples showing the nature of PCR-based molecular markers (SSLP, CAPS) commonly used for positional cloning. SSLP: a primer pair (small arrows) is used to PCR-amplify a 150 bp fragment containing a GA_{15} dinucleotide repeat at a certain locus in the genome of *Arabidopsis* accession Columbia (Col). PCR with the same primer pair and genomic DNA from *Arabidopsis* accession Landsberg erecta (Ler) as template yields a 140 bp fragment, because the GA-repeat (GA_{10}) is 5 dinucleotide units shorter. The size difference between Col and Ler is visualized by agarose gel-electrophoresis (GE) with a standard (Std) as comparison. Genomic DNA from a recombinant plant (Het), which is heterozygous at the locus, will yield both fragments. CAPS: a single nucleotide polymorphism (SNP), detected by sequence comparison of Col and Ler genomic DNA, results in the disappearance of a restriction enzyme site (e.g. *Eco*RI site) in Ler DNA. Using a primer pair 600 bp DNA fragments are PCR-amplified from both DNAs and subsequently digested with restriction enzyme *Eco*RI, yielding 400 bp and 200 bp fragments for Col, whereas the 600 bp Ler fragment remains undigested. Identically treated recombinant heterozygous plant DNA (Het) will give all three bands when visualized on an agarose gel.

CAPS marker (Michaels and Amasino, 1998; Neff et al., 1998). DNA fragments are first PCR amplified and then cleaved with restriction endonucleases. The presence or absence of a restriction site results in different sized fragments from each accession. In the rare case where no such PCR based marker can be designed, it is still possible to exploit differential hybridization or FRET-melting curves of fluorescently labeled hybridization probes covering the SNP for genotyping recombinants and alleles (Bernard et al., 1998; de Silva et al., 1998).

As an adjunct to the complete genome sequencing of Columbia, Cereon Genomics (Cambridge, MA) has completed a low pass (i.e. 3×) nucleotide sequence of the commonly used Landsberg *erecta* (Ler) accession (Rounsley et al, 1999), using a whole genome shotgun sequencing approach (Fleischmann et al. 1995). This resource will soon be made available to academic scientists through TAIR (Table 1, #8). Also, the Institute for Genomic Research (TIGR) has sequenced around 15,000 random fragments of approximately 500 bp from Ler, and these sequences can be accessed and searched with a BLAST-tool via the Internet (Table 1, #15). These efforts provide a wealth of information regarding the extent of genetic diversity between *Arabidopsis* accessions, as well as generating a large pool of molecular markers (Table 1, #14). This large pool of markers combined with new technologies such as 'Genotyping Chips' (Wang et al., 1998; Cho et al., 1999; Hacia, 1999) should greatly simplify genetic fine mapping. It is reasonable to assume that the number and density of molecular markers present in the *Arabidopsis* genome will enable chromosome 'landing' on a genetically defined region that is so small it encompasses only a single or a few open reading frames. These candidate genes can then be independently cloned and transformed into mutants to determine the gene of interest via functional complementation of the mutant phenotype.

With the availability of over 80% (December 1999) and soon the complete *Arabidopsis* genome sequence, positional cloning of most *Arabidopsis* genes will be limited only by the ease of scoring the phenotype and by the production of the recombinant mapping populations. Genetic fine mapping itself can now be considered an inexpensive and straightforward standard technique (cp. Figure 2; Lukowitz et al., 2000), which also puts it in the reach of small labs. An example underlining this comes from our own laboratory. Starting with the F2 segregating mapping population, a herbicide-resistance gene was cloned in less than four weeks. This time included the development of several new PCR markers for fine mapping and the identification of the point mutations in different alleles (W.-R. Scheible and C.R. Somerville, unpublished).

Finally, the ability to quickly generate large numbers of transformants in *Arabidopsis* (Section III.B) and the availability of transformation-competent cosmid- and artificial chromosome vectors (Meyer et al., 1994; Hamilton, 1997; Liu et al., 1999) should also allow accelerated cloning of point mutations by whole-genome functional complementation.

B. Insertion Mutagenesis and Retrieval of Tagged Genes

Two main approaches exist for tagging genes in *Arabidopsis* with mutagenic DNA elements: namely, tagging with T-DNA from *Agrobacterium tumefaciens* (Koncz et al., 1989; Marks and Feldmann, 1989; Feldmann, 1991; Koncz et al., 1992; Azpiroz-Leehan and Feldmann, 1997; Krysan et al., 1999), and tagging with transposons (such as Ac/Ds, En/Spm, or Mu) (Dean et al., 1991; Walbot, 1992; Aarts et al. 1993; Hehl 1994; Fedoroff et al., 1995; Bhatt et al., 1996; Long and Coupland, 1998; Martienssen, 1998; Wisman et al., 1998). Both kinds of mutagenic DNA elements insert more-or-less randomly into chromosomes and can therefore be used to create 'loss-of-function' mutants for any gene.

In addition, certain tailor-made transposons and T-DNAs, that carry strong enhancers or a promoter directing transcription into the region flanking the insertion, can lead to 'gain of function' or 'activation tagged' mutants (Walden et al., 1994; Martienssen, 1998; Weigel et al., unpublished). These mutations in which ectopic activation of a flanking gene promotes a mutant phenotype, are usually dominant or semi-dominant. Thus they are visible in the T1-generation, though somewhat rare as they require the activating DNA elements to insert in the vicinity of a gene without disrupting the transcription unit. Insertion mutants can also be designed to contain genes encoding beta-glucuronidase or green fluorescent protein, that report the expression of the chromosomal gene at the insertion site (gene- and enhancer traps, Topping and Lindsey, 1995; Martienssen 1998 and references therein; Campisi et al., 1999). Lines with very specific expression patterns (cell type, developmental stage, environmental condition) and lines that might have been missed in conventional insertion mutant screens (e.g. homozygous-lethal insertions in heterozygous mutants) can be identified and the responsible genes and promoters isolated (Sundaresan et al. 1995).

T-DNA tagging involves Agrobacterium-mediated plant transformation using vectors based on native Ti-plasmids. DNA present within two imperfect repeats, the left and right borders, is transferred, and integrates into the plant genome (Zupan and Zambryski, 1995, Sheng and Citovsky, 1996; and references therein). A key feature for the production of insertion mutants is the availability of an efficient transformation system, like the *Agrobacterium tumefaciens* based vacuum-infiltration or *in planta* inoculation procedures for *Arabidopsis* (Bechtold et al., 1993; Chang et al., 1994). More recently high throughput transformation protocols for *Arabidopsis* were developed, such as the 'floral-dip' (Clough and Bent, 1998) or 'floral spray' method (W.-R. Scheible, unpublished). These methods enable a single person to conveniently produce hundreds of thousands independent T1-transformants. These in planta methods bypass the tissue culture and regeneration phase and the related problem of somaclonal variations or reduced fertility common to previous transformation methods (Scholl et al., 1981; Valvekens et al., 1988; van den Bulk et al., 1990). Recent research shows that the basis of the in planta methods is that the oocyte of the female gametophyte is transformed, after meiosis and presumably during fertilization (Bouchez et al., 1999; Ye et al., 1999). Transformed T1-plants are selected using selectable markers (e.g., herbicide resistance) encoded within the T-DNA. It seems likely that these high throughput methods will eventually be adapted to other plant species, like cereals (Komari et al, 1998; Chin et al., 1999; Enoki et al., 1999), and will permit the creation of large populations of insertion mutants.

Transposon tagging has some advantages over T-DNA tagging. For plant species that are difficult to transform, transposons are more efficient because a single transformation event can be used to generate thousands of mutations. Transposons can move around in the genome, and many independent insertions (up to 200 with Mutator) can be generated in one line (Chandler and Hardeman, 1992; Das and Martienssen, 1995), whereas T-DNA tagging generates only an average of ~1.5 insertions per line. (Feldmann and Marks 1987; Martienssen 1998; Bouchez et al., 1999). However T-DNA tagging results in stable insertions while transposon tagging requires the removal of transposase activity for insertion stability. Hence the smaller number of transposon tagged lines that need to be screened to find a mutant of interest is offset by the difficulty of the subsequent analysis to determine the responsible insertion event. Another advantage/disadvantage of transposons is their tendency to jump into locations near their original integration site (Jones et al., 1990; Bancroft and Dean, 1993; Cardon et al., 1993). While this limits the effectiveness of whole genome mutagenesis from a single transposon donor site, it raises the probability of knocking out tandem-repeat genes or clustered gene families, both of which are common in the *Arabidopsis* genome (Bevan et al., 1998; Lin et al., 1999; Mayer et al., 1999). Finally, transposon insertions in a gene undergo somatic and germinal reversion, which can provide evidence of a gene's function without the need for functional complementation (Bhatt et al., 1996; Liu and Crawford 1998 and references therein; Uwer et al., 1998).

Large T-DNA- and transposon insertion and activation tagged mutant collections for *Arabidopsis* have been generated by various labs (e.g. J. Ecker, K. Feldmann, INRA Versailles, T. Jack, R. Martienssen, M. Sussman, D. Weigel, see also Scholl et al., 1999; Wisman et al. 1998; Bouchez et al., 1999), altogether comprising several hundred thousand lines. Many of these lines have been made available to the *Arabidopsis* community through the Arabidopsis Biological Resource Center (ABRC, Table 1, #16) and the Nottingham Arabidopsis Stock Center (NASC, Table 1, #17). In collections of this size it is theoretically possible to find knockout mutants for any given gene. Large collections of insertion mutants (T-DNA or transposon) also exist for maize, tomato, petunia, rice and snapdragon and additional collections will probably be created in several other species.

Once an interesting phenotypic insertion mutant has been identified in the course of a screening-experiment, the identification and retrieval of the disrupted gene, which is tagged by a DNA tag of known sequence, can be very straightforward. Using methods like adapter-mediated PCR (Riley et al., 1990; Rosenthal and Jones, 1990; Siebert et al., 1995), plasmid-rescue (Koncz et al., 1992; Mandal et al., 1993) or thermal-asymmetric-interlaced (TAIL-) PCR (Liu et al., 1995b) the plant genomic DNA flanking the T-DNA tag on both sides is cloned and sequenced. However, T-DNA insertions can sometimes be of complex nature (e.g. tandem or inverted repeats, truncated T-DNA or ragged borders), making molecular analysis difficult (Gheysen et al., 1990; Nacry et al., 1998). In activation-tagged mutants the targeted gene is not tagged by the DNA-insertion element; therefore, it is not immediately clear which gene is affected by the enhancer elements. However, the existing evidence in *Arabidopsis* shows that the activation tag is usually located close to the affected gene, with distances ranging from ~0.4–3.6 kb upstream or downstream (Kakimoto, 1996; Weigel et al., 2000).

With the approaching completion and availability of the *Arabidopsis* genome sequence, flanking DNA information is sufficient to immediately identify the insertion by a simple database search. Confirmation that the gene is disrupted by the DNA tag and is responsible for the corresponding phenotype is accomplished by functional complementation analysis of the mutant line with a wild-type copy of the gene. In *Arabidopsis*, this is routinely done using *Agrobacterium tumefaciens* mediated T-DNA transformation. However in other plant species this step is often more problematic.

IV. Assigning Gene Functions Using the Tools of Functional Genomics

The term 'functional genomics' refers to the development and application of high throughput or large-scale (genome-wide or system-wide) experimental approaches to assess gene function by making use of the information and reagents provided by structural genomics (EST- and genome sequencing) (Hieter and Boguski, 1997). A tight connection normally exists between the function of a gene product and its expression pattern. As a rule, each gene is expressed in the specific cells and under the specific conditions in which its product makes a contribution to fitness (Brown and Botstein, 1999). Hence, analysis of the spatial (e.g. tissue, cell type, subcellular), temporal (e.g. developmental, diurnal) or environmental (e.g. biotic and abiotic stress, elevated CO_2) effects on transcript- and protein-expression patterns should give useful clues about gene functions. This is particularly desirable for understanding the biological functions of the thousands of *Arabidopsis* genes that presently have no assigned function or just a general function (e.g. genes encoding transcription factors or protein phosphatases). With the major tools of functional genomics and reverse genetics at hand it is anticipated that the task of assigning some degree of function to all *Arabidopsis* genes will be finished within the next decade (Somerville and Somerville, 1999).

Various competing technologies have been developed in recent years for the large scale detection and quantification of the transcriptome, the transcript complement of a genome. These include high-throughput-sequencing of cDNA libraries (Newman et al., 1994; Lee et al., 1995), serial analysis of gene expression (Velculescu et al., 1995; Adams, 1996; Velculescu, 1997), enhanced concatemer cloning (Powell, 1998), and parallel approaches like high-density cDNA arrays on nylon membranes (for review see Baldwin et al., 1999), DNA microarrays and gene chips. The latter two will be focus of the discussion here, due to their emerging predominance in the field. For more details the interested reader is also referred to 'The Chipping Forecast,' a collection of excellent and comprehensive reviews on different aspects of the microarray and gene chip technologies, written by leading experts in the field. "The Chipping Forecast" was published in January 1999 as a supplement to 'Nature Genetics' and is publicly available (Table 1, #18).

A. Microarrays

In principle, microarrays are reverse Northern blots, with thousands of probes hybridizing to single target sequences. Microarrays are constructed by arraying thousands of single-stranded target DNA fragments in a grid-format onto miniature solid supports, such as poly-lysine coated glass slides. These target clones are either partial cDNA sequences (ESTs, Section II.A) or PCR products amplified from genomic DNA. Ideally the fragments are gene specific and represent all genes of an organism with a minimum of redundancy. However, this criterion is difficult to meet with EST sequences, especially for the considerable number of *Arabidopsis* genes that exist in multigene families (Section II.B 3). It is estimated that sequences with >70% identity over >200 nucleotides in length are likely to exhibit some degree of cross-hybridization under standard conditions (Richmond et al., 1999).

Microarrays are probed with a complex mixture of fluorescently labeled fragments made from polyA-mRNA or total RNA. Total RNA (50–100 μg) or polyA-RNA (1–2 μg) from two different samples is

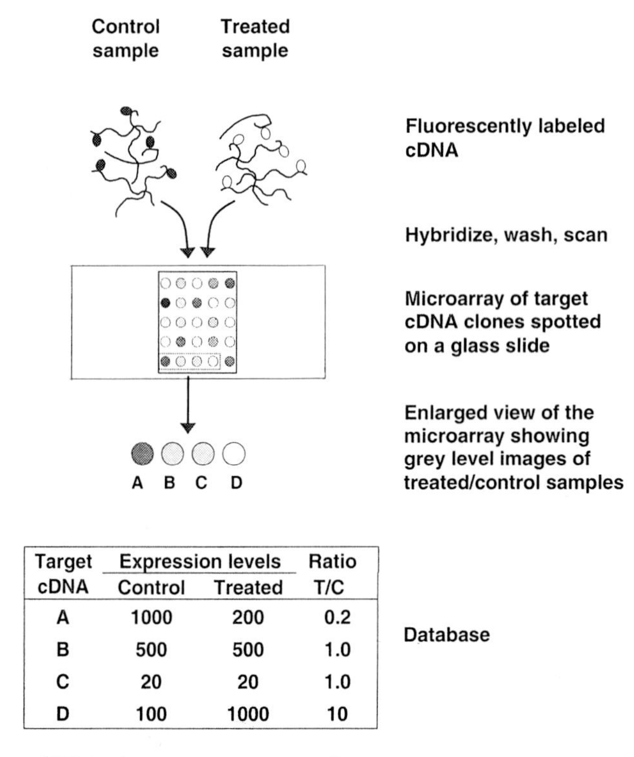

Target cDNA	Expression levels Control	Treated	Ratio T/C
A	1000	200	0.2
B	500	500	1.0
C	20	20	1.0
D	100	1000	10

Fig. 4. Diagram outlining the use of DNA microarrays to assess steady-state mRNA levels in two mRNA populations. The grey levels of the spots represent the ratio of the absolute expression values derived from mRNA from treated versus control tissues. The grey-level-saturation coding is as follows: Black, mRNA levels are much higher in the control than in the treated sample. White, mRNA levels are much higher in the treated than in the control sample. Intermediate grey-level (as in spot B and C), mRNA levels are similar in the two samples compared.

labeled with a fluorochrome, a different color for each sample, using a reverse transcription step and the two labeled cDNA populations are hybridized to the target fragments, which are present in excess (Fig. 4). For RNA samples derived from small amounts of tissue or even single cells, PCR methods can be used to produce sufficient material for fluorescent probes.

All probe fragments are assayed simultaneously in a very small hybridization volume and hence at high probe concentrations. These features along with the low background hybridization on glass slides improve the detection of rare mRNA species, and detection-limits of 1 mRNA molecule in 100,000 (= 1 molecule in 5–10 cells) and lower have been reported (Ruan et al., 1998; Gerhold et al., 1999; Table 1, #19). The fluorescence hybridization signal for each probe is quantitated with an instrument designed for the purpose (e.g. a special laser microscope) and used to assess mRNA abundance in a given tissue. Fluorescence-based detection offers an important benefit: two or more target mixtures, each labeled with a different fluorochrome (e.g. green Cy3, red Cy5; Yu et al., 1994) can be hybridized simultaneously to a given microarray slide, permitting direct comparisons between samples. Emission signals of these fluorochromes are linear over a broad range, which makes it possible to evaluate abundant and rare probe species in the same experiment. Due to the limitations of hybridization in very small volumes (e.g. limited mixing and edge effects) we believe that DNA microarrays might not always be very accurate for measuring the absolute abundance of a message, but they can reliably measure the relative change in expression between two samples.

The range of applications for DNA based microarrays is very diverse (Eisen and Brown, 1999). For example, microarrays are applicable to all organisms for which cDNA can be prepared; and even fragments of genomic DNA can be used as targets. In principle there is no need for sequencing of any of the target sequences unless interesting results are seen during experimentation. Microarrays also allow for simultaneous gene expression studies in biochemically interacting species (such as plants and their pathogens), and arrays containing all known plant viruses should be very useful in determining viral infections. Furthermore the microarray technique can also be adapted for use in 'reverse Southerns,' in which complex DNA probes (e.g. BAC, YAC or even total genomic DNA) are used in place of cDNA (Shalon et al., 1996). This technique can be useful for defining large insertions or deletions in the genome. In addition, insertion sites of transposons or T-DNAs can be verified using microarrays by hybridizing with DNA flanking the insertion elements, produced using inverse PCR (Lemieux et al., 1998). Last but not least microarrays constructed of genomic fragments might be used in combination with labeled DNA binding proteins to determine candidate target genes (Bulyk et al., 1999; Kehoe et al., 1999).

One of the major drawbacks of the microarray technology in the plant field at this point is the limited availability of high-density quality arrays. Sophisticated gene expression experiments, which involve comparisons of two or more genotypes at multiple time points and/or encompassing more than one experimental condition are therefore limited, especially since current microarrays are non-reusable.

The microarray technology is also very expensive to implement and requires major investments (>US$ 100,000) in equipment plus operating expenses and reagents, which place it outside the means of many researchers. As a result, many universities and research institutes are establishing centers where expensive equipment of this type is available to researchers for reasonable fees.

Microarray projects have many of the same problems as other large-scale genome projects. At the top of this list is the problem of dealing with the organization and handling of thousands or tens of thousands of ESTs or genomic fragments. There is a tremendous amount of labor and expense associated with selecting, growing, maintaining and arraying a large number of clones. For maximal effectiveness, each of the clones on the microarray should be sequenced as well. For a new microarray project, the selection and sequencing of clones for the production of microarrays can be the major expense. The effort necessary to analyze the results of the experiments is discussed in Section IV.C.

There are technical problems with microarrays that must be addressed as well. There has not been extensive study of the sensitivity of microarrays. Nor has the extent to which cross hybridization between similar genes on the microarray occurs been explored in detail. Also, the reproducibility between experiments has sometimes been found to be unsatisfactory (V. Haake, personal communication). This problem might be related to differences between copies of the same array and/or inconsistencies in plant sampling (wounding responses, age, time of day, changing environmental conditions). Extensive quality controls of microarray slides and standardization of plant growth and sampling procedures are therefore crucial. The use of technical replicates (same sample, different slides) and biological replicates (different samples, different slides) is one way that has been used to measure reproducibility. Only changes in expression that occur in several replicates are viewed with confidence. Finally, the general lack of appropriate software, standard statistical approaches and expertise in the scientific communities for the collection and analysis of expression data, data management and storage represent yet another major limitation to the microarray and gene chip technologies (Section IV.C).

It seems likely that these many limitations will be resolved by the development of more robust methods and, possibly, by the emergence of public or private service providers. Indeed, Monsanto Life Sciences Company, in collaboration with Incyte's Microarray Division has established an *Arabidopsis* microarray project (Table 1, #20) to enhance plant genomics research in academia and non-profit organizations. The program provides researchers with the opportunity to access greater than 10,000 non-redundant *Arabidopsis* ESTs on a microarray. A consortium of scientists from the Carnegie Institution, Stanford University, Michigan State University, the University of Wisconsin and Yale University is also beginning to offer access to *Arabidopsis* arrays to the scientific community (Table 1, #21).

B. Gene Chips

The other technology for large-scale transcript analysis is the DNA chip technology pioneered by Affimetrix (Santa Clara, CA, Lockhart et al., 1996; Lipshutz et al., 1999). This technology is based on high-density oligonucleotide arrays synthesized directly on derivatized glass slides using a combination of photolithography and oligonucleotide chemistry (Fodor et al., 1991; Lipshutz et al., 1999). As with microarrays fluorescently labeled nucleic acid can be hybridized under the same favorable conditions to the chip and quantitative hybridization information from each oligonucleotide is also obtained using a specialized scanning laser microscope. The oligonucleotide targets are designed in pairs, one perfectly complementary to the probe, and a companion that is identical except for a single base difference in the central position. Typically twenty of these pairs are designed for each gene to be represented on the chip (Wodicka et al., 1997). The mismatches serve as sensitive internal controls for hybridization specificity, and comparison of the perfect match signals with the mismatch signals allows low-intensity hybridization patterns from rare RNAs to be obtained. DNA chips can contain several hundreds of thousands of different 20–25 mer oligonucleotides of defined sequence and it will, therefore, be possible to produce chips with sufficient complexity to represent an entire plant genome.

Unlike the microarrays, however, chip-technology is dependent on prior gene sequence information and is thus limited to organisms for which high quality EST or genomic sequence exists. For photosynthetic organisms, at present, these include *Rhodobacter capsulatus*, *Synechocystis* PCC6803, *Arabidopsis* and others (Table 1, #5, Table 2). Another drawback associated with chip technology is the additional high cost of producing the photolithographic masks used to direct the computerized synthesis of the oligonucleotides. With the completion of the *Arabidopsis* genome sequence and improvements in manufacturing, like a new, cheaper and faster maskless fabrication method using micromirror arrays (Singh-Gasson et al., 1999), the chip-technology is likely to be used extensively in the future to obtain genome-wide expression information in *Arabidopsis* and many other organisms. Besides their use in gene expression studies, gene chips can also be used for genotypic analysis (e.g. mutation analysis), resequencing, using variant detector arrays (Hacia, 1999), and large scale genetic mapping of single nucleotide polymorphisms (SNPs) (Wang et al., 1998; Cho et al., 1999). This latter use may revolutionize genetic mapping and marker-assisted breeding in plants (Somerville and Somerville, 1999).

C. Analysis and Mining of Large-Scale mRNA-Expression Experiments

The highly regular arrangement of probe elements and sharply defined signals that result from robotic printing on glass matrices and laser imaging of fluorescence-based arrays renders image data amenable to extraction by digital image processing procedures. A perfect image analysis program is able to automatically detect spots, sample the local background and extract a mean signal-intensity above background for each target on the array. For proper comparison of expression data from a double-label array, signal-intensities have to be normalized. There are several normalization techniques that can be used. Genes that are unlikely to change in expression under the conditions being tested, like constitutive house-keeping genes, can be used to adjust one fluorescence channel to another. Or one can assume that under the conditions being tested, most genes will not change in expression and the mean or median fluorescence for each channel for the slide as whole can be used as means of normalization. Another simple method is to convert the fluorescence of each target spot into a normalized score, based on the standard deviation (Student normalization) or the mean and standard deviation (Z score normalization) of the fluorescence for each channel as a whole. The two normalized signal intensities gained from each spot of a double-labeled microarray can then be plotted as XY-data pair in a double-logarithmic graph. Genes whose change in expression is >2 and that show the equivalent changes in replicate arrays or replicate samples are considered to be differentially expressed in the two conditions compared.

Examining a wild-type control and an experimental sample and determining which genes change in expression constitutes the simplest use of microarrays (see experiment described in Fig. 5). The range of conceivable comparisons is endless, comprising different environmental or experimental conditions (e.g. light quality, photoperiod, nutrient availability, CO_2, cold, drought, salt, phytohormones, herbicides, pathogens, pests), developmental processes and stages (e.g. flowering, germination), tissue types or genotypes (wild type versus mutant). Initial experiments with microarrays have been focused on gene discovery, attempting to find small numbers of genes that greatly change in expression under the conditions tested. For example, Schena et al. (1996) used microarrays containing 1046 human cDNAs to explore differential expression patterns in response to heat shock or protein kinase C activation by phorbol esters, and Heller et al. (1997) used the same microarray to discover inflammatory disease-related genes. Ruan et al. (1998) published the first large-scale microarray experiment in the plant field,

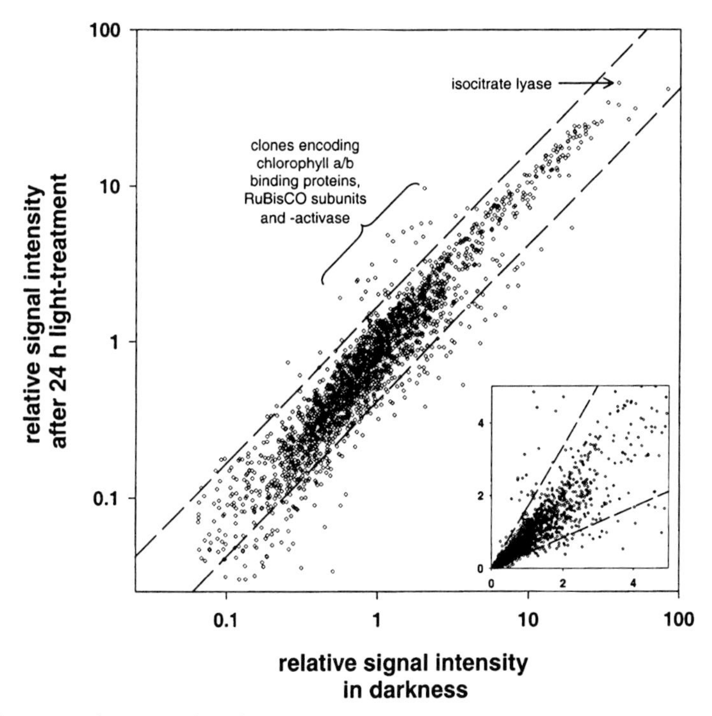

Fig. 5. Double-logarithmic scatter-plot comparing the mean hybridization signals of ~2300 target-clones for etiolated versus de-etiolating *Arabidopsis* seedlings. The insert is a linear-scaled cut-out of the same data set. Dashed lines represent the margins for two-fold changes in expression. Etiolated seedlings grew in darkness, whereas de-etiolating seedlings were transferred into light (120 μmoles photons m^{-2} s^{-1}) after 3 days in darkness. Seedlings from both conditions were harvested after 4 days and used for isolation of RNA. The vast majority of genes are expressed at similar levels in the two samples. As expected, some of the most significantly upregulated genes in de-etiolating seedlings are photosynthetic genes encoding chlorophyll *a/b* binding proteins, subunits of ribulose-1,5-bisphosphate carboxylase/oxygenase (Rubisco) and also Rubisco activase. A gene encoding isocitrate lyase, an enzyme involved in beta-oxidation of storage lipids, is heavily but not differentially expressed. Some of the target clones that show significantly higher hybridization signals in etiolated seedlings correspond to unknown genes. RNA samples were kindly provided by Eric Schaller (University of New Hampshire, Durham).

studying the expression of 1443 genes in major organs of *Arabidopsis* (roots, leaves, flowers).

As microarrays and gene chips become more common and inexpensive, experiments become more complex and comprehensive. The next logical experiments are comparisons between multiple treatments, multiple specimens, dose-response series or time course experiments. The first publications of this kind came from P. Brown's lab at Stanford University: DeRisi et al. (1997), for example, analyzed the dynamic temporal expression pattern of all yeast genes in wild type and two transcription factor mutants during the metabolic shift from fermentation to respiration. Chu et al. (1998) analyzed the transcriptional program during sporulation in budding yeast, and using a microarray representing 8,600 human genes Iyer et al. (1999) subsequently explored the transcriptional changes in the response of human fibroplasts to serum. These types of experiments quickly yield overwhelming amounts of raw data. Powerful software is, therefore, needed to sort, organize and intuitively visualize the huge data sets, according to gene expression profiles and/or experimental parameters. A number of freely available analysis programs, of varying quality and utility, and expensive commercial software packages are available or still in development. Most of them deal with the concept of 'clustering' - grouping together genes based on orderly features in their temporal expression patterns throughout the experiment (Eisen et al., 1998).

Clustering analysis of time course microarray experiments has been shown to precisely group together co-expressed genes encoding components of molecular complexes like the ribosome, the proteasome or the nucleosome during the yeast cell-cycle (Spellman et al., 1998), or genes whose products work in concert as in metabolic pathways (DeRisi et al., 1997; Iyer et al., 1999). For plants like *Arabidopsis* it will be exciting and instructive to see similar genome-wide temporal gene expression analyses for example during the environmentally and developmentally influenced formation, change or degradation of photosynthetic multi-component complexes (Photosystems I and II, cytochrome b_6f, ATPase) and biosynthetic pathways (Calvin cycle, chloroplastic starch turnover, sucrose-production and export). Such experiments will also help to understand and identify the interplay, regulatory mechanisms and elements, like promotors or transcription factors, necessary for coordinated expression of nuclear- and/or organellar-encoded genes and targeting of their products (Martin and Herrmann, 1998). Likewise, differential temporal expression patterns of multigene family members will help to understand which member participates in the process under study.

Comparison of gene expression between multiple treatments and specimens or time course experiments combined with clustering analysis is an especially powerful tool for predicting gene function. Shoemaker et al. (1996), for example, identified many previously uncharacterized genes with significant expression changes during the different phases of yeast spore morphogenesis. When they tried three of those genes by deletion analysis, all three knockout mutants were found to have sporulation defects. Thus, the additional information gained from such experiments when compared to simple 'gene discovery' experiments (see above) is more likely to yield causal relationships (i.e. to pinpoint a more precise biological function for a new gene). The additional effort required for planning, executing and analyzing an experiment might be well worth it, since the time needed for characterization of corresponding knockout mutants might be shortened (Section V.C). Time course experiments with plants will require very careful planning and maximal control of plant growth parameters, since trivial factors, like day-time or nutrient availability can alter gene expression and steady-state transcript levels tremendously (W.-R. Scheible, A. Krapp and M. Stitt, unpublished). For longer time-courses it will also be important to include all the necessary internal controls, which account for changes of gene expression during plant growth and development.

The final level of gene chip and microarray experimentation will be achieved once the expression data of many (hundreds and thousands) experiments from individual researchers are being collected in an ever-expanding public database, which can be compared, queried and cross-searched by researchers. For *Arabidopsis* such a database is currently under construction at Carnegie Institution Department of Plant Biology at Stanford University. When the complete genome of *Arabidopsis* is finished and placed on a microarray or gene chip, theoretically a gene expression experiment for a given cell type, tissue or developmental stage will only have to be done a single time, given experiments are conducted in comparable and standardized conditions. Once the amount of expression data is large enough researchers will then be able to extract meaningful information without ever approaching the bench. This could include the discovery of new regulatory patterns and pathways or the determination of the specific biological roles of multigene family members. Brown and Botstein (1999) propose that the set of genes expressed in a cell will tell us what biochemical and regulatory mechanisms are operative, how the cell is built and what it can and cannot do.

D. Proteomics

Large-scale transcript expression patterns in conjunction with bioinformatics are very valuable tools for plant functional genomics with increasing impact in the future. However, mRNA-levels do not necessarily reflect protein-levels (Anderson and Seilhamer, 1997), and this is also well known in the field of plant photosynthesis (Gruissem, 1989; Reinbothe et al., 1993; Flachmann and Kühlbrandt, 1995). Hence, knowledge of the expression level and expression pattern of proteins, as well as knowledge of their interacting partners can be very informative for understanding their biological functions. Knowledge of protein-protein interactions or post-translational modifications may also be useful if protein expression is unchanged or has no physiological consequence. The technology to address these points at a global scale is proteomics (Wilkins et al., 1997; Blackstock and Weir, 1999; also see Table 1, #19, #22).

Proteomics has a number of concerns not shared by DNA/RNA technologies. Sample handling and sensitivity are critical issues for proteomics, since there is no equivalent to PCR amplification. Problems with low abundance proteins, membrane proteins or basic proteins exist, but can be at least partially overcome by special preparation methods (Ben, 1999;

Santoni et al., 1999; Seigneurin-Berny et al, 1999). At the present time proteomics is technically more laborious and challenging to perform and hence does not yet offer the same high throughput as DNA microarrays.

For the reproducible separation of complex protein mixtures the method of choice is two-dimensional (2D) polyacrylamide gel electrophoresis (Dunn and Corbett, 1996), which separates proteins by molecular weight and isoelectric point. Similar to microarrays, fluorescent dyes can be used to stain 2D-gels, allowing the visualization and quantification of thousands of protein-spots over a broad dynamic range (Steinberg et al., 1996; Patton et al., 1999). Two-dimensional mobility data might suffice in the future to reliably identify many proteins and their modified forms. If a large enough body of mobility data for a given experimental and electrophoretic system is not available, or if the proteins cannot be clearly separated on 2D-gels, proteomics still requires the individual identification (and quantification) of each protein.

The rapid identification of proteins has greatly benefited from the development of highly sensitive mass-spectrometric techniques, like matrix-assisted-laser-desorption-ionization-mass-spectrometry (MALDI-MS) (Beavis and Chait, 1996; Jungblut and Thiede, 1997), and the availability of EST, genome and protein sequence databases. The process of identification typically involves several steps. First, an in-gel digest of proteins is performed with a protease (e.g. trypsin). Then each fragmented protein-spot is picked by a robot and delivered on a matrix to the laser-ionization chamber of a mass-spectrometer. The times needed for the various ionized fragments of a protein to reach the detector are monitored and give a highly precise measure of their molecular weights. The peptide masses are subsequently matched with computer-generated lists of expected peptide weights formed from the simulated digestion of a protein database with the same protease. The combination of the different fragment masses from the same protein, the 'peptide mass fingerprint,' reveals its identity (Henzel et al., 1993). The technique is capable of dealing with peptide fragment mixtures from two and more proteins (e.g. proteins with identical mobility on 2D-gels) and is powerful enough to identify all proteins from a fully sequenced organism (Shevchenko et al., 1996). However if a protein is degraded, modified or simply unknown then the composition and partial sequence information of its proteolytic fragments can be obtained by using methods like tandem- or electrospray-ionization-mass spectrometry (Banks and Whitehouse, 1996; Gillece-Castro and Stults, 1996; Jensen et al., 1999). The partial peptide sequence (a peptide-sequence tag) can then be used for highly specific searches of protein and EST databases (Shevchenko et al., 1996).

In addition to this 2D-gel MS-based technology, more sensitive peptide-antibody-array approaches are envisaged and promise to allow proteomics to achieve similar ease-of-use and high-throughput as large-scale transcript-measurements (Blackstock and Weir, 1999). These antibody-arrays could also be equipped with peptide-antibodies that discriminate between different modified forms of the same protein (Weiner, 1995). Managing and mining the obtained protein-expression data poses the same challenge as for microarrays and gene-chips, requiring custom-made software, biocomputing skills and organization in databases. Large 2D-gel databases exist for a variety of organisms like *Escherichia coli*, yeast and mammals (human, mouse, rat) and to a more limited extent for plants (*Arabidopsis*, rice, maize, wheat and pine) (Table 1, #23).

E. Metabolic Profiling

Establishing metabolic profiles for mutants with e.g. known gene-knockouts (Section V) is another large-scale approach to rapidly gain functional information of many *Arabidopsis* genes, as well as being the most direct way for beginning the metabolic engineering of plants. Understanding and redesigning the metabolic pathways and regulatory networks, which determine the chemical composition of a plant, is the central goal of plant biotechnology. Metabolic engineering of plants by targeted genetic manipulation of certain 'key-enzymes' has not met with a great deal of success (see reviews of Stitt and Sonnewald, 1995; Kinney, 1998; Nuccio et al., 1999; Ohlrogge, 1999). Such attempts have highlighted the complexity of the regulatory networks underlying plant metabolism, where control is exerted at, and shared between, different levels. For example, metabolites and nutrients not only control the activity and stability of proteins (e.g. allosteric regulators), they also act as signals and can trigger the expression of many genes and complete biosynthetic pathways (Koch ,1996; Stitt, 1999; Stitt and Krapp, 1999). Since our knowledge is too rudimentary to predict the metabolic effects of targeted genetic manipulations, the most straightforward way for metabolic engineering is the non-targeted chemical analysis of mutant collections (EMS- or DNA insertion populations, Section III.B) by high-throughput metabolic profiling. Although this technology is still in its infancy, it promises the simultaneous determination of a large range of metabolically relevant compounds in a single or a few measurement(s), by making use of powerful analytical methods like GC-MS or electrospray tandem mass spectrometry (Rashed et al., 1997; Avery et al., 1998; Tretheway et al., 1999).

Untargeted metabolic profiling per se can be considered as a potent multiparallel screening procedure for fast classical genetics. Metabolic profiling however is also a tremendous tool for 'reverse genetics' and 'functional genomics'. It has a reasonably high throughput (Trethewey et al, 1999) and allows for the assignment of more specific functions to previously identified genes, if changed metabolite patterns are obvious in their corresponding gain- or loss-of-function mutants. For example, a lower level of seed storage lipids in a transcription factor knockout-mutant may suggest that the transcription factor has a function in controlling the expression of lipid-biosynthetic enzymes.

Approximately 15–20% of all *Arabidopsis* genes with known or putative function are involved in metabolism and about the same percentage have a role in signal transduction and transcription (Bevan et al., 1998; Lin et al, 1999; Mayer et al., 1999). Given the apparent diversity and complexity of plant metabolism and its regulation networks, it is conceivable that metabolic profiling will play an important role in the assignment of functions to many of these genes. When the genes that control the chemical composition in *Arabidopsis* are identified, it will be possible to genetically target their orthologs in agriculturally relevant plant species.

V. Reverse *Arabidopsis* Genetics

Once differentially expressed anonymous genes have been discovered in the conditions of interest using DNA microarrays or proteomics, the next step is to collect information regarding their biological role in vivo. This typically requires the isolation and characterization of loss- or gain-of-function mutants for each gene of interest. This strategy, from the gene to the mutant and the biological function, is commonly referred to as reverse genetics (Fig. 6).

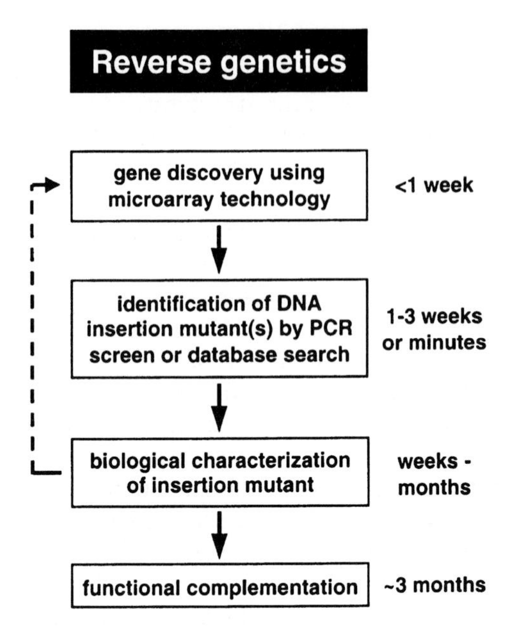

Fig. 6. Flowchart for reverse genetics depicting the steps and time-estimates for assigning a biological function to a previously uncharacterized gene. The dashed arrow reflects the eventual necessity of crossing mutants with insertions in different discovered candidate genes as described in Section V.C.

A. Finding a Specific Mutant in DNA-Insertion Collections

As described in Section III.B, many T-DNA and transposon insertion mutant collections already exist for *Arabidopsis* and also other plant species. The saturation of an insertion mutant collection depends strongly on different factors like the number of independent lines, the size and organization of the genes and the genome, the randomness of insertion of the mutagen, as well as the number of insertions per genome. Considering these factors it is estimated that for *Arabidopsis* it is necessary to screen 120,000 independent mutant T-DNA insertion lines (1.5 insertions/line) to find at least one insertion mutant for a given gene with 95% probability (Bouchez and Höfte, 1998; Krysan et al., 1999).

To make such an undertaking possible it is crucial to use appropriate pooling strategies and powerful PCR screening tools, since this method requires detection of single insertions in very complex DNA mixtures. The pooling strategy for one of the existing *Arabidopsis* collections containing approximately 61,000 independent T-DNA insertion lines (Krysan et al., 1999 and Table 1, #24) is outlined here briefly: The entire collection is divided in 270 pools containing 225 (25 × 9) T1-plants each. Tissue samples from all 225 plants in each pool were harvested and combined to prepare 270 genomic DNA pools. Aliquots from 9 of these DNA pools were combined to give 30 DNA superpools, each representing 2025 plants. Seeds from nine mature plants were harvested for one seed pool, thus giving 25 seed pools per plant pool or 25 × 270 = 6750 seed pools for the entire collection.

The DNA superpools can be rapidly screened for an insertion in any gene of interest by using PCR with oligonucleotide primers based on the sequences of the target gene and the insertion element (Ballinger and Benzer, 1989; Zwaal et al., 1993; McKinney et al., 1995; Krysan et al., 1996; Winkler et al., 1998). The presence of an insertion in the target gene is indicated by the presence of a PCR product, the specificity of which is verified by DNA gel blot using probes from the gene of interest and from the insertion element. Another method of screening is 'inverse display of insertions' (IDI) (Tissier et al., 1999). IDI involves spotting inverse PCR products of insertions from the DNA pools on a membrane that is then hybridized with the probe of interest.

Once a hit is found and confirmed, a second round of PCR screening is performed with the DNA pools constituting the positive superpool. DNA is prepared from the plants from the T2-seed pools representing the identified DNA pool and re-screened by PCR. This will identify the seed pool that should contain both homozygous and heterozygous insertion mutants for the given gene. Homozygous-lethal mutations can still be represented as heterozygotes, and maintained in this way. Ideally this combination of pooling and screening can identify a desired mutant with less than 100 PCR reactions.

Systematic sequencing of insertion sites in much the same way as described in Section III.B is another approach to find corresponding insertion mutants to any gene of interest. Several databases that contain the genomic sequences flanking the insertions in many thousand lines are under construction (for example Martienssen, 1998; Tissier et al., 1999; Parinov et al., 1999; Speulman et al., 1999; also Table 1, #25). In the near future a simple database search should be sufficient to quickly identify insertion lines that disrupt any given gene of interest. DNA microarray experiments will greatly help to discover potentially interesting genes, based on changes in their expression patterns. Gene discovery plus the identification of corresponding knockout mutant will possibly require less than a week. These loss-of-function alleles can then be examined separately or in combination using DNA microarrays, metabolic profiling, or other more conventional techniques to quickly gain an understanding of the process under study.

B. New Approaches for Targeted Gene Disruption and Gene Silencing in Plants

A major limitation to the analysis of gene function by insertion mutation is the high degree of gene duplication that is apparent in the *Arabidopsis* genome (Bevan et al., 1998; Lin et al., 1999; Mayer et al., 1999) and is probably a common feature in plant genomes (Somerville and Somerville, 1999). Since many of these gene duplications are tightly linked, it will usually not be feasible to produce double mutants by genetic recombination.

A labor-intensive approach using transposons might help to generate such double mutants. Since transposons tend to jump into locations near their original integration site, a 'launch-pad' line, containing an insertion in one of the two gene copies, could be used to generate a new transposon tagged population, thereby increasing the probability of finding the double mutant of interest (Das and Martienssen, 1995; Martienssen, 1998).

Targeted gene disruption using homologous recombination (HR) may be the better solution to eliminate tandem genes simultaneously by gene replacement. Homologous recombination is the primary tool for gene knockouts and allele substitutions in bacteria and yeast. More recently it has also become a routine method of gene inactivation in mammals and in the moss *Physcomitrella patens* (Schaefer and Zryd, 1997; Reski, 1998) as a model system for plants, allowing the study of targeted disruption of e.g. photosynthetic genes (Hofmann et al., 1999). In higher plants like *Arabidopsis* the unresolved problem of illegitimate recombination (Puchta and Hohn, 1996) makes targeted gene disruption by HR difficult, though there has been limited success on a gene-by-gene basis (Miao and Lam, 1995; Kempin et al., 1997). Homologous recombination in plants can be stimulated by overexpression of bacterial proteins involved in HR (Shalev et al., 1999). Fine tuning the HR system through genetic modification and recent improvements like high efficiency *Arabidopsis* transformation and high throughput PCR screening techniques may allow homologous recombination to become a routine tool for molecular and genetic studies of *Arabidopsis* (Kempin et al., 1997).

Another gene-by-gene approach for targeted heritable gene destruction is the use of RNA-DNA hybrids to introduce site-directed point mutations (Cole-Strauss et al., 1996; Kren et al., 1998; Beetham et al., 1999). By introducing mutagenic sequences that produce stop codons in regions that are conserved among all duplicates it may be possible to generate all combinations of null mutations in all members of a plant multigene family from one experiment (Somerville and Somerville, 1999). In maize the efficiency of gene conversion mediated by chimeric oligonucleotides was reported to be up to 1000-fold higher than homologous recombination (Zhu et al., 1999). Without an appropriate screen for the desired change, however, it is very tedious to identify lines in which the site-directed changes have been introduced. This technique is also hampered by the limitation that changes are introduced into protoplasts and adult plants are regenerated via tissue culture. For plant species without a working tissue culture system, other methods must be found.

Strategies to downregulate the expression level of a given gene by transcriptional or post-transcriptional gene silencing, e.g. antisense approach or cosuppression (Matzke and Matzke, 1995; Baulcombe, 1996; Meyer and Saedler, 1996; Depicker and van Montagu, 1997; Smyth, 1997) have been widely used in the past ten years, but are not ideal. Transgenic constructs integrate into the genome at unpredictable sites and possibly create new mutations. The expression of randomly inserted transgenes suffers from position effects, and the loss of gene function in antisense-experiments is often just partial rather than complete, and usually a significant number of transgenics have to be produced in order to identify one which is considerably silenced. Perhaps more importantly, if the target gene is a member of a multigene family, it may be impossible to specifically affect the activity of only the target gene.

A method of gene silencing based on producing double-stranded RNA from bidirectional transcription of genes in transgenic plants may be more useful for high-throughput gene inactivation (Waterhouse et al., 1998). Use of cell-type-specific, developmentally regulated or inducible promotors in combination with this approach could also significantly obviate problems associated with the lethality of some mutations. Although the mechanism is not properly understood, it resembles double-stranded RNA-mediated gene silencing in nematodes (Fire et al., 1998, Fire, 1999).

For many applications, particularly in species other than *Arabidopsis* where production of thousands of transformants is difficult, virus-induced gene silencing may be the easiest method for suppressing gene function (Baulcombe, 1999). This method exploits the fact that some or all plants have a surveillance system that can specifically recognize viral nucleic acids and mount a sequence-specific suppression of viral RNA accumulation. By inoculating plants with a recombinant virus containing part of a plant gene, it is possible to rapidly silence the endogenous plant gene (Kjemtrup et al., 1998; Voinnet et al., 1998).

Over-expression can be a potentially useful complement to targeted gene disruption and gene silencing. Strong constitutive, inducible (by e.g. hormones, antibiotics, heavy metals or environmental cues) or tissue- and developmentally specific promoters can be used to this effect (Reynolds, 1999), and can lead to strong and/or ectopic expression of genes and their products. Over-expression of a given gene can also lead to cosuppression, the homology-dependent gene-silencing of the introduced trans-gene(s) and the endogenous gene (for reviews see above). Untargeted over-expression of genes is achieved by activation tagging with T-DNA and transposons (Section III.B). The issues presented by over-expression, however, are different than unraveling the normal biological function for a given, yet uncharacterized gene. Over-expression and ectopic expression of gene products in tissues and conditions where they are not properly regulated can alter the biological function of the gene product and lead to additional and new phenotypes. For example, the ectopic expression of the maize transcription factor Lc-gene under the control of a 35S promotor leads to anthocyanin pigmentation in various tissues including roots, which are not normally pigmented (Ludwig et al., 1990). The ectopic expression of the *Arabidopsis LEC*1 transcription factor gene on the other hand is sufficient to induce embryo development in vegetative cells (Lotan et al., 1998). Although such information does not unravel the normal biological function of a given gene, it might nevertheless be helpful when dealing with uncharacterized genes, genes that act redundantly, or genes that are required during multiple stages of the plant life cycle and whose loss of function results in early embryonic lethality (Weigel et al., 2000). In addition, over-expression of genes encoding putative flux-limiting enzymes in metabolic pathways has been widely used to evaluate their importance in vivo as well as for metabolic engineering (Stitt and Sonnewald, 1995).

C. The Gauntlet

A gene is expressed in specific cells and under the specific conditions in which its product makes a contribution to fitness (Brown and Botstein, 1999). However, this connection can be difficult to demonstrate and may be the most time consuming step in a reverse genetic approach towards the determination of the biological function of a gene (Fig. 6). Careful study of the 'reverse genetic' mutant in a wide range of experimental, environmental and developmental conditions, including selective conditions at high plant density and competition for resources such as light and nutrients, may be required to reveal a phenotype. This is reflected by the observation that only ~2% of loss- or gain-of-function T-DNA insertion mutants show a clear visible phenotype when grown in ambient conditions (Azpiroz-Leehan and Feldmann, 1997; W.-R. Scheible, unpublished). The *Arabidopsis* potassium-channel T-DNA insertion mutant *akt*1-1 is a good example of the need for careful evaluation of possible phenotypes. This mutant is phenotypically indistinguishable from wild-type plants when grown on many nutrient media. However, growth of *akt*1-1 plants is significantly inhibited compared with wild type on media containing 100 μM potassium in the presence of ammonium (Hirsch et al., 1998; Spalding et al., 1999). Mutations can also result in very weak, and thus easily overlooked phenotypes or can have low penetrance (Ogas et al., 1997), and might be revealed only by multigenerational population studies. For example, adult *Arabidopsis* plants homozygous for various mutant actin alleles appear to be morphologically normal and fully fertile. However, when grown as populations descended from a single heterozygous parent, the mutant alleles are found at extremely low frequencies relative to the wild type in the F2- and following generations and thus appear to be deleterious (Gilliland et al., 1998). In addition, gene functions that seem to be nonessential or redundant (see above) might require the isolation or generation of double, triple, etc. mutants with mutations in different family members before phenotypic changes can be found in specific conditions. Likewise, for knockout mutants of a candidate gene discovered in microarray experiments or by other means, it may be necessary to cross these mutants to other candidate genes discovered in the same experiment, as indicated in Fig. 6.

VI. Conclusions and Outlook

Rapidly growing amounts of EST- and genome sequence for *Arabidopsis* and other plants, together with the new tools of functional genomics and reverse genetics are profoundly changing and accelerating research in all fields of plant biology. Linking mutant phenotypes to their corresponding genes by map-based cloning, or identifying mutants and phenotypes for new genes—discovered for example during large-scale gene expression experiments—is becoming progressively straightforward. Hence we expect that some degree of biological function for most if not all genes in a typical angiosperm genome can be deciphered within the next decade. Public databases and powerful software permitting convenient access, archiving and efficient mining of the new wealth of information such as genome sequence, map position of genes and alleles, or expression data, are being developed and will yield unprecedented insight into all aspects of plant growth and development. The genomic sequences of *Arabidopsis* and rice in conjunction with large-scale EST sequencing and mapping of a few major crop species should suffice to exploit the new knowledge for improving many crop plants and will be of key-importance to produce enough food for the world's growing human population.

Acknowledgments

We are grateful to Shauna Somerville for providing an original version of Figure 4 and Detlef Weigel for making results available prior to publication. W.-R. Scheible was a recipient of a fellowship (Sche 548-1/1) from the Deutsche Forschungsgemeinschaft.

References

Aarts MGM, Dirkse WG, Stiekema WJ and Pereira A (1993) Transposon tagging of a male sterility gene in *Arabidopsis*. Nature 363: 715–717

Adams MD (1996) Serial analysis of gene expression: ESTs get smaller. Bioessays 18: 261–262

Ahn S, Anderson JA, Sorrells ME and Tanksley SD (1993) Homoeologous relationships of rice, wheat and maize chromosomes. Mol Gen Genet 241: 483–490

Anderson L and Seilhamer J (1997) A comparison of selected mRNA and protein abundances in human liver. Electrophoresis 18: 533–537

Avery EL, Dunstran RH and Nell JA (1998) The use of lipid metabolic profiling to assess the biological impact of marine sewage pollution. Arch Env Cont Tox 35: 229–235

Avramova Z, Tikhonov A, Chen M and Bennetzen JL (1998) Matrix attachment regions and structural colinearity in the genomes of two grass species. Nucl Acids Res 26: 761–767

Azpiroz-Leehan R and Feldmann KA (1997) T-DNA insertion mutagenesis in *Arabidopsis*: Going back and forth. Trends Genet 13: 152–156

Baldwin D, Crane V and Rice D (1999) A comparison of gel-based, nylon filter and microarray techniques to detect differential RNA expression in plants. Curr Opin Plant Biol 2: 96–103

Ballinger DG and Benzer S (1989) Targeted gene mutations in *Drosophila*. Proc Natl Acad Sci USA 86: 9402–9406

Bancroft I and Dean C (1993) Transposition pattern of the maize element Ds in *Arabidopsis thaliana*. Genetics 134: 1221–1229

Banks JF and Whitehouse CM (1996) Electrospray ionization mass spectrometry. In: Karger BL and Hancock WS (eds) Methods in Enzymology, Vol 270, pp 486–518. Academic Press, San Diego

Baulcombe DC (1996) RNA as a target and an initiator of post-transcriptional gene silencing in transgenic plants. Plant Mol Biol 32: 79–88

Baulcombe DC (1999) Fast forward genetics based on virus-induced gene silencing. Curr Opin Plant Biol 2: 109–113

Beavis RC and Chait BT (1996) Matrix-assisted laser desorption ionization mass-spectrometry of proteins. In: Karger BL and Hancock WS (eds) Methods in Enzymology, Vol 270, pp 519–551. Academic Press, San Diego

Bechtold N, Ellis J and Pelletier G (1993) *In planta* Agrobacterium-mediated gene transfer by infiltration of adult *Arabidopsis thaliana* plants. CR Acad Sci Paris 316: 1194–1199

Beetham PR, Kipp PB, Sawycky XL, Arntzen CJ and May GD (1999) A tool for functional plant genomics: Chimeric RNA/DNA oligonucleotides cause in vivo gene-specific mutations. Proc Natl Acad Sci USA 96: 8774–8778

Bell CJ and Ecker JR (1994) Assignment of 30 microsatellite loci to the linkage map of *Arabidopsis*. Genomics 19: 137–144

Ben H (1999) Advances in protein solubilisation for two-dimensional electrophoresis. Electrophoresis 20: 660–663

Bennetzen JL and Freeling M (1993) Grasses as a single genetic system: Genome composition, colinearity and compatibility. Trends Genet 9: 259–261

Bennetzen JL and Freeling M (1997) The unified grass genome: Synergy in synteny. Genome Res 7: 301–306

Bernard PS, Lay MJ and Wittwer CT (1998) Integrated amplification and detection of the C677T point mutation in the methylenetetrahydrofolate reductase gene by fluorescence resonance energy transfer and probe melting curves. Anal Biochem 255:101–107

Bevan M (1997) Objective: The complete sequence of a plant genome. Plant Cell 9: 476–478

Bevan M, Bancroft I, Bent E, Love K, Goodman H, Dean C, Bergkamp R, Dirkse W, Van Staveren M, Stiekema W, Drost L, Ridley P, Hudson S-A, Patel K, Murphy G, Piffanelli P, Wedler H, Wedler E, Wambutt R, Weitzenegger T, Pohl TM, Terryn N, Gielen J, Villarroel R, De Clerck R, Van Montagu M, Lecharny A, Auborg S, Gy I, Kreis M, Lao N, Kavanagh T, Hempel S, Kotter P, Entian K-D, Rieger M, Schaeffer M, Funk B, Mueller-Auer S, Silvey M, James R, Montfort A, Pons A, Puigdomenech P, Douka A, Voukelatou E, Milioni D, Hatzopoulos P, Piravandi E, Obermaier B, Hilbert H, Duesterhoeft A, Moores T, Jones JDG, Eneva T, Palme K, Benes V, Rechman S, Ansorge W, Delseny M, Voet M, Volckaert G, Mewes H-W, Klosterman S, Schueller C and Chalwatzis N (1998) Analysis of 1.9 Mb of contiguous sequence from chromosome 4 of *Arabidopsis thaliana*. Nature 391: 485–488

Bhatt AM, Page T, Lawson EJR, Lister C and Dean C (1996) Use of Ac as an insertional mutagen in *Arabidopsis*. Plant J 9: 935–945

Blackstock WP and Weir MP (1999) Proteomics: Qualitative and physical mapping of cellular proteins. Trends Biotech 17: 121–127

Bouchez D and Höfte H (1998) Functional genomics in plants. Plant Physiol 118: 725–732

Bouchez D, Granier F, Bouché N, Caboche M, Bechtold N and Pelletier G (1999) T-DNA insertional mutagenesis for reverse genetics in *Arabidopsis*, Abstract 4-1. In: Proceedings Book of the 10th International Conference on *Arabidopsis* Research, University of Melbourne, Australia

Brown PO and Botstein D (1999) Exploring the new world of the genome with DNA microarrays. Nature Genet 21: 33—37

Bulyk ML, Gentalen E, Lockhart DJ and Church GM (1999) Quantifying DNA-protein interactions by double-stranded DNA arrays. Nature Biotech 17: 573–578

Camilleri C, Lafleuriel J, Macadré C, Varoquaux F, Parmentier Y, Picard G, Caboche M and Bouchez D (1998) A YAC contig map of *Arabidopsis thaliana* chromosome 3. Plant J 14: 633–642

Campisi L, Yang Y, Yi Y, Heilig E, Herman B, Cassista AJ, Allen DW, Xiang H and Jack T (1999) Generation of enhancer trap lines in *Arabidopsis* and characterization of expression patterns in the inflorescence. Plant J 17: 699–707

Cardon GH, Frey M, Saedler H and Gierl A (1993) Definition and characterization of an artificial En/Spm-based transposon tagging system in transgenic tobacco. Plant Mol Biol 23: 157–178

Cavell AC, Lydiate DJ, Parkin IAP, Dean C and Trick M (1998) Collinearity between a 30-centimorgan segment of *Arabidopsis thaliana* chromosome 4 and duplicated regions within the Brassica napus genome. Genome 41: 62–69

Chandler VL and Hardeman KJ (1992) The Mu elements of *Zea mays*. Adv Genet 30: 77–122

Chang SS, Park SK, Kim BC, Kang BJ, Kim DU and Nam HG (1994) Stable genetic transformation of *Arabidopsis thaliana* by *Agrobacterium* inoculation *in planta*. Plant J 5: 551–558

Chen M, San Miguel P, de Oliveira AC, Woo SS, Zhang H, Wing RA and Bennetzen JL (1997) Microcolinearity in sh2-homologous regions of the maize, rice, and sorghum genomes. Proc Natl Acad Sci USA 94: 3431–3435

Chen M, San Miguel P and Bennetzen JL (1998) Sequence organization and conservation in sh2/a1-homologous regions of sorghum and rice. Genetics 148: 435–443

Chin HG, Choe MS, Lee SH, Park SH, Park SH, Koo JC, Kim NY, Lee JJ, Oh BG, Yi GH, Kim SC, Choi HC; Cho MJ and Han C (1999) Molecular analysis of rice plants harboring an *Ac/Ds* transposable element-mediated gene trapping system. Plant J 19: 615–623

Cho RJ, Mindrinos M, Daniel R. Richards DR, Ronald J. Sapolsky RJ, Anderson M, Drenkard E, Dewdney J, Reuber TL, Stammers M, Federspiel N, Theologis A, Yang W-H, Hubbell E, Au M, Chung EY, Lashkari D, Lemieux B, Dean C, Lipshutz RJ, Ausubel FM, Davis RW and Oefner PJ (1999) Genome-wide mapping with biallelic markers in *Arabidopsis thaliana*. Nature Genet 23: 203—207

Choi SD, Creelman R, Mullet J and Wing RA (1995) Construction and characterisation of a bacterial artificial chromosome library from *Arabidopsis thaliana*. Weeds World 2: 17–20

Chu S, DeRisi J, Eisen M, Mulholland J, Botstein D, Brown PO and Herskowitz I (1998) The transcriptional program of sporulation in budding yeast. Science 282: 699–705

Clough SJ and Bent A (1998) Floral-dip: A simplified method for *Agrobacterium*-mediated transformation of *Arabidopsis thaliana*. Plant J 16: 735–743

Cole-Strauss A, Yoon K, Xiang Y, Byrne BC, Rice MC, Gryn J, Holloman WK and Kmiec EB (1996) Correction of the mutation responsible for sickle cell anemia by an RNA-DNA oligonucleotide. Science 273: 1386–1389

Cooke R, Raynal M, Laudie M and Delseny M (1997) Identification of members of gene families in *Arabidopsis thaliana* by contig construction from partial cDNA sequences: 106 genes encoding 50 cytoplasmic ribosomal proteins. Plant J 11: 1127–1140

Cutler S and Somerville C (1997) Cellulose synthesis: Cloning in silico. Curr Biol 7: R108–R111

Das L and Martienssen R (1995) Site-selected transposon mutagenesis at the hcf106 locus in maize. Plant Cell 7: 287–294

Dean C, Sjodin C, Bancroft I, Lawson E, Lister C, Scofield S and Jones J (1991) Development of an efficient transposon tagging system in *Arabidopsis thaliana*. Symp Soc Exp Biol 45: 63–75

Depicker A and van Montagu M (1997) Post-transcriptional gene silencing in plants. Curr Opin Cell Biol 9: 373–382

DeRisi JL, Iyer VR and Brown PO (1997) Exploring the metabolic and genetic control of gene expression on a genomic scale. Science 278: 680–686

DeSilva D, Reiser A, Herrmann M, Tabiti K and Wittwer C (1998) Rapid genotyping and quantification using the LightCycler with hybridization probes. Biochemica 2: 12–15

Dunn MJ and Corbett JM (1996) Two-dimensional polyacrylamide gel electrophoresis. In: Karger BL and Hancock WS (eds) Methods in Enzymology, Vol 271, pp 177–202. Academic Press, San Diego

Eisen MB and Brown PO (1999) DNA arrays for analysis of gene expression. In: Weissman S (ed) Methods in Enzymology, Vol 303, pp 179–205, Academic Press, San Diego

Eisen MB, Spellman PT, Brown PO and Botstein D (1998) Cluster analysis and display of genome-wide expression patterns. Proc Natl Acad Sci USA 95: 14863–14868

Enoki H, Izawa T, Kawahara M, Komatsu M, Koh S, Kyozuka J and Shimamoto K (1999) *Ac* as a tool for functional genomics of rice. Plant J 19: 605–613

ESSA, European Scientists Sequencing Arabidopsis (1999) *Arabidopsis* chromosome IV. Abstract 3-12. In: Proceedings Book of the 10th International Conference on *Arabidopsis* Research, University of Melbourne, Australia

Fedoroff NV, Chen F, Eckardt NA, Gomez-Buitrago AM and Raina R (1999) Global analysis of environmental stress and developmentally regulated genes in *Arabidopsis*. Abstract 4-4. In: Proceedings Book of the 10th International Conference on *Arabidopsis* Research, University of Melbourne, Australia

Feldmann KA (1991) T-DNA insertion mutagenesis in *Arabidopsis*: Mutational spectrum. Plant J 1: 71–82

Feldmann KA and Marks MD (1987) *Agrobacterium*-mediated transformation of germinating seeds of *Arabidopsis thaliana*: A non-tissue culture approach. Mol Gen Genet 208: 1–9

Fire A (1999) RNA-triggered gene silencing. Trends Genet 15: 358–363

Fire A, Xu S, Montgomery MK, Kostas SA, Driver SE and Mello CC (1998) Potent and specific genetic interference by double-stranded RNA in *Caenorhabditis elegans*. Nature 391: 806–811

Flachmann R and Kühlbrandt W (1995) Accumulation of plant antenna complexes Is regulated by post-transcriptional mechanisms in tobacco. Plant Cell 7: 149–160

Flavell RB, Bennett MD, Smith JB and Smith DB (1974) Genome size and proportion of repeated nucleotide sequence DNA in plants. Biochem Genet 12: 257–269

Fleischmann RD, Adams MD, White O, Clayton RA, Kirkness EF, Kerlavage AR, Bult CJ, Tomb J-F, Dougherty BA, Merrick JM, McKenney K, Sutton G, Fitzhugh W, Fields C, Gocayne JD, Scott J, Shirley R, Liu L-I, Glodek A, Kelley JM, Weidman JF, Phillips CA, Spriggs T, Hedblom E, Cotton MD, Utterback TR, Hanna MC, Nguyen DT, Saudek DM, Brandon RC, Fine LD, Fritchman JL, Fuhrmann JL, Geoghagen NSM, Gnehm CL, McDonald LA, Small KV, Fraser CM, Smith HO and Venter JC (1995) Whole-genome random sequencing and assembly of *Haemophilus influenzae* Rd. Science 269: 507–512

Fodor SPA, Read JL, Pirrung MC, Stryer L, Lu AT and Solas D (1991) Light-directed, spatially addressable parallel chemical synthesis. Science 251: 767–773

Gale MD and Devos KM (1998a) Plant comparative genetics after 10 years. Science 282: 656–659

Gale MD and Devos KM (1998b) Comparative genomics in the grasses. Proc Natl Acad Sci USA 95: 1971–1974

Gerhold D, Rushmore T and Caskey CT (1999) DNA chips: promising toys have become powerful tools. Trends Biochem Sci 24: 168–173

Gheysen G, Herman L, Breyne P, Gielen J, Van Montagu M and Depicker A (1990) Cloning and sequence analysis of truncated T-DNA inserts from *Nicotiana tabacum*. Gene 94: 155–163

Gibson S and Somerville C (1993) Isolating plant genes. Trends Biochem Sci 11: 306–313

Gillece-Castro BL and Stults JT (1996) Peptide characterization by mass spectrometry. In: Karger BL and Hancock WS (eds) Methods in Enzymology, Vol 271, pp 427–448. Academic Press, San Diego

Gilliland LU, McKinney EC, Asmussen MA and Meagher RB (1998) Detection of deleterious genotypes in multigenerational studies. I. Disruptions in individual *Arabidopsis* actin genes. Genetics 149: 717–725

Goff SA (1999) Rice as a model for cereal genomics. Curr Opin Plant Biol 2: 86–89

Goodman HM, Ecker JR and Dean C (1995) The genome of

Arabidopsis thaliana. Proc Natl Acad Sci USA 92: 10831–10835

Gruissem W (1989) Chloroplast gene expression, how plants turn their plastids on. Cell 56: 161–170

Hacia JG (1999) Resequencing and mutational analysis using oligonucleotide microarrays. Nature Genet 21: 42–47

Hake S and Walbot V (1980) The genome of *Zea mays*, its organization and homology to related grasses. Chromosoma 79: 251–270

Hamilton CM (1997) A binary-BAC system for plant transformation with high-molecular-weight DNA. Gene 200: 107–116

Hehl R (1994) Transposon tagging in heterologous host plants. Trends Genet 10: 385–386

Heller RA, Schena M, Chai A, Shalon D, Bedilion T, Gilmore J, Woolley DE and Davis RW (1997) Discovery and analysis of inflammatory disease-related genes using cDNA microarrays. Proc Natl Acad Sci USA 94: 2150–2155

Henzel WJ, Billeci TM, Stults JT, Wong SC, Grimley C and Watanabe C (1993) Identifying proteins from two-dimensional gels by molecular mass searching of peptide fragments in protein sequence databases. Proc Natl Acad Sci USA 93: 5011–5015

Heslop Harrison JS, Murata M, Ogura Y, Schwarzacher T and Motoyoshi F (1999) Polymorphisms and genomic organization of repetitive DNA from centromeric regions of *Arabidopsis* chromosomes. Plant Cell 11: 31–41

Hieter P and Boguski M (1997) Functional genomics: it's all how you read it. Science 278: 601–602

Hirsch RE, Lewis BD, Spalding EP and Sussman MR (1998) A role for the AKT1 potassium channel in plant nutrition. Science 280: 918–921

Hofmann AH, Codon AC, Ivascu C, Russo VEA, Knight C, Cove D, Schäfer DG, Chakhparonian M and Zryd J-P (1999) A specific member of the cab multigene family can be efficiently targeted and disrupted in the moss *Physcomitrella patens*. Mol Gen Genet 261: 92–99

Höfte H, Desprez T, Amselem J, Chiapello H, Caboche M, Moisan A, Jourjon M-F, Charpenteau J-L, Berthomieu P, Guerrier D, Giraudat J, Quigley F, Thomas F, Yu D-Y, Mache R, Raynal M, Cooke R, Grellet F, Delseny M, Parmentier Y, de Marcillac G, Gigogt C, Fleck J, Philipps G, Axelos M, Bardet C, Tremousaygue D and Lescure B (1993) An inventory of 1152 expressed sequence tags obtained by partial sequencing of cDNAs from *Arabidopsis thaliana*. Plant J 4: 1051–1061

Iyer VR, Eisen MB, Ross DT, Schuler G, Moore T, Lee JCF, Trent JM, Staudt LM, Hudson J Jr, Boguski MS, Lashkari D, Shalon D, Botstein D, Brown PO (1999) The transcriptional program in the response of human fibroplasts to serum. Science 283: 83–87

Jensen ON, Wilm M, Shevchenko A and Mann M (1999) Peptide sequencing of 2-DE gel-isolated proteins by nanoelectrospray tandem mass spectrometry. Meth Mol Biol 112: 571–588

Jones JDG, Carland FC, Lim E, Ralston E and Dooner HK (1990) Preferential transposition of the maize element Activator to linked chromosomal locations in tobacco. Plant Cell 2: 701–707

Jungblut P and Thiede B (1997) Protein identification from 2-DE gels by MALDI mass spectrometry. Mass Spectrom Rev 16: 145–162

Kakimoto T (1996) CKI1, a histidine kinase homolog implicated in cytokinin signal transduction. Science 274: 982–985

Kaneko T, Katoh T, Sato S, Nakamura Y, Asamizu E, Kotani H, Miyahima N and Tabata S (1999) Structural analysis of *Arabidopsis thaliana* chromosome 5. IX. Sequence features of the regions of 1,011,550 bp covered by seventeen P1 and TAC clones. DNA Research 6: 183–195

Kehoe DM, Villand P and Somerville S (1999) Technical focus: DNA microarrays for studies of higher plants and other photosynthetic organisms. Trends Plant Sci 4: 38–41

Kellogg EA (1998) Relationships of cereal crops and other grasses. Proc Natl Acad Sci USA 95: 2005–2010

Kempin SA, Liljegren SJ, Block LM, Rounsley SD and Yanofsky MF (1997) Targeted disruption in *Arabidopsis*. Nature 389: 802–803

Kinney AJ (1998) Manipulating flux through plant metabolic pathways. Curr Opin Plant Biol 1: 173–178

Kjemtrup S, Sampson KS, Peele CG, Nguyen LV, Conkling MA, Thompson WF and Robertson D (1998) Gene silencing from plant DNA carried by a geminivirus. Plant J 14: 91–100

Klein M, Eckert-Ossenkopp U, Schmiedeberg I, Brandt P, Unseld M, Brennicke A and Schuster W (1994) Physical mapping of the mitochondrial genome of *Arabidopsis thaliana* by cosmid and YAC clones. Plant J 6: 447–455

Klimyuk VI, Carroll BJ, Thomas CM and Jones JDG (1993) Alkali treatment for rapid preparation of plant material for reliable PCR analysis. Technical Advance. Plant J 3: 493–494

Koch KE (1996) Carbohydrate-modulated gene expression in plants. Ann Rev Plant Physiol Plant Mol Biol 47: 509–540

Kohchi T, Fujishige K and Ohyama K (1995) Construction of an equalized cDNA library from *Arabidopsis thaliana*. Plant J 8: 771–776

Komari T, Hiei Y, Ishida Y, Kumashiro T and Kubo T (1998) Advances in cereal gene transfer. Curr Opin Plant Biol 1: 161–165

Koncz C, Martini N, Mayerhofer R, Koncz KZ, Koerber H, Redei GP and Schell J (1989) High-frequency T-DNA-mediated gene tagging in plants. Proc Natl Acad Sci USA 86: 8467–8471

Koncz C, Nemeth K, Redei GP and Schell J (1992) T-DNA insertional mutagenesis in *Arabidopsis*. Plant Mol Biol 20: 963–976

Konieczny A and Ausubel F (1993) A procedure for mapping *Arabidopsis* mutations using co-dominant ecotype-specific PCR-based markers. Plant J 4: 403–410

Kotani H, Sato S, Fukami M, Hosouchi T, Nakazaki N, Okumura S, Wada T, Liu Y-G, Shibata D and Tabata S (1997) A fine physical map of *Arabidopsis thaliana* chromosome 5: Construction of a sequence-ready contig map. DNA Res 4: 371–378

Kren BT, Bandtopadhyay P and Steer CJ (1998) In vivo site-directed mutagenesis of the factor IX gene by chimeric RNA/DNA oligonucleotides. Nature Med 4: 285–290

Krysan PJ, Young JC, Tax F and Sussman MR (1996) Identification of transferred DNA insertions within *Arabidopsis* genes involved in signal transduction and ion transport. Proc Natl Acad Sci USA 93: 8145–8150

Krysan PJ, Young JC and Sussman MR (1999) T-DNA as an Insertional Mutagen in *Arabidopsis*. Plant Cell 11: 2283–2290

Lagercrantz U (1998) Comparative mapping between *Arabidopsis thaliana* and *Brassica nigra* indicates that *Brassica* genomes have evolved through extensive genome replication accompanied by chromosome fusions and frequent rearrangements. Genetics 150: 1217–1228

Lagercrantz U, Putterill J, Coupland G and Lydiate D (1996) Comparative mapping in *Arabidopsis* and *Brassica*, fine scale genome collinearity and congruence of genes controlling flowering time. Plant J 9: 13–20

Lee NH, Weinstock KG, Kirkness EF, Earle-Hughes JA, Fuldner RA, Marmaros S, Glodek A, Gocayne JD, Adams MD, Kerlavage AR, Fraser CM and Venter JC (1995) Comparative expressed-sequence-tag analysis of differential gene expression profiles in PC-12 cells before and after nerve growth factor treatment. Proc Natl Acad Sci USA 92: 8303–8309

Lemieux B, Aharoni A and Schena M (1998) DNA chip technology. Mol Breed 4: 277–289

Lin X, Kaul S, Rounsley S, Shea TP, Benito M-I, Town CD, Fujii CY, Mason T, Bowman CL, Barnstead M, Feldblyum TV, Buell R, Ketchum KA, Lee J, Ronning CM, Koo HL, Moffat KS, Cronin LA, Shen M, Pai G, Van Aken S, Umayam L, Tallon LK, Gill JE, Adams MD, Carrera AJ, Creasy TH, Goodman HM, Somerville CR, Copenhaver GP, Preuss D, Nioerman WC, White O, Eisen JA, Salzberg SL, Fraser CM and Venter JC (1999) Sequence and analysis of chromosome 2 of the plant *Arabidopsis thaliana*. Nature 402: 761–768

Lipshutz RJ, Fodor SPA, Gingeras TR and Lockhart DJ (1999) High density synthetic oligonucleotide arrays. Nature Genet 21: 20–24

Liu D and Crawford NM (1998) Characterization of the germinal and somatic activity of the *Arabidopsis* transposable element Tag1. Genetics 148: 445–456

Liu YG, Mitsukawa N, Vazquez-Tello A and Whittier RF (1995a) Generation of a high quality P1 library of *Arabidopsis thaliana* suitable for chromosome walking. Plant J 7: 351–358

Liu YG, Mitsukawa N, Oosumi T and Whittier RF (1995b) Efficient isolation and mapping of *Arabidopsis thaliana* T-DNA insert junctions by thermal asymmetric interlaced PCR. Plant J 8: 457–463

Liu YG, Shirano Y, Fukaki H, Yanai Y, Tasaka M, Tabata S and Shibata D (1999) Complementation of plant mutants with large genomic DNA fragments by a transformation-competent artificial chromosome vector accelerates positional cloning. Proc Natl Acad Sci USA 96: 6535–6540

Lockhart DJ, Dong H, Byrne MC, Follettie MT, Gallo MV, Chee MS, Mittmann M, Wang C, Kobayashi M, Horton H and Brown EL (1996) Expression monitoring by hybridization to high-density oligonucleotide arrays. Nature Biotech 14: 1675–1680

Long D and Coupland G (1998) Transposon tagging with Ac/Ds in *Arabidopsis*. Methods Mol Biol 82: 315–328

Lotan T, Ohto M-A, Yee KM, West MAL, Lo R, Kwong RW, Yamagishi K, Fischer RL, Goldberg RB and Harada JJ (1998) *Arabidopsis* LEAFY COTYLEDON1 is sufficient to induce embryo development in vegetative cells. Cell 93: 1195–1205

Ludwig SR, Bowen B, Beach L and Wessler SR (1990) A regulatory gene as a novel visible marker for maize transformation. Science 247: 449–450

Lukowitz W, Gillmor CS and Scheible W-R (2000) Why it feels good to have a genome initiative working for you. Plant Physiol 123: 795–805

Mandal A, Lang V, Orczyk W and Palva ET (1993) Improved efficiency for T-DNA-mediated transformation and plasmid rescue in *Arabidopsis thaliana*. Theor Appl Genet 86: 621–628

Marks MD and Feldmann KA (1989) Trichome development in *Arabidopsis thaliana*: I. T-DNA tagging of the GLABROUS 1 gene. Plant Cell 1: 1043–1050

Martienssen RA (1998) Functional genomics: probing plant gene function and expression with transposons. Proc Natl Acad Sci USA 95: 2021–2026

Martin W and Herrmann RG (1998) Gene transfer from organelles to the nucleus: how much, what happens, and why? Plant Physiol 118: 9–17

Matzke MA and Matzke AJM (1995) How and why do plants inactivate homologous (trans)genes? Plant Physiol 107: 679–685

Mayer K, Schüller C, Wambutt R, Murphy G, Volckaert G, Pohl T, Düsterhöft A, Stiekema W, Entian K-D, Terryn N, Harris B, Ansorge W, Brandt P, Grivell L, Rieger M, Weichselgartner M, de Simone V, Obermaier B, Mache R, Müller M, Kreis M, Delseny M, Puigdomenech P, Watson M, Schmidtheini T, Reichert B, Portatelle D, Perez-Alonso M, Boutry M, Bancroft I, Vos P, Hoheisel J, Zimmermann W, Wedler H, Ridley P, Langham S-A, McCullagh B, Bilham L, Robben J, Van der Schueren J, Grymonprez B, Chuang Y-J, Vandenbussche F, Braeken M, Weltjens I, Voet M, Bastiaens I, Aert R, Defoor E, Weitzenegger T, Bothe G, Ramsperger U, Hilbert H, Braun M, Holzer E, Brandt A, Peters S, van Staveren M, Dirkse W, Mooijman P, Klein Lankhorst R, Rose M, Hauf J, Kötter P, Berneiser S, Hempel S, Feldpausch M, Lamberth S, Van den Daele H, De Keyser A, Buysshaert C, Gielen J, Villarroel R, De Clercq R, Van Montagu M, Rogers J, Cronin A, Quail M, Bray-Allen S, Clark L, Doggett J, Hall S, Kay M, Lennard N, McLay K, Mayes R, Pettett A, Rajandream M-A, Lyne M, Benes V, Rechmann S, Borkova D, Blöcker H, Scharfe M, Grimm M, Löhnert T-H, Dose S, de Haan M, Maarse A, Schäfer M, Müller-Auer S, Gabel C, Fuchs M, Fartmann B, Granderath K, Dauner D, Herzl A, Neumann S, Argiriou A, Vitale D, Liguori R, Piravandi E, Massenet O, Quigley F, Clabauld G, Mündlein A, Felber R, Schnabl S, Hiller R, Schmidt W, Lecharny A, Aubourg S, Chefdor F, Cooke R, Berger C, Montfort A, Casacuberta E, Gibbons T, Weber N, Vandenbol M, Bargues M, Terol J, Torres A, Perez-Perez A, Purnelle B, Bent E, Johnson S, Tacon D, Jesse T, Heijnen L, Schwarz S, Scholler P, Heber S, Francs P, Bielke C, Frishman D, Haase D, Lemcke K, Mewes HW, Stocker S, Zaccaria P, Bevan M, Wilson RK, de la Bastide M, Habermann K, Parnell L, Dedhia N, Gnoj L, Schutz K, Huang E, Spiegel L, Sehkon M, Murray J, Sheet P, Cordes M, Abu-Threideh J, Stoneking T, Kalicki J, Graves T, Harmon G, Edwards J, Latreille P, Courtney L, Cloud J, Abbott A, Scott K, Johnson D, Minx P, Bentley D, Fulton B, Miller N, Greco T, Kemp K, Kramer J, Fulton L, Mardis E, Dante M, Pepin K, Hillier L, Nelson J, Spieth J, Ryan E, Andrews S, Geisel C, Layman D, Du H, Ali J, Berghoff A, Jones K, Drone K, Cotton M, Joshu C, Antonoiu B, Zidanic M, Strong C, Sun H, Lamar B, Yordan C, Ma P, Zhong J, Preston R, Vil D, Shekher M, Matero A, Shah R, Swaby I'K, O'Shaughnessy A, Rodriguez M, Hoffman J, Till S, Granat S, Shohdy N, Hasegawa A, Hameed A, Lodhi M, Johnson A, Chen E, Marra M, Martienssen R and McCombie WR (1999) Sequence and analysis of chromosome 4 of the plant *Arabidopsis thaliana*. Nature 402: 769–777

McKinney EC, Aali N, Traut A, Feldmann KA, Belostotsky DA, McDowell JM and Meagher RB (1995) Sequence-based identification of T-DNA insertion mutations in *Arabidopsis*: Actin mutants act2-1 and act4-1. Plant J 8: 613–622

Meinke DW, Cherry JM, Dean C, Rounsley SD and Koornneef M (1998) *Arabidopsis thaliana*: A model plant for genome analysis. Science 282: 662–682

Meyer K, Leube MP and Grill E (1994) A protein phosphatase 2C involved in ABA signal transduction in *Arabidopsis thaliana*. Science 264: 1452–1455

Meyer P and Saedler H (1996) Homology-dependent gene silencing in plants. Annu Rev Plant Physiol Plant Mol Biol 47: 23–48

Meyerowitz EM (1992) Introduction to the *Arabidopsis* genome. In: Koncz C, Chua N and Schell J (eds) Methods in *Arabidopsis* Research, pp 100–118. World Scientific Publishing, Singapore

Meyerowitz EM (1994) Plant developmental biology: Green genes for the 21st Century. Bioessays 16: 621–625

Meyerowitz EM (1999) Today we have naming of parts. Nature 402: 731–732

Meyerowitz EM and Somerville CR (1994) Arabidopsis. Cold Spring Harbor Laboratory Press, Cold Spring Harbor, New York

Miao ZH and Lam E (1995) Targeted disruption of the TGA3 locus in *Arabidopsis thaliana*. Plant J 7: 359–365

Michaels SD and Amasino RM (1998) A robust method for detecting single-nucleotide changes as polymorphic markers by PCR. Plant J 14: 381–385

Mozo T, Fischer S, Meier-Ewert S, Lehrach H and Altmann TSO (1998) Use of the IGF BAC library for physical mapping of the *Arabidopsis thaliana* genome. Plant J 16: 377—384

Mozo T, Dewar K, Dunn P, Ecker JR, Fischer S, Kloska S, Lehrach H, Marra M, Martienssen R, Meier-Ewert S and Altmann T (1999) A complete BAC-based physical map of the *Arabidopsis thaliana* genome. Nature Genet 22:271–275

Nacry P, Camilleri C, Courtial B, Caboche M and Bouchez D (1998) Major chromosomal rearrangements induced by T-DNA transformation in *Arabidopsis*. Genetics 149: 641–650

Neff MM, Neff JD, Chory J and Pepper AE (1998) dCAPS, a simple technique for the genetic analysis of single nucleotide polymorphisms: Experimental applications in *Arabidopsis thaliana* genetics. Plant J 14: 387–392

Newman T, De Bruijn FJ, Green P, Keegstra K, Kende H, McIntosh L, Ohlrogge J, Raikhel N, Somerville S, Thomashow M, Retzel E, Somerville C (1994) Genes galore: A summary of methods for accessing results from large-scale partial sequencing of anonymous *Arabidopsis* cDNA clones. Plant Physiol 106: 1241–1255

Nuccio ML, Rhodes D, McNeil SD and Hanson AD (1999) Metabolic engineering of plants for osmotic stress resistance. Curr Opin Plant Biol 2: 128–134

Ogas J, Cheng J-C, Sung ZR and Somerville C (1997) Cellular differentiation regulated by gibberellin in the *Arabidopsis thaliana* pickle mutant. Science 277: 91–94

Ohlrogge J (1999) Plant metabolic engineering: are we ready for phase two? Curr Opin Plant Biol 2: 121–122

Parinov S, Sevugan M, Ye D, Yang W-C, Kumaran M and Sundaresan V (1999) Analysis of Flanking Sequences from Dissociation Insertion Lines: A Database for Reverse Genetics in *Arabidopsis*. Plant Cell 11: 2263–2270

Parnell L, Schueller C, Zaccaria P and Mayer K (1999) The nature of the EST: Distribution across chromosome IV. Abstract 3-4. In: Proceedings Book of the 10th International Conference on *Arabidopsis* Research, University of Melbourne, Australia

Patton WF, Steinberg TH, Berggren KN, Diwu Z and Haugland RP (1999) Combining ultrasensitive fluorescence detection with 2-D gel electrophoresis in high-throughput proteomics. FASEB J 13: A1455

Pearson WR and Lipman DJ (1988) Improved tools for biological sequence comparison. Proc Natl Acad Sci USA 85: 2444–2448

Ponce MR, Robles P and Micol JL (1999) High throughput mapping in *Arabidopsis thaliana*. Mol Gen Genet 261: 408–415

Powell J (1998) Enhanced concatemer cloning-a modification to the SAGE (Serial Analysis of Gene Expression) technique. Nucl Acids Res 26: 3445–3446

Pruitt RE and Meyerowitz EM (1986) Characterization of the genome of *Arabidopsis thaliana*. J Mol Biol 187: 169–183

Puchta H and Hohn B (1996) From centimorgans to base pairs—homologous recombination in plants. Trends Plant Sci 1: 340–348

Rashed MS, Bucknall MP, Little D, Awad A, Jacob M, Alamoudi M, Alwattar M and Ozand PT (1997) Screening blood spots for inborn errors of metabolism by electrospray tandem mass spectrometry with a microplate batch process and a computer algorithm for automated flagging of abnormal profiles. Clin Chem 43: 1129–1141

Reinbothe S, Reinbothe C and Parthier B (1993) Methyl jasmonate-regulated translation of nuclear-encoded chloroplast proteins in barley (*Hordeum vulgare* L. cv. Salome). J Biol Chem 268: 10606–10611

Reski R (1998) *Physcomitrella* and *Arabidopsis*: The David and Goliath of reverse genetics. Trends Plant Sci 3: 209–210

Reynolds PHS (1999) Inducible gene expression in plants. CABI Publishing, New York

Richmond CS, Glasner JD, Mau R, Jin H, Blattner FR (1999) Genome-wide expression profiling in *Escherichia coli* K-12. Nucl Acids Res 27: 3821–3835

Riley J, Butler R, Ogilvie D, Finniear R, Jenner D, Powell S, Anand R, Smith JC and Markham AF (1990) A novel, rapid method for the isolation of terminal sequences from yeast artificial chromosome (YAC) clones. Nucl Acids Res 18: 2887–2890

Rosenthal A and Jones DSC (1990) Genomic walking and sequencing by oligo-cassette mediated polymerase chain reaction. Nucl Acids Res 18: 3095–3096

Round EK, Flowers SK and Richards EJ (1997) *Arabidopsis thaliana* centromere regions: genetic map positions and repetitive DNA structure. Genome Res 11: 1045–1053

Rounsley SD, Glodek A, Sutton G, Adams MD, Somerville CR, Venter JC and Kerlavage AR (1996) The construction of *Arabidopsis* EST assemblies: A new resource to facilitate gene identification. Plant Physiol 112: 1179–1183

Rounsley S, Lin X and Ketchum KA (1998) Large-scale sequencing of plant genomes. Curr Opin Plant Biol 1: 136–141

Rounsley S, Subramaniam S, Cao Y, Bush D, DeLoughery C, Field C, Philipps C, Iartchouk O, Last R, Wiegand R, Fischhoff D and Timberlake B (1999) Whole genome shotgun sequencing of *Arabidopsis*—analysis and insight. Abstract 3-1. In: Proceedings book of the 10th International Conference on *Arabidopsis* Research, University of Melbourne, Australia

Rouzé P, Pavy N and Rombauts S (1999) Genome annotation: which tools do we have for it ? Curr Opin Plant Biol 2: 90–95

Ruan Y, Gilmore J and Conner T (1998) Towards *Arabidopsis* genome analysis: monitoring expression profiles of 1400 genes using cDNA microarrays. Plant J 15: 821–833

Saier MH Jr (1998) Genome sequencing and informatics: New tools for biochemical discoveries. Plant Physiol 117: 1129–1133

Sankhavaram RP, Parimoo S and Weissman SM (1991) Construction of a uniform abundance (normalized) cDNA library. Proc Natl Acad Sci USA 88: 1943–1947

San Miguel P, Tikhonov A, Jin Y-K, Motchoulskaia N, Zakharov D, Melake-Berhan A, Springer PS, Edwards KJ, Lee M, Avramova Z and Bennetzen JL (1996) Nested retrotransposons in the intergenic regions of the maize genome. Science 274: 765–768

Santoni V, Doumas P, Rouquie D, Mansion M, Rabilloud T and Rossignol M (1999) Large scale characterization of plant plasma membrane proteins. Biochimie 81: 655–661

Sasaki T (1998) The rice genome project in Japan. Proc Natl Acad Sci USA 95: 2027–2028

Sato S, Kotani H, Hayashi R, Liu Y-G, Shibata D and Tabata S (1998) A physical map of *Arabidopsis thaliana* chromosome 3 represented by two contigs of CIC YAC, P1, TAC and BAC clones. DNA Res 5: 163–168

Schaefer DG and Zryd JP (1997) Efficient gene targeting in the moss *Physcomitrella patens*. Plant J 11: 1195–1206

Schena M, Shalon D, Heller R, Chai A, Brown PO and Davis RW (1996) Parallel human genome analysis: Microarray-based expression monitoring of 1000 genes. Proc Natl Acad Sci USA: 10614–10619

Schmidt R, West J, Love K, Lenehan Z, Lister C, Thompson H, Bouchez D and Dean C (1995) Physical map and organization of *Arabidopsis thaliana* chromosome 4. Science 270: 480–483

Schmidt R, Love K, West J, Lenehan Z and Dean C (1997) Description of 31 YAC contigs spanning the majority of *Arabidopsis thaliana* chromosome 5. Plant J 11: 563–573

Schmidt R, Acarkan A, Boivin K, Koch M and Rossberg M (1999) Analysis of microsynteny in *Arabidopsis thaliana*, *Brassica oleracea* and *Capsella rubella*. Abstract 3-14. In: Proceedings Book of the 10th International Conference on *Arabidopsis* Research, University of Melbourne, Australia

Scholl RL, Keathley DE and Baribault TJ (1981) Enhancement of root formation and fertility of shoots regenerated anther- and seedling-derived callus cultures of *Arabidopsis thaliana*. Z Pflanzenphysiol 104: 225–231

Scholl R, Rivero L, Crist D, Ware D and Davis K (1999) Progress of ABRC in obtaining new T-DNA stocks and preparing DNA from T-DNA populations for PCR screening. Abstract 4-15. In: Proceedings Book of the 10th International Conference on *Arabidopsis* Research, University of Melbourne, Australia

Seigneurin-Berny D, Rolland N, Garin J and Joyard J (1999) Differential extraction of hydrophobic proteins from chloroplast envelope membranes: A subcellular-specific proteomic approach to identify rare intrinsic membrane proteins. Plant J 19: 217–228

Shalev G, Sitrit Y, Avivi Ragolski N, Lichtenstein C and Levy AA (1999) Stimulation of homologous recombination in plants by expression of the bacterial resolvase RuvC. Proc Natl Acad Sci USA 96: 7398–7402

Shalon D, Smith SJ and Brown PO (1996) A DNA-microarray system for analyzing complex DNA samples using two-color fluorescent probe hybridization. Genome Res 6: 639–645

Sheng J and Citovsky V (1996) *Agrobacterium*-plant cell DNA transport: Have virulence proteins, will travel. Plant Cell 8: 1699–1710

Shevchenko A, Jensen ON, Podtelejnikov AV, Sagliocco F, Wilm M, Vorm O, Mortensen P, Shevchenko A, Boucherie H and Mann M (1996) Linking genome and proteome by mass spectrometry: Large-scale identification of yeast proteins from two dimensional gels. Proc Natl Acad Sci USA 93: 14440–14445

Shoemaker DD, Lashkari DA, Morris D, Mittmann M and Davis RW (1996) Quantitative phenotypic analysis of yeast deletion mutants using a highly parallel molecular bar-coding strategy. Nature Genet 14: 450–456

Siebert PD, Chenchik A, Kellogg DE, Lukyanov KA and Lukyanov SA (1995) An improved PCR method for walking in uncloned genomic DNA. Nucl Acids Res 23: 1087–1088

Singh-Gasson S, Green RD, Yue Y, Nelson C, Blattner F, Cerrina F and Sussman MR (1999) Maskless fabrication of light-directed oligonucleotide microarrays using a digital micromirror array. Nature Biotech 17: 974–978

Smyth DR (1997) Gene silencing: Cosuppression at a distance. Curr Biol 7: R793-R795

Somerville C and Somerville S (1999) Plant functional genomics. Science 285: 380–383

Spalding EP, Hirsch RE, Lewis DR, Qi Z, Sussman MR and Lewis BD (1999) Potassium uptake supporting plant growth in the absence of AKT1 channel activity: Inhibition by ammonium and stimulation by sodium. J Gen Phys 113: 909–918

Spellman PT, Sherlock G, Zhang MQ, Iyer VR, Anders K, Eisen MB, Brown PO, Botstein D, Futcher B (1998) Comprehensive identification of cell cycle-regulated genes of the yeast *Saccharomyces cerevisiae* by microarray hybridization. Mol Biol Cell 9: 3273–3297

Speulman E, Metz PLJ, van Arkel G, Lintel Hekkert B, Stiekema WJ and Pereira A (1999) A two-component *enhancer-inhibitor* transposon mutagenesis system for functional analysis of the *Arabidopsis* Genome. Plant Cell 11: 1853–1866

Steinberg TH, Jones LJ, Haugland RP and Singer VL (1996) SYPRO orange and SYPRO red protein gel stains: One-step fluorescent staining of denaturing gels for detection of nanogram levels of protein. Anal Biochem 239: 223–237

Stitt M (1999) Nitrate regulation of metabolism and growth. Curr Opin Plant Biol 2: 178–186

Stitt M and Krapp A (1999) The interaction between elevated carbon dioxide and nitrogen nutrition: the physiological and molecular background. Plant Cell Environ 22: 583–621

Stitt M and Sonnewald U (1995) Regulation of metabolism in transgenic plants. Annu Rev Plant Phys Plant Mol Biol 46: 341–368

Sundaresan V, Springer P, Volpe T, Haward S and Jones JDG (1995) Patterns of gene action in plant development revealed by enhancer trap and gene trap transposable elements. Genes Dev 9: 1797–1810

Sutton G, White O, Adams M and Kerlavage A (1995) TIGR assembler: a new tool for assembling large shotgun sequencing projects. Genome Sci Technol 1: 9–19

Tanksley SD, Bernatzky R, Lapitan NL and Prince JP (1988) Conservation of gene repertoire but not gene order in pepper and tomato. Proc Natl Acad Sci USA 85: 6419–6423

Tautz D (1989) Hypervariability of simple sequences as a general source of polymorphic DNA markers. Nucl Acids Res 17: 6463–6471

The C. elegans Sequencing Consortium (1998) Genome sequence of the nematode *C. elegans*: A platform for investigating biology. Science 282: 2012–2018

Thompson HL, Schmidt R and Dean C (1996a) Analysis of the occurrence and nature of repeated DNA in an 850 kb region of *Arabidopsis thaliana* chromosome 4. Plant Mol Biol 32: 553–557

Thompson HL, Schmidt R and Dean C (1996b) Identification and distribution of seven classes of middle-repetitive DNA in the *Arabidopsis thaliana* genome. Nucl Acids Res 24: 3017–3022

Tissier AF, Marillonnet S, Klimyuk V, Patel K, Angel Torres M, Murphy G and Jones JDG (1999) Multiple independent defective suppressor-mutator transposon insertions in *Arabidopsis*: A tool for functional genomics. Plant Cell 11: 1841–1852

Topping JF and Lindsey K (1995) Insertional mutagenesis and promoter trapping in plants for the isolation of genes and the study of development. Transgenic Res 4: 291–305

Trethewey RN, Krotzky AJ and Willmitzer L (1999) Metabolic profiling: A rosetta stone for genomics? Curr Opin Plant Biol 2: 83–85

Unseld M, Marienfeld JR, Brandt P and Brennicke A (1997) The mitochondrial genome of *Arabidopsis thaliana* contains 57 genes in 366,924 nucleotides. Nature Genet 15: 57–61

Uwer U, Willmitzer L and Altmann T (1998) Inactivation of a glycyl-tRNA synthetase leads to an arrest in plant embryo development. Plant Cell 10: 1277–1294

Valvekens D, Montagu, M and Lijsebettens MV (1988) *Agrobacterium tumefaciens*-mediated transformation of *Arabidopsis thaliana* root explants by using kanamycin selection. Proc Natl Acad Sci USA 85: 5536–5540

Van den Bulk RW, Löffler HJM, Lindhaut WH and Koornneef M (1990) Somaclonal variation in tomato: effect of explant source and a comparison with chemical mutagenesis. Theor Appl Genet 80: 817–825

Velculescu VE, Zhang L, Vogelstein B and Kinzler KW (1995) Serial analysis of gene expression. Science 270: 484–487

Velculescu VE, Zhang L, Zhou W, Vogelstein Jacob, Basrai MA, Bassett DE Jr, Hieter P, Vogelstein B and Kinzler KW (1997) Characterization of the yeast transcriptome. Cell 88: 243–251

Voinnet O, Vain P, Angell S and Baulcombe DC (1998) Systemic spread of sequence-specific transgene RNA degradation in plants is initiated by localized introduction of ectopic promoterless DNA. Cell 95: 177–187

Walbot V (1992) Strategies for mutagenesis and gene cloning using transposon tagging and T-DNA insertional mutagenesis. Annu Rev Plant Physiol Plant Mol Biol 43: 49–82

Walden R, Fritze H, Hayashi H, Miklashevichs E, Harling H and Schell J (1994) Activation tagging: A means of isolating genes implicated as playing a role in plant growth and development. Plant Mol Biol 26: 1521–1528

Wang DG, Fan J-B, Siao C-J, Berno A, Young P, Sapolsky R, Ghandour G, Perkins N, Winchester E, Spencer J, Kruglyak L, Stein L, Hsie L, Topaloglou T, Hubbell E, Robinson E, Mittmann M, Morris MS, Shen N, Kilburn D, Rioux J, Nusbaum C, Rozen S, Hudson TJ, Lipshutz R, Chee M and Lander ES (1998) Large-scale identification, mapping, and genotyping of single-nucleotide polymorphisms in the human genome. Science 280: 1077–1082

Waterhouse PM, Graham MW and Wang MB (1998) Virus resistance and gene silencing in plants can be induced by simultaneous expression of sense and antisense RNA. Proc Natl Acad Sci USA 95: 13959–13964

Weigel D, Ahn JH, Blázquez MA, Borevitz J, Christensen SK, Fankhauser C, Ferrándiz C, Kardailsky I, Neff MM, Nguyen JT, Sato S, Wang Z, Xia Y, Dixon RA, Harrison MJ, Lamb CJ, Yanofsky MF and Chory J (2000) Activation tagging in *Arabidopsis*. Plant Physiol 122: 1003–1013

Weiner H (1995) Antibodies that distinguish between the serine-158 phospho- and dephospho-form of spinach leaf sucrose-phosphate synthase. Plant Physiol 108: 219—225

Wilkins MR, Williams KL, Appel RD and Hochstrasser DF (1997) Proteome Research: New Frontiers in Functional Genomics. Springer-Verlag, Berlin Heidelberg New York

Winkler RG, Frank MR, Galbraight DW, Feyereisen R and Feldmann KA (1998) Systematic reverse genetics of transfer-DNA-tagged lines of *Arabidopsis*: Isolation of mutations in the cytochrome P450 gene superfamily. Plant Physiol 118: 743–750

Wisman E, Hartmann U, Sagasser M, Baumann E, Palme K, Hahlbrock K, Saedler H and Weisshaar B (1998) Knock-out mutants from an En-1 mutagenized *Arabidopsis thaliana* population generated phenylpropanoid biosynthesis phenotypes. Proc Natl Acad Sci USA 95: 12432–12437

Wodicka L, Dong H, Mittmann M, Ho M-H and Lockhart DJ (1997) Genome-wide expression monitoring in *Saccharomyces cerevisiae*. Nature Biotech 15: 1359–1367

Yamamoto K and Sasaki T (1997) Large-scale EST sequencing in rice. Plant Mol Biol 35: 135–144

Ye G-N, Stone D, Pang S-Z, Creely W, Gonzalez K and Hinchee M (1999) *Arabidopsis* ovule is the target for *Agrobacterium in planta* vacuum infiltration transformation. Plant J 19: 249–257

Yu H, Chao J, Patek D, Mujumdar R, Mujumdar S and Waggoner AS (1994) Cyanine dye dUTP analogs for enzymatic labeling of DNA probes. Nucl Acids Res 22: 3226–3232

Zachgo EA, Wang ML, Dewdney J, Bouchez D, Camilleri C, Belmonte S, Huang L, Dolan M and Goodman HM (1996) A physical map of chromosome 2 of *Arabidopsis thaliana*. Genome Res 6: 19–25

Zhu T, Peterson DJ, Tagliani L, St Clair G, Baszczynski CL and Bowen B (1999) Targeted manipulation of maize genes in vivo using chimeric RNA/DNA oligonucleotides. Proc Natl Acad Sci USA 96: 8768–8773

Zupan JR and Zambryski P (1995) Transfer of T-DNA from *Agrobacterium* to the plant cell. Plant Physiol 107: 1041–1047

Zwaal RR, Broeks A, van Meurs J, Groenin JTM and Plasterk RHA (1993) Target-selected gene inactivation in *Caenorhabditis elegans* by using a frozen transposon insertion mutant bank. Proc Natl Acad Sci USA 90: 7431–7435

Index

Numerical

A

B

C

D

E

F

H

I

J

K

L

N

Q

R

S

T

U

V

W

X

Y

Z